ISBN 978-0-483-60800-9
PIBN 10650324

FB &c Ltd, Dalton House, 60 Windsor Avenue, London, SW19 2RR.
Company number 08720141. Registered in England and Wales.

For support please visit www.forgottenbooks.com

THE WORLD BOOK

ORGANIZED KNOWLEDGE IN STORY AND PICTURE

TRADE MARK REGISTERED

EDITOR-IN-CHIEF

M. V. O'SHEA

DEPARTMENT OF EDUCATION, UNIVERSITY OF WISCONSIN
MADISON, WISCONSIN

EDITOR

ELLSWORTH D. FOSTER

ASSOCIATE EDITOR NEW PRACTICAL REFERENCE LIBRARY; AUTHOR OF CYCLOPEDIA OF CIVIL GOVERNMENT

EDITOR FOR CANADA

GEORGE H. LOCKE

LIBRARIAN, TORONTO PUBLIC LIBRARY, TORONTO, ONTARIO

ASSISTED BY ONE HUNDRED FIFTY DISTINGUISHED SCIENTISTS, EDUCATORS, ARTISTS AND LEADERS OF THOUGHT IN THE UNITED STATES AND CANADA

1920

W. F. QUARRIE & COMPANY

TORONTO CHICAGO NEW YORK

THE WORLD BOOK
ORGANIZED KNOWLEDGE
IN STORY AND PICTURE

BEECH, a forest tree, beautiful in summer because of its spreading symmetrical branches and thin, silky leaves; in autumn because of the rich gold to which its leaves turn with the frost; and in winter by reason of its smooth gray trunk and its wealth of shining brown twigs with their polished leaf buds. Europe has one species and North America another, the two differing chiefly in size, for while the American beech attains a height of from fifty to seventy-five feet, that of Europe may be 100 or 120 feet. The beech has one advantage over many other trees—its seedlings grow well in dense shade and are thus able to spring up in sufficient numbers to choke out other trees.

The flowers of the beech are of two kinds, both small and inconspicuous, and it is by those hidden near the ends of the twigs that the bur-sheathed nuts are produced.

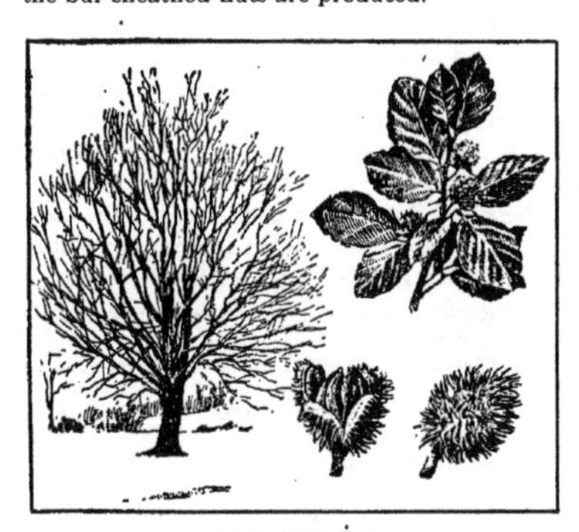

THE BEECH
The form of the tree, as seen in winter, the prominently-ribbed leaf, and beechnuts, the fruit of the tree.

The hard, brittle wood is liable soon to decay if exposed to the air, but under water it lasts well, and is accordingly much used for piles or dams. It is an excellent firewood.

BEECHER, Henry Ward (1813-1887), one of the most effective and powerful pulpit orators that the United States has produced. He was born in Litchfield, Conn., the third son of Lyman Beecher, also a great preacher, and he was the brother of Harriet Beecher Stowe (which see). As a child he was diffident and sensitive, loved the ocean and was only prevented from going to sea by his admission to the Church in 1826. He studied theology under his father's instruction in Lane Seminary, for a time was pastor of a Presbyterian church in Lawrenceburg, Ind. (1837-1839), and from 1839 to 1847 preached in Indianapolis. In 1847 he took charge of Plymouth Church, Brooklyn; he remained here until his death.

HENRY WARD BEECHER

Beecher was original in choice and treatment of subjects for his sermons, and his delivery was eloquent, dramatic, pathetic and witty. Tender-hearted and charitable himself, any form of injustice called from him bitter denunciations. He was a Republican and aided the cause of that party by pen and speech, taking a specially active part in the campaign of 1856. During the War of Secession he visited England, and there showed his wonderful power over men by controlling clamoring mobs and forcing them to listen to him. He did much to enlist British sympathy for the Northern rather than the Southern cause. During his life in Brooklyn he contributed much to the *Independent,* of which he became editor in 1861; he also edited the *Christian Union* and was a frequent contributor to the *Ledger*. Among his works are a novel entitled *Norwood; Lectures to Young Men* and *A Circuit of the Continent.*

BEEF, the world's favorite meat, derived from cattle. That it was an important article of man's diet in very early times we know from the hero-stories in the *Odyssey* and the *Iliad,* and the Bible tells us that it was occasionally eaten even before the days of Abraham. Though to-day it is a staple food in Canada and the United States and the more prosperous European countries, it is a luxury and quite rare in many others. As the great cattle ranges of the world are quite rapidly being converted into grain fields, beef may soon become too expensive for every-day consumption.

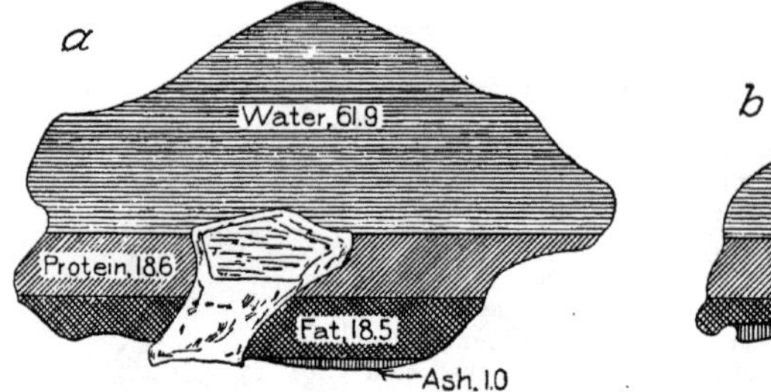

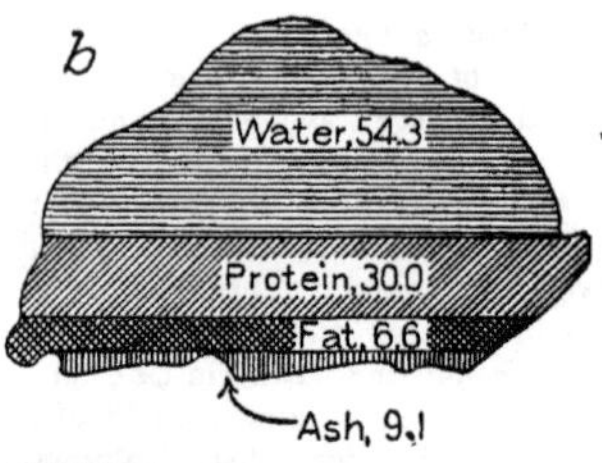

CROSS-SECTION OF A CUT OF BEEF
(*a*) Fresh beef constituents; (*b*) the composition of dried beef.

THE CUTS OF BEEF

(1) Neck	(9) Navel
(2) Chuck	(10) Loin
(3) Ribs	(11) Flank
(4) Shoulder	(12) Rump
(5) Fore shank	(13) Round
(6) Brisket	(14) Second cut, round
(7) Cross ribs	
(8) Plate	(15) Hind shank

The above cuts are so named in Bulletin 28 of the United States Department of Agriculture.

Beef is very nourishing, as the first illustration shows. It is eaten in a number of forms. *Fresh beef* must be consumed soon after it is killed, and is of course the best. *Dried beef* (or *jerked beef* as it is called in England) will keep much longer. *Corned beef* is cured by salting. *Canned beef* is eaten principally in out-of-the-way corners of the world, where fresh beef cannot be sent; it frequently consists of otherwise unmarketable scraps. *Extract of beef,* if properly made of those elements of the meat which dissolve in hot water, is a stimulant, but not very nourishing. *Beef tea,* composed of juices extracted by pressure, may be, on the other hand, quite nourishing.

Steaks, Roasts, and Other Cuts of Beef. In preparing a carcass the butcher first dresses it, then splits it along the backbone into halves, and usually into hindquarters and forequarters. The second part of the illustration shows a half-beef as you might see it hanging in the butcher shop, the lines showing where the butcher will afterward cut it. The numbers are here explained; where the use of a part is not told it is an inferior cut suitable for stews, gravies, soups, corned beef, etc.

(1) Neck. (The tongue is the only eatable part of the head.)

(2) Chuck, an inferior quality of roast or steak.

(3) Ribs. The prime roasts are from the ribs nearest the loin.

(4) Shoulder clod.

(5) Fore shank.

(6) Brisket, (7) Cross ribs, (8) Plate and (9) Navel, are most frequently eaten as corned beef.

(10) Loin, steaks. Short steak is nearest the ribs, flank steak next the flank, and rump steak next the rump. Sirloin steak is between the short steak and the rump. The tenderloin is underneath all.

(11) Flank, and (12) Rump, usually corned.

(13) Round, steak. The inside of the leg is more tender than the outside.

(14) Second cut round.

(15) Hind shank, soup bones.

For detailed description of the world's meat industry, see MEAT AND MEAT-PACKING. See, also, DIET; FOOD. C.H.H.

BEELZEBUB, *be el' ze bub,* a word meaning *the god of flies,* is the name by which the Philistines of Ekron worshipped their chief god. The word was originally written Baalzebub and probably referred to Baal (which see). As used in the New Testament, Beelzebub means the worst of the evil spirits. Jesus said, "And if I by Beelzebub cast out devils, by whom do your children cast them out?" (*Matt.* XII, 27.)

BEER, a malt liquor used as a beverage. The liquor may be made from nearly all kinds of mealy grains that will ferment. However, barley is the one most used. The process of making beer is fully described in these volumes under the heading BREWING.

Kinds of Beer. As beer is a name applied to all malt liquors, it includes several different kinds. The one most generally used in Germany and the United States is *lager beer.* This is a light beer which is stored for a time before being used, in order to allow it to ripen and mellow. *Ale* is a carefully-made kind of beer much used in England; it is stronger than lager and lighter in color. *Porter* is a dark, strong English beer, while *stout* is a strong porter. Beer is sometimes brewed from other substances, as the names *ginger beer, root beer,* etc., indicate.

History of Beer. There is authority for saying that beer was brewed and drunk in very ancient times. The Egyptians must have been familiar with it several thousand years ago, as the process of making it is shown in detail on their wonderful monuments. The Greeks obtained their knowledge of it from the Egyptians, and from them it spread to other peoples along the Mediterranean. When the Romans invaded Britain about the time of the Christian Era, they found the inhabitants brewing a sort of ale from barley. In the thirteenth century the Normans had laws regulating its sale. To-day large quantities are drunk all over the civilized world. Perhaps the people of Germany and the United States are the greatest beer drinkers. Nearly two billion gallons were consumed in the latter country in the year 1916. See ALCOHOLIC DRINKS. J.F.S.

BEERSHEBA, *be ur she' bah,* a city on the southern boundary of the Promised Land of the Hebrews. Since Dan was on the northern boundary, from Dan to Beersheba was a common form of expressing the extent of the country. The expression is still in common use, but with an entirely different application. "From Dan to Beersheba" is now a sort of slang phrase meaning within the outermost limitations, meant to exaggerate an effort that has been or must be made. The ancient city is now but a heap of ruins. The word means *well of the oath,* so named because there Abraham made a covenant with Abimelech.

BEESWAX, *beez' wax,* a solid, fatty substance secreted by bees and obtained from the honeycomb. Before swarming bees eat honey freely, for it is needed to produce wax for comb-building. For every pound of wax produced ten to twenty pounds of honey must be consumed. The process by which the wax is prepared for commercial use is not generally understood. The comb is boiled, and the melted wax, rising to the surface, is dipped off. On cooling, it becomes solid. As thus obtained, beeswax is of a dark yellow or brownish color, containing numerous impurities, which can be removed by melting and filtering. Most of the beeswax placed upon the market is bleached. It is used in small quantities by seamstresses, and by housewives for their flatirons, also in the manufacture of candles, the preparation of ointments and cements and in modeling. See BEE; WAX.

BEET, a garden vegetable of which the entire plant may be eaten, both roots and leaves. Originally only the leaves were eaten, as "greens," but gradually cultivation increased the size and sweetness of the roots until now

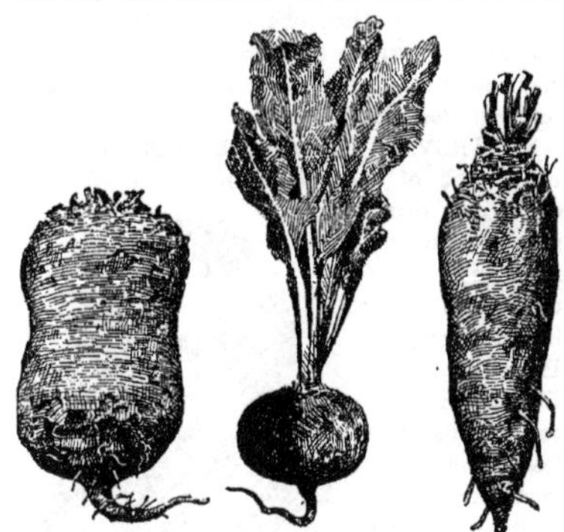

THREE VARIETIES OF BEETS
Mangel, common beet and sugar beet.

it is rather as a root vegetable than as a leaf vegetable that it is grown. The edible parts, usually turnip-shaped, vary in color from white to dark red, the red variety being preferred, and they are prepared for the table by several hours' boiling. They are then served hot, with butter, or are pickled or used as a salad. Beets have a high water percentage and a correspondingly low food value, though the starch and sugar which they contain make them more valuable than cabbage, for instance.

During recent years renewed attention has been given to cultivating and refining the leafy tops of beets, and the excellent Swiss chard has resulted. The coarse varieties of beets, known as mangels, are extensively grown as food for cattle. See CHARD.

Sugar Beets. The most important use of beets, however, is in the manufacture of sugar, about three-fifths of all the sugar produced in the world coming from this source. Germany, Austria, Russia and France are the leading countries in the industry, but the cultivation of the sugar beet is rapidly spreading in the United States. This beet closely resembles the varieties ordinarily raised in gardens, except that it is usually of light color; and it thrives best in a cool temperate climate, having a reasonable supply of moisture. It has been successfully raised in California, Michigan and Utah, but Colorado is the leading state in its production; while in Canada, Ontario far surpasses any other province. In the United States the annual production is about 4,000,000 tons, valued at close to $20,000,000, and in Canada about 148,000 tons, valued at $906,000. Practically all of this crop is made into sugar, not more than one or two per cent being used as forage; and one ton of beets will make about 240 pounds of sugar. See SUGAR, subhead *Beet Sugar*.

BEETHOVEN, *ba' to ven,* LUDWIG VON (1770-1827), one of the most notable of German musical composers, and the supreme master of modern music for the orchestra. He was born at Bonn, and received his first lessons in music from his father, a tenor singer, who is said to have beaten music into the boy. Under teachers of great name and ability he made rapid progress, and his talent was so marked that he was sent in 1792 to Vienna to study under Hadyn.

BEETHOVEN

In or near that city the greater part of his life was spent, and there he composed the works that have given him immortal fame. He was a great pianist as well as a great composer, winning recognition for his playing before his age fully appreciated his genius in musical composition. The latter, indeed, was the result of long years of patient, careful study and perseverance.

The life of the great musician was not one of the happiest. He had probably the greatest misfortune to bear that can come to a composer, the loss of hearing. Yet he bore this affliction with admirable courage, and, though totally deaf before reaching middle age, he continued to give to the world the benefits of his genius. His compositions, begun when he was eleven years of age, consist of nine symphonies, a number of concertos, sonatas for the piano and for the piano and violin, the opera *Fidelio,* an oratorio, *The Mount of Olives,* string quartettes, masses and songs. With him the development of orchestra music reached its highest point; he also surpassed any of his predecessors in giving warmth and poetic color to the sonata and the symphony. The latter form he brought almost to perfection, and his *Fifth* and *Ninth* symphonies are among the most beautiful works ever composed. His own statement, "Music is the only spiritual entrance to a higher world of knowledge," expressed the high ideals he held of his art.

BEETHOVEN'S BIRTHPLACE AT BONN

The Moonlight Sonata. One evening as Beethoven and a companion were hurrying through the streets of Bonn they heard, through the open window of a humble dwelling, the lovely strains of the *Sonata in F.* As they stopped to listen there came a cry of despair, "Oh! If I could but hear some really good musician play this wonderful piece." Entering the cottage, they found the player to be a young blind girl. Beethoven sat down at the old piano and played as if inspired, but the blind girl, who listened with awed delight, knew not that it was the great Beethoven until he struck the opening bars of the *Sonata in F.* Suddenly the candle flickered and went out and the music ceased. Beethoven's friend stepped to the window and opened the shutters, and as the beautiful moonlight flooded the room he began to improvise a tender melody that seemed the embodiment of the silvery light that transfigured everything it touched. The great composer then hastened home to put the music into form. He named it the *Sonata in C Sharp Minor*, but lovers of its exquisite melody will always call it the *Moonlight Sonata.* B.M.W.

BEETLE, *bee' t'l,* the common name of the largest order of insects in the world, of which no fewer than 150,000 species have been studied and described, over 12,000 of them in America, north of Mexico. The name *beetle*

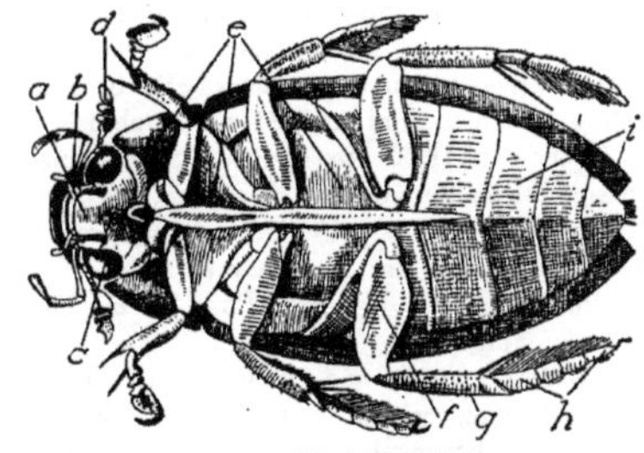

PARTS OF A BEETLE

(*a*) Mouth parts
(*b*) Head
(*c*) Compound eye
(*d*) Antenna
(*e*) First, second, third thorax
(*f*) Femur
(*g*) Tibia
(*h*) Tarsal segments
(*i*) Abdomen

means *biter,* and has reference to the strong mouth parts, which are well adapted to seizing and tearing their prey, but it is the scientific name *Coleoptera,* meaning *sheath-wings,* which indicates their distinctive characteristic. Some beetles are so tiny that they can scarcely be seen, and some are startling-looking creatures as much as four inches in length; some are very flat and others almost as round as balls; some are long and slim, others circular; and there are the greatest variations in color, from sober black or brown to brilliant reds and greens and metallic blues. All are distinguished, however, by the horny forewings which cover and hide the membranous hind wings when the insect is at rest, and protect the soft back. The under part of the body, too, has a thick, horny armor, and there is no doubt that this "shell," easily crushed though it is, has had much to do with making the beetles the most numerous of insects; for the stiff and prickly substance is not especially agreeable to birds, which would otherwise greedily devour them.

Life History. No other insects have been studied as widely and as carefully as have the adult beetles, and they have always been favorite objects with collectors because their hard outer covering makes it possible to preserve them in perfect shape with very little trouble; but the earlier stages of beetle-life are comparatively little known. It is certain, however, that all beetles have what is known as a "complete metamorphosis"—that is, there are four stages in their development. First comes the egg laid by the female in such a location that the newly-hatched young will have food ready at hand. The next form is the *larva,* usually a disgusting-looking, soft, white grub, which differs from the caterpillar, or butterfly larva, in that it lives for three or four years in that state, before it becomes a dormant *pupa.* Either in a rude cocoon of its own making or in a burrow, the pupa rests quietly for a period which varies from two to three weeks to as many years, and when it emerges it is a full-grown beetle.

Habits. Few places on the earth are free from beetles, which live in the water, either salt or fresh, and on the land, underground and at the surface, while some species live as parasites on other animals. Of course there are modifications of structure to meet all these different modes of life, as well as to adapt the beetles for securing the particular food which they prefer. Some live on vegetable matter, and certain of these species, both as larvae and as full-grown beetles are very harmful to crops (see POTATO BUG). Others live on animal food, some of them catching insects alive, others preferring decaying matter, and these latter, in their office of scavengers, render valuable service to man. Two of the most helpful

Outline and Questions on the Beetle

I. Variety of Species

(1) Most numerous of insects; 150,000 species
(2) Over 12,000 species in North America
(3) Various sizes, shapes and colors

II. Distinctive Characteristics

(1) Strong mouth parts
(2) Sheath-wings
(3) Protective, horny covering

III. Life History

(1) Complete metamorphosis
  (a) Egg
  (b) Larva
  (c) Pupa
  (d) Fully developed beetle
(2) Habits
  (a) Food
  (b) Harmful beetles
  (c) Helpful beetles

Questions

Why do young collectors find beetles attractive objects of study?

Why should children be taught not to kill ladybirds?

What characteristic of beetles is shown in their common name? In their scientific name?

Sketch the life history of a beetle.

Why do not birds find beetles desirable food?

What shows the adaptability of these insects?

What is meant by a "complete metamorphosis"?

What European beetle has been brought to America and carefully cultivated here? Why?

About how many species are there in the United States and Canada?

What great difference is there between the larva existence of a beetle and that of a butterfly?

Are beetles on the whole harmful or helpful?

How long does the pupa stay in its cocoon?

Name four important species of beetle.

With how many of those listed here are you familiar?

Do all species of beetles eat the same kind of food?

In what country was the scarab looked upon as sacred?

What peculiar habit has given the burying beetle its name?

beetles are the ladybird, which feeds upon plant lice and thus saves many a garden plant from destruction, and the European Calosoma beetle, which both as larva and as winged adult preys on the destructive larvae of the gypsy moth. Into certain sections of North America which are particularly troubled with gypsy moths this European beetle has been introduced, and has proved a most effective preventive measure. On the whole, however, beetles do more harm than good. M.S.

Related Subjects. The best-known species of beetles are treated in these volumes under the following headings:

Blister Beetle
Burying Beetle
Carpet Beetle
Click Beetle
Cockchafer
Deathwatch
Grain Beetle
Hercules Beetle
June Bug
Ladybird
Potato Bug
Scarab
Stag Beetle
Water Beetle
Weevil

BEGIN, *bay' zhaN,* LOUIS NAZAIRE, Cardinal (1840-), a Canadian Roman Catholic prelate, archbishop of Quebec, and created cardinal on May 25, 1914, at the last consistory held by Pope Pius X. Cardinal Begin was born at Lewis, Que., attended the seminary of Quebec, was graduated from Laval University in 1862, and after two years of study at the French Seminary in Rome was ordained a priest in 1865. He was for sixteen years professor of theology at Laval University, and then for four years was the head of the university's normal school. He became bishop of Chicoutimi in 1881 and archbishop of Quebec in 1898. In addition to the distinction which his position gives him, Cardinal Begin is known as a writer on religious topics. Among his books are *The Rule of Faith, The Infallibility of the Sovereign Pontiffs* and *The Catholic Faith.*

BEGONIA, *be go' ni a,* a group of juicy-stemmed herbs which in temperate climates are favorite house plants. Their fleshy, waxy leaves, oddly one-sided, sometimes display beautiful shades of color, and the flowers are often showy. They have a waxy look, also, and those of the common varieties range in color from pink to deep red, one flower often showing several shades. The carefully cultivated kinds grown in greenhouses sometimes have wonderfully developed flowers, from four to six inches in length and of the deepest orange color. One advantage of begonias as house plants is that they are easily grown from "slips," either leaf or stem cuttings tak-

Brown False Long-horn

Stag Beetle

Red Milkweed Beetle

Milkweed Leaf Beetle

Goldsmith Leaf Beetle

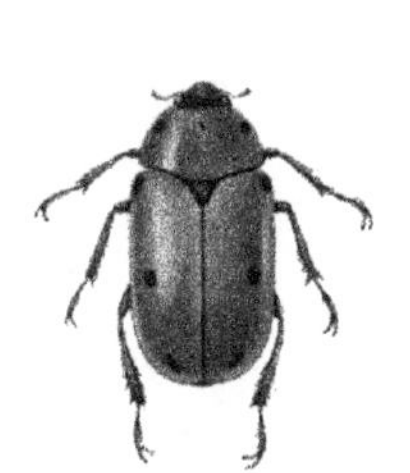

Spotted Vine-chafer

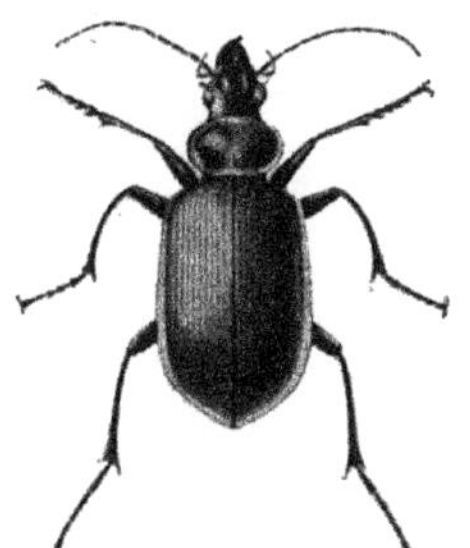

The Searcher

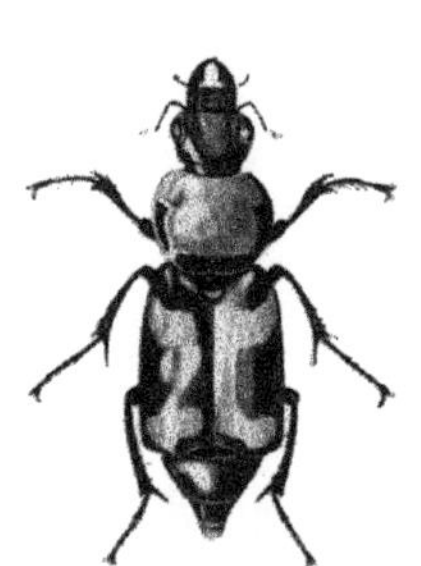

Burying Beetle

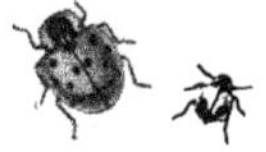

Ladybird Beetles

Bombardier Beetle

Helmet Beetle

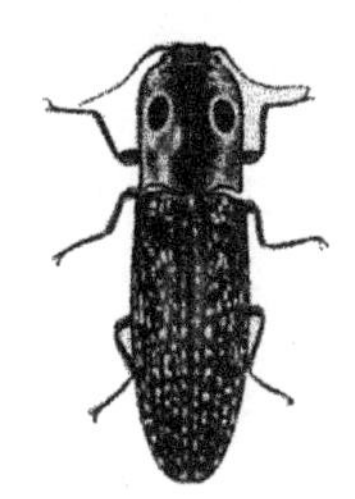

Eyed Elater

Purple Long-horn

Carl F. Gronemann

ing root without difficulty and producing vigorous new plants.

BEGONIAS

BEIRUT, or **BEYROUT,** *ba' root,* has surpassed all other cities in Asiatic Turkey in commercial and industrial growth. It is the capital of a province of the same name and the chief seaport of Syria, and is located sixty miles northwest of Damascus, at the base of the mountains of Lebanon. The term *briar root,* applied to pipes, is generally supposed to be a corruption of *Beirut,* which at one time exported great quantities of wood suitable for pipe making. Its chief exports now are olive oil, cereals, sesame, tobacco and wood; its manufactures are silk and cotton, and articles of gold and silver. In ancient times Beirut was a large and important Phoenician city, but for centuries the city was passed from ruler to ruler until 1763, when the Turks took possession. It was bombarded and taken by the British in 1840, but was again restored to Turkey. Population, 150,000.

BELASCO, *be las' ko,* DAVID (1862-), one of the best-known American theatrical managers of modern times, and a playwright of some distinction. He was born in San Francisco, and in that city began his career as an actor at the age of fourteen. Finding that he had rather unusual talent in adapting plays for the stage, he turned his attention to dramatic writing, and throughout a long period has won success in three fields—as a playwright, theatrical manager and producer E. H. Sothern, Blanche Bates, Mrs Leslie Carter and David Warfield are among the famous people of the stage who have been under his management, and the Republic and several Belasco theaters of New York City are important playhouses that passed into his control during a period of less than a dozen years.

The name of Belasco is especially associated with stage realism, for it is his belief that a stage setting should correspond in the smallest details with the scene which is being represented, whether it be a restaurant, an opium den, a second-hand clothing shop or a drawing room. In the spring of 1915 he joined with Charles Frohman in the production of an old success, *The Celebrated Case,* several of the best players of the world being engaged for the revival of this play. Plans for future productions of a similar strong character were interrupted by the death of Mr. Frohman on the *Lusitania,* the great ship that was torpedoed and destroyed by a German submarine in the War of the Nations, in June, 1915. Among the successful plays written by Belasco are *The Girl I Left Behind Me* (with Franklin Fyles), *The Heart of Maryland, Zaza, Naughty Anthony* and *The Girl of the Golden West.* The latter was one of his most successful efforts.

BELEM, *ba leN',* a name sometimes used for Pará, a city of Brazil, under which title it is described in these volumes.

BELFAST, IRELAND, the center of the linen industries, is the capital of the province of Ulster, on the borders of the counties of Down and Antrim. It is the first city in Ireland in population, manufactures and trade, because it is advantageously situated on Belfast Lough where the River Layan flows into the lake. It is 113 miles north of Dublin, with a harbor which is one of the best in the United Kingdom. Although there are some unsightly slums, the city is for the most part well laid out. The town hall is a notable building, erected in 1906 at a cost of $1,500,000. The ship-building industry is steadily increasing in importance, and the city has manufactures of machinery, distilleries, flour mills, tan yards, chemical works and more power looms and spinning mills than are contained in all the rest of Ireland. The linen made in Belfast has been a superior article for many years.

Belfast has been the center of resistance to Home Rule for Ireland, and clashes between political and religious factions have caused frequent riots, sometimes resulting in bloodshed. Population in 1911, 386,947. See LINEN; HOME RULE.

BELGIUM, one of the smallest countries in Europe, but one that has stood in the very front rank among the states of the world by its achievements in all the arts of peace, progress and civilization. Belgium attained this position in a very short time, for not until 1830 did it become an independent country. But the Belgian *nation* is not a new nation. The people that inhabit this country have a long history and have played a prominent part in the development of the civilization of Europe. Caesar, in commenting on the valor of his various enemies, gave first place to the Belgae (Belgians). Its noblemen, such as Godfrey of Bouillon and Baldwin of Flanders, were leaders of several of the Crusades (which see). The first industrial activity in Northern Europe was developed in the cities of this country. The guild system of the medieval towns originated first in Flanders, and was introduced afterwards into England, France and Germany. Belgium had long been famous for its historic cities, its beautiful medieval buildings and its priceless art treasures. Its art, the expression of the ideals of the people, is indeed unique.

After a successful revolution in 1830 Belgium was recognized as an independent country by the Great Powers. The extent and boundaries of the country to 1914 were determined by the Treaty of London, signed April 19, 1839, in which the Great Powers, namely, Great Britain, France, Prussia, Austria, and Russia guaranteed the independence and neutrality of Belgium. Before the Franco-German War of 1870, both France and Prussia, at the request of England, pledged themselves to respect this neutrality.

Since Belgium gained its independence it has developed an extensive manufacturing industry and an enormous trade. From being at that time a poverty-stricken nation, it developed into one of the lending nations of the world, joining in that capacity such old banker-nations as the English, French and Dutch. For many years its surplus capital was invested in enterprises all over the world.

The progress of the country had been steady and continuous until the fateful day in August, 1914, when the War of the Nations broke out. Suddenly Belgium, that had no part in any European quarrel and had absolutely nothing to gain by taking up arms, was confronted with the choice of either granting the demand of Germany to pass through the country, as an easy way of invading France, or of protecting its independence and neutrality. Although a small nation, Belgium, true to the lofty political idealism and the spiritual traditions of its people, chose the latter course.

Under the guidance of its ruler, King Albert, Belgium decided to fight in defence of its honor, its historic heritage of freedom, and the right to be mistress in its own house (see ALBERT). But the country had to pay a heavy price for this decision, for it became once more the battle-ground of Europe, the early scene of the greatest and most destructive war in human history. Its beautiful country has been devastated, its villages razed to the ground, many of its historic monuments and art treasures destroyed, its ancient seats of learning rendered desolate, and its people driven from their homes to seek refuge in foreign lands. No greater tragedy has ever

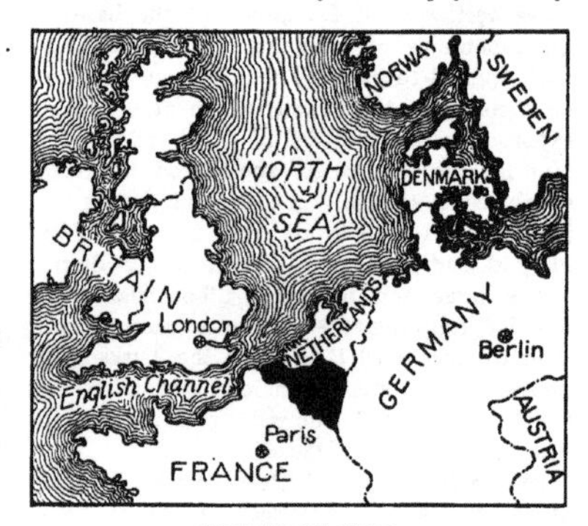

LOCATION MAP

Windmill near Brussels

Belgian milk-cart

befallen a nation. And just as we value the portrait of a man or woman taken at some critical period of life, so it would be of special interest to have the picture of this nation on the eve of the greatest crisis in its existence.

Location. Belgium, together with Holland, lies across the central plain of Europe near its narrowest point. It has a sea coast on the North Sea of forty-two miles, and the land frontiers measure 793 miles. Of this distance 384 miles lie on the border of France, 269 miles join Holland, eighty miles border the grand duchy of Luxemburg, and sixty miles adjoin Germany. Geographical situation has shaped the character and destiny of the people of Belgium more than those of any other people in the world. For trade purposes it forms the natural meeting ground of the nations of Western Europe, but, unfortunately, on account of its position, it has been again and again their battle ground. Belgium has been called "the cockpit of Europe," and history has shown that any conflicts, in which the leadership of the continent was sought, whether by Spain or France, by the Hapsburgs or the Hohenzollerns, were fought on its territory. Among the great contests fought on its soil prior to the twentieth century may be mentioned the famous and decisive battles of Neerwinden (1693), Fleurus (1690), Ramillies (1706), Oudenarde (1708), Fontenoy (1745), Jemappes (1792), Waterloo (1815). For recent battles see WAR OF THE NATIONS.

Area and Population. Belgium, with an area of 11,373 square miles and a population of 7,423,784 inhabitants in 1910, is the most

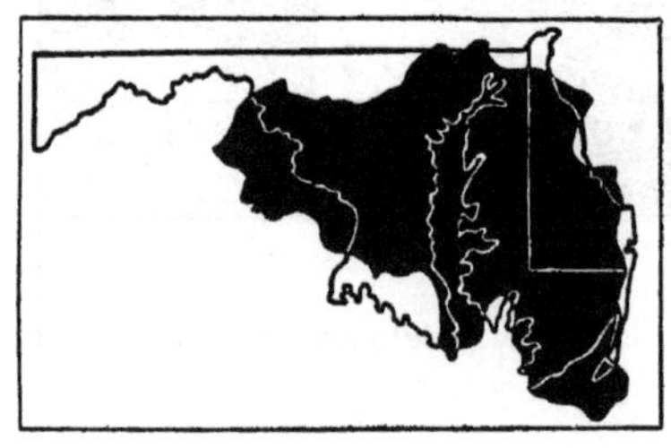

COMPARATIVE AREAS
Belgium compared with Maryland and Delaware.

densely-populated country in Europe. It had in the year named about the same population as Pennsylvania, which, however, has an area nearly four times as large. Belgium contained, before the War of the Nations, 653 persons to the square mile, and is therefore credited with an average density of population twenty-one times larger than that of the United States, where the average density is 30.9 persons to the square mile. The most densely-peopled provinces are Brabant and East Flanders, where the average density is 1,158 and 957 persons to the square mile, respectively. The neighboring state of Holland has an area eleven per cent larger and a population twenty-one per cent smaller. The treaty which closed the War of the Nations gave to Belgium 382 square miles more territory, in two small districts between Holland and Luxembourg. Its area is now 11,755 square miles.

Inhabitants and Languages. The inhabitants of Belgium belong to two different races, known as Flemings and Walloons, and they speak two different languages. The Flemings, occupying the northern part, are of Teutonic origin and speak Flemish, which is a Dutch dialect. The Walloons occupy the southern part and speak French, or rather Walloon, a dialect of French. According to the language spoken, the population in 1910 was divided as follows: Flemish, 3,220,662; French, 2,833,324; French and Flemish, 871,299; French and German, 74,993; German only, 31,435; those who speak all the three languages, 52,547. The fact that the people are of two different ethnical stocks and speak two different languages has had an important bearing upon the political problems that have agitated Belgium since 1831, and Germany, after its occupancy of the country in 1914, took all possible political advantage of this condition. French was the official language until 1870, when Flemish, as the result of a vigorous agitation on the part of the Flemings, was put on a footing of equality with French. Belgium is now a bi-lingual (two-language) country, and nearly all the cities and provinces have two names, one Flemish and the other French.

Religion. The constitution of Belgium guarantees full religious freedom. With the exception of about 30,000 Protestants, mostly foreigners, and about 15,000 Jews, the people are Roman Catholics. There is no state Church, but the state grants a small contribution towards the pay of the clergy of all denominations. Belgium is divided into six dioceses, presided over by the archbishop of Malines, who is the primate of Belgium, and the bishops of Liége, Namur, Tournai, Bruges and Ghent. The archbishop of Malines, the old and venerable Cardinal Mercier, pluckily defended the rights of his flock following the dark days of the German invasion in 1914.

Physical Features. The surface of the country resembles an inclined plane, and the land rises by a succession of stages from the sea coast to the low mountains of the Ardennes on the southeast. Looking inwards from the sea across Belgium, there is seen a narrow belt of dunes on the coast, then a strip of reclaimed territory, almost level with the sea and protected by dykes, known as *polders*. Following this comes a broad central region, composed of sandy soil and extending from west to east almost across Belgium. The eastern portion, including parts of the provinces of Antwerp and Limburg, consists of sterile heaths and arid wastes, and is known as the *Campine*. The western portion is the celebrated Flemish plain, which is described below. The surface then rises gradually until the valleys of the Meuse and Sambre are reached. This region contains one of the best military routes leading from Holland and Germany into

France, and here were situated the fortresses and important strategic points of Liége, Namur, Mons and others, that have been the object of numerous and famous sieges. South of these rivers extends the plateau of Ardennes, reaching an average altitude of 1,400 feet and containing fine and extensive woods, mostly of beech.

The Flemish Plain. Its Historic Cities. The most interesting part of Belgium is the Flemish plain, that represents to-day a wonderful triumph of human industry over nature. The sandy soil is nearly barren, yet the incessant labor of its inhabitants has transformed it into one of the most populous, best cultivated and most productive areas in the whole world. The Flemish plain has played an important rôle in the civilization of Europe, and here are situated many old cities with their relics of ancient fame and their records of past greatness. The old city walls are usually gone, but the town halls, the guild halls and the belfry, the possession of which was for a medieval town the greatest of chartered privileges, remain. Some of these cities are but the ghosts of their former greatness. But the belfry still rises in their midst, chiming the hours and quarters over streets that are now quiet and pigeon-haunted, but that were formerly filled with bustling crowds surging to the sound of its tocsin. Such a city is Bruges, the belfry of which is immortalized by Longfellow in the well-known lines:

> In the market place of Bruges
> Stands the belfry old and brown;
> Thrice consumed and thrice rebuilded,
> Still it watches o'er the town.

Such cities also are Malines, Termonde, Ypres. The famous hall of the clothworkers' guild at Ypres, built early in the thirteenth century, was one of the finest examples of Gothic architecture at its best. But it was something more; it was the earliest and noblest piece of architecture designed for a civic purpose in Belgium, and perhaps in Europe. It was here that the guilds were first organized (see GUILD). The Germans burned this building in 1914. In these cities of Belgium the first free municipal life was established, and here the first industrial activity in the north of Europe was developed. In other cases modern industrial revival has been effected upon the ancient sites. Such cities are Ghent and Antwerp; and, in other parts of Belgium, Brussels and Liége.

Rivers. Belgium is watered by the Scheldt with its chief tributary, the Lys, in the north, and by the Meuse and its chief tributary, the Sambre, in the south. All these rivers rise in France, while the main rivers, the Scheldt and the Meuse, have to pass through Holland before reaching the sea. This fact has greatly retarded the development of Antwerp and of other towns. In West Flanders, the river Yser, which receives the little Yperlee, runs into the sea at Nieupoort. Besides these there are other small streams flowing in all parts of the country, so that Belgium is exceedingly well supplied with rivers.

Agriculture. Belgium is typically, but not wholly, a country of very small farms, especially when compared with the United States and Canada. But the intensive system of agriculture has attained a high degree of perfection, and the production per acre, regardless of the fact of a thousand years of cultivation, is the highest in Europe. Of its total area of 7,369,000 acres, 4,340,000 acres were under cultivation, 1,303,000 acres were forests and 476,000 acres were fallow and uncultivated, in 1913. Belgium yields per acre more oats, barley, potatoes, flax and tobacco than any of its neighbors, Great Britain, France, Germany, Holland or Denmark. Except Denmark and Holland none of the above countries has as many cattle and pigs to the square mile. In proportion to its size it feeds a greater number of persons than any of its neighbors. All this agricultural wealth is raised on small farms ranging from one and one-half to twelve and fourteen acres in size, which are not owned by the men who till them, but are rented at a high price. For this reason the Belgian agricultural classes do not enjoy any such prosperity as falls to the lot of the peasant proprietors of Holland and Denmark. Of later years the inhabitants, with their accustomed painstaking industry, have reclaimed for agricultural purposes tens of thousands of acres in the sandy Campine district, while fertile pasture lands now replace the old marshes in the northwestern part of the country.

Mining. The mineral wealth of Belgium is very great; size for size, it is greater than that of any other country in the world. The chief products are coal, zinc, lead and iron. The coal fields cover an area of over 550 square miles and extend across the country along the rivers Sambre and Meuse. The chief coal-bearing districts are around Mons, Charleroi and Liége. A new and extensive coal-field has been discovered in the Campine district, and has been worked since 1906. The output of coal, which is of excellent quality, averages about 23,000,000 tons a year, valued at $76,000,000. Other minerals are zinc, with a production of 206,000 tons, valued at $27,000,000; lead, with a production of 50,000 tons, valued at $9,000,000; and silver from lead, with a production of 580,000 pounds, valued at $6,500,000. These figures represent production in normal times; by comparing them with the similar products in the United States we are better able to realize their magnitude. In the United States the yearly production of zinc averages 337,000 tons; of lead, 436,000 tons; and of silver, 70,000,000 ounces. The metals are found chiefly in the Ardennes region.

Numerous quarries of marble, granite and slate are worked, and their products reach an average value of $12,500,000 yearly.

Manufactures. Belgium is normally one of the greatest manufacturing countries in Europe. Coal and iron are the bases of modern industry and, as we have seen, Belgium possesses them in abundance. The iron and steel manufactures are on a large scale. The production of pig iron in 1913 amounted to 2,500,000 tons; of manufactured iron, to 335,000 tons; of steel ingots, to 2,500,000 tons; and of steel rails, to 1,903,000 tons. When one notes that the United States produces about 30,000,000 tons of pig iron, a better idea of the enormous production of this little country is gained. Machinery of all kinds, firearms, tinware, nails, wire, brass, are some of the chief metal manufactures, and these are produced at Liége, Malines, Namur, Ghent, Charleroi, Mons and Brussels. Brussels and Ghent are the centers of an important jewelry manufacture.

Flanders, the center of the flax industry, has for centuries been noted for the superior quality of its linens. The manufacture of woolen and cotton goods is very extensive. It is carried on in all parts of the country, but the chief seats of the industry are at Liége, Verviers and Ghent. Lace is one of the most widely known of Belgian manufactures. Much of this is made by hand in the homes of the people, and the same can be said about weaving and the manufacture of gloves. One of the characteristics of the industry of Belgium is the great number of articles made in the homes of the people or in small shops, which employ only a few hands. These subsist side by side with large establishments where the latest machinery is used.

Belgium is one of the leading glass manufacturing countries in the world. Porcelain and other varieties of pottery ware of high grade are manufactured at Tournai, Mons, Ghent and Brussels. Belgium has also become of late years one of the principal beet sugar manufacturing countries of Europe. Other manufactured products are leather, chemicals, paper, beer, tobacco and furniture.

Commerce. This little kingdom carries on in normal times an immense foreign trade. In fact, only the United States, Great Britain, Germany, France and Holland export and import greater totals. Belgium nearly quadrupled its trade between 1880 and 1914; its trade in 1913 was twice as large as that in 1900. The total value of the trade of Belgium in 1912 was 2,270 million dollars, of which nearly 1,000 million dollars were imports for home consumption, 600 million dollars exports of Belgian produce; 470 million dollars represented transit trade. In order to realize the magnitude of these figures we will again compare them with those of the United States, which were 1,788 million dollars imports and 2,071 million dollars exports in 1914. Germany, France, Great Britain, Holland and the United States, in the order named, are the leading countries connected with the foreign trade of Belgium. During the fiscal year 1913-1914 the United States exported to Belgium merchandise worth $61,219,000, and imported from that country merchandise worth $41,000,000.

Transportation. Belgium has an unrivaled system of transport facilities, consisting of railways, navigable rivers, canals and good roads. With its 5,400 miles of railroads, Belgium has a greater railway mileage compared to its size than any other country. The railways are owned by the government, and the fares are the cheapest in the world. There are in Belgium 1,240 miles of navigable waterways, rivers and canals, which reach all the principal towns of the country, and are also connected with the canal and river systems of Holland, Germany and France. The Flemish plain especially is intersected by a network of canals.

The chief seaports are Antwerp and Ostend. Nearly seventeen million tons of shipping entered and left Antwerp yearly, before 1914. Good roads, most of them macadamized or paved with stone, traverse all parts of the country.

Education. Official elementary education is created and organized by the communes (see COMMUNE); there must be at least one school in every locality. Elementary education, which has been mostly in the hands of the Church, has been almost as backward as in Spain or

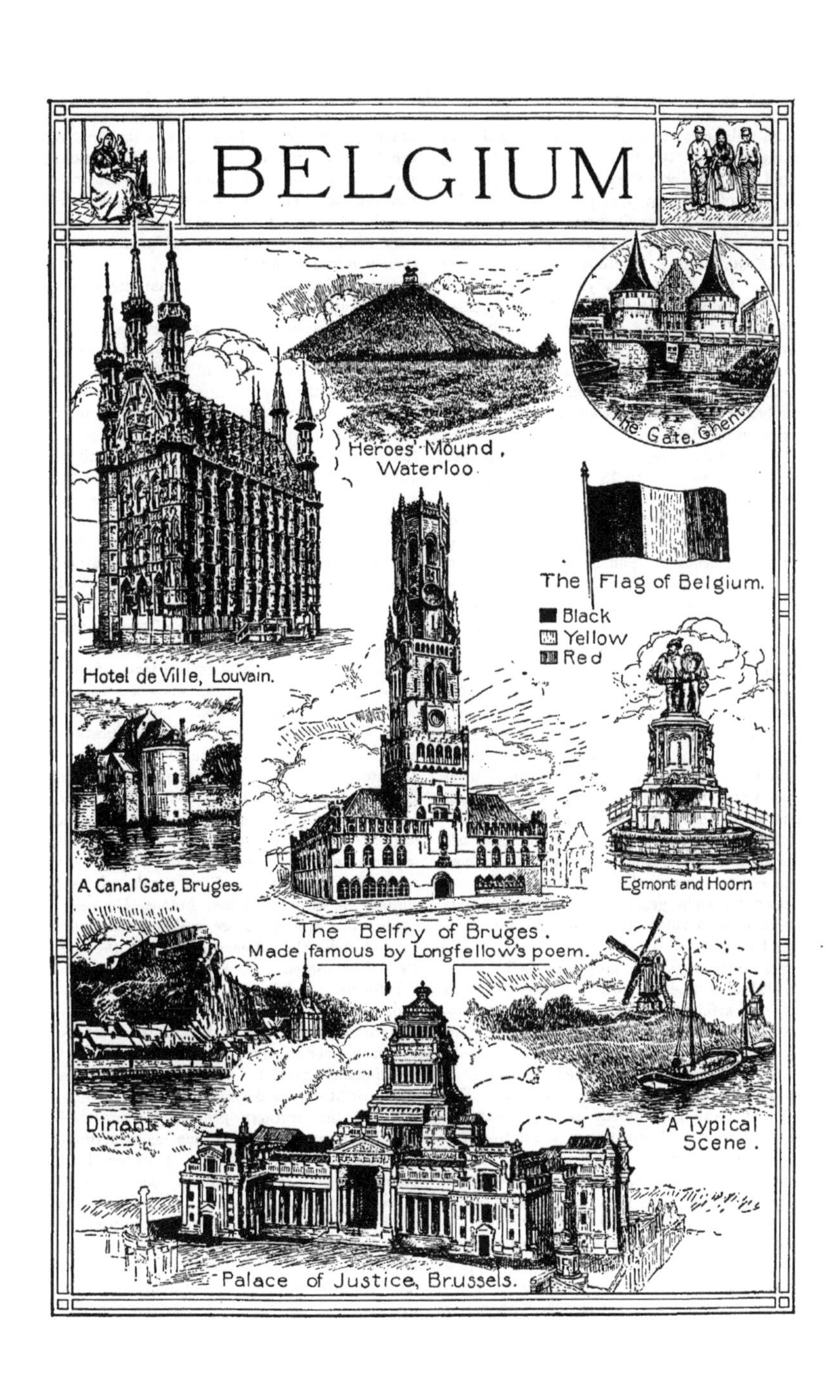
BELGIUM
Heroes' Mound, Waterloo.
The Gate, Ghent
Hotel de Ville, Louvain.
The Flag of Belgium.
Black
Yellow
Red
A Canal Gate, Bruges.
Egmont and Hoorn
The Belfry of Bruges.
Made famous by Longfellow's poem.
Dinant
A Typical Scene.
Palace of Justice, Brussels.

Italy. Illiteracy, although still great, is diminishing rapidly; the number of persons who could not read or write formed over thirteen per cent of the total population in 1910. The struggle of the Roman Catholic clergy to obtain control over the elementary schools has been one of the burning questions in the politics of Belgium for a great many years. Higher education is provided by the state universities at Ghent and Liége, and the free universities at Brussels and Louvain; each of these contains good technical schools for arts, manufactures, engineering and mines. Belgium has a good number of high schools, as well as a great number of special schools.

Government. Belgium is governed under the constitution adopted in 1831, which has justly been admired. This constitution has been taken as a model by several European nations that have gained their independence during the nineteenth century. The Belgian constitution guarantees to the citizen equality before the law, security of person and property, freedom of the press, religious freedom and the right of association, all liberal principles that many European nations gained much later than 1831.

Belgium is a constitutional monarchy, the crown being hereditary in the direct male line; the king must be a Roman Catholic. The executive power is vested in the king, assisted by a Cabinet appointed by him, which must possess the confidence of the legislative chambers, which consist of a Senate and a Chamber of Deputies. The deputies, 186 in 1913, are elected directly by the people for four years. The senate consists of 120 members, who are elected for eight years. All registered voters are obliged to vote; failure to vote is punishable by law.

For administrative purposes Belgium is divided into nine provinces. These provinces are not artificial divisions, like the departments of France, but are based to a great extent on historical grounds. They are administered by governors appointed by the king, assisted by elected councils. But the chief unit of local government is the *commune* (which see). Each commune is administered by an elective council and a mayor. In a country like Belgium, where the communal life was developed early and where the glorious traditions of the towns are so cherished, the power and prestige of the mayor is very great. Very often in national crises the joint appeal of the mayors of the chief cities has exercised a greater influence upon the people than any other agency. When Belgium was invaded by the Germans in 1914 the mayors of the cities showed that the trust placed upon them was not mislaid. They all made a plucky fight for the defense of the rights of the people, very often under conditions of great risk to their personal safety.

Army. Belgium had only a small army, that was recruited by voluntary enlistment until 1909, when a light form of conscription was introduced. The peace strength of the army in 1913 was 44,000 men and 3,540 officers. In addition to the regular army there is a gendarmerie of about 3,800 men, and a voluntary civil guard of about 180,000. It was estimated that the army in war time would reach about 350,000 men, but there were only about 100,000 available for the national defense at the beginning of the War of the Nations.

Belgium possessed several fortresses that were considered among the strongest in Europe. Among them were Liége and Namur, which guarded the passage of the Meuse at important strategic points. But the pivot of the national defense was Antwerp, with its great fortified camp and its number of detached forts. Before the War of the Nations, Antwerp was considered an impregnable fortress, but all former ideas about the value and use of fortifications have since been greatly altered.

History. From the time of the ancient Belgae, whom Caesar in his *Gallic Wars* called his most valiant foes, and from whom Belgium took its name, the territory comprising the modern country shared the fortunes of the other provinces of the Netherlands. It was subject in turn to Rome, France, Germany, Spain and Austria; these earlier periods are treated under the subhead *History* in the article NETHERLANDS.

Differences did exist, however, between the people of the northern and those of the southern part of the Netherlands, and these tended to become constantly more and more marked. The inhabitants of the southern country were an agricultural and manufacturing people; those of the northern primarily commercial. The doctrines of the Reformation penetrated the entire Netherlands, but they found no permanent hold in the southern country, which remained true to the Catholic Church. In the north, however, the Reformation triumphed.

In 1579 a formal division occurred, the Northern, or Dutch, Netherlands declaring its independence, while the southern provinces, those comprising modern Belgium, remained

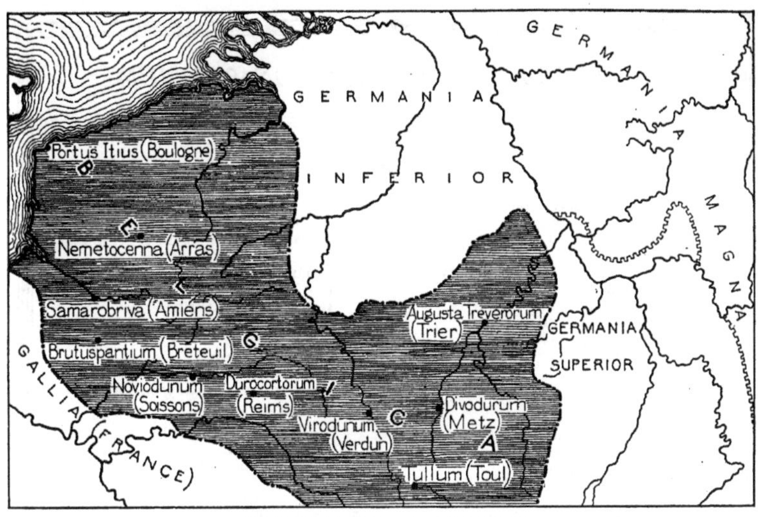

BELGIUM UNDER THE ROMANS

First conquered by Julius Caesar, 57-52 B. C., it remained a part of the Roman Empire for about five hundred years.

under Spanish rule. This Spanish domination was far from favorable to the progress of the country, for industry was stifled and many of the most enterprising citizens left for other countries. In 1598 Philip II of Spain made Belgium a separate state under the rule of his daughter Isabel and her husband, the Archduke Alfred, and during its brief period of independence the country prospered. On the death of Albert without heirs in 1621, Belgium reverted to Spain, and from that time on was known as the Spanish Netherlands.

For many years Belgium was in a most unfortunate position. Spain, which was steadily declining in power, was continually at war with France, and of many of these struggles Belgium was the battle ground. Belgian territory, too, was constantly changing hands, as one or another of the countries was victorious. Naturally, this troubled condition made internal prosperity and progress impossible. In 1713 the War of the Spanish Succession was closed by the Treaty of Utrecht, by the terms of which Belgium passed to Austria (see SUCCESSION WARS). The exhausted country did not immediately recover from the effects of Spanish misrule, but during the governorship of Charles of Lorraine, brother-in-law of Maria Theresa, which lasted from 1741 to 1780, it became really prosperous. For a few years during this period France gained control of almost all Belgium, which was, however, restored to Austria in 1748 by the Treaty of Aix-la-Chapelle.

Independence Achieved. During the French Revolution and the Napoleonic era the Netherlands fell into French hands, and in the readjustment of European affairs which took place at the Congress of Vienna (1815) Belgium and Holland were united as a single state. The old differences seemed to have increased rather than diminished, and the Belgians particularly objected to the union because more concessions were made to the Dutch than to them. In 1830, therefore, when the spirit of revolution was strong in Europe, Belgium revolted and set up an independent government, choosing as king a son of Louis Philippe, the King of France. He declined, and Prince Leopold of Saxe-Coburg was elected king in 1831 as Leopold I. His policies were wise, and the new state prospered, passing without upheaval through the revolutionary period of 1848. Leopold II came to the throne in 1865, and won for himself a place in history by his part in founding the Congo Free State. At his death in 1909 he was succeeded by his nephew, Albert, a man of serious purpose and progress-

OUTLINE AND QUESTIONS ON BELGIUM

Outline

I. Location
(1) Latitude, 49° 30′ to 51° 30′ north
(2) Longitude, 2° 33′ to 6° 6′ east
(3) Boundaries
(4) Effects of position on history
(a) Trade
(b) Warfare

II. Size
(1) Comparative
(2) Actual area

III. Physical Features
(1) Coast
(2) Surface
(a) Coastal lowlands
(b) The *Campine*
(c) The Flemish plain
(d) Highlands
(3) Drainage
(a) Abundance of rivers

IV. The People
(1) Races
(a) Flemings
(b) Walloons
(c) Other peoples
1. Germans
2. French
(2) Languages
(a) Two official tongues
(3) Education
(a) Decreasing illiteracy
(b) Elementary schools
(c) Higher institutions
(4) Religion

V. Industries and Communication
(1) Agriculture
(a) Intensive farming
(b) Chief crops
(c) Live stock
2) Mining
(a) Chief minerals
(b) Production compared with that of other countries
(3) Manufacturing
(a) Iron and steel manufactures
(b) Textile industries
(c) Other manufactures
(4) Commerce
(a) Recent growth
(5) Transportation
(a) Excellent railway facilities

VI. Government
(1) Its constitution
(2) The ruler and his cabinet
(3) Legislature
(4) Local government
(5) Chief cities
(6) Public debt

VII. Defense
(1) Army
(2) Fortresses

VIII. History
(1) In ancient times
(2) Part in Crusades
(3) The rise of the guilds
(4) Long connection with Netherlands
(5) Separation in 1579
(6) Unfortunate conditions under Spanish domination
(7) Napoleonic era
(8) Reunion with Netherlands
(9) Independence
(10) Steady progress
(11) Belgium's part in the War of the Nations

Questions

Which state in the American Union does Belgium most closely resemble in size?

If the populations of the two were exchanged, how much more densely would the American state be populated than it is at present?

How does Belgium rank among the countries of Europe as regards crop production per acre?

What phase of the educational question has entered largely into politics?

How did the ancient inhabitants of Belgium rank among the Gauls as to bravery?

How does the language situation in Belgium differ from that in England, for example?

Outline and Questions on Belgium—Continued

What is the most valuable output of the mines?

How did the country prove its progressiveness in making its constitution?

Why could not the inhabitants of Belgium and Holland be contented under one government?

Name five important battles that have been fought on the soil of Belgium.

What effect has the location of the rivers had on the development of Antwerp?

How many countries have a greater railroad mileage in comparison with their size?

Where was the government of the country established during the War of the Nations?

What relation was there between ancient Flanders and modern Belgium?

What were the guilds, and what part did Belgium play in their development?

How many nations surpass Belgium in bulk of commerce during normal times?

What was the most strongly-fortified city in the country?

What attitude did the country wish to maintain in the War of the Nations?

How does it happen that the Flemish plain, naturally an unfertile region, is one of the most densely-populated territories in the world?

What ornamental wares are manufactured in Belgium?

What was the strength of the army in time of peace?

What did the Congress of Vienna have to do with the history of the country?

What phase of financial activity proved the prosperity of the country?

What are *polders,* and why are they necessary?

In what two ways are a number of industries carried on in Belgium?

How do you account for the fact that the mayors are so much more important here than in most countries?

What was the effect of Spanish rule on Belgium?

When was the neutrality of the country guaranteed, and by what nations?

To what Church do most of the people belong?

What two mineral resources have made possible the development of great manufactures?

When before 1830 was the country independent?

How did the population in 1912 compare with that in 1830, when the country became independent?

Why are not the Belgian farmers as prosperous as those of Holland?

Who are the Flemings? The Walloons? To which race of people is each most nearly allied?

Compare the production of zinc and lead with that of the United States. Which is larger in proportion to area?

Why is the country called "the cockpit of Europe"?

What has been the difference in the history of Bruges and that of Antwerp?

How many men was Belgium able to put into the field to defend herself at the outbreak of the War of the Nations?

Who was the first king of the new country? What was the most important achievement of his successor?

Where is the Campine, and of what does it consist?

To how many countries has the territory composing Belgium belonged in the course of its history?

How many kings has the country had since its final declaration of independence?

Who was king at the outbreak of the War of the Nations?

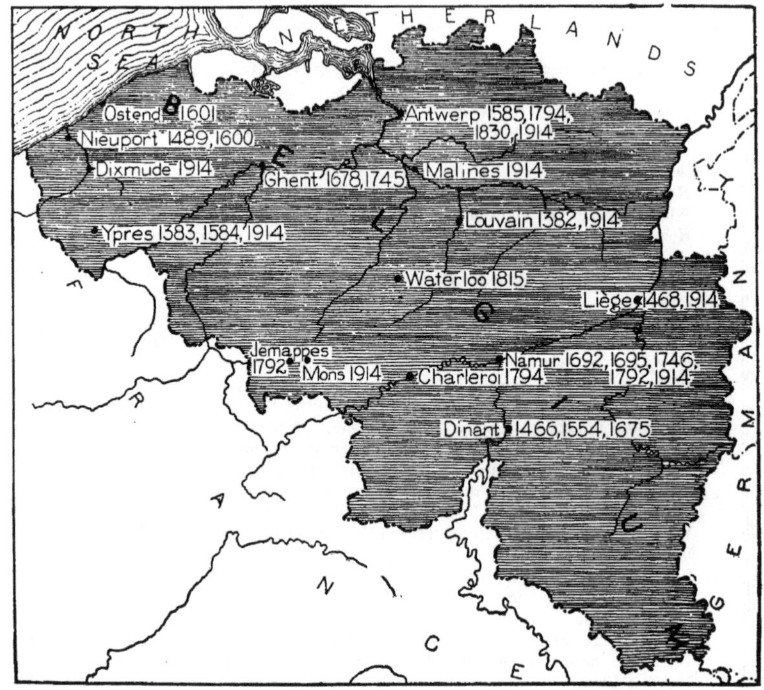

BATTLEFIELDS OF BELGIUM

On the soil of Belgium French, British, Austrian, German and Spanish armies at various times have met in historic encounters. Of these the battle of most far-reaching consequence was probably that of Waterloo, which decided Napoleon's fate.

ive ideas, destined to become the idolized leader of his people in their greatest crisis.

The War of the Nations. In the great war beginning in August, 1914, which involved nearly all the powers of civilization, Belgium expected to remain neutral, as its neutrality had been guaranteed by treaty in 1839. However, at the outset of the struggle Germany demanded permission to pass through the country, for strategic reasons, in its campaign against France. Belgium disputed the passage of the German army and fought at Liége to turn back the invaders. Its defense of its rights was heroic, but the little Belgian army was pushed back by the force of numbers.

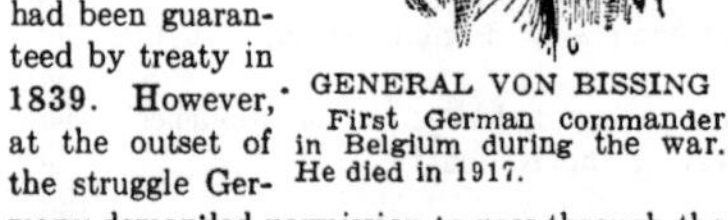

GENERAL VON BISSING
First German commander in Belgium during the war. He died in 1917.

The heroic defense of Liége retarded the German advance, but within a month the Germans had occupied practically the whole of Belgium, except Antwerp and the coast. In October Antwerp, which had been considered an impregnable fortress, also was taken, after a ten-day bombardment. The government of Belgium, which had first been transferred from Brussels to Antwerp, was then transferred to Ostend, and then to Havre, in France.

The Germans could not clear Belgian soil entirely of Belgian soldiers. The extreme northwestern corner remained in the hands of its defenders throughout the war, and though only a few square miles remained to them their stubborn resistance elicited the admiration of all civilized peoples. Ypres, near the French boundary, was pounded repeatedly by heavy artillery and "ground to powder"; some of the heaviest fighting of the war was here, in which Canadians and Australians particularly distinguished themselves

Not only did Belgians suffer unspeakable personal horrors, but the Germans, determined to destroy the country which had so offended them by resistance in 1914, systematically plundered the industries of the people. Machinery was taken out of factories to the value of many millions of dollars and shipped to Germany; what could not well be transported was destroyed.

The peace conference could not hope to force Germany to pay full cash value for the billions of dollars lost through wanton destruction of civilian property, but it provided that the first 2,500,000,000 francs ($500,000,000) exacted should be paid to Belgium.

Other Items of Interest. In the sixteenth century Antwerp was the richest and most splendid city in the world, its harbors often sheltering more than two thousand ships.

The battle ground of Waterloo is about nine and one-half miles from Brussels. It is marked by a Heroes' Mound two hundred feet in height, on the top of which there is a huge lion.

Two of the really great painters of Europe, Rubens and Van Dyck, and a number of minor ones belong to little Belgium.

In normal years the deep-sea fisheries of Belgium are valued at about $1,000,000, about one-seventh as much as those of Massachusetts.

If there be no male heir to the throne, the king has the privilege of nominating his successor, with the consent of the legislative chambers.

There are no publications relating to Belgium from American presses, but English books are available. Consult Holland's *The Belgians at Home;* Smyth's *The Story of Belgium;* Scudamore's *Belgium and the Belgians.*

Related Subjects. The reader who is interested in Belgium will find the following articles helpful:

CITIES AND TOWNS

Antwerp	Louvain
Bruges	Malines
Brussels	Namur
Ghent	Ostend
Liége	Ypres

LEADING PRODUCTS

Coal	Linen
Glass	Pottery
Iron	Silver
Lace	Steel
Lead	Zinc

RIVERS

Meuse	Scheldt

HISTORY

Aix-la-Chapelle, Treaties of	Utrecht, Peace of
Congo	Vienna, Congress of
Gaul	Walloons
Succession Wars	War of the Nations

BELGRADE, *bel grayd',* the capital of Serbia, situated on a promontory formed by the confluence of the Danube and the Save rivers. It is a strongly fortified city, and no other fortress in Europe has witnessed more battles than have been waged round the walls of Belgrade. In ancient days it formed a strong outpost, protecting the West against the advancing Turks who named Belgrade "the home of wars for the faith." With varying fortunes the city was a buffer, constantly changing owners, now captured by Turks, now relieved by Austria, and it was not until 1866 that the city was finally handed over to the Serbians.

Modern Belgrade dates from its evacuation by the Turks. An ancient mixture of East and West has quickly grown into a modern city with wide, clean streets, fine buildings, electric lighting, and electric street cars. The educational institutions are of the best, and there are a royal palace, many churches, a cathedral, national and public libraries and modern business houses. It carries on extensive commerce and has manufactures of machinery, cloth, boots, cigarettes, carpets, silk stuffs, pottery and cutlery. During the preliminary stages of the War of the Nations, beginning in 1914, Belgrade was captured by the Austrians, but was quickly retaken by the Serbians. For some months thereafter Belgrade was almost forgotten in the rush of events in Belgium, France and Russia. Late in 1915, however, an Austro-German invasion of Serbia occurred, and the Serbians were again driven out of their capital, October 9, 1915. Population, previous to the war, 90,890. See SERBIA.

BELIZE, *be leez',* a city that has been noted for more than three hundred years as the center of the mahogany and logwood trade of the world. It is the chief seaport and the capital of British Honduras, in Central America, and it is situated at the mouth of the southern arm of the Belize River. It has a harbor available for small vessels, but ocean-going steamers have to anchor a mile or more from the river mouth and land their cargoes by lighters. Besides mahogany and logwood the city exports rosewood, cedar, cocoanuts and sugar. Population in 1911, 10,478. See BRITISH HONDURAS.

BELL. In modern life there is perhaps no sound more familiar than the ringing of a bell. Those who trust to the morning alarm clock may be awakened by the tinkling of a bell; during business hours a bell calls messengers,

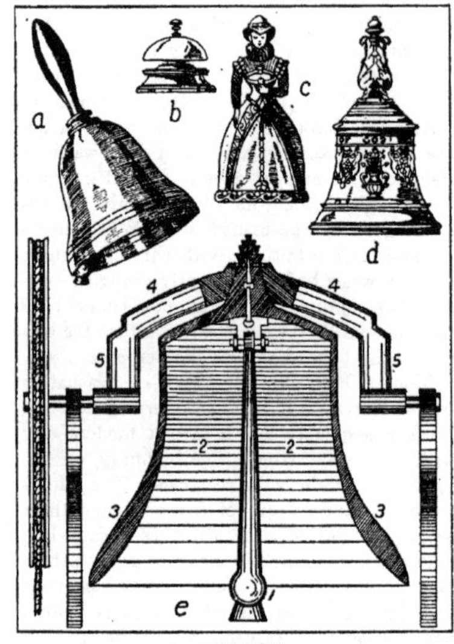

BELLS

(*a*) Ordinary hand bell; (*b*) modern, of the latest style; (*c*) brass bell, after an old model; (*d*) a bell of the period of the Renaissance; (*e*) church or school bell of the present day. Its parts are named as follows: (1) Tongue, or clapper; (2) barrel; (3) sound bow; (4) yoke; (5) shoulder.

and in some cases signals the time to commence and to stop work. The deep-sounding bell summons worshippers to church; the "curfew tolls the knell of parting day." So familiar is a bell that many articles, such as flowers, which would be hard to describe otherwise, are plainly visualized by the expression "bell-shaped."

Ancient **Bells.** Before the Christian Era, bells as we know them now probably did not exist. Bars of metal and instruments shaped like cymbals were used for the purposes of announcing meetings. The modern cup-shaped instrument dates from the fourth century, and apparently was first used for the purpose of summoning the devout to worship. That the bell has always been closely associated with religious exercises is shown by its use in the pronunciation of excommunication from the Church by *bell, book* and *candle,* and by the tolling of bells to announce death. Bells were introduced into France in 550, and into England about a century later. One of the oldest bells existing in Great Britain is preserved at Belfast, and is known as "the bell of Saint Patrick's will." It is six inches high, five inches broad and four deep, and is thought to date from about 652.

The original Latin word for bell is *campana,* from which we derive the modern *campanile,* meaning *bell tower* (see CAMPANILE). In the thirteenth century it became customary to make large bells and place them in special towers, where they were sounded by being struck with a metal rod or by being swung so as to be struck by a metal clapper suspended within the hollow of the bell. In 1400 a notable bell weighing six and one-half tons was made in Paris, and from that time the size of bells increased.

Famous Bells. The largest bell in the world is the "Tsar Kolokol" of Moscow, weighing 193 tons. It was never rung, however, as it was cracked in the making and a piece eleven tons in weight broke off (see illustration). The most famous bell in the United States is the Liberty Bell, which pealed forth the news of the signing of the Declaration of Independence and the birth of a nation (see LIBERTY BELL). "Big Ben," in the Westminster clock tower of the Houses of Parliament in London, weighs thirteen and one-half tons and was made in 1858. It is recorded that on one occasion during its fifty-seven years of staid time-sounding Big Ben boomed forth thirteen strokes at the hour of midnight, and numerous pieces of fiction have centered their plot around the alleged incident. The cathedral of Notre Dame, Montreal, Canada, has a bell weighing fourteen and one-half tons. The great bell of Cologne cathedral, cast from the metal of cannon captured from the French in 1870, weighs twenty-seven and one-half tons. In Burma there is a pagoda with a bell weighing eighty tons. All of these bells of great size and weight are sounded by being struck; they are not movable.

How Bells Are Made. Ancient bells were made in pieces, riveted together, but it is now customary to mold them in a single piece from molten metal. Bell metal, as it is called, consists of a mixture of copper and tin in proportions of four parts of copper to one of tin. A mold is made of baked clay, the melted metal is poured into it and allowed to cool. The clay mold is then destroyed and the bell is complete. A very large bell may require several weeks to cool thoroughly.

Uses of Bells. In addition to its original use as a summons to worship, the swinging bell has for centuries been employed for numerous important purposes. In the feudal days

THE GREAT BELL OF MOSCOW

of England a bell at evening signified it was time to "cover the fire" and was called the *curfew;* this custom is immortalized in Rose Hartwick Thorpe's poem, *Curfew Must Not Ring To-night,* a tragic incident of the Cromwell era. In pioneer days of America a bell pealed warning of Indian attacks. At sea, time is marked by the sounding of a bell, and buoys sometimes carry bells to warn ships off dangerous coasts. In England even at the present day a "town crier" is sent around in small communities to announce sales or other important events, and he rings a bell to attract attention. F.ST.A.

Bell, Nautical. On board ship time is marked by the striking of a bell every hour and half hour, and the term *bells* is used exactly as one says *o'clock* on land. The day at sea is divided into six periods of four hours each, termed *watches,* commencing at midnight. Half an hour after midnight *one bell* is struck; at one o'clock it is *two bells;* at two, *four bells;* at three, *six bells;* at four, *eight bells.* Then the round commences again, *eight bells* being sounded every four hours. The odd number of strokes denotes the half hour.

BELL, Alexander Graham (1847-), an American scientist, celebrated as the inventor of the first successful telephone. Everyone realizes the importance of this instrument, but only the business man, who may wonder how business was ever carried on in the days before the telephone, appreciates what a miracle-worker its inventor really was. Bell's first telephone was patented in 1876 and was exhibited at the Centennial Exposition. Its possibilities were evident, but it was far from being a perfect instrument. The inventor never ceased his experiments upon it, and by successive improvements brought it to such a point that in January, 1915, conversation was easily carried on between New York and San Francisco. Bell was not the only experimenter with long-distance conversation. Elisha Gray (which see) also invented a telephone and applied for a patent on the same day on which Bell sent in his application, but Bell was adjudged to have the prior claim, and his rights were later sustained against all claimants. The Bell Telephone Company has held practically the monopoly of telephone business in the United States, and it is stated on good authority that a thousand million conversations are held over the Bell telephones of the United States and Canada annually. See TELEPHONE.

ALEXANDER GRAHAM BELL

Alexander Graham Bell was born in Edinburgh, Scotland, received his education there and in London, and in 1870 moved to Canada with his father, who was the inventor of the "visible speech" method of teaching deaf-mutes to talk. Of the thousands who have profited by this system, Helen Keller (which see) is a notable example. In 1872 the younger Bell became professor of vocal physiology at Boston University, and there introduced his father's system. Even after his inventions demanded most of his time, he never lost interest in the education of the deaf, and wrote various works which entitled him to rank as an authority on that subject. His inventions, besides that for which he is chiefly noted, include the photophone, in which sounds are conveyed by a vibratory beam of light; the graphophone, the forerunner of the phonograph; and the telephone probe, which detects bullets in the human body.

BELL, JOHN (1797-1869), an American statesman, prominent in all the vital movements that preceded the War of Secession. He was born near Nashville, Tenn., was graduated from what is now the University of Nashville in 1814, was admitted to the bar at the age of nineteen, and when but twenty was elected to the state senate. So able did he prove himself in that body that in 1827 he was sent to Congress, where he served until 1841, becoming known for his outspoken support of a protective tariff. In 1832 he vigorously supported Jackson in his campaign for the Presidency, and two years later was made Speaker of the House. President Harrison appointed him to his Cabinet in 1841 as Secretary of War, but at Harrison's death and the succession of Tyler, Bell and the others of the Cabinet resigned.

From 1847 to 1859 he was in the United States Senate, where he favored Clay's compromise measure of 1850, voted against the Kansas-Nebraska Bill and opposed the repeal of the Missouri Compromise. The Constitutional Union party (which see) nominated him for President in 1860, and in the ensuing election Tennessee, Virginia and Kentucky gave him their electoral votes (see electoral map, in article LINCOLN, ABRAHAM). For a time he opposed the secession of Tennessee, favoring a policy of armed neutrality for that state, but later he actively supported the Confederate cause, though he took no part in the war.

BELL, ROBERT (1841-), a Canadian geologist who made the first surveys of many of the rivers and lakes of Western Canada, including Great Slave Lake, Lake Nipigon and part of Lake of the Woods, Lake Winnipeg, and the Athabaska, Slave, Nelson and Churchill rivers. After studying at McGill University, he joined the staff of the Geological Survey in 1857, when he was only sixteen years old. For half a century thereafter the reports of his geological and topographical surveys were the chief sources of information about the West. He was geologist on an exploring expedition to Hudson Bay in 1884, and in 1897 surveyed the south coast of Baffin Land and reached the great lakes in the interior. The value of Bell's work was recognized by his appointment as director of the Dominion Geological Survey, by the award of honorary degrees from McGill, Cambridge and other universities, and by membership and office in many honorary scientific societies.

BELLACOOLA, *bel' la koo' la,* a tribe of Indians which has suffered much from contact with white settlers. Their home is now in British Columbia. They were once a strong and important tribe, extending their range into the Northern United States, displaying unusual skill as fishermen and building great houses which were marvels of strength and ingenuity. Diseases introduced by the white men, however, have killed them off until now but a few hundred remain.

BELLADONNA, *bel' la don' a,* or **DEADLY NIGHTSHADE,** a bushy herb of the nightshade family and a native of Europe and Asia, poisonous in all its parts, but cultivated extensively because of its healing properties. In

BELLADONNA
(*a*) Branch; (*b*) fruit.

the United States the plant is grown in gardens because of its showy flowers as well as for use in medicine. The roots and leaves yield a drug known as belladonna, that is widely used for the relief of asthma, bronchitis, and whooping cough, colic and other intestinal troubles, while plasters or liniments made from belladonna are applied externally to soothe the pains of inflammation and neuralgia. Belladonna is also valued highly by oculists because of its paralyzing effect on certain muscles of the eye, which causes the pupil to expand and greatly aids the doctor in his examination and treatment of the eye. No one but a competent oculist should ever apply belladonna to the eye, and the use of this drug by persons who hope to add beauty to the eye by increasing its brilliance cannot be too strongly condemned. In fact, no belladonna preparation should ever be used for any purpose except under the direction of a physician.

The plant grows to a height of from two to six feet, and bears drooping, bell-shaped purple flowers, broad, oval leaves and large black berries. Could it speak it would probably say, "Touch me not," for persons are known to have died from eating its berries, and even breaking off a spray may cause eye trouble. Children, especially, should be warned against touching the plant. Vinegar is a simple remedy in cases of poisoning, but a physician should, if possible, be summoned immediately.

Once upon a time Italian ladies used the juice of belladonna as a face ointment, and from this custom, as well as from its use in beautifying the eye, came its name, which is the Italian for *fair lady.*

BELLAIRE, *bel air',* OHIO, a manufacturing and trading city located in the valley of the Ohio River. It is in Belmont County, on the eastern state boundary, which is the Ohio River. The city covers an area of nearly three square miles. In 1910 the population was 12,946; in 1914 it was 13,896. Americans predominate; foreign-born include principally Germans, Bohemians, Hungarians and Slavs.

Bellaire is four miles south of Wheeling, W. Va.; Pittsburgh, Pa., is ninety-seven miles northeast, and Columbus, the state capital, is 137 miles west. The Baltimore & Ohio Railroad, built to the city in 1854, crosses the river over a fine iron bridge. Other lines running into the city are the Pennsylvania, built in 1849, the Ohio River & Western, built in 1880, and the Wabash, built in 1907. River commerce, handling coal, stock and freight of all kinds, is extensive. The important manufactures include machinery, glass, steel, iron, enamel ware, farm implements and stoves. The monthly pay roll of the Carnegie Steel Company at Bellaire averages $85,000. Near the city are large deposits of coal, iron, limestone, fire clay and cement. Natural gas and coal, easily obtained from the surrounding country, make the cost of manufacturing exceptionally low.

Among the many handsome buildings of the city are the post office, built in 1914 at a cost of $125,000, four large banks and a city hospital. The city library is conducted in connection with the public schools, which have in addition to the regular courses a commercial department. Bellaire was settled about 1795, incorporated as a village in 1858 and received a city charter in 1874. W.L.M.

BELLBIRD, the name of several species of birds whose notes sound like the tolling of a bell. The most famous of these is a South American songster of glossy white plumage, which dwells among the dense forests of the Amazon region. Its clear, melodious notes, which carry for a long distance, may be heard at mid-day. To some travelers these seem like the blow of the blacksmith's hammer on the anvil. A remarkable feature of this bellbird is a fleshy projection about an inch in length, growing from the base of the beak, black in color and dotted here and there with starlike tufts of small feathers. It hangs down loosely at the side of the beak, except when the bird becomes excited and when it utters its bell-like tones. Then the projection slowly extends sometimes to a length of five inches, but does not, as was formerly supposed, assume an erect position.

THE AMAZON BELLBIRD

The name is applied in Australia and New Zealand to two members of the honey-sucker group, birds which feed on the nectar of flowers. The New Zealand bellbird utters notes that sound like the tinkling of a silver bell.

BELLEFONTAINE, *bel fon' tain,* OHIO, the county seat of Logan County, is situated in the western part of the state about midway between the northern and southern state lines. Columbus, the capital, is fifty-six miles southeast, Lima is thirty-three miles northwest and Cincinnati is 113 miles southwest. The Toledo & Ohio Central, two divisions of the Cleveland, Cincinnati, Chicago & Saint Louis (Big Four) railways and an electric line serve the city. Bellefontaine was settled in 1806, was organized as a town and became the county seat in 1820 and was incorporated as a city in 1835. The name is taken from the French, and means *beautiful fountain;* it was suggested by several springs of clear water, spurting at the base of the elevation on which the city is built. In 1910 the population was 8,238; it increased to 8,915 in 1914. The area is less than three square miles.

The site of Bellefontaine is on the highest point of land of the state, 1,550 feet above sea level. In the vicinity is Lewiston reservoir, with an area exceeding 13,000 acres and a capacity of four and a half billion cubic feet. Drinking water is supplied the city by artesian wells drilled into limestone to a depth of about 200 feet. The leading manufactures are iron and steel bridges, automobiles, mattresses and comforts. The large shops of the Big Four railroad are located here. Bellefontaine has a Carnegie Library and three parks. C. OF C.

BELLEROPHON, *be ler' o fon,* a mythological Greek hero, the son of Glucus, king of Corinth, and the slayer of the dreadful, three-headed Chimaera. He was sent by the king of Argos to the Lycian king with a sealed message asking his death, and to accomplish this he was sent to kill the fire-breathing monster. He was assisted by the goddess Minerva, who gave him the golden bridle with which he secured the winged horse Pegasus (which see). Mounted upon this steed, he was able to attack the monster from above and slay him.

Seeing that Bellerophon was a favorite of the gods, the king gave him his daughter in marriage and made him heir to his throne. Legend says that in his later years Bellerophon, made proud by his good fortune, attempted to mount upon Pegasus to the home of the gods, and that Jupiter, angered by his boldness, caused him to be dashed to the earth. Lame and blind, he dragged out his few remaining years in misery. Hawthorne has told the story of Bellerophon's great victory delightfully in *The Chimaera,* omitting the later mournful part of the tale:

> He whose blind thought futurity denies,
> Unconscious bears, Bellerophon, like thee
> His own indictment; he condemns himself.

BELLES-LETTRES, *bel let' r,* an expression applied in a somewhat indefinite way to those forms of literature that appeal to the emotions and the imagination. *Belles* and *lettres* are two French words meaning *beautiful* and *letters,* which, taken together, mean *fine literature* or *polite literature.* Poetry, fiction, literary criticism, dramatic writings and certain forms of history would be included in the term, as distinguished from the more practical or prosaic forms of literature, such as scientific treatises. There is, however, no hard and fast rule governing the use of the term. The first English writer to employ this expression was Jonathan Swift, who in 1710, made use of it in the periodical called *The Tatler.*

BELLEVILLE, *bel' vil,* the county town of Hastings County, Ontario, on the Bay of Quinte, an arm of Lake Ontario, and at the mouth of the Moira River. It has a fine harbor which accommodates the largest steamers which ply on the Great Lakes. The city is on the Grand Trunk, Canadian Northern and Canadian Pacific railways, 112 miles northeast of Toronto and fifty-one miles west of Kingston. It is a noted dairying center, and manufactures butter in large quantities. There are also cement works, hardware factories, foundries and boiler works, flour and paper mills, potteries, shirt, furniture and soda-water factories. Belleville has excellent public schools, and is the seat of the provincial institute for the deaf and dumb; of Saint Agnes College, for women; and Albert College, with separate departments for men and women. The latter is one of the oldest educational institutions in the province. It was founded by the Methodists in 1757, and was for many years a university, but now provides only secondary instruction. Belleville is also known for its beautiful churches and as the home of Sir Mackenzie Bowell. Population in 1911, 9,876; in 1916, estimated, 12,000.

BELLEVILLE, ILL., an important manufacturing city with a population, chiefly of German descent, of 21,139 in 1914. It is the county seat of Saint Clair County, situated in the southwestern part of the state, about fourteen miles southeast of Saint Louis. Excellent railway accommodations are provided by the Illinois Central Railway, constructed to the city in 1860, and the Louisville & Nashville Railway, built in 1870. Electric lines connect with Saint Louis and adjacent towns. The city was founded in 1814, and was incorporated as a city in 1850. The name "Belleville" is of French origin and means *beautiful city.* The area of the city is about nine square miles.

Belleville is situated in the midst of an agricultural and coal-mining region, and has a large trade in flour and general produce. Its industries include foundries, machine works, nail mills, stencil works and carbon work. Here is located one of the largest rolling mills in the West. Besides the public school system, a convent, two business colleges and a Carnegie Library serve the educational needs. There are three banks, two hospitals, a Roman Catholic cathedral, an Elk's hall and a post office, the latter erected in 1913 at a cost of $60,000.

Falling Springs, in the vicinity, is a favorite resort for excursionists; it has a waterfall and presents a delightful view of high bluffs. F.S.B.

BELLINGHAM, *bel' ing am,* WASH., the county seat of Whatcom County and the nearest United States shipping point to Alaska. It is on the east shore of Bellingham Bay, which is part of Puget Sound, in the extreme northwestern part of the state and is on the Nooksack River and on the Great Northern and Northern Pacific railways. It is also the terminus of the Bellingham & Northern Railway (now a part of the Chicago, Milwaukee & Saint Paul system), a line sixty-two miles long, connecting with Mount Baker goldfields and the agricultural country of Nooksack Valley. Bellingham is the home port of a number of coastwise and trans-Pacific steamship lines. The area of the city is about twenty square miles. The population, largely American, was 24,299 in 1910 and 29,937 in 1914.

Bellingham has an excellent harbor. The great natural wealth of the surrounding country consists of salmon, lumber, minerals, clay, limestone and sandstone beds. A gravity plant installed by the city at a cost of $1,000,000 brings water from Lake Whatcom, a high natural reservoir two and a half miles southeast. Electric power is obtained from the falls of the Nooksack River. In Bellingham are some of the largest salmon canneries in the world. These companies operate their own ships to Alaska and by way of the Panama Canal to New York City. Lumber interests, saw and shingle (especially cedar) mills, logging companies, wood-working factories, salmon, fruit and vegetable canning, milk condensing and the manufacture of cans and cement, represent the important commercial interests of Bellingham.

Among the more prominent buildings are the city hall, a Federal building, erected in 1913 at a cost of $280,000, the Y. M. C. A., two Carnegie libraries and two hospitals. The largest state normal school, an industrial school and several business colleges are located here. There are four city parks, Elizabeth, Larrabee, Cornwall and Whatcom Falls. Snow-capped peaks, Mount Baker, the largest, to the south and east, the cascades and falls of the river, magnificent pine forests and the ocean islands, all contribute to the beautiful scenery of this region.

Bellingham Bay is thought to have been given its name by Vancouver in 1792, in honor of Sir Henry Bellingham. Along the shore there were once four separate settlements which have united in the present city. In 1903 Bellingham, named from the bay, was founded by the consolidation of Whatcom and Fairhaven. A city charter was obtained in 1904. W.F.C.

BELLINI, *bel le' ne,* GIOVANNI (1430 or 1431-1516), the greatest figure in early Venetian painting, and the most important member of a family of Italian painters who flourished during the early Renaissance. Giovanni received his first lessons in painting from his father, Jacopo Bellini, a distinguished portrait painter. His fame rests chiefly on his religious paintings, many important examples of which survive. One of his numerous Madonnas hangs in the Metropolitan Museum of New York, and there are others in private collections; his *Christ on the Cross* and *Transfiguration* are in Venice; and the National Gallery of London possesses his *Agony in the Garden.*

His early works seem to have been painted under the influence of profound religious feeling, and show great severity of treatment, but his later canvases exhibit a blending of noble form and soft, rich color that is characteristic of Venetian painting at its best. He was commissioned to paint several great altarpieces for churches in Venice, and he assisted his brother Gentile in the decoration of the ducal palace at Constantinople. All of his paintings in the latter city were destroyed by fire, and this was the fate also of his work on the decoration of the great council hall at Venice. Bellini was also distinguished as a painter of portraits; one of his masterpieces, the portrait of Doge Loredan, is now in the London Gallery. His influence on the art of his time was profound, and among the throngs of pupils who studied with him were Giorgione and Titian.

BELLE ISLE, *bel' ile,* STRAIT OF, the body of water separating the island of Newfoundland from the mainland to the north and west. This strait, which is about eighty miles long and about twelve miles wide, is the northern entrance from the Atlantic Ocean into the Gulf of Saint Lawrence. The passage through the strait is made not without danger, but as the route is the shortest from Montreal and Quebec to England, it is usually followed by ocean liners.

At the eastern end of the strait is Belle Isle, the island, which lies nearly midway between the Newfoundland and Labrador shores. The island has an area of fifteen square miles. At its southern end, which steamships pass, is a great lighthouse whose signal light is visible for nearly thirty miles.

BELLOWS, a wind-making machine, used to fan flames to intensify their heat, or to operate a reed organ or pipe organ. The bellows of a blacksmith shop has two compartments formed by three boards and soft but airtight leather sides. By a simple arrangement of weights and levers which move the upper and lower board, air is first drawn into the under compartment, then into the upper and then forcibly expelled through a nozzle facing the forge. The advantage of two compartments is a continuous current of air, but small bellows, consisting of one chamber, with a valve at the bottom and a nozzle through which the air is expelled by pressure, are operated by hand to quicken the burning of fires in open hearths and grates. See BLOWING MACHINE

HAND BELLOWS

BELL-SMITH, FREDERICK MARLETT (1846-), a Canadian painter, equally successful in portraits and landscapes. Among his principal paintings are *Queen Victoria's Tribute to Canada,* for which Her Majesty gave personal sittings, *Landing of the Blenheim,* in the national collection at Ottawa, and *Lights of a City,* in the Ontario collection at Toronto. For many years he has made annual visits to the Rocky Mountains, and his sketches of mountain scenery are among his best work. Though born and educated in London, England, he made his home in Canada after 1867, when he settled at London, Ont. He was for seven years teacher of drawing in the schools of that city, and then from 1889 to 1891 was director of the Toronto Art School. For thirty years he was also director of fine arts at Alma College, Saint Thomas, Ont. Bell-Smith was a charter member (1867) of the Society of Canadian Artists, and in 1886 became an associate of the Royal Canadian Academy of Artists.

BELMONT, the name of a family of American financiers, two of whose members, father and son, achieved distinction in national life.

August Belmont (1816-1890) was born in Germany and was sent to the United States in 1837 as representative of the banking house of the Rothschilds. He was Austria's consul-general at New York from 1844 to 1850 and in 1854 he became American minister to Holland. In addition to a business career which would usually demand all of one's time, he took an active interest in politics, being chairman of the national Democratic committee for twelve years, and he was also an eager sportsman and a collector of paintings.

August Belmont (1853-), son of the above, also became a prominent capitalist, rising to posts of officer and director in many large railway, banking and manufacturing corporations, including the consolidated traction lines of New York City. Like his father, he became interested in Democratic politics. In 1910 he married Eleanor Robson, a leading actress.

BELOIT, *be loit',* WIS., a manufacturing center in Rock County, situated on the Rock River, close to the southern state line, about midway between the eastern and western state borders. Chicago is ninety miles southeast. Railway transportation is provided by the Chicago, Milwaukee & Saint Paul and the Chicago & North Western railways; an electric line is in operation to Janesville. Beloit was settled in 1824 and was chartered as a city in 1856. The population increased from 15,125 in 1910 to 17,122 in 1914. The area is less than five square miles.

The city occupies both banks of the river, at the point where it receives the waters of Turtle Creek. In its territory it is noted as the seat of Beloit College (Congregational), founded in 1846. Abundant water power is derived from the river for manufactures, the largest plants being engaged in making windmills, gasoline engines, paper and paper-mill machinery and scales. It has one of the largest wood-working plants and the oldest rye-flour mill in the United States.

BELSHAZZAR, *bel shaz' ar,* the king of Babylon who saw the "handwriting on the wall" which was interpreted by the prophet Daniel (*Dan.* V). Belshazzar was the last of the Babylonian kings, and he reigned with his father Nabonidus. He was killed in the storming of Babylon by Cyrus, according to the account in *Daniel,* on the night in which he saw the writing on the wall.

The following stanza contains the opening lines from a favorite song of Sunday school children, which graphically describes this episode:

At the feast of Balshazzar and a thousand of his lords,
As they drank from golden vessels, as the Book of Truth records,
In the night, as they reveled in the royal palace hall,
They were seized with consternation
'Twas the hand upon the wall.

BELT or **BELTING,** a flexible leather or rubber band, or like material, passing around wheels, pulleys or drums for the purpose of transmitting motion or power from one to another. The term *belt* is commonly applied to the broad, flat bands or woven material. Chains and ropes may also be employed, but these are generally classed as *chain drives* and *rope drives*. The two ends of the material of which a belt is to be made are joined securely together to produce what is called an endless

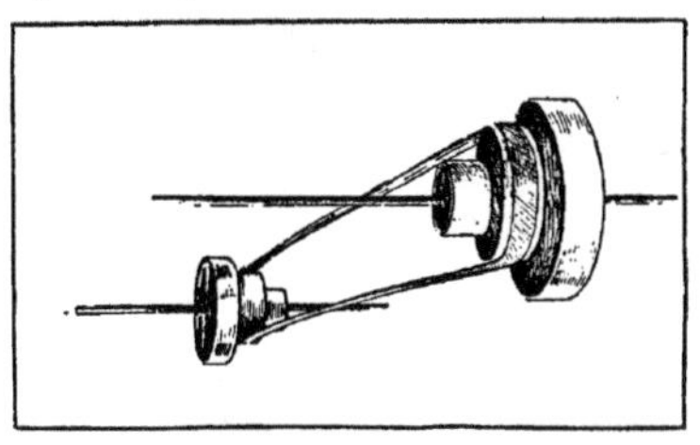

BELT AND SHAFTING
The belt may be shifted to larger or smaller wheels, to increase or decrease speed.

belt. Each manufacturer and machinist may have his own favorite way of effecting the joint, but the principle is the same in all.

In lacing, the joint may be made by running the lace through a series of holes as in lacing a shoe, care being taken when crossing the lace to do so on the top, or the side farthest from the surface of the wheel. Another method is to lace through two series of holes without crossing, the two ends of the lace being tied together on the upper surface. Metal rivets present an even surface, but are more liable to breakage than lacing. A combination of lacing and cement has grown in favor, for it produces a more permanent joint than any other method. Imperfect lacing is often the cause of trouble and power waste. A well-adjusted, properly-jointed belt may be kept in use for thirty years, if treated with proper care. Joints of belts are usually tested to stand a certain pressure per square inch, but in actual operation a belt should not be called upon to bear more than one-third of the stated maximum pressure.

BELUCHISTAN, *be lu chi stahn'*, another form of the word BALUCHISTAN (which see).

BENARES, *ben ah' rez*, the headquarters of the Hindu religion in India, situated on the right bank of the holy Ganges River, 390 miles northwest of Calcutta. The Hindus regard it as the holiest place on earth, created at the beginning of the world, eternal; those who are sufficiently blessed to die there are sure of instant admittance to paradise.

In reality, Benares is a very busy, flourishing city, capital of the Benares district of the United Provinces of Agra and Oudh, and decidedly modern for an Eastern city. But, as in all Indian cities, there is a deep undercurrent of oriental religious life, and the city contains 1,500 temples which are visited by vast numbers of pilgrims. Sacred cattle are found roaming at will through many parts of the city, and many of the temples are haunted by numbers of sacred monkeys. The inhabitants eye askance all Europeans, fearing and quick to resent any imaginary slight to their religious or racial prejudices.

Benares carries on a large trade in cotton, silk and woolen goods, and is famous for ornamental brass work known as Benares-ware. Its stores contain the fine shawls, embroidery and jewelry of the East and the most modern of Western inventions. The streets are narrow and crowded, but during recent years many sanitary improvements have been made. Population, 203,800.

BENEDICT, *ben' e dikt*, the name of fifteen Popes, of whom the last will probably be ranked by future historians as most important.

Benedict XV (1854-), GIACOMO DELLA CHIESA, became Pope at a troublesome time, shortly after the outbreak of the gigantic War of the Nations in 1914. He was born at Pegli, Italy, on November 21, 1854, was ordained priest in 1878, and in 1887 became secretary to Cardinal Rampolla, then the Papal secretary of state. In 1907 he became one of the Advisers to the Holy Office, and later in the same year was appointed Bishop of Bologna. On May 30, 1914, he was created cardinal, and a few months later, after the death of Pope Pius X, was chosen Pope on September 3, in a conclave which lasted only four days. This was the shortest conclave in the history of the Papacy, and no other Pope has been chosen after so short a service in the office of cardinal.

Pope Benedict came to his high office after a brief but thorough training. While secretary to Cardinal Rampolla he was intimately connected with the negotiations between the Papacy and the European powers, thus acquiring a knowledge of facts and diplomatic methods which stood him in good stead in the delicate situations growing out of the War of the Nations. Repeatedly during that great conflict he endeavored to bring peace to the war-torn countries, but his kind offices were rejected. Similarly, the Pope's administration of the see of Bologna, one of the most important in

Italy, proved invaluable experience for the administration of the supreme office which he was later called upon to fill. A man of aristocratic birth and training, a noted scholar, famous for his fearlessness and moral courage, Pope Benedict is certain to occupy a prominent place in the history of his time.

HIS HOLINESS POPE BENEDICT XV

Other Popes of the Name. Of the other Popes of the name, from Benedict I, Pope from 574 to 578, to Benedict XIV, Pope from 1740 to 1758, several are worthy of mention.

Benedict VIII, who was raised to the Papal chair in 1012 and ruled until 1024, had to contend for his right to the office with the antipope Gregory VI, but was confirmed in his possession of it by the Emperor Henry II. His later reign was disturbed by contests with the Saracens, but Henry remained his friend and helped him to gain possession of the island of Sardinia. One of his interdicts was directed against marriage of the clergy.

Benedict IX, nephew of the preceding, was first chosen Pope in 1033, and lived until 1056. Within that time, however, he was never Pope for more than eight months at a time, for he was constantly being deposed by one power or another. Four times he was reinstated, but in 1048 was permanently superseded and retired to a convent, where he died.

Benedict XIV (1675-1758) is worthy to rank with the very ablest holders of the Papal throne. After having served as bishop of Ancona, archbishop of Bologna and cardinal, he became Pope on the death of Clement XII in 1740. He distinguished himself especially by the interest he took in education and archaeological matters, establishing several chairs in the University of Rome, founding academies and directing the excavation of various Roman antiquities. Himself a man of uprightness and sincerity, he labored for reform among the clergy. His chief work is *On the Beatification and Canonization of Saints,* considered an authority on the subject. G.W.M.

BENEDICTINES, *ben e dik' tinz,* a religious Order of men, so named because of their adherence to the rule of life dictated by Saint Benedict. His idea was that each monastery should be a separate organization and should, for the monk, take the place of the family. The first monastery of the Order was established by Benedict at Monte Cassino, in the Appenines, in the year 529. The Order spread rapidly, and in the next century the Benedictines were foremost in implanting Christianity and civilization in the West. During the Dark Ages these monasteries were the only places where the followers of the Church could find meeting-places of retreat from the social classes, and the Order at this time was very influential in preserving some of the traditions which the bishops had succeeded in keeping alive.

The Benedictines have always been noted for their piety and for their encouragement of learning. Within their monasteries no branch of art or industry known at that time was neglected, and many of the books written before the invention of printing were made there.

BENEFIT ASSOCIATIONS. See FRATERNAL SOCIETIES.

BENEFIT OF CLERGY, the privilege enjoyed by the clergy of the medieval Church by which they were exempted from the jurisdiction of the ordinary courts and were responsible only to their bishops or Church courts. Originally, this exemption applied only to the clergy, but it was later extended to all clerks; and since everyone who could read and write even a little was considered a clerk, the privilege was much abused. A layman could receive the benefit of clergy only once, however, and before obtaining his liberty was branded on the thumb, a punishment which later was commuted to whipping, imprisonment or banishment. The benefit of clergy was abolished in 1827. One of Kipling's finest stories is entitled *Without Benefit of Clergy,* but he uses the phrase in an entirely different sense from the form above.

ELECTION OF POPE BENEDICT XV. Announcing successor of Pope Pius X from balcony of Saint Peter's.

BENGAL, *ben gawl'*, one of the three presidencies or large governing units of India, the others being Bombay and Madras. Few other regions in the world are more densely-populated, for Bengal has in its relatively small territory of 78,669 square miles 45,500,000 people, or 578 to the square mile. If half of the total population of the United States or twelve times the number of people in all Canada were compelled to live within the single state of Nebraska, conditions would be almost exactly the same. Certainly it would not seem possible to the crowded North Americans under such circumstances that they could support themselves by farming, for it is generally considered that wide areas are necessary for that industry, yet Bengal is distinctively agricultural. Over half of its area is under cultivation and as the soil in the Ganges and Brahmaputra valleys is very fertile, large returns are received for labor, even though methods are most primitive. The most important crop is rice, to which almost three-fourths of the cultivated area is given over, and rice is the principal food of the people. The other prodnets include wheat, barley and millet, oil seeds, opium, indigo, sugar cane and several varieties of fiber plants.

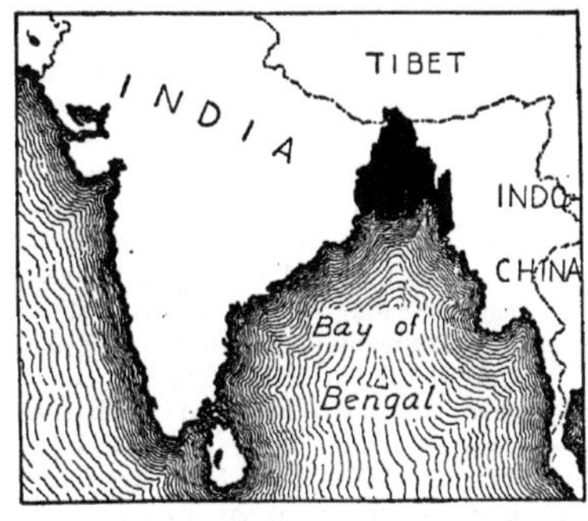

LOCATION MAP
Bengal (in black) and the Bay of Bengal.

Formerly, when Bengal had poor transportation facilities and knew little of machine-made articles from beyond the sea, there were important native manufactures. In particular, delicate cotton and silk fabrics were woven. But to-day cheaper textiles from Great Britain can be brought over the everywhere-present railroads or procured from the factories in the larger Bengalese towns, and hand-weaving has largely decreased.

Geographical Features. Bengal lies at the head of the Bay of Bengal and comprises the choicest land of India. Just as Egypt is indebted to the Nile for its fertility, so Bengal depends on the Ganges and the Brahmaputra, which spread the silt they bring from their high mountain sources over the plain of which Bengal is chiefly composed. Sometimes the Ganges is too generous with its floodings, and in places it is confined within embankments.

Three lines of railways meet at Calcutta, the capital of Bengal and the largest city of India; and from that center a huge commerce is carried on, largely with the United Kingdom.

History and Government. Bengal had a long independent existence before it began to be ruled by the Mogul emperors, from which it was wrested in 1757 by the English under Clive. British rule in Bengal was confirmed by Warren Hastings, and the province was early recognized as one of the most valuable parts of India. The name has been applied to widely-varying stretches of territory, and before 1905 included far more than it does to-day. The province was so large as to be unwieldy, however, and in that year the eastern part was united with Assam to form a new province of Eastern Bengal and Assam. Discontent resulted, for the people felt that the British government was trying to discourage any national feeling; and in 1912 a new adjustment was made. Eastern Bengal was reunited with Bengal proper, and on the west Bihar, Chota, Nagpur and Orissa were erected into a new province, while Bengal was raised to the rank of a presidency.

The chief executive of Bengal is a governor, appointed by the government in London, who has as his aids two members of the India Civil Service. The legislative department consists of an assembly of fifty-two members. Calcutta, once the capital of all India, remains the capital of Bengal, while Darjiling, the "city above the clouds," is the summer residence of the governor. See INDIA. E.D.F.

BENGAL, BAY OF, the northern portion of the Indian Ocean, lying between India and Burma. It receives the waters of the Ganges, Brahmaputra and Irrawaddy rivers, and on or near its shores are the important towns of Calcutta, Madras and Rangoon. In the eastern part of the bay are the Andaman Islands, on which a prison colony is maintained by the Indian government. Off the coast of Burma is a chain of islands containing active volcanoes which frequently erupt great quantities of mud, sometimes sufficient to form fresh islands, which after a time are washed away by the sea.

BENGALI, *ben gah' le,* one of the modern languages of India, spoken by about 50,000,000 people in Bengal. Related to the Sanskrit, it is written in characters derived from that language, and it also possesses many words borrowed from the Sanskrit. The name Bengali is an English form, its native form being *Banga-Bhasa.* Its grammar bears a resemblance to the modern Persian and to some degree to the English. Of the several Bengali dialects, that spoken at Calcutta is the standard. Bengali literature dates from the beginning of the fifteenth century; the greatest name in its history is that of Rabindranath Tagore, the poet who received the Nobel prize for literature in 1913. He writes in English, however, as well as in Bengali. See SANSKRIT LANGUAGE AND LITERATURE; TAGORE, RABINDRANATH.

BENGOUGH, *ben goff',* JOHN WILSON (1851-), a Canadian caricaturist, journalist and poet, and also known as one of the leaders of the single tax movement. He was born in Toronto, and after receiving a grammar school education began the study of law, which he soon dropped for journalism. In 1873 he established in Toronto a humorous weekly called *The Grip,* most of whose illustrations were drawn by himself. For nearly twenty years this paper held a unique place in Canadian life, and Bengough's good-humored, yet sometimes sharp, wit exposed the weaknesses of many Canadian statesmen. After 1892 he contributed to various Canadian and English newspapers, and also devoted much of his time to lecturing in Canada, Great Britain, Australia and the United States. He has written some excellent verse, and is the author of the famous election song, *Ontario, Ontario.* Among his publications are *Popular Readings, Original and Selected; Caricature History of Canadian Politics; Motley: Verses Grave and Gay. The Up-to-Date Primer, A First Book of Lessons for Little Political Economists,* is a humorous exposition of the single tax theory. He was appointed an associate of the Royal Canadian Academy of Arts upon the formation of that institution in 1880.

BEN-HUR, a novel by General Lew Wallace, of which over 1,000,000 copies have been sold. The dramatic version produced in 1899, nineteen years after the book first appeared, was also very popular, and the spectacular features, particularly the magnificent chariot race which Ben-Hur won against terrible odds, later made it a striking and successful moving-picture play.

The scene of the story is laid in the time of Christ, and its hero is a young Jew, Ben-Hur, who by reason of the enmity of Messala is wrongly convicted of a crime and sentenced to slave in the Roman galleys. His adoption by the Roman commander Arrius; his return to Palestine to seek his mother and sister; the furious contest with his old enemy in the chariot race, and his final conversion to Christianity make a powerful dramatic story.

BEN'JAMIN, the youngest and favorite son of Jacob, and a brother of Joseph. No finer story was ever written than that of Joseph and Benjamin, beginning with *Genesis* XLII. The name means *son of my right hand.*

He became the founder of the Benjamites, who in the march through the Wilderness had the honor of being placed next to the Tabernacle. The land assigned the Benjamites was between that of Ephraim on the north and Judah on the south. The territory was small but important, since it was the key to the entrance to the Promised Land from the east. The Benjamites were famous warriors, and the tribe contained many men noted for their skill in the use of the left hand. The Benjamites were nearly destroyed in a war which they waged against all the other tribes of Israel (see *Judges* XIX-XXI), but later they recovered their strength and at the division of the kingdom united with Judah.

BENJAMIN, JUDAH PHILIP (1811-1884), an American lawyer and statesman, commonly known during the War of Secession as "the brains of the Confederacy." His parents were English Jews and he was born in the West Indies, but when still a child was taken to Wilmington, N. C. After studying at Yale he was admitted to the bar in New Orleans in 1832, and soon rose to a high rank in his profession. Elected to the United States Senate in 1852, he was active in the Southern cause and with the secession of Louisiana in 1861 withdrew from the Senate. His pronounced ability as Attorney-General and as Secretary of State in the Cabinet of the Confederacy won him the name referred to above. At the close of the war he went to London, where he built up a large practice and acquired an enviable reputation.

BEN LOMOND, *ben lo' mund,* a mountain in the west-central part of Scotland, made famous by Sir Walter Scott's *Lady of the Lake.* It is just east of Loch Lomond and is the southern extremity of the Grampian Mountains, the Central Scottish Highlands. On the north side of Ben Lomond is a precipice 2,000 feet high.

BENNETT, the family name of two of the most prominent and successful American journalists, father and son.

James Gordon Bennett (1795-1872) was the founder and editor of the New York *Herald* and originator of many of the modern devices of journalism. Foreign correspondents, financial articles, full reports of important speeches, prompt and lively accounts of every-day events—these were unknown before his time, but other editors immediately adopted them.

Bennett was born at Newmills, in Scotland, and studied for the Roman Catholic priesthood, but the reading of Franklin's *Autobiography* interested him in America, and he emigrated in 1819. In Halifax and later in Boston he earned a scanty living on various journals; in 1832 he went to New York and at intervals attempted to found a paper of his own, always without success until 1835, when on May 6 there appeared the first number of the New York *Herald.* It was issued from a cellar, this little four-page penny paper, but its editor was a born journalist and knew how to interest and hold the public, whether he always pleased or not. At first he wrote the entire paper, and sold it as well, but soon began to employ reporters and newsboys, and to introduce the features mentioned above. Financially the paper was a great success, earning for its proprietor, in his later years, almost $750,000 annually. If Bennett himself was not always liked, if his opinions were not always respected, it was because he deliberately chose to increase the prominence and the circulation of his paper at the expense of his personal influence.

James Gordon Bennett, Jr. (1841-1918), his son, succeeded to the proprietorship of the *Herald,* but directed the affairs of the paper largely by cable, as he preferred to live in Paris. The chief innovation which he introduced was the publication of London and Paris editions of the *Herald.* Although he lives in Europe, Mr. Bennett keeps in intimate touch with his New York paper, sometimes cabling editorials on vital American topics. From his youth he has been intensely interested in yachting, and has won various races. Bennett, at his own expense, sent Henry M. Stanley to Africa to hunt for Livingstone, and equipped the *Jeannette* for its Polar expedition. He was also one of the founders of the Commercial Cable Company.

BENNETT, [ENOCH] ARNOLD (1867-), an English writer whose realistic novels have won him a popularity which few present-day novelists have succeeded in gaining. Most famous are his stories of the "Five Towns," which include *Anna of the Five Towns, The Old Wives' Tale, Clayhanger, Hilda Lessways* and *The Matador of the Five Towns.* These depend for their interest not on exciting events or romantic situations but on the realism with which they set forth the every-day lives of commonplace people and the insight into human nature which they show. Longer than most novels of the day, they yet succeed in holding their reader's interest.

Bennett was born at Hanley, one of the Five Towns, studied at the University of London, and worked as a magazine editor until the increasing popularity of his writings allowed him to give up other work. Some of his essays, as *The Human Machine* and *How to Live on Twenty-four Hours a Day* were widely read. *The Pretty Lady* (1918) is a revelation of the social awakening under the stress of the great war.

BEN NEVIS, *ben ne' vis,* which in Gaelic means *mountain of snow,* is the tallest peak in Great Britain. It rises 4,406 feet above the sea, at the south end of the Caledonian Canal in Inverness, and at its summit on clear days practiced observers can see nearly to the North Sea, one hundred miles away.

BENNINGTON, VT., a village in the southwestern part of the state, the name of which has been made famous by an important battle of the Revolutionary War, fought about five miles distant, on August 16, 1777. On that day nearly 2,000 "Green Mountain Boys" under General John Stark defeated in succession two divisions of the British army commanded by Burgoyne, who had dispatched the troops to Bennington to seize a store of supplies. The English suffered a loss of over 200 killed and 700 captured; the Americans, only fourteen killed and forty-two wounded. The battle had results of first importance. It not only seriously weakened Burgoyne but encouraged the colonial troops to continue their campaign against him, and his surrender two months later at Saratoga is generally regarded as the turning point in the war.

Bennington township, consisting of the three incorporated villages of Bennington, North Bennington and Bennington Centre, is the county seat of Bennington County. It is situated thirty-five miles northeast of Albany, N. Y., and fifty-two miles southwest of Rutland, Vt., on the New York Central Railroad. Special interest attaches to the place because of its picturesque location at the foot of the Green Mountains and its association with colonial and Revolutionary history. Seth Warner and Ethan Allen made Bennington their home, and its citizens had much to do with the formation of Vermont as a separate state. The state soldiers' home is located at Bennington village, and in Bennington Centre a monument over 300 feet high has been erected to commemorate the famous battle. The township is a manufacturing center, with a large output of knit goods, woolens, hosiery, collars and cuffs. Township matters are under the control of a board of selectmen, while each village manages its own local affairs. Population of the township in 1910, 8,698. E.S.H.

BENTON, THOMAS HART (1782-1858), an American statesman, for thirty years United States Senator for Missouri and an influential factor in every important public question. He was born near Hillsborough, N. C., began to study at the University of North Carolina but removed to Tennessee, and in 1811 was admitted to the bar in that state. While in the Tennessee legislature, to which he was elected in 1809, he became acquainted with Andrew Jackson, and in the War of 1812 he joined Jackson's staff. In 1813 a quarrel between these friends led to one of those shooting affrays so common in the frontier country, and both were injured. Not until years afterward were the two reunited in friendship.

THOMAS HART BENTON

After attaining the rank of lieutenant-colonel in the war, Benton removed to Saint Louis, where he practiced law and founded the Missouri *Inquirer*, a newspaper of pronounced proslavery temper. When Missouri was admitted to the Union in 1820 he was elected to the United States Senate, and then began his generation-long term. Intensely loyal to the West and its needs, he worked for the construction of a transcontinental railway and for the opening of the mineral lands to settlement. It was in connection with Jackson's fight against the United States Bank that Benton came most prominently before the public, winning by his ardent advocacy of a gold and silver currency the nickname of "Old Bullion." He also took an active part in the Oregon boundary discussion and the disputes over the annexation of Texas, and was entirely in favor of the Mexican War. Having lost his seat in the Senate in 1850 through his opposition to the compromise proposed by Clay, he was elected two years later to the House of Representatives, but in 1854 he retired from public life and completed his great book *Thirty Years' View*. This has become a valuable historical record.

BENTON HARBOR, MICH., a city in Berrien County, in the southwestern part of the state, sixty miles northeast across Lake Michigan from Chicago. It is inland one and a half miles from the lake, with which it is connected by ship canal, and is at the junction of the Saint Joseph and Paw Paw rivers. The city has the advantages of Saint Joseph harbor at the mouth of the canal and has steamboat lines to Chicago and Milwaukee. It is served by the Cleveland, Cincinnati, Chicago & Saint Louis, the Pere Marquette and the Michigan Central railroads. The population in 1910 was 9,185; it was 10,302 in 1914. The area is nearly three square miles.

Benton Harbor is in the Michigan fruit belt and is one of the largest peach markets in the world. In 1915 the total boat and rail fruit shipments amounted to 10,000 car loads. The exportation of fruit, grain and lumber, fruit packing, cider and vinegar making, fruit evaporating and manufactories of fruit baskets, spray pumps, lumber and furniture are the important industries. Medicinal water from mineral springs found in the vicinity is bottled and shipped. Many visitors come directly to the springs.

The city was settled about 1860. It has several small parks, several noteworthy bank buildings, a Carnegie Library and a public hospital. H.G.K.

BENZENE, *ben' zene*, or **BENZOL**, *ben' zahl*, a colorless liquid lighter than water and having a pleasant odor, which burns with a smoky flame. The lighting power of illuminating gas is due in a measure to the benzene it contains. Benzene readily dissolves fats, resins, rubber, sulphur and iodine, and is very important because of the great number of compounds obtained from it. One of these, nitrobenzene, has the odor of bitter almonds and is used to produce the flavor of almonds in essences and perfumery, but its chief use is in the manufacture of aniline. Benzene is sometimes called benzol. It should not be confused with *benzine,* which is obtained from the distillation of petroleum. See ANILINE; BENZINE.

BENZINE, *ben' zin* or *ben' zeen,* a light colorless liquid obtained in refining crude petroleum. It has an odor resembling that of kerosene, is very inflammable, and evaporates rapidly when exposed to the air. When mixed with air its vapor is highly explosive, and serious accidents may occur if a lighted pipe or cigar, or a lighted match or lamp is brought near benzine exposed to the air. Bringing it near a hot stove may also cause an explosion. Benzine is used for cleaning type and rollers of presses in printing offices, for removing spots from clothing, for dissolving fats, oils and resins, and in the manufacture of illuminating gas. See PETROLEUM. J.F.S.

BENZOATE, *ben' zo ate,* **OF SODA,** a compound of soda and benzoic acid, used to a considerable extent in preserving food substances that are liable to decay or ferment from exposure to the air. The Federal government allows benzoate of soda to be used in quantities not larger than one-tenth of one per cent, careful investigation by the Referee Board of the Department of Agriculture having shown that such small quantities are not injurious to health. On account of the prejudice existing against it, however, many manufacturers of foods prefer to avoid its use. J.F.S.

BEOWULF, *ba' o woolf,* an Anglo-Saxon epic poem, the most important relic of Old English literature. The only existing manuscript of this was written about A. D. 1000 and may now be seen in the British Museum. The poem recounts the adventures of the hero Beowulf, especially his delivery of the Danish kingdom from the half-human monster Grendel and his equally formidable mother, the slaughter of a fiery dragon and the death of the hero from wounds received in the conflict. The character of the hero is attractive through his noble simplicity and disregard of self. In imaginative quality and strength this poem compares well with the epics of Homer.

BERBERS, *bur' burs,* the people of the Mediterranean coast of Africa. Their name and the word *Barbary* are probably derived from the Greek *Barbaros,* which meant *one who babbles*—hence, anyone who did not talk Greek, that is, a foreigner. Our word *barbarian* is from the same source.

Berbers are often nearly black, with shiny brown hair. They are sparely built, not tall, but strong and graceful. They cultivate the land after a primitive fashion, and raise many sheep, goats and camels. They live usually in tents, but are finding their way into towns in increasing numbers. There are three distinct types, known as Tuaregs, Kabyles and Shilluhs. The Tuaregs are desert wanderers, dreaded by all peaceful tribes, and call themselves "Amazirg," meaning *the free.* The Kabyles and Shilluhs have been brought more under the influence of civilization.

BEREA, *be re' a,* **COLLEGE,** a school at Berea, Ky., founded in 1855 for the purpose of educating both white and negro youths of both sexes. It is situated in the Cumberland Mountains and has gained a wide reputation by providing courses of study and methods of training especially suited to the needs of the mountain people of this region. Most of its students come from the mountain sections of Kentucky, Tennessee, North Carolina, Virginia and West Virginia. It has a model school, and industrial, academic, normal, collegiate, commercial and music departments. Since 1904 the work for whites and negroes has been carried on in separate institutions, by requirement of the state law.

Berea College is non-sectarian, has about 55 professors, over 1,700 students, and a library of 25,000 volumes.

BERESFORD, *ber' es ferd,* LORD CHARLES WILLIAM DE LA POER (1846-), a British admiral who rendered great assistance in raising the British navy to a high state of efficiency. Entering the navy in 1868, he gave indication at once of great ability and proved gallantry in action. At the bombardment of Alexandria in 1882, while in command of H. M. S. *Condor,* he distinguished himself and received special recognition. Entering Parliament soon afterwards, he devoted his energies to forcing on the government a complete reorganization of the naval program. In this he was only partially successful, and he brought on himself the enmity of the Admiralty, or navy department. In 1906, however, he was appointed commander-in-chief of the Channel Fleet. In 1910, his term of command having expired, he was again returned to Parliament as member for Portsmouth. Owing to his advanced age he did not take a command in the War of the Nations.

BERGAMOT, *bur' ga mot*, a small evergreen tree whose leaves and flowers resemble those of the bitter orange, and whose pear-shaped fruit has a smooth rind the color of a lemon. The rind yields a greenish-yellow fluid known as *oil of bergamot*, which is chiefly used in making perfumery. The tree is cultivated in Calabria (Italy) and in France. The name is also applied to a number of different pears, and in America to several pleasingly-fragrant plants of the mint family.

BERGEN, one of the chief seaports of Norway. It is beautifully situated amidst picturesque scenery, at the head of a deep bay, with a background of lofty mountains, 186 miles northwest of Christiania. The older portion of the city, founded in 1070, is built of wood and has several times been ravaged by fire and pestilence. The modern town is well built, with wide streets and houses of brick and stone. It has some fine buildings, among which are the cathedral, museum, library and public schools. The principal industries are shipbuilding and barrel making, and the town carries on an extensive trade in dried fish, herring, tar, grain and flour. Although only a little north of the latitude of Petrograd, Bergen has a much milder climate than that city, owing to its sheltered position and the ocean currents from warmer latitudes. The rainfall is unusual, however, averaging annually about seventy-three inches.

Early in the fourteenth century Bergen became an important member of the Hanseatic League (which see). The city was the birthplace of a number of notable men, among them being Ole Bull and Edward Grieg, the musicians, and the poet Holberg. Statues of Holberg and Bull have been erected by the city. In January, 1916, a great fire destroyed a large part of the city. Population in 1910, 76,867.

BERGH, *burg*, HENRY (1820-1888), founder of the two American societies that strive to prevent cruelty to children and to animals. He was born in New York City and educated at Columbia University. In 1864 he resigned from a position with the American legation at Petrograd, then known as Saint Petersburg, and soon after his return to America became interested in the work of mercy to which he devoted the rest of his life.

Though ridiculed and opposed, he succeeded in 1866 in having incorporated, under the laws of New York State, the Society for the Prevention of Cruelty to Animals, and two years before his death he saw this society established in thirty-nine states of the Union, and in Canada, Brazil and Argentina. In 1874 he rescued a little girl from brutal treatment, which led to the founding of the Society for the Prevention of Cruelty to Children, now established in nearly every state in the Union and the Canadian provinces. These societies have back of them the powerful arm of the law, and can bring about the arrest and punishment of those persons who are inhuman in their treatment of children or animals. Bergh also invented artificial pigeons for the sportsman's gun, and was the first to suggest the use of an ambulance for removing injured animals from the street.

BERGSON, HENRI LOUIS (1859-), a French philosopher whose theories have gained for him a wide following. Intuition and not intellect is to him the trustworthy guide; all former philosophies he rejects, at least partially, because intellect had too large a part in their construction. Time to him is the great reality, but he means by time not just what is usually understood. In the ordinary sense of yesterday, to-day and to-morrow time does not exist, according to Bergson, for the past is gone, the present vanishes before one can say "It is here," and the future no one can state positively will ever be. But time in the sense of pure duration, as one feels it in dreams, is the only sure foundation upon which life rests.

Bergson was born in Paris, of Jewish parentage, studied in that city, and by 1900 had become so well known as a teacher that he was appointed to the chair of philosophy in the College de France. Though he has never striven for general popularity, his doctrines have gained such a vogue that his lecture rooms have been crowded constantly, and people have been turned away. In 1913 he visited the United States, lecturing at Columbia University, and in 1914 was elected to the French Academy. His publications include *Time and Free Will*, a setting forth of the main principles of his philosophy; *Laughter*, an essay on the value of the comic; and *Matter and Memory*, a discussion of the relation between body and mind.

BERING, *be' ring*, **SEA**, that part of the North Pacific Ocean which lies between Alaska and the eastern coast of Asia. Bering Strait connects it on the north with the Arctic Ocean; on the south the narrow Alaskan peninsula and the Aleutian Islands separate it from the open Pacific. From north to south it is about 1,000 miles in extent, from east to west 1,500 miles.

During winter it contains floating and pack ice, and most of the year its waters are covered with a dense fog. Of its islands by far the most

IN HONOR OF VITUS BERING
The man, and the sea and strait named for him.

important are the Pribilof, the home of the largest fur seal colony in the world.

Bering Island, on which the explorer for whom it was named died, is the most westerly of the Aleutian chain and lies off the east coast of Kamchatka. It is barren, rocky and uninhabited. *Bering Strait* is the narrow channel which separates North America from Asia at their nearest points. At its narrowest place, between East Cape and Cape Prince of Wales, it is thirty-six miles wide, and were it not for the fog which always shrouds the water the highlands of Asia might be visible from America. Though Bering explored this strait, it was Captain Cook, in 1778, who first thoroughly studied it.

Vitus **Bering (1680-1741)** was a famous Dutch navigator who gave his name to these northern waters and islands. The courage which he displayed as captain in the navy of Peter the Great, during the Swedish wars, led to his being chosen to command a voyage of discovery in the Sea of Kamchatka. In 1728, and later, he examined the coasts of Kamchatka, Okhotsk and the north of Siberia, discovering the relation between the northeastern Asiatic and northwestern American coasts. Returning from America in 1741, he was wrecked upon the island which bears his name, and died there.

Related Topics. Though not intimately connected with the geography of this section, Bering Sea has a close relation to the following topics:

Alaska	Pribilof Islands
Bering Sea Controversy	Seal

BERING SEA CONTROVERSY, a dispute between Great Britain and the United States which centered about the valuable seal fisheries on the Pribilof Islands. Since 1867 the United States had carefully regulated by license the killing of seals on the Pribilof Islands, where there is located the largest seal rookery in the world (see FUR SEAL), and had received a bounty of ten dollars for each skin; but after 1886 unlicensed fleets of Canadians and Americans were organized to kill the seals during feeding time, when they were more than three miles from shore, or beyond the jurisdiction of the United States government. Many of the animals killed were females, and it became apparent that the herd under such treatment would soon be exterminated.

In order to restrict unlicensed killing, the United States set up a claim that Bering Sea was a closed sea, that is, subject to the exclusive jurisdiction of the United States. This was protested by Great Britain, and by a treaty in 1892 the question was referred to arbitration. The tribunal, which consisted of one Englishman, one Canadian, two Americans and one representative each of France, Italy and Sweden and Norway, reported August 15, 1893, a decision which was generally unfavorable to the United States. However, it established a closed season from May 1 to July 31, forbade the killing of seals in the open sea within sixty miles of the Pribilofs and prohibited explosive weapons. These restrictions proved ineffectual, and in spite of almost constant negotiations since that time no satisfactory solution of the problem has been found, and in consequence there is danger that the valuable fur seals will be exterminated.

BERKELEY, *burk' li,* SIR WILLIAM (1610-1677), a colonial governor of Virginia whose faithlessness and obstinacy in dealing with the uprisings of the Indians in 1676 caused the revolt known as Bacon's Rebellion (which see). He was born near London, studied at Oxford and became governor of Virginia in 1641. When Cromwell gained control of the British government Governor Berkeley offered an asylum in Virginia to loyalists and kept the colony loyal to the king until 1652. In that year he was compelled to resign, but he was again chosen governor in 1660.

His second term of office was marked throughout with dissatisfaction, and his harshness in punishing the leaders in Bacon's rebellion displeased even Charles II, who said, "The old fool has taken more lives in that naked country than I did for the murder of my father." Berkeley was recalled to England by the king in 1677, and died soon after reaching home.

BERKELEY, CALIF., a residential suburb of San Francisco, situated in Alameda County, near the eastern coast of San Francisco Bay. Adjoining it on the south is Oakland, the county seat. The Southern Pacific and Santa Fe roads provide excellent railway accommodations, and electric and ferry lines connect with San Francisco, seven miles across the bay; five electric lines join Berkeley and Oakland. The population, of which three-fourths are native Americans, shows a rapidly growing city; in 1910 it numbered 40,434, and in 1914 had increased to 52,105. The area of the city is eight and a quarter square miles.

A bird's-eye view of Berkeley gives one the impression of an immense park dotted with attractive homes of modern architecture. It has a fine location on the heights, commanding a beautiful view of San Francisco Bay and of the Golden Gate. There is a water front of three miles, offering inducements to manufacturers; a new wharf was constructed at a cost of $100,000. The city has manufactures of refined petroleum, soaps, health foods, elevators, aeroplane motors and pumps. One of its fruit establishments has an annual output of $1,100,000; that of a cocoanut-oil plant is $750,000, and that of an ink company, $500,000. The fisheries are also important. Berkelev has a United States post office which was completed in 1915 at a cost of $180,000; a $300,000 high school building, a $150,000 city hall, a Masonic Hall, Newman Hall and an armory building.

The city is noted as the seat of the University of California, with an enrollment of over 8,500 students (see CALIFORNIA, UNIVERSITY OF). Three theological seminaries, the State Agricultural and Mechanical College, the California School of Arts and Crafts, the State Institution for the Deaf, Dumb and Blind, Boone University Academy, Snell Seminary for Girls and Saint Joseph's Academy are also located here. The library was the joint gift of Mrs. Rosa Shattuck and Andrew Carnegie.

The site which Berkeley now occupies was a farming district until it was chosen as the seat of the university. It was settled in 1868, was incorporated as a town in 1878 and was named in honor of George Berkeley, bishop, philosopher and author, on account of his famous line, "Westward the course of empire takes its way." In 1909 the city adopted the commission form of government. W.D.

BERKSHIRE, *burk' shir,* **HILLS,** one of the most beautiful resort regions of the Eastern United States. These hills are not a separate range but are a continuation into Berkshire County, Mass., of the Green Mountains of Vermont. They attain in Greylock a height of 3,535 feet, the greatest altitude in the state. The wooded slopes are cut by mountain streams, which find their way to the Hoosac, the Housatonic and the Deerfield rivers. Quaint inns are located in picturesque spots. The towns of Pittsfield, Stockbridge, Lenox and Great Barrington, besides various summer resorts, are situated in this region. The Hills provide one of the most delightful motoring trips in the eastern part of America, and are visited by thousands of tourists every year.

Unter-den-linden — Bismarck Memorial — The royal palace

BERLIN, the capital city of the former kingdom of Prussia and from 1871 to 1918 of the German Empire. Greater Berlin, which includes numerous suburbs, stands third in size among the world's great cities. Its population according to the last census (1910) is 3,974,300. Only Greater London and Greater New York are larger. The city proper, with a population of 2,071,257, ranks fifth among the world's cities. Greater Berlin has an area of 1,376 square miles, but Berlin proper covers only twenty-four and one-half square miles.

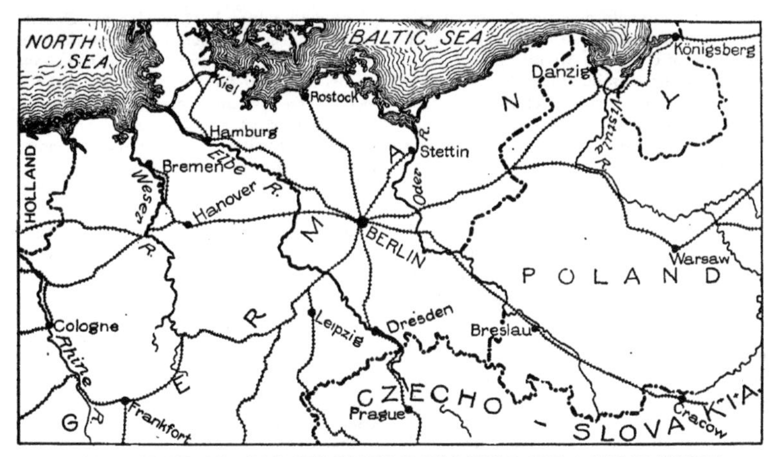

BERLIN, CENTRALLY LOCATED IN WHAT WAS UNTIL 1918 A GREAT EMPIRE

The city lies in the same latitude as Edmonton, Alberta, and Southern Labrador, but its climate and average temperature correspond to that of New York. It is as far north of the equator as the Strait of Magellan is south of it. It is 180 miles from Hamburg and eighty-four miles from Stettin, and has water communication with both ports through the Elbe and the Oder rivers and the canal connecting them. It is 427 miles from Vienna, 674 miles from Paris, 746 miles from London, 1,048 miles from Rome, 1,098 miles from Petrograd and 1,699 miles from Constantinople.

Famous Buildings and Monuments. Berlin is famous for its imposing buildings, beautiful parks and splendid avenues. The center of the social and political life is Unter-den-Linden, one of the most famous streets of the world. At the eastern end of this promenade is the palace formerly occupied by Emperor William II, a rectangular brown sandstone building with over 600 rooms. At the western end is the famous Brandenburg Gate surmounted by the bronze quadriga, or chariot of victory, which Napoleon carried off to Paris in 1807. Between the palace and the Brandenburg Gate, a distance of less than a mile, are the French and Russian embassies, the University of Berlin, the royal library, the opera house, the palaces of Emperor William I and Emperor Frederick III, the finest hotels and the most elegant shops.

Of all the public buildings, by far the most magnificent is the Parliament building (Reichstagsgebaude), in a modified classic style. Though unattractive in some of its details, it is a strikingly-powerful architectural conception. Also notable is the new cathedral, dedicated in 1905. It is in the Italian Renaissance style, with a dome rising to a height of 380 feet, the loftiest building in Berlin. Almost equally conspicuous is the Emperor William Memorial Church, completed in 1895 as a memorial to the "Old Kaiser," as the Germans still affectionately call the grandfather of the last of the Hohenzollerns.

Berlin has hundreds of monuments, in all parts of the city. Perhaps the most famous is Rauch's equestrian statue of Frederick the Great, which is on Unter-den-Linden, near the palace. Near by is the national monument of Emperor William I. In the Königs Platz, the great square in front of the Reichstag building, is the Victory Column, erected to celebrate the victories of the Franco-German War of 1870-1871. Statues of Bismarck, Von Moltke and Von Roon stand not far away. From the Victory Column it is but a step to the Avenue of Victory (Sieges-Allee), a promenade adorned with statues of the thirty-two Hohenzollerns who ruled in Prussia before the formation of the new German Empire. This statuary was a gift of Emperor William II to the city.

Intellectual and Artistic Life. While the city, the kingdom of Prussia and the Empire have all been spending money freely to make Berlin outwardly a great capital, people of ability, of brains and of genius, have been drawn to it to make it a capital in other senses. Its university and schools, its art galleries, its orchestras and its theaters, are newer than those of cities like Munich, Dresden and Leipzig, but are crowding their older rivals for first place. The University of Berlin (which see), though founded as late as 1810, is already the first in attendance among all German universities.

In painting and sculpture Berlin holds a leading place. Its former royal museums and the national gallery are filled with ancient and modern art treasures, providing excellent opportunities for study. Museums and galleries in other cities are richer in special limited fields, but few offer a greater variety of excellence. In music Berlin does not yet rival Leipzig, but its conservatory and philharmonic orchestra have great influence. In theatrical affairs Berlin takes first place, not only in the production of modern German dramas, but also in translations from Moliere, Shakespeare, Ibsen and George Bernard Shaw.

Commerce and Industry. The geographical position of the city makes it a natural market and distributing point for the agricultural regions of Prussia, Austria and Russia. It is the hub of the railway systems of North Germany, the great trunk lines (shown on the accompanying map) furnishing communication in every direction. Commerce is in cattle, wool, lumber, coal and grain. Berlin is the center for trade in grains, its position corresponding in this respect to Chicago in the United States. It is also the chief market for speculation and legitimate transactions in stocks and bonds, and may fairly claim to be the financial center of the Empire.

More than one-half of the city's working population is engaged in manufactures. The transformation of iron and steel into useful machinery is, perhaps, the greatest industry, no fewer than a hundred large firms being engaged in this branch alone. Munitions of war, railway supplies, bicycles, steel pens, electrical supplies and sewing machines are important products. The production of Berlin's breweries now rival those of Munich. The city is a publishing center, and also is a leader in the manufacture of soaps, lamps, china, pianos, furniture and ladies' cloaks.

Government. The government of the city under the empire was partly under the control of the Prussian Minister of the Interior, and partly of the local civil authorities. All police functions, including the regulation of building, crime and passports, and the prevention of fire, belonged to a branch of the Interior Department. The civil authorities had jurisdiction over the water supply, street lighting and cleaning, drainage, and the care of the sick and the poor. There was an elected common council of 144 members; the heads of the departments formed the Stadtrat; its thirty-two members were elected by the council, but had to be approved by the king. The common council elected the burgomaster (mayor) and the chief burgomaster, who did not have to be residents of Berlin at the time of their election. They were usually, in fact, men who had distinguished themselves as mayors of other cities. Under the new and untried republic of 1919 the city government has not been put on a permanent basis.

History. Of Berlin's beginnings nothing definite is known, but by the fifteenth century it was an important community, in the mark of Brandenburg. About 1500 it became the official residence of the margraves of Brandenburg, but it was practically wiped out during the Thirty Years' War. Its renewed prosperity was due to the Great Elector, Frederick William. By his order public buildings and private dwellings were rebuilt, new suburbs were laid out and strong fortifications were added. He also laid the foundation for its future greatness by constructing a canal to connect the Spree and the Oder rivers. At his death, in 1688, Berlin had a population of 20,000 people, compared with 6,000 in 1648.

Under the kings of Prussia, the successors of the Great Elector, the growth and beautifying of the city continued. Frederick the Great gave special encouragement to the manufactures of silk and cotton; and Berlin, formerly little more than a small garrison town, began to take on the character of an industrial center. During the nineteenth century it made enormous strides, especially after it became the capital of the German Empire in 1871. Most of Berlin, in fact, is of nineteenth century construction; the important buildings which are old total fewer than a dozen.

As the capital of the new Empire, Berlin presented a sorry spectacle in 1871. Its sanitary conditions were worse than those of any other large city in Europe. Many of its streets were unpaved, and the cobble-stones on the remainder were but a slight improvement. Open sewers and drains, public pumps and underground tenements were characteristic. To-day this is all changed, partly because the state and the people have worked together to

Outline and Questions on Berlin

I. Size and Location

(1) Rank among world's cities
 (a) City proper
 (b) Greater Berlin
(2) Population
(3) Position with reference to other large cities

II. Description

(1) Streets
 (a) Unter-den-Linden
(2) Public buildings
(3) Monuments
(4) Educational institutions
 (a) University of Berlin
(5) Museums and galleries
(6) Drama

III. Commerce and Industry

(1) Communication
(2) Trade
(3) Manufactures
 (a) Machinery
 (b) Brewing
 (c) Other products

IV. Government

(1) Functions of the Department of the Interior
(2) Local officials
 (a) Mayors
 (b) Council

V. History

(1) Early history
(2) The Great Elector
(3) The city's newness

Questions

What is there unusual about the history of the University of Berlin?

During what century did the city have its greatest growth?

What is the most noted street in Berlin?

How large a proportion of the people earn their living from manufactures?

Why can it be said that Berlin ranks both third and fifth among the cities of the world?

In what English dramatists are the people especially interested?

What change has there been in the sanitary conditions of the city since it became the capital?

What place is as far from the equator on the south as Berlin is on the north?

With what American city may it be compared in respect to its trade in a great commodity?

What is the most imposing building?

How did the city commemorate the German victories in the Franco-German War?

What ruler laid the foundation for the greatness of Berlin?

What city of North America is in the same latitude?

What is the tallest building in the city?

How does Berlin rank among the cities of Germany as regards art and literature?

make it a great capital. More than anything else, the new national spirit of the German nation made this result possible in a short time. Berlin has been one of the cleanest and most healthful cities in Europe, but after the flight of the emperor in November, 1918, it entered upon a period of stress. Fighting among factions for control took its heavy toll of deaths. Gunfire damaged stately government buildings. Order was certain to follow after hysteria passed, but should the new republic survive, Berlin will miss for many years the pomp and splendor of the court of its Hohenzollerns.

University of Berlin, an institution of learning, which, though it does not date back to the Middle Ages as do so many of the great European universities, has attained a commanding position among the universities of the world. Plans for founding it were begun during the latter part of the eighteenth century, but the Napoleonic struggle interfered, and not until 1810 was it really established. Its full name is the Royal Frederick Wilhelm University of Berlin, for it was founded during the reign of Frederick Wilhelm III of Prussia. After somewhat more than a century of existence, it stands as the most prominent university of Germany, and during its history it has numbered many illustrious scholars among its instructors, of whom may be mentioned the Grimm brothers, Hegel, Fichte, Ranke, Virchow, Mommsen, Niebuhr, Treitschke, Helmholtz and Harnack. In its departments of theology, law, medicine and philosophy, there was, before the outbreak of the War of the Nations (1914), a combined enrollment of almost 11,000 students. Several American universities have exchanged professorships with the University of Berlin. Attached to the university are the Royal Library, one of the largest libraries in the world, housed in a beautiful new palace; the technological, agricultural and veterinary institutes, and the so-called "institute for research." W.F.Z.

Consult Siepen's *Berlin;* Shaw's *Municipal Government in Continental Europe.* There are no books printed in America relating particularly to Berlin, but the above English books are available.

BERLIN, a town in Ontario, since 1916 known as KITCHENER (which see).

BERLIN, CONGRESS OF, an assembly of representatives of the powers of Europe which emphasized the fact that Turkey in Europe was not to become the prey of any one nation, and that questions relating to it were to be settled in conference. By victories in the

Russo-Turkish War of 1878 Russia gained the power practically to dictate terms of peace to Turkey, and the Treaty of San Stefano plainly looked toward the annihilation of Turkey in Europe. This roused the jealousy of the other powers, and in June, 1878, representatives from Germany, Austria, France, England, Italy, Russia and Turkey met at Berlin to modify the objectionable terms. Seldom has a more illustrious body of statesmen met for any purpose, and they did their work thoroughly, denying Russia much which the Treaty of San Stefano had conceded.

That treaty had created Bulgaria and Eastern Rumelia an independent state; however, the Congress of Berlin allowed self-government to Bulgaria, under Turkish auspices, but made Eastern Rumelia merely a Turkish province. Bosnia and Herzegovina remained nominally under Turkish sovereignty, but Austria-Hungary was made administrator of their affairs; Montenegro, Serbia and Rumania were confirmed in their independence, and Great Britain was made practical owner of Cypress. All in all, the Congress of Berlin did much to make the map of Europe what it remained throughout the rest of the century. It was disturbed but little until the Balkan Wars of 1912 and 1913, although in 1908 Bulgaria proclaimed itself a sovereign state.

BERLIN DECREE. See CONTINENTAL SYSTEM.

BERLIN, N. H., noted for its large paper mills, is situated in Coos County, about midway between the geographical center and the northern border of the state, on the Androscoggin River. Portland, Maine, is ninety-eight miles southeast, and Manchester, N. H., is 175 miles south. Railway transportation is provided by the Boston & Maine and the Grand Trunk lines. The first settlement was made in 1827, and the city was incorporated in 1897. The population increased from 11,780 in 1910 to 13,013 in 1914; more than 7,500 are French-Canadians and 5,000 are Americans. The city has an area of three and one-half square miles.

Berlin is located in the heart of the White Mountains, a region noted for wonderful mountain scenery. The Androscoggin River has a rapid descent here of nearly 200 feet in the course of a mile, and while the falls contribute greatly to the picturesque scenic effect they also furnish power for manufacture. The city has the largest sulphite-pulp mill in the United States, the daily output being over 400 tons; this product is a wood pulp used in making paper, the manufacture of which is so extensive that the city is known locally as the *Paper City*. There are also large lumber mills, knitting mills, foundries and chemical plants. Berlin has a city hall, recently built at a cost of $100,000, a $125,000 Y. M. C. A. building, a high school, a business college and a Carnegie Library.

BERLINER, *ber' le ner,* EMILE (1851-), the inventor of the telephone transmitter, or microphone, and of the method of reproducing sound utilized in one style of talking machine. He was born and educated in Germany, emigrated to the United States in 1870, and nine years later became chief instrument inspector for the Bell Telephone Company. In addition to the microphone, invented in 1877, he discovered several other means of improving the telephone service, the patents for which are held by the Bell Company.

Berliner's inventions in connection with the talking machine date from 1887. In the old-style machine the sharp-pointed instrument, known as the stylus, moved through a groove of varying depth, and the reproduction of sound was inaccurate and unsatisfactory. Berliner conceived the idea of using for the record, instead of a cylinder, a disk on which a horizontal record of uniform depth but varying direction should be cut. In reproducing the sound the stylus, as the invention developed, is guided only by the groove through which it moves, and not by a feed screw as formerly. He also invented and perfected the modern method of making duplicates of disk records and saw his prediction fulfilled that famous singers would some day earn great sums of money by having their voices recorded. The talking machine company which owns his devices has spent $500,000 in sustaining the patents.

The air-cooled engine with revolving cylinders, now used extensively on aeroplanes, is also the invention of Berliner.

BERLIOZ, *bare le ose',* HECTOR (1803-1869), one of the most intellectual and poetical composers of music that France has ever produced. He studied at the Paris Conservatory, where, in 1830, his *Sardanapalus* won first prize—expenses for foreign travel. Thereafter he gained a wide reputation for the composition of so-called *program music,* in which a story is realistically expressed by the music; and for his masterful work in the arrangement of orchestra music, in which he laid the foundation for Wagner, Liszt and Richard Strauss. His symphonies, *Harold in Italy* and *Romeo and*

Juliet; his opera, *The Trojans,* and his celebrated *Te Deum* are now considered masterpieces, though the composer was little appreciated in his own day.

BERMUDA or **SOMERS ISLANDS,** a group of about 350 islands and islets in the Atlantic Ocean, midway between the West Indies and the North American coast. By steamer they are forty-eight hours distant from New York

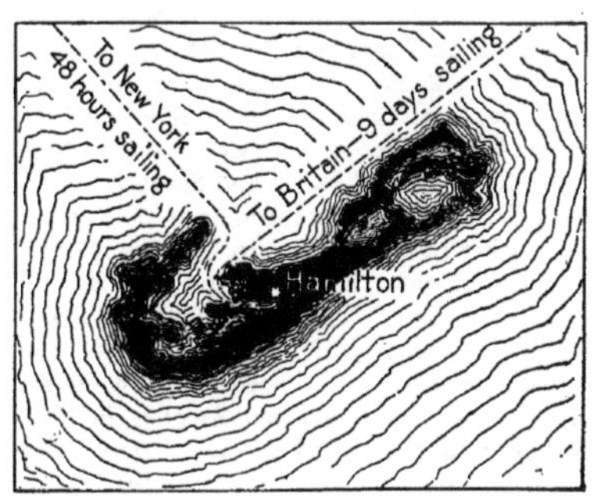

THE FORM OF THE BERMUDAS
Hamilton, the capital, and sailing distances to New York and Europe.

City. Twenty of the group are inhabited, forming one of the most important of British possessions in the Atlantic. They are the most northerly islands in the world entirely of coral formation, and unlike other coral islands they contain hills and ridges which rise to a height of nearly 300 feet. The largest islands are Bermuda, named after Juan Bermudez, who discovered the group in 1515, Saint George, Ireland and Somerset. The total area is 12,000 acres, of which Bermuda contains 9,000 acres.

The climate is especially pleasant and healthful. In winter frost is unknown, and the temperature seldom falls below 50° F. The sharp contrast between summer and winter on the mainland is here entirely absent, for in summer the heat is seldom greater than 87° F., and delightfully cooling breezes blow from the sea. The Bermudas have long been regarded as almost an ideal health resort, and are visited both summer and winter by great numbers of Americans and Englishmen. The scenery is magnificent and many enjoyable drives may be taken, but so conservative are the people that no automobiles are allowed on any of the islands.

Ireland Island is an important British naval station, forming the winter headquarters of the North Atlantic fleet. A garrison of 2,500 men is permanently maintained. The capital, Hamilton, on Bermuda, with a population of 2,627, is described elsewhere in these volumes. The islands were first settled in 1609 by Sir George Somers, supported by a number of colonists from Virginia. They are now administered by a governor, assisted by two councils of appointed members and an assembly of thirty-six members elected by the people. Population in 1913, 19,935, of whom 7,060 were whites.

BERN, *burn,* the capital of the Swiss canton of Bern, and since 1848 the capital of the republic of Switzerland. Its name is taken from the German word *Bären,* meaning *bears,* and was adopted, according to legend, because many bears were killed on the day the city was founded. In consequence of this tradition the bear is almost a sacred animal in Bern, and the municipality keeps a Bears' Den, on the right bank of the Aar, in which a number of fat, brown bears are always on view. Bears in wood and stone are everywhere to be seen throughout the city as ornaments on foundations and buildings.

Bern is nearly 1,800 feet above sea level, and is surrounded on three sides by the river Aar. No other city in Switzerland excels it in beauty, and it is among the most regularly-built towns in Europe. Its fountains, its arcade-covered walks, its fine bridges and quaint old shops are a joy to the tourist. Among the public buildings are the great Gothic cathedral, built between 1421 and 1502; the Church of the Holy Spirit; the Federal-council buildings, or Parliament house, commanding a splendid view of the Alps; the university; the town house, a Gothic edifice of the fifteenth century, and the mint. Bern has an academy and an excellent public library. The manufactures are woolens, linens, silk stuffs, stockings, watches, clocks and toys. The city was founded in 1191, became a free city of the empire in 1218 and in 1353 entered the Swiss Confederacy. Population in 1910, 85,264. See SWITZERLAND, subhead *History*.

BERNARD, *bur'nard* or *bur nahrd',* SAINT (1091-1153), one of the most influential Roman Catholics of the Middle Ages, whose life and character, to the people of his own day, were the expression of their highest ideals. He was born in Burgundy, France, and at the age of twenty-two became a monk of the Cistercian Order. Soon after, he founded the famous Cistercian monastery at Clairvaux, becoming its first abbot, and it was to him that the Order owed its wonderful growth and influence.

Saint Bernard dominated the affairs of the Church for the rest of his life and in 1128 gave the rules of government to the newly-founded Order of Knights Templars. His stirring eloquence inspired the Christians to undertake the second Crusade (see CRUSADES), but of the innumerable host that marched to the holy war only a remnant ever returned, and his disappointment over this failure was a blow from which he never recovered.

He wrote a large number of epistles, sermons and treatises, and his beautiful hymns, *Jesus, the Very Thought of Thee* and *O, Sacred Head, Now Wounded,* are sung in Christian churches to-day. Martin Luther's words concerning him are famous: "If there ever lived a God-fearing and holy monk, it was Saint Bernard of Clairvaux."

BERNHARDT, ROSINE, known as SARAH (1845-), one of the world's greatest actresses, who through fifty years of almost uninterrupted work retained unimpaired her vitality and her artistic gifts. The impression that her art has made on people is expressed in the name so frequently applied to her—"the Divine Sarah."

SARAH BERNHARDT

She was born in Paris, and though of Jewish descent was baptized with Christian rites, in accordance with her father's wish, and spent the early years of her life in a convent. At the Paris Conservatory, where she studied from 1858 to 1862, she won second prizes for tragedy and comedy, but her first appearance on the stage in 1862, in a small part in Racine's *Iphigenia,* was in no way exceptional. After an unsuccessful trial of burlesque she turned to serious parts, and as Cordelia in *King Lear,* as the queen in Hugo's *Ruy Blas* and as Zanetto in *The Passer-by,* she made it known to the theatrical world that a new actress of rare promise had appeared.

Joining the company at the French Comedy Theater in 1872, she achieved a series of triumphs, one of her most remarkable performances being the rôle of Dona Sol in Hugo's *Hernani* (1877). In 1879 she acted with great success at the London Gaiety Theater. On her return to Paris she suddenly terminated an engagement with the management of the French Comedy Theater, a breach of contract which cost her $20,000. During 1880 and 1881 she toured Denmark, Russia and America, including in her repertoire the famous *Camille* of Dumas. In 1882, in London, she married a Greek actor, Jaques Damala, from whom she was separated a year later. Her next appearances on the stage were in a number of plays by Sardou, who wrote for her especial use *Theodora, La Tosca* and *Cleopatra.*

American Tours. Bernhardt's first American tour began in 1886, and she was everywhere received with great enthusiasm. Between 1891 and 1893 she visited North and South America, Australia, and the chief countries of Europe, and on her return to Paris in 1893 she became manager of the Theater of the Renaissance. Five years later she established the Sarah Bernhardt Theater, of which she is still the manager, and which she opened with a revival of *La Tosca.* She revisited America in 1900, 1911 and 1913, the first of these tours being devoted to Rostand's *L'Aiglon,* with the famous Coquelin in the leading male rôle. Her 1913 tour consisted of vaudeville performances of single acts from a number of plays, and the presentation of a new one-act play entitled *A Christmas Night During the Reign of Terror.*

During the American engagement of 1913 she suffered from an accident which later developed into blood poisoning and made necessary the amputation of a leg in February, 1915. Yet she learned to walk on an artificial leg and resolutely returned to activity, although not to the stage. In 1914 she appeared in a moving-picture production of *Queen Elizabeth,* which she said gave her great joy because it would make her live a thousand years.

In October, 1916, this great actress returned to the United States to appear during the following winter in the principal cities. She was accorded the greatest reception which ever marked her American experiences.

Her services to America have been aptly summarized by a writer in these words: "Of French literature we knew nothing. She opened that great treasure-house to us. She made living realities of great dramatists and created an intellectual sympathy between France and America." The countries of Europe which the great Bernhardt has visited owe her a similar tribute.

Estimate of the Actress. Bernhardt is also a gifted painter and sculptor, and has written a volume of *Memoirs* and two plays. In recognition of her great talents she was elected to membership in the French Legion of Honor in 1914. As an actress she represents the perfection of the art of acting in all its forms, and it is said with truth that "where inspiration fails her she triumphs by sheer technique."

A number of peculiar mannerisms and oddities are associated with her acting, but these give her individuality without detracting from the flawlessness of her art. B.M.W.

Consult Jules Huret's *Sarah Bernhardt*, with a preface by Edmond Rostand.

BERNSTORFF, Johann Heinrich, Count (1862-), a German diplomat, in 1908 appointed ambassador to the United States. His father, Count Albrecht Bernstorff (1809-1873), was also prominent in German diplomacy, and was for many years ambassador at London, where his son was born. After serving in the army for eight years, the son entered the German diplomatic service in 1889 as an attaché of the embassy at Constantinople. Thereafter he served in various diplomatic capacities at Belgrade, Dresden, Munich, Petrograd and London, and was German consul-general in Egypt for two years prior to his appointment as ambassador to the United States. In this position he conducted many delicate negotiations with America after the beginning of the great war in Europe, and, especially as he had an American wife, he was believed to be honestly friendly toward the United States. However, it soon developed that he was the virtual head of the intricate spy system maintained in America by his government, and therefore he was grossly abusing the privileges of his honorable office. The American government sent him home in February, 1917. He was then ambassador to Turkey until the Empire crumbled.

COUNT VON BERNSTORFF

BERSAGLIERI, *ber sa lya' re,* the famous fast-marching sharpshooters of Italy, whose picturesque uniform of dark green is distinguished by the full plume of cock feathers surmounting a broad-brimmed felt hat. They are said to be able to march farther in a day than any other body of troops in the world, and their normal gait is a rapid, swinging walk by which good pedestrians are soon outdistanced. The Bersaglieri won distinction as fighters in the Crimean War, especially at the battle of Tchernaya (August 16, 1855), and upheld their traditions in the War of the Nations.

BERTILLON, *ber te yoN',* **SYSTEM,** a system for the identification of criminals, named after its inventor, Alphonse Bertillon. The object of the system was to arrive at so complete a description of a criminal that, no matter how he might disguise himself in future, his recognition and identification would be certain when the Bertillon tests were applied. The means of identification are based on measurements of certain parts of the body and description of general appearance. It is known that the bones of adults over twenty years of age do not change, neither are the measurements of any two persons exactly similar. In the Bertillon system the following measurements are taken:

Body: Height standing; height sitting; inches from finger tips to finger tips, with arms outstretched.

Head: Length and width; length and width of right ear.

Limbs: Length of foot, left middle finger, little finger and forearm.

The system has proved successful and its efficiency has been greatly increased by the inclusion of thumb and finger prints, which by themselves are considered an almost infallible means of identification. See Finger Print Identification.

Alphonse Bertillon (1853-1914), the inventor of the system named after him, was born in Paris, and after years of study devoted to criminology became the head of the identification department of the police of Paris. In recognition of his services he was made chevalier of the Legion of Honor. He was a prolific writer on criminology and kindred subjects, on which he is regarded as the world's chief authority.

BERYL, *ber' il,* a mineral which occurs sometimes in great six-sided prisms and sometimes in smaller crystals, which are among the most valuable of the precious stones. Some beryls are bright green, and are called *emeralds;* some are a bluish sea-green, when they are known as *aquamarines;* and some are light blue or yellowish. The best beryls are found in Brazil, in Siberia and in Alexander County, North Carolina, but various other parts of the United States produce the coarser, less valuable varieties. The scholars of the Middle Ages credited the beryl with many virtues, one learned writer declaring it "rendered the wearer unconquerable and at the same time amiable, while his intellect was quickened and he was cured of laziness." For a long time the beryl was the birthstone for October, but it has been replaced in that connection by the opal.

BESANT, *be zant'*, SIR WALTER (1836-1901), an English novelist and critic. His most notable work, *All Sorts and Conditions of Men,* giving a clear picture of the sordid life and surroundings of the people of East London, was the inspiration of the People's Palace in Mile End Road, London, and placed the author among the social reformers of his day. Like Dickens, he tried to arouse public sympathy for the poor and oppressed. He was born at Portsmouth and educated at King's College, London, and Christ's College, Cambridge.

After teaching mathematics for a few years he formed a literary partnership with James Rice. Among the novels which they produced together are *Ready Money Mortiboy, The Golden Butterfly* and *The Seamy Side.* After the death of Rice, Besant wrote alone a number of popular novels, including, besides his masterpiece mentioned above, *Dorothy Forster, Armorel of Lyonesse* and *Beyond the Dreams of Avarice.* He was the founder of the Society of Authors, a trade-union of writers, and the editor of the *Author*, the publication of the society. He was knighted by Queen Victoria in 1895.

BESSARABIA, *bes a ray' bi a,* a province formerly in the extreme southwest of Russia, but which voted late in the year 1918 to join the enlarged country of Rumania, with permission of the peace conference. It covers an area of

LOCATION OF BESSARABIA

16,181 square miles. The province is for the most part flat, but the soil is extremely fertile, producing an abundance of grain and raising great numbers of sheep, cattle and horses, which are among the finest of Europe. Vine growing is also an important industry, and wine of good quality is exported. Manufactures are in a backward state, the inhabitants being more devoted to out-of-door pursuits than to confining industries.

In the sixteenth century Bessarabia was seized by the Turks, but it was taken by Russia in 1770, to which it was formally ceded in 1812. In 1856 the southeast portion was restored to Turkey, but by the Treaty of Berlin in 1878 it was again declared Russian territory. The population is very much mixed, consisting of Russians, Poles, Bulgarians, Rumanians, Greeks, Tartars, Jews, Turks, Armenians and numerous tribes of gypsies. The capital is Kishinev, the scene of a massacre of the Jews in 1907 (see KISHINEV). Population, 2,400,000.

BESSEMER, *bes' e mer,* SIR HENRY (1813-1898), an English inventor and engineer, celebrated for his discovery of the Bessemer process for making steel. This discovery made it possible to roll steel into shape without hammering it (see STEEL, subhead *Bessemer Process*). It is true that the process was discovered by an American named William Kelly eight or nine years earlier than by Bessemer, and the American patents were granted to Kelly, yet it was from Bessemer that the world learned of it. Both inventors were first ridiculed by iron-makers, but Bessemer's superior resources and business ability enabled him to put his ideas into successful operation. His experiments and discoveries relating to steel and iron are best known, but his work in many other fields also brought him fame. He discovered a new process of making bronze powder, made several important improvements in type-casting machinery, and invented a method, still in use, for compressing the graphite used in making lead pencils. He was knighted by Queen Victoria in 1879.

BESSEMER, ALA., a city famous for its iron, steel and coal industries. It is situated in Jefferson County, on the Valley River, a little north of the geographical center of the state, and about eleven miles southwest of Birmingham. The large tonnage produced in the city and vicinity make it the objective point of seven railroad lines. The city was founded in 1887, incorporated in 1888 and named for Sir Henry Bessemer, the English inventor of the Bessemer process of steel manufacture. Its population is principally American, but with a large percentage of negroes and Italians, and increased from 10,864 in 1910 to

15,360 in 1914. Since the last Federal census (1910) Jonesboro has been included within the city limits. The area is about four and one-half square miles.

Bessemer occupies the physical center of the great iron ore and coal district of Alabama, located high above Valley River, with a border of cedar, chestnut and pine forests. To the northwest are the extensive Dolomite and the Black Warrior coal fields. From southeast to southwest, twelve miles from the city, extends the famous Cahaba coal field, with iron ore seams intervening. The city's industries are concerned chiefly with the products of the natural resources of the region. About 10,000 people are employed in the three principal iron industries and in the Louisville & Nashville Railroad shops. The manufacture of fire and building brick is important, and the lumber products are considerable. Valuable limestone quarries also occur in the vicinity.

Educational advantages are provided by a high school, a business college and a library. There are three banks, and a post office, erected at a cost of $37,000. The Elizabeth Duncan Hospital, the Presbyterian church and the Methodist Episcopal church are the most notable buildings. G.H.S.

BETEL or **BETLE,** *be' t'l,* the name of two different plants common in Asia—the betel palm and the betel vine. The betel palm, the commonest and most important of the areca palms (see PALM), is a graceful tree, usually forty to fifty feet high. Its fruit, the betel nut, is about the size of a small hen's egg, and has a fibrous shell, within which is enclosed the betel nut which is chewed by the natives of the Orient. But the name *betel* has been given to this nut and tree which bears it, only because of the association with the *betel vine,* to which the name originally belonged. This is a creeping plant of the pepper family, the leaves of which have a sharp, stinging taste. A number of different plants nearly related to the peppers, the leaves of which have similar properties, are also extensively cultivated in the East, and are sometimes called betel.

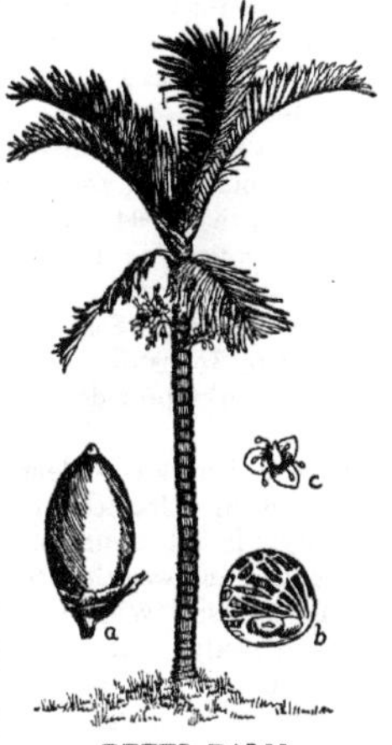

BETEL PALM
(*a*) Nut; (*b*) cross-section of nut; (*c*) flower.

Betel Chewing. In the East Indies, practically every native, man and woman, young and old, carries a betel-box. It is rare indeed to find a Malay of either sex who is not addicted to the habit, but it has never spread to any extent among other peoples. The seeds of the betel nut are boiled, cut into slices and dried in the sun and are then grated upon fresh betel leaves which have been smeared with quicklime. The whole is rolled into a pellet and chewed, and, while it is far too biting for pleasure to a person not used to it, the Malays find so much satisfaction in it and chew it so continuously that the proper handling of betel is an important part of the etiquette in every ceremonial meeting. Betel tinges the saliva, gums and lips brick red, blackens the teeth and causes them to decay rapidly.

BETH'EL, the place where Abraham pitched his tent, where Jacob saw in his dream the ladder reaching to heaven, upon which angels were ascending and descending, and where he wrestled with the angel and had his name changed to Israel. There are various opinions about the location of Bethel, but most authorities consider it to be about twelve miles north of Jerusalem. The name means *house of God.* The present town is of no importance.

BETHLEHEM, *beth' le hem,* the town in which Jesus Christ was born. The name means *house of bread.* Bethlehem is now a small village about five miles from Jerusalem, with but one street of low, flat-roofed houses. Our interest in it is due to its importance in Bible history. David, Boaz and Obed were also born there. As the birthplace of Israel's great men and the Savior of the world, Bethlehem will always be held sacred in the memory of Christians.

BETHLEHEM, PA., a borough in Northampton County, fifty miles northwest of Philadelphia, near the New Jersey boundary line. It is on the Lehigh River and Canal, and is served by the Central of New Jersey, the Lehigh & New England and the Lehigh Valley railroads and by electric interurban lines to nearby cities and towns. The population in 1910 was 12,837; in 1914 it was 13,721.

Bethlehem was founded by Moravians in 1741 and is the chief Moravian center in the United States. A college and theological seminary and a seminary for women are maintained there by that denomination. South Bethlehem (which see), connected with Bethlehem by two fine bridges across the Lehigh River, is the seat of Lehigh University. Among the institutions of Bethlehem are Saint Luke's Hospital and a public library. The important industrial plants include great steel and iron works, zinc and graphite works, silk and knitting mills and manufactories of hosiery, paint and other commodities.

West Bethlehem was annexed to Bethlehem in 1904. At the former place was located a general hospital of the Continental army, 1776-1778, and 500 soldiers are buried there. Bethlehem became a borough in 1845. It is gaining an enviable musical reputation as a result of the Moravian musical festivities.

The area of the borough is a little more than one square mile.

BETHMANN-HOLLWEG, *bate'mahn hol' vayK,* THEOBALD THEODORE VON (1856-), a German jurist and statesman, fifth Chancellor of the German Empire and the leader of the German government during part of the War of the Nations, which began in 1914. Bethmann-Hollweg acquired a peculiar interest for Americans in 1915, when it became known that he was opposed to the policy of torpedoing, without warning, passenger ships of the nations with which Germany was at war. It is said to have been largely through his influence with Emperor William II that the submarine warfare was modified and that diplomatic relations were preserved for so long between Germany and the United States.

THE WAR CHANCELLOR

Bethmann-Hollweg is by nature quiet, reserved and studious. It has been said that "gravity is the essence of him." To this man, apparently more at home in the study than in the councils of state, it fell to pilot the Empire through its greatest crisis. The son and grandson of famous men who had held high offices, and born of a wealthy family, he rose rapidly in the ranks of officialdom, becoming governor of the province of Brandenburg in 1899, and in 1905 Prussian Minister of the Interior. In 1907 he left the Prussian ministry and became Secretary of State for the Interior in the German Empire and Vice-Chancellor under Prince von Bulow. In 1909 he succeeded to the chancellorship, but resigned in 1917. In July, 1919, he asked the allied nations to spare the former kaiser from trial and, instead, put him on trial for Germany's crimes.

BEVERLY, *bev' er li,* MASS., a progressive manufacturing center and a popular summer resort, situated in the northeastern part of the state on Salem Bay, an inlet of the Atlantic Ocean. A bridge connects Beverly with the city of Salem, two miles distant, on the opposite shore of the bay. Boston is eighteen miles southwest. Railway accommodations are provided by the Boston & Maine and by electric lines. Beverly was founded in 1668 and was incorporated as a city in 1895. In 1915 it had a population of 22,959, an increase of 4,309 since 1910. The city's area is about fourteen and one-half square miles.

Beverly is the center of the shoe-machinery industry, about 4,500 people being employed in one factory. There are also manufactories of shoes, boots, morocco, belting and oil clothing. Beverly is a distributing point for the products of the Texas oil region, steamers operating regularly between this point and Port Arthur, Tex. The city has a number of vessels engaged in cod-fishery, and many of the inhabitants gain a livelihood in this industry. The most notable buildings are the $110,000 public library, with 12,000 volumes, and a Federal building erected in 1910. In addition to its public school system, the city has the New England Industrial School for Deaf Mutes and the Beverly Industrial School. Six playgrounds and two parks provide recreation and amusement.

Beverly Farms is a delightful summer residential district, located in the eastern part of the city. It is the summer residence of many Bostonians and was the "summer capital" during Ex-President William H. Taft's administration. Beverly Farms has a branch public library, erected at a cost of $25,000.

BHUTAN, *bhoo tahn',* is the name of an independent state, covering an area of about 16,800 square miles in the southern part of Asia. It lies between Tibet on the north and Bengal on the south, in the eastern Himalayas, where the mountains abound in sublime scenery and raise their snowclad peaks to a height of over 24,000 feet. Some parts of the territory are fertile, the chief crops being millet, wheat, and rice. The chief manufactures are coarse cloths and silks. The inhabitants are allied to the Tibetans, and are skilled agriculturists. They are Buddhists and have two rulers—a spiritual ruler, the Dharm Raja, and a secular ruler, Deb Raja. The capital is Punakha, or Dosen. The government of Bhutan is subsidized by the English government in India, receiving £3,333 (nearly $17,000) annually in return for a portion of the state annexed by the British in 1865. Population, estimated, 200,000. Only a very few of the people are of the white race.

BHUTAN

Peter Preaching at Athens

THE STORY OF THE BIBLE

Old Walls of Damascus

BIBLE. The man who sets out on a journey around the world finds that it does not take long to reach a place where most of the things which assume importance in his own life seem unimportant or unknown. The automobile or the works of Shakespeare; the telephone or the dramas of Goethe—these have found their way over a comparatively small part of the earth's area. But there is one book which the traveler finds in the remotest part of the earth. Farther than the railroad has penetrated or the most daring gold-seeker has ventured, the Bible has found its way, until it is almost literally true that there is no country in which it has not tried to make for itself a home.

This does not mean that the knowledge of the Bible has spread throughout the world; that the Eskimo hut or the African *kraal* has its Bible as has the Ontario or Texas farmhouse; but it does mean that it has been so frequently translated and so widely distributed that almost everybody can have at least parts of it in his own language. In all, there are 108 complete translations and over 500 partial translations, and more than 14,000,000 copies of the Bible or of the New Testament are distributed each year.

What It Is. *The Book,* the Bible is frequently called, but it could with greater accuracy be called *the library,* for it is in reality a library of sixty-six books which are usually bound together as one. Even the name to-day has the singular form, but originally it was a Greek plural meaning *books.* These sixty-six books have a greater unity than most libraries, for while they include essays, stories, love-songs, dramas and legal documents, they constitute altogether the sacred writings of Judaism and Christianity, and they have running through them a single thread of purpose which makes of them a whole. The Jews felt that to them, and to them alone, had been granted the revelation of the true God, and their sacred writings had as their purpose the setting forth of this revelation for future generations—for future generations of *Jews,* for it was too wonderful and too sacred to be wasted on Gentiles, or "heathen."

This was the older part of the Bible—the Old Testament, as it is commonly known. The New Testament aimed with no less singleness of purpose at giving a revelation of God, but it spoke to a wider audience. Its teachings were for "all the world," "to the Jew first, and also to the Gentile." Thus it is that the Jews of to-day accept only the Old Testament, while for the Christians the two parts have equal authority.

The Story of Its Long Life. Long ago, when the sacred writings received their name of *The Books,* that word meant a very different thing from what it does to-day. It referred to the rolls of papyrus, on which were inscribed by hand those writings which were counted worthy of publication (see MANUSCRIPT). On such rolls were inscribed the earliest sacred writings of the Jews, and it was probably such a manuscript which was delivered to Jesus when "He went into the synagogue and stood up for to read." At that time there existed only the Old Testament, but as the founders of the Christian Church wrote out their records of its origin or sent to individual churches their advice in letters, these, too, became a part of the Holy Scriptures.

As Christianity spread to Europe the influence of the sacred writings spread also. Interest in secular learning died out in the troubled times of the later Roman Empire and in the Dark Ages which followed, but love for "the book" never flagged, and it seems not too much to say that knowledge of the arts of reading and writing might have died out had it not been for this concern for the sacred writings. There were no printing presses in those days, and every new copy of the Scriptures which was made had to be laboriously written out by hand. Each monastery had its writing room, in which the monks toiled over their precious manuscripts, keeping heart and hand and pen clean, that they might be worthy of the high task. In his *Golden Legend* Longfellow put into the mouth of a monk the words—

> It is growing dark! Yet one line more,
> And then my work for to-day is o'er.
> I come again to the name of the Lord!
> Ere I that awful name record,
> That is spoken so lightly among men,
> Let me pause awhile, and wash my pen;
> Pure from blemish and blot must it be
> When it writes that word of mystery.

Very beautiful were some of these manuscripts, for they were frequently embellished or illuminated with gold and silver and glowing colors. Longfellow's monk continues:

> There, now, is an initial letter!
> Saint Uric himself never made a better!
> Finished down to the leaf and the snail,
> Down to the eyes on the peacock's tail!

So the Bible was kept alive by the monks' labor of love, until in the fifteenth century there came that wonderful invention which revolutionized learning—the invention of printing from movable types. So far as is known the first book printed by Gutenberg (which see) was a Bible; and from that time on the number of Bibles turned out by the presses has steadily increased. In many instances it has been necessary to design and make special types for the printing of the Bible in languages which had never before known a printed book, but no difficulty has been so great as to daunt those who have determined to make it possible for everyone, everywhere, to read the Bible in his own language.

From One Language to Many. In a sense, the writings of any great author make up a library, but the Bible differs from such a collection in that it is the work of about forty authors. All of its books, with the exception of *Daniel* and *Ezra*, which were in Aramaic, were written in Hebrew or in Greek; but the

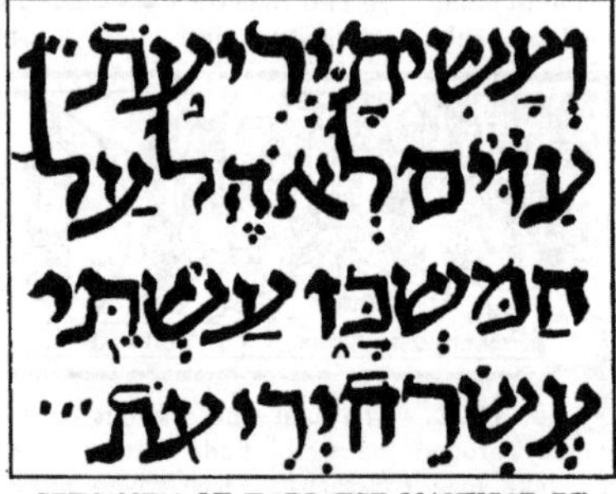

SPECIMEN OF EARLIEST MANUSCRIPT
A portion of *Exodus* (chapter XXVI, verse 7), written in square Hebrew. This was translated into Syriac, Greek, Latin, Anglo-Saxon, Old English, and finally into present-day English.

Hebrew books, those which comprised the Scriptures in the time of Christ, had all been translated into Aramaic, for Hebrew had at that time become a dead language. Further translations, too, were necessary, for many of the Jews had settled in different countries and had adopted the languages of the native inhabitants.

Of these ancient translations into other tongues by far the most valuable was that known as the *Septuagint*, which was completed in 285 B.C. It is a translation from Hebrew into Greek of such parts of the Scriptures as then existed, and received its name, which means *seventy*, from the fact that about that many scholars took part in the work of translating.

The time came, however, when Christianity became the religion of the Roman world, and Latin-speaking Christians demanded the Bible in their own tongue. Jerome, one of the most famous of the Church fathers, completed such a translation in A.D. 405, and this *Vulgate*, as it was called, became the authorized Bible of the Western Church. Down to the present time the Bible in use in the Roman Catholic Church is based on the Vulgate.

In Modern Tongues. Long after Christianity spread to countries where no Latin was spoken, that language continued to be used in the churches, and the earliest suggestions that the Bible actually be translated into the vernacular met with violent opposition. Express the sacred truths which had stood supreme in the Church for centuries in the vulgar language of the people? It was not to be thought of! But in every enlightened country there were brave men who dared to fight against tradition and to bring the Bible within reach of all. In Germany, Luther's translation was epoch-making, not only in religion but in the literature and language of the country, for it helped to crystallize the all-too-fluid dialect forms. In England there were several worthy of note—Wycliffe's translation, finished about 1380; Tyndale's, upon which modern versions of the English Bible are based; and Coverdale's, which was printed in 1535 and was the first complete English Bible published.

The Accepted English Versions. Several editions of the Bible followed Coverdale's, each of which attempted to correct the errors and improve the language of previous editions, but none proved wholly satisfactory. In 1604, therefore, at the petition of the leading clergymen of the country, James I of England appointed fifty-four eminent scholars to make a new translation. For seven years they worked at their task, and the *King James,* or *Authorized, Version* which they produced, so far surpassed in accuracy, arrangement and language all those versions which had gone before that within a few years it supplanted them all. English-speaking Protestant churches throughout the world adopted it, and until the latter part of the nineteenth century it had no rival.

Biblical knowledge had not stood still, however, and the English language had changed so that some parts of the Bible had acquired different meanings from those which they had had when the translation was made; and in 1870 a company of English scholars, aided by a group of distinguished Americans, set about making a new translation. So far as possible they kept the stately language of the King James Version which had become so firmly fixed in the affections of the people; but where changes in diction were necessary for the clearing up of obscurities they did not hesitate to make them. Many a reader to-day, who is grateful for the new light thrown by this *Revised Version,* still prefers to read the older version, familiar through long use.

45

The *American Revised Version,* which embodies all the results of recent investigations, was made still later.

The Divisions of the Bible. The Jews looked upon their religion as a compact between themselves and God, and applied to their sacred writings the title of *Covenant* or *Testament.* This latter form attaining wider use, the later books were called the New Testament, and as the Old and New Testaments the two great divisions of the Bible are known to-day.

The Old Testament. This division, which deals with the Jews and their history under the old Mosaic laws, consists of thirty-nine books, which admit of a very definite classification into law, historical books, prophecy and poetical books. The law includes the first five books, *Genesis, Exodus, Leviticus, Numbers* and *Deuteronomy;* the historical division takes in all the books from *Joshua* to *Esther;* the poetry embraces *Job, Psalms, Proverbs, Ecclesiastes, Song of Solomon* and *Lamentations;* and the prophetic division comprises the remaining sixteen books. Of some of these books the authors are definitely known; of others the authorship is traditional; while as to the origin of a number of them, absolutely nothing is known. Throughout century after century the first five books were ascribed to Moses, and the daring person who ventured the question, "But how could Moses have described his own death?" was looked upon as a dangerous skeptic. The *Psalms,* too, were credited without question to David. But present-day criticism is by no means so sure of its ground, and hesitates to declare with certainty the authorship of either of these groups of writings.

The New Testament. The twenty-seven books which make up this division fall into three natural groups. First, there are the historical books, the four Gospels and the *Acts of the Apostles;* then the epistles, and finally the curious, prophetic, visionary *Revelation.* Even the earliest of the Gospels was not written until years after the Ascension; the story of Jesus was handed down by word of mouth until there grew up the fear that parts of it might be lost, and then Mark wrote his narrative, the first of the Gospel accounts. The Gospel of John was the latest book, and was probably not written until the beginning of the second century of the Christian Era.

The Canon. This is a curious word with an interesting history, as used in this connection.

Originally it meant a *rod,* or by derivation, a *carpenter's rule,* and so in time it came to have reference to the standard by which the authority of the sacred writings was measured. Now the term "canon of the Scriptures" denotes those books which are believed to be inspired and are therefore authorized by the Christian Church. Just those books which at present make up the Bible, and no more, were not always looked upon as the canon; over certain specific books long controversies raged. Could the book of *Esther*, in which the name of God is not once mentioned, possibly be one of the "holy" writings? Were the books of the Apocrypha (which see) to be regarded as inspired, or were they merely secular? Before the time of Christ. the Jews had determined the canon of the Old Testament, but that of the New was naturally not entirely fixed until some centuries after Christ's death. To-day the Roman Catholic Church accepts as canonical the Apocrypha, which Protestant churches reject.

The Bible as Literature. The writers of the Bible were not learned men, as the world counts learning to-day, and yet their works have lived and multiplied as have no others ever written. In large measure, to be sure, this is because of the subject matter with which they deal, but in a lesser degree it is an outgrowth of the literary character of the Bible. Time was, and not so very long ago, when to have spoken of the "literary character" of the Bible would have been looked upon as sacrilege. It was a sacred book, designed to appeal to man's spiritual nature and taking no account of the feeling for beauty which is no less an inborn sense. Indeed, in those days when the Bible was perhaps more in men's thoughts than at any time since the days of the Church fathers—in the Puritan age—the spiritual sense and the sense of beauty were looked upon as antagonistic.

Then, too, the Book was looked upon as too sacred to be studied as any other book might be; people actually held that its holy character extended to the paper upon which it was printed and the boards in which it was bound. To ponder over it spiritually was right and necessary—"in His law doth he meditate day and night"; but to attempt to solve its problems with the aid of cold reason—such an act was to lay profane hands upon the Ark of the Covenant or to enter with unhallowed feet the Holy of Holies. One eminent writer upon things Biblical phrased the matter picturesquely when he called the Bible "a literature smothered by reverence."

In recent years students of the Bible have been paying more and more attention to its literary quality, but yet students of literature have placed upon it too little stress. Many a person who accounts himself well read, and is ashamed to confess his ignorance of the works of some minor poet, almost boasts of his lack of knowledge of the Bible. And yet no one can really understand English literature or grasp its full significance who has not a background of Biblical knowledge. What can Byron's stirring *Destruction of Sennacherib* mean to the person who has never read the account in II *Kings,* 25? We are told that—

> The Assyrian came down like the wolf on the fold,
> And his cohorts were gleaming in purple and gold,

and that later—

> The Angel of Death spread his wings on the blast,
> And breathed in the face of the foe as he pass'd,

but no statement is made as to who Sennacherib was or as to why such dreadful woe came upon the Assyrians. Or how can the person who had not read the story of Moses' experience on Sinai grasp the full significance of Lowell's—

> Daily, with souls that cringe and plot,
> We Sinais climb and know it not.

But it is not only as a background for other literature that the Bible merits attention. It is itself a literature unmatched in certain literary forms. Epigram, epic, thrilling tale, exquisite pastoral romance—all these and many more are to be found in its pages, and in lyric poetry, especially, critics agree that it equals if it does not surpass all other literatures in the world. The translators, too, have been so wise in their choice of words that the English reader has little or none of that disappointed feeling which usually accompanies the reading of translations. So wonderfully phrased are most parts of the Scriptures that the person who cares nothing for the spiritual significance of the Bible cannot fail to be impressed with the fact that if Biblical quotations are contrasted with those from almost any other work, they take from such quotations much of their dignity and stateliness.

The Bible for Children. This topic is closely related to the preceding, since it means to take no account of the religious teaching of children, but to emphasize the "story" side of the

Bible. Any truth which is presented to children must of necessity be in story form, and the Bible is at its best in this field. It was written in the childhood of the race, by men who were striving not to attain literary distinction, not to embellish their works highly, but merely to tell their stories and sing their songs simply and directly, that they might not fail of their appeal. Tolstoi phrased most forcibly the relation between this fact and the appeal of the Bible to children. "It seems to me," he wrote, "that the Book of the childhood of the race will always be the best book for the childhood of each man." And he goes on to tell of his experience in teaching peasant children, whom he had tried to interest through other means:

The Finding of Moses

> Then I tried reading the Bible to them, and quite took possession of them. . . . The corner of the veil was lifted, and they yielded themselves to me completely. They fell in love with the Book and with learning and with me.

The vivid stories of men and women who seem alive and who have to do with just those fascinating things which interest children most cannot fail to attract and hold attention.

The Handwriting on the Wall

The Wise Men of the East

The Good Samaritan

Joseph, with his marvelous coat, with his adventures in the pit and in prison and with his final triumph, is a hero after their own hearts, and David's giant, Jonathan's bow and arrows and Daniel's lions are just the exciting touches which they love. The finding of Moses is an old story which to children is ever new. As they grow older, there is the pastoral of Ruth, with its "happy ending," and the no less charming but more stirring prose epic of Esther. There are books of Bible stories without number, some excellent, some mediocre and some poor; but as soon as the child is old enough he should be given his Bible stories in Bible language. Not only does this enhance their charm, but it gives the boy or girl a taste of well-chosen, virile English which most of the harmless but carelessly constructed children's books of the present day do not afford. The story of the Three Wise Men, the moral from the tale of the Good Samaritan and the warning in the handwriting on the wall cannot be excelled in this twentieth century.

Biblical Chronology. The general reader of the Bible, as well as the student, cannot fail to be interested in the location of the events in time. Does anyone know when Abraham went out from Ur of the Chaldees? Did David live before or after the Trojan War? The Jews reckoned time from the Creation of the world, and by computing the generations before and after the Deluge, worked out a complete system of chronology. According to them, the creation of the world took place about 4,000 years before the birth of Christ, and the outstanding events occurred in the years here set down:

Event	(B. C.)
Creation of the world	4000
The Flood	3200
Call of Abraham	2000
Israelites enter Egypt	1700
The Exodus	1300
Saul anointed king	1095
David began to reign	1055
Solomon began to reign	1015
Division of the kingdom	975
Fall of the kingdom of Israel	722
Fall of the kingdom of Judah	586
Return from the Captivity	536
Conquest of Palestine by Alexander	330
Capture of Jerusalem by Pompey	63

The correctness of these dates previous to the reign of Saul, it should be stated, is open to serious criticism, and among historians and Bible students Hebrew chronology is considered to stand by itself until such time as the dates given coincide with dates of events in secular history.

Books of the Bible. The table which follows lists the books of the Bible and gives their authors, so far as they are known, but it must be borne in mind that often the authorship cited here is merely traditional. The books marked with a star are those about which there has been most controversy and uncertainty. Separate articles on the more important of the books listed here will be found in their alphabetical order in this work:

THE OLD TESTAMENT

Book	*Author*
Genesis	Moses
Exodus	Moses
Leviticus	Moses
Numbers	Moses
Deuteronomy	Moses
Joshua	Joshua
Judges	*Samuel
Ruth	Unknown
I and II Samuel	Unknown
I and II Kings	Unknown
I and II Chronicles	*Ezra
Ezra	Ezra
Nehemiah	Ezra
Esther	*Unknown
Job	Uncertain
Psalms	David, Asaph, Levites, Hezekiah, Solomon, Prophets
Proverbs	Solomon, Agur
Ecclesiastes	Unknown
Song of Solomon	Solomon
Isaiah	Isaiah
Jeremiah	Jeremiah
Lamentations	Jeremiah
Ezekiel	Ezekiel
Daniel	Daniel
Hosea	Hosea
Joel	Joel
Amos	Amos
Obadiah	Obadiah
Jonah	Jonah
Micah	Micah
Nahum	*Nahum
Habakkuk	Habakkuk
Zephaniah	Zephaniah
Haggai	Haggai
Zechariah	Zechariah
Malachi	Malachi

THE NEW TESTAMENT

Book	*Author*
Matthew	Matthew
Mark	Mark
Luke	Luke
John	John
Acts of the Apostles	Luke
Pauline Epistles	Saint Paul
Romans	
I and II Corinthians	
Galatians	
Ephesians	
Philippians	
Colossians	
I and II Thessalonians	
I and II Timothy	
Titus	
Philemon	
James	James
I and II Peter	Peter

OUTLINE AND QUESTIONS ON THE BIBLE

Outline

I. What It Is

(1) Its name
(2) Its place in the world
(3) A library, not a book

II. The Canon of the Scriptures

(1) Old Testament.
(2) New Testament
(3) Apocrypha

III. Origin

(1) Dates of writing
  (a) The oldest book
  (b) The youngest book
(2) Authors
(3) Language

IV. Translations

(1) The Septuagint
(2) Vulgate
(3) The Bible in English
  (a) Early versions
    1. Wycliffe's
    2. Tyndale's
    3. Coverdale's
  (b) The Authorized Version
  (c) The Revised Version
  (d) American Revised Version

V. Contents

1) Old Testament
  (a) Law
  (b) History
  (c) Prophecy
  (d) Poetry
(2) New Testament
  (a) History
  (b) Letters
  (c) Prophecy

VI. Chronology of the Bible

1) Important dates
(2) Its uncertainty

VII. The Bible as Literature

(1) The new attitude
(2) Reasons for its charm
(3) Its value for children

Questions

What event, according to Bible reckoning, took place almost as long before the beginning of the Christian Era as the Norman Conquest did after it?

What was the first complete Bible published in English?

What was the form of the sacred writings possessed by the Jews?

How many letters of Paul are part of the New Testament?

What change has taken place in comparatively recent times in the attitude toward the Bible?

How many complete translations of the Bible have been made?

By what other name is the Authorized Version known? Why is it so called?

What quality have many of the Bible stories that makes them suitable for children?

According to tradition, who wrote the first five books of the Old Testament?

In what language were *Daniel* and *Ezra* originally written? Why was the entire Old Testament later translated into this tongue?

What reasons were there for making a revised version of the Scriptures?

What is the meaning of the name *Bible?*

How many books of history are there in the New Testament? Name them.

Why was the Septuagint so called? Is the title strictly accurate?

What are the books of law? The prophetic books?

About how many Bibles and New Testaments are distributed each year?

What is meant by the "canon"? What books form part of it?

About how early were the first translations from the Hebrew?

I, II and III John........John
Jude ....Jude
Hebrews ....*Uncertain
Revelation ....John

Consult Brown's *History of the Bible;* Jacobus's *Roman Catholic and Protestant Bibles Compared;* Selleck's *The New Appreciation of the Bible.*

Related Subjects. These volumes contain a great number of articles on Biblical topics—the books of the Bible, the characters, and kindred topics. For convenience of reference these are listed here, with the exception of the books, which are given above:

Aaron	Haggai
Abel	Hallelujah
Abraham	Ham
Absalom	Herod
Adam and Eve	Hezekiah
Ahab	Hittites
Ahasuerus	Hosea
Ahaz	Isaac
Amalekites	Isaiah
Ananias	Ishmael
Apocalypse	Jacob
Apocrypha	James, Saint
Ark	Japheth
Baal	Jehoshaphat
Babel, Tower of	Jehu
Balaam	Jeremiah
Barabbas	Jeroboam
Barnabas	Jezebel
Bartholomew	John, the Baptist
Beelzebub	John, Saint
Benjamin	Jonah
Caiaphas	Joseph
Cain	Joseph
Calvary	Joseph of Arimathea
Canaanites	Joshua
Cherub	Josiah
Cities of Refuge	Jubilee
Daniel	Judas
David	Jude
Deborah	Judith
Decalogue	Lazarus
Delilah	Leviathan
Deluge	Levites
Douai Bible	Lucifer
Eden	Luke, Saint
Edom	Magdalen, Mary
Eli	Malachi
Elijah	Manna
Elisha	Mark
Esau	Mary, The Virgin
Esther	Matthew, Saint
Ezekiel	Michael, Saint
Ezra	Miracle
Felix, Antonius	Moabites
Festus, Porcius	Moses
Gabriel	Nahum
Gamaliel	Nazarite
Gath	Nehemiah
Gehenna	Nimrod
Gideon	Noah
Goliath	Ophir
Goshen	Passover
Gospels	Patriarchs
Habakkuk	Paul
Pentateuch	Scribes
Pentecost	Selah
Peter	Septuagint
Pharisees	Shem
Philip	Sinai
Philip	Sodom
Philistines	Solomon
Pilate, Pontius	Tabernacle
Rachel	Tabernacles, Feast of
Ruth	Targum
Sadducees	Thomas, Saint
Samaritans	Timothy
Samson	Titus
Samuel	Vulgate
Sanhedrin	Zebulum
Saul	Zedekiah
Scapegoat	Zephaniah

In addition to these, there is given under PALESTINE a list of the Biblical places treated in these volumes. A.MC C.

BIBLIOGRAPHY, *bib li og' ra fi,* a term which has changed from its original meaning—the copying of books—and has come to denote writings about books, whether these concern themselves with the externals of books or with their contents. Collectors of rare books are especially interested in the former branch, and will pore long over bibliographies which give descriptions of old or rare books, their bindings, and the comparative value of different editions; but the general reader is far more interested in a bibliography which treats of the content of books.

Bibliographers have never quite relinquished the ambition of compiling one huge bibliography which shall include and classify all the books that have ever been published, but this remains as yet an unrealized dream. *Special* bibliographies, however, which either limit themselves to the books of one country or to those treating of some one subject, are numerous and are constantly increasing in number. To the unsystematic reader whose choice of books is guided by what he happens to see or hear, such works are of little value, but to the person who wants to pursue a certain line of reading or find out all that has been written on some topic, they are indispensable. Scribner's *Biographical Guide to American Literature* and the *Reader's Guide to Periodical Literature* are probably the best-known of the general book lists.

At the end of important articles in these volumes are names of books which give broader treatment of the topic just discussed. These names are selected with a view to their popular character and their appeal to the general reader. See BIBLIOMANIA, below.

BIBLIOMANIA, *bib li o may' ni a,* a term derived from two Greek words meaning *book* and *madness*. It is used to denote the passion a person may display for collecting books, not for the interest or helpfulness of their contents but rather for their rarity. Andrew Lang describes it as the "love of books for their own sake, for their paper, print, binding, and for their associations, as distinct from the love of literature." Thus the books of famous printers or binders, early editions or *de luxe* editions, uncut copies, specially illustrated copies, or even books which contain certain misprints, are eagerly sought after.

There seems to be almost no limit to the price which a real bibliophile, or book-lover, will pay for a certain choice volume, if only other book-lovers want it also. The largest sum on record paid for a single volume is $50,000, the price of a Bible printed on vellum by Gutenberg about 1450 and sold for that sum in 1911. A copy of the same edition, but printed on paper, was sold in 1912 for $27,500. The only known perfect copy of Malory's *Mort d'Arthur*, printed by William Caxton, was sold in 1885 for $9,750 and in 1912 for $42,800; other books from Caxton's press bring from $1,000 to $15,000. Books from the presses of other early printers also bring high prices, notably those of Jenson and Aldus at Venice, Sweynheym and Pannartz at Rome, and Ulric Gering at Paris. The many books printed by the Plantin and the Elzevir families at Antwerp, Amsterdam and Leyden are also eagerly sought.

A special field of great interest is the collection of books which have belonged to famous people and contain bookplates, autographs or other marks of ownership. Of recent years there has been a growing interest in early books on America, and in the first editions or obscure works of famous modern writers, notably Poe, Thackeray, Dickens, Stevenson and Kipling. A copy of the original 1827 edition of Poe's *Al Aaraaf, Tamerlane, and Other Poems* sold at auction in 1900 for $2,050, and another copy, with an inscription in Poe's handwriting, in 1909 brought $2,900.

BIBLIOTHEQUE NATIONALE, *be ble o tek' nah syo nal'*, the largest library, not only in France, but on the Continent. It is located in Paris, and is the national library of the French people. Its splendid collections include over 2,600,000 printed volumes and maps, about 102,000 manuscripts, more than 250,000 engravings and 150,000 coins and cameos. In 1536 it was decreed that one copy of every book printed in France should be deposited in the national library; this explains the large number of printed volumes it contains.

The sources of this library are, in the main, the library that Charles V arranged in the Louvre in 1367, and that of the royal Orleans family, at Blois. Francis I united the two libraries, and Charles IX brought them to Paris. The library is now located on Richelieu Street.

BICEPS, *bi' seps,* the large muscle in the front of the upper arm. Its upper end is attached to the shoulder blade, or scapula, and its lower end by a tendon to the large bone, or radius, of the fore arm. This is the muscle usually proudly exhibited as evidence of muscular development. It bends the elbow and turns the hand outward, and is the largest and strongest muscle of the arm. The muscle opposed to the biceps, and by which the arm is straightened, is the triceps. See MUSCLES.

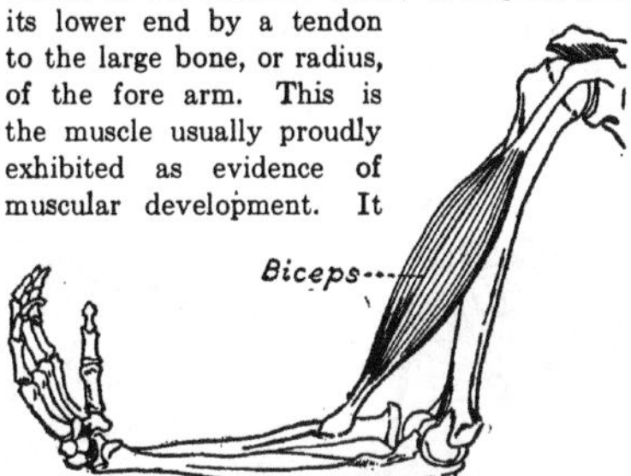

LOCATION AND FORM OF BICEPS

BI'CYCLE. The story of the modern bicycle furnishes an interesting example of the manner in which the delight of the public in various forms of recreation rises and declines. The familiar "safety" bicycle entered upon its career of wide popularity in America about 1889, and during the next ten years it enjoyed an astonishing vogue. It was ridden by old and young, was as popular with women as with men, and bicycle enthusiasts found it equally satisfactory for recreation and for business purposes. In 1899 there were 312 bicycle establishments in the United States, and their combined output for the year reached a total of 1,112,880 machines. Then the tide turned, and after 1900 there was a gradual decline in the demand for these machines. Manufacturers of bicycles ceased advertising their wares, and many of them transferred their capital and equipment to the production of the motorcycle (a bicycle equipped with a gas engine) and the automobile. The bicycle is still used to a limited extent as a business convenience, but it will probably never again be as popular as the motorcycle (which see), because of the speed of the latter and the fact that it can take the rider long distances without causing him undue fatigue.

The essential features of the bicycle of latest construction are as follows: It consists of two wheels, arranged tandem and attached to a frame upon which the rider's seat is mounted. The vehicle is propelled by the rider's feet, which act upon pedals connected with the axle of the rear wheel. A handle bar, guiding the direction of the front wheel, is the steering apparatus. The mechanism for driving the vehicle is usually a chain whose links fit over a sprocket wheel turned by the crank shaft and running over another sprocket on the rear wheel; but some chainless bicycles are made, having levers with bevel gear which transmit the motion of the crank shaft to the rear wheel.

The wheels are of equal size and have steel wire spokes, ball bearings, wooden rims and pneumatic tires. The latter may be single-tube or double-tube, but the double-tube are generally regarded as the more satisfactory. The frame consists of hollow steel tubing, and the weight of the vehicle is from twenty-three to thirty pounds. A recent appliance is the coaster brake, attached to the rear wheel, which enables the rider to check the speed of the bicycle by pressing back on the pedals. In the older type of machine the brake was attached to the handle bar.

In the early stages of the bicycle's vogue a high grade machine, more than twice as heavy as the present-day vehicle, cost from $100 to $150. An excellent bicycle, with the latest appliances, can be purchased to-day for twenty-five or thirty dollars. About 169,000 machines, valued at approximately $2,437,000, are manufactured in a year in the United States. The province of Ontario is the center of the Canadian industry, the yearly output of the entire Dominion being valued at about $445,000.

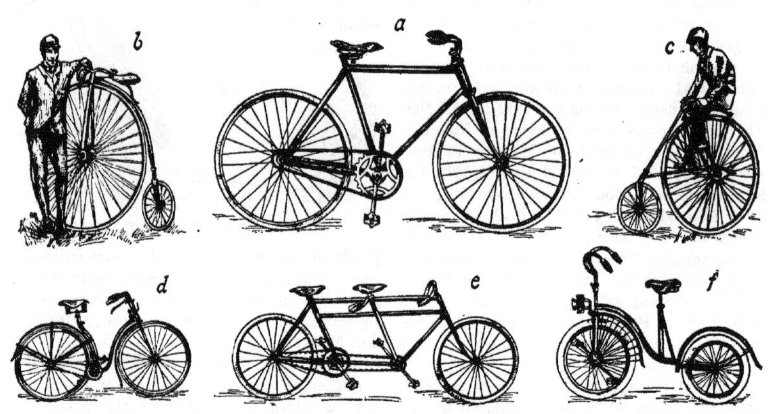

THIRTY YEARS OF BICYCLE DEVELOPMENT

(*a*) Modern bicycle; (*b*) the so-called "ordinary," the immediate predecessor of (*f*), the first "safety"; (*c*) the "Star" bicycle, with small wheel in front; (*d*) the "safety" adapted for the use of women; (*e*) the "tandem," for two, which could be developed on the same principle for three, four or even six people.

Evolution of the Bicycle. A crude, two-wheeled vehicle propelled by the feet was known to the ancients, and machines of this character were in use in England, France and Germany many years before the real forerunner of the modern bicycle was devised. This machine, invented in 1816 by Baron von Drais, chief forester to the Grand Duke of Baden, and called *draisine* in honor of him, consisted of two wheels of equal size, arranged tandem and connected by a perch. The rider, who rested part of his weight on the perch, propelled the machine by striking the ground with his feet. A padded arm-rest on a bar attached to the front wheel served as a handle bar, and the whole formed a steering apparatus.

A machine modeled on the draisine was

patented in England in 1818, where it enjoyed a brief period of popularity under various names—hobby-horse, dandy-horse, swift-walker and others. Its manufacturer, however, named it the *pedestrian curricule*. This machine and others modeled on it failed to win permanent favor. In 1855 a French carriage-maker brought out a vehicle resembling the modern bicycle in that it had cranks and pedals fitted to the front wheels, and ten years later an improved form of two-wheeled *velocipede* appeared in Paris, greatly stimulating interest in cycling and resulting in the introduction of the French velocipede into the United States.

This velocipede, like the high-wheeled bicycle which followed it, had the front wheel much larger than the rear, and was propelled by cranks attached to the hubs of the front wheel. Its successor, known as the *ordinary*, which was developed in England about 1873, was in general use in Europe and America until replaced by the low-wheeled bicycle, or *safety*. The first vehicle of this type had the front wheel about three times the size of the back one, but the tendency was to increase the size of the former and to reduce that of the rear wheel, until the diameters were respectively sixty and sixteen inches. The average roadster of 1875 weighed sixty-five pounds and had a steel frame and solid rubber tires.

During the next ten years various improvements were introduced, including adjustable ball bearings and cushion tires. The "ordinary" was popular because it afforded exhilarating exercise, but it had serious defects. The high seat, more than five feet above the ground, and nearly over the center of the large wheel, made falls, or "headers," as they were called, somewhat frequent and dangerous, and the vehicle was difficult to mount. It is not surprising, therefore, that a low-wheel vehicle embodying the principal features of the modern bicycle was finally brought out in 1884. Of the subsequent improvements, including the coaster brake, adjustable handle bars, cushion saddles and the drop frame for ladies' bicycles, the most important was the pneumatic tire. Though invented in England in 1843, this device was not applied to the bicycle until 1889, and it was the greatest single factor in stimulating the popularity of the "wheel." The cyclist might be subjected to the trouble of mending punctured tires, but whatever annoyance he suffered on this account was more than offset by the freedom from jolts, greater ease in running, and the increased lightness of his machine.

The cycling era was not unproductive of good results, for in moderation it provided very healthful exercise; and the extensive use of the bicycle, together with the influence of the League of American Wheelmen, contributed much to the good-roads movement that has been so greatly advanced by the advent of the automobile. E.D.F.

Consult Garratt's *Modern Safety Bicycle;* Henry's *Cycle Building and Repairing.*

BIDDEFORD, ME., an old city in York County, which was for a long time the chief settlement in the province of Maine. It is situated in the extreme southwestern part of the state, on the right bank of the Saco River, six miles from the sea. Portland is fifteen miles northeast, and Portsmouth, N. H., is forty miles southwest. Railway transportation is provided by the Boston & Maine Railroad, and electric lines extend north and south. For eight months in the year there is water transportation for freight that can be shipped by schooner or barge. The first settlement was made as early as 1616 by people from Biddeford, England, who named the place for their home town. Biddeford was settled under a patent in 1630, and Saco, on the opposite bank of the river, was included within its limits until 1718, when Biddeford was incorporated separately. In 1855 it received a city charter. In 1910 the population was 17,079; it increased to 17,475 in 1914. The foreign element is largely French. The area exceeds fifteen square miles.

The development of Biddeford was favored by the abundant water power of the Saco River, which at this point has a fall of forty feet. The manufacture of textiles is its leading industry, about 16,000 people being employed in the two large cotton mills, which rank with the foremost in New England. There are machine shops, making cotton-mill machinery and accessories for cotton mills, and two big saw mills which have an annual output of 15,000,000 feet. The post office, erected in 1914, the city hall, Masonic Temple, Webber Hospital and McArthur Library are the noteworthy buildings.

Biddeford is located in the heart of a summer resort region and on account of its fine boating, fishing and bathing facilities is a popular resort. Old Orchard Beach, which comprises several small beaches, is not excelled by any resort on the New England coast. B.M.A.

BIENNIALS, *by en' i alz,* in botany, the name given those plants that put forth leaves and roots the first season, remain quiet through the winter, and the next spring or summer bear blossoms, fruit and seed, and then die. That is, their life-span is two years, or practically so. Familiar examples of biennials are the turnip, beet and carrot, which store up food in their roots the first season for the plants to feed upon during the period of flowering. Plants that live for one year are known as *annuals,* while *perennials* live on year after year for an indefinite period. See ANNUALS; PERENNIALS.

BIENVILLE, *byaN veel',* JEAN BAPTISTE LE MOYNE, Sieur de (1680-1758), a well-known explorer who was prominent in the settlement of the French province of Louisiana. As a boy he accompanied his brother Iberville (which see) on his early explorations of the Mississippi, and assisted him in founding Biloxi in 1699. He was appointed governor of Louisiana in 1701, at the age of twenty-one, and founded the city of Mobile, which he made the seat of government, but was removed from office in 1707. He continued to be active in the upbuilding of the province, however, and in 1718 was again made governor, in which year, with the aid of Law's Mississippi Company, he founded the city of New Orleans. This place became the capital of Louisiana province in 1722. Four years later Bienville was again dismissed, but was reinstated in 1733, and for the next ten years worked with untiring zeal for the prosperity of his colony. From 1743 until his death he lived in France. See MISSISSIPPI SCHEME.

BIERSTADT, *beer' staht,* ALBERT (1830-1902), a landscape painter whose popularity rests on his pictures of scenes in the Western United States, especially in the Rocky Mountains. He was born in Germany, but spent most of his life in America. Though his canvases do not have the highest artistic merit, they possess a grandeur and impressiveness that make them wonderfully effective. His *Sierra Nevada* may be viewed in the Corcoran Art Gallery in Washington, D. C.; his *Valley of Yosemite* is in the James Lenox collection in New York. The Capitol at Washington contains two historical pictures painted by him—*The Discovery of the Hudson River* and *The Settlement of California.* Bierstadt was awarded several German and Austrian decorations and was a member of the National Academy of Design of New York.

BIGAMY, *big' a mi,* in law, the crime of marrying a second time while a first husband or wife is living and not legally divorced. The laws of the Roman Catholic Church formerly regarded a second marriage as bigamous even though the first spouse were dead, the first marriage being a sacrament that was binding on each of the contracting parties for life. That Church does not even now admit the right of civil courts to annul marriages by divorce, but recognizes the right of one to marry again whose first marriage has been broken by death. In most countries the government recognizes a properly-granted divorce as sufficient authority to remarry. If a person remarrying can prove that he or she is unaware of the fact that a former husband or wife is living, and has had no such knowledge for seven years, the law does not now consider remarriage under such circumstances as bigamous.

In England, until the reign of William III, bigamy was a crime punishable by death. The penalty was reduced in the time of George I to a minimum of two years' imprisonment, with hard labor. In the United States the punishment is usually a term of imprisonment, varying from two to five years, accompanied by a fine. Canadian laws concerning bigamy are based on those of England.

BIGHORN, the name given a wild sheep, native to the Rocky Mountains, and named from the massive horns of the ram, which

THE BIGHORN

curve back from the forehead, down and then forward, frequently measuring forty inches. This is also the height of the animal at the shoulder. The horns of the ewe are short. Bighorns, which are not to be confused with the much less wary Rocky Mountain goat (which see), formerly ranged in the highest mountains from New Mexico to the Arctic Circle. They are grayish brown, with face of a lighter shade, a dark line down the spine and the under parts and two patches on the rump, white. They live in herds of thirty or forty, are quick and nimble, jumping and climbing easily in the most dangerous places. Their meat has an excellent flavor and they are considered fine game, but in most states are protected by law. No other animal on earth, not even the chamois, has the agility of the bighorn. In the Rocky Mountain National Park, in Colorado, tourists sometimes see herds of them, even the lambs, plunge head downward from précipices hundreds of feet high, then landing on their four feet held tight together, break their fall only for an instant, repeating their leaps until safe from their enemies, in the valley below.

BIGHORN RIVER, the largest southern tributary of the Yellowstone River. It rises in the Wind River Mountains of Central Wyoming and flows in a northeasterly direction for a distance of 450 miles, through Wyoming into Montana. Its course is through scenic, mountainous country, along the former home of the Sioux Indians. The stream is navigable as far as Fort Custer, near the scene of the Custer massacre in 1876.

BIGLOW PAPERS, two series of humorous and satirical poems which at two critical times in the nation's history made James Russell Lowell not only a popular poet but an important figure in the life of the country. The first, relating to the Mexican War, appeared in 1848; the second, dealing with the War of Secession and reconstruction, in 1867. Both were signed with the fictitious name of Hosea Biglow. Hosea, who wrote in Yankee dialect, possessed all of New England's clear-headedness and sharpness. "Editorial matter" by a fictitious Homer Wilbur reinforced the poems. The following quotations, among many others, have made a permanent place for themselves in American literature:

This goin' ware glory waits ye haint one agreeable featur.

I don't believe in princerple
But oh I du in interest.
An' you've gut to git up airly
Ef you want to take in God.

BIGNONIA, *big no' ni a,* a family of plants, consisting both of shrubs and trees, which bear showy, trumpet-shaped flowers and are native to warm climates. One of the most interesting members of this family is the *trumpet-flower,* or *trumpet-creeper,* which is found in the United States from New Jersey and Pennsylvania westward to Illinois and south to the Gulf States. This plant loves the moist woodlands and fields, where its stem creeps along the ground or climbs over fences and bushes and often finds its way up the trunks of small trees. Because of its excellent climbing qualities the trumpet-creeper is cultivated as a porch vine. The beautiful, trumpet-shaped flowers are red within and tawny or orange outside, and are the favorite blossom of the humming bird; in order to reach the nectar this little bird must thrust its tiny head and shoulders well into the long tube.

Another species of bignonia found commonly in the woodlands of Southern United States is the *cross-vine,* a shrub with a high-climbing stem and numerous flowers which are yellow within and reddish-brown without. This plant takes its name from a curious formation in the stem, a cross-section of which shows a conspicuous cross. The familiar *catalpa* tree is also a member of the bignonia family.

BILBAO, *bil bah' o,* an important commercial center and port of Spain, on the River Nervion, about eight miles inland from the Bay of Biscay. It is a well-built city, capital of the province of Vizcaya, or Biscay, and lies on both banks of the river, which is crossed by five bridges. Bilbao has always been famous for the manufacture of swords, which in the days of Shakespeare were commonly called *bilbos.* The river Nervion is navigable. The city exports wool, iron, fruit, oil, grain and wines; its imports are cotton and woolen goods, fish, spirits and machinery. Copper and iron are mined in the vicinity. Population in 1910, 92,514.

BILE, a yellowish, bitter fluid secreted by the liver. The bile is collected in the gall bladder, to which it passes through the biliary ducts. The amount formed varies from twenty-five to fifty ounces in twenty-four hours. From the gall bladder the bile passes to the small intestine, entering the duodenum by the cystic duct. The bile acts as an aid to digestion in the following ways:

(1) It emulsifies the fats; that is, it breaks them up into small globules, so that they can pass into the lacteals.

(2) It moistens the mucous membrane of the intestines so that it will absorb fats.

(3) It prevents a decomposition of food while it remains in the intestines.

(4) It stimulates the muscles of the intestines, and so aids in the movement of their contents and prevents constipation.

When the regular flow of bile is stopped the coloring matter is absorbed into the blood and causes jaundice, a condition in which the whites of the eyes and the skin turn yellow. Gall stones result from a chronic mild infection of the contents of the gall bladder; they are sometimes caused by the bacillus of typhoid fever. W.A.E.

Related Subjects. The articles on the following topics contain information that will be helpful in this connection:

Digestion	Jaundice
Gall Bladder	Liver
Intestines	

BILL. When the men of the Middle Ages wrote an important document they closed it with a leaden or other seal called a *bulla*. Gradually the word came to be applied to the paper itself (see BULL), and in England and France the similar words *bill* and *billet* were applied to less formal writing as well. As a result Anglo-Saxons have at the present time several widely different sorts of written or printed papers which are called bills. Besides the bills of attainder, of exchange, of health, of lading, of rights and of sale, each of which is described in the succeeding pages of this volume, the most important forms are the following.

In Bookkeeping. When a promissory note is received by the proprietors of a business, or a draft or bill of exchange in their favor is accepted, the bookkeeper enters them on his page for *Bills Receivable*. Notes given or drafts accepted by the firm are *Bills Payable*. See PROMISSORY NOTE; BILL OF EXCHANGE; BOOKKEEPING.

In Commerce. Besides the formal statement of money due, a bill with which everyone is familiar, there is the invoice or bill of goods, which contains a list of goods sold, with the price of each item, and is commonly sent out at the time of a sale. The usual form of invoice is here shown:

Kitchener, Ont.,
Sept. 22, 19—

T. Campbell, Esq.,
Bay City, Mich.
Bought of MATTHEWS Co. Ltd.
Terms 60 days.

100 bbls. Baldwin apples	@	$2.04	$204.00
50 bbls. Northern Spies	@	$2.86	143.00
			$347.00

Shipped Grand Trunk.

In Law. When anyone is accused formally and in accordance with legal procedure, the written statement is known as a true bill or a bill of indictment (see JURY, subhead *Grand Jury*). There are several other types of formal statements in legal practice which are called bills.

In Law-Making. When we read of the Jones Bill or the Smith Bill being passed by the House, we know that a statute has been proposed by a Representative or Member of that name. Later, if it becomes law, it will be termed an *act*. See LAW.

BILLBOARD, a board erected for the display of posters or other advertising matter, It is used to advertise nearly everything the public buys, from shoe polish and hooks and eyes to theaters, automobiles and hotels. Billboards are found in every village, town and city, and in the rural districts they are frequently prominent along railroads and at crossroads. In many cities nearly every vacant lot has billboards. They are usually built of wood, eight to twelve feet high, and of any length from a few feet to an entire block.

The billboard is almost always unsightly, but sometimes the posters displayed have real artistic merit. This is especially true in England, France and Germany, where the art of designing large posters has received much attention. Many boards are now constructed so that they may be illuminated at night, either by colored electric lights which form a part of the design or by reflection from powerful lamps overhead.

The erection of a billboard requires that the property-owner give his consent, but this is not always sufficient. In many large cities the boards must conform to a standard size set by law, and frequently the consent of a majority of the property-owners in the block is necessary. Usually the regulation of billboards is left to the local authorities, both in the United States and Canada as well as in Europe, but there are a few exceptions. Massachusetts, for example, allows no billboards along public highways within the state.

There is developing a consistent crusade against the billboard, with the hope of abolishing it. Some communities are almost a unit in their agreement not to purchase anything advertised in this manner in their vicinity. The theory on which such a crusade is based is that business interests will cease to disfigure the landscape when it is no longer profitable to do so. See ADVERTISING.

BILLIARDS, *bil' yardz,* a popular indoor game usually played on a rectangular table with three ivory balls, which are driven against one another by means of a rod called a *cue.* Billiards, as now played, is a scientific game, requiring much skill and practice, and like many other games has won the attention of enthusiasts who have become professional players. So skilful are many of them that special rules have been devised as handicaps. The average amateur, however, finds ordinary billiards sufficiently difficult and interesting.

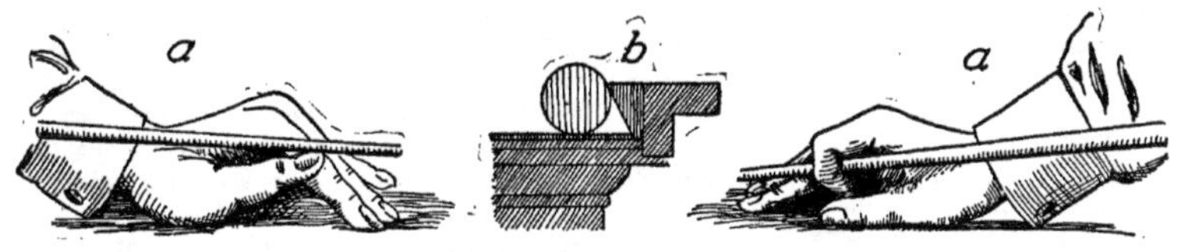

IN THE GAME OF BILLIARDS
(*a*) Position of left or right hand, showing bridge; (*b*) cushion and ball

Before the game begins, the players *bank for the lead;* this is done by driving the balls against the end cushion so that they rebound toward the other end. The player whose ball approaches nearer to the head cushion on the rebound has the choice of playing first or of allowing his opponent to play first. He may also pick out one of the three balls as his *cue ball,* this being the one which he drives with his cue against the other two. One of the balls is red and two are white, but one of the white is distinguished by a black spot. The two cue balls are used in rotation throughout the game. If any player uses the wrong ball he is usually penalized by the loss of his turn.

To begin the game the red ball and one of the white balls must be placed on the two spots near the ends of the table. The red ball is placed on the spot at the *foot* of the table, and one of the white balls at the *head.* The first player places his cue ball anywhere on the table back of or to the side of the white ball of his opponent. He then tries to hit the red ball at the other end of the table in such a way that his cue ball will rebound and hit the third ball. If he succeeds he makes a *carom,* or *billiard,* and scores one point. He may then drive his ball at either of the others, and may continue to play until he misses a carom. Then his opponent, driving his own ball at either of the others, plays until he misses. The players shoot alternately until the game is ended. Any number of points may be agreed upon as constituting a game.

The Table and the Cues. Billiard tables are made in many sizes, but the size generally used is four feet six inches wide and nine feet long. In championship games between professionals, a table five feet wide and ten feet long is sometimes used. The bed of the table is a tablet of slate or marble, covered with a fine green broadcloth. At the sides and ends of the table are cloth-covered rubber cushions, from which the balls rebound. Billiards may also be played on tables with pockets at the corners and at the middle of the sides, but this style is usually reserved for pool (which see).

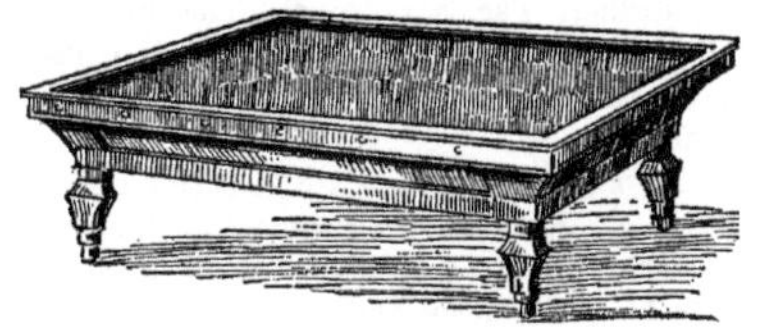

A BILLIARD TABLE

A good billiard cue, made of ash or maple, should weigh about a pound, though the heavier cues run to twenty or twenty-two ounces. The handle, or butt end, is about an inch and one-half in diameter, and the cue tapers down to a tip from one-fourth to one-half inch in thickness. Glued to the small end is a button or tip, made of two pieces of leather; the under piece is hard and flat, while the upper or outer piece is springy and spongy, thus giving a "jump" to the ball.

Billiard Balls. The best billiard balls are made of solid ivory, but cheaper ones are of various imitations. When an ivory tusk reaches the manufacturer it is carefully examined for flaws. If it is perfect it is then sawed into blocks two and one-half to three inches square. The blocks are then cut into a rough spherical shape, and laid aside for six months to dry. When the ivory is perfectly dry it is chiseled smooth and spherical. Lastly it is polished by machinery which gives it a rubbing with chalk, chamois skin and soft leather. See IVORY.

All particles of ivory sawdust and shavings are carefully saved. They are treated with various chemicals, and are molded into any shape desired under enormous hydraulic pressure. Cheap grades of billiard balls are sometimes made of such a composition, as are many other small articles. The process of manufacture is so nearly perfect that only an expert can distinguish the composition from the solid ivory. W.F.Z.

Consult *Modern Billiards;* Daly and Harris's *Daly's Billiard Book.* For the English game, see Roberts's *The Game of Billiards.*

BILLINGS, JOSH. See SHAW, HENRY WHEELER.

BILLINGS, *bil' ingz,* MONT., the county seat of Yellowstone County, is in the south-central part of the state, 238 miles east and south of Helena, on the Yellowstone River and on the Burlington, the Great Northern and Northern Pacific railroads. The area exceeds two square miles. In 1910 the population was 10,031; in 1914 it had increased to 13,020.

Cattle and sheep raising are extensive industries in a large territory surrounding Billings, and from the city are shipped great consignments of cattle, sheep and wool. It is one of the most important inland wool markets in the United States. In the vicinity are coal, marble and limestone deposits, and among the industrial establishments are railroad shops, a beet-sugar factory, flour and lumber mills and brickyards. The city has a county courthouse, city hall and public library.

BILL OF ATTAINDER, a bill or statute declaring a person to be guilty of an offense named and his property appropriated or confiscated. It is an act of a law-making body which pronounces a verdict of guilt upon a man for some alleged offense without giving him a legal trial, or even a hearing. Under such a bill a person found guilty could neither receive nor transmit by inheritance, neither could he testify in any court or claim any rights or legal protection. The Constitution of the United States declares that no state shall pass any bill of attainder, and furthermore explicitly assures any accused person the right to a fair and speedy trial. Bills of attainder were abolished in England and its possessions in 1870. See ATTAINDER.

BILL OF EXCHANGE, a written order by one person directed to another person, requesting him to pay to a third a certain sum of money. The word *person,* as used in this case, may refer to an individual, a bank or any other corporation. The writer of the bill of exchange is called the *drawer,* or *maker;* the person who is asked to pay the fixed sum is the *drawee;* the person to whom the money is to be paid is the *payee.*

Technically, there is not the slightest difference between a bill of exchange and that commercial paper called a *draft.* Common usage, however, has in a measure established slightly-different meanings; so to-day when a bill of exchange is correctly referred to, it means a draft drawn by a person of one country upon a drawee in another country; in such case the word *exchange,* while its ordinary meaning of transfer of money remains, is also significant of the value of money of one country in terms of the other country's monetary unit. In explanation of the latter statement we speak of the *rate of exchange* between the United States and England as being $4.86 for £1. Accepting this view of a bill of exchange, a *draft* then becomes an order for transfer of money between people of the same country. In the days before the term *draft* became current, bills of exchange were termed either *foreign* or *domestic.*

Form of a Bill of Exchange. Only in one respect does the form of a bill of exchange differ from the ordinary draft. Triplicate copies are made in interest of security because of long distances required for transmission and dangers incident thereto. The first, or original copy, is sent to the payee; the second is for the purpose of presentation to the drawee, and the third is retained in the hands of the drawer. The wording of the first, or original copy, may be as follows:

$500

New York, N. Y., Dec. 22, 1916.

At sight of this first of exchange (second and third unpaid) pay to William Jones, or order, five hundred dollars ($500), for value received, and charge to the account of

JOHN KNIGHT.

To LESTER MORGAN, London, England.

As soon as the drawee has honored the request made upon him by paying the amount demanded, the other two copies of the bill of exchange automatically become null and void.

Further particulars of the laws relating to bills of exchange will be found in the article DRAFT.

BILL OF HEALTH, a certificate signed by port authorities, certifying the state of health of a ship's crew and passengers at the time the ship sails. If no infectious disease is known to exist the master of the vessel receives what is called a "clean bill of health." A "suspected bill" is given when disease is feared, and a "foul bill" when it is known to exist. At the port of arrival the vessel is held in quarantine (which see) until the state of health of those on board is ascertained.

A "clean bill of health" is now often used to indicate that a person suspected of some wrong doing has been proven to be innocent.

BILL OF LADING, a written receipt for goods consigned to a common carrier for shipment, either on land or sea. A bill of lading is not merely a receipt for the goods to be transported, but it is a contract between the shipper and the carrier by which the latter agrees, for a stated consideration and under specified conditions, to deliver the goods to their destination. The term was originally applied only to agreements made with shipowners or captains, but now it is also given without distinction to receipts issued by other common carriers. Such receipts issued by railways are sometimes called *way bills.*

A complete bill of lading should mention the name and address of the shipper or consignor, and the name and address of the consignee, the person to whom delivery is to be made. It should name the railway or railways over which shipment is made; on the sea the name of the ship and of the company which owns it are necessary. The goods should be described as accurately as possible, the number of cases or packages and the character of the contents being noted. If there are numbers or other marks of identification these should be recorded. It is further customary to state the value of the shipment, as the charge for transportation is based partly on value and partly on weight or bulk. The charges for transportation, whether paid in advance or payable on delivery, must be mentioned.

A bill of lading has some of the characteristics of a negotiable paper. While the goods are in transit the owner may endorse the bill to a purchaser, who assumes, in law, all the rights and responsibilities of the original consignee. Goods in transit cannot themselves be transferred, but ownership in them may be changed by a transfer of the bill of lading. Only the owner of the goods, however, may endorse it. A finder or thief of a bill, even though it is endorsed in blank or to bearer, cannot claim title. See COMMON CARRIER.

BILL OF RIGHTS, a phrase used rather loosely to mean an enactment or agreement setting forth some fundamental right of the people. When the term is used without modification or explanation it is usually the English Bill of Rights of 1689 which is referred to. When William and Mary came to the throne in 1688 they were forced to assent to a Declaration of Rights, which guaranteed to all the principles of political liberty now established in the English system of government. It is one of the great instruments on which the English constitution rests, the others being the Magna Charter and the Petition of Rights (which see).

Most of the states of the United States have in their constitutions bills of rights, which make clear just what the state may or may not do, and define the inalienable rights of the citizens. Because the first ten Amendments to the Constitution of the United States, proposed by the first Congress which met after the adoption of that instrument, had as their object a more definite statement of the rights of the citizens, they have received collectively the name of "bill of rights." The enactment of these Amendments had been promised at the hands of the First Congress; it is believed the new American states would not have ratified the Constitution in the form in which it was presented to them if such an understanding had not been reached.

BILL OF SALE, a formal statement of the sale or transfer of personal property. There is a recognized legal form in which a bill of sale should be written, which may be purchased at any stationer's store. However, any clear and definitely-written statement of transfer of rights and interest in fully described property is accepted as legal evidence of sale. A bill of sale is often given to a creditor as security for borrowed money, and empowers the receiver to sell the goods named in it if the loan is not repaid at the appointed time. The following is a form of bill of sale used in connection with the sale of ordinary property:

Know all Men by These Presents, That I, James Brown, of Springfield, Ill., in consideration of Eight Hundred Seventy Dollars ($870), the receipt of which is hereby acknowledged, do hereby grant, sell, transfer and deliver unto John Howard the following property, to-wit:

Six Horses	@ $100	$600
Two Buggies	@ 90	180
Two Harness	@ 25	50
Two Plows	@ 20	40
Total		$870

To have and to hold the said goods and chattels unto the said John Howard, his executors, administrators and assigns, to his own proper use and benefit forever. And I, the said James Brown, do avow myself to be the true and lawful owner of said goods and chattels; that I have full power, good right and lawful authority to dispose of said goods and chattels in manner aforesaid; and that I will warrant and defend the same against the lawful claims and demands of all persons whomsoever.

In witness whereof, I, the said James Brown, have hereunto set my hand this 12th day of June, 1916.

JAMES BROWN.

(Witness) ——————————

BILOXI, *bi lok' si,* MISS., lays claim to distinction as a favorite watering place and as a great market for sea foods. It is in Harrison County, in the southeastern corner of Mississippi, and is situated on a peninsula formed by the Black Bay of Biloxi, north and east of the city, and the Mississippi Sound, both parts of the Gulf of Mexico. A chain of islands in the Gulf creates the Sound, a quiet, protected strip of water for all manner of craft sailing between Biloxi and Mobile, sixty-one miles northeast, and New Orleans, eighty miles west. Biloxi covers an area of six square miles. It is served by the Louisville & Nashville Railroad and by an interurban line for a distance of twenty-five miles along the coast. The population was 8,049 in 1910; in 1914 it was 9,147.

More shrimps are caught and canned in the twelve factories in Biloxi than at any other city in America. The oyster production, both raw and canned, is fast taking a leading place. The combined oyster, fish and shrimp business amounts to $2,500,000 annually. The sea food industry has required three others of importance, shipbuilding, artificial ice plants and box-making. Biloxi is an attractive city. Magnificent trees, live oaks, pines, magnolias and palms, are part of an abundant vegetation. Wide streets extend in shell-paved drives fifty miles in length. A marble Federal building, costing $175,000, was completed in 1907. Beauvoir, the former house of Jefferson Davis, near Biloxi, is now a home for Confederate veterans.

Across the Bay from the present city, d'Iberville established in 1699 Fort Maurepas, the first French post in the Gulf region, and soon after a settlement near the fort named Biloxi, from the Biloxi Indians. After a fire, the first site was abandoned and a permanent settlement made where the present city stands. For a number of years it was the seat of government of the Colony of Louisiana. In 1872 it was incorporated as a village and in 1896 received a city charter. Water is supplied by artesian wells, owned and operated by the city. H.H.R.

BIMETALLISM, *bi met' al iz'm,* a monetary system in which two metals, gold and silver, are legal tender for any amount. The ratio of value between the two metals, under such a system, is fixed either by law, if the system

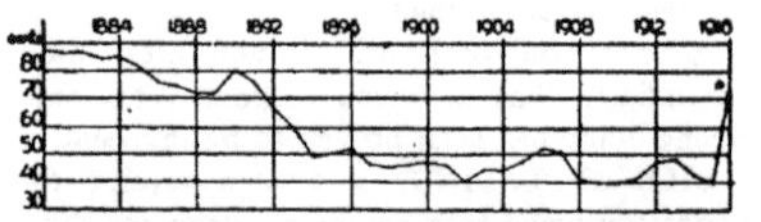

CHART OF SILVER VALUES

In 1873 the silver in a silver dollar of United States coinage was worth 100 cents in gold. The line in the chart shows its depreciation until 1915. Later its value increased to nearly 75 cents. The dot indicates value on June 1, 1916.

is confined to a single country, or by international agreement if it applies to several nations. The supporters of bimetallism argue that a fixed legal ratio will prevent nearly all fluctuation in the prices or market values of the two metals. If this were true, the prices of commodities would become practically stable and foreign exchange would be greatly simplified, whether other countries used a single or a double standard.

The opponents of bimetallism, who now include most economists, assert that the arguments for such a system are false. They claim, apparently with justice, that the cheaper metal will drive the dearer from use (see GRESHAM'S LAW). An example taken from history will make their point clear. In the United States in 1896 the Democratic party platform demanded the "free and unlimited coinage of silver at the ratio of 16 to 1." As a matter of fact, silver in 1896 was not worth 16 to 1, and its market price later fell to 25 and even 30 to 1; that is, it required thirty ounces of silver to buy one ounce of gold. If the Democrats had won the election of 1896, free silver coinage would probably have been established. Before many years all the money of the country would have been silver, for fifty cents worth of silver, bought in the open market, could then have been exchanged at the mint for a coin stamped one dollar. Gold would have been withdrawn from circulation, for a dollar's worth of gold could have been exchanged intrinsically for two dollars in silver. The opponents of bimetallism further argued that even if the system were logical, and if both metals remained in circulation, there was no proof that the fluctuating prices of commodities would become stable. The total output of the precious metals might fluctuate as much as the output of one alone.

Bimetallism was for a century a political issue in France and other European countries. In the United States it was a burning issue only during the last quarter of the nineteenth century. This was an era of falling prices, of industrial depression and lack of work. The industrial and agricultural classes seized on bimetallism as a remedy, not because they were sure it was right but because they felt that the gold standard was wrong. The discovery of new sources of gold supply in various parts of the world was followed by a gradual rise in prices which removed the causes of complaints. Bimetallism is now a dead issue, not because its theories were entirely disproved, but because there is no need for it. E.D.F.

BINDER TWINE, the string with which grain in the harvest fields is fastened into bundles by mechanical binders, is an article of an importance which few people realize. Because of the almost universal employment of machinery in the wheat fields of the United States and Canada and its growing popularity in Russia and Argentina, a failure in the supply of twine at any harvest would mean the loss of millions of bushels of the grain from which the world's bread is made.

About ninety per cent of all binder twine is made from the leaf fiber of the *sisal,* a plant described in another volume of this work, which grows mainly in Yucatan. Most of the rest is spun from the fiber of the manila tree, a banana-like plant of the Philippines. Though twine manufacturing companies have spent time and money on agricultural experiments in the attempt to find a substitute for manila or sisal which will grow in the United States or Canada, they have not yet found one which resists insects. About 200,000 tons of sisal are needed each year for the twine of Canada and the United States.

Almost all binder twine is made entirely by machinery. Combs traveling on a belt pick out the foreign matter and straighten the fibers, delivering them in a loose but continuous coil. A length of this coil is placed in a receptacle in front of each spinning machine, to which it is automatically fed, and which takes just enough fibers to form the proper size strand, and twists them into a firm twine. The balling machines then finish the work.

BINDWEED, the name of a group of plants of the morning glory family, the creeping, twining stems of which cling so tightly to other plants as often to cause their death through suffocation. Thus they are aptly named. One of the most familiar bindweeds is the hedge bindweed, or wild morning glory, found commonly along the roadside, trailing over walls or on the edge of the woods, both in Europe and throughout the Middle and Eastern United States and lower Canada. The stem is somewhat hairy and grows from three to ten feet long. The blossoms are beautiful in color, pink with white stripes, and have the shape of a funnel with a flaring mouth. The tubes of these flowers are so long that only long-tongued bees and other insects can reach the stores of nectar within, and the blossoms remain open only in the sunshine or on bright moonlight nights.

FLOWER AND FRUIT

The field bindweed is a troublesome weed that came from Europe and now grows from Nova Scotia and Ontario southward to New Jersey and westward to Kansas. Its stem is rarely more than two feet in length, but it trails along the ground in the fields of the farmer with a persistency that makes it anything but a welcome visitor. If the weed is kept from going to seed and the land is cultivated in the late fall, the plant may be controlled in a few seasons. Kerosene oil applied to the roots will kill the plant.

BINGEN, *bing' en,* a town of Germany which has attained a place in literature and in legend. Mrs. Caroline Norton's world-famous ballad has made the name a household word; it begins as follows:

A soldier of the legion lay dying in Algiers,
There was lack of woman's nursing, there was dearth of woman's tears,

and closes with the refrain—

For I was born at Bingen, fair Bingen on the Rhine.

Also the legend of the Mouse Tower, on a rock in the Rhine, is one of the most famous tales which has come down from the Middle Ages. According to this tale the hard-hearted Bishop Hatto, having caused hundreds of poor people to be burned to death in a barn to avoid feeding them, fled to the Mouse Tower and was there devoured by rats and mice. This tower, built before the year 1000, was restored in 1856 and now serves as a signal station for ships.

Bingen is in the Duchy of Hesse, at the point where the Nahe joins the Rhine. The district is noted for the culture of the vine, and there is a large trade in wine. There are also important manufactures of starch and tobacco products. Opposite the city, on a bluff above the Rhine, stands a great monument commemorating the founding of the German Empire in 1871. Population in 1910, 10,200.

BINGHAMTON, *bing' am tun,* N. Y., the county seat of Broome County, is in the south-central part of the state, 215 miles northwest of New York City and eighty miles south and east of Syracuse. It is attractively situated at an altitude of 867 feet, at the junction of the Susquehanna and Chenango rivers, which are crossed by several bridges within the city limits. It is on the Erie, the Delaware & Hudson and the Delaware, Lackawanna & Western railroads; interurban electric lines connect with various suburbs, some of which have important manufacturing interests. The area is nearly ten square miles. In 1910 the population was 48,443; in 1915 it was 53,082.

Prominent features of Binghamton include Ross Park (100 acres) and Ely Park (134 acres), O'Neil Park, Industrial Exposition grounds, a Federal building, city hall, public library, supreme court law library, state armory, an opera house and the armory theater. The Binghamton State Hospital for the Insane, and Susquehanna Valley Orphans' Home and Saint Mary's Home are among the leading institutions of the city.

Binghamton is an important manufacturing center; among the chief factory products are boots and shoes, tobacco and cigars, cameras and photographic supplies, carriages and carriage trimmings, spices, leather goods, motors, clocks, furniture (Morris chairs a specialty), electrical apparatus, clothing, cotton fabrics, medicine, paper bags and envelopes, glass, pottery, etc. The total value of factory products is about $18,000,000 a year, and is increasing at a rapid rate.

Settled in 1787, and known as Chenango Point until 1800, Binghamton was then platted and named in honor of William Binghamton, an owner of land in the neighborhood. The place was incorporated as a village in 1834 and became a city in 1867.

BINOMIAL, *by no' mi al,* in algebra, a quantity consisting of two terms or members, connected by the sign + or — ; thus $a+b$ or $a-b$ is a binomial. Either term of a binomial may itself contain two or more numbers, as $(x+y)-(y-x-z)$. One of the most interesting and important operations in all algebra is the *binomial theorum,* perfected by Newton, by which a binomial may be raised to any power. See ALGEBRA.

BIOGENESIS, *by o jen' e sis,* the theory that all living things come from living things similar to themselves, as the chick from the egg and the plant from the seed. This theory accounts for all life from the lowest to the highest forms, both animal and vegetable. It is now the theory generally accepted by biologists, and is opposed to the old theory which accounted for each form of life by supposing it to have originated from a separate creation. See BIOLOGY.

BIOGRAPHY, which is derived from two Greek words meaning *life* and *to write,* means literally the written lives of men and women. It is a very important department of literature. Biographies range in length from what is probably the shortest one ever written, the Biblical "And Enoch walked with God; and he was not, for God took him," to lengthy works of many volumes which touch upon philosophy, upon science, upon history.

Development of Biographic Writing. It seems natural that biographical writing should be one of the oldest branches of literature, for there has never been a time when men have not been interested in the lives of other men; but the earliest writings of this type were very simple, like the accounts of the patriarchs in the Bible. The myths and legends of the Greeks and Romans were but brief biographies of their gods and heroes, usually written to bring out some one point of character. The more formal biography was then, as it remained for centuries, little more than an account of the happenings in the life of a man; but nevertheless ancient Greece and Rome produced some examples of biographical writing which in their way have never been surpassed. The *Agricola* of Tacitus, for instance, is beyond criticism, while the parallel *Lives of Plutarch* have had an influence on modern literature which it is difficult to overestimate.

The Middle Ages had little biography except the lives of saints and martyrs. These works did not strive for historic accuracy, and it was not until the seventeenth century that biography in the modern sense really appeared. That an appreciation of the influence of such writings did not take an earlier hold upon the people is to be much deplored, for had the sixteenth century felt the keen interest that the twentieth does, in the lives of its poets, for instance, the world would have much valuable information which now it lacks. To-day every "idle singer of an empty day" has the minute details of his life chronicled in the magazines, and of all writers of any note exhaustive biographies are sure to appear; but of Shakespeare, the world's greatest poet, there is no biography—not even so much as to make absolutely certain that he was the author of the plays ascribed to him.

Important Biographies. Preëminent among English biographies is Boswell's *Life of Samuel Johnson,* and. second perhaps is Lockhart's *Scott.* Forster's *Dickens,* Mrs. Gaskell's *Charlotte Bronte,* Cross's *George Eliot* and *Tennyson's Life* by his son are all excellent examples of biographical writing. The most interesting and famous American biography is of the type known as *autobiography*—the life of a person written by himself. This is Benjamin Franklin's *Autobiography.*

Biographical material has also entered largely into the writing of fiction. Sometimes a slight sketch of some real person is introduced by an author into his novels, as when Dickens in *Bleak House* pictures Walter Savage Landor and Leight Hunt as Mr. Boythorne and Harold Skimpole; sometimes an entire novel is woven about the career either of the writer himself or of some person prominent in public life. Thus Goethe wrote of himself in *The Sorrows of Werther* and Dickens made use of many of the events of his own life in *David Copperfield;* Meredith described in *Diana of the Crossways* the celebrated Carolyn Norton, a granddaughter of Sheridan; Gertrude Atherton wrote of Alexander Hamilton in *The Conqueror,* and Maurice Hewlett gave a more or less distorted picture of Lord Byron in his *Bendix.* In reading such works allowance must always be made for the fact that after all the author is writing fiction and not biography, and therefore may feel free to take certain liberties with his material.

Biography for Children. If the teacher or parent of a child ever hears him say, "I don't like to read biography—I don't care for 'lives' of people," that teacher or parent may be sure that the lives have simply been presented to him in the wrong way. For everyone, young or old, is naturally interested in "lives"—if they are shown him from the right angle. What, indeed, are most of the stories which so delight children but biography, presented from the point of view which appeals to a child? *Bluebeard, Dick Whittington and His Cat, The Ugly Duckling,* "fairy tales" though they may be, are but biographies which the child feels should be true, even if they are not; while as regards less legendary heroes—Joseph, Daniel, David, King Arthur, the Cid, Roland—any child will listen to stories of them told over and over again, and then ask to hear them once more.

We expect a child to like stories of these heroes; we pick out the points that will strike the child's fancy, fire his imagination, hold his interest. But our attitude changes when we come to consider other men whom tradition has not marked as children's heroes. "Why," we say, "should a child be interested in the Apostle Paul? A boy or girl does not care particularly for preaching and for missionary work." And we forget that Paul had, if ever a man had, just those experiences that children love to hear about; that he was "in deaths oft, . . . in journeyings often, in perils of waters, in perils of robbers, . . . in perils of the wilderness, in perils of the sea." Or we think again, "Of course a child doesn't care to read about Dickens or Longfellow or Hawthorne. Of what particular interest is it to him that one man wrote *The Tale of Two Cities* and another man wrote *The Marble Faun?*"

But even in those biographies which at first

sight seem adapted only to older people, a little study will almost always reveal much that will interest a child. The "things that happen" must of course be emphasized, especially childhood events, while all moralizing must be omitted. There may be in such a biography an apparent lack of proportion, but it is simply an exaggeration of some points, not a distortion. As a child grows older and gradually widens his interests, he will learn other facts to fit on to and fill out those he has already learned; but he will not need to unlearn anything of what he has remembered.

Biography in the School. The teacher finds many uses for biography besides the merely intellectual one, for there is nothing so helpful in character-building as well-selected, well-presented biographical material. This does not mean that the admonition, "Do thou likewise," is to be given every time a forceful act or an attractive character is presented; in fact, it means quite the opposite. If the factors that made a man great or good are put impressively before him, the child will have an instinctive desire to imitate them. A knowledge of the early hardships of Dickens, for instance, and of his struggles to educate himself is almost certain to waken in a child some appreciation of his greater opportunities; the story of Wolfe or of Washington contains a never-dulled spur toward achievement; the wonderful career of Florence Nightingale is a striking lesson in the benefits of preparation for life work. These illustrations might be multiplied almost endlessly, but every teacher and parent will have in mind heroes or heroines of his own whom he can present sympathetically to the children.

The Biography Method of Studying History. "History is but the essence of innumerable biographies," wrote Carlyle, and the deeper a student delves into history, the more thoroughly is he convinced that this is true. Some men seem to gather unto themselves the entire history of their times; the story of Rome in its late republican days was the story of Caesar; France for almost a generation had almost no history save that of Napoleon; while the trend of affairs in England during a momentons period is best understood through a study of the life of Cromwell. In other eras no one man stands out as clearly, and it is the reaction of many less dominant figures upon each other which works out the history of such periods.

To come to concrete examples, a student might possess himself very satisfactorily of the history of the early national period in the United States by studying the lives of Washington, John Adams, Hamilton and Jefferson. If the history of Germany from 1860 to 1880, for instance, be the topic, it may be approached through the lives of Emperor William I, Bismarck and Napoleon III of France. To most students this biographical method is very interesting, for it introduces that personal touch which makes history a living subject.

How to Write a Brief Biography. No hard and fast rules can be laid down for this, for every person looks at the subject from a different angle, but certain general directions can be given which will be of use to school children in writing such brief biographies as they may be called upon to produce. First of all, an outline is most helpful. It need not be elaborate, but it should contain every important point which the writer intends to touch upon. The natural method in writing a biography is to proceed chronologically, and, simple as that direction may seem, it is one that is frequently overlooked by young writers. One apparent exception should be made to this rule: the great achievement of the man under consideration should be briefly noted very near the beginning of the biography, that the one who reads may know why he is important enough to deserve study. Later it may be dwelt upon more at length.

One rule cannot be too strongly insisted upon—a biography must not be a mere catalogue of dates. No one finds interesting a story of a man's life which proceeds thus: "He was born in 1847. In 1853 he started to school and in 1865 to college. He graduated in 1869 and two years later moved to Philadelphia." That is the mere skeleton, the framework about which interesting events and anecdotes are to be grouped. Above all, one thing should be kept in mind, and that is the crowning achievement; and every happening, however trivial, which helped to shape the man's character or ability toward that end should be pointed out. In Nathaniel Hawthorne's life, for instance, the years spent in the Maine woods were quiet, unproductive years, but no one could understand Hawthorne's later life, his intense reserve and his unsociable habits, without a knowledge of that formative period.

Some men's lives, from the point of view of a biographer, do not end with their death. Those classical biographies known as the *Gos-*

pels chronicle the entire career of Jesus of Nazareth, but the account of his life is not complete without that supplementary book called the *Acts of the Apostles* which gives the account of the carrying on of his work by those whom he inspired. Thus the life of Dickens cannot be regarded as complete without a discussion of those humane reforms which have taken place since his death, but which he initiated. A.MC C.

Related Subjects. These volumes are particularly rich in biography, and to make reference to it easier and more effective, the biographical articles are all indexed. Many of them, as indicated below, are listed under special departments; the others are given here.

ACTORS

See *Drama*.

ASTRONOMERS

See *Astronomy*.

AUTHORS—MISCELLANEOUS

Adams, Charles Francis
Asbjörnsen, Peter Christen
Baedeker, Karl
Boswell, James
Brandes, Georg M. C.
Burke, Edmund
Butterworth, Hezekiah
Caedmon
Carpenter, Frank George
Chesterfield, Earl of
Collier, John Payne
Curtis, George William
Custis, George Washington P.
Dana, Richard Henry
Diderot, Denis
Dodge, Mary Abigail
Dole, Nathan Haskell
Dowden, Edward
France, Anatole
Franklin, Benjamin
Frechette, Louis H.
Furness, Horace Howard
Gosse, Edmund
Grady, Henry Woodfin
Hale, Edward Everett
Heine, Heinrich
Johnson, Samuel
Lafontaine, Jean de
Landor, Walter Savage
Lang, Andrew
Le Moine, James MacPherson
Lewes, George
Malory, Sir Thomas
Mandeville, Sir John
Martineau, Harriet
Matthews, James Brander
Mitchell, Donald Grant
Morley, Henry
Morley, John
Oliphant, Laurence
O'Rell, Max
Ossoli, Sarah Margaret Fuller
Paine, Thomas
Pepys, Samuel
Pliny the Elder
Pliny the Younger
Proudhon, Pierre Joseph
Quintilian
Richter, Johann Paul F.
Riis, Jacob A.
Roosevelt, Theodore
Rousseau, Jean Jacques
Sainte-Beuve, Charles Augustin
Scudder, Henry Elisha
Sheldon, Charles M.
Smiles, Samuel
Smith, Goldwin
Smollett, Tobias George
Southey, Robert
Sparks, Jared
Stael-Holstein, Baroness de
Steele, Sir Richard
Stoddard, Richard Henry
Strabo
Swift, Jonathan
Taine, Hippolyte Adolphe
Tarbell, Ida Minerva
Taylor, Bayard
Thomas à Kempis
Tieck, Ludwig
Voltaire
Walpole, Sir Robert
Walton, Izaak
Warner, Charles Dudley
Wilde, Oscar
Wilson, John
Winter, William

BUSINESS MEN AND FINANCIERS

Allan, Sir Hugh
Armour, Philip D.
Ashburton, Alexander Baring, Lord
Astor, John Jacob
Astor, William B.
Astor, William Waldorf
Belmont, August
Carnegie, Andrew
Cooke, Jay
Field, Marshall
Field, Cyrus West
Fish, Stuyvesant
Gould, George Jay
Gould, Jay
Green, Hetty H.
Harriman, Edward H.
Hill, James J.
Hirsch, Maurice, Baron de
Law, John
Mackay, John William
Morgan, John Pierpont
Plant, Morton F.
Pullman, George M.
Rockefeller, John D.
Rothschild, Mayer A.
Rothschild, Lionel
Sage, Russell
Schwab, Charles M.
Stanford, Leland
Vanderbilt, Cornelius
Vanderbilt, William Henry
Vanderbilt, William K.
Wanamaker, John

CARTOONISTS

See *Cartoon*.

CHEMISTS

See *Chemistry*.

CHURCHMEN

See *Protestants; Roman Catholic Church.*

DRAMATISTS

See *Drama*.

ECONOMISTS

See *Political Economy*.

EDITORS

Bennett, James Gordon
Bennett, James Gordon, Jr.
Bok, Edward W
Bourassa, Henri
Brisbane, Arthur
Brown, George
Bryan, William Jennings
Carter-Cotton, Francis L.
Dana, Charles A.
Douglass, Frederick
Garrison, William Lloyd
Greeley, Horace
Hapgood, Norman
Hearst, William R.
Hincks, Sir Francis
Howe, Joseph
Howell, Clark
Labouchere, Henry D.
Macdonald, James A.
Mackenzie, William L.
Northcliffe, Alfred Charles Harmsworth, First Baron
Pulitzer, Joseph
Robertson, John R.
Rochefort, Victor H.
Scott, Walter
Smith, Goldwin
Stead, William T.
Watterson, Henry
Weed, Thurlow
White, William Allen
Willison, Sir John

EDUCATORS

See *Education*.

ENGINEERS

See *Engineering*.

ESSAYISTS

See *Essay*.

EXPLORERS

Abruzzi, Luigi Amadeo, Duke of the
Alvarado, Pedro de
Amundsen, Roald
Andree, Salomon
Baker, Sir Samuel White
Balboa, Vasco Nunez de
Bienville, Jean Baptiste, Sieur de
Cabot, John
Cabot, Sebastian
Cabral, Pedro Alvarez
Carson, Christopher
Cartier, Jacques
Champlain, Samuel
Clark, William
Columbus, Christopher
Cook, James
Cook, Frederick
Coronado, Francisco Vasquez
Cortez, Hernando
Dawson, George M.

De Soto, Fernando
Dias, Bartholomeu
Drake, Sir Francis
Du Chaillu, Paul Belloni
Emin Pasha
Eric the Red
Franklin, Sir John
Frémont, John Charles
Frobisher, Sir Martin
Cama, Vasco da
Gilbert, Sir Humphrey
Gosnold, Bartholomew
Hawkins, Sir John
Hedin, Sven
Hennepin, Louis
Henry the Navigator
Hudson, Henry
Joliet, Louis
Kennan, George
La Salle, Sieur de
Lewis, Meriwether
Livingstone, David
Mackenzie, Sir Alexander
Magellan, Ferdinand
Marquette, Jacques
Nansen, Fridtjof
Narvaez, Panfilo de
Nordenskjöld, Nils Adolf Erik, Baron
Park, Mungo
Parry, Sir William Edward
Peary, Robert E.
Pike, Zebulon Montgomery
Pizarro, Francisco
Polo, Marco
Ponce de Leon, Juan
Raleigh, Sir Walter
Rohlfs, Gerhard
Ross, James Clark
Schwatka, Frederick
Scott, Robert Falcon
Shackelton, Sir Ernest
Speke, John Hanning
Stanley, Sir Henry Morton
Stefansson, Vilhjálmur
Thompson, David
Tonty, Henry de
Verrazano, Giovanni da
Vespucius, Americus

GEOLOGISTS

See *Geology*.

HISTORIANS

See *History*.

HUMORISTS

Ade, George
Bangs, John Kendrick
Browne, Charles F.
Burdette, Robert J.
Clark, Charles H.
Clemens, Samuel L.
Cobb, Irvin S.
Dunne, Finley P.
Haliburton, Thomas C.
Jerome, Jerome K.
Jerrold, Douglas W.
Locke, David R.
Nye, Edgar Wilson
Shaw, Henry Wheeler
Shillaber, Benjamin P.
Smith, Sydney

INVENTORS

See *Invention*.

JOURNALISTS

See *Newspaper*.

JURISTS

See *Law*.

LABOR LEADERS

See *Labor Organizations*.

MATHEMATICIANS

See *Mathematics*.

MISSIONARIES

See *Missions*.

MUSICIANS

See *Music*.

MILITARY AND NAVAL LEADERS

See subhead *History* under each country.

NOVELISTS

See *Novel*.

ORATORS

See *Oration*.

PAINTERS

See *Painting*.

PATRIOTS

See *Patriotism*.

PHILANTHROPISTS

Addams, Jane
Armour, Philip D.
Barton, Clara
Carnegie, Andrew
Cooper, Peter
Corcoran, William W.
Cornell, Ezra
Drexel, Anthony
Durant, Henry F.
Gerard, Stephen
Gould, Helen Miller
Hirsch, Maurice
Lenox, James
Lick, James
Nightingale, Florence
Peabody, George
Raikes, Robert
Rockefeller, John D.
Smithson, James
Vassar, Matthew
Yale, Elihu

PHILOSOPHERS

See *Philosophy*.

PHYSICIANS

See *Medicine and Surgery*.

PHYSICISTS

See *Physics*.

PIONEERS

Austin, Stephen F.
Boone, Daniel
Carson, Christopher
Clark, George Rogers
Crockett, David
Cutler, Manasseh
Ross, Alexander

POETS

See *Poetry*.

POPES

See *Pope*.

PRESIDENTS

See *President of the United States*.

PSYCHOLOGISTS

See *Psychology*.

REFORMERS

Addams, Jane
Anthony, Susan B.
Bergh, Henry
Booth, Maud B.
Booth, Bramwell
Booth, Ballington
Booth, William
Booth, Tucker
Brown, John
Calvin, John
Catt, Carrie Chapman
Dow, Neal
Fry, Elizabeth
Garrison, William L.
Gough, John B.
Hunt, Mary H.
Huss, John
Knox, John
Jerome of Prague
Livermore, Mary A. R.
Lockwood, Belva Ann
Luther, Martin
Melancthon, Philip
Mott, Lucretia Coffin
Owen, Robert
Phillips, Wendell
Stanton, Elizabeth Cady
Tyndale, William
Willard, Frances E.
Wycliffe, John
Zwingli, Ulric

RELIGIOUS LEADERS

See *Religion*.

RULERS

See subhead *History* under each country.

SAINTS

See *Canonization*.

SCIENTISTS

See *Science*.

SCULPTORS

See *Sculpture*.

SOCIALISTS

See *Socialism*.

STATESMEN

See subhead *History* under each country.

MISCELLANEOUS

Agrippina
Alden, John
Aspasia
Beatrice Portinari
Beard, Daniel C.
Blennerhassett, Harman
Blondel
Boabdil
Boleyn, Anne
Bothwell, James Hepburn
Bridgman, Laura D.
Brummell, George Bryan
Buckingham, George Villiers, Duke of
Cade, John
Camp, Walter
Carteret, Sir George
Carver, John
Cassius Longinus
Catiline
Cenci, Beatrice
Corday d'Armont, Marie Anne Charlotte
Coriolanus
Cornelia
Coverdale, Miles
Crassus, Marcus Licinius
Crichton, James
Damocles
Damon and Pythias
Darling, Grace H.
Dinwiddie, Robert
Du Barry, Marie Jeanne Becu
Duns, John
Eaton, Margaret O'Neill
Erasmus, Desiderius
Eugenie-Marie de Montijo
Eulenspiegel, Till
Faust, Johann
Fawkes, Guy
Force, Peter
Fra Diavolo
Frohman, Charles and Daniel
Furnivall, F. J.
Gary, E. H.
Grey, Lady Jane
Grotius, Hugo
Hays, Charles M.
Jahn, Frederich L.
John of Gaunt
Josephine, Marie Rose
Keller, Helen Adams
Kidd, William
Kneipp, Sebastian
Kropotkin, Peter A., Prince
Lathrop, Julia C.
Leicester, Robert Dudley, Earl of
Lycurgus
Maintenon, Françoise Marquise de
Maria Louisa
Marie Antoinette
Mercator, Gerard
Morgan, Sir Henry
Mother Shipton
Müller, Friedrich Max
Münchhausen, Baron
Nana Sahib
Octavia
Oglethorpe, James E.
Olmsted, Frederick L.
Origen
Orleans, Dukes of
Pankhurst, Emmeline
Paris, Count of
Peter the Hermit
Petrarch, Francesco
Petrie, William M. F.
Pinchot, Gifford
Pinkerton, Allan
Pitman, Sir Isaac
Pompadour, Madame
Queensberry, Marquis of
Reichstadt, Duke of
Ridley, Nicholas
Riel, Louis
Rienzi, Cola di
Robin Hood
Rob Roy
Roland de la Platiere, Madame
Rolfe, John
Rosamond
Schliemann, Heinrich
Selkirk, Alexander
Servetus, Michael
Seymour, Horatio
Shaw, Anna
Skeat, Walter William
Smith, John
Spartacus
Standish, Miles
Stradivarius, Antonio
Stuart, Charles Edward
Tetzel, Johann
Turner, Nat
Van Horne, Sir William C.
Walker, William
Warbeck, Perkin
Warwick, Richard Neville, Earl of
Washington, Martha
Webster, Noah
White, Richard Grant
Xanthippe

BIOLOGY, *by ol' o ji,* from two Greek words meaning *life* and *speech* or *discourse,* is an old subject, but a modern word, for it seems to have been used only since the beginning of the nineteenth century. It means, as is clear enough from its origin, the *study of life,* whether plant or animal, but the word is used in somewhat varying senses. Thus botany, zoölogy, ethnology, physiology, as well as other branches —all those sciences which deal with living things as distinguished from the non-living things in which geology, astronomy and physics interest themselves—are a part of biology.

Related Subjects. All the articles listed under botany, zoölogy, physiology and kindred sciences really fall within the scope of biology; but the following list contains those topics which are too broad to be classified under any one branch, and so belong to biology as a whole:

Acclimatization
Albino
Assimilation
Atavism
Biogenesis
Cell
Death
Degeneration
Dwarf
Evolution
Fibrin
Genus
Hybrid
Man
Metamorphosis
Morphology
Natural Selection
Order
Parasites
Protoplasm
Reproduction
Species
Spontaneous Generation
Variety

BIRCH, the name of all trees belonging to the birch family, found widely spread in North America, Europe and Northern Asia. In general, birches may be distinguished by their smooth bark arranged in horizontal layers and thin, delicate, triangular leaves, which usually turn yellow late in autumn. The flowers appear before the leaves, and are in long clusters which hang downwards. The fruit is a small, scaly cone, and the seed is flat and winged (see SEEDS, subhead *Seed Dispersal*). The birch is a hardy tree, and only one or two other trees can live in so cold a climate as it can endure.

Canoe Birch, or **Paper Birch,** also called WHITE BIRCH. This tree is easily recognized by its yellowish-white bark, which is easily separated into thin layers. It reaches a height of sixty to eighty feet, has a few erect, large branches and many small horizontal ones. It is the bark of this tree that was used by the Indians in making canoes, and they still employ it for making ornaments and small baskets which they sell to tourists. Ladies visiting the northern woods collect it for stationery, for writing can be put upon the thin bark. This birch is found throughout North America from the Arctic Circle southward as far as Long Island. A variety of the white birch, known as the *weeping birch,* is common throughout Scotland and England.

Yellow Birch. The yellow birch, sometimes called the *gray birch,* is a tree from fifty to seventy-five feet in height, with a broad, round top. The bark may be yellow or dark gray, but on young twigs it is silvery yellow. The thin layers often break, forming loose ends which give the tree a ragged appearance. The yellow birch is found from Newfoundland south as far as Delaware, east to North Carolina and Tennessee, and west to Minnesota. The wood is valuable for furniture, sleigh frames and a large number of small articles. It is one of the best of timber trees.

Red Birch, or **River Birch.** The bark of this tree is a dark reddish-brown, varying on the different branches from cinnamon to silver. The red birch is a tall graceful tree, reaching a height of sixty to ninety feet. It is the birch of the South, being found along rivers and in ponds and marshes from Massachusetts to Florida and west as far as Texas. When transplanted, it flourishes equally well in the Northern states.

Cherry Birch, or **Sweet Birch.** This species is found throughout the Northern United States and Canada, as far west as the Great Plains. Its branches are slender, with delicate twigs. The tree reaches a height of sixty to eighty feet, and has a beautiful, rounded top. The wood is dark brown, hard and close-grained, and is valuable especially for furniture and interior decorations. Boys chew the twigs for the flavor of wintergreen in the sap, and from the sap the genuine birch beer is made. Much of the latter, however, is made from chemicals.

White Birch, or Aspen-leaved **Birch.** This is a small, graceful tree, found along the Atlantic coast from Nova Scotia to Delaware and northwest to Lake Ontario. It rarely grows higher than forty feet, and has slender horizontal branches and small, dainty, tremulous leaves that in shape and poise resemble those of the aspen. Its bark is chalky white or grayish, is hard and close and not easily separated into layers. Dark-colored V-shaped patches appear on the bark wherever there has been a bud or branch. The wood of the white birch is used for making spools, shoe pegs and wood-pulp and is also valued as a fuel. This tree is hardy, and will thrive in almost any soil, though it is generally found in dry, gravelly places. W.F.R.

BIRDS, FROM NORTH TO SOUTH
Auk
Eider Duck
Canada Goose
Snowy Owl
Stork
Eagle
Female
Bobolink
Pelican
Prairie Chicken
Ostrich
Parrot
Condor
Bird of Paradise
Albatross
Black Swan
Penguin
KATE ABELMANN

BIRD, an animal with feathers—that is the briefest description that can be given, and at the same time one which will enable a person to recognize a bird wherever seen. Of course many more facts must be known if one is to have even the most general knowledge of birds—that they are warm-blooded, that each has a backbone and but two legs, and that most of them can fly. This last feature might at first seem the truly distinguishing one, but a little thought brings to mind the fact that some birds, as the ostrich, the penguin and the curious apteryx, cannot fly, while bats, which are not birds, flying squirrels and flying fishes do have the power of flight.

That birds have backbones show that they belong in the great class of vertebrates, for *vertebrate* means *backboned.* But it is very curious to an untrained person to learn that they are much more closely related to the reptiles than to any other living animals. There seems to be absolutely no resemblance between the brilliantly-colored bird, with its exquisite song, poised lightly on tree or bush, and the venomous snake which slips stealthily through the underbrush, perhaps intent on robbing the nest of that very bird. One is perhaps the most dearly-loved by man of all the animal world; the other the most hated and feared. Yet scientists discover resemblances in their structure, and students of fossil forms can state positively that ages ago there were bird-reptiles or reptile-birds from which existing birds developed.

Framework of the Body. The most natural thing in the world is to think, "Since a bird can fly, its body must be light," and so indeed it is. For the bones are thin and hollow, and in the thorax and abdomen are air cavities which can be inflated when the bird wants to fly. Everyone who has watched a housewife prepare a fowl for the table knows the shape of the breastbone, which is really a keel to which are fastened the strong wing-muscles. All birds that fly have this keel-like breastbone, and the lowest classes of birds, as the ostrich, which do not have it, are called *ratitae,* meaning *keelless* or *flat-bottomed boat.*

Flight. It is the modification of the fore limbs or "arms" of birds into wings which gives them their most wonderful distinction—the power of flight. From the "arms and hands" as they may be called, grow strong heavy feathers, making a broad surface with which the bird can beat the air; while above and below these heavy quills there are short feathers which prevent the air from passing through. The great condor has a wing-stretch of from ten to twelve feet, while that of the humming bird is from two to about five inches, but each is perfectly adapted to bearing up the body which it must support. The tail does not help much in flight, but is rather a rudder by which the bird steers itself and holds its body level.

And the power of flight to which all these special structures contribute is marvelous indeed. The grace of a bird in soaring, hovering, sailing and swooping, everyone has remarked, and poets have often found in it their inspirations. Thus Leland has—

> Great albatross!—the meanest birds
> Spring up and flit away,
> While thou must toil to gain a flight,
> And spread those pinions gray;
> But when they once are fairly poised,
> Far o'er each chirping thing
> Thou sailest wide to other lands,
> E'en sleeping on the wing;

and Shelley sings—

> Around, around, in ceaseless circles wheeling
> With clang of wings and scream, the Eagle sailed
> Incessantly.

The flight of smaller birds, too, has had its strain of praise, for Riley tells of—

> . . . the humming-bird that hung
> Like a jewel up among
> The tilted honeysuckle horns.

and Christina Rossetti sings to the skylark—

> O happy skylark springing
> Up to the broad, blue sky,
> Too fearless in thy winging.

It does not seem strange that the giant birds that seem to hang motionless for hours in the upper air should be able to take long, swift journeys in search of food or a new abiding place, but the endurance of some of the smaller birds is surprising. Apparently feeble birds, that usually confine their flight to short dashes from bush to bush, may during the migrating season cover in a single flight distances ranging from 500 to 2,000 miles. Some birds, as the yellowlegs, migrate in the fall from the Arctic regions to Southern South America, and return in the spring, thus making each year journeys amounting to 16,000 miles. See subtitle, *Migration of Birds,* below.

Just how rapidly they fly during these journeys there is wide difference of opinion. It may safely be stated that no land animal can move nearly as rapidly as these dwellers in the air, but it seems likely that their speed has been exaggerated by some writers. Carrier pigeons, not the very fleetest of birds, can keep up for hours an average of fifty-five miles an hour; an eagle can rise out of sight in less than three minutes, and thus, according to calculations, must be able to fly at least sixty miles an hour. But it seems probable that the average speed of birds is not more than thirty or forty miles an hour.

The flight of birds, their ability to live in an element peculiarly their own, has been an inspiration to inventors as well as to poets. For their flight clearly proved that solid bodies of considerable weight could be held suspended in the air if only the right sort of "wings" could be constructed, and all the attempts at building flying machines have grown out of this perception. From Daedalus (which see), who made wings of feathers cemented with wax and flew with them over the sea, to the successful maker of the most complete aeroplane—all have had constantly before them the example of the birds.

Feathers. That birds are among the most beautiful objects in nature is due not only to their grace, but to their clothing of feathers. These feathers, while they take the place of the hair that covers the higher animals, do not grow from practically every part of the body, as does a cat's fur, for instance. They grow in certain definite areas or patches, and the spaces between may be bare or covered with down. Usually, however, the feathers overlap enough to furnish a complete coat. The wing and tail feathers are always the largest and strongest.

The wonderful adaptations which nature is capable of are plainly to be seen in the feathery coats or plumage of birds. Thus the birds which live in a warm climate and do not fly, like the ostrich or the cassowary, have fewer and thinner feathers than do the birds of colder regions. A thick covering is not needed as protection; strong quills are not needed for beating the air—and the bird therefore has neither. In any water fowl is seen the same wise provision—a special gland exists for the secretion of oil, with which the bird oils its plumage and thus prevents the water from reaching the body and the feathers from becoming wet and heavy. "Like water from a duck's back" is the common proverb which has grown out of this peculiarity.

The difference in color which often exists between the male and the female bird of the same species is another example of this same wonderful adaptation. At the mating season the male must be attractive, if at no other time, and therefore it is true of many birds as of the bobolink that—

Robert of Lincoln is gayly drest,
  Wearing a bright, black, wedding coat;
White are his shoulders and white his crest;

while—

Robert of Lincoln's Quaker wife,
  Pretty and quiet, with plain brown wings,
Passing at home a patient life,
  Broods in the grass while her husband sings.

Some of these mating-season coats of the male birds are gorgeous in the extreme, but many of them are shed when the nesting season is over.

Molting. Almost all birds shed their feathers at least once a year. Anyone who has had a canary or other cage-bird has watched the process often, and has noticed how the best songster fails at that time to sing. Nor will the barnyard fowl, however well fed, lay eggs during this shedding or *molting* time, as it is called. The feathers never all drop off at once, as do the leaves from a tree, to make room for the new plumage, but the old, worn feathers gradually give place to new. In some birds the molting is partial, the wing and tail feathers remaining in place; in others complete, those strong feathers falling also. Some birds, like the bobolink mentioned above, have a gay plumage for the mating season and a duller one thereafter—

Robert of Lincoln at length is made
  Sober with work and silent with care;
Off is his holiday garment laid;

and such birds have of necessity two molting seasons in a year.

Senses. As compared with other animals

birds have a very keen sense of sight. Indeed, in some species this is little less than marvelous. A kite which has soared so high that the human eye cannot see it, can often spy far below on the ground a tiny mouse, upon which it will drop with unerring aim. Sometimes, as in the owl, the sensitiveness to light is so great that the bird is practically blind in the daytime, but at night he is truly "lord of the dark green wood." The eye of a bird is much like that of a human being, but it possesses certain advantages all its own. For it has an extra eyelid, called by naturalists the *nictitating* or *winking* membrane, which it may draw at will.

The sense of hearing, too, is very well developed in birds—not better than in human beings, as is the sense of sight, but probably as well. They have no external ears, but the opening to the internal ears is fairly large. No better proof of this acute sense of hearing is necessary than the ease with which certain birds mimic sounds, as does the mocking bird, or learn to repeat words, as does the parrot.

The senses of touch and taste are not highly developed in birds, nor, it is believed, is that of smell, though some birds seem to make use of this last-named sense in seeking out their food. The condor and the common buzzard, however, find their food by sight and not by smell.

Special Senses. In addition to these senses possessed by man, birds, according to some authorities, have also a so-called "sixth sense" —the sense of direction. Whatever this sense be called, it is exhibited in a marked degree by the carrier pigeon, which will fly swiftly back to its home, even when it has been sent away by train in a wicker case. Even more remarkable is the ability of a flock of migrating birds to fly straight to the spot where they had their homes the year before, even when to reach it they must cross tracts of water which afford no landmarks whatever. A person who has interested himself in providing homes for certain kinds of birds, as the bluebird or martin, may feel certain that each spring they will return to him. One observer writes:

> My neighbor was very busy one Saturday morning putting up his ornamental bird house—"For my martins," he explained, "will arrive about the twelfth." The completed house was promptly taken over by the sparrows, but the bird-lover did not worry. On the morning of the twelfth we heard a commotion, and sure enough, the martins had arrived and were driving the sparrows from their premises.

It is not only in this special sense or instinct that birds show their adaptation to the necessities of their life; there are peculiarities of structure, too. Thus there is no danger of a perching bird falling from its twig while sleeping, for the muscles in its legs are so arranged that, when the bird perches, its toes are bent and cannot open until the bird rises again. Some birds' feet are specially made for scratching in the ground, some for holding prey, some for swimming, and the bills are equally well adapted to just the services they must perform.

Song. Without their power of song, birds would not be so greatly loved, however beautiful many of them are. Yet only a comparatively small number can really sing, and these are often the ones which have the plainest plumage. Almost all birds make some sound—many seem to have a well-developed language by which they make themselves understood; but no one welcomes especially the harsh *caw* of the crow, the scream of the jay or the hoot of the owl. However, the nightingale, the mocking bird, the thrush, the linnet, the warblers, the skylark and certain others—these are welcome everywhere; for the song of the male in the mating or nesting season affords one of the greatest pleasures which all out-of-doors has to offer. The lark, of whom Shelley sings—

> Sound of vernal showers
> On the twinkling grass,
> Rain-awakened flowers,
> All that ever was
> Joyous, and clear, and fresh, thy music
> doth surpass,

and the nightingale with its "eternal passion, eternal pain" have been most celebrated in poetry, but there are other birds which give equal pleasure to innumerable listeners.

The organ which produces the voice of birds, whether it be melodious song or discordant cry, is not the larynx, as in man, but a special organ which only birds have, called the *syrinx*. Man uses his tongue in his speech; a bird does not.

"Home Life" of Birds. Birds usually live in pairs and in some species, as the eagles and hawks, a union lasts for life. With the majority of birds, however, a new mating takes place each spring, and the first act of the pair is to seek out a suitable place for a home. Most birds build nests, all those of one species making just the same kind, but there are some curious exceptions to the nest-building rule (see Nest). In the nest the eggs are laid, for all birds without exception are hatched from eggs; and in some way or other these must be kept warm until the little ones appear. Sometimes the eggs are laid in warm sand or rotting vegetation, but the nesting birds *brood* their eggs, and keep them warm by their bodies. Usually, the mother bird performs this act of service, while her mate brings her food or cheers her with his song, but in some species the father bird takes his turn at brooding the eggs. With most of the smaller birds it takes from two to three weeks for the little chicks to mature and break from their shells. The number of eggs varies from two to a score, seeming to be in proportion to the dangers which the young are to meet; the number is practically the same at every sitting of each species. See EGG.

With the lower orders of birds, such as the chickens and geese, and other water fowl, the young come from the shells clothed with a complete suit of down, and are able to take partial care of themselves. Within a few hours after their appearance the fluffy chicks run about searching for bugs or grain, though they still have to be kept warm by the mother at night. The chicks of the higher orders of birds, however, are very helpless when they first come from their shells—as helpless as is a tiny kitten or a baby. They are naked of feathers, scrawny and apparently "all mouth"; and the frequent saying that "all young things are beautiful" certainly does not hold true of young birds. The parent birds show the most unselfish devotion to their helpless little ones, guarding the nest from enemies at the risk of their own lives, keeping the nestlings warm at night, and above all things providing them with the incredible amount of food which they need for growth (see subhead *Food*, below). When the time comes for the little birds to fly, the parents teach them to do so by calling to them and withholding food from them, thus inducing the young to leave the nest. Few members of the animal kingdom show more parental affection than do the birds. In some cases it is but a few weeks that the little ones are so dependent, but with some of the larger birds this period may be almost a year.

Food. The food of birds varies according to the species, for like the four-footed animals they are divided into two classes—those which eat animal food and those which eat vegetable. Many birds, however, feed on grain and insects indiscriminately. No bird has teeth, but the beak of each species is fitted to handle the kind of food which it eats, and its digestive organs are peculiarly adapted to treating food that has not been chewed. When the food is swallowed it finds lodgment first in the *crop*, a large sack at the bottom of the gullet. Here it is soaked and softened for some time, and is then passed on to the *gizzard*, a kind of stomach, with exceedingly strong, muscular walls and tough, wrinkled lining. Here the food is ground fine by vigorous rubbing, aided by small stones which have been swallowed by the bird. The flesh-eating birds have smaller gizzards, with thinner muscular coats, than have the grain-eaters, for the softer food does not need nearly so much grinding as do the hard-skinned seeds and grains; and in some species of flesh-eaters there is no gizzard at all.

Like larger animals which live upon flesh, the carnivorous, or flesh-eating, birds are strong, fierce and unsociable. They do not live in friendly colonies, as do the cheerful and harmless grain-eaters, but withdraw for the most part into some secluded region of forest or rock.

It is seldom that birds, during their waking hours, are to be seen at rest. They are constantly flying about, darting from one place to another, not at random but with a very definite purpose—to find food. For so much food do their active bodies, high temperature and rapid circulation demand that they must be eating most of the time. To anyone who has watched fledglings being fed it must have seemed incredible that such tiny bodies could possibly absorb so much nourishment, for it is an actual fact that these little birds regularly eat more than their own weight of food in the course of a day. Most young birds, even those which later prefer seeds and grain, are fed on insect food, and as they appear when insect pests are at the very worst, man is usually much benefited by their hearty appetites.

Relation to Man. It is this enormous quantity of food required by birds which makes the various wild species of such value to the farmer. Not many years ago the farmers were inclined to look upon nearly all birds as pests, because a few of them stole grain or nipped the cherries and blackberries, but to-day there exists a far truer appreciation of the part they really play in industrial life. For the insects upon which birds feed so voraciously are in many instances the very worst plant enemies—the enemies which each year result in a loss of hundreds of millions of dollars to the farmers. This loss would be much greater without the services of the birds, and careful scientists who have observed and collected facts for years go so far as to say that successful agriculture would probably be impossible without the birds. The seeds, too, which furnish so large a part of the food of many birds, are most frequently weed seeds; it is estimated that the sparrows alone save to the farmers of the United States and Canada millions of dollars each year by eating these seeds. Some birds there are, it is true, who do help themselves occasionally to the farmer's grain, but there are very few which do not more than pay for such thefts. Even the crow, the owl and most of the hawks, generally classed as robbers, by their wholesale destruction of insects and small rodents prove themselves friends rather than foes to the farmer.

The safe thing for a farmer to do is to acquaint himself so far as possible with the birds of his neighborhood and to find out their favorite diet. Then such as do really prey on the crops without making adequate return may be driven away or killed, while the others, far more numerous, may be attracted in every possible way. Protection of such birds from enemies, especially cats, providing them with food at times when it is not otherwise available, and of water for drinking and bathing, and the supplying of artificial nest boxes or other safe places for building will do much to attract the useful birds to spots where they are needed. Even the village or city dweller, who does not need the help of the birds as does the farmer, may by similar methods bring about his house a colony of birds which will give him continuous pleasure.

Cage Birds. Certain birds have for a long time held a peculiar relation to man in that they have lived in his home and been, in a sense, domesticated. The gracefulness, beauty and cheerful songs of birds have always made their strong appeal, and centuries before the beginning of the Christian Era the custom of keeping birds in cages was common. To-day the number of such pets is greater than ever before; most of them have not been captured and confined to a cage after having known the joys of a free life, but have been bred and raised in captivity. In such a case the keeping of caged birds cannot be called cruel, as in the case of any free, wild thing.

Canaries are by far the best known of all the birds which are kept for their song, but caged nightingales, finches, thrushes and mocking birds are by no means uncommon. Parrots are the favorite talking birds, while parrakeets, cockatoos and certain finches are often kept by reason of their beautiful plumage. It is almost always the grain-eating birds which are trained as cage birds, both because it is far easier to feed them and because they are not used to so active a life as are the insect-eating birds. No person who is not willing to take excellent care of a bird should ever have one, for one of the absolutely necessary things is that the cage and all it contains should be kept strictly clean, and that the bird itself should have opportunity for frequent baths.

Classification. Many systems of classifying birds have been offered, no one of which seems entirely to please scientists. The outline at the end of this article gives perhaps as simple a method of grouping as any which is now in use.

J.B.

Study of Birds

In recent years there has been a growing interest in birds—their appearance, their habits, their value to man—and many "bird books" have been published to comply with the demand for accurate information. Some of these are simple little books which list and describe the birds of a certain locality, so that dwellers there may recognize them on sight, while others are more elaborate publications with beautifully-colored illustrations. More general treatises also abound, which give charming accounts of the manner of life of the birds. But no one, no matter how carefully he has studied such books, can really know birds unless he has studied them firsthand, at first in the laboratory or schoolroom and later in the field. The following lessons on birds are designed to aid the teacher or parent in interesting the boys and girls and showing them how to begin the study of the most fascinating of all their wild neighbors.

Lessons on Birds. *General Suggestions.* 1. Birds in their natural state are so hard to approach that they must be tamed before one can successfully study them. This is accomplished only by feeding them, for, as one writer has said, "The way to a bird's heart is through its crop." Instructions as to feeding and taming birds should be given during the first lessons.

2. It is exceedingly important that the teacher's knowledge of her subject be much broader than the work actually assigned to the class.

she must be familiar with the size and color of the bird and the quality of its singing note; she must be able to recognize the features that distinguish male and female, and in addition know the habits and haunts of the bird. What it feeds upon, how its food is obtained, when and where it makes its nest, when the young are hatched, how long they remain in the nest and the dangers that beset them after they have learned to fly are some of the facts of which the successful teacher will make herself mistress. The children must be taught how to help the young birds, many of which die from lack of care.

3. The greater part of the work on bird study should be devoted to outdoor exercises.

4. Classroom work should consist chiefly of hearing reports and giving instruction for future observations. On completing the study of a bird assign the review of the work for a class exercise. Finally, the review should be written; this affords an opportunity for drill in composition.

Suggestions for Observations. 1. The return of the birds in the spring is a subject that never fails to interest the children. Have them report the first birds they see and keep a record of these reports in some such form as this:

The first crow, March 1.

The first robin, March 10.

Besides a general school record, there should be separate records, kept by the pupils individually. Instructions for observing the birds should be complete and clear. Observations are most successfully made by one pupil alone, or by groups of not more than two or three, as birds are easily frightened by noises and by the approach of people, whom they are likely to regard as enemies. The directions should include these points:

a. In making trips to study birds wear dull-colored clothing, preferably shades of brown which tone in with the ground and trunks of trees. White and bright colors are not practicable because they attract attention.

b. Move gently and quietly so as not to frighten the bird. As a rule, the observer can approach the bird more successfully if he pretends to be searching for something else.

2. Train the pupils to describe only what they actually see, and insist on their making careful observations. Caution them about imagining that they see the thing they are looking for.

3. An opera glass will be found very convenient, but it can be dispensed with much

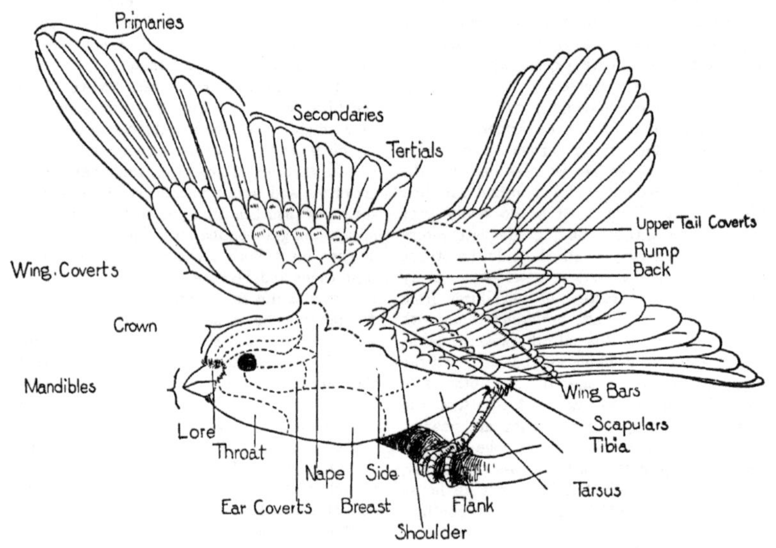

PARTS OF A BIRD

more easily than a few good books. Such works as Chapman and Reed's *Bird Guide,* or Chapman's *Handbook of the Birds of Eastern*

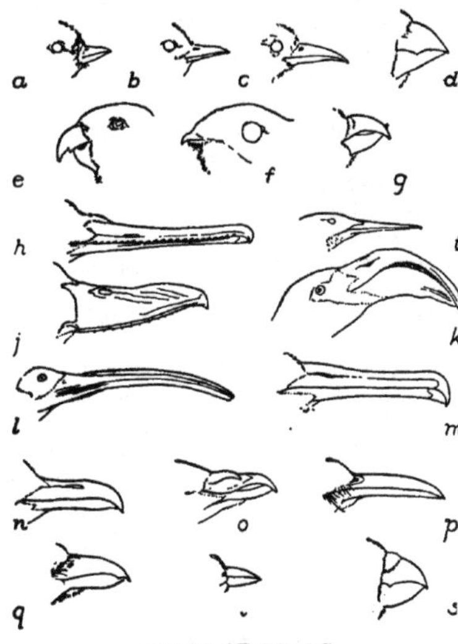

TYPES OF BEAKS

(*a*) Kinglet
(*b*) Gnatcatcher
(*c*) Wren
(*d*) Sparrow
(*e*) Parrot
(*f*) Whip-poor-will
(*g*) Grosbeak
(*h*) Mersanger
(*i*) Humming bird
(*j*) Duck
(*k*) Flamingo
(*l*) Ibis
(*m*) Cormorant
(*n*) Gull
(*o*) Pigeon
(*p*) Mocking bird
(*q*) Shrike
(*r*) Nuthatch
(*s*) Grosbeak

North America, should be found in every school library, and be used on observational trips.

4. Each pupil should carry a pocket note book in which to jot down observations as they are made. If the memory alone is relied upon valuable points may be omitted.

Parts of a Bird. The parts of a bird, with names attached, are shown in the diagram on page 734. All but the very young pupils should learn these terms, as they occur in all books about birds. The little ones should learn to recognize the most prominent parts, as the head, wings, tail, etc.

1. For a practical exercise have the children compare a live bird, a mounted specimen or the skin of the bird with the diagram, and ask them to name the corresponding parts. The first lessons should be the easier ones, the naming of the head, mandibles, wings, legs and tail. Have the pupils measure the specimen from the point of the beak to the end of the tail.

How many inches long is it?

Spread out the wings, measuring them from tip to tip. What is the distance?

Compare this distance with the length.

2. To make sure that the pupils remember what they have observed, begin the second lesson by reviewing the first. Then proceed to the study of the smaller parts, taking up this part of the work systematically, as follows: first study the parts of the body, the head, nape, breast, beak and rump. When these parts have been learned study in like manner the wings, tail and legs. Classes above the fifth grade should have no difficulty in distinguishing and naming all these parts. An occasional exercise on them in connection with the other lessons will help to fix them in the memory.

3. The way in which the structure of a bird is adapted to its mode of life is a topic of absorbing interest. The older classes should be led to see that a bird of prey has a different beak from one which feeds on fruit and insects, and that their feet also are not alike. Variations in the structure of different birds are clearly illustrated in the accompanying diagrams of beaks and feet.

Local Protection of Birds. The lessons on bird protection are designed to increase the

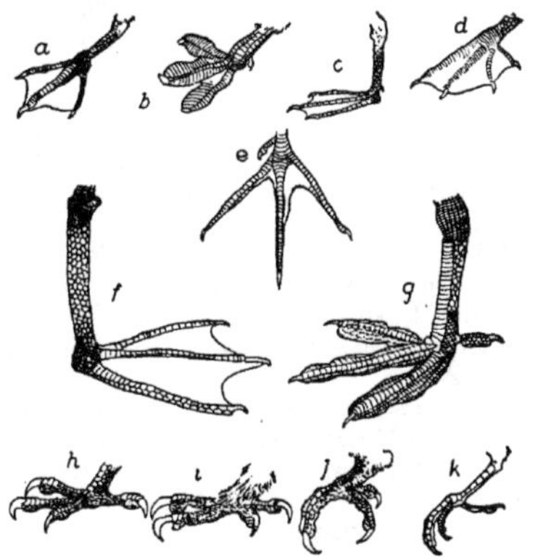

PRINCIPAL TYPES OF FEET

(*a*) Auk
(*b*) Grebe
(*c*) Petrel
(*d*) Pelican
(*e*) Sandpiper
(*f*) Albatross
(*g*) Coot
(*h*) Owl
(*i*) Hawk
(*j*) Parrot
(*k*) Swallow

interest of the children in birds and their love for them, and to impress upon them how valuable the birds are to the farmer and gardener. They should be encouraged to give their services in protecting the birds and in making it easy for their feathered friends to return to the same nesting places year after year. There are several ways in which this can be done. To illustrate:

a. Suggest that the children supply the birds with food when the spring first brings them back, and, in addition, that they provide them with fresh water throughout the season. The birds are always attracted to a running fountain in the garden or yard, but if there is none on the premises they will be grateful for pans or other vessels from which they may drink. Birds which stay in the locality through the winter can be induced to remain about the yard and buildings if regularly fed, and in time they will feel quite at home.

b. Nesting places should be provided. A small structure which affords a place of shelter from the sun and the rain, and which is conveniently located, will appeal to the birds as a suitable nesting place (see subtitle *A Home for Birds and a Decoration for the Garden*).

c. Young birds should be protected from cats and other enemies. Little fledglings which fall from the nest before they can fly are especially in danger; these should be placed back in the little home as soon as they are found, for otherwise they will be killed by cats or die from starvation. Nearly all the young must learn how to feed themselves after leaving the nest, and if such birds are approached carefully and fed, they will become quite tame. An older bird that has broken a leg or wing can also be helped. The broken member should be tightly bandaged so that it will heal properly, and protecting care be given the injured creature until it is able to fly again. Such works of mercy are invaluable in teaching the children daily lessons in kindness and gentleness.

In addition to the books mentioned above, Chapman's *Color Key to North American Birds*, Olive Thorne Miller's *Bird Ways*, *In Nesting Time* and *Our Home Pets*, and Mabel Osgood Wright's *Bird Craft* and *Citizen Birds* will be found interesting as well as helpful. Besides, this set of books contains descriptions of all the most important birds of the world, each appearing in its alphabetical order. J.B.

A Home for Birds and a Decoration for the Garden

Mr. and Mrs. Wren were house-hunting. They had recently returned from a long southern trip, and they found it necessary to look up new quarters, for the home they had occupied the season before was sadly out of repair. The roof leaked, the inside of the house was very dirty, and the premises generally looked as though the caretaker had neglected his duty.

"Well, I suppose we'll have to build a new one," said Mr. Wren, heaving a sigh. "I confess that I had hoped to make the old house do for another season, but this seems pretty hopeless."

"I was much attracted by those new houses on Garden Terrace," said Mrs. Wren. "Let's take another look at them before we start to build. They certainly are different from the style of house that we have lived in so long, and I for one would be glad of a change."

"Do you mean those fancy cottages with bright-colored roofs?" said Mr. Wren, doubtfully. "They did look rather well on the posts of the garden fence, but I never supposed they were intended for modest people like ourselves. I thought they were built for aristocrats!"

"Nonsense!" said Mrs. Wren. "Those are the houses that were advertised in the papers, as being especially adapted for the homes of wrens, bluebirds, chickadees and all kinds of middle-class birds. Why shouldn't we have a beautiful house, if we can find one that is within our means, and at the same time is practical? Come along!"

Mr. Wren knew from experience that it was useless to argue against his wife, when her mind was thoroughly made up; so he flew meekly after her, across the fields and over the streets until they came to a beautiful garden. The fence was quite high, and every post held a tiny wooden house, gaily painted in bright colors. One house with white and green decorations and an orange-colored roof seemed to please Mrs. Wren particularly. It was not too large, its entrance was of the right size, and there was a neat and stylish porch. It was the prettiest house in Garden Terrace. The Wrens decided to take it.

Building Bird Houses. Every friend of the birds will appreciate the "moral" back of this little story. There is no better way to encourage the pretty songsters to make the yards and gardens their permanent abiding places than to build attractive homes for

HOMES FOR BIRD FRIENDS

them. Some birds will return season after season to the same nesting place after they learn that comfort and shelter can there be enjoyed. Children should be urged to build homes for the little summer visitors, for in this way they help to protect the birds as well as to make the premises about the home delightful throughout the season. Below are given directions for building a type of bird house that the children will enjoy working upon. The two pages of drawings and the page of colored pictures accompanying this article show various kinds of bird homes that can be constructed by any children who are familiar with simple tools.

The children who built the house occupied by the Wrens used ⅝" stock of white-wood, and it and the other houses on the Terrace were made so they could be nailed against a barn, a post or a tree (see page 739). The roof provided a protective overhang. Six pieces of wood, the shapes and sizes of which are shown in Fig. 1, were used in making the house selected by the Wrens. These pieces were nailed together with ⅞" brads, making what are known as butt joints, as shown in the working drawing (Fig. 2). The top edge of the front piece and the two end edges of the roof were carefully beveled, in order to secure the proper pitch or slant of the roof.

To protect the house from the weather, and also to make it an attractive feature of the garden, two coats of outside paint were applied when the house was built. The four sides and the bottom were painted white. The roof received two coats of orange coach paint. On the sides of the house were painted simple tree-shapes, in green. The shapes of these decorations were first cut from paper, as shown in Fig. 3. The unit developed in this way was placed three times on the sides, and twice on the front of the house. The shapes were traced with a pencil on the white paint, as shown in Fig. 4. The tree shapes were then carefully filled in with green paint, and the trunks and bases with orange paint.

Fig. 5 shows two completed wren houses, with the color schemes indicated. B.S.

Nests of the Birds

Of all the wonderful things which the bird-lover discovers about his favorites, there is nothing more wonderful than their nest-building power. How can young birds in their very first spring know how to set about building a nest exactly like the one out of which they were forced, as awkward fledglings, in the summer before? The parent birds are not near to instruct them, and yet the nests are in practically all instances as perfect as were those built by their parents or their grandparents. It cannot be explained, this curious nest-building instinct; it must ever remain a mystery.

Purpose of Nests. In one sense, a nest is not a home. Two robins building their first nest do not pick out the spot that suits them best and fashion a structure which will afford them most comfort, but they build with one purpose in mind—to provide a place in which eggs may be laid and brooded until they hatch, and in which the young may be protected until they are ready to take care of themselves. The nest is in a sense the little birds' nursery, in which they live and are happy until they are ready to start out into the world; but very, very few of them ever come near the nest again once they have learned to fly, and even the parent-birds abandon it when it has served its one purpose. But for that one purpose it is remarkably well adapted, and each bird builds just the type of nest which fits its own need.

Variations in Nests. In nothing do the various birds differ more widely than in the kinds of nests they build, and it is very interesting to observe the growing perfection of the nests as the birds rise higher and higher in development. In part this is because the higher orders of birds are more intelligent, but largely it is because the young of these higher forms are more helpless. The "chicks" of the domestic fowl, of the ostrich and of various other birds have a full covering of down when they are hatched, and are very soon able to run about and seek food for themselves, but the young of the birds of flight, and especially of the song-birds, are utterly helpless and require the protection of a carefully-built nest and the watchful care of the parents. So an ostrich merely makes a hole in the warm sand and there deposits its great thick-shelled eggs, while an oriole, on the other hand, fashions a most curious and wonderful structure, in which its eggs and its helpless young are comparatively safe from reptiles and other preying animals. In general, the nests of the smaller birds are the more beautiful.

FOR OUR BIRD NEIGHBORS
KATE ABELMANN

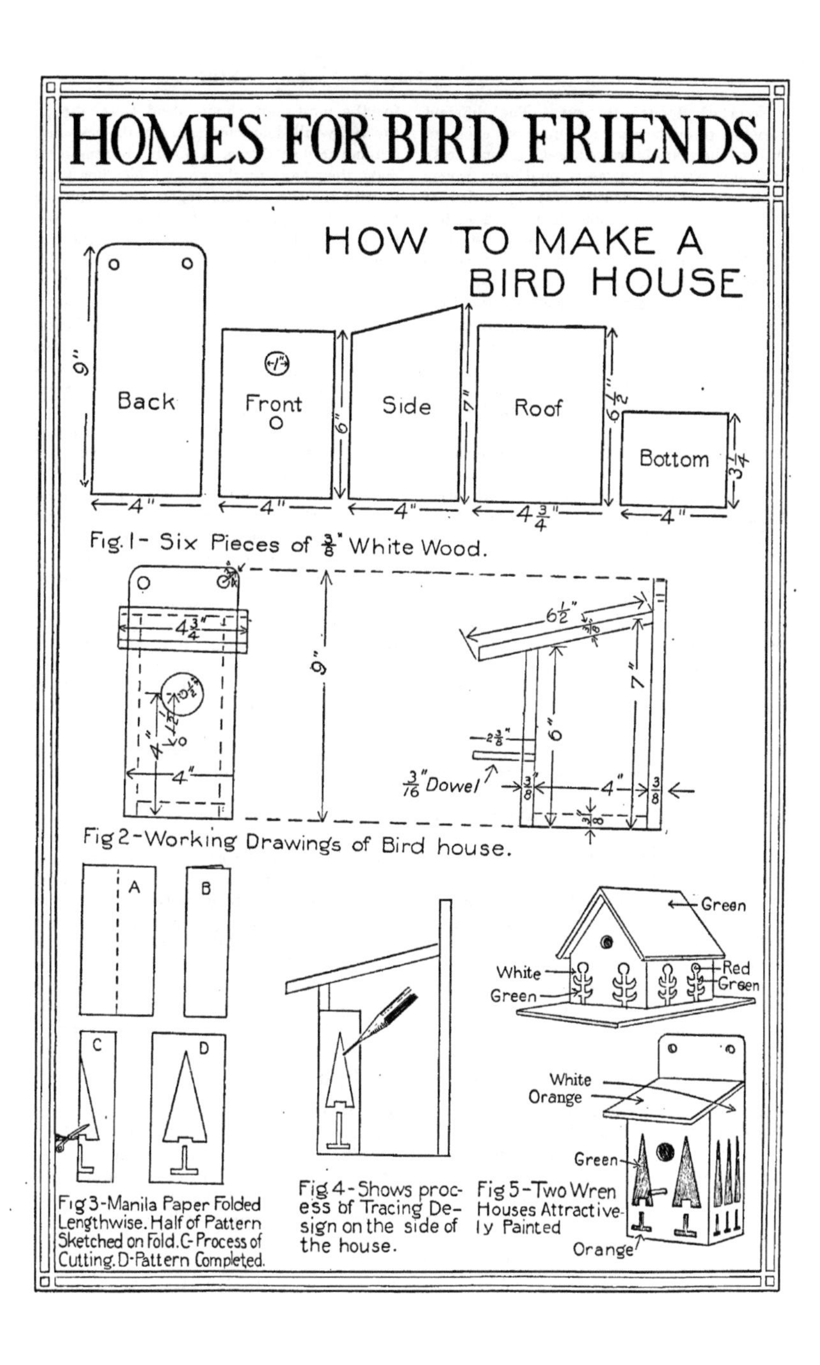
HOMES FOR BIRD FRIENDS
HOW TO MAKE A BIRD HOUSE
Back
Front
Side
Roof
Bottom
Fig. 1 - Six Pieces of 3/8" White Wood.
Dowel
Fig 2 - Working Drawings of Bird house.
A
B
C
D
Fig 3 - Manila Paper Folded Lengthwise. Half of Pattern Sketched on Fold. C - Process of Cutting. D - Pattern Completed.
Fig 4 - Shows process of Tracing Design on the side of the house.
Green
White
Green
Red
Green
White
Orange
Green
Orange
Fig 5 - Two Wren Houses Attractively Painted

Nests of Special Interest. The different varieties of nests are too numerous even for listing—there are burrowed nests, ground nests, carpenter nests, basket nests, mud nests, woven nests and ever so many more; but a few of the more interesting kinds may be described here. Hunting and studying birds'-nests is a fascinating pursuit, but it should not be indulged in while the nests are still in use. The bird-lover may discover the whereabouts of the nests of various birds, and later, when the young have flown, may examine them at his leisure; but it is not safe to bring nests into the house until they have been carefully looked over, as they are likely to be swarming with vermin.

Beginning at the very bottom, there is the king-penquin, which simply deposits its eggs on the bare rock and pays little further attention to them. Then there are the burrowing birds, such as the sand-martin and the kingfisher, which dig deep tunnels into a bank, usually facing the water, and lay their eggs at the end; and the ground-nesters, as the turkey or the goose, which scrape together on the ground shelters of leaves and grass. Very curious is the habit of the tropic mound-birds, which build up great heaps of vegetable matter, lay their eggs therein and trust to the heat of the decaying pile to hatch them. Somewhat higher in the scale are the mud or clay nests, made by some of the commonest American birds. The swallows' nests are entirely of clay, and are of varying shapes, some looking much like gourds with their bent necks, in the end of which is the "door," but the robin uses mud merely as a cement to hold together a nest of twigs and leaves, which it lines with cotton or wool or hair—any soft substance it can find.

Then there are the carpenter nests, which are in reality but holes in stumps or trees, with a very little soft material laid down for a bed. Of the birds with such wooden nests the commonest examples are the woodpeckers, which for the most part choose dead trees, though the ivory-billed species can cut its nest into a living hardwood tree. A number of song-birds make their nests on the ground, blending them in with the surrounding vegetation so cleverly that only the practiced eye can find them, but by far the larger number choose shrubs or trees, and the majority of them build what are known as basket nests—loosely woven structures of twigs, stems, grass or any other material that may be near at hand. Not all of these are equally well made, the crows and eagles, for example, constructing very rough, coarse nests, while the thrushes and the warblers are very neat and skilful.

Probably the most beautiful and interesting of the comparatively common bird homes are the woven nests of the humming birds, the orioles and a few other species. To do their best work these birds must live near dwellings where they can get wool or hair or twine; and their eyes are very sharp in spying any bit which may have been left about for them. The humming bird's nest is so closely and smoothly woven that it looks like felt, and so tiny that it may swing from the top of a branch, almost hidden from the eyes of the curious by the greenish lichens with which its little makers have adorned it; the oriole's nest, larger but no less exquisite in structure, usually hangs as a pouch from the tip of a long, drooping elm branch (see ORIOLE, for illustration).

A few birds, as the cowbird, are too lazy or too stupid to build nests for themselves, and lay their eggs in the nests of other birds; but for the most part, no trouble is too great for the parent-birds to take for the suitable housing of their eggs and young, and the heroism which they show in defending their nests from enemies far larger and stronger than themselves entitles them to rank among the bravest of all the animals. We usually fail to credit them with this noble characteristic. A.MC C.

Migration of Birds

This is one of the great marvels of nature—the advance and retreat of an army of birds, so great that one might imagine the sky at times would be dark with them. But few people ever see them in their flight. Astronomers, viewing the full moon through their large telescopes in August and September, frequently see the small migrating birds, high in air, cross their field of vision, and are often able to distinguish the finches, thrushes and others. The Canada geese, in their honking, orderly ranks, are a familiar enough sight in the fall to those who live in Central North America, and hunters can predict very closely when the "duck season" will begin. But why does no one ever see bobolinks and the plovers, the swallows, the sparrows and the warblers sweeping over the hills and valleys and

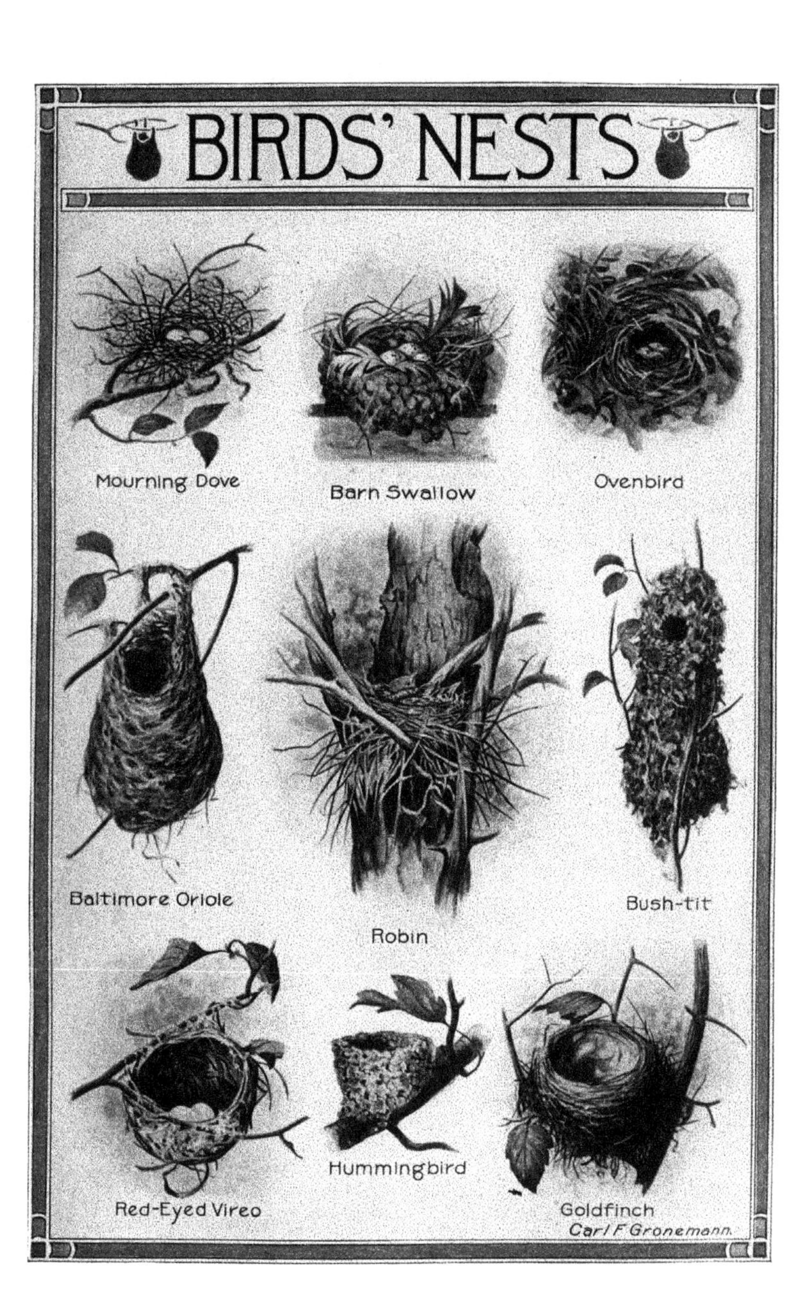
BIRDS' NESTS
Mourning Dove
Barn Swallow
Ovenbird
Baltimore Oriole
Robin
Bush-tit
Red-Eyed Vireo
Hummingbird
Goldfinch
Carl F. Gronemann

over the roofs of the cities? One day there are no martins to be seen, no warblers to be heard; the next day they seem to be everywhere, making themselves as much at home as though they were steady residents and not mere summer tourists settling for a brief sojourn. How far have they flown? Where have they spent the winter? How long did it take them to come back? Why do they migrate? These questions and many more are certain to occur to any student of this most fascinating subject.

Why Birds Migrate. The little sparrows that hop about in the deepest snow of winter and seem to enjoy themselves prove clearly that some birds are not affected by the cold, and careful study has made it practically conclusive that this is true of all birds. A mocking bird, which everyone is accustomed to think of as a dweller in warm climates, was observed to live happily in New York during an unusually cold winter, and frequently bluebirds, looked upon as the surest harbingers of spring, will winter in a northern wood. The condition of residence depends on their food supply. The vagrant mocking bird has found a privet tree; the bluebirds hover near a berry-covered cedar tree. But those birds which feed upon insects or worms would starve if they could not seek their winter food in less severe climates than the temperate zones. There is thus a rough division of birds into those which make long migrations and those which do not, strictly according to the demands of their appetites.

But the need for food cannot alone account for the marvelous departure of the birds from home. If it did, insect-eating birds would remain the year round in the warm parts of the earth, where the air is full of their prey. Moreover, some of the birds start south in August, when insect life is still plentiful in the higher latitudes. Behind the great movements there is some primal cause, some strong urge which man cannot comprehend. Scientists explain that ages ago, when climatic conditions were much more extreme, birds were actually forced to migrate by the slowly pushing ice cap which came down from the Pole, crowding out vegetable and animal life before it. As this receded, the birds began to press northward from their enforced home, and somehow this planted in them an instinct, a heredity of habit, which resulted in the seasonal migrations. However that may be, it is certain that, as spring approaches, something says to certain birds, "Fly northward, and there build your nests and rear your young;" and in a few days the northern woods and fields begin to swarm with welcome visitors. Bird-lovers who find the domestic life of the pretty creatures that nest in their yards a source of unceasing interest, report that pairs of birds sometimes return season after season to the same tree or bird house.

How They Migrate. When they have felt the "mysterious call" the birds respond at once, each species organizing its flight as have its ancestors from time immemorial. Some species are most systematic; others are more happy-go-lucky, traveling as the spirit moves, and covering no great distance at a time. Almost invariably, however, there is a certain order. The old males, strong of wing, fly first, next come the females, and at the close of the procession the young birds, who are re-visiting their northern birthplace for the first time. Perhaps as they advance northward they may find that they have started a little too soon. Then they slacken their pace, and tarry in the intermediate regions for a time. Some kinds of birds fly openly day by day, but

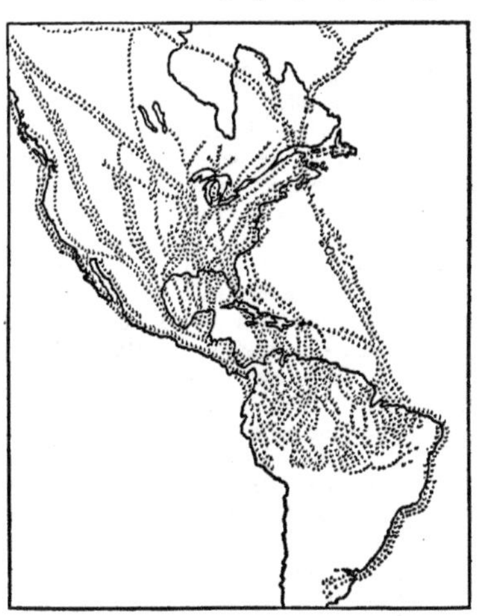

PRINCIPAL ROUTES OF MIGRATIONS, SHOWING EXTENT OF FLIGHT

many of the ordinary night-fliers and the more timid species migrate only by night, sheltering themselves in secluded places during the day.

There is a difference observable in the spring and fall flights. In the former, the birds are fresh and strong from their long winter's feeding without family cares, and the trip may be made easily. The males don their brilliant courting dress, and their arrival at any given place is a conspicuous event. In the fall, however, the birds move more slowly. They have worked hard during the summer feeding their young, for this is a tremendous task, and have as yet scarcely recovered from their molting season, and there are, besides, the little birds which have not yet tried their wings in any very long flight. Stop-overs are frequent, and sometimes the whole flock tarries in a rich feeding ground for weeks, gathering strength for continuation of the long journey. The bobolink, for instance, stays for a time in the Carolina rice fields.

Migration Routes. Some of the birds make what would be a very long journey even for larger travelers. The golden plover, one of the most ambitious birds, breeds far north on the shores of the Arctic Ocean, and with the approach of fall moves into Nova Scotia and from there lays a direct course southward to the north coast of South America, flying for hundreds of miles across the open Atlantic, far out of sight of the shore (see diagram herewith). The yellowlegs are the world's most famous travelers, for they summer as far north as the Arctic Circle and take an 8,000-mile flight to Central Argentina every autumn, returning in the spring. Sixteen thousand miles a year on a serious adventure, borne onward by one slender pair of wings! Other birds there are, too, which cross great stretches of trackless water, and the mind of man cannot even conjecture what guides them. Observation has been extensive enough to show that there are a number of well-marked migration routes along which most of the birds travel—in America no fewer than seven "trunk lines." In part these are determined by the land surface features, and in part by the presence along the way of food supplies, but nothing has as yet been discovered which will account in the least for the route of the "ocean-going" birds. The valleys of the great rivers running from north to south are favorite highways of the migrating birds. J.B.

Government Protection of Birds

Bird Reservations. Tracts of land or water—islands and marshy places along rivers and shores, wild stretches in mountainous districts and lakes—which are set apart by national, state or provincial governments as permanent and safe retreats for the native wild birds, are known as *bird reservations.* There they rest in peace and security from the hunter the year round. Without such abiding places many species of birds would before many years become extinct. The first United States reservation was established by President Roosevelt in March, 1903, when, by special proclamation, he set aside Pelican Island, in Indian River, Fla., as a home for the pelicans that nested there. In 1909, when he retired from office, he had established fifty-three different bird reservations. This great work was continued by his successors; four new national reserves were created in 1913, and two in 1915.

These places of refuge are located in all parts of the national domain, from Porto Rico on the south and east to the chain of Aleutian Islands off the coast of Alaska; along the Gulf and Atlantic states, midland in Nebraska and South Dakota, and westward in Oregon and California; while in the Pacific is the great Hawaiian Island Reservation. The two reservations in Oregon and California, including Lower Klamath Lake, Malheur and Harney lakes and the great marshy stretches around them, are the largest wild-fowl nurseries on the Pacific coast.

In bird reservations the feathered creatures seem instinctively to know they are safe from the gun of the huntsman; admission to one of these sanctuaries is only through the kindness of officials in charge, and indeed on some reservations human beings are not allowed under any circumstances.

In 1916 a treaty was ratified by the United States and Canada providing that no bird which helps the farmer by destroying insects shall be shot at any time, and that the open season for game laws may be restricted to three months and a half.

Game Laws. The wanton and careless destruction of wild birds by sportsmen and plumage-hunters, which in the past has caused the birds to diminish at a deplorable rate, has awakened the people of both the United States and Canada to the need of protective legislation. The loss to farmers from preying insects would be much greater than it is but for the birds, and it would be a calamity not to preserve for mankind the many beautiful and sweet-singing creatures of the woodland

BIRD DAY

A Suggestive Program

Do you ne'er think what wondrous beings these?
Do you ne'er think who made them, and who taught
The dialect they speak, where melodies
Alone are the interpreters of thought?

—*Longfellow*

Song, *The Skylark*........*James Hogg*
Roll Call, answered with names of birds
Who Stole the Bird's Nest?..................*Lydia M. Child*
(By a Group of Pupils)
Essay, *The Meaning of Bird Day*
The Building of a Nest...*Margaret Sangster*
To a Water-Fowl.........*William Cullen Bryant*
The Blue-Bird............*Eben E. Rexford*
Quotations about Birds
The Sandpiper............*Celia Thaxter*
Remorse ....*Sydney Dagre*
Essay, *Birds' Eggs I Can Recognize*
Song, *The Birds' Ball*
The Whitethroat..........*Theodore H. Rand*
In April..................*Emily G. Arnold*
Essay, *How Birds Help the Farmers*
Robert of Lincoln.........*William Cullen Bryant*
What Robin Told.........*George Cooper*
Robin's Return...........*Edith Thomas*
A Dialogue between Mr. and Mrs. Wren
Song, *The Brown Thrush*..*Lucy Larcom*
Birds in Summer..........*Mary Howitt*

OUTLINE AND QUESTIONS ON BIRDS

Outline

I. Important Characteristics

(1) A covering of feathers
(2) Power of flight
(3) Possession of a backbone and warm blood
(4) All have two legs

II. Framework of the Body

(1) Hollow bones
(2) Air cavities
(3) Keel-like breastbone

III. How the Birds Fly

(1) Wings
  (a) Modifications of forelimbs
  (b) Uses of the feathers
(2) Tail
  (a) The bird's rudder
(3) Long-distance flights
(4) Rapidity of flight
(5) Examples for airship-builders

IV. The Coat of Feathers

(1) Where the feathers grow
(2) Why they vary for different birds
(3) Purpose of oil glands
(4) Why the male is brightly dressed
(5) Molting

V. Senses

(1) Sight
  (a) Compared with that of other animals
  (b) Sensitiveness to light
(2) Hearing
  (a) How birds hear
  (b) Acuteness of this sense
(3) Touch, taste and smell
(4) Sense of direction
  (a) The carrier pigeon
  (b) Birds that return to last-year's nests

VI. The Song of Birds

(1) Their "organ of speech"
(2) Birds most gifted in song
(3) Birds which cannot sing

VII. Their Home Life

(1) The mating instinct
(2) Hatching of eggs
(3) Care of the young
  (a) Feeding
  (b) Teaching them how to fly

VIII. Food

(1) Animal and vegetable food
(2) How birds eat
(3) Methods of finding food

IX. Birds and Mankind

(1) Value to farmer
(2) Protection of birds
(3) Birds as pets

X. Classification

See lists, fully classified, under the heading *Related Subjects*

XI. Nests of Birds

(1) Purposes of nests
(2) How nests vary
(3) Nests of special interest
  (a) Burrowed nests
  (b) Mounds
  (c) Nests of clay
  (d) Carpenter nests
  (e) Wooden nests
  (f) Woven nests.
(4) Birds that steal nests

XII. Migration of Birds

(1) Why birds migrate
(2) How they migrate
(3) The routes they follow

XIII. Bird Reservations

(1) Their purpose
(2) Location of important reserves

XIV. Bird Day

(1) What it means
(2) Its influence
(3) How observed

Questions

How are birds related to snakes?
What resemblance is there between a flying bird and a boat? An aeroplane?
What is the origin of the old saying, "Like water from a duck's back"?
Why does the bobolink wear different plumage for different seasons?

Outline and Questions on Birds—Continued

What instinct brings the carrier pigeon back home even when it has been sent away by train?

How fast do birds fly?

How do birds and mankind differ in their method of producing sounds?

Do birds talk to one another?

Are the same number of eggs laid at each sitting?

How do the newly-hatched birds of the lower orders differ from those of the higher orders?

How do baby birds learn to fly?

Why can birds digest their food without chewing it?

In what respect are the birds an aid to agriculture?

What services does mankind owe the birds?

How ancient is the custom of keeping birds in cages?

Would you choose a grain-eating bird for a pet, or one that feeds on insects? Why?

What are the principal classes into which birds are divided?

Why do some birds build elaborate nests, and others such crude ones?

What birds dig tunnels for nesting places?

How are the eggs of the mound-bird hatched?

What sort of cement does the robin use to hold its nest together?

What bird travels 16,000 miles a year?

What are the chief reasons for the periodical flights of birds?

What is the usual order of arrangement when a flock of birds flies off in a procession?

Do the birds travel at the same rate in the spring and fall flights?

What determine the "trunk lines" followed by migrating birds?

Why do governments set apart reservations for birds?

When birds molt do their feathers drop off all at once?

Do they shed all of their feathers?

What proof is there that the sense of hearing in birds is acute?

What senses do they use in seeking out their food?

Why do birds eat during most of their waking hours?

How can wild birds be tamed?

What sort of clothing should one wear on a bird-study excursion?

If you should find a bird with a broken wing what ought you to do?

What bird never builds a nest of its own?

Do mated birds remain together season after season?

Are birds the only creatures that fly? Do all birds fly?

Where do their largest and strongest feathers grow?

Do birds return to their last-year's nests?

How many miles can a carrier pigeon fly in an hour?

How does the manner in which feathers grow on a bird differ from the manner in which hair grows on a dog, for instance?

What peculiarity of birds is indicated in the lines "Robert of Lincoln is gayly drest" and "Robert of Lincoln's Quaker wife, pretty and quiet, with plain brown wings"?

What difference is there between the manner in which a bird produces its song and the manner in which a human being speaks?

What difference is noticeable between the young of the lower orders of birds and those of the more highly developed orders?

so greatly in danger of becoming extinct. The most far-reaching law in force in the United States, the McLean Act of 1913, gives the Federal government control of all migratory birds and game birds which do not remain permanently within the limits of one state. The Department of Agriculture is empowered to prescribe regulations in regard to closed seasons, zones and similar matters. The various states have also passed protective laws, no fewer than 240 having been enacted in 1915 by legislatures of forty states. There are special laws pertaining to the hunting of birds in force in nearly all of the Canadian provinces, and in 1915 a law prohibiting the importation of wild bird skins for commercial purposes went into effect over the entire Dominion. The importation of skins or feathers of birds into the United States is forbidden by the Underwood tariff law of 1913. J.B.

Bird Day is a special day observed by the school children of the United States and Canada. It was set apart to teach them the importance of protecting the birds, "the winged wardens of the farms." Not only are they taught that the birds save the farmers of the country millions of dollars each year by their destruction of harmful insects, but they learn the equally important lesson expressed so beautifully by Coleridge in his *Rime of the Ancient Mariner:*

> He prayeth best who loveth best
> All things both great and small;
> For the dear God who loveth us,
> He made and loveth all.

Bird Day was first celebrated in the public schools of Oil City, Pa., in May, 1894, its founder being C. A. Babcock, superintendent of schools. Two years later the United States Department of Agriculture issued a circular urging all public schools to devote a special day to the cause of bird protection. The observance of the day is left to the schools of each state and province, and the movement is growing each year. The exercises are similar to those used in the celebration of Arbor Day (which see), and the two programs are frequently combined. Bird Day in Louisiana is observed on May 5, the birthday of that great friend of the birds, John James Audubon. J.B.

Consult the books referred to in the body of the above article; also, *Check-list of North American Birds*, published by the North American Ornithologists' Union; Nuttall's *Manual of the Ornithology of the United States and Canada.*

Related Subjects. The following classified lists will show how fully birds are treated in these volumes:

BIRDS OF PREY

Buzzard	Lammergeier
Condor	Marsh Hawk
Eagle	Owl
Falcon	Secretary Bird
Goshawk	Shrike
Hawk	Sparrow Hawk
Kestrel	Turkey Buzzard
Kite	Vulture

CREEPERS AND CLIMBERS

Cockatoo	Quetzal
Creeper	Sapsucker
Flicker	Toucan
Lory	Woodpecker
Macaw	Wren
Parrakeet	Wryneck
Parrot	Yellow-hammer

FISHERS

Booby	Frigate Bird
Cormorant	Hornbill
Darter	Pelican
Fish Hawk	

PERCHERS

American Goldfinch	Lark
Baltimore Oriole	Linnet
Bellbird	Magpie
Bird of Paradise	Martin
Blackbird	Meadow Lark
Bluebird	Mocking Bird
Bobolink	Motmots
Bower-bird	Nightingale
Brown Thrasher	Nutcracker
Bullfinch	Nuthatch
Bunting	Oriole
Canary	Ortolan
Cardinal Bird	Ouzel
Catbird	Oven Bird
Chaffinch	Pipit
Chat	Raven
Cowbird	Redbird
Crossbill	Redstart
Crow	Robin
Crow Blackbird	Rook
Cuckoo	Snowbird
Curassow	Sparrow
Dickcissel	Starlings
Dipper	Stone Chat
Finch	Sunbird
Flycatcher	Swallow
Goldfinch	Tailor Bird
Grackle	Tanager
Grosbeak	Thrush
Halcyon	Titmouse
Hoopoe	Umbrella Bird
Indigo Bird	Vireo
Jackdaw	Wagtail
Jay	Warblers
Junco	Waxwing
Kingfisher	Weaver Bird
Kinglet	Wood Pewee

PIGEONS

Carrier Pigeon	Pigeon
Dove	Turtle Dove
Passenger Pigeon	

RUNNERS

Apteryx	Ostrich
Cassowary	Rhea
Emu	Road Runner

SCRATCHERS

Bustard	Partridge
Francolin	Peacock
Grouse	Pheasant
Guan	Prairie Chicken
Guinea Fowl	Ptarmigan
Jungle Fowl	Quail
Lyre Bird	Tragopan
Mound Bird	Turkey

SEA BIRDS

Albatross	Petrel
Fulmar	Scissorsbill
Gannet	Skua
Gulls	Tern
Kittiwake	Tropic Bird

SWIMMERS

Auk	Goose
Brant Goose	Grebe
Canada Goose	Guillemot
Canvasback	Merganser
Coot	Penquin
Diver	Puffin
Duck	Shoveler
Eider Duck	Swan
Gadwall	Widgeon
Gallinule	

WADERS

Adjutant	Oyster Catcher
Avocet	Plover
Bittern	Rail
Crane	Ruff
Curlew	Sanderling
Egret	Sandpiper
Flamingo	Snipe
Heron	Spoonbill
Ibis	Stilt
Jabiru	Stork
Jacana	Turnstone
Killdeer	Woodcock
Lapwing	Yellowlegs
Night Heron	

WEAK-FOOTED BIRDS

Goatsucker	Swift
Humming Bird	Whip-poor-will
Night Hawk	

BIRD'S-EYE MAPLE, a beautiful form of the wood of the sugar maple, which shows a variation from the straight grain by having numerous small, round spots that resemble the eyes of a bird. When smoothed and polished, this wood is highly valued in furniture making. The hard and red maples are the varieties that most frequently yield bird's-eye wood. It is especially prized in the manufacture of choice writing desks and bedroom furniture.

This peculiar formation is the result of an irregularity in growth. When the bark of the maple is injured, the trunk usually starts to sprout, and a multitude of little twigs appear that have just about enough vitality to keep alive. Each twig is the center of a series of wood rings which give to the wood its characteristic appearance under the skilled hand of the finisher. Special methods of sawing are necessary, that the beauty of the grain may be preserved; a typical method consists in taking short lengths of log to a saw which pares off a thin layer from the surface as the log revolves, a layer 100 feet in length sometimes being thus cut off before the heart of the wood is reached.

Curled Maple. This is another variation from the straight-grained wood, and it is quite as beautiful as bird's-eye. Here the wood fibers lie upon each other in ripples, and are somewhat longer than those of the ordinary wood. When polished, curled maple presents most beautiful effects in light and shade. The reason for the formation of curled maple is not known. Beeches and birches also are subject to this irregularity. Aside from its use for furniture, curled maple is prized as a wood for stocks of guns, as it combines lightness and beauty with great strength and durability. See MAPLE.

BIRDS OF PARADISE, a family of birds that are equaled only by the humming birds in their splendor. There are between forty and fifty species, in forests of Australia, New Guinea and other Pacific islands. Strangely enough, they are related to the family of crows, which are as remarkable for plainness as their famous cousins are for beauty. The name was given them by early travelers, who wrongly supposed that they were without feet and lived in the air, always keeping their eyes turned toward the sun, and never touching the earth till they died. Europeans first saw specimens of them in the sixteenth century, when they were carried to Europe by the sailors who made the voyage around the world in the Magellan expedition (1519-1522). In their native haunts they live almost entirely in the tree-tops. There they build their simple nests, and run and play about the branches, a graceful and agile company; the gorgeously-attired males sit and plume themselves to attract the more soberly-clad females.

The brilliant plumage of the males presents a bewildering variety of form and color, reminding one of Thoreau's statement that Nature made their feathers to show what she could do. Among the larger species is the twelve-wired bird of paradise, so called because out of its short, square tail twelve long, wire-

like feathers grow, which curve around towards the sides of the wings. This bird is a foot long, and truly has a coat of many colors, with purple-bronze on the head, green, black and purple on the neck, bronze-green on the back and shoulders, and emerald-green on the edges of its violet-purple wings. The tail is also violet-purple and the breast a rich yellow.

Not less splendidly arrayed is the long-tailed bird of New Guinea, over a yard in length and having as a distinctive feature a fan-like arrangement of feathers on the sides of the breast, which can be raised to form two half circles. The ends of these feathers are bright blue and green, and the tail feathers opal-blue. The great bird of paradise, about half the size of the one just described, is a superb creature whose chief glory is a spray of feathers growing out from under each wing. These rise from the body and fall backward in a graceful curve, a showery mass of brilliant color that seems to envelop the bird like a magic fountain. The smallest of the family is known as the *king bird of paradise*, a little creature whose two middle tail feathers terminate in spiral disks of beautiful emerald-green.

The lovely birds of paradise have to pay a bitter price for their glorious plumage, for thousands upon thousands have been sacrificed to man's greed and woman's vanity. The birds are captured for the millinery trade during the mating season, when the males assemble in the trees in groups of from twelve to twenty to display their splendid garments. The bird-catcher shoots them with blunt arrows which stun them, thus permitting their capture without injury to the plumage. The United States, according to a provision of the Underwood Tariff Bill of 1913, now prohibits the importation of the plumage of these birds, and the movement to check their cruel slaughter is also progressing rapidly in other countries. See Audubon Societies; Bird. B.M.W.

BIRKENHEAD, *bir' ken hed,* a town and parliamentary borough of Cheshire, England, on the Mersey River, opposite Liverpool, of which town it is practically a suburb. Its growth has been rapid, caused by the increasing importance of Liverpool, with which it is connected by ferry and by a railroad tunnel underneath the river. This tunnel is four and one-half miles long and cost $6,000,000.

The town has many fine buildings, and all public utilities such as water works, electric light plant, electric street car service and gas plants are municipally owned and managed at a substantial profit. It has large shipbuilding yards, wagon factories, machine and engineering works. In the shipbuilding yards of Birkenhead the famous *Alabama* (which see) was constructed for the Confederate States of America. Population in 1915, 130,794.

BIRMINGHAM, *bir' ming am* or *bir' ming ham,* ALA., one of the most rapidly-developing cities of the South, the largest city of the state and the county seat of Jefferson County, famous for its steel, iron and coal industries. It is situated north of the geographical center of the state, ninety-five miles northwest of Montgomery, the capital, and 168 miles west and south of Atlanta, Ga. Nine railroad lines enter the city and provide exceptional transportation facilities; they are the Louisville & Nashville; Mobile & Ohio; Alabama Great Southern; Seaboard Air Line; Central of Georgia; Illinois Central; Atlanta, Birmingham & Atlantic; Southern, and Frisco. Interurban lines reach the rapidly-growing suburbs and near-by cities. An increase of population from 132,685 in 1910 to 174,108 in 1915 is evidence of the rapid growth of the city. The area exceeds forty-eight square miles.

Birmingham is located 608 feet above sea level, in a valley abundantly rich in coal, iron and limestone; around it lie three famous coal fields, the Warrior, the Cahaba and the Coosa. Iron Mountain, six miles distant, has almost inexhaustible deposits of hematite, a valuable iron ore. The city is partly on the sloping side of Red Mountain, and is attractively laid out, with wide streets, beautiful residences, winding parkways and numerous parks, the latter comprising 600 acres; of these, Capitol, Avondale, Lake View and East Lake parks are especially attractive.

Buildings and Institutions. The most notable of the city's public buildings are the Federal building, the county courthouse, a city hall, recently completed at a cost of $200,000, the Auditorium, Saint Vincent's hospital, First National Bank building, the $2,000,000 terminal station, a Y. M. C. A. building, two excellent hotels costing more than $1,000,000 each, and several ten-story "skyscraper" office buildings. The benevolent institutions include Hillman's hospital, Mercy Home and Jefferson County almshouse. Besides the public schools, with five high schools, there are Howard College (Baptist) at East Lake, Birmingham College (Methodist) at Owenton Heights, the Boys' Industrial School, also at East Lake, a normal training school (colored) and medical, dental and business colleges.

Industry and Manufacture. Owing to the extent of its iron and coal industry, Birmingham is sometimes called *The Pittsburgh of the South;* the annual output of iron is estimated at 2,000,000 tons. The first steel plant in the South was established at Birmingham in 1897, with two open-hearth furnaces; at Ensley, in the vicinity, the Tennessee Coal and Iron Company now owns the largest basic open-hearth plant in the United States, excepting that of the Carnegie mills in Homestead, Pa.; a rail mill and casting plant are operated in connection with it. Alabama's largest yield of iron ore is obtained from the Birmingham district, and more than one-half of the United States export of pig iron is made from this city. While coal mining and the manufacture of iron and steel products are the chief industries, the cotton, cottonseed and lumber interests are important, Birmingham is also one of the leading yellow-pine markets of the South. Besides all of the above, the city has cement factories, fertilizer factories, clay-pipe and brick plants, clay being one of the important resources of the district. Here, also, are located the repair shops of the Southern Railway.

History. Early in 1871, following the discovery of the rich natural resources of this section, Birmingham was founded by the Elyton Land Company, who named it for the celebrated steel and iron manufacturing city of England. A small iron furnace was built and mining was begun. In the same year the town was incorporated. In 1873 it was scourged by cholera; it recovered and was again prostrated by a panic in 1889. Following this period growth was rapid. The Greater Birmingham law became effective January 1, 1910, at which time the suburbs of North Birmingham, East Birmingham, Avondale, Woodlawn, East Lake, Wylan, Ensley, Pratt City, West End and Elyton became part of the greater city. In March, 1911, the commission form of government was adopted. C. OF C.

BIRMINGHAM, *bir' ming am,* ENGLAND, one of the most important manufacturing cities in the world, in the county of Warwick, 103 miles northwest of London and seventy-eight miles southeast of Liverpool. In addition to enjoying the position of acknowledged center of the world's hardware industries, Birmingham is a striking example in favor of municipal ownership of utilities. Its water supply, street railways, gas works, electric lighting plants, markets and slaughter houses are the property of the municipality and are run at substantial profit. Its water supply is derived from two lakes in Wales and is conveyed seventy-three miles by pipes and aqueducts. A recently-installed sewage system has made this the best-drained city in England. In 1876 the municipality also acquired a larger district in the most populous part of the city at a cost of $8,000,000, and after condemning it as insanitary rebuilt and made it the leading business section.

The manufactures consist of steam and gas engines, motors, hydraulic presses, machinery, cutlery, screws, nails, railway cars and firearms, in addition to which there are numerous glass works, breweries and chemical works. There are many fine buildings, the chief of which is the town hall, in Greek style of architecture, completed in 1850. Here is annually held the Birmingham Musical Festival, one of the most notable events of the English musical world. In 1846 Mendelssohn's oratorio *Elijah* was here presented to the public for the first time. As much as it is possible for a city to owe its prosperity to one man, Birmingham is indebted for its present position to Joseph Chamberlain (which see), who became mayor in 1874 and inaugurated an era of progress. It is now the fourth city of England, with a population of 525,830 in 1911.

BIRTHDAYS OF FAMOUS PEOPLE. Lists of birthdays of hundreds of the world's men and women will be found in the articles relating to the months of the year. To these the reader is directed.

BIRTHSTONES, certain precious gems that through custom, imagination and sentiment are associated with the twelve months of the year, each month having dedicated to it a special stone. Thus the stone of any month is the birthstone of every person whose birthday falls in that month, and the belief that a natal stone is more intimately associated with one's personality than any other stone is very widespread. This belief may be traced to the writings of Josephus, a Jewish historian of the first century of the Christian Era, who found a connection between the twelve stones of the high priest's breastplate (see *Exodus* XVIII, 15-20, also article HIGH PRIEST), and the yearly circle of months. Yet the custom of wearing birthstones is comparatively of recent date, as it is supposed to have originated in the eighteenth century, in Poland.

The list of birthstones given below was adopted by the American National Retail Jewelers' Association at their convention held in August, 1913. A comparison of this list with the gems mentioned in the adornment of the high priest's breastplate will suggest the changes that the centuries have produced in the popular beliefs about birthstones:

January .....Garnet
February .....Amethyst
March ....Bloodstone or aquamarine
April ....Diamond
May ....Emerald
June ....Pearl or moonstone
July ....Ruby
August ....Sardonyx or peridot
September ....Sapphire
October ......Opal or tourmaline
November ....Topaz
December ....Turquoise or lapis-lazuli

Each gem named in the above list is described in its place in these volumes.

BISBEE, ARIZ., a prosperous copper-mining and smelting center, with a mixed American, English, Mexican and Slavonian population, which in 1910 numbered 9,019. It is situated in the extreme southeastern part of the state, in Cochise County, about eight miles from the Mexican border, thirty miles from Tombstone, the county seat, and ninety miles southeast of Tucson. Bisbee is the terminus of the El Paso & Southwestern Railway, which connects with the Southern Pacific Railroad at Benson. This branch line was constructed to Bisbee in 1902. The city was founded in 1877 and incorporated as a city in 1901.

Bisbee is picturesquely located in the heart of the Mule Pass Mountains, in a famous copper region; its Copper Queen and Calumet and Arizona mines are among the richest in the world. The industries are dependent upon the extensive copper, gold, silver and lead resources of this district; in a period of three months, 18,000,000 pounds of copper were produced from its ores. About 5,000 people are engaged in mining, and the annual output of this industry is valued at $18,000,000. An excellent school system for a city of its size and a public library offer educational advantages, and the Young Men's Christian Association and the Young Women's Christian Association have branches here. The buildings of these associations, with the Elks Club, comprise the most notable structures of the city. Shattuck Crystal Cave and Tombstone Mountain Boulevard are scenic points in the vicinity. J.H.G.

BISCAY, *bis' kay,* a great bay indenting France, is an eastern arm of the Atlantic Ocean, called by the old Romans the Cantabrian Sea. It is bounded on the east and northeast by France, and on the south as far as Cape Ortegal by Spain, and forms a fairly-regular curve about 400 miles long and 300 miles wide. The rugged Spanish coast is indented by bays, but the low and sandy French coast is broken by the great mouths of the Loire, Garonne, Adour and Charente rivers. Bordeaux, Bayonne, Nantes, Rochefort and Brest are the principal French ports; the cities on the Spanish coast include San Sebastian, Santander, Bilbao and La Rochelle. The people who live on the Spanish shore are called Basques, and are industrious and active, raising crops of nectarines, barley and maize in spite of unfertile soil along the coast. From the Basques the bay takes its name.

On account of its exposed position and diverse currents the bay is noted for storms and is especially trying to voyagers. In 1588 the great Spanish Armada, as it was starting on its career of conquest, encountered a terrible storm on the Bay of Biscay, in which the unwieldy vessels were scattered in all directions, and were assembled again only after several days. Byron in *Childe Harold* says of the bay:

On, on, the vessel flies, the land is gone,
And winds are rude in Biscay's sleepless bay.

BISHOP, *bish'up.* In the Roman Catholic, Anglican, Greek and some other Eastern churches, the title is given to one who has jurisdiction over the local churches which constitute his diocese. The office of bishop in the churches named is considered as descending without interruption from the Twelve Apostles, but this order of succession is not accepted by most Protestant churches. In the Methodist Episcopal Church the authority of a bishop is recognized; in this Church the bishops are elected by the General Conference, and after reaching a certain age they are placed upon the retired list.

The duties of the bishop vary with different denominations. In general, he has authority over the clergy and various church interests within his diocese. He may call conventions of the clergy, at which he presides, and he may appoint clergymen to churches and for cause may remove them from their positions.

BISMARCK, *biz' mark,* N. D., the state capital and county seat of Burleigh County. It is south and west of the center of the state, and is 194 miles east of Fargo, 446 miles northwest of Saint Paul and 685 miles east of Helena, Mont. The city is situated on the east bank of the Missouri River, a wide stream at this point and navigable 1,500 miles farther north, to Fort Benton, Mont. River commerce is extensive. Bismarck is on the Northern Pacific, which crosses the river on a steel and iron bridge near the city, and on the Minneapolis, Saint Paul & Sault Sainte Marie and the Chicago, Milwaukee & Saint Paul railroads. The population, largely American, in 1910 was 5,443; it increased to 6,344 in 1915. The foreign-born inhabitants are largely Germans and Scandinavians.

Bismarck is an important trade center for a large part of the great wheat country of its section of the United States and Canada. It has more than fifty large jobbing and wholesale houses. Manufactories are increasing rapidly; grain elevators, a twine plant, flour mills, foundries and assembly plants for farm implements and automobiles are the chief industrial establishments. Near the city are rich clay fields and the largest coal mine in the state.

Fort Lincoln, an army post four miles from town, has a plant valued at $1,000,000. The property of the United States Indian School, located in this city, is worth $150,000. These two institutions and other posts and Indian agencies are supplied through Bismarck. The North Dakota state capitol, a building costing $750,000, the Federal building, costing $150,000, Bismarck hospital, Saint Alexius hospital, the municipal auditorium, the National Guard Armory and the Masonic Temple are noteworthy buildings. A Methodist college, a business college and a school of music supplement the public school system. There are two parks which together contain nearly 200 acres.

In 1804-1805 Lewis and Clark spent the winter on or near the site of Bismarck. A prominent settlement was made in 1873, and three years later a city charter was granted. Bismarck, named for the great German statesman of that name, was made the capital of Dakota Territory in 1883 and in 1889 became the capital of the state of North Dakota. The city adopted the commission form of government in 1912. G.L.P.

BISMARCK ARCHIPELAGO, *biz' mark ar ki pel' a go,* the name given to New Britain, New Ireland and other islands in the Pacific in honor of the first chancellor of the German Empire, in 1885, when they were declared German territory. At the commencement of the War of the Nations in 1914 they were occupied by a combined force of Australian and New Zealand troops and annexed to Great Britain. They are situated about sixty miles east of New Guinea and cover a total area of 18,200 square miles. The principal products are copra, coffee, cotton, rubber and copper. Population, consisting chiefly of Papuans, about 189,000.

BISMARCK - SCHÖNHAUSEN, *bis' mark shoen' how zen,* KARL OTTO EDUARD LEOPOLD VON, Prince (1815-1898), the greatest German statesman of the nineteenth century and one of the most commanding figures in all history. Through his genius the German people, after centuries of disunion and oppression, were brought together to begin their history anew under the government of a united Empire. He was born of a noble Prussian family of Schönhausen, in the district of Magdeburg. At the age of seventeen he began the study of law and political science at the University of Göttingen, completed his studies at Berlin and was admitted to the bar in 1835. After serving his term in the army as lieutenant of the Life Guards, he began to take an interest in local affairs and in 1846 became a member of the provincial diet, or legislative assembly, of Saxony; in 1847 he was elected to the Prussian diet.

PRINCE BISMARCK

The following year was the time of a great revolutionary outbreak that swept over all Europe. In Prussia peace was secured only when King Frederick William IV granted the people a constitution and promised to rule according to its provisions. During this critical period Bismarck had been coming to the front as a strong advocate of increased power for the king, and his speeches in the new Prussian Parliament elected in 1849 brought him favorably to the attention of Frederick William. Accordingly, in 1851 his sovereign appointed him representative of Prussia in the Germanic diet at Frankfort, the most important event thus far in his career.

During eight years of service at Frankfort, Bismarck established in his mind the policy which was later to bring about such tremeudous changes. He clearly saw that Prussia and Austria, rival leading states in the Germanic Confederation, could never remain in that league on equal terms, and the only hope for German unity and freedom was to form a new confederation, with Prussia at the head and Austria excluded from it. How he accomplished this belongs to the story of the birth of the German Empire.

The Making of an Empire. Between 1858 and 1861 Bismarck represented Prussia at the court of Alexander II, czar of Russia. In the latter year he was transferred to Paris by William I, who had just succeeded to the throne of Prussia, and in 1862 he was summoned to Berlin by that monarch to become his Prime Minister and Secretary of Foreign Affairs. "With that day," wrote one historian, "a new era did in truth begin for Prussia and Germany, and so for Europe." From that time Bismarck worked with one end in view—the unification of the German fatherland. When the Prussian diet refused to work with him in the reorganization of the army he dissolved that body and carried out his policy without parliamentary authority. In a speech made in 1862 he said, "Not by speeches and resolutions of majorities are the mighty problems of the age to be solved, but by blood and iron." This often-quoted expression was his way of saying that only by war could the jealous German states be brought together, and, in truth, three wars were fought within the next ten years before his great purpose was realized.

In 1864 Prussia and Austria united against Christian IX of Denmark, forcing him to resign all claim to the duchies of Schleswig and Holstein. No sooner was this war ended than Prussia and Austria began to quarrel over the provinces wrested from Denmark, a dispute that gave Bismarck just the opportunity he craved. Having secured the neutrality of France and made an ally of Italy, he sent the army of Prussia, disciplined to the highest point of efficiency, against the Austrians, and in a brief war of seven weeks Austria was completely defeated.

The next step in Bismarck's program was the establishment, in 1867, of the North German Confederation, a league of the German states north of the Main River, under the presidency of Prussia. In this union Austria had no place. Bismarck realized, however, that the states south of the Main must come into the Confederation before German unity could be accomplished. He foresaw, too, that a war with France must come sooner or later, for Napoleon III, emperor of the French, fearful of losing his leadership in European affairs, was jealously watching every move in the great

BISMARCK MEMORIAL IN HAMBURG
This is one of the most massive and most impressive monuments on the continent of Europe.

historic drama across the Rhine, and was openly opposed to the unification of the German states.

In 1869 France and Prussia became involved in a dispute over the succession to the throne of Spain, and in 1870 they began war. The states north and south of the Main joined together under the standard of William I, and in the triumph of the German arms in that war Bismarck saw at last the realization of the greatest dream of his life.

The Iron Chancellor of the Empire. Just as Bismarck's personality had dominated affairs throughout the preceding period, so his was the guiding hand in the organization of the Empire, of which he became first chancellor, with the title of *prince*. In this office he revealed the same strength of will and purpose that had characterized him as Prime Minister, fittingly expressed in the title that historians give him—"The Iron Chancellor."

His genius was conspicuous in home and foreign affairs. He originated those measures in behalf of the working classes that are the foundation of modern state socialism in Germany, the system known as paternalism. He inaugurated Germany's colonial policy and did much to extend his country's trade throughout the world; the conference which met at Berlin in 1884 to arrange for the recognition of the Congo Free State in Africa was suggested by him. His one great mistake was his opposition to the Catholic party in Germany, for in this he took an extreme position from which he had to recede.

In foreign relations his policy was one of peace. He kept Germany from becoming involved in the troublesome Eastern Question, and endeavored to promote its security by forming the Triple Alliance, which guaranteed him the support of Austria and Italy in case of aggression on the part of Russia or France. The Triple Alliance remained in force until the great conflict that set all Europe aflame in 1914, for in the War of the Nations Italy joined the allies and fought against Germany and Austria.

Bismarck remained at the head of affairs throughout the reign of William I and during the short period when Emperor Frederick III ruled. In 1888 William II came to the throne, a ruler possessed of as dominating a personality as Bismarck himself. Numerous disagreements followed, and in 1890 Bismarck tendered his resignation and retired to his estates at Friedrichsruh. Later the two became reconciled, and when in 1895 the great chancellor's eightieth birthday was celebrated throughout Germany, the emperor visited him. Bismarck is one of the great characters of history, for the results of his achievements have been lasting. The simple epitaph which he wrote for himself reflects his own estimate of his life and character—"a faithful German servant of the Emperor William I." There are many statues of him in the cities of the Empire.

B.M.W.

Related Subjects. The following articles in these volumes will throw much light on the life and work of Bismarck:

Franco-German War
Germany, subhead *History*
Schleswig-Holstein
Seven Weeks' War
Triple Alliance
William I
William II

BISMUTH, *biz' muth*, a grayish-white metal with a slightly red tint. Small quantities are found in the pure state, but most of that used in the arts is obtained from the ores bismite and bismuthinite and from cobalt. Bismuth is purified by heating on an inclined shelf, so that the metal will drain away from the impurities. It is brittle, harder than lead, and 9.9 times heavier than water. One of its chief uses is in the manufacture of "fusible metals," or mixtures that melt at a low temperature. For example, fusible metal, formed of eight parts bismuth, five parts lead and three parts tin, melts at 202° F., or ten degrees below the boiling point of water. This metal can be melted in a piece of stiff paper held over the flame of a candle or lamp without burning the paper. Fusible metals are used for plugs in steam boilers to prevent explosions, in automatic sprinkler systems designed for fire protection in large buildings and for various other purposes.

Some compounds of bismuth are used in the manufacture of paint, and the subnitrate, a white powder, is sometimes used as a remedy for dyspepsia.

BISON. See BUFFALO.

BITHYNIA, *bi thin' i ah,* an ancient country of Asia Minor, bounded on the north by the Euxine, the old name for the Black Sea, and separated from Europe by the Sea of Marmora and the Bosporus. In early times the

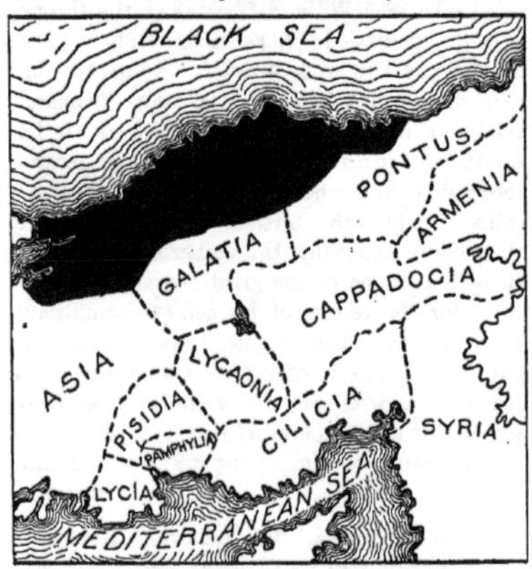

LOCATION OF BITHYNIA

Greeks established in this region the colonies of Chalcedon and Heraclea. Later, Nicaea, Brusa and Nicomedia were flourishing cities, the last-named being the royal residence of the Emperor Diocletian. Bithynia became a Roman province in 74 B.C., and under Trajan was governed by Pliny the Younger. The latter wrote a famous letter to Trajan respecting the treatment to be given the Christians of the province, which shows that Christianity gained a strong foothold there.

BIT'TERN, a marsh bird of the heron family, the most familiar species being the common bittern of North America, whose remarkable, dismal cry, sounding like the blow of an axe on a stake, has given it vari-

THE BITTERN

ous local names, as *stake-driver, mire-drum, bog-pumper, thunder-pump,* etc. This bird nests in lonely marshes and swamps, and in summer is found north of Virginia as far as the fur-bearing sections of Canada; in winter it ranges from Virginia southward to the West Indies. It is from twenty-five to thirty inches in length, having a shorter neck and shorter legs than the heron, but longer toes. The upper parts of its body are brownish-buff, spotted with reddish-brown and black, whence its occasional name of *freckled heron.* Its under parts are pale buff striped with brown, and its legs are yellowish-green. The three to five brownish-drab eggs are laid in a crude nest that is merely a thick mat of coarse grass placed on the ground.

The bird is a solitary creature of many peenliar habits. In the daytime it sometimes stands motionless for hours, on the lookout for frogs, lizards, large-winged insects and meadow mice, on which it feeds, and at night it becomes most active. Its gruesome call of *pump-er-lunk, pump-er-lunk,* which comes sounding over the marshes with solemn regularity each evening, has given rise to many absurd stories concerning it. Though its flesh is prized by some, the bittern is not an important game bird. In Utah, where wild game is not found in abundance, this bird is protected by law throughout the year, but most states do not find it necessary to enact laws regulating the hunting of the bird. In the Canadian provinces of British Columbia and Nova Scotia, the bittern may be hunted from September to March and from August to March, respectively.

BIT'TERNUT, a tall, handsome variety of the hickory group, so called because of its bitter-tasting kernel. This tree is usually found in low, wet woods and swamps, from which fact its other name, *swamp hickory,* is derived. It is found in latitudes of Ontario and Maine to Florida, and west to Minnesota, Nebraska and Texas.

The bitternut grows to a height of from sixty to 100 feet, and has several marked peculiarities that set it apart from other hickories. Its flattened, tapering, yellow buds are borne the year round, and it has the smallest leaflets and the slenderest twigs of all the hickory family. The nut is smooth and round, its thin shell enclosing a plump, white kernel, which is so bitter that no creature of the woodland will touch it, but it is relished by some people. The wood of the tree is hard, tough and close-grained, of use in making ox yokes and hoops, and as a fuel. The attractive appearance of the bitternut and its value as a shade tree are bringing it more and more into use in parks and landscape gardens. See HICKORY.

BIT'TERROOT, a plant of the dogbane family which grows in Canada and Northwestern United States. The name bitterroot comes from its long, fleshy, tapering root, which, though bitter, is a nutritious article

BITTERROOT

of food, esteemed both by the whites and the Indians. It is locally known as *tobacco root,* because of its tobacco-like odor while cooking. The plant has juicy, green leaves and a fleshy stalk bearing a single rose-colored blossom that remains open only in the sunshine.

This plant has given its name to a range of mountains between Montana and Idaho, to a forest and a river in Montana and to a fertile and beautiful valley in Montana east of the Bitterroot range, ninety miles long and seven miles in width. The bitterroot is also the state flower of Montana.

BITTERS, a term usually applied to liquid compounds containing tonic properties, taken to aid digestion or as appetizers. The most generally used bitters are angostura, quassia, cinchona, orange and wild cherry. They are usually taken with spirits or wine—a few drops of the bitters in a small glass of the liquor. Many so-called bitters are composed mainly of alcohol, and have little or no medicinal value. They should never be used except on the advice of a physician.

BITTERSWEET, or **WOODY NIGHT-SHADE,** an interesting member of the potato family, native to Europe and Asia, but now grown generally throughout Eastern United States and in Canada. Its stem is a weakling, choosing for its dwelling place moist thickets or edges of ponds, where it may climb upon the surrounding vegetation or lazily creep along the ground. The dark green leaves show a variety of form, some being heart-shaped, others having ear-like leaflets at the base, and still others having wing-like lobes. The blossom much resembles that of the potato, though it is smaller, and is blue or purple, with a yellow center.

The special attraction of the plant, however, is its fruit, or egg-shaped berries that change, as they ripen, from green to yellow and then to ruby-red. Berries in all stages of growth appear in the same cluster, and the mingling of the different colors seen against the deep green of the leaves is delightful to the eye. No one should be tempted by the beauty of the fruit, however, as the berries are poisonous. The twigs of bittersweet yield a fluid that helps to deaden pain; this is used in medicine as a remedy for certain skin diseases.

BITUMEN, *bi tu' men,* a name given a number of mineral substances which are composed chiefly of hydrogen and carbon, such as naphtha, petroleum, mineral pitch or mineral tar, and asphalt. Bituminous coal, or "soft coal," contains a large proportion of bitumen. All forms of bitumen burn, producing a great volume of smoke unless abundantly supplied with air. Bitumen, in the various forms mentioned, is widely distributed over the earth. See ASPHALT; COAL, subhead *Bituminous Coal;* PETROLEUM.

BIZET, *be za',* GEORGES (1838-1875), a French music composer whose real name was ALEXANDRE CÉSAR LÉOPOLD. He is remembered almost entirely for his brilliant and popular opera,

Carmen. Bizet studied at the Paris Conservatory, where he distinguished himself in 1857 by winning the Prize of Rome, the highest honor the institution could award. After further study in Italy he began the composition of operatic music, but *Carmen,* produced in 1875, shortly before his death, was the only work that received permanent recognition. This bright and melodious opera has enjoyed uninterrupted popularity to the present time. See CARMEN.

BJÖRNSON, *byurn son,* BJÖRNSTJERNE (1832-1910), one of the most distinctively national of all Norwegian writers, famous especially for his tales of Norse peasant life. For him his countrymen have a respect that is little less than worship, for he is to them the personification of their nation. Critics of other lands recognize in him the greatest novelist of Norway, one of its greatest poets, if not its very greatest, and its foremost dramatist except Ibsen.

BJÖRNSON
The greatest of Norway's novelists.

He has none of the pessimism of Ibsen, but shows in all his works a persistent, though not a sentimental, optimism. A patriotic politician as well as a literary artist, he took an active part in every movement for the advancement of his country, and was very influential in bringing about the separation of Norway from Sweden, in 1905.

He was born at Kvikne, studied at the University of Christiania, and after leaving that institution earned a reputation as a journalist before the appearance of his first play, *Between the Battles,* and his first peasant novel, *Synnove Solbakken.* From 1857 to 1859 he was director of the Bergen theater; from 1860 to 1863 he traveled in Europe, and in 1880-1881 he visited the United States on a lecture tour. He wrote, besides the works mentioned above, the novels *Arue, The Fishermaiden, A Happy Boy, The Bridal March, Dust, The Heritage of the Kurts* and *In God's Way;* the dramas *Mary Stuart in Scotland, The Newly Wedded Pair, The New System, A Glove* and *Dagelannet;* and many poems. For his works, many of which have been translated into English, he was awarded the Nobel prize for literature in 1903.

Nearly every child finds in his school readers *The Tree,* which begins—

The Tree's early leaf-buds were bursting their brown:
"Shall I take them away?" said the Frost, sweeping down.
"No, leave them alone
Till the blossoms have grown,"
Prayed the Tree, while he trembled from rootlet to crown.

BLACK, commonly referred to as the darkest of all the colors. In the theory of color, white is produced by all the colors mixed in the proportion in which they are found in the solar spectrum and the rainbow, and black is produced by the absorption of all colors; in other words, then, black is really the absence of color. Of course, in practice this does not hold; all objects absorb some color and reflect some, but black objects reflect the smallest proportion. Black is the emblem of sorrow and mourning in Europe and America, but not throughout the world, for in some Asiatic countries, particularly in Chosen (Korea), white is so used. See COLOR; SOLAR SPECTRUM.

BLACK ART. See NECROMANCY.

BLACKBERRY
Leaves, fruit and flower.

BLACKBERRY, a vine-like shrub, one of the most important and profitable of small fruits. It is cultivated in most fruit-growing localities of America for jams, jellies, preserves, wine and dessert, but it also grows wild in cool climates. The leaves are three or five, round-pointed; the blossoms, pink or white. The berries are plump and juicy, somewhat longer and more solid than a raspberry. Most varieties are black, but a new Burbank creation is a beautiful transparent white. And in his thornless blackberry Burbank has developed a blackberry stalk as smooth as velvet.

Blackberries thrive on well-drained soil. The weak and diseased stems of the shrubs should be cut off each spring, and the outer stems cut back about one-third of the length. The early Harvest, Snyder and Agawam are three favorites.

Blackberry Jam. Although fresh berries can be obtained only during a short season, yet the delicious flavor of blackberries may be enjoyed all the year. Pick over and wash a quantity of berries. Weigh the fruit and measure an equal weight of sugar. Put the berries into a preserving kettle, mash them as they are being heated, and when considerable juice has been pressed out add the sugar gradually. Let the mixture boil up well, then skim out the fruit or strain it. Boil the juice again until it is thick and will *jelly*. Put the fruit back and boil once more. Then pour into jars and seal.

BLACKBIRD, the family name of several species of birds whose distinguishing feature is the glossy black coat of the male birds. Of the North American blackbirds none is more familiar than the red-winged variety, known also as the *swamp blackbird, red-winged oriole* and *red-winged starling.* The red wings are from seven and one-half to ten inches in length, and receive their name from the scarlet, yellow-tipped shoulders of the male, which otherwise is as black as coal. The female, as is true of most birds, is somewhat commonplace in appearance, with blackish-brown upper parts streaked with rusty black and gray, and dusty-white under parts streaked with reddish-brown.

These birds breed throughout the United States, and in the summer are found in Eastern Canada, west to the Saskatchewan valley and north to Great Slave Lake. They begin their southward migration in October, or early in November, traveling in large flocks, and are often winter residents of Mexico. Some, however, are brave enough to endure the hardship of a New England winter, and refuse to join the autumn migration parties.

The favorite nesting place of blackbirds is a low bush on the edge of a pond or the moist grass of the marshes. In the nest, which is made of grass, leaves and mud, are deposited from three to five eggs, pale blue in color, with

A slender young Blackbird built in a thorn tree:
A spruce little fellow as ever could be;
His bill was so yellow, his feathers so black,
So long was his tail, and so glossy his back,
That good Mrs. B., who sat hatching her eggs,
And only just left them to stretch her poor legs,
And pick for a minute the worm she preferred,
Thought there never was seen such a beautiful bird.
—D. M. MULOCK.

streaks and spots of black or purple. They are cheerful, sociable birds, given much to noisy chattering, and for this reason are sometimes regarded as a nuisance. Yet their liquid warbling would be very delightful if they did not perpetually interrupt their singing with discordant notes which sound like fretful complainings. They feed upon worms, insects, fruit and grain, and their frequent raids upon the unripe corn of the farmer have given them an undesirable reputation. On the other hand, they more than offset the damage done to the corn by the service they render in destroying immense numbers of harmful insects.

For other familiar species of blackbirds, see COWBIRD; CROW BLACKBIRD. See, also, BIRD, subtitle *Migration of Birds.*

BLACKBURN, next to Manchester the most important center of the cotton industries of England. It is twenty-four and one-half miles northwest of the latter city, and 210 miles northwest of London. It is an ancient city, at one time the capital of a district named Blackburnshire but now incorporated in Lancashire. The cotton industries give employment to many thousands of men, women and girls, and the town also has important manufactures of hardware and machinery. The mill operatives of Blackburn earn higher wages than any others of the laboring classes in England, and they form large clubs among themselves, to which they subscribe weekly; the total sum subscribed is drawn annually in order to spend a week at the seaside.

Blackpool, one of the finest watering places in England, receives annually several hundred thousand pounds sterling in this way from Blackburn operatives. In 1557 a grammar school was founded here by Queen Elizabeth, and it now occupies extensive modern buildings. Population in 1911, 133,064.

BLACK DEATH. See PLAGUE.

BLACK-EYED SUSAN, or YELLOW DAISY, is a sunshiny wild flower found in dry fields and along the byroads and highroads almost everywhere. The flowers have orange-yellow rays, or petals, so gaily they advertise to the bees and butterflies the wealth of nectar and pollen in their purple-black centers. Just one flower grows at the top of each rough, hairy stem. The leaves, too, are stiff and hairy, and placed alternately along the stems. It is difficult to pick these inviting flowers without pulling up the whole plant, because the stems are tough and rigid. Cut them, and you will be rewarded for your care, as the roots are perennial and they will come up year after year. See PERENNIAL.

BLACK-EYED SUSAN

BLACKFOOT, a tribe of Indians who are supposed to have received their name from white men, who noticed that the tops of their moccasins were blackened by the burnt grass and brush through which they passed. The Blackfoot are a branch of the great Algonquian family, and formerly occupied all the region from the North Saskatchewan River in Canada to the head waters of the Missouri in Montana. They were restless, aggressive, and frequently at war with other tribes, but never with the United States or Canada. They became noted for their large herds of horses. They now live on reservations in Alberta and Montana, where they are acquiring the customs of civilization. In all they number about 4,700. Tourists in Glacier National Park and in Rocky Mountains National Park, north of the former, in Alberta, may see many camps of these Indians.

Illustrations will be found with the articles INDIANS and GALCIER NATIONAL PARK.

BLACK FOREST, a mountainous district in the southwestern part of Germany, covering an area of 1,844 square miles. It runs almost parallel with the River Rhine for eighty-five miles and forms an elevated chain of plateaus rather than a series of isolated peaks. The highest summit is Feldberg, 4,900 feet above sea level. The slopes contain many lakes and streams, in which the Danube, Neckar, Kinzig and many smaller rivers have their sources. The principal mineral found is iron. The region is noted for its mineral springs, which have led to the establishment of numerous watering-places and health resorts, of which Baden-Baden (which see) is the most famous. The forests yield much timber, especially the pine of the black fir, from which the forest takes its name. The region is being continually reforested by the best-known, thorough German methods.

The manufacture of wooden toys, clocks, and musical instruments is the most important industry of the section, employing about 40,000 persons. The inhabitants of the forest preserve a quaint simplicity in habits, and the district is rich in old legendary associations. The trade is centered in the towns of Freiburg, Rastatt, Lahr and Offenburg.

BLACK FRIDAY, in the United States the name given to two days that ushered in many panics. The first Friday was on September 24, 1869, when Jay Gould and James Fisk, Jr. attempted to create a "corner" by buying all the gold contained in the New York City banks. The value of gold had been steadily rising for several days, and speculators were aiming to carry it still higher. On Friday the whole city was in a state of tremendous excitement, when gold rose to 162½ and was still rising; a possibility seemed imminent that business houses would be closed, as no one knew what prices goods should bring. At this exciting time $4,000,000 was taken from the United States Treasury and placed on the market by Secretary Boutwell to break the "corner," and the value of gold immediately fell, not, however, without leaving the speculators richer by almost $11,000,000.

The second Black Friday was on September 19, 1873, when the New York Stock Exchange reported numerous failures, which precipitated what is known as the panic of 1873.

BLACK GUM, an American tree of which there are two well-known species, one yielding a tough, close-grained wood, used for hubs of wheels; the other yielding a light, soft wood employed chiefly for such articles as berry crates and fruit boxes. Both are handsome trees from eighty to ninety feet in height, with twisted branches and bright green leaves which turn crimson in autumn. The black gum is native to America, but has been introduced into Europe as an ornamental tree. Just why it should be called a gum tree is uncertain, for there is no gum connection in any way with it.

BLACK HAW, or **STAG-BUSH,** a shrub or small tree belonging to a family of most valuable ornamental plants. The buds, reddish and downy, appear in winter. This shrub grows sometimes to a height of fifteen feet and has spreading, rather stout branches. The flowers are attractive and pure white; the fruit, nearly half an inch long, is oval, bluish-black and covered with a soft bloom. The black haw grows well in dry places.

BLACK HAW

BLACK HAWK (1767-1838), a chief of the Sac Indians, one of the most persistent enemies of the white men in their westward progress. He was born at Kaskaskia, Ill., became chief of his tribe in 1788, and from the first showed himself strongly opposed to any concessions to the whites. Despite his influence, the Sacs and Foxes in 1804 agreed to give up to the United States their lands east of the Mississippi River, but Black Hawk repudiated the contract, declaring that the chiefs had been made drunk before they signed the documents.

During the War of 1812 Black Hawk, tempted by British agents, joined them with about 500 warriors, but soon retired from British service. In 1823 most of the Sacs and Foxes, under the leadership of Keokuk, removed to their reservation beyond the Mississippi River; but Black Hawk, with part of his tribe, refused to emigrate and fought with the whites what is known as the Black Hawk War. After several encounters, the Indians were defeated, and Black Hawk and his two sons were taken captive. The three were confined in Fortress Monroe until 1833, and afterward joined the tribe in the reservation near Fort Des Moines.

BLACK HAWK

In 1911 there was erected near Oregon, Ill., on a bluff which overlooks the picturesque valley of Rock River, a massive reinforced concrete statue to this indomitable Indian chief. The figure is fifty feet in height and was constructed by Lorado Taft. It is not a portrait statue, for it presents features from more than one Indian tribe, but its simple strength and majesty and its prophetic gaze down the valley once dominated by this grim warrior gives it a very real personality.

THE GREAT STATUE
This majestic figure, the work of Lorado Taft, is of concrete and stands fifty feet in height.

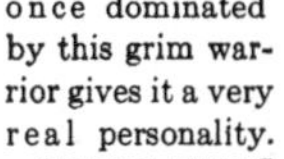

BLACK HILLS, a name that describes one of the richest gold-mining sections in the United States. It is a low, mountainous district, covering an area of 6,000 square miles in South Dakota and extending into Wyoming. The territory was purchased from the Indians in 1876, and mining operations were begun the year following. Gold, silver, copper, lead, iron and a number of valuable building stones are obtained in the region. The great Homestake mine, one of the largest in the United States, is located at Lead, in the Black Hills. See SOUTH DAKOTA.

BLACK HOLE OF CALCUTTA, a small room, 14 feet 10 inches wide and 18 feet long, in the old English fort of Calcutta, India. On June 20, 1756, Surajah Dowlah and his men, after forcefully robbing the city, captured the fort. On that intensely hot night in June, 146 men were thrust as prisoners into that one small room with but two tiny windows, and by morning only twenty-three were alive to tell of their terrible experience. See INDIA, subhead *History*.

BLACKING, for shoes, is commonly formed of ivory black, bone black or lamp black, in a liquid mixture of oil, vinegar, molasses or sugar dissolved in water, and sulphuric acid. The chemical combination of the acid, the sugar element and the black gives the blacking its power to adhere to the leather. The difference between liquid and paste blackings is in the amount of vinegar contained. After blacking is applied to leather a high polish may be imparted by rubbing briskly with a cloth or soft brush.

BLACK LAKE, QUE., a town in Megantic County, four and one-half miles south of Thetford Mines and forty-six miles south of Quebec, on the Quebec Central Railway. The town is of importance chiefly for its valuable mines, which yield asbestos, chrome and iron. The mines of Black Lake and the vicinity furnish about eighty per cent of the world's supply of asbestos (which see). The town was incorporated in 1908. Population in 1911, 2,645; in 1916, estimated, 4,000.

BLACK'LIST, a list of names of persons thought deserving of censure or punishment, or considered as untrustworthy as workmen, or classed as delinquent debtors. Lists of the latter are sometimes published by mercantile agencies and others for the protection of employers or tradesmen; to a very general extent general blacklists were formerly used by employers who wished to warn others against the employment of persons whom they considered objectionable. As used in connection with labor problems the term refers to lists of persons considered undesirable by labor unions or by employers as workmen. Activity in the cause of unionism by a prospective employee is frequently objected to by an employer, while the unions object to men for exactly the opposite reasons, namely, that such persons have refused to join the union or obey its orders or have assisted as strikebreakers. Blacklists are usually distributed secretely, because the persons responsible for such lists do not care to be known.

The Dominion of Canada and more than one-half of the states of the American Union have passed laws against the use of blacklists, but it has been found difficult to enforce these because employers often discharge workmen without stating any reason other than that their services are no longer needed, and also because it is easy to conceal the exchange of information on which these blacklists are based. Better understanding between employers and laborers of late years has decreased the temptation to maintain such lists, even without the pressure of law. See LABOR ORGANIZATIONS.

Trading with the Enemy. The term blacklist also refers to enemies at war. In the War of the Nations, for example, Great Britain and others of the allies published lists of persons and firms who were "by reason of enemy nationality, sympathy or association, found to be zealous to advance the cause of the enemy and to make their trading or profits in trade a means to this end." Trading with the enemy is illegal under English common law, and the effect of these lists was merely to add certain persons nominally neutral to the class of those with whom trade was prohibited. The publication of these lists, which included some firms in the United States, was regarded by many as an invasion of neutral rights, and called forth a protest from the United States government.

BLACK'MAIL, a term legally applied to an attempt to extort money from a person under threat of exposure of an alleged past offense or of some damaging secret. Whether successful or not, blackmail is an offense punishable in the United States and Canada by imprisonment or fine, or both. The laws of Great Britain are exceptionally severe on blackmailers, a sentence of twenty years' imprisonment having been pronounced on a person proved guilty of even an attempt to blackmail another. In olden times blackmail was a certain amount of money, corn, cattle or some other thing of value, paid to men allied to robbers for protection from pillage; from that ancient usage the present meaning of the term was derived.

BLACKMORE, RICHARD DODDRIDGE (1825-1900), an English novelist whose name would now be forgotten were it not for his *Lorna Doone,* a vigorous and beautifully-written story of Exmoor and the neighboring district, the home of the Doone family. He was born at Longworth, in Berkshire, educated at Oxford and began the practice of law in 1852. Failing health forced him to abandon his chosen profession, and settling on a fruit farm up the Thames from London he devoted his time to writing. Among the several novels from his pen are *Cradock Nowell, The Maid of Sker, Alice Lorraine* and *Kit and Kitty,* none of which has the charm or literary merit of *Lorna Doone.*

BLACK MOUNTAINS, a range of mountains extending across North Carolina into the northern parts of Georgia and North Carolina. Their direction is nearly east and west, and they contain the highest peaks in the Appalachian system, among which are Mount Mitchell, 6,710 feet, and Clingman's Peak and Guyot's Peak, the latter two exceeding 6,500 feet in altitude. Their sides are covered with evergreen forests, from which they take their name. Many of the valleys are highly fertile. See APPALACHIAN MOUNTAINS.

BLACK PRINCE, THE. See EDWARD, THE BLACK PRINCE.

BLACK SEA, called by the ancients PONTUS EUXINUS, is a sea situated between Europe and Asia, bounded by Russia, Bulgaria, Rumania and Turkish dominions in Europe and Asia. It is connected with the Mediterranean by the

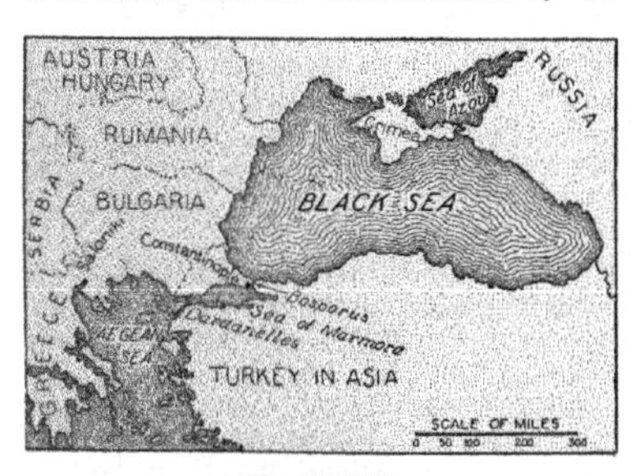

BLACK SEA
Very important commercially, though subject to violent storms and dangerous breakers.

historic Bosporus, the Sea of Marmora and the Dardanelles, and by the Strait of Kertch with the Sea of Azov, which is, in fact, only a bay of the Black Sea. It has a length of 750 miles, a greatest width of 380 miles, a maximum depth of 7,000 feet and covers an area of 180,000 square miles, more than the combined area of Illinois, Iowa and Wisconsin. The water is not so clear as that of the Mediterranean, and it is less salt, on account of the many large rivers flowing into it, among which are the Danube, Dniester, Dnieper and Don.

During January and February the shores from Odessa to the Crimea are ice-bound. The Black Sea contains few islands and those of small extent. The most important ports are Odessa, Kherson, Sebastopol, Batum, Trebizond, Sinope and Varna. The sea is of great commercial importance to Russia, as it furnishes an outlet for the agricultural region of the south, but its exit to the Mediterranean Sea and the ocean has long been closed by the Great Powers of Europe to the battle fleet of Russia. At the outbreak of the War of the Nations this great sea assumed first importance, and on it many battles occurred between Russian and Turkish vessels. The attempted forcing of the Dardanelles by the allies to allow free access to the Mediterranean for Russian warships failed utterly, though it was one of the most gigantic undertakings in the history of military and naval operations.

Related Subjects. Connected more or less intimately with the Black Sea are the following topics:

Azov	Marmora
Bosporus	War of the Nations
Dardanelles	

BLACK SNAKE, or **BLUE RACER,** the most agile and swift of all snakes. It is common in North America, and often reaches a length of five or six feet. It is comparatively harmless, as it has no poison fangs, but it possesses the power of destroying its prey by the contraction of its folds—in other words, by squeezing it to death. It is a deadly enemy of the rattlesnake, in destroying which it shows great skill. It is said to follow its prey by scent, and, being quicker in its movements, is able to seize the rattlesnake by the back of the neck and quickly crush it to death. The black snake is usually bluish above and slate color beneath, though in South America it is more often an olive-green. Birds' eggs and small animals, like mice, frogs and birds, comprise its principal food.

BLACKSTONE, SIR WILLIAM (1723-1780), a distinguished English judge and writer on law, whose most famous work, *Commentaries on the Laws of England,* has had a wider influence than any other treatise on law in the English language. He was educated at Oxford University and was admitted to the bar in 1746. He first attracted attention by a course of lectures delivered at Oxford in 1753, and when five years later a new professorship of law was established at the university he was given the appointment. The lectures delivered at this time established his fame, and his progress thereafter was rapid. He resigned his professorship in 1766, and four years later was knighted and appointed justice of the court of common pleas. Until his death he was occupied with his duties as judge and as an advocate of prison reform.

Blackstone's *Commentaries,* made up of his Oxford lectures, furnished the model for all later English and American commentaries. For a century at least after Blackstone's death it was practically the only text-book on law, and to-day is regarded as a leading authority.

BLACKWELL, ELIZABETH (1821-1910), the first woman to receive a medical diploma and take up the profession of medicine in the United States. When she was ten years of age her parents moved from Bristol, England, her birthplace, to New York, and later settled in Cincinnati, where she was engaged for several years in teaching. After numerous difficulties, due to prejudice against women in the profession, she was graduated in medicine with the highest honor, in 1849. She continued her studies in Paris and London, and on her return to America settled in New York, where she acquired an excellent practice. In 1854 she opened a hospital for women and children, and in 1868, with her sister Emily, also a physician, she founded the Woman's Medical College of the New York Infirmary. Miss Blackwell lectured to women on health and physical development, and published several books relating to women and medical education.

BLACKWELL'S ISLAND, a narrow island in East River, between Manhattan and Long Island, a part of the territory covered by New York City. It is about one and one-half miles long and one-eighth of a mile wide. It is devoted exclusively to penal institutions and hospitals of New York City. Among these are the workhouse, to which petty criminals are sentenced and for which "the Island" is most noted; the city hospitals, asylums for the insane and blind, the almshouse, the home for the aged and infirm and the hospital for consumptives. The island in summer is carpeted with grass and the buildings are well kept. All vessels which sail out upon Long Island Sound pass Blackwell's. On the rocks at the north end of the island a lighthouse rises fifty-four feet above the water.

BLADDER. See KIDNEYS.

BLADDERWORT, *blad' er wert,* a curious, slender water plant, found in ditches, pools and marshes throughout the world. About a dozen kinds, yellow and blue in color, are natives of the United States or Canada. The leaves have little bladders that fill with air at flowering time and raise the plant in the water sufficiently to bring the blossoms above the surface. From this fact the name is derived. These bladders have a valve-like door, through which insects enter when looking for food or shelter and so are trapped. If the ditch or pool dries up, the bladders hold moisture enough to keep the plant alive for some time.

BLAINE, *blane,* JAMES GILLESPIE (1830-1893), an American statesman whose personality, keenness of intellect, power in debate and profound knowledge of history and of human nature made him one of the most influential political leaders for nearly two decades. He was born at West Brownsville, Pa., of Scotch-Irish ancestry, at the age of seventeen was graduated from Washington College, in Washington, Pa., and during the next few years taught school and engaged in the study of law. After settling in Maine in 1854, he became editor of the Kennebec *Journal,* soon accepted a more influential position on the Portland *Daily Advertiser,* and in 1858 was elected to the lower house of the state legislature.

JAMES G. BLAINE
Named the "Plumed Knight" by Robert G. Ingersoll in his speech in the 1876 nominating convention at Cincinnati. The name electrified the audience and clung to Blaine as long as he lived.

Blaine's Public Career. Four years later, upon election to Congress, Blaine began his long career as a national Republican leader. Between 1869 and 1874 he was Speaker of the House, and his constructive ability and genius in debate were conspicuously revealed in the consideration of the various measures for the reconstruction of the Southern states (see RECONSTRUCTION). As the Presidential election of 1876 drew near it was apparent that Blaine would be a candidate for the Republican nomination, but he failed to secure that honor by twenty-eight votes. Charges of corrupt practices in securing legislation in favor of certain railroad projects had been made against him by his political opponents, and his defense was not satisfactory to many of his own political faith. Therefore he never again had the undivided support of the Republicans, and the great ambition of his life was never realized.

In 1877 Blaine entered the United States Senate, where he zealously championed the cause of protective tariff, and labored for the advancement of American shipping and for subsidies for American industries (see SUBSIDY). In the Republican convention of 1880 his friends fought for his nomination through thirty-six ballots, finally giving their support to the "dark horse" of the convention, James A. Garfield. After the latter's inauguration Blaine was appointed Secretary of State, but he held office only eight months because of the death of the President and the consequent reorganization of the Cabinet by the new President, Chester A. Arthur.

After three years of retirement Blaine returned to public life as the Republican candidate for the Presidency, but was defeated in the election by Grover Cleveland, after a campaign unequaled up to that time in bitterness of personal attack. He thereupon devoted himself to literary work, publishing in 1886 the second volume of his *Twenty Years in Congress,* the first volume having appeared in 1884. It is a valuable historical record.

Refusing to permit his friends to press his candidacy for the nomination of 1888, he entered President Harrison's Cabinet as Secretary of State in 1889 and served with distinction until his resignation in 1892. With his administration of the State Department are connected the treaty with Germany concerning the Samoan Islands (see SAMOA) and the assembling of the first Pan-American Congress (which see). He also negotiated a large number of reciprocity treaties for the encouragement of American commerce (see RECIPROCITY), and vigorously upheld the dignity and honor of his country in its foreign relations. A last and unsuccessful attempt was made to secure his nomination for the Presidency in the Republican convention of 1892 which renominated Harrison. Six months later he died. B.M.W.

Consult Blaine's *Twenty Years in Congress;* Peck's *American Party Leaders.*

BLAIRMORE, a village in Alberta, on the Crow's Nest River and the Canadian Pacific Railway, thirteen miles east of Crow's Nest Pass and eighty-three miles west of Lethbridge. Blairmore has a number of coal mines, and is also the distributing point for the coal fields of the region. Cement and brick clay are found in the neighborhood, and bricks, lime and lumber are shipped. The population increased from 239 in 1901 to 1,137 in 1911, and to about 1,800 in 1916.

BLAKE, EDWARD (1833-1912), a Canadian lawyer and statesman, for twenty years a member of the Dominion House of Commons, one of the foremost orators who ever addressed that body, and predecessor of Sir Wilfrid Laurier as leader of the Liberal party. Blake was born at Adelaide, Ontario, on October 13, 1833, but to the day of his death he was undeniably Irish. His father had emigrated from County Galway to Ontario in 1832, and the son exactly sixty years later reversed the process. After winning fame in the Dominion he removed to Ireland and was elected a member of the British Parliament, where he sat as an Irish Nationalist until 1907. Failing health then led him to resign and return to Canada, and the scene of his early triumphs.

EDWARD BLAKE

After graduation from Upper Canada College and the University of Toronto, he was called to the bar in 1856, and in a few years won a large practice. He was elected to the first Dominion Parliament in 1867, and at the same time sat in the Ontario legislature. He led the Liberal opposition in the legislature until 1871, when he was prime minister for a few months. He declined the leadership of the Liberals in Parliament, but in 1873 accepted a ministership without portfolio in Alexander Mackenzie's Cabinet. His health was uncertain during the next few years, but he held the portfolio of Minister of Justice from 1875 to 1877, long enough to take the chief part in planning the organization of the Dominion Supreme Court. From 1880 to 1887 he was leader of the Liberal Opposition in the Dominion House of Commons. Though he resigned this leadership in 1887 he still sat in the House until 1891, when he withdrew from Canadian public life.

In the field of imperial politics, which he entered in 1892 at the request of the Irish Nationalists, he rendered good service to the cause of Home Rule. He was conspicuous in 1896 as one of the committee to investigate South African affairs. In early life Blake was a strong supporter of imperial federation, but he later became less outspoken and even questioned the wisdom of such a plan. In addition to his political activities he practiced law most of his life and appeared in many important cases, both in England and in Canada. He was vitally interested in education, founded several scholarships in political science in the University of Toronto and served a term as chancellor of that institution. U.H.L.

BLAKE, Robert (1599-1657), a famous British admiral who distinguished himself in the naval battles of the seventeenth century that wrested the supremacy of the seas from the Dutch. During the civil war between Charles I and Parliament, in which he aided Cromwell, he destroyed the squadron of the Royalist general, Prince Rupert, and as a reward for his services was made sole admiral of the English fleet. Between 1652 and 1653 he won a series of victories over the Dutch Admiral Van Tromp, and forever ended Holland's claim to being mistress of the seas.

In 1654 Cromwell sent Blake to the Mediterranean, where he succeeded in upholding the dignity of the British flag in contest with the Dutch, the Spanish and the French. He attacked Tunis, the dey of which had insulted the British flag, routed an army of 3,000 Turks, and at Algiers and at Tripoli set free all the English held there as slaves.

BLAKE, William (1757-1827), an English engraver and poet who occupies a unique place among English poets by reason of the unusual character of his writings. His poetry, which he himself illustrated with drawings of great beauty and originality, has the imaginative quality of Spenser's work, the spirit and ring of the Elizabethan lyrics, and also many elements of the writings of the Romantic poets who followed him—love of children and animals, and an appreciation of the beauty that lies in ordinary life. His best-known poems are found under the titles *Songs of Innocence* and *Songs of Experience*.

Charles Lamb regarded him as one of the most extraordinary men of his age, and Swinburne has characterized him as "the single Englishman of supreme and simple poetic genius of his time."

BLANC-MANGE, *blah manzh'*, a popular dessert having the appearance of jelly. It is whitish in color and is made of Irish moss, cornstarch, arrowroot and other starchy substances, boiled with milk and flavored. Frequently chocolate and various fruit juices are added. Served with cream or sauce, blanc-mange is a wholesome and agreeable dessert, pleasing to the eye as well as the taste. The name comes from the French *blanc*, meaning *white*, and *manger*, meaning *to eat*.

BLANC, Mont. See Mont Blanc.

BLAND, Richard Parks (1835-1899), for many years the leader of free coinage sentiment in the national House of Representatives and the author of the famous Bland-Allison bill of 1878, which was passed over President Hayes' veto. He was born near Hartford, Ky., and after wandering as far west as California finally returned to Missouri to practice law. He was elected to the lower house of Congress in 1872, where he served until his death, with the exception of the years between 1895 and 1897.

The *Bland-Allison bill*, through which his name became famous, provided for the purchase by the government of not less than $2,000,000 nor more than $4,000,000 worth of silver bullion a month.

This bill was in effect until repealed by the passage of the Sherman Law in 1890. Bland was a prominent candidate for President at the Democratic convention in Chicago in 1896, but he withdrew in favor of William Jennings Bryan, who was nominated because of a brilliant appeal on the silver issue, known as the "cross of gold" speech. However, Byran was defeated in the ensuing election by William McKinley, after which the coinage of silver ceased to be a political issue.

BLANK VERSE, poetry written without rhyme, a form employed in some of the noblest and sweetest poems. The name *blank* here refers to the lack of the rhyme at the ends of lines of poetry, which the ear expects to hear. The first English poet to employ blank verse was Henry Howard, earl of Surrey, who in the sixteenth century translated into unrhymed verse two books of Virgil's *Aeneid*. Marlowe and Shakespeare brought this form of verse to perfection, and since their time it has been used by practically all the English poetic dramatists except Dryden, and by Milton, Tennyson, Browning, Wordsworth and other writers in many different forms of poetry.

A familiar example of blank verse is Bryant's *Thanatopsis*, in which is used the typical blank verse line—a ten-syllable line having five feet of two syllables each, the second of which is accented. Such a line is an example of iambic pentameter, and is divided and accented thus:

To him' | who in' | the love' | of na' | ture holds' |

The concluding stanza of *Thanatopsis* very well illustrates the effects produced by blank verse, which critics agree is admirably adapted to use in poems that combine harmony of sound and dignity of music with nobility of thought:

So live, that when thy summons comes to join
The innumerable caravan, that moves
To that mysterious realm, where each shall take
His chamber in the silent halls of death,
Thou go not, like the quarry slave at night,
Scourged to his dungeon, but, sustained and soothed
By an unfaltering trust, approach thy grave,
Like one who wraps the drapery of his couch
About him, and lies down to pleasant dreams.

BLARNEY, *blahr'ni,* **STONE,** a stone much kissed because of the superstitious belief that it will give to those who kiss it the power of saying easily things which flatter, compliment or persuade. From this has come the expres-

BLARNEY CASTLE

The stone on the highest point of the corner in center of picture is the Blarney stone, and bears date of 1703. It is held in place by two iron bars.

sion *blarneying.* This stone is in a wall near the top of Blarney Castle, Blarney, Ireland, a village of about 1,000 people, near the city of Cork. Tourists from every part of the globe visit the castle just to kiss the stone. It is commonly believed that the Blarney Stone legend originated because the first owner of the castle delayed its surrender in medieval times by promises and flattering speech.

BLASHFIELD, Edwin Howland (1848-), an American artist whose wall paintings, adorning some of the finest buildings in the United States, have brought him into the front rank of decorative painters. He was born in New York City. After studying in Paris under the famous French painter, Leon Bonnat, he spent several years in France, Italy, Greece and Egypt, returning to the United States in 1881 and beginning work as a figure painter. Of his early canvases, the best known are *Christmas Bells* and *The Angel with the Flaming Sword.* Since 1892 he has given his time entirely to decorative painting. Some of his splendid achievements in this field are the central dome of the Library of Congress in Washington, picturing the *Development of Civilization;* in the Baltimore courthouse, *Washington Resigning His Commission* and *Lord Baltimore's Edict of Toleration;* and the ceiling of the great ballroom of the Waldorf Astoria Hotel, New York, representing *Dance and Music.*

Blashfield's work is characterized by delicate and beautiful coloring. He is a member of the National Academy of Design, and with his wife has written *Italian Cities* and edited Vasari's *Lives of the Painters.*

BLASPHEMY, *blass' fee mi,* a term, originally meaning *profanity,* but now applied to a spoken or written insult to the Diety. From the earliest days blasphemy has been an offense against man's laws, and once was punishable with death or other severe penalty. In England until 1547 blasphemous persons were whipped and imprisoned; until 1825 the legal punishment in Scotland was death, and by the present laws of that country the offense is punishable by imprisonment. In the United States and Canada the punishment for blasphemy was formerly imprisonment and whipping, but, as in England, the law has fallen gradually into disuse. The Bible declares that at the final day there shall be a strict accounting for blasphemy.

BLAST FURNACE, the name given to the common smelting-furnace, used for obtaining iron from its ores with the aid of a powerful blast of air. This device is fully described in the article Iron, under the subhead *The Blast Furnace.*

BLASTING, the operation of breaking up masses of rock or other hard substances, by means of explosives. Previous to the invention of gunpowder the usual method of blasting was by heat, followed by quick cooling. Hannibal in forcing his way over the Alps is supposed to have employed this method by lighting fires against rocks and then dashing cold water on them. This caused them to crack, when they could be moved or further broken by the tools at his command. In ordinary blasting operations, such as occur in railroad construction and mining, holes are bored to the requisite depth by means of drills, the explosive is introduced, the hole is *tamped* or filled up with broken stone, clay or sand, and the charge is exploded by means of a fuse or by electricity. In larger operations, mines or shafts of considerable diameter take the place of the holes above described.

In the construction of the Panama Canal blasting was resorted to on a scale which surpassed all previous operations. The largest single blast was used at the blowing up of the Gamboa dyke at the northern end of the Culebra (now Gaillard) Cut. Here forty tons of dynamite were used, placed in 1,000 holes each containing about eighty pounds of the explosive. When all was ready for blasting, on October 10, 1913, President Wilson, in Washington, touched an electric button which completed a circuit with the explosive, more than 2,000 miles distant. Instantly a huge explosion took place, tearing down the dyke and removing the last obstacle to the joining of the waters of the east and west.

BLAVATSKY, *bla vahts' ke,* HELENA PETROVNA HAHN-HAHN (1831-1891), a famous Russian spiritual leader whose teachings embody the doctrines held by modern theosophists (see THEOSOPHY). During twenty years of travel in various parts of the world she made a special study of the mystic factor in religion, and in 1858 became a famous spiritualistic medium in Russia. Later she moved to the United States, where in 1875 she founded the Theosophical Society. Four years afterward a branch society was organized in Bombay, India. Though her claims as a worker of miracles were disproved, she is regarded by theosophists as their greatest leader, and when she died she had nearly 100,000 followers in England, France, the United States and Canada. Her most important work, *Isis Unveiled,* is the textbook of the theosophists.

BLEACHING, *bleech' ing,* from the German *bleichen,* meaning *to whiten,* is the process of making cotton, linen, wool, silk and other fabrics white by removing from them their natural coloring matters. Bleaching is practiced in its simplest form by the housewife who spreads her washing on the grass to whiten in the sunlight, a custom that originated many centuries ago. The process of bleaching is supposed to have been known to the Egyptians, Babylonians and Hebrews. The ancient method, which was in use until the eighteenth century, consisted in spreading the cloth on a stretch of grass, and leaving it exposed to the air and sunlight for several months, with sprinklings of water each day.

In the eighteenth century the Dutch discovered a new method, and Holland became a very important center of the bleaching industry. The fabrics were steeped repeatedly in potash lye, soaked in buttermilk for about a week and then washed and spread on the ground to whiten. The Hollanders obtained such good results that the name *hollands,* still in use, was given the excellent fabrics bleached in this manner; also, a very desirable quality of linen, which was spread on plots of grass, came to be known as *lawn.* The Scotch and Irish still bleach their fabrics by spreading them on the grass, a process called *crofting,* from the Scotch word *croft,* meaning a small tract of meadow land.

Bleaching as carried on at the present time is a complicated process requiring the special machinery of the modern factory. It consists of steepings, boilings, washings and dryings and the use of various chemicals, particularly the bleaching powder called chloride of lime. The operations vary according to the materials of which the fabric is composed, and according to the fineness or coarseness of the yarn. Cotton bleaches more quickly than linen and requires fewer operations, for the latter must be subjected first to a series of alkaline boilings to dissolve the impurities that are present in the flax fiber. Linen fabrics are often exposed to the action of air, light and moisture for several days, as this is supposed to make the fiber retain its strength, and it adds to the life of the cloth. This step of the process is called *grassing.*

Wool and silk goods in the process of bleaching are subjected to the fumes of burning sulphur; the sulphur combines with the coloring matters in these fabrics to form a colorless compound without destroying the coloring matters. Washing bleached silk or wool goods several times with soap containing potash makes them turn yellowish, for the soap destroys the colorless compound. Hydrogen peroxide is coming into general use as a bleaching agent for silk.

Straw, beeswax, feathers, hair, ivory, oils, sponges and the rags and paper used in paper making are also bleached. B.M.W.

BLEEDING, or HEMORRHAGE, *hem' o rayj,* is the escape of blood from the body, or from one part of the body into another. In popular use the term sometimes signifies a more sudden and severe flow of blood than ordinary bleeding, but there is no real distinction. Bleeding may be caused by a cut or tear in a vein or artery, or it may be the result of bruising a new surface or a mucous membrane.

Under abnormal conditions blood may also escape into the lungs, the stomach and other cavities and remain concealed for a considerable length of time. This is called *internal bleeding.*

Arterial Bleeding. This may be recognized by the fresh red color of the blood and by the way it issues from the wound—that is, in jets or spurts. The cutting of a large artery sometimes results fatally with the first gush of blood, and in other cases the shock is sufficient to cause unconsciousness. Fainting, in which the supply of blood to the brain is greatly diminished and the blood pressure reduced, favors the formation of a blood clot, one of the best possible checks to bleeding (see BLOOD, subhead *Clotting*). In case of a superficial wound, too, the cut usually closes quickly by the formation of a clot, but where no such clot is formed artificial methods must be used to arrest the flow of blood. In many cases this can be done by applying clean hot water or clean ice water to the wound, or by using direct pressure with a small piece of clean cotton. If the bleeding seems to be excessive, place one or both thumbs over the place from which the blood issues, and press firmly. This will check bleeding until a better remedy can be employed. Wounds about the face and neck can be treated especially well with thumb pressure.

In case of serious cuts it is best to tie a handkerchief or a strip of strong cloth loosely around the limb above the cut, and to slip a short stick into the loop; by twisting the stick sufficient pressure is secured to check the bleeding. This device has become known as a *tourniquet.* A tourniquet should not be left twisted for more than a half hour at a time, as it may cause congestion. In many cases a blood clot will have formed by the time the doctor has arrived. Tying the artery, that is, applying a ligature to each end of the cut vessel, is a reliable method of stopping hemorrhage.

Venous Bleeding. A hemorrhage from the veins is ordinarily less dangerous than arterial bleeding. It may be recognized by the dark red color of the blood and the even flow. Pressure over the wound is usually enough to stop the flow, but in extreme cases a ligature or other method may be necessary.

Other Forms. All forms of bleeding, if they continue, cause death. A serious hemorrhage demands the attention of a physician. Some persons, known among physicians as *bleeders,* have a natural tendency to bleed. The blood clots with difficulty, and a slight wound may cause death. The normal person, however, need not fear death from this cause. Internal hemorrhage should have immediate medical attention. Chiefly in young children hemorrhages sometimes occur through the skin. The only remedy in severe cases seems to be the transfusion of blood from some other person, and this is not a certain cure. Bleeding of the nose may sometimes be stopped by applying cold water to the back of the neck, forehead and bridge of the nose. W.A.E.

BLEEDING HEART, a hardy, late spring flower, rich with home associations and memories of old-fashioned gardens. The flowers are irregularly heart-shaped, deep rosy red, with the inner petals white. The structure of stem and flower is so delicate that it seems the little rosy blossoms are dripping toward the ground. With its wealth of fresh foliage and interesting flowers, the bleeding heart makes an attractive border. This plant was brought from Japan and introduced into England about 1850, after which it spread to all home gardens and is now a favorite everywhere. It is easily cultivated,

BLEEDING HEART

and if lifted in the fall and potted will grow very successfully with gentle heat.

BLENHEIM, *blen' im,* a village in Bavaria, on the Danube, twenty-three miles northwest of Augsburg, which gave its name to a celebrated battle of the War of the Spanish Succession, fought near the place on August 13, 1704. In this battle the allied forces of England and Germany, under the Duke of Marlborough and Prince Eugene, gained a decisive victory over the French and Bavarians. The residence of the dukes of Marlborough at Woodstock, Oxfordshire, known as Blenheim House, was erected at public expense as a token of gratitude to the English hero of Blenheim.

In Robert Southey's poem, *The Battle of Blenheim,* three stanzas of which are here given, an old man is supposed to be telling his two grandchildren the story of the battle:

"They say it was a shocking sight,
After the field was won,
For many thousand bodies here
Lay rotting in the sun;
But things like that, you know, must be,
After a famous victory.

"Great praise the Duke of Marlbro' won,
And our good Prince Eugene."
"Why, 'twas a very wicked thing!"
Said little Wilhelmine.
"Nay, nay, my little girl," quoth he,
"It was a famous victory.

"And everybody praised the Duke,
Who such a fight did win."
"But what good came of it, at last?"
Quoth little Peterkin.
"Why, that I cannot tell," said he,
"But 'twas a famous victory."

BLENNERHASSETT, *blen' er hass et,* HARMAN (1764-1831), an English emigrant to the United States who gave aid and encouragement to Aaron Burr (which see) in the latter's plan to establish an empire in the Southwest. He was born in Hampshire, England, and was educated in London and Dublin. In 1797 he emigrated to America, settling on an island in the Ohio River below the present city of Parkersburg, W. Va. It was there that Aaron Burr visited him and persuaded him to become a fellow-conspirator. When the plan failed Blennerhassett was arrested on the charge of treason. Though not convicted of this crime he lost the greater part of an ample fortune. He died on the island of Guernsey, in the English Channel.

The experiences of Blennerhassett form the basis of a novel by Charles Felton Pidgin, entitled *Blennerhassett, or Decrees of Fate.*

BLEWETT, *blew' et,* JEAN MCKISHNEY (1862-), a Canadian poet, sometimes called the sweetest of Canada's poets. She was born at Scotia, Ont., and was educated at the Saint Thomas Collegiate Institute. At seventeen she wrote *Out of the Depths,* a book of some merit but inferior to her later works. Her success in literature began with the publication of *Cabinet Articles,* a series of pen pictures which appeared in various magazines and newspapers. These sketches were unique and attracted wide attention. Among Mrs. Blewett's poems, which are her chief claim to distinction, are *Margaret, Spring* and *She Just Keeps House for Me.*

BLIGHT, *blite,* a term commonly given to the effects of disease upon plants, or the effect of any other conditions which cause them to wither or decay. Rust and smut are frequently known as blight among farmers. Sometimes the term is applied to the effects produced by insects, but its use by botanists is restricted to effects produced by fungi and germ diseases within the plant. Remedies and preventive measures are given in the article INSECTICIDES AND FUNGICIDES. See, also, FUNGI; RUST; SMUT, and references there suggested.

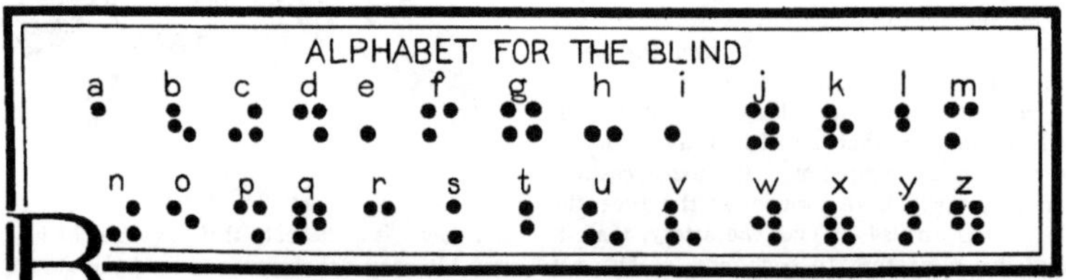

ALPHABET FOR THE BLIND

a b c d e f g h i j k l m

n o p q r s t u v w x y z

BLINDNESS, partial or total loss of the sense of sight, may be due to a wide variety of causes. In some cases it is brought about by imperfect development of the visual apparatus or defects of certain parts of the delicate mechanism of the eye. Such cases are said to be examples of *congenital* blindness. *Acquired* blindness is caused principally by accidents, operations, various eye diseases, certain diseases of the body, poor lighting, protracted eye-strain and misuse of certain poisons, notably wood alcohol.

Common Eye Diseases. *Conjunctivitis,* inflammation of the membrane that covers the lids and the eyeball, is an infectious disease that is very often transmitted by that relic of barbarism, the public towel. Attacks of this malady, if not promptly treated, are liable to result in the chronic form of the disease, in which the membrane becomes thickened and reddened. Victims of chronic conjunctivitis suffer from heaviness of the lids and eye-fatigue, and bright illumination dazzles them. Acute and chronic forms of the disease should have the attention of a competent oculist until they are cured.

Trachoma, or granulated lids, is another infection of the conjunctiva, but of a more serious character. Infection in a single family is easily transmitted by the common use of towels, wash basins, etc., and the utmost precaution should be taken to prevent the spread of the disease. Children should not even be allowed to play with toys that have been handled by victims. Trachoma is responsible for so large a number of cases of blindness, and is known to be so serious an ailment that the United States government prohibits immigrants suffering from it from entering the country. The disease tends to produce inflammation in the interior of the eye, ulcerations of the cornea and deformities of the lid. It usually requires several months of medical treatment to cure a case of trachoma.

Ophthalmia neonatorum, infection of the conjunctiva of new-born infants, is a terrible disease that usually makes its appearance on the second or third day after birth. It produces blindness by causing ulceration of the cornea. Characteristic symptoms are badly-swollen lids and profuse discharge of pus. When it is realized that one-third of the blindness in children is due to this disease one is not surprised to learn that societies have been formed for the prevention of infant blindness. If a new-born baby's eyes show the slightest signs of inflammation the nurse or doctor should cleanse the lids thoroughly with pure cotton and water, and then drop into each eye a few drops of a two per cent solution of silver nitrate. This treatment is a sure preventive.

Iritis is an inflammation of the iris and of the muscular body which controls the shape of the lens (the ciliary body). It is either the result of disease elsewhere in the body or of infection which has entered the body through some other channel than the eye. Rheumatism and syphilis are common causes, and a form known as *traumatic iritis* is the result of injury to the eye. Symptoms include pain in the eye, over the forehead or in the temple, contraction of the pupil, sensitiveness to light and redness of the eyeball. Neglect of this serious disease may cause shrinkage of the eyeball and complete loss of sight, but proper treatment taken in time usually effects a cure. In case of injury to one eye precautions should be taken to keep the infection from the uninjured organ. One of the most serious features of traumatic iritis is the tendency of the uninjured eye to become diseased through what oculists call "sympathy."

Glaucoma (hardening of the eyeball) results from an interference with the drainage of the fluids from the eyeball. Only a competent physician can correctly diagnose this ailment when it is inflammatory in character, for then its symptoms are like those of iritis. No one should attempt to prescribe his own treatment or to take drugs without advice. Belladonna, which is helpful in iritis, would cause blindness if used as a remedy for glaucoma. Another common eye disease is cataract, which causes the crystalline lens to become opaque. This ailment is fully described in this volume under the heading CATARACT. Other grave eye disorders are *retinitis,* or inflammation of the retina, *atrophy of the optic nerve* and tobacco and alcohol *amblyopia.* The latter, inflammation of the optic nerve between the eyeball and the brain, is the result of chronic poisoning from the use of tobacco or alcoholic beverages. The victim at first complains of a cloud before his eyes, which always appears in the direction in which he is looking. Interference with the ability to distinguish between colors is another early symptom. Abstinence from liquor and narcotics is absolutely essential to effect a cure.

There are many cases of defective vision, due to nearsightedness, astigmatism, etc., which can be corrected by wearing the right kind of glasses. Any evidence of eye trouble, whatever its nature, should be diagnosed without delay. Educational authorities are coming more and more to realize the dangers that result from neglect of the eyes, and it is customary in many towns and cities to have systematic physical examination of pupils by reliable physicians. Such precautions have been the means of saving or conserving the sight of many children.

Accidents. Eye injuries in industrial plants occur in largest numbers among workmen in the iron and steel industries. Small slivers of steel, struck off from large pieces under the blow of the hammer, are a fruitful source of trouble. If these tiny particles fly into the eye and are left there they may cause serious trouble, even loss of sight. Workmen thus injured should have the particle located by means of the X-ray, and see that it is immediately extracted. In many cases these tiny pieces of metal can be drawn out with a magnet. Those who have made a study of the subject believe that the wearing of protective glasses should be made obligatory upon all workmen exposed to injuries of this nature.

In blasting operations the premature explosion of charges is a common cause of injury to the eye. Such accidents may be due to too short fuses, also to delay in explosions. In the latter case the workman goes to the charge to find the source of the trouble and the blast meets him square in the face. Similar accidents occur in hunting and in Fourth of July celebrations, though the latter are each year becoming more infrequent. Forethought and ordinary care will prevent nearly all disasters of this nature.

Everyone is liable to eye injuries from dust particles in the air, cinders, etc. The great lesson for all is not to neglect the organ that gives the priceless possession of sight. Here, as in so many cases, "an ounce of prevention is worth a pound of cure." Prompt measures of relief should be the rule whether the eye is attacked by disease or is harmed. E.E.A.

Education of the Blind

Sad as is the condition of the blind to-day, it is happiness itself when compared with that of even a hundred years ago. Up to that time almost no attempt had been made to educate the blind. It was assumed that they must go through life dependent, unoccupied and restless; while they may have resented the fact of their blindness they could see no way to fight against its consequences.

Beginning of Movement for Education. But all people everywhere could not be content thus to take for granted the misfortune of the blind, and in 1646 an Italian writer published a book which brought to the front the question whether something could not be done for them. The interest aroused by this book led for a long time to nothing practical and to no really systematic attempt to give instruction, but in 1784 a Frenchman, Valentin Haüy, opened in Paris the first school for blind youth. He also invented books with raised letters, which could be read by the sense of touch, for investigations had long before revealed the fact that this sense was likely to be highly developed in the blind. England soon took up the work, though the English schools aimed at first to give manual rather than literary instruction, and this is still a prominent aim.

The movement spread rapidly over Europe during the early years of the nineteenth century, until now practically every country has its schools for the blind, nearly all of them residential. To-day there are over 150 such institutions on the continent of Europe, largely in France and Germany.

Growth in the United States. The United States schools have from the first been on a somewhat different basis from those of Europe, forming a part of the regular educational system provided by the states, though instruction is also given in special residential schools. The blind have come to look upon themselves not as a class apart, but as simply one division of the great school-going public. Some authorities believe that whenever possible the blind should be taught with the seeing; that it is advantageous both to the blind and the other children to be brought together at an early age. Accordingly, several cities conduct day school classes for their blind children.

Education has been for the most part along three lines: *literary,* including the branches taught in most grammar and secondary schools, and some not commonly taught there, such as typewriting and simple business; *musical,* including voice training and instruction on the piano or other instrument, with special training toward composition or teaching if the talent of the pupil seems to warrant this; and *industrial,* beginning with general manual training and including training in those occupations in which the blind can successfully engage, such as cane-seating, broom-making, basket-making, knitting, crocheting, housework, carpet-weaving and piano-tuning. Since the sense of hearing is unusually acute in the blind, they often become expert in this last operation; for years the pianos in the public schools of Boston have been kept in tune by graduates of the Perkins Institution, a school for the blind.

This, the oldest as well as the most famous institution of the kind in the United States, was established in Boston in 1829 and incorporated as the New England Asylum for the Blind. From the start it received help from the state, and the other New England states took advantage of the opportunities it offered by sending their blind to it at state expense, as they continue to do. In honor of a generous benefactor the institution was renamed the Perkins Institution and Massachusetts Asylum (later, School) for the Blind, and under the direction or Dr. Samuel G. Howe it rapidly attained the high rank which it still holds. The exhibitions which its pupils gave before different state legislatures led to the founding of like institutions in many parts of the country, and in 1910 there were forty-eight schools for the blind (all except five of them state schools). After the Perkins Institution the most noted of such schools is the Pennsylvania Institution at Philadelphia, which ranks among the foremost in the world. Only the residential schools in Boston, New York City, Philadelphia, Baltimore and Pittsburgh are incorporated.

A very special triumph for Dr. Howe and the Perkins Institution was the education of Laura Bridgman, a blind deaf-mute who was brought to the school in 1837, at the age of seven. Despite very general predictions of failure, Dr. Howe and his assistants succeeded in opening up the world to the child, hitherto so entirely shut in, and in making of her a busy, useful woman. Authorities the world over took an intense interest in her and in other cases, and in the published reports of the methods used in dealing with them.

A comparatively new development has been the attempt to educate the adult blind in their homes. In several states teachers are provided at government expense, who give instructions not only in reading but in certain of the crafts. All of the institutions gladly permit the use of the embossed books in their libraries by the adult blind. What is particularly needed is the creation of more state commissions to deal with the adult blind, those who have lost their sight when over school age, for about three-fourths of the blind population is adult.

In Canada. The training of the blind has received careful attention in the Dominion, there being five residential schools, two of which are supported by the government. The new developments outlined above have been studied and in some cases adopted, but some of the Canadian schools for the blind are organized on the European principle and are looked upon as charitable institutions rather than as a part of the educational system.

Books and Apparatus. Naturally, oral reading is made use of even more in institutions for the blind than in ordinary schools, but much must be done through the sense of touch, and special books and apparatus are therefore necessary. The first attempts at teaching the blind to read were made by means of raised letters, in form similar to the ordinary letters of the alphabet, which most of the pupils learned to recognize by running their fingers over them.

The effort to improve this method to the end that every pupil might be able really to read led to the devising of various arbitrary systems, of which by far the best known are the different modifications of groups of dots or "points." By all means the most widely used of these modifications is the *braille,* which takes as its basis six points, or dots, arranged in two vertical parallel columns, and shifts them into different combinations to stand for the letters. Those who have advocated the Roman-letter system based their arguments on the conviction that the education of the blind should be as nearly as possible like that of people with good eyesight, and that everything which tends to make differences between the two helps to deprive the blind of a normal view of life.

Adherents of a point system, on the other hand, assert as its chief advantage the fact that it enables the blind to write as well as read. By means of a grooved board, a perforated metal rule and a stiletto, any blind person may indeed learn to write, with a fair degree of rapidity, notes in words, figures or music, the writing being done from right to left, and the paper reversed for reading. Most schools for the blind now use only one of these methods, but a few still make use of both, combining their advantages.

Geography is taught by the aid of relief maps, in which the towns are indicated by metallic points, the boundaries by raised lines, and the mountains, valleys and rivers in the ordinary manner of relief maps. Dissected maps, cut along state and county boundary lines, such as those which are given to children as puzzles, are a help in teaching outlines. A person who has his sight can with difficulty appreciate the delicacy of the trained touch of the blind, and the ease with which they can master details through their fingers.

Natural history is taught by the use of life-size models and mounted specimens of animals and birds, and models of papier-mache also help in the study of anatomy. The models used in teaching botany must sometimes be much more than life size, that the different parts may be perceived by touch.

In most residential schools kindergartens for the blind have been established. The work is very successful, since all those occupations which do not call for blending of colors or for drawing may be taught. Froebel's *gifts* (see KINDERGARTEN), in use in all kindergarten work, require but little adaptation for use with sightless children.

At Louisville, Ky., and in connection with the Perkins Institution in Boston, there are special printing establishments which put out works for the blind, and thousands of volumes, both of school text-books and of the choicest works, are now available in embossed type. The schools, all of which have libraries, are generally anxious to loan these to any blind person, and the United States Postoffice Department carries them free. The Congressional Library at Washington has a special reading room in which books for the blind are kept, and elsewhere several of the large public libraries have embossed books for circulation. E.E.A.

Related Subjects. Those interested in this subject will find the following articles helpful:

Astigmatism	Eye, Subhead *Care of the Eye*
Bridgman, Laura	Keller, Helen A.
Cataract	

BLINDFISH, the name given to several kinds of fish inhabiting the waters of caves. Those in Mammoth Cave, Kentucky, especially, are objects of curiosity to tourists. Places for the eyes are indicated on the head, but the fish have no organs resembling eyes. The head and body, however, are covered with rows of small projecting bodies, or papillae, that are very sensitive to the touch. These fish never exceed five inches in length; the body is colorless and when held between the eye and a light shows the light through it dimly. Scientists have never definitely settled the question whether or not these fish formerly had eyes, but the general supposition is that the eyes have been lost through living for ages in waters from which light was excluded. See MAMMOTH CAVE.

BLINDWORM, sometimes called BLINDSNAKE, because of the very small size of their eyes. There are three principal species, but actual blindness is not present in any of them. The most important is a group of serpent-like lizards, of which the common blindworm, or slowworm, gives the name to the type. They are found west of the Mississippi River. Another is a family of true serpents known as blindsnakes, inhabiting warm climates. These are most nearly blind, for their eyes are very small and weak and occasionally almost invisible. The third group comprises a family of degenerate amphibians (see AMPHIBIAN) which inhabit South America and Mexico principally.

BLISS, TASKER HOWARD (1853-), an American army officer of varied experience who in 1917 was sent to Paris as the military representative of the United States on the allied war board. He was born at Lewisburg, Pa., finished half of the course of study at Buchnell University and then was appointed a cadet at West Point, from which he was graduated in 1875. He served three years as professor of military science in the Naval War College, and in 1888 went to Spain as military attache of the American legation. During the Spanish-American War he served in the Porto Rican campaign, and at its close was appointed collector of customs at the port of Havana, during American occupation.

By this time he had risen to the rank of brigadier-general. Upon release from Havana he became a member of Army War College Board, then commandant of the Army War College. In 1903 he was sent to the Philippine Islands as a department commander. When he returned home in 1909 he became president of the Army War College and assistant chief of staff of the army. During the early part of the Mexican uprising he commanded a brigade on the border, and in 1915 returned to Washington as assistant chief of staff. In November, 1915, he was raised to the rank of major-general. President Wilson assigned him to the Paris post as soon as American soldiers reached France in large numbers.

BLIZZARD, *bliz' ard,* a severe winter storm characterized by violent, cold wind filled with tiny particles of ice and snow. These storms, known by different names in different northern countries, are caused in the great polar regions by high barometric pressure forcing out currents of cold air. They are usually preceded by a short period of warm weather, and Weather Bureau officials are able to forecast their course with great accuracy. They are common in Central and Eastern Canada, in the northern part of the Mississippi basin in the United States, and in Russia and Siberia. Because of the cold, exhausting wind and blinding snow, human beings and live stock frequently lose their lives in these storms. Numerous instances are recorded of people being lost and frozen to death as a result of trying to find their way between house and barn in a blizzard.

BLOCKADE, *block aid'*, the patrolling by warships of coasts belonging to an enemy, to prevent the passage of forbidden vessels, in or out. To keep all vessels away from an enemy's country and thus to cripple the foe by shutting out arms, munitions and even food is held to be a legitimate act in warfare. Most nations hold that notice must be given of any blockade, that neutral vessels within the blockaded district must be given a reasonable time to leave, that ships attempting to *run* the blockade (pass through it) may be captured, and that their cargoes are also liable to seizure unless the owners prove themselves not responsible for the attempt.

Paper Blockade. In 1806 Napoleon declared a blockade against the British Isles (see CONTINENTAL SYSTEM) and England retaliated by a similar measure against France. Neither country had sufficient ships really to enforce such extensive blockades. They existed only on paper, and the proclamations became merely excuses for capturing ships—a mild form of piracy. A "paper blockade" has therefore come to be known as a blockade declared only by publication, or merely a warning to neutral vessels to remain away from a forbidden zone. After the Crimean War the representatives of the powers gathered in Paris declared that "a blockade to be binding must be effective," that is, the blockaders must be able to endanger all ships that may attempt to pass. The German submarine blockade of the British Isles in the War of the Nations was considered a "paper blockade" by other countries.

Pacific Blockade. Since 1814 there have been a number of instances when one country has exerted force against another without resorting to war, by means of a *pacific*, or *peaceful, blockade*, to secure redress of grievances. Such a measure cannot be enforced against neutral vessels, but only against those of the nation blockaded, and ships cannot be confiscated, but merely held until the conclusion of the blockade. The English-German-Italian blockade of Chinese waters in 1902 was at first a "pacific" blockade. E.D.F.

BLOCK AND TACKLE, a mechanical appliance consisting of a combination of pulleys and ropes. It is a *machine*, for it is designed to perform work. *Block* refers to the casing for the pulleys, *tackle*, to the ropes. A single block contains one pulley, a double block, two

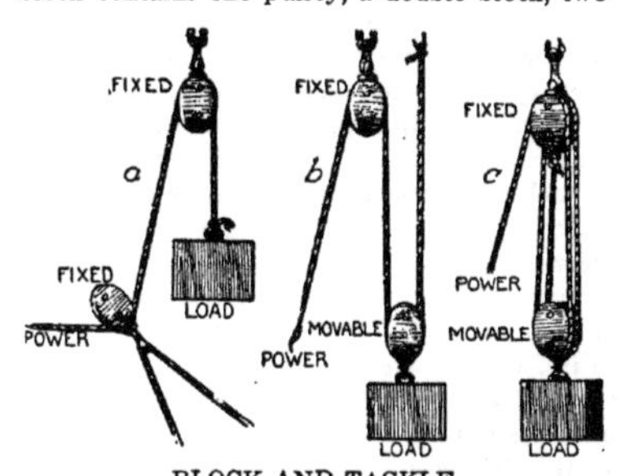

BLOCK AND TACKLE
Fig. 1. (*a*) No mechanical advantage; (*b*) mechanical advantage of *two;* (*c*) mechanical advantage of *four.*

pulleys, and so on. Each block usually has a hook with which to fasten it to its support or to the object to be moved.

In the article PULLEY it is shown that a simple movable pulley, such as is contained in a *single* block and tackle, has a *mechanical advantage of two;* that is, with its help a force will move practically two times the weight that

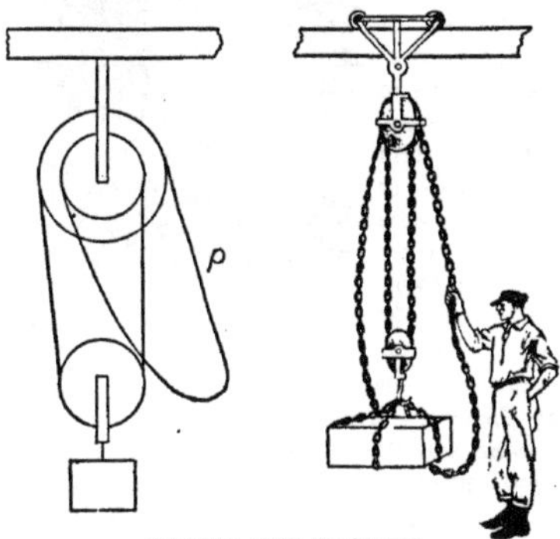

BLOCK AND TACKLE
Fig. 2. The endless chain.

it can move without it. The mechanical ad vantage of a double movable block is *four,* for, as shown in Fig. 1 c, there are four ropes, each of which bears one-fourth the weight; therefore any pull on the end in excess of one-fourth the weight will lift the object. Similarly, the advantage of a triple movable block is *six.*

Endless Chain. The apparatus shown in Fig. 2 is technically known as the differential pulley. It can be made to possess a very large mechanical advantage. The lower block is single, but the upper block has two pulleys of different sizes, fastened so that neither one can turn without the other. When the chain is pulled at *p* it is wound on the larger pulley faster than it unwinds from the smaller. If the circumference of the larger pulley is thirty-six inches and that of the smaller is thirty-two inches, when the wheels have been turned once the weight will have been raised the difference, or four inches. Meanwhile, the hand pulling at the rope moves thirty-six inches, nine times as far, and has exerted a force only one-ninth that of the weight. The mechanical advantage in this case is thus seen to be *nine*, but increases as the difference in the size of the pulleys decreases. C.H.H.

BLOCK'HOUSE, a military fortification, formerly erected on frontiers and in pioneer settlements, usually to serve as a place of last resort in case of attack by hostile forces. Such houses were generally built of heavy logs or blocks of hewn timber, banked with earth. They were fitted with loopholes for musketry at the sides and in overhanging floors of upper stories. Built in the form of a square or cross, they were made large enough for twenty-five to 100 men. Such houses saved many lives in the early wars with Indians. Even later, in the Spanish-American and Boer wars they were employed to good effect. Against such modern artillery as was developed in the War of the Nations, beginning in 1914, blockhouses were useless, and fighting from trenches became the chief defense of all armies.

BLOCKHOUSE

BLÖEMFONTEIN, *bloom' fon tane,* a Dutch word meaning *fountain of flowers,* is the name of the capital of Orange Free State in the Union of South Africa, founded in 1846, about 600 miles northeast of Cape Town. It is well located on a plateau 5,000 feet above sea level, and has a healthful climate. A small stream, named the Bloemspruit, flows through the town and supplies a portion of the city's water, the remainder being obtained from the Modder River and conveyed a distance of twenty-four miles to the city.

It is on the main line of the Cape-to-Cairo Railway, and is in direct communication with all the principal South African towns. In educational facilities Blöemfontein is far more advanced than most South African towns. It has good schools and a university, the buildings of which were erected in 1906 at a cost of $625,000. In the South African War Blöemfontein was occupied by Lord Roberts without opposition. In 1910 the city was chosen as the capital of the Orange Free State Province, then incorporated into the Union of South Africa, and became the seat of the Supreme Court of the Union. Population in 1911, 26,925, of whom 14,720 were white and 12,205 colored, the latter mostly of the Bechuana and Basuto tribes.

BLONDEL, *bloN del',* a French minstrel who figures in a romantic tale of the twelfth century. He was the trusted attendant of Richard the Lion-hearted, king of England, as well as his instructor in music. During Richard's journey homeward from the Crusades he was captured by the Duke of Austria and confined in a castle on the Danube. Blondel, as the story goes, wandered all over Germany in search of his royal master. Hearing that a distinguished captive lay within the Castle of Dürrenstein, he stood before that fortress and began to sing a song which he and the king had written together. Joyfully the minstrel heard the loved voice of his master take up the second stanza, and then he hastened home to England to secure the king's ransom. Blondel is mentioned in Sir Walter Scott's *Talisman.* See RICHARD I.

BLONDIN, *bloN daN',* (1824-1897), the assumed name of a famous French tight-rope walker, JEAN FRANCOIS GRAVELET, whose remarkable feats in his dangerous profession included the crossing of Niagara Falls on a tight rope. After a brilliant career in France he sailed to America with the Ravel family of acrobats, and while visiting Niagara conceived the idea of making the thrilling trip above the seething waters. Having bridged the distance with a rope 1,100 feet long and 160 feet above the water, he made the trip on August 17, 1859, in the presence of 50,000 spectators. Not once but many times did he walk across; some trips were made blindfolded and more than once he carried a man on his back over the roaring cataract.

BLONDIN, Pierre Edouard (1874-), a Canadian legislator and political leader, member of the House of Commons since 1908, and since 1914 a member of the Conservative Ministry headed by Sir Robert L. Borden. Blondin was born at Saint François du Lac, Que., and attended Nicolet College and Laval University. After holding several local offices he entered Parliament in 1908. In 1914 he was appointed Minister of Inland Revenue, a position which he exchanged in the next year for that of Secretary of State and Minister of Mines.

BLOOD. Nature has provided for the distribution of dissolved food materials throughout the human body by a wonderful system of tubes, called arteries and veins, through which circulates a life-giving substance we call *blood.* This collection of tubes may be compared to the water-works system of a great city, by which the water is pumped into pipe-lines and carried to the different homes, office buildings and factories.

Composition. Blood itself is a liquid like water, for it consists of a substance called *plasma,* which is about nine-tenths water. In this are floating millions of minute bodies called *corpuscles.* Blood also contains another substance, a ferment which causes it to clot when drawn from the body or exposed to the air at the surface of a wound (see subhead *Clotting,* below).

Corpuscles. There are two kinds of blood corpuscles, red and white, and because the former are by far the more numerous, they give the blood its color. In a drop of blood the size of a pinhead there are about 5,000,000 red corpuscles and 6,000 white ones. The red are shaped like a coin, except that they are thinner

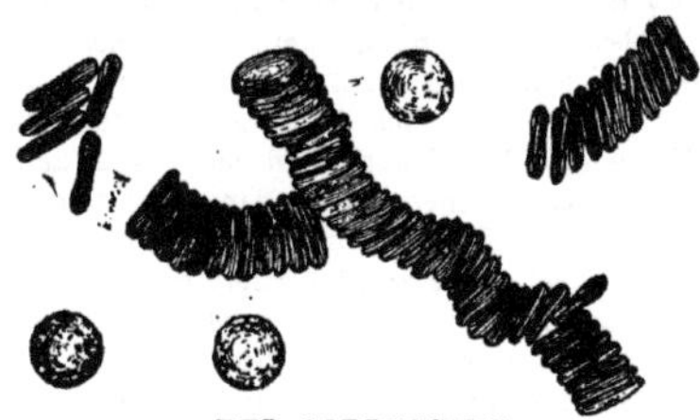

RED CORPUSCLES
Magnified about one thousand diameters.

in the center than at the edges, and they are only 1/3200 of an inch in diameter. In them is an important substance containing iron, which absorbs oxygen from the air in the lungs and carries it to the tissues of the body. This substance is bright red when charged with oxygen, but turns to a dark red or purple when the oxygen has been given off. It is called *hemoglobin.*

The white corpuscles, though fewer in number than the red, are a little larger. They are shaped like tiny balls and one of their functions is to protect the body against certain diseases. To illustrate: If a wound is made in such a manner that bacteria are left on the torn surfaces, the white corpuscles rapidly multiply on these surfaces and tend to destroy the bacteria. This they do by absorbing them and carrying them out of the system, or destroying them.

Clotting. When blood is drawn from the body or exposed to the air on the surface of a wound, minute threads soon extend through it in all directions, and these hold the corpuscles in their meshes, forming a solid mass, or clot. The clotting is caused by *fibrinogen,* which manufactures the threads. It seems to be a wise provision of nature to stop the flow of blood from a wound. A clot will form more quickly over a wound with rough, ragged edges. Clean sand placed on a wound assists in the formation of clot, and may be the means of saving life.

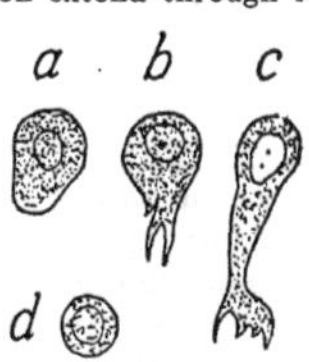

WHITE CORPUSCLES
Showing movements, similar to those of the amoeba (which see). (*a*) Beginning of movement; (*b*) formation of foot-like projection; (*c*) nucleus changing form; (*d*) dead corpuscle.

Quantity of Blood. The quantity of blood in the body is about one-thirteenth the weight of the body. A man of average weight has from twelve to fifteen pounds of blood; allowing a pint to the pound, this amounts to about six quarts. As the normal quantity in any human body is almost perfectly proportioned to its needs, the loss of only a little through bleeding or disease induces weakness and dizziness. If not replaced by natural processes serious results may follow (see Anaemia).

Functions. One of the chief functions or uses of the blood is to carry the nutriment received from food to all parts of the body. Those portions of food absorbed through the membrane of the intestines are conveyed by the thoracic duct to the large vein. Those portions that are absorbed through the liver are conveyed directly from that organ into the blood current. The function of carrying oxygen to all parts of the body has been explained under *Corpuscles*. The third function is to gather up the waste throughout the body and convey it to the kidneys and other organs of excretion, by which it is expelled from the system.

The Blood in Circulation. The various functions referred to in the preceding paragraph could not be carried on by the blood had there been no provision made for its circulation through the body. How it passes through the trunk-lines and branches of the marvelous circulatory system—the network of veins, arteries and capillaries—is fully described under the heading CIRCULATION OF THE BLOOD. A drop of blood traveling from the heart through the body and back again makes the complete circuit in about half a minute.

Highly-magnified crystals of oxy-hemoglobin from human blood.

Blood Poisoning. Blood poisoning is caused by the presence in the blood of germs, or bacteria (which see). Blood poisoning sometimes results as a complication of or as secondary to such diseases as appendicitis or typhoid fever. It is occasionally a result of infection by instruments on which bacteria have found lodgment. Even a scratch with an infected needle or knife may cause blood poisoning that will terminate fatally. Any object that has been brought in contact with decaying animal matter or with diseased organs is especially dangerous. Surgeons have been known to contract blood poisoning while performing operations, and to die from the effects. Formerly the loss of life due to blood poisoning contracted during operations was very great, but modern methods of sterilization have worked a revolution in the practice of surgery. Now the instruments, the hands of doctors and nurses and the body of the patient are thoroughly scrubbed and treated with antiseptics, and the danger from infection is reduced to a minimum. Cuts and bruises that are seemingly trivial may be a source of danger if dirt finds lodgment in them, and it is a wise plan to pour a harmless antiseptic, such as iodine, over the wounds to prevent possible infection. Blood poisoning, formerly exceedingly difficult to combat, is now successfully fought with an antitoxin.

Transfusion of Blood. This is an operation for the relief of those suffering from certain diseases or from the effects of hemorrhage. Transfusion is the injection into one person of blood taken from another. If the transference is from vein to vein the operation is known as *direct* or *immediate* transfusion; if the blood is first freed from fibrin and injected from a receptacle the operation is called *indirect* transfusion. In recent years physicians have shown a tendency to substitute for blood-injection that of a hot salt-solution having a temperature of about 115° F. The solution, containing six drams of sterilized salt to one gallon of sterilized water, is injected into the veins of the patient or into the tissue beneath the skin. This operation is considered superior to the older method in that it is more efficient, less dangerous and less difficult to perform. W.A.E.

Related Subjects. The above article cannot be fully understood without reference to some or all of the topics that follow:

Anaemia	Circulation
Antitoxin	Heart
Aorta	Lungs
Arteries	Veins
Capillaries	

BLOOD, AVENGER OF. In most primitive societies when a man was intentionally killed or seriously injured by another his nearest relative felt it to be not only his right but his solemn duty to take vengeance. This next of kin was known as the *avenger of blood*. The act of vengeance, even if it went so far as murder, was not looked upon as a crime, and failure to execute it brought disgrace not only upon the avenger of blood himself, but upon the dead as well. As society developed, these "blood feuds," as they were called, became subject to stricter regulation, and among many peoples cities of refuge were established, to which a manslayer might flee to be safe from the avenger until his case was investigated. The law of Moses appointed such cities, and the Greeks had them in abundance (see CITIES OF REFUGE). Still later it was provided that the criminal might gain his safety by paying a fine known as *blood money*, which the avenger was compelled to accept (see BLOOD MONEY).

The *vendetta* of Corsica and the *feud* of Kentucky, which play so large a part in literature dealing with those sections, are but survivals of the old avenger of blood idea.

BLOODHOUND, a breed of hound distinguished by a remarkably keen scent. All hounds are primarily hunting dogs. The bloodhound, which of all the many breeds is probably nearest the original, is a powerful dog about two feet high at the shoulder, having a short-haired black-and-tan coat, large, but not broad head, with wrinkled skin, deep-set hazel eyes, a deep, square muzzle and long silky ears. These dogs were once trained to hunt large game, such as the boar, bear and stag, and also to hunt man, but now are chiefly valuable for tracking fleeing or missing persons. In the days of slavery in America they were used to hunt fugitive slaves. Their name came from their ability to track blood, but so acute

BLOODHOUND

is their sense of smell that they are able to pick out a trail that has been crossed and recrossed by many others, and are baffled only by running water.

BLOOD MONEY. In the Middle Ages and well into the more modern period, this name was applied to the money paid for bloodshed. It might be either the compensation paid by a manslayer to the nearest relatives of the victim (see BLOOD, AVENGER OF), to secure himself and his kin from vengeance, or the money paid as a reward for bringing about the death of another, directly or through evidence. It was once common among the Scandinavian and Teutonic peoples, who called this money payment *wergild*, and the custom is still practiced in Arabia. The price varied with the nature of the crime and the rank of the victim. Certain crimes, such as the slaying of a sleeping person, could not be compensated by a money payment; such criminals were declared outlaws and could be slain with impunity. The term is now applied to the reward or bribe paid for giving up a criminal to justice.

BLOODROOT, sometimes called INDIAN PAINT or RED PUCCOON, one of the earliest spring flowers found in Canada and the United States. Its root and sap are rich orange-red, hence the

BLOODROOT

name. The leaves are heart-shaped, deeply-lobed, and, folded around the flower stalk, they come from the ground singly. Each stalk bears one dainty white or rose-tinted blossom. The plant is rich in tannin and has been used as an astringent. The juice, at one time prized by the Indians as war paint, is now used by them for dyeing baskets, quills and moose hair. It is a member of the poppy family.

BLOODSTONE. See HELIOTROPE.

BLOOMFIELD, N. J., a city in Essex County, adjoining Newark on the northwest, twelve miles from New York City and a residential suburb of both cities. It is on the Morris Canal and on the Erie and the Delaware, Lackawanna & Western railroads. Electric interurban lines extend to a number of neighboring cities. The population, which in 1910 was 15,070, was 17,306 by the state census of 1915. The area is five and a half square miles.

Bloomfield is the seat of the German Theological Seminary of Newark (Presbyterian). Noteworthy features of the city include a large park which was a military training ground during the War of Independence, Jarvie Memorial Library, Knox Hall and Job Haines Home for Aged People. The principal manufactured products are railroad brake shoes, paper, pins, rubber goods, woolen cloth, silk, electric elevators, lamps, brushes, plumbers' supplies, cod-liver oil, strawboard and hats.

Bloomfield was settled about 1675 and was a part of Newark until 1812, when it was incorporated as a separate township. It received its present name in 1796 in honor of General Joseph Bloomfield, an officer in the War of Independence and afterward governor of New Jersey.

BLOOMFIELD-ZEISLER, FANNIE. See ZEISLER, FANNIE BLOOMFIELD.

BLOOMINGTON, ILL., an important industrial city and the county seat of McLean County, one of the richest agricultural sections of the United States. It is situated north of the geographical center of the state, on the Illinois Central, Chicago & Alton, Lake Erie & Western and the Cleveland, Cincinnati, Chicago & Saint Louis railways. Electric lines connect with Peoria, Decatur, Springfield and Saint Louis. Chicago is 126 miles northeast, Springfield, the capital, is fifty-nine miles southwest and Peoria is forty-five miles northwest. The city was founded in 1824 and was incorporated as a city in 1850. Its population, mostly American, increased from 25,768 in 1910 to 26,850 in 1914. The area is over four square miles.

Bloomington is styled the *Evergreen City,* as it presents the appearance of an immense park of shrubbery and trees, dotted with attractive homes. The largest nurseries of the state, covering more than 1,000 acres, are located here. Hundreds of ornamental lights illumine the well-paved streets, and numerous small parks and playgrounds are as gardens within a garden. The Illinois State Normal University, two miles north, at Normal, Ill., an institution for the education of teachers; the Illinois Wesleyan University; a Roman Catholic academy; a business college; a high school, which was erected at a cost of $400,000, and a library with 30,000 volumes, offer educational advantages; besides, there are colleges of music and of oratory. Among the notable structures are a $125,000 Y. M. C. A. building, a $75,000 Federal building, erected in 1895, and a marble courthouse. The city also contains two hospitals and about forty churches.

The car works and repair shops of the Chicago & Alton Railroad are located here, and the pork-packing and canning industries are important. Most prominent among the manufactures are farm implements, flour, stoves, bricks, tiles, silos and portable elevators, and there are several tanneries, cigar factories and a glass plant. Extensive limestone quarries and coal mines are located in the vicinity; one coal shaft lifts several hundred tons a day. The city is also an important horse market, the sale of horses averaging $300,000 yearly.

Since 1915 Bloomington has had the commission form of government, with a mayor and four elective officers. In 1900 the city sustained a loss of several millions of dollars by fire in the business district. Bloomington has been the home of such distinguished men as former Vice-President Adlai E. Stevenson, David Davis of the Supreme Court, and former governors Joseph W. Fifer and John M. Hamilton. J.H.H.

BLOOMINGTON, IND., popularly styled *The University City,* because it is the seat of the Indiana State University. Its population, chiefly American, increased from 8,833 in 1910 to 9,850 in 1914, exclusive of students. The city, the county seat of Monroe County, is situated southwest of the geographical center of the state, about sixty miles southwest of Indianapolis, 103 miles northwest of Louisville and 222 miles southeast of Chicago. The Chicago, Indianapolis & Louisville Railway (Monon Route), constructed through the city in 1854, and the Illinois Central (Indianapolis Southern), built in 1906, provide railway accommodations. The city was founded in 1818 and was incorporated in 1876. Its area is over two square miles.

Bloomington is located on a ridge between the east and west forks of the White River, in a vast limestone quarry district. Besides quarrying, the city is actively engaged in the manufacture of furniture; about 1,000 people are employed in this industry and the annual output averages $1,500,000. In addition to the university, with its library of 5,000 volumes, the city has two high schools and a Carnegie Library. The most conspicuous structures are the university buildings, representing a cost of about $1,100,000; the post office, erected in 1912 at an outlay of $70,000, and a $150,000 Methodist church, built in 1913. The Indiana University Park, with an area of ten acres, is the principal recreation spot. R.G.B.

BLOWFLY, a large blue and black fly, about twice the size of the ordinary house fly, which

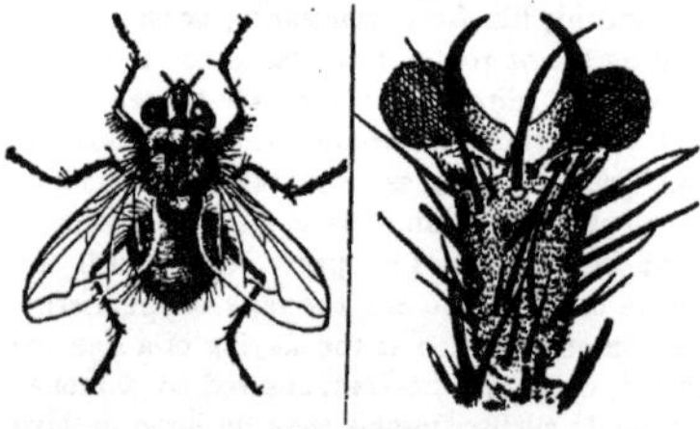

BLOWFLY

The figure at the left is about twice actual size. At the right is a highly-magnified foot of the insect.

lays its eggs upon meat or dead animals. These eggs are called *fly blows,* and hatch very quickly into maggots, which destroy the meat. In some places these flies and maggots play an important and useful part in quickly disposing of carrion, but in the house they are a menace, carrying germs of disease and spreading infection. Great care should be taken to keep all food beyond their reach, and garbage cans, their favorite breeding places, should be securely covered. The blowfly breeds so quickly that the offspring of one fly may number many thousands in a few days.

BLOWING MACHINES, a name given to various devices which are used to produce, supply and direct a strong artificial current of air under pressure. One of the oldest, simplest and most common of blowing machines is the bellows (which see), which has been used by blacksmiths and workers of metals since time immemorial. Blowing machines are of several kinds, according to the uses to which they are put. Some are used to produce a forced draught in order to assist the burning of fire in boilers and furnaces; others for ventilating purposes, that is, for extracting the foul air from buildings, mines or ships and forcing in pure air; while other big special machines are employed for supplying air to blast furnaces and in the Bessemer process for manufacturing steel. The chief varieties of blowing machines are known as *disk* blowers, *fan* blowers and *jet* blowers.

Disk Blower. This consists of an axle which has attached to it several blades so arranged together as to form a rimless wheel, which is enclosed and moves within a cylindrical casing open at both ends. The ordinary electric fan used in offices and homes, which is a form of disk blower, gives a good idea of the form of this wheel and the way it rotates. As this machine is used for ventilating purposes, it is usually set in the wall of the building, with one opening of the cylindrical casing towards the building and the other to the outside air. The axle and blades are rotated by means of an electric motor, and according to the way the machine is set it either sucks the foul air from the building or forces fresh air into it.

Fan Blower. The fan blower is the most common of modern blowing machines. It is like the disk blower, except that the axle of the fan blowers is provided with radial spokes, as in a rimless wheel, to the ends of which are fastened blades arranged parallel with the axis. This wheel is situated inside of a circular casing of steel or cast iron, within which it is made to revolve rapidly. When this wheel revolves it causes the air within the casing to revolve also. In this way a centrifugal action is set up by which there is a diminution of pressure at the center of the machine and an increase of pressure against the sides of the casing. The air is then sucked in through circular orifices at the central side of the casing and is forced out through an outlet in the side called the delivery tube.

Jet Blower. Another kind of blower in common use is the jet blower, in which a jet of steam is used to produce a current of air. The principle upon which it is based is the following: When a jet of steam is allowed to escape through a small opening or through a tube of small diameter which is inserted into a larger tube open at both ends, it creates a current of air. The air is pushed in the direction of the escaping jet and a fresh supply is drawn in through the other opening, so that a continuous stream of air passes along the tube. The exhaust nozzles used in the smokestacks of locomotives and fire engines are examples of jet blowers. O.B.

BLOWPIPE, a small tapering tube used to direct a current of air upon the flame of a lamp, candle or gas jet, forcing it in any desired direction and causing it to burn very rapidly, thus intensifying its heat. In its simplest form the blowpipe consists of a funnel-shaped tube of brass or glass, usually seven inches long and one-half inch in diameter at the larger end and tapering so as to have a very small opening at the lower end. Within about two inches of the smaller end the pipe is bent nearly to a right angle, so that the fine current of air may be directed sidewise to the operator.

The current of air may be produced by a bellows instead of the breath. The blowpipe is used by jewelers for soldering, and by glassblowers in making thermometers. See BLOWING MACHINES.

BLUBBER, *blub' er,* the fatty strip which lies just beneath the skin of whales and other warm-blooded sea animals, furnishing a protection against the cold. When refined it yields oils for lubrication, fuel and soapmaking. Whale oil was extensively used in lamps until petroleum was discovered, but it is now employed in such a manner only by the Eskimos. Blubber is an important part of Eskimo diet, and is also eaten by the Ainos and other inhabitants of northern Japanese territory. Whalers remove blubber from the body of the animal in large strips by means of spades and shovels, and secure two or three tons from every catch.

BLÜCHER, *blü'Kur,* GERHARD LEBRECHT VON, Prince of Wahlstadt (1742-1819), the famous Prussian general whose timely arrival with reinforcements saved the day for the allies at the Battle of Waterloo and crushed Napoleon Bonaparte. He began his military career in the Swedish army at the age of sixteen, but soon entered the Prussian service and fought under Frederick the Great in the Seven Years' War, and later in the wars of the allied nations, against the French Revolutionists. Throughout the wars with Napoleon he was one of the most bitter and untiring foes of that conqueror, and even after the Peace of Tilsit, when Prussia was made a dependent state of Napoleon's empire, he was never shaken in his belief that his country could be liberated.

When the Prussians renewed the war against the French in 1813, Blücher was given an important command and shared in the glory of Napoleon's defeat at Leipzig. After Napoleon's return from Elba, Blücher was placed at the head of the Prussian troops and led his army to Belgium. On June 16, 1815, Napoleon defeated the Prussians at Ligny, and then turned swiftly to attack the English, concentrated near Waterloo under the Duke of Wellington.

On June 18 the great Battle of Waterloo was fought. Throughout the day the French hurled themselves in vain against the squares of British troops, while Wellington, knowing that his soldiers were growing tired, anxiously waited for reinforcements and prayed for either "Blücher or night!" At last, as the French were making a final desperate charge, the Prussians swept upon the field, and the Battle of Waterloo was won.

That he might suitably reward Blücher for his services, Frederick William III of Prussia created in his honor the Order of the Iron Cross, and in 1819 a colossal bronze statue of the great general was erected in his native town of Rostock. See IRON CROSS; WATERLOO, BATTLE OF.

BLUE, one of the primary colors. It appears in nature most permanently and brilliantly in stones, such as the turquoise, sapphire, lapis lazuli and labradorite, and in the clear sky and sea, in flowers and in the feathers of fowls, the peacock and blue bird. There are many blue paints and dyes prepared from minerals or from plants. Only a few of them are permanent colors; of these ultramarine, prepared from the mineral lapis lazuli, and cobalt blue, made by mixing aluminum and cobalt salt, are the best known. Indigo is the most common vegetable source of blue.

Blue is used in painting and printing and dyeing. It is the second of the three colors used in the three-color process of reproducing colored pictures. Mixed with yellow it produces green; with red, purple. See COLOR; PHOTOGRAPHY, subhead *Color Photography*.

As an Emblem. Blue has been adopted at various times as a badge or symbol. The Scotch Covenanters of the seventeenth century chose it as their emblem in opposition to the royal red, and from this circumstance came the expression *true blue,* originally applied to a loyal Presbyterian. The winner of the Derby in the English races is said to carry off the *blue ribbon,* this expression having originated in the use of a blue ribbon as the distinguishing badge of the Knights of the Garter, an order founded by Edward III. The phrase *boys in blue,* applied to soldiers on the Union side in the War of Secession, to distinguish them from the Confederates, the *boys in gray,* has become current. There is a well-known Memorial Day poem by Francis Miles Finch which has as its refrain—

> Under the sod and the dew,
> Waiting the Judgment Day;
> Under the one the blue;
> Under the other, the gray.

BLUEBEARD, *blu'beerd,* the bloodthirsty hero of a famous legend, who murdered in succession six wives and was himself killed by the brothers of the seventh. The story first appeared in a book of fairy tales written in the seventeenth century by a French author, Charles Perrault. The chief character is supposed to have been suggested by an historic personage of the fifteenth century, Gilles de Laval, who was remembered chiefly on account of his inhuman acts of cruelty.

In Perrault's story, Bluebeard, so called because of the tint of his beard, entrusted to his wife Fatima the keys of his castle, warning her not to open the door of a certain room while he should be away. Lured by her curiosity, Fatima opened the forbidden door and discovered a chamber in which lay the bodies of the six wives who had preceded her. On his return home Bluebeard learned of her act of disobedience by the blood on the key, and only the timely arrival of her brothers, who speedily put an end to the brutal husband, saved her from sharing the fate of her predecessors. The essential details of this story are found in the folk lore of various peoples. F.J.C.

BLUE'BELL, the popular name for several blue, bell-shaped flowers—the bellflower, hairbell, or harebell, and others. When the name is heard one thinks first of the lovely Scotland—

> Hang-head bluebell,
> Bending like Moses' sister over Moses,
> Full of a secret that thou dar'st not tell!

BLUEBELL

This little wonder of Nature melts its way through snow and ice by the heat it generates. In England, a delicate hyacinth-like flower is known as the *heather-bell,* or *harebell.* . Hanging downward as it does, the bluebell forms a roof to shield its pollen from the rains, and keeps out unwelcome insect-visitors. But the bees cling to the pistils and dip far into the bells for their nectar. Bluebells make hardy flower borders and are easy to cultivate.

BLUEBERRY. See HUCKLEBERRY, subhead *The Blueberry.*

BLUEBIRD, a beautiful wild bird of the thrush family, whose soft, pretty warble is one of the earliest signs of approaching spring, as Bryant tells us in his *Yellow Violet:*

> When beechen buds begin to swell,
> And woods the bluebird's warble know,
> The yellow violet's modest bell
> Peeps from the last year's leaves below.

The bluebird is a North American songster whose range is from Nova Scotia and Manitoba to Bermuda and the West Indies. The birds, which spend the summer in Canada and Northern New England, visit Virginia and the Carolinas in the winter, while those of the Middle States migrate farther south. From March to November in the Middle and Northern states they are among the most interesting and lovable of the wild birds that nest in the house yard, orchard or garden. The male, smartly dressed in a bright blue coat with cinnamon-red trimmings, is a most devoted husband, though he permits his active little mate to do the bulk of the work of nest building, preferring to show his admiration by outbursts of song in her honor and by bringing her choice insects for her bill of fare.

THE BLUEBIRD

The nest of the bluebird, which is placed in a hollow tree, a deserted woodpecker's hole or other crevice, or in a bird house, if one is at hand, is not an elaborate structure, being thinly lined with grass or feathers. The pale blue or nearly white eggs number from four to six, and two or three broods of little ones are raised each season. The baby birds are black at first, the blue feathers appearing by the time the young wings are ready for the first flight. The birds are seven inches long when full grown. They feed upon spiders, grasshoppers, beetles and other insects, and, as winter approaches, on various wild berries of the autumn woods. The bluebird is the friend of the farmer, and is welcomed as a destroyer of harmful insects no less than for its cheery singing and agreeable manners. See BIRD; THRUSH.

BLUE BOOKS, the name given to official reports, papers and documents printed by the British government for the information of Parliament, which are so called because they are bound in dark blue paper covers. They include bills presented to Parliament and passed by that body; reports and papers called for by members, or authorized by the government, and reports of committees. In the United States the name is officially applied to the published lists of people in the employ of the government and to the manual containing the navy rules and regulations.

In May, 1915, the United States government published what was known as the *White Book,* giving the text of the important diplomatic correspondence with the governments of Europe then engaged in the War of the Nations. A second installment appeared in October of the same year. The White Book is a valuable public record of the position taken by the United States in that struggle, and it also reveals the difficulties that confront a nation attempting to preserve a strictly neutral attitude in a world-wide war.

The nations involved in the great European conflict also issued from time to time statements of their position and the diplomatic correspondence pertaining to the war. These volumes were named according to the color of their binding, and included the Belgium *Gray Book*, the British *White Paper*, the German *White Book*, the French *Green Book* and the Russian *Orange Book*.

BLUEFIELD, W. Va., the distributing point for the large coal region in which it is located. It is in Mercer County, close to the southern border of the state, 100 miles west of Roanoke, Va. Transportation is provided by the Norfolk & Western Railroad. The place was settled in 1888 and was incorporated in 1893; the commission form of government is in operation. In 1914 the population was estimated at 13,974, an increase of 2,786 since 1910. The area exceeds five square miles.

Bluefield has extensive coal and coke industries and is the shipping point for vast Pocahontas coal fields. There are large wholesale houses and railway shops here. The Federal building, state normal school and two sanitariums are the principal buildings.

BLUEFIELDS, an important seaport, the capital of the department of Zelaga, on the Mosquito Coast in Nicaragua. It has a landlocked harbor, and is connected with Galveston and New Orleans by direct lines of steamers. Large quantities of bananas and other tropical fruits are exported to the United States. From 1655 to 1850 Bluefields and the surrounding territory formed a British protectorate. A form of government by a native chief was then established, but in 1894, under the name of Zelaga, the district was incorporated with Nicaragua. Population, about 5,000. See Mosquito Coast.

BLUEFISH, a sea fish, common on the eastern coasts of America and the most destructive fish in northern seas, for it preys incessantly on smaller fishes. It is allied to the mackerel, but is larger, growing to the length of three feet or more, and is much esteemed for the table. Bluefish are taken in nets and by hook, furnishing great sport by the latter method. New York City alone uses $250,000 to $300,000 worth of bluefish in a year.

BLUE FLAG. See Iris.

BLUE GRASS, the grass that has made the pastures of Kentucky famous and which gave that state its popular name of "Blue Grass State." The name is applied because of the bluish tinge of its seed pods in the month of June. It is also called spear grass, or meadow grass. In favorable soil blue grass will reach a height of two feet. It has many long, narrow root leaves, and is one of the best pasture grasses known, though it is not as valuable as some other grasses for hay because its yield is less. It is an excellent park and lawn grass, and thrives best on clay soils containing lime.

BLUE LAWS, the name applied to a set of laws regulating the conduct of the members of the colony of New Haven, Conn. These were once supposed to be genuine and binding upon the deeply-religious people, but now are known to have been the product, in large part, of the imagination of Rev. Samuel Peters, a minister who was driven from the colony to England, and who thereafter devoted himself to ridiculing the Americans. Among those laws which he declared had been passed were the following:

No one shall be a freeman or have a vote unless he is converted and a member of one of the churches allowed in the dominion.

No one shall cross a river on Sunday but an authorized clergyman.

No one shall run on the Sabbath day, or walk in his garden, except reverently to and from meeting.

No woman shall kiss her child on the Sabbath or fasting day.

No one shall travel, cook victuals, make beds, sweep houses, cut hair or shave on the Sabbath day.

Whoever wears clothes trimmed with gold, silver, or bone lace above two shillings by the yard, shall be presented by the grand jurors, and the selectmen shall tax the offender at 300 pounds estate.

No one shall read common prayer, keep Christmas or saint-days, make minced pies, dance, play cards, or play on any instrument of music, except the drum, trumpet and Jew's-harp.

Every male shall have his hair cut round according to a cap.

No gospel minister shall join people in marriage. The magistrate may join them, as he may do it with less scandal to Christ's Church.

A man who strikes his wife shall be fined £10.

A woman who strikes her husband shall be punished as the law directs.

When parents refuse their children convenient marriages, the magistrate shall decide the point.

A drunkard shall have a master appointed by the selectmen, who are to debar him from the liberty of buying and selling.

Whoever publishes a lie to the prejudice of his neighbor shall sit in the stocks, or be whipped fifteen stripes.

The name *blue laws* came into use when the colony of Connecticut printed a set of rules and bound them in a blue paper cover.

Modern Blue Laws. The term is applied at the present time to legal restrictions in matters of conduct that are usually left to the conscience of the individual. Laws regarding the observance of Sunday are very often known as blue laws, and those who oppose the restriction or abolition of liquor selling quite generally regard prohibition and local option measures as belonging to the same classification. B.M.W.

BLUE PRINT, the photographic print used instead of the original plans by architects and engineers to guide them in the construction of buildings, bridges and other engineering works. A blue print is a photograph obtained by preparing a sensitive paper by brushing it over with a solution of oxalic acid and iron and then treating it with a solution of potassium ferrocyanide. The drawing, which is on vellum or some other material that will allow light to pass through it, is placed over the paper and exposed to light. After exposure the paper is washed in pure water. The lines in the drawing protect the cyanide from the action of the light, and when the paper is washed it dissolves, leaving white lines where the black lines were on the drawings. The completed print is therefore composed of white lines only, on a solid background of blue. Blue photographic prints from ordinary negatives can be made in the same way. The prepared paper can be procured of dealers in photographic and architect's supplies.

BLUE RIDGE, the most easterly range of the Appalachian Mountain system, in North America, extending across New Jersey, Pennsylvania and Virginia into the northern parts of Georgia and Alabama. In the southern portion this range is crossed by the Black, the Nantahala and the South mountains. In the strictest sense the name applies only to that portion of the range which crosses Virginia and separates the Piedmont region from the Great Valley. The highest peaks are about 4,000 feet in altitude. The Hudson, the Potomac and several other rivers have cut their way through these mountains, and thus form narrow, picturesque valleys. See APPALACHIAN MOUNTAINS; BLACK MOUNTAINS; PIEDMONT REGION.

"BLUE SKY" LAWS, statutes which regulate the issue and sale of stocks and bonds by corporations. The origin of the term is uncertain, but it is generally supposed to have been from the remark of a Kansas bank commissioner, who said that certain companies were capitalizing the blue sky. Some of the stocks and bonds offered continually to the public have little more than "blue sky" as security, and it is to prevent such frauds on the public that laws have been passed to protect the investor. The importance of such protection may be understood when it is known that each year, in the United States alone, the investing public is defrauded of $100,000,000 to $125,000,000 by dishonest promoters.

The lead in this form of legislation was taken by Kansas in 1911, as the result of the activity of a large number of companies claiming to own oil lands in Oklahoma. About half of the states of the Union now have similar laws. Usually the law requires dealers in stocks and bonds to have a state license and to file with the proper authorities all information about the securities they offer for sale, the amount of the issue, actual value of the property in hand, etc. The Kansas law is very strict and requires specific information on many points, but in some states the information required is more general. The enforcement of the law is sometimes placed in the hands of one of the regular state officers, such as the bank commissioner, but usually a special board is created.

BLUE VITRIOL, *vit'ri ul,* or **BLUE STONE,** is a compound of copper and sulphuric acid, in the form of dark blue crystals. Its chemical name is *copper sulphate.* Blue vitriol is the most useful compound of copper. It is used in calico printing and dyeing, in making electrotypes and in copperplating, for preserving timber, in electric batteries and in the production of other compounds of copper. It is poisonous, and is extensively used in spraying mixtures to destroy insects (see INSECTICIDES). When exposed to the air the crystals lose their water, turn white and crumble to a powder. The blue vitriol of commerce is obtained as a by-product in refining gold and silver with sulphuric acid.

BLU'ING, a blue-colored mixture used in the laundry to whiten clothes. White clothes have a tendency to turn yellowish when laundered, and the bluing overcomes this tendency. Bluing can be purchased in liquid form; bluing paddles and bluing balls are also in common use. The balls are placed in a bag which is whirled through the rinsing water in the same manner as the paddle. The liquid preparations are so strong that a teaspoonful is usually all that is needed to blue the water used in rinsing a tub of clothes. The bluing must be thoroughly stirred in the water before the clothes are placed in the tub, otherwise they will be streaked. Bluing is also put in starch. The substances used most commonly in making bluing are soluble Prussian blue and coal-tar blue.

BLUN'DERBUSS, a corruption of a Dutch word meaning *thunder box,* is applied to an

THE BLUNDERBUSS

old-fashioned smooth-bore, muzzle-loading gun. The barrel terminated in a somewhat bell-shaped muzzle, and several bullets could be put in at one load. It made an effective weapon at short range, as the charge scattered widely and some of the bullets were almost certain to take effect. It became entirely obsolete with the introduction of breech-loading weapons, but the word is much used in a figurative sense to apply to anything which does not concentrate effort.

BLUSHING, which causes the face and neck to redden, is brought about by the stimulation of certain nerves called *vaso-dilator.* As a result the arteries become larger and more blood flows through them. Blushing is also accompanied by a sensation of warmth in the face. Children blush less frequently than older persons because they are too young to have the mental states that cause blushing—that is, modesty, shame and similar sensations. *Unblushing* is often applied to a person who is so hardened that he has lost the sense of shame. Mark Twain once said, "Man is the only animal that blushes—or needs to."

Terror produces a physical state the opposite of the one described above. The vaso-constrictor nerves are in this case stimulated, which causes the tiny blood vessels to contract. Thus the amount of blood is lessened, and the skin grows cold and pale. W.A.E.

BOA, *bo'a,* a variety of South American serpents of great size and enormous strength, but without poison fangs. Many fantastic stories are told of the terribly destructive power of boas. In the still-popular boy's book ***Swiss Family Robinson*** is a vivid description of the destruction and swallowing of a donkey by a boa constrictor. They are said to be able to swallow horses, oxen and other large animals whole. Of course this is not true, but it is well known that they can swallow animals much larger than their own heads. This is due to the formation of their mouths; the jaws are joined by an elastic substance which allows them to stretch both vertically and horizontally.

The boas are generally found in dense forests, where they suspend themselves from tree branches and seize animals passing beneath. With their strong teeth, pointing backward towards the throat, they easily hold a small animal, round which they wrap a fold of the body and quickly crush it to death. Before the animal is swallowed it is covered with a thick coating of saliva. After a meal a long sleep is taken, lasting sometimes a week, until the food is thoroughly digested; bones, horns and other indigestible substances are disgorged and the serpent is ready for another meal. However, it can live in comfort several weeks without eating.

The most common species is the *boa constrictor.* It inhabits the tropical regions of

THE BOA CONSTRICTOR

Mexico and Brazil, and seldom exceeds twelve feet in length. Stories of boas thirty and more feet in length are fabrications, though the *anaconda,* sometimes called a boa, attains a length of nearly thirty feet. In Arizona and other southern parts of the United States two very small species of boas are found. They live chiefly on small animals and insects.

BOABDIL, *bo ahb deel',* or **ABU-ABDULLAH,** *ah boo' ab dil'ah,* also called *the little* and *the unfortunate,* was the last Moorish king of Granada, which is now a province of Southern Spain. Boabdil claimed the throne in 1482 and expelled his father, who died of a broken heart. Because of his tyranny, his subjects were not loyal. Taking advantage of such unsettled conditions, the Castilian army of Ferdinand and Isabella, king and queen of Castile and Aragon, was sent to besiege Granada. Its surrender was obtained in January, 1492, only a few months before Queen Isabella assisted the enterprise of Christopher Columbus. Riding away from his lost kingdom, Boabdil turned for a farewell look, and that spot is still shown to tourists as "the last sigh of the Moor." Later Boabdil went to Africa, where he is said to have been killed while fighting for the ruler of Fez. In Irving's *Alhambra,* a literary masterpiece, he is a prominent character.

BOAR, *bohr,* WILD, the wild hog of Europe, North Africa and Asia Minor, far exceeding in size the largest of domestic hogs. From very early times hunting the wild boar has been a favorite pastime of kings and nobles,

WILD BOAR

and in most countries the animals have been carefully preserved for this purpose. In England in the days of the Norman kings a person killing a wild boar without royal permission was liable to have his eyes put out. In most parts of Europe this animal in its free state is extinct, but is preserved on some large estates. It is grayish-black in color, with short hair and coarse bristles. It has formidable tusks, and is a dangerous enemy when wounded. The boar hunt, on foot, with spears and hounds of a strong and fierce breed, was an exciting and dangerous sport. In India at the present time a boar of larger species, forty inches high at the shoulder, is hunted on horseback, and "pig sticking" is the most popular sport among British military and civil residents.

The wild boar feeds at night on roots and grain, though it will sometimes eat small animals and birds' eggs. The meat of the wild boar was once highly esteemed, and considered far superior in flavor to the flesh of the domestic hog. The boar's head was a great delicacy and was brought to table at feasts with great ceremony in ancient times; in parts of Europe the ceremony of heading the procession to the table with a boar's head held aloft is still maintained.

BOARD OF HEALTH, an organization formed to protect the health of the community over which it has jurisdiction. Boards of health are established by city and state governments. A city board of health is provided for in the charter under which the city government is organized; a state or provincial board is provided for by the legislature. In townships and counties the boards of trustees and supervisors usually act as the board of health for the territory under their respective jurisdictions. The powers and duties of every board of health are prescribed by law.

Duties. The duties of the board of health in a large city are more numerous than those of one in a small town or village, though they are of similar nature. The most important of these duties are to prevent the spread of contagious diseases by enforcing quarantine regulations and requiring vaccination of children before they enter school; to prevent the sale of unwholesome food; to see that garbage is collected and removed regularly; to see that dead animals are removed and properly disposed of; to prescribe and oversee the duties of coroners, and to perform such other duties as the city council may direct.

The duties of state boards of health are of a more general nature than those of a city board. In many cases their work is advisory, and they render important service by investigating causes of disease and publishing bulletins giving information that will prevent the spread of any particular malady. An illustration of this kind of general service is the series of bulletins issued by the board of health of Arkansas on the hookworm disease. Recommendations to the state legislature often secure the enactment of laws in the interest of public health. In some states pure food commissions perform the duties of a state board of health.

There is no national board of health in the United States. The duties of such a board are performed by inspectors of foods and drugs under the direction of the Bureau of Animal Industry in the Department of Agriculture and by the Marine Hospital Service. See SANITARY SCIENCE.

BOARD OF TRADE. The farmer raises wheat, oats, rye, corn, etc., expecting to sell in the highest market and reap a profit to which his hard labor entitles him. Does he fix the price for his grain, and can he demand that he shall receive that price? If he had the only good crop in the land he probably could get whatever price per bushel he demanded, but thousands of others have like crops and not one of all those farmers has the means of knowing whether the combined crops of all will meet or exceed the world's demands. And upon the demand the price per bushel is based. Moreover, the farmer acting independently would have trouble to learn where there was need of his wheat or corn and therefore would find difficulty in securing a market.

We know that he may take his grain to the nearest village any day and sell it. His daily paper tells him the price that prevailed the day before. If he is a careful reader he knows the possibilities as to higher or lower prices in the future, and he may determine to hold his crop until he can get more money for it.

Who fixes the price per bushel, and how is the price determined? Is it controlled by a group of men, and is the farmer at their mercy? Many people believe that powerful men with immense capital control the situation, and that they can raise or lower prices at will—in other words, that they gamble on the upward or downward turn of the market and that the weight of their dollars turns the scale in the direction of higher or lower prices. This is almost entirely a mistaken idea. If such a condition ever should exist there would be instant demand for legislation against it.

Prices are fixed by men, it is true, but they are forced in the beginning to recognize what is a just and reasonable price for all grains. It is a well-established principle that wheat should average in price, year after year, at least one dollar per bushel. A group of men in Chicago and another in Liverpool are secondarily responsible for any changes from an average price. The dominant influences which control price are the *supply* of a product and the *demand* for it. The two groups of responsible men who dictate prices are guided primarily by these two conditions. They comprise organizations known as *boards of trade*.

The reason for the location of the most important boards of trade in Chicago and Liverpool is that Liverpool is the world's greatest grain market and Chicago is the great market in the western hemisphere, close to the vast grain-producing sections, and the largest grain-shipping center in the western world. Therefore Liverpool and Chicago men are naturally able to determine more accurately than other people what the grain supply for any year is to be. Their agents, located all over the grain-producing areas, make daily reports of crop conditions and prospects. The world's demand for cereals is quite accurately known from year to year, therefore it is not difficult to determine whether there will be enough grain to meet the demand. Prospects of a shortage is sure to raise the price; evidence of overproduction will naturally make every bushel less valuable.

The board of trade in Chicago reflects every day the world's crop conditions. Wheat may sell at $1.30 per bushel as a result of dryness over a vast grain area, thus making an average crop doubtful. One night over the wires may flash a report of heavy rains extending the length and breadth of that area. The next morning operators on the board of trade will hammer the price downward, for the wheat crop prospects have changed for the better, and the cereal will clearly bring less money per bushel. Thus do prices fluctuate. Boards of trade in other cities receive telegraphic quotations from Chicago and Liverpool and almost instantaneously prices are equalized throughout the world.

In addition to the above necessary service performed by boards of trade they have established uniform standards of quality, or grades, of grains. The term *number one hard* applied to wheat means the same throughout the wheat-growing world.

The above is a brief sketch of the beneficial aspects of the board of trade—those activities which in the nature of things some agency must control. There is another and a distinctly different and unfavorable view held by many people, and at times it would appear there is evidence to support it. That there is gambling in the fluctuating prices of cereals is admitted; the gambling instinct is ever present in some men, and they will place bets regardless of law or moral considerations upon whatever offers the possibility of gain. The board of trade seems unable to suppress the speculative tendencies of men determined to venture money upon the rise or fall of prices. Beneath their activity, however, is the inevitable law of production and demand, and seldom do they seriously change stable conditions by their onslaughts upon the markets.

Occasionally, when very unusual conditions have made such operations possible, the speculative instinct has led men to attempt to "corner" the supply of a commodity. A "corner" is an *artificial scarcity* of a grain, created by a combination of men with large capital for the purpose of holding the article affected off the market by buying practically all the visible supply. The object is to extort abnormally high prices. Such efforts seldom succeed. The most memorable attempt to "corner" wheat occurred on the Chicago Board of Trade in 1867, when the price reached $2.85 per bushel; in 1898 another attempt sent the price to $1.85 but cost the operator, Joseph Leiter, several millions of dollars in losses.

Transaction of Business. In the midst of all the noise and confusion which the outsider observes on the floor of the board of trade during the hours when it is in session, there is a vast and thoroughly-systematized volume of business being transacted with a speed which amazes the onlooker. The brokers on each board of trade have a sign language peculiar to themselves, by which they can make themselves understood above the din constantly prevailing. A sign made with the open hand of the broker toward the person with whom he is in communication signifies "sell"; if he shows the back of his hand, it means "buy"; one finger raised means 5,000 bushels or other unit of the article dealt in; two fingers raised signifies 10,000 bushels, and so on. The circular platform or depression where the business is transacted is called the "pit."

Margins. The practice of dealing in margins has become a leading feature of the business of all boards of trade. According to this method of dealing, the trader deposits with his broker a sufficient amount to cover the ordinary fluctuations in price, and the broker furnishes the rest of the necessary capital. For instance, in January the trader wishes to buy 5,000 bushels of wheat for delivery in February. If the present price is $1 a bushel, he may advance his broker $250, which is a margin of five cents a bushel; he may margin it for even two or three cents per bushel, if he prefers. If the price of wheat advances, he can order the broker to sell it, and if he chooses, withdraw his margin as well as a profit, according to the extent of the rise. In case the margin is five cents per bushel, if the price recedes below $0.95 or below the point where his margin will cover the loss, he must either deposit enough margin with his broker to cover the falling off or lose what he has advanced.

Settlement. Most boards of trade have their own clearing houses, and at the end of each business day all parties who have been trading on the board must send reports of sales and purchases to the clearing house. Those whose reports show net loss must send certified checks for the amount, and those who have made net gains are paid.

Rules and Regulations. The most stringent regulations are made to prevent fraudulent practice on the board. The smallest fraud on the part of any member, however prominent he may be, is punished by immediate suspension, and his trial is prosecuted with a rigid impartiality not surpassed by the courts of law. A board of trade contract matures on the last day of the term mentioned in it, and all transactions between members for purchases or sales on the floor of the board are strictly contracts under its rules.

The distinction between so-called *long* and *short* transactions is as follows: In the former, the trader buys, expecting a later advance in price to net him a profit; in the latter, he sells, expecting a subsequent decline.

Caution. While boards of trade are reputable commercial organizations, the manner of transacting business is such that they are no place for the inexperienced trader, whose judgment of the market may be exceedingly faulty. Because of this lack of good judgment very frequently he loses his investment. See CHAMBER OF COMMERCE. E.D.F.

Two novels dealing realistically with the production and distribution of wheat are Frank Norris's *The Pit* and *The Octopus*.

BOAT, as generally referred to, is a small vessel or water craft usually built of wood. It is propelled by oars, by sails set to catch the wind or by a screw turned by mechanical means. There are many special names applied to the different kinds of boats; among these are the punt and dory—flat-bottomed boats used in shallow water; cutters, dinghys, launches, gigs and barges; whaleboats and lifeboats, sharp at both ends; racing shells, long and narrow, with sliding seats to give the oarsmen great power; and skiffs, small, light boats usually fitted with a centerboard, a weight suspended below the keel, to give stability and lessen drifting when sailing.

Every ship is required by law to carry a number of boats proportional to its size and to the number of passengers. Among such boats must be fully-equipped lifeboats, in addition to which are numerous collapsible boats and rafts. Vessels carrying no passengers must provide sufficient boats to accommodate the crew, if they should be forced to leave the vessel. Ship's boats when not in use are hung on *davits* at the sides of the vessel and are lowered and raised by hand with block and tackle, or by steam or electric power. The seats in boats are called *thwarts*, and the spaces for oars are *rowlocks*. For description of large vessels, see SHIP; also CANOE; YACHT; MOTOR BOAT. Under the title SAILBOAT AND SAILING directions for handling a sailboat are given.

BOATSWAIN, *bote' swane,* called by sailors *bo's'n,* is a petty officer on board a ship. In the navy he is responsible for the sails, rigging, colors, anchors, cables and cordage, and he calls the members of the crew to their work. The pay of a boatswain in the United States navy is from $1,200 to $1,500 a year. After ten years of service he is made *chief boatswain,* a commissioned officer ranking with the *ensigns.* On merchant ships, the boatswain is next in importance to the mates; he is foreman of the crew and is in charge of the rigging. The "bo's'n" is a character known to all readers of sea stories, but perhaps the most remarkable specimen is in W. S. Gilbert's *Yarn of the Nancy Bell,* whose hero is—

> "A cook and a captain bold,
> And the mate of the Nancy brig,
> And a bo'sun tight, and a midshipmite,
> And the crew of the captain's gig."

BOBOLINK, *bob' o lingk,* one of the most interesting of the North American wild birds, named in imitation of its gay and sprightly song, in which the sounds of *bob-o-lee, bob-o-link,* can be heard distinctly. The bobolinks

BOBOLINKS
Above the male; below, the female.

are related to the blackbirds and orioles, and are about seven inches long. In the spring the male wears very handsome attire—black, with markings of buff and ashy-white on the back of the head, shoulders, back and wings. In the autumn the male and female have the same plumage—yellowish-brown above, paler beneath and a light stripe on the crown.

These birds are found from Labrador to Mexico and the West Indies. In March or April they appear in the Southern United States and work their way northward, nesting in May in the cool, grassy meadows of the Middle states, New England and Canada. The nest is built on the ground, hidden in the tall grass, and contains from four to six dull white eggs, with irregular markings of lilac and brown. In their northern abode the bobolinks destroy great numbers of crickets, grasshoppers, beetles and spiders, and are also fond of dandelion and grass seeds, thus earning their right to the friendship and protection of man.

Far different is the history of the gay-singing bobolink after the southward migrations begin. In August great flocks appear in the reedy places and marshes of the seacoast and inland waters in the vicinity of New York and Pennsylvania, where they grow fat on the wild rice found there and are eagerly sought by the huntsmen. Known at this period as the *reedbird,* or *ricebird,* they are much-prized game, for their flesh is delicious. About October those which have escaped the hunter fly to the rice fields of the Southern states, where they do considerable damage to the growing rice crops. Winter finds them located in the West Indies. The wholesale destruction of the ricebirds has caused the United States government to appoint a closed season for them in Maryland, the District of Columbia, Virginia and South Carolina, continuing from November 1 to August 31. See BIRD.

Bryant's happy description of the bobolink in his poem *Robert of Lincoln* is a charming bit of natural history. The opening stanza is here given:

> Merrily swinging on brier and weed,
> Near to the nest of his little dame,
> Over the mountain-side or mead,
> Robert of Lincoln is telling his name:
> Bob-o'-link, bob-o'-link,
> Spink, spank, spink.
> Snug and safe is that nest of ours,
> Hidden among the summer flowers
> Chee, chee, chee.

BOCCACCIO, *bok kah' cho,* GIOVANNI **(1313-1375),** an Italian novelist and poet, the earliest writer of classic Italian prose and the author of some of the most famous stories in the world's literature. He was the son of a merchant of Florence and early showed unusual talent, writing verses before he was seven years of age. Nevertheless, he spent several years in the study of law before he was able to devote himself entirely to literature.

Early in his career Boccaccio fell in love with Maria, daughter of King Robert of Naples, and his first work, a romantic love tale in prose, *Filocopo,* was written at her command. The *Decameron,* or *Ten Days' Entertainments,* on which his fame rests, is a collection of one hundred tales, supposed to have been related in ten days by a party of ladies and gentlemen who had withdrawn to a country house near Florence while the plague was raging in that city. In these tales Boccaccio reveals himself as a master of the art of story-telling, and their influence on later writers was tremendous. Vividly told and beautifully written, they are marred only by the coarseness that was characteristic of the age in which their author lived.

BOEHMERIA, *bo me' ri ah,* a group of plants closely resembling the stinging nettle, several species of which are of great service to man because of their strong fiber. One of the most important species, known as Chinese grass, yields most of the fiber used by the Chinese in making China-grass cloth, a glossy, transparent fabric of great beauty. Chinese grass is a non-stinging, herb-like plant with broad, oval leaves which are white and downy beneath. It is grown by seeds and by root division and thrives in shady and moist ground, three crops being raised each year. The fiber is removed by hand stripping, by boiling the stalks in water or by means of chemicals or by machinery, but hand labor has been found most successful. Therefore most of the Chinese-grass fiber is obtained from China and India, where labor is cheap. The plant is cultivated to a limited extent in the Southern United States.

BOEHMERIA
Stalk, leaves, flower and seed pods.

Another species of boehmeria, commonly known as *ramie,* or *rhea,* is an East Indian shrub whose upright stem yields a firm, durable fiber that is stronger than hemp and almost as fine and lustrous as silk. This fiber is used in making cordage, nets, fabrics and the paper employed in the manufacture of banknotes. The plant is grown by seeds, cuttings or root division, and requires a hot, moist, even climate and soil that is rich and damp.

BOEOTIA, *be o' shi a,* in ancient times a division of Central Greece lying between Attica and Phocis, having an area of 1,100 square miles. Its shores were washed by two arms of the sea, the Corinthian Gulf to the west and the Strait of Euboea to the east. The northern portion of Boeotia formed the basin of Lake Copaïs, into which the Cephissus River flowed, and south of the lake rose the Helicon Mountains, famed in ancient legends as the seat of worship of the Muses. The atmosphere of ancient Boeotia was heavy, the summers sultry and the winters foggy, and the Boeotians seemed to have been affected by the dullness of their native air, for they were stolid, coarse and unimaginative. In this respect they were a striking contrast to their brilliant neighbors of Attica, whose lofty achievements in art and literature were in part inspired by the clear air and exhilarating climate of their land.

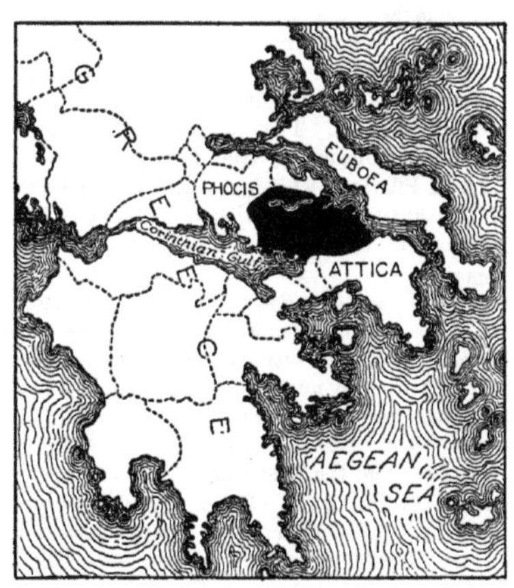

BOEOTIA (IN BLACK)
North of ancient Attica, thus near Athens, the center of old Greek culture.

Thebes, the chief city of Boeotia, was in ancient times the head of an important confederacy of cities known as the *Boeotian League*. The League assisted the Persians during their invasions, and in the Peloponnesian War fought on the side of Sparta against Athens. The confederacy was at the height of its power under the Theban generals, Epaminondas and Pelopidas, to whose military genius Thebes owed its brief period of supremacy in Greece.

Modern Boeotia is a political division of Greece, with a population of about 65,800. See THEBES; GREECE, subhead *History*.

BOER, *boor*, a Dutch word meaning *farmer*, and generally applied to settlers of Dutch descent in South Africa. In 1652 a party of Dutch left Holland to found a colony in South Africa, where they might obtain religious and political freedom. They were influenced to a considerable degree by the migration of the English Pilgrims, who had sailed to America about thirty years previously, after a brief residence in Holland to escape persecution at home.

In 1796 the territory where they had made their new African homes was seized by the British. The Boers bitterly resented the British rule and in 1836 numbers of them boldly set forth into the unknown and hostile lands to the north, which afterwards became the Orange Free State and the Transvaal. History has no parallel to the record of danger and hardship faced by the Boers on what is known as the "Great Trek" in search of freedom. See TRANSVAAL; SOUTH AFRICAN WAR.

BOGOTA, *bo go tah'*, the capital of the republic of Colombia, South America, and of the state of Cundinamarca. It is pleasantly situated on a plateau 8,700 feet above sea level, in the eastern section of the Andes Mountains. Although the city lies only 4° 37′ north of the equator, its elevation gives it the climate of perpetual spring, the air being moist and fresh and not too warm for comfort. The San Francisco and San Augustine rivers flow from the mountains through the city in cool streams and divide it into four parts, each with broad streets, well shaded with trees. The principal buildings, grouped round a fine central square, include a grand cathedral in Corinthian style and the government buildings.

The city, which is a center of culture and education, has numerous churches, schools and public libraries, a national university, three endowed colleges, a botanic garden and a museum. Those buildings which are not massive are usually one story in height in order to withstand earthquakes, which are of frequent occurrence. Bogota has manufactures of carpets, matches, glass, cordage, porcelain, soap, cloth and leather, but its location is unfavorable to great expansion in trade and manufacture. The city was founded in 1538 and soon became the capital of the province of New Granada. When the republic of Colombia was established in 1819, Bogota became the capital of the new state. Population in 1915, 122,369. See COLOMBIA.

BOHEMIA, *bo he' mi a*, a former kingdom and crownland of Austria-Hungary, once an independent kingdom, and now the northwestern corner of Czecho-Slovakia. It formed the northwestern part of the old monarchy, has an

THE ABSORPTION OF BOHEMIA
The ancient kingdom is merged into the republic of the Czecho-Slovaks.

area of 20,064 square miles, or about half that of Kentucky, and a population in 1910 of 6,769,548, or nearly three times that of the American state. Primarily an agricultural province, it produces large stores of rye, wheat, oats, barley, potatoes and sugar beets; but it was also first among the Austrian provinces in the possession of mineral wealth. Its resources, its industrial life and its geographic features are treated in the article AUSTRIA-HUNGARY, for reasons there stated, and need not be repeated here.

Government. Under the old Austro-Hungarian rule, Bohemia had a diet of about 250 members, but it was not thoroughly representative of the people, for most of the membership was chosen by the aristocracy or large land owners. Before the end of the War of the Nations the province began to plan for ultimate independence, and when the Czecho-Slovak republic became a possibility Bohemia earnestly joined the movement. See CZECHO-SLOVAKIA.

The chief cities are Prague, the former capital, which has two great universities, one German and one Czech; and Pilsen, famous for Pilsener beer (see PILSEN; PRAGUE).

History. Its first inhabitants were the Boii, a Celtic people from whom the country took its name, but they were driven out by the Germans during the first century B. C. A Slavic race called the Czechs appeared during the sixth century, and by the ninth they had been converted to Christianity by the Germans. After forming for a time a part of the Moravian kingdom of Svatopluk, Bohemia became in the tenth century a duchy which paid tribute to Germany, but so ambitious were its dukes and so enlightened its policies that the German rulers at length gave it recognition as a kingdom. Its period of greatest glory was from 1253 to 1278, when it came forward as one of the strongest European powers.

The House of Luxemburg ruled in Bohemia from 1310 to 1437, and of this line of kings three were also emperors of Germany. It was under Sigismund, last of the Luxemburg sovereigns, that the religious movement which centered about John Huss (which see) reached its height and culminated in actual warfare. Religious in its origin, this movement soon took on a political aspect, for Czech set himself up against German, and for a time stayed the Germanization of the kingdom.

The independence of Bohemia, however, was not much longer to endure, for in 1526 Ferdinand of Hapsburg was chosen its king, and he proved to have the Hapsburg tendency to sacrifice all interests to the advancement of Austria, the hereditary Hapsburg domain. The religious question again came to the fore with the Reformation, which the Bohemians were much inclined to accept, and the determined repression of Protestantism by the Catholic Hapsburgs did much to bring on the Thirty Years' War (which see).

From that struggle Bohemia emerged broken in spirit and helpless, with two-thirds of its population dead and its civilization ruined. Protestantism was utterly crushed, as was all active opposition to the Hapsburg rulers; no really national feeling seemed to awaken until almost a century and a half later, near the end of the eighteenth century. From that time on the Czech element of the kingdom never ceased to agitate for independence, or at least for official recognition of the Czech language, as opposed to the German. Bohemia became fully as restless a member of the Austro-Hungarian monarchy as was Hungary. Long before the close of the War of the Nations Bohemian soldiers of Czech blood openly deserted from the armies of the dual monarchy and joined the allied forces, that they might more quickly achieve their dream of independence. J.B.

BOIL, a small but very painful eruption or swelling on the body or face. Not only is it dreaded for its unsightliness but also on account of the suffering which it occasions, an amount of pain which appears out of all proportion to the cause. It begins as a small hard point of a dark red color, which throbs painfully and feels hot. As these symptoms increase in severity the boil grows larger, finally reaching a cone-shaped form having a broad, firm base and a whitish blister on the apex. The latter contains a little pus. A few days after the blister opens a core of cellular tissue is discharged, following which the cavity heals rapidly. Boils are the result of infection with a pus germ which gains entrance through a hair follicle or sweat gland. The most efficacious form of treatment is lancing the swelling after the appearance of pus. The use of laxatives to clean out the intestinal tract and washing the boil with mild disinfectants are advisable measures. Care should be taken to keep the discharge away from the skin, as it carries infection.

In its first stages a carbuncle somewhat resembles a boil, but the former is by far the more serious. W.A.E.

BOILER, a vessel in which steam is generated by the boiling of water. Boilers for steam engines are formed of riveted plates of metal, are supplied with a furnace in which a fire is made, a grate on which the fire is laid and an ash pit into which the ashes fall. On top of that part of the boiler containing water is a *steam dome* into which the steam passes as generated and from which it is conducted by pipes to the points at which it is to exert its power.

The uses to which the steam is put when developed are very numerous, and include the supplying of engines with power for industrial purposes, for driving railroad and marine engines, for heating, and in fact wherever mechanical power is more efficient or more economical than man power. When steam is produced by boiling water in an open vessel, that steam has no power that could be utilized; it simply rises and mixes with the atmosphere.

If steam is confined it endeavors to escape, and in doing so exercises *pressure*. The higher the temperature is raised the greater the pressure exerted.

Strength of Boilers. The first essential of a boiler is strength to withstand the pressure of the steam without bursting. Boilers are therefore carefully constructed with a strength sufficient to resist a certain amount of power or pressure to the square inch. Each boiler is fitted with a gauge which indicates the pressure of steam. There is also an ingenious contrivance called a safety valve which, if the pressure rises too high, automatically opens and allows sufficient steam to escape to bring the pressure within the safety limit. Powerful modern boilers are constructed to stand a pressure of as much as 200 to 225 pounds per square inch.

VERTICAL FIRE-TUBE BOILER

The tubes are like chimneys for the fire, and by passing through the water give a large heating surface.

Methods of Heating. The main object of all boilers is the same, but that object is arrived at by different methods. It is desirable quickly to heat the water, which is supplied by pipes leading into the boiler, and to maintain that heat as economically as possible. *Fire-tube boilers* are provided with flues, or passages by which the heated gases generated by the fire pass through the chamber containing water. *Water-tube boilers* have the water in a series of tubes instead of in one compartment and are considered the safest and most effective style of boiler.

Upright and Horizontal Models. Boilers may be divided into two classes, namely, *upright* and *horizontal*. The principles of operation are the same in both cases, the difference in shape being simply a matter of convenience. On board steamships upright boilers are obviously more convenient than the horizontal boilers used in locomotives on land would be. Small upright boilers are used on land for industrial purposes, and in some cases for heating large buildings. The cylindrical shape has been universally adopted as being most efficient and most economically maintained in a state of cleanliness.

Horse Power of Boilers. In stating that a boiler is of a certain *horse power* it is not intended to convey the idea that the power exerted is equal to the power of the same number of horses. Horse power is a term indicating a definite unit of force, established by the American Society of Mechanical Engineers. The horse-power unit as applied to steam boilers is thirty pounds of water evaporated per hour from a temperature of 100° F. under a pressure of seventy pounds by gauge. See STEAM ENGINE.

C.H.H.

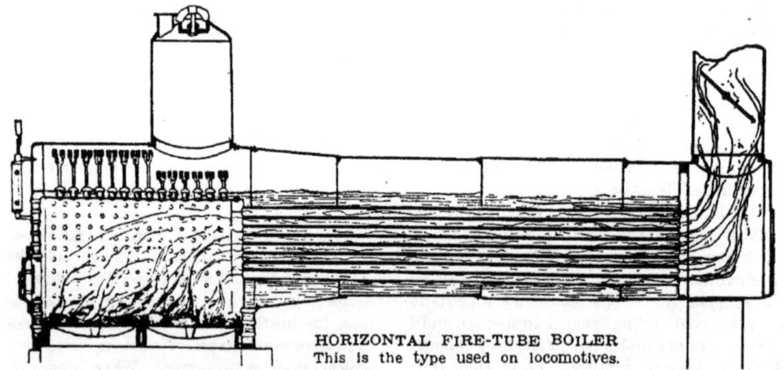

HORIZONTAL FIRE-TUBE BOILER

This is the type used on locomotives.

BOILING POINT. At certain temperatures liquids bubble up and give off vapor. The temperature at which this occurs is in each case the *boiling point* of the liquid. This depends upon the amount of atmospheric pressure to be overcome, because the boiling point occurs where the tension of the vapor is equal to the pressure of the atmosphere. Pressure is heaviest at sea level—14.7 pounds to the square inch—where the boiling point of water is 212° F. In ascending to higher elevations the pressure diminishes and the temperature of the boiling point falls in the ratio of 1° F. to every 550 feet of altitude. At lofty heights the boiling point of water is so low that food cannot be cooked in open vessels. This fact is sometimes utilized in calculating altitudes at different stages in the ascent of mountains. Thus the boiling point on the summit of Mont Blanc, 15,781 feet above sea level, is 185° F.

Not all liquids boil at the same temperature. Thus, while water boils at 212° F. at sea level, ether boils at 96°, alcohol at 173° and mercury at 662°. In finding the boiling point the thermometer is not immersed in the liquid but is held in the vapor just above the surface. The atmospheric pressure of fifteen pounds to the square inch is used as the unit of measure in mechanics; for instance, in making steam, the pressure of *ten atmospheres,* 150 pounds, raises the boiling point to 356° F. The injection of gases into water lowers the boiling point. See STEAM. C.R.M.

BOISE, *boi' za,* IDAHO, the state capital, the largest and most important city of the state and the county seat of Ada County. It is situated in Southwestern Idaho, on the Boise River, 265 miles west and north of Pocatello. Transportation is provided by the Oregon Short Line and the Boise Valley Interurban Railway. The first settlement was made in 1834 by French-Canadian explorers. In 1863 Major Lugabill of the United States army established here the military post of Fort Boise, now Boise Barracks, and the city was organized the following year. It was the territorial capital and has been the state capital since 1890, when Idaho was admitted to the Union. Since 1912 the city government has been on the commission plan. In 1910 the population was 17,358; by 1915 it had increased to 29,637. The area exceeds four square miles.

Boise is a city of broad streets, stately trees and attractive homes; it is modern in every respect. Probably its most striking feature is an abundant supply of naturally hot water for domestic and public use, supplied by boiling wells. The most notable building is the massive state capitol, constructed of native sandstone at a cost of nearly $2,000,000. The city hall, Federal building, Carnegie Library and the Empire building are important structures. The Natatorium, in the Moorish type of architecture, is one of the largest and finest resorts of its kind in the United States; the supply of hot water in the swimming tank is cooled to the proper temperature.

Boise is the seat of a Roman Catholic archbishop and an Episcopal bishop. The high school is one of the finest in the West; Saint Theresa's and Saint Margaret's academies offer opportunities for advanced education. The state penitentiary, a United States assay office, a United States reclamation office and a soldiers' home are located here. There are two hospitals, three theaters and the "White City," a large amusement park.

By means of irrigation this section of the country has become a rich agricultural, dairying and stock-raising district, of which Boise is the commercial center. From this point are shipped large quantities of fruit, wool, hides and cattle. Boise is one of the greatest horse markets of the Northwest and one of the leading inland wool markets of the United States. The city does a large wholesale business, the value of the annual trade being estimated at $13,000,000. Great deposits of gold, silver, copper, lead and zinc ores are found in this locality and mining is an important industry. The Boise River furnishes water for irrigation through canals, and water power for manufacture through the construction of the Arrowrock Dam, twenty-six miles above the city (see IDAHO). This is the highest dam in the world (351 feet). Ample electric power also offers inducements to manufacturers. Boise has a large number of factories which have not yet attained their greatest proportions. R.W.C.

BOK, *bahk,* EDWARD WILLIAM (1863-), an American journalist, who as editor of the *Ladies' Home Journal* has been called "a lay preacher to the largest congregation in the United States." Under his editorial direction that publication has won a larger circulation than any other standard monthly magazine in America, if not in the world, and through its editorial columns he has wielded tremendous power in support of many good causes. The better babies' movement, the education of boys and girls in sex hygiene, the campaigns against useless patent medicines and in favor of a sane Fourth of July are some of the causes which he has consistently supported.

Although Bok has a keen understanding of the problems of American life, he himself is a native of the Netherlands. He was only six years old when his parents emigrated to the United States, as the result of business troubles through which his father lost his entire fortune. The death of the father soon after arrival in the United States left the widow and two sons in sore need. Bok himself has told of those days, when he and his brother used to go out on the streets at night and pick up bits of coal and pieces of wood because the family did not have five cents to spare for a bundle of kindling.

EDWARD W. BOK

Until he was thirteen Bok went to the public schools in Brooklyn, N. Y. Then he secured work as an office boy, and at seventeen he edited *The Brooklyn Magazine*, which flourished for two years with the encouragement and material help of Henry Ward Beecher, whose church Bok attended. After several years' work as a stenographer for publishers in New York he founded the Bok Syndicate Press, which furnished special articles to newspapers throughout the United States. These articles, though signed by William Bok, were mostly written by Edward, and won the attention of Cyrus H. K. Curtis (born 1850), the publisher of the *Ladies' Home Journal* and the *Saturday Evening Post*. In 1889 Bok became editor-in-chief of the former periodical, and in 1891 was made vice-president of the Curtis Publishing Company. In 1896 he married Miss Mary Curtis, the daughter of Cyrus H. K. Curtis.

BOKHARA, *bo kah' rah,* a Russian protectorate in Central Asia, situated between Turkestan and Afghanistan, covering an area variously estimated at from 80,000 to 90,000 square miles. One is accustomed to associate Bokhara only with the carpets for which it is famous, but there are many things combining to make the country important and interesting. It is under the influence of Russia to the extent that it can deal independently only upon local matters, yet is recognized as a kingdom and is ruled by an *ameer* whose local power is absolute. This ameer appoints officials to rule over certain districts, each of which must pay money and send gifts of cattle to the ruler every year. The government does not pay its officials, who are expected to take from the people of their districts what they need for themselves. It is not safe in Bokhara to question the methods of rulers and government officials. Naturally industries are in a backward state, except agriculture, on which people, officials and rulers alike must depend.

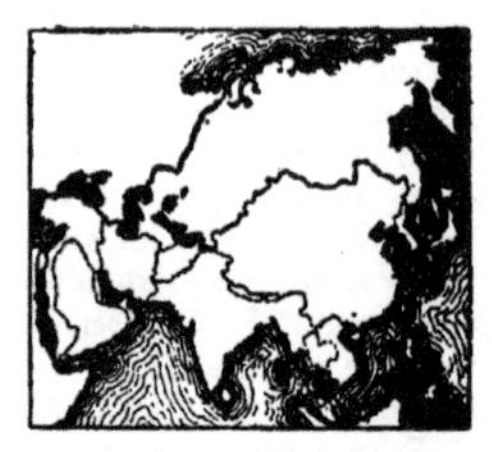

LOCATION OF BOKHARA
The small space in black represents this Russian dependency.

Horses, cattle, sheep, camels and goats are raised in great numbers, and cereals and fruit are produced abundantly. The only important industries are the manufacture of swords, knives and gold and silver ornaments. Carpets of excellent quality, textiles of silk, wool and cotton are made in native houses on primitive looms and find a ready market. The population is of the most varied kind, but not mixed, each nationality keeping to its own communities. Kirghiz, Turkomans, Uzbegs, Tadjiks, Afghans, Arabs, Persians, Jews and Russians, with many nomadic tribes, live in their settlements and plunder each other and are in turn robbed by those who are supposed to govern them. The only important towns are Bokhara, the capital, and Karshi, both cities carrying on extensive commerce with India, Persia and Russia. Internal communication is still restricted to camel caravans, such roads as exist being almost too rough for any wheeled vehicles. Population, estimated at about 2,500,000.

BOLEYN, *bool' in,* ANNE (1507-1536), second wife of Henry VIII of England, of importance in English history not for anything which she did herself, but because Henry's desire to marry her led to the separation of the English Church from the Church of Rome. As lady of honor to Queen Catharine she was conspicuous at court, and the king soon fell in love with her. She refused to listen to his addresses, however, unless he would divorce Catharine. Because the Pope would not permit the divorce

Henry denied his supremacy over the Church in England and forced Archbishop Cranmer to pronounce a divorce legal (see CRANMER).

Meanwhile, however, he married Anne in January, 1533, and in the following May she was crowned at Westminster with great splendor. In September she became the mother of Elizabeth, who was later one of England's great rulers. Henry soon tired of Anne, and to make possible his marriage with Jane Seymour, he had her thrown into prison on a charge of infidelity. Though she vigorously protested her innocence and prayed for release, she was condemned to death and on May 19, 1536, was beheaded. Historians have never settled to their own satisfaction the question of her guilt or innocence, but all agree that she was at times indiscreet. See HENRY VIII; CATHARINE OF ARAGON; SEYMOUR, JANE.

ANNE BOLEYN
The second of the six wives of King Henry VIII of England.

BOLIVAR, *bol' i vahr,* SIMON (1783-1830), a South American patriot whose services in behalf of the struggling republics of the southern American continent won for him the title of "Liberator." He completed his education by law studies in Madrid and extensive traveling on the continent of Europe, returning to South America just before the revolutionary uprising of 1810 in his native country of Venezuela. For years he led his people in a heroic struggle against the mother country, Spain, and when, in 1819, New Granada and Venezuela were consolidated into the republic of Colombia, Bolivar was chosen its President. By 1822 the Spanish troops were completely driven out of Colombia, and in that year Bolivar was called upon to help the revolutionists of Peru. In 1824 he was made dictator of that country, resigning his office a year later when independence had been secured.

The southern provinces of Peru joined together in a separate state in 1825, which was named Bolivia in honor of the great Liberator. To him was entrusted the task of framing a constitution for the new republic, which was adopted by the Bolivians, but it gave such power to the chief executive that Bolivar was accused of desiring to make himself perpetual dictator of a South American empire. The last four years of his life were full of discord, and though he was chosen President of Colombia in 1826 and 1828 he was forced to retire in 1829, when Venezuela separated from the republic of Colombia.

Bolivar is justly regarded as the "Washington of South America." Necessity forced him to use arbitrary methods, but he gave himself with supreme devotion and courage to the great object of his life—to unite the South American republics into a strong federation. Statues in his honor have been erected in Caracas, Bogota and Lima; in 1884 one was placed in Central Park, N. Y., and this was replaced by a new one in 1917.

Related Subjects. The reader is referred, for additional facts respecting the life of Bolivar, to the following articles in these volumes:

Bolivia	South America
Colombia	Venezuela
Peru	

BOLIVIA, *bo liv' i a,* one of the two South American countries which are entirely inland; it lies south of the equator, from latitude 9° 44′ to 22° 50′. It was named for Bolivar, the Great Liberator, who helped it to become what it is to-day—an independent republic (see BOLIVAR, SIMON). Surrounded by Peru, Brazil, Paraguay, Argentina and Chile, it comes into contact with the leading states of the continent, and itself bids fair to become, with the development of its resources, one of the foremost among them.

As to the size of Bolivia, there are various estimates ranging from 510,000 to 730,000 square miles, for the boundary lines on the south and east remain uncertain. But the best authorities at the present day give the area as approximately 512,000 square miles, and Bolivia thus ranks as the third largest South American state. It is almost as large as all the states bordering both sides of the Mississippi, with California added, but whereas the combined population of those states is over 27,440,000, the population of Bolivia is only about 2,267,000, or less than that of California alone. The great Canadian provinces of Ontario and Quebec are only a little larger. The average density of population is less than four to the square mile, and so unevenly are the inhabitants distributed that over vast tracts they average less than one to the square mile.

LOCATION MAP

The People. About twenty per cent of the inhabitants are of Spanish origin; the rest are of pure Indian and of mixed blood, the latter called *mestizos*. The Indians are the descendants of the old Inca Empire; those of the northern part of the high plateau speak "aymara," and those of the southern section, "quichua." The descendants of the Spaniards fill the leading professional and commercial positions and hold the public offices of importance.

Education and Religion. Great impulse has been given in late years to public instruction. Primary education is imparted to more than 80,000 children. The schools are well provided with desks and other accessories imported from the United States. High schools for training in commerce, mining and agriculture, and normal colleges for teachers have been established. In the American institutes at La Paz and at Cochabamba English is included in the courses of study. Special schools for Indians are also in operation, where manual training is taught.

Religiously, Bolivia is Roman Catholic, and most of the Indians as well as the whites belong to this Church, but other faiths are not molested.

Government and Cities. The government is republican in form, with the executive power vested in a President elected by the people for four years. He may not succeed himself, nor may the two Vice-Presidents succeed themselves. The Congress has a Senate of eighteen members and a Chamber of Deputies of seventy-five members, all elected for four years. Division into nine departments facilitates local government, and these are in their turn subdivided. There is a Supreme Court and a system of inferior courts.

Military service is compulsory, and the country has an army of about 80,000, all but about 3,150 of whom are reservists and guards. The chief cities are La Paz, the capital; Cochabamba, Potosi, Oruro and Sucre. See LA PAZ.

The Land and Its Rivers. The western part of Bolivia is called, not without reason, the "South American Switzerland," for here two great ranges of the Andes stretch for hundreds of miles, and with their lofty peaks, covered with eternal snow, afford some of the most picturesque scenery to be found in all the western hemisphere. The eastern range, the Cordillera Real, is much the loftier of the two, and contains some giant peaks—Illampu and Illimani, each over 21,000 feet, and Chorolque, 18,500 feet. By far the tallest of the western peaks is Sajama, 21,000 feet high. Between these ranges lies the Bolivian plateau, the region in which the largest cities are located and the most progressive people are to be found. The altitude of this plateau is from 12,000 to 13,000 feet, and a number of small mountain ranges cross it. At its northwestern end, part in Bolivia and part in Peru, is Lake Titicaca, the most elevated large lake in the world. Deep, clear and icy cold, it lies at an altitude but 1,500 feet below that of Pike's Peak. See TITICACA.

Stretching east and northeast from the mountains are the great plains, which slope from a height of 3,000 feet in the foothills to 300 feet at the Brazilian boundary. Very flat and well-watered, these plains contain large, grassy tracts, and along the river margins are dense tropical forests. When the heavy rains set in much of this eastern region is a great morass, but it is this very overflowing of the rivers which makes the soil so fertile and gives promise of future grazing and agricultural developments. Many thousands of wild cattle are found in the rich lowlands in the southeastern sections.

Among the principal rivers of the country are the Beni and the Mamore, which unite to form the Madeira, the largest tributary of the Amazon. Others are the Itenez, the Madre de Dios, the Pilcomayo and the Bermejo, the two latter forming part of the River Plata system.

Climate. Lying entirely within the tropics, Bolivia is saved by its varying altitudes from a monotonous tropical climate. The plains of the east, it is true, afford just what one might expect in that latitude—intense heat and excessive moisture; and they are therefore very unhealthful, especially for white men. In the high mountains is the other extreme of comparatively cold weather, with a yearly average of about 50°, but between the two, in the plateau and mountain basins, lies a temperate region with moderate rainfall and little menace to health. Thus we can account for the fact that the eastern plains are sparsely settled and several large districts are yet unexplored.

Animal Life. One of the most characteristic animals of Bolivia is the llama. This valuable beast roams the mountain regions, and its importance to the people can scarcely be overestimated. It is the beast of burden which finds its way over precipitous mountain passes; it yields long, strong wool, and its flesh, though not palatable to Europeans, is eaten freely. The armadillo, peccary, tapir, puma, jaguar and various monkeys flourish, while the swampy river banks are the home of huge alligators, the high mountains, of the vulture and condor, and the plains, of the rhea, the South American ostrich. The forests swarm with brilliant tropic birds, harsh-voiced but beautiful. Each animal named above is described in these volumes.

Agriculture. The gradations in the climate cause distinct zones in the plant life, for what flourishes in the hot lowlands will not grow in the cold, clear air of the mountains. In these highest regions little will grow—no trees, and only such grains as mature very rapidly, but the intermediate districts, in the mountain basins and on the eastern slopes of the Andes, produce most of the temperate and sub-tropic growths. These are almost the only regions which have been really cultivated, and they yield lemons, oranges, sugar cane, bananas, pineapples, coffee and some cotton.

The great eastern plain, it is believed, contains much exceedingly fertile land, but agriculture is as yet almost entirely neglected and the methods used in cultivating the soil are of the most primitive sort. The great wealth of this lowland region lies in its forests. Rubber trees are plentiful, and the occupation of many of the Indians of the forest region is the gathering and transporting of rubber. The peon system prevails here as in Brazil (see BRAZIL, subhead *The Great Forests and the Rubber Industry*). Every year thousands of tons of crude rubber, worth almost $8,000,000, are exported. See, also, PEONAGE.

Mineral Wealth. Bolivia is rich in minerals, but the development of the mining industry was long hindered by poor transportation facilities and lack of capital. In recent years rail-

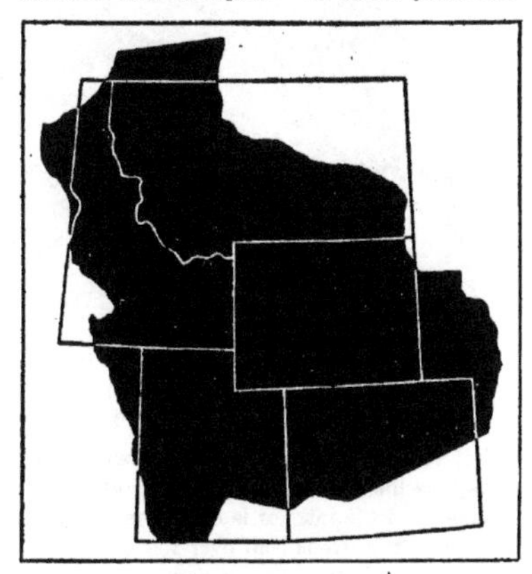

COMPARATIVE AREAS

Bolivia is of uncertain area, but the best authority ranks it nearly equal in size to Idaho, Montana, Wyoming, Utah and Colorado.

road construction through the Bolivian high plateau has somewhat remedied this condition; the tin, silver and copper mines in operation are worked with the latest mining machinery, and the country produces each year about 40,000 tons of tin, 3,000 tons of copper and silver worth millions of dollars. Once Bolivia was noted for its gold. A part of that which the Spaniards found decorating the buildings and the apparel of the Incas came from that country, but after the conquest the mines were not worked until the Spaniards made slaves of the natives and compelled them to labor. Since the country became independent comparatively little attention has been paid to the mining of gold.

Transportation. The mountainous character of the most thickly-settled part of Bolivia

A TYPE OF OPEN-AIR MARKET
Such markets are characteristic of Spanish-American countries.

makes excellent roads almost impossible, and to-day, as hundreds of years ago, most of the commerce of the interior is carried on by pack animals. Steep trails lead over the mountains, and long trains of mules or llamas wind their way along them, bringing goods from the central provinces to points that are in touch with the railroads or cart roads that lead to the Pacific. For near the cities there are cart roads, over which heavy, two-wheeled carts rumble; and in the western part of the country there are railways with a total length of 1,500 miles.

With the country's recent development has come a recognition of the fact that there can be no great progress without railways; many new lines are being built and others are projected. Among those under construction is a line connecting the Bolivian town of Tiepiza with La Quiaca, on the Argentina frontier. When this road is finished a continuous all-rail route will be open between La Paz and Buenos Aires. Because of its central location in the South American continent, Bolivia is destined to become the connecting link for all the railroads between Peru, Chile, Argentina and Brazil. The rivers communicating with the Atlantic furnish a possible outlet for the eastern sections, but the distance and consequently the cost of transportation is so great that little use is made of them.

In grand total, Bolivia exports about $32,000,000 worth of products each year, and imports $23,000,000 worth, even with inadequate facilities.

History. The opening up of Bolivia to Europeans was as romantic as that of Peru, for the former country was a part of the ancient empire of the Incas, about which legend and mystery long centered. In 1538 it was conquered by Pizarro, and Spaniards flocked in to enrich themselves from its mines. They enslaved the natives so thoroughly that not until 1780 was there a serious disturbance.

Bolivia gained its liberty after fifteen years of fighting (see BOLIVAR, SIMON), and in 1825 the first Congress that met at Sucre proclaimed the independence of Bolivia. The first President was General Sucré, one of the most illustrious South American generals and statesmen of the time. Santa Cruz, who succeeded Sucre, initiated the plan of uniting Peru and Bolivia

OUTLINE AND QUESTIONS ON BOLIVIA

Outline

I. Location and Size

(1) Latitude, 9° 44′ to 22° 50′ south
(2) Longitude, 58° to 73° 20′ west
(3) Comparative area
(4) Actual size
  (a) Reason for uncertainty

II. Physical Features

(1) "The Mountain Republic"
(2) Plateau
(3) Great plains
(4) Rivers and lakes
(5) Close relation between surface and climate

III. Animal Life

(1) Importance of llama
(2) Other animals and birds

IV. Industries and Communication

(1) Agriculture
  (a) Relation of crops to altitude
  (b) Possibilities of future development
  (c) The peonage system
(2) Mining
  (a) Silver and tin
(3) Transportation
  (a) Absence of good roads
  (b) Growth of railroads

V. The People

(1) Proportion of Indians
(2) Costumes of women
(3) Educational and religious condition
(4) Government
  (a) Executive
  (b) Legislature
  (c) Judiciary
  (d) Military system

VI. History

(1) Under Spain
(2) Independence achieved
(3) Modern development

Questions

Why has not the country a tropical climate everywhere?

Why would you not care to spend a winter in a Bolivian home?

Why cannot the area of the country be stated exactly?

What effect has the compulsory military system had on the Indians?

What article that the school children use every day comes from the United States?

If the Bolivians could keep only one of their native animals, which would they choose? Why?

Would you like to be a stylish Bolivian woman? Why?

What South American country resembles Bolivia in its lack of seacoast?

If you were transported to the market place of La Paz, what are some of the interesting things that you would look for?

What section of the country promises great development as a grazing region?

In what respect are the transportation facilities in the most populous section just as they were hundreds of years ago?

Give three popular nicknames of Bolivia, and account for each.

Why can there never be in Bolivia an exciting campaign for the reëlection of the President?

Why will not oranges, which are easily grown in Florida, grow in Western Bolivia, which is nearer the equator?

Express picturesquely the fact that the capital of the country has been by no means stationary.

In which part of the country are the people most progressive?

Native Woman Weaving

Home of Indian Family

Street in La Paz

under the name of the "Confederacion Peru-Boliviana." He succeeded in occupying Lima, the capital of Peru, but the Chileans and the Peruvian faction opposed to the confederation were successful in deposing Santa Cruz, and the whole scheme collapsed. In 1880 Chile made war against Peru and Bolivia and took from them all the coast where are found the saltpetre deposits. As Bolivia was deprived of its seacoast, trade is carried on through Peru, Chile and Argentina, and through Brazil by way of the Amazon. Though the country is peaceful and being gradually developed, it is greatly in need of the help of foreign capital and immigration. D.I.C.

Other Items of Interest. One of the popular names for the country is the "Mountain Republic."

Bolivia's output of antimony is of growing importance. In 1911 only 312 tons were produced, but by 1915 this had increased to 17,923. Great Britain takes by far the largest part of this supply.

Though the climate in July and August is like that of the United States in November and December, the houses are unheated, the price of fuel being practically prohibitive. As the rooms are very large, often thirty feet long and sixteen feet high, they are far from comfortable.

Bolivia is a land of contrasts. In the market place of La Paz are to be seen llamas from the north with their loads of ice, and mules from the south laden with oranges and tropical fruits.

The electric car line which leads to La Paz was built by Americans.

An Englishman once asked, "What is really the capital of Bolivia?" And a Bolivian replied, "The capital of Bolivia is the back of the horse which the President of the republic rides."

The Indians between nineteen and twenty-one who are serving in the army are given not only military drill but a general education, and thus compulsory service is advancing this part of the population very rapidly.

Bolivia is proud of its title, "The Land of Ten Thousand Silver Mines." At one time the entire world was dependent for its silver upon this one country.

So gay are the costumes generally worn that someone has compared the market place of La Paz to a "field of poppies in the month of June."

A recent law provides for the closing of saloons on Sunday, heavy fines being the penalty for violation.

Bolivia has a Boy Scouts organization, well-uniformed and well-equipped.

At Tiahuanacu may be seen the ruins of a city that flourished possibly 3,000 years ago.

Among the countries of the world, only the Straits Settlements produce more tin than Bolivia.

The woman of Bolivia, if she would be in style, must wear as many skirts as she can carry. As each is of a different color, and each is somewhat shorter than the one below it, the effect is rainbow-like.

The automobile-truck has been introduced into Bolivia, and is solving in some sections the problem of transportation.

No other country in the world except Northern India has so many mountains of towering heights as has Bolivia.

Related Subjects. The reader interested in Bolivia will find the following articles helpful:

Andes	Bolivar, Simon
Inca	Pizarro, Francisco
La Paz	Silver
Llama	Sucré
Madeira River	Tin
Pilcomayo	Titicaca, Lake

BOLL WEEVIL, *bole wee' v'l.* One of the chief enemies of the cotton plant is the insect known as the boll weevil. It is estimated that the damage done by it to the cotton crop of the United States decreases the yield about 400,000 bales yearly.

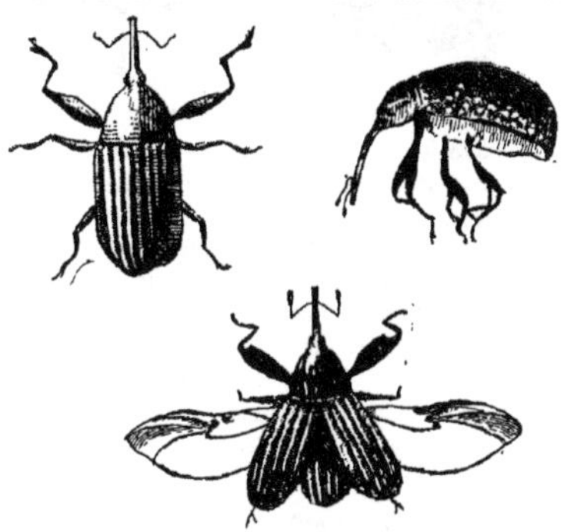

BOLL WEEVIL

Three views of this Southern pest. The weevil is somewhat smaller than an ordinary house fly. The views above are considerably enlarged.

History and Extent. The boll weevil was found for the first time in 1892 around Brownsville, Texas, where it was introduced from Mexico. That is why it is sometimes called the *Mexican cotton boll weevil.* Since that time it has continuously spread northward and eastward until it has reached the cotton fields of Georgia, where it made its first appearance during 1915.

Description. The cotton boll weevil is a small, grayish beetle, a little less than a quarter of an inch long. The insect lays its eggs in punctures made by the female weevil in the buds of the cotton plant, which are called *squares,* and in the bolls. The larvae (young), which are a little longer than the adults, live within the squares and bolls and feed upon their contents. As a rule the square in which the weevil has laid an egg drops to the ground, but most of the damaged bolls remain on the plant, and become stunted or dwarfed. The cotton fiber in such bolls becomes useless.

The insect passes the winter in the adult state. When the plants are cut in the fall or when the bolls that have fallen on the ground dry or rot as a result of the frost, the adult insects, called weevils, leave them and seek shelter under the rubbish of the field or among

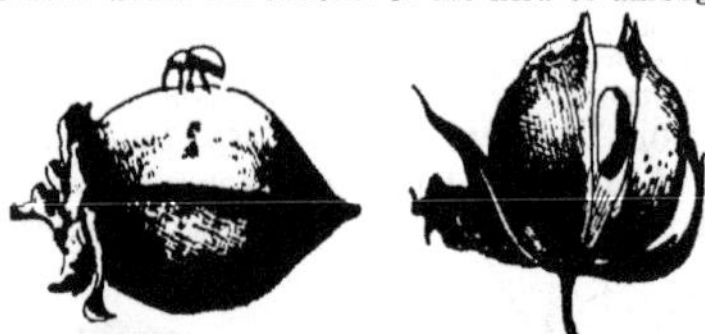

THE ATTACK OF THE BOLL WEEVIL

the vegetation of the roadside or in the adjoining woods, where they remain until the following spring.

Methods of Combating Its Ravages. The United States Department of Agriculture has conducted experiments to discover the best means of ridding the country of the boll weevil or checking its ravages. Below are given some of the chief methods recommended:

> The field ought to be cleaned in the fall by uprooting the stalks of the old plant, collecting with them the fallen bolls and burning them. This is a very important step, for it destroys all the insects and larvae that have accumulated there. Then the field ought to be plowed deep in the fall and prepared during the winter for an early crop. This can be done by planting early maturing varieties, and by fertilizing when necessary.

Recent Researches. In 1913 a new variety of the typical boll weevil was discovered. This attacks the so-called Arizona wild cotton, a cotton-like plant, which grows in a number of mountain ranges in Southeastern Arizona, and also in parts of Mexico and New Mexico. This western variety of the boll weevil has adapted itself to life under extremely arid conditions, in which respect it differs greatly from the typical boll weevil. It therefore appears that the western weevil might thrive in the drier portions of Texas, where the typical weevil has not been able to establish itself.

It has also been discovered that the boll weevil is able to feed and develop on plants other than cotton, especially on all the numerous species belonging to the mallow family. This discovery may have important results as regards the methods to be used for the extermination of these pests. See COTTON. W.F.R.

BOLOGNA, *bo lohn' yah,* an ancient seat of learning in Italy, possessing one of the oldest and most famous universities in the world (see below), now also an important modern industrial center, capital of the province of the same name. On account of its university, founded as early as 1088, Bologna was formerly spoken of as *The Learned City;* later, on account of its democratic principles, it was styled *The Free,* and in more modern times its beautiful climate and fruitful soil earned for it the name of *The Fertile.* Although the inhabitants are progressive and industrious, the city, with narrow and crooked streets, still maintains a medieval appearance. One of its most striking features is the system of arcades, extending far in all directions on each side of the street. On these arcades the shops open.

Bologna might well be called the city of churches, as it has 130 dating from the early part of the eleventh century down to recent years. Of the 180 leaning towers which ornamented the city in the Middle Ages, only two, dating from about 1110, now remain; these are among the most conspicuous objects of the city. The art treasures of Bologna are world renowned; the most famous of its pictures is Raphael's *Saint Cecilia,* painted for the Church of Saint Giovanni in 1515.

The city has important manufactures of silks, velvet, chemicals, paper, and a sausage widely known as Bologna. Population in 1911, 172,639.

University of Bologna. This is one of the oldest and most famous universities of the world. Among its most celebrated students were Dante, Petrarch and Tasso. Its continuous existence is known to date from the early part of the eleventh century, but it had had a periodic existence for several hundred years prior to that time. The influence of the university upon the civil and ecclesiastical organization of Europe during the Middle Ages was marked, and it also became prominent at an early day in scientific investigations and discoveries. It was here that Galvani made his discovery, which led to what was later known as *galvanic electricity* (see ELECTRICITY). The university admits women as students and has a number of women on its faculty.

BOLTON, *bole' ton,* or **BOLTON-LE-MOORS,** a manufacturing town of Lancashire, England, ten miles northwest of Manchester, on the River Croal. It is one of the most important centers of the cotton industry, employing more than 20,000 workmen in its factories and containing some of the largest cotton mills in the world. Bolton was the home of Arkwright (which see), whose inventions revolutionized the spinning industries. All public utilities, including street railways, electric light, gas works, water works, markets and slaughter houses are municipally owned and are operated at a substantial profit, which is devoted to the reduction of taxation. The most important public buildings are the town hall, the market hall and Saint Peter's Church. There are six free public libraries and four public parks. The city is one of the oldest in England; it was designated as a market town as early as 1256. Population in 1911, 180,885.

BOMB, *bom* or *bum,* a hollow ball, usually of cast iron or steel, filled with shot and explosive chemicals which are ignited by a fuse, or by a percussion cap when thrown. As applied to a projectile fired from a cannon the word has been entirely superseded by *shell.*

A bomb is popularly regarded as the peculiar weapon of anarchists and other dissatisfied classes, and the name has acquired a sinister meaning as that of a dreaded destroyer. An ordinary tube such as a gaspipe may be made into a bomb by filling it with explosives and attaching to it a fuse or percussion cap to explode it. A more ingenious bomb is made with a clock-work attachment which will cause an explosion at a certain time, by completing a circuit between two small electric batteries, or by allowing a weight to fall on a percussion cap.

In the War of the Nations, bombs to be thrown by hand, called *grenades,* were extensively used in trench warfare. Others, sometimes of primitive manufacture and consisting of tin cans filled with powder ignited by a fuse, were hurled by catapults. Such a bomb is made somewhat on the principle of a shrapnel shell (see SHRAPNEL). A metal case contains the explosive, on the top of which bullets, pieces of metal, or even stones are placed. The powder or explosive bursts the metal case, fragments of which, together with the other articles it contains, are scattered powerfully in all directions.

BOMBARD'MENT, an attack on a fortress, city or field position by the continued and concentrated fire of big guns. Previous to the War of the Nations a bombardment of strongly-fortified positions had usually resulted in favor of the defenders, the guns then employed not doing sufficient damage to warrant the expense and time of the bombardment. The bombardment of Port Arthur by the Japanese in the Russo-Japanese War, although doubtless of considerable moral effect on the defenders, entirely failed to win the town until the Japanese troops had been pushed forward sufficiently to dominate the interior of all the Russian positions. In the South African War the bombardment of Ladysmith and Kimberley had only trivial effects. The inhabitants escaped injury by digging underground chambers, to which they retired when necessary.

An entire change of opinion as to what bombardment can accomplish occurred at the very beginning of the War of the Nations in 1914, when the forts of Liége, though considered impregnable and among the most modern in the world, were easily battered down by German artillery. The forts of Namur, Antwerp, Przemysl and other towns shared the same fate, and it has been convincingly shown that no matter how strong a fort may be made, guns can be produced to reduce it. Bombardment of trenches was also more effective than in any previous war, and more than sixty per cent of the wounds inflicted were caused by artillery fire. This is due in a great measure to the fact that in previous wars guns employed were counted by tens, whereas in the gigantic European struggle the fire of hundreds of guns was concentrated on the positions attacked. In the last attack on Przemysl by the Austro-German forces, 240,000 shells were fired at the Russian defenders within twenty-four hours.

Aerial Bombardment. The War of the Nations brought about many alterations in the conduct of war, and none of the new methods of destruction was more dreaded than bombardment from the air. The dirigible airships of the Germans and the aeroplanes of other nations hovered over important points and dropped bombs of great power. Cities in England were repeatedly attacked and great damage to buildings and considerable loss of life resulted. This plan of attack being entirely new, it called for new methods of defense. Guns were mounted on hills, on high buildings, and wherever they could be utilized to greatest advantage. Pointing almost directly upward, these guns fired shells into the air in the attempt to destroy or drive away the aerial invaders. The damage done by air raids was considerably less than might have been expected, owing chiefly to the difficulty of ascertaining when the airship was in proper position above the object at which bombs were to be thrown. See WAR OF THE NATIONS. F.ST.A.

BOMBAY, *bom bay',* a presidency, or province, of British India, extending from Baluchistan in a long strip down the western coast, nearly to the southern end of the peninsula. The surface is mountainous, the Western Ghats running parallel to the coast for nearly the entire length of the territory. The principal rivers are the Indus, the Tapti and the Nerbudda, all flowing into the Arabian Sea. The climate is

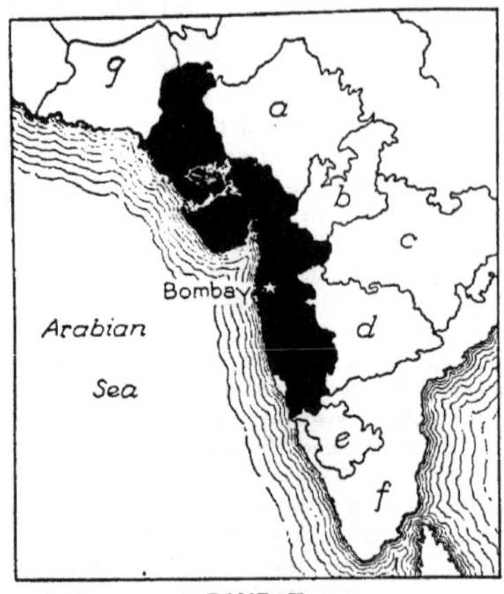

BOMBAY

Location of Bombay presidency, the city of Bombay, and the surrounding presidencies.
(a) Rajputana
(b) Central India Agency
(c) Berar Provinces
(d) Hyderabad
(e) Mysore
(f) Madras
(g) Baluchistan (not Indian)

hot for six months in the year, averaging 95° F. in the shade; during the other six months it is about 10° cooler. The leading agricultural products are cotton, rice, wheat and millet. Bombay is the largest cotton-producing district of India and furnishes nearly one-fourth of the entire crop of the peninsula.

The manufactures are cotton and silk fabrics, leather and brassware. The commerce is more extensive than that of any other Indian province, large quantities of cotton, tea, sugar and wool being exported to Great Britain.

The government is in the hands of a governor and an executive council, and for local administration the presidency is divided into four divisions, the Northern, Central, Southern and Sind. There are several native dependencies within the territory, each of which is controlled by a chief, who is subject to the governor of the presidency and is assisted by a British agent residing at his court. Population in 1911, 19,672,642. See INDIA.

BOMBAY, the name of the most important seaport on the west coast of India, and the capital of the presidency, or province, of the same name. The city is one of England's choicest Eastern centers of population, and is recognized as the "gateway of India." It is sometimes stated that the name is derived from the Portuguese *boom bahia,* meaning *good harbor,* but it really comes from *Bambai Mumba,* the name of a Hindu goddess. Bombay is the second city of India, having fallen behind Calcutta in the race for position, chiefly on account of famines and plagues which have from time to time caused great losses in population and money. Within recent years, however, the city has made great progress, materially assisted by its monopoly of the Indian cotton trade.

Approached from the sea, Bombay is said to afford a spectacle unsurpassed for beauty except by the bay and city of Naples. What was originally the island of Bombay, on which the city stands, is now connected with the mainland, forming a peninsula, of which twenty-two square miles comprise the city area. The streets are wide and in the business section are lined by modern and substantial buildings, rivaling those of many European cities and unequaled in any other Asiatic city under British rule. The city has two waterfronts, one facing the outer harbor, called Back Bay, the inner fronting on a magnificent land-locked harbor five miles broad and fourteen miles long.

The population of Bombay is of a very mixed character, including representatives of every Eastern race. The European inhabitants occupy a suburb apart from the native quarter and the business section. The most powerful community, yet the smallest in numbers, is that of the Parsees, who control by far the greater number of the native business establishments (see PARSEE). The high-caste Mahrattas dominate the political situation, next to the British element, which is in authority. Arabs, Afghans, Bengalis, Rajputus, Chinese, Malays, Tibetans, Japanese, Siamese and Singalese are included in these races; they are too far apart in racial instincts to have common interests, and each race is too insignificant to form a strong party. To tourists the most impressive native sight, and one of the most grewsome in the world, is the so-called Towers of Silence, where the Parsee dead are placed, to be devoured by vultures. See TOWERS OF SILENCE.

Apart from the cotton industries there are manufactories of metal goods, gold and silver work, dyeing and tanning establishments and numerous flour mills. Bombay possesses one of the finest railroad terminal buildings in the world, has excellent docks, good internal communications, and its coastal trade is more than double that of Calcutta. Next to Madras, the city is the oldest of British Indian possessions, having been ceded in 1661 by Portugal, under whose dominion the city never flourished. Population, 1911, 979,445. F.ST.A.

BONA FIDE, *bo'nah fi'de,* a Latin term meaning *in good faith,* a term used in law in the sense of *honestly, genuinely* or *without deception.* A *bona fide* purchaser is a person who buys property and pays for it, believing that the title to such property is clear and that the owner has the right to sell it. A contract entered into in good faith cannot be annulled without the consent of both parties, but when a contract is shown to be not *bona fide,* the injured person may lawfully cancel it. The term is also used in connection with suits for libel, where a distinction is made between malicious acts and acts done in good faith.

BONANZA, *bo nan' za,* a Spanish word meaning *fair weather* or a *favoring wind,* used in mining districts of various countries to signify an abundance of precious metal or rich ore. The word was first applied in this way by the miners of the Comstock Lode, a wonderful gold and silver mine in Nevada which yielded $340,000,000 worth of ore in thirty years. It is now also used to signify any good fortune or successful enterprise.

BONAPARTE, *bo' na pahrt,* the family name of one of the greatest characters in history, Napoleon I, emperor of the French. Bonaparte is the French form of the original Italian name *Buonaparte,* borne by Napoleon's family in Corsica, and by several Italian families prominent in the early Middle Ages. The French spelling was used entirely by Napoleon after 1796. The Corsican Buonapartes were descendants of Francesco Buonaparte, who came to Corsica from Italy in the middle of the sixteenth century. In the eighteenth century three male representatives of the family were residing in Ajaccio, capital of Corsica; these were the Archdeacon Lucien Bonaparte, his brother Napoleon, and the nephew of both, Carlo, the father of Napoleon I.

Carlo, or **Charles,** Bonaparte (1746-1785), Napoleon's father, completed a law course at the University of Pisa, and after the French conquest of Corsica accepted a government position at Ajaccio. Then he married Letizia Ramolino, a beautiful girl of noble character (in 1767). He fought under Paoli, the Corsican patriot who carried on the vain struggle for independence from French rule, and when he saw that the cause was hopeless he went over to the side of France. In 1771 Louis XV included the Bonaparte family in the list of those which were to enjoy rank among the French nobility. Of a restless and adventurous disposition, Carlo went to Paris in 1777, where he resided for several years, obtaining free admission for his second son Napoleon to the military school of Brienne, where the boy laid the foundation for the career that was to change the destiny of Europe.

Eight children survived him: Joseph, later the king of Spain; Napoleon I, emperor of the French; Lucien, who became prince of Canino; Maria Anna, afterward called Elisa, Princess of Lucca and Piombino and later wife of Prince Bacciocchi; Louis, whom Napoleon made king of Holland; Carlotta, afterward named Marie Pauline, who married Prince Camillo Borghese; Annunciata, afterward called Caroline, wife of Murat, for a time king of Naples; and Jerome, by Napoleon's decree king of Westphalia.

Jerome Bonaparte (1784-1860), the youngest brother of Napoleon, after a short European career began an American line of the Bonaparte family by his marriage to Miss Elizabeth Patterson, of Baltimore, in 1803. Having become a lieutenant in the French navy, he was sent on an expedition to the West Indies, and on the outbreak of the war between France and England in 1803 was forced to run his ship into New York harbor to escape the pursuit of English cruisers. While traveling in the United States, Jerome met Miss Patterson and married her in spite of his august brother's protests. Two years after their marriage he separated from his wife at Napoleon's command, and in 1807 was created king of Westphalia, the kingdom erected by Napoleon from conquered German territory. The emperor also forced his brother to marry Catherine, Princess of Württemberg, for he decreed that the queen of Westphalia must be of royal blood.

Jerome's short and troubled reign, in which the state was all but ruined financially, came to an end with the defeat of Napoleon at the Battle of the Nations (Leipzig) in 1813. He was loyal to his brother's cause through all the events that followed, and fought bravely for him at the Battle of Waterloo. Thereafter, until 1847, he lived in various European cities. When his nephew, Louis Napoleon, became President of the French republic established in 1848, Jerome was given charge of the home for disabled soldiers in Paris, and later became marshal of France and president of the Senate.

By his early Baltimore marriage, Jerome had one son, likewise called Jerome, who became the father of Charles Joseph Bonaparte, who was Secretary of War and later Attorney-General in President Roosevelt's Cabinet. Two sons and a daughter were born of the second marriage. The younger son, afterwards known as Prince Napoleon, became one of the pretenders to the French throne.

Joseph Bonaparte (1768-1844), the eldest son of Carlo Bonaparte, was closely associated with his brother, the Emperor Napoleon, throughout the latter's period of triumph. He was born in Corsica and educated in France at the College of Autun. Returning to Corsica in 1785, he studied law and in 1792 became a member of the administration of Corsica during its brief period of independence. In 1793 he emigrated to Marseilles and married the daughter of a wealthy banker; later, with the rise of his brother to fame after his brilliant campaign in Italy, Joseph began an important public career. Napoleon selected him to conclude a friendly treaty with the United States in 1800, and he later signed the historically famous Treaty of Lunéville and that of Amiens.

JOSEPH BONAPARTE
Once an exile in America.

In 1806 Napoleon made him king of Naples, and two years after transferred him to Madrid as king of Spain. His position there, entirely dependent on the support of French armies, became almost intolerable; he was twice driven from his capital by the approach of hostile armies, and the third time, in 1813, he fled, not to return. After the Battle of Waterloo he went to the United States and lived for a time near Philadelphia, assuming the title of Count of Survilliers. He afterwards went to England, and from there to Italy, where he died.

Louis Bonaparte (1778-1846), the second brother of the Emperor Napoleon and the father of Napoleon III. On completing a course at the artillery school of Châlons, France, he served under Napoleon in the Italian and Egyptian campaigns. In 1802 he married Hortense Beauharnais, Napoleon's stepdaughter, and four years later yielded very reluctantly to his brother's demand that he accept the Dutch crown. It is to the credit of Louis that he tried to rule in the interests of his subjects, and he gave up the throne in 1810 because he thoroughly disapproved of Napoleon's "Continental System," which was most injurious to Dutch commerce. After his abdication, Holland was annexed to France. From that time on Louis lived chiefly in Rome and in Florence. He was a writer of considerable ability and published a novel and several historical works. See CONTINENTAL SYSTEM.

LOUIS BONAPARTE

Lucien Bonaparte (1775-1840), Prince of Canino, next younger brother of Napoleon, was an enthusiastic supporter of the people's cause throughout the French Revolution. Having been a disciple of Robespierre, he was imprisoned for a time after the fall of that leader, but was released through Napoleon's influence, and in 1798 was settled in Paris as a member of the newly-elected Council of Five Hundred. Shortly after Napoleon's return from Egypt, Lucien was elected president of the Council and in this position saved his brother by his high-handed dismissal of that body when an attempt was made to pass a vote of outlawry against him. As Napoleon began to develop his system of military despotism, Lucien, who still held to his republican principles and candidly expressed his disapproval of his brother's conduct, fell into disfavor and was sent out of France as ambassador to Spain. Settling finally in Rome, he devoted himself to the arts and sciences and lived in apparent indifference to the growth of Napoleon's power. He came to France, however, and exerted himself on his brother's behalf, both before and after the Battle of Waterloo. A man of literary tastes, his published works include two long poems and an autobiography. B.M.W.

Napoleon Bonaparte. See NAPOLEON I.

BO'NAR, JAMES (1852-), a Canadian economist, deputy master of the Canadian branch of the royal mint. He was born at

Collace, Perthshire, Scotland, and was educated at the universities of Glasgow, Tübingen and Leipzig and at Balliol College, Oxford. From 1881 to 1895 he was junior examiner in H. M. Civil Service Commission, and from 1895 to 1907 senior examiner. He went to Canada in 1907 to assume direction of the royal mint, established in that year. Dr. Bonar made his first contribution to economic literature in 1881, with a book entitled *Parson Malthus*. Of his numerous later books the best are *Malthus and His Work; Philosophy and Political Economy; Political Economy,* an elementary treatise; and *Disturbing Elements in Political Economy.*

BOND, a word of many meanings, but most commonly applied to a special form of contract—a written or printed evidence of debt issued by a government or a corporation, usually for the purpose of borrowing money. The name is given to any obligations issued in a group and bearing a fixed rate of interest, provided they are under seal. If the evidence of debt is merely a promise, not under seal, it is a promissory note, not a bond.

Mortgage and Debenture Bonds. Bonds are of two kinds, *mortgage* and *debenture.* A *mortgage* bond, as the name indicates, is a direct lien on the assets of the company, or on a specially designated part of the assets. A railroad, for example, may issue general mortgage bonds based on its entire assets, or it may issue bonds whose security is only a branch line or a certain division. A *debenture* bond, on the other hand, is merely a promissory note under seal, and has no characteristics of a mortgage. If the corporation fails to pay the interest or principal, the holders of mortgage bonds may bring foreclosure suit; the holders of debenture bonds are merely creditors who may share in the assets.

Theoretically and legally there is no limit to the number of bond issues which a corporation may issue. It may have first and second mortgage bonds, or even third mortgage bonds. It may have all of these and also several issues of debenture bonds. Mortgage bonds are usually for longer periods, varying from twenty to ninety-nine years, and bring a moderate return—four to six per cent is the average. Debenture bonds, being merely promissory notes, require fewer legal preliminaries, and are usually issued when a corporation desires to borrow money for a short period of years. The rate of interest on such short-time loans is usually larger than on long-time mortgages.

Bonds issued by governments are not based on a mortgage as security, but merely on the apparent ability and intention of the government to repay their face value. If the government of the United States or of Canada should refuse to redeem its bonds, that is, should *repudiate* them, the holders would have no redress. In the past, a number of the states have repudiated their bonded debts, or have compromised with their creditors. While repudiation is still possible, in the United States and Canada the issue of bonds is so carefully controlled by statute that the bonds of a state or province, or any of its divisions, are as safe as any investment ever can be.

Registered and Coupon Bonds. Advertisements of bond issues frequently state that "these bonds are sold with the privilege of registry." This means that the owner's name, the serial number of the bonds and the total amount held by each owner are registered on the company's books. The interest and principal are payable only to the person whose name is registered; in this way the holder is protected from loss or theft. Registered bonds may be transferred by giving proper notice to the secretary, who makes the necessary changes in the register.

Coupon bonds have certificates of interest, or *coupons,* attached to the bond and giving the exact amounts of interest due, and dates when due. These coupons are cut off on the interest date and are presented for payment. Any bank, on receipt of a coupon, will usually cash the coupon for its customers, and will in turn collect from the corporation which issued the bonds.

The difference between registered and coupon bonds is merely in the form. It does not in any way affect the character of the security. A mortgage bond may be either registered or coupon; so, too, may a debenture bond.

Bonds as Investments. The popularity of bonds is easily explained. In the first place they give a corporation an opportunity to borrow money without giving outsiders a voice in the management, for the owner of a bond does not become a stockholder, but merely a creditor. If the corporation is being mismanaged, the bond-holders have the power to protect their interests through legal processes. In the second place the safety of bonds appeals to investors. Not all bonds, of course, are safe; the bonds of a company which has no assets are as worthless as its stock. In all cases, however, the bond-holders are preferred creditors, and bonds are the first obligations of a company. No dividends can be paid on either common or preferred stock until the interest on the outstanding bonds has been earned and paid. In good business years the capital stock of a company may earn large dividends, but if business is poor there may be no dividends. Bonds pay a smaller, fixed rate of interest, but a company must be close to bankruptcy if it cannot pay interest on its bonds. If the bond interest is not paid it is said to be *defaulted.*

To say which class of bonds is the safest is a difficult, if not impossible, task. As a rule, the bonds of national governments are the safest, and for that reason pay the lowest rate of interest. The bonds of Haiti, Nicaragua or any other country in which the government is unstable, are unsafe; they have occasionally been repudiated, and may be repudiated again. The bonds of states, provinces, counties, townships and school districts form a second class, normally just as safe as the first class. Third in preference should come the bonds issued by large corporations, whose conservative management is known and whose semi-public character makes secret manipulation practically impossible. Such corporations include the large railway systems and most public service corporations. Their bonds are usually listed on the stock exchange (which see), but the market prices must not always be accepted as a fair index of the value of the bond. A fourth class of bonds would include those issued by smaller companies, many of which are probably as sound financially as the larger ones. Independent investigation, however, is advisable in every case. But if the intending investor is not in a position to learn the facts for himself he should at least ask the advice of a reputable banker or broker.

Market Prices. Many bond issues, particularly if the issue or the company is a large one, are listed on the stock exchange. In this case the quotations of sales furnish a fairly adequate test of the value of a bond. Bonds are quoted on a basis of $100 par value. For example, if the quotation is 97½, this means that a bond whose face value is $100 is being sold for $97.50. If the market price falls far below par it is a fairly sure sign that the bond is not a good one for the conservative investor to hold. It is customary in the market to designate each bond issue by a short abbreviation which identifies the bond; for example, Bethlehem Steel 1st 5s, means first mortgage bonds, drawing five per cent interest, issued by the Bethlehem Steel Company; Panama 2s 36 means a bond drawing two per cent interest, principal due in 1936, issued by the Republic of Panama. If the quotation reads Panama 2s '36, 99 bid, it means that somebody has offered $99 for a $100 bond issued by Panama and due in 1936.

Other Meanings of the Word "Bond." The words *bond, bind* and *band* all come from the same root, and originally had the same meaning, that is, a *fastening.* In a figurative sense there was a fastening between two people who made any kind of a contract; and in law today a bond is any contract under seal by which a person agrees to do or not to do a thing. More specifically, a bond involves the payment of money; any other contract is usually called a *covenant.*

Bonds may be *simple* or *conditional.* A simple bond is a definite promise to pay. On the other hand, a bond may involve a condition. A bank cashier, for example, furnishes a bond to his employers; that is, an individual or a company agrees to make good the bank's loss if the cashier steals any of its money. A bondsman may agree to pay the court a certain amount if some accused prisoner is not on hand when he is wanted; instead of staying in jail the prisoner may then be temporarily released. He is said to be out on *bail* (which see), and the bond given to insure his appearance is a *bail bond.* Goods which are liable to customs duties or internal revenue taxes are said to be *in bond* when they are placed in storage under a bond that they will not be removed until the duty or tax on them is paid. This is the meaning of the phrases "bottled in bond" and "imported in bond." Such storage houses are called *bonded warehouses.* W.F.Z.

Related Subjects. The following topics will be found helpful by the reader interested in bonds:

Bail	Interest
Bonded Warehouse	Mortgage
Commercial Paper	Stock Exchange

BOND, Sir Robert (1857-), a Newfoundland statesman, Premier from 1900 to 1909, and for many years the leading Liberal in the political life of that colony. He was educated for the bar, but the call of politics was stronger than that of the law. In 1882 he was elected to the Newfoundland Assembly, and in two years so established his leadership that he was chosen Speaker. From 1889 to 1897 he was Colonial Secretary in the executive council, or Cabinet.

During these years he was conspicuous in diplomacy. In 1890 he negotiated the Bond-Blaine reciprocity agreement with the United States, but the treaty was rejected by the British government at the request of Sir John A. Macdonald, Premier of Canada. He was one of the Newfoundland delegates who met Sir John J. S. Thompson and Sir Mackenzie Bowell in 1895 to discuss the control of the Atlantic fisheries, and in 1895 was chairman of the delegates who went to Ottawa to discuss union with Canada, to which he was personally opposed. In 1902 he again negotiated a reciprocity treaty with the United States, but this one was rejected by the United States Senate.

As Premier and Colonial Secretary from 1900 to 1909, Sir Robert necessarily played an important part in the settlement of the disputes with France, in 1902, and with the United States, from 1905 to 1910. He opposed any commercial concessions to the United States, and in 1907 went to England to protest against the policy adopted by the British government. Sir Robert's visit failed to secure any modification, and to a large extent he was personally blamed for the hard times which existed in Newfoundland. At the general elections of 1909 his ministry was ousted, and he then became leader of the opposition. In January, 1914, however, he charged that the Fisherman's Union, his political ally, was betraying him, and he retired to private life. Sir Robert's knighthood dates from 1901, when he was created Knight Commander of the Order of Saint Michael and Saint George (K. C. M. G.).

BONDED WAREHOUSE, a warehouse maintained or controlled by a government for storage of goods on which duty is levied, to be held until such time as they are required. This system has been found of great convenience to merchants, as it enables them to pay duty by installments, removing the bonded goods in such quantities as may be required from time to time. Goods of domestic manufacture, on which internal revenue taxes are levied, may be stored in the same way. While in bond the goods may be made ready for sale by the manufacturer or merchant, under supervision of government inspectors, who must see that nothing dutiable is removed from the warehouse until due payments have been made.

BONE, the hard substance that forms the skeleton or framework of the bodies of man and all other animals that have backbones. Someone has called the bones the "stiffening rods of the body." Many of the lower orders of animals, such as oysters, clams, jelly-fish and insects, have no bones.

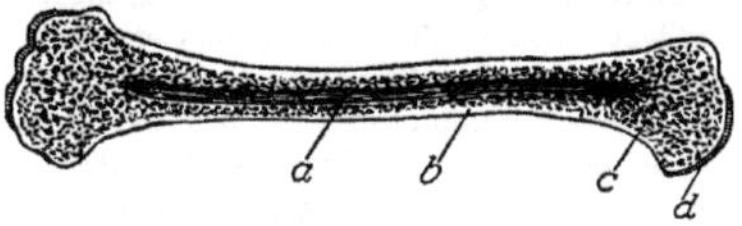

BONE AND CROSS-SECTION
(*a*) Marrow cavity; (*b*) hard substance; (*c*) spongy bone; (*d*) cartilage.

Bones contain both animal matter and mineral matter—a larger proportion of the latter than any other organs of the body. About one-third of their composition is animal matter, and two-thirds mineral matter. If a bone is soaked for two or three days in a weak acid the mineral matter is dissolved, and the animal matter remaining can be bent into almost any shape. If bones are burned the animal matter is destroyed, leaving the mineral matter, which is chiefly a compound of lime, containing phosphorus. Burnt bone is very brittle and easily ground into powder. Bones are sometimes called the "petrified" organs of the body, because they have been slowly changed into a sort of porous limestone as the body reaches maturity. In childhood they contain only a small portion of mineral matter, and are not easily broken or injured by falls; this is why children suffer few severe injuries. On the other hand, children's bones may easily be bent or otherwise deformed by lack of exercise or by keeping the body too long confined in one position.

Structure. On the outside bones are hard and compact, but the interior is porous; the long bones like the humerus and femur are hollow, and these hollow spaces are filled with yellow marrow. Numerous small nerve canals traverse the bones, and through these nerves blood vessels and lymphatics find entrance and convey nourishment. The bone is covered with a close-fitting membrane called the *periosteum.*

Broken Bones. When a bone is broken it should at once be "set", that is, put in its proper position; then the joined ends should be so fastened that they cannot move upon each other. They will then grow together, and in a few weeks, ordinarily, the bone will be as good as ever. Dr. Alexis Carrel, in his successful experiments in limb grafting, has shown that bones can be grafted upon each other. The bones of the aged are more brittle than those of the young and are more easily broken. They are also more difficult to heal.

Used as Fertilizer. The mineral matter in bones, comprising lime and phosphates, makes excellent fertilizer. The animal matter is first removed by burning the bones or boiling them; either process accomplishes the desired result, but by boiling less of the mineral matter is lost. The bones are then ground to powder, and the latter is treated by chemical processes to make all the contained minerals soluble in water. W.F.R.

Related Subjects. The following articles will be found interesting and helpful in connection with this topic:

Arm	Hand
Carrel, Alexis	Joints
Cartilage	Lymphatics
Fertilizer	Pelvis
Foot	Skeleton

BONEBLACK, IVORY BLACK, or **ANIMAL CHARCOAL,** is the charcoal obtained by heating bones in closed iron retorts until the animal matter is burned. The bones are reduced to small, coarse grains; these are then ground to a fine powder, which forms the boneblack of commerce. Boneblack will absorb the coloring of liquids that pass through it, consequently it is extensively used in refining sugar and in removing the coloring matter from syrup (see SUGAR). It will also absorb odors and is sometimes used to remove disagreeable smells from clothing and houses.

BONHEUR, *bo nur',* ROSA (1822-1899), a famous French artist, whose spirited reproductions of animal life have given her repute as the greatest woman painter of animals. She was baptized Marie Rosalie, but has always been known as Rosa. She painted from early girlhood. When she was nineteen years of age her *Rabbits Eating Carrots* was displayed in the annual Paris exhibition known as the *Salon.* Thereafter until 1855 her pictures were exhibited each year, and in 1848 she was awarded a medal of the first class. In 1865 Empress Eugenie honored her with the cross of the Legion of Honor.

ROSA BONHEUR

In order to study animals at close range she practiced for a time in various slaughterhouses on the outskirts of Paris. Her canvases are remarkable for the truthfulness shown in the representation not only of the animals painted, but of the landscape setting of the pictures. The drawing and composition are likewise excellent, but the color is somewhat hard and she was only moderately successful in giving her pictures an effective atmosphere.

Possibly the most admired of her canvases is the great *Horse Fair,* which gives an extraordinary representation of struggle and action. This was bought by Cornelius Vanderbilt for $55,000 and presented to the Metropolitan Art Museum of New York City, where it now hangs. *Plowing in the Nivernais,* one of her best pictures, is now in the Louvre, Paris, which also possesses her *Haymaking Season in Auvergne.* The *Deer in the Forest, A Limier-Briquet Hound* and *Weaning the Calves* are in the Metropolitan Museum, and her *Deer Drinking* is in the New York Lenox Library. The *Shepherd of the Pyrenees* is reproduced here.

BONESET, *bone' set,* or **THOROUGHWORT,** *tho' ro wert,* a useful annual plant, found in meadows and lowlands of the United States and Canada. It is easily recognized by its tall stem, four or five feet in height, passing through the middle of a large, double, hairy leaf; and especially by its rayless flower-heads in loose, flat tops. It is about the only white blossom appearing among the gay, late summer colors. It is much used in medicine as a tonic and for causing perspiration. The name *thoroughwort* refers to the appearance of the union of the exactly opposite leaves.

BON'IFACE, the name of nine Popes, three of whom are of special historic importance.

Boniface II (530-532) was the first Pope to assume the title of Universal Bishop of Christendom, the head of the Church having been known up to that time as Bishop of Rome.

Boniface VIII (1294-1303) was the most famous Pope of the name. Two edicts showed clearly his opinion as to the place the Church should hold in the life of the day. The first of these was the *Clericis Laicis,* issued in 1296, which forbade the payment or collection of taxes on Church property without the consent

THE SHEPHERD OF THE PYRENEES

"THE SHEPHERD OF THE PYRENEES," Rosa Bonheur called this picture, but the lover of her work knows well that it was not the shepherd who was the central figure in her mind. For she loved animals, and brought to the delineation of them a sympathy and an understanding almost unmatched in the history of art. She has not made of her sheep and cattle mere soulless brutes, nor has she erred on the other side and given them too many human characteristics; but she has caught the very nature of the animals and transferred it to the canvas, so that every animal-lover may see it. The care and affection with which she painted her favorite subjects are evident in every detail. The sheen on a horse's coat, the tangled roughness of a lion's mane, the softness and thickness of a sheep's curled wool—these she loved to bring out as accurately as possible. All in all, so successful was this greatest of woman animal painters that the person who would learn more of animals could not do better than to use her pictures as his text-book.

In this picture she has made it evident that the shepherd loves his sheep just as she does. He must spend his days high in the mountains, where week after week he sees no human being, but he is not lonesome, for he has the companionship of his flock. Even the glory of the softly tinted mountain peaks cannot draw his attention from the sheep, which in turn show their affection by clustering about him as he sits to rest.

L. J. B.

of the Holy See; the second was the *Unam Sanctam,* which declared that those who did not hold the spiritual power of the Church superior to any merely temporal power were maintaining false doctrine, dangerous to salvation. His attempts to assert the Papal supremacy in France led to contests with King Philip the Fair, during the course of which Boniface was arrested and imprisoned.

Boniface IX (1389-1404) bent all his energy to strengthening Papal authority and succeeded in making himself almost absolute in Rome. Without the city, however, he met with firm opposition and was obliged to make concessions. G.W.M.

BONIFACE, SAINT (680-755), a celebrated English missionary, whose wonderful labors among the heathen tribes of Germany have been perpetuated in his title, the *Apostle of Germany.* He was born at Kirton, Devonshire, of a noble Anglo-Saxon family, and received in baptism the name Winfrid. In 718 Pope Gregory II called him to preach the Gospel to the German tribes, and for three years he labored among them, seeing multitudes converted through his preaching. The Pope appointed him bishop in 722 and gave him the name of Boniface. About 743 he founded the Abbey of Fulda, now one of the most famous monasteries in Germany, and for ten years, beginning in 744, he was archbishop of Mainz. He is said to have enforced his missionary teaching by cutting down, with his own hands, the oak at Geismar, sacred to the pagan gods. The festival of Saint Boniface is celebrated in both the Roman and Anglican churches on June 5, the anniversary of the day on which he met his death at the hands of a heathen mob. G.W.M.

BONN, a town of Germany, the birthplace of Beethoven and the home of an important modern German university. It is situated in a region noted for its beautiful scenery, on the left bank of the River Rhine, fifteen miles southeast of Cologne. As a commercial center Bonn is not of great importance, but breweries and jute spinning and weaving afford employment for a considerable number of inhabitants. The chief buildings, in addition to the university, are the Münster Church, in late Romanesque style, the Rathaus, or town hall, and a fine court of law. The city passed into the possession of Prussia by the terms of the Congress of Vienna in 1815, and its renewed prosperity dates from that year. Population in 1910, 87,978.

Bonn University. This is one of the most famous German universities, established in 1818, by Frederick William III, king of Prussia. Its faculties include those of theology, law, medicine and philosophy. It had over 2,400 students at the outbreak of the War of the Nations in 1914, but the number was very small during the continuance of that struggle. The library contains 360,000 volumes, besides a large number of manuscripts. The university also has a celebrated observatory. Among its most famous students was the crown prince of Germany, who in 1888 became Emperor William II.

BOOBY, *boo'bi,* a swimming bird which is a tropical species of gannet (which see). Because it alights on ships and allows itself to be caught easily, the sailors gave it its name, the word originating from the Spanish word meaning *idiot.* The booby is really a fearless bird, and its seeming stupidity may be merely the difficulty of setting its long, heavy wings in motion or from living in desolate places, not knowing man and thus having no instinctive reason to fear him. The booby has a naked throat and lower jaw, in one species colored blue. It lives on fish which swim near the surface of the water; this food its tyrant enemy, the frigate bird, often steals out of its bill.

BOOKKEEPING. Did you ever keep the score of a ball game by cutting notches on a stick? If you did, you did not think of it as bookkeeping, and yet it was in a very similar manner that the first business accounts were kept. Indeed, it is only a few hundred years since the Chancellor of the Exchequer, the treasurer of the funds of the British government, recorded all his transactions on tallies, or notched sticks, and it is less than a century since the tally ceased to be part of the official bookkeeping of the British nation.

Everyone should have some knowledge of bookkeeping. Most people would like to be able to keep accounts for themselves or to understand the accounts of others, but many are deterred by the belief that the subject is complicated and the work difficult. In reality modern bookkeeping is nearly as simple as notching sticks, though the process is longer. This does not mean that it is easy for a farmer to comprehend the account books of a factory, or for a housekeeper to understand the records of a large corporation. But it does mean that a farmer may easily keep books for his farm and a housekeeper for her household. Of course, the books can be made complicated, but they do not have to be.

One valuable service of good bookkeeping is the help it gives in planning for the future. Is it wise for the storekeeper to put in a larger stock of overcoats this winter, or to invest his money in spring suits? Will it pay the farmer to buy a gasoline tractor, or to continue using horses? Can the man who owns a home afford to buy new furniture now, or will he do better to save his money toward future expenses, such as painting the house? Will it be worth while for the housekeeper to continue taking boarders, or would it be more profitable for her to spend her extra time in baking goods for sale? Does the manufacturer make more or less money by owning his factory building than he could make if he rented? All such questions can never be properly answered by people who do not keep accounts, or who keep them with little regard to scientific principles.

Single-Entry Bookkeeping

Let us suppose you have written the following in your diary:

I sold Harry my old camera to-day. He will pay me five dollars later.

You have *entered* in your book a record of the transaction, or, in bookkeeping language, have *made an entry*.

In the single-entry system of bookkeeping just such a diary is kept, called a *day book*. It usually contains two double columns at the right of the page, for the figures representing money, and a single *checking* column at the left, the purpose of which will be explained later. Sometimes the first of the two money columns contains the figures for each item in a list of goods bought or sold and the second column shows only the total of the list. For example, if you sold Harry both a fishing rod and a camera, you might write this:

	June 8 Harry Wilson, sold him				
	Camera	5	00		
	Fishing rod	1	25	6	25

This is seldom the best use to make of the two columns, however. What will usually be found a more satisfactory method is to write in the first column the figures for merchandise or money which people get from us, and in the second column the figures for merchandise or money which we get from others. The first column is called the *debit* column; the second, the *credit* column. Written in this fashion, the entry for Harry would appear as follows:

	Harry Wilson, sold him Camera 5.00 Fishing rod 1.25	6	25		

Now suppose that after a time you wanted to make out a bill of what Harry owes you. He has bought several things from you, and has made two or three small payments. If you had to look through all the items in the day book it would take you a long time. That is why you keep a second book called a *ledger*, in which all the items of business with each person are gathered together on one page. You should copy, or as bookkeepers say, *post* your entries from the day book into the ledger as soon as they are made, so that your ledger will always be up to date. Then when you want to see how much any person owes you, only a moment is needed.

In a large ledger the right half of each page is for credits, the left half for debits. In a smaller ledger it is generally necessary to give to each account two pages that face each other. Each page or division of a page has at the left a double column for the date, then a space for the explanation, next a single column for checking (explained in the next paragraph), and at the right a double column for dollars and cents. If the day book has the debit and credit columns, as advised above, items with figures in the debit column are posted to the debit half of a ledger account, and those with figures in the credit column to the credit half. If the columns in the day book are not so employed, it is necessary to separate the debits from the credits at the time of posting, a task apt to be troublesome.

When a posting is made, the number of the page in the day book from which it is copied is written in the checking column; the posting to Harry's account is then like this:

Harry Wilson

June	8	Camera 5.00 Fishing rod 1.25	1	6	25

This number serves as an index if it is ever necessary to refer to the original entry. Immediately after the posting is made, the number of the ledger page is written in the day book, in the column at the left of the page. This, too, serves as an index, and also shows the bookkeeper that he has completed his posting. For this reason great care should be taken always to check an item the moment the posting is finished, but never sooner. Otherwise, if the bookkeeper were interrupted an item might accidentally be posted twice, or not at all.

The Cash Book. The day book and the ledger are the only two books necessary in single-entry bookkeeping, but others are often helpful. The commonest of these *auxiliary* books is the *cash book*. It is really a day book for transactions in which cash is concerned. Sometimes it is made subordinate to the day book; that is, its items must be posted to the day book. Except in special circumstances, however, this is a waste of time, and the cash book may be considered as a part of the day book; that is, its items may be posted directly to the ledger.

The cash book may confuse a beginner in bookkeeping, for its left-hand column is for receipts of cash, which must be credited to people, and its right-hand column for disbursements, which must be debited to people —just the opposite of the day book. This is a feature copied from double-entry bookkeeping, and aside from custom there is no reason why it should be adopted in single-entry. But it is always wise to follow custom in such matters, so that accounts can be understood by other people than the bookkeeper.

The advantage of the cash book is ease in *balancing* the cash. If the sum of the column for receipts is $3.00 greater than the sum of the column for disbursements, this amount is said to be the *balance*. It should exactly equal the cash on hand. All receipts or payments of money should therefore be recorded in the cash book, even if, because no accounts with persons are affected, they will not be posted to the ledger. The word *balance* really stands for the *sum necessary to balance,* or make equal, the two columns. It is also used in speaking of accounts in the ledger. Thus, in the Harry Wilson account below shown, the amount $2.25 (the difference between $6.25+ $1.00 and $2.00+$3.00) is the balance; it is called a *debit balance,* because the debit figures have the larger total:

HARRY WILSON

June	8	Camera 5.00				July	1	Cash	2	2	
		Fishing rod 1.25	1	6	25			Adventure books	2	3	
	29	Postage stamps	2	1							

It will be noticed that the amounts $2.00, $1.00 and $3.00 have been written with the zeros omitted. It is never necessary to write zeros in the column for cents when the sum is in even dollars, but of course five cents must be written *05,* and fifty cents *50,* in the cents column.

Determining Profits. To find out, in single-entry bookkeeping, how much money a business has made in the course of a year, it is necessary to find the sum of all balances owed to the business, the sum of all balances owed by it, the value of property which it owns and the amount of cash on hand. Suppose the state of a small merchant's affairs to be as shown below:

Due from customers	$ 326.15	
Merchandise on hand	1540.94	
Cash on hand	74.80	
Total	$1941.89	
Due to others		$237.82

The figure $1941.89 represents *total assets.* The figure $237.82 is for *liabilities,* and if it is subtracted from the other the difference is the merchant's *net worth,* or the amount of property he would have after paying his debts. The difference between the present net worth and that of a year ago will in a general way show the profits for the period, as follows:

Total assets	$1941.89
Liabilities	237.82
Net worth (by subtraction)	1704.07
Net worth a year ago	1000.00
Profit for the year (by subtraction)	704.07

Practice Examples. Single-entry bookkeeping is no longer employed except in very simple accounts or by people who have not learned the more efficient double-entry system.

Here are ten items such as a boy might have to record if he were keeping single-entry books for himself. After studying the explanations given above anyone should be able to make the proper entries in the day book, cash book and ledger.

July

17. **Mowed Mr. H. C. Smith's lawn. Charge him 50¢.**
18. **Received $1.00 for work done to-day for Mrs. Potter.**
    Spent 25¢ for a baseball.
19. **Bought Tom Pearson's bat for 75¢. Pay him next week.**
    Weeded Mr. Smith's garden. Charge him 35¢.
20. **Picked and sold 6 quarts blueberries, at 7¢ a quart, cash.**
22. **Received payment in full from Mr. Smith.**
    Spent 55¢ at the circus.
23. **Received 50¢ for mowing Mrs. Potter's lawn.**
    Paid Tom Pearson in full.

When he is making his postings to the ledger from the cash book the student should remember that items which do not concern debts do not appear in a single-entry ledger. Where there is an entry in the cash book which will not be posted, a *blank check* (√) should be put in the checking column so that it will not appear that a posting has been omitted. When a ledger account is settled in full the figures of each side should be added and the totals written opposite each other below a single line and underscored with a double line.

After the work is done it may be compared with the following illustrations:

Day Book (page 1)

	July 17				
1	H. C. Smith. Mowed his lawn		50		
	19				
2	Tom Pearson. Bought his bat				75
1	H. C. Smith. Weeded his garden		35		

Cash Book (page 1)

		Receipts		Dis-burse'ts	
	July 18				
√	Earned at Mrs. Potter's	1			
√	Baseball				25
	20				
√	Picked and sold 6 qts. blueberries @ 7¢		42		
	22				
1	H. C. Smith in full		85		
√	Circus				55
	23				
√	Earned mowing Mrs. Potter's lawn		50		
2	Tom Pearson in full				75

Ledger (page 1)

H. C. Smith

July	17	Mowing lawn	D1		50	July	22	Cash	C1		85
	19	Weeding garden	D1		35						
					85						85

(page 2)

Tom Pearson

July	23	Cash	C1		75	July	19	Baseball bat	D1		75

Double-Entry Bookkeeping

The Venetians are usually credited with having invented the double-entry system, in the fifteenth century, but it is quite probable that they merely elaborated and made popular methods that had been used by the ancient Romans. From the name anyone might judge that the double-entry system requires twice as much labor as the single-entry. In some cases it does, giving in return twice as much information, but in any active business it actually saves labor, besides telling the proprietors much about their affairs which they never could learn from the older method.

The principle of the double-entry system is not that there are two entries for every transaction, but that there are two phases, or sides, of every entry. For example, when a merchant sells ten yards of cloth he increases his cash or the amount of money due him, but at the same time he decreases his assets of merchandise. In strictly single-entry you merely record the increased asset; in double entry you note also the corresponding decrease. Similarly, if a ten-dollar bill is found on the floor, single-entry shows only the increased cash. At the end of the year this ten dollars will appear in the profits of the business, though the latter obviously deserves no credit for it. Double entry, on the contrary, records the source of every profit or loss.

Debit and *credit* are the two words which designate the opposite sides of each transaction. Debit in Latin means *he owes.* In the sale of the camera it is plain that Harry owes, so our entry in Harry's account is called a *debit entry* or just a *debit,* and in making it we are said to *debit* him. Credit is Latin for *he entrusts.* If Harry were to lend us money we should *credit* him, for he would be entrusting us with it. In these two instances we have shown the words in their original meanings; in actual practice, since they cover transactions of every description, both *debit* and *credit* have a much wider significance. For the present we shall just note that as we debit Harry for the camera the opposite half of the entry is a credit to the list of goods in which the camera was included, and that as we credit Harry when he lends us money we must debit some other account at the same time. Farther on in this article will be found thorough instructions what to debit and what to credit in any instance. The main thing to be remembered at this stage is that *the vital point of double-entry bookkeeping is that there must be both a debit and a credit in every entry, or for every transaction.*

The Day Book, The Journal and The Ledger. These are the three books used in the old-fashioned system of double-entry bookkeeping. To-day the day book is almost never seen, the journal often receives only a few of the entries, and there is a multitude of different ledgers and auxiliary books, but an understanding of the three original types is the easiest and surest road to success with the newer ones.

The day book, as in single-entry, is a sort of diary. But transactions can be recorded in a simpler fashion than in single-entry, because the journal later separates the debits from the credits. The day book of a country storekeeper might read something like this:

Feb. 27

Sold Mrs. Ericsson on account 3 sacks of flour at $3.20.

Hiram Watts paid $10.00 on account.

George Henderson brought in 10 dozen eggs at 24¢. Sold him a pair of shoes for $3.00 on account.

Bought 2 tons of coal for cash, $17.00.

Alex. McKenzie brought in 20 lbs. butter at 30¢. and paid $5.00, both on account.

The journal is the book in which the debit items of each transaction are separated from the credits. If they were copied into the ledger directly from such a number of facts as the day book contains, many mistakes would be made. So they are first entered in the journal in an orderly manner, and are said to be *journalized.*

The ledger has a page, or part of one, for accounts *with* each person and *of* each important class of asset, liability, expense or source of profit. Entries in it are all posted from the journal.

Journalizing. Let us see how, if we were keeping books for the storekeeper, we should enter in the journal the items shown above in the day book.

The first transaction was a sale. It is marked *on account,* which means that Mrs. Ericsson did not pay for the flour, and it is to be charged to her account in the ledger. Plainly we shall debit her. But whom or what shall we credit?

In answer let us suppose for a moment that we did not own the flour, but it was entrusted to us by the person who did. In such a case we should certainly credit him. In double-entry bookkeeping every source of loss or gain, every asset or liability, is treated as though it were owned by a person or as though it were *itself a person.* So in the present instance we credit flour. To put this matter in another way, let us assume that the storekeeper hires several boys and makes one responsible for all the flour, another for all the dry goods, a third for all the shoes, and so on. To make sure that they are not careless with their trusts he charges them with the value of the goods, and makes them pay him if they lose any of them. When he buys the merchandise and they assume responsibility for it, he debits them. Now since a credit is the opposite of a debit, if he debits them when they assume responsibility, he must credit them when they cease to be responsible for any part of the goods. So the flour account in the ledger may be imagined to represent an imaginary person who accepts responsibility for all flour; when we buy flour we debit flour account, and when we sell flour we credit flour account.

But if a country merchant kept a special account in his ledger for every kind of article in his store from pins to paint and from calico to cocoanut he would need so many pages that it would take more time to keep his books than to run his business. So he generally includes all his stock under the heading *Merchandise.* If there is any one article or class of goods that forms a large part of his trade, or any one that he wishes to learn his exact profits in, he may open separate accounts for them. Thus, for instance, he may keep his groceries and his hardware apart from the rest. In the present instance we shall assume that everything is merchandise.

Our first entry in the journal will appear as follows:

Feb. 27

	Dr.		Cr.	
Mrs. Ericsson to Merchandise For 3 sacks flour at $3.20	9	60	9	60

The first column is headed *Dr.*, an abbreviation for *debtor*, and the second *Cr.*, which signifies *creditor*. The columns in a journal are in reality seldom lettered, but the first one is always to be used for the debit figures, the second for the credits.

Our next item is a receipt of cash. In this case *on account* shows that the money is to be applied to the account of goods charged some time ago to Mr. Watts. He ceases to be responsible to us for $10, and the fictitious boy in charge of the cash, or the *cash account*, accepts the responsibility. So we have:

	Dr.		Cr.	
Cash to Hiram Watts On Account	10		10	

The third item is really two. We bought eggs from Mr. Henderson and he bought shoes from us. No cash is recorded in either sale. So we write:

	Dr.		Cr.	
Merchandise to George Henderson 10 doz. eggs @ 24¢	2	40	2	40
George Henderson to Merchandise Pair shoes	3		3	

Why can we not save time and space by consolidating the two items into one, *George Henderson to Merchandise $0.60*, since the net result of the transaction is that he owes us 60 cents and we have 60 cents worth of merchandise less than before? We can, but there are a great many reasons why we should not. Suppose we do, and at the end of a few months Mr. Henderson comes in to pay his bill. We copy the ledger page, or show it to him, and he says, "But you have not given me any credit for those eggs I brought in." Then we have to take time to turn to the journal and show him that we have, and if he does not know much about bookkeeping he will still think we are cheating him. So the extra step now saves many later. Furthermore, at the end of the year we shall want to know how many dollars' worth of goods we have sold. If we only credit Merchandise with 60 cents, our report of total sales as taken from the credit columns of the Merchandise account will be $2.40 too small.

So far each of the items we have journalized has concerned only assets and liabilities. Now we come to a different kind. When the storekeeper bought eggs he intended to sell them later. Now he buys coal for heating his store. Though the coal is an asset in a sense, as he could at any time sell it again, in fact he does not intend to do so, but will soon use it up, and will get no financial return. It is one of the *expenses* of conducting his business. In a larger business, where it is important to keep count of the cost of heating, coal would have an account of its own, but in this case we class it as expense.

We *credit* cash in this instance. The imaginary cash boy to whom we entrusted our money is now relieved of the responsibility of caring for $17.00. Our journal entry reads

	Dr.		Cr.	
Expense to Cash Two tons of coal for store	17		17	

In the last item on our day book page we have two credits to Mr. McKenzie, one for butter and one for cash. It is plain that we can combine the two in this case without affecting our totals. In the entry for Mr. Henderson we could not combine without subtracting, and in double-entry bookkeeping it is never permissible to subtract. We write in the journal, with two debits balancing the one credit:

	Dr.		Cr.	
Cash	5			
Merchandise to Alex. McKenzie 20 lbs. butter, 6.00			11	

For a bookkeeper thoroughly familiar with the principles of debiting and crediting it is just as easy to record a transaction in the journalized form as in the cruder day book style. That is why the day book has practically ceased to exist. Anyone keeping books for the first time would be wise to make simple memorandums of each transaction in order to have plenty of time to reason out the proper form for the journal entries, but if the following distinctions be kept in mind, it soon becomes as natural to think of transactions in journalized form as to put words into sentences:

To debit an account means:

(1) The account assumes responsibility to the business; or

(2) The account has taken something from the business.

To credit an account means:

(1) The account is released from responsibility to the business; or

(2) The account has conferred a benefit on the business.

If any debits and credits have been wrongly made, they should never be crossed out or erased, but corrected by a reversing entry. That is, if Cash has been credited and Merchandise debited for a sale, an entry of the same amount debiting Cash and crediting Merchandise will offset the error and the proper entry can then be made. The explanation of the reversing entry should be written with it, and the word error added to the entry which needs correction. The most important reason for not making erasures in books is that the latter may some day be used as legal evidence, and erasures would give the impression that the bookkeeper had attempted to conceal facts.

Posting. The instructions given for single-entry posting, including the careful use of the checking column, apply to double-entry also. The first item in our journal is a debit to Mrs. Ericsson, which will be posted exactly like those to Mr. Smith in the single-entry examples.

The next posting is to the Merchandise account, and as it is a credit, it goes on the right side. It is not necessary to write in the explanation column, for we shall probably never care to know anything more about the Merchandise account than the total figures for the year. The date and journal page number must of course be put in as usual. For the debit cash posting, immediately following, explanations are also unnecessary. When we post the credit to Hiram Watts' account we can write the word *Cash* in the explanation column, if desired, but as most of our customers will pay their bills with money it is just as clear to assume that all unexplained postings on the credit side of a personal account represent cash. In the credit posting to George Henderson that follows, which is not cash, we shall write:

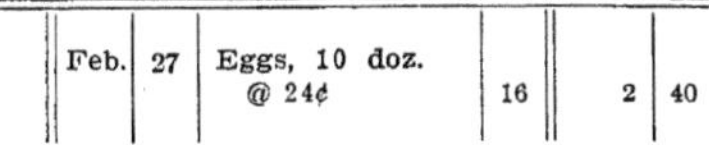

	Feb.	27	Eggs, 10 doz. @ 24¢	16		2	40

A Labor-Saving Journal. Every one of the entries we have so far made involves at least one of the accounts, Cash and Merchandise. In country storekeeping this will be true, with very few exceptions, of all the records of the year's dealings, and in all businesses these two or similar items are constantly occurring. In the course of a day a firm might have hundreds of postings to Cash, each without any explanation. If you have in mind the precept that bookkeeping should give the maximum of explanation with the minimum of labor you will wonder if there is not some way in which this huge number of postings can be combined into one.

The answer to your question is found in a *special column* journal. In such a book the debit columns are placed at the left of the page, the description of the entry in the center, the credit columns at the right. For each account occurring often, there is a special column.

In the store from which we have been taking examples the commonest debit posting will be for receipts of cash and the most frequent credit for sales of merchandise. Therefore a special debit column for cash and a special credit column for merchandise will save us much time. In journalizing the first of the entries given above, we shall put the debit figure in the general column at the left, the credit figure in the special merchandise column at the right. The first we post as before, but not the second. In the checking column opposite it we make a *check* (√) so that it will not appear that a posting has been neglected. At the end of a day the figures in the special column are added and the total posted to Merchandise account in the ledger. The cash items are treated similarly, and our entries in the journal will look as shown at the top of the next page.

When the special columns are added, it will be seen, the totals are transferred into the general columns. If the last then have equal sums we have proof that the amounts posted

Cash		General				General		Merchandise	
					Feb. 27				
		9	60	34	Mrs. Ericsson				
				✓	3 sacks flour at 3.20			9	60
10				✓	On Account				
				25	Hiram Watts	10			
		2	40	12	Merchandise				
				47	George Henderson	2	40		
					10 doz. eggs @ 24¢				
		3		46	George Henderson				
				✓	Pair shoes				
		17		8	Expense				
				3	Cash	17			
					2 tons coal for store				
5		6		12	20 lbs. butter @ 30¢				
				49	Alex. McKenzie	11			
15		15	00	2	Cash, dr.				
				13	Merchandise, cr.	12	60	12	60
		53	00			53			

to Cash and Merchandise as the same as though each item had been posted separately.

As we have shown only two debits to Cash and two credits to Merchandise in the above examples not much labor in posting has been saved, but if in the course of a day's business there were several dozen of each there would be an enormous gain in time to the bookkeeper. It will be noticed, too, that we have been able to make adequate explanations with a dozen fewer words than in the older form. Other advantages of the special column journal will occur to anyone who uses it, as, for instance, that the proprietor can tell almost at a glance how much cash has come in during the day and how much merchandise has been sold. Further modifications of this form of journal are found in the cash book and sales book explained below, and in many special books devised for unusual forms of business. In deciding what special columns to have for any particular journal it should be remembered that unless items of one sort are frequent an extra column for them will add to labor rather than reduce it.

How to Open a Set of Books. To *open books* means to start recording the affairs of a person or business which has not previously kept accounts. The first step in the process is always to list the assets and liabilities; the second, to journalize them.

For illustration let us make the entries necessary to open the books of two partners who are starting a small factory. The first, A. J. Steele, is to invest $5,000 in the business. The second, W. F. Gordon, puts in some machinery worth $8,000 which he owns, and a note in his favor dated November 7, 1916, for $2,500 and bearing interest at six per cent per annum. The firm is to assume responsibility for a debt of $633.14 which Gordon owes on the machinery, and he will furnish the amount of cash necessary to bring his total investment to $10,000. The two commence business on May 3, 1917.

As the note has been bearing interest it is worth more than its face value (see NOTE; INTEREST). Therefore to find how much cash Gordon must invest we must first calculate the amount of the interest which has been earned. Then we make our first entries in the journal, as shown on the opposite page.

The entry which concerns Steele scarcely needs explanation. He has conferred a benefit on the business, therefore is credited, while the account which accepts responsibility is debited. The description *Capital Account* is adopted to distinguish the invested funds from the money which the partners will draw out of the business from time to time for personal use.

Machinery is an asset which will not be sold, like merchandise, nor used up, like coal and other items of expense. Though it will wear

May 3, 1917

			Dr.		Cr.	
		W. F. Gordon and A. J. Steele commence business this day under the firm name of Gordon and Steele.				
		Cash	5000			
		A. J. Steele, Capital Account			5000	
		Machinery	8000			
		Bills Receivable	2500			
		(Note of Q. A. Shaw, Nov. 7, 1916, with Int. 6%. Due May 7.)				
		Interest on the above	73	33		
		Cash	59	81		
		W. F. Gordon, Capital Account			10000	
		Accounts Payable (Gunn and Co.)			633	14

out in time, the process will be slow and at the end of the year it will be nearly as valuable as now. The account is debited because it assumes responsibility for a certain amount of the firm's capital, though no actual money is paid to Gordon for it (see CAPITAL).

Bills Receivable always refers to promissory notes owed to the business by outsiders, and its opposite, *Bills Payable,* to notes which the business owes. Because people speak of owing a bill whenever they have purchased on account, beginners at bookkeeping sometimes confuse Bills Payable and Bills Receivable with Accounts Payable and Accounts Receivable. To avoid this error it is only necessary to remember that simple debts are *accounts* and only notes and accepted drafts or bills of exchange are *bills.*

It is sometimes difficult for beginners to remember when to debit Bills and Accounts Receivable and their opposites, and when to credit them. Accounts Receivable bears the same relation to accounts with individuals who owe money to the business that Merchandise does to the different varieties of goods, such as shoes and groceries. That is, unless it is worth while to open a separate account for an individual, his debts are grouped with those of other customers under the general heading Accounts Receivable, which is to be debited exactly as the individual accounts would be. Bills Receivable, on the contrary, represents an asset which resembles Cash more closely than it does a personal account. This is because in law an account paid with a promissory note is settled, so far as the showing on the books is concerned, just as completely as though it had been paid with money. Therefore, when a note is received Bills Receivable is *debited* and the payer credited, exactly as if he had given cash. When the note is paid, Bills Receivable is *credited,* for it is released from its responsibility for assets of the business.

Bills and Accounts Payable are treated similarly. Thus if we pay an account with a note we debit the account, and *credit* Bills Payable for having conferred a benefit on the business. When we pay the note we *debit* Bills Payable and credit the account which settled it. For example, if an old note is paid partly in cash and partly with a new note, the journal entry might be as follows:

			Dr.		Cr.	
		Bills Payable (note of Dec. 7 to N. W. Hillis)	4000			
		Interest (on above)	120			
		Cash			1500	
		Bills Payable (new note of this date)			2620	

The whole matter may be summed up by saying that we can never debit an account which *we* owe or a note which *we* owe, because the word means *he owes,* and we can never credit an account or a note representing money for which *we* trust another, because credit means *he entrusts.* When an account or bills payable ceases to represent what we owe, then we debit it, and when an account or bills receivable no longer shows that for which we have trusted others, then we credit it.

A bill receivable or a bill payable should never be entered in the name of the *person* from or to whom it is due, because of the legal provision mentioned above. Of course it would be possible to open an account called Q. A. Shaw's Notes, but as it is very seldom that a firm holds many notes from or owes many notes to one party, it is easiest to group all under Bills Receivable and Bills Payable.

Interest is the next item in the entry which we made for Gordon and Steele. Shaw's note has been earning interest since November 7. This interest is not due until the note itself is payable, but part of it has been earned, and the paper is worth more than its face value of $2,500. Therefore the firm must credit Gordon its full worth, including the accrued interest of $73.33 (for five months and twenty-six days). Bills Receivable is made responsible for the face of the note and Interest for the remainder. When the note is paid, only the amount of $2,500 will be credited to Bills Receivable. The balance will be a credit to Interest. If the note is paid on time the interest credit will be $75. As the account has already been debited $73.33, the net result will be a credit of $1.67, the amount of interest earned by the note after it came into the firm's hands.

Suppose, for further illustration, a case in which the note had been drawn without interest, so that on May 7 Shaw will pay $2,500 and no more. If we took the note to the bank on May 3, we could not sell it for its face value, for the bank must make interest on its money in the meantime. Deducting $1.67 (see DISCOUNT) we find the note to be worth $2,498.33. If we debit Bills Receivable only this amount and when the note is paid credit Bills Receivable the full $2,500, there will be a discrepancy, for as the debit side represents assets and the credit side payments on them, it will appear that we have been overpaid. To avoid this we debit Bills Receivable $2,500 in our first entry and credit Interest $1.67, the amount of the discount. Then when the note is paid Bills Receivable will balance and interest will have received credit for earning $1.67; the exact state of affairs then appears in the pages of the ledger. Our entry under this supposition would read as follows:

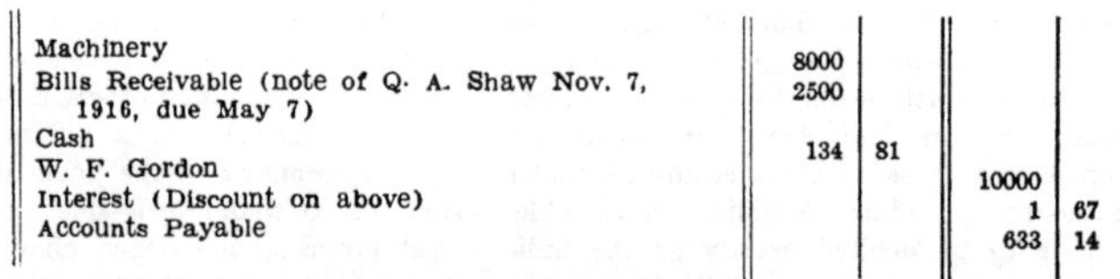

Machinery	8000			
Bills Receivable (note of Q. A. Shaw Nov. 7, 1916, due May 7)	2500			
Cash	134	81		
W. F. Gordon			10000	
Interest (Discount on above)			1	67
Accounts Payable			633	14

Until one has the principles of debiting and crediting interest thoroughly in hand, the easiest way to remember the proper treatment of interest and discounts of this nature is to keep in mind that bills must always be debited or credited their face value, any differences being assigned to Interest.

Interest on Bills Payable is similarly handled. If you discount at the bank a note signed by your own firm you will credit Bills Payable the face value, and debit Interest for the amount which the bank deducts. If you pay the note before it is due, you then debit Bills Payable for the face, and credit Interest for the difference between the face and the sum you actually pay.

Some Common Accounts. To illustrate the accounts which most frequently occur in the average business let us continue with the entries of Gordon and Steele.

Rent. As it is customary to pay rent in advance this would probably be the firm's first entry. It is clearly a debit, for the expense has taken something from the business. With the first entry it is well to write an explanation including the description of the rented premises and the length of the time paid for. Unless there is a written lease the terms of occupation are often added. In all subsequent entries only the time need be recorded. The first entry might be:

Rent 27 High St. to June 3	50	
Cash		50

A credit to rent may appear in two ways. If the firm owns premises which it lets to others, rent will be a source of revenue instead of a cause of expense. Secondly, if a portion of rent already paid is refunded either actually, or, in the case of closing the books as explained below, theoretically, it will be income.

Bank. Practically all payments of a business firm, including rent, are nowadays made by check instead of in cash. But it is seldom found worth while in keeping books to distinguish cash in the bank from cash in the till, or checks from currency. The account named Cash covers all money transactions. In many businesses, however, the Bank account is kept separate from the Cash account. In such event Bank is debited and Cash is credited when a deposit is made, and the Bank is credited for all withdrawals by check.

Furniture and Fixtures. If the firm purchases office furniture, a typewriter, sales counters or permanent fixtures of any sort for the office or shop, they are entered in this account, which is exactly similar to Machinery. Things which will be used up, however, such as stationery, pens and ink or paper towels, belong under the head of Expense or of a special account called Office Expense.

Insurance. One of the first acts of the new firm will be to insure its property. Insurance is a form of expense but is often recorded in a separate account. Besides the reason of its importance there is the one that policies are sometimes written for more than one year, so that at the end of that time some of the insurance will remain to be used up.

Material. Before commencing to manufacture, the firm will have to purchase raw materials. These will be charged to an appropriately-named account, or to separate accounts for each important class of material. After the raw material has been manufactured into the article to be sold it becomes Merchandise. Usually, therefore, all entries to Material will be debits, and all those to Merchandise, credits. At the end of a month or year, when an inventory is taken of the amount of material remaining on hand, there will be credits to the account, as explained below under *How to Close the Books.*

Wages. At the end of the first week the firm will have to pay its employees. The usual entry will be a debit to Wages and a credit to Cash. Unless an employee has the privilege of drawing money from his account at any time it is unusual to keep a separate page for him. The details may be written under the entry, or, if there are a large number of employees, kept in a separate book, usually the Time Book.

Personal Accounts. In a partnership such as this there will probably be two accounts, *A. J. Steele, Personal,* and *W. F. Gordon, Personal.* If the partnership articles arrange for the proprietors to draw monthly salaries, the amount due them at the end of a month will be credited to these accounts, and any sums from time to time drawn by either one of them, or paid for the interests of either, will be debited to his personal account.

Expense includes items of expenditure which can not be posted to any of the special accounts like Insurance and Rent, nor to any of those representing assets. Pins for the office, for instance, or the mending of a broken window would be charged to it. Sometimes special accounts are opened for Office Expense, Selling Expense, and the like.

Other accounts which are apt to occur are similar to those already described. Real Estate, for instance, represents assets and resembles Machinery; Mortgage Payable is a liability to be handled like Bills Payable; Taxes must be treated like Insurance. Commissions to salesmen are somewhat like Wages, though often more difficult to handle; there are so many ways to treat them that a full discussion cannot be given here.

The Trial Balance. Since for every debit there is an equal credit, and vice versa, it is evident that if books are correctly kept all the debits in them will equal the sum of all the credits. In bookkeeping language, the books will *balance.* Once a month, or oftener for a new bookkeeper, it is well to take a *trial balance,* or list of the total debits and credits for each account in the ledger, to prove that in posting from the journal there have been no errors or omissions which will affect the balance. Of course errors in posting to the wrong account or others which do not affect the balance cannot be discovered in this way. In taking a trial balance the totals of accounts should not be written in the books, unless in pencil marks easily erased.

A trial balance for Gordon and Steele might be as follows:

Trial Balance Dec. 31, 1917

	Dr.		Cr.	
Accts. Payable	13001	71	17652	53
Accts. Receivable	20322	17	15690	12
Bills Payable	2934	17	4951	36
Bills Receivable	4188	33	2980	21
Cash	23610	30	22338	99
Expense	57	60		
Furniture and Fixtures	346			
Gordon, Capital			10000	
Gordon, Personal	813	28	1200	
Insurance	360			
Interest	136	91	89	40
Machinery	8340			
Material	22166	90		
Merchandise			21801	26
Rent	400			
Steele, Capital			5000	
Steele, Personal	760		800	
Taxes	225	98		
Wages	4840	52		
	102503	87	102503	87

If the two columns of the trial balance do not have equal sums the error to be sought for may be either in the work of taking the trial balance itself, or a previous mistake in the books. If the latter, it may be an error in journalizing, a posting omitted, a posting to the wrong side of the ledger, a posting made twice, or a posting of the wrong amount. To discover the error find the difference between the sums of the two columns. If Gordon and Steele's bookkeeper had found that his debits were $225.98 less than his credits, he would probably have recalled that he had given a check for this amount of taxes, and looked at once to see if he had omitted to post it or to copy it from the ledger to the trial balance. If the discrepancy had been $451.96, an unfamiliar number, he might have divided it by two and finding the result to be $225.98, would would have guessed that the amount was posted to the credit side of the ledger instead of to the debit side. If the difference of the two columns were a number divisible by nine, say $2700, the error might be due to a transposition; that is, $23610 might have been written $26310, with the figures 3 and 6 transposed. (It will be noted that 27 divided by 9 gives 3, the difference between the two figures transposed.) In the same way that one who drives an automobile learns to locate the trouble in his engine, a bookkeeper soon comes to know the short cuts to the correction of his errors. The fact that almost any sort of error in the books will be revealed in the trial balance is perhaps the greatest advantage of the double-entry system.

Taking the accounts in the order in which they appear in the trial balance, the first is Accounts Payable. The credits to this account represent sums the firm has owed, the debits the amount paid on them. The excesses of credits over debits must therefore be debts still unpaid, or liabilities. This balance is written (usually in red ink to show that it is a figure arbitrarily introduced into the books instead of a posting) on the side which has the smaller total. All subtracting necessary to discover the balance should be performed outside of the books. Beneath the figures of the last written line of the longer column a single line is drawn in either black or red ink, then the total is written and beneath it a double line drawn. A single line in bookkeeping always signifies that the figures above it are to be added, a double line that the addition is completed. In the other column the single line, the total and the double line are written directly opposite the first. The red ink entry is given the date of closing, in this case December 31, 1917. Then below the double line of the other column is written in black ink a copy of the entry and the date January 1, 1918. This restores the equality of debits and credits as shown in the trial balance, and gives to next year's books the total amount owed on accounts.

Balanced in this fashion, the close of the account will now appear as follows. *Italics* are used to represent red ink in all the examples hereafter given:

Dec.	28	J. C. Moakley	47	500		Dec.	21	Cook Co. Invoice 265	47	207	
Dec.	*31*	*Balance*		*4650*	*82*		24	Bronson Bros.	47	1164	09
							29	O. R. O'Gorman	48	368	92
				17652	53					17652	53
						1918 Jan.	1	Balance		4650	82

How to Close the Books. Closing the books refers not only to bringing accounts to a conclusion when a firm ceases to transact business, but also—and in most instances—to arranging them at the close of a year or other period in order to reckon profits or losses.

In the work of closing, entries are usually made directly into the ledger. Care must be taken not to disturb the balance of debits and credits; whenever figures are entered on one side, corresponding ones must be posted somewhere on the opposite side. For this reason a few bookkeepers make their closing entries through the journal.

Accounts Receivable, Bills Payable, Bills Receivable and Cash are handled in the same manner, and sums due to or from individuals.

In Furniture and Fixtures we consider a different class of account. It represents an asset, but one which is presumably worth less than when it was entered in the books at the time of purchase. There has therefore been some loss to the business; this must be shown, and the proper value must be entered for the start of the new year. If the furnishings are now considered worth $300, this figure written in red ink together with the word *Inventory* on the credit side of the account will indicate that Furniture and Fixtures is relieved of responsibility for that amount. As the trial balance shows that the account has debits of $346 and no credits, there has been a loss of $46. An account called Profit and Loss is now opened and this loss transferred to it. This may be done by a journal entry:

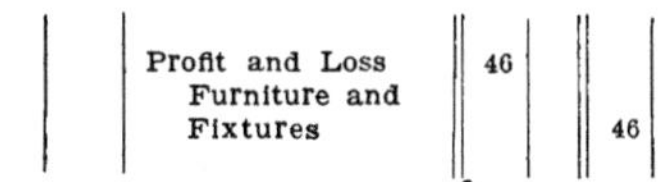

	Item	Dr.	Cr.
	Profit and Loss	46	
	Furniture and Fixtures		46

The more favored procedure, however, is to write the balance of loss on the credit side of Furniture and Fixtures in red ink and then to write it on the debit side of Profit and Loss in black ink. The difference in inks shows that in the one case the entry is arbitrarily introduced to balance the accounts, but in the other is in effect a posting. In both, the figure in the checking or index column is preceded by the letter *L* to show that it refers to a page in the ledger instead of the journal. The totals of Furniture and Fixtures are now found. The balance of the books has been destroyed by the introduction of the inventory figure on the credit side with no corresponding debit entry. But this fault is corrected by writing a debit to the account for the new year. This will be in black ink, because it is practically posted from above. The whole account then appears as shown below:

Date		Item	Fol.	Amount	Date		Item	Fol.	Amount
1917 May	3	Office equipment	1	346	*Dec.*	*31*	*Inventory*		*300*
							Profit and Loss	*L42*	*46*
				346					346
1918 Jan.	1	Inventory		300					

each proprietor earns and draws. The credits to them represent debits to the Wages account, and the debits to them show the money which each has drawn from the business. Neither partner has drawn his full salary, so the accounts will be closed exactly like Accounts Payable, to indicate that the firm owes them a balance.

Insurance presents a different type of account from any we have examined. Insurance premiums are generally paid in advance for one or more years. In the present figures let it be assumed that Gordon and Steele have insured for three years. But they have only been in business eight months of this time. To charge the profits of the present year with the expense of insurance for the twenty-eight months to follow would obviously misrepresent the facts. So an inventory is taken of insurance just as of furniture and fixtures; that is, a red ink entry is made to show that next year's accounts relieve this year's accounts of responsibility for part of the money paid for insurance. The result is as in Example A, on the following page.

Rent and taxes are handled in the same manner as Insurance.

Interest is just the reverse of insurance, in one respect. It is paid after it is earned, instead of before. Therefore a certain amount of interest on Bills Payable and Bills Receivable has accrued during the present year but will not fall due until the next. To make the books tell the proper story for each year, adjusting entries must be made. This year's account must assume the responsibility for interest which will have to be paid next year on this year's borrowings, and it must get the credit for interest which will be received next year on this year's loans. Therefore a red ink debit is made for accrued interest on Bills Payable, and a red ink credit for that on Bills Receivable. The close of the account will then appear as shown in Example B.

Wages account is treated like interest on Bills Payable.

Material account, if its balance is transferred directly to Profit and Loss, will appear

Expense is exactly like Furniture and Fixtures. Ordinarily, of course, its items represent total losses. But suppose, for instance, that the stock of postage stamps charged to it has not been used up, or that magazine advertising has been paid for that will not be published until next year. In either of these cases there will be an inventory and a transfer of the balance to Profit and Loss, just as explained above.

Machinery account is also treated in the same manner as Furniture and Fixtures.

For reasons which will appear later, the Capital accounts are the last ones to be closed. The Personal accounts, as used by Gordon and Steele, are merely to show the salaries which

EXAMPLE A, INSURANCE

1917											
May	3	Niagara Policy No. 7392	1	249	12	*Dec.*	*31*	*Unexpired*		*281*	*20*
	7	Globe No. 301982	2	110	88			*Profit and Loss*	L42	*78*	*80*
				360						360	
1918											
Jan.	1	Unexpired		281	20						

EXAMPLE B, INTEREST

Dec.	21	Smith note	47	8	25	Dec.	28	V. J. Nelson note	47	8	76
	28	Henderson note	47	13	50		*31*	*Accrued on B. R.*		*21*	*73*
	31	*Accrued on B. P.*		*47*	*21*			*Profit and Loss*	*L42*	*72*	*99*
				184	12					184	12
1918											
Jan.	1	Accrued		21	73	Jan.	1	Accrued		47	21

EXAMPLE C, MATERIAL

Dec.	21	Cook shipment	47	207	60	*Dec.*	*31*	*Inventory*		*4378*	*22*
	24	Bronson "	47	1164	09			*Merchandise*	*L6*	*17788*	*68*
	29	O'Gorman	48	368	92					22166	90
				22166	90						
1918											
Jan.	1	Inventory		4378	22						

EXAMPLE D, MERCHANDISE

Dec.	31	Material	L4	17788	68	Dec.	21	Brown Bros.	47	267	10
		Wages	L8	4840	52		28	J. T. McDonald	47	610	41
		Profit and Loss	*L42*	*3158*	*25*		*31*	*Inventory*		*3986*	*19*
				25787	45					25787	45
1918											
Jan.	1	Inventory		3986	19						

to have been responsible for a large loss. This is of course not the true state of affairs, for that part of it which has been made into Merchandise has been a source of profit to the firm. In a business which buys its merchandise already manufactured, debits for purchases appear in the Merchandise account, and it is of course possible to combine Material and Merchandise in manufacturing accounts. But it is generally more advisable to keep them apart until the books are closed. If the expenses of freight, express and cartage on material have been kept in a separate account they should also be transferred to Merchandise, for they are part of the cost of the material. The closing entries of Material account appear in Example C.

Merchandise, before its balance is transferred to Profit and Loss, should be charged not only with the cost of material, but with Wages. In the article ACCOUNTING is an explanation of what is called *cost accounting,* a system by which manufacturing costs are made to include a share of interest, rent, insurance, taxes and depreciation. In the present instance, Merchandise will be closed as in Example D.

Profit and Loss has now received all the balances from accounts which affect it, and appears as shown in Example E.

There are several ways in which this account may now be closed. The business policies of the proprietors will determine which shall be chosen. If they wish to spend the money they will divide and carry it to their personal accounts. If they wish to leave it in the business to increase their working capital they can carry it to their capital accounts. A third method, less frequently adopted in private partnerships than in corporations, is to close Profit and Loss account exactly like Ac-

EXAMPLE E, PROFIT AND LOSS

Dec.	31	Expense	L10	53	48	Dec.	31	Merchandise	L6	3158	25
		Furniture and Fixtures	L5	46							
		Insurance	L7	78	80						
		Interest	L9	72	99						
		Machinery	L2	840							
		Rent	L3	396	78						
		Taxes	L18	104	76						

counts Payable, in order to show that the business owes the profit balance to its proprietors.

The Balance Sheet. Providing that the work of closing has been correctly done a list of the debits and credits will again balance. But the accounts included in the list will differ from those in the trial balance in that they will represent only assets and liabilities. Thus Insurance, which before indicated merely a source of expense, now designates an asset—money entrusted to the insurance companies for protection which they have not yet given. Gordon and Steele's Balance Sheet follows:

Balance Sheet January 1, 1918.

Assets		Liabilities	
Accounts Receivable	4632.05	Accounts Payable	4650.82
Bills Receivable	1208.12	Bills Payable	2017.19
Cash	1271.31	W. F. Gordon, Capital	11043.63
Expense	4.12	W. F. Gordon, Personal	386.72
Furniture and Fixtures	300.	Interest	47.21
Insurance	281.20	A. J. Steele, Capital	5521.81
Interest	21.73	A. J. Steele, Personal	40.
Machinery	7500.		
Material	4378.22		
Merchandise	3986.19		
Rent	3.22		
Taxes	121.22		
	23707.38		23707.38

Corporation Bookkeeping

The principal differences between the account books of corporations and those of private businesses occur in the treatment of proprietorship and profit and loss accounts. The owners of a corporation are its *stockholders,* and these may be constantly changing. It is not convenient, therefore, to have their names appear in the ledger, and they are recorded in a special set of books. The ledger then deals with the stockholders as a whole, in an account called Capital Stock.

Opening Corporation Books. The entries necessary in opening the books of a corporation vary greatly according to the conditions under which the company is organized. For the simplest case let us suppose a company which starts business with all its authorized capital stock sold and fully paid for. The first journal entry would be:

	Cash				
	Capital Stock				

If a corporation were organized to take over the business of Gordon and Steele on the date of the balance sheet given above, paying each partner for his interest in the business with shares of the new stock and selling enough more stock to bring the capitalization to $25,000, the entries could be made, in the old journal, as follows:

	W. F. Gordon, Capital	11043	63		
	A. J. Steele, Capital	5521	81		
	Cash	8434	56		
	Capital Stock			25000	

When these were posted the result would be to cancel the claims of the old proprietors, giving them an interest only through their holdings of stock as shown in the stock-record books, and to substitute the account called Capital Stock for the old Capital accounts.

If not all the stock of a company is sold and

paid for at the outset, the capital account is usually called Capital Stock Subscribed. When all the authorized capital has been issued and paid for, a journal entry debiting Capital Stock Subscribed and crediting Capital Stock will wipe out the former account, leaving the state of the books the same as in the first two examples.

If stock subscribed is not paid for at the time of its sale, this, too, must be shown by the books. In such cases it is usually the practice to collect money from the stockholders in installments. The original journal entry would be something like this:

Account	Debit	Credit
Installment Subscription No. 1	5000	
Installment Subscription No. 2	5000	
Capital Stock Subscribed		10000

As fast as money was paid to the company by the subscribers, the proper installments would be credited and Cash debited. When all the installments had been paid the installment accounts would balance and could be closed.

If the proprietors of a business are paid more for it than the value of their actual assets (see GOOD WILL), an entry recording this must be made before the transfer of stock to them is shown. If Gordon and Steele were to be given $1500 more than the net assets of their balance sheet, the entry might be this:

Account	Debit	Credit
Good Will	1500	
W. F. Gordon, Capital		000
A. J. Steele, Capital		500

Profit and Loss. Except in the case of certain corporations in the state of New York (see STOCK, CAPITAL) the account for Capital Stock is never affected by profits and losses. When the gains of a company are sufficiently large, a *dividend* is declared. The entry at such a time is a debit to Profit and Loss and a credit to Dividend. Later, when the dividend is paid, there is a debit to Dividend and a credit to Cash. The balance of profit not given out in dividends is left as a credit to Profit and Loss or transferred to an account called *Surplus* or to *Undivided Profits*. If the company loses so much money that its original capital is lessened, there will be a debit balance under Profit and Loss, or an account called *Deficit*.

Farm Bookkeeping

There is no class of business man to whom the keeping of good accounts will be more helpful than to the farmer. There are many farm-owners and farm-renters in the world who continue raising unprofitable crops year after year because they are unaware that they are doing so. For instance, a farmer may be making money on corn and losing on wheat, but because he has a profit from his operations as a whole he does not discover this. Or it may be that he is deriving less income from the farm which he owns than he would from investing his money elsewhere and renting a farm. In fact, a pamphlet issued by the United States Department of Agriculture says there is reason to believe that the majority of farmers are really living on the interest of their investments rather than on the profits from their farms.

Farm bookkeeping is not difficult. Only a few minutes a day are necessary on the average farm to preserve thorough records. But the difficulty has been that there has been nobody to teach a farmer how to keep his books, for the reason that accounts of the style kept in cities are usually quite unsuited to the country. Recently, however, the United States government has studied the matter, and several bulletins on the subject have been issued by the Department of Agriculture, which it will pay any farmer to read. To comprehend them thoroughly he should have an understanding of the principles explained in the earlier part of this article and under the head ACCOUNTING.

To open books for the farm an inventory must be taken of all property. Some of the accounts to be opened will be exactly similar to those described above; an instance of this is Machinery. In addition there will be such accounts as Live Stock and Poultry. The account for Real Estate (including buildings) will be charged for rent, taxes, insurance, repairs and depreciation and credited at the end of the year for its service to the different parts of the farm. Cash, bills and accounts are the same as for city bookkeeping. Reasonable wages should be credited to the farmer and any members of his family who do work which would otherwise have to be paid for. Household Expense should be kept separate from Farm Expense, and any vegetables, milk or other products of the farm used at home that could be sold should be charged to the household and credited to the proper division of the farm. Similarly, in order to learn the cost of keeping cattle, any of their fodder which is raised on the farm should be charged to the cattle expense account just as though it had been bought. To the accounts which represent the products of the farm (corresponding to the manufacturer's Merchandise account) should be charged the labor, seed, fertilizers and other elements of cost. This is explained more fully under ACCOUNTING.

At the end of a year, under such a system, the books will show with considerable exactness the profits of the farm. If to these are added the household, personal and other expenses which the farm has paid for the farmer, and a reasonable amount for the rent of the farmhouse, the resulting sum will be the actual return to the farmer on his investment. If this return is not as high a percentage as could be secured through investment elsewhere, the farmer may consider that he has lost money. This applies to the man who rents his land and owns only his machinery and stock, as much as to the one who owns both. C.H.H.

Labor-Saving Books

Besides special books for unusual types of business, there are in common use a number of *auxiliary books* and variations of the simple journal and ledger. One of them, a labor-saving journal, has already been described.

The *Cash Book* is a specialized journal. In its usual form the left-hand page is for cash received, the right-hand for cash paid out. Every item on the left-hand page represents a debit to cash and a credit to some other account. It is obvious, then, that the cash need not be posted to the ledger until the foot of the page is reached, when the total of all figures can be debited to it. As a matter of fact it need not be posted at all, but, since the other page includes all the credits to the account, the cash book can be taken to represent the cash page in the ledger. In this case the cash book is usually balanced every day, exactly as such an account is balanced in the ledger when the books are closed. The balance at any time indicates the amount of cash which should be on hand.

A *Sales Book*, like a cash book, is a form of journal. All sales of Merchandise are entered in it. It is often made with two columns, one for cash sales, the other for sales on account. At the end of a day the *footing*, or total, of the cash column is entered in the cash book. The sum of the two columns is posted to the ledger as a credit to Merchandise, the debits being posted singly to the proper persons.

A *Sales Ledger*, or *Accounts Receivable Ledger*, is helpful to any firm which has a large number of accounts. It contains an account for each customer, to which entries are posted instead of to similar accounts in the general ledger. But in the latter there is a *controlling account* called Accounts Receivable, the balance of which should at any time represent the difference between the total debits and the total credits in the sales ledger. Therefore care must always be taken to see that postings of a sale are made to both ledgers. To effect this the usual method is to post the footing of the On Account column in the sales book to Accounts Receivable in the general ledger, and the individual items to the accounts in the sales ledger.

Purchase Books and *Purchase Ledgers* are similar in principle to sales books and ledgers.

Bookkeeping for Children

Many parents find it pleasant as well as profitable, when a child has reached the age of ten years, to instruct it in the simpler principles of keeping accounts. Most children grow up without learning even the rudiments of bookkeeping, and when they have to face the handling of their own affairs suddenly find themselves helpless. Boys may acquire the necessary knowledge after they enter business, but girls ordinarily have no occasion to unless they are left alone in the world, in the very situation in which they need it most. It is easier to teach money-handling to a young child than to an older one, because for the former it can be made an interesting part of play, but the latter is apt to regard the necessary work as drudgery.

The First Steps. Here is the method by which one father taught his little girl the fundamentals of account-keeping. She was still in the penny stage, but had already learned that it was wise not to spend her pennies for candy as fast as she acquired them, but to save a few for other childish wants that might arise. One day her father gave her a small cash box. They counted her money and found that she had fourteen cents. Then he showed her how to make out a little slip to put in place of any pennies she took from the box. The first one read, *Candy, 2¢*, and the next day, when her mother borrowed from her, a second slip said *Lent to mamma, 5¢*. When her mother returned the money, the slip was torn up. Later, after she had mastered these simple steps, the little girl was shown how to take a receipt from her mother for money borrowed and how to return it when the slip was destroyed. On another day, when given more pennies, she was told to write a slip, *Given me by mamma, 10¢*, so that this transaction could not be confused with money paid back after a loan.

When the first of the month came, the father showed his young pupil how easy it was to *balance* her cash. First of all, they counted the pennies in the box and found sixteen. They knew that at first she had fourteen cents, and the slips showed that she had been given ten cents, so there was a total of twenty-four cents for which to account. Since sixteen cents was on hand, eight cents must have been spent. But the slips told only of seven cents having been taken out of the box, so her father made it clear to her that she must have forgotten a slip for one cent. And then the little girl suddenly remembered that she had hurried to get a penny to give to the organ grinder's monkey, and so had neglected her bookkeeping.

To make things easier for the next month all the slips were torn up and a new one put in which said *Balance on hand February 1, 16¢*. Of course if there had been any slips representing money lent, they would have been preserved.

The Simplest Books. From the cash box and its slips it was a natural advance to a *cash book*. The child could see, especially when she began to have more slips, that it would be an advantage to write her records in a book. In the first place, this would prevent slips being lost; in the second, it would save the time of copying all the figures upon one sheet of paper when the cash was to be balanced, and best of all, it would give her an opportunity to compare one month's affairs with another. For five cents her father bought a little book with a single column on each page. He showed the little bookkeeper how to write on the left-hand page the explanations for money received and on the right-hand page those for money taken out of the box, and how to put the figures directly under each other in the columns, so that addition was easy. He told her not to cross out items representing loans when the sums were repaid, as she had torn up slips, but to make an independent entry on the *Cash Received* side of the page.

Little folks do not have enough to do with lending and borrowing to need a ledger. The signed receipts show money that is to come in, and if a debt is contracted a temporary slip can be made out. Later the advantage of a ledger will appear, and then parents may easily explain its principles, or even start a child in a simple system of double-entry by studying the first part of this article, which explains the more advanced steps in bookkeeping. J.O'S.

BOOKPLATE, a printed or engraved label pasted in a book as a mark of ownership. It is usually made of paper, but vellum, morocco leather and other materials are sometimes used. The best bookplates are impressions from engravings on copper or wood, but zinc etchings and other cheaper processes are also employed. A printed label, bearing only the name and address of the owner, is an inexpensive and dignified substitute for the more elaborate and decorative designs.

Bookplates were first used in Germany during the last quarter of the fifteenth century. These were crude affairs, usually colored by hand. The earliest known printed copy was used about 1480, and the earliest dated bookplate was designed by Albert Dürer in 1516. From Dürer's day to our own the lover of good and rare books has lavished care and skill on the ornamentation of his bookplate. For many years its most prominent feature was the owner's coat-of-arms, but this style has been partly displaced by allegorical or pictorial designs. One of the greatest book collectors of all time was Robert Hoe, the son of the man who invented the rotary printing press; the central feature of his bookplate was an old hand press. In place of the idea that a bookplate must show the owner's lineage, it now usually indicates something of the character and tastes of its owner.

The collection of bookplates and of books containing bookplates is a fascinating pastime. Many of the plates designed by great artists, such as Dürer, Holbein, Hogarth, Bartolozzi and Piranesi, are worthy of a place in an art

BOOKPLATES OF GEORGE WASHINGTON AND WALTER CRANE
In modern designs the tendency is away from the classical.

collection. There are the bookplates of famous men, many of them designed by artists of note. The bookplates of George Washington, Daniel Webster, Horace Walpole, David Garrick, Gladstone, Dickens, Carlyle, Tennyson and Victor Hugo are among the most sought-for examples. The modern tendency in decoration is shown in the copy of the plate of Walter Crane, with its very obvious play on the name of the owner.

BOOKS AND BOOKBINDING. In an essay called *Heroes and Hero Worship,* Thomas Carlyle says: "All that mankind has done, thought, gained or been, is lying as in magic preservation in the pages of books." By books Carlyle meant all written records, whether preserved in manuscript or in printed form. The present article deals only with printed books; the written records of civilization are discussed in the article MANUSCRIPT.

Modern Book-Making. Modern book-making comprises three distinct processes, type setting, printing and binding. The invention of printing from movable types is usually credited to John Gutenberg, and the date assigned is 1450. The first book printed from movable types was a Latin Bible in two large volumes; a few copies of it are still in existence, and an exceptionally fine one was sold in New York at public auction in 1911 for $50,000. The history of printing and the steps which make up the process are given in the article PRINTING.

EVOLUTION OF THE BOOK

1. The cairn. 2. Story-telling.

At first the only books printed were copies of the Greek and Roman classics and religious works. They were bound without title page, and there was no statement anywhere as to when the books were printed or where or by whom they were produced. Occasionally the printer put in a paragraph on the last page of the book containing this information and used the seal of the town in which he lived, or his own coat-of-arms, as a trade mark. The first dated title page was used in 1470, but title pages did not become common until after 1500. Most of the books were very large. The Bibles especially were immensely thick and heavy, and the paper on which they were printed was very thick and strong.

About the year 1500 smaller volumes began to appear. Books were read more and more, and they had to be made so that they could be conveniently handled. Thinner paper was used, and pasteboard was substituted for wooden boards for stiffening the binding. During the seventeenth century the art of printing was at its worst, but the badly-printed pages were often most beautifully bound in ornamented leather or in velvet gaily embroidered in gold and silver and bright-colored threads.

At the end of the eighteenth century and the beginning of the nineteenth, book-making made a conspicuous advance. The outward appearance of books was greatly changed by the introduction of glazed cloth as a covering for the pasteboard sides. Better paper was used, and printing presses were greatly improved. The most famous artists of the day were doing book-illustrations, principally engravings on copper and wood-cuts, but also etchings and lithographs. These methods have been almost entirely superseded by zinc etchings and other photo-mechanical processes. A noteworthy feature of the last decade of the century was the general revival of printing as an art, due chiefly to William Morris, the English poet, painter and craftsman. The most noteworthy advance of recent years has been the introduction of very thin paper, called *India paper*, in standard books, resulting in volumes an inch thick, yet containing a thousand pages. Such paper is hard to handle, and has been largely superseded by a paper somewhat heavier but still much lighter than the average of the last century.

Book Publishing and Selling. There was a time when the author of a book was often also the publisher and bookseller, and not so many years ago books were sold only at the printer's or the author's house. Practically all of the early printers acted as their own booksellers, and later some of them began to print books for other printers and carry in stock books printed by others. To-day, however, few authors have a more intimate connection with the publishing and selling of books than to make their terms with the publishing company and receive their royalties, the royalty being a percentage of the price at which the book is sold. The publisher makes all arrangements for printing and binding, and also distributes the book through the retail booksellers.

American Book Trade. In 1672 the first book store was opened in Boston by a man named Hezekiah Usher. Benjamin Franklin, too, was one of the early American printers and booksellers and because of his many other achievements he probably did more than anyone else to make the new trade famous. Many of the books sold by the early booksellers were imported from Europe, but the printing trade

EVOLUTION OF THE BOOK

3. Egyptian hieroglyphics. 4. Picture writing.

prospered, for the American printers, instead of importing the books they wished to sell, would simply reprint as many copies as they wished. As for paying the foreign author or publisher, no one dreamed of such a thing. After the War of Secession, although there was still no copyright law, there was so much rivalry among the American publishers as to who should get out some new book that had been published in England, that they began offering large sums of money for advance sheets. Boston, New York and Philadelphia became and have remained the leading publishing centers of the United States, but Chicago has gained on them of late years.

Some Statistics. Between the years 1825 and 1840, about 1,115 books were published in the United States. In 1880, with a population of 50,000,000 people, not more than 2,000 books were published. In 1910 there were published about 13,360 books; in 1915, about 11,250. More books of fiction were published than any other kind. The production of school books during the entire nineteenth century and since the beginning of the twentieth has been remarkable. Webster's speller sold at the rate of more than a million copies a year for many successive years. School books, prayer books, Bibles and hymn books have sold in enormous numbers. In 1910 the total number of copies of books and pamphlets published in the United States was more than 161,000,000.

Bookbinding

This name is applied to the process of putting the printed pages of a book between covers. Once an art, or more properly, a handicraft, it has grown into a mammoth industry. Instead of the old-time patient workman who spent days on a single book, we have to-day machines which turn out 10,000 bound books in a day. Books are still bound, and beautifully bound, by hand, but the real triumph of our age is to be able to print and bind books not by the dozen or hundred, but by the thousand or hundred thousand, so that by increasing output and decreasing cost they may come within the reach of every one. America leads the world in the invention and use of machinery for binding books. In an up-to-date bindery nearly all the work is done by machines of American invention, and some of the machines do not even require an operator. Bookbinding includes the entire process of putting the book between covers—folding, gathering and stitching the sheets, making and decorating the cover, and fastening it to the book.

Folding. The book pages come from the printers in great sheets, sometimes as many as sixty-four pages to the sheet, and so arranged that when the sheet is folded and cut, the pages will appear in the proper order. The first step is to fold these sheets, and cut them into signatures, which usually contain sixteen pages. This is done by an ingenious machine which requires no operator. All that human hands do is to place on a sort of shelf at one side of the machine a great pile of the printed sheets, to turn on the electricity, and to remove the piles of folded sheets, or "signatures," which are dropped into a rack at the other end.

Gathering. If there are to be *inserts* of any sort in the book they are now put in by hand.

EVOLUTION OF THE BOOK

5. Manuscript.

6. The printing press.

Most books have rather a large number of inserts because modern processes for reproducing pictures usually require a more highly finished and higher grade paper than is needed for the printed pages. When all the inserts have been pasted in, the signatures are distributed by an operator into the boxes of a "gathering" machine. There are as many boxes as there are signatures in the book. An endless belt, moving underneath them, carries the signatures along so that one drops on top of the other in the proper order and the number of signatures required for the book is complete when the belt gets to the end of the rows of boxes.

Stitching. Sewing used to be one of the slowest and most tedious processes in binding a book. It is never done by hand now, except in the binding of costly or rare books. All sewing in a big shop is done by machinery. Two operators are required. One operator puts the signatures, one at a time, astride an arm which has a row of sharp teeth along its upper edge. These points puncture the paper so as to make the task of sewing easier for the curved needles which flash in and out. The stitching is done around a cord or tape which is later pasted to the covers of the book. A second operator stands at the back of the machine and cuts apart the books as they appear, fed to her in a continuous stream. There are other types of sewing machine—one, for instance, which wire stitches pamphlets or very cheap books not over an inch thick.

Smashing. After stitching the books are put into a machine which "smashes" the book where it has been folded, so that it will be no bulkier at the back than at the front. This also makes the books more substantial.

End Papers and Head Bands. The *end sheets* are now pasted into place. These sheets are of very strong paper, sometimes ornamentally stamped, and serve the purpose of lining the cover of the book when it is put on. Sometimes a head band, so called, is put on also, as in the case of this present volume. If you will examine any well-bound book you will discover that a fold of cloth has been pasted in at the top and bottom as a finish for the edge of the sheets. This is done for appearances only and does not materially strengthen the binding, although it does protect the top of the back from injury when the book is pulled out of a bookcase.

Trimming. The edges of the book are still uncut. In expensive books they are very often left uncut until the book is in the hands of a reader, but this requires such very exact folding that it is more common to trim all of the edges in a machine called a *guillotine*. Ten or a dozen books are stacked together and clamped on a moving platform. The operator moves them into position and a huge knife descends diagonally and cuts through the thousands of tightly-pressed pages.

Gilding and Marbling. Here are two processes for which no machine has yet been invented. They are performed to-day exactly as they were three hundred years ago. If the edges of the pages are to be gilded, a number of books are clamped together and put on a table. Trimming has not left the edges smooth enough for gilding, so the man who does the gilding first rubs them down with a steel knife, just as if he were working on a surface of wood. He coats this surface with white of egg and then lays on thin sheets of gold leaf until the entire surface is covered. After drying for twenty minutes, the gold is burnished by hand. The cost of gilding books is very great compared with that of other processes. Real gold is used and the process is very slow.

The man who marbles the edges of books stands before a tank filled with liquid gum in a thin solution. On the surface of this tank he sprinkles colors with a round whisk broom. These colors float on the surface and mingle in the patterns you see on "marbled" paper edges. He takes as many as a dozen books, tightly clamped together, and dips their edges into the tank, then puts them on a rack to dry. He can dip them only once because this immediately disturbs the colors, and he has then to skim the tank and start over again.

Plain-colored edges are put on with a sponge, just as stain is put on wood. Sometimes the edges are stained red before gilding. All of the surfaces, whether stained or marbled, are burnished after they dry, with emery or agate wheels.

Rounding and Backing. As soon as the edges are finished, the books are backed with hot glue; this is well rubbed in, so the spaces between the signatures will not tend to break apart. When the glue is dry, the books are put into another machine which rounds out the back and makes the front curve in a little. The same machine creases the book at the back where the cover hinges on. This process is called *backing.*

Lining. The next operation is to paste the cloth on the back edge of the book, which, in addition to the sewing, holds the leaves together, and is fastened to the cover of the book. This lining is usually of crinoline, but occasionally, if it is desired to make the book exceptionally strong, canton flannel is used. This flannel is very fuzzy and, once it is glued into place, can hardly be loosened. Lining the back is the last operation before the book is put into the cover.

The Cover. Covers are of leather, cloth or paper, usually stiffened with pasteboard; if no pasteboard is used the cover is *limp.* Cloth and paper covers are cut by machinery, but leather covers are cut by hand to avoid waste of material. The edges of leather are beveled in a machine to avoid bulkiness when the material is folded over the pasteboard. The covers are pasted on the boards by another machine. An operator stands before it and feeds it the pieces of cloth or leather, and the machine does the rest of the work. The glue is heated by machinery, the cloth is coated with it, the heavy pasteboard sides are moved up, a roller presses the two together and the finished covers drop out.

Another machine stamps the lettering on the cover. If the lettering is to be in gold, the covers first go through the hands of girls who cover the spaces with gold leaf. After the title has been stamped other girls rub off the surplus gold with balls of soft rubber.

Perhaps the most interesting machine of all is the one which pastes the cover and book together. This machine has two great paste boxes which move up to the book and coat the end sheets with glue. The book then drops to meet the cover, which is pressed firmly against it. This process is accomplished so accurately that the end sheets serve to line the covers of the book. The books, now complete, are put into great presses and a cover is screwed down very tightly; then they are left several hours to dry.

Expensive books, particularly those bound in leather, often have the covers put on by hand. A book is said to be bound in *full leather,* if the back and sides are fully covered with a single piece of leather. A book is *half bound,* or bound in *half leather,* if the back and a narrow strip on the sides are of leather, leaving the rest of the sides to be covered with paper or cloth. Small pieces of leather are usually added on the corners to keep them from breaking and to add to the attractiveness of the volume; this style of binding is called *three-quarter leather.* In binding by hand the hemp strings to which the sheets are sewed are run through holes pierced in the heavy pasteboard cover, the leather back and corners are pasted on, and the paper or cloth cover and inside lining are pasted down by hand.

One of the marvels in modern invention is a machine which is used in binding the cheapest books. This one machine performs all of the processes. It folds, gathers, wire stitches and covers a book without the aid of human hands.

The Art of Bookbinding. If we consider binding as a method of preserving documents, the Assyrian tablets in the British Museum, dating from the sixteenth century B.C., were bound, for they were placed in a sort of envelope made of terra cotta. The earliest Oriental books, which were written on separate leaves, were placed between two flat pieces of wood or copper or richly-carved ivory and bound tightly together with a leather thong. The Greeks and Romans wrapped a leather cover around their waxed tablets of wood.

Papyrus and parchment books, however, were usually written on a long strip of the material and rolled up just as modern maps are rolled, to preserve them. Gradually the binding of books became a craft that equaled the making of rare jewelry, and books were cherished for their beauty and not for their contents. Enormous books bound in bright-colored leathers were sometimes carried in public processions. Marvels of beauty and workmanship adorned the altars of cathedrals, books with vellum leaves, bound between plates of silver studded with precious stones. Some of the old books had covers of enamel, some had carved ivory covers, others had covers studded with crystals, gems and cameos. The book of Gospels on which the English kings took their coronation oath was bound between boards an inch thick with leather thongs for fastening them together, with ornaments of hammered brass on the corners and a huge brazen crucifix on the side.

Chinese and Japanese Bookbinding. The art of making paper was invented by the Chinese, probably in the first century A.D., although it did not reach Europe until nearly a thousand years later. The Chinese, too, were using block printing long before it was invented in the West. But whereas we have to-day countless machines for printing and binding, the Chinese and Japanese are still printing and binding as they did a thousand years ago. They print their books on very thin paper, so thin that only one side can be used. All the sheets are double, with a fold at the edge of a leaf instead of a cut edge, and sometimes, if the paper is very thin, there is an interlining of paper as well, making three thicknesses in all. The books are put between covers of pasteboard covered with cloth, and the leaves are sewed together and the covers sewed on at the same time. Even the latest Japanese *Who's Who* (although that is not what it is called in Japan) is bound in this primitive way. A. C.

Related Subjects. Under the following headings will be found material of interest to the reader of this article:

Bookplate
Caxton, William
Faust, Johann
Gutenberg, John
Manuscript
Morris, William
Palimpsest
Printing

BOOKWORM, a book-ruining insect or grub, which feeds on the leather and paste of book bindings, and sometimes even eats holes in the paper. A very common form of bookworm is the grub of a kind of beetle. All bookworms prefer old books; at one time a cigarette beetle was found breeding in a copy of Dante's *Divine Comedy* which had been printed in 1536. The best preventive is frequent overhauling of books; the grubs can be eradicated by exposing books in a tight box to the fumes of carbon bisulphide.

A person who spends more time with books than with people is called a "bookworm."

BOOMERANG, *boom' er ang,* the famous missile of the Australian aborigines, the best known form of which, when thrown, returns to its owner. The return boomerang is shaped like a sickle or a rude and very open V, and

FORMS OF BOOMERANGS

it is a deadly weapon in the hands of experts. It is about three feet in length and weighs from eight ounces upward. It can be thrown accurately at least a hundred yards. The non-return boomerang is straighter and is thrown as nearly as possible in a straight line. The use of the boomerang has much decreased with the introduction of more effective weapons, and it has become largely a somewhat dangerous toy, though the Australians sometimes use it to kill birds. Some of the natives of India also make boomerangs.

A false statement made by a person, coming back to embarrass him, is called a boomerang.

BOONE, *boon,* DANIEL (1734-1820), the most famous pioneer and backwoodsman of early American history, was born in Pennsylvania, near the present city of Reading, on November 2, 1734. In 1751 the family settled at the forks of the Yadkin in Davie County, then on the North Carolina frontier. Daniel learned to read and write and acquired enough knowledge of arithmetic to enable him to become a surveyor, but in the lore of the woods and in sagacity and fearlessness not even the craftiest Indian could outdo him, and he became a great hunter and trapper.

From the time he was twenty his life was full of adventure. He took part in the disastrous Braddock expedition against Fort Duquesne in 1755 (see BRADDOCK, EDWARD), was a leader in protecting the frontier against Indian attacks, and fought in Lord Dunmore's War, an Indian conflict occurring just before the outbreak of the Revolution.

In 1767 he had made his first expedition to the wilds of Kentucky. Eight years later he led a party of settlers to that region and built a fort on the Kentucky River, calling it Boonesboro. Thither he brought his family, his wife and daughters being the first white women to stand on the banks of the Kentucky. Early in 1778 Boone and thirty companions left the settlement to procure a supply of salt, and all were captured by a band of Shawnee Indians. Boone was carried to Detroit by his captors and finally adopted by a Shawnee chief. Pretending to be highly pleased with Indian life, he was given considerable liberty, and so was able to escape when he learned of an intended raid on Boonesboro.

He made the perilous journey of 160 miles back home in four days, reached there in time to give warning to the settlers, and bore a conspicuous part in repelling the attack. Boone then removed with his family to North Carolina, but returned to Kentucky in 1780. His services as guide and surveyor were in great demand, and he represented the settlers in the legislature and acted as sheriff and county lieutenant of Fayette County. His carelessness in securing clear titles for his land holdings caused him to lose all that he possessed, and in about 1790 he took his family to Point Pleasant, Va. (now W. Va.).

Five years later they settled in Spanish territory about forty-five miles west of Saint Louis, where Boone was appointed commander of the Femme Osage district. The Spanish authorities made him a land grant of about 845 acres, but when the territory passed to the United States through the Louisiana Purchase he lost it all. In 1810, however, Congress confirmed the grant in recognition of the services of the man who had "opened the way for millions of his fellow men."

He died September 21, 1820, and was buried in Missouri, but in 1845 his remains, with those of his wife, were transferred to Frankfort, Ky., where a splendid monument has been erected to honor his memory. In 1915 the Daughters of the American Revolution of North Carolina, Tennessee, Kentucky and Virginia completed a marking of the Boone trail through the four states mentioned. The first marker is placed at Boone's home on the Yadkin in North Carolina, and the trail passes over several hundred miles of most picturesque scenery, ending at Boonesboro. The work of marking the trail, a task of four years, is representative of the place that Daniel Boone occupies in the hearts of his countrymen—that of the typical American pioneer. See illustration of Boone monument, in article KENTUCKY.

An excellent biography of Boone has been written by R. G. Thwaites. For young people, an interesting life has been prepared by Lucile Gulliver (in the "True Stories of Great Americans" series). E.C.B.

BOONE, IOWA, an important coal center, with a mixed population of Americans, Swedes, Germans, Irish and Scotch, the number of whom increased from 10,347 in 1910 to 12,253 in 1915. Boone is the county seat of Boone County, and is situated in the central part of the state, near the Des Moines River. Des Moines is fifty miles southeast and Fort Dodge is fifty miles northwest. The city is served by the Chicago & North Western, constructed to this point in 1865; the Chicago, Milwaukee & Saint Paul, built to the city in 1875, and the Fort Dodge, Des Moines & Southern (interurban); the last-named was constructed in 1906 and has its headquarters at Boone. The city was founded in 1848, was incorporated as a city in 1865, and named for Daniel Boone, the American pioneer. Boonesboro was annexed in 1887. The area is a little less than five square miles.

The industries of Boone are largely dependent on the natural resources of the vicinity, where extensive deposits of coal and potter's clay are found. This section is also rich in agricultural products. Live-stock raising and pork and beef packing industries are important; the principal manufactures are bricks, tiles, machines, flour, tobacco products, hosiery and hardware specialties. The car shops of the Chicago & North Western Railway are located here. The buildings of note are the post office, constructed in 1903 at a cost of $50,000; a $200,000 courthouse, erected in 1916; the high school building, erected in 1914 at a cost of $180,000; an armory, a Masonic Temple and the Eleanor Moore Hospital. In addition to the public school system, the city has a theological seminary and a library donated by former Senator C. J. A. Ericson.

BOÖTES, *bo oh' teez,* "THE HUNTSMAN," is a constellation in the northern hemisphere, close to the Great Bear, or Dipper. It contains Arcturus, one of the brightest stars visible in the northern heavens, its light being exceeded only by that of Sirius, Vega and perhaps Capella. In ancient mythology Boötes represents Arcas, the son of Callisto. With his dogs he hunted and would have killed his mother, the Great Bear, but for the intervention of Jupiter.

BOOTH, the name of a family originally English, which produced several well-known actors. One of them was for years the acknowledged leader of the American stage; another, the misguided slayer of the greatest man of his age.

Junius Brutus Booth (1796-1852) made his first appearance on the London stage in 1815, at the Covent Garden Theater. There he won great popularity in the rôle of Richard III and later as Iago to the Othello of Edmund Kean, at the Drury Lane Theater. After 1821 he acted chiefly in the United States, where he was enthusiastically received and held in high esteem as a tragedian, especially in the rôles of Richard III, Lear, Shylock, Hamlet and Iago.

Edwin Thomas Booth (1833-1893), son of Junius Brutus, became one of the most distinguished actors on the American stage. He was born in Belair, Md., and his acting career began in 1849 in Boston. Two years later he played his father's rôle of Richard III, in New York City. His numerous tours of the United States, Australia and Europe were uniformly profitable, and he became known as the leading American tragedian. His most-admired acting was in Shakespearean plays, especially in the rôles of Hamlet, Shylock, Richard III, Lear and Othello. Though not imposing in appearance he was dignified and graceful, and his marvelous voice could express the finest shades of feeling. Booth was a man of intellectual distinction and fine artistic gifts.

EDWIN BOOTH

John Wilkes Booth (1839-1865), the younger brother of Edwin, is remembered only as the assassin of Abraham Lincoln. He gave up the stage in 1863, after several years of acting in which he was only moderately successful, and his intense sympathy for the Southern cause led him to form a conspiracy, with others, that brought about the murder of the President. On the evening of April 14, 1865, he entered Ford's Theater, in Washington, where Lincoln was sitting in a private box, and shot him. Shouting *"Sic semper tyrannis"* ("So be it ever to tyrants"), he leaped to the stage below, and, though he had broken his leg in the effort, made his escape through a back door, mounted a horse that was held in waiting and fled to Virginia. At Bowling Green he was overtaken, and, hiding in a barn, was shot after he refused to surrender. His mad act shocked the South almost as much as it did the North.

BOOTH, the name of a family that has become widely known for its activity in religious work, especially in connection with the world-famous Salvation Army and Volunteers of America (which see).

William Booth (1829-1912), the best known of the family and the founder of the Salvation Army, was born at Nottingham, England. Adopting the faith of the Wesleyans at the age of fifteen, he later joined the Methodists and was ordained a minister, but left that body in 1861 in order to carry on the preaching of the Gospel in his own way. In the meantime, in 1855, he had married Miss Catherine Mumford, who was heartily in sympathy with his methods. In 1864 they began evangelistic work in London, holding open-air meetings and striving to brighten the lives of the forlorn dwellers in the East End of the city. The converts were organized by Booth into a mission band in 1865, and from this developed the great semi-military organization that in 1878 received its present name of Salvation Army.

Since that time the Army has made its way all over the world, and its open-air meetings, processions, stirring music and the zeal and self-denial of its workers are known to all. General Booth was wholly devoted to the work. He wrote many of the hymns sung in the Army meetings, and the weekly paper of the organization, the *War Cry,* was founded by him. *In Darkest England,* the best known of his books, presents his idea of how to deal with poverty and vice. Though ridiculed and violently opposed at the beginning of his labors, General Booth came to enjoy the highest respect, and in 1902 was honored by an invitation from Edward VII to be present at the coronation ceremonies. His sons and daughters were his trained assistants in the Army work, and on his death in 1912 the leadership passed to his son Bramwell.

Bramwell Booth (1856-), son and successor of General William Booth, was born in Halifax, Nova Scotia, and educated in private schools. He took an active part in the work of the Army from the time he was eighteen, and in 1880 he was appointed chief of the staff. In 1896 his younger brother, Ballington, left the Salvation Army to form a separate organization, and thereafter General Booth relied upon his son Bramwell more and more for advice and coöperation, and named him in his will as his choice for the position of commanding officer.

Ballington Booth (1859-), second son of William Booth, and the eventual organizer of the Volunteers of America, was born in London. In 1886 he married Maud Charlesworth, and the following year they were sent to America to take charge of the Salvation Army work in the United States and Canada. Because of a disagreement with General Booth concerning methods to be followed, they left the Army in 1896 and founded the new organization above named. The general methods of the Volunteers of America are similar to those of the Salvation Army, but the Volunteers make a special effort to work with the various churches. Ballington Booth is an eloquent speaker, and is the author of *From Ocean to Ocean.*

Maud Ballington Charlesworth Booth (1865-), wife of Ballington Booth, was born near London, the daughter of a wealthy clergyman of the Church of England. At the age of seventeen she went to Paris with Miss Catherine Booth, a daughter of General Booth, and they organized a branch of the Salvation Army in that city. Two years later she accompanied a band of workers to Switzerland, where her zeal and activity caused her to suffer imprisonment. After her marriage to Ballington Booth she took her husband's rank of marshal and was his co-worker in the organization of the Volunteers of America. She is known throughout the United States for her helpful service in behalf of prisoners, both before and after their release. Several homes, each called Hope Hall, are maintained in large cities as first refuge of released convicts. Like her husband, she is a forceful public speaker. Her writings include, *Branded, Look Up and Hope, Sleepy Time Stories, After Prison—What?* and *Twilight Fairy Tales.*

Frederick Saint George de Latour Booth-Tucker (1853-), who succeeded Ballington Booth in the leadership of the Salvation Army in the United States, was born in Bengal, India. He was educated for the civil service of British India, and held positions in the Punjab until 1881, when he resigned to join the Salvation Army. In 1882 he established the Army work in India and took charge of it until 1891, in that year becoming foreign secretary of the Salvation Army headquarters in London. In 1888 he married Emma Ross Booth, a daughter of William Booth, and at that time added the name of Booth to his own name, Tucker.

On the withdrawal of Ballington Booth from the Salvation Army in 1896, Booth-Tucker took charge of the work in the United States, serving with conspicuous ability until 1904, in which year he resumed his work at the London headquarters of the Army. In 1907 he returned to India as special commissioner for India and Ceylon. His writings include *Life of General William Booth, Favorite Songs of the Salvation Army* and *Farm Colonies of the Salvation Army.* B.M.W.

BOOTS AND SHOES. In the far-distant past of human history our ancestors became aware that travel with unprotected feet was a painful matter. Probably as soon as they reached the stage of using weapons or tools of

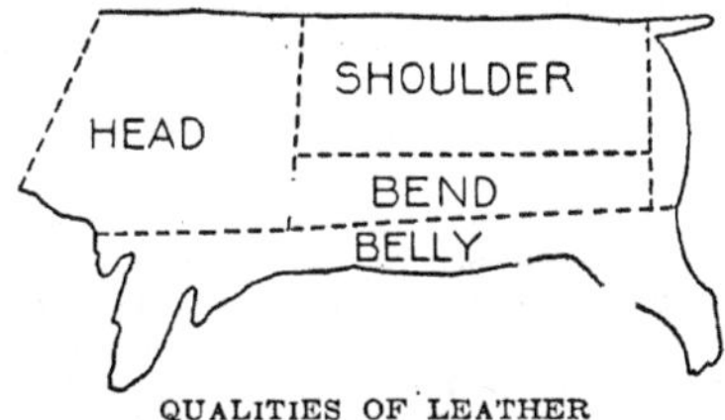

QUALITIES OF LEATHER
The skin on the bend is the firmest part; the shoulder, head and belly next, in order.

any kind they looked about for means of covering their feet. The animals which provided them with food were made to render still further service. From their skins, garments to protect the body were made, and it was but a very natural step forward when, from skins of the same kind, strips were torn and wrapped around the feet. Such was probably the origin of the first foot covering, from which was evolved the sandal, the earliest form of shoe of which we have knowledge. The sandal was merely a piece of untanned skin, covering the sole of the foot only and tied above by thongs. As the sandal became a thoroughly established article of wear it was naturally improved, and fashion, even in remote times, decreed the manner of its construction (see illustration in article FOOT). Covering for the sole was not found sufficient for all purposes, so the boot was gradually evolved.

From Ancient to Modern Forms. The Egyptians of ancient days wore sandals of woven papyrus or grass, a custom which yet prevails in many parts of China, India and other Eastern countries. The Romans wore sandals, the soles of which were sometimes protected by metal plates. Shoes shaped out of a piece of wood are now worn in parts of France, where they are called *sabots*. The same kind of shoe, known as a *clog*, is the favorite footwear of the mill workers of Yorkshire and Lancashire, England. In the Middle Ages boots began to assume most fantastic shapes. With wide flaps coming well above the knee, the leather was embroidered and highly ornamented, and at one time the upward length of a man's footwear signified his rank.

In modern days men and women often take pride in small feet. This was not so in France in the latter part of the seventeenth century. At that time a member of a royal family wore shoes two feet and a half in length; a baron could not have more than two feet of shoe; a mere knight had to be content with eighteen inches of foot covering. The long, pointed toes of such boots were usually looped up and tied at the knee. In some parts of Ireland a primitive form of boot survives, and a few peasants still wear *brogues*, or foot coverings. many sizes too large, stuffed with hay or moss. The boot in all its shapes is fast giving place to the more rational shoe of to-day, reaching just above the ankle. In the lumber camps of the United States and Canada a boot reaching to just below the knee and often called a *shoepack*, is worn. The British horseguards still wear the *jackboot*, which comes above the knee. This boot may be regarded as the father of all modern forms of boot In England it is still customary to call a shoe reaching just above the ankle, a boot, a shoe being one which is of a lower cut.

The Industry in America. The shoemaking industry was introduced into America in 1629 when Thomas Beard, a shoemaker, arrived on the *Mayflower* with a supply of cowhides. Within a few years from that time the town of Lynn, Mass., then a primitive settlement, became the center of the shoe trade. For many years, however, journeying shoemakers continued to travel from farm to farm making shoes out of the hides provided by the settlers. Comfort and utility were more sought in those days than style and appearance.

The boot and shoe industry in America had its beginning in Massachusetts, and to-day that state leads so greatly in production that no other state produces one-fourth as great an output. The value of Massachusetts footwear of all kinds is $240,000,000 a year (factory prices); Lynn makes more women's shoes than any other city in the world, and Brockton enjoys the same supremacy in the manufacture of men's shoes. After Massachusetts, in order of rank, is Missouri, with a product of $50,000,000, Saint Louis being the producing center. The next five states in order of importance are New York, New Hampshire, Ohio, Pennsylvania and Illinois.

Over 250,000,000 pairs of boots, shoes and slippers are made in the United States every year.

Shoemaking in Canada. The shoemaking industry is of great importance in all Canadian cities, but it has not yet been sufficiently developed to meet all the requirements of the market. Boots and shoes are imported from the United States and England. As a general rule, shoes are sold in Canada for a slightly higher price than is received for the same quality in the United States.

CUTTING LEATHER
Showing how patterns are laid to prevent waste in stock.

The 1911 census of Canadian manufactures gives the following particulars of the boot and shoe industries: There are in all Canadian provinces 180 establishments manufacturing boots and shoes. The total capital invested is $23,630,649. The wages and salaries paid to employees reach an annual total of nearly $8,000,000, and the value of the product is estimated at $34,000,000. This output would be sufficient to provide every inhabitant of Canada with one pair of shoes costing $4.75.

How Modern Shoes Are Made. In the making

of a modern shoe the hide undergoes more than 100 processes before it comes out of the factory as a finished product. The first process is that of *tanning*. This is done by soaking the hide in chemical solutions which change the nature of the skin and turn it into leather (see TANNING). The modern shoe consists of a *sole* and *upper;* these are also divided into several component parts. Usually twenty-six pieces of leather are used, with the addition of fourteen pieces of cloth, twenty-eight nails, eighty tacks, two tips, two heels, two box toes, two steel shanks and twenty yards of thread, and eyes of brass or other metal for lace holes, or perhaps twenty-four buttons. By far the greater number of shoes sold now are made almost entirely by machinery, and so perfect has that machinery become that a pair of shoes can be completed in less than a quarter of an hour.

Bootmaking Machinery. In modern boot and shoemaking almost the only work done by hand is the assembling of the various pieces. Many machines are employed, each doing its own particular portion of the work. One makes button holes, while another is used only for fastening on the heel. Certain seams are sewn by one machine; others, by different machines. From the year 1810, when the first shoemaking machinery was invented, constant improvements have been made. F.ST.A.

BORACIC, *bo ras' ik,* **ACID,** or **BORIC ACID,** is a substance used in medicine and surgery to prevent infection of wounds, for sprays for the nose and throat and for an eye wash. In manufactures and the arts it is valuable for preserving meats, for making glazes for pottery and in enamel work.

The chief supply of boracic acid comes from Tuscany, where the steam that issues from crevices in the ground is charged with it. The steam is passed through water which collects the boracic acid; the water is then evaporated by the use of volcanic heat, and the acid forms into crystals. It is purified by reheating, and when again cooled appears as white flaky crystals. Boracic acid is one of the chief constituents of borax, which is found so extensively in Death Valley, in Southern California. See BORAX; DEATH VALLEY.

BORAX, *bo' raks,* a borate of sodium, most familiar in the form of white crystals which dissolve in water, giving a solution of alkaline reaction. Borax is used for many purposes, as in the manufacture of colored enamels and glazes for porcelain, and in soldering and welding, since it dissolves the oxide on the surface and thus gives clean surfaces of the metal to stick together. It is employed to some extent in preserving butter, soft cheese, canned meat, fish and other foods. Whether in the quantities used for such purposes it is injurious to health is a disputed question. Some soaps and washing powders contain borax, which, like soda, softens hard water. In the laundry borax is used to soften water and to enhance the gloss of starch in ironing. It is employed to some extent in the textile industries and in medicine.

When borax crystals are heated in a loop of platinum wire they swell up greatly, owing to the boiling out of the water. When the bubbling ceases a clear, glassy liquid remains, which combines with many metallic oxides, giving "borax beads" of various colors. In mineralogical and chemical analysis these beads are used to detect the presence of certain metals. Thus cobalt gives a blue bead, copper and nickel, green, and manganese, amethyst.

Borax is found in large quantities in the Death Valley region in Southern California, and an impure variety called *tincal* comes from Tibet. Large quantities are manufactured from borio acid (which is found in Tuscany, Italy) by boiling the acid with carbonate of soda. See DEATH VALLEY.

BORDEAUX, *bawr doh',* one of the chief commercial cities of France, noted as the center of the wine export trade. It is situated on the River Garonne, about seventy miles from the Bay of Biscay and 358 miles southwest of Paris. It has a spacious harbor. During the early stages of the War of the Nations, beginning in 1914, it was made the temporary capital of France, its situation, far removed from the scene of hostilities, being admirable for the preservation of government documents and property. The red wine of Bordeaux, generally known as claret, is extremely popular in England and all European countries. In addition to wine and brandy, the city exports sugar, vinegar, earthenware, glass bottles, carpets, paper and great quantities of dried fruits. Shipbuilding is an important industry. The fisheries are important, and a large fleet is sent every year to the cod-fishing grounds off Newfoundland.

Bordeaux came into the possession of England through the marriage of Henry II to Eleanor of France, but it was regained by France in 1451. Its great Cathedral of Saint Andre dates from the middle of the eleventh century, although the building was not entirely constructed until the fourteenth century. Population in 1911, 261,678.

BORDEN, ***bawr' den,*** SIR FREDERICK WILLIAM **(1847-1917),** a Canadian statesman, for over thirty years a Liberal member of the Dominion House of Commons, and for fifteen years Minister of Militia and Defense. He was born at Cornwallis, N. S., and was educated at King's College, Windsor, and the Harvard Medical School. Immediately after his graduation from the latter he practiced medicine for six years at Canning, N. S. At the same time he took an active interest in the Canadian militia, in which he was appointed assistant-surgeon. Some years later he was appointed honorary colonel of the Canadian army medical corps, and in 1911 was made honorary surgeon-general in the imperial army.

Medicine was his chosen profession, and the militia was his hobby, but Borden was still young when his interests were transferred to a wider field. In 1874, when he was twenty-seven, he was first elected to the House of Commons. Except for a break from 1883 to 1886 he served in Parliament without interruption until 1911. There he quickly became a recognized leader, and in 1896, when the Liberals were returned to power, was appointed Minister of Militia and Defense. In 1899, on the outbreak of the South African War, he was active in raising and equipping troops to be sent to South Africa, and in 1904 reorganized the Canadian militia, which was placed under the direction of a council. Borden was a conspicuous advocate of Canadian participation in imperial defense, and was a delegate to several imperial conferences at which this question was discussed. His services were recognized in 1902 by the honor of knighthood.

BORDEN, SIR ROBERT LAIRD (1854-), a Canadian statesman who became Premier of the Dominion in 1911. He was born in the village of Grand Pre, N. S., June 26, 1854. After graduating from an academy in a neighboring town, he taught school for several years, then decided to adopt the law as a profession. He was called to the bar, and for nearly twenty years devoted himself to his practice. It is interesting to note that he became partner in the law firm of which another Conservative Premier, Sir John S. D. Thompson, was formerly senior partner. The junior partner for several years was Sir Charles Hibbert Tupper, whose father, also a Premier, was succeeded by Borden as leader of the Conservative party.

Borden was the recognized leader of the Nova Scotia bar, when he was persuaded, in 1896, to offer himself as a candidate for the Dominion House of Commons. As a young man he had been affiliated with the Liberal party, of which his cousin, Sir Frederick Borden, was also a member, but long before 1896 he had become a Conservative. He was elected one of the members for Halifax.

ROBERT LAIRD BORDEN

In Parliament at first he made few speeches, and these were usually confined to the legal aspect of public questions, but his ability was quickly recognized, and in 1901 he was chosen to succeed Sir Charles Tupper as leader of the Conservative opposition. This honor had come unsought to him, indeed somewhat against his wishes, for he realized the difficulties of the position. His political career up to this point had been so brief that it is no injustice to say that he was practically unknown to the public outside of the Maritime Provinces.

After he had accepted the position, however, he threw himself actively into the work of leadership. The first important issue he had to face arose out of the question of establishing provincial governments in the western territories. Under the territorial government the Roman Catholics had secured the right to maintain separate schools. Sir Wilfrid Laurier undertook to preserve this right in the new provincial constitutions. Borden, on the other hand, took the position that the control of education was a provincial matter. This attitude cost him for a time the support of the French-Canadians. He was also severely criticized for accepting a salary as leader of the opposition, an innovation introduced by the Laurier government. The result of this and other criticism was the loss of his seat in the House of Commons at the general elections of 1904. The blow was severe, and Borden was inclined to withdraw from public life, but he was persuaded to accept the representation of an Ontario district, and to retain his party leadership.

In the general election of 1908 Borden regained his seat for Halifax, and the Conservative representation from Nova Scotia, which had disappeared in 1904, now numbered six. Thus strengthened in his native province, Borden continued to lead the opposition in the House of Commons until 1911. In the elections of that year all other issues were gradually subordinated to that of reciprocity with the United States, a policy which Borden vigorously opposed, for patriotic, rather than economic, reasons. As the elections resulted in a decisive victory for the Conservatives, Borden was naturally called upon to form a ministry; which took office on October 10, 1911. In 1912 Borden was appointed to the Imperial Privy Council, and in 1914 was created Knight Grand Cross of the Order of Saint Michael and Saint George (G. C. M. G.). He is a forceful, convincing, logical speaker, but lacking in the fire of the orator and the lightness of the politician. G.H.L.

For the record of the Borden ministry see CANADA, subtitle *History of Canada.*

BORE, the name of a tidal wave which rushes up the estuary of a river or a narrow arm of the sea at spring tide. It varies from three to ten feet in height, moves with irresistible force and wears away the banks and bed of a stream. Vessels sometimes find it impossible to make headway against its strong current. Some of the places where this wave is especially prominent are the Bay of Fundy, the Amazon River, the estuaries of the Seine, and the Severn and the Dee rivers. See TIDES.

BOREAS, *bo' re as,* in Greek and Roman myths, one of the six sons of Aeolus, god of the storms and winds, and of Eos (Aurora), goddess of the dawn. Boreas personified the north wind; in classic writings he is called boisterous and blustering, and is regarded as the type of rudeness, in contrast to his youngest brother Zephyrus, the west wind, who is the type of gentleness. The origin of the term *zephyr*, applied to mild breezes, is thus explained. Aeolus sometimes sent his sons forth with orders to stir up terrible storms on the sea; the part that Boreas had in this gigantic frolic of the gods is picturesquely described by Lucan, a Roman poet:

The curling surges loud conflicting meet,
Dash their proud heads, and bellow as they beat;
While piercing Boreas, from the Scythian strand,
Plows up the waves and scoops the lowest sand.

It is told by the myth writers that Boreas loved the nymph Orithyia, but that he could not woo her successfully, because it was so difficult to breathe gently and to sigh was quite out of the question. Finally, in despair, he seized her and bore her away to far-distant regions of snow and ice, where she became his wife. They were the parents of two sons and two daughters. The former, Zetes and Calaïs, were winged warriors who took part in the expedition of the Argonauts (which see).

The use of the term *Boreas* as a symbol of the north wind occurs frequently in modern literature, as in the familiar line of Burns: "Cauld Boreas, wi' his boisterous crew." See AEOLUS.

BORGHESE, *bor ga' ze,* a celebrated aristocratic family of Italy, originally of Sienna and later of Rome. In the latter city was born Camillo Borghese (1550-1620), who was elected Pope as Paul V in the year 1605.

Prince **Camillo** (1775-1832) was another well-known member of the family. In 1803 he married Pauline, the sister of Napoleon, and three years later, having received the title Duke of Guastalla, was appointed governor of the provinces of Piedmont and Genoa. After the fall of Napoleon he broke off all connection with the Bonaparte family, separated from his wife and retired to Florence, where he died.

The Borghese Palace, the town residence of the Borghese family, is one of the finest buildings in Rome. It was begun in 1590, completed in 1607, and at one time contained a valuable collection of art treasures. These were sold at public auction in 1892 by Prince Paolo Borghese; at the same time the important family records were acquired for the Vatican by Pope Leo XIII. The palace still possesses, however, a famous collection of paintings, numbering nearly six hundred. Among these are a *Madonna* by Botticelli; four paintings by Raphael, including the *Burial of Jesus;* Titian's *Sacred and Profane Love,* and Correggio's *Danae.*

BORGIA, *bor' ja,* an Italian family of Spanish origin, prominent in the fifteenth and sixteenth centuries, whose most powerful members established a reputation for wickedness and treachery. The first to gain prominence was ALFONSO BORGIA, who became Pope as Calixtus III in 1455. His nephew, RODRIGO BORGIA, was elevated to the Papacy in 1492 as Alexander VI. While Cardinal he became the father of a large family whose most notorious members were Cesare and Lucrezia.

Cesare, or Caesar, Borgia (1457-1507) was the favorite son of his father, whose power aided him in his career. He was clever, but seemed without conscience. He attempted to establish an hereditary monarchy in Central Italy, but his methods were so unscrupulous and gained him so many powerful enemies that upon the death of his father, in 1503, his ambition failed completely. After two years' imprisonment in Spain, he escaped to the king of Navarre, in whose service he was killed in battle in 1507.

Lucrezia **Borgia** (1480-1519), the sister of Cesare, was a woman of great beauty and charm, but in her early years she was as wicked as beautiful. Her first marriage was annulled; her second husband was murdered by Cesare. After becoming the wife of the Duke of Terrara her life was above reproach. She was a patron of learning and the arts.

BORGLUM, *bawr'glum,* GUTZON (1867-), one of the foremost American sculptors, whose work is noted for its power, its simplicity and its half-concealed disregard for convention. He shows strongly the influence of Rodin (which see), and, like Rodin, reveals imagination through a splendid technique. He has also won distinction as a painter, but in later years has deserted the palette for the chisel.

GUTZON BORGLUM

Borglum was born of Danish parents who had settled in Idaho. After a brief period of studying in San Francisco, he spent a number of years in Paris and London. Since 1902 New York has been his home, and his most important pieces of sculpture are in the United States. Among them are *Mares of Diomedes* in the Metropolitan Museum, New York; a series of statues for the Cathedral of Saint John the Divine in the same city; the colossal head of Lincoln in the rotunda of the Capitol at Washington, D. C.; the imposing seated figure of Lincoln in Newark, N. J.; the statues of Philip Sheridan and James Smithson in Washington and the statue of John P. Altgeld in Chicago.

Borglum's great work is as yet unfinished. Plans were completed in 1915 for the transformation of Stone Mountain, sixteen miles from Atlanta, Ga., into a great Confederate Memorial—a "memorial to a movement," in Borglum's own words (see STONE MOUNTAIN). Across the face of the mountain will be carved a frieze 2,000 feet long and fifty feet high. The great leaders of the South will appear as a group of fifty or sixty horsemen, before whom march the main body of the army. The figures will all be cut in full relief, in such a way as to give the impression of an army moving over the mountain. The plans also call for a colonnade of thirteen columns, one for each of the Confederate states, and for a great hall, all to be cut in the solid rock of the mountain. The work will probably be completed in 1924 and will cost about $2,000,000.

His brother, SOLON BORGLUM (1868-), also a distinguished sculptor, has perpetuated in stone and bronze the life of the western frontier. Cowboys, ranchmen, Indians and especially horses are his favorite subjects. Unlike that of his brother, his work shows practically no foreign influence.

BORING MACHINES, for making holes in all sorts of materials, vary from simple hand instruments to complex, power-driven devices. Among the former the *awl* acts on the principle of a nail, pushing its way directly in, while the *auger* and *gimlet* resemble the screw. A *bit and brace* is provided with a handle called the *brace,* which may be held upright with one hand and turned with the other so that the *bit,* or removable cutter, bites into the material to be pierced.

Boring instruments for piercing stone and metal are called *drills.* These are of many kinds, and are usually operated by steam, electricity or compressed air. In boring small holes in rocks for blasting purposes a steel drill somewhat like a chisel is used, the head of the drill being struck by a heavy hammer. In all boring operations requiring a deep hole a diamond drill is used. This consists of a hollow steel cylinder on the end of a shaft which is turned by steam or electric power. The actual cutting is done by black diamond teeth, which bite into the rock and remove a core, which is brought to the surface in the cylinder.

For boring holes in steel plates, bridge girders and other iron and steel structures, tools operated by compressed air are much employed. The air is directed from a supply tank through rubber tubes to the tool, which works with great rapidity and with but little waste of power. See COMPRESSED AIR.

BORNEO, *bawr'ne o,* the third largest island in the world, only Greenland and New Guinea exceeding it in size, is situated southeast of Asia in the Malay Archipelago. It is about 400 miles east of Singapore, surrounded by the China, Java and Celebes seas. Its area is 289,496 square miles, or a little less than half the area of Alaska, and is a little larger than

LOCATION OF BORNEO

the province of Alberta. By far the largest and most valuable portion of the island is under Dutch rule, the remainder being divided into three small states under British protection. For the most part the island is mountainous, the highest peak being 13,698 feet above sea level. The rivers are a most important feature, as they form the main arteries of internal commerce, several being navigable for more than 100 miles.

The climate is hot and moist and the humid lowlands are very unhealthful for Europeans The soil is very fertile and produces spices, potatoes, yams, cotton, tobacco and many kinds of tropical fruits. Sugar cane is extensively cultivated, and groves of sago palms abound, furnishing a valuable source of food to the natives. There are extensive forests of teak and ironwood and almost inexhaustible supplies of other valuable timber. The water buffalo, the usual beast of burden, is used for plowing and for drawing native carriages, and it is also ridden; horses are owned only by the European inhabitants and a few of the richest natives.

The mineral wealth of the island is very great, but mining operations are carried on under difficulties. Gold, quicksilver, copper, iron, tin, sulphur and mineral oils are found, and diamonds have been discovered in many places. The diamonds of Borneo are inferior to those of Africa, for they are of a yellowish tinge. Edible birds' nests (which see), found in great numbers in the caves and cliffs on the coasts, form an important article of commerce. Gutta percha, trepang, rattan and timber are exported in large quantities.

The most-highly civilized of the native inhabitants are Malays from Java, Sumatra and other East Indian Islands, and Bugis from Celebes. In the northern part there are many Sulu types from the Philippine Islands. The head-hunting Dyaks of the interior still believe in nature worship and idolatry (see DYAKS). The Malays are mostly Mohammedans and are the principal merchants and traders; they are also noted as bold sailors. Chinese are numerous, especially in mining districts, where they toil incessantly to accumulate enough money to allow them to return home to live in comfort. The natives wear costumes of bright and gaudy colors. The women excel in weaving cotton fabrics and in making mats of beautiful designs and coloring. Many different languages are spoken on the island; the Dyaks are divided into numerous tribes, each with its own dialect, and the Chinese and Malays retain their own language and customs. Population, about 1,700,000. F.ST.A.

BOSNIA, formerly the most southerly province of the Austro-Hungarian monarchy. The name had a different significance after the prov-

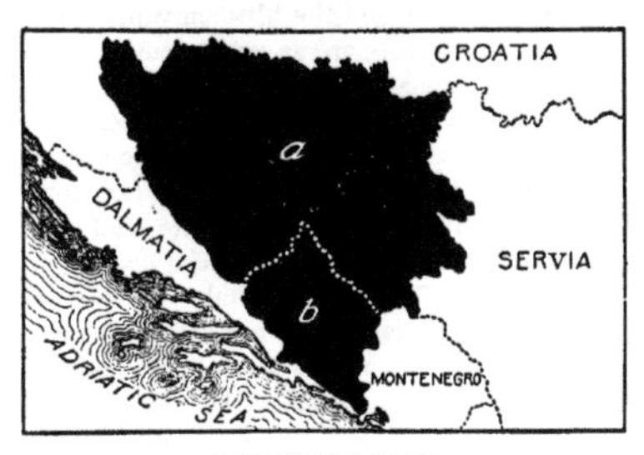

LOCATION MAP
(*a*) Bosnia; (*b*) Herzegovina.

ince was made part of Austria-Hungary in 1908, for it then included not only Bosnia proper, but Herzegovina, as well. Its area, with Herzegovina, is 19,768 square miles, and its population in 1910 was 1,931,802. It is therefore about twice the size of Vermont, and has over five times as many people. About one-third of the population are Mohammedans, for Bosnia belonged to Turkey from the beginning of the fifteenth century until the Treaty of Berlin at the close of the Russo-Turkish War of 1877-1878 made it a protectorate of Austria. Because Austria had done much to advance civilization in the province the powers of Europe offered no objection to its being made a part of the monarchy in 1908. When the Austro-Hungarian Empire crumbled in November, 1918, and new independent states rose from its ruins, Bosnia and Herzegovina became a part of the new Jugo-Slavia (which see), together with other southwestern provinces. (For industrial and geographic features, see AUSTRIA-HUNGARY.)

The People. The inhabitants of the region are known as Bosniaks, but the religious distinctions are so sharply drawn that the people themselves adopt no such general name. The Mohammedans prefer to be called *Turks;* the Roman Catholics, *Latins;* and the members of the Greek Church, *Serbs;* but they all have certain characteristics which show their kinship with the Serbs. They are a romantic people, loving poetry, music and the heroic tales of their ancestors, but they are not too proud to work thriftily on their small farms. Women as well as men are expected to work hard in the fields, but on the whole they are not unkindly treated, as in most countries where Mohammedanism has gained a strong hold. In dress the Bosniaks much resemble the Serbs (see SERBIA), but the Oriental fez and turban are very common, and the Moslem women who flit about the streets are as mysteriously veiled as are their sisters of the Orient. Indeed, some of the "Turkish" women hide even their eyes whenever they step beyond the boundaries of their own homes. As education is not compulsory, many people cannot write and read the curious language with its special characters.

The Capital City. The capital of Bosnia was *Serajevo*, or Bosnia-Serai, a particularly beautiful city. Serajevo means the *city of palaces,* and was applied in honor of the palace built in the fifteenth century by one of its Mohammedan rulers. Though in the last half century the city has been to a certain extent modernized, the old cypress groves, the curious wooden houses, the scores of mosques with their glittering minarets, and above all, the great bazaar, make it still seem more Oriental than European. It is a busy trading center, and the dark, twisting lanes of its bazaar are thronged with merchants. Silks, metal filigree work, rugs and embroideries are produced. It was in Serajevo that Prince Ferdinand, heir to the Austrian throne, and his wife were assassinated in June, 1914, a tragic event which speedily culminated in war. Population in 1910, 51,919. A.MC C.

BOSPORUS, *bahs' po rus,* or **BOSPHORUS,** a strongly-fortified strait connecting the Black Sea with the Sea of Marmora. The name is derived from Greek words meaning *ox ford,* probably given because it is so narrow in parts as to be easily crossed by an ox; according

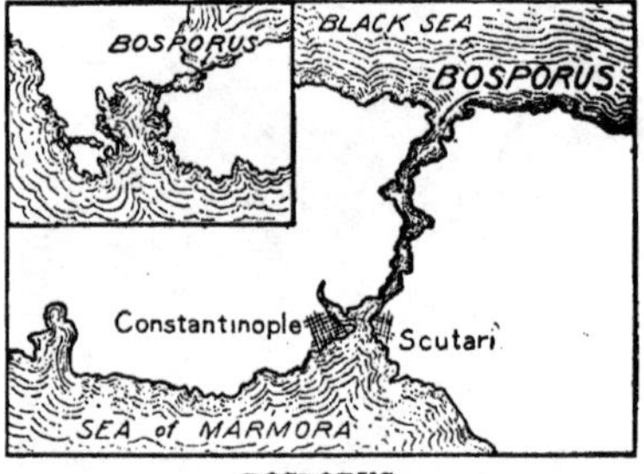

BOSPORUS

to legend it was crossed by Io after she was changed into a cow (see Io). It is nineteen miles long and from one-half to two miles wide. A strong current usually flows from the Black Sea towards the Mediterranean. The strait is an important commercial route and is frequented by the vessels of all nations. Only Turkish warships were to be seen there, however, during the late nineteenth and early twentieth century, for a treaty among the Great Powers, made in 1841 and confirmed in 1878, guaranteed that no war vessels of other nations should pass either the Bosporus or the Dardanelles without Turkey's consent.

The strategical value of the Bosporus is almost equal to that of the Dardanelles (which see). During the War of the Nations the allied fleets of Britain, France and Italy found in the Dardanelles an impregnable barrier, keeping them from Constantinople. On the east, the Russian warships in the Black Sea could not even attempt to force the passage of the Bosporus. The greatly-coveted city of Constantinople, on the Golden Horn, was practically safe from attack by water.

Over the middle of the channel, in that place about 3,000 feet wide, Darius constructed a bridge of boats on his Scythian expedition.

BOSTON, Mass., popularly called The Hub, or The Hub of the Universe, is the capital of the state, the county seat of Suffolk County, and the metropolis of New England. Its population in 1910 was 670,585, and in 1915, according to government estimates, was 745,139. It covers an area of 30,295 acres, or about 47.30 square miles. Boston is in the east-central part of Massachusetts, on an arm of Massachusetts Bay. It is 232 miles northeast of New York, and lies about as far north of the equator as Rome, Constantinople and Vladivostok.

The city has been for generations a commercial and manufacturing center, but its chief claim to fame is in its historical and literary associations, its libraries and educational institutions—in short, its position, more or less freely acknowledged, as the chief center of culture in the United States. It may be said that the things of the mind and spirit—books, pictures, music, practical religion, the love of nature and the healthy sports which bring mind and body and spirit together—all these are characteristic interests of Boston, and they are characteristic because they are so vitally interesting to so large a portion of the population. Oliver Wendell Holmes makes one of the characters in *The Professor at the Breakfast Table* say that Boston is full of crooked little streets, but it—

"has opened, and kept open, more turnpikes that lead straight to free thought and free speech and free deeds than any other city of live men or dead men—I don't care how broad their streets are, nor how high their steeples."

General Description. The site of Boston was originally a small peninsula, with its shores deeply indented by inlets and coves, surrounded by marshes, and connected with the mainland by a narrow neck, which was so low that it was often submerged at high tide. The area of the peninsula was 783 acres, but in the nineteenth century this was increased to 1,829 acres by filling in the inlets, the tidal marshes and the large area called Back Bay. This district, which now includes the finest residential section of the city, was once a part of the wide mouth of the Charles River and formed an inner harbor. The work of filling in required thirty years (1856 to 1886), and whole forests, quarries and hills were used in the process. The narrow neck was gradually widened until it became the widest part of the peninsula.

Until the beginning of the nineteenth century the most conspicuous features of the landscape were the three hills—Beacon, Copp's and Fort. Fort Hill has long since been leveled, but Beacon and Copp's hills, though considerably

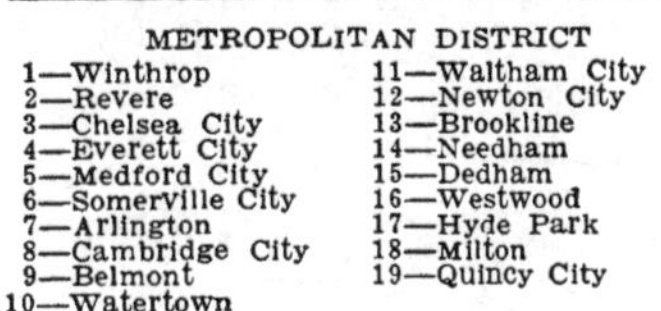
METROPOLITAN DISTRICT

1—Winthrop
2—Revere
3—Chelsea City
4—Everett City
5—Medford City
6—Somerville City
7—Arlington
8—Cambridge City
9—Belmont
10—Watertown
11—Waltham City
12—Newton City
13—Brookline
14—Needham
15—Dedham
16—Westwood
17—Hyde Park
18—Milton
19—Quincy City

cut down, still remain. Beacon Hill, so called because it was used as a signal station, is about 110 feet high, and is topped by the State House, whose gilded dome is visible for many miles. The oldest part of the city, the North End, still retains to some degree its eighteenth-century appearance. The narrow, winding streets seem to follow the cow-paths of early days, and their direction is determined by the irregularities of the surface. Many of them are so narrow that vehicles are allowed to traverse them only in one direction. Even Washington Street, the main thoroughfare in the retail district, is so narrow that sufficient space cannot be allowed for reasonably-wide sidewalks, and during the busy hours crowds of pedestrians regularly use the paved street. Tremont Street, another important avenue for retail shops, is wider, and for a considerable part of its length runs along the Common, a large and famous park in the heart of the city. In striking contrast to the narrow streets of the retail district are the many broad, regular avenues in the newer portions of the city (see, below, *Parks and Boulevards*).

Historic Buildings. In the gradual rearrangement and reconstruction which have been almost continuous since the beginning of the nineteenth century, Boston has suffered heavily in the destruction of many historic buildings. Fortunately, however, the most important structures have been preserved with a care and reverence which is seldom found in America.

FANEUIL HALL
As it appears to-day.

Faneuil Hall. Probably first in interest is Faneuil Hall, popularly known as the "Cradle of Liberty." It was the gift of Peter Faneuil (1700-1743), a prosperous merchant, who offered to build a market house if the town would agree to maintain it. As completed in 1743 it was 150 feet long and forty feet wide, and was substantially built of bricks imported from England. In 1761 it was almost destroyed by fire, only the four walls remaining. It was at once rebuilt, considerably enlarged, and was dedicated to the cause of liberty in a fiery speech by James Otis.

Then followed the exciting times of the Revolutionary period. In Faneuil Hall the local Committee of Correspondence established its headquarters in 1772. There, on the night preceding the Battle of Bunker Hill, General Howe was watching the performance of a farce, *The Blockade of Boston,* presented by some of the English officers and their wives. The audience was being amused by a dialogue between a caricatured Washington, bearing a rusty sword, and his grotesque squire, when a messenger burst into the hall with the shout, "The Yankees are attacking our works on Bunker's Hill." Many in the audience were highly amused, thinking that the interruption was a part of the play. But they were undeceived by Howe's sharp command, "Officers, to your alarm posts," and amid shouts and screams the performance came to an abrupt end. For a description of the battle and the monument, see BUNKER HILL.

After the colonies had won their independence, Faneuil Hall became the favorite meeting-place for the discussion of public affairs. The hall could not be rented or sold, but was open to the public by permission of the aldermen. This semi-public character gave a peculiar authority to Faneuil Hall meetings, and in other parts of the country the speeches and resolutions at such meetings were considered to represent public opinion in Boston. There were held many blood-stirring assemblies at the height of the anti-slavery movement. There, in 1837, Wendell Phillips delivered his first great oration as a protest against the murder of Elijah Lovejoy, at Alton, Ill. Webster, Choate, Sumner and Theodore Parker were a few of the others who made history within those four walls. Although much enlarged in 1805 and carefully restored in 1898 Faneuil Hall has long since been too small for great public meetings, and it is now a museum of colonial and Revolutionary days. The lower floor is still used as a market.

Old South Meeting-House. "Old South," as it is familiarly and affectionately known, is intimately connected with the stirring times before the Revolutionary War. The present structure was built in 1730 to replace an older church which had stood on the same site since 1669. During the pre-Revolutionary agitation meetings were held there and addresses were delivered which won for it the title of the "Sanctuary of Freedom." There was held the great public meeting which preceded the Boston Tea Party (which see). During the siege of Boston it was despoiled of its library and was turned into a riding school for the British soldiers, and for a short time after the great fire of 1872 it was used as a post office, but it now serves as an historical museum.

Other Historic Features. The oldest church in the city is Christ Church, built in 1723; it is the "Old North Church" from which were hung the signal lanterns for Paul Revere. Revere's house, in the North End, is still stand-

OLD SOUTH MEETING-HOUSE
As it now appears, surrounded by towering business blocks of steel and stone.

ing (see REVERE, PAUL). King's Chapel was built in 1754 on the site of the first Episcopal Church in Boston. The adjoining burial-ground is the oldest in the city; in it is the grave of Governor Winthrop. Granary Burying Ground, dating from 1660, has the graves of Samuel Adams, James Otis, Peter Faneuil, the Hancock family and Paul Revere. In Copp's Hill Burying Ground, also dating from 1660, are the graves of Increase and Cotton Mather. All of these cemeteries are in the heart of the old city. The Old State House (1748), on Washington Street, has been restored to its original appearance, and has many interesting relics and paintings. Of it, Oliver Wendell Holmes said, in the *Autocrat of the Breakfast Table:* "Boston Statehouse is the hub of the solar system. You couldn't pry that out of a Boston man if you had the tire of all creation straightened out for a crowbar."

Noteworthy Modern Buildings. Architecturally the most interesting part of the city is Copley Square, in the Back Bay district. Grouped around a large triangular grassy plot are the public library, Trinity Church, New Old South Church and the Copley Plaza Hotel. The library, completed in 1895, is a stately structure of gray granite, in the Italian Renaissance style. It is nearly square, and encloses an open court surrounded by an arcade. Over the main entrance are reliefs by Augustus Saint Gaudens, and the interior is richly decorated with mural paintings by Puvis de Chavannes, John S. Sargent and Edwin A. Abbey. Directly opposite the library is Trinity Church (Protestant Episcopal), generally considered as the finest American example of the French Romanesque style. Its beautiful interior decorations and stained-glass windows by William Morris, Burne-Jones and La Farge are particularly praiseworthy. Other notable churches are the Roman Catholic Cathedral on Washington Street, the Arlington Street Church and the First Church of Christ, Scientist, the Mother Church (for illustration, see CHRISTIAN SCIENCE).

The buildings devoted to the arts and to education are among the most conspicuous in the city. On Huntington Avenue is a splendid group, including Symphony Hall, the New England Conservatory of Music, the Opera House

PAUL REVERE'S HOUSE
It stands to-day in an unattractive, unsightly part of the city.

and the Museum of Fine Arts. Not far away are the buildings of the Young Men's Christian Association and the costly marble group of the Harvard Medical School.

Owing to a limit of 125 feet placed upon the height of buildings, Boston has few large office buildings. It has, however, a number of conspicuous structures devoted to government purposes. First of these is the State House, which was begun in 1795 and completed in 1798 after designs by Charles Bulfinch (1763-1844), later the architect of the Capitol at Washington. The building was greatly enlarged in 1890, the extension maintaining the colonial style of the original part. The customhouse, erected in 1837-1848, was originally a low building with a large central dome, but was transformed by the addition of a lofty, central tower over the dome; the work of remodeling was completed in 1915. Not far distant is the United States Government building, covering an entire block and erected at a cost of $6,000,000. The county courthouse, the city hall and the city hall annex, completed in 1914, are other large structures.

Parks and Boulevards. The pride of the city is Boston Common, the oldest public park in America. It was set off in 1634 and has ever since been cherished because of its close connection with the city's history. Originally the Common ran to the edge of the Charles River, but the reclamation of Back Bay put it in the heart of the city. Adjoining it on the west are the Public Gardens, tastefully laid out with an artificial lake and brilliant masses of flowers in season. Both the Common and the Public Gardens are ornamented with numerous monuments, the most important being the statues of Washington, Sumner, William E. Channing, the monument commemorating the Boston Massacre and the Ether Monument, celebrating the discovery of ether. Most famous of them all is the great Shaw Memorial, by Saint Gaudens, in memory of Robert Gould Shaw, the commander of the first regiment of negro soldiers to serve in the Federal army during the War of Secession.

Extending westward from the Public Gardens is Commonwealth Avenue, one of the finest boulevards in America. It is 240 feet wide, and down the center has a shaded parkway or promenade. Many of the finest residences and apartment buildings are on this avenue and on Beacon Street, two blocks north. Commonwealth Avenue is a link in Boston's great park system. The parks really form two separate systems, consisting of two concentric rings. The inner ring, belonging to the city proper, has been laid out since 1875, and includes about 2,300 acres. Its central point is Franklin Park. The outer, or metropolitan, system, comprises over 10,000 acres within a radius of ten or twelve miles from the city. It includes the famous Blue Hill Reservation, the highest land in Eastern Massachusetts, and the lovely Middlesex Fells, a half-wild, half-cultivated area of forest and pond. The Metropolitan Parks are administered by a special commission of five members appointed by the governor of Massachusetts.

Educational and Other Institutions. Perhaps first among the many educational influences in the city is the public library, which is the second largest library in the United States and has a larger circulation than any other free public library in the world. It has over a million bound volumes, including one of the greatest existing Shakespeare collections and the best collection on music, according to report, in America. Other famous libraries in Boston are those of the Boston Athenaeum (250,000 volumes) and the Boston Medical Library (80,000 volumes). It is estimated that within a radius of fifteen miles from the State House are 5,000,000 volumes available for public use.

Boston's public schools, especially those of secondary grade, are famous for their excellence. In addition to the kindergarten, grammar and usual high schools, there are special high schools of the mechanic arts, the practical arts and commerce, and special trade schools for boys and girls. The Boston Latin School, founded in 1635, is the oldest and one of the most famous schools in America. Among higher institutions of learning are Boston University (Methodist Episcopal), Boston College (Roman Catholic), Simmons College (for women), College of Physicians and Surgeons, and the Massachusetts College of Pharmacy. Massachusetts Institute of Technology, known the world over as "Boston Tech," has since 1916 been located in a magnificent group of buildings on the Cambridge side of the Charles River. The medical school of Tufts College and the medical and dental schools of Harvard University are also in Boston. Besides these collegiate institutions Boston is noted for many free lectures, foremost among which are those given under the auspices of the Lowell Institute. It was endowed by John Lowell, Jr. (1802-1836), a first cousin of James Russell Lowell, and since its foundation in 1839 has been noted for the high standard of its lectures and the eager response which the people at large have made to the opportunity to hear them.

Although these institutions exist for the benefit of the more fortunate citizens, for those less fortunate there is also adequate provision. The Perkins Institution for the Blind, made famous by the work of Dr. Samuel G. Howe (see HOWE, JULIA WARD), is the best-known institution of its kind in the United States. The Perkins Institution was removed from South Boston to Watertown, a suburb, in 1913, but its name will always be associated with Boston. The Massachusetts General Hospital, established in 1799, has an excellent training school for nurses. Noteworthy, too, are the Peter Bent Brigham Hospital, in connection with the Harvard Medical School, the Children's and Infants' hospitals, and the Boston Insane Hospital. Most of the city's penal institutions are located on islands in Boston Harbor.

Literature and the Arts. In literature and the fine arts Boston has always held a high place, and it is to-day the home of many well-known authors and artists. Its literary ascendancy, however, began in the middle decades of the nineteenth century and continued nearly to the end of the century. Longfellow, Lowell, Holmes, Hawthorne, Emerson, Parkman, Motley and Henry James are a few of the most famous writers who made their homes in or near Boston. As a music center Boston rivals New York. The Boston Symphony Orchestra, founded in 1881 and long maintained by the generosity of Henry Lee Higginson, is one of the great orchestras of the world. Scarcely less noted are such organizations as the Handel and Haydn Society, the Cecilia Society and the Apollo Society. The New England Conservatory of Music, with about 3,000 students, is the largest institution of its kind in the United States.

In the theater Boston at first was behind other cities—undoubtedly because of its Puritan atmosphere. It is now, however, a rival of New York and Chicago as a producing center, and has many beautiful theaters. Among them are the Boston, the largest theater in New England, the Colonial, Tremont, Hollis, Cort, Plymouth, Wilbur, Majestic and Keith's. The Museum of Fine Arts has one of the greatest art collections in the world, and is the natural center for the artistic life of the city. The museum school, the Massachusetts Normal Art School and the Lowell School of Design are favorably known.

Commerce and Industry. A glance at the map shows that Boston is somewhat away from the main lines of communication. This fact explains the change which gradually came over the city's business interests during the nineteenth century. During the eighteenth century Boston was larger than New York or Philadelphia, and had a greater trade. But in the nineteenth century, when the railroads opened the West, Boston's commercial field remained practically the same, while New York and other cities found new territories from which to draw trade. This does not mean that Boston has no commercial interests, but it does mean that Boston has become a manufacturing center as well. Though many of the factories are in the suburbs or even at considerable distances, they are the largest business interests of the city. Boston is also a great financial center, and on or near State Street, which is the Wall Street of Boston, are the home offices of many great mining, railway and insurance corporations and of powerful banking interests.

As a market Boston is noted for its trade in wool and fish, in both of which it stands second only to London among the cities of the world. It is second only to New York in its foreign trade, which amounts to more than $200,000,000 a year. The exports are chiefly meats, meat and dairy products, breadstuffs, leather and leather goods, cotton and woolen goods, and various products of iron and steel. The imports are mostly wool, hides, sugar, chemicals, drugs, india-rubber and fish. The manufactured articles, which reach an estimated total value of $250,000,000 a year, include refined sugar, malt liquors, boots and shoes and men's and women's clothing in addition to the chief exports mentioned above. The Waltham watch and the Singer sewing-machine were first made in Boston in 1850; the manufacture of Chickering pianos dates from 1824, and that of Mason & Hamlin pianos from 1854. Shipbuilding, including repairs to warships at the Charlestown navy yard, is a distinctive industry.

To care for its trade and to distribute its manufactures Boston has excellent transportation facilities. It is a large railway center, being the terminus of the Boston & Maine, the New York, New Haven & Hartford, the Boston & Albany and a number of smaller railways. It has rail connection with every part of the United States and Canada. There are two large union stations, both on water fronts. The North Station is used by the Boston & Maine; the South Station serves the Boston & Albany and the New York, New Haven & Hartford lines. The latter, at the time of its completion in 1898, was the largest railway terminal in the world; it covers thirteen acres and has thirty-two tracks.

The harbor is large and safe, though the entrance channel is rendered somewhat difficult by many small islands. Since 1840, when Boston was chosen as the American terminus of the Cunard Line, it has had direct steamship lines to Europe, and in recent years has greatly extended its range of shipping connections to other parts of the world. In 1911 the Massachusetts state legislature appropriated $25,000,000 for the improvement of the port of Boston. The first large expenditure, of $2,500,000, was for the enlargement of the great Commonwealth Docks in South Boston. Another large expenditure, of $3,000,000, was for a drydock 1,200 feet long and 149 feet wide; the great Liverpool Dock, completed in 1913 and at that time the largest in the world, is only 1,050 feet by 120 feet.

Government. Until 1822 Boston had no corporate existence, and its government, like that of every New England community, was controlled by the town-meeting. In 1822 it was incorporated as a city. The present charter, adopted in 1909, provides a quasi-commission form of government. There is a mayor and a city council, but there are also several commissions over which the mayor has no control. The mayor's term is four years, but he may be recalled at the end of two years by a vote of a majority of all the registered voters. The city council has one chamber of nine members, three elected each year for terms of three years. The school committee, of five members, is also elected. All nominations are made on petition of not fewer than 5,000 voters. A remarkable feature is the finance commission, of five members, all residents of the city but appointed by the governor of the state for five years, one member being appointed each year. This commission is free to investigate any matters relating to appropriations, expenditures, accounts or methods of administration which it feels may need investigation, and may report to the mayor, the council, the governor or the legislature. While Boston has home rule, therefore, in all essentials, the state provides a possible check in case of maladministration.

The People. In 1850, of the four elements of the population, those born in other parts of the United States ranked first, those born in Boston of native parents ranked second; the foreign-born were third, and the children of foreign-born were last. Half a century later the foreign-born were first in numbers, the children of foreigners second, those born elsewhere in the United States were third, and the native Bostonians were last. No better summary could be given of the change which has occurred in the character of the population. In 1910, out of a total population of 670,585, only 23.5 per cent were native-born of native parents, 38.3 per cent were native-born of foreign or mixed parentage, and 35.9 per cent were foreign-born. The Irish constitute by far the largest foreign element, the number of Irish-born and native-born of Irish parentage being nearly twenty-five per cent of the total population. The Scotch, English and Germans are represented in smaller numbers, and in recent years there has been a steady influx of Italians and Russian-Jews. Although it is apparent that the old or real Bostonians are now only a small part of the city's residents, they have given it its character and its reputation among the cities of America.

History. "This town of Boston," Emerson once said, "has a history. It is not an accident, not a windmill, or a railroad station, or cross-roads tavern, or an army-barracks grown up by time and luck to a place of wealth; but a seat of humanity, of men of principle, obeying a sentiment and marching whither that should lead them; so that its annals are great historical lines, inextricably national; part of the history of political liberty." In the preceding paragraphs enough references have been made to events in Boston's history to show the truth of Emerson's remarks. Before the Revolutionary War the history of Boston was largely that of Massachusetts, and after the Union was formed the important events in its story were nearly all connected with national movements.

The first man to enter Boston Harbor was Captain John Smith, who came in 1614. In 1630 John Winthrop's colony settled at Charlestown, across the river, but before the end of the summer decided to move to the peninsula then known as Trimountaine, from its three hills. This name is more familiar in its modern form, *Tremont.* On September 17, 1630, it was officially ordered that the settlement should be called Boston, after the city of that name in England, from which some of the leading colonists had come. In 1632 Boston was chosen as the capital of the colony. During the seventeenth and eighteenth centuries the town prospered, and at the beginning of the War of Independence, with a population of 20,000, it was

OUTLINE AND QUESTIONS ON BOSTON

Outline

I. Location and Size

(1) Latitude—comparative and actual
(2) Location
(3) Area
(4) Population

II. Description

(1) Its site
(2) The three hills
(3) Business district
(4) Residence districts
(5) Historic buildings
- (a) Faneuil Hall
- (b) Old South Meeting-House
- (c) Old North Church
- (d) King's Chapel

(6) Modern buildings
- (a) Public Library
- (b) Symphony Hall
- (c) Museum of Fine Arts
- (d) State House
- (e) Customhouse
- (f) Government building
- (g) Trinity Church

(7) Parks and boulevards
- (a) The Common
- (b) The outer and inner systems
- (c) Monuments

(8) Burying grounds
(9) Institutions

III. Industrial Life

(1) Manufactures
- (a) Food products
- (b) Leather goods
- (c) Fabrics
- (d) Iron and steel products

(2) Commerce
- (a) Bulk
- (b) Chief imports and exports

(3) Transportation

IV. Population

(1) Proportion of native-born to foreigners
(2) Nationalities of foreign-born
(3) Boston's place in literature and art

V. Government

(1) Commission features
(2) Officials
(3) Finance commission

VI. History

(1) Settlement
(2) Part in Revolution
(3) Later events
(4) Importance in history of the country

Questions

What was Boston originally called, and why? What street in the city recalls the old name?

What did Holmes mean when he spoke of the city as having opened "turnpikes that lead straight to free thought and free speech and free deeds"?

Why has Boston no great "skyscrapers"?

If an early settler could return now, what would he find different about the site of the city?

What drawbacks have the streets in the "downtown" section? How do you account for this?

Under what circumstances was the announcement of an epoch-making event taken for part of a play?

What building is known as the "Sanctuary of Freedom," and why?

How would it affect the literature of the United States if Boston and its literary history were blotted out?

What building is intimately associated with one of the best-known of American poems?

How did the opening-up the West affect the industrial life of Boston?

What great religious organization has its home here?

the most important town in the colonies. The Stamp Act, the Intolerable Acts, the Boston Massacre and Boston's part in the Revolutionary War are discussed elsewhere.

During the first half-century of independence Boston was the home port of hundreds of ships which traded as far as Rio de Janeiro and Calcutta. This period was also noteworthy for the spread of Unitarianism. Then came an era of anti-slavery agitation, and then after the great fire of 1872, which destroyed $75,000,000 worth of property, an era of reconstruction and development which has left its mark in the splendid buildings and other public works of the city. The construction of the subway, from 1895 to 1898, its enlargement a little more than a decade later, and the construction of the great dam across the mouth of the Charles River are the most important of these newer improvements. W.F.Z.

Consult Sullivan's *Boston, New and Old;* Howe's *Boston Common—Scenes from Four Centuries.*

BOSTON MASSACRE, *mas' a ker,* a fight between a mob of Boston citizens and a squad of British soldiers, on March 5, 1770. It was one of the earliest and most serious disturbances of the critical period before the Revolutionary War. The opposition of the colonists to the various acts of oppression on the part of Parliament and the king had led to the quartering of two regiments of soldiers upon the people of Boston, and this the latter bitterly resented as an unwarranted invasion of their rights as British subjects. A squad of soldiers stationed on King, now State, Street, angered by the taunts and stone-throwing of a mob of men and boys, fired into the throng, killing three and wounding seven, two of whom died later. The officers and men responsible for the firing were tried for murder, and were acquitted of this charge after being defended by John Adams and Josiah Quincy, who later were to become famous among American Revolutionists. Two soldiers found guilty of manslaughter were branded on the hand, and the entire garrison was removed to Castle Island, out in the harbor. The colonists regarded the affair as a triumph for their principles. A monument in memory of the victims of the Boston Massacre was erected in Boston in 1888.

BOSTON PORT BILL, a bill passed by the Parliament of England and signed by King George III, on March 31, 1774, which was occasioned by the wholesale destruction of tea in Boston Harbor on December 16, 1773 (see BOSTON TEA PARTY). This bill was to go into effect on June 1, 1774, and was virtually to close the port of Boston to trade. Salem was to be the seat of government and Marblehead the port of entry until such time as the people of Boston should pay for the property destroyed and meet other imposed conditions. Such an attack upon the liberties of the colonists aroused a deep feeling of indignation, and the people were given strong assurances of support by the legislatures of other colonies. The Boston Port Bill was one of five laws passed by Parliament to punish the people of Massachusetts for their defiant attitude (see INTOLERABLE ACTS, FIVE). As these acts aroused the colonists to further resistance, they helped to bring on the Revolutionary War (which see).

BOSTON TEA PARTY, in American history, the name applied to a famous raid on a number of English tea ships anchored in Boston Harbor. This occurred on the night of December 16, 1773, when American indignation over the tax imposed on imports of teas was at its height. When ships laden with tea, sent out by the English East India Company, arrived at Boston, the people assembled in a town meeting and passed resolutions urging Governor Hutchinson to demand the immediate return of the vessels. His message of refusal was conveyed to a mass-meeting held in the Old South Meeting-House, and was followed immediately by the famous "Tea Party." A band of citizens disguised as Indians and armed with hatchets hurried to the wharf, boarded one of the vessels and broke open 342 chests of tea, throwing their contents into the harbor. In this forceful and picturesque fashion the citizens of Boston announced their conviction that "Taxation without representation is tyranny."

BOSTON UNIVERSITY, established in Boston, Mass., in 1869, by the Methodist Episcopal Church. The university is open to both men and women on equal terms. It includes college and graduate departments, has schools of theology, law, medicine and science, and provides post-graduate work in science, language, history and philosophy. The agricultural college is allied with the Massachusetts Agricultural College at Amherst. There are about 170 members of the faculty, and the institution has over 1,800 students.

BOSWELL, *boz' wel,* JAMES (1740-1795), a Scottish writer whose masterpiece, *The Life of Samuel Johnson,* is one of the greatest biographies in English literature. He was born at Edinburgh, and educated there and in Glasgow and at the University of Utrecht. The most important event in Boswell's life was his introduction in 1763 to Johnson, whose writings he ardently admired. The acquaintance ripened into a friendship that shaped Boswell's whole career. He was admitted in 1773 to the Literary Club founded by Johnson, which included among its members Burke, Goldsmith, Sir Joshua Reynolds and David Garrick. From that time on Boswell made careful notes of all that his idol said and did. The two made a famous visit to Scotland and the Hebrides in 1773, Boswell publishing a description of their travels in 1785 under the name *Journal of a Tour to the Hebrides.* In 1791, seven years after Johnson's death, the great *Life* appeared. This masterly work, with its wealth of detail and vivid portrayal of Johnson's personality, is ranked by Macaulay as the first among biographies.

BOTANICAL, *bo tan' i kal,* **GARDEN,** a garden in which plants are cultivated for scientific, educational, artistic or economic purposes. Until modern times their sole aim was the cultivation of plants for the needs of medical science. Modern botanical gardens are usually connected with universities or are situated in parks under local government control, and are like museums in that they aim to show as far as possible the principal types of plant life of the earth. Very frequently in parks they are given the name of *conservatories.*

In the United States there are many fine collections of plants, but only a few bear the name of botanical garden. The New York Botanical Garden, occupying 250 acres in Bronx Park, New York City, has the finest greenhouses on the continent. The Missouri Botanical Gardens of Saint Louis, the botanic gardens of Cambridge, Mass., and the Arnold Arboretum at Brookline, in connection with Harvard University, are well known. In Canada, the principal botanical garden is at McGill University, Montreal, still in the course of development.

Of the numerous ones in France, the Jardin des Plantes is the most noteworthy. It is one of the oldest and largest in the world, growing over 15,000 species of plants. The Royal Gardens at Kew, near London, are world-famed, as are also those at Edinburgh, Oxford and Dublin. Other famous European gardens are located at Bologna, Strassburg, Munich and Leipzig. (See illustration of typical building in the article CONSERVATORY.)

BOTANY, the science which deals with plants—the description of them, their relationships, their habits, their distribution and their uses. Directly or indirectly man and all the lower animals are dependent upon the plant world for food, for shelter, for clothing (see PLANT), and since plants can be made to yield their best treasures only by means of some such systematic study as botany affords, the value of the science is extremely great, not only to the scholar, but to everybody.

> "Where's the second boy?" asked Mr. Squeers.
> "Please, sir, he's weeding the garden," replied a small voice.
> "To be sure," said Squeers, by no means disconcerted. "So, he is. B-o-t, bot, t-i-n, tin, bottin, n-e-y, ney, bottinney; noun substantive, a knowledge of plants. When he has learned that bottinney means a knowledge of plants, he goes and knows 'em. That's our system, Nickleby."

So spoke the schoolmaster in Dickens' *Nicholas Nickleby,* defining the word *botany* better than he spelled it. Though his working out of the method was absurd in the extreme, he described it correctly enough, for the only way of mastering the science of plants is simply to "go and know 'em" by actual experience. It is rather the fashion of many who have never made any systematic study of botany to affect to despise it—to declare "I love plants, but I hate botany"; but this is not the attitude of a true plant lover. He will want to know as much as possible about the plants which he loves; and while there are occasional botanists who, as Emerson says,

> Love not the flower they pluck and know it not,
> And all their botany is Latin names,

most of them have intelligent delight in the objects of their study.

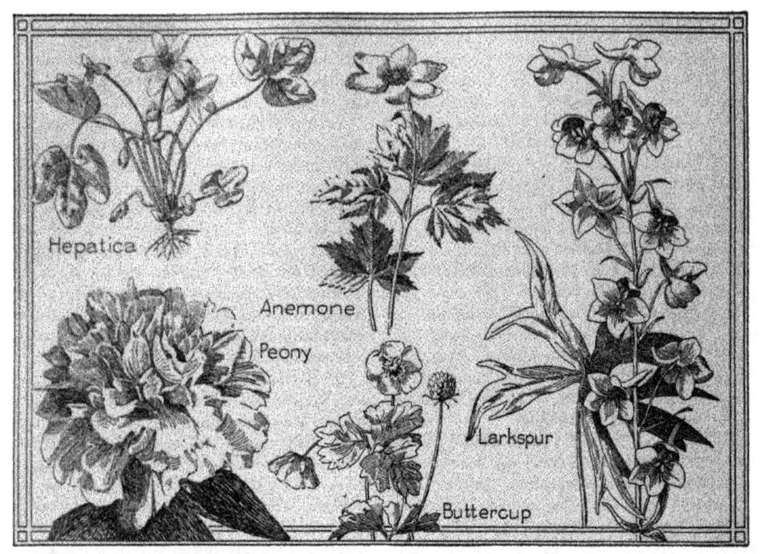

SOME MEMBERS OF THE BUTTERCUP FAMILY

Story of the Growth of Botany. With some sciences it is difficult enough to imagine just how they began, but this is not the case with botany. All about, wherever men lived, there plants lived, too, and so conspicuous were many of them that they could not help attracting attention. In the earliest ages, doubtless, this attention was somewhat casual—man had done nothing to bring the plants there, and he troubled himself not at all about their growth or development. If they bore berries or fruits which might be eaten, he ate them, and probably until distinctions were established many persons died from eating poisonous fruits.

As men advanced and came to have a little leisure for other things besides gaining a bare living and fighting off their enemies, they began to take more interest in the things about them, and the plants, apparently useful for many things, drew their attention. There were numerous curious differences in these plants which demanded notice. Some were always green; some spent the winter with bare, dead-looking branches, but came to beautiful life in the spring; some really died, and never became green again with the coming of warm weather. Some bore wonderful flowers, but were good for nothing save to look at, while others, with no flowers to attract attention, had luscious fruits, rich nuts, leaf-buds that served as food, or roots that might be made into bread. Gradually there came to the knowledge of men one most important fact—that in certain plants there were qualities which made them good for various sicknesses. And it was in the study of these medicine plants that botany really began, physicians being the first botanists. Writers of Greece and Rome occasionally produced works on plants, which dealt almost entirely with such as could be used as medicines.

These students of plant life, who called their slowly-growing science *botany*, from a Greek word meaning *plant*, had little thought of a systematic classification of plants, nor could they, with their limited knowledge, have attempted anything like the modern classifications. The Middle Ages saw no advancement in this plant-science, but when, about the sixteenth century, there was a new interest in learning of all sorts, botany came in for its share. Some of the earliest illustrated books were about plants, and many of the old woodcuts used would be, in point of beauty, a credit to modern books. Still the chief interest was in disease-healing plants, and only gradually did the more general phases of the subject make their appeal. Attempts at

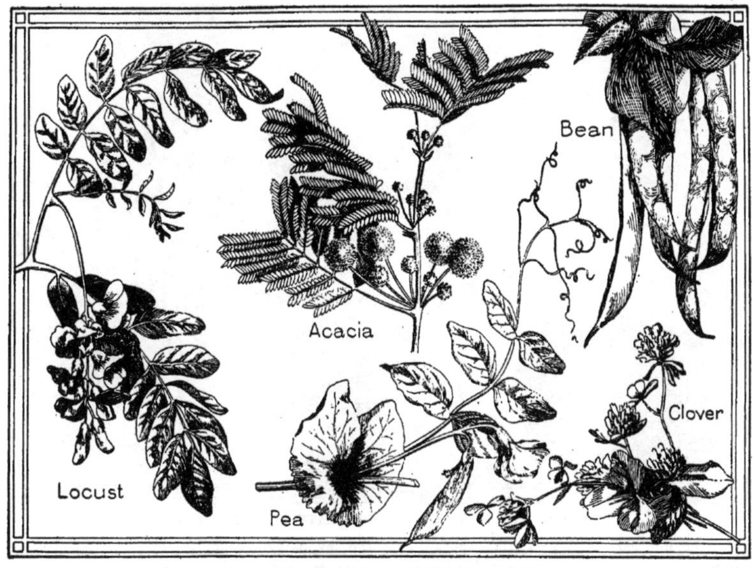

SOME MEMBERS OF THE PULSE FAMILY

classification were made, until, with Linnaeus, in the eighteenth century, the period of wavering and uncertainty came to an end, and botany began to be considered an exact science. From the time of that great naturalist, looked upon as the founder of modern botany, classification has grown constantly more elaborate and more exact (see LINNAEUS).

Place of Botany in Schools. To-day botany has a regular place among the sciences studied, but under its technical name is to be found only in high schools and colleges, for the classifications, the close examination of plant parts, the long-sounding Latin names are too difficult for young children. But these do not make up the whole of botany, and children are by no means cut off from the enjoyment of this subjcet which seems of all the sciences the one designed to appeal to them. The simplest elements of botany are being grasped by any child who learns to know a buttercup by name, to tell a petunia from a morning-glory, or to distinguish poison ivy that he may avoid it in the woods, and few are the schools which do not include some study of plant life. In the lower grades this is called nature study, but it is the real beginning of botany (see NATURE STUDY).

Of Interest to Boys and Girls. Very many things which come under the head of botany are so curious and interesting that no child who approaches them correctly can fail to find in them as much pleasure as in a story. For instance, there is the marvelous way in which each plant knows how to draw from the soil just the elements it needs to make it what it is. A story every child loves is that old *Arabian Nights* tale of the gigantic genie which came out of the jar, causing the man who watched the process to wonder and to doubt the evidence of his senses that anything so huge could emerge from anything so small. But all over the land, in every garden, a more wonderful thing happens hundreds of times every year. One of the tiniest of all seeds is the poppy seed—little more than a grain of dust it seems. But when it is planted in the dark ground, it chooses just the right kind of food, until presently there stands in the garden above the spot where it was placed one of the most beautiful of all plants—dusty-looking, gray-green leaves, straight, slender flower-stems, and crowning all, the gorgeous blossoms with their crinkled satin petals—all of which must have been present, in some form, in the tiny seed.

And close beside the poppy, perhaps, if the

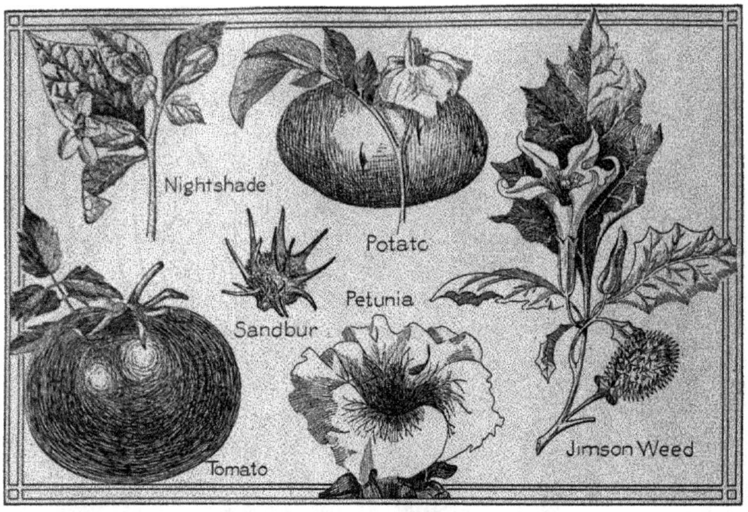

SOME MEMBERS OF THE NIGHTSHADE FAMILY

garden be not very well kept, there grows an ugly weed—a cocklebur. It has nothing beautiful about it, nor is it of any use. Nobody planted it or knows how it came there—perhaps the dog brought home the burry seeds in his coat and shook them off here; and now everyone calls the plant a nuisance and wishes it out of the way. But the wonderful thing is that these two, the poppy and the weed, can grow there within a few inches of each other, in exactly the same soil, and while the poppy takes water and food and turns them into soft green leaves and delicate flame-colored flowers, the cocklebur takes up water and food and makes them into harsh, rough leaves and troublesome burs. It is one of the unsolvable mysteries of nature, which even the wisest of scientists do not fully understand.

Plant Families. Very curious, too, are the family relationships which exist among the plants. In the animal world the cat, the tiger, the lion and the panther all belong to the same family, but that does not seem strange, for they all look very much alike. A child, seeing a tiger for the first time, will naturally call it "a great, big kitty." No one is surprised to learn that the dog and the wolf are of one family, for some kinds of dogs look enough like wolves to be cousins, if not brothers. In the plant world, too, there are some relationships that cause no surprise—that the blackberry and the raspberry should be close connections, for instance, is perfectly natural. But in some of the families there are members which do not look in the least like each other, and only a botanist would dare class them together. This grouping into families, indeed, is one of the most difficult and skill-requiring tasks of the botanists, and no child can hope to master the technical classifications and the hard-sounding Latin names—the *Liliaceae*, the *Amaryllidaceae*, and so on. But the facts about many of the best-known families are simple enough.

The Lily Family. There is the lily family, for instance. The name calls up at once the Easter lily, the day lily, the lily of the valley, the Chinese lily, and many other beautiful flowers, but it does not suggest the tulips and the hyacinths, the dog-tooth violets and the trilliums, all of which are just as truly members of the lily family. Two other members are thought of as vegetables rather than as flowers; these are the onion, with its unpleasant scent, so different from that of a true lily, and the asparagus, which certainly does not look like a lily in any way, but has resemblances that only a botanist can see. (See illustration, in article LILY.)

The Rose Family. There is also the interesting rose family. "O yes," exclaims the child who is fortunate enough to have a garden, "that's a big family. There's the Cherokee

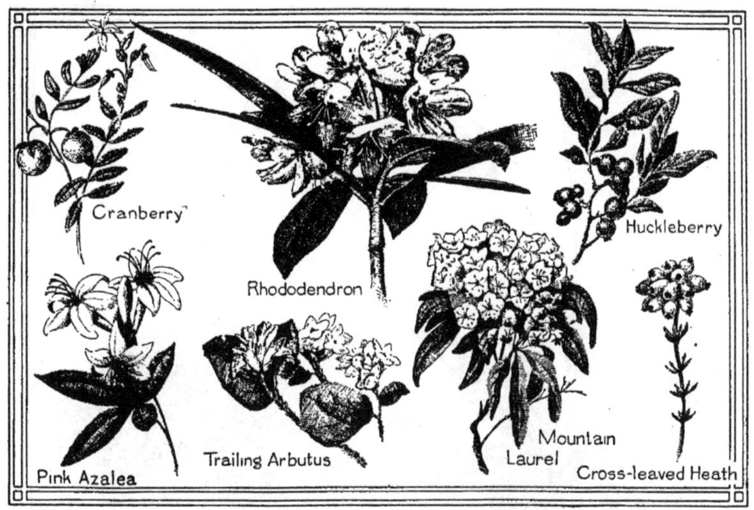

SOME MEMBERS OF THE HEATH FAMILY

rose and the yellow rose and the moss rose and the tea rose and the American Beauty and O, ever so many more." But that is not the end, and that same child might well be surprised to learn that, if the entire rose family were all to be destroyed, he would miss not only those beautiful ornaments of the garden and the florist's window, but the apples, pears, peaches, cherries, blackberries, raspberries, strawberries, plums, quinces and almonds, as well as the sweet-brier and the bridal-wreath. In this great family, one of the most useful in the plant world, there are resemblances which a child can easily distinguish. The blossoms of most of the fruits mentioned above are, when examined, much like wild roses, after all, and the wild rose is the one true "natural" rose; all the other beautiful varieties have been produced from it. (See illustration, in article ROSE.)

Other Families. Another family which betrays its relationships to the careful observer is the pulse family, to which belong the peas and beans, the clover, the locust, the alfalfa and the acacia. If a red clover blossom is examined closely it will be seen that each of the tiny flowerets of which it is made up looks very much like a sweet pea, or like one of the flowers of the scented locust cluster. But suppose a big red peony, a bluish-purple larkspur, a pink hepatica, a white anemone and a yellow buttercup be grouped together—would anyone ever dream that they were near relatives? And yet they are, and the buttercup has given its name to the family. The poppy family includes the bleeding heart, the Dutchman's breeches, the bloodroot and all the various kinds and colors of poppies; and the heath family is made up of the cranberry, the huckleberry, the exquisite trailing arbutus and the rhododendron; but perhaps the strangest of all families is one which includes some of the most useful vegetables, a widely-grown, much-loved flower, several troublesome weeds, and a poisonous plant which should be carefully shunned. This is the nightshade family, of which the potato, the tomato, the petunia, the sandbur, the jimson weed and the deadly nightshade are members.

Explanations as to how botanists have found out these relationships, some of them almost incredible, are impossible in language which anyone not a scientist could understand. It took centuries to work them out, and even now there are differences of opinion as to where certain plants belong.

Why Plants Need Insects. A very interesting phase in the study of plant life concerns their connection with insects. Many insects are pests of the worst kind, doing immeasurable harm to plants, but there are some without which a great many plants could not fulfill their life history satisfactorily. In order to bear

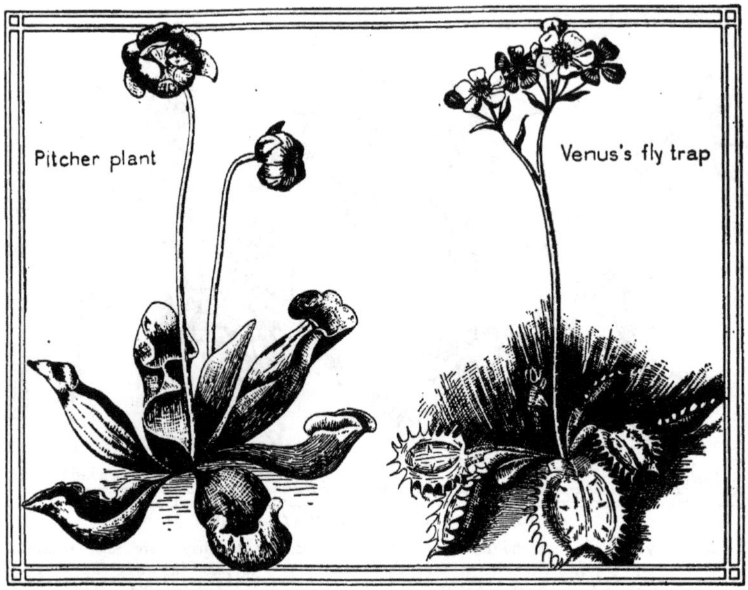

TWO OF THE CARNIVOROUS (INSECT-EATING) PLANTS

fruit and produce seed—the final purpose of every plant, useful or troublesome alike, that springs from the ground—a plant must have its blossoms *fertilized*. That is, the dusty, yellow or brown pollen must reach the part of the flower where the seeds are to be formed. The whole process may take place within a single flower, but almost all plants, especially those which have showy or sweet-smelling blossoms, thrive better if the pollen comes from another plant (see CROSS-FERTILIZATION). For carrying this pollen the plants are dependent upon the wind or upon insects, and wonderful indeed are the attractions which the flowers have developed to entice the insects into doing them this service. The bright colors, the curious shapes, the sweet scents, above all the delicious nectar or honey, are all but lures for the roving insects, and some blossoms have special features which make it impossible for any insect but that which best fulfils their purpose to enter them (see BEE).

This cross-fertilization boys and girls cannot observe for themselves in its more technical phases, but much that is interesting about it they can trace. For instance, they may watch a patch of red and white clover and discover what insects visit it. How many blossoms can a honey-seeking insect visit in five minutes? Does an insect visit first one flower and then another of an entirely different kind—first a clover, then a nasturtium in a near-by garden —or does it take only flowers of one kind?

Interesting and Curious Plants. Plants are not "alive" in the sense that animals are; that is, they cannot move about and they have no intelligence to help adapt themselves to their circumstances; but some of them display traits which might almost lead a plant-lover to declare that they actually have sense. There are the insect-eating plants, for instance, with their various ways of catching and holding insects until the digestible parts have been absorbed; they are quite as adept at catching their insect food as is the whip-poor-will, which flies through the air with its mouth open, simply swallowing all that fly into its mouth (see CARNIVOROUS PLANTS). Then there are the uninvited guests, or parasites, plants which are not willing to do the work necessary to the preparation of food, and so fasten themselves upon some other hard-working plant and draw their nourishment from it until in the majority of cases the host dies (see PARASITES).

It has to be a very broad-minded plant-lover who can love weeds as well as the beautiful garden plants, and yet to a botanist a weed may be a far more interesting plant than the most carefully nurtured flower. For it is no weakling—it does not demand care and congenial surroundings, but grows wherever it finds itself. And if, at times, it is too self-assertive, and steals food, moisture and air from weaker plants about it, why that is what makes it a weed. The name *weed* is but a relative term, and almost any plant might be so called under certain conditions. See WEED.

Another very interesting class of plants consists of those that store food. Just as truly as the squirrels, which during the fall bear off to their tree-homes nuts and acorns, store up more than they can use at the time, do these thrifty plants provide for a time that is coming. They are not always allowed to keep this surplus food and use it as they intended, however, for man likes it, too, and it is frequently to be seen on his table. For it is the part of the beet, the onion, the carrot, the parsnip or the turnip which man eats that constitutes the stored-up food. These are *biennial*, or *two-year*, plants—that is, they do not produce seed the first year they are in the ground. During the first summer they are busy manufacturing food and laying it away either in their roots, as in the case of the turnip or carrot, or in that part of the stem known as the bulb, as in the case of the onion. The second year this stored food is made use of to build a tall stem, on the top of which the flowers and finally the seeds appear.

Everything in those plants which have blossoms works toward that one end—the production of seed; but after these have ripened something remains to be done. How are they to be carried to some place where they may take root and grow? So interesting and important is this question that an entire article has been devoted to it in these volumes, and in the article SEEDS, under the subtitle *Seed Dispersal*, will be found descriptions of the devices by which seeds go on long journeys.

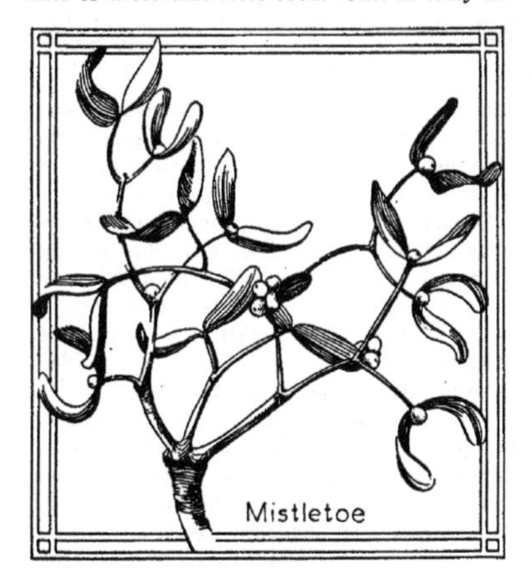

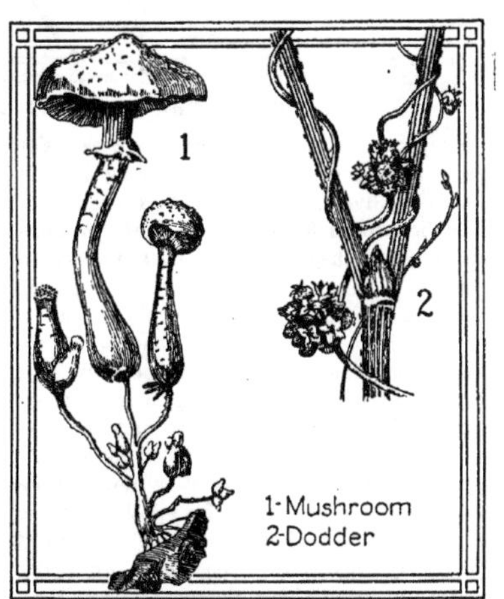

THREE FAMILIAR EXAMPLES OF PARASITES

The Struggle for Existence. There are many, many people who go through life and never see a foxglove; it is not one of the very commonest flowers. And yet a knowledge of the seed-producing ability of that plant makes it seem strange that there are not foxgloves to be pushed aside wherever one steps. For a single foxglove plant may bear in a year over 1,250,000 seeds, and if every one of these were planted and if each grew and bore seeds, there would be at the end of the second season 1,562,500,-000,000 seeds. Continuing the process, it may be seen that the descendants of the one foxglove plant would within a comparatively few years spread all over the earth. But all of this hinges on a very large *if—if* every seed could be planted, grow and produce seeds. It is here that the "struggle for existence" and the "survival of the fittest," as the great scientist Dar-

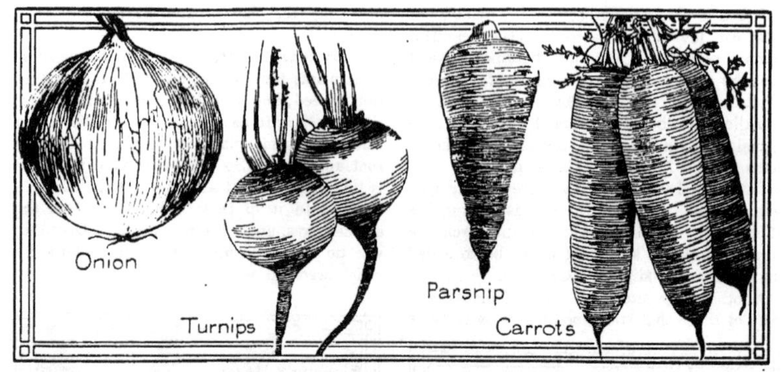

FOUR OF THE VALUABLE PLANTS THAT STORE FOOD

win expressed it, must be reckoned with. For plants as well as animals have to fight for a chance to live, and only the strongest and those best fitted for life under the conditions among which they find themselves manage to exist.

One of the things plants have to fight is overcrowding. So many little plants spring up close together that all cannot possibly find room to grow and food enough to make them strong. Every farmer knows this and is guided by it, for if he sows his seed too thickly the plants must be thinned out as soon as they show above the ground, or few of them are as strong as they should be. Everyone has noticed how many tiny trees often spring up under the parent tree where the seeds have been dropped, but comparatively few of these, and those the strongest, ever grow up.

Changes in the climate or moisture conditions are other enemies of plant growth. Certain plants need much moisture, and if for any reason the water in a locality grows less and less and finally is gone entirely, these moisture-loving plants must die; swamp vegetation is a definite type which is destroyed when a marsh or swamp is drained. On the other hand, land plants frequently suffer from flood conditions. In the late spring untimely frosts kill off millions and millions of little plants which are still too feeble to resist the cold, and the first frosts of autumn prevent many a plant from ripening its seed.

Animal enemies, too, are numerous and include not only cattle, sheep and other grazing animals, but smaller and still more deadly foes —the locust, chinch bug and other insects. Whole crops in various parts of a country are destroyed by these insects every year. (See INSECTS, and each pest named above.)

Through these and other agencies it comes to pass that the earth is not a jungle like that of the Amazon—a jungle so dense that men and animals could not find their way through. In the long run, though vegetation may be luxuriant one year and sparse the next, the growth and the dying-out just about balance, so that if left to itself vegetation remains fairly constant. Man, by his deliberate efforts, can do much to change this; he can cut down centuries-old forests and so cause all the surrounding region to suffer, or he can till and plant barren stretches until "the desert shall rejoice, and blossom as the rose." See FORESTS.

For a discussion of the structure of plants, the functions of their different parts, and the way they grow, see BUD; FLOWERS; LEAVES; SEEDS; STEM.

Botany for Older People

The botanist or the serious, mature student finds in the science far more than a series of loosely-connected, interesting facts about flowers and trees and weeds. His botany includes the study of every plant form that grows, from the tiniest bacteria that can be seen only with the aid of a microscope, to the gigantic trees whose "slender tops are close against the sky." In fact, so wide is the study that it might be described as a group of sciences rather than as a single one. There is *morphology,* perhaps the fundamental branch, which deals with the development and mature form of plant structures, tracing the life history from the single cell to the final complex stage; *taxonomy,* or classification (see subhead below), which depends upon morphology for the facts upon which the arrangement of plants into groups is based; plant *anatomy,* the study of the tissues of which the plant is composed; *physiology,* which deals with the functions of the various organs, leaf, stem, root, and the parts of the flower; and *ecology,* a somewhat recent phase of the science and an extremely interesting one. It deals with plants in relation to their surroundings—the changes made in them by changing environments, and the special structures that adapt certain plants to conditions which others could not endure.

Botanical Terms. Anyone in North America who knows anything about trees knows an oak, but the trees which are called by this name differ in size, in manner of growth, in the shape of leaves and acorns, and in other particulars. Some are live oaks, others white oaks, red oaks or bur oaks, but all have an equal right to the name oak. In his scientific naming of plants the botanist takes into account these variations, and calls each different kind a *species* (which see). That is, the live oak is a species, and the bur oak is a species; but all the oaks together constitute a larger group, called a *genus* (which see). The Latin name bestowed upon a plant consists of two parts, the first indicating the genus to which it belongs, the second its particular species. *Quercus,* for instance, is the Latin word for *oak,* and is borne by all trees belonging to that genus, but *alba,* meaning *white,* is borne only by one species, the white oak. The botanical name for the white oak, therefore, is *Quercus alba.*

But all plants which are related are not so closely connected as is one oak with another, and there must therefore be larger groups than the genus. The next higher is the *family,* which is composed of several or many similar genera (plural of *genus*). As pointed out above, members of a family frequently bear no closer resemblance to each other than do members of a human family, and only a botanist can trace their similarities (see FAMILY). Above the family is the *order,* into which families that have certain resemblances are grouped, and still higher is the *class.* At the summit of the classification is the division into great groups, of which there are but four in the entire plant kingdom, and these are described below. Very definite differences in structure or in habit divide these groups from each other.

Classification. As botany has been a steadily-growing science, it is natural that the classification should not have remained stationary; thus to-day all botanists do not use the same terms or even the same divisions into groups. In the end, however, they all amount to exactly the same thing, and the classification here given is the one acceptable to most scientists. Some of the names are difficult, but in every case explanations or alternative terms are given.

I. *Thallophytes (thal' o fites),* very simple plants without roots, stems or leaves. Some are tiny, one-celled structures, others are large and showy, but none of them produces seed or flowers. To this lowest of plant groups belong the following forms:

(a) Bacteria are the simplest of plants. Most people who speak of these as microbes, germs or bacilli do not realize that they are plants as truly as is the greatest oak of the forest. (See BACTERIA.)

(b) Algae, or as more commonly named, seaweeds. These include not merely the seaweeds, properly so called, but the scum which forms on stagnant pools. (See ALGAE.)

(c) Fungi, including such differing forms as bread mold, water mold, mushrooms and puffballs. (See FUNGI.)

(d) Lichens, which partake of the character of both algae and fungi. (See LICHEN.)

II. *Bryophytes (bri' o fites)*, or, to use a simpler name which means the same, "moss plants." This group includes two forms:

(a) Liverworts.
(b) Mosses.

III. *Pteridophytes (ter' id o fites)*, or "fern plants." Though the name means literally *fern plants*, other plants belong to the group, as listed below. Some of these pteridophytes are large and beautiful plants and to the untaught observer might seem to be as high in the scale as the grasses or garden plants, but they do not have flowers nor produce seed.

(a) True ferns.
(b) Horsetails or scouring rushes.
(c) Club mosses. The name must not lead to confusion with the true mosses lower in the scale.

IV. *Spermatophytes (spurm' a to fites)*. The name means literally *seed plants*, and it is of this group people usually are thinking when they talk of plants, for by all means the most conspicuous, beautiful and useful members of the plant kingdom belong to it. Not all the seed plants of this great group belong to one class, the method of bearing the seed giving rise to a division into *gymnosperms* and *angiosperms* (see these titles in these volumes).

1. Gymnosperms *(jim' no spurmz)*, or "naked-seed plants," which include—

(a) Pines
(b) Cycads
(c) Ginkgo or maidenhair trees.

2. Angiosperms *(an' je o spurmz)*, or plants with enclosed seeds, which include—

All of the common trees except the pine and all of the conspicuous garden plants.

They are divided into two groups, the divisions being based on the number of seed-leaves. Those plants which have but one are classed as *monocotyledons*, those which have two as *dicotyledons*, the term *cotyledon* here meaning *seed-leaf*, and the *mono* and *di*, respectively, *one* and *two*.

(a) Monocotyledons *(mon o kot i le' donz)*. To this order or subclass belong about forty plant families, some of which are of great importance. The chief of these are the following:

(1) Grass Family. This is probably the most useful to man of all plant families, for it includes not only the grasses ordinarily so called, which are valuable for hay and for grazing, but the bamboo, the sugar cane, and all the cereals.

(2) Lily Family, which has about 2,600 species scattered all over the world. The section above on PLANT FAMILIES gives an idea of the importance of this family.

(3) Palm Family. There are parts of the world where certain trees of this family are the mainstay of the people, and some of its members, as the cocoa palm and date palm, are of widespread importance.

(4) Amaryllis Family.
(5) Iris Family.
(6) Orchis Family.
(7) Pineapple Family.
(8) Banana Family.

(b) Dicotyledons *(di kot i le' donz)*. To this subclass belong more than 200 families, some of them small and obscure, many of them very large and important, like the rose family. It is impossible to list all of these families here, but a glance at the few which follow will make plain the great variety of dicotyledonous plants:

(1) Willow Family
  (a) Poplar
  (b) Willow
(2) Beech Family
  (a) Beech
  (b) Chestnut
  (c) Oak
(3) Buttercup Family (see above)
(4) Poppy Family (see above)
(5) Mustard Family
(6) Rose Family (see above)
(7) Pulse Family (see above)
(8) Maple Family
  (a) Maple
  (b) Box Elder
(9) Vine Family
  (a) Grape
  (b) Virginia creeper
  (c) Boston ivy

Practical Aspects of Botany. It must not be inferred that botany is a mere theoretical science, a pursuit of facts for facts' sake. If all that resulted from the researches of botanists were a classification, however complete and exact, the subject would be a barren one, indeed. But it has its very practical phases. For one thing, though the medical side is no longer dominant as it once was, it is still of great importance, and a study of certain plants and their effects is one of the necessary parts of a physician's education. Closely related to this is the study, of late so absorbing to many scientists of distinction, of those simple plant forms, the bacteria, which play so large a part in many diseases.

Many plants, too, are useful in the arts, and the study of these is known as *economic* botany, while *agricultural* botany, or the study of farm crops, is of an importance which demands no emphasis. A knowledge of botany is essential to forestry, and connected with this is the lumber industry in many of its phases. In fact, if a careful examination is made, it is difficult to find an industry which does not touch botany at some point. A certain phase of botany

OUTLINE AND QUESTIONS ON BOTANY

Outline

I. Definition

(1) Science of plants
(2) A field study—not a mere book study

II. What It Deals With

(1) The composition of plants
(2) Their structure
(3) Functions of their parts

III. Structure of Plants

(1) Cells
(2) Root
 (a) Structure
 (b) Functions
 (c) Kind of root
  1. Fibrous
  2. Fleshy
 (d) Storage of food in root
(3) Stem
 (a) Structure
 (b) Functions
 (c) Movement of water in stem
 (d) Storage of food
(4) Bud
 (a) Structure
 (b) Kinds of buds
 (c) Position
(5) Leaf
 (a) Venation
 (b) Arrangement
 (c) Movements
 (d) Structure
  1. Chlorophyll
 (e) Functions
(6) Flower
 (a) Structure
  1. Calyx
  2. Corolla
  3. Stamens
   a. Pollination
  4. Pistils
 (b) Functions
(7) Fruit
 (a) Fleshy fruits
 (b) Dry fruits
 (c) Aggregate fruits
(8) Seed
 (a) Storage of food in seed
 (b) Germination
 (c) Seed dispersal

IV. Classification of Plants

See article above

V. History of the Science

(1) Its early connection with medicine
(2) Gradual development
(3) Becomes an exact science

Questions

How do the insects pay for the sweets they take from the flowers?

What plant family contains a beautiful flower, a deadly poison and two favorite vegetables?

What are the simplest and tiniest of all plants?

What favorite children's story is illustrated in the growth of every plant from its seed?

Why do not all the acorns that fall to the ground produce great oak trees?

Describe three things which plant-breeding has accomplished.

How important are plants in the life of men and of animals?

Why do flowers have beautifully-colored petals and sweet-smelling nectar?

How do the seeds of the maple differ from those of the pine, and what use is made of this in their classification?

What is the name of the great group of plants which is distinguished by having two seed-leaves?

Where is the pollen in the flower to be found?

Mention six things which you would miss if the rose family were taken out of the world.

If you see the name *Quercus alba,* what does it tell you?

which has grown very rapidly since the beginning of the twentieth century, with results which look to the uninitiated like miracle-working, is plant-breeding. By it the cultivation of many plants has been practically revolutionized, and others which were once of no commercial value whatever have been made useful members of the plant world (see BURBANK, LUTHER). See, also, PLANT. M.S.

Consult Bergen's *Botany;* Bergen and Caldwell's *Botany;* Gray's *How Plants Grow;* Barnes' *Outlines of Plant Life;* Hodge's *Nature Study and Life;* Coulter's *Elementary Studies in Botany.*

Related Subjects. These volumes contain articles on hundreds of plants, and these are all classified and indexed under the articles on the following topics:

Carnivorous Plants	Plant—subheads
Dyeing	Creepers
Ferns	Desert Plants
Fiber	Forage Plants
Flowers	Shrubs
Fruits	Tropical Plants
Fungi	Unclassified
Grains	Seaweed
Grasses	Spice
Herbs	Trees
Medicine and Drugs	Vegetables
Nuts	Water Plants
Parasites	Weeds

There are also more general articles which will prove very helpful to the reader interested in botany:

Air Cells	Ecology
Air Plants	Etiolation
Alburnum	Evergreen
Angiosperms	Exotic
Annuals	Flora
Biennials	Galls
Boehmeria	Germination
Botanical Garden	Gymnosperms
Bract	Herbarium
Breeding	Inflorescence
Bryophytes	Kew Gardens
Bud	Leaves
Bulb	Leguminous Plants
Catkin	Osmosis
Cell	Perennials
Cellulose	Phanerogamous Plants
Chlorophyll	Pollination
Citrus	Protoplasm
Colchicum	Pteridophytes
Composite Family	Puffball
Coniferae	Ranunculus
Corm	Roots
Cotyledon	Rose Family
Cross-Fertilization	Sap
Cryptogams	Seeds
Cycads	Spore
Diatom	Spurge Family
Disease	Stems
Subhead *Diseases of Plants*	Umbelliferae
	Venation

BOTFLY, or **HEEL-FLY,** a common, hairy parasite-fly, which lays its eggs upon the hairs of cattle, usually above the hoof. Licking their legs, the animals pick up the eggs on their tongues. There they hatch quickly, and in sheep they move to the nostrils, in horses to the stomach and in cows to the back. When ready the larvae (young) push their way out, drop to the ground, burrow and soon come forth as flies.

BOTFLY
About twice natural size.

The lumps seen on the backs of cattle in the late winter and spring show the presence of botfly larvae, or grubs, and the pests should be removed, or kerosene should be injected into the spots. Much suffering is caused to "grubby" cattle, even loss of life, and it is estimated that the total yearly loss to the United States and Canada through the botfly is nearly $50,000,000.

BOTHA, *bo' tah,* LOUIS (1862-), a statesman, general of the British army, and the first Premier of the Union of South Africa. He was born at Greytown, Natal, at a time when native affairs were in a very unsettled state, and early served in wars against native tribes. During the South African War he was one of the ablest of the Boer leaders, and prolonged their resistance by skilful organization of guerrilla warfare, until he was forced to admit that British success was inevitable. In the period of reorganization following the war he worked for the best interests of his countrymen, and loyally supported the British government. Becoming Premier in 1907, when the South African provinces were given self-government, he proved himself a leader worthy of admiration for his broad-minded policy.

In 1915, during the War of the Nations, leading the forces of the Union of South Africa, General Botha invaded German Southwest Africa and after a most remarkable campaign received the surrender of the territory from the leader of the German forces.

BOTHNIA, *bahth' ni a,* GULF OF, a landlocked sea, extending from the island of Aland northward between Sweden and Finland, forming an arm of the Baltic Sea. It covers an area nearly as large as the state of New York. Commercially and politically it is of little value. On the east it is bordered by land devoid of all agricultural advantages, on the north are the barren and Arctic regions of

Lapland, and it is only on the southeastern coast of Sweden that land of real value is to be found. Within comparatively recent times there appears to have been a general rising of the coast line of the gulf and a sinking of the level of the sea. Towns which were at one time seaports are now several miles inland. Navigation is difficult, owing to the presence of many small islands and sandbanks; on account of its shallowness, never exceeding 300 feet in depth, the frequent storms quickly cause a rough sea. In the winter the gulf freezes over and traffic between Sweden and Finland is carried on by sleighs.

BOTHWELL, *bahth' wel,* JAMES HEPBURN, Earl of (1536?-1578), a Scottish nobleman who is important in history only because of his marriage to Mary Queen of Scots and for the events which that union brought about. He won the confidence and regard of the queen after her marriage to Lord Darnley in 1565, and when the latter was murdered in 1567 he was accused of having had a hand in the affair. Being summoned to answer the charge of murder, he appeared at the trial with 4,000 of his followers, and was speedily acquitted. He was then in high favor with the queen, and, with or without her consent, he seized her at Edinburgh, carried her a prisoner to Dunbar Castle and prevailed upon her to marry him after he had divorced his own wife. A confederacy was formed against him, and in a short time Mary was brought to Edinburgh a prisoner. Bothwell fled to Norway, from which country he was sent under arrest to Denmark, where he died. See MARY STUART.

BOTTICELLI, *bot te chel' le,* SANDRO (1447-1515), an Italian painter, one of the greatest representatives of the Florentine school during the early Renaissance (see RENAISSANCE). His work, distinguished for breadth of culture, variety of subject, richness and delicacy of coloring and high imaginative quality, is at all times an expression of his individual moods and ideas. He excelled in painting Madonnas (see MADONNA), and important examples of these may be seen in a private collection in Boston, in the Uffizi Palace, Florence, and in the Berlin Museum. The masterpiece of his early career, a panel representing the *Adoration of the Magi* (in the Uffizi), shows a few traces of the influence of one of his first teachers, Fra Filippo Lippi.

Botticelli was commissioned by Pope Sixtus IV, in 1481, to take charge of the decoration of the latter's new chapel in the Vatican. Three of the frescoes in that chapel—*The Life of Moses, The Temptation of Christ* and *The Punishment of Korah, Dathan and Abiram*—and several portraits of Popes were painted by Botticelli himself. On his return to Florence he executed commissions for Lorenzo the Magnificent and other Florentine notables, and for a cousin of Lorenzo he painted some of his greatest canvases portraying scenes from mythology. In the Academy of the Fine Arts, Florence, hangs his most celebrated mythological picture—*Spring,* or *The Realm of Venus.* Of equal beauty is its companion picture, *Birth of Venus,* now in the Uffizi. Both of these canvases are characterized by delicate coloring, grace and lightness of touch, and richness of imagination. They illustrate also the artist's greatest weakness, his inability to represent correctly the human figure. Among other well-known works are three panels representing episodes in the life of Saint Zenobius, one of which is in the Metropolitan Museum, New York; and a decorative panel portraying the story of Lucretia, in the Gardner collection, Boston.

BOTTLE, *bot' 'l,* a vessel for liquids, usually of glass or earthenware, and generally made with a narrow neck and small opening which can be closed with a cork. In ancient times bottles were made of skins and in many parts of Asia and in Southern Europe such primitive vessels are still in use. The bottles mentioned in the Bible were probably made of goat skin, sewn as nearly as possible in a natural shape, with one leg forming the neck. The chief disadvantage of skin bottles lay in the fact of their affecting and being affected by their contents. The same objection applied to metal bottles. It was not until glass was utilized that a bottle was obtained which was practically impervious to its contents, no matter how injurious they might be to other materials. Acids that would quickly eat their way through leather or metal vessels produce no effect upon glass bottles. Wines may be kept for hundreds of years in glass bottles without acquiring any unnatural taste or in any way affecting the glass.

How Bottles Are Made. In making a glass bottle the operator takes a mass of molten glass from the smelting furnace. This is placed on the end of a hollow metal tube through which air is blown, either from the lungs of the workman or by machinery. When blown out into a pear-shaped hollow, the glass is placed in a red hot mold the size and shape of the bottle required. The blowing is continued until the glass is forced by pressure of the air into all parts of the mold. It is then cooled and removed, the neck and mouth being worked to the desired size and smoothed by special tools. Lettering and other marks on the bottle are made by dies placed in the mold. Bottle making is one of the most important branches of the glass industries and has grown to vast proportions. It is estimated that in the United States alone more than $41,000,000 worth of glass bottles are annually used by bottlers of wine, spirits and beer; this is one of the allied industries which is declining with the spread of prohibition.

BOTTLE-TREE, a curious Australian tree which has dense foliage and a short trunk resembling a bottle with bulging sides. The leaves of this tree are long and narrow, smooth-edged and pointed at the tip. The flowers grow

AUSTRALIAN BOTTLE-TREE

in short clusters on slender stems. The fruit is a pod with six seeds. The natives make nets of the fibers and use the sap in the stems as a drink, this latter often having been found useful in times of drought.

BOUCICAULT, *boo' se ko,* DION (1822-1890), an Irish comedian and playwright who won extraordinary success as an author of romantic Irish plays and as an actor of eccentric Irish parts. Though he studied to become an architect he was far more interested in the stage than in his intended profession, and the success of his famous comedy, *London Assurance,* produced when he was only nineteen years of age, made him resolve to seek his fortune in the theater. Other plays followed rapidly, and in 1852 he began to act. The following year he visited America, where his popularity was assured from his first appearance on the stage. Several successful plays were produced in America, among them his famous *Octoroon.*

Returning to London in 1860, Boucicault produced *The Colleen Bawn.* To this play he gave the name "Sensation Drama," and it was the first of those dramas, still popular, that depend upon thrilling effects and situations for their interest. *The Colleen Bawn* earned for its author a fortune. Among others of this type he wrote *Arrah-na-Pogue* and *The Shaughrau*. With Charles Reade he wrote the novel *Foul Play,* which was afterwards dramatized, and with Joseph Jefferson he dramatized Irving's story of *Rip Van Winkle,* in which Jefferson became world-famous as an actor. In all Boucicault wrote over 300 dramatic pieces. As an actor he was not highly gifted, but his keen sense of humor and likable personality made him immensely popular.

BOUGHTON, *baw' t'n,* GEORGE HENRY (1834-1905), an English-American painter of Puritan and Dutch life, whose work shows remarkable fidelity to the subjects portrayed. He was born near Norwich, England, but was taken to the United States by his parents in 1839, where they settled near Albany, N. Y. At an early age and without a master he took up the study of art, but later went abroad and studied in Paris. In 1862 he opened a studio in London. His skill and fine sense of color soon gained for him associate membership (1879) in the Royal Academy, and in 1896 he became a full member. In 1871 he was made a member of the National Academy of Design, New York City. His most popular pictures are the *Return of the Mayflower, Puritans Going to Church* and *The Scarlet Letter.* His picture of quaint little Dutch girls *Weeding the Pavement* is in the Tate Gallery, London. His *Edict of William the Testy* is in Washington, D. C.

BOUILLON, *boo yoN',* GODFREY DE. See GODFREY DE BOUILLON.

BOULDER, *bole' dur,* COLO., a city noted for its beautiful setting in the Rocky Mountains and for its mineral and agricultural resources. Its population increased from 9,539 in 1910 to 10,983 in 1914. It is situated in Boulder County, of which it is the county seat, about midway between the geographical center and the northern border of the state. Denver is twenty-nine miles southeast. The city is served by the Colorado & Southern Railroad, constructed to the city in 1870; by the Union Pacific, built in 1873; the Denver, Boulder & Western (the Switzerland Trail of America), built in 1899; and the Interurban Electric, constructed in 1908. Boulder was founded in 1859, was incorporated as a city in 1871 and named for the immense boulders which occur in the vicinity. The area is about two and one-half square miles.

Boulder lies on the eastern slope of the foothills of the Rocky Mountains, in the midst of some of the most wonderful scenery of the range, at an elevation of 5,835 feet above sea level. It is the south gateway to Rocky Mountain National Park and Estes Park. Long's Peak, one of the highest in the state, with an altitude of 14,271 feet, and Royal Arch are interesting features of the vicinity. Pure mountain water comes from a chain of eight lakes, fed by the Arapahoe Glacier. The surrounding country is rich in agricultural products; fruits, grain and garden truck are raised in great variety, and there are abundant crops of alfalfa. Mining and smelting are prominent industries of the locality, which produces gold, silver, lead, copper and zinc ores, and an abundance of tungsten.

Boulder is the seat of the University of Colorado (see COLORADO, UNIVERSITY OF). In addition to its public school system, it has a business college, Saint Gertrude's Academy and a Carnegie Library. F.E.E.

BOULDER, a rounded stone which has been worn to its shape by water and is too large to be called a pebble. The name is also given to large surface rocks that have been partially smoothed by the action of ice in the Glacial Period. Boulders are found on the surface, also imbedded in clay and gravel and in the beds of streams. They are usually of different composition from the rocks in the vicinity, which is evidence that they must have been transported a long distance by ice. When lying on the surface they are known as *erratic blocks*. See GLACIER; GLACIAL PERIOD.

BOUNTY, a reward or premium paid from public funds to encourage a certain kind of labor or production calculated to be of benefit to the whole community. The term is also applied to money paid for the extermination of certain destructive wild animals. In the United States this form of bounty is now practically obsolete. In Canada such a bounty is not uncommon; as an instance, the sum of $15 is paid by the government for every female wolf destroyed. In Australia, rabbits have become so destructive that a small sum is paid for every one killed. In South Africa, India, and many Eastern countries subject to visits of bubonic plague a bounty is also offered for the destruction of rats, for by these filthy animals the dreadful disease is spread. Locusts at times inflict such terrible damage on crops in Africa that the provincial governments have put aside large sums to be utilized in taking measures for their destruction.

Industrial Bounties in Canada. To encourage certain industries in Canada, the Dominion government annually appropriates a sum of money to be distributed as a bounty to all owners of fishing vessels and all men engaged in the fisheries. A bounty is also paid on all crude petroleum produced in Canada, on iron, steel and lead smelted in the Dominion and, in order to encourage the manufacture of binder twine, a rebate of duty is allowed on all Manila hemp imported.

BOURASSA, *boo ras' sa,* HENRI (1868-), a Canadian journalist, orator and political leader, the chief of the French-Canadian Nationalists. Bourassa was born in Montreal, and his fellow French-Canadians have repeatedly elected him to public office. He was mayor of Montebello, Que., from 1890 to 1894, and of Papineauville in 1897. He sat in the Dominion House of Commons from 1896 to 1908, and in the Quebec assembly from 1908 to 1912. But these offices do not indicate the importance of his career, though they did give him an opportunity to voice his opinions. His influence is now felt chiefly through his public speeches and through *Le Devoir,* the newspaper which he founded in 1910. He believes that French-Canadians ought not to endanger their interests by interfering in any complications caused by British rule. In other words, the interests of the British Empire, he says, are not the interests of the French-Canadians. For this reason he opposed Canadian participation in the South African War in 1899 and in the War of the Nations, which began in 1914. The French-Canadian Nationalists, believing that they exist in fact, and should be recognized in government, as a separate nation, have carried to its logical extreme the doctrine that small nations have a right to live and govern themselves as they wish. With this wish the majority of Canadians may sympathize, but they do not regard the French-Canadians as a separate nation.

BOURBON, *boor' bon,* an ancient family which gave many kings to France, Spain and Naples. Originally the Bourbons were lords of the old province of Bourbonnais, in France, but by marriage they became connected with the royal family, and in 1589 Henry of Navarre, a Bourbon, came to the throne of France as Henry IV (which see). Louis XIII, Louis XIV, Louis XV and Louis XVI were all of this dynasty. This line of rulers was deposed at the outset of the French Revolution, but was restored to the throne in 1815 in the person of Louis XVIII. Charles X succeeded to the crown in 1830, but was forced to abdicate in favor of Louis Philippe, a member of a younger branch, the Bourbon-Orleans. He, too, was deposed in 1848. His heirs, however, have never given up their claim to the throne of France, and until the War of the Nations (1914) cemented all factions there yet persisted in the country a political party which centered around these Bourbon claimants. Now the republic is extremely popular.

In Spain and Naples. In Spain the Bourbons came to power when Louis XIV of France placed his grandson, Philip of Anjou, on the Spanish throne in 1700. There are to-day in Spain two branches of this famous family, one represented by the present king, Alfonso XIII, the other by Don Carlos, commonly known as *The Pretender*. During the two hundred years and more after the Bourbons began to reign in France the only break in their rule occurred during the wars with Napoleon Bonaparte.

In Naples the Bourbons came to the throne in the person of Charles III, son of Philip V of Spain, who gained the crown in 1735. His descendants reigned in Naples until 1806, when Napoleon took over the kingdom for his brother. After the downfall of Napoleon the Bourbon Ferdinand I became king of the Two Sicilies, and the Bourbon line continued to hold the throne until Sicily and Naples were made part of United Italy in 1860.

BOURGET, *boor zheh'*, Paul (1852-), the author of *The Disciple* and *The Promised Land*, a present-day novelist and critic whose reputation is likely to endure, because of his insight into human nature. He was born at Amiens, France, and attended school at Clermont-Ferrand, where his father was professor of mathematics. When he entered the famous College de Sainte-Barbe he showed marked ability as a student. After graduating from that school he entered journalism and traveled in many countries, meeting people of different classes. In this way he studied character and saw life among the rich and poor. His writings are serious and deal with the motives which control the action of people. Some have said that his work is pessimistic, but that is doubtless because he shows every-day life as it exists.

BOURINOT, *boo re no'*, Sir John George (1837-1902), one of the foremost Canadian historians, for twenty-two years clerk of the Dominion House of Commons. His official position gave him an intimate acquaintance with the leading men of his day, a privilege fully as great as the access to the records of the government. He became an authority on parliamentary procedure, which was the subject of one of his first books. Of his other books the most important are *Manual of Constitutional History; Parliamentary Government in Canada; How Canada is Governed; Canada Under British Rule; Intellectual Development of the Canadian People,* and a life of *Lord Elgin.*

He was especially interested in Nova Scotia, for he was born at Sydney, in that province. Two of his best books are about it and bear the titles *Cape Breton and Its Memorials of the French Regime* and *Builders of Nova Scotia.* He was for many years the editor of the Halifax *Reporter,* a journal which he established in 1860, and was at one time president of the Royal Society of Canada. From 1882 until his death he was the Society's honorary secretary. Bourinot's services to Canadian scholarship won general recognition, and he was knighted by Queen Victoria in 1898.

BOW, *bo,* **AND ARROW,** the world's most destructive weapon of offense before the invention of gunpowder. The bow was usually made of steel, wood, bone, horn or other elastic substance, and bent into a slight curve, the two ends being joined by a cord. The use of the bow is practically obsolete except in the pastime called *archery,* a word derived from the Latin *arcus,* meaning *a bow* (see Archery). The long bow, which was used by English archers, the most-dreaded fighting men of Europe for centuries, was usually the height of the archer, or about six feet in length, and the arrow was a metal-tipped shaft three feet in length. Three feathers near the blunt end of this shaft, doing service as a rudder, kept it in a straight course, point first, and prevented it from turning in the air. In the blunt end also was a notch which fitted the string of the bow. When the archer made ready for the shot he placed the arrow across the bow, notch on string, grasped the shaft with the fingers of the right hand and held the bow with the left arm, firmly extended. With a steady pressure the bow was bent until the hand holding the arrow on the string was drawn to the right ear. Suddenly released, the bow sprung back into its former position and the shaft sped on its mission. The weapons used in modern archery are made on the same principles, but are smaller.

The usual wood employed for bows is yew; in medieval times the importation of yew staves into England was demanded of every merchant, to insure a sufficient supply, and bow-making was an important industry. Laws were passed compelling towns to provide grounds for archery and in the thirteenth and fourteenth centuries all other sports were forbidden on Sundays and holidays. The crossbow was a bow placed in a rest, similar to the barrel of a gun, with a "stock" which was held against the shoulder. The bow was bent, and the string was placed over a trigger which when pulled released it and discharged the arrow, which lay in a groove along the barrel. The crossbow (which see) never acquired the popularity of the longbow, which was in every way superior.

The bow and arrow constituted the typical weapon of the chase employed by the North American Indians before they learned the use of firearms. Their arrowheads were of two kinds, sharp, and blunt or top-shaped. The

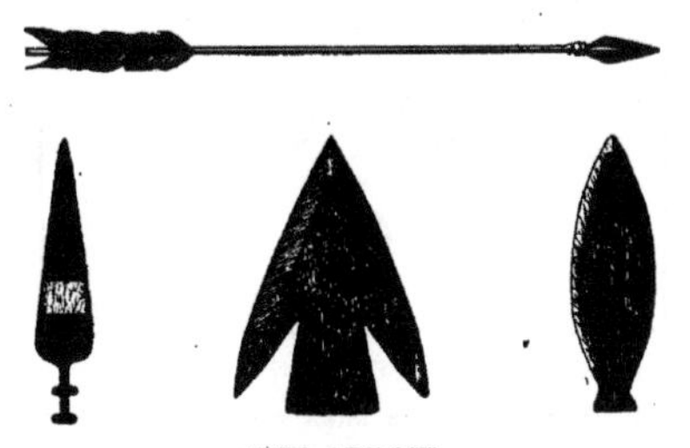

THE ARROW
At top, the Indian war arrow. Below: the first and second arrowheads were for use in war; the third was the usual form employed in hunting.

latter were for the purpose of stunning the prey. The arrow points were made of various materials, including ivory, bone, wood, copper and stone, and were attached to the shaft by lashing with sinew or riveting with gum. Reeds, canes, ivory, bone and wood were utilized in making the body part of the arrow. Several varieties of bow were in use, differing in size, material and shape. In the Arctic regions, where material was scarce, whales' ribs and driftwood were utilized. The Northern Athapascan tribes made long, straight bows of willow or birch, the tribes of the Saint Lawrence region used ash, hickory, oak and other hard wood, and so on. Sometimes the bows were beautified with painted decorations. F.ST.A.

BOWDOIN, *bo' d'n,* **COLLEGE,** the best-known college in Maine, famed as the oldest institution of learning in the state and numbering among its graduates such distinguished men as Nathaniel Hawthorne, Henry W. Longfellow, Franklin Pierce, Melville W. Fuller, Thomas B. Reed and Robert E. Peary. The college was chartered in 1794 under the laws of Massachusetts, of which Maine was then a district, and was named for James Bowdoin, a prominent statesman of the period (see below). It was opened at Brunswick in 1802, now comprises a group of buildings valued at $1,000,000, and holds rank with the best of the smaller colleges of the United States. The Medical School of Maine, opened in 1820, is connected with it. The college has an average of eighty-five professors and instructors and 450 students. The library contains nearly 110,000 volumes.

Bowdoin, James (1726-1790) was born in Boston and educated at Harvard College. An eager patriot throughout the pre-Revolutionary struggle, he served Massachusetts as a representative in the general court, as state senator and as president of the colonial council. In 1779 he presided with distinction over the Massachusetts constitutional convention, and succeeded John Hancock as governor in 1785. His administration was notable for the energy he showed in putting down Shays's Rebellion (which see). At the close of his term of office he took part in the proceedings of the Massachusetts convention that ratified the Constitution of the United States. Bowdoin was a man of scholarly and artistic tastes, and was one of the founders of the American Academy of Arts and Sciences (which see).

BOWELL, *bou' el,* SIR MACKENZIE (1823-1917), a Canadian journalist and statesman, Premier of the Dominion from 1894 to 1896. He was at one time Grand Master of The Orange Association of British America, and for years its spokesman in the Dominion Parliament. Sir Mackenzie was born at Rickinghall, England, on December 23, 1823, but his parents emigrated to Canada in 1833, bringing their ten-year-old son with them. At Belleville, where they settled, the boy received a common school education and won success as the editor and proprietor of the *Intelligencer*, which he made one of the most influential newspapers in Ontario.

With the organization of the Dominion in 1867, Sir Mackenzie was elected to the House of Commons, where he served without interruption for twenty-five years. During this period he became one of the leaders of the Conservative party. He was Sir John Macdonald's Minister of Customs from 1878 to 1891, and held the same post for a year in the Cabinet of Sir John J. C. Abbott. He was then in turn Minister of Militia and Minister of Trade and Commerce, and on December 21, 1894, himself became Premier. Meanwhile, in 1892, after a service of twenty-five years in the House of Commons, he accepted an appointment to the Senate, of which he is still a member. His brief ministry was overthrown chiefly because of his attitude towards the Manitoba schools (see MANITOBA, subhead *History*); though an Orangeman, he favored support of separate schools for Roman Catholics. He resigned as Premier when his policy failed, but from 1896 to 1906 remained the Senate leader of the Conservatives. He was created Knight Commander of the Order of Saint Michael and Saint George in 1895.

SIR MACKENZIE BOWELL

THE BOWER-BIRD

BOWER-BIRD, a name given to several different birds about the size of the jay, living in Australia or the Pacific islands. They are so called because in the nesting season the male birds build remarkable bowers in which to woo their mates. The female bird is of a grayish-green, while the male is dark brown or purplish-black, with buff spots on the upper and gray and yellow on the under side. The only touch of bright color which he has is a band of rose-pink feathers around his neck. So, when other birds are wooing with beautiful song or display of brilliant plumage, the bower-bird, lacking these attractions, invites his chosen mate to his bower. In some secluded part of the forest, under the overhanging branches, he has built a wonderful little room with twigs and leaves, decorated with shells, or feathers, or orchids. And there he dances and bows, like a cavalier in a minuet, until his mate is won.

BOWLING, *bole' ing,* sometimes called TEN-PINS, is a modern improvement upon the old Dutch game of *skittles,* which Rip Van Winkle encountered in his wanderings in the mountains. It is an indoor game, in which the player tries to roll a large wooden ball down a long, narrow alley in such a way that it will upset ten wooden pins placed in position at the other end. It is also played with a less number of pins. The game is an excellent winter sport, requiring the use of practically every muscle in the body, yet it does not put too severe a strain on any one part.

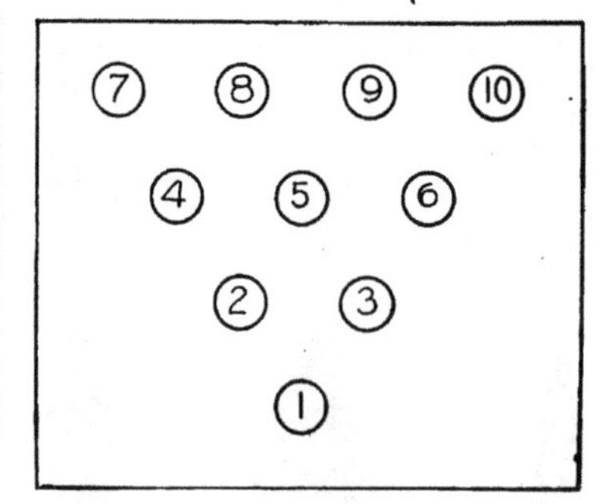

POSITION OF PINS IN BOWLING

The Alley. The long wooden platform down which the ball is rolled is called the *alley.* It is made of hardwood strips, usually set on edge. The surface is very smooth and highly polished, and slopes slightly from the center. A regulation alley must not be less than forty-one inches or more than forty-two inches wide. At one end a line is marked across the alley;

this is the *foul line,* beyond which the bowler is not allowed to step or slide as he rolls the ball. The foul line should be sixty feet from the first pin. Beyond the alley is a small pit, two and one-half feet wide, and back of this is a swinging padded cushion to stop the force of the balls. Alongside the alley are shallow gutters, into which the balls drop if they are not properly rolled.

The Pins and Balls. The pins are set up in triangular formation, each pin being exactly twelve inches from the next one. The last row, of four pins, must be three inches from the end of the alley. Each pin of regulation size is two and one-quarter inches in diameter at the bottom and fifteen inches long, and has a maximum circumference of eleven and five-eighths inches. The weight of each pin must be at least three pounds and two ounces.

The balls are usually of wood, but sometimes of a hard-rubber composition. They are made in several sizes, the largest being twenty-seven inches in circumference and weighing sixteen and one-half pounds. In each ball there must be two small holes, into which the thumb and a finger may be inserted to give a firm grip; sometimes, however, balls with three finger-holes are used.

Scoring. A game consists of ten *frames;* each player is allowed to bowl two balls in each frame. If he knocks down all the pins with the first ball, it is called a *strike;* if he knocks them all down with the two balls, it is a *spare.* If the bowler makes a strike, he adds to the ten thus scored the number of points scored on the next two balls that he rolls. If he rolls three strikes in succession he scores thirty in the first frame. A perfect score, comprising nothing but strikes, is 300; this is a feat seldom accomplished. If the player makes a spare, he adds to the ten thus scored the number of pins bowled over on his next ball. The example herewith will show how the scoring is done.

	NAME	1	2	3	4	5	6	7	8	9	10	ST.	SP.	TOTAL
1	W.F.Z.	20	39	48	65	84	93	122	142	162	182	4	5	182
2	M.E.C.	9	25	32	51	60	77	93	102	110	119	1	3	119
3	J.C.H.	30	60	90	120	150	180	210	240	270	300	12	0	300
4														

SCORING A GAME

This is the record of a "three-cornered" match. In the first frame two of the players made strikes, indicated by the x's. As this entitled them to 10 points plus the number of pins knocked over with the next two balls, no figures were written till after the next turns. In the second frame W. F. Z. made ten points to add to his strike, writing 20 as the score. As this also constituted a spare, an oblique mark was made, no figures being recorded till after the first ball of the third frame. Then 9 was added to the 10 earned by the spare, making the total 19 plus 20, or 39. As he failed to hit the remaining pin with his second ball, only 9 was scored for the third frame. J. C. H. made another strike in the second frame and so was obliged to wait till after the first ball of the third frame before scoring his first frame. As this was also a strike, the score was 30. Both W. F. Z. and J. C. H. made a strike in their last frame, so that each had to roll two extra balls. As shown by the score and the marks, one made a spare and the other rolled two strikes. The score of J. C. H. was perfect.

Bowls, or Lawn Bowling. This game has been popular in the British Isles since the thirteenth century, and may have been played there earlier. It has little resemblance to the indoor pin bowling just described, but is similar in principle to curling and to shuffleboard, both of which are explained elsewhere in this work.

A bowling green is a smooth grass plot about forty yards square. Usually it is level, but sometimes built with a *crown,* the center being a foot or more higher than the edges. *Rinks* six or seven yards wide are marked off, so that six games may be played at the same time. The word *rink* also refers to a team, just as in curling. Each green is surrounded by a ditch.

The object of a rink is to roll its balls as near as possible to the *jack,* a white earthenware ball about two and one-half inches in diameter. The bowls are made of wood, about five inches in diameter, and weigh not more than three and one-half pounds. They are made with a *bias;* that is, one side of a bowl is more convex than the other, so that unless thrown with great speed it will curve as it rolls. On a good green a bowl may curve six feet from a straight line. It is this bias which is the excuse for Shakespeare's pun in *King Richard II,* act III, scene IV:

Queen. What sport shall we devise here in this garden,
To drive away the heavy thought of care?
Lady. Madam, we'll play at bowls.
Queen. 'Twill make me think
The world is full of rubs, and that my fortune runs against the bias.

To start the game the first player of one side, called the *lead,* stands with one foot on a rubber mat and throws the jack twenty-five yards or more. He then throws his first bowl as directed by his *skip,* or captain. His first opponent follows, then the second man of his own side, and so on, the skips throwing last. Each player endeavors to place his own bowl in a good position, or to dislodge an opponent's bowl or the jack, or to guard a previously thrown bowl of his own team. In a match game each player throws two bowls, but in a friendly game four. Then the rink which owns the bowl resting nearest the jack scores one. Twenty-one points are a match. Sometimes a point is given for each bowl nearer the jack than the nearest of the opponent's bowls. When one *end* has been played the mat is moved beyond and the latter thrown out in the opposite direction for the next end.

Scotch and English rules differ as to the treatment of bowls or a jack knocked into the ditch. The Scotch, too, have a game of points which is entirely distinct from the regular game except that it is played with the same objects on the same greens.

Lawn bowling is better known in Canada than in the United States. In Great Britain there are many clubs. One in Southampton was founded in 1299 and has played on the same green for centuries. Bowlers from Canada and Great Britain have met in international matches in both countries, and Australian and New Zealand players have visited England. W.C.

For other recreations see lists of titles at end of articles AMUSEMENTS and GAMES AND PLAYS. Consult Spalding's *Guide to Bowling.*

BOWLING GREEN, KY., an important shipping point and horse market, with a population of 9,175 in 1910, which had increased to 9,597 in 1914. It is the county seat of Warren County, and is situated in the northwestern part of the state, at the head of navigation of the Barren River and on the Louisville & Nashville Railroad, which was constructed to the city in 1859. Louisville is 114 miles northeast and Nashville is seventy-two miles south. The city was founded about 1805, was incorporated in 1810 and named for the character of the land, *sloping and green.* The area is nearly two and a half square miles.

Locally, Bowling Green is known as the PARK CITY. It is located in a fertile agricultural region, which produces alfalfa, tobacco, grains, vegetables and fruits. Extensive quarries of building stone (oolite) and rock asphalt are found in the vicinity. Buildings of note are the Federal building, erected in 1914 at a cost of $160,000, and the courthouse. The most important industrial establishment is a dressmaking house which employs 300 seamstresses and makes apparel to order. The state normal school is located here; and Ogden College, Patten College, the Bowling Green Business College, a library and a good system of public schools offer unusual educational advantages. L.G.D.

BOWMANVILLE, *bo' man vil,* ONT., a town in Durham County, situated on Lake Ontario, forty-two miles northeast of Toronto. It is on the main lines of the Canadian Pacific, the Canadian Northern and the Grand Trunk railways, and is the terminus of the Toronto & Eastern Electric Railway. The harbor is deep and will accommodate the largest steamers on the lakes; the water commerce is considerable. Bowmanville is the center of a rich agricultural district, noted for apples and dairy products, and it is also an important manufacturing center. The largest establishment, which produces tires and other rubber goods, has about 750 employees, and a piano and organ factory employs about 200. Foundries, a glove factory, a flour and barley mill and canning works are also worthy of mention. Population in 1911, 2,814; in 1916, about 4,000.

BOWSER, *bou' zer,* WILLIAM JOHN (1867-), a Canadian barrister and legislator, long one of the leading Conservatives in British Columbia and in 1915-1916 premier of the province. After his graduation from Dalhousie University he practiced law for a year in New Brunswick, his native province, but in 1891 removed to Vancouver, B. C., which remains his home. He was an unsuccessful candidate for the Dominion House of Commons in 1896, and for the provincial assembly in 1898, but was elected to the latter in 1903. Sir Richard McBride in 1907 appointed him Attorney-General and Commissioner of Fisheries. In 1909 to 1910 and again in 1915 he temporarily performed the duties of Minister of Finance and Minister of Agriculture, in addition to his other offices. In December, 1915, on the resignation of Sir Richard McBride, Bowser was appointed premier, but his ministry lasted less than a year, for in the general elections of September, 1916, the Conservatives were defeated. Bowser was succeeded as premier by H. C. Brewster, the Liberal leader.

BOX-ELDER. See subhead, under ELDER.

BOXER REBELLION. The defeat of China by Japan in the war of 1894 to 1895 was the signal for aggressive action in China by a number of the European powers. By one means or another Great Britain, Russia, Germany and France obtained "concessions" of territory or commercial privileges from China. By 1900 it was clear that continued aggression by the foreigners would mean the division of China into spheres in which European influence would predominate, resulting in the practical loss of its independence.

The most obvious method of Chinese resistance was to drive out the foreigners. Anti-foreign demonstrations and attacks on foreigners became common in many parts of China, and early in 1900 foreigners began to hear of an organization called *Boxers*. The Chinese name really means "The Fist of Righteous Harmony," which was incorrectly translated *Boxers*.

The Boxers, though originally a patriotic society much like a volunteer American militia, came under the control of a group of fanatics whose cry was "Exterminate the foreigners." Their attacks on foreign residents became bolder, until Peking itself was in the hands of a mob which threatened the foreign legations. On June 10, 1901, a relief force of 2,000 marines left the foreign warships at Taku, with Peking as their destination, but it was driven back.

The situation in Peking was then desperate. The secretary of the Japanese legation and the German ambassador were murdered in the streets, and there remained little doubt that the Boxers were encouraged by the Chinese government. The foreign diplomats and their families, together with a number of missionaries and Christian Chinese, took refuge in the British legation. They fortified it as best they could, and here for nearly two months they were besieged, without any communication with the outside world. Meanwhile, Taku had been occupied by the allies, who prepared an army of 18,000 men to march to the relief of the besieged. On August 14, after a ten-days' march, the allied army entered Peking. The empress dowager, the boy emperor and the court fled into the interior. Exactly two weeks later, the allies—Japanese, Russians, British, Americans, French and Germans—marched through the Forbidden City, in whose sacred royal precincts no foreigner had ever stepped, as a "symbol of the humiliation of China."

Thus ended the Boxer Rebellion. It cost China and the foreign nations a few lives and much money. China agreed to pay an indemnity of $330,000,000, to be distributed among the allies. The United States, a few years later, formally renounced its share in this amount, and returned it. China also agreed not to import arms, ammunition or the materials for making them; it prohibited, under penalty of death, membership of any Chinaman in anti-foreign organizations of any kind, and it made several concessions in favor of foreign commerce. To prevent a repetition of such danger to the residents of the legations a district in Peking was set aside for foreigners. This section is now fortified. The peace agreement also provided that certain points might be occupied by the foreign powers in order to keep open communication between the sea and the legation quarter in Peking. See CHINA, subhead *History*.

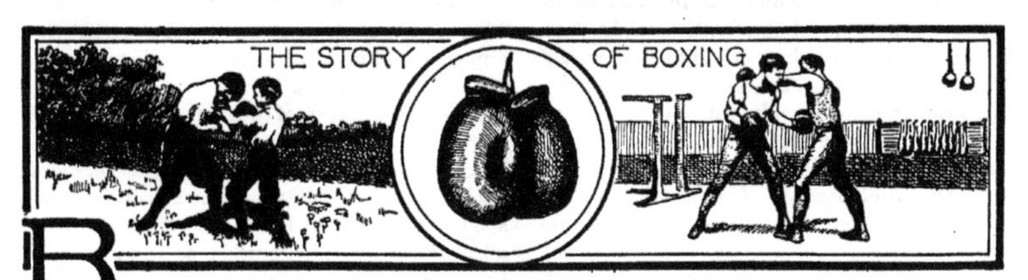

BOXING, called the "manly art of self-defense," is the art of hitting an opponent without getting hit. In popular use the term boxing is associated with friendly exhibitions of skill rather than with public prize-fights between professionals. Professional boxing is discussed in the article PRIZE-FIGHTING (which see). All boxers now use padded leather gloves, usually weighing six to eight ounces, partly to protect their hands and partly to prevent serious injury to each other.

Marquis of Queensberry Rules. These rules, as now in force throughout the world, are not quite the same as those originally adopted in 1867. Their first purpose was to eliminate the brutality of professional prize-fighting and make boxing a sport. Under the original rules a match included three five-minute rounds, with one minute's rest between rounds. If one of the boxers was knocked down or fell from weakness he was allowed any reasonable time to recover.

These rules, slightly changed, now apply both to professional and to amateur boxing. They require a "fair, stand-up" match, no wrestling or clinching allowed. Each round is three minutes long, and the intermission between rounds is one minute. If a boxer falls, either from weakness or from a blow, he is allowed ten seconds to get up without assistance. If he is not on his feet as the referee counts ten, the boxer is said to "take the count" and is "knocked out." His opponent is the winner. The American Amateur Athletic Association adopted the Queensberry rules with only a few minor changes. It fixed the maximum weight of gloves at eight ounces, and the size of the "ring" in which the match takes place at sixteen to twenty-four feet square. Under English rules the ring may be as small as twelve feet square.

Boxers are divided by weight into six different classes, the maximum weight for each class being as follows: Bantam weight, 105 pounds; feather weight, 115 pounds; light weight, 135 pounds; welter weight, 145 pounds; middle weight, 158 pounds; heavy weight, over 158 pounds. For example, if a man's weight increases from 105 pounds to 107 pounds he becomes a member of the feather-weight class, remaining in that division until his weight exceeds 115 pounds. Boxers in each class usually arrange matches only with others in their own class, but it is not unusual for a man to advance from a lighter to a heavier class.

Technique of Boxing. The object of boxing is to deliver blows and at the same time protect oneself from attack by one's opponent. Every experienced boxer develops an individual, characteristic style, certain defensive positions which allow him to use his strength and skill to best advantage. The beginner, however, should adhere to the accepted principles and not try unnatural poses until he has mastered at least the rudiments of the art.

The boxers face each other, just out of reach. Most boxers stand with the left foot in advance and slightly turn the left side of the body toward the opponent. Beginners should assume an upright position. Professional boxers, especially in the United States, assume a position called the *crouch*, in which the body is bent towards the right, while the left arm is stretched out toward the opponent. The left arm is used to make the most leads; the right delivers the heavier blows.

The attack may be made in several ways. First is the *straight lead*, a hard blow straight from the shoulder without preliminary feinting. A feint is an attempt to throw the opponent off his guard, as, for example, by apparently aiming with the right at his head while really planning to strike him with the left in the stomach. A *counter* is just like a lead, but is delivered at the same time that the opponent attempts to strike; it is really a *counter-lead*. A counter may be, and usually is, delivered in connection with a rapid change in position, such as a step to one side or a drop from a standing to a crouching position. A *cross counter* is a blow usually delivered with the right hand, which goes over the opponent's left arm as he counters a left lead or as he leads his left.

Guarding may be done in various ways, usually with the arm or hand. If the blow is stopped by the guard it is said to be *blocked*, otherwise it *lands*. A blow may be avoided by a quick duck of the head, thus allowing it to pass over. It may also be pushed to one side, so that its force is delivered only against the body. This is a valuable device, for if the opponent's blow is partly blocked and glances off, he is nearly always slightly off his balance and himself exposed to a stiff blow. There is always the further possibility that the opponent's blow will fall short if a boxer bends his back or his head.

Some Advantages of Boxing. As a form of physical exercise boxing is of undeniable value. Not only are the muscles of the arms, legs and back developed, but the various positions assumed in attack and defense bring into play all of those movements which serve to increase the agility of the boxer, make him light on his feet and develop control of his body. Moreover, he receives training in mental alertness as he practices the art of "hitting without getting hit." Boxing is excellent for mental discipline, and it also has a part in the development of character, for under right conditions it brings out the finest qualities of sportsmanship—courage, honesty and the ability to be a "good loser," and to control the temper in all situations.

Some Technical Terms. Below is a list of some special terms used by boxers and not already mentioned.

Break Away. Usually the call of the referee when the two are in a clinch.

Breaking Ground. To retreat diagonally from right to left.

Clinching. To catch hold of an opponent in such a way that he cannot swing his arms. This is a common device when a boxer is slightly dazed from a blow and wants a moment's rest without danger.

Corners. The corners of the ring. Each boxer has his own corner, diagonally opposite that of his opponent, to which he retires between rounds.

Cross-Counter. A blow in which the arm crosses the opponent's lead.

Drawing. Apparently leaving an opening for the opponent, but really preparing a counter-attack.

Fiddling. The preliminary motions at the beginning of a round, each boxer maneuvering for an opening.

Foul. Hitting below the belt, or with the palm or back of the hand. Kicking, tripping, wrestling and unnecessary roughness also constitute fouls. If the foul is a serious one the referee usually awards the match to the boxer against whom the foul was committed.

Hook. A sharp blow from the side, with the arm bent.

In-Fighting. Close quarters, so that full-body blows or blows straight from the shoulder are prevented. The boxers are also too close to swing.

Lead-Off. The first blow; or sometimes a straight lead.

Return. A blow after the opponent's blow has been blocked or avoided.

Second. Each boxer may have one or two assistants, who wrap him in blankets, fan him with towels, or otherwise help him between rounds.

Solar Plexus. The pit of the stomach. A powerful blow landed here almost always means a knock-out.

Swing. A swinging blow from the side. Usually the boxer jerks his body in the same direction, to give the blow greater force.

Throw Up the Sponge. When a boxer or his second tosses his sponge into the ring it is an acknowledgment of defeat. The boxer gives up the fight.

Upper-Cut. A short, sharp blow from below aimed at the opponent's chin. If successfully delivered it knocks him off his balance and may even knock him out. W.C.

BOXING THE COMPASS. See subhead, under COMPASS.

BOX TORTOISE, *tor' tis,* or **BOX TURTLE,** a species of tortoise or turtle found in North America from Long Island to New Mexico. It has a hard shell above and below, into which it can completely shut itself, locking itself in by means of hinged joints in the lower shell. The animal lives chiefly on mushrooms and berries, sometimes eating earthworms, slugs and eggs of insects. It lives to a great age, sometimes as long as 300 years, but its exact span of life has not yet been determined. The tortoise shell used for ornaments is not obtained from the box tortoise, but comes from the sea tortoise, especially the hawksbill.

BOX TREE, or **BOXWOOD,** a shrubby evergreen tree twelve or fifteen feet high. The leaves are small, oval, leathery and deep green. The flowers are inconspicuous, male and female on the same tree. It is a native of England, Southern Europe and parts of Asia. Formerly very common in England, it has given its name to several places—Boxhill, in Surrey, for instance, and Boxley, in Kent. The wood is of a yellowish color, close-grained, very hard and heavy, and it takes a beautiful polish. It is therefore much used by wood turners and carvers, engravers on wood and makers of mathematical instruments. As far back as the times of Pliny it was used to make flutes and other wind instruments.

The boxwood of commerce comes mostly from the regions adjoining the Black and Caspian seas, and is said to be diminishing in quantity. In gardens and shrubberies box trees may often be seen clipped into various formal shapes. There is also a dwarf variety reared as a hedge for garden walks and lawn borders.

BOY'COTT, the name given to an organized movement to injure or ruin the business of a person by refusing to deal or associate with him. The name is taken from that of an English land agent, Captain Charles Boycott, whose harsh measures against the Irish tenants of his employer, Lord Erne, caused the people for miles around to refuse to have anything to do with himself or family. In trade disputes the boycott is a favorite weapon of combinations of trade-unionists or workingmen, who seek thereby to injure an employer's business so he will accede to his employees' demands, or suffer punishment for refusing to do so.

Such a form of boycott is generally regarded lawful by English courts if it is employed to secure an increase in wages or improved working conditions, and if it is carried on without violence or other unlawful acts. In the United States a different view prevails in some localities, Alabama, Colorado, Illinois, Indiana and Texas expressly forbidding the boycott, even when it is unaccompanied by acts of violence.

BOYLE'S, *boil's*, **LAW**, also called MARIOTTE'S LAW, the law governing the elasticity of gases, as follows: Under the same temperature the volume of a gas will decrease in proportion to the increase of pressure on it. To illustrate, take two glass cylinders, *a* and *b*, each holding one pint. Fit each cylinder with an air-tight piston. Let the cylinders be filled with any gas, as an illuminating gas or hydrogen. On the piston of *a* place a weight of four pounds, and on that of *b* place a weight of eight pounds. The temperature of the gas in each cylinder being 70° F., the volume of gas in *b* will be one-half that in *a*. See GAS; ELASTICITY.

BOYNE, *boin*, BATTLE OF THE, in English history, the battle which decided that England's rule should be dominant in Ireland. It was fought in 1690, on the banks of the weedy River Boyne, near Drogheda. There the army of William of Orange (William III of England) met the forces of James II, last of the Stuart kings, and over them gained a decisive victory (see STUART, HOUSE OF). The reign of the Protestant Queen Elizabeth had forced the Protestant faith upon all Ireland, though the majority of the people were Roman Catholics. During the years that followed there had been several unsuccessful attempts to throw off the English yoke, and when James II, himself a Roman Catholic, came to the throne, he helped the Irish cause by putting into the hands of the Roman Catholics the civil government and military power of the island.

The Revolution in 1688, which dethroned James and gave the English crown to William of Orange, was the signal for another Irish rebellion, and James was invited by its leaders to join cause with them. Accordingly, he landed in Ireland in the spring of 1689 with a small French force loaned by Louis XIV, and established his headquarters at Dublin. The year following occurred his overwhelming defeat at the Battle of the Boyne. In 1691 the unfortunate war was terminated by the Treaty of Limerick. An obelisk 150 feet high marks the site of the battle, and Irish Protestants celebrate July 12 as the anniversary. See WILLIAM III; IRELAND, subhead *History*.

BOYS' AND GIRLS' CLUBS. The soil is the basis of life itself; mother's kitchen is the basis and business center of organized domestic life. The people are directly dependent upon these two interests of life for their very existence, and the entire human race constitutes a common brotherhood through our twin benefactors, *crop production* and *housekeeping*. It is no wonder then that we have developed through these two interests a type of education commonly known as "Boys' and Girls' Club Work," an agency for the perfecting of real democracy. Club work makes education possible to all; it takes facts from school, class room, laboratory and experiment station, and gathers a variety of information from books, bulletins and institutions. It carries to every home, regardless of its remoteness from class room or college, by means of this extension agency and through a program of follow-up work, printed instructions, field meetings, demonstrations and personal direction. Paid leaders or itinerant teachers through this means make education universal, and common knowledge more readily becomes common practice.

Beginnings of the Movement. The boys' and girls' club work had its inception in a few agricultural counties of the central states of the American Union, and was promoted by a few county superintendents of schools in Iowa, Illinois, Indiana and Ohio, as early as 1896. The movement, however, made little or no permanent progress within these states until the colleges of agriculture, through their extension departments, assumed definite leadership of the movement and made the project statewide in scope. The early development of the work with young people was fraught with considerable difficulty, in that the leaders looked upon the movement as a temporary agency through which to exploit the benefits of a college, or to make it possible to advertise and also to popularize more rapidly the work for adult farmers. Consequently much of the work was outlined on a basis of temporary contests in crop production, the making of big yields and the exhibiting of corn, vegetables and chickens at county and state fairs. No adequate follow-up system was made available for the work. State appropriations for extension work were used in the promotion of the work for men, and so the work for young people was not provided with the paid leadership needed, and very little definite direction was given to the movement.

It remained for the United States Department of Agriculture in the year 1908 to launch the club work as a permanent and worth-while coöperative educational movement. This was first undertaken in the state of Mississippi, in coöperation with the county superintendents of schools; later, during the period from 1910 to 1916, the movement spread to all the states of the Union and was supported by Federal and state legislative enactment. Under the terms of the Smith-Lever Act it became a definite agency with a fixed policy of extension education, the work being carried on coöperatively by the United States Department of Agriculture and the state colleges of agriculture through the States' Relations Service. In the year 1916 over half a million dollars was spent in the support of the work.

Every state in the Union has now a corps of paid coöperative leaders, and a total of from four to forty club leaders are now coöperatively employed in each of the states of the Union. In addition to this body of workers there are four county agricultural agents, both men and women, in most of the counties of the Union, who serve the people in a very practical way in the interests of better agriculture and home-making. A corps of specialists, supervisors and directors in various phases of extension work are also engaged in this extension service. All of these contribute a great deal of their time to the promotion of the boys' and girls' club work.

The Scope of Club Work. The scope of the work carried on with young people is limited only by the amount of funds and leadership available for the proper support of the movement in any given state or locality. The real motive for the club organization is industrial and has to do with the growing of gardens, trees and farm crops, the raising of poultry and farm animals, or the carrying out of definite home projects in connection with the

A CLUB GIRL
Making use of a neglected yard.

household arts, such as home canning, cooking, bread making, garment making and kitchen management. The time of work covers a season, a year or a series of four years, according to project and program outlined and the detail with which it is carried out.

Club members, in addition to doing the home projects, also keep records of costs, receipts and observations. and they follow a carefully-prepared system of directions which come in small installments and always in season. Leaders conduct field meetings and demonstrations, hold contests, play at holding festivals, and fairs, and hold banquets in honor of all members who have entered, "stayed in the game, and finished the job"—or, in other words, who have *achieved.*

Definite Objects. Some of the more direct and important objects of this type of extension work are to offer to young people careful guidance and direction in agricultural and home economic activities, in the hope that there may be retained for the farming business and for rural life many of the best young people; to demonstrate through the boys and girls what is efficient and worth while for the general practice of farm and home management; to offer to the boys and girls through the group or club definite training for leadership and the advantages of coöperative effort, and to make available the kind of team work essential to better community ideals; to furnish a plan of study available for immediate practice, to help the rural and village schools to teach the subjects of agriculture and home economics in a more vital and practical way; to offer to the isolated boy and girl of rural communities the educational advantages so essential to better, all-around social development; to teach habits of thrift, economy, industry and a positive liking for the farm and home.

This is made possible through the club group and constant coöperation with others in the same community. Out of all this it is believed that a deeper interest and a greater efficiency in all of the activities and enterprises of farm and home life will be forthcoming. Indeed, results are already apparent.

The Club Programs. In order to give a better notion of what the club leaders mean by club work, it will be well to submit here a complete program of the different activities which are part of a year's work in a completed project; for the sake of brevity we will outline the club work of a Home Garden and Canning Club, and show the steps of development for the year:

First, a campaign for the enrollment through the schools and the selection of a local leader who can look after a club group for the entire year.

Second, members are then organized into a group called a club, with its own officers, such as president, vice-president, secretary, program committee, custodian of equipment or tools, bookkeepers, superintendent of gardens, sergeant-at-arms, publicity agent, superintendent of markets, garden surveyor, etc.

> Illinois, Douglas.
> (state) (county)
> [Date] Jan. 4 1917.
> I hereby make application for membership in the national Boys' and Girls' *Home Garden* Club, and if admitted I shall endeavor to follow all instructions, attend meetings, and exhibit products from my crop at a local or district fair. I will keep an accurate record of my work, expenses, and receipts, and will make a complete crop report at the close of the season and forward same to the State Club Leader, State College of Agriculture.
> (Signed) JAMES R. KNOWLES.
> P. O. Address Argo R. D. 46
> Illinois.

SAMPLE ENROLLMENT CARD

The state leader instructs officers and local leaders as to the manner of conducting the club meetings, gives definite suggestions as to the business part of the club work, and then outlines programs that will have to do with the agriculture or home activities such as the project represents.

In this particular project plans should be made so that both boys and girls may be members and all made to feel at home while sitting in council at club meetings, where they will discuss together the important problems of gardening, home canning, sprays, soils, fertilizers, weeds, insect pests, methods and devices. They will also make demonstrations and discuss other problems related to the farm, such as the use of their food supplies, where and how to secure seed, markets, labels, etc. In addition to this, literary and musical selections may be enjoyed at each program, the program hour to be preceded by a short business session and parliamentary drill for the members, and followed by a social meeting at which the young people through the club agency offer the best possible combination for the social development of an entire community. This social side often attracts young people, who later become regular workers.

TYPE PROGRAM FOR HOME GARDEN CLUB MEETING

Prairie Dell Home Garden Club

Meeting Friday 3:30 to 5:00 P. M., Prairie Dell School.

Official Program.

3:30 to 3:50 P. M. Business Session.

Order of Business:

- Call to order by president.
- Reciting in concert club motto; all members.
- Reading of minutes by secretary of club and approval of same by members.
- Unfinished business.
- New business.
- Discussion by some member of a phase of parliamentary practice.
- Club records and definite achievements recorded by secretary of club.

3:50 to 4:40 P. M. Literary Program.

Subjects:

- Club Plot Work; Farm Water Supply.
- Music—Club quartette.
- Parody—"Coming Thru the Corn"—Club member.
- Addresses—Club Plot Work.
  - (a) How I prepared my Garden—Club member.
  - (b) Best Methods of Plowing Garden Soil—Club member.
  - (c) Best Fertilizers for Home Garden —Club member.
- Special Club Interests.
  - (a) Life History of a Tomato—Club member.
  - (b) Address—How to Plan and Provide a Better Water Supply for Our Homes—Some successful neighbor who has succeeded in his own home.
  - (c) Debate—Resolved: That the Home Garden is the Most Profitable Enterprise of a Farm.
- Instrumental Music—Club member.

4:40 to 5:00 P. M. Social Period.

Carry out a definite social program, with guessing games, plays, contests, etc., including one guessing or play contest that will develop interest in the garden work; have games or plays that will encourage social intermingling.

The program committee should outline and plan definitely all parts of the program, including the business and social sessions.

Club members receive follow-up instructions in all of their work. These instructions are sent to them in small installments and directions are given in plain, simple language, so that the members may successfully interpret, and, together with members of the family, put the subject-matter into effectual practice. This method of instruction continues from the time the club members have enrolled until they have completed the entire year's work. The following illustrates one lesson of follow-up instructions to club members:

Nasturtiums or Asters for Border

First Planting.	*Second Planting.*	*Third Planting*
Tomatoes		
Carrots	Late Icicle Radish	
Stringless Beans		
Carrots		
Tomatoes		
Carrots		
Wax Beans	Late Icicle Radish	
Carrots		
Tomatoes		
Beets		
Stringless Beans	Late Head Lettuce	
Beets		
Tomatoes		
Onions		
Onions	Turnips	
Onions	Early Beans	Spinach
Radishes	Late Beans	
Radishes	Early Beans	Spinach
Lettuce	Late Beans	
Lettuce	Early Beans	Spinach
Lettuce	Late Beans	

Parsley and Mint (half of each) for Border

PLAN FOR A CLUB MEMBER'S SMALL HOME GARDEN

This garden plan is based upon very intensive cultivation, very fertile soil and plenty of moisture throughout the entire growing season. In order to mature the beans, carrots and tomatoes in midsummer it will probably be necessary to irrigate when the rainfall is not sufficient for the garden. It will be noted that the plan calls for a second and a third planting in the case of a number of vegetables. These should be planted in the same rows after maturity of the previous crop.

The eastern and western borders may be of vegetable greens, such as spinach, Swiss chard, etc. Distance between rows, nine inches. In place of two rows of carrots as shown in the plan, it may be well to substitute peas. In the Eastern states the early varieties will keep the ground occupied throughout the season. Instead of one or two rows of tomatoes, use a row of salsify and a row of parsnips. The first row of onions should be onion sets, with the idea of harvesting them early and getting them out of the way before the tomato plants are large. Bulbs of the early blooming flowers, such as tulips, crocuses, etc., may be placed in the ground late in the fall and much of their blooming be completed before planting time for the garden. Fall greens, such as spinach and kale, should be planted as other crops mature. Plant asters and other fall-blooming plants, wherever possible. Leave no vacant space in your garden.

Instructions on keeping records and cost accounting are furnished early in the season, with a view to teaching cost accounting and securing a complete record of the year's work at the close of the year.

In the home projects members are visited by teachers, county leaders, county agents and sometimes district and state agents in charge of the work, for the purpose of commending them for well-doing, encouraging them to manage better the project they have undertaken and to give definite direction. A carefully-planned system of follow-up work contemplates a visit from a local leader not less than two times a month. A brief letter giving timely advice should be sent by the leader to each member. Like the public school teacher, the local leader should not have too many club members to direct during the summer. They should be near a common center and capable of being reached without too much travel.

A BOYS' CLUB

Members of an apple club receiving instruction.

Club members should be reinforced in their cultural instructions by field meetings and field demonstrations, for the purpose of definitely instructing them in how to do things. Then, in turn, these members should be called upon to demonstrate privately to others in the community and at public gatherings the proper way of doing things; if in home canning, they should demonstrate how to can fruits and vegetables in the most economical and efficient way.

In the fall of the year, for the purpose of educating club members in standards of industrial achievements and quality of products, and incidentally giving them the inspiration that one always gets from seeing what other people do, a club fair or festival will be held, at which both fresh and canned products will be exhibited. At these fairs public demonstrations will be conducted by the children for the purpose of showing the people of the neighborhood systematic methods of pruning, grading, crating and canning, the preparation of the canned products for the table and proper methods of serving. Lecturers who are successful farmers and home builders, and extension workers from the college extension departments usually give lectures on various phases of home-garden and canning work.

In connection with the fair or annual club festival, a series of play contests is definitely carried out, such as contests in potato-paring, in labeling cans, in giving recipes and in judging, and other related plays possible with vegetables, cans and equipment.

TYPE PROGRAM

Club Fair or Festival:

Saturday, July 22, at Hickory Grove.

10:00 A. M. Club pageant or parade, club members bearing club colors and banners.

10:45 A. M. Awards announced by leaders.

11:00 A. M. Baseball game—North Side vs. South Side.

12:00 M. Picnic dinner by club groups.

1:15 P. M. 50-yard dash—Boys under 13.
50-yard dash—Girls under 13.
50-yard dash—Boys over 13.
50-yard dash—Girls over 13.
Potato relay—4 men team, ¼ mile.
Potato paring contest—6 potatoes. (Speed, 25; skill 25; waste, 50.)
Needle-threading contest—7 needles. (Remove thread from spool and knot. Speed, 30; skill, 30; quality, 40.)

2:30 P. M. Bird House contest.
Best house
Most original house.

3:30 P. M. Canning demonstration contest.

4:30 P. M. Spelling contest. 300 words selected from gardening and canning literature. One trial. No hesitation. Two entries from each club.

Photographs from U. S. Dept. of Agriculture

BOYS' AND GIRLS' CLUBS

A permanent garden record. Field instruction for father and son. Exhibiting their first efforts at canning. Regular Saturday market conducted by a Lincoln (Neb.) boys' and gir's' club.

WORLD BOOK

Photograph from U. S. Dept. of Agriculture

A CLASS IN HOME CANNING.

A rural teacher and her canning club conducting a demonstration, assisted by state and Federal club workers.

A four-H banner will be awarded to the club winning the greatest number of points. A pennant will be awarded for second place, and a book for third place.

SPECIAL PROGRAM FOR ANNUAL EXHIBIT OF CLUB WORK:

Club Exhibit Hall—September 22-27.

Club Exhibits:

1. All members of clubs are required to exhibit.
2. Only *bona fide* club members may exhibit.
3. All exhibits must be in the hands of local committee before 9:30 A. M., September 22.
4. Cards will be furnished on which will be written the names and addresses of exhibitors and numbers of entries for which they intend to compete.
5. No exhibit may be removed before 10 A. M., September 28.
6. All products not removed before 10 P. M., September 28, will be sold and receipts used to defray the cost of the exhibition.
7. An exhibit may be entered in but one class, record books and stories excepted.
8. Ribbons or medals will be granted first-, second- and third-prize winners in each class.
9. All articles must have been grown or produced by exhibitor in whose name they are entered.

Exhibit List:

1. Four ears sweet corn.
2. Two ears sweet corn and ½ gal. jar canned sweet corn on cob.
3. Two ears sweet corn, 3 qts. canned corn cut from cob.
4. Plate of seven potatoes.
5. Half-bushel box of tomatoes.
6. Two 1-qt. jars of tomatoes, canned whole, and plate of seven tomatoes, and record book.
7. Two 1-qt. jars of tomatoes, canned whole.
8. Pound basket string beans.
9. Two 1-qt. jars string beans and pound basket of string beans.
10. Pound basket string beans, two 1-qt. jars of canned string beans, and record book.
11. Best record book at exhibit.
12. 10 qts. of products—all different—canned in glass.
13. 12 No. 3 cans of products—one to be opened for test.
14. Greatest variety of canned vegetables and fruit by club member.
15. Illustrated story of experience in home canning of surplus.
16. Illustrated story of "How I Made My Garden."
17. Illustrated story of "Experience in Marketing."
18. Canning contest: 1 bushel of tomatoes to be canned by each team. Judged on speed, skill and appearance of pack. Each team to furnish all equipment except stove and fuel.
19. Best drawing of garden plan.
20. Best exhibit of garden equipment by member.

Committee on Awards:

The committee on awards will meet at the office of the superintendent of schools on December 5, to examine reports and grade club members. The following shall be the basis:

	Per Cent
Cost of production........	30
Yield or quantity produced	30
Exhibit	20
Club report and story..	20

No system of education designed for the training of boys and girls is complete until it properly recognizes the *place of play,* and offers an opportunity for children to measure themselves by *contest methods* with their fellows. No sound system of education has yet been devised or offered to the public which does not properly recognize the dual powers for child development, namely, *play* and *contest,* as a means for training in efficiency and as a motive for working out standards or industrial measurements with young people.

The next stage in the club program is to secure the proper measurements of the club achievements in yields or products, and have the club members submit signed and properly-certified reports to the county, district or state leaders. Anyone who has followed up this particular phase of the work will understand its importance and will appreciate that the training a child gets in cost accounting alone is worth all the time and money which is being spent for boys' and girls' club work; for in the home as well as upon the farm the work of keeping books, records and cost accounting has been greatly neglected, and some "first aid" to this phase of farming can easily be administered through the club work. (See ACCOUNTING.)

After the measures have been taken and children graded upon the basis of their achievements by a competent committee, the winners are then called together and banqueted or given a reception by the business men, farmers, grangers, federations of women's clubs or some other organization of the community.

In a number of the states *club achievement days* are held, on which a carefully-planned achievement program is carried out in the interest of the young people. Members are called upon to give reports, illustrate methods of winning, and to tell their own stories of "How I Made My Crop," "How I Raised My Chickens," "How I Kept My Records," "How I Canned and Marketed My Surplus," etc. Leaders also make their awards of diplomas, medals, scholarships, educational trips, pennants, pure-bred farm animals, bank deposits, etc., to those who have reached the standard of highest achievements. In this way common work is dignified and made both interesting and attractive to all the young people. Heroes of the industrial army are thus honored, and they become "guiding lines" for others of their class.

At the conclusion of a program of this type the leaders usually outline the club program for the new year and thus take advantage of all the inspiration and enthusiasm of the day. They utilize all the advantage for the next year's work and secure the support of friends and members because of the worth-while results.

TYPE PROGRAM

Central School, 2:00 P. M., December 10.

Program:

Invocation—By Pastor of local church.
Summary of club work—Chairman county committee.
Giving reports—Club members. (No reading.)
Music—Club band.
Report of Committee on Awards.
Awarding of 4-H Brand Medal and Prizes to winners.
Plans for next year's club work—Club leader or county superintendent of schools.

APPLICATION OF TERMS USED IN THE BOYS' AND GIRLS' EXTENSION WORK

Club Work; an organized system of extension teaching for young people through demonstrations in the field and home. It contemplates the organizing of members into groups called clubs, for the purpose of definite work under carefully-prepared projects and with adequate local leadership.

Experiment; an effort designed to discover principles or facts and the methods of their application.

Test; effort made to prove or disprove the practical, local application of established principles and facts under a given set of conditions.

Demonstration; an effort designed to show by example the practical application of an established fact. Demonstrations may be of methods or of definite results.

Field Meeting; a gathering of young people for the purpose of observing, discussing and studying the progress or results of a definite field or home garden demonstration.

Club Visit; a visit by state, district or local club leader to the club meeting, which may be held in the home, school, club room, courthouse or other convenient place.

Club Project; the definitely outlined work of the club group for a given year, including the entire program of a season or year.

Home Project; the individual work undertaken by a club member at home, as part of the club project or work of the group.

Contest; a competitive phase of the club work, in which club members measure themselves and their work by a common standard, called the *basis of award.*

Basis of Award; a standard of measurement outlined for the convenience of leaders in determining the achievement, grade or standing of a club member, and used for both the awarding of school credits for club work and the awarding of prizes, medals, diplomas, etc.

Club Festival; an organized part of the club work having reference to the activities of a single day, on which members exhibit products, demonstrate methods, hold related plays and other program features, for the purpose of arousing interest, giving instruction and creating enthusiasm for the work.

Score Card; a standard of measurement employed for use in judging work to determine the quality of particular products, animals, grains and articles produced in connection with the boys' and girls' club work.

Agricultural Club; a club of young people organized for more general instruction, for the purpose of developing the efficiency of its members in all agricultural subjects and practices and not requiring the same kind of home work or a common motive.

Follow-up Instructions; the especially prepared typewritten, multigraphed or printed directions and subject-matter prepared for club members. They deal with both the club group work and the subject-matter of the individual projects.

Follow-up Work; the term used to designate the different efforts, activities and organization work which seek to give such aid, from time to time, to the club group and the individual member that definite results at the close of the year may be shown.

Specialist in Club Work; an individual who has been assigned to extension teaching in boys' and girls' work on a particular subject, and who, as a rule, deals with subject-matter.

Local Leader; the one who has been selected by the state, district or county leader for the purpose of looking after the local group of club members, meeting with them and inspecting their club plots and home project work, and in other ways representing the state leader in the conduct of the work. This person may be a paid or a volunteer leader.

Project Program; a definitely-outlined series of activities required of club members, club groups and club leaders, for a given season or year.

Club Plot; a piece of ground required as a unit or acreage upon which the home project and field work is to be performed.

Club Unit; the outlined unit of work of a given project. It refers to quantity measurement of projects that cannot be measured on an acreage.

Completed Project; an expression used in boys' and girls' club work to indicate that a club leader, club group or club member has worked out the entire program and fulfilled all requirements of those in charge of the work, including attendance at meetings, growing of the crop, harvesting, marketing, making exhibits and submitting a properly filled out, signed and attested crop report to the leader in charge.

Results. During the calendar year ending December 31, 1915, 209,178 boys and girls in the United States were organized into club groups and enlisted in the regular boys' and girls' work. Of this number 127,882 did regular home project work as required by the leaders; 62,882 made complete and attested reports on the regular blanks furnished for the purpose and completed all the work of the season or year, which was forty per cent of the total enrollment and sixty-four per cent of those who undertook the work. Out of the total number that completed the work, 24,299 boys and girls were engaged in what is termed profit-making projects, from which they produced $509,325 worth of food products (gross receipts). The total cost of the work from all sources—local, state and national—was $166,-405.67, thus showing a *per capita* cost for the work of eighty cents, while the production value per member was $20.96. Basing the *per capita* cost of the work only upon those who completed *all* the work and made the crop reports required, that cost would be $2.02. Most of the boys and girls who were enrolled in the club work and who were not able to do the home project work were members of the club groups, met monthly or bi-monthly, studied the lessons, participated in the discussions and received the instructions from the state leaders in connection with the work, and most of them did part of the work, but for various reasons—lack of land, objection on the part of parents, etc.—they failed to complete the work and render reports.

The state leaders, in addition to their own direction of details of the work, secured 11,478 local volunteer leaders who assumed coöperative leadership with them and helped the paid leaders in local follow-up work, such as holding of group meetings, visiting club plots, conducting demonstrations and assisting in keeping up a live active interest in the work during the hot summer months. The coöperative leaders of twenty-eight states conducted 1,670 canning demonstrations for the training of club members in the art of home canning and for saving surplus fruits and vegetables; the total attendance of these canning demonstrations was 156,580. In addition to this the same leaders conducted 3,829 field meetings and personally visited 27,733 club plots. During the same calendar year they prepared and distributed 2,108,456 pieces of follow-up instructions or directions in the support of the work.

The United States Department of Agriculture, through the Extension Office of the North and West, supplemented this amount. with 1,140,146 circulars, sheets of instructions, etc. An illustration of the influence of boys' and girls' club work on adults may well be shown from one single project called the home-canning club work. During the twelve months of the calendar year 1915, 26,534 adults wrote to the office and requested that they be given the privilege of using the home canning instructions which were available to the boys' and girls' club work; 3,156 of these adults reported to the office at the close of the season that they had canned 275,836 quarts of fruits, and 270,659 quarts of vegetables, or a total of 546,495 quarts of canned products, an average of over 109 quarts per family represented in the work; one-half of these products were made up of the inexpensive home-grown vegetables. It would doubtless be safe to say that much of the fruits and vegetables thus

FROM HOME GARDEN TO HOME CANNING

canned would have gone to waste had it not been for the instructions and definite directions received.

Club members who have won state championships or made unusual records in connection with their projects become members of the National All-Star Achievement Club. To such members advanced work is assigned in crop rotation, farm management work and duties as leaders. Forty-two champions of the Northern states have furnished reports, and the following data may prove of interest: The average age of these forty-two champions was seventeen years; average number of years in club work, two and one-third years; longest time of any club member in the work, six years; twenty-nine of these champions are still attending school, twenty-six are making plans to attend college and twenty-two of them will go to colleges of agriculture. When questioned with reference to their selection of a vocation for life these champions reported as follows: Fifteen will take up farming and stock raising; three will take up domestic science teaching; seventeen, veterinary medicine; one, engineering; and two, teaching or club-leadership. Eight of these forty-two were champions in club work for 1915. O.F.B.

Information relating to all phases of the Boys' and Girls' Club movement may be obtained from the Department of Agriculture, Washington, D. C.

BOY SCOUTS, an organization of boys, international in its scope, which has won for itself this commendation: "No movement of our time towards child betterment has been more practical than the Boy Scout movement." All boys between the ages of twelve and eighteen may join, if they fulfil certain requirements, and it offers to them an opportunity for training in resourcefulness, self-control, thrift, courage—in fact, almost all the virtues that make for efficient manhood and good citizenship.

History. The first organization of this kind in the world was founded by Daniel Carter Beard in May, 1905. It was called the Sons of Daniel Boone, "a society of tenderfeet and boy scouts." Each officer bore as a title the name of a famous scout, Daniel Boone (president), Davy Crockett (secretary), Kit Carson (treasurer) and Simon Kenton (keeper of the tally gun). About the same time Ernest Thompson Seton, the naturalist, formed an organization called the Woodcraft Indians, his purpose being to glorify the American Indian. Beard's idea, on the other hand, was to inculcate the manly qualities of the early American scouts in the characters of the modern American boys. In 1910 the Woodcraft Indians decided to unite with the Sons of Daniel Boone, to form a larger organization, the Boy Scouts of America. Meanwhile, in 1908, Sir Robert Baden-Powell had formed the Boy Scouts of England, modeled on the Sons of Daniel Boone, and at the first banquet of the executive board of the Boy Scouts of America, Baden-Powell acknowledged the source of his inspiration:

> "I am not," he said, "the father of the Boy Scouts movement. I might be called its uncle. . I looked at what the United States were doing, read some of Beard's books on various plans that you had under way over here, cribbed from them right and left, and started the Boy Scouts of England."

So exactly did the aims and activities of the new society meet the demands of the boys themselves, as well as the ideals of their parents or guardians, that the movement spread rapidly, until at present not only England and the United States, but Canada, Germany, France, Italy, Australia, China and several South American republics have branches. In the United States alone there are about 200,000 Boy Scouts, and in all other countries there are about 800,000 more.

Purposes and Methods. The method is summed up in the term *scoutcraft,* which includes first aid, life-saving, tracking, signaling, cycling, nature study, swimming, rowing and other accomplishments. Much of this scoutcraft is gained through games and team work, and is thus pleasure, not work, to the boy. Especially attractive are the long walks, or "hikes," as the boys themselves call them, on which they are taken frequently by their leaders. These not only give excellent physical exercise, but afford a splendid opportunity for nature study. The summer camps, which are a feature of many of the organizations, make possible drills in fire-building, tent-pitching and cooking, and teach the boys how to be comfortable and content without the luxuries of home.

Before he becomes a scout of the lowest rank—that is, a *tenderfoot*—a boy must take the scout's oath, which declares:

"On my honor I promise that I will do my best (1) to do my duty to God and my country; (2) to help other people at all times; (3) to obey the Scout Law."

This scout law demands honor, loyalty, helpfulness, friendliness without snobbishness, courtesy, kindness to animals, obedience to scoutmaster and parents, cheerfulness and thrift. In order to become a second-class or first-class scout a boy must meet certain definite requirements in scoutcraft which prove not only that he has made a definite effort, but that he can help himself and others and feels confident of himself in an emergency.

In form, the organization is semi-military. Eight boys constitute a *patrol*, which chooses one of its own members as patrol leader; three patrols form a *troop*, which has an adult scoutmaster. At the head is the *chief scout*. But it is only in form that it is military. Drill with rifles is forbidden; no military marching tactics are taught, and the simple uniform is advised merely because it makes for democracy. Since the outbreak of the War of the Nations in 1914, the Boy Scouts in the European countries, and particularly in England, have shown an ardent patriotism. At police duty the older ones have proved most effective, and the training which they have had in first-aid work has stood them in good stead in many a time of need.

In Canada the head of the organization is the Governor-General, with title of chief scout. In the United States Ernest Thompson Seton was chief scout from the beginning of the organization until 1915, when the office was abolished. The present head is Daniel Carter Beard, National Scout Commissioner. D.C.B.

Related Subjects. In connection with the Boy Scout movement, the following articles in these volumes should be read:

Baden-Powell — Campfire Girls
Beard, Daniel Carter — Seton, Ernest Thompson

BOZEMAN, *boze' man,* MONT., is the county seat of Gallatin County, in the southwestern part of the state, with a population of 5,107 in 1910. The city is situated on a branch of the Gallatin River, twenty-six miles west of Livingston, ninety-six miles southeast of Butte and ninety-eight miles southeast of Helena. Transportation is provided by the Northern Pacific and the Chicago, Milwaukee & Saint Paul railways. The city was settled in 1864 and named for its first settler.

Gallatin valley, in which Bozeman is situated, is so fertile that it is locally called the "Egypt of America." A number of streams flow into this valley and supply abundant water for irrigation; consequently great quantities of grain, vegetables, fruits and live stock are annually shipped from the city. Bozeman is largely engaged in the manufacture of flour, one of the largest mills in the Northwest being located here. The lumber industry, too, is important. A Federal building, erected in 1916 at a cost of $80,000, an $80,000 Y. M. C. A. building, a $25,000 public library and a $60,000 Elks' Home are evidences of the city's growth. Bozeman is the seat of the State College of Agriculture and Mechanic Arts; this with a business college and a good public school system constitute the educational advantages.

The vicinity of Bozeman abounds in mountainous scenery of unusual beauty, and it is on the trail to Yellowstone Park. In the neighborhood are found gold, silver, lead, copper, iron and coal. A government fish hatchery is located near the city. J.M.R.

BOZZARIS, *bo tsah' rees,* MARCO (about 1790-1823), a noted patriot and hero in the Greek struggle for independence. He was born in Albania of a family famous for its bravery. Through his skilful defense of Missilonghi he gained especial renown. After many victories over the Turks he finally lost his life in a daring night attack upon the camp of the pasha of Scutari. In the poem *Marco Bozzaris*, Fitz-Greene Halleck has glowingly told the story of this last attack, and he ranks Bozzaris in history in these words:

One of the few, the immortal names,
That were not born to die.

BRABANT, *brah bant',* the central district of the lowlands of Holland and Belgium. The northern portion comprises the present Dutch province of North Brabant, and the southern portion the Belgian provinces of Brabant and Antwerp. This district extends from the Waal to the sources of the Dyle, and from the Meuse and the plain of Limburg to the lower Scheldt. In the time of Caesar, Brabant was inhabited by a mixed race of Germans and Celts; in the fifth century the Franks took possession of it.

During the centuries which followed, this region was at various times a part of the dominions of France and of Germany; in 1477 it passed to the Austrian House of the Hapsburgs, and so to the Emperor Charles V, becoming thereby the inheritance of his son, Philip II of Spain. The people of Northern Brabant joined the other Netherlanders in their revolt against that cruel monarch, and their province became a part of the independent Dutch Republic. At the close of the wars of Napoleon, when the kingdom of the Netherlands was established, all of Brabant was included in the realm, and it was divided into the provinces of North Brabant, South Brabant and Antwerp. The two latter provinces became parts of Belgium in 1830, when that country declared its independence and separated from the kingdom of the Netherlands, South Brabant being known from that time as Brabant. The oldest son of the king of Belgium bears the title Duke of Brabant. See BELGIUM, for fate of Brabant during the War of the Nations.

BRACE'BRIDGE, ONT., the county town of Muskoka County, a town in the heart of the Muskoka Lakes district, famous for its beautiful scenery and as the resort for hundreds of summer visitors. Bracebridge is on the Muskoka River, which carries a heavy tourist traffic in summer, and on the Grand Trunk Railway. The town was a pioneer in developing electric power from waterfalls and has become an important industrial community. Especially noteworthy among the manufacturing establishments are large tanneries, woolen mills, lumber mills and motor-boat factories. Bracebridge was incorporated in 1871; it adopted the commission form of government in 1911. The Dominion post office, erected in 1914 at a cost of $50,000, is a conspicuous building. Population in 1911, 2,776. G.H.O.T.

BRACT, *brackt,* a small form of leaf near the flower of a plant. From the point where the bract joins the stem, the flower or flower-stalk always develops. Bracts are often very much like the ordinary leaf of the plant and are then called *leaf-bracts.* Usually, however, they are unlike the leaf, often being merely scales, or hair-like. Sometimes, too, they are

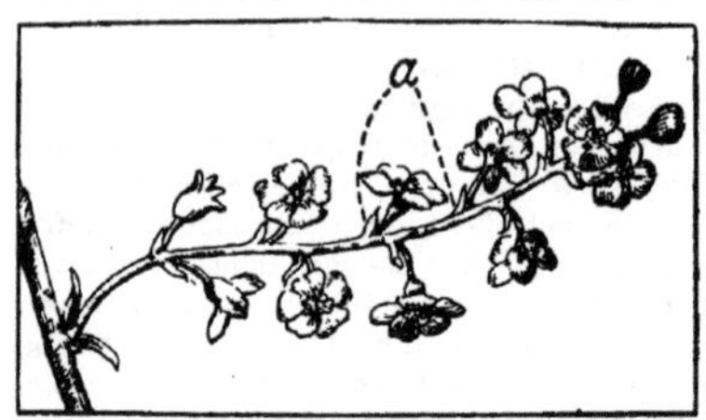

BRACTS
a indicates location.

colored and are mistaken for part of the flower.

BRADDOCK, *brad'ock,* EDWARD (1698-1755), a famous English general who led the British and American troops in the disastrous expedition against Fort Duquesne, during the French and Indian War. Having been appointed commander of all the British forces in America in 1754, he planned as the first event of the campaign against the French the capture of Fort Duquesne, on the site of the present city of Pittsburgh. In 1755 he was joined by Virginia troops near Alexandria, and at Frederick, Md., he added to his staff George Washington and Benjamin Franklin.

Scorning the advice of both these experienced Colonials regarding the danger of exposing himself to the Indians, who were accustomed to fight from ambush, he set out from Fort Cumberland by the path marked out by Washington two years before. On July 9 the advance guard was attacked by a band of French and Indians. The British, frightened by the war whoop of the red men and confused by the Indian method of fighting from behind trees, were defeated. Braddock showed conspicuous personal bravery, but fell mortally wounded after five horses had been shot from under him. Washington led the survivors, less than half the force, to a place of safety.

BRADDOCK, PA., a borough in Allegheny County, noted for its great steel mills, which rank with the largest in the United States. It is on the right bank of the Monongahela River, in the southwestern part of the state, ten miles southeast of Pittsburgh. The Pennsylvania, Baltimore & Ohio and Pittsburgh & Lake Erie railways provide steam transportation, and electric lines are in operation to near-by cities and towns. The first settlement was made in 1795 on the site of *Braddock's Field,* where, in July, 1755, the British general was defeated and fatally wounded by the French and Indians. The borough was incorporated in 1867. In 1910 the population was 19,357; in 1914 it was 20,936. The area is less than one square mile.

Braddock has extensive manufactures of steel and pig-iron, especially of steel rails and railroad and car supplies; the production of cement and plaster is considerable. Possibly because of the large steel interests in Braddock, the first Carnegie Library in the United States was established here. There are fine public schools, many churches and a Federal building. Kenneywood Park is the principal recreation ground.

BRAD'FORD, an important center of the woolen and cotton industries of England, situated on the River Aire, eight miles west of Leeds in Yorkshire. The name is derived from the Anglo-Saxon *broad-ford.* It stands in the center of a rich coal and iron mining district, is served by three main lines of railway and is connected by canal with Liverpool and with important ports on the Humber. The smoke-blackened buildings give the city a very gloomy appearance, though the streets are broad and well laid out. Its textile manufactures are among the most extensive in England, the worsted mills alone employing 36,000 persons. The cotton industries are not so important, giving employment to only about 8,000.

All public utilities are municipally owned, and the town is progressive and quick to adopt measures calculated to benefit the community. In 1907 the mayor was raised to the dignity of lord mayor, a title at one time conferred only on the chief executive of London. There are numerous public parks and gardens, but the smoke-filled atmosphere prevents trees, flowers and grass from flourishing. Bradford has always taken a prominent part in the temperance movement, and the first Temperance Hall in England was erected there in 1837. It is an ancient city, records showing that it existed previous to 1066, the year of the Norman Conquest, and that it was a flourishing market town in 1251. Population in 1911, 288,458.

BRADFORD, PA., a city in McKean County, in the northwestern part of the state, situated near the New York state line, sixty-seven miles south of Buffalo. Railway transportation is provided by the Buffalo, Rochester & Pittsburgh, the Pennsylvania and the Erie railroads; electric lines operate to towns north and south. Bradford was settled in 1823, was chartered as a city in 1879, and since 1914 its government has been administered on the commission plan. In 1910 the population was 14,544. The area exceeds two square miles.

Industry in Bradford is associated with the natural resources of the region in which it is located, namely, coal, petroleum and natural gas. There is a large trade in oil, gasoline, wood acid and alkalis, and extensive manufactures of tanks, well supplies, refined oil, nitroglycerine, building and paving brick, cutlery, gas engines, torpedoes and air compressors. The large oil refineries have pipe lines to the seacoast. Large railway shops are also located here. Bradford has a fine library, a well-equipped hospital and a driving park. It is lighted and heated by natural gas.

BRADFORD, WILLIAM (about 1590-1657), an American of the colonial period, second governor of Plymouth Colony, whose work as an historian of the Massachusetts Pilgrims makes him the father of American history. He was born in Yorkshire, England, joined the Separatists at Scrooby, and suffered imprisonment for trying to escape from England. Later he succeeded in reaching the Pilgrims in Holland, whither many of them had removed to secure freedom of worship, and there he became a tradesman. In 1620 he sailed with them on the *Mayflower,* and helped to found Plymouth Colony. Chosen governor of the colony in 1621 to succeed John Carver, he held that office until his death, except for five years when by "importunity he got off." Throughout this period his tact, good judgment and high executive ability were important factors in making the colonizing experiment a success.

WILLIAM BRADFORD

Bradford's historical work, *The History of Plymouth Plantation,* is a day-by-day account of the colony from 1631 to 1646, upon which all later histories of Plymouth have been based. The author left the work in manuscript. It found its way in the course of time into the archives of the Old South Church, Boston, but disappeared during the Revolution. In 1855 it was discovered in the Fulham Library, England, and was soon after sent to the United States and published two hundred and fifty years after Bradford wrote it. It is one of the most valuable of all the original documents dealing with American history. The original work is now among the archives of Massachusetts.

BRADSTREET, Anne (1612-1672), the first woman writer America produced, was the wife of Simon Bradstreet (1603-1697), who became governor of Massachusetts Colony in 1679. She was born in England, a daughter of Thomas Dudley, the second colonial governor of Massachusetts, and came to America in 1630, two years after her marriage. There, in addition to bringing up a family of eight children, she found time to write a long list of poems that were the delight and admiration of her friends in the colony, who called her "The Tenth Muse." Her poetry is made up, in general, of discourses on the universe, its history and its phenomena, and to the modern reader it seems to be nothing more than rhymed prose. Nevertheless, though she lacked the real poetic gift, she deserves a place in the history of American literature, for she devoted herself to literature for its own sake, and was one of the first writers to do so.

BRADSTREET COMMERCIAL AGENCY. See Commercial Agencies.

BRADY, *bray'di,* Cyrus Townsend (1861-), an American clergyman and author whose writings, though clean and wholesome, have in them little of a ministerial character. *Under Tops'ls and Tents, Hohenzollern, In the Wasps' Nest, A Little Traitor to the South, The Love Test, The Island of Regeneration, The Fetters of Freedom* and the *Bob Dashaway* series are for the most part stories of the masculine, warlike type, and Brady cannot be accused, as are many clerical story-writers, of "helping in the great work of feminizing the world."

Born in Allegheny, Pa., Brady graduated from the United States Naval Academy, and then went West in the service of a railroad. After studying theology he became a priest of the Protestant Episcopal Church, and later was made archdeacon of Kansas. He was rector successively of churches in Philadelphia, Toledo, Kansas City and Mount Vernon, N. Y., removing to the last-named place in 1913. During the Spanish-American War he served as chaplain in a Pennsylvania regiment. His non-fictional books, including lives of Decatur and Paul Jones, *Border Fights and Fighters* and *Sir Henry Morgan, Buccaneer,* are wholesome and yet as full of thrills and adventure as are his stories.

BRAGG, Braxton (1817-1876), an American soldier who won distinction in the Confederate army during the War of Secession. He was born in North Carolina, was graduated at West Point in 1837, and in that same year saw active service in the Seminole War. During the Mexican War he was brevetted captain-major and lieutenant-colonel for gallant conduct, but in 1856 he resigned from the army and engaged in sugar-planting in Louisiana. When the War of Secession broke out he was placed in command of Southern forces at Pensacola, Fla., with the rank of brigadier-general, and in the following year was made major-general. At the Battle of Shiloh he became the commanding general on the death of A. S. Johnston, and later in the same year succeeded Beauregard in command of the Army of the West. His defeat at Perryville and at Murfreesboro brought upon him severe criticism in the South, but he regained his popularity by his defeat of Rosecrans at Chickamauga in September, 1863. Defeated by Grant at Chattanooga, he asked to be relieved of his command and then became military adviser to Jefferson Davis, but resumed active service to conduct an unsuccessful expedition against Sherman in Georgia. After the close of the war his only public office was that of chief engineer of the state of Alabama.

BRAHE, *brah* or *brah'ay,* Tycho (1546-1601), a celebrated Danish astronomer, whose tireless study of the heavens added much to what was known in his day of the facts of astronomy. He studied philosophy and rhetoric at the University of Copenhagen and entered the University of Leipzig as a law student, but was interested only in astronomy, which he regarded as "something divine." From 1571 until his death he was able to devote the greater part of his time to his chosen field of labor. On November 11, 1572, he sighted a new star in the constellation Cassiopeia, and, after carefully studying its position, published an account of his observations.

Frederick II of Denmark became interested in his investigations, and in 1576 fitted up for him on the Island of Hveen, in the Sound between Denmark and Sweden, the magnificent observatory of Uraniborg, "the fortress of the heavens." There Brahe worked and studied for twenty years, testing and improving upon the theories of Copernicus, discovering new laws governing the motion of the moon, and throwing new light on the subject of comets. After the death of Frederick, Brahe suffered from opposition and persecution, and was obliged to leave Uraniborg. In 1599 he went to Prague, in Bohemia, where he was able to render valuable assistance to a young man destined to be a greater astronomer than himself, Johann Kepler (which see).

BRAHMA, *brah' ma,* the Supreme Being in the religion of the orthodox Hindus. He is conceived to be the Creator of the world, which is to endure for 2,160,000,000 years and then be destroyed, to be recreated by him after the same number of years. He is called the Self-existing, the Great Father, the Lord of Creatures and the Ruler of the World, but he has little part in actual religious worship (see BRAHMANISM), and has never been worshiped by the common people. Only one temple sacred to him is known.

In Hindu mythology Brahma is represented with four heads and four arms. Originally he had five heads, but one was destroyed by the god Siva. His color is red, and he rides upon the Swan, which is sacred to him. Sarasvati, the goddess of eloquence, is his consort.

BRAHMANISM, *brah' man iz'm,* one of the great religions of the world, the faith of over 200,000,000 Hindus, which takes its name from the priestly class of India, known as the Brahmans. It is based on ancient religious writings called the Vedas (which see), the sacred revelations of which are interpreted by the Brahmans. Unlike Christianity, Mohammedanism and Buddhism, Brahmanism has no definite founder. The word *Brahma,* so importantly connected with this system, is used in two ways. In its neuter form it refers to the Universal Power, the all-pervading essence of the universe, of which the human soul is a part; in its masculine form it is the name of the Supreme Being in Brahmanism (see BRAHMA).

Brahmanism is both a religious and a social system. As a religion its chief doctrine is that all existence is bound up in sorrow and evil, and that the one way for the human soul to be saved is to become merged with the Universal Power, Brahma. To return to Brahma the soul must be purified by meditation, self-control and self-denial. Growing out of this belief is the doctrine of the rebirth of the soul (see TRANSMIGRATION OF THE SOUL). As only a few reach the state of perfection in a single lifetime, the great majority must be born again and again. Those who are righteous live again in a higher spiritual state, but those who die in sin pass to a lower condition, and their souls may even enter the bodies of unclean beasts.

Brahman Castes. Socially Brahmanism is a system of castes, by which the Hindus are divided into distinct classes. According to an ancient tradition, Brahma created four castes—Brahmans, soldiers, laborers and serfs. These sprang from his mouth, his arms, his thighs and his feet. The changing conditions of the passing centuries have brought about a great modification of this system, and at the present time it is based largely on industrial conditions, with the members of every trade and occupation forming a separate caste. In the modern Hindu society there are about 2,000 castes. The Brahman himself may now engage in commercial occupations and hold office. The priestly class represents the highest culture of India, and as the result of centuries of education they have produced a type of distinct superiority. See CASTE.

Stages Towards Perfection. To attain the ideal state of perfection the Brahman is supposed to pass through four stages. In the first he begins to study the Vedas, and learns all about the privileges of his caste; he is taught that he has a right to ask alms, and to be free from paying taxes or from suffering capital punishment. He is not allowed to eat flesh and eggs, and must not touch leather or the skins of animals. Above all, he is taught to abhor sin, and he receives instruction in the rules and ceremonies prescribed for his purification.

The second stage of the Brahman begins with his marriage, and he is then called upon to observe many new rules regarding fasting, washing, etc. When he has trained a son for the holy calling, and has seen the son of that son, he is ready for the third stage. The Brahman who attains to this retires to the forest, where he prays and meditates and studies the sacred Vedas. Bathing is practiced morning, noon and night, and many severe penances are imposed. In the fourth stage he inflicts still more cruel tortures upon himself, and in the end his soul is ready to become a part of the divine nature. The third and fourth stages are an ideal which is rarely attained.

The Triad. Brahmanism is an outgrowth of early nature worship, practiced in the period when the oldest hymns of the Vedas were being written. Brahma, the Supreme Creator, is the head of the Triad of three gods, the other two being Vishnu, the Preserver, and Siva, the Destroyer and Reproducer. The Triad is a later development in Hindu theology, and in modern religious practice Brahma has almost no part, for the worship of one Supreme Being is too abstract an idea to appeal to the Hindu intellect. The worship of Brahma has therefore given way to that of Vishnu and Siva. There is, however, a reform spiritual movement going on in India in which God is worshiped under the form of Brahma in the three aspects of Creator, Preserver and Destroyer. B.M.W.

Consult Hopkins' *Religions of India;* Oman's *The Brahmans.*

BRAHMAPUTRA, *brah ma poo' tra,* one of the most important rivers of Asia, rising in the snowy slopes of the Himalaya Mountains, and after a course of nearly 1,800 miles swelling the waters of the sacred Ganges. In its northern reaches the river receives various names, being called Tsanpo, or the *pure one,* by the inhabitants of certain districts of Tibet, and Nari Chu and Maghang in other parts. For 170 miles it follows a southeasterly direction, then turns eastward for 500 miles and then southward to the plains of Assam, through which it keeps a straight course. At a height of 13,800 feet above sea level the river is navigable and forms an important link in internal commerce. In certain parts navigation is rendered impossible on account of narrow gorges and rapids, but large boats may ascend the river a distance of 800 miles from the sea.

The Brahmaputra retains the distinction of being unbridged throughout its length, communication across it being maintained entirely by boats and rafts. The principal tributaries, themselves mighty rivers, are the Lobit, Dibong, Dihong and Subansiri, each receiving the drainage of a large area. The valley of the Brahmaputra is fertile and extensively cultivated, tea, rice and jute being produced in immense quantities. In times of rain the river floods many hundreds of square miles along its banks, a natural irrigation which is of great advantage to rice growers.

BRAHMS, Johannes (1833-1897), the last of that famous line of German musical composers of first rank, which began with Bach. Brahms was born at Hamburg, where his musical education was begun by his father. At the age of ten he was sent to the best teacher in Hamburg, who was so impressed by the boy's talent that he prophesied he would win a greater name in music than did Mendelssohn. He early came in touch with the greatest musicians of his day, including the famous violinist, Joseph Joachim, and Liszt and Schumann. The latter wrote enthusiastically in praise of the young musician, after he had heard Brahms play some of his own compositions. Brahms composed and played thereafter in several of the music centers of Germany, and in addition held positions as music director and concert conductor. The later years of his life were spent mostly in Vienna, where his great masterpieces, appearing from year to year, won for him the highest honors the music world could bestow.

JOHANNES BRAHMS

The principal compositions of Brahms include four symphonies; two serenades; the *Tragic Overture;* the majestic *German Requiem,* a choral work suggested by the death of his mother; variations on one of Hadyn's themes, known as the *Chorale Saint Antoine;* the Hungarian Dances for the piano; trios, quartets and quintets; several concertos; and nearly 200 songs, the best known of which is the beautiful *How Art Thou, My Queen?* Brahms was one of the most intellectual of the master composers, and though he is not yet fully understood his fame continues to grow as the world is learning better to appreciate him. His work is characterized by dignity and grandeur of idea, at all times expressing the deep earnestness and sincerity that were among his strongest traits.

BRAIN, *brane,* the organ of the body which controls all thought, action and feeling, and the center of the nervous system in man and the higher animals. An eminent authority has called it the "great central exchange of our telephone system." Nothing is accomplished in the world that is not first conceived in the brain of man. The steam engine first assumed shape in the mind of Watt; in the wonderful brain of Edison were built the electric light machine, the phonograph and the moving picture apparatus; the brain of Napoleon was so gigantic that it required nearly all of Europe to move it to Saint Helena that it might no longer demolish thrones and crumble empires. The human brain is relatively larger than that of the lower animals. The average weight of the brain for men is about three pounds; for women, it is a little less.

Structure. The brain fills all the cavity enclosed by the skull and commonly known as the *cranium*. It is divided into four parts, the *cerebrum*, or large brain; the *cerebellum*, or small brain; the *pons*, and the *medulla oblongata*, which forms the upper end of the spinal cord.

Membranes. The brain is surrounded by the membranes. The *dura mater* is strong and tough. It forms the lining of the skull and the partitions which separate the hemispheres of the cerebrum and the cerebrum from the cerebellum. The *pia mater* is more delicate and lies next the brain. Between these is the *arachnoid* (spider's web), which is formed from the inner layer of the dura mater and the outer layer of the pia mater.

Cavities. Under the cerebrum, and almost in the center of the head, are two cavities known as *lateral ventricles*. They are separated by a thin membrane and connect with

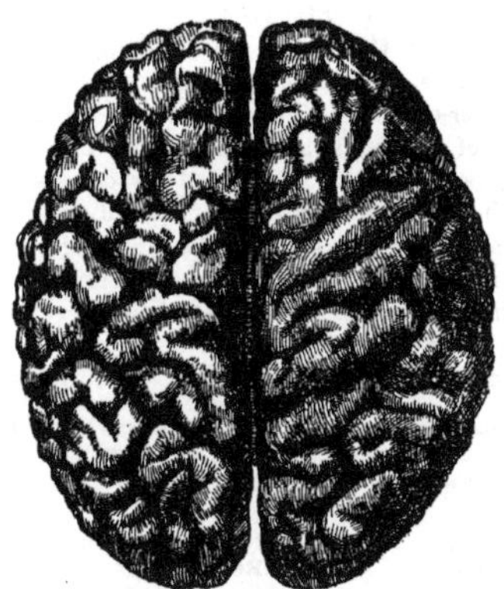

UPPER SURFACE OF CEREBRUM

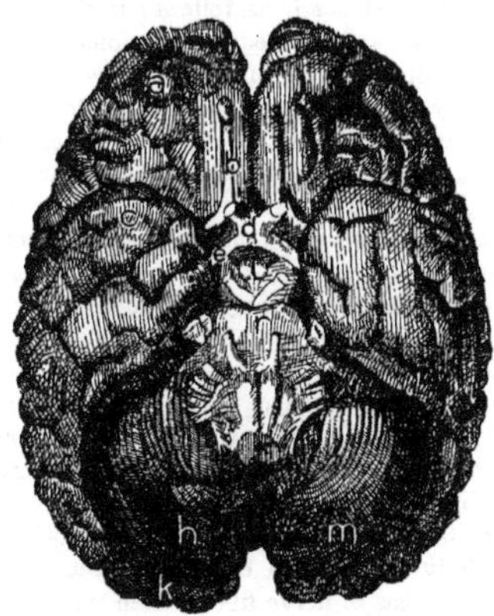

THE BASE OF THE BRAIN

TWO VIEWS OF THE BRAIN

(*a*) Anterior lobe of the cerebrum
(*b*) Olfactory nerve
(*c*) Portion of posterior lobe
(*d*) Optic chasm
(*e*) Optic tract
(*h-m*) Hemispheres of the cerebellum
(*r*) Pons Varolii

Gray and White Matter. Brain substance is of two kinds, gray matter and white matter. The gray matter is on the outside, and forms that portion of the brain which is the center of all mental and voluntary physical acts. It follows the folds, or convolutions, in a thin layer, composed of minute cells, so small they cannot be seen except by a powerful microscope. It is estimated that the gray matter of the cerebrum contains more than nine billion of these cells. The white matter is more compact and forms the greater portion of the brain. It unites with the membrane which holds the parts of the brain together, and with the nerves. Extending from the gray matter through the white to the various centers in the brain are numberless nerve fibers, which serve to connect these centers with each other.

the third ventricle, which is just below them. Below the third, and almost in front of the medulla oblongata, is the fourth ventricle, which connects with the canal in the spinal cord.

The Cerebrum. The cerebrum occupies all the upper part of the cranium, and forms over seven-eighths of the brain. It is divided into two equal parts called *hemispheres* by a deep fissure extending from front to back. In shape each hemisphere closely resembles a very large coffee-bean. A band of *fibers* unites the hemispheres on their under side and forms the roof of the lateral ventricles. Each hemisphere is divided into parts known as the frontal, the parietal, the temporal and the occipital lobes, each named from the bone of the skull under which it lies. The brain substance in these contains many folds or convolutions, which greatly increase the surface of the gray matter. It is estimated that the surface of the cerebrum is equal to that of the human trunk, but by means of these folds it is compacted into a small space.

The cerebrum is the controlling organ of the body, the seat of mental activities and voluntary action. It also contains centers of sight, hearing, taste, touch and smell. It is divided into numerous centers, each of which controls certain movements, or functions. Injury to one of these centers results in the loss of the function over which it presides. If the visual centers are injured blindness follows; if the auditory centers are affected one becomes deaf. Paralysis of any part of the body is usually caused by the formation of a blood clot over the brain center controlling the motor, or motion, nerves that extend to that part of the body. While many centers have been located, there are many more whose locations are not yet known. In this respect the brain is like an unknown country, of which no complete maps have been made.

The Cerebellum. The cerebellum is situated in the back of the head, below the cerebrum. It is partially divided into hemispheres, and each hemisphere is connected with the cerebrum by three sets of nerve fibers. The gray and white matter intermingle in the cerebellum more fully than in the cerebrum. The white matter consists of nerve fibers which unite into a system of nerves that extend through the spinal cord to all parts of the body. The chief function of the cerebellum seems to be to harmonize those muscular movements necessary to maintain the body in an upright position when standing, walking and running. If this part of the brain is injured one staggers like a drunken man when one attempts to walk.

The Pons. The pons is a passage way for nerves from various parts of the nervous system. It seems to be used chiefly as a means of communication, but it is supposed to be the seat of other functions not yet understood.

The Medulla Oblongata. We have already remarked that the medulla, at the base of the brain, is really a part of the spinal cord. The nerve fibers passing from the brain to the spinal cord cross in the medulla to the other side, so that the nerve centers in the brain control the movements on the opposite side of the body; paralysis of the motor nerves in the right side of the brain results in paralysis of the left side of the body. The motor centers of the right hemisphere of the cerebrum control the movements of the left hand and leg, and those of the left hemisphere control the right arm and leg.

The first function of the medulla is to transmit the nerve impulse to and from the brain. Its second function is reflex action, that is, sending a motor impulse to any part of the body, before the sensation resulting in that impulse reaches the brain. The third function of the medulla is to preside over those movements necessary to the maintenance of life which are carried on without the action of the will, such as respiration, circulation and digestion. Because of this function the medulla is sometimes called the "vital knot."

Health of the Brain. The brain, like the muscles, is strengthened by use and fatigued by over-exertion. Sleep is merely a resting time of the brain and nerves, and plenty of sleep is essential to a healthful nervous system. Fresh air, nourishing food and bodily exercise are also necessary to clear thinking and vigorous mental action, which can come only from a healthy brain. Narcotics, alcoholic beverages and drugs all work more or less injury to the brain and should be avoided. W.A.E.

Related Subjects. In connection with the above article, the following topics will be found interesting and helpful:

Aphasia	Nerves
Apoplexy	Nervous System
Epilepsy	Reflex Action
Meningitis	Spinal Cord

BRAINERD, MINN., the county seat of Crow Wing County. It is on the east bank of the Mississippi River, which is dammed and provides water power for the city's mills and factories. It is on the Northern Pacific and the Minnesota & International Falls railways, 138 miles northwest of Saint Paul and 118 miles southwest of Duluth. Brainerd is a shipping center for grain, iron ore and lumber, and has large railway shops. It was chartered as a city in 1883. Population in 1910, 8,526.

BRAKE, or **BRACKEN,** a species of fern very common in North America, Europe and Asia, often covering large areas of hillside and untilled ground. It has a black, creeping rootstock, from which fronds grow often to the height of several feet and divide into three branches. As the plants do not fall when the frost kills the tops, they form a good cover for small game throughout the year. The rootstock is bitter, but has been eaten in times of famine. It has also been used in place of hops in beer making. In dressing chamois and kid leather it is still occasionally employed. The term *bracken* also means a rough, tangled undergrowth.

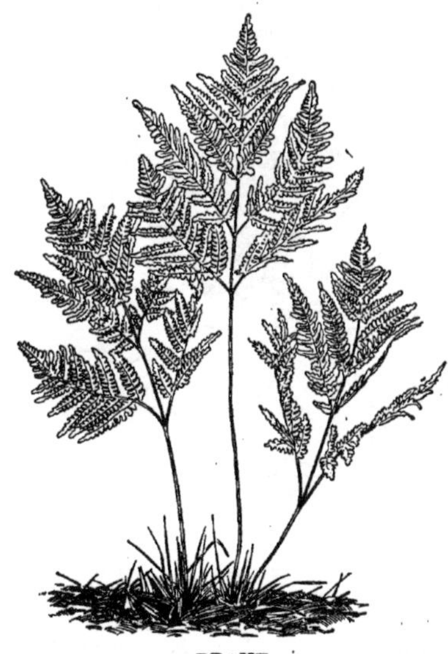

BRAKE

BRAMANTE, *bra mahn' ta,* DONATO (1444-1514), one of the greatest Italian architects of the Renaissance. During the first twenty-eight years of his career, beginning in 1472, he worked in Milan, his masterpiece being the choir and dome of the Church of Santa Maria delle Grazie. In 1499 he went to Rome, where he was so profoundly impressed by the splendid examples of Roman architecture that he adopted a new style in his art, becoming thereby the founder and leader of the Middle Renaissance school of architecture. Not only did his work show his appreciation of beauty, but also a thorough understanding of the laws of perspective and of engineering principles. Pope Julius II commissioned him to design new galleries for the Vatican, and then to plan the greatest architectural work of the Renaissance, the rebuilding of Saint Peter's Church. His death in 1514 interrupted this great work, and his plans were considerably altered by his successors, including Raphael and Michelangelo. Nevertheless, the drawings that he made for the reconstruction of the famous church were studied by later architects.

BRAMBLE, *bram'b'l,* the name commonly given to any shrub of the blackberry family, with trailing, prickly stems. It is rarely cultivated, but as a wild plant it grows in great abundance. The flowers do not appear till late in the summer, and the fruit, which is deep purple or almost black in color, ripens in the autumn. Says the English poet Elliott, in the poem *To the Bramble Flower:*

> Thy fruit full well the schoolboy knows,
> Wild bramble of the brake!
> So, put thou forth thy small white rose;
> I love it for his sake.

BRAMP'TON, the county town of Peel County, Ontario, twenty-one miles west of Toronto, on the Grand Trunk and Canadian Pacific railways. Brampton has the largest factory in Canada for making loose-leaf books. There are several boot and shoe factories, a paper box factory, a foundry, planing, flour and woolen mills and a pressed-brick yard. One of Brampton's chief industries is to supply the Toronto market with cut flowers; one of the firms employs over 250 people. There is also a considerable trade in dairy products, apples and other produce of the surrounding farm districts. Population in 1911, 3,412; in 1916, estimated, nearly 4,000.

BRAN, the coarse outer coat of wheat, rye and other cereal grains, which by sifting is separated from the flour in the process of milling. In ordinary speech, when bran is referred to without any qualifying word, wheat bran is meant; the other brans are spoken of as rye bran, corn bran, etc. Wheat bran contains protein, 15.4 per cent; nitrogen, 53.9; fat, 4.0; fiber, 9.0; ash, 5.8; water, 11.9. It is an excellent food for all kinds of farm animals, and when mixed with corn meal is especially prized by the dairy farmer because of its milk-producing qualities. The other varieties of bran are also used in feeding stock, but less extensively than wheat bran because smaller amounts are obtained from the other cereal grains.

Although bran is not a nutritious food for human beings, it has a certain value because of its laxative effects, and therefore wheat-bran preparations for making bread and muffins are extensively sold. See FOOD, subhead *Chemistry of Food.*

BRANDEIS, LOUIS DEMBITZ (1856-), an American lawyer and publicist, a conspicuous figure in the struggle for economic, social and political justice, and since 1916 an Associate Justice of the United States Supreme Court. He was born in Louisville, Ky., but since his graduation from the Harvard Law School in 1877 has made Boston his home. He began the practice of his profession in 1879, and in few years became one of the leaders of the local bar. His advanced political and sociological tendencies, however, grew with his practice, and eventually cost him many powerful clients, who discouraged his activities against "vested interests." Socially he was later made unwelcome in many homes which were at first open to him. But no ostracism, social or otherwise, made him swerve in the course of conduct he believed to be right.

He was one of the earliest advocates of conservation and in 1910 was legal adviser of Glavis in the Ballinger-Pinchot controversy (see ALASKA, subhead *History*). In the same year he appeared before the Interstate Commerce Commission as counsel for several associations of shippers, and made the startling claim that the railways of the United States were wasting $1,000,000 a day by retaining methods in management which up-to-date manufacturers discarded long ago in favor of more scientific systems. Railroad managers vigorously denied such a possibility, but at once began to effect economies that seemed in large measure to have justified the charge. He has frequently appeared in trials and public hearings involving such subjects of reform as a minimum wage and shorter hours for working women and children, and in 1910 was chairman of the arbitration board which settled the strike of the New York garment workers. Brandeis was for many years a bitter critic of the New York, New Haven & Hartford Railway, and he foretold the difficulties which beset that company in 1912 and 1913 long before the general public had any warning of disaster. His appointment in January, 1916, to membership in the Supreme Court was contested for nearly five months before confirmation was won in the Senate. See SUPREME COURT OF THE UNITED STATES, for illustration.

BRAN'DENBURG, the central province of Prussia, which was the greatest kingdom of the former German Empire. It has Berlin as its capital, and that city is now the capital of the new republic. Brandenburg covers an area of 15,381 square miles. The surface is flat and well watered, and the industries and resources of the province are most highly developed. Internal communications by road, railway, river and canal are excellent; the railroads are owned by the state. Agricultural pursuits have suffered during recent years owing to the inclination of rural inhabitants to move to the cities. The principal crops are barley, rye, potatoes, tobacco, hemp, flax, hops and sugar beets. Cattle raising also gives employment to a large number of people.

As a manufacturing center the province is of the greatest importance. Wool, silk, linen,

BRANDENBURG

In the heart of Prussia, with Berlin as its center.

paper and leather goods form the bulk of the products, and there are numerous breweries and distilleries. The principal towns, next to the capital, are Potsdam, Königsberg, Brandenburg and Frankfort-on-the-Oder. The famous Brandenburg Gate, described in the article BERLIN, is named after the province. Brandenburg's industrial life suffered severely during the War of the Nations; situated near the center of Germany and containing the capital city, it was keenly alive to every phase of the tremendous conflict. Population in 1910, exclusive of Berlin, 4,092,616.

BRANDES, *brahn'des,* GEORG MORRIS COHEN (1842-), the greatest Danish literary critic of his day and one of the greatest of modern times. He was born in Copenhagen, of Jewish parents, and was educated at the university in that city. After extensive travel in Europe he taught for five years in the University of Copenhagen and then removed to Berlin. Since 1882 he has made his home in his native city, devoting himself to study and lecturing. In 1914 he visited the chief cities of the United States, everywhere arousing keen interest by his lectures and his personality.

Brandes has not only brought new life to Danish letters, but he is a stimulating critic of the world's literature. To him, literature itself is a "criticism of life," and his viewpoint is that of a philosopher—a disciple of the French critic Taine, and of John Stuart Mill and Herbert Spencer. In his greatest work, *Main Literary Currents of the Nineteenth Century,* he discusses clearly the chief tendencies of the literatures of the European nations since 1800. He has also written *Eminent Authors of the Nineteenth Century, Poland,* and *Men and Works in European Literature, Recollections of My Childhood and My Youth.*

GEORG BRANDES

BRANDON, a city in Manitoba, well situated at an altitude of 1,180 feet, on a hill overlooking the Assiniboine River, 133 miles west of Winnipeg. It is a division point on the Canadian Pacific and the Canadian Northern railways, and is also served by the Great Northern Railway from the United States. Brandon is growing rapidly, both as a manufacturing center and as a distributing point. Its population, only 5,620 in 1901, increased to 13,839 in 1911 and to 17,450 in 1916.

Brandon is the distributing and shipping center for one of the richest agricultural sections in Canada. It has about twenty wholesale houses, also branches of practically all the agricultural-implement makers of Canada and the United States. It has large grain elevators and produces over $2,000,000 worth of flour a year. Of its other manufactures the leaders are car repairs, gasoline engines, stoves, pumps and windmills, store fixtures, saddles, harness and other leather goods, school desks, tents, lightning rods, butter and cheese. Brandon's manufactures have an average annual value of $5,000,000, about one-tenth of the total for the province.

The city is also known as an educational center. It has Brandon College, affiliated with McMaster University (which see) at Toronto, the provincial normal school, an Indian industrial school, a ladies' college, a business college and a collegiate institute, besides the usual public and Roman Catholic separate schools. It is the seat of the provincial insane asylum. The electric street railway is owned by the city. Brandon was founded in 1881, and was incorporated as a city in the next year.

BRANDY, *bran'di,* the name generally applied to the liquor obtained by distilling the fermented juice of the grape. If no qualifying word is used it is understood that the grape is the basis of the brandy. When other fruits are used the liquor is known as apple brandy, cherry brandy, etc., according to the fruit used. All of these brandies contain about fifty per cent of alcohol, and differ from each other only in the essential oil which gives to each its particular flavor and aroma.

Brandy is clear and colorless when distilled, and retains these characteristics if kept in glass vessels. Placed in wooden casks, it takes on the color of pale amber, for the coloring matter in the wood is dissolved out by the spirit. It is then sold under the trade name of *pale brandy.* Some dealers darken the liquor by means of caramel and sell it as *brown brandy.* The flavor of brandy improves with age, but its strength declines.

Genuine Cognac, the best brandy in the world, made in the southwestern part of France, is not easily obtained in the United States, because the high tariff on imported liquors makes it a costly beverage; the price is sometimes as high as $20 a gallon. American manufacturers now produce a brandy that in quality is said to be a rival of Cognac. Because of its stimulating properties, brandy is sometimes used as a medicine, being given occasionally to persons suffering from shock or exhaustion. However, its use in this manner is decreasing, for physicians use other and more desirable stimulating medicines. See ALCOHOLIC DRINKS.

BRANDYWINE, BATTLE OF THE, one of the important battles of the Revolutionary War, fought on Brandywine Creek, near Dilworth, N. J., September 11, 1777. At Chad's Ford, commanded by General Washington, the American force of 11,000 met the British army of 18,000, under General Howe. The British took the offensive, and after a stubborn fight, by a brilliant flank movement on the part of Cornwallis, the Americans were forced to retreat. The losses were about equal, but the victory enabled Howe to enter Philadelphia.

BRANGWYN, *brang'win,* FRANK (1867-), one of the foremost of present-day English painters and etchers. Brangwyn was born at Bruges, Belgium, where his father, a Welshman, was a manufacturer of church embroideries and vestments, and one of his early teachers was William Morris, whose ideal in art was Gothic. But in Brangwyn's mature work the predominating characteristic is not Gothic purity but an Oriental magnificence of color. His paintings have the decorative quality which is associated with the names of Morris and Rossetti and their friends, but this quality is obtained by emphasizing color and mass. Similarly, in his etchings he obtains brilliant contrasts in lights and shadows by treating his subjects as masses rather than as lines. This effect naturally involves the neglect or suppression of architectural and other details, but the loss of detail is more than balanced by the universal character which is given to the design.

All of Brangwyn's work is vigorous and dignified, and he is notably successful in handling large paintings or etchings. His mural paintings include *Modern Commerce,* in the Royal Exchange, London; *King John Signing the Magna Charta,* in the courthouse at Cleveland, Ohio; and a series of four paintings for the Panama-Pacific Exposition at San Francisco. *The Convict Ship, Trade on the Beach, Saint Simon Stylites* and *Venetian Funeral* are among his best paintings. His etchings include *The Sawyers* and *The Paper Mill.*

BRANT, Joseph (about 1742-1807), a brave and diplomatic Mohawk Indian chief, devoted to the welfare of his people. Brant was the name of his foster-father, his Indian name being Thayendanega. At the age of thirteen he took part with his two elder brothers in Johnson's campaign against the French at Lake George. He was sent to the Rev. Eleazar Wheelock's Indian school at Lebanon, Conn., from which Dartmouth College grew, became interpreter to a missionary and taught religion to the Mohawks. During the Revolutionary War the Mohawks adhered to the British, and because of his ability Brant soon attained the rank of colonel. He fought in the Battle of Oriskany, one of the bloodiest engagements of the war, and led the Indians in many raids, but he was not present at the Wyoming Valley Massacre. Later he lived in Canada on an estate granted him by the British government, and a bronze statue has been erected at Brantford, Ontario, in his honor (see below).

JOSEPH BRANT

BRANTFORD, Ont., the chief city and county town of Brant County, and the fifth city in size in the province. It is popularly called the *Telephone City,* because here Alexander Graham Bell perfected his invention.

STATUE TO JOSEPH BRANT

It is on the Grand Trunk, the Canadian Pacific and the Toronto, Hamilton & Buffalo railroads, and is an important manufacturing and exporting center. Electric interurban railways also connect with Galt and other cities. The Grand River, whose valley is noted for its scenic beauty, flows through the city, but is of no commercial importance. Brantford is sixty-five miles southwest of Toronto, twenty-five miles west of Hamilton and seventy-six miles west of Buffalo. Population in 1911, 23,132; in 1916, 25,420, not including suburbs.

The city was founded in 1823, and was named in honor of Joseph Brant, the famous Mohawk chief, to whose memory a fine monument was erected in Victoria Square in 1887. Originally the center of an excellent farming region, the city is now noted for its manufactures, the most important of which are agricultural implements, engines, wagons, paper boxes and envelopes, electrical supplies, bricks and other clay products and various iron, steel and wood products. The annual output from Brantford's seventy manufacturing establishments has a total value of more than $15,000,000. About twenty wholesale houses also contribute largely to the prosperity of the city. It is supplied with electrical power from Niagara Falls and also uses natural gas. The city has many fine buildings, including the home of the Young Men's Christian Association, built at a cost of $160,000, many churches, the post office and the Ontario Institute for the Blind.

BRANT GOOSE, also called BRENT GOOSE, a high-flying, sea-loving bird of the northern hemisphere, about twenty-six inches in length. It nests within the Arctic Circle and migrates in winter as far south as the Carolinas. Although at home especially on the Atlantic shores of Canada and the United States, it is occasionally seen inland. The male is distinguished by its blackish head, throat and shoulders, brownish-gray back and white patches on the sides of the neck. The brant is a popular game bird, but is protected from the hunter's gun in most states and provinces for about nine months of the year. See DUCK; GOOSE; GAME, subhead *Game Laws.*

BRAS D'OR LAKE, *brah dohr',* a tideless salt-water lake or lagoon which divides Cape Breton Island, N. S., nearly into two parts. This division is actually made by a ship canal which connects the southern end of the lake with Saint Peter's Bay, on the southwest coast. The entrance to the lake, which is on the northeast side of the island, is through two long channels separated by Boulardeire Island. The north channel is not navigable for large vessels, but the south channel, which is twenty-two miles long and about a mile wide, has an average depth of 350 feet. The two channels open into a small basin, called Little Bras d'Or, which in turn opens into a basin twice as large, usually called Great Bras d'Or. These two basins have a combined area of 360 square miles. The shores of the two basins are well-wooded and picturesque, and attract many tourists and summer residents. The waters swarm with salt-water fish, and the fisheries are of considerable commercial importance. *Bras d'Or* means *arm of gold,* in French, but it is supposed that the word is really of Indian origin and was corrupted by the early French colonists to its present form.

BRASS, a bright-yellow or reddish alloy produced by melting copper and zinc together, usually in the proportion of two parts of copper to one part of zinc. Brass is harder and stronger than either metal from which it is made, a condition true of any alloy. It can be cast in molds, rolled into thin sheets and drawn into fine wire, and it takes a high polish. See ALLOY.

Varieties. Different varieties of brass are made by varying the proportions of copper and zinc. Red brass contains four parts copper and one part zinc; yellow brass, the variety most often seen, is two parts copper and one part zinc. Muntz metal is three parts copper to two parts zinc, and what is known as spelter solder is one part copper to one part zinc, with a little silver added when intended for soldering articles of gold and silver. Brass intended for engraving purposes contains a little tin, and that to be turned or filed contains a small portion of lead, which makes it harder than the ordinary variety.

Uses. We are familiar with brass buttons, brass wire, brass beds and numerous other articles made from this metal, but we probably know little of the large quantities of brass wire, woven into screening and used for numerous other purposes, for its color is less characteristic. Another extensive use of brass is found in the manufacture of gas and electric light fixtures.

Some varieties bear such a close resemblance to gold that they are mistaken for it by those unacquainted with the two metals. These varieties are employed in making watch cases, cheap jewelry and other ornaments, most of which are often sold at prices far above their real value.

Manufacture. The manufacture of brass and articles made from it requires skilled workmen. The most common method of manufacture is by heating carbonate of zinc, charcoal and thin pieces of copper in crucibles. The molten metal is then cast into bars or ingots, which are again melted and recast to free the metal from impurities.

Brass castings are made by pouring the molten brass into molds of sand, in much the same manner as iron castings are produced. Sheet brass is made by rolling the refined ingots in mills especially designed for that purpose (see ROLLING MILL) and brass wire is made in the same manner as other wire (see WIRE). The most extensive brass works in the world are in Western Connecticut. (See BRONZE.)

History. We do not know when brass first came into use. Some authorities think it was discovered in the thirteenth century, but others are of the opinion that it was not known until about 200 years later. The brass mentioned in the Bible is believed to be some variety of bronze which was probably produced by smelting an ore that contained two or more metals. The word is also used in Scripture to denote strength. In modern slang, "brass" has meant for many years undue boldness or shamelessness. W.F.R.

BRAZIL, *bra zil'*, UNITED STATES OF, in area one of the greatest republics in the world, comprising almost half of the continent of South America, and now one of the most rapidly developing nations. It is almost as large as Canada or Europe, including European Russia, and larger than the United States and its island possessions, exclusive of Alaska. Of the other twelve countries of South America, all but Ecuador and Chile touch Brazil on the north, west or south; it is bordered on the northeast, east and southeast by the Atlantic Ocean. Its greatest extent north and south approximately equals that east and west, the former being 2,660 miles, the latter 2,700 miles, while the total area is about 3,258,200 square miles.

Brazil has regular steamship connection with the principal American and European ports. Rio de Janeiro, the capital, is 5,204 (nautical) miles from London and 4,748 miles from New York; Para, on the northern coast, is 4,153 miles from London and 2,915 miles from New York; it is 4,144 miles from Pernambuco, the easternmost port of Brazil, to London, and 3,678 miles to New York. Ocean vessels have an average speed of from twenty to twenty-one nautical miles an hour, and cover approximately 500 miles a day. It would therefore take about ten and a half days to make the trip from Rio de Janeiro to London, and about nine and a half from Rio de Janeiro to New York.

The People and Their Cities. While all the other republics of South America are of Spanish origin, Brazil was first peopled by the Portuguese; and the Portuguese language, ennobled by the great works of Camoens and Lobeira

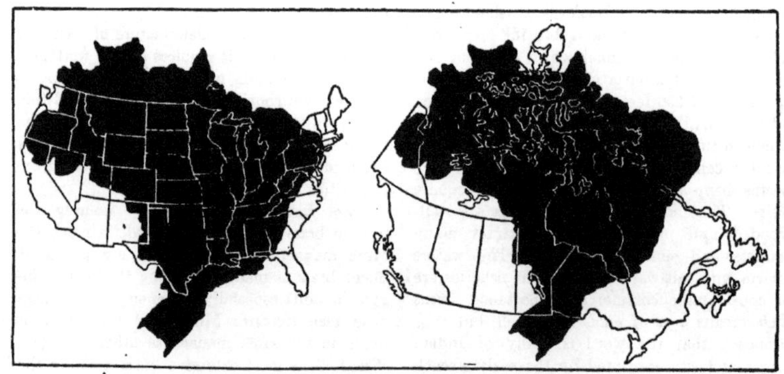

COMPARING THE AREAS OF BRAZIL, THE UNITED STATES AND CANADA

and by the lyrics of many gifted men and women of Portugal, is the official language of this great republic. The Brazilians, like the Portuguese, are poetically designated as *Lusitanians* (from the ancient Latin designation of their ancestors). Brazil was named from the Brazil-wood, chips of which, falling from the woodman's ax, seemed like coals of fire to the first explorers. *Brazil* is a modification of the old English word *brazier*, meaning a metal basket of coals for heating rooms, and the interchange of *l* and r is not uncommon in such words.

Of the population of Brazil, estimated at 21,500,000, less than one-half are whites of unmixed blood, and one-third are of mixed blood. The remainder are negroes and Indians, many of the latter being little influenced by civilization. The negroes of Pernambuco played so remarkable a part in the early history of Brazil that they claim a respect seldom accorded to their race in other lands. Robert Southey sought to do for Brazil what W. H. Prescott accomplished for Peru and Mexico; and his great history of the country in past centuries relates the story of the Brazilian negro republic of the seventeenth century.

BRAZIL
The proportion of the South American continent it occupies.

Despite the warm personal attachment of many American authors to the last emperor, Pedro II, his two visits to the United States and the historic friendship existing between the British and the Portuguese, there has never been a large number of British or American residents in Brazil. A large majority of the white population are of Portuguese descent, and while there has been much Italian immigration within recent years, it has been speedily assimilated. The Portuguese language, which has much of the beauty of the Spanish, but some features to remind us of the French, is not lacking in tender and beautiful poems, such as those of Violante do Ceo. In translations these are not unfamiliar to Northern readers. And the women of Brazil may take pride in the noble careers of Catherine of Braganca—wife of Charles II of Great Britain—of their late empresses and of that marvelous scholar and writer, Dona Bernarda Ferreyra, as types of Lusitanian character and achievement. Brazilian women of the upper class are very carefully trained in the proprieties of deportment, and to some their *dueñas* (chaperons) seem to be over-strict.

The population of Brazil is densest in the coast regions. In the state of Alagoas, on the coast, it averages 76.6 to the square mile; in the state of Rio de Janeiro, 57.6; in Amazonas, 0.5; and in Matto Grosso, 0.3.

The most important cities of Brazil are Rio de Janeiro, the capital, São Paulo, Bahia, Pernambuco, Para or Belem, and Manaos. Of these, Manaos alone is far from the coast. Each city is described in these volumes.

Education and Religion. There is no compulsory education law in Brazil, and in cousequence the free schools which exist in every parish are very poorly attended. The illiteracy percentage, counting the entire central population, is eighty, but changes in the school system introduced in 1911 are expected to improve conditions. The more thickly populated coast states are far in advance of the interior educationally, for they have secondary and technical schools, while the large cities possess libraries, museums and professional schools. Rio de Janeiro is one of the world's most progressive cities. In the entire country there is nothing which can correctly be called a university.

Theoretically, freedom of worship prevails in Brazil, but there seems little need for such legal regulation, since over ninety-nine per cent of the population are Roman Catholics. This does not mean, as in most of the South American states, that there is an established Church, for Church and State are entirely separate.

Highlands and Lowlands. The two great regions into which Brazil is divided according to its surface features are sharply distinguished. To the south and east lies the vast tableland known as the Brazilian Highlands, which geologists consider the first part of the continent to have been lifted above the sea. Everywhere this is in the neighborhood of 2,000 feet above sea level, and through much of its extent 4,000 feet. Little rivers carrying the drainage of this plateau region to the great river systems have cut gorges in the surface, and at intervals mountain ranges rise, though they nowhere attain great height. Most important of these mountain systems is the Serra do Mar, or Sea Mountains, which border the southeastern coast, leaving but a narrow plain between mountains and sea. The highest peak in these mountains is Itatiaya (8,900 feet), west of Rio de Janeiro. The very spot where Rio stands was ages ago in the midst of a mountain range which has been drowned or washed over by the sea; this is shown by the wonderful harbor of Rio, with its steep, clear-cut sides and nearly 300 islands. Parallel with the Sea Mountains run lesser ranges, most of which reach almost to the northeastern coast.

To the north and west of this plateau is the vast Amazon River basin, which forms the lowland portion of Brazil and with the valleys of its tributaries comprises over half of the country. This great plain, much of which is flooded each year, is for the most part less than 500 feet above sea level. It is in this basin that there still exists the largest unexplored area on the western hemisphere (see subhead *Forests*, below).

The Amazonian lowland is bordered on the north by a plateau, less important than the southern tableland, which forms a natural line between Brazil, the Guianas and Venezuela.

A RAILROAD THAT REDUCED A VOYAGE
Corumbá is an important city in Matto Grosso. Before 1914 it was reached by a sea voyage and a boat trip up the Paraná and Paraguay rivers —a six weeks' journey from Rio. Now it is joined to the coast by a railroad little longer than one from Boston to Chicago.

Rivers and Transportation. Brazil has not only the greatest river system of the world, the Amazon, which drains two-thirds of the country, but it has also several other large rivers. Chief of these are the Paraguay and Paraná, together draining one-fourth of the country, and the São Francisco, which is the waterway for the eastern plateau region. This last-named river, together with several of the streams which enter the Amazon in its lower course, have sharp falls at the edge of the plateau, and are thus practically useless for navigation, but, as a whole, the Amazon system affords a navigable waterway which only the Mississippi-Missouri system can approach in length. Of the 19,000 miles of the Amazon system which lie within Brazil, over two-thirds, or 13,000 miles, is navigable. The importance of this is hard to over-estimate, for throughout much of their course these rivers flow through a valuable forest area, and are the only means of transportation. At the best, in a region of so vast a size, river travel is unsatisfactory, because the up-stream voyage must be very slow, weeks of journeying by canoe along the Amazon scarce equaling a day's travel by rail along the shores of the Mississippi; but the time is not yet in sight when roads or even trails can be opened up through the forests, so dense is the undergrowth and so heavy the rainfall. See AMAZON.

All of Brazil is not dependent upon the rivers for travel, however, as the more open parts of the country have well-developed railroad systems. In 1915 there were 15,120 miles of railway in operation, and several thousands more under construction or planned. Of the present mileage, the government owns about one-fifth and various states over one-fourth.

The river systems of Brazil received new popular attention in 1914 when Theodore Roosevelt, during his exploration of the wild central portion of the country, announced the discovery of a new river—a tributary of the Madeira, almost 1,000 miles in length. A chorus of skeptical comment arose at once—there could be no such river; the "River of Doubt" it was called in the press. The government of Brazil, however, sent out explorers to the region, with the result that Roosevelt's findings were confirmed, and a new river, the Rio Téodoro (Theodore), was placed on official maps.

Climate. Save for three little states in the extreme south, Brazil lies entirely within the tropics, with comparatively even climate. In the Amazon valley, the hottest part of the country, the average temperature is about 90°, but in the higher, dryer regions there is a somewhat greater range; in some large areas it is

The Harbor at Rio De Janeiro

Smoking Rubber

Approach to the President's Palace

Where the American Ambassador Lives

The Harbor Customs at Rio De Janeiro

Drying Coffee

not only healthful but delightful. There are, as in most tropic countries, a wet season and a dry season, but these are not very well marked. Most of the rain, however, falls between January and June, while from June to October the weather is comparatively clear and dry.

In general, Brazil has a very heavy rainfall, only a small region in the interior of the highlands having too little moisture for successful agriculture. The eastern plateau is well watered, but by far the heaviest rainfall is in the Amazon basin. Near the coast the annual precipitation is from seventy-five to 100 inches, but inland, in the dense forest region, it is estimated that it must be from 300 to 400 inches.

Throughout much of the river country white men find it difficult to live, for the heat and the excessive moisture are fever-breeding, and distances are so great that a man stricken with disease has small chance of getting to a more healthful locality before he dies.

The Great Forests and the Rubber Industry. There has always hung about the forests of the Amazon a haze of romance, largely because they are so little known, and might, therefore, contain almost anything. Along the Amazon itself, and its chief tributaries, people have traveled for generations, but away from these water highways few adventurers have been daring enough to journey. Great trees of many species, among which palms are prominent, grow to heights of from sixty to one hundred feet, and from their trunks and branches hang tropic vines, with stems as big as a man's arm. Heavy underbrush rises to meet these overhanging vines, and so makes a growth that is practically impassable. The enterprising traveler may cut a trail, but in an amazingly short time it is again overgrown with luxuriant vegetation, and all traces of it are washed away by the pouring rains.

In the forest there are monkeys of many kinds, sloths, opossums, pumas and jaguars, while the trees swarm with birds of brilliant plumage. Great snakes, among them the dreaded boa (which see), glide through the underbrush, but are not so numerous as early writers believed. In the rivers are to be found sea-cows and great turtles, the latter valued for their flesh as well as for their eggs.

The Brazilian forest is by no means all waste, however, for some of the trees are useful and valuable. Dye-wood and Brazil nuts are produced in abundance from the part of the jungle accessible from the rivers; and above all else, the famous Para rubber. Rubber production has long been the chief industry of this part of Brazil, and until the last decade the greater part of the india rubber of the world came from that section. At the present time, however, Africa and the East Indies are producing fully as much, and it seems possible that with all its natural supply Brazil may not be able to meet the competition of other countries because of the scarcity of laborers and the absence of transportation facilities in the forest section. See RUBBER AND RUBBER MANUFACTURE; also subhead *Business Opportunities*, below.

Brazil's Great Source of Wealth. Brazil has the largest foreign commerce of any of the South American republics, exporting each year over $300,000,000 worth of products; of these, coffee forms over sixty per cent. For Brazil is the great coffee-growing country of the world, producing two-thirds of the world's supply. The industry centers in the state of São Paulo, in the southern part, and is of such great importance to the region that the government has taken all possible measures to advance it. The southern states of Rio de Janeiro, Minas Geraes and São Paulo, are very different in their general prosperity from the backward, undeveloped states of the interior, and the difference is due to coffee. Practically all the coffee used in the United States and Canada comes from Brazil.

Other Agricultural Products. There is much fertile land in Brazil, but only a very small portion of it has been put under cultivation. Almost everywhere the most primitive methods of tillage are used. Second to coffee in importance is sugar, which is produced in the Atlantic coast states, and next comes cotton, which every state produces to a greater or less extent. Tobacco is also grown, but is of rather inferior quality.

Brazil has scattered throughout its vast area much excellent grazing land which would make possible the raising of enormous herds of cattle. Three or four of the states have developed the cattle industry, but for the most part the transportation facilities are so limited that it has not been profitable.

Mining. Once Brazil was the leading country of the world in the production of gold and diamonds, Minas Geraes being especially rich in gold. The source of this earlier gold was chiefly, however, the sand and gravel along the rivers, and such a surface supply could not last long; while the opening of the fabulously-rich diamond mines of South Africa caused those of Brazil to decline. Gold still exists in large quantities, but the lack of capital and of energetic legislation in favor of mining has retarded the industry. Iron, lead, copper, silver and a few of the precious stones are also present, but the scarcity of fuel prevents the development of these riches, for Brazil has little coal, and what it has is of poor quality.

Manufactures. This same scarcity of coal has interfered with the development of manufactures, but of recent years there has been a decided increase in these, the waterfalls being utilized in many cases to furnish power. Chief of the manufacturing industries is the spinning and weaving of cotton, and next is the making of woolen goods. Sugar is also refined, but much of this work is yet done by primitive methods. No goods are manufactured in sufficient quantities for export; the exports consist mainly of rubber, coffee and cotton.

Business Opportunities. In a land so rich in natural resources and so little developed, the opportunities for the investment of capital and labor are so varied that it would be idle to attempt to enumerate them. The production of rubber in the past has been performed chiefly by cheap laborers who have gathered the sap from incisions made in the bark of the "up-river *hevea*" trees, generally growing wild. This is by far the best rubber known. In the vicinity of Ceará the *manihot* tree is cultivated for its rubber sap; and there is also the *castilloa* of Mexico and Central America to be taken into account. The rubber tree of Northern conservatories is not one of these, and is merely ornamental; and *castilloa* and *manihot* rubbers are not of the best. The cultivation of the *hevea* in its natural habitat is highly productive.

Medicinal plants and dye woods abound in Brazil. The use of nuts—except the chestnut—as a food for diabetic persons has made of nut products a food staple; and Brazil abounds in nut-bearing trees. The late Henry T. Blow, formerly United States minister to Brazil, constructed a reception cabinet, or chamber, of Brazilian woods for his home in Saint Louis; and at every international exposition there are exhibits of the inexhaustible timber of the Brazilian forests, desirable for use in North America. The people of the United States buy Brazilian products to the amount of more than a hundred millions of dollars annually, and sell to Brazilians merchandise valued at but little more than a quarter of this amount.

The trade should be reciprocal between Brazil and its best customer. Commercial diplomacy must be enlisted to correct the existing lack of balance in trade. Why do Brazilian orders for staple manufactures and Brazilian contracts for public works go so largely to European competitors of North Americans? In a witty Brazilian proverb, one "hears with the ears of a merchant" when he pays little attention to what is said. North Americans hitherto have spoken to Brazilians perhaps, only with the voice of a merchant, without mutual sympathy and appreciation. Social qualities are essential in salesmanship, and the American house which would establish Brazilian trade must extend terms of credit of three to six months, to meet like courtesies from European business firms.

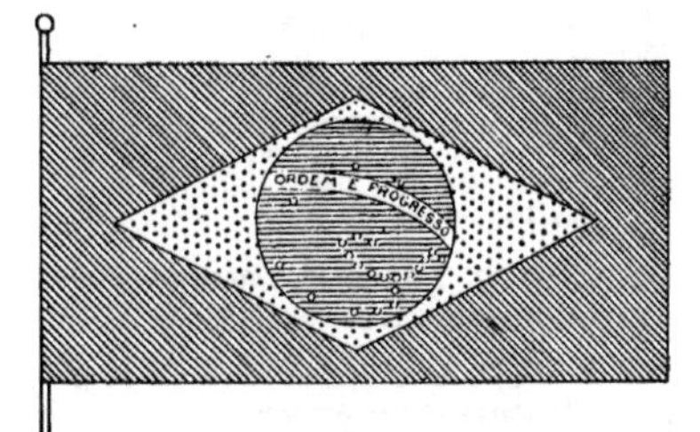

THE NATIONAL FLAG
The rectangle is green; the dotted center, yellow; the globe, blue, with band and stars in white. See FLAG, color plate.

Brazilian money—like the Portuguese—is counted in milreis, and the milreis, equal to a fraction less than fifty-five cents, is theoretically divided into a thousand parts; for *milreis* means *a thousand mills.* But counting by thousands is not difficult to acquire, when one is familiar with the centesimal accounting of the United States and Canada.

Government. Brazil is a republic, with a constitution modeled on that of the United States, but allowing a far greater measure of states' rights. Indeed, each of the twenty states is almost independent, accepting the interference of the Federal government in but few matters.

At the head of the republic is the President, elected by direct vote for a period of four years. His Cabinet of seven ministers is appointed by him and is not responsible to the legislative body. This latter consists of two houses, a Senate and a House of Deputies. Each of the twenty states, as well as the Federal District, elects three senators, who remain in office for

OUTLINE AND QUESTIONS ON BRAZIL

Outline

I. Location and Size

(1) Latitude, 34° south to 5° north
(2) Longitude, 35° to 74° west
(3) Boundaries
(4) Length, 2,660 miles
(5) Breadth, 2,700 miles
(6) Actual area, 3,258,200 square miles
(7) Comparative size
(8) Population

II. Surface and Drainage

(1) Brazilian Highlands
  (a) River gorges
  (b) Sea Mountains and other ranges
(2) Basin of the Amazon
  (a) Largest unexplored area on western hemisphere
  (b) Vast forests
(3) Northern plateau
(4) Rivers
  (a) Amazon
  (b) Paraguay
  (c) Paraná
  (d) São Francisco
  (e) "River of Doubt"

III. Climate

(1) Tropical character
(2) Variations due to altitude
(3) Rainfall
(4) Unhealthfulness of certain localities

IV. Resources

(1) Forests
(2) Agriculture
  (a) Coffee
  (b) Sugar
  (c) Cotton
(3) Stock-raising
(4) Mines
(5) Manufactures
(6) Possibilities for industrial improvement

V. People

(1) Portuguese origin
(2) Mixture of whites, Indians and negroes
(3) Influx of Italians
(4) Language
(5) Education
  (a) Lack of compulsory school laws
  (b) Greater progress of coast states
(6) Religion
(7) Cities

VI. Transportation and Commerce

(1) River travel
(2) Railroads
(3) Steamship connection
(4) Largest foreign commerce of any South American country

VII. Government

(1) Republican form
(2) "States' rights" features
(3) Executive department
(4) Legislature
(5) Local government

VIII. History

(1) Discovery and settlement
(2) Portuguese rule
(3) The negro revolt
(4) Period of Spanish domination
(5) Brazil the seat of Portuguese government
(6) Independence achieved
(7) Reign of Pedro II
(8) Establishment of republic
(9) Later development

Questions

In what peculiarity of location do practically all of the large cities of Brazil resemble each other?

What is the "River of Doubt," and why was it so called?

Where is the deepest mine in the world, and what is taken from it?

What great change took place in the appearance of the capital city during the administration of the third President?

If difficulties arose between Brazil and the United States, what procedure would be followed?

Outline and Questions on Brazil—Continued

Who tried to do for Brazil what Prescott did for another South American country?

Where is the largest unexplored area in the western hemisphere?

Is it an absurdity to speak of a "cotton tree"? Why?

If a trail were cut through the Amazon jungle in the spring, would it be easily followed in the autumn? Why?

With what great European ruler was Brazil's most famous emperor connected by marriage?

How would the United States suffer if commerce with Brazil were cut off?

Compare the country as to size with the United States and Canada.

If the inhabitants of the United States were transferred to Brazil, how much more densely populated would that country be than it is at present?

What advantage have the children in the coast states over those in the inland regions?

How does the smallest Brazilian state compare in size and population with the smallest state in the American Union?

How does it happen that Brazil is Portuguese rather than Spanish, like the rest of South America?

Do all locomotives here have coal cars? Why?

How long would it take you to go from New York to Rio de Janeiro? From Para to London? From New York to Pernambuco?

Why are laws relating to freedom of worship unnecessary?

How many times as much rain falls in the Amazon basin each year as in the best-watered part of the United States?

Four men in a restaurant in Bahia were charged for their dinner ten thousand *reis,* a reis being the thousandth part of a milreis. How much, in American money, did each have to pay?

Why has it been more profitable to allow excellent grazing lands to lie idle in most parts of the country than to raise cattle on them?

Why would it not be good advertising to state that cigars were made of Brazilian tobacco?

How did the country get its name?

Why would you not care to live in the Amazon region?

What is meant by the expression, "He hears with the ears of a merchant"?

What is the country's great source of wealth? How large a part of the industries centers about it?

How large a part of the Amazon system lies within this one country?

What food besides fish does the Amazon furnish?

What valuable product is received from the "up-river *hevea*" tree?

How does the density of population in the state of Rio de Janeiro compare with that of Kentucky? How does that of Amazonas compare with that of British Columbia?

What effect has the fuel supply had on manufactures?

Name three things which the great forests yield.

When was the capital of Brazil the capital of another country as well?

When did a subject people establish and control a state of their own? What did they call it?

If a young man of Brazil determines to go to a university, what must he do?

Describe briefly the harbor of Rio de Janeiro.

nine years. The lower house consists of 212 members, elected for three years. Each state has its own governor, legislatures and judicial system, and is divided into districts for the purpose of local government.

History. The discovery of America by Columbus, and of the Cape of Good Hope by Da Gama six years later, were both the result of maritime enterprise fostered in the period of Portugal's glory by Henry, the Navigator, called by Thomson—

> The Lusitanian prince who, Heaven-inspired
> To love of useful glory, roused mankind,
> And in unbounded commerce mixed the world.

The coast of Brazil was first reached in 1500, by Vicente Pinzon, a Spaniard, who claimed it for his king. But since it proved to be east of the line of demarcation fixed by arbitration of the Pope between the fields for exploitation by the Portuguese and by the Spanish, Spain yielded its claim to the region (see DEMARCATION, LINE OF). The first settlement was made in 1501, at Rio de Janeiro. Bahia was founded in 1549. Between 1532 and 1545 there were formed twelve proprietary divisions of the coast lying between latitudes 30° S. and the equator, but the proprietors failed to take advantage of the opportunities offered them. An attempt was made to enslave the Indians, but this was abandoned in 1680 in deference to the priests, whose labors among the natives were bearing fruit. African slavery followed, great numbers of negroes being imported for the purpose.

The story of the negro revolt against their taskmasters and of the state of Palmares, which they established, fortified, and maintained for more than sixty years, ending in 1695, is one of the most romantic in American history. Southey compares it with the story of early Rome, and describes its capital city of more than 20,000 people, with its palace having a "rude kind of magnificence." An incentive to the African race forever will be the memory of Palmares. It was for them more than a state. They used to call it "Republica des Palmares."

From 1580 to 1640 the Spanish royal house controlled both Spain and Portugal. Gold was discovered in 1691 and diamonds about a score of years later. When Portugal was invaded by the French, in 1807, the Portuguese royal family fled from Lisbon to Brazil, which remained for fourteen years the seat of the Portuguese government. In 1821 the king returned to Lisbon, leaving his son Pedro as regent of Brazil; but in the following year the great colony proclaimed its independence and chose the regent to be its emperor, under the title of Pedro I. He proved a worthless ruler and after nine years was induced to abdicate, leaving his crown and scepter to his son, who succeeded at the age of fifteen years, under the title of Pedro II. The long reign of this ruler was beneficent, and early won the respect of the world.

Dom Pedro II represented three of the great royal houses of European history—the Braganeas, the Hapsburgs and the Bourbons. His mother was a sister of Napoleon's second wife, and his stepmother a granddaughter of Napoleon's first wife. His public and private life was free from blemish. There were few wars in his reign. In 1825 there was a short contest with the new republic of La Plata (Argentina) for the possession of Uruguay, which fell to neither power.

From 1860 to 1865 Brazil was compelled to wage a war in concert with Argentina and Uruguay, against Francisco S. Lopez, the dictator of Paraguay, who preferred death to surrender. Dom Pedro issued in 1871 an imperial decree for the gradual abolition of slavery, deeming this the best method of ridding the nation of that system. In 1888 he proclaimed total emancipation. In all the years of his long reign he was the patron of art, letters, science, production, commerce and public benevolence.

In 1874 and again in 1876 Dom Pedro visited the United States, where he enjoyed unbounded popularity, and where he chose for his most intimate friends the men of letters with whose works he was familiar. The feelings of the people were happily expressed in the words of Whittier:

> Wear unashamed a crown by thy desert
> More than by birth thy own,
> Careless of watch and war: thou art begirt
> By faithful hearts alone.

The visit of the North American scientist, Louis Agassiz, to Brazil in the preceding decade was a matter of general as well as scientific interest.

In 1889 Dom Pedro was deposed, his opponents thus anticipating the republic for which he was seeking to prepare his people. He died in Paris, two years later. In the year of his deposition Deodoro da Fonseca, the first President of the new republic, was impeached and he resigned.

Dr. Prudente de Moraes, elected President in 1894, displayed ability and firmness. His successor, Dr. Campos Salles, chosen in 1898, inaugurated a great improvement in the government finances. Within the next decade the capital city took on a new appearance, being modernized in many ways, and became a delightful city alike for permanent residence or for temporary sojourn (see RIO DE JANEIRO).

In 1902 Rodrigues Alves was chosen for the Presidency, and in 1906 the choice fell upon Affonso Penna, who died in office, three years later. The International Exposition held at Rio de Janeiro for six months in 1909 was notable for the plan then conceived to maintain a continuous Brazilian-American exhibit of samples of goods offered for trade.

Hermes da Fonseca was elected President in 1910, and Wenceslao Braz in 1914. The visit of Dr. Lauro Müller, Minister of Foreign Affairs, to the United States in 1913 was appreciated, for Americans credited him largely with the marvelous improvement of the Brazilian capital and the advancement of his nation within recent years, and recognized in him a promoter both of trade and of good feeling.

On October 26, 1917, Brazil declared war upon Germany, and threw its moral force with the allies. A.DE F.M.

Other Items of Interest. On the northern coast of Brazil there is a native variety of cotton which grows in the form of a little tree. This often attains a height of eight feet and produces a long, strong fiber.

With a view to bringing about closer relations between the United States and Brazil, an American Chamber of Commerce has been established in Rio de Janeiro. Its first quarterly magazine, printed in English and Portuguese, appeared in July, 1916.

Santos-Dumont, the famous inventor and aëronaut, was born at São Paulo, Brazil.

There are various species of ants in Brazil which build huge nests. On the way from São Paulo to Rio de Janeiro may be seen hundreds of hills each taller than a man.

The production of coffee and of rubber comprises about sixty per cent of the industries of Brazil, and practically all of the product is exported, the country using no rubber in its raw state and very little coffee.

About seventy per cent of the coffee of the United States, over 700,000,000 pounds annually, comes from Brazil.

Of the states of Brazil, the smallest, Sergipe, has an area about one-fifth greater than that of Maryland. Its population, however, is only about one-third that of the latter state.

"Long distance" telephone lines are being rapidly extended in Brazil, and many of the large cities now communicate with each other across hundreds of miles.

Because of the scarcity of coal, most of the locomotive and steamboat engines use wood and oil as fuel.

New industries are constantly coming into prominence. In the not far distant future, for instance, it is believed that Brazil will be exporting and not importing rice.

There exists between Brazil and the United States a treaty which provides for amicable settlement by arbitration of any disputes which may arise between the two countries.

The deepest mine in the world is a gold mine in the state of Minas Geraes, Brazil. It is sunk in the hard rock, and has a shaft over 5,800 feet in depth, while its lowest point is 3,056 feet below sea level.

Consult *Brazilian Year Book*, published in Rio de Janeiro, London and New York (inquire of Brazilian embassy, Washington); Winter's *Brazil and Her People To-day;* Roosevelt's *Through The Brazilian Wilderness.*

Related Subjects. The reader who is interested in Brazil will find helpful the articles on the following topics:

CITIES AND TOWNS

Bahia	Rio de Janeiro
Manaos	Santos
Para	São Paulo
Pernambuco	

LEADING PRODUCTS

Coffee	Rubber
Cotton	Sugar
Gold	

RIVERS

Amazon	São Francisco
Madeira	Tapajos
Paraguay	Uruguay
Paraná	

BRAZIL, IND., an important manufacturing city of the state and the county seat of Clay County, with a population of 10,001 in 1914, an increase of 661 since 1901. English, Scotch, Irish and Poles comprise about forty per cent of the inhabitants. Brazil is situated in the western part of the state, sixteen miles northeast of Terre Haute and fifty-seven miles southwest of Indianapolis. The city is served by the Vandalia Line, constructed to this point in 1850; the Chicago & Eastern Illinois Railroad, built in 1885, and the Central Indiana Railway, completed in 1890. The city was founded in 1844, was incorporated in 1873 and named for Brazil, South America. The area is about two and one-half square miles.

Brazil is located in the center of the black-coal field of the state. Extensive deposits of clay, shales and bituminous coal are also found here, and the prosperity of the city is largely dependent upon these natural resources. The clay plants employ about 1,000 men, and the annual output exceeds $2,000,000. The machine shops, foundries, boiler works and planing mills all produce large supplies. The surrounding country is rich in agricultural products, chief among which are corn, wheat, oats, alfalfa and vegetables. About 300,000 bushels of tomatoes are annually used by one industry, in the manufacture of products of this vegetable.

The buildings of note are the Federal building, erected in 1913 at a cost of $74,000, and a $275,000 courthouse, constructed in the same year. The city has a hospital, a Carnegie Library and a business college. J.B.

BRAZIL NUT, or **PARA NUT,** the edible seeds of two species of Brazilian trees, in America commonly called *nigger-toe* or *cream-nut.* The tree grows as high as 150 feet, and is very abundant along the Amazon and Orinoco rivers.

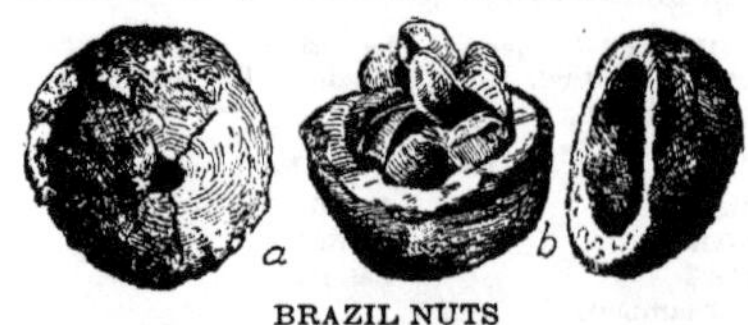

BRAZIL NUTS

(*a*) A seed vessel of Brazil nut; (*b*) same, opened, showing the nuts as they are known in the markets of the world.

The leaves are bright green and leathery, two feet long and six inches wide. The flowers are cream colored; the fruits are round, very hard shelled, about the color of a cocoanut and nearly six inches in diameter. Each fruit contains about twenty of the well-known three-sided, wrinkled seeds or nuts, tightly fitted into the shell like the slices of an orange. These nuts are eaten as delicacies, are used for dessert and candies and yield an oil used in oil painting, oiling of delicate machinery, lighting and sometimes for cooking.

BRAZOS, *brah'zose,* the largest river of Texas, formed by the junction of two streams called the Clear and Salt forks. It flows southeast by a winding course and empties into the Gulf of Mexico, forty miles southwest of Galveston. It has a length of 900 miles and is navigable for 300 miles from the Gulf during periods of unusually high water and at all seasons for forty miles from the Gulf. The commerce it carries is only local.

BREAD, *bred,* the most widely-used food of civilized man. For centuries it has been of such overshadowing importance that it has become commonly known as the "staff of life." Figuratively, the name is often applied to food in general, as in the Biblical passages, "Give us this day our daily *bread,*" or, "Man shall not live by *bread* alone," or in the old couplet—

Seven wealthy towns contend for Homer dead
Through which the living Homer begged his
bread.

What Bread Is. People in different ages have meant very different things by the word *bread,* just as to-day people the world over vary in their ideas as to what this "staff of life" is. But always there is a general resemblance, in that the principal ingredients are water and some sort of meal or cereal. The loaves may be big and light, brown without and white and flaky within, or they may be little, tough, flat cakes, grayish-white and gritty with ashes; but they are all *bread* to the people who eat them.

In the most advanced countries wheat is the common grain for bread, because it makes lighter, better-tasting food, and is more easily digested. The peasants of some countries of Europe, particularly Russia, Austria-Hungary and the Balkan states, still live largely upon a black bread made from rye, while others make their bread of oats. The people of the United States and Canada think that wheat is the world's most important cereal, because it is so to them; but more people are fed on rice than on any other grain, and throughout much of the Orient it is not only boiled whole, but ground up into flour and made into bread. In the United States the people of the South use much corn meal, and in many places when bread is spoken of, it is the corn *pone* or *dodger* that is meant, wheat bread going under the name of *light bread.*

Many countries have their characteristic or "national" kinds of bread, which are used in addition to the regular wheaten loaves. Thus Scotland has its *oat cakes* and its *bannocks* of barley meal; the Central American countries have their *tortillas,* which are cakes made of crushed and parboiled corn; and the United States and Canada have their *hot biscuits.* Different conditions, too, call for various kinds of bread. Ships which are to make long voyages must have a bread which will keep indefinitely, and to meet this demand, ship's biscuit, or *pilot bread,* is made in large quanti-

WHITE BREAD

Water, 35.3
Protein, 9.2
Fat 1.3
Carbohydrates 53.1
Ash, 1.1

WHOLE WHEAT BREAD

Water, 38.4
Protein, 9.7
Fat 0.9
Carbohydrates 49.7
Ash, 1.3

ties. It is simply flour mixed with water and baked slowly, and it is so hard that excellent teeth are required for chewing it.

Every child who has read *The Swiss Family Robinson* remembers the fascinating account of the making of bread from manioc roots. The roots were grated, and the pulp was then placed under heavy weights and squeezed dry, for the wise head of the family knew that the juice was poisonous. This was not a mere fairy tale. There are many people in the West Indies and in Africa who use the roots of various plants for making bread, and in some cases these roots are poisonous unless properly prepared. The West Indians use cassava, which is manioc under another name, and make their little cakes by hand just as the Swiss Family Robinson did.

Kinds of Bread. All these breads, no matter how different they may look or taste, fall into two classes. They are either *leavened*—that is, fermented or "raised" with some sort of yeast or baking powder—or *unleavened,* baked without "rising." Undoubtedly the latter form, which is the simpler, was the earlier, but the process of making leavened or fermented bread must have been understood for thousands of years, for the book of *Exodus* distinguishes between the two kinds. The yeast plant used to-day was of course not known, but leaven, consisting of a portion of dough which had been allowed to ferment, answered the purpose. The Egyptians, doubtless by accident, seem to have discovered the possibility of making bread light in this way, and the Jews as well as the Greeks probably learned it from them. Yeast as "rising" has not everywhere taken the place of leaven to this day.

Of breads not raised with yeast or baking powder, the most common are *salt-rising* bread, in which the ferment necessary to lightness is brought about by a sour batter of corn meal and milk; *aerated* bread, made with water which has been charged with carbon dioxide; *gluten* bread, made from flour which has been freed of much of its starch; the oaten cakes, bannocks and corn pone mentioned above; and the various kinds of *crackers,* or biscuits, as they are called in England. These latter may contain many ingredients besides cereal and water—may have shortening, flavoring, fruits or meat juices introduced to improve their taste or food value. *Pancakes* made from specially prepared self-raising flours contain the elements which, when moistened, produce the necessary fermentation.

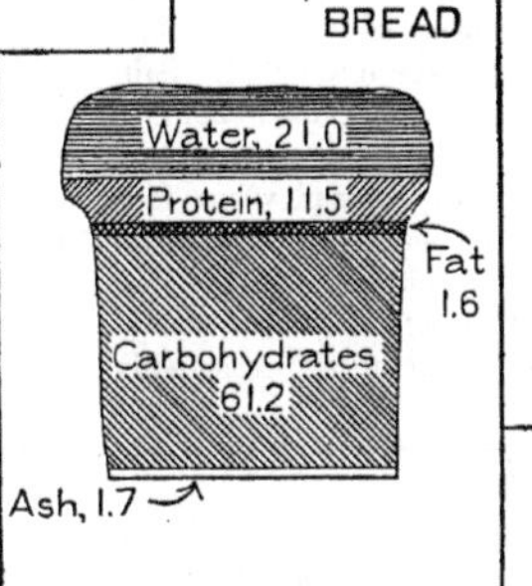

Of leavened breads by far the most familiar and important to the people of America is the white wheat bread, made with yeast. Other flours may be combined with or substituted for the wheat, but the principle remains unchanged. In some forms of bread, however, as in biscuits, muffins, corn bread and many of the popular brown breads, baking powder or soda is used as the leavening agent. Against the chemical mix-

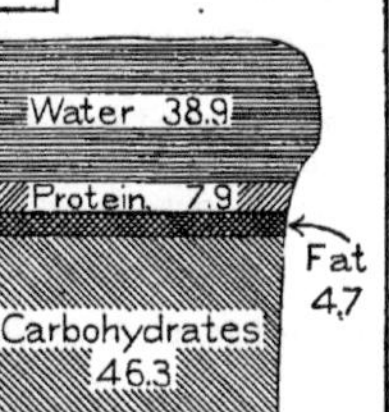

tures known as baking powders there has been much prejudice, but under the pure food laws the possible harmful effects have been largely done away with. The housewife should be careful, however, to use only a standard make, of whose purity she feels sure.

How Yeast Bread Is Made. Many housekeepers, especially in country districts, still make their own bread, but in recent years the bakeries have done a large proportion of the work. In most cities the day has passed when dirty, unsanitary underground bakeries are permitted to exist, and many of the large bakery companies have put up great factories which are marvels of cleanliness. The public is invited to inspect the process of manufacture in all its stages, so it may realize that the advertising slogan, "Yours are the first hands that touch our bread" is no empty boast. In such sanitary bakeries all the work is done by machinery, even to the wrapping of the loaves in oiled paper.

The process is much the same whether bread is made in the home or in a great bakery, but it is rather more interesting to watch in the latter place. Most bakers use compressed yeast cakes, which are dissolved in warm water and mixed with enough flour to make a thin paste, or *sponge*. This is left to ferment for two or three hours in a temperature of from 70° to 75°, and is then ready to be made into dough. Salt, milk, lard, a little sugar, and enough flour to make a good stiff dough are then added, and the whole is dumped into a mixer, where it is thoroughly stirred and kneaded by iron arms, that the ferment may reach every part of the mass. The whole is then left to rise for several hours, and as it gets light it is occasionally beaten down and kneaded, that the large gas bubbles which have formed may be broken into little ones.

When the practiced baker sees that the dough is light enough it is separated by a dividing machine into small pieces, rolled into loaves and placed in pans, which are again allowed to stand for a short time before being put into the oven. During all these "rising" stages care must be taken that the dough does not get too warm, or sour bread will result.

An ordinary bakers' oven is circular, and will hold from 300 to 500 loaves at one time. The pans are slid in and lifted out with long wooden paddles, and all the men who attend to it in this as in its other stages are scrupulously clean and dressed in white. The temperature of the oven is 400° to 450° F., but the interior of the loaf rises only two or three degrees above the boiling point of water. Steam is sometimes admitted to the oven. This dextrinizes the starch of the crust, producing a glazed surface, which prevents evaporation of moisture from the interior and so produces a moister loaf. On the average, nine pounds of flour yield nine pounds of bread. Flour made largely from the hard spring wheats of the Northwest are what the bakers call *strong* flours. These take up more water than weak flours and give a higher yield of bread. On the average white bread contains thirty-five per cent of water.

Test of Purity. Bakers are sometimes tempted to make their bread heavy by putting into it a great deal of salt, which cannot be detected by taste. The salt makes the bread weigh heavy because of the moisture it retains.

TEST FOR ALUM

To determine the presence of alum in bread, take a sample of the suspected article, place it in a saucer and pour over it a solution of carbonate of ammonia. If alum is present the bread will turn black, but no change will occur if the bread is pure.

To test the relative food value, dry pieces of both a heavy and a light loaf in a slow oven. The heavier of the two will throw off more moisture and be lighter after heating, therefore less nutritious than the other.

Small quantities of alum, which are very harmful to the system, are sometimes added to bread to make it white and cover up the use of a dark, cheap flour. Its presence may be detected by pouring a solution of carbonate of ammonia over the bread; if it contains alum it will turn black, if it does not, there will be no change of color.

Food Value. Well-made bread is wholesome and nourishing, but does not contain all of the food elements in sufficient quantities to make by itself a perfect food. Its deficiency in fats is commonly made up by eating it with butter. Bread that is soggy and heavy is liable to be harmful, the digestibility increasing with its lightness and with thorough baking. There has been much controversy as to whether white-wheat bread or one of the darker breads which

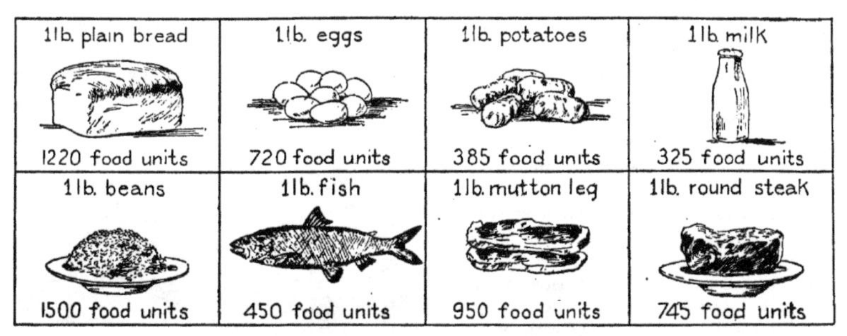

A COMPARISON OF FOOD VALUES

Meat is sixty per cent water and costs from twenty to thirty cents per pound. Bread is nearly forty per cent water and costs five cents per pound. There is more energy stored in a pound of bread than in a pound of meat or in any other foods illustrated above, except beans.

contains the bran as well as the flour is the better, but the weight of opinion is now in favor of the latter. Physicians are advising that everyone eat some bread every day in which bran is included. Bread should be very well chewed, however, because it is a starch food and the saliva helps to digest the starch. The only harm that hot breads, long looked upon as extremely unwholesome, can do, arises from the fact that the inside, or "crumb," is easily compressed into a solid mass and swallowed before the saliva has a chance to mix with it. Crusty breads or rolls are better than softer breads, because they call for more chewing. Breads become stale by losing water, about fourteen per cent being lost in the course of a week. Stale bread can be refreshened by heating to about 300° F. All in all, no other food probably returns better value for money expended than does bread, and Benjamin Franklin, on his first arrival in Philadelphia, acted with characteristic wisdom when he spent his last penny for "three wheaten rolls." See FOOD, subhead *Chemistry of Foods.* J.F.S.

BREADFRUIT, *bred'frute,* one of the most important food staples of tropical islands in the Pacific Ocean. The tree that produces it grows about forty feet high, often limbless half its height, with large, spreading upper branches and glossy dark-green leaves over a foot in length. The fruit, usually seedless, is green at first, later brown, and lastly yellow. Six or eight inches in diameter, the fruit hangs singly by short, thick stems, or in clusters of two or three. There is almost a constant supply throughout the year.

The fruit is generally eaten immediately after being gathered. It is also often prepared to keep for some time, either by baking it whole in close, underground pits, or by beating it into paste and storing it underground, where a slight fermentation takes place. The eatable part lies between the skin and the core and is somewhat of the consistency of new bread. Mixed with cocoanut milk it makes an excellent pudding. The inner bark of the tree is made into a kind of cloth. The wood, when seasoned, closely resembles mahogany, and is used for the building of boats and for furniture.

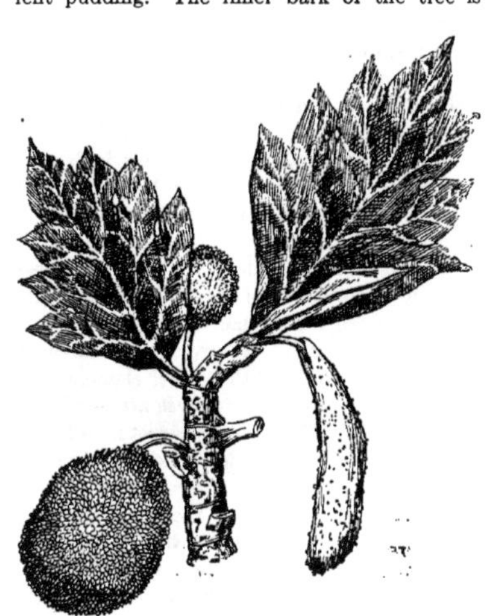

BREADFRUIT
Stalk, fruit, leaves and flower.

The breadfruit tree has been cultivated in Southern Florida, but the fruits are not seen in Northern markets as they do not bear shipment well.

BREAKFAST, *brek'fast,* **FOODS** are cereal preparations, placed on the market either in ready-cooked or in uncooked form. There has been within recent years an ever-increasing demand for such foods, which has been met by the constant development of new varieties and combinations. Wheat, rice, corn, oats, and to a much smaller extent barley, make up the breakfast foods. These are treated in various ways and sold in many forms, so the possible number seems far greater; for these grains are crushed or rolled, shredded or toasted, steamed or ground. Moreover they come to the housewife in forms that may be served without cooking, direct from the package, in forms that require but little cooking, or in others that require hours of cooking.

In general it may be said that these cereals form a nourishing and digestible food, if they are properly cooked. Long boiling at a high temperature is necessary if the highest food value as well as the most pleasing taste is to be secured, and housewives often err in cooking their cereals too little. Wheat has the highest value of any of the cereals, and is the most digestible, though Scotland has proved through long centuries that oats form a food well suited to build brain and brawn. To those not accustomed to eating oats, they are, however, somewhat heating, because of their high percentage of fat. Rice is much less nutritious than any of the other grains, consisting as it does almost entirely of starch, but the starch appears in a form which is very easy of digestion. Rice loses less of its valuable elements if it is steamed rather than boiled in water.

Breakfast foods are usually eaten with sugar and milk or cream, and thus other elements not present in the cereals themselves are added.

BREAK'WATER, a solid structure of masonry or other construction placed between a harbor and the open sea to serve as a protection against the violence of the waves. Breakwaters are usually constructed by sinking loads of unwrought stone along the line selected and allowing them to settle under the action of the waves. When a thorough foundation of stone and gravel has been secured and the mass rises toward the surface it is surmounted by a pile of masonry, sloped in such a manner as will best enable it to resist the action of the waves. The greatest breakwaters are those of Cherbourg in France; Plymouth in England; Delaware Bay and Buffalo in the United States, and Valparaiso in Chile. In less important localities floating breakwaters are occasionally used. These are built of strong, open woodwork, partly above and partly under water, divided into several sections and secured by

BREAKWATER PROTECTING A HARBOR

chains attached to fixed bodies. The breakers lose nearly all their force in passing through the beams of such a structure.

BREATH AND BREATHING. Breathing, or the taking in and expelling of air by the body, is so necessary a function that "he has stopped breathing" is equivalent to saying "he is dead"; yet it is so simple a process that no thought is given to it. The moment a person starts to watch his breathing or count the intakes of air he becomes self-conscious and cannot breathe regularly or naturally.

Reasons for Breathing. Food and drink are necessary to the body, and yet under extreme conditions life may exist for hours or even days without them; but the body must have a constant supply of oxygen or it will die. The act of breathing, or *respiration,* as it is more scientifically called, supplies this oxygen and carries off carbon dioxide which has been produced by the various changes taking place in the body. The lungs constitute the special organ of breathing.

Acts of Breathing. Breathing consists of two acts, *inspiration,* or breathing in, and *expiration,* or breathing out. These words suggest in their very form interesting connections—to *inspire* is to breathe into someone a special power of achievement; to *expire* is to draw the last breath—to die. Between inspiration and expiration are definite pauses which form part of the breathing movements, that after the expiration being considerably longer than that after the inspiration. With the taking in of air the chest cavity expands, for the diaphragm flattens its dome and the ribs are raised; with the expelling of air the diaphragm again arches upward and the ribs draw back into their places. The increase in the chest cavity is not very great in ordinary breathing, but forced breaths make it considerable. When the chest is expanded, the air rushes in to fill the vacuum caused by the expanding lungs. The act of respiration is usually mechanical, except when a person deliberately tries to expel as much of the air from the lungs as possible.

Frequency of Breathing. No definite statement can be made as to how many times a person breathes in a minute, for the variations are considerable. In general, however, the number of respirations is in a fairly definite proportion to the number of heart beats—one to four and one-half or five. A full-grown man or woman in good health may breathe normally from sixteen to twenty times in a minute—the average is eighteen—but exercise, illness or certain personal characteristics may increase this range. There have been known cases of hysteria during which the patient breathed over 100 times a minute.

Lung Capacity. The quantity of air that is changed in each act of breathing, known as *breathing*, or *tidal*, air, is from twenty to thirty cubic inches, but this is increased during physical exertion of any kind. Over and above this normal thirty cubic inches, *complemental* air to the amount of about 100 cubic inches may be drawn into the lungs by the deepest possible inspiration, and about the same amount, known as *reserve* air, may be driven out by a forced expiration. But no matter how hard a person tries, he cannot expel all the air from his lungs. There remains always about 100 cubic inches, known as *residual* air. All in all, about 686,000 cubic inches of air pass through the lungs of an adult person every twenty-four hours; this may be doubled by continued exertion of any sort.

What Breathing Accomplishes. In passing through the lungs the air gives up oxygen to the blood and receives in exchange carbon dioxide, or carbonic acid gas, a waste, the exchange taking place in the tiny, thin-walled blood vessels of the lungs. The amount of carbon dioxide carried out varies with the temperature and moisture of the air, the age, sex, muscular development, state of health and the nature and quantity of food and exercise. In both sexes it increases to about the thirtieth year, and is greater in strong, healthy persons than in frail, slender ones. If this gas were not carried off, asphyxiation would result, just as surely as it does from escaping illuminating gas. By a remarkable provision of nature the proportions and supply of oxygen and carbonic acid gas are always assured. Plants breathe out, or exhale, oxygen, and they inhale carbonic acid gas, to them a life-giving principle. Oxygen is the fuel needed by man to keep the fires of life going. Thus do plants and animals help each other.

Breathing in Relation to Health. Breathing is an instinctive act—that is, a person does not have to think or plan in order to breathe; and if conditions were always the best this natural breathing would supply plenty of pure air. But many people find themselves under artificial conditions which are harmful—in crowded cities where smoke makes the air impure, or in heated rooms where the ventilation is not good. First of all, care should be taken to have the air as pure as possible. Out-of-door air is always better than that which has grown heavy and dead within a room, even when the former is damp or perhaps dusty. In the second place, deep breathing should be practiced. If as much air as possible is drawn into the lungs, and is then completely exhaled, the whole circulatory system is benefited. Five minutes of deep breathing each day will be found helpful, for such exercises warm up the body and exercise certain accessory breathing muscles. The air so breathed in should be fresh air, however. No matter how cold a room may be, such exercises should never be taken with the windows closed.

Bad Breath. This most troublesome condition may arise from any one of various causes —either from decayed teeth, catarrh, adenoids or bad tonsils, diseases of the stomach, or constipation. In any case breath perfumes or even mouth washes do little good. If it is the teeth that are at fault, a dentist should be consulted; if any of the other organs, a physician should be visited. Occasionally, where the case is not extreme or of long standing, a simple laxative and a gargle, as listerine or hydrogen peroxide, will be effective. W.A.E.

Related Subjects. All of the articles in the following list will be found to contain information related to some phase of the question of breath or breathing:

Adenoids	Lungs
Catarrh	Pulmotor
Circulation	Stomach
Drowning	Teeth

BRECK'INRIDGE, JOHN CABELL (1821-1875), an American soldier, statesman and Vice-President of the United States during Buchanan's term as President. He was born near Lexington, Ky., and was a descendant of a long line of notable men. After graduating at Center College, Ky., he practiced law for a time, then served in the Mexican War. On his return he was elected to the state legislature, then to Congress in 1851 and 1853. Near the close of his term as Vice-President, in 1860, he was nominated for President by the extreme Southern Democrats, but was defeated by Lincoln. He was elected to the United States Senate in 1861, but resigned to enter the Confederate army, serving prominently as a major-general throughout the war, particularly in the battles of Shiloh, Murfreesboro, Chickamauga, Chattanooga and Cold Harbor. In Jefferson Davis's Cabinet he was Secretary of War from January to April, 1865. After the downfall of the Confederacy he went to Europe, returning to Kentucky in 1868 to practice law.

BREEDING, the science of improving breeds of domestic animals and varieties of plants, by constant attention to their pairing, in case of animals, and by cross-fertilization in plants. Breeding has for its purpose the development of some particular quality or characteristic, as the production of milk in the dairy cow, of speed in the race horse, of strength in the draft horse, of quality of wool in the sheep, or to increase the size or richness of all vegetable growths designed for food or for ornament.

Plant Breeding. New varieties of plants are produced by fertilizing the ovules of one variety with the pollen of another variety. Plants grown from the seeds thus obtained are known as *hybrids.* It is reasonable to suppose that the hybrid will possess some of the characteristics of the two varieties from whose cross-fertilization the seeds were obtained, but the plant breeder cannot foretell what the characteristics of the hybrid will be, and he is often disappointed in his results. When the hybrid fails to reach the desired standard it is crossed with some other variety, and this process is repeated until a plant possessing the desired qualities is secured. From this we see that it may require several years to produce a variety of corn, wheat or barley especially adapted to a given climate and soil, or to the production of a given product, as a wheat that contains enough protein to make macaroni.

Much of plant breeding, however, has for its purpose the improvement of one particular strain of wheat, corn or other plant. This is accomplished by selecting the most perfect plants and planting the seed obtained from them, and continuing the process for several seasons, until a seed is obtained that assures a crop in the desired abundance and quality or that will grow successfully in a given climate, such as the varieties of spring wheat grown in Minnesota, North Dakota and Canada. See CROSS-FERTILIZATION.

The man who has possibly attained the most conspicuous success in plant development is Luther Burbank. His experiments and some of the most startling results secured are related in the article bearing his name.

Breeding Domestic Animals. The numerous varieties or "breeds" of cattle, horses, sheep and swine are proof of the wonderful changes that can be produced by careful breeding. Every successful breeder must first form distinct ideas as to the results he wishes to produce and then work carefully and persistently towards their realization. The most successful breeders seek to improve their animals only in one quality; one cannot develop a successful dairy cow and a beef cow in the same animal. The most important principle in breeding is that *like produces like,* on the principle of heredity; therefore the breeder selects his animals with the greatest care, giving more attention to the selection of the male, because his characteristics are more prominent in the offspring. The old saying, "Blood will tell," is especially applicable to stock breeding. Characteristics that have been developed through a long line of ancestors are more permanent than those of recent origin. Because of this the successful breeder places a high value on pedigree, or ancestry, and those unacquainted with the facts usually consider the high prices paid for blood-animals a foolish expenditure of money. For the breeder, however, such an outlay of money is a wise investment.

The highest degree of perfection in domestic animals is reached only under the most favorable conditions of climate, food and shelter, and when animals that have been raised under these conditions are removed where such influences no longer exist, they usually fail to produce the results expected. Breeding domestic animals is both a science and an art. It requires first, a love for the work, and secondly, years of training and experience on the part of those who would engage in it.

The same general principles which apply to the breeding of domestic animals may be applied more or less fully to the improvement of the human race. See EUGENICS. W.F.R.

BREMEN, *brem'en,* the most important port of Germany, next to Hamburg. It is the capital of the free state of Bremen, and is situated on both banks of the River Weser, forty-six miles from the North Sea. The older portion of the town, forming the business section, has narrow and crooked streets, but the modern part has fine boulevards, gardens and handsome residences. Bremen is the commercial center of Northwest Germany and has four harbors capable of accommodating the largest vessels afloat. The exports consist chiefly of woolen goods, linens, glass, hemp, hides, oil, wooden toys and other manufactured goods. More than eighty per cent of the imports consist of foodstuffs. All public utilities, in-

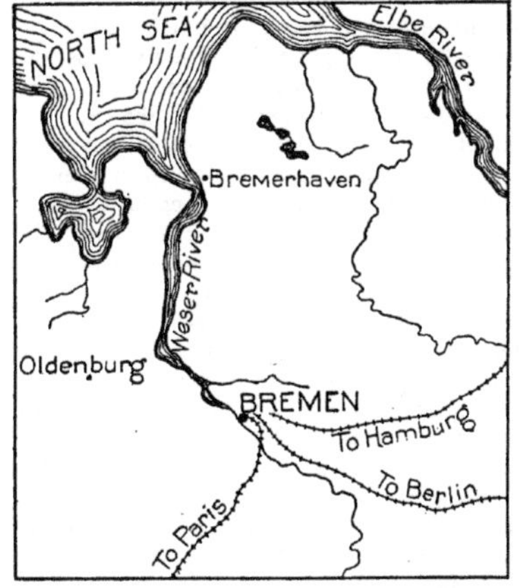

LOCATION OF BREMEN

cluding gas, electric light and water works, are municipally owned and are profitably conducted. Previous to the War of the Nations more immigrants to the United States and Canada sailed from Bremen than from any other European port. In June, 1916, the first merchant submarine in the history of the world, the *Deutschland,* sailed from Bremen and arrived in Baltimore, Md. By the employment of such vessels the Germans hoped to break the blockade established against them by the allied powers.

The schools of Bremen are excellent; the city spends twice as much money annually on education as it does on its police force. Population in 1910, 246,827.

BREMERHAVEN, *brem' er hay v'n,* an important seaport of Germany, in the state of Bremen on the estuary of the Weser River, thirty-eight miles north of the city of Bremen. It has a good harbor and here are situated the drydocks and repair shops of the North German Lloyd Steamship Company. During the War of the Nations the harbor was the headquarters of the German submarine fleet. Vessels bound for the port of Bremen sometimes wait at Bremerhaven until the tide is sufficiently high for them to proceed up the river. The chief industry is shipbuilding and the manufacture of articles needed in that industry. It is connected with Geestemünde by a drawbridge over the River Geeste, the two towns forming practically one municipality. Population in 1910, 24,165.

In September, 1916, a German submarine vessel, fully war-equipped, sailed from Bremerhaven to the shores of the United States, a distance of 3,400 miles, eluding hundreds of English and French traps set for vessels of its class.

BRESLAU, *bres' lou,* capital of the province of Silesia, in Germany, beautifully situated on both banks of the Oder River, 350 miles from its mouth and 202 miles southeast of Berlin. The city is divided into an old and a new portion, the old retaining the characteristics of a medieval city, the new being a modern manufacturing town. The cathedral, built in the twelfth century, is an imposing structure containing very valuable art treasures and an altar of beaten silver. The flourishing university occupies a fine Gothic building overlooking the river; it has a library of 400,000 volumes and an observatory. The trade of the city is very extensive, as it is situated close to large iron and coal fields and is a forwarding and receiving depot for the products of the surrounding territory. The manufactures consist of machinery, tools, railway cars, gold and silver work, carpets, furs, cloth, glass and china. It is also an important military center, because of its location near the Polish frontier, and has a large permanent garrison. Its ancient ramparts have, however, been converted into promenades. Population in 1911, 512,105.

BREST, an important seaport and one of the chief naval stations of France, in the department of Finistère, 389 miles west of Paris. It has a fine harbor, protected by powerful forts which overlook the narrow entrance. The city is ancient, and many of its streets are narrow, crooked and dirty; in some parts they are so steep that the second story of a house may be on a level with the ground floor of the one next above it. Steps are cut in the street to facilitate ascent and descent. Although extensive commerce is carried on, the city derives its importance chiefly from the naval docks, and the principal industries are connected with the manufacture of naval supplies. In the War of the Nations the city was the principal port in France for the use of the Americans. In ancient times Brest was the most important town of Brittany (which see), and it changed hands many times, sometimes being held by the English, sometimes by the French. There was an old saying to the effect that "He is not lord of Brittany who is not lord of Brest." Population in 1911, 90,540.

BRETON, *bre toN'*, JULES ADOLPH (1827-1906), a French painter of peasant life, born at Courrières. He began as a painter of historical subjects, but soon discovered that his genius lay in depicting the life of the peasants among whom he was born. His works are characterized by tender feeling, but they lack that strength and power which mark Millet's work. Among Breton's principal paintings are *Blessing the Grain,* now in the Luxembourg, in which the rendering of sunlight is so admirable; *Return of the Gleaners,* his most celebrated work; *Erecting a Calvary;* and *Song of the Lark;* the original of this last is one of the most popular pictures in the Chicago Art Institute. Breton also wrote both poetry and prose. Among his literary works are *Jeanne, The Life of an Artist, A Peasant Painter* and *The Fields and the Sea.* The *Song of the Lark,* one of his finest paintings, is faithfully reproduced here.

BREVET, *bre vet'*, a military title conferred on a commissioned officer who receives honorary rank higher than that which he holds in his regiment. It does not carry with it any increase of pay and does not entitle the recipient to seniority over officers of equal rank except when in the field. In the United States army officers are addressed by the title of their brevet rank. In the British army, officially both titles are used. For instance, a captain receiving a brevet majority would be styled "Captain and Brevet Major ———."

BREVIARY, *bre' vi a ri,* a book containing the prayers, mostly Psalms, which in all sacred Orders of the Roman Catholic Church must recite daily, unless excused by Papal permission. The breviary is printed in four volumes, one for each season of the year. The *Missal* contains the various prayers which must be read or sung by the celebrant in the celebration of mass. The *Ritual* contains the prayers and various blessings used in funerals, baptisms and marriages.

BREWING, *broo'ing,* the process of making from barley and other cereals such beverages as beer and ale, which are fermented but not distilled. In popular usage the term *brewing* includes all the steps in the process, but technically this is untrue. The manufacturers distinguish between *malting* and *brewing,* which are really two separate processes. In order to make the explanation as simple as possible, the distinction will be kept in this article.

Malting. The first step is to generate the ferment, *diastase,* which will change the starch in the barley or other grain to maltose and dextrin. This is done by malting, or causing the grain to germinate. It is first steeped in water for forty-eight to seventy-six hours, the water being drawn off and renewed at least once a day during this period. When the grain has absorbed enough water to soften it, it is taken from the steeping tanks and piled in a nicely leveled heap about two feet deep on the "germinating floor." This is called *couching.* The grain then begins to sprout, and must be carefully watched. As soon as the rootlets appear it is *floored;* that is, spread out to a depth of ten inches or less. The sprouting must be regular, not too fast or too slow; the temperature must be varied for different grains, and germination must be checked when the sprouts are two-thirds the length of the grain. The checking is done by heating the grain in a dry kiln. The temperature is raised by stages to 150° F. for light beer and to 220° F. for dark beer. The malt is then left dry and crisp.

Brewing. When the malt is thoroughly dry it is ready for brewing, and this is the second process. It is crushed (*bruised* is the technical term) between iron rollers, and is then mixed with warm water, forming a *mash,* which in the German process has the consistency of a breakfast porridge, but in the English is much thinner. Both processes are used in America. The mixture is placed in mash tubs, and sufficient hot water added to bring the mixture to the desired temperature. *Diastase* is most active at 145° F. and for some time the mash is kept at about that temperature. Eventually, however, it is further heated, reaching 158° in the English and 167° in the German process. During the heating the mash is stirred by a mechanical device. At last the liquid is drawn off from the grain, or grist. The liquor in this stage is called the *wort.*

The next step is to boil the wort with hops, from one to twelve pounds of hops being used

BRETON loved the fields, and in the peasants and their work there found his happiest subjects. The *Song of the Lark*, one of his best-loved paintings, is typical of the finest qualities of his genius. His peasant girl is not beautiful, but she is strong and thoroughly alive to the joy and beauty all about her. The work-day before her will be long, for the sun is just pushing above the horizon; and the labor of gleaning is by no means easy. But that has not deadened her to the fact that there is a shimmer of dew on the fields, a rosy glow in the sky, and above all, that there is somewhere in the air above her a singing lark.

In the intensity of her interest she stands still for a moment, but not as one of Millet's work-worn peasants would stand. There is an uplifted look about the whole figure and face. One heel is raised from the ground; the hand that holds the sickle is slightly lifted and clenched; the shoulders are thrown back, and the lips are parted as if to echo the exquisite notes of the morning songster. The observer finds himself holding his breath that he, too, may catch the sounds that have brightened this peasant girl's day.

L. J. B.

to 100 gallons of wort. The boiling lasts from one to six hours. The boiled wort may be cooled in a number of different ways—by running through pipes immersed in cold water, or by trickling over pipes through which cold water or liquefied carbon dioxide is flowing, or by using a surface cooler. The last method involves the use of a shallow vessel which allows a large surface to cool rapidly from exposure to the air.

From the cooler the wort is run into fermenting vats or tubs; about five pounds of yeast are used for every hundred gallons of wort. There are two varieties of yeast used, one of which is most active at the ordinary temperature (60° to 68° F.), the other at a lower temperature (43° to 46° F.). Both convert the malt sugar and dextrin into alcohol and carbon dioxide. In the case of the high-temperature yeast, the escaping gas carries the yeast to the surface of the liquid, producing what is called *top fermentation.* The low-temperature yeast remains at the bottom, giving bottom fermentation. Bottom fermentation is used for lager beer, top fermentation for ale, porter and stout. After several days the fermented liquid is run into settling vats, where any remnants of the yeast rise to the surface in a scum. The beer is then ready to be drawn off and stored in casks or barrels until it matures.

Chemistry of Brewing. Malting and brewing are highly technical subjects, demanding an expert knowledge of chemistry and microbiology. The success of every step in the process, from crushing the grain to storing the "green" beer, depends on many factors. The length of time devoted to every process, the temperature of the mixture and the kind of water used all affect the beer. The amount of alcohol is dependent almost entirely on the temperature in making the wort, a high temperature yielding little alcohol. If the temperature is not kept right while the wort is being made or while fermentation is taking place, the entire brew will be spoiled. The water must be analyzed, and if necessary suitable salts must be added to it. Thus water for mild ale must be rich in sodium chloride (common salt) and that for pale ale must contain an adequate quantity of calcium and magnesium salts. The water used in malting must be moderately hard.

In making beer and other malt liquors the starch in the grain is changed into sugars and dextrin during the mashing. The hops clarify, preserve and flavor the liquor. Fermentation decomposes the sugars, such as *maltose,* and some of the dextrins, into alcohol and carbonic-acid gas. It is estimated that 100 parts of sugar when fermented yield fifty parts of alcohol and forty-seven parts of carbonic-acid gas, besides a little glycerine and other products. In many modern breweries the carbonic-acid gas is collected and converted into a liquid by cooling and compression. In the brewery itself this carbon dioxide is used in the cooling pipes (see above), in charging the finished beer with gas and, in place of pumps, to force the beer from the storage cellar into the bottling room. J.F.S.

Related Subjects. In the following articles will be found much information of interest in connection with this topic:

Alcoholic Drinks	Hop
Ale	Malt
Beer	Yeast
Fermentation	

BREWSTER, *broo'ster,* WILLIAM (1560-1644), one of the best-known of the early American colonists, the leader of the Pilgrims who came to America in 1620 in the *Mayflower.* In his home in Scrooby, England, the Dissenters from the Established Church were accustomed to meet for worship each Sunday, and he was one of the company who went to Holland, in 1608, to escape persecution. Until the emigration to America in 1620 he supported himself at Leyden by teaching and book publishing. For twenty-four years a leader of the Plymouth colonists, he helped to make the colony prosperous by his energy and cheerfulness. He was generally known as Elder Brewster, and for several years was the only preacher among these colonists.

BRIAND, *bre aNd,* ARISTIDE (1863-), a French statesman, the first Socialist to become Prime Minister of France. Briand was trained for the law, but even as a young man devoted all his time to journalism and politics. Entering the Chamber of Deputies as a Socialist Radical in 1902, he won prominence by his keen powers of analysis and exposition of complex subjects. He was chairman of the committee which drafted the bill separating Church and State, and in 1906, as Minister of Public Instruction and Worship, with great tact put the law into effect.

In 1909 Briand became Prime Minister. Though he was a professed Socialist, under the responsibilities of office he became more and more conservative, and was finally expelled from the Socialist party. The chief event of his Ministry was the great railway strike of

1910, which he broke by the unusual course of calling the strikers to the colors for military service. They were ordered to protect the railroads, and in fact became their own strike-breakers. Briand resigned in 1911, but was again Prime Minister in 1913 for a short time. In August, 1914, he was appointed Minister of Justice in the Viviani ministry, which fell in October, 1915, as a result of the diplomatic failures of the Allies in the Balkans (see WAR OF THE NATIONS). Briand once again became Prime Minister, a position in which he soon showed himself an abler statesman, a greater leader than ever before, and to a remarkable degree he won the confidence of the whole French people.

BRIBERY, *bribe' er i,* the giving or offering of something of value to one who in return violates his duty or the law in order to benefit the giver. In the eyes of the law the giver and receiver of a bribe are equally guilty. The gift, to constitute a bribe, need not consist of money, but may take the form of property, or position, or undue influence. Large fines and sometimes long terms of imprisonment are the punishments inflicted on those convicted.

BRICK AND BRICK-LAYING. The story of brick carries one back to days of the greatest antiquity. The most familiar early reference to brick-making relates to the struggles of the captive Israelites in Egypt, over 2,000 years

BRICK-MAKING IN ANCIENT EGYPT

before the birth of Christ. According to Biblical history this was their principal employment. There was a plentiful supply of clay and sand on the banks of the Nile, also water with which to mix them and the intense heat of the sun to bake the bricks. When the captives clamored for straw, it was not for burning the bricks but to chop up and use as a binding material in the same way that horsehair is used in modern plaster.

Brick was probably the first material used for buildings of a permanent nature. The art of brick-making was well known in Babylon over 6,000 years ago and in that country there were neither trees for lumber nor quarries for stone. The Chinese employed bricks for building many centuries before the Christian Era. The Romans introduced the industry into Britain and other conquered territories. At the present time bricks exist in England stamped with the initials of Roman brick-makers who lived many centuries ago. The first brick building in America was built in 1633 on Manhattan Island with material imported from Holland. At the present time wherever there is suitable clay and building is to be done there springs up a brick-making industry.

Brick-Making. The first necessity in the making of good bricks is a clay free from fossil remains and containing little iron or lime. If sand is not already present in the clay in the proportion of one part of sand to two of clay, sufficient sand must be added to secure these proportions. The clay and sand are first mixed into a pliable mass by the addition of water. From this mass the bricks may be molded by hand or they may be molded and cut by machinery. There are now in general use machines which will make over one hundred thousand bricks a day. From the trough in which the sand and clay are mixed the material is forced through tunnel-like openings the size of the required brick. As the column of clay comes from the machine, like meat from a mincing machine, it is cut into proper lengths by wires.

The pieces drop on the endless belts on which they are carried to drying sheds. After the bricks are dried they must be hardened by burning. This is done in kilns in which the bricks are stacked so that the heat of a fire may readily penetrate to all of them. Kilns are of various sizes, the average being about thirty feet in diameter and twelve feet in height. The firing takes from six to ten days. Bricks for ordinary building are kept at a cherry-red heat. Others for finer purposes are raised almost to white heat.

Varieties of Brick. In addition to those used for building purposes, *pavement* bricks are made in great quantities. To make these hard enough to withstand the wear of heavy traffic, lime is added to the clay and sand. During the burning the lime fuses and renders the bricks extremely firm and durable. The introduction of steel frames into buildings has greatly increased the use of bricks. In parts that are not seen, where strength only is needed and finish is not a matter of importance, a brick of somewhat rough appearance is used. *Facing bricks,* or those which occupy prominent, exposed positions, are more elaborately finished, more uniform in color and sometimes glazed by a special process. Bricks are also used for foundations, for linings of sewers, tunnels, chimneys, cisterns and for numerous other purposes. Tiles and pipes baked in the same way as bricks are extensively used for drainage purposes. Enameled tiles are also used for decorative purposes, for floors, for fireplaces and for fancy wall linings and wainscots.

Brick-Making Industry. The great centers of this industry in the United States are along the Hudson River from Troy to New York City, in Philadelphia County, Pa., and in Cook County, Ill. Over 30,000,000,000 bricks are annually produced in these districts, with a value of about one hundred million dollars. Throughout the United States and in Canada every town of importance has brick-making plants and the output is steadily increasing.

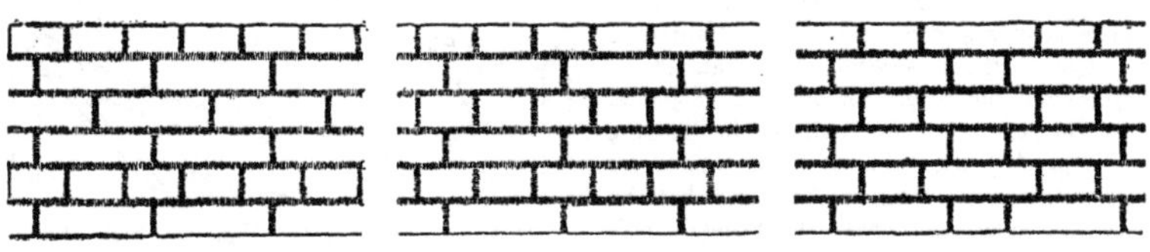

HOW BRICKS ARE LAID

The three styles pictured are, from left to right, American, English and Flemish bonds. If brick walls were built entirely of *stretchers*, or bricks whose long sides appear on the outer edge, they would not be strong. If the inner and outer rows are bonded with *headers*, the weight upon any point is borne partly by each row.

Brick-laying. Bricks are laid in horizontal rows called *courses* and held together by a lime mortar. The latter also gives elasticity to a wall and makes it dryer. For the sake of strength the weight of each brick should be borne by more than one in the course below; therefore no vertical joint is permitted directly over another. On the other hand, a pleasing appearance is gained by having every joint in line with one in the second course below it. Bricks in America are usually eight inches long, four inches wide and two inches thick; in England they are slightly larger. The ancient Roman brick, sometimes copied to-day for architectural reasons, were twelve inches long and less than two inches thick.

Brick walls are of several sorts—solid, hollow, veneer and curtain. A solid wall is from one to three feet thick, according to the weight it must bear. Hollow walls give coolness in summer and warmth in winter and keep out moisture, but are expensive. Brick veneer four inches thick is built around wooden buildings to improve their appearance; the veneer should be separated from the wood by an air space. Curtain walls are those in steel or concrete skeletons; they carry no weight but their own. More information about brick work will be found in the articles BUILDING; ARCH; BRIDGE.

In the United States and Canada bricklayers are formed into powerful unions which regulate the scale of pay and conditions of labor. The bricklayer who belongs to a labor union is among the best-paid of all skilled workmen and he usually works but eight hours a day. In building, each bricklayer is accompanied by a laborer whose duty it is to keep the skilled workman supplied with bricks and mortar. In large buildings the materials are hoisted from floor to floor by machinery and carried to the bricklayer in what is called a "hod," a scuttle-shaped wooden box carried on a pole over the shoulder. F.ST.A.

BRIDAL WREATH, *bride'al reeth,* a small, graceful, flowering shrub, common in gardens or on lawns in north temperate regions. The branches are slender and curving, the leaves smooth, small and oblong, sometimes with cut edges. In spring, especially, there is a great profusion of small rounded clusters of dainty white flowers. The willowy blossoming branches make very attractive decoration. The various shrubs belonging to the bridal wreath family, the *spirea,* are adapted to many soils, from swamps to hillsides. Some thrive in the bright sunshine, while others grow better in the shade.

BRIDGE, *brij,* a structure to carry a road, railway, or waterway across water or over a valley. Bridge-building is, perhaps, the oldest branch of engineering, though until the invention of the railway made necessary more bridges, and longer and stronger bridges, the science had advanced but little since the days of the ancient Romans. Nature built the first bridges—logs fallen over streams, giant grapevines growing across ravines, or arches of rock carved by the wind and rain. (See the article NATURAL BRIDGE for pictures.) Timber foot-crossings were probably the first made by man, though possibly suspension bridges of the grape-vine type, such as those constructed in modern times by the Indians of British Columbia and other primitive peoples, are even older.

The Romans were the first great bridge builders. Besides braced timber structures like that defended by Horatius, they constructed magnificent arches of stone, some of which remain to this day (see AQUEDUCT). After the fall of Rome little progress was made until the era of iron and steel, though the length of both timber and arch bridges was increased.

Timber Bridges. The simplest form of timber bridge, a log or plank with one end on each bank of a stream, is of course limited in span (the distance between the points of its support) by the length of the log or plank. Moreover, everyone has noticed, in crossing such a plank bridge, that it bends more and more as the center is approached, and that the longer the span the weaker the bridge. This is because each point of support acts as the fulcrum of a lever (which see). It is obvious, then, that to lengthen or to strengthen a wooden bridge it is necessary to have additional points of support. But the stream or valley may be too deep for a pier, or the current too swift for one to be built with ease. In such a case added support must be a part of the structure itself.

In an ordinary wagon bridge such as that represented in Fig. 1, the strength, of course, depends not on the planks but upon the framework which supports them. If the stream to be crossed is seven yards wide, a frame of tim-

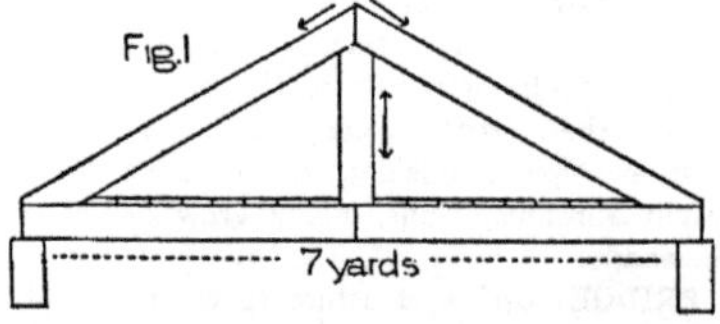

bers unsupported except at the ends would perhaps break under heavy loads. In the simple truss shown in Fig. 1, any weight upon the bridge exerts a downward pull upon the post at the center, which in turn distributes the load between the two slanting supports in an outward thrust upon the foundations. The post is said to be subject to *tension,* or stretching, and the slanting supports to *compression,* or squeezing. It will be shown later that the horizontal timbers are subject to both tension and compression. In Fig. 1 the span of seven yards has been safely bridged with timbers none of which is over four yards in length. The exact strain which each one will be called upon to bear under a given load

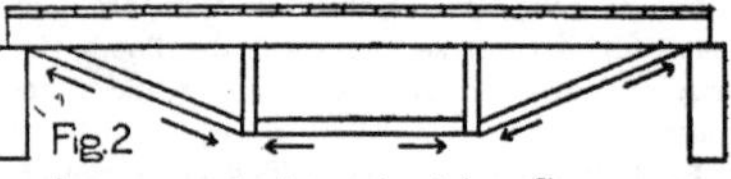

can be accurately determined (see COMPOSITION OF FORCES).

In many bridges the truss is underneath the roadway, as in Fig. 2. Here the upright posts are subject to compression and the horizontal and two diagonal braces to tension. That this

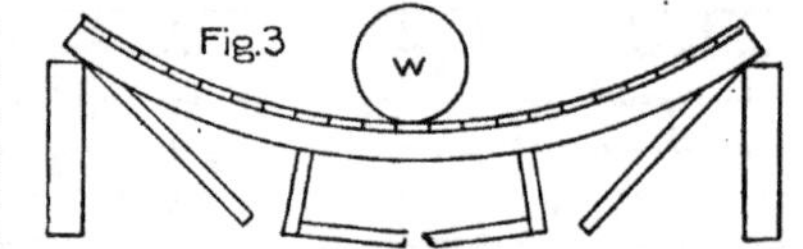

is so can be seen from Fig. 3, which shows that if a weight is placed on the bridge heavy enough to bend it, the distance between the points connected by the truss timbers is increased.

By the construction of more elaborate trusses, timber bridges may be greatly extended in length. One erected at Wittingen, Germany,

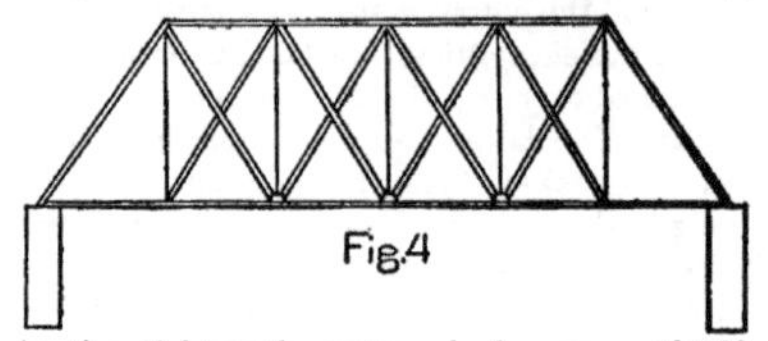

in the eighteenth century had a span of 390 feet, probably the longest in any wooden bridge. A truss frequently seen on timber bridges is the Howe truss, shown in Fig. 4, in which the diagonals are of timber and the vertical tie-rods (so called because they tie together the parts which weight tends to pull apart) are of iron.

Wooden bridges have the great disadvantage of being rapidly worn by weather and easily burned. They are serviceable only where those made of other materials would be at the time too expensive. Railroad *trestles,* bridges of many small spans which cross valleys, are commonly built of wood, and seldom resist decay for more than fifteen years.

Arch Bridges. A century and a half ago the only bridge materials other than wood were stone and brick. They were employed in the arch, which had been a favorite form

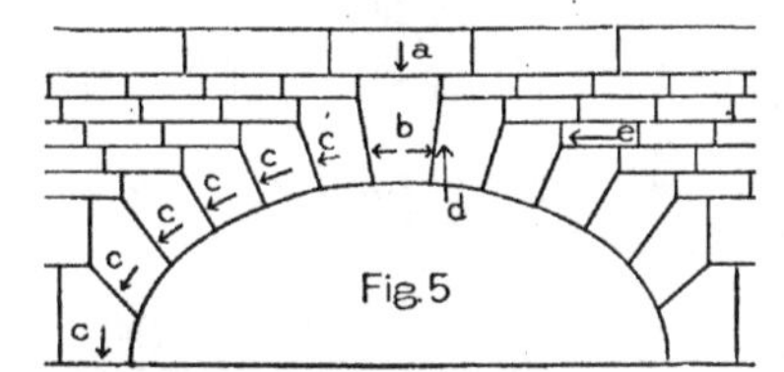

from very ancient times. Figs. 5 and 6 illustrate the points of weakness and strength in this type of bridge. The weight of the *keystone* (the central stone crowning the arch) exerts a downward pull represented in Fig. 5 by the arrow *a*. But in effect this pull is resolved into thrusts against each of the next stones, called *voussoirs* (see COMPOSITION OF FORCES). The voussoirs in turn thrust against their neighbors until in the *skewback*, the lowest stone, the force is exerted directly downward (see ARCH). It is a law of physics that when bodies are in equilibrium there is an equal and opposite force for every force exerted by them or upon them. Thus when an arch is properly built the downward element of the thrust *b* is exactly equaled by the upward thrust *d* of the voussoir, and the outward thrust of a voussoir is counterbalanced by the inward thrust *e* of the pier. If an arch is too broad, its top too heavy or the piers too light,

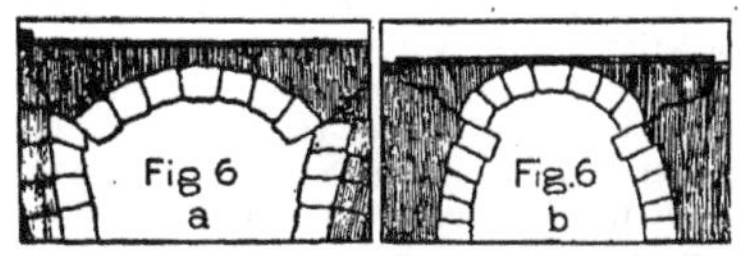

it will break, as shown in Fig. 6*a*, because of the excessive outward thrust. If, on the other hand, an arch is too light at the top, or too heavy at the sides, the upward thrust will be too powerful, and the result will be as illustrated in the Fig. 6*b*.

The Romans did not build arches of more than one hundred feet span, but modern engineers have learned to construct them nearly three times as long with stones, and even longer with concrete. A stone arch bridge at Plauen, Germany, has a span of 295.3 feet, and a concrete bridge in Rome, Italy, a span of 328 feet. The principles of the arch bridge combined with the truss are seen in some iron bridges, instances of which are the cast-steel Eads Bridge (which see) over the Mississippi at Saint Louis, Mo. (520 feet span), the beautiful wrought-steel bridge over the upper Niagara between the United States and Canada (840 feet span), and the new Hell Gate Bridge at New York, with the remarkable span of 1,016 feet (see HELL GATE, for illustration).

Perhaps the most usual reason for the choice of the arch form of construction is its beauty, which is well illustrated in the accompanying picture of the Tunkhannock viaduct, and in

TUNKHANNOCK VIADUCT

A great concrete and steel structure, on the Lackawanna Railroad, in New York. It is a half mile in length, is 240 feet high, and cost $12,000,000. This vast expenditure was undertaken to shorten the running distance between New York City and Buffalo three and one-half miles and reduce heavy grades.

the picture of the Los Angeles viaduct which will be found with the article CONCRETE.

Suspension Bridges. A short time before the year 1800 a form of bridge hung on chains or wire cables was found to be strong enough for long spans. The suspension bridge has the merit of being less expensive than other long bridges, and can be made an object of great beauty, as the picture of the Manhattan Bridge shows. But its great fault is its flexibility; it sways with every wind and an army marching over it with uniform step would cause it to swing so violently that it would collapse. The prominent steel work at the roadway of the Manhattan Bridge is merely stiffening and wind-bracing; the whole support of the structure comes from the four steel cables that pass over the high towers. These cables were spun by drawing a steel wire back and forth across the river, over the towers; each cable has a diameter of about 21¼ inches, is composed of 2,368 wires, and is fastened to anchorages at both ends. The longest suspension bridge is the Williamsburgh Bridge which crosses the East River of New York with a span of 1,600 feet. See also, BROOKLYN BRIDGE.

MANHATTAN BRIDGE. NEW YORK CITY
One of the newer of the suspension bridges connecting Manhattan Island with the Borough of Brooklyn and New York suburbs on Long Island.

Iron Beam and Truss Bridges. The iron beam bridge is of course an evolution of the simple log bridge. Fig. 7, *a,* shows what takes place when a wooden beam bends beneath weight. The upper fibers are crushed together, the lower ones pulled apart; that is, the beam is subject to both compression and tension. The intensity of the forces is least near the

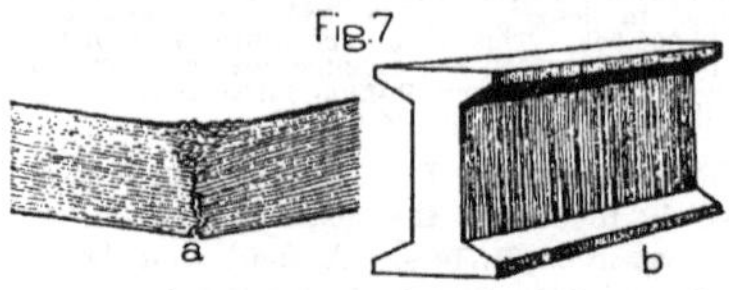

center and greatest at the edges, hence the usual steel beam, shown in Fig. 7, *b,* is made stronger at top and bottom. From its shape it is called an I-beam. Steel beam bridges are suitable for railway spans of not more than twenty-five feet, and for roadway spans of less than forty feet.

The iron truss bridge is constructed on the same principles as the wooden truss, but with much greater variety of design to suit its particular purpose. In the Saint Louis Municipal Bridge a simple truss has a span of 668 feet, and the central section of the new Quebec Bridge (Fig. 8) is a simple truss of 640 feet length.

Cantilever Bridges. A bracket fastened on the wall to hold a shelf is a cantilever. In the forms of bridge already described it is plain that if any section were removed the structure would collapse. If in a cantilever bridge, however, the central section were removed, as shown in Fig. 8, the remainder of the bridge would be unharmed. At each of the two piers of the Quebec Bridge an independent truss structure was erected, which is in perfect balance at all times, though the cantilever arm (*a*) extends outward 580 feet from the pier and the anchor arm (*b*) 515 feet. In most cantilever bridges the central part (c) is built from the cantilever arms, and to offset its weight, which might tip the whole strueture inward at the center, the anchor arms are fastened to the anchor piers (*d*). For the Quebec Bridge, however, the engineers planned to lift the 5,000-ton center section into place from six scows on which it had been floated to the bridge site. This was attempted on September 11, 1916, but through defects in the hoisting gear the heavy load fell into the river. The Quebec Bridge has the longest span (1,800 feet) of any bridge in the world. An

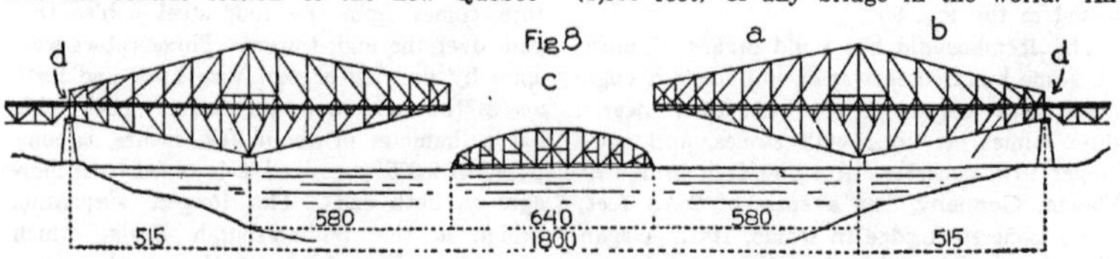

THE GREAT CANTILEVER BRIDGE AT QUEBEC
It was the central portion of the 1,800-foot span, itself 640 feet in length, which collapsed in 1916.

earlier one on the same site collapsed while under construction in 1907, presumably because the under side of the cantilever arm was too weak and it *buckled,* that is, it bent under the enormous compression of the upper half.

The great advantage of the cantilever bridge is that it can be erected without temporary supports, which might block the channel, or which, as in the case of the huge bridge over the Firth of Forth in Scotland, could not be constructed at all because of the depth of the channel and the swiftness of the current. While this is also true of the suspension bridge, the latter, on account of its tendency to sway, is not strong enough for heavy railroad traffic.

OVER THE FIRTH OF FORTH
In this graceful cantilever bridge are two spans of 1,710 feet. From 1889 until the completion of the Quebec Bridge these were the longest spans in any structure in the world.

Movable Bridges. Where river bridges cannot have the roadway high enough for boats to pass beneath them it is frequently necessary to make them movable. The most usual type of this class of bridge is the swinging drawbridge, which turns upon a vertical axis, like a merry-go-round. Sometimes, however, the central pier necessary for a swinging bridge would occupy space in the stream that is needed for ships. In this case a *lift* bridge which can be raised like an elevator, a *bascule* bridge opening like a jack-knife, or a ferry bridge consisting of a car traveling back and forth suspended from a track high overhead may be suitable. In both the lift and the bascule bridges massive counterweights move down when the bridge goes up so that, like a window and its weights, the whole structure is always balanced and requires very little power to move it. In a bascule bridge the circular swing of the two lever-like halves makes it very difficult to maintain perfect balance in all positions. Thus it will be seen from the accompanying picture that in the Canadian Pacific Railway's bridge over the Soo Canal

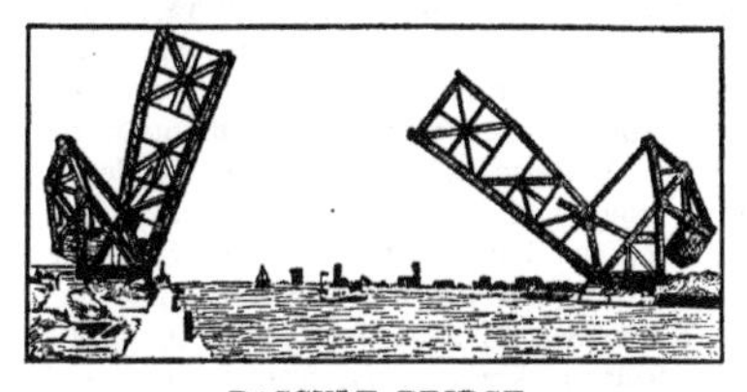

BASCULE BRIDGE
Canadian Pacific Railway Bridge over the canal at Sault Ste. Marie. When raised, it provides a navigation clearance of 279 feet. Over all, the bridge is 426 feet in length and is credited with establishing a record in this type of construction.

the counterweights swing inward as the bridge opens.

Pontoon Bridges. A bridge built on boats is called a pontoon bridge. It is usually a temporary structure erected by an army. Perhaps the most famous pontoon bridges were those by which the hordes of Xerxes crossed the Hellespont. C.H.H.

BRIDGE, a game of cards played with the full pack of fifty-two cards, and usually by four people. It is derived from whist, and was formerly called bridge whist. The game is of uncertain origin; similar games were played a century ago in Russia, Turkey and Denmark, but its popularity dates from 1894 and 1895, when it was introduced in London. Since then various new rules and styles of play have been introduced from time to time. The most important of these variations is the game of *auction bridge,* whose popularity has been steadily increasing since its introduction in 1907. It has now completely supplanted ordinary bridge in public favor. Both in regular bridge and in auction bridge the tendency at first was to adhere rigidly to certain rules, but as the games developed greater freedom was allowed to the players. Bridge, however, is still an exacting game, and no beginner can expect to master it easily. Power of coneentration and a retentive memory are required to a considerable degree, but the best equipment of the bridge player is experience. Praetice is the best teacher, but the beginner is advised also to accept advice from better players and to consult at least one of the standard books mentioned at the end of this article.

Regular Bridge. The play in bridge is similar to that in whist, except that the dealer or his partner must declare the trump. The dealer may declare any suit to be trump, or he may prefer to play without trumps. If he feels that his own hand does not permit a satisfactory declaration he may ask his partner to declare the trump. This is called the "bridge," or "bridging the declaration," and the dealer usually signifies his intention by saying, "I bridge it." In any case the dealer plays the hand without assistance from his partner; the latter, after the player to the dealer's left has led the first card, lays his cards, arranged by suits, face up on the table. The dealer makes all plays from his own and his partner's hand, the latter being known as the *dummy*. Partners always sit opposite each other, and the order of play is from left to right, the player to the left of the dealer having the first lead. Thereafter the winner of each trick leads.

Scoring. The method of scoring is somewhat complicated. The score is kept in two parts, the *game* score and the *honor* score. The game score is determined entirely by the number of tricks taken by the dealer; the honor score includes special amounts for high cards held at the beginning of play, and certain special penalties and special rewards. These two scores, though kept separately, are added together at the conclusion of play.

The *game* consists of thirty points. Points for game are credited only to the side making the declaration. This side wins a certain number of points for each trick over six, as follows: if "no trumps" is the declaration, each trick over six counts twelve points; if hearts, eight points; if diamonds, six points; if clubs, four points; if spades, two points. Every hand must be played out, and if the declarant makes more than thirty points the total is credited to his score. The winner of two games out of three wins a *rubber*. If one side wins two games in succession, the third game is not played.

The *honor* score is more complicated than in whist. If the declaration is "no trumps" the four aces are the *honors*, and for each ace held at the beginning of play the side holding it counts ten, provided that it holds at least three of the aces. If each side has two aces, the honors are *divided*, or *easy*, and are not counted. If one of the players holds the four aces in his hand, his side is credited 100 points in the honor score. If a trump has been declared, the honor cards are the ace, king, queen, jack or knave, and ten, the value of the honors being determined by the value of a trick.

Three honors, either divided or in one hand; twice the value of a trick.

Four honors, divided; four times the value of a trick.

Four honors, in one hand; eight times the value of a trick.

Five honors, three in one hand and two in the other; five times the value of a trick.

Five honors, four in one hand and one in the other; nine times the value of a trick.

Five honors, in one hand; ten times the value of a trick.

For example, if a player holds the ace, king and ten of diamonds, when diamonds are trumps, he adds twice six, or twelve, to his honor score. If hearts are trumps and he has three honors, he adds twice eight, or sixteen, to the honor score.

Slam. When one side takes all thirteen tricks it counts as a *grand slam*, for which forty points should be added to the honor score. If one side takes twelve tricks it counts as a *little slam*, for which twenty points should be added.

Rubber. The side winning the rubber adds 100 to its honor score.

Doubling. Either opponent of the dealer and his partner may *double*, which means that each trick shall have double its usual value. It signifies, of course, that he believes that he can prevent the dealer from taking more than six tricks. The dealer or his partner may *redouble*, in which case each trick counts four times its usual value. It is customary to limit the value of a trick to 100. Doubling does not affect the value of honors, grand slam or little slam.

Auction Bridge. In auction bridge, which is played almost to the total exclusion of the earlier form, the method of play is the same. The dealer, however, instead of making a declaration, is allowed to bid for that privilege, and each of the other players in turn may make a higher bid. The bidding continues until three of the players in succession have refused to bid further. The player who contracts to take the largest number of tricks of the highest value is given the privilege of playing the hand; provided, however, that as between partners who have bid on the same suit, the player who first made such a declaration shall play the hand. His partner becomes the *dummy*, as in ordinary bridge.

The value of a bid depends on the value of a trick "in trumps" or in the suit. Under the present rules the tricks count as follows: if clubs are trumps, each trick over six counts six; if diamonds, seven; if hearts, eight; if spades, nine; if no trumps, ten. In the bidding, therefore, a bid of "one heart" is higher than one diamond, but a bid of "two diamonds" (fourteen points) is higher than "one heart" (eight points). Any bid for a higher number of tricks takes precedence, even though the value of the bid is the same; for example, "four clubs" (twenty-four) is better than "three hearts" (also twenty-four). A bid of one heart signifies that the bidder, if the bid is not raised, makes hearts trumps and undertakes to win one trick more than six. If he is successful he scores not merely what he bid but all he makes; for example, if he bids one heart and takes ten tricks, or four more than six, he scores four times eight, or thirty-two. If he fails to fulfil his contract the opposing side adds fifty points to its honor score for each trick which the declarant loses. For example, if the declarant bids "two hearts" he undertakes two more than six, or eight, tricks; if he succeeds in taking only six tricks in all, he has taken two less than his contract, and the opponents add twice fifty or 100 points to their honor score. The additions to the game score are made only when the side which has declared the trump has fulfilled its contract.

Honors. The rule for honors in auction bridge is the same as given above for ordinary bridge. The same cards are counted for honors, but as the value of the tricks is different, the value of the honors is changed. A *grand slam* in auction bridge adds 100 to the honor score, and a *little slam* adds 50. The winning side in a rubber adds 250 to its honor score. Doubling, as in ordinary bridge, does not affeet the value of honors or either slam.

Other Varieties. Bridge, either regular or auction, may also be played by two or three people. In every case, however, four hands are dealt. In two-handed bridge, or double-dummy, each player plays two hands, his own and his dummy's. In three-handed bridge the players alternate in playing with the dummy, but in three-handed auction they bid for the privilege, and the two unsuccessful bidders become partners for that deal. Points are scored in the same way as in the other forms, but in three-handed bridge, each player has an individual score. W.F.Z.

Consult *Elwell on Auction Bridge; Elwell on Bridge;* Foster's *Complete Bridge;* Dalton's *Complete Bridge;* Dalton's *Auction Bridge.*

BRIDGE OF SIGHS, a beautiful bridge in Venice spanning the canal between the Doge's Palace and the state prison, so named because prisoners passed over it from the hall of judgment to the place of execution. It was built in the closing years of the sixteenth century by Antonio Contino, who was also the builder

I stood in Venice, on the Bridge of Sighs;
A palace and a prison on each hand.
—BYRON: *Childe Harold.*

of the famous Rialto Bridge and was therefore known as "Antonio of the Bridge." The Bridge of Sighs, about whose name much sentiment has circled, is a lofty structure, arched at the top and closed at the sides. It contains two separate passages; through one of these prisoners passed to the palace, and through the other were led back to the prison.

Bridge of Sighs is also applied to a covered passage-way in New York City between the Tombs and the Criminal Courts Building.

BRIDGEPORT, CONN., one of the county seats of Fairfield County, ranking among the chief manufacturing cities of the state, and the second city in size, following New Haven. Its population was 102,054 in 1910 and 118,434 in 1915. Bridgeport is located in the southwestern part of the state, on Bridgeport harbor, a small inlet of Long Island Sound, at the mouth of the Pequonnock River. New York is fifty-eight miles southwest. Excellent transportation facilities are afforded by the New York, New Haven & Hartford Railroad. There are also steamer lines to New York, Boston and cities of the Sound. The fine transportation facilities make the city the home of many people whose business interests are in New York. The area is about fourteen and one-half square miles.

Bridgeport is divided into three sections by two arms of its harbor. The city proper and business section are situated west of the harbor. The finest residential districts are Golden Hill, Fairfield Avenue, Brooklawn, Mill Hill and North Main Street. Black Rock, a suburb of the city, is a noted summer resort, its harbor affording ample protection for yachts.

Parks and Boulevards. Bridgeport is locally called *The Park City,* from the number of its recreation grounds, the most notable being Beardsley Park, Washington Park and Pembroke Park, a broad expansion of the old King's Highway, which led from Boston to New York. Seaside Park, on the shore, with a sea wall and drive two miles long, contains a soldiers' and sailors' monument and monuments erected to the memory of Elias Howe and P. T. Barnum, the industry and enterprise of these two men having been great factors in the development of the city.

Manufactures. Owing to the variety of its products, the city is sometimes styled the *Industrial Capital.* Since 1856 the name of Bridgeport has been associated with sewing machines. The works of the Singer Sewing Machine Company cover about ten acres. The Union Metallic Cartridge Company has one of the largest cartridge factories in the world, its ammunition park covering several hundred acres. The American Tube and Stamping Company, the Holmes and Edwards Silver-Plating Company, the Crane Company and the Warner Corset Company are also located here. Besides these, there are extensive manufactories of carriages, locomobiles, machinery, hardware, hats, woolen goods, aluminum, bronze and brass. There is considerable coasting trade, as the harbor is safe for small vessels.

Public Buildings and Institutions. The most notable buildings are the Federal building, the Barnum Memorial Institute, the Young Men's Christian Association, the Burroughs Home for Women, the Sterling Widows' Home, the Saint Vincent Hospital and a city hospital and the railway station.

History and Progress. The first settlement, in 1665, was known as Pequonnock. In 1694 it was renamed Fairfield Village, in 1701 it became Stratfield and in 1800 was incorporated as the borough of Bridgeport. It was incorporated as a city in 1836; Fairfield was included in 1870, and in 1899 Summerfield and West Stratford were annexed. In 1832 the first Bridgeport steamer, *The Citizen,* made its first trip. The Housatonic Railroad was constructed as far as New Milford in 1840; in 1848 the New York, New Haven & Hartford Railroad was opened, and the development of the city began. F.E.M.

BRIDGES, *brij'ez,* Robert (1844-), a scholarly English physician and poet whose name and writings were unknown to the majority of readers until 1913, when he was chosen to succeed Alfred Austin as poet laureate of England (see Poet Laureate). He distinguished himself as a student at Eton and at Oxford, and after graduation from the university traveled for several years in Europe and the Far East. Following this he studied at Saint Bartholomew's Hospital in London and on receiving his degree in medicine began life as a physician, becoming a member of the staffs of two London hospitals. In 1892 Dr. Bridges sought the seclusion of his beautiful rural estate in Berkshire, where he has since lived as a student of literature and as a writer of poetry that has always charmed the critics.

ROBERT BRIDGES
Poet Laureate of England.

The poetry of Dr. Bridges is characterized by noble serenity, high finish of style, and pure lyric beauty. These qualities make him the poet of the critic and the scholar, but he has never stirred the multitude in the manner of his popular contemporary, Rudyard Kipling, who many people thought would be named laureate instead of Bridges. He has used both the familiar English rhythms and the verse forms of the Greek and Latin poets, always with faultless technique, restraint and delicacy.

The edition of his poetical works issued by the Oxford University Press in 1913 has greatly increased the number of his readers, and made known many beautiful lyrics that anyone with a taste for poetry can enjoy. The following stanza from *A Passer-by* is a good example of Dr. Bridges' style:

Whither, O splendid ship, thy white sails crowding,
Leaning across the bosom of the urgent west,
That fearest nor sea rising, nor sky clouding,
Whither away, fair rover, and what thy quest?

Ah, soon, when winter has all our vales oppressed,
When skies are cold and misty and hail is hurling,
Wilt thou glide on the blue Pacific, or rest
In a summer haven asleep, thy white sails furling.

BRIDGET, *brij'et,* SAINT, the name of two saints of the Roman Catholic Church. The first is a patron saint of Ireland, commonly known as Saint Bride. She was a woman of unusual ability and beauty. Not wanting to marry or be troubled with suitors, she prayed to become ugly. Her prayer was answered, and she went into seclusion in a cell built under a large oak. Here the Monastery of Kildare was founded—"the church of the oak." Many wonderful stories are woven about this saint. Her feast is celebrated on the day of her death, February 1.

Another Saint Bridget, or more properly, *Brigitta* or *Birgitta,* of Sweden, is the most celebrated saint of the Northern kingdoms. She married and bore eight children, one of whom was afterwards honored as Saint Catherine of Sweden. Saint Brigitta's charitable, saintly life made her widely loved. Her feast is celebrated on October 9. D.J.D.

BRIDGE'TON, N. J., the county seat of Cumberland County, situated in the southwestern part of the state at the head of navigation on Cohansey Creek. Philadelphia is thirty-eight miles north; New York is 126 miles northeast. Railway transportation is provided by the New Jersey Central and the West Jersey and Seashore railroads, and electric lines extend east and southeast. Bridgeton is an old place, settled long before the Revolutionary War, but it was not incorporated until 1865. It is controlled by a modified form of commission government. The Federal census shows an increase in population from 14,209 in 1910 to 14,335 in 1914.

Bridgeton is located in a fertile and well-cultivated district which supplies produce for the large fruit and vegetable canning industry of the city. The manufactories produce hosiery, shirts, dresses, gaspipe, castings, carriages, wagons and boilers. Bridgeton has fine county and city government buildings, county and city hospitals, an insane asylum and a park. For higher education there are Ivy Hall Seminary, West Jersey Academy, Seven Gables Seminary and South Jersey Institute. There is a good public library. The climate and scenic beauty of the locality attract large numbers of summer visitors.

BRIDGE'WATER, a town in Lunenburg County, Nova Scotia, at the head of an estuary formed by La Have River, eighty miles southwest of Halifax. It is on the Halifax & Southwestern Railway, whose repair shops and main offices are located here. Lumbering is the chief industry of the vicinity, and Bridgewater exports each year over 50,000,000 board feet of lumber. Naturally there are large saw, planing and shingle mills and several wagon and wood-working factories. Foundries, tanneries, shipbuilding yards and granite and marble works are also important. The neighborhood is a favorite resort for trout and salmon fishing, and for hunters for duck, woodcock and grouse. Population in 1911, 2,775; in 1916, 2,900.

BRIDGMAN, *brij'man,* LAURA DEWEY (1829-1889), a remarkable blind deaf-mute, born at Hanover, N. H. At the age of two a severe illness deprived her of sight, hearing and speech, and to some extent, also, of smell and taste. Learning of her case, Dr. Samuel G. Howe, of the Perkins Institution for the Blind at Boston, undertook her education when she was eight years old. Never before had an attempt been made to teach one who had been so deprived of her senses. She made rapid progress, however, and learned to read and write, to reason and to think well. She also learned to do household work and to sew, both by hand and on the machine. After receiving her education, Miss Bridgman taught in the Perkins Institution. Through the success of this experiment many other cases were brought to Dr. Howe. See BLINDNESS.

BRIGADE, *bri gaid',* a term applied to a body of troops consisting usually of two regiments, although the term is elastic and is often applied to a unit of four or even more regiments. In the United States a brigade usually consists of three regiments. The German army is divided as follows: three battalions equal one regiment (usually 1,000 men); two regiments make a brigade, two brigades equal a division; two divisions make one army corps. A brigade may consist of cavalry, artillery or infantry. See ARMY.

Brigadier-General, a field officer who commands a brigade. In rank he is above the highest regimental officer, the colonel, and below the major-general. In the United States army the salary of this officer is $6,000; in Great Britain, $4,866; in France, $2,433; in Germany, $2,441, and in Italy, $1,900. See RANK IN ARMY AND NAVY.

BRIGAND, *brig'and.* See BANDIT.

BRIGGS, CLARE (1875-), a cartoonist and the originator of the popular series *The Days of Real Sport, Oh Skinnay!, When a Feller Needs a Friend, Friend Wife* and *Kelly Pool,* which have endeared him to everybody who likes to see the follies and foibles of life, robbed of their sting, and who believes "a smile is worth a guinea in any market." Briggs was born at Reedsburg, Wis. He first won recognition in 1896 as a newspaper artist with the Saint Louis *Globe-Democrat,* and since then his creations have become a daily feature of newspapers throughout the United States. He became nationally famous as one of the Chicago *Tribune* cartoonists, and a wide demand for his work resulted in 1915 in an arrangement whereby his cartoons should be syndicated. One of his *When a Feller Needs a Friend* series appears in these volumes in connection with articles on EDUCATION.

BRIGHT, *brite,* JOHN (1811-1889), an English orator, reformer and statesman, for nearly half a century the personification of his country's conscience. Bright was an earnest Quaker, and he made his religion a part of his daily life. There was in him something of the austerity and righteousness of the ancient Hebrew prophets. With Cobden he fought for free trade when mere suspicion that a man was a free-trader made him unwelcome among "respectable" people. He opposed England's participation in the Crimean War, though people called him traitor. During the War of Secession in America he was one of the solitary figures in England who opposed recognition of the Confederacy. His letters to Lincoln, Sumner, Greeley and other Americans were rays of hope to them, and it is known that their personal letters to him were read by Bright at meetings of the British Cabinet. Late in life he opposed Home Rule for Ireland, although his action cost him the friendship of Gladstone, his old leader, and saddened his last days.

In early life Bright gave little promise of his future greatness. He was born near Rochdale, where his father was a leading cotton manufacturer, and until his thirtieth year Bright was only a local leader. He was prosperous in business, happy in his home life and was always ready to lead helpful movements in the town. He was known locally as favoring the repeal of the Corn Laws, but it was not until 1841, after the death of his wife, that he became a national figure. Three days after her death Richard Cobden came to see him, and appealed to him in the name of his dead wife: "There are thousands of homes in England at this moment where wives, mothers and children are dying of hunger. Now, when the first paroxysm of your grief is past, I would advise you to come with me, and we will never rest until the Corn Laws are repealed." See CORN LAWS.

This invitation Bright accepted. He was elected to Parliament in the same year, and in the next two years spoke throughout England and Scotland for free trade. In 1841 Bright was "Cobden's chief ally" and "John Bright of Rochdale"; in 1843 the posters proclaimed him merely "John Bright"—this was enough. From this time until his death, except two periods caused by ill health, Bright was one of the conspicuous figures in the House of Commons. Under Gladstone he was in the Cabinet from 1868 to 1870 as President of the Board of Trade, and from 1873 to 1874 and again from 1880 to 1882 he was Chancellor of the Duchy of Lancaster. He accepted these offices, however, rather to lend the support of his name to Gladstone, and he was glad when the chance came to resign from the routine of office-holding. W.F.Z.

BRIGHTON, *bry'tun,* called the "queen of watering places," is the most fashionable seaside resort in England. It is fifty-one miles south of London, in the county of Sussex, and is so much visited by Londoners that it is often spoken of as London-by-the-Sea. It has a magnificent sandy beach and one of the finest promenades in Europe. Brighton Aquarium has a world-wide reputation, and its museum contains the world's most complete collection of British birds. The town has no manufactures and depends entirely on its visitors for its prosperity. All public utilities are municipally owned and operated at a profit. Its rise into favor dates from the time of George III, when the Prince of Wales showed great partiality for it and had a beautiful residence erected there. What were then the royal stables have been converted into a magnificent concert hall containing one of the largest organs in the world. Population in 1911, 131,250.

BRIGHT'S DISEASE, a name given to various forms of kidney disease, so named because they were first recognized and described by Dr. Richard Bright, in 1827. Acute Bright's disease is acute inflammation of the kidneys. It is frequently a complication of scarlet fever, diphtheria and other infectious diseases, but may result from exposure to wet and cold, from diseases that have interfered with the skin's function of excreting waste matter, or from the use of arsenic, carbolic acid, iodoform, lead, phosphorus, mercury and other poisonous substances.

The symptoms vary somewhat. The patient may suffer from slight headache, pain in the back and legs, loss of appetite and nausea, or there may be more serious symptoms—fever, prostration, stupor, shortness of breath and convulsions. Dropsy is a common accompaniment, and the urine usually contains albumin, casts and blood corpuscles. The patient should be kept in bed and be given saline purgatives daily. Sweating should be induced by hot packs or similar devices, and there should be a reduction of starches and sugars in the diet.

Chronic Bright's disease occurs in several forms, among which are inflammation of the substance proper of the kidney and inflammation of the kidney connective tissue. Anaemia, dropsy and the presence of albuminous deposits in the urine are marked symptoms of the first type. The urine is voided in reduced quantities and is dark and contains a heavy sediment. Uraemic poisoning sometimes occurs, often resulting in death. Victims of this form of chronic Bright's disease often live for years. A warm, genial climate and out-of-door life are recommended as safeguards for such patients. In the other form of chronic Bright's disease there is an increased amount of urine excreted, but albumin is not found in it in such large amounts. Dropsy of the lungs is liable to occur and attacks of uraemic poisoning are not infrequent. All cases should be in the care of a skilled physician. W.A.E.

BRIM'STONE, the name for a commercial form of sulphur. In order to purify it, which means to free it from foreign matter, sulphur is generally melted in a closed vessel and allowed to settle. Then it is poured into cylindrical molds, in which it becomes hard and brittle, and is known in commerce as *roll sulphur,* or *brimstone.* When ground into a fine powder it becomes the well-known sulphur of commerce. See SULPHUR.

BRISBANE, *briz'bane,* the capital of Queensland, Australia, a city which grew from a convict settlement established in 1825, but abandoned in 1839. It is well situated on the Brisbane River and is remarkably well built. Its commercial importance has increased rapidly, owing to the quick settlement of the surrounding country. The chief industries are boot and shoe making and soap boiling; it has numerous breweries and distilleries and is an important center of the wool trade. It was named after Sir Thomas Brisbane, a former governor of New South Wales, who was instrumental in removing the convict settlements and materially assisted in the development of the country. Population in 1911, 141,342.

BRISBANE, ARTHUR (1864-), an American newspaper man who as editor of one of the daily papers in New York has found a wide audience and acquired an influence on the popular mind possessed by few editorial writers of his day. Other papers, too, whether they approve or condemn his policy, have been influenced by his methods to the extent of patterning their editorials after his—making them simple in language, replete with illustrations and printed in short paragraphs.

Brisbane was born in Buffalo, N. Y., studied for a time in Europe, and in 1882 became a reporter for the New York *Sun.* After serving this paper for a time in London, he became editor successively of the *Evening Sun,* the *World* and the *Evening Journal,* beginning his connections with the last named in 1897. It is this paper for which he has done his most effective work. Many of his editorials were published in 1906 as *Editorials from the Hearst Newspapers.* He has also written a biography, *Mary Baker Glover Eddy.*

BRISTLES, *bris''ls,* the stiff, coarse hairs of the hog or the wild boar, especially those which grow on the back. Various kinds of brushes are made of them, and the waxed threads used by shoemakers and saddlers are often tipped with them. Though the United States is one of the greatest hog-growing countries of the world it imports most of its bristles, for the well-fed, fat hogs which are slaughtered in the packing houses produce but soft, inferior bristles. It is the lean, underfed hogs of the cold regions of Russia which furnish the very best, and the method of procuring them is curious. The hogs are not killed, but are led through the forests that they may rub themselves against the trees and so scatter their bristles. These are collected from the ground, tied into bundles, and sold at from $2.00 to $3.00 a pound. China also furnishes many bristles, though not quite as high in quality.

BRIS'TOL, one of the oldest cities of England, known to have existed before the Roman invasion in 55 B.C, and a town of considerable importance when William the Conqueror fought the Battle of Hastings (1066). The name is derived from Anglo-Saxon words meaning *the place at the bridge*. It is situated partly in Somersetshire and partly in Gloucestershire, 118 miles west of London, and in population is the seventh city in England, having 357,059 people in 1911. The original town was built on the triangular piece of land at the junction of the rivers Frome and Avon, but it has now far outgrown those limitations. It is one of the most important ports of England, and it was from here that John Cabot sailed in 1497 on his voyage to America. It has beautiful suburbs, including Clifton, which is the favorite residential district of the wealthy classes. Here is a famous suspension bridge, at the time of completion the largest in the world, with a single span of 676 feet, and having an elevation of 245 feet above high water.

The cathedral founded in 1142 is a notable building, and the Church of Saint Mary Redcliffe, dating from 1293, is the most beautiful parish church in England. The industries include glass works, potteries, soap works, tanneries, sugar refineries, chemical works, machinery works and shipbuilding yards, and the city is the greatest cattle market in the kingdom. All public utilities are municipally owned, the profits derived from their operation being devoted to the reduction of taxation.

BRISTOL, CONN., a manufacturing city in Hartford County, named for the great English trade center, Bristol. It is situated in the middle western part of the state, on the Pequabuck River, whose name recalls the old native Pequot Indian inhabitants. Bristol is one of the old-new cities of Connecticut; it was settled about 1728 and received its city charter in 1911. It was incorporated as a town in 1785 and as a borough in 1893. Before and during the War of Independence it was a stronghold for English sympathizers. If walls could speak, those of "Tory Den," a cave in Chippen's Hill, three miles distant from the center of the town, could tell many stories of Tory plots and planning. The population, mostly Americans, with some Italians and Poles, increased 1,560 during the four years from 1910 to 1914, when it reached 15,145. The area is about twenty-eight square miles.

The surrounding country is hilly. Eighteen miles northeast is the city of Hartford, the state capital; about fifteen miles southwest is Waterbury, and New York is 100 miles south and west. Since 1849 the New York, New Haven & Hartford Railway has connected Bristol with all the largest cities of New England and with New York; over forty steam trains enter and leave the city daily. Electric interurban lines run to Hartford, Waterbury and Terryville. The town, preëminently a manufacturing center, containing about fifty industries, is locally known as the *Bell City*, because of one of its early products. Ball bearings, coaster brakes, clocks, fishing rods, woolen and knit goods, engines, brass goods and tableware are the principal articles made, whose manufacture, together with the industry of silver plating, employs nearly 6,000 people. A Federal building, erected in 1909 at a cost of $90,000; an Emergency Hospital, conducted by the Bristol Visiting Nurses' Association, and a public library are among the most important public buildings of the place. Rockwell Park, the largest in the city, is one-half square mile in area. All of the city's elective officers are subject to recall (see RECALL). D.W.N.

BRISTOL, PA., a manufacturing town in Bucks County, in the southeastern part of the state. It is on the Delaware River and the Pennsylvania Railroad, and is the terminus of the Delaware and Lehigh Canal. It is ten miles northeast of the city limits of Philadelphia, ten miles southwest of Trenton, N. J., and across the river from Burlington, N. J., with which it is connected by ferry. The population of Bristol in 1910 was 9,256; in 1914 it was 10,172. The greater proportion of the inhabitants are American; among the foreign-born, Italians, Austrians and Greeks predominate.

Bristol is attractively situated along the river, and is surrounded by a productive agricultural country. It has machine shops, rolling and worsted mills and large manufactories of hosiery, cast-iron pipe, carpets and patent leather. The town has two banks, a Federal building, erected in 1913 at a cost of $50,000, and a library.

The town was originally called Buckingham. It was settled in 1681, the same year William Penn received his charter for the province. Until 1725 it was the seat of government of the county. It was first incorporated as a town in 1720. The present charter is a revised form (1905) of the charter of 1851. E.L.M.

BRISTOL, R. I., one of the old colonial towns of New England, a port of entry and an important industrial center of the state. Owing to its charming location, its wide, elm-shaded streets and beautiful residences, Bristol is locally known as *Beautiful Bristol.* The population, which is largely American, with a mixture of Italians and Portuguese, increased from 8,936 in 1910 to 10,302 in 1915. Bristol is situated on Narragansett Bay, in the eastern part of the state, and in Bristol County, of which it is the county seat. Fall River is twelve miles northeast, Newport is twelve miles south and Providence is fifteen miles northwest. The town was founded in 1680, incorporated in 1747 and named for the English city of the same name. The area is over nine and one-half square miles.

An extensive passenger and freight service is carried on between Bristol and Fall River and Providence. Connection with the latter city is by two railway lines, the New York, New Haven & Hartford and a suburban line, passenger trains on both lines being operated by electricity. The town is also served by the New York Steamship Company and by ferry lines. The principal industry is the manufacture of rubber goods, in which the National India Rubber Company alone employs more than 2,000 people. Here also is located the plant of the Herreshoff Manufacturing Company, which has built a number of noteworthy yachts and torpedo boats. Some of the Herreshoff racing yachts have successfully defended America's Cup against the world for twenty years. The most notable buildings are the white marble Colt Memorial High School, constructed in 1911 at a cost of $250,000, and the Federal building, erected in 1876. The school system of the town is its special pride, and its public library contains 12,000 volumes.

The Narragansett village of Mount Hope, the residence of Massasoit and King Philip, was located within the limits of Bristol, and near here the latter was killed in the Indian War of 1676. Some historians believe that the dwellings of the Northmen, mentioned in the Sagas of Iceland, were built near the present site of Bristol. J.F.F.

BRISTOL, TENN., and VA., called the BORDER CITY, or TWIN CITY, since it is a dual city situated on the state line, partly in Washington County, Va., and partly in Sullivan County, Tenn. The population, composed of seventy per cent American, eighteen per cent negro and twelve per cent German and Greek, increased from 13,395 in 1910 to 14,906 in 1914, and is about equally divided between the two communities. The principal street of the city is the state line, and two separate city governments are necessary. In other respects Bristol is one city. Johnson City is twenty-five miles southwest, Knoxville, is 131 miles northeast, and Washington, D. C., is 400 miles, also, northeast. The city is the terminus of five railroads which connect with the Norfolk & Western, the Southern and the Virginia & Southwestern railways.

The valley in which Bristol is located is about 1,700 feet above sea level and is sheltered by mountains. The section is rich in timber and mineral deposits of coal and iron; consequently it is the headquarters of a number of large companies, the most important of which is the Virginia Iron, Coal and Coke Company. Nearly all the industries are dependent on the resources of the surrounding country, manufactures of car wheels, sheet-metal, foundry products, wood-pulp, bricks and tobacco products being the most prominent. The city is also a distributing center for iron, coal and coke. Its buildings of note are a $60,000 Federal building, erected in 1900, with a more recent addition costing $50,000, and a city hall. About $750,000 have been expended on its schools and colleges. King's College (Presbyterian), founded in 1868, the Southwest Virginia Institute, Sullins College and two business colleges provide higher and specialized education. The city has a library, a hospital and a number of costly churches. Virginia Park (thirty-five acres) and Anderson Park (two acres) are the recreation grounds.

Bristol was settled in 1851, and incorporated in 1856. The present town charter was adopted in 1898 and revised in 1901. It was named for the English city of Bristol. Bristol, Tenn., has adopted the commission form of government, with a mayor and two commissioners. Bristol, Va., is governed by a mayor and council elected every two years. Previous to 1890 the latter was known as Goodson. N.B.R.

BRITISH AMERICA. This phrase is used with somewhat varying meaning. In its narrow sense it refers to that portion of North America north of the United States which belongs to Great Britain, and includes Canada and Newfoundland. More broadly, it takes in as well certain other British possessions in North America or close to it, as British Guiana, British Honduras, the Bermuda Islands, the British West Indies and the Falkland Islands. Each of these colonies or possessions is described under its own title in these volumes, while in the article GREAT BRITAIN, subhead *Colonial Possessions of Great Britain,* is given a list of all the British outlying territories.

BRITISH COLUMBIA, the westernmost province of the Dominion of Canada. Though it has many pleasant valleys, whose fertility is one of its greatest assets, the province is essentially a mountainous region. It has been called a "sea of mountains" and the "Switzerland of America," though it is twenty-two times as large as the latter country. Almost the entire width of the great Cordilleras, the backbone of the continent, lies within the province, and the whole province, except the northeast corner, is included in the mountain system. Mining, fishing and lumbering are the main industries. Trapping and fur-trading, which originally brought the white man to this region, are no longer of first importance, but mining, which led to permanent settlements, is still the most profitable industry.

Area and Population. British Columbia has a total area of 355,855 square miles, nearly one-tenth of the entire Dominion. Until 1912 it was the largest province, but in that year the extension of the boundaries of Ontario and Quebec placed it third in size. It has five times the area of the state of Washington, which adjoins it on the south, and about five and one-half times the area of all the New England states. If it were placed bodily on top of the United States it would extend from Milwaukee to New Orleans, and at the northern boundary, its widest point, it would extend from Detroit, Mich., to Sioux City, Iowa. British Columbia is nearly 1,000 miles long from north to south, and at its widest part is about 650 miles from east to west.

LOCATION MAP
Comparative size with respect to the entire Dominion.

So much of the province is mountainous and not fit for permanent settlement that the average population is only 1.09 per square mile of surface. This gives a total population of 392,480, according to the census of 1911. Over sixty per cent of the population is male, and a little more than half is urban, that is, living in cities, towns and villages. The large cities, in the order of their size, are Vancouver, Victoria (the capital), New Westminster, Nanaimo, Nelson, Prince Rupert and Kamloops. Each of these is discussed in a separate article in these volumes.

The People. Of the total population, one-third is of English birth or descent. The Scotch are second, with about one-fifth, and the Irish include one-tenth. There are about 20,000 native Indians; the number of Chinese is about 20,000, but is not likely to increase materially because of the heavy tax levied on Chinese immigrants. Of the other foreigners or persons of foreign descent the most numerous are the Scandinavians, Germans, Italians, Japanese and French, in the order named. The Anglican Church has the largest

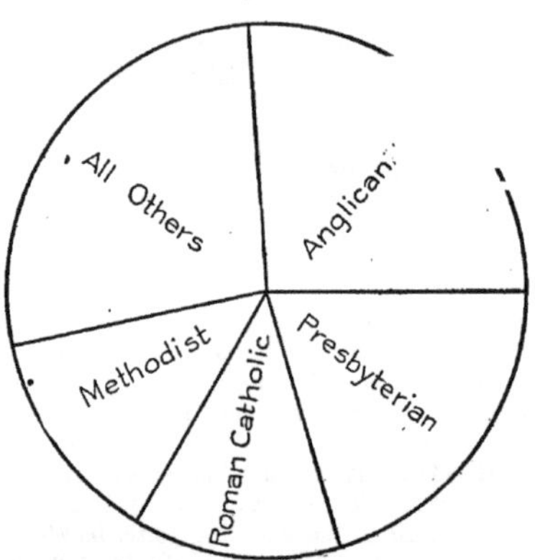

RELIGIONS IN THE PROVINCE

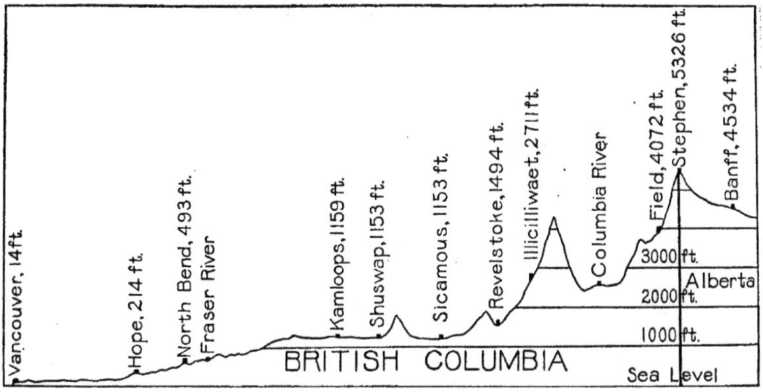

ELEVATION OF THE LAND, ACROSS THE PROVINCE
On the line of the Canadian Pacific Railroad

number of communicants, closely followed by the Presbyterians and the Methodists.

Characteristics of the Surface. The predominating physical feature is the great mountain mass called the Cordillera. This belt of parallel ranges is about 450 miles wide at the southern boundary of the province, and 350 miles wide at the northern boundary. The entire mass may be divided into the Rocky Mountains and the Coast Range, between which lies the Great Basin, or Central Plateau. Most geographers also speak of the Island Range, which is now almost completely submerged; the only remnants are Vancouver Island and the Queen Charlotte Islands. All of these ranges have a general direction from northwest to southeast.

The Rocky Mountains. From the northern boundary to latitude 54° N. the whole of the Rocky Mountain system lies within the province, but from this point southward to the United States line the crest of the main chain is the boundary between Alberta and British Columbia. From an average altitude of 10,000 feet at the south these mountains decrease to 5,000 feet at the north, and the northern boundary of the province practically coincides with the end of the Rockies as a distinct chain. The highest mountain in British Columbia is Mount Robson (13,068 feet). The range is broken by the Peace River Valley and by a number of gaps, or passes. These include Crow's Nest Pass and Kicking Horse Pass, which are crossed by the Canadian Pacific Railway, and Yellowhead Pass, crossed by the Grand Trunk Pacific.

The minor ranges of the Rocky Mountain system, in order from east to west, are the Purcell Range, the Selkirk Mountains and the Gold Range. The Selkirk Mountains, though usually regarded as a part of the Rocky Mountain system, belong to an older geological period. Separating these ranges from the main chain is a valley 900 miles long, and varying from one to eight miles in width. In this valley, called the Rocky Mountain Trench, rise the chief rivers of the province, the Columbia, Fraser, Kootenay and Finlay.

The Central Plateau. This region is a continuation of the Cascade Range in the United States. It has an average elevation of 3,500 feet, and lies between the Rocky Mountains and the Coast Range. About midway from north to south are several small ranges which divide it into two sections, the southern half of which has a special name, the Interior Plateau of British Columbia. Here the Fraser, the Columbia and other rivers have worn deep valleys, whose rocky walls form miniature Grand Canyons. The whole of the Central Plateau is dotted with lakes, many of them merely expansions of the larger rivers; chief among them are the Babine, Tacla, Stuart, Quesnel, Shuswap, Okanagan and the two Arrow lakes.

The Coast Range and the Islands. The Coast Range extends practically the whole length of the province from north to south. North of the Portland Canal its western slope lies in Alaska. On the ocean side the mountains rise abruptly, sometimes from the water's edge, and the coast resembles that of Norway.

The shore line is indented by hundreds of fiords and smaller inlets, which have been caused by the partial submergence of the range. This coast is one of the most remarkable in the world, and with all its irregularities has a total shore-line of about 7,000 miles. If

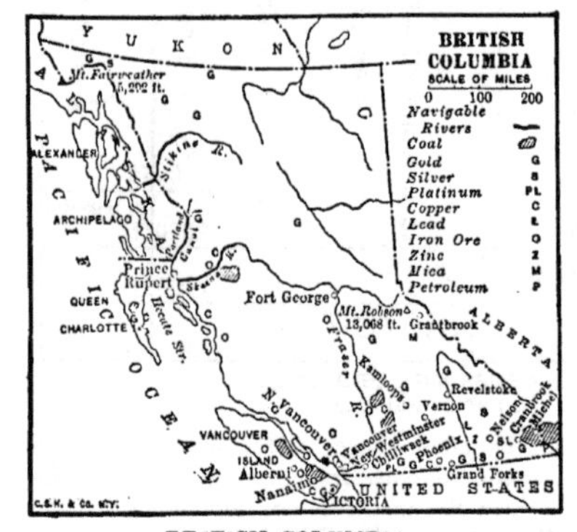

BRITISH COLUMBIA

The map locates the principal cities, chief rivers, coal measures now being worked, boundaries and highest point of land.

this coast-line were straightened out it would reach from New York westward to Yokohama, Japan, or from Vancouver to Yokohama and back again as far as the Hawaiian Islands.

Queen Charlotte Islands and Vancouver Island are the only remains of a mountain range which ages ago was a part of the mainland. The Pacific side of the islands has many inlets, but not as large or as numerous as those on the mainland. Mount Victoria, on Vancouver Island, has an altitude of 7,484 feet, the highest peak of the Island Range.

Influence of the Mountains on Climate. If British Columbia had no mountains, if it were flat, like Saskatchewan, its climate would probably be nearly uniform in every part. As it is, however, the parallel mountain ranges run at right angles to the prevailing winds from the west. This makes a marked difference in the climate of the east and west, and creates a number of longitudinal zones. Thus the traveler who journeys along the coast from Skagway, Alaska, to Vancouver will have about the same weather during his entire trip, but if he should move eastward either from Skagway or Vancouver he would encounter new conditions as soon as he crossed the mountains.

The climate of the coast is like that of Western Europe. The prevailing westerly winds are warm and are loaded with moisture absorbed in their passage over the ocean. When they strike the cool summits of the Coast Range, condensation of the moisture takes place rapidly, causing a large amount of rainfall. The average annual rainfall is from eighty to ninety inches, about the same as that of Southern England. At Victoria, which is farther north than the city of Quebec, where the cold becomes intense, the temperature during January very seldom falls to the freezing point, and flowers often bloom in the gardens all the year.

After the warm winds have lost most of their moisture and have become cooled, they rise to a height of 7,000 to 9,000 feet to cross the Coast Range. As they pass over the valleys and the great Central Plateau they are kept at this height by the warm currents rising from the plains. Because there is no obstruction in the path of the winds, nothing to cause them to lose their remaining moisture, the interior has little rain in summer and only light snow in winter. Droughts and extremes of heat and cold are common. At Kamloops, which is only 200 miles from the coast, the annual rainfall seldom exceeds twelve inches, and the temperature ranges from 25° to 30° below zero to 100° above.

The prevailing westerlies retain some of their heat and moisture as they cross the interior plateau, but when they strike the snowy peaks of the Rocky Mountains they lose the last of these. The western slopes of these ranges, especially the Selkirk Mountains, have an annual rainfall nearly as heavy as that of the Pacific coast, and in the mountain passes it it is not unusual for snow to reach a depth of twenty-five to thirty feet. The lower parts of the western slopes are covered with extensive forests, and the heavy snows feed many great glaciers. It is this section, more than any other, which has received the name of "the Switzerland of America."

Plant and Animal Life. The wild life of British Columbia is not extremely varied, and all of it belongs to the type called northern or northwestern. Wherever there is rain there is abundant vegetation, from mosses to great forests. The western slopes of the Coast Range, the Selkirks and the Gold Range constitute probably the greatest areas of virgin timber in North America. It is estimated that the great forests of the province, excluding sparsely-timbered areas, cover 30,000,000 acres, of which 2,500,000 acres have been set aside as Dominion forest reserves. There are also

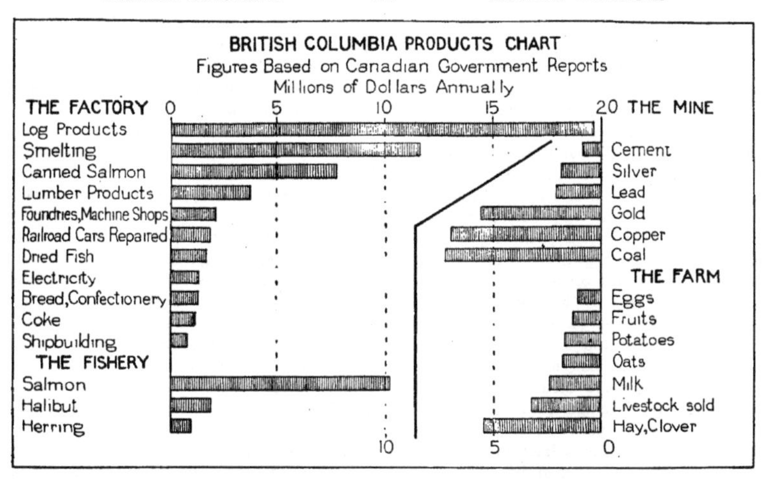

two large provincial reserves, Strathcona Park, on Vancouver Island, and Mount Robson Park, protecting the headwaters of the Fraser River. These two parks include about 1,000,000 acres.

The Douglas fir, a magnificent tree which frequently exceeds a height of 300 feet, is the most valuable and most abundant species, but white and yellow cedar are also important. On the mainland, directly north of Queen Charlotte Sound, are great stands of spruce and hemlock, which are especially valuable for making wood-pulp. The importance of lumbering has only recently been recognized; between 1900 and 1910 the average annual cut increased nearly threefold, afterwards decreasing. It is now 1,000,000,000 board feet or more, valued at about $30,000,000.

Among the animals the moose, black-tailed deer and caribou are common in the valleys and wooded sections; the wapiti, once numerous, is now extinct except in the foothills of the Rocky Mountains. Bears, wolves and wild cats are so common that in some sections they are considered a nuisance. In the mountains are the haunts of the bighorn and the goat, the pursuit of which furnishes thrills for even the most seasoned hunters. In the north are many fur-bearing animals, and trapping and trading are still important. The birds are especially numerous, more than 300 different species making their homes here. One of the characteristic birds is the burrowing owl of the interior plateau. Jays and magpies are conspicuous; among the game birds are grouse, partridge, teal, mallard, pin-tail and canvas-back. Fish, especially trout, are plentiful in nearly all the streams. The province, in short, has an abundance of game, and is a sportsman's paradise.

Mineral Wealth and Mining. The coal deposits on Vancouver Island and the gold along the Fraser River first drew attention to the section which now forms British Columbia, and mining is still the industry of first importance. The annual output of minerals fluctuates considerably, but $25,000,000 is a fair average. In 1912 it was $32,000,000, but this was a high record, and there was a slight falling-off in the succeeding years. The average is about one-fifth of the total for the Dominion.

As early as 1857 placer mining was carried on along the Fraser River, and by 1862 and 1863 the annual output had reached $3,000,000. The production then slowly declined until it was about $400,000 in 1893, but since then new hydraulic methods and new gold fields have revived the industry. The placer mines are of comparatively little importance, but vein-mining brings the annual total to an average of more than $5,000,000. The year 1913 set a high record of 297,450 ounces, valued at more than $6,100,000.

In the southeastern part of the province are large deposits of silver-lead ores, and the production of each of these minerals averages $2,000,000 a year. Copper has been mined commercially only since 1894, when the output was 324,000 pounds. It now averages 45,000,000 to 50,000,000 pounds, valued at $7,000,000 to $8,000,000. Of the non-metallic minerals coal is most important. It is mined chiefly on Vancouver Island, but there are known to be rich deposits in the Rocky Mountains, especially in the southern third of the province. In the production of coal British Columbia is second to Nova Scotia among the provinces, but its annual average of 2,500,000 to 3,000,000 short tons is less than half that of the leader.

Fisheries. For many years Nova Scotia held first rank among the provinces for its fisheries, but in 1912 British Columbia forged ahead; the fisheries of this province are now the most important in Canada and yield about forty per cent of the annual total for the Dominion. The salmon fisheries are by far the most valuable, and their products, including canned fish, average about $10,000,000 a year. Seventy-five per cent of the catch is canned each year. The salmon are of several kinds and ascend the rivers at different seasons. The spring salmon, or quinnat, is the largest, and the best for use when fresh. The sockeye, which follows in the summer, is smaller, though of more uniform size, and is preferred for canning because when the run has once begun it invariably continues steadily in enormous numbers. The dog-salmon are caught by the Japanese, who salt them for export to the Far East. During the season from 2,000 to 2,500 boats are used at the mouth of the Fraser River, from which the largest part of the catch is taken. About two-thirds of the canneries are on the Fraser River; most of the remainder are on the Skeena River and Rivers Inlet.

The only other fish of much importance is halibut, the catch of which has an annual average value of $1,750,000. Cod, herring, shad, sturgeon, clams and crabs add about $2,000,000 each year, bringing the total value of the fisheries to $13,000,000 or $14,000,000 a year. The fur-seal fisheries were formerly important, but are now almost extinct (see SEAL).

Agriculture. Mining and fishing are the greatest industries, but not the only ones. There are many fertile sections where grains and other field crops are being successfully raised, and the lower valley of the Fraser River is one of the regions where fruits and vegetables of all kinds reward the farmer's work. Practically all of the valleys are fertile, and require only irrigation to make them productive. In the Okanagan, Thompson and Columbia valleys are more than 100,000 acres tributary to irrigating canals, and there are several hundred thousand more acres available. The irrigated land will support many kinds of crops, but the expense of constructing irrigation systems makes intensive cultivation of fruits and vegetables most profitable.

Wheat is raised only in small quantities, but the production of oats averages 3,000,000 bushels. The potato crop is about the same size. Live-stock raising is of some importance, and there are perhaps 300,000 head, including horses, in the province. The production of butter and cheese is important locally. Apples, peaches, pears, plums, cherries and various small fruits, including strawberries, are raised with great success, especially in the warm delta of the Fraser River. Agriculture in British Columbia represents a total investment of perhaps $200,000,000, and the total annual yield of all products is about one-tenth of that amount.

Transformation of Raw Products. British Columbia has three great industries, which are dependent on its own raw products. These industries, in the order of their importance, are the transformation of logs into lumber and dozens of other products, the smelting of ores, and the canning and preserving of fish. These three industries produce more than one-half of the total manufactures of the province. The annual grand total is now over $80,000,000; in 1900 it was $19,000,000. Log and lumber products are worth $20,000,000 to $25,000,000; smelted ores represent a value of $15,000,000, and canned fish $8,000,000 more. Vancouver is the center of manufacturing, and alone produces about one-fourth of the total. The many rapid streams which descend the west slope of the Coast Range furnish abundant water power where it is most needed—on or near the coast—and there is no visible reason why the manufacturing industries should not continue to grow.

Waterways and Railways. Owing to the mountainous character of a large part of the province, the means of transportation have not yet been fully developed. The construction of wagon-roads and even trails is a task involving great skill and expense. The construction of railways is even more costly, but the province is now served by the three great transcontinental lines, the Canadian Pacific, the Canadian Northern and the Grand Trunk Pacific. The total railway mileage in 1915 was about 3,000, and other lines already surveyed or under contract will add at least 4,500 miles more.

In the Kootenay district the steamers on the lakes furnish connections between some of the points not reached by railroads, and on the Fraser River steamers ascend as far as Yale. There is also steamship connection between Vancouver, Victoria and Prince Rupert. Most of the interior, however, is dependent on highways and trails, and of each of these there are over 15,000 miles.

Trade and Commerce. The ports of British Columbia—Vancouver, Victoria, Nanaimo, New Westminster and Prince Rupert—are the natural outlets for Canadian trade with the Far East. To some extent they compete with ports in the United States not only for Canadian but for American trade. Vancouver, for example, is 500 miles nearer Yokohama than is San Francisco, and this is no slight advantage. Prince Rupert is 400 miles nearer Japan than is any other port in North America. From Vancouver and Victoria the Canadian Pacific operates two lines of steamships, the *Empress* to Japan and China, and the *Australian* to Honolulu, Fiji Islands and Sydney. There are a number of lines which operate between the various Pacific ports in Canada and the United States. Several other lines, both Canadian and foreign, operate between British Columbian ports and Alaska, Australia, Hawaiian Islands, China and Japan. Over 12,000 ships, with a total tonnage of 12,000,000, enter and clear each year from all these ports. The foreign commerce of the province amounts to about $100,000,000, of which sixty per cent is imports.

Education. The provincial ministry or executive council is also a council of public instruction, and the provincial secretary also acts as minister of education. Subject to the approval of the council, the superintendent of education has entire charge of the school system. All public schools are free, and no religious instruction or control is permitted. Education is compulsory for all children between the ages of seven and fourteen, although some exceptions are allowed. The school enrollment is about 50,000, and the teachers number 2,000. As in the other provinces, the schools are supported partly by the provincial government and partly by local taxation. The total expenditure for educational purposes is about $5,000,000 a year, of which the province provides $2,000,000.

High schools, or superior schools, may be formed in connection with the common schools whenever ten pupils are qualified to carry on the studies. Admission to high schools and promotion through the four-year course is regulated by examinations conducted by the provincial department of education. Normal schools for teachers have been established at Vancouver and Victoria, and in 1912 a charter was given to the University of British Columbia, whose endowment consists of 2,000,000 acres of crown lands. This school was ready to receive students in 1915; it absorbed the McGill University College of British Columbia, which had been affiliated for ten years with the great McGill University at Montreal.

Government. The government is exactly like that of the other provinces of the Dominion. The lieutenant-governor, who receives an annual salary of $9,000, is appointed by the Governor-General of the Dominion for a five-year term. He is the direct representative of the Crown. The ministry, or executive council, is composed of six members of the legislative assembly. Their appointments are made by the lieutenant-governor, but they are responsible to the assembly. Ministers receive $6,000 a year, except that the premier receives an additional $3,000. The assembly comprises forty-two members, each of whom is elected for four years. Three senators and eleven members of the House of Commons represent British Columbia in the Dominion Parliament at Ottawa.

Justice is administered by the police magistrates, the county courts, the supreme court and the court of appeal. The police magistrates have jurisdiction in minor cases, such as actions for debt involving less than $100. The intermediate courts, ten in number, have a wider original jurisdiction and also some appellate jurisdiction. The supreme court, composed of a chief justice and five associate, or *puisne,* judges, has full criminal and civil jurisdiction. The court of appeal is the court of last resort, except that some important cases may be further appealed to the Privy Council (which see); it is composed of a chief justice and four associate judges. The judges of county courts receive $3,000 a year; of the supreme court, $6,000; of the court of appeal, $7,000. A chief justice receives $1,000 additional.

Provincial Revenues. British Columbia, like the other provinces, receives an annual subsidy, whose size is determined by the population, the debt, the wealth of the province and other factors. This subsidy now amounts to about $750,000, about seven per cent of the total provincial revenue. The largest source of income is from timber, royalties and licenses, which bring in about $2,500,000, and the proceeds from the sale of public lands amount to more than $1,000,000 a year. A large number of miscellaneous taxes, including a poll tax on Chinamen and a graduated income tax, bring the total revenue nearly to $11,000,000.

History. Until the last quarter of the eighteenth century the existence of the area now included in British Columbia was unknown to the civilized world. The first white men who saw the Pacific coast so far north were Spaniards. In 1778, four years after the Spanish discovery, the famous English navigator, Captain James Cook (which see), began the exploration and accurate description of the coast. In 1788 the first white settlement was made at Nootka, on the west shore of Vancouver Island, by a party of Englishmen, but it was broken up almost immediately by the Spaniards, who claimed by right of discovery the entire coast northward so far as the Russian possessions. For a year or more it seemed as if Spain and England would go to war over this distant land, but in 1793 they divided the territory by arbitration. Spain took the area south of Nootka Sound, including the present states of Washington and Oregon; England took the coast northward to the Russian territory, now Alaska.

The dispute over ownership had not deterred the British from exploring the territory. Between 1792 and 1794 Captain George Vancouver, under orders from the British government, surveyed the coast about as far north as Milbank Sound. He was the first to circumnavigate the island which is named for him. Meanwhile other explorers, in the employ of the Northwest Company, had reached the coast after a long and dangerous trip from the interior. The most famous of these men was Sir Alexander Mackenzie, whose achievements are recorded in his biography in these volumes.

Under the Rule of the Hudson's Bay Company. For half a century the great Northwest, including British Columbia, was ruled as the private property of the great fur-trading companies, the Northwest Company until 1821, and thereafter the Hudson's Bay Company (which see). In 1846 the latter built a fort where Victoria now stands. This action created great excitement in the United States, which claimed the entire coast north to the line of 54° 40′. There was a shadow of justice in this claim; it was based on the indefinite character of the boundaries of the Louisiana Purchase (which see). President Polk, though he was elected on the issue of "fifty-four forty or fight," found it wiser to compromise. (For further details, see OREGON, subhead *History*.)

During these years the Northwest was constantly growing in population, and the despotic, though just, government of the Hudson's Bay Company gradually became unsuited to new conditions. Vancouver Island was finally created a crown colony in 1859, and the discovery of gold on the Fraser River in 1856 led to the organization of a separate government for the mainland two years later. The name New Caledonia, by which the region was previously known, was changed to British Columbia. For a number of years the royal government was more a matter of form than of fact. The governor, both of Vancouver Island and of British Columbia, was Sir James Douglas, who was also chief factor of the Hudson's Bay Company. Douglas was an able administrator, and he is justly regarded as the founder of British Columbia; it is no discredit to his fame to admit that he ruled with an iron hand. In 1866 the rule of the Hudson's Bay Company came to an end when the two colonies were united under a single government as British Columbia, and five years later it became one of the provinces of the Dominion of Canada.

A Province of the Dominion. The adherence of British Columbia to the Dominion was not secured without difficulty. It was argued that the Pacific coast colony could never have interests in common with the eastern provinces. Communication between them was slow, and for practical purposes Ottawa was as far from Vancouver as from London. After much discussion British Columbia voted to join the confederation of provinces, and the Dominion government in return promised to build, or have built, a railroad to connect the Pacific coast with the railways of Ontario. This new railroad, the Canadian Pacific, was not completed until 1885, and more than once British Columbia was all but ready to leave the Dominion, because the agreement had not been kept. But with the driving of the last spike the province was firmly bound to the Dominion, and the question of separation no longer disturbed the country.

The autocratic nature of the government under the Hudson's Bay Company's régime left its mark on British Columbia until 1903. It is not unfair to say that until that year the government was more or less personal.

BRITISH COLUMBIA
Landing a Salmon
Mighty Cedars
Taku Indian
Canadian Rockies
Along the Coast
A Smelter in the Mining District
Parliament Buildings, Victoria

The premier and the other members of the council were not strictly responsible to the legislative assembly, and frequently disagreed with it on matters of policy. Since 1903 responsible government has been unquestioned both in practice and in principle. The lieutenant-governor at that time was Sir Henri Joly de Lotbinière, who was instrumental in securing the change, and the first premier under the new system was Sir Richard McBride.

Immigration and Other Problems. Since 1903 the chief questions of public interest have involved the development and conservation of the mining, fishing and agricultural resources. There has also been a powerful demand for a larger subsidy from the Dominion government. A question which has several times threatened foreign complications is the restriction of immigration. A provincial act of 1900 prohibited the immigration of Asiatics, but was disallowed or vetoed by the Dominion government as being contrary to the national policy. So far as regards the Japanese, the question was settled when the Japanese government forbade emigration to Canada (see CANADA, subhead *History*). Chinese immigration was practically stopped in 1903 when the Dominion imposed a poll tax of $500 on every Chinese immigrant.

A striking event of 1914 was the attempt of a shipload of Hindus to secure admission to Canada. The problem of Hindu immigration has long perplexed Canadian authorities, and particularly British Columbia. The situation was relieved temporarily by a Dominion statute which forbade the entry of Orientals into Canada except by direct passage from the land of their birth. As there is no direct steamship line between India and Canada, Hindus have been barred by this law. Hindu resentment at this treatment came to a head in 1915, when Gurdit Singh, a wealthy Hindu, specially chartered a Japanese steamer, the *Komagata Maru,* and with 379 other Hindus sailed directly from Calcutta to Vancouver. The port officials refused the intending immigrants permission to land, and after various judicial preliminaries sent a hundred Vancouver policemen to the ship to compel it to put back to India. The Hindus, thoroughly enraged, beat off the police, and it required the threat of a Canadian cruiser's guns to persuade them to leave.

Labor problems, aside from the influence of immigration, have also received considerable attention. A law of 1905 provided an eight-hour day for all miners working underground, and a law of 1907 made the same provision for workers in smelters. There has been, nevertheless, considerable discontent among the miners, and strikes have several times interrupted the progress of the industry. In 1912 the Industrial Workers of the World made a great effort to establish their organization, and succeeded in causing a strike of the railway employees. There were riots in Vancouver, and several strikers and agitators were given prison sentences. In the next year Vancouver Island was the scene of a great strike among the coal miners. Chinese and Italians were brought in as strike-breakers, and there were riots during the summer at Nanaimo, Ladysmith and other places. Many of the strikers were arrested and severely handled, and much criticism was directed against the government because all the prisoners, even before their trials, were treated as convicted criminals.

Elections and Political Changes. On December 15, 1915, the resignation of Premier McBride was announced. McBride had been premier since 1903, and his administration was marked by the ever-increasing prosperity of the province. His successor, the Hon. William J. Bowser, had been his chief lieutenant for many years, and was well known throughout the province. The Conservative Bowser administration, however, held office only for a few months, for at the general elections in September, 1916, the Liberals won a large majority in the legislative assembly. On the resignation of the Conservative ministry, the Liberal leader, Hon. H. C. Brewster, became premier and formed a ministry. At the same time the voters gave a decisive verdict in favor of prohibition and woman suffrage.

Other Items of Interest. It is estimated that the coal measures of British Columbia contain about one hundred billion tons of coal.

The Douglas fir, in the coast regions where it thrives best, frequently attains a base circumference of from thirty to fifty feet.

The province has no fewer than thirty-six native varieties of trees.

Certain small streams in Vancouver Island and in Northern British Columbia are at times literally choked with salmon, and the natives simply toss them out upon the banks with pitchforks.

Though in general a region of high altitudes, the province has "river-bottoms" along the Lower Fraser and the Upper Columbia where dykes are a necessity.

QUESTIONS ON BRITISH COLUMBIA

An outline suitable for British Columbia will be found with the article "Province."

What is the nearest Pacific coast port to the Orient?

If you could control one of British Columbia's natural resources, which would you choose?

What is the "Rocky Mountain Trench"? What relation has it to the drainage of the province?

What section is known as "the Switzerland of America"? Why?

What is a quinnat? A sockeye? For what is each best adapted?

Is Vancouver nearer Japan or farther from it than San Francisco? How much?

What first attracted white men to this far western region?

Which part of the province has many miniature "Grand Canyons"?

What have the Dominion and the provincial governments done to protect the forests?

How does the province rank with other parts of the country in the possession of timber land?

What industry, once of prime importance, is now nearly extinct? Why?

How large a part of the school expenses is met by the provincial government and how much by local communities?

If British Columbia were placed on top of the United States, how large an area would it cover?

How many times as long as the coastline of California is that of British Columbia?

What was meant by the slogan "Fifty-four forty or fight"?

How large does the most valuable tree of the province grow?

Why are fruits and vegetables the most profitable crops on irrigated land?

What is the smallest number of pupils for whom the government will establish a high school?

What was the British Columbia region originally called?

In what unexplored region is there a road over a half-century old?

Are there more males or females in the province?

Is the province more or less densely populated than the Dominion as a whole?

Where do people live on mountain tops which are no longer mountain tops?

Name six reasons why, if you were a hunter, you would like to visit British Columbia.

If all the people in Canada were removed to this one province, would the density of population be greater or less than that of the United States?

Why is there little danger of Chinamen flocking to the province in great numbers?

Show the close relation between the natural resources and the manufacturing industries.

If you brought a lawsuit in British Columbia for a debt of $79.47, before what court would it be tried?

Is the loftiest peak in the province taller than Pike's Peak or not so tall?

Why is the snowfall so light in the Central Plateau and so heavy in the mountain region farther east?

How long was it after the coast of Labrador was sighted that this western coast of Canada was first visited by white men?

Why is railroad building so difficult and so expensive in this province?

Victoria is the oldest city in the province, dating back to 1846, when a trading post known as Camosun was established on its site.

A provincial prohibition law was passed in 1917.

With the exception of the Colorado, all the great rivers which empty into the Pacific take their rise in this province. The largest of them are the Yukon, Columbia, Fraser and Skeena.

The province seems to be, in popular parlance, "mineralized all over," for there is scarcely a region which has not its share of underground wealth.

Sockeye salmon appear in unusually great numbers every fourth year, and fishermen and canning factories make special preparation to catch them.

One of the northern districts of the province is the Cariboo—a great unexplored, undeveloped region, into which leads the Cariboo Trail, a government post road three hundred miles long. It was built in the early "sixties," when there was a "gold rush" to these northern parts, and is still used by the British Columbia Express Company, which operates stage lines the entire distance.

Strange as it may seem, tobacco, usually looked upon as a southland crop, is grown with profit in British Columbia.

Near Victoria is an observatory with a seventy-two inch reflector. This region offers peculiar advantages to observers because of the small daily range in temperature and certain atmospheric peculiarities.

The Indians call Mount Robson, which is one of the most picturesque peaks in the Dominion, *Yuh-hai-has-kun*—"the road winding upward"; for to their eyes the hard rock left standing when the softer rock was eroded has the appearance of an ascending spiral road.

Consult Fairford's *British Columbia;* Thornhill's *British Columbia in the Making.* These books may be purchased by sending to bookstores in the cities of the province.

Related Subjects. A more detailed knowledge of British Columbia may be gained from the following articles:

CITIES AND TOWNS

Chilliwack	Penticton
Cranbrook	Prince George
Esquimalt	Prince Rupert
Fernie	Revelstoke
Grand Forks	Rossland
Kamloops	Summerland
Kelowna	Trail
Nanaimo	Vancouver
Nelson	Vernon
New Westminster	Victoria
North Vancouver	

ISLANDS

Vancouver	Queen Charlotte

LEADING PRODUCTS

Coal	Lumber
Copper	Oats
Gold	Salmon
Halibut	Silver

MOUNTAINS

Assiniboine, Mount	Rocky Mountains
Cascade Range	Saint Elias Mountains
Robson, Mount	Selkirk Mountains

RIVERS

Columbia	Peace
Fraser	Skeena
Kootenay	

UNCLASSIFIED

Hudson's Bay Company	Queen Charlotte Sound
Juan de Fuca, Strait of	

BRITISH COLUMBIA, UNIVERSITY OF, at Vancouver, B. C., a non-denominational, coeducational institution for higher education, an integral part of the public educational system of the province. The university undertakes to "furnish instruction in the various branches of a liberal education, and in the technical branches that have a bearing upon the life and industries of the province." The act creating the university reserves for it the sole right to confer degrees, except in theology, in the entire province. The institution opened on September 29, 1915, and during its first year had an enrollment of 435 students, of whom about fifty were in active service with the Canadian troops in the War of the Nations. In May, 1916, the first class, comprising forty students, was graduated.

History. The need of a university in British Columbia was first publicly emphasized in 1877. Not until 1890, however, was a serious attempt made to found one, which in that year failed for lack of interest. Finally, in 1907, the provincial legislature passed a University Endowment Act which authorized the government to set aside 2,000,000 acres of land as an endowment. In 1908 an act was passed establishing and incorporating the University of British Columbia. The work of organization proceeded slowly, and a site was not definitely selected until three years later. The location chosen is a tract of 250 acres on Point Grey, a short distance west of Vancouver.

The plans for the first group of four buildings were drawn by a Vancouver firm of architects, whose designs were chosen in competition. Construction was begun on the science building in 1914, and at the same time the legislature voted $500,000 for further construction. Upon the outbreak of the War of the Nations, however, the board of governors, feeling that it would be shortsighted and unpatriotic to commit the public to a large expenditure and heavy fixed charges when every dollar in the country might be required in the struggle, decided to postpone the completion of the science building and to avoid large expenditures of every kind. In the meantime, by vote of the legislature, sufficient funds were provided to allow the university to begin its work in temporary quarters and to take over the work of the McGill University of British Columbia, which had been giving university instruction since 1906. The transfer of the student body from the old institution, which suspended operations in 1915, to the new university, made it possible for the latter to graduate a class in the first year of its existence.

BRITISH GUIANA, *ge ah' na,* a colony of Northern South America, whose area, 90,277 square miles, is greater than the combined area of its neighbors, Dutch Guiana and French Guiana (which see). To the north of it is the Atlantic Ocean; to the east, Dutch Guiana; to the south, Brazil, and to the west, Venezuela. It lies between 2° and 8° north of the equator. With an area greater than that of Minnesota, it had by the census of 1911 but 296,041 inhabitants, or little more than one-tenth as many as Minnesota, but a census in Guiana is apt to be misleading. The interior parts of the country have never been explored, and the native savages who live there have never been numbered. It is estimated, however, that there are not more than 10,000 or 15,000 of them.

LOCATION MAP
The small black area compares the size of British Guiana with the entire continent.

The Land. British Guiana is divided into three settlements, Essequibo, Demerara and Berbice. In general, the surface is a plateau, higher toward the northern part, and so far as is known the highest mountain is the peak Roraima, on the western boundary. This is a flat-topped mountain, almost inaccessible, which rises to a height of 8,600 feet. The longest river in Guiana, the Essequibo, is in this British colony, as is also the Berbice, while the Corentyn forms the boundary between it and Dutch Guiana. These rivers are of no great importance for navigation, as falls and rapids interrupt their courses.

The climate is hot, with an excess of moisture, and like practically all countries where such conditions occur British Guiana is unhealthful for Europeans, of whom there are fewer than 5,000. Most of the inhabitants live in the coast region, the only part that has been put under cultivation.

Resources. As in most tropical countries with heavy rainfall, the forest growth of British Guiana is very heavy, and valuable woods abound. Most of the forest area is as yet unexplored, however, and the mineral resources are almost equally undeveloped. Some gold is found in the river bottoms, and a few diamonds have been mined in the interior.

But the chief wealth of the country lies in its agricultural possibilities. The soil which has been cultivated in the coast region is very rich, and sugar, rice, sea-island cotton and coffee are grown in considerable quantities. Much of the work is done by East Indian and Chinese laborers, called *coolies.* Each year British Guiana exports about $10,000,000 worth of

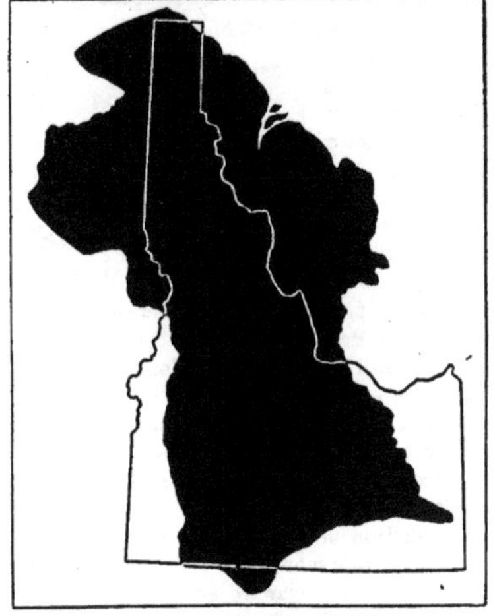

COMPARATIVE AREAS
British Guiana is nearly 7,000 square miles larger than the North American state of Idaho.

products, and imports approximately the same amount.

Government and History. The capital is Georgetown, on the coast, a curious city of 53,000 inhabitants, which has its houses built on great piles to avoid the high tides. A governor, appointed by the British Crown, a council and a court of policy constitute the executive department, and the legislature consists of a single house of elected members.

The Dutch made the first settlements in Guiana in 1613, and not until 1815 did the British make good their claim to the territory they now hold, though they had previously attempted its conquest. The latter part of the nineteenth century saw the country much disturbed by boundary disputes with Venezuela and Dutch Guiana, but these were settled in 1899 by an international commission. H.M.S.

BRITISH HONDURAS, *hon doo' ras,* or **BELIZE,** *be leez',* a colony of Great Britain, in the northeastern part of Central America. From 1636 it was frequently visited by logcutters, but the first permanent settlement was made early in the eighteenth century by a Scotchman. The Spaniards, holders of the adjacent territory, frequently tried to drive out the settlers, but in 1783 a treaty formally recognized the right of Great Britain to develop the section, provided all British subjects in the Spanish parts of Central America removed at once to Belize, as it was then called. It was not until 1836 that the colony was recognized as a permanent possession of Great Britain. Since that time the country has progressed steadily, and is in many ways more advanced than the independent Latin countries of Central America.

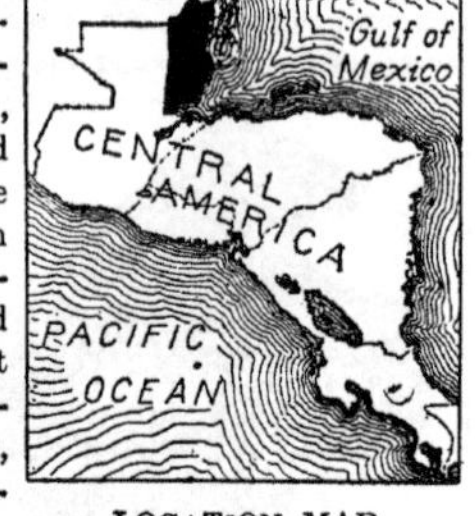

LOCATION MAP

British Honduras has an area of 7,562 square miles, or a little more than the combined areas of Connecticut, Delaware and the District of Columbia, but its population of 40,458 (1911) is less than one-fortieth of that of the states named. It is bounded on the east and northeast by the Bay of Honduras, on the north and northwest by Yucatan, and on the south and west by Guatemala. The coast is low and swampy, the climate sub-tropical, and, in spite of dampness and occasional epidemics of yellow fever and cholera, the colony is not as unhealthful as are most of the Central American states. Fewer than a thousand of the inhabitants are whites, however, the rest being negroes, Indians and half-breeds.

It was the timber, especially the mahogany, which originally led Great Britain to seize the colony, and timber has remained the chief product. Numerous palms, some of them valuable, grow wild, and bananas and sugar cane are cultivated, but in all the colony there are fewer than 100 square miles of land under tillage. The trade, about one-half of which is with Great Britain, amounts to nearly $6,000,000 annually.

The government of the colony is in the hands of a governor, assisted by an executive and a legislative council, all appointed by the sovereign of Great Britain. The capital and principal seaport is Belize, a thriving town with a population of about 10,500. In the neighborhood of the cities there are fairly good roads, but elsewhere they are very poor, and there are no railways. See CENTRAL AMERICA.

BRITISH ISLES, *iles,* the most important archipelago in the world, bounded by the English Channel, Strait of Dover, North Sea and the Atlantic Ocean. They include Great Britain, consisting of England, Scotland and Wales; Ireland, Isle of Man, the Hebrides, the Orkneys and Shetland Islands and numerous small and unimportant islands and islets. For full description and history, see GREAT BRITAIN; ENGLAND; SCOTLAND, etc.

BRITISH MUSEUM, *mu ze' um,* a great national institution in London, which includes one of the largest libraries in the world, the largest reading room, wonderful collections of antiquities and an exhibit of rare drawings. In 1753, Sir Hans Sloane bequeathed to the nation his various collections, including 50,000 books and manuscripts, on the condition of $100,000 being paid to his heirs. Montague House was appropriated for the museum, and it was first opened on January 15, 1759. Additions to the collection poured in, and the original building soon proved too small; a new building in Great Russell Street was planned, but it was not completed till 1847. In 1857 a second new building was completed at a cost of $750,000.

More recently, the accommodation having become again inadequate, it was resolved to separate the objects belonging to the natural

THE BRITISH MUSEUM

history department from the rest and to lodge them in a building by themselves. Accordingly, a large natural history museum was built at South Kensington, and the specimens pertaining to natural history, including geology and mineralogy, were transferred to it.

General Features. The museum is under the management of forty-eight trustees. It is open daily, free of charge, and there are official guards whose duty it is to point out to visitors the most interesting points. Stupendous as are the collections, they are so arranged and catalogued that they are easy of access, and a large staff of attendants is always ready to furnish assistance. That people avail themselves of the advantages of this great institution which the government keeps up at a cost of approximately $1,000,000 annually, may be seen from the fact that the reading rooms are used each year by about 200,000 persons, while over 700,000 people visit the various departments.

Divisions. Of the eight departments into which the museum is divided, that most used is the department of *printed books*. The vast library, which has almost 4,000,000 volumes and is gaining new ones at the rate of 50,000 a year, occupies forty-five miles of shelves and requires about 600 volumes for its catalogue alone. The circular reading room is 140 feet in diameter and accommodates 200 readers. Admission is by ticket, but anybody may obtain one by complying with a few simple conditions. As in the United States, where a copy of every book published and copyrighted must be sent to the Congressional Library, so in England a free copy of every publication must be forwarded to the British Museum. Frequently, too, it receives valuable gifts.

The *manuscript* collection consists of almost 60,000 volumes, in addition to Oriental manuscripts. Many documents important in state history are to be found there, as well as unpublished memoirs which throw much light on the history of the times in which they were written. *Greek and Roman antiquities,* another department, attracts many visitors, for it contains the famous Elgin Marbles (which see), representing the very highest period of Greek art.

Then, too, there are coins and medals, ancient and modern; British antiquities, which include memorials of the time when Roman civilization was carried to Britain; prints and drawings, and a remarkably complete collection of the remains of Egyptian and Assyrian civilization. Sculptures, papyri, scarabs, mummies, the famous Rosetta Stone (which see), carved inscriptions from thousands of years before the Christian Era—all these are to be found in orderly array in the Egyptian and Assyrian rooms.

The South Kensington natural history collections are as rich as are those in the main museum, and school children by thousands flock there to see the stuffed animals from all over the world, the collections of nests and eggs, and the plants, the gorgeous butterflies and the birds, placed in such natural attitudes and surroundings that they seem to be actually alive. A.MC C.

For a description of the largest library on the Continent, the national library of France, see BIBLIOTHEQUE NATIONALE.

BRITISH NORTH AMERICA ACT, the official title of an act passed by the British Parliament in March, 1867, providing for the formation of the Dominion of Canada. On May 22 the queen issued a proclamation "for uniting the provinces of Canada, Nova Scotia, and New Brunswick, into one Dominion, under the name of Canada," and on July 1, 1867, the act went into effect. This day is now celebrated each year as Dominion Day, the birthday of the Dominion of Canada. See DOMINION DAY.

By the terms of the act, the province of Canada, which had been formed in 1841 by the union of Upper and Lower Canada, was again divided, and the provinces were given the names of Ontario and Quebec. New Brunswick and Nova Scotia retained their names, and provision was also made for the future admission of other provinces.

The British North America Act has been several times amended. The first time was in 1871, when the British Parliament removed doubt as to the power of the Canadian Parliament to establish provinces in the territories acquired after the passage of the original act. Again in 1875 it was found necessary to define more clearly one of the sections regarding the powers of Parliament, and in 1886 the Dominion Parliament was given authority to admit representatives from "any territory which for the time being forms part of the Dominion of Canada but is not included in any province."

For details of the governmental organization established by the British North America Act, see CANADA, subhead *Government.*

BRITISH WEST INDIES, *in'diz,* a term embracing all the islands in the West Indies, sometimes called THE ANTILLES, now in the possession of Great Britain. They include the Bahamas Barbados, Jamaica, Leeward Islands, Windward Islands, Trinidad, Tobago and numerous smaller islands and islets. For purposes of administration they are divided into Crown colouies, governed by the Crown of Great Britain through governors appointed by the king, and colonies with a limited amount of self-government. The federation of all of these British islands into one self-governing colony has been proposed, but has not been found practicable. The climate is healthful, and, although at times the heat is excessive, refreshingly cool breezes blow from the sea. Agriculture is the principal occupation of the inhabitants, most of whom are of mixed races. Sugar, fruit, vegetables and cereals are grown, and the islands produce much valuable timber. Locust trees grow to enormous size, and there are many specimens said to be 4,000 years old. A kind of rice which, unlike the usual species, requires little irrigation, is extensively cultivated. Population, estimated at 1,680,651. Each group of islands named above is described elsewhere in these volumes. For a list of all the outlying possessions of the Empire, see GREAT BRITAIN, subhead *Colonial Possessions of Great Britain.*

BRITISH WEST INDIES (In Black)

BRITTANY, or **BRETAGNE,** *bre tahn'i,* formerly one of the largest provinces in France, now subdivided into five departments, or states, of the republic. It is a peninsula projecting into the Atlantic, between the British Channel on the north and the Bay of Biscay on the south, forming the extreme northwestern and most picturesque portion of France. Brittany is supposed to have taken its name from the ancient Britons, who sought refuge there when driven from the island of Britain between the fifth and seventh centuries. Agriculture is backward, but good crops of corn, grapes and other fruits are raised. The inhabitants along the coast engage in the manufacture of salt; coal, lead and iron are found in small quantities in the interior. The fisheries are quite important. Many remains of works of the ancient inhabitants are found throughout the country, and the native peasants retain the ancient language, which closely resembles the Welsh.

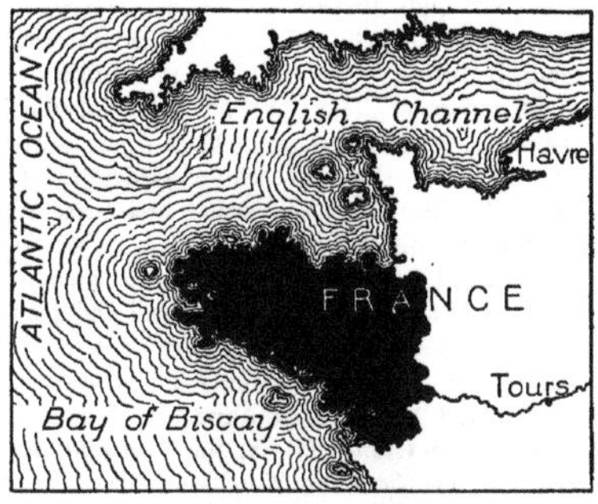

LOCATION OF BRITTANY

BROAD'CLOTH, a wide, wool cloth of superior quality, plain or twilled, and dyed. Owing to the introduction of a standard of widths long ago, the distinction arose between *broad cloth* (cloth of two yards) and *straight* (cloth of one yard wide or narrower). The name now, however, merely signifies a certain cloth of especial smoothness and excellence. From the sheep to garments of broadcloth is a story of many careful processes—shearing of the sheep; selection of best parts; perhaps dyeing; oiling, carding, spinning into yarn; then weaving, shrinking, felting, stretching, napping and shearing; and finally hot pressing, to give it an attractive polish. Then at last it is ready for display, sale and use.

For list of cloths of all kinds see *Related Subjects* list, in article TEXTILES.

BROADSWORD, *brawd'sohrd,* a broad-bladed, single-edged short sword designed for cutting but not for stabbing. The broadsword was formerly used by regiments of cavalry and Highland infantry in the British service, but was not the best weapon for defense. Its use is mentioned in Scott's *Marmion:*

> O Young Lochinvar is come out of the west,
> Through all the wide Border his steed was the best;
> And, save his good broadsword, he weapon had none,
> He rode all unarmed, and he rode all alone.

Swords used in warfare now are more flexible than broadswords and are designed for both cutting and stabbing. The *claymore,* a type of broadsword, but double-edged and longer, was the national weapon of the Highlanders.

BROCADE, *bro kade',* a cloth of silk, enriched with raised flowers, foliage or other ornaments, this pattern often being in gold and silver threads. The term is restricted to silks which are figured in the loom, thus distinguished from those which are embroidered after being woven. Brocade was manufactured in Oriental countries at an early date and in Europe as early as the thirteenth century. It is used on furniture, in hangings and in gowns, being a favored fabric for royal garments. The invention of the Jacquard loom attachment by the French weaver, Jacquard, has made possible many beautiful effects in the weaving of brocade and other cloths.

For list of cloths of all kinds, see *Related Subjects,* in article TEXTILES.

BROCK, SIR ISAAC (1769-1812), a British soldier whose distinguished service in America won for him the title "Hero of Upper Canada." Early in his career he served in the West Indies, in Holland and with Nelson at the Battle of Copenhagen, and in 1802 was sent to Canada to suppress a rebellion of the troops. In 1806, when war between England and the United States was threatening, he was placed in command of the garrison at Quebec, and soon attained the rank of major-general. On the outbreak of the War of 1812 Brock made active preparations for the defense of the Canadian frontier; on August 16, 1812, he captured Detroit, where General William Hull was in command, receiving the ignominious surrender of the entire American force. Brock was mortally wounded on October 13 while leading his men against an American attack on Queenstown (now Queenston), near Niagara

Falls. On Queenston Heights, the site of the battle, a magnificent monument has been erected to honor his memory.

TOWERING ABOVE QUEENSTON HEIGHTS

BROCK'TON, MASS., famous as a shoe-manufacturing center, especially of men's high-grade shoes, in which it leads the world. It is the county seat of Plymouth County, in the southeastern part of the state, twenty miles south of Boston and twenty-nine miles north of Fall River. The New York, New Haven & Hartford Railroad serves the city, and there is trolley connection with fifteen large surrounding towns. In 1915 the population, which is a mixture of Americans, Irish, English-Canadians and Swedes, numbered 62,288, an increase of 5,410 since 1914. Brockton was settled in 1700, and was incorporated as the town of North Bridgewater in 1821. Its present name was adopted in 1874, and the charter was granted in 1881. The area exceeds twenty-one square miles.

The city is the industrial center for all the people living in the surrounding towns of East and West Bridgewater, Avon, North Easton, Randolph, Whitman and Holbrook. About 17,000 people are employed in the thirty-five shoe factories of the city; the annual output amounts to $65,000,000. The manufacture of shoe accessories is also extensive, and of rubber goods, wooden and paper boxes, sewing machines, pianos, automobiles, gloves, razors and sporting goods the output is large.

Brockton has a Federal building, a million dollar city hall, three hospitals, a business college and a fine public library with an historical room, lecture hall, art gallery and a library of 70,000 volumes. The city had 2,700 individual home and school gardens in 1915, when it won first prize for the best school garden in the state. In 1916 it was awarded first prize for the purest milk supply. E.C.J.

BROCKVILLE, ONT., the county town of Leeds County, an important railroad and manufacturing center, on the northern bank of the Saint Lawrence River, seventy miles south of Ottawa and 126 miles southwest of Montreal. It is on the Grand Trunk, the Canadian Pacific and the Canadian Northern railways, and is a port of call for Saint Lawrence steamers. The town was named in honor of Sir Isaac Brock (see above), the hero of the Battle of Queenstown Heights. Population in 1911, 9,374; in 1916, about 10,000.

The industrial importance of the town is due partly to its manufactures, of which the most important are stoves, hardware, steam engines, agricultural implements, hats and gloves, carriages and automobiles. It is also an important dairying center, and has a large trade in butter and cheese. It is the headquarters of the Eastern Ontario Dairymen's Association. Brockville is also a resort for tourists and sportsmen, due to the proximity of the Thousand Islands on the southwest and of many small lakes, abundantly stocked with black bass, salmon, trout, pike and pickerel, from twenty to forty miles to the north and west. Within the town itself are several beautiful spots, of which Saint Lawrence Park, covering fifty acres, is noteworthy. G.A.K.

BRODEUR, *bro dur'*, LOUIS PHILIPPE (1862-), a Canadian statesman and jurist, since 1911 a judge of the Supreme Court of the Dominion. He was born at Beloeil, Que., and was educated at Saint Hyacinthe College and Laval University. He was admitted to the bar in 1884 and from 1891 to 1911 sat in the House of Commons. In 1900 he was chosen Speaker of the House, but resigned in 1904 to become Minister of Inland Revenue; in 1907 he became Minister of Marine and Fisheries. With W. S. Fielding he negotiated the first treaty ever put into effect by a British colony—the

French-Canadian treaty of 1907. He accompanied Sir Wilfrid Laurier to England as a delegate to the Colonial Conference of 1907 and was a delegate to the Imperial Defense Conference in 1909. In 1910 he introduced the first naval bill ever considered in the Canadian Parliament, and in the following year, when the naval service was organized, was appointed Minister of the new department. Before the end of 1911 he resigned from Parliament and from the Cabinet to become a judge of the Dominion Supreme Court.

BROKER, one who acts as an agent for another in a business transaction, and charges a certain sum for his services. Such charge is called *commission,* or *brokerage,* and is always based on a definitely-arranged percentage of the sum involved in the transaction. Brokers usually confine themselves to one line of business, being called accordingly *stock brokers,* those who deal in stocks and bonds, *bill brokers, insurance brokers, ship brokers,* etc.

Although both conduct business in practically the same manner, there is a difference between a commission merchant and a broker. Both are paid by commission, but while the commission merchant sells and makes delivery of certain goods while acting as an agent, the broker does not necessarily have possession of the goods or stock bought and sold. The broker forms the connection between two principals and hands to each a written statement of the bargain.

BROMIDES, *bro'midz,* the name given compounds of bromine with potassium, silver and other metals. Bromide of potassium is extensively used in medicine and photography. It is found in the form of white crystals, shaped like those of common salt. As a medicine it sometimes produces a soothing effect on the nerves, but it is dangerous to use except on a physician's prescription. It is used in photography in making silver bromide, which is very sensitive to the light, and is used in preparing sensitized plates and films (see PHOTOGRAPHY).

Bromine, *bro' min,* one of the two chemical elements which is a liquid at room temperature, the other being mercury. Bromine takes its name from its offensive odor (Greek *bromos,* meaning a *stench*). It is a heavy, red-brown liquid which produces painful burns when spilt upon the skin. It resembles chlorine in its chemical behavior towards other elements. It is obtained as a by-product of the salt industry. J.F.S.

BRONCHITIS, *bron ki' tis,* inflammation of the mucous membrane lining the bronchial tubes, a disease which is characterized by a distressing cough and discharge of mucus from the air passages. It occurs in two forms—*acute* and *chronic* bronchitis. The former may be caused by exposure to cold and wet or by the inhalation of dust or irritating gases, or it may occur as a complication of certain other diseases.

Acute bronchitis frequently develops into pneumonia. Besides coughing and expectorating mucus, the patient suffers from sore throat, fever, pain in the chest and shortness of breath on exertion. The sputum is scanty in the beginning, but later becomes abundant, and there is sometimes pus present in it. The fever may rise to 103°, but does not usually last as long as the other symptoms. Poultices to relieve the cough and steam inhalations are helpful remedies. On the approach of an attack the patient should take a hot bath, go to bed and take a laxative. Cold sponge-baths taken every morning and careful attention to the ventilation of the sleeping room will help ward off attacks.

Chronic bronchitis, with symptoms similar to those described above, but not so severe, is a common affliction of the aged. Many cases supposed to be this form of bronchitis are really consumption, and their diagnosis and treatment should be placed in the proper hands. W.A.E.

BRONTE, *bron'te,* CHARLOTTE (1816-1855), an English novelist whose writings, even her popular *Jane Eyre,* possess no more fascination and no greater appeal than does her life, gloomy and almost eventless as it was. Her clergyman father removed in 1820 from Thornton, where Charlotte was born, to Haworth, and there in the dreary parsonage, with the stern Yorkshire moors on every side, the remarkable Bronte children grew up. Charlotte was the eldest of four who survived childhood. For a time in their childhood Charlotte and her sister Emily attended a school to which the former brought later an unenviable fame by her descriptions in *Jane Eyre,* but their education was continued elsewhere. After teaching and serving as governess, Charlotte went with Emily in 1842 to Brussels, with a view to learning French and German.

She taught for a year in the school she had attended there, but returned to Haworth in 1844, convinced that whatever happened she must remain home with her family. Her father was rapidly becoming blind, her brother was drinking himself to death, and her two sisters were feeling the first touches of a disease which was shortly to cause their death. To support themselves the three sisters turned to writing, and in 1846 published a volume of poems under the names of Currer, Ellis and Acton Bell. It was issued at their own risk and attracted little attention, so the sisters turned to fiction and each produced a novel. Charlotte wrote *The Professor,* but it was refused by publishers everywhere and did not appear until after her death. *Jane Eyre,* her next novel, had a different fate, for on its appearance in 1847 it took the world by storm. It had faults of style, but its realism, and above all its passion, were a new note in literature. A second novel, *Shirley,* appeared in 1849, and *Villette,* based on her experiences in Brussels, in 1852. Meanwhile her sisters and brother had died, and an unbroken gloom settled over her life. In 1854 she married her father's curate, the Rev. Arthur Nicholls, and she had a few happy months. Worn out, however, by the tenseness of her nature and the violence of her inner protests against her lot, no less than by the cold, cheerless home and bleak climate, she died in 1855. Mrs. Gaskell, in her *Life of Charlotte Bronte,* shows to the reader very clearly the dreariness of the "home among the graves" and the unquelled spirit of the Bronte sisters.

BRONZE, *bronz,* a hard, durable, sounding metal, which melts easily and is capable of enduring exposure to the weather and of taking a fine, smooth finish. These qualities, combined with its possibilities for artistic coloring, make it valuable for statuary, lighting fixtures and ornamental work for both the exterior and interior of buildings. It is an alloy of copper and tin in varying proportions, with occasionally the addition of small quantities of lead, zinc and silver (see ALLOY). The most common varieties of bronze in use are *gun metal,* used in making ordnance; *bell metal; specular metal,* used for making mirrors and reflectors in telescopes; *statuary bronze,* used in sculpture; *aluminum bronze,* a composition of copper and aluminum, closely resembling gold; and *manganese bronze,* often called white bronze, a composition of iron and manganese with other bronzes. Japanese bronzes contain quite a large proportion of lead, which makes them softer.

Bronze has been known from a very early period of history. The Chinese and ancient Egyptians were familiar with it centuries before the Christian Era, and it is supposed that their early bronzes were produced by smelting the ores of the metals. It was also used by the Assyrians and Romans. See BRONZE AGE, below.

BRONZE AGE, a term describing that period in the development of mankind when bronze, made from a mixture of copper and tin, as to-day, was used as the material for weapons, implements and ornaments. It stood for a distinct advance in civilization, for before that time people had contented themselves with stone, a material which lay ready to their hand. Iron had not yet been discovered, or at least had not been reduced to such form to be useful.

The Bronze Age is not an absolute division of time, but a relative condition of culture, which in some places may have been reached early, in others late; in some it may have been prolonged, and in others brief, or even nonexistent, the people passing directly from the use of stone to that of iron. It is believed, however, that over much of Europe there was this stage approximately eighteen centuries B. C., though it frequently overlapped the ages before and after it, bronze being used side by side with the earlier stone or the later iron. In North America, too, there was undoubtedly a Bronze Age, though the use of this metal compound seems never to have been very extensive. See STONE AGE; IRON AGE.

BROOCH, *brohch,* an article of jewelry used for fastening the dress, or for adornment only. It has a pin passing across it, which is fastened at one end with a joint and at the other with a hook. Following the thorns used in primeval

BROOCHES
(*a, b*) Modern brooches; (*c*) brooch of the sixteenth century; (*d*) Anglo-Saxon specimen; (*e*) modern cameo brooch.

times to hold together the fig-leaf garments, came the use of wood and bone. Then came the use of metals in the form of brooches, worn by both men and women in Greece and Rome and in Europe from the time of Homer to the fall of the Western Empire. Typical of the times and the people, they often bore inscriptions, and it is an interesting fact that the oldest example of Latin now in existence is inscribed on a brooch. Brooches were often used, also, as a kind of amulet or talisman. They are now made chiefly of gold, silver or platinum, some set with jewels or decorated with enamel, while others are merely engraved.

BROOK FARM, an experiment in "brotherly coöperation" which owes much of its continued fame to Hawthorne's *Blithedale Romance.* In 1841 a number of the New England Transcendentalists (see TRANSCENDENTALISM) decided to put some of their theories into practice, and accordingly bought at West Roxbury, Mass., 200 acres of land, where they organized the Brook Farm community, under the direction of George Ripley. The object was to prove that people might live happily and inexpensively on the product of their labor, and have abundance of time for intellectual pursuits. Hawthorne, Emerson, Margaret Fuller, Charles A. Dana, George W. Curtis and Amos B. Alcott were at various times members of the community, working with the rest at their allotted tasks for a certain period each day. All were paid equally for their labor; all had the same claim on the property and products of the establishment. Of this interesting experiment Holmes said that everything was common there but common sense.

At one time there were as many as seventy members, and visitors from all over America came to view the experiment; but financial difficulties, together with the destruction of the chief building by fire, led some of the members to withdraw, and in October, 1847, the association was dissolved.

BROOKLINE, MASS., the wealthiest suburb of Boston, also claims to be the richest and most beautiful town of its size in the world. The population is largely American, with a mixture of Hebrews; it increased from 27,792 in 1910 to 33,490 in 1915. The town is situated in Norfolk County, about three miles southwest of the State House in Boston, on the Boston & Albany Railroad. It is also connected with Boston by an electric road. The area is more than six square miles.

Brookline has long been regarded as a model suburb. It is laid out as a vast park, with elegant villas and country seats, beautiful gardens and shrubbery. On the northeastern border of the town flows the Charles River. Frederick Law Olmsted, the landscape gardener, resided here, and the floral beauty of the town bears testimony to his art. Brookline is connected with Boston Common by the boulevards of the Metropolitan park system. At Clyde Park are located the clubhouse and grounds of the Boston Country Club, and Corey Hill affords a fine view of Boston and the surrounding country. The town hall is built of granite. The library contains about 64,000 volumes. The wealth of the town is estimated at $150,000,000; of this sum $17,000,000 is deposited in one bank. Though chiefly a fashionable residential district, it has manufactories of electric appliances and motors.

Brookline was settled in 1635 under the name of Muddy River, and was incorporated as a town under its present name in 1705; up to 1793 it belonged to Suffolk County, in which Boston is situated. The growth of the latter city has almost surrounded Brookline, and on account of its great wealth many attempts have been made to annex it, which so far have not succeeded. The town limits include the villages of Cottage Farm, Longwood and Reservoir Station. S.C.J.

BROOKLYN, N. Y., the largest city in America ever absorbed by another, was at the census years of 1860, 1870 and 1880 the third community in size on the continent. It lost this latter distinction to Chicago ten years afterward, and in 1898 gave up its individual government and became a part of New York City. Were it still independent it would again be the third city, for it now contains more people than Philadelphia. In 1910 the census reported 1,634,351 population, but there are now nearly 2,000,000 within its area.

Though it is generally thought of as "The Sleeping Room of New York," Brooklyn is also a great manufacturing city, exceeding in the value of its products all others but New York (Manhattan), Chicago and Philadelphia. Especially famous are the mammoth sugar refineries and the Brooklyn Navy Yard, where warships are built.

With the island of Manhattan, Brooklyn is connected by four suspension bridges, by subway tubes, railway tunnels and ferries. It is situated on Long Island, and occupies the whole of King's County. Its name survives from the Dutch hamlet of Breuckelen, founded in 1636. The chief influences which have kept the name alive since incorporation with New York are the world-famous Brooklyn Bridge and its newspapers, one of which, the *Brooklyn Daily Eagle,* was once edited by Walt Whitman.

Henry Ward Beecher, who preached in the city forty years, was perhaps its most famous citizen. See NEW YORK (City) suburban map.

BROOKLYN BRIDGE, a bridge over the East River, connecting New York with Brooklyn, famous because at the time of its construction it was considered the greatest suspension bridge in the world, and still one

BROOKLYN BRIDGE
The original suspension bridge across East River.

of the most notable. It was fourteen years in building, having been begun in 1869 and finished in 1883, at a cost of $15,000,000. The center span, between the towers, is 1,595½ feet; the side spans at either end are 930 feet; and as there is an approach of 1,562½ feet on the New York side and of 971 feet on the Brooklyn side, the total length of the bridge is 5,989 feet, or somewhat more than a mile. Four great cables nearly sixteen inches in diameter support the bridge, their resting place on the towers being 329 feet above high water. With a total width of eighty-five feet, the bridge carries a roadway, a double line of electric railway, and a broad promenade.

As the New York metropolitan district has grown the one bridge has been found inadequate, and three others of like proportions, the Manhattan, Williamsburgh and Greensboro, have been built farther up East River. The latter was opened for traffic during 1915. These also are suspension bridges, but of a somewhat improved type.

See BRIDGE, for another illustration of one of the New York-Brooklyn bridges.

BROOKS, PHILLIPS (1835-1893), one of the great pulpit orators of America, for twenty-two years the rector of Trinity Church, Boston, and for the last two years of his life bishop of the Protestant Episcopal Church in Massachusetts. Brooks was born in Boston; he traced his ancestry on his father's side to the Rev. John Cotton of Puritan fame, and on his mother's side to the founder of the Phillips academies. He entered Harvard when Lowell, Holmes, Agassiz and Longfellow were teaching there, was graduated with high honors in 1855, and then studied for the ministry at the Alexandria (Va.) Protestant Episcopal seminary. Following his ordination in 1859, he became rector of the Church of the Advent in Philadelphia, and between 1862 and 1869 had charge of the Holy Trinity Church of that city.

PHILLIPS BROOKS

During his long period of service as rector of Trinity Church in Boston, Brooks became one of the best-known men in America, distinguished alike for his broad, liberal views, intellectual gifts, eloquence and winning personality. He wrote widely on religious subjects, important titles of his works including *Yale Lectures on Preaching, The Influence of Jesus* and *The Light of the World.* Among his several Christmas and Easter carols the best-loved is *O Little Town of Bethlehem,* a stanza of which is here given:

> O little town of Bethlehem
> How still we see thee lie!
> Above thy deep and dreamless sleep
> The silent stars go by;
> Yet in thy dark streets shineth
> The everlasting light;
> The hopes and fears of all the years
> Are met in thee to-night.

BROOM, the name of an ornamental shrub of the pea family. The common broom of Europe is bushy, with straight, angular dark-green branches and deep, golden-yellow, butterfly-shaped flowers. From the botanical name, *planta genista,* came the name of the royal English family Plantagenet, and one of this family used the broom for his crest. In Europe the broom is used for tanning and dyeing, and the fibers are made into cloth. The tops and seeds have been used in medicine in cases of dropsy. This plant must not be confused with broom corn, which is described below. See PLANTAGENET.

BROOM CORN, or **BROOM GRASS,** a member of the grass family, native of the East

Indies, but extensively cultivated in the United States for the single purpose of furnishing material for making brooms. The plant is of the same family as sorghum and kafir corn.

BROOM CORN

There are two kinds, standard and dwarf. The standard grows to a height of eight or ten feet; the dwarf, four and a half to six feet. The stalk is pithy, with long, pointed leaves which enclose it, as in corn. The stem is jointed and topped by branched clusters of seed heads, which are used in making carpet brooms and clothes brushes. Before the plant is fully matured the stem is broken over about eighteen inches from the top. This part is cut off, dried quickly in an airy, shady spot, and the seeds are removed. It is then ready for broom-making.

The ground for a broom-corn crop should be thoroughly prepared before seed planting, and during the growth of the plant should be continually cultivated and kept free from weeds. An acre will produce from 500 to 600 pounds of broom-making material. Oklahoma leads the American states in production, with about fifty per cent of the total yield of nearly 80,000,000 pounds; the 42,000,000 pounds from that state annually are worth over $2,560,000. Illinois is second with twenty-eight per cent of the crop, and Kansas third, with twelve per cent.

Brooms and Broom-Making. At one time (and in many places in Europe even now) brooms for rough sweeping were made of twigs; long-haired brushes were used in housework. In 1850, however, Americans discovered the value of broom corn for sweeping, and in 1859 Ebenezer Howard started a broom-making factory at Fort Hunter, N. Y.; this city is now the Eastern center of that industry. The corn is sorted as to size and quality, the green, tough, springy fibers, free from seed, being the most desirable. It is then sent to the factory in bales. The required number of the long fibers are bound by wire to a turned stick or handle about four feet long, enlarged at the end to which the brush is fastened. The broom is then flattened in a vise and sewed. The ends are trimmed evenly, and the broom is then ready for the market. Whisk brooms, or small brooms for brushing clothing, are made in a similar manner, but are of the finer fibers and often have ornamental handles. Most of the work of broom-making is by hand, and is often done in penitentiaries and by the blind, for sightless people easily learn the comparatively-simple operations. Quantities of broom corn are exported to Europe each year. The annual value of brooms from factories in the United States is nearly $20,000,000. M.S.

BROTHER JONATHAN, a popular personification of the United States, or rather, of its people, collectively. Various accounts of its origin are given, but the most likely one traces it back to George Washington and his friend and adviser, Jonathan Trumbull. Trumbull, governor of Connecticut during the Revolutionary War, was so wise a counselor that Washington was frequently heard to say in times of uncertainty, "We must ask Brother Jonathan." The remark at length became a proverb, the name losing its first meaning and broadening to include the whole people. It always kept more than a hint of its most favorable significance, however, for *Brother Jonathan* represents only the broad-minded, trustworthy phase of the national life. Other very familiar personifications, which differ in that they represent not the people but the nation as a political whole, are *John Bull* for England, *Uncle Sam* for the United States and *Johnnie Crapaud* for France. See BULL, JOHN.

BROWN, a color obtained by the mixture of black and the primary colors red and yellow. A large number of shades and tints may be produced by mixing in varying proportions, and by adding other colors a still greater variety is obtainable. There are various brown coloring matters, most of them being mineral pigments; umber and sepia are good examples.

BROWN, ELMER ELLSWORTH (1861-), a distinguished American educator, university president and former United States Commissioner of Education. He received his education at the Illinois State Normal University, at the University of Michigan and in Germany. After filling several public school positions, Dr. Brown was chosen assistant professor of the science and art of teaching in the University of Michigan in 1891. From there he went to the University of California as associate professor of pedagogy, and in 1893 was appointed head of the department. In June,

1906, he succeeded William T. Harris as Commissioner of Education, resigning in 1911 to become chancellor of New York University. He has written *Origin of American State Universities* and other educational books.

BROWN, George (1818-1880), a Canadian journalist and statesman, at once one of the most honored and most opposed of the men who have figured in the political history of Canada. His public career covered a period of bitter political strife, when Canada was faced by great questions which meant failure or success to the country. All these questions, the relation of the Church to the State, the problems of education, and especially of Confederation, Brown studied with a zeal which brushed aside formalities and surface appearances. Once he had made up his mind, his convictions were permanent. He was never inclined to accept discipline or criticism, and his opinions were always forcibly expressed, even though they made him unpopular. Such a man could not compromise, and in the stormy days preceding Confederation only compromise could give a man high public office. Yet it is a tribute to his ability and to his keen and accurate vision that his influence was so great. He correctly gauged and not infrequently led public opinion.

GEORGE BROWN

Brown was born in Edinburgh, Scotland, where he received a high school and academy education. At the age of twenty he came to America with his parents, settling in New York. Here the father and son conducted a weekly paper which was the organ of the Scotch Free Church adherents, but after 1843 George Brown made his home in Toronto. He founded the Toronto *Globe,* still one of the famous papers in the Dominion, and made it the organ of reform. He became the friend of Baldwin, Lafontaine and other statesmen, and in 1851 was elected to the Assembly of Canada. On July 31, 1858, he was appointed premier, and formed the Brown-Dorion ministry. Four days later he resigned, the governor having refused to dissolve the Assembly and call a new general election.

With the exception of two years (1861-1863), Brown continued to sit in the Assembly until 1867. He was an ardent advocate of Confederation, was a member of the Charlottetown and Quebec conferences, and was President of the Council in the coalition ministry of Sir Etienne Taché. Until his death he retained the active management and editorship of the *Globe.* In 1873 he accepted a nomination to the Dominion Senate, but subsequently refused the lieutenant-governorship of Ontario and the honor of knighthood.

BROWN, Henry Kirke (1814-1886), an American sculptor, noteworthy in the days when the United States was seeking and finding for itself a place among the producers of art. Especially well known are his *Indian and Panther,* the first bronze sculpture cast in the United States; statues of Winfield Scott, Nathanael Greene, De Witt Clinton and Lincoln; and, finest of all, the equestrian statue of Washington, in Union Square, New York. Few of the equestrian statues produced by later sculptors outrank this. Brown was born at Leyden, Mass., studied first in Cincinnati and from 1842 to 1846 in Italy; but though it was there that he became master of his art he never ceased to oppose Italian influence and to strive for something more truly national. At the outbreak of the War of Secession he was engaged on a group of figures for the state house in Columbia, S. C., which he was obliged to leave unfinished.

BROWN, John (1800-1859), one of the most extreme of American abolitionists, whose name still lives in the widely sung—

John Brown's body lies a-mouldering in the grave
But his soul goes marching on.

Brown was born at Torrington, Conn., of *Mayflower* ancestry, and spent the years of his young manhood in aimless wanderings, living at various times in Connecticut, Ohio and New York. Unwilling to learn any trade, he earned but a scanty living for his twenty children. When the Kansas-Nebraska Bill was passed, allowing the Kansas

JOHN BROWN

settlers to decide whether the state should be slave or free territory, enthusiasts from both factions thronged to the state, and among the strongest of the free-state men was Brown. In the fierce warfare which was carried on for several years in Kansas and Missouri he proved an expert fighter, and his victory over a band of Missourians at Osawatomie won him the popular name of "Osawatomie Brown."

But he was not content with resistance of this nature; he had formed a more aggressive plan to free the slaves, and on the night of October 16, 1859, put it into effect. With about a score of followers he descended upon

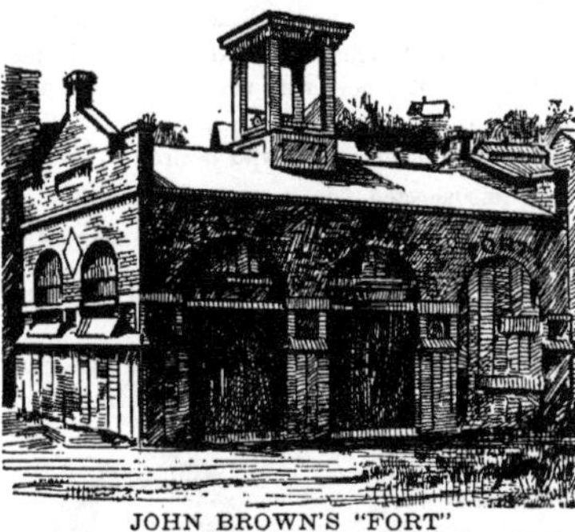

JOHN BROWN'S "FORT"

An engine house at Harper's Ferry which Brown held and later lost to Colonel Robert E. Lee, commanding government forces.

Harper's Ferry, in West Virginia, and seized the national arsenal there, believing that the result would be an immediate rising of the slaves. No such thing occurred, however, and within a day or two United States troops under Robert E. Lee, later the great leader of the Confederate armies, regained possession of the arsenal and took prisoners Brown and such of his followers as had not fled. The leader was tried at Charlestown, convicted of treason, and on December 2, 1859, was hanged.

Intense was the excitement which the insurrection kindled; the South naturally looked upon Brown as nothing less than a criminal, while many in the North, even including some of the more temperate abolitionists, regarded him as a martyr to conscientious, if mistaken, enthusiasm. The Harper's Ferry insurrection had no immediate consequences of great importance, but the fact that it was condoned in the North made the South more determined in its defense of slavery, and it was one of the indirect causes of the War of Secession (which see).

BROWNE, Charles Farrar (1834-1867), an American humorist who, under the name of Artemus Ward, became one of the most popular lecturers of his day. He was born in Waterford, Me., and had to begin his career with very little schooling, as he was his mother's only support. His first position was that of typesetter in a printing office; then he became a reporter, and, finding that the anecdotes he printed were widely quoted, decided to make use of them on the lecture platform. His lectures took him west to the Rocky Mountains, and he also visited England, where his unusual style of fun-making was very popular.

He was accustomed to say the most absurd things with an air of great solemnity, and his unexpected turns and ridiculous puns used to set his audiences into unbounded laughter. "Africa is famed for its roses," he would say. "It has the red rose, the white rose and the neg-roes"; or, "If spring is some, June is summer." Often he would give such sound advice as "Always live within your income, if you have to borrow money to do it"; and he admonished the Prince of Wales to be as "good a man as his mother was." The lectures of Artemus Ward, in book form, with their impossible spelling and grammar, preserve for the modern reader the best sayings of this humorist, but do not excite the mirth they did when the lecturer's personality made them the best of their kind.

BROWNIE, a fairy-like creature in Scotland's superstitions, formerly believed to haunt houses, particularly farmhouses. It was assumed he was very useful to the family, particularly to good servants, for whom he merrily did many acts of drudgery while they slept. If offered food or pay for his tasks, he disappeared and never came again. The brownie bears a close resemblance to the Robin Goodfellow of England and the Kobold of Germany. Many stories have been woven about this little elf. The best known and most popular are the *Brownie Books* of Palmer Cox, the American artist and writer for children. See Cox, Palmer.

BROWNING, Elizabeth Barrett (1806-1861), considered by competent critics the greatest woman poet that England has produced. Her marriage to one of the most eminent poets of his day did not lead to the eclipse of her genius by his, but rather to the strengthening of both (see Browning, Robert).

Elizabeth Barrett, born at Coxhoe Hall, in

Durham, grew up at Hope End, in Herefordshire, where she spent a most happy childhood. Some of her poems, written years later, show that her joy in the beautiful out-of-doors about her in her girlhood remained with her all her life. She was never very strong, but her mind was alert and vigorous, and she found pleasure in reading which would be far too difficult for most children. Very early she began to write poetry, but of these early poems she was ashamed in her later years. When she was about twenty, however, she published a little volume, *An Essay on Mind, and Other Poems,* and from that time on her rise to recognition was steady.

ELIZABETH BARRETT BROWNING

After about 1835, when the family moved to London, she was an invalid, confined to her room, but her letters show her to have been possessed of a cheerful, gallant spirit. In 1838, however, there occurred a tragedy in her life, from the shock of which she never fully recovered. Her favorite brother was drowned while at Torquay alone with her, and for a time it seemed as if her life, too, would be sacrificed. But she rallied, and began to produce poems again, some of her best-known works, as *The Cry of the Children* and *Lady Geraldine's Courtship,* coming from her sickroom. A reference to Robert Browning in this latter poem led to an acquaintance which grew into mutual love, and in 1846 the two were married. The union was unusually happy, her only grief being that her father was opposed to it, and never forgave her. From the time of their marriage the poets lived in Italy, where Mrs. Browning's health was far better than in England. She died and was buried in Florence, a city which she loved.

Her greatest work, in the opinion of most critics, is the *Sonnets from the Portuguese,* which bear comparison with the finest sonnets in the English language and perhaps surpass all others which deal with the same subject; for they are love-sonnets, recording the growth of her love for Browning and his for her. Written during her engagement, they were not shown even to Browning until after their marriage, and then he insisted upon their publication, allowing the use of the words *From the Portuguese* simply as a disguise. Mrs. Browning's longest work, and the one she herself ranked highest, is the narrative poem *Aurora Leigh.* While not autobiographical in its story features, it aimed to present a picture of Mrs. Browning's ideals and beliefs. C.W.K.

BROWNING, Robert (1812-1889), one of the most distinguished and original thinkers of England. Some critics hold that he is more philosopher than poet, but such an opinion fails to do justice to the musical quality of much of his verse, as well as to his truly inspired poetic touches. Could it be said that the man is not a poet who wrote such lines as—

Oh lyric love, half angel and half bird,
And all a wonder and a wild desire,

Some unsuspected isle in far-off seas.

That's the wise thrush; he sings each song twice over
Lest you should think he never could recapture
The first fine careless rapture.

And yet it is not for smooth, lilting lines that Browning is best known and best loved, but for the strength and optimism that show through his rugged verse—an optimism that is far from being mere placid acceptance, but sees the good in man despite what is evil.

ROBERT BROWNING

Early Life. He was born in Camberwell, a suburb of London, on May 7, 1812, and grew up amid pleasant surroundings. His father and his mother were in sympathy with his aspirations, and seem to have known how to direct his education so as to bring out the best that was in him. The fact, too, that he inherited perfect health had much to do with the pure physical enjoyment of life which he so often expressed in his poems, as in the lines from *Saul*—

How good is man's life, the mere living! how fit to employ
All the heart and the soul and the senses forever in joy!

He studied under tutors and for a brief time at University College in London, but most of his education came from the books which surrounded him from his childhood. In all his works there is evident a very wide acquaintance not only with the literature that everyone reads but with obscure works that come within the reach of comparatively few. This accounts for the difficulty many experience in reading some of his poems. They are filled with allusions to facts and fancies which only a person as widely read as Browning himself could hope fully to understand. Travel on the Continent did much to broaden his outlook on life and to convince him that his early attempts at writing poetry were feeble and immature. The works of Keats and Shelley were a genuine inspiration to him, and confirmed him in the desire to devote his life to poetry.

Marriage with Elizabeth Barrett. In 1844 Browning became acquainted with Elizabeth Barrett through calling on her to thank her for a compliment she had paid him in one of her poems. Friendship grew into love, and in 1846 they were married. Their life together was very beautiful, and her death in 1861 was a shock from which Browning never completely recovered (see BROWNING, ELIZABETH BARRETT). He removed from Italy, where all his married life had been spent, to England, that he might educate his son. In England he was very popular socially. Later, however, he returned to Venice, where he died. His body was taken to England and buried in Westminster Abbey.

As a Poet. From the time his first poem appeared, in 1832, he wrote rapidly, revising little. He seemed unable to revise his work, and this probably kept him from attaining the faultless form which distinguishes Tennyson's poetry. The form which he adopted, however, rugged as it is, fits far better the content of his poems than would more smoothly-flowing measures. And at times his lines have all the swing desired by the most music-loving reader, as for instance, these lines:

> Just for a handful of silver he left us,
> Just for a riband to stick in his coat.

To Browning the most fascinating of all studies was the human mind, and he was able to analyze it and to describe its experiences as perhaps no other English poet except Shakespeare has ever been able to do. His genius was distinctly dramatic, and had he lived in an age when the drama was the chief form of literary expression, he might have done his best work in that field. As it was, however, he brought the dramatic monologue to a high point of perfection, such poems as *My Last Duchess, Andrea del Sarto, The Bishop Orders His Tomb at Saint Praxed's, Fra Lippo Lippi* and *A Forgiveness* showing him at his best. *The Ring and the Book,* considered by many critics his masterpiece, is a series of monologues forming one great poem. Besides the poems mentioned above, his best-known works are the dramas, *A Blot in the 'Scutcheon, In a Balcony, Pippa Passes* and *Colombe's Birthday; Saul, Rabbi Ben Ezra,* and the poems comprised in the collection called *Men and Women.* Browning is for the most part distinctly not a children's poet, but such ringing dramatic poems as *An Incident of the French Camp, How They Brought the Good News from Ghent to Aix* and *Hervé Riel* will appeal to any boy who likes war and action. C.W.K.

BROWNSVILLE, TEXAS, county seat of Cameron County, is the southernmost city of the state. It is situated on the north bank of the Rio Grande, opposite Matamoras, Mexico, and about thirty-five miles from the Gulf of Mexico. Galveston is 372 miles northeast, Corpus Christi 157 miles northeast and Laredo 234 miles northwest, up the river. The population of Brownsville is more than fifty per cent Mexican. In 1910 there were 10,517 inhabitants; in 1914 there were 12,310. The area of the town is nearly eight square miles.

The value of exports and imports between the United States and Mexico, handled by the United States customhouse at Brownsville, exceeds $2,150,000 annually. The city is the market for a large, developing agricultural and cattle country. The important products are stock, garden vegetables, corn, cotton, fruits and sugar cane. A large sugar mill and oil mills are the chief commercial establishments.

The prominent buildings of the town are the United States customhouse; a Federal building costing $50,000, erected in 1889; Cameron County courthouse, costing $250,000; two large bank buildings, a hospital, library and the Roman Catholic cathedral. Brownsville is the seat of a Roman Catholic college and convent. On Point Isabel, on the coast, is a government wireless station.

As a border town Brownsville has been the scene of a number of minor battles. A Mexican settlement occupied the site before the Mexican War. It was at that time fortified by General Zachary Taylor. Major Brown, whose name was later given to the town, was put in command of the fort and lost his life in its defense. The Battle of Resaca de la Palma was fought near here in 1846. In 1859 the town was besieged and occupied by Mexican raiders. It was made a Federal blockade port in 1863. The last engagement of the War of Secession took place at Palmetto Ranch, near the battlefield of Palo Alto, May 13, 1865.

The place was settled by Americans in 1848, and incorporated in 1853. The commission form of government was adopted in 1916. The water works, electric light plant and street railway are owned by the city. L.B.

BROWN-TAIL MOTH, originally a European moth, whose caterpillars are very destructive to orchard, forest and shade trees. It was accidentally brought into Massachusetts about 1890, where it has done much damage, and has gradually spread through New England. The wings of this moth are pure white, the name brown-tail being given it on account of a bunch of brown hair at the tip of the abdomen of the female. With wings outspread the female moth is one and one-half inches across, the male being slightly smaller. On mornings during the flying season hundreds of these moths can be seen collected on poles or posts near electric lights, whence they scatter to trees.

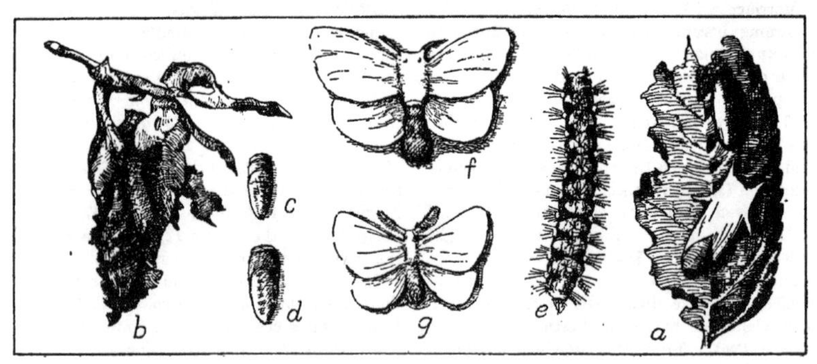

BROWN-TAIL MOTH
(a) Egg mass and moth laying egg
(b) Winter nest
(c) Male pupa
(d) Female pupa
(e) Full-grown caterpillar
(f) Female moth
(g) Male moth

The first three weeks in July the female deposits her eggs—about 250 of them—under the tip of a leaf. About fifteen or twenty days later they develop into the destructive caterpillar, about two-thirds of an inch long, with a wavy line of light spots on each side of its back and two red spots at the end. Its barbed hairs contain a poison which causes a burning itch when it touches the skin.

The young caterpillars feed on the outer coat of leaves, and they also eat into apples and pears. In October they spin their winter webs —grayish silk nests—and so, attached to twigs, they stay until April. The greatest damage is done in the spring by the hungry caterpillars emerging from the nests. The destruction of these nests in the winter by removal and burning is the best means of exterminating the moths. Spraying with kerosene emulsion or a strong solution of arsenate of lead or even strong soapsuds destroys the caterpillar (see INSECTICIDES). See GYPSY MOTH.

BROWN THRASHER, often incorrectly called a brown thrush, is a handsome, reddish-brown bird longer than the robin. It has a long tail, which it thrashes about to show its emotions, hence the name. Its breast is speckled with white; its bill is long and curved at the tip; its eyes are yellow. In all parts of the Eastern United States and Lower Canada it sings in gardens and orchards and from roadside fences, in notes not much inferior to those of the mocking bird. It is a good mimic. In the early morning or evening time it perches in the top of a tree and sings sometimes for an hour or more. It nests in shrubbery and brush piles and lays four or five eggs, bluish-white and dotted with reddish-brown. Not only is the brown thrasher a joy to the ear and eye, but it is a help to the gardener, for in return for the few berries it takes it feeds on many insects which would be harmful to plants.

BROWN UNIVERSITY, one of the comparatively few institutions of higher learning in the United States which date back to colonial days. It was chartered in 1764 at Warren under the name of Rhode Island College, but was transferred in 1770 to its present location at Providence. In 1804 it was rechristened Brown University, in honor of Nicholas Brown, a generous benefactor. Though non-sectarian in spirit, the university owes its origin to the Baptists, and the president as well as a majority of the board of fellows and the board of trustees must be Baptists. As a matter of fact, all the nine presidents have been clergymen of that denomination.

During the early and middle part of the nineteenth century the scope of the institution was greatly enlarged. President Francis Wayland (1827-1855) introduced the elective system, laying stress on the hitherto neglected sciences and emphasizing graduate work. The university does not have colleges of law, medicine, engineering or agriculture. A great increase in the number of students took place in the last decade of the nineteenth century and the early years of the twentieth; the annual attendance is about 1,000, while the faculty numbers about one hundred. The libraries contain about 225,000 volumes, and the endowment is over $4,250,000.

In 1891 a Woman's College was founded, which six years later was accepted by the trustees as the Woman's College in Brown University. This branch of the university has its own buildings and its own courses, and the women comprise about one-fourth of the total enrollment of the university.

BRUCE, *broos,* Robert (1274-1329), the greatest of the kings of Scotland, famed in legend and poetry as well as in history. A large part of his life was spent in trying to wrest Scotland from the English, and it was during the very darkest days of that struggle that a famous incident known as the story of Robert Bruce and the spider is said to have occurred. While a fugitive, Bruce lay one morning on his hard bed in a wretched hut and saw on the roof above him a spider swinging by a thread of its own spinning. It was trying to swing itself from one beam to another, and again and again made the attempt in vain. When it had tried six times, Bruce realized that that was just the number of battles which he had vainly fought against the English, and he made a vow that if the spider tried a seventh time and succeeded he would renew his courage and try again. The spider's seventh attempt was successful, and Bruce took heart and went forth to victory.

Early in his career Bruce, then Earl of Carrick, swore allegiance to Edward I, king of England, and though he occasionally changed sides and aided the patriot William Wallace (which see), he managed to maintain friendly relations with Edward until 1306. In that year in a quarrel he killed "Red Comyn," a claimant to the Scottish throne, and immediately afterward assembled his vassals and had himself crowned king at Scone. Defeat at the hands of the English followed, and late in the year he dismissed his troops, retired to the Irish coast and let his enemies think him dead. In the following spring, however, he landed in Carrick, defeated the English and within two years had almost all Scotland in his hands.

He then advanced into England, laying waste the country, and in 1314 defeated the English in a famous battle at Bannockburn as the latter were advancing under Edward II to the relief of the garrison at Stirling. For years hostilities continued at intervals, with occasional treaties which did not really establish peace, and it was not until 1328 that England finally recognized Scotland's independence and the right of Bruce to the throne. Bruce did not live long after the completion of his great work, but died in 1329 of leprosy. His son David succeeded him.

BRUCHESI, *broo ka'se,* Louis Paul Napoleon (1855-), Roman Catholic archbishop of Montreal since 1897. He was born at Montreal and educated at Saint Sulpice College in that city, and in Paris and Rome; in the latter city he was ordained priest in 1878. Upon his return to Canada he was appointed to a professorship in Laval University. In 1887 he became canon of the cathedral of Montreal and in the same year professor of Christian apologetics at Laval. He prepared the educational exhibits of his native province for the World's Fair in Chicago, and was for a time chairman of the Roman Catholic school board of Montreal. He was appointed archbishop of Montreal in 1897. Archbishop Bruchesi has been conspicuous in many public movements not purely religious. His efforts for temperance resulted in the organization of the Anti-Alcoholic League in 1907, and he was at one time vice-president of the Dominion Forestry Association. Movements for social and industrial betterment have always received his hearty support.

BRUGES, *broo'jez,* the "city of bridges," is an ancient walled city of Belgium, capital of the province of West Flanders, fifty-five miles northwest of Brussels. It contains a regular network of canals crossed by fifty bridges, all opening in the middle to permit the passage of vessels. Architecturally it is one of the most interesting cities in Europe, for it is full of remarkable buildings dating from the Middle Ages. The Market Hall is a fine old building with a tower 354 feet high and containing forty-eight bells said to be among the finest in Europe. The town hall is a Gothic structure dating from the fourteenth century; the Palace of Justice is noted for its beautifully-carved chimney piece, and its almost priceless art treasures. The Church of Notre Dame contains tombs of Charles the Bold and Mary of Burgundy and a life-sized statue of the Virgin and Child, attributed to Michelangelo.

From 1240 to 1426 Bruges was one of the most important cities of the Hanseatic League (which see), and its commerce was far more extensive than at the present time. The manufactures include lace, textiles and tobacco; there are shipbuilding yards and numerous breweries and distilleries. Early in the War of the Nations (1914) Bruges was occupied by the Germans. Population in 1910, 53,285.

BRUMMELL, *brum'el,* GEORGE BRYAN (1778-1840), generally known as BEAU BRUMMELL, an English man of fashion who for twenty-one years set the London taste in dress and manners, and yet died miserably in an asylum for the poor. At the age of sixteen he gained the friendship of the Prince of Wales, afterward George IV, who made him a cornetist in his own regiment of the Tenth Hussars, and showed him flattering attention during the period when Brummell, prosperous and admired, lived handsomely on his father's fortune. Brummell was not a fop; he was fastidious about his appearance, but dressed, as Lord Byron said, "with exquisite propriety." A long course of extravagant living, however, brought him heavily into debt, and in 1816 he fled to Calais, in Northern France, to escape his creditors. Thereafter he was dependent on the generosity of his friends, growing poorer from year to year and dying wretchedly in Caen, France, where for a brief period he had held the position of consul.

BRUNELLESCHI, *broo nel les' ke,* FILIPPO (1377-1446), the real founder of the architecture of the Renaissance (which see). He was born in Florence, but went to Rome with Donatello to study his chosen art. While there he evolved the idea of bringing architecture back from the Gothic style to the principles of Greece and Rome. In this he was successful, as his work opened the way for Bramantè and others, but he himself never freed himself entirely from the traditions of medieval art. In 1417 he removed to Florence, where he lived the rest of his life. His greatest work was the dome of the Cathedral of Saint Mary, which he erected despite warnings from other architects as to its impossibility. It has remained unsurpassed, for the dome of Saint Peter's, though greater in height, is inferior to it in massiveness of effect. Among other important works by him were the Pitti Palace at Florence and the Pazzi Chapel at Santa Croce.

BRUNHILDE, *broon hil' de.* See NIBELUNGENLIED.

BRÜNN, the capital of the Austrian province of Moravia. It is a beautiful city, situated at the junction of the rivers Schwarzawa and Zwittawa, eighty-nine miles north of Vienna. The name is derived from the Hungarian word *bruno,* meaning *mud* or *clay,* conferred on account of the nature of the soil on which it stands. The city is noted for its manufacture of woolen goods, which have earned for it the name of the "Austrian Leeds," the English city of Leeds being the world's greatest wool-manufacturing center. There are also manufactures of leather, cotton, silk, chemicals, hardware and machinery. Its extensive commerce is chiefly promoted by fairs which attract merchants from all parts of Europe. Until 1860 the town was surrounded by fortifications, but the ramparts were then converted into beautiful promenades. Population in 1910, 125,737.

BRUNSWICK, *brunz'wik,* an irregularly-built medieval city, capital of the duchy of the same name, situated in a very fertile region on the Oker River, in Germany. It is thirty-five miles southeast of Hanover by rail. The name is derived from old Latin words meaning the *village of Bruno,* Bruno being the ruler of the land in the latter part of the ninth century. Its ancient ramparts have been converted into promenades and parks, but the great age of the city is clearly shown in many of its fine old buildings. The cathedral, palace, town hall and the ancient council house are remarkable examples of Romanesque and Gothic architecture. It is now an important industrial center, with manufactures of woolen and linen goods, jute, machinery and chemical products. The city is governed by a municipal council, consisting of thirty-six members, with an executive board of eight members. All public utilities are municipally owned. Population in 1910, 143,552.

BRUNSWICK, a former duchy of the German Empire, which gave to Great Britain its present reigning line (see BRUNSWICK, FAMILY OF). The duchy had its hereditary ruler, its constitution, its diet, or legislative body, of

THE FORMER DUCHY OF BRUNSWICK
In black.

forty-eight members, and the right to send three delegates to the imperial Parliament at Berlin. All these have disappeared in the new republic.

Brunswick, with an area of 1,418 square miles, a little greater than that of Rhode Island, is surrounded by the Prussian provinces of Hanover, Saxony and Westphalia. The northern part, though hilly or rolling, nowhere reaches any considerable altitude, but the southern part contains a portion of the Harz Mountains system and rises in places to heights of more than 3,000 feet. Deposits of iron ore, lead, copper, asphalt and lignite are found, but the mining industry is far surpassed in importance by agriculture. About one-half of the land is capable of tillage, and the leading crops are grain, potatoes, sugar beets and fruit. The manufacturing industries include brewing, distilling and the making of linens and woolens, hats, chemicals and beet sugar. The population in 1910 was 494,339.

BRUNSWICK, FAMILY OF, a distinguished family of which a younger branch furnished to Great Britain its present line of rulers. The House of Brunswick was founded in the twelfth century by the famous Henry the Lion, a rebellious vassal of Frederick Barbarossa (which see). Much of his territory was lost to him in conflict with his emperor, but he continued to hold Brunswick and Lüneburg, and it was his grandson, Otto the Child, who in 1235 was given the title of first Duke of Brunswick.

By the two sons of Ernst the Confessor, who became duke in 1532, the family was divided into the two branches of Brunswick-Wolfenbüttel and Brunswick-Lüneburg (House of Hanover), and it was a representative of this latter branch who became king of Great Britain as George I in 1714, his claim being based on the fact that he was the son of a granddaughter of James I of England. The Brunswick-Wolfenbüttel was the family in possession of the duchy of Brunswick until the death of the last duke in 1884, and after that years of conflict and of regency followed. In 1913 the difficulties were settled and Ernst August, son of the Duke of Cumberland and son-in-law of the emperor of Germany, was made Duke of Brunswick. He is a member of the Brunswick-Lüneburg line.

BRUNSWICK, GA., the county seat of Glynn County, in the southeastern part of the state, located eight miles from the Atlantic Ocean, on a broad curve of Saint Simons Sound. Savannah is nearly sixty miles north, and Jacksonville, Fla., is nearly as far south. The state capital, Atlanta, is 275 miles west. The nearest fresh-water rivers are the Altamaha, twelve miles north, and the Greater Satilla, twelve miles south. The numerous streams along the coast are tidal estuaries of the sea, rising and falling with the tide among the marshes, about which Sidney Lanier has written. The city gets drinking water from artesian wells. The area is more than five square miles. The population was 10,182 in 1910; in 1914, 10,645.

Steamship lines operate between Brunswick and Europe, the West Indies and South America, and all the important Atlantic seaboard towns. The four railroads of the city, the Atlantic, Birmingham & Atlantic; the Southern; the Atlantic Coast Line, and the Georgia Coast & Piedmont, own extensive terminals on the harbor. They also own large warehouses and refrigerating plants. On the docks are handled exports of locally-manufactured products, such as ties, lumber, naval stores, turpentine, rosin, cotton, phosphates, canned oysters and vegetables, and a vast amount of imports for distribution in the South.

Large hotels, accommodating many winter tourists, clubs, banks and churches, a Federal building costing $150,000, a city hall costing $75,000 and a $150,000 county courthouse are among the noteworthy public buildings. Near the city are the two pleasure resorts, Saint

Simons Island and Jekyl Island; the latter is exclusively the home of American capitalists. The Seminole Trail, the automobile highway from New York to Florida, and many miles of shell roads lead to points of historical interest. Brunswick was founded by General Oglethorpe in 1735, and named for the Duke of Brunswick. The city charter was obtained in 1856 and revised in 1890.

BRUSH, Charles Francis (1849-), an American electrician, known especially as the inventor of the Brush dynamo for arc lighting. He also invented an electric lamp and a large number of devices which have greatly developed the electric light. He was born at Euclid, Ohio, and was graduated at the University of Michigan in 1869, with the degree of mining engineer. For his achievements in electrical science he has been awarded the Rumford medal by the American Academy of Arts and Sciences, and has been honored by election to membership in the French Legion of Honor. See Electric Light.

BRUSSELS, *brus'elz,* the capital of Belgium and one of the finest cities in Europe. It is situated near the middle of the country, twenty-seven miles by rail south of Antwerp. On account of its wide boulevards, its beautiful buildings, the animated appearance of its streets and its artistic and intellectual life, under normal conditions, Brussels has been nicknamed a "miniature Paris."

Special Features. The city consists of a lower town and an upper town. The lower town, containing the older parts and devoted now almost entirely to commerce and industry, is surrounded with a circle of wide boulevards which have been built on the site of the old walls of the city. The upper town, which is partly inside and partly outside the boulevards, is the finest part and contains the king's palace, the government offices, wide streets, beautiful parks and modern residential quarters. The chief point of interest in the lower town is the famous Grande Place, which is probably the most splendid example of a medieval market-square still left in Europe.

On one side of this square is situated the beautiful town hall (Hôtel de Ville), dating in part from the fifteenth century; it is an imposing Gothic structure, with a spire 364 feet in height. Here also are situated the corporation houses of various medieval guilds, which lend a special charm to that square on account of their medieval architecture and ornament. Another important ancient building is the Church of Sainte Gudule, begun in 1220, and considered one of the finest specimens of pointed Gothic architecture. It is richly adorned with sculpture, paintings and beautiful stained-glass windows, and has a famous carved pulpit.

The Palace of Justice. One of the finest buildings in Europe is undoubtedly the Palace of Justice, built between 1866 and 1883, which crowns the highest point in the city. It impresses one by its magnificent proportions, for it occupies a larger area than that of Saint Peter's Church in Rome. Its style reminds one of the great massive buildings of ancient Egypt or Nineveh, from which it differs by its dome and by its columns in Greco-Roman style.

Institutions. Among the numerous educational institutions are a university, a polytechnic school, an academy of science and fine arts, one of the best conservatories of music in Europe, a fine astronomical observatory and a large public library. Brussels possesses a celebrated picture gallery, which contains some of the finest specimens of Flemish art, and also a good museum of modern paintings. Many of the most-valued pieces of art were hurriedly removed from the city in 1914 upon the approach of the Germans (see below).

Industries and Transportation. The industries of Brussels have for centuries been varied and important. It is celebrated for its manufacture of lace, which is an old-established industry. It has manufactories of cotton and woolen goods, curtains, paper, carriages and articles of bronze, as well as breweries, distilleries, sugar refineries and foundries.

Brussels is the center of the well-developed network of the railways of Belgium and is also connected by canals with all parts of the country. The town has now direct communication with the sea through the recently-constructed Willibroek Canal, which connects it with the Rupel River not far from its confluence with the Scheldt. The length of the canal to the Rupel is twenty miles.

History. Brussels appears to have been founded in the sixth century. From the tenth century onwards it began to develop its trade and industry, and various trade guilds similar to those of Ghent were formed here (see Guild). It became in 1477 the capital of the Austrian Netherlands, and under the fostering care of several of the Hapsburg governors was for a long time one of the pleasantest capitals in Europe. It suffered severely during the bombardment by the French under Marshal

Villeroi in 1695. It was captured in 1794 by the French, who retained it till 1814. Near Brussels the Battle of Waterloo was fought, which decided the fate of Europe and brought about the exile of Napoleon. From 1815 to 1830 it was one of the capitals of the kingdom of the Netherlands, and in 1830 it was the center of the revolt which separated Belgium from Holland.

In September, 1914, the city was occupied by the Germans when they swept over Belgium; it surrendered without a battle in order to save its beautiful buildings from bombardment (see WAR OF THE NATIONS). The conquering forces at once levied a tribute of $40,000,000 upon the city, and later other demands increased this tax to nearly $100,000,000. See, also, BELGIUM. O.B.

BRUSSELS SPROUTS, a cultivated variety of cabbage originating in Belgium. It is long-stemmed, four or five feet high, with small clustering green heads an inch or two thick. Brussels sprouts are an autumn crop, cultivated

BRUSSELS SPROUTS

and served for the table much the same as cabbage and cauliflower. The heads need plenty of room, therefore when they begin to crowd each other the leaves should be broken away, leaving but a few at the top where new heads form.

BRUTUS, *broo'tus,* MARCUS JUNIUS (85-42 B.C.), a distinguished Roman, one of the conspirators against the life of Caesar. He fought with Pompey against Caesar, but when Pompey was defeated at Pharsalia he was pardoned by Caesar, who made him governor of that part of Gaul lying south of the Alps. Although Caesar had thus befriended him, Brutus allowed himself to be drawn into the great conspiracy, and actually was one of those who raised a dagger against Caesar. See CAESAR, CAIUS JULIUS.

When Mark Antony, with his ironic "And Brutus is an honorable man," had roused the people to fury in his oration over the dead body of Caesar, Brutus took refuge in the East and raised a large force in Greece and Macedonia. With Cassius, he met Antony and Octavius at Philippi, but when the battle went against him he committed suicide by falling on his sword. Shakespeare in his *Julius Caesar*, of which Brutus is really the hero, has given a more favorable picture of him than most historians sanction. His defense of his action in helping to kill Caesar is one of the most famous speeches in all Shakespeare's works. It concludes with the words:

> As Caesar loved me, I weep for him; as he was fortunate, I rejoice at it; as he was valiant, I honor him; but as he was ambitious, I slew him. There is tears for his love; joy for his fortune; honor for his valor; and death for his ambition.

BRY'AN, WILLIAM JENNINGS (1860-), an American orator, statesman, journalist and political leader, three times the unsuccessful Democratic candidate for President of the United States, and yet, in spite of defeat, the man whom a large number of voters of his party looked upon for years thereafter as the greatest of their leaders.

In the Democratic national convention at Baltimore in 1912 the most conspicuous figure was a middle-aged man of medium height. His hair was much thinner than it was sixteen years before, and it was fringed with gray, but he was easily recognized as the same man who was Democratic candidate for President in 1896. In the intervening sixteen years he had met one defeat after another, and his party in his own state had refused his leadership as recently as 1910. Yet here he was, in spite of all, the leader of his party, bitterly attacking several of the prominent candidates as the agents of "reaction" and "predatory interests." Amid great confusion he finally dominated the convention, forced the nomination of the Presidential candidate before the adoption of the platform—an action without precedent—and finally wrote into the platform the planks he wanted. This picturesque fighting figure was William Jennings Bryan, and the candidate to whom he then gave his support was Woodrow Wilson.

When Bryan first became a national figure he was thirty-one years old. He had just been elected to the House of Representatives from a Nebraska district, and his fame as the "boy orator of the Platte" accompanied him to Washington. Here he was given the unprecedented honor of membership on the Committee on Ways and Means during his first term.

Early Career. The question arose in many minds at the same time—"Who is this 'boy orator of the Platte'?" Investigation showed that he was born on March 19, 1860, at Salem, Illinois. He was valedictorian of his class when he was graduated from Illinois College, at Jacksonville, in 1881, and he attended a Chicago law school from 1881 to 1883, at the same time studying in the office of Lyman Trumbull. In the next year he married Miss Mary Baird, who later graduated in law, and to whom he publicly gave credit for frequent advice on both legal and political questions. After he had practiced law in Jacksonville for four years, the family moved to Lincoln, Neb., where he quickly became a leader of the bar and a popular Democratic campaign orator.

WILLIAM JENNINGS BRYAN
Three times an aspirant for the Presidency of the United States, the second so-called "Great Commoner" in American history.

When the first Nebraska Congressional District, normally a Republican stronghold, sent him to Washington in 1891, there was considerable good-natured comment at his expense. But Bryan was a hard worker, and he forced recognition for himself in the discussions on the tariff and frée silver. His advocacy of the unlimited coinage of silver at the ratio of 16 to 1 (see BIMETALLISM) found no favor with his Republican constituents, who refused to elect him for a third term and also defeated him as candidate for United States Senator. He then became editor of the Omaha *World-Herald,* and continued to advocate free silver both in his paper and on the public platform.

Three Campaigns for the Presidency. In 1896 Bryan was an alternate delegate to the Democratic national convention at Chicago, and became a member upon the withdrawal of a regular delegate. He wrote the plank of the platform declaring for free silver, and during a heated debate which lasted for seven hours he swept the convention off its feet by a great oration, closing with these words:

> We shall answer their demand for a gold standard by saying to them: You shall not press down upon the brow of labor this crown of thorns! You shall not crucify mankind upon a cross of gold.

The speech won him the nomination for President. In the campaign that followed Bryan traveled over 18,000 miles and delivered over 600 speeches in twenty-seven states—a record number. The story of the campaign was told in his book, *The First Battle.* The election resulted in the choice of McKinley, the Republican candidate (see MCKINLEY, WILLIAM, subhead *Administration*). Again in 1900 Bryan was defeated by McKinley after a campaign almost as exciting as that of 1896. After this second defeat Bryan founded *The Commoner*, a weekly journal published at Lincoln (later changed to a monthly). In 1904 he did not seek the nomination, but bitterly opposed the conservative stand taken by the convention and its repudiation of the principles for which he stood.

For the next two years the political world saw nothing of him. In 1905 and 1906 he was on a tour of the world, and was everywhere received with many honors. The decisive defeat of Parker in the 1904 election turned the Democrats again to Bryan, who was mentioned as a candidate for President as soon as he returned to the United States. He was nominated in 1908, but was defeated by William H. Taft. In spite of these three defeats Bryan was still the leader of his party, although there were many who tried to lessen his influence. In 1910 he was all but repudiated by the Democratic convention of his own state; every candidate whom he supported was defeated for nomination, and the county local option plank which he proposed was also beaten.

Recent Political Activity. His Democratic opponents freely predicted that this defeat meant the end of Bryan's leadership, but two years later, at Baltimore, he appeared stronger than ever before and established his power without question. His influence in Woodrow Wilson's nomination and election were recognized by his appointment as Secretary of State in the new President's Cabinet. Bryan was not trained in diplomacy, and his administration of this office met with considerable eriticism, especially because he was absent from Washington at several critical periods. He continued in office until June 8, 1915, when he resigned because he was unable to agree with the President on the policies to be pursued toward Germany and the other European nations at war. For further details, see WILSON, WOODROW, subhead *Administration.*

Temperance Advocate and Lecturer. Bryan's courage in advocating policies which seem to him right and just has been strikingly illustrated by his outspoken stand for prohibition of the liquor traffic. In this he has assumed no halfway position, but, seemingly indifferent to the effect on his political career, has definitely taken his place in the ranks of those who are working to legislate out of existence the sale of intoxicating liquors. It was a source of great satisfaction to him that his state of Nebraska joined the ranks of the prohibition states in the fall elections of 1916, and he announced at that time that he would thenceforth devote himself to the work of making the entire United States dry.

As a lecturer he is probably known to more people than any other public man of his time, and no speaker has won more enduring popularity on the Chautauqua circuits. Of the Winona (Ind.) Chautauqua Assembly he was in 1915 elected president. The quality of his oratory is often described as "silver-tongued." Never at a loss for a telling phrase or for words to express his thoughts, he captivates his audiences by his splendid delivery, his graceful flow of language and his earnestness and sincerity. His greatest effort, *The Prince of Peace,* is a masterpiece of American oratory. W.F.Z.

BRY'ANT, WILLIAM CULLEN (1794-1878), the first great American poet, known as the "Father of American poets." Though he has never been popular and beloved in just the way that Longfellow has been, yet there are certain of his poems, as *Thanatopsis, To a Waterfowl, The Death of the Flowers* and *To the Fringed Gentian,* which are as well known as anything else in American literature. They have given inspiration as well as pleasure to innumerable people.

Events of His Life. Bryant was born at Cummington, Mass., and was the son of a country doctor. From his childhood his inclinations were toward literature, and he spent in reading and study much of the time which most boys devote to play. This does not mean

WILLIAM CULLEN BRYANT

that he had no pleasures, for he has left an account of snowball fights, of dam-building and of the races in which he used to join the other children. At the age of ten he had a poem published in a country newspaper, and three years later attracted much attention on the appearance of *The Embargo,* a satiric poem, based on the celebrated Embargo Act (which see) and addressed to President Jefferson. In 1810 he entered Williams College, but after a year gave up the idea of a college course and studied law. He was admitted to the bar in 1815, and practiced for ten years, most of the time at Great Barrington, Mass. Before he left that town in 1825 he was married to Miss Fairchild, with whom he had a very happy life.

Meanwhile, when he was but seventeen, he had written *Thanatopsis,* the first great poem that America had produced. He left it carelessly among some papers, but six years later his father discovered it and sent it to the editor of the *North American Review.* Its publication commanded immediate attention and was much discussed, some European critics going so far as to assert they did not believe anyone in the United States could have written it. Possibly the most frequently quoted words of the poem are these:

So live, that when thy summons comes to join
The innumerable caravan, which moves
To that mysterious realm, where each shall take
His chamber in the silent halls of death,
Thou go not, like the quarry-slave at night,
Scourged to his dungeon, but, sustained and soothed
By an unfaltering trust, approach thy grave,
Like one who wraps the drapery of his couch
About him, and lies down to pleasant dreams.

The same periodical afterward published others of his writings, and he was well known as a poet and literary critic before he removed in 1825 to New York. There he became associate editor of the *New York Evening Post,* of which three years later he became editor-in-chief. For over a half century he held this position, for his mind was vigorous and active to the time of his death. He became a noteworthy figure in the life of New York, and through all his years of prominence there was no breath of criticism against his character.

Place as a Poet. As has been emphasized above, Bryant's place in American literature is unique. There had been one great American prose writer before him—Washington Irving; but America had as yet produced no poet, and Bryant's verse was the model until the appearance of Longfellow. Frequently those whose genius flowers as early as did Bryant's show a decided decline in the work of their later lives, but Bryant remained productive and progressive to the last. Before all, he was the poet of nature, which he described with wonderful sympathy, but he never failed to see in the beautiful scenes he pictured some suggestion which had its effect on his life. Thus his poems usually close with a moral, but that is always so truly an outgrowth of what has preceded that it does not affect the artistic tone of the poem. If Bryant is cold and removed from human sympathy, as Lowell charged in his *Fable for Critics,* it is due merely to his dignity and serenity, and not to any lack of ability to feel.

Besides the short poems mentioned above, for which he is chiefly famous, Bryant published translations of the *Iliad* and the *Odyssey, Letters of a Traveler, Letters from the East* and *Orations and Addresses.*

Place as an Editor. As an editor of one of the great papers of the largest city in the country, Bryant attained distinction. His editorials were plain, straightforward and convincing, and if they had no permanent place in literature they exercised a strong influence in their day. Many of the reforms which he had advocated Bryant lived to see firmly established, and he rejoiced particularly in the downfall of slavery, against which he had been very active. A.MC C.

BRYCE, George (1844-), a Canadian Presbyterian clergyman, educator and author, an authority on the Canadian Northwest and its history. Although a native of Brantford, Ont., Dr. Bryce's name will always be associated with the development of Manitoba, particularly in education. In 1871, a few months after the organization of the province, the General Assembly of the Presbyterian Church of Canada decided to organize a college and church at Winnipeg. Bryce, although young and only recently graduated from the University of Toronto, was sent to Winnipeg, where he organized Manitoba College in 1871, and Knox Church in the next year. A few years later he took a leading part in organizing the University of Manitoba (which see), of which he was a councillor and examiner and for fourteen years head of the faculty of science. From its organization in 1871 until 1909 Dr. Bryce was also professor of English literature in Manitoba College.

In addition to his activities as a teacher, Dr. Bryce has taken an active interest in other fields. He was president of the Royal Society of Canada in 1909, was moderator-general of the Assembly of the Presbyterian Church in 1902-1903, and is a member of the Dominion conservation commission. He is a voluminous and popular writer of biography and history, and many of his books are unexcelled in their field. The most important among them are *Manitoba: Infancy, Progress and Present Condition; Short History of the Canadian People; Remarkable History of the Hudson's Bay Company,* and biographies of the Earl of Selkirk, Sir Alexander Mackenzie and Sir George Simpson.

BRYCE, James, Viscount (1838-), a British historian, publicist and statesman, author of *The American Commonwealth,* generally accepted as the best existing interpretation of political institutions in the United States. Not without justice Bryce has been called the "unofficial interpreter of the United States to Great Britain." For five years, from 1907 to 1912, as British ambassador at Washington, he was as well the official interpreter of Great Britain to the United States. No other man has contributed more to the perpetuation of friendly relations between these two countries than has James Bryce.

This unique service is but a small part of Bryce's activities. He was born on May 10, 1838, at Belfast, Ireland, where his father was for many years head master of a school. Inclination led the son back to Glasgow, his father's home. After completing his course in the high school and the University of Glasgow, he went to Trinity College, Oxford, where he took his degree in 1862. In 1867 Bryce was called to the bar, and only three years later, in 1870, was given the important position of Regius Professor of Civil Law at Oxford. He had already won distinction in 1862 by the publication of his history, *The Holy Roman Empire;* this study, written when he was only twenty-four years old, is as authoritative in its field as *The American Commonwealth.* His later books, all maintaining a high standard, include *Impressions of South Africa; Studies in History and Jurisprudence; Studies in Contemporary Biography,* and *South America.* The last volume is a keen, critical appreciation of conditions in Brazil, Argentina and other countries.

Bryce continued to lecture on civil law at Oxford until 1893, when the pressure of public and semi-public duties compelled him to resign. He was elected to the House of Commons in 1880, and served continuously until 1906. In Gladstone's third and fourth ministries Bryce held several positions, and again in 1905 and 1906 he sat in the Cabinet as Chief Secretary for Ireland. Home Rule had one of its strongest supporters in him, but it could not be introduced during his term of office (see HOME RULE).

In recognition of his many services to Great Britain he was raised to the peerage as Viscount Bryce in 1914. In the course of his long life many honors have been awarded him, including honorary degrees from Edinburgh, Aberdeen, Glasgow, Saint Andrew's, Cambridge, Oxford, Harvard, Princeton, Buenos Aires and other universities. He was at one time president of the British Academy, and holds memberships in many foreign learned societies.

BRYMNER, WILLIAM (1855-), a Canadian painter, born at Greenock, Scotland, educated at Saint Francis College, Richmond, Que., and Saint Therese College, Saint Therese, Que. Later he studied art in Paris under Bouguereau and T. Robert Fleury. He received a gold medal for painting at the Pan-American Exposition at Buffalo and at the Louisiana Purchase Exposition at Saint Louis. Since 1886 he has conducted the advanced art classes of the Art Association of Montreal. He has been president of the Royal Canadian Academy of Art since 1909.

His father, DOUGLAS BRYMNER (1823-1902), was a Scotchman who settled in Canada in 1857 and became one of the foremost journalists of his day. In 1872 he was appointed the first Dominion archivist, a position in which he rendered great service to American and Canadian historians by bringing order into the archives and by publishing extracts from many of the precious manuscripts stored there.

BRYN MAWR, *brin mar,* **COLLEGE,** one of the few distinguished institutions for the higher education of women in America, located at Bryn Mawr, Pa., a few miles from Philadelphia. It was founded in 1880 by Joseph W. Taylor, who was a member of the Society of Friends, or Quakers. The college is characterized by its high requirements for admission and the general culture and scholarship of its students. The buildings are of gray stone, in Gothic style. These include a library, containing about 55,000 volumes, a science hall, a lecture hall, gymnasium, hospital and six dormitories for students. A faculty of sixty members is maintained, and the college has about 450 students. The endowment now exceeds $1,000,000.

The aim of Bryn Mawr is to make the collegiate education of women very much like that of men in the best men's colleges. Courses in domestic science, art and the like are not offered. The students are encouraged to be independent, and to prepare themselves for professional, literary or public, rather than domestic, careers.

BRYOPHYTES, *bri'o fites.* The plants of the world which do not produce flowers are divided into four great groups, or orders, and of these the *bryophytes* are one. The word means *mosslike plants,* and the bryophytes include the *mosses* and the related *liverworts.* Some of these plants have leaves, while others are leafless, but none of them has true roots. Perhaps the most outstanding fact in connection with the bryophytes is what is known as *alternation of generations;* that is, the plants have different phases which do not resemble each other any more than do the caterpillar and the butterfly which develops from it. See LIVERWORTS; MOSSES.

BUBONIC PLAGUE, *bew bon'ik playg.* See PLAGUE.

BUCCANEER, *buk'a neer,* one who robs, plunders and murders. The name really applies, however, in its original sense to the adventurers of the sixteenth and seventeenth centuries who preyed upon vessels in the Caribbean Sea and on neighboring coasts. It originated from the French *boucan,* meaning *place for curing meat,* because the earliest of these adventurers stole cattle, smoked the meat and sold it to passing vessels. In time they captured such vessels and went to sea.

Religious wars between Britain and Spain produced the daring buccaneers Drake and Hawkins; and Sir Henry Morgan, the Welshman, is among the most famous leaders of early buccaneers. In the eighteenth century, the government no longer assisting in or consenting to such robbery, the methods of pirates were adopted. Among pirates perhaps the most famous is Captain Kidd, whose treasure-trove makes part of the story of Edgar Allan Poe's *Gold Bug.* Later, *marooning* was practiced; that is, putting ashore on desert islands those whom the pirates robbed. By the end of the eighteenth century, practically all these practices had been abandoned, or were carried on in the most out-of-the-way places. (See articles on each man named above.)

BUCHANAN, *bu kan'an,* JAMES (1791-1868), the fifteenth President of the United States. At the time of his inauguration he was in his sixty-sixth year; with the exception of William Henry Harrison he was the oldest man ever chosen to that office. Unlike Franklin Pierce, his predecessor, and Abraham Lincoln, his successor, Buchanan entered office after he had passed the prime of life. He had long been known as a statesman of ability and experience; he had won honor in both houses of Congress, sat in President Polk's Cabinet as Secretary of State and had been his country's representative at the courts of Great Britain and Russia. He was without question one of the leaders, not merely of the Democratic party, but of the nation. His reward was the highest office in the gift of the people, yet he retired to private life under a cloud of dislike and indignation.

His Youth. James Buchanan was born near Mercersburg, Pa., on April 23, 1791. His parents were Scotch-Irish Presbyterians, who worked hard for a living on their farm. The son was sent to Dickinson College at Carlisle, Pa., where he was graduated in 1809. He then studied in a law office for several years, and in 1812 began to practice his profession. In politics he was at first a Federalist. He was opposed to the second war with Great Britain, but when it came he said that it was "the duty of every patriot to defend the country," and in 1814 he volunteered as a private for the defense of Baltimore. Buchanan was already known locally as an orator, and in the autumn elections was chosen to the legislature, where he served two terms.

Political Career. It was then his intention to remain in private life, but the sudden death of his fiancée altered his resolve and he became active politically. He was first elected to the national House of Representatives in 1820, and served from 1821 to 1831 without a break. President Jackson then appointed him minister to Russia, where he negotiated the first commercial treaty between the United States and that country. This treaty of 1832 remained in force for eighty years, for it was not abrogated until President Taft's administration.

On his return to the United States, Buchanan lived quietly for a year, but in 1834 was elected to fill a vacancy in the United States Senate. The Pennsylvania legislature twice reëlected him, but he resigned before the end of his third term in order to become Secretary of State under Polk. In the Senate, as previously in the House, Buchanan was a leading supporter of Jackson, especially on the issue of the President's right to remove executive officers without explaining his reasons to the

Senate. During the many discussions over the right of petition, Buchanan held, with John Quincy Adams, that any citizen possessed this

JAMES BUCHANAN

His reason for accepting the Presidential nomination he stated thus:

They tell me that the use of my name will still the agitated waters, restore public harmony, by banishing sectionalism, and remove all apprehension of disunion. For these objects I would not only surrender my own ease and comfort, but cheerfully lay down my life.

privilege; but on the slavery question, to which most of the petitions referred, he maintained that Congress had no control over slavery in the states.

In 1844 he was Pennsylvania's "favorite son" for the Presidential nomination, but he withdrew his name in favor of James K. Polk. The latter was elected, and he appointed Buchanan Secretary of State. In this office he conducted the negotiations concerning the Oregon country, and the friendly settlement was largely due to his efforts. Buchanan was heartily in favor of the annexation of Texas, and during the war with Mexico succeeded in keeping the United States out of disputes with other countries. It was on Buchanan's advice that Polk, in his first message to Congress, reaffirmed the Monroe Doctrine in a declaration aimed at British schemes of colonization in California. (See MONROE DOCTRINE.)

At the end of Polk's term in 1849, Buchanan was succeeded as Secretary of State by John M. Clayton, who negotiated the Clayton-Bulwer Treaty (which see). This treaty marked important departures from the policy of Polk and Buchanan, who had been uneasy over British attempts to establish a protectorate over part of Central America. As the interpretation of the treaty became almost immediately a cause of controversy, President Pierce in 1853 sent Buchanan to London, where he remained as United States minister for three years, most of his time being spent in an attempt to find a basis for the settlement of Central American problems.

One of the incidents of Buchanan's mission was his share in the Ostend Manifesto of 1854, which won the unqualified approval of many Southerners (see OSTEND MANIFESTO), but classed Buchanan with the pro-slavery men. He was fortunate in being at a distance while the struggle over the Kansas-Nebraska Act (which see) was taking place, for he was free from the bitterness which descended on some of the leaders of the Democratic party. He was therefore nominated for President by the Democrats in 1856, with John C. Breckinridge of Kentucky for Vice-President. "Buck and Breck," as they were called, were elected by a large electoral majority; 174 votes to 114 for Fremont, the Republican candidate, and eight for Fillmore. The popular vote was much closer. Buchanan, like Lincoln and some others, was a minority President; he lacked about 372,000 votes of an absolute majority.

Buchanan's Administration. Buchanan's experience as Secretary of State and as diplomatic representative aided him in establishing more friendly relations with Great Britain.

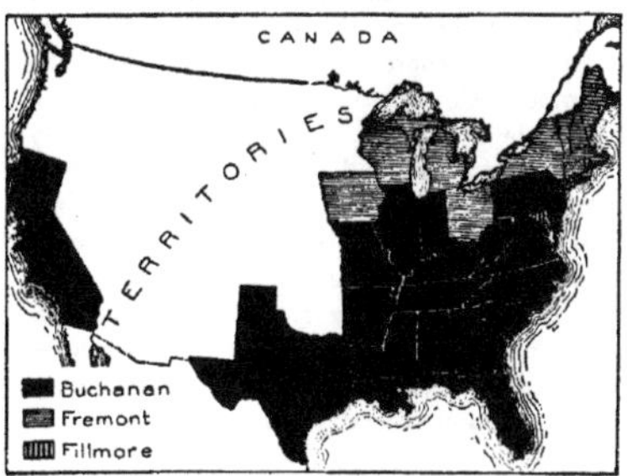

ELECTION RESULT IN 1856

The Central American controversy, which he failed to settle while he was minister at London, was finally disposed of by treaties between Great Britain, Nicaragua and Honduras, largely through his intervention.

In his American policy, Buchanan worked hard for the annexation of Cuba, parts of Central America and possibly Mexico, and even went to the point of urging Congress to give him authority to send troops into Mexico in order to dispose of one of the rival factions there. He encouraged William Walker, the filibuster who for a time was dictator of Nicaragua. The Republicans insisted that his purpose was to bring more slave-holding territory into the Union. The Senate, although Democratic, refused to sanction any of these schemes.

In his domestic policy, Buchanan was equally unfortunate. He foresaw the danger of disunion, but he took no steps to prevent it, and even gave a sort of passive encouragement to those who urged a separation. He was by nature a compromiser. Two days after his inauguration the Supreme Court of the United States announced its decision in the Dred Scott case (which see); the Court held that Congress had no right to interfere with slavery in the territories. Meanwhile, the struggle in Kansas went on with increasing bitterness. At first Buchanan had agreed to the principle of popular sovereignty, but under the influence of the Southerners in his Cabinet he seems to have been led to the conclusion that the only way to prevent secession of the Southern states was to secure the adoption of the Lecompton Constitution. Although it was once rejected by the people of Kansas, the President gave it his support, declared that Kansas was "already a slave state, as much as Georgia or South Carolina," and urged Congress to admit it. Largely through the influence of Stephen A. Douglas, Congress refused to consider the Lecompton Constitution, and sent it back to the people of Kansas, where it was decisively defeated for the second time. In 1859 occurred John Brown's raid at Harper's Ferry (see BROWN, JOHN). In spite of this evidence of violent hostility to slavery, Buchanan did nothing to quell the rising storm. He was still much influenced in his attitude by the Southern members of the Cabinet, particularly Howell Cobb of Georgia and John B. Floyd of Virginia.

Buchanan was by this time identified in the popular mind with the extreme pro-slavery Democrats, yet the fact that he was not considered for renomination either by the Northern or Southern Democrats is some evidence that he was still trying to steer a middle course. Vice-President Breckinridge was nominated for President by the extreme pro-slavery Democrats, and Douglas by the moderate Democrats. The Constitutional Union party nominated John Bell, while the Republicans named Abraham Lincoln. Although the Republicans disclaimed any intention of interfering with slavery in the states, the election of Lincoln was followed by the secession of the Southern states, led by South Carolina. In February, 1861, the seceded states formed a new government which they called the Confederate States of America (which see). During this period, between Lincoln's election in November, 1860, and his inauguration in March, 1861, a heavy responsibility rested on Buchanan. In his annual message to Congress in December, 1860, he argued that there was no right of secession, but on the other hand, he could see no way of preventing secession because any interference would involve war upon a state. He declared that the North was responsible for the break in the Union because it would not cease its criticism of slavery.

The whole issue came to a head when, on December 26, Major Anderson removed his little garrison to Fort Sumter from Fort Moultrie, which was almost defenseless against new batteries constructed by the South Carolina government. When South Carolina's commissioners came to him to offer "peace and amity between that commonwealth and the government at Washington," Buchanan refused to see them, except as "private gentlemen of the highest character." Yet he was inclined to yield to South Carolina's demand that he withdraw the garrison from Fort Sumter. Lewis Cass, Secretary of State, resigned because he thought that the President was not properly defending his country, and Floyd, Secretary of War, resigned after being discredited in a money scandal. Jeremiah S. Black and Edwin M. Stanton, who had succeeded Cass and Floyd, respectively, threatened to resign if the President surrendered Fort Sumter. Under their influence and that of John A. Dix of New York and Horatio King of Maine, two staunch Unionists who came into the Cabinet when Howell Cobb and Jacob Thompson resigned to join their states in secession, Buchanan was persuaded to take a firmer stand, and even consented to the attempt to relieve Fort Sumter (which see) by the steamer *Star of the West.* (See also, WAR OF SECESSION.)

Buchanan did not want war, but after its beginning he wrote to John A. Dix that Lin-

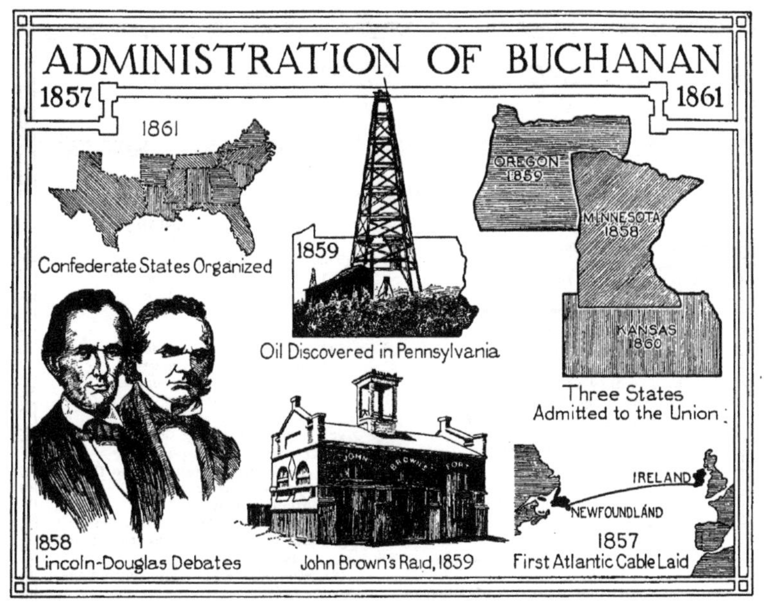

coln "had no alternative but to accept the war initiated by South Carolina or the Southern Confederacy." When his term ended he was within a few weeks of being seventy years old, and it was with great relief that he retired to Wheatlands, his little estate a mile from Lancaster, Pa. There he died, on June 1, 1868, and was buried in the local cemetery.

Summary of His Career. Under normal conditions Buchanan might have been a President of distinction. But in 1860 he allied himself with the extremists of his party and when the crisis came failed to show self-confidence and energy. The mistakes and weakness of his last year in office have overshadowed his earlier achievements. The President tried to reflect the divided sentiment of the country. On one side he believed that a state had no right to secede; on the other, he said that the United States had no right to force a state to remain in the Union. His moral scruples prevented him from taking any decisive steps, and brought on him general disapproval. The most bitter of his Southern critics charged him with treachery, and some of his Northern opponents accused him of treason. Like most men who seek to compromise, he was disliked by extremists on both sides. A.B.H.

Other Items of Interest. Every slave-holding state except Maryland gave Buchanan its electoral vote in 1856.

When, on the eve of his return from Russia to the United States, Buchanan paid his farewell visit to the imperial palace, the emperor asked him to request the President to "send another minister exactly like himself."

After his retirement from office he published *Mr. Buchanan's Administration on the Eve of the Rebellion,* a defense of his policies.

Buchanan never married, but he was not a lonely man, for he took into his household a niece and a nephew, to whom he gave a beautiful devotion. When his niece was absent from him he wrote to her nearly every day.

A man of very simple tastes, the formalities and etiquette of court life were very disagreeable to him, and only his sense of duty to his country caused him to accept his appointment to Saint Petersburg and later to London.

Authorities hold that of all the American Presidents only Jefferson and John Quincy Adams have equaled Buchanan in the administration of foreign affairs.

It was during this administration that the first Atlantic cable, from Newfoundland to Ireland, was laid.

OUTLINE AND QUESTIONS ON JAMES BUCHANAN

Outline

I. Early Years

(1) Birth and education
(2) Law study
(3) In the War of 1812

II. Political Career

(1) In the House of Representatives
(2) As minister to Russia
(3) In the Senate
(4) As Secretary of State
 (a) The Oregon dispute
(5) As minister to Great Britain

III. Administration

(1) The slavery issue
 (a) Dred Scott decision
  1. The questions at issue
  2. The decision
  3. The result
 (b) The struggle for Kansas
  1. "Bleeding Kansas"
  2. Attempts to secure statehood
  3. Lecompton Constitution
 (c) The underground railway
 (d) John Brown's raid
(2) The election of 1860
 (a) Democratic convention
 (b) Republican convention
 (c) Result
(3) Secession
 (a) South Carolina secedes
 (b) Efforts at compromise
  1. Crittenden Compromise
  2. Peace conference
  3. Corwin Amendment
 (c) Formation of the Confederacy
(4) Other events
 (a) Admission of states to the Union
 (b) Financial panic of 1857
 (c) Atlantic cable, 1857
 (d) Lincoln-Douglas debates, 1858
 (e) Oil discovered in Pennsylvania, 1859
 (f) William Walker's filibustering expedition
 (g) Death of Washington Irving

Questions

When and why was "Buck and Breck" a popular slogan?

In what sense was North America nearer to Europe at the close of Buchanan's administration than at the beginning?

What stand did Buchanan take on the question of the right of a citizen to petition Congress?

In what way was the President connected with the history of certain Central American states?

When did he refuse to recognize a formal delegation save as "private gentlemen of the highest character"?

How many Presidents were older than Buchanan at the time of their nomination?

Does Canada owe anything to him?

If he had had his way, how much larger would the United States be than it is at present?

What was his attitude at the close of his term of office on the subject of the coming war?

Why did not Buchanan care to be sent as minister to Great Britain? What led him to accept the post?

What decision, epoch-making in its effects, was pronounced by the Supreme Court during his term of office?

In what way did the death of a girl once affect the history of the United States?

What was Buchanan's attitude toward secession?

Washington Irving, called in his own day the "Prince of American Letters," died during this administration.

It was during Buchanan's term of office that the Lincoln-Douglas debates, among the most famous in the history of the country, took place.

When Buchanan was nominated for the Presidency a paper in his home town which was opposed to him politically and could not therefore be accused of partiality wrote as follows:

> We know him as a friend of the poor—as a perpetual benefactor of the poor widows of this city, who, when the piercing blasts of each successive winter brought shrieks of cold, and hunger, and want, in the frail tenements of Poverty, could apply to the Buchanan Relief Donation for their annual supply of wood.

Mr. Buchanan was "a large, muscular man, who enjoyed the most perfect health," and at the age of sixty-five was "capable of enduring as much labor as a young man."

Minnesota, Oregon and Kansas were admitted to the Union during this administration.

Consult Wilson's *Division and Reunion;* Buchanan's own book, *Mr. Buchanan's Administration on the Eve of the Rebellion;* Grant's *Personal Memoirs.*

BUCHAREST, or **BUKHAREST,** *boo ka-rest',* the capital of Rumania, a picturesque city on both banks of the River Dimbovitza, about thirty-three miles north of the Danube. The city is ordinarily one of the gayest of European capitals and has acquired the popular name of "Little Paris." It is famous alike for its bohemian atmosphere and for its fashion.

Although there are still portions which retain their Oriental appearance, with narrow, crooked and somewhat dirty streets, the city is for the most part modern and well-planned. Twelve fine bridges across the river connect the two portions of the city. There are many beautiful buildings, and the city is noted for its great churches. The manufactures are not highly developed, and before Rumania entered the War of the Nations in 1916 the industries were chiefly in the hands of Germans. Bucharest is an important center of trade in petroleum, cereals, timber, hides, honey and wax. The population is composed of many nationalities, including Greeks, Turks, Jews, Russians, Poles, Germans and Hungarians. The city was made the capital of Rumania in 1862 when that kingdom was created by the union of Wallachia and Moldavia. Bucharest has been visited by plague several times, and in 1813 this dreadful scourge claimed 70,000 victims in six weeks. In December, 1916, after a campaign of but a few weeks, the Austro-German forces captured the city and 15,000 square miles of Rumanian territory. Population in 1913, 338,109. See RUMANIA.

BUCK, DUDLEY (1839-1909), a distinguished organist and composer, holding first rank among modern American composers of church music. He was born at Hartford, Conn., began his musical studies at the age of sixteen, and was an organist in his home city until 1858. Thereafter he studied in Europe for five years, and on his return to America began a long and honored public career. He held important organ positions in Chicago, New York and Boston, engaged in concert tours, assisted Theodore Thomas in conducting his orchestra concerts in New York City, and for several years was organist and conductor for the Brooklyn Apollo Club. Occupying a conspicuous place among musicians, he rendered a lasting service to his fellow Americans by developing in them an appreciation for good music. By his church music, especially, he helped to elevate musical taste in his own country.

Buck's best-known compositions include numerous songs, hymns, anthems and other church music; the music for Sidney Lanier's cantata sung at the opening of the Philadelphia Centennial in 1876; the cantata based on Longfellow's *Golden Legend,* which won a thousand-dollar prize offered by the Cincinnati Musical Festival of 1880; the cantatas entitled *King Olaf's Christmas, The Light of Asia* and *The Voyage of Columbus;* the overture to *Marmion;* two operas; and the organ sonatas in E flat and G minor.

BUCKETSHOP, an establishment which appears to transact a regular, legitimate business in buying and selling grain or securities, but is in reality merely gambling with its customers. A customer, for example, buys 100 shares of stock on "margin." The bucketshop broker, instead of buying the stock, merely makes the necessary entry in his books. If the stock rises in price the customer gets his margin back and the extra profit; if it falls, the broker keeps the margin.

The margins are usually small, sometimes as low as one per cent, and a slight fluctuation is enough to put them in the pockets of the bucketshop operator. Bucketshops are nothing more or less than gambling houses, and laws prohibiting them are practically universal.

Some operators in the past have literally stolen the margins from the customer by quoting false prices, but as the entire transaction was illegal, the customer had no chance of recovering his money. See BOARD OF TRADE.

BUCKEYE, an American name for certain species of horse-chestnuts. Ohio is called the *Buckeye State*, because at one time the sweet or yellow buckeye grew very abundantly throughout the state; that species is now called the Ohio buckeye. See CHESTNUT; HORSE-CHESTNUT.

BUCKINGHAM, *buck'ing am,* the county town of Labelle County, Que., twenty miles east of Ottawa, the capital of the Dominion. It is on the Canadian Pacific Railway and on the Rivière du Lièvre, four miles above its junction with the Ottawa River. Buckingham is a lumbering center, and has pulp, saw, shingle and planing mills and a sash and door factory. There is also some dairying in the neighborhood, as is shown by the presence of a cheese factory and a creamery. Mica, plumbago and phosphate are mined near-by, and there is an electric reduction plant in the town. Population in 1911, 3,854; in 1916, estimated, 4,100.

BUCKINGHAM, GEORGE VILLIERS, Duke of (1592-1628), an English nobleman, known as the "power behind the throne" during the reigns of James I and Charles I. In 1623, when a marriage was being arranged between Prince Charles and the Infanta of Spain, Buckingham went with the prince to Madrid to carry on the suit in person. The result, however, was the breaking off of the marriage and the declaration of war with Spain. After the death of James, Buckingham was sent to France, as proxy for Charles I, to marry Henrietta Maria.

In 1626, after the failure of an expedition against Cadiz, he was impeached but was saved by the favor of the king. Despite the difficulty in obtaining supplies, Buckingham took upon himself the conduct of a war with France, but the attempt proved a failure. His incapacity, no less than the injustice of his having received such high honors, titles and preferments at the hands of the king, made him extremely unpopular, and a second impeachment was prevented only by the dissolution of Parliament. He then set out on another expedition to Rochelle, but was assassinated while embarking. It has been declared that few men more unworthy have found a high place in English history. It was entirely to his handsome and graceful person and to his manners, affable enough to those who might benefit him, that he owed the royal favor. See ENGLAND, subtitle *History*.

BUCKLE, *buck''l,* HENRY THOMAS (1821-1862), an English historical writer who will always be remembered for his *History of Civilization in England*. When he was twenty years old he was reputed to be one of the greatest chess players in the world, but he gave up this game, of which he was very fond, to devote his time to the writing of his great history. Though for seventeen years, from 1844 until the year before his death, he labored ten hours a day to accomplish his purpose, he completed only two volumes of the work.

These volumes are themselves merely an introduction to the work he planned. The first states the general principles of his method and the laws by which human progress is governed; the second illustrates these laws and principles by references to the histories of Spain, Scotland, Germany and the United States. Buckle's arguments are frequently open to criticism, but his work has been of value in stimulating historical research and discussion.

BUCKSKIN, a soft, yellowish or grayish leather, formerly used by Indians and frontiersmen for clothing. It is now used mostly in the manufacture of gloves. In the early days it was made of the skin of deer, from which fact its name was derived, but it is now usually made from sheepskin. To obtain the softness which is the chief quality of buckskin, oil is used in the dressing. A kind of twilled woolen cloth, now largely used for riding breeches, is also called buckskin. In tales of frontier life and of the Indians, buckskin is often mentioned.

BUCKTHORN, the name of a large genus of useful trees and shrubs, natives of Europe and Asia, several species of which belong to North America. The buckthorn common in North America is a spiny shrub, growing to seven or eight feet in height. The leaves are oval, usually rounded at the base. The flowers, which appear in May, are of an inconspicuous green, male and female flowers being produced on different shrubs. The fruits are about one-quarter of an inch across, berry-like and black, bearing four seeds. The juice of the ripe berries, mixed with alum, is used by artists as sap-green. Medicinally, the berries produce a powerful purgative, but one which is not often used. The bark yields a beautiful yellow dye, and one species in the Pacific states furnishes the medicinal cascara bark.

BUCK'WHEAT, a plant producing a three-sided seed or grain, at present considered of secondary importance but well worthy of cultivation. The origin of buckwheat is not known, but it is supposed to be a native of Asia, and was therefore named *Saracen wheat* by the French. It takes its present name from a German word meaning *beech wheat,* because of the resemblance of the seeds to the beechnut.

The plant has smooth, branching stems, green leaves with dark veins and white flowers.

BUCKWHEAT
Top of stalk, flower and fruit.

Rather light, well-drained soils are best suited to buckwheat, but the crop is hardy and is less affected by soil than by frost. Because it grows well on poor soil, with little cultivation, it has been called the poor farmer's crop, and so the term *buckwheaters* has been applied in some localities to the unskilled farmers. For the best results, plant buckwheat as late as possible to secure a sufficient crop before the severe frosts. It begins to bloom early and continues to blossom until harvest, so at that time all of the grain is not fully matured; but the farmer soon learns to judge the best time for harvest. Buckwheat is a useful plant in another sense, for by its shade it stifles many weeds which appear and leaves a clean field for the following year. Its blossoms are a favorite forage for bees.

While cultivated in China and other Eastern countries as a food plant, in Europe buckwheat is used principally as feed for stock and poultry,

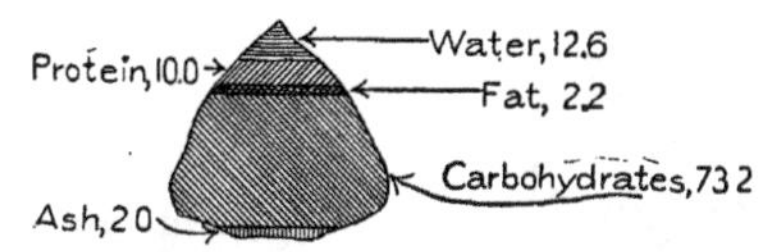

FOOD QUALITIES OF BUCKWHEAT

being excellent for pheasants. It is sometimes given to horses with bran, chaff or grain. In the United States and Canada it is extensively used to make flour from which breakfast cakes are prepared. Beer may be brewed from it, and it has been used in preparing cordials. The blossoms have been used for brown dye.

Buckwheat is quite extensively grown in the United States east of the Mississippi River and from Pennsylvania northward, averaging about 15,000,000 bushels a year, bringing nearly nine and a half million dollars. New York contributes one-third of the amount, Pennsylvania nearly as much, Michigan about one-fifteenth. In Canada the average yield is over 9,000,000 bushels, valued at over five million dollars; Ontario's crop is about one-half

PRODUCTION AREA
Where dots are most numerous the production of buckwheat is greatest.

of Canada's total, Quebec's about a quarter, and New Brunswick's one-sixth. The legal weight of buckwheat is from forty-eight to fifty-two pounds per bushel in the various states and provinces. M.S.

BUD, as the term is most commonly used, means an unopened flower, but it applies just as truly to an undeveloped leaf.

Leaf Buds are a provision of nature for allowing leaves to exist safely through a winter, and they furnish one of the most remarkable instances of the way plant parts are adapted to their purpose. On the stem, either at the end or at the sides, appears a little swelling. If it is to remain in this condition through a hard winter this bud has a thick, strong husk,

VARIOUS BUDS
(a) Wild rose
(b) Shagbark hickory
(c) Lilac
(d) Cherry
(e) Horse-chestnut
(f) Iris
(g) Meadow lily
(h) Cottonwood
(i) Water lily

as shiny oftentimes as if it had been varnished. Within may be a warm, woolly coat which protects the sensitive little leaf from the cold. However tiny, the leaflet in this bud is in form the fully-developed leaf, only it is folded and packed away so closely that not the slightest space goes to waste. Every plant always has its little leaves folded or rolled in just the same manner.

Everyone has doubtless watched the leaves of a fern unroll from the tip, but it is not so easy to see just how all leaf buds develop. The sorrel, or sour grass, has its three leaflets, each folded smoothly in the middle, pressed closely together; the magnolia leaf is folded along its central vein, with the dull inner surface outward; the currant leaf is plaited like a fan, and the violet leaf has its two margins rolled inward toward the center. Careful watching in the spring when green things are just beginning to appear will reveal many more interesting methods of close packing.

Flower Buds, too, are for protection, but their coverings are rarely as thick and strong as are those of some leaf buds. Often the green outer part of the flower, the calyx, is folded about the colored corolla, and the beautiful tints do not begin to appear until the bud opens. It is easy to watch flower buds in the process of unfolding; everybody has seen the quickly-expanding morning glory, for instance, unfold from a tight bud, furled like a rolled umbrella, into a glowing bell; or a rose loosen deliberately its overlapping petals. Violets, daffodils, hollyhocks, lilies—all have their own individual methods of opening, and all of these are well worth observing. Of recent years moving pictures have been taken of many flower buds expanding into flowers. When shown on the screen the process is quickened, a series being shown in two or three minutes which it took the camera days to obtain, in exposures at regular intervals of several hours. A.MC C.

BUDAPEST, *boo da pest'*, a city on both banks of the Danube, and the capital of the new state of Hungary, is composed of two towns, Buda and Pest, which were united as one municipality in 1872. *Buda* was named after Bleda, brother of Attila the Hun. *Pest* is derived from a Russian word meaning *oven*, and the name was given to the town probably on account of its great lime kilns. It is admirably situated to be the central terminus from which all Hungarian railways radiate, and receives nearly all the products of the surrounding territory. Its commerce is very extensive, grain, wines, tobacco, hides, hemp, cattle, sheep and pigs being exported.

There are numerous mineral springs in the vicinity, and the waters, especially that known as Hunyadi Janos, have achieved a wide reputation for medicinal properties. Among these springs is one of the deepest artesian wells in

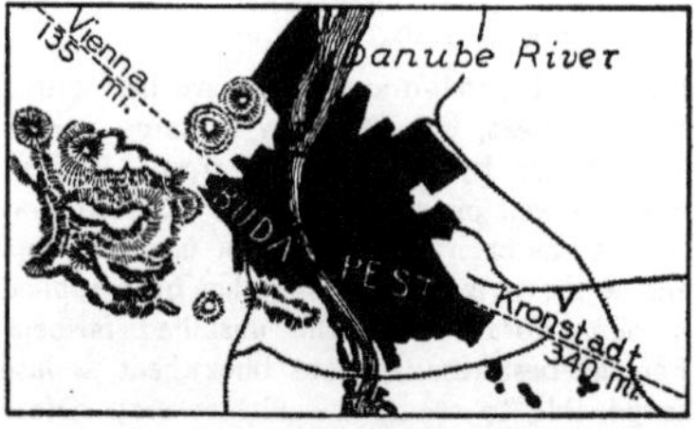

LOCATION OF BUDAPEST

the world, descending over 3,000 feet and discharging over 260,000 gallons of water daily at a temperature of 165° F. Budapest is best known commercially, however, as a city of model flour mills and one of the greatest milling centers of the world. Grain is brought to its elevators by rail and by huge barges on the Danube, carrying over 600 tons each. The river is an even more important highway of commerce than the railways.

Pest, although more modern than Buda, has far outgrown the latter city in importance, both commercially and as a seat of learning and culture. One of its streets, the Andrassy Strasse (Street), nearly two miles in length and lined by fine buildings, is among the most beautiful in Europe. Even as late as 1870 the towns now united as Budapest were backward in all matters relative to public health. The unfiltered water of the Danube was used for all domestic purposes, many people lived in damp cellars and unsanitary tenements, the little drainage was discharged into the river, and the death rate was abnormally high. But the Budapest of to-day is well drained, has practical sewage systems and filtration plants and possesses all municipally-owned public utilities demanded by a progressive city. It was the first city in the world to establish an underground electric trolley system.

The two portions of the town are joined by six bridges across the Danube, one of them a suspension bridge 1,230 feet in length. The growth of the combined cities has been remarkable. In 1841 the population was 107,240, but by 1910 this had increased to 863,735. E.ST.A.

BUDDHISM, *bood'iz'm,* the religious system founded in the sixth century B. C., by the Hindu sage, Buddha, "the Enlightened," who, in his life and teachings, was more like Christ than any other of the great teachers of mankind. When the founder of Buddhism began to teach, the prevailing religion in India was that taught by the Brahmans (see BRAHMANISM). Buddha, like the Brahmans, believed that existence was a sorrow and an evil, but he taught the people that salvation should be sought through a change of heart, not by sacrifices, ceremonies and self-torture. Buddha declared that his followers were released from the restraints of caste, and the poorest outcast was encouraged to seek the benefits of his teachings (see CASTE). Thus the new faith represented a revolt against the teachings of the Brahmans, and in its spirit Buddhism bore somewhat the same relation to Brahmanism as Christianity did to Judaism in the days of the early Church.

The central idea in Buddhism is that *Nirvana,* or release from existence, is the chief good, and that Nirvana can be attained only by crushing all desire. The Buddhist who follows the "path to the other shore" must hold fast to eight conditions: right view, right judgment, right language, right purpose, right profession, right application, right memory and right meditation. The five great commandments of Buddhism forbid killing, stealing, the

THE GREAT BUDDHA

committing of adultery, lying and drunkenness. Almsgiving, purity, patience, courage, contemplation and knowledge are the virtues especially cultivated, and charity is so broadly interpreted that it includes acts of kindness toward even the lowest animals.

When all the conditions have been met, the individual may hope to have his reward in ceasing to exist. But the path to Nirvana is too difficult to travel over in one lifetime, so when the body dies the soul must find another abode. This doctrine, the transmigration of the soul (which see), is found both in Buddhism and in Brahmanism.

It has been said of the moral code of Buddhism that for pureness, excellence and wisdom it is second only to Christianity, but Buddhism and Christianity are far apart in the great end which the followers of these religions seek. For the Christian, the end of striving is eternal life, and the thought of ceasing to exist is repugnant.

Buddha taught that there were no gods, and that priests, ceremonies, prayers and sacraments were unnecessary. His teachings, however, have been greatly changed during the passing centuries, and modern Buddhists rear temples, offer prayers and hold formal religious services. Though Buddha is not worshiped as a god, the adoration of his statues and of his relics is a very important feature of the worship of his followers. In the modern Buddhist temple the central object is an image of the great teacher or a shrine containing his relics, and before this flowers, fruit and incense are offered each day.

In Hindustan, the birthplace of the Buddhist faith, this religion has now but a feeble hold, but it prevails in Ceylon, Burma, Java, Cochin-China, Laos, Nepal, Tibet, where it takes the form of Lamaism (which see), Mongolia, China and Japan, and its followers are estimated to number 500,000,000. See RELIGIONS, subhead *Religions of the World.*

Buddha, *bood'ah,* the sacred name of a great reformer and teacher of early India, who became the founder of the religion known as Buddhism. His name was Siddhartha, and his family name Gautama; *Buddha,* acquired by him as the founder of a great religious system, is the Sanskrit for *the Wise* or *the Enlightened.* He was born in the sixth century before Christ, in the town of Kapilavastu, a few days' journey north of Benares. Tradition says that he was a prince of royal blood. Of his youth little is known except what has come down in legendary form; the many tales that sprang up about his early life and achievements have been picturesquely woven together by Sir Edwin

At left: Chinese many-armed Buddha. At right: Buddha of India.

Arnold in his romantic poem, *The Light of Asia.*

It is told that he married, and that in his thirtieth year, shortly after the birth of his son, he left his father's court and his wife and child, and wandered forth to seek the path of salvation, as taught by the Brahmans. He then entered upon a period of meditation, fasting and self-torture, and while seated under the sacred bo-tree, the light of the truth was declared to have dawned upon his troubled spirit. To him it was revealed that the one

THE GREATEST BUDDHIST TEMPLE

The great Temple of Maha Bodha, at Buddha Gaya, 100 miles from Benares, is Buddhism's most sacred shrine. The tree in the right foreground is the sacred bo-tree, lineal descendant of the bo-tree under which Buddha sat when the principles of his faith were unfolding.

way to find deliverance from suffering was by crushing all desires of the heart. Commencing at Benares, Buddha began to teach his new faith, in opposition to Brahmanism, winning thousands of converts by his pure life and gentle and earnest spirit. During his lifetime he saw his doctrines carried to all parts of India. B.M.W.

Consult Carus's *Buddhism and Its Christian Critics;* Hopkins's *The Religions of India.*

BUDGET, *buj'et,* as usually defined, is the official summary of a country's finances, including a forecast of the receipts and expenditures of the coming year. But this is a comparatively new meaning. The word was derived from *bougette,* an old French word meaning *bag, pouch* or *wallet.* In England the name was applied to a wallet or box in which legal and official documents were carried, and especially to the tiny trunk in which the Chancellor of the Exchequer kept his papers. Therefore when he appeared in the House of Commons he "opened his budget" to read the statement of the country's financial condition. About

1760 the term was first commonly applied to the financial statement itself.

The procedure in the British Parliament is typical of all countries in which a budget is prepared. The Chancellor of the Exchequer submits to Parliament a yearly statement of the expenses for the coming year and the means for raising revenue. The figures are usually compared with those of the preceding year. If the estimated revenue is smaller than the expenditures planned, the budget includes suggestions for increasing the revenue. In Canada the same procedure is followed, the budget being presented to the Dominion Parliament by the Minister of Finance.

The theory on which a budget is based is that the executive officers of a government are best qualified to determine how much money is needed and how it shall be spent. The people's representatives in the legislative department merely give their consent. The budget appropriates a grand total, and indicates approximately how much each department of the government will receive, but just how this share shall be disposed of is an executive, not a legislative, function.

In the United States this process is reversed. The United States, properly speaking, has no budget. Congress makes all appropriations and determines how every dollar shall be spent. The executive departments may make suggestions, but they have no voice in the decisions. Congress is at liberty to spend all it wants to spend, then it must find the revenue to pay the bills. If the United States had a scientific budget system, Congressional appropriations would be restricted. The theory of the American system is that the elected representatives of the people know how much they want to spend, and that the executives are hired to spend it.

The American system, on the whole, is more wasteful. It gives more opportunities for appropriations which are not vitally necessary but are due to political expediency. Economists and financiers, and some recent Presidents of the United States, as well, have recommended the adoption of a budget system, but Congress has taken no action.

BUENA VISTA, *bwa'na vees'ta,* BATTLE OF, an important battle of the Mexican War, through which the United States gained control of Northeastern Mexico. It was fought on February 22 and 23, 1847, near Buena Vista, at the upper end of a long, narrow, mountain pass called Angostura Heights. There the American general, Zachary Taylor, had stationed his force of 5,000 men, who were protected on one side by high cliffs and on the other by deep ravines. For two days a Mexican army of 20,000 under Santa Anna made desperate attempts to drive the Americans from the heights, but they were beaten back in every charge and finally driven from the field. The Americans lost about 750 in killed and wounded; the Mexicans, about 2,000. Taylor's victory at Buena Vista made him the hero of the hour, and the next year he was elected President of the United States. See MEXICAN WAR.

BUENOS AIRES, *bway'nohs i'raz,* or *bo'nus a'riz,* the capital of Argentina, known by reason of its beauty, its fashion and its cosmopolitan character as the "Paris of America." It enjoys several distinctions, for its

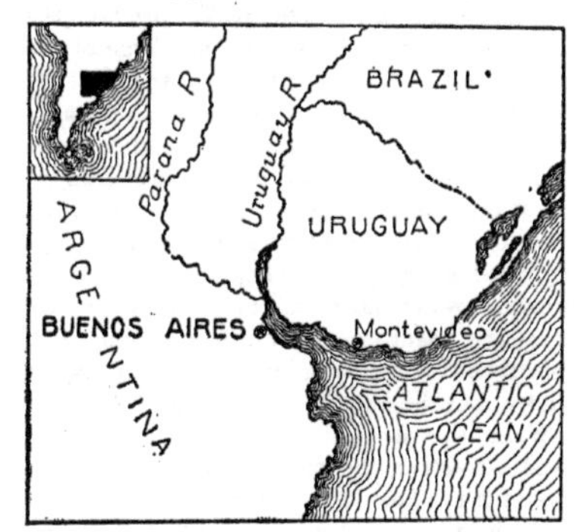

LOCATION MAP

In the small corner map the black space represents that part of Southern South America shown in the larger map.

population of 1,329,697 in 1911 not only makes it the largest city in the world south of the equator but the second largest Latin city in the world, Paris alone surpassing it. In 1914 the number of people was estimated at 1,560,163. Rome, with its thousands of years of history and with the civilization of Europe all about it, is not half so large as this comparatively-modern city in a continent which until recent times has scarcely been ranked with the progressive parts of the earth.

Location and Appearance. Though Buenos Aires is situated 175 miles from the mouth of the Rio de la Plata, it is in effect a seaboard city, for the river at that point is thirty miles wide and the silt has been dredged out so that large vessels may approach the city. The name Buenos Aires means *good air,* but the locality is not naturally very healthful, though improved sanitation has done much to make the death rate low. On a broad, spreading plain but fifteen or twenty-five feet above sea level, it stretches over about 72.8 square miles, and has all the characteristics—the business section, residence section, factory section, slums, parks and show places—of a Northern city.

The streets are for the most part broad, though in certain of the busy downtown districts they are too narrow for the throngs that crowd through them, and congestion is avoided by allowing vehicles of all kinds to move in but one direction on one street. Trees line many of the streets, such as the Avenida de Mayo and the Avenida Alvear; handsome homes and magnificent business blocks are common, and attractive parks display the plants and animals of the region. Some of the city's most beautiful buildings, the hall of Congress, the municipal building, the palace of justice, the government palace, the cathedral and the Episcopal palace, are grouped about the Plaza de Mayo. Buenos Aires prides itself on having the finest and most perfectly-equipped newspaper building in the world—a building which contains, in addition to the necessary plant, a library, a museum, club rooms and offices for physicians whose expert advice may be had for the asking.

While this modern part of the city is the most attractive, the old section is even more interesting. Low, flat-roofed houses, built in Spanish style about an open court and presenting to the street heavily-barred windows, are most in evidence.

The People. Like most great cities of America, Buenos Aires has drawn its inhabitants from all over the world. Native Argentinians, descended from old Spanish settlers, make up about half the population, but the other half comprises Italians in large numbers, Spaniards, Frenchmen, Englishmen, Germans and a few Americans. True to its Latin origin and character, the city is pleasure-loving, and life for the well-to-do there has the gayety, the brightness and the excitement which are usually associated with Paris. It is far from being a cheap place to live, and the authorities encourage immigrants to find their way inward to the rich agricultural land and not to settle in the city.

Progressiveness. The city has excellent communications with other towns, and a

Outline and Questions on Buenos Aires

I. Location and Size

(1) Latitude, 34° 36′ 21″ south
(2) Longitude, 58° 21′ 33″ west
(3) Latitude compared with that of Northern cities
(4) Area
(5) Population

II. Description

(1) Streets
  (a) How overcrowding is avoided
(2) Public buildings
  (a) Government palace
  (b) Cathedral
  (c) Finest newspaper building in the world
(3) Parks
( ) The old section
( ) Schools
(6) Public hygiene

III. Commerce

(1) Exports
  (a) Value
  (b) Character
(2) Imports
  (a) Value
  (b) Character

IV. The People

(1) Native Argentinians
(2) Foreigners
  (a) Origins
(3) Character of the people
  (a) Gay and pleasure-loving

V. History

(1) Settlement
(2) Troubled times
(3) Recent progress

Questions

Why is there little danger of collisions in the narrow crowded streets of the downtown section?

What has been the attitude of the city toward the question of public hygiene?

How many cities larger than Buenos Aires are there in which a Romance language is the ordinary language of the people?

What picturesque buildings are to be seen in the old section of the city?

Has Buenos Aires always been the capital of Argentina?

What does the name of the city mean? The name of the river on which it is situated?

What attitude does the government take toward emigration?

How does one newspaper in the city provide for its employees?

How many cities in the western hemisphere have a larger commerce?

What is meant by calling Buenos Aires the "Paris of America"?

What encouraged the Argentinians to declare themselves independent of Spain in 1810?

How can a city which is 175 miles inland be reckoned a coast city?

Besides the native Argentinians, what nationalities make up the largest part of the population?

How many cities in the United States surpass Buenos Aires in population?

thoroughly-adequate system of street railways within its own boundaries. In its schools and other educational institutions it is most progressive, and many Northern cities might take lessons from its careful supervision of the city hygiene. But it is in its commerce that the city shows most clearly its up-to-date character, for it is the second port in America, only New York carrying on a larger foreign trade. Its imports, consisting largely of manufactured goods, amount in normal times to almost $1,000,000 daily, and its exports of grain, wool, live stock and cattle products slightly exceed that sum. Most of the trade has always been with Europe, but of late years that with the United States has been increasing steadily.

History. Attempts were made in 1535 and in 1542 to found a Spanish colony on the site of Buenos Aires, but without success; the present city dates from 1580. Its growth, though slow, was steady, and its importance as a port was such as to lead to its choice in 1776 as the capital of the province of Rio de la Plata. Having discovered their strength by defeating in 1806 and 1807 attempts of the English to seize the colony, the people in 1810 declared their independence from Spain. From 1851 to 1859 Buenos Aires, with the province of the same name, was a separate state, but the difficulties which had led to its secession were finally adjusted, and in 1880 the city was made the capital of the republic. Since that time its history has been one of steady growth, for it has suffered but little from the revolutions which have prevented progress in most South American states. H.M.S.

AMERICAN BISON

WATER BUFFALO

For books relating to the city see list at end of the article ARGENTINA.

BUFFALO, *buf'a lo,* the name given to several species of wild ox, the best known of which is the common black buffalo of India, now found domesticated in nearly all the warmer countries of Asia and Africa. This buffalo is of an ashy-black color, frequently with white feet. The horns are triangular and covered with wrinkles that run around them; they curve outward and backward. This buffalo is smaller than the African buffalo, but larger than the domestic ox, the carabao or water buffalo of the Philippines; the latter is a smaller variety of the same species. The Indian buffalo has long been used in the rice fields of Asia, and is now found in Egypt and several countries of Southern Europe. The hide is tough and thick and makes excellent leather. The milk of the cow is excellent for food, and is used in India for making a sort of fluid butter.

The African, or Cape, buffalo is found throughout the southern part of Africa, and is the largest and fiercest buffalo known. The color is black; the horns are short, and they unite on the forehead, forming a sort of helmet. A smaller species having a brown color is found in the Congo region.

American Buffalo, or Bison. The bison, or American buffalo, is not a buffalo at all from the viewpoint of the zoölogist, because the structure of this animal does not correspond to that of the buffalo. For instance, the buffalo has thirteen ribs and the bison fourteen; the shoulders, head and neck of the bison are much heavier, and the withers are much lighter in proportion than the corresponding parts in the buffalo. Nevertheless, the bison has been called a buffalo for so many years that it is now commonly known by this name.

The American buffalo is of a dark, reddish-brown color, and the head, neck and shoulders of the male are covered with a thick growth of coarse hair which in some instances is almost black. This hair forms a great beard on the throat and chin. The head is very large and is carried low. A full-grown bull is about six feet high at the shoulders and when in good flesh will weigh about 2,000 pounds.

Formerly these animals were found by the thousands in all that portion of North America between the Appalachians and the Rocky Mountains and from Texas to the Peace River in Canada. The Indians used their flesh for food and their skins for clothing. The value of the skins soon became known to white men, who used them for robes and coats. The slaughter of the buffalo, which began soon after the coming of the whites, continued until they have practically disappeared, and are no longer found in the wild state. There are a few small private herds, and some of these are found in city parks. Several hundred are protected by the American government in Yellowstone Park, and over 2,000 by the Canadian government in Buffalo Park. An attempt has been made to produce a new, useful, domestic animal by crossing the bison and the domestic cattle. The product, called the "cattalo," has not, however, proved very satisfactory, and interest in the experiment is lessening. V.L.K.

BUFFALO, N. Y., the county seat of Erie County, next to New York City the largest city of the state, and one of the most important cities in the United States. It is situated at the eastern end of Lake Erie and at the head of Niagara River, twenty miles southeast of Niagara Falls. New York is 439 miles southeast, and Chicago is 523 miles slightly southwest. Buffalo is one of the most rapidly-growing cities of the Union; in 1900 the population was 352,387; in 1910 it was 423,715, and during the next five years it increased over 37,000, reaching 461,335 in 1915. Thirty per cent of the inhabitants are foreign-born, and among these Germans, Poles and Italians are found in greater numbers.

Buffalo is connected with the Hudson River by the Erie Canal, which, together with several other existing canals, has been enlarged and improved, forming the New York State Barge Canal, a waterway accommodating boats of 1,500 to 3,000 tons capacity (see NEW YORK STATE BARGE CANAL; ERIE CANAL). Through the Welland Canal the city has direct boat connection with the ports of Lake Ontario and the Saint Lawrence River. Buffalo is the home port of more than a dozen steamship lines, has several lines of ferries to the Canadian side, and many independent boats ply between this and other lake ports. Among the twelve trunk lines entering the city are the Buffalo, Rochester & Pittsburgh; Delaware, Lackawanna & Western; Grand Trunk; New York Central Lines; West Shore; Lehigh Valley; Michigan Central; New York, Chicago & Saint Louis; Pennsylvania, and the Wabash railroads. Two belt lines operate within the city, connecting the trunk lines with the various industries. The street car service is extended by interurban electric lines.

Location and Description. The early city was at the mouth of Buffalo Creek. It has grown northward and now covers an area of forty square miles, on a gradual slope of land which rises from fifty to eighty feet above the lake and 600 feet above sea level. The streets, with few exceptions, are broad, well-shaded and paved for more than half of their total extent, 750 miles. The business section is on or near the lake front, and from here the principal business thoroughfares radiate north and northeast; Main Street extends north and northeast to the city limits; Delaware Avenue, a parallel street, is crossed about a mile from the business center by North, Summer and Ferry streets, and by Lincoln Parkway. These streets are in the chief residence section. Niagara Street, branching from the foot of Main, follows the lake and river to Tonawanda, a suburb north of the city. From Niagara and

Lafayette squares extend Broadway, Genesee and Sycamore streets, each several miles long.

The shallow mouth of Buffalo Creek was the original harbor. It has been deepened, and a ship canal for wharfing has been built from this harbor southward, parallel with the lake shore. Inner and outer harbors in the lake have been created by a series of breakwaters. One nearly five miles long, constructed by the United States government, is among the largest breakwaters in the world. The government has also built a large lock in the Black Rock harbor, between Squaw Island and the mainland, to accommodate boats to and from Tonawanda. Buffalo has a total wharf frontage of ten miles. The great international bridge, completed in 1873 at a cost of $1,500,000, spans the river from Squaw Island to Bridgeburgh, Canada.

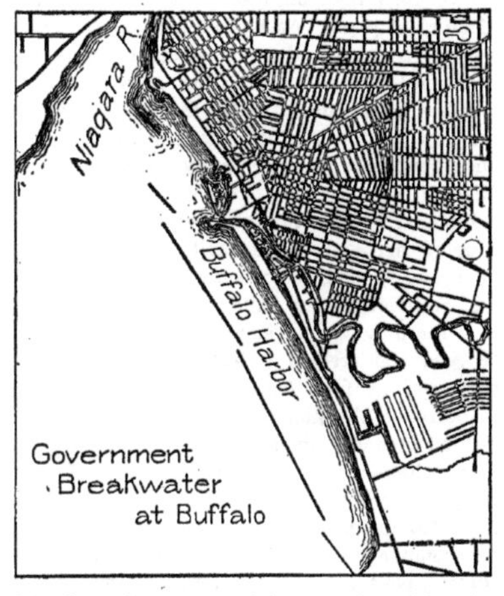

Government Breakwater at Buffalo

Parks and Boulevards. The total park area of the city is nearly 1,200 acres, and the larger parks are connected by boulevards and parkways which almost encircle the city. Delaware Park, the site of the Pan-American Exposition in 1901, is on the north side; it covers 365 acres, including a lake of forty-six acres. Contiguous to this park on the south is beautiful Forest Lawn Cemetery, with 239 acres, in which are monuments to President Fillmore and the Indian chief, Red Jacket; west are the grounds of the state insane hospital, 230 acres in extent. North of "The Front," a park of forty-five acres along the cliffs overlooking the lake at the head of Niagara River, is a small United States military post, Fort Porter, whose parade grounds, a favorite promenade, command a fine view of the lake. Humboldt Park, fifty-six acres, is on the east side of the city. The larger parks of the south side are South Park, 155 acres, which has a notable conservatory containing a great variety of plants, and Cazenovia Park, with 106 acres. There are a number of small parks and squares, among them Lafayette Square with a soldiers' and sailors' monument; Niagara Square, containing a monument to William McKinley, in the business section; and Franklin, Washington and Delaware squares. Niagara Falls, Crystal Beach and Fort Erie Beach (on the Canadian shore) are the popular resorts of the community.

Public Buildings. Among the principal buildings are the Federal building, completed in 1902 at a cost of $2,000,000; the county and city hall, constructed of granite at a cost of $1,400,000; the Sixty-fifth and Seventy-fourth regiment armories, each costing $2,000,000; the state arsenal, the Historical Building and Albright Art Gallery in Delaware Park, preserved from the Pan-American Exposition; the Music Hall, Chamber of Commerce, Merchants' Exchange, Masonic Temple, Y. M. C. A. building, four high school buildings, each costing $1,000,000; Buffalo Library, Grosvenor Library and the Protestant Episcopal and Roman Catholic cathedrals. Among the prominent club buildings are those of the Buffalo, University Park, Saturn and Country clubs. Ellicott Square, one of the largest office buildings in the world, covers an entire block.

Institutions. Institutions of higher education are the University of Buffalo, with schools of law, medicine and dentistry; a state normal school, Niagara University, the German Martin Luther Seminary (Evangelical Lutheran), Saint Joseph's and Canisius colleges and Sacred Heart and Holy Angels' academies (Roman Catholic), and the Buffalo College of Pharmacy. The Fine Arts Academy and the Society of Natural Sciences occupy a part of the Buffalo Library building. There are numerous private schools and night schools. In addition to the two municipal libraries are those of the Historical Society, Society of Natural Sciences, Law (Eighth Judicial District), Merchants' Exchange, Erie Railroad, German Young Men's Association, Lutheran Young Men's Association and Young Men's Christian Association.

Social service has been thoroughly and wisely developed. Buffalo has a large number of benevolent organizations, including about twenty hospitals, several of which conduct nurses' training schools. The Charity Organization Society (the first of its kind in the United States, organized in 1877), with headquarters in the building of Fitch Institute, a combined day nursery for children of poor working mothers and a training school for nursemaids, conducts charitable activities. In the city are also the Buffalo Orphan Asylum, Saint John's Orphan Home, the Home for the Friendless, Saint Vincent's and Saint Joseph's orphanages (Roman Catholic), the Church Home for Aged Women, Ingleside Home for Erring Women, Saint Mary's Asylum for Widows and Foundlings, Saint Mary's Institute for Deaf Mutes, the state insane hospital and the Erie County penitentiary.

Trade and Industry. In extent of traffic, Buffalo is one of the principal ports of the world, although open for commerce but eight months of the year. The total water tonnage of the port during 1915 was 19,488,427, an increase of more than 5,000,000 tons over that for the year 1910. A large volume of business is done with Canada, and the annual export trade amounts to about $80,000,000. The extensive water-borne commerce of the Great Lakes, consisting principally of grain and flour, lumber, coal, iron ore and fish, is here reshipped to be sent by smaller craft and by rail to other cities and to the ports of the Atlantic seaboard. Buffalo handles more wheat, flour and coal than any other city in the world. The total grain receipts in 1915 were 258,404,000 bushels, of which nearly one-half was wheat. It is handled by huge fixed and floating grain elevators and transfer towers. The elevators have a combined storage capacity of 22,000,000 bushels and can handle 5,000,000 bushels daily. Large quantities of the lumber and iron ore are docked at Tonawanda. Buffalo is one of the largest live-stock markets for horses, cattle, sheep and hogs in the United States, and annually distributes more than 15,000,000 pounds of fish over territory from Boston to Denver. The railroad stockyards are in the southeastern part of the city, and south are extensive coal trestles, the permanent supply stations of the railroads.

In manufacture Buffalo ranks second to New York City in the state. The industries include sixty per cent of all those recognized by the United States Census Bureau, the value of the products being about $251,000,000 annually. The capital invested in these is $243,311,000, and nearly 68,000 men are employed. The greatest number, 22,000, are engaged in the manufacture of foundry and machine-shop products, in which the city ranks high. It exceeds all other United States cities as a market for linseed oil, used largely in the manufacture of paint. At Lackawanna, a southern suburb, is one of the largest steel plants in the world. Manufacturing has grown rapidly and has been stimulated by the unusual shipping opportunities of the city and by a cheap and abundant power supply furnished by Niagara Falls.

History. When this was yet the red man's country, herds of buffalo visited the salt licks here, and their name was associated with the first settlement. La Salle, in 1679, built near here the first ship navigated on Lake Erie, the *Griffon,* and erected a fort, which was soon destroyed by fire. Cornelius Winney, an Indian trader, the first permanent white settler, arrived probably about 1788. A large tract of land, which included the present site of the city, was purchased in 1792 by the Holland Land Company. Joseph Ellicott, the agent of the company, later known as the founder of Buffalo, platted the city in 1801 and 1802, after the plan of the city of Washington. The place was called New Amsterdam until 1801.

During the second war with Great Britain the territory about Niagara Falls was the scene of active military operations, and in 1813 Buffalo was almost completely destroyed by a force of British and Indians. The settlement, which had been incorporated as a village in 1813, was rebuilt after the conclusion of peace, and its importance as a trade center was apparent as commerce developed on the Great Lakes. The Erie Canal was completed in 1825 and from that time the growth of Buffalo has been rapid. It became a city in 1832. The first grain elevator in the world was erected here in 1843. In 1852, Black Rock, a village to the north, and Buffalo's trade rival, was annexed. Millard Fillmore and Grover Cleveland made Buffalo their home, and Cleveland was mayor of the city when elected governor of New York in 1882. The Pan-American Exposition was held in Buffalo from May 1 to November 1, 1901. President William McKinley, while attending the exhibition, was assassinated on September 6. T.E.F.

BUFFALO BILL. See CODY, WILLIAM FREDERICK.

BUG, a name generally applied to all small insects which are hard to catch, as well as to tiny crawling ones. The true bugs, or *Hemiptera,* as the scientist calls them as a group, have beaks bent toward the breast, adapted for sucking or piercing, but not for biting. The so-called Junebugs, ladybugs, tumblebugs and potato bugs, however, are really *beetles.* Among the most common and troublesome bugs are the bedbug, chinch bug and louse. Most bugs are injurious to plant life, but their lives and habits are interesting to study. See INSECT; INSECTICIDES; BEETLE.

BUGLE, *bu'g'l,* a treble wind instrument, of brass or copper, but originally made of horn. It resembles the trumpet but has a shorter tube and a smaller bell-shaped opening. The tone of the trumpet is brilliantly blaring, while that of the bugle is softer and has a penetrating quality. The bugle is used chiefly for sounding the calls to the infantry; the trumpet, for the cavalry. In peace the soldier is reminded of every routine duty by a special call from a bugler, while in war his marches and movements are directed and guided by bugle calls.

Bugle Calls. The routine of life in the army barracks is marked by bugle or trumpet calls, from the *reveille,* which calls the soldiers from their slumbers, to *taps,* when the day is done. There are warning calls, formation calls, alarm calls and service calls. The accompanying calls are from the *Infantry Drill Regulations* of the United States army.

BUILDING, *bild'ing.* Since the days when men dwelt in caves, and like Polyphemus in the *Odyssey* shared them with their flocks, mankind's knowledge of the science of building has kept pace with advance in other fields. Though the rude peasants of some parts of Europe still live in sod hovels, Eskimos in huts of ice, Malays in tree-top shelters and present-day Indians of the American Southwest in houses of adobe, their more civilized brothers have learned how to erect beautiful homes and magnificent workplaces of stone, brick, concrete, steel or timber. The stages through which structural skill has passed and the principal types of buildings are described in the article ARCHITECTURE; this is the story of how modern buildings are constructed.

Building Trades. To erect a skyscraper the work of hundreds of men is required. Each has a particular kind of work—digging, shoveling sand, carrying hods of plaster, putting in place the hot steel rivets, laying the stone, bricks or blocks of terra cotta, plumbing, wiring for electricity, finishing walls or floors, or one of a hundred other tasks. On smaller buildings the work is divided into trades in the same way, and even to build a small wooden frame house, excavators, masons or cement workers, carpenters and bricklayers are required.

TAPS

Slow

RETREAT

Moderate

BOOTS AND SADDLES

Quick

REVEILLE

Quick

End

D.C.

Contracting. The men who supervise all these workers are usually called *contractors.* Sometimes a contractor agrees to erect a building for a stated amount of money, and if through careful direction of his men and wise buying of materials he is able to do it with less expense than anticipated, the saving is his profit. In other cases the owner of the building pays all expenses and a commission to the contractor. Frequently a contractor will in turn engage others to do part of the work; thus plumbing, wiring, plastering and similar tasks are often handled by *sub-contractors.* To be a successful contractor a man must possess a thorough knowledge of building principles, be able to estimate costs accurately and have the ability to govern others and to direct their operations so that good work will be done without waste of time. Many contractors have risen from the ranks of laborers.

The Substructure. The two parts of a building, the substructure and the superstructure, might theoretically be termed the below-ground and the above-ground sections. But the substructure includes the basement walls, which in actual practice extend sometimes many feet above the ground level.

Footings. The narrow end of a shingle can be pushed into the earth and the edge of a board can be driven in, but the flat side of a plank will support a very large weight without sinking. If the foundations of a heavy building are set directly on the soil they will penetrate it like the shingle, but if they are built on *footings*—which, like the plank, are sustained by a greater area of ground—they can be made nearly as solid as though on rock. Footings may be of concrete, reinforced-concrete, flat stone, stone and brick, concrete and brick, steel or timber. In very soft ground they are sometimes set on piles, as are the tall buildings of many large cities.

For any except a light frame structure the builders are always very careful to determine the proper size of footings according to the character of the soil and the weight of the building. An illustration of what will happen if the load is not evenly distributed to the footings is given in the illustration. Most large buildings have *isolated footings,* with the size of each one proportioned according to the average load it will be called upon to bear. The reason for making the average load the basis of the plans will be clear if one will suppose the case of a store building in the back of which are kept the tons of stock, but the front of which is lightly loaded except when crowds fill it on special sale days. If the footings are proportioned according to the greatest load at any one time, those in front will be stronger and the average load will cause those at the back to settle more rapidly. On the other hand, each footing must be large enough

EXAMPLES OF FOOTINGS

1. *A continuous footing* under an opening will cause cracks, because, as shown by the arrows, the downward pressure is all at the sides, and there is in effect an upward thrust at the center, which will "break the back" of the arch.

2. *Isolated footings,* formed by omitting that part of the footing beneath the opening, allow uniform settlement, because all upward force is directly under the downward pressure.

to support its share of the greatest possible load, which includes (1) the *dead load,* or weight of the building itself; (2) the *live load,* or weight of the people and objects which may at any one time be in the building; (3) the *snow load, wind load* and other emergency strains.

Footings must always be below the point to which frost penetrates the ground, or they will move to and fro and cause cracks in the walls above.

Foundations. Besides supporting the superstructure, foundation walls must withstand the inward thrust of the earth and, if there is a cellar, keep out dampness. Stone or concrete foundations are built at least eight inches thicker than the wall next above them. Stone is to some extent sponge-like, brick is more so, and concrete walls admit water readily unless covered with tar, asphalt or other waterproofing substance.

The Superstructure. The details of this part of a building vary widely, according to the materials of which its frame is composed. How a wooden structure is put together is told in the article CARPENTRY.

Brick buildings are of two sorts, those which have brick walls from twelve to forty inches thick, and those with wooden walls and a brick veneer only four inches thick. Both types are often constructed with air spaces in the walls as an aid to dryness and warmth. Sometimes stone is used on the outside and brick on the inside, the two materials being bonded together.

The more windows there are in a brick or stone wall the more danger there is that cracks will appear; the principle of this is exactly that illustrated above concerning footings. If window and door openings in upper stories are directly over those below, as in colonial architecture, the danger is less. The distribution of weight is also to be considered in placing stone steps, sills or lintels. Sills are of two types, *slip* sills and *lug* sills. The former are just the width of the opening which they border; the latter enter the wall a little at each end. Lug sills and stone steps are supported by mortar at the ends only; in other words, they are given isolated footings. A lintel is a beam supporting the wall over an opening; it is inserted into the brickwork only a few inches at each end.

Steel Frames. The highest buildings in the world have skeletons of steel which carry the weight of the whole structure. The walls of each story are called *curtain walls,* because they bear no weight but their own; they are usually of brick or tile or terra cotta. The floors are of hollow tile, which, though flat on their surfaces, are built on the principle of an arch, with a keystone in the center and all the weight thrusting against the I-beams at the side, as in the illustration. An explanation of the use of I-shaped beams will be found in the article BRIDGE.

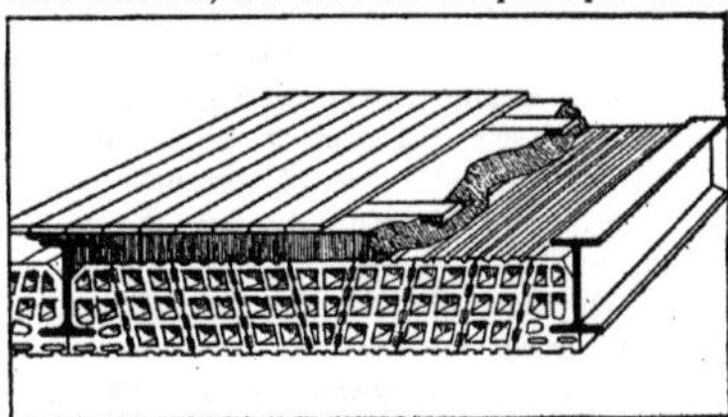

FLOOR OF FIREPROOF BUILDING
The hollow tiles, which are supported like an arch, are covered with concrete before the wood floor is laid.

Reinforced Concrete. Except for the frame itself, the making of which is described in the article CONCRETE, buildings of this material are constructed like steel frame structures. C.H.H.

Related Subjects. The following article will give a more detailed knowledge of the materials and principles in use in building:

Adobe	Calcimine
Brick	Carpentry
Building Stone	Carrara Marble
Cement	Pile
Concrete	Plastering
Fireproofing	Plumbing
Girder	Roof
Glass	Sandstone
Granite	Shingles
Iron	Staff
Limestone	Strength of Materials
Lumber	Steel
Marble	Stucco
Masonry	Terra Cotta
Mortar	Tiles
Nails	Wire Glass

See, also, the articles listed under ARCHITECTURE.

BUILDING AND LOAN ASSOCIATIONS, or **BUILDING SOCIETIES,** are private organizations whose primary object is to make it possible for people of small financial resources to own their homes. This is best accomplished by lending the savings of all the members to the few who at any one time wish to build or buy a home. All such societies may be divided into two classes, private corporations and mutual companies.

Private Corporations. Private associations are operated for the profit of the shareholders. The corporation receives money on deposit, like an ordinary savings deposit, and pays interest therefor. These deposits it lends to people who wish to build homes. The borrower mortgages the new home as security, and agrees to repay the loan in installments. The borrower pays a higher rate of interest than the depositor receives, thus giving the corporation its profits.

Mutual Companies. In a mutual company every depositor is a stockholder. A stockholder who wishes to borrow money subscribes for the number of shares which equal in par value the amount he wishes to borrow. He pays for this stock in small installments, and meanwhile he is also paying interest on the loan. If he is unable to complete the payments, he must surrender possession of the house and land, which becomes the association's property. If he completes the payments on the stock the mortgage on the property is automatically cancelled.

A mutual company has no capital except the weekly or monthly savings of the members. This capital naturally grows from year to year. It sometimes happens, however, that the amount available for loans is less than that required by a number of the members. In such a case the money is usually loaned to the one who offers the highest bonus over the regular interest. The disappointed members must wait until more money is available.

As only a few of the members can borrow the accumulated capital at any one time, the others must continue their payments, with the idea that they will borrow at some future time. Members are always allowed to withdraw, but few of them do so, as they then lose a part of the savings already invested in the shares. Objection has been raised to such organizations because they charge high rates of interest on their loans. But it is true that they lend money to people who would not otherwise be able to own their homes. In most of the states and provinces, as well as in many European countries, the operations of building and loan associations are very strictly controlled by special laws. W.F.Z.

BUILDING STONE. There are a number of varieties of stone used in the construction of piers and bridges, in the foundations and walls of buildings and in the finishing of interiors. These stones are known to the building trades

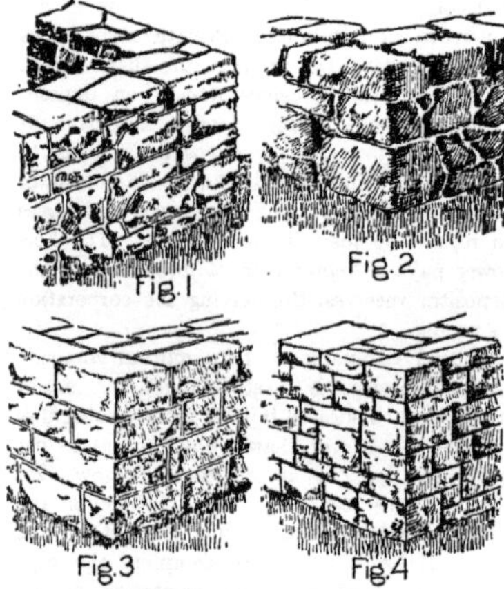

HOW BUILDING STONE IS LAID
(1) A rubble wall.
(2) Field stone
(3) Ashlar, in which all stones are squared.
(4) Random-coursed ashlar.

as *building stones.* Those most extensively used in the United States and Canada are granite, limestone, marble and sandstone. In addition, onyx, marble and slate are used in finishing interiors. Each of the stones named above is described under its title.

In selecting the stone for a building or other structure, the architect or engineer is governed by the following considerations:

1. Strength required.
2. Durability.
3. Convenience of access.
4. Expense of working.
5. Architectural effect.

Granite is the strongest and most durable stone. It is usually selected for foundations of heavy structures and for piers, and it is frequently used for the walls of large buildings.

Limestone in great variety is also used for foundations, walls and trimmings. It is not so strong as granite, but it is an excellent stone for all ordinary foundations and walls. *Marble* is softer than ordinary limestone, and is now practically restricted in its use to stairways, interior finishings and ornamental work.

Sandstone of hard quality is used for expensive dwellings such as the "brownstone" structures so famous in all large cities. It is more brittle than limestone and will not withstand so great a pressure. *Slate* is used for interior work in sinks, mantels and other furnishings. It is very durable.

Durability. The causes affecting the durability of building stone are the action of air and rain (weathering), change in temperature and chemical changes within the stone. Weathering is the chief cause of change. If the stone is porous it may absorb small quantities of moisture from the air and from rain. When the temperature falls below the freezing point this water freezes and expands; the next season more water enters, effecting a further change, until finally the surface begins to crumble. The presence of iron or sulphur in the stone is likewise a source of weakness. The iron, on exposure to the air, discolors the stone, and the sulphur is liable to unite with oxygen, forming sulphuric acid, which attacks the structure of the rock.

Building stone usually withstands the climate better if used near the locality where quarried. Many fine public buildings have rapidly lost their beauty because constructed of stone brought from a considerable distance.

The following table shows the life of the stones described, before they begin to show deterioration:

KIND OF STONE	LIFE IN YEARS
Coarse brownstone	5 to 10
Fine brownstone	20 to 50
Coarse fossiliferous limestone	20 to 40
Marble, coarse dolomitic	40 to 50
Marble, fine	50 to 100
Granite	75 to 200
Best Ohio limestone	100 to 200
Nova Scotia sandstone	50 to 200

The above variations in years are due not only to quality but to location of the building of which the stone is a part.

BUKOWINA, *boo ko ve'nah,* a former Austrian duchy, with an area of 4,031 square miles, between Hungary, Russia and Rumania, in the region of the Carpathian Mountains. It is well watered by the Pruth, Sereth and Dniester rivers, and in their valleys the soil is extremely fertile. Cattle raising is an important industry, and crops of cereals, fruits and vegetables are raised. Brewing, distilling and milling are extensively carried on, but manufactures are not numerous and few minerals are produced. The capital is Czernowitz, with a population of about 88,000. The inhabitants are of mixed races, consisting chiefly of Ruthenians, Russians, Rumanians, Germans and Poles. Population, about 750,000.

BULB, the underground storehouse of a plant—a kind of bud whose very compact leaves, through storage of food, have become thick and fleshy. Like a seed, a bulb holds within it the undeveloped plant of the future—leaves, flower and all. But, unlike the seed, a bulb has much stored-up nourishment, and a plant grows more quickly from the bulb than from the seed. Some bulbs are formed in rings or layers, like onion and hyacinth bulbs; others are scaly, like those of the lily. Put in the ground at the right time, which depends entirely upon climate and season conditions, and given the proper attention and continued nourishment, a bulb will soon send forth roots from the bottom. And from the center, up through the ground, will come shooting the leaves, stems and flowers.

The bulb of a plant is formed either above-ground or beneath the surface. Being usually brownish or earth-colored, it does not advertise for outside help, as do some of the flowers to bees, for pollination (which see). So each

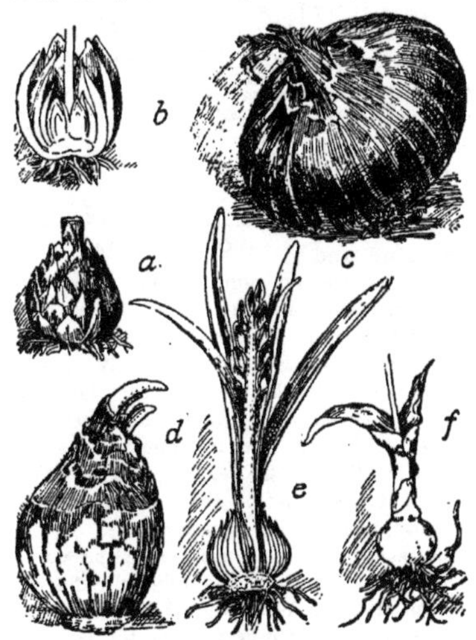

SOME COMMON BULBS
(*a, b*) Lily, entire and cross-section; (*c*) early red onion; (*d*) narcissus; (*e*) hyacinth; (*f*) tulip.

bulb produces decade after decade a plant exactly like the parent. Some bulbs are edible, like the onion, and most of them produce fragrant flowers of wonderful charm and interest.

BULGARIA, *bul ga'ri a,* a kingdom in the eastern part of the Balkan Peninsula of Europe. Long ground under the heel of Turkey, in the last four or five decades it found its strength and asserted its independence. The Balkan War of 1912-1913 won it a sixteen per cent increase in territory. Its future was promising until it was defeated with its Germanic allies in the War of the Nations, in 1918. Area, 43,310 square miles; the population in 1910 was 4,467,000. Many thousands perished in the war. It would take six Bulgarias to equal in size the province of Alberta, Canada.

The Land and Its Resources. The Danube River runs along the northern boundary, and it is the valley of that stream which forms the northernmost of the three physical divisions of Bulgaria. South of this stretch the Balkan Mountains, with peaks which reach at times heights of 10,000 feet, and beyond these is a lowland region which slopes to the Aegean Sea.

Differences in climate are as well marked as those in surface. The territory north of the Balkans is colder than that of the south, as the Balkan Mountains protect the latter from the north winds. Forests of oak, pine and beech grow on the northern slopes of the mountains, which are also known to be rich in minerals. South of the mountains a fine, temperate climate is found, but only in the extreme south, in the Aegean coast land, is there such a climate as, for instance, prevails in Italy. Bulgaria lies in about the same latitude as the middle regions of Italy, but much of it is shut off by the mountains from the warm, moist Mediterranean winds.

LOCATION MAP

The country in late years has witnessed a steady development of the mining industry. Between 1892 and 1915 over 2,200 permits for mine-prospecting were granted by the government. Three coal mines, two copper mines, one lead and one zinc mine were in operation in 1916. Agriculture, however, is the chief industry, almost three-fourths of the population maintaining themselves by it. Modern farming implements have taken the place of crude wooden plows, and reaping, threshing and winnowing are now performed with the aid of efficient machines instead of by hand. The land in Bulgaria belongs to the farmers and not to the government; there are no large land proprietors, and out of nearly 934,000 farms, 905,965 have less than fifty acres each.

By far the most important crops are the grains, which are grown in quantities large enough for export. Potatoes are also grown extensively, and in Eastern Rumelia rice and cotton flourish. Tobacco-raising, the production of cocoons for silk, and apiculture (bee-raising) are also important. The most interesting industry to a visitor is the raising of roses for the perfume "attar of roses." In the southern valleys of the Balkans roses grow everywhere in profusion, scenting the air long before the traveler comes close enough to see them. Great quantities of them are needed, for it takes over 6,000 pounds of rose leaves to make one pound of attar, and Bulgaria produces 9,000 pounds of attar annually. See ATTAR OF ROSES.

Transportation and Trade. Had Bulgaria been off the direct line of trans-European travel, it might have remained obscure and unimportant; however, the shortest route from Western Europe to Asia lies through its territory. Branch roads run from this, but it is not on these or on the main line that Bulgaria depends for transporting its products. A large part of its exports go by way of the Danube, and much of the foreign trade is with Austria-Hungary and Germany.

Cereals make up over two-thirds of the exports, but attar of roses, fruits and animal products are also exported, while the chief imports are textiles, metal goods, petroleum and coal. Although the bulk of the population is busy with agriculture, the Bulgarians have shown themselves proficient also in commerce and trade, as is proven by the many and various commercial and industrial enterprises and the number of its banks.

The People and Their Condition. Bulgarians constitute about three-fourths of the population of the country, and the second race in point of numbers is the Turkish, of which there are about 466,000. The native Bulgarians are not a quick, brilliant race, but they are determined and are willing to work very hard for anything that seems to them desirable. One of the things which has seemed most desirable to them is education. Even while the Turks were crushing them, schools were maintained, and with freedom these increased and improved rapidly. Primary education is free and compulsory, and there are normal schools and colleges of excellent grade, commercial, industrial and trade schools and a number of agricultural schools, including stations where practical instruction in agriculture is given.

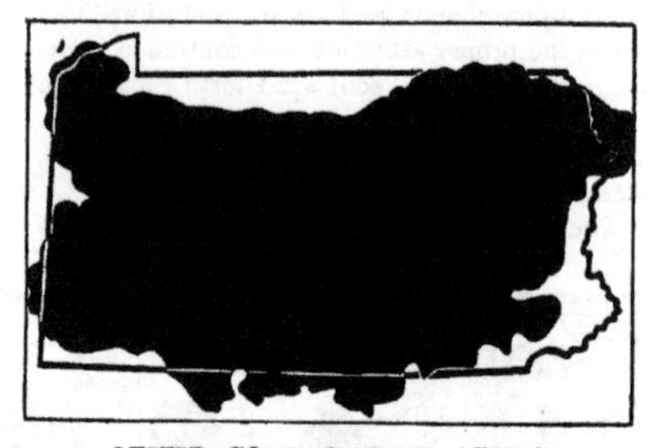

OTHER COMPARATIVE AREAS
Pennsylvania is slightly larger, with 44,832 square miles, while Bulgaria has 43,310.

The Bulgarian Church, a branch of the Greek Orthodox, is the church of over eighty per cent of the people. Other religions are not molested, however, and many of the cities have a numerous population of Mohammedans.

Government. Bulgaria is a constitutional monarchy, under a new constitution which dates from 1908. The ruler has in his hands practically all of the executive power, and his sanction is necessary to put into effect laws passed by the legislative assembly. This body, known as the *Narodno Sobranye* (National Assembly), is made up of representatives elected by universal male suffrage, and there is one representative for every 20,000 inhabitants. For the administration of local government Bulgaria is divided into seventy-one districts.

The chief cities are Sofia and Philippopolis, both on the railway line which connects Europe with Asia; and Varna and Bourgas, the principal seaports. See SOFIA.

History. During Roman times present-day Bulgaria was Roman territory, partly Moesia, partly Thrace. Over the plains to the eastward came first the Slavs and later the Bulgars, relatives of the Huns, and since neither

FORMER KING FERDINAND
He cast Bulgaria's lot with Germany in the second year of the war. He guessed wrong as to the outcome and lost his throne.

was strong enough to drive out the other, a gradual blending took place which resulted in the Bulgarians of to-day. In 864 they accepted Christianity, and from them the Church language and Church, as well as secular, literature were introduced into Russia. For a time the Bulgarian state was a strong enemy of the Byzantine Empire (which see), but in the fourteenth century it fell a prey to the Turks. Five centuries passed before the Bulgarians were strong enough to rise against the Turks; even then the first insurrection, in 1876, was comparatively slight.

The Turks, however, punished the subject people severely, committing those unspeakable "Bulgarian atrocities" which roused statesmen everywhere to denunciations of the Turk. Russia, whose people are akin in race and in religion to those of Bulgaria, undertook to right Bulgarian wrongs, though not with a purely unselfish motive, and the result was the Russo-Turkish War (which see). Turkey was defeated, and the Congress of Berlin made Bulgaria a semi-independent principality under the nominal suzerainty of Turkey. The country progressed steadily. It chose as its prince the German Alexander of Battenberg, and when he abdicated in 1886, Ferdinand of Coburg was selected. Meanwhile, in 1886, Eastern Rumelia had united with Bulgaria, and the result was a sharp clash with Serbia, in which, to the amazement of Europe, Bulgarian armies were victorious.

In 1908 Bulgaria took advantage of the internal troubles which were occupying the attention of Turkey to declare itself independent. Prince Ferdinand then assumed the title of king. But a greater crisis in Bulgarian affairs occurred in connection with the Balkan War of 1912-1913. With the other Balkan states—Serbia, Montenegro and Greece—Bulgaria was victorious over Turkey, its armies proving themselves well disciplined and most courageous (see BALKAN WARS). Turkey was forced to give up almost all its European territory, but a dispute arose between Bulgaria and its allies, because the latter, especially Serbia, refused to stand by the treaty which regulated the division of the conquered territory. The result was another war, in which Greece, Serbia, Rumania and Montenegro joined against Bulgaria, with entire success. Exhausted, Bulgaria consented to a peace which gave to Rumania a portion of its northeast territory, and permitted it to annex but a fraction of the land which had been won from Turkey. Bitter was the disappointment caused in the young, self-confident nation, and had not its resources been well-nigh exhausted, it would probably have renewed the struggle.

In October, 1915, Bulgaria entered the War of the Nations as an ally of the Central Powers, Germany and Austria-Hungary. For

Outline and Questions on Bulgaria

I. Position and Size

(1) Latitude, 40° 45′ to 44° north
(2) Longitude, 22° 20′ to 28° east
(3) Area
  (a) Comparative
  (b) Actual
(4) Population

II. Physical Features

(1) Balkan Mountains
(2) Lowland region
(3) Rivers
  (a) Danube

III. Climate and Vegetation

(1) Distinct zones
(2) Effect of mountains
(3) Forests
(4) Roses

IV. Industries and Commerce

(1) Agriculture
  (a) Grains
  (b) Potatoes
  (c) Rice, cotton and tobacco
(2) Mining
(3) Manufacturing
  (a) Attar of roses
(4) Railway routes
(5) Exports and imports

V. The People

(1 Proportion of Bulgarians
(2) Other races
(3 Characteristics of Bulgarians
(4 Education, religion and government

VI. History

(1 Its antiquity
(2 Coming of the Bulgars
(3) Acceptance of Christianity
(4 Under Turkish rule
(5 Independence achieved
(6) Balkan Wars
(7) War of the Nations

Questions

Why has not Bulgaria as warm a climate as Italy?

What is meant by "Bulgarian atrocities"?

If the populations of Alberta and Bulgaria were exchanged, would the latter country be more or less densely populated than at present?

What industry helps to make the country regions delightful?

Is the executive or the legislature the more powerful?

What river and what mountains dominate the northern and the central physical divisions?

What fortunate fact of location has made Bulgaria important?

What great empire got its religion from Bulgaria?

What sort of agricultural implements have been used until very recently?

For what peculiar reason did the Turkish government permit the destruction of Bulgaria's magnificent forests?

What attitude have the people taken toward education?

Was the territory of Bulgaria greater at the end of the first Balkan War or the second?

the course of that gigantic struggle, see WAR OF THE NATIONS. S P.

Other Items of Interest. The name *Bulgaria,* according to popular etymology, is derived from that of the *Volga,* the river on which the original Bulgar kingdom was located.

Bulgaria once had magnificent forests, but the Turkish government permitted their destruction and even encouraged it, that robbers and highwaymen might not find shelter in them.

In the mountains, the little brown Bulgarian bear is still found in great numbers, and some of the outlying villages suffer in the winter from the ravages of wolves.

Autumn is called "the clear time," and is the most delightful of the seasons. Frequently it lasts well into the month of December.

The clergymen of the national Bulgarian Church are paid by the government.

Morals are high among all classes, and religion is a very real part of the life of the people, but they are by no means fanatical.

A compulsory military system prevails, and of the able-bodied men only Mohammedans are exempt from service.

Every summer thousands of wandering shepherds come from Macedonia and Greece to pasture their flocks on the mountains of Bulgaria.

In general, the Bulgarians are slightly below average height, and are rather stocky. The women early lose what good looks they possess, and by middle life look far older than they are. This is due to their life of toil.

Almost the only large estates in the country are those held by the monasteries.

Birds of prey are everywhere abundant, and in consequence singing birds are not numerous.

In its mountain regions the country has scores of mineral springs, many of which have been in high repute since Roman times for their medicinal qualities.

The money unit is the *lev,* which has a value of about nineteen and one-fourth cents.

Related Subjects. The articles under the following titles will give added information on the subject of Bulgaria:

Balkan Peninsula	Slavs
Balkan Wars	Sofia
Danube River	War of the Nations

BULL, a letter or order from the Pope, acting as the head of the Church, and published or sent to the Roman Catholic churches. It is written in Latin, usually with elaborate lettering, dated from "the day of Incarnation," and named after the first word or phrase. The name *bull* comes from the seal used, which is a *bulla,* a round piece of lead impressed on one side with the heads of Saint Peter and Saint Paul, on the other with the name of the reigning Pope. If the bull be a "Bull of Justice," the seal is attached by a cord of hemp; if a "Bull of Grace," the cord is of red or yellow silk. Pope Leo XIII ordered the use of ordinary instead of Gothic characters on the less important bulls. D.J.D.

BULL, JOHN, the name popularly used to typify England and the English people, in the same sense that "Uncle Sam" typifies the United States and its people. It was first used by the witty Scottish doctor and writer, John Arbuthnot, in *The History of John Bull,* in which, in a discussion of the political affairs of Europe at that time, John Bull, representing England, appears as a jolly, honest, plain-dealing but hot-tempered farmer. Arbuthnot's word-picture was later reproduced in a drawing by Sir Francis Carruthers Gould, and now the name and the picture through long use have grown familiar to all. John Bull wears a "tile" hat, a swallow-tail coat, trousers tucked in boots, and across his ample waistcoat usually appear the outlines of the British flag.

JOHN BULL

BULL'DOG, a species of dog with fierce, savage eyes, leering features and affectionate regard for its master—the best watchdog and one of the most faithful canine companions. It has been said that the bulldog is "so homely as to be positively beautiful."

Many years ago, to get just the right animal for the barbarous sport of bull-baiting, the British mastiff was crossed with the pug of Eastern Asia. And through careful breeding and selection we have the bulldog of to-day—with massive head, short, wrinkled muzzle and little, rounded nose, loose-hanging lips and protruding lower jaw, with lower front teeth showing. The ears are drooping, the neck thick and short. From the strong chest, held by slightly-bowed legs, the compact, short-haired body tapers to higher, straight hind legs. The prize bulldog of the dog shows weighs about fifty pounds, and is brindle, red, fawn, white or piebald. The close, tight grip of the jaws of the bulldog makes it a terror to thieves and others bent on mischief. Its faithful affection for its master and his family makes it usually a safe companion for children. The *bull terrier* came originally from a cross between the bulldog and the terrier. It is smaller than the bulldog, lively and very courageous. See DOG; TERRIER.

BULLDOGS

At left, French bulldog; at right, English bulldog.

BULLET, *bul'et,* the small projectile, from one to three inches in length, discharged from a rifle, pistol, revolver, machine gun or similar firearm. The bullet used in modern rifles is conical in shape and consists of a core of copper covered with steel or nickel. In size it varies according to the caliber of the weapon used. Revolver bullets are heavier and shorter and the wounds they inflict are more dangerous than those inflicted by rifle or machine gun fire. It has been found that unless struck in a vital organ, such as heart or brain, sixty-five per cent of those wounded in warfare to-day make speedy recovery. Bullets used for hunting big game usually have soft lead points, or are hollowed to insure spreading when they reach their mark. The use of such missiles, called *dum-dum* bullets, is condemned in present-day, so-called civilized warfare. Modern bullets are always held in a cartridge case, usually of brass. See AMMUNITION; PROJECTILE; RIFLE; MACHINE GUN.

BULLFIGHTING, a contest between men and bulls, the latter tormented and goaded to fury. This sport, frowned upon by all peoples except certain of the Latin races, was very popular among the Greeks and Romans and was introduced into Spain by the Moors. It at once captured the fancy of the Spaniards and became their national sport. In Spain and Mexico bullfights are still of regular occurrence on holidays and feast days, and skilled bullfighters are popular heroes. The fights are usually held in an amphitheater having circular tiers of seats rising one above another, and are attended by vast crowds.

The combatants, who make bullfighting their profession, march into the arena in procession. They are of various grades—the *picadores,* combatants on horseback, in the old Spanish knightly garb; the *banderilleros,* combatants on foot, in gay dresses, with colored cloaks or banners; and lastly, the *matador,* who deals the death blow to the defeated bull. As soon as the signal is given, the bull is let into the arena. The *picadores,* who have stationed themselves near him, commence the attack with their lances, and the bull is soon infuriated. Often a horse is wounded or killed, and the rider is obliged to run for his life; sometimes men are seriously gored by the horns of the distracted animal. The *banderilleros* assist the horsemen by drawing the attention of the bull with their red cloaks, and try to fasten on the bull their *banderillas*—barbed darts ornamented with colored paper, and often having explosive crackers attached. If they succeed, the crackers are discharged, and the bull races yet more madly around the arena. In case of danger the *banderilleros* and *picadores* save themselves by leaping over the wooden fence which surrounds the scene of conflict.

The *matador* enters the arena with a naked sword and red flag; when the animal has afforded sufficient sport or is too weak to offer further resistance, he dispatches it with one skilful thrust. The slaughtered bull is dragged away and another is turned into the arena. Eight or more bulls may be sacrificed in a single afternoon. See illustration, in full-page illustration, in article SPAIN.

BULLFINCH,' a singing bird which can be taught to repeat musical airs. In Germany great care is given to training bullfinches, and good singers command high prices. The bullfinch's name is no doubt derived from its thick, bulging bill, which makes its little head faintly resemble that of a bull. It is about the size of an English sparrow, and has a blue-gray body with a bright red breast and a black beak and crown. It is found in well-wooded places in Britain, Southern Europe and Asia, where it lives on berries, small fruits and the buds of trees.

BULLFROG, the frog of the deep bass voice, at home in weedy lakes and ponds. In Canada and the United States east of the Rocky Mountains, including Florida and Texas, the bullfrog is found. In ponds or lakes where water is hidden from the shore by low trees and bushes, weeds and reeds and water lilies

BULLFROG

he makes his home. There he feasts on the bugs, snails, shrimps, or toads and fish—all the dainties which he relishes; this food is found near the roots and stems, and under the leaves of the sheltering water plants. When he is not searching for food or sleeping one can see him basking in the sunshine on some half-hidden log, revealing his large yellow-green body, which, in the shade, is olive-green or reddish-brown, with large brown or black spots, and

BULLFROG TADPOLES
(*a*) A developing bullfrog; (*b*) the same four days later, showing beginning of absorption of the tail.

yellow lines across the back. And all the day, or in the still summer evening, above the chorus of other little creatures, comes the loud, hollow "jug-o-rum" of the bullfrog. A powerful swimmer, he loves the water and is rarely seen on land, excepting occasionally after long-continued rains. The hind legs of the frog are often used as food, and also as bait for fish. See FROG.

BULLHEAD. See CATFISH.

BULLION, *bul'yun,* the name given gold and silver in any form except legal-tender coin. We usually think of bullion as bars of gold and silver brought to a mint to be made into coin. When anyone deposits bullion at a mint for coinage it is carefully weighed and its degree of purity is ascertained. A report is then made to the depositor setting forth these facts, together with a statement of the net value of the bullion deposited. If there are any charges or deductions the amount of these is also stated. When the coins are ready for delivery they are given to the depositor, on his order, by the superintendent of the mint. The name *bullion* is also applied to gold dust, gold nuggets, gold and silver plate and foreign coins. See COINAGE; MINT.

BULL RUN, BATTLES OF, two engagements of the War of Secession, fought near Bull Run, a small creek in Northeastern Virginia, both of which gave the Confederates reason to hope for the ultimate triumph of their cause. The name *Manassas* instead of Bull Run is frequently applied to these battles, that being the name of a small town in the vicinity.

The First Battle of Bull Run, the first serious conflict of the war, was fought on Sunday, July 21, 1861, between Federal troops commanded by General McDowell, and Confederates under generals Joseph E. Johnston and Beauregard. About 18,000 men were engaged on each side, though the entire Federal army numbered about 30,000, and the Confederate about 28,000. Both armies were made up of raw, poorly-trained volunteers. McDowell began the battle in the morning by sending Tyler, Heintzelman and Hunter to turn the left wing of the Confederate army, which was posted along Bull Run Creek. This movement was so successfully carried out that by three o'clock victory was apparently with the Federals. Jackson's brigade, however, which was stationed on a small hill, held its position unflinchingly throughout the fight, causing one of the Confederate generals to exclaim, "There stands Jackson like a stone wall," and giving that gallant officer his familiar title of "Stonewall."

With the aid of reinforcements, generals Johnston, Beauregard, Jackson and Kirby Smith made a new attack on the Federals and drove them from the field in such disorder that the retreat became a rout. The total Federal loss was about 2,800; the Confederate, about 2,000. The effects of this battle were momentous. The North, realizing for the first time that the conflict was more than a three-months' fight, began preparations for a serious war; the Confederate cause was greatly strengthened in Europe, and the South, enthusiastic over the result, became over-confident.

The Second Battle of Bull Run, fought on August 29 and 30, 1862, marked the close of Pope's campaign in Virginia. His army of about 64,000 men lay encamped along the Rappahannock, engaged in the defense of Washington. On August 26, "Stonewall" Jackson, with a force of about 25,000, destroyed the Union supplies at Manassas and Bristow, and three days later Pope, who had formed his lines near Bull Run Creek, ordered an attack on the Confederates. In the meantime Lee and Longstreet had succeeded in effecting a union with Jackson.

At the close of the first day's fight both sides claimed a victory. At noon on the following day Pope ordered Porter, McDowell and Heintzelman to make a united attack, but the movement was unsuccessful, and a counter attack by Longstreet's troops drove the Federals from the field. The Confederates lost about 9,500 out of a total force of 54,000, while the Union loss was fully 14,500. Pope then led his army back to Washington, and Lee, with the enthusiasm born of a decisive victory, took the aggressive and began his first invasion of the North (see ANTIETAM, BATTLE OF). See, also, WAR OF SECESSION.

BÜ'LOW, BERNHARD HEINRICH, Prince von (1849-), a German diplomat and statesman, Prime Minister of Prussia and Chancellor of the German Empire from 1900 to 1909. After five years of retirement he again became conspicuous in world politics late in 1914, when he was appointed ambassador extraordinary to Italy. The object of this appointment, which was due partly to von Bülow's personal popularity in Italy and partly to the fact that his wife was an Italian, was to use every possible influence to keep Italy from joining the allies against the Teutonic powers. Although von Bülow's skilful diplomacy is given credit for delaying Italy's declaration of war, it could not prevent it (see WAR OF THE NATIONS).

PRINCE VON BÜLOW

Von Bülow was born in Holstein, at that time still under Danish rule. His family had been politically prominent for several generations, and his father, though originally in the Danish service, later entered the service of Prussia, and from 1873 until his death in 1879 was Secretary of State for Foreign Affairs. His son was naturally destined for a public career, and after serving in the army during the Franco-German War, entered the German Foreign Office. He was in turn secretary of legation at Rome, Petrograd and Vienna, and was chargé d'affaires at Athens during the Russo-Turkish War of 1877-78. He was a secretary at the Congress of Berlin, at which his father was one of the three Prussian representatives. After further service in Petrograd and Paris he became minister to Rumania in 1888 and ambassador to Italy in 1893. Four years later he was recalled to become Secretary of State for Foreign Affairs, and in 1900 succeeded Hohenlohe as Chancellor of the Empire.

In the former position he was active in carrying out Germany's desire for colonial expansion, and as Chancellor he was particularly skilful in holding together the heterogeneous parties in the Reichstag and adroit in interpreting to the world the policies of the government. In 1905 von Bülow won a notable triumph in the negotiations with France concerning Morocco, and three days after the resignation of Delcassé he was raised to the rank of prince (Fürst). Although the Chancellor is not legally responsible to the German Parliament, von Bülow resigned in 1909 when his budget, including proposals for tax reforms, was defeated by the Reichstag.

BÜ'LOW, Hans Guido von (1830-1894), a famous German musician, one of the greatest pianists and orchestra conductors of his century. He was born in Dresden and began to study music at the age of nine. It was many years later, however, before he made up his mind to follow music as a profession, for at one time he expected to become a lawyer. In the year 1850 he attended a magnificent production of Wagner's *Lohengrin,* at Weimar, and at that time resolved himself to become a musician. Soon after this he went to Zurich to study under Wagner, whose genius he worshiped all his life, and later he was taught by Liszt at Weimar.

Bülow's public career began with a concert tour of Germany and Austria, in the year 1853. For many years afterward he traveled widely, giving orchestra concerts in the various European countries and in America. He also held several important positions, being at different times head professor in a leading conservatory of Berlin, conductor of the royal opera in Munich, and court conductor to the Duke of Meiningen. The Meiningen orchestra, which he took with him on numerous tours, became the most famous in all Germany. His work as a conductor was remarkable because of his attention to details and his masterly interpretation of the music. His memory, too, was marvelous, and he used notes neither in conducting nor in playing.

Bülow was one of the best authorities on Beethoven, and published an edition of the latter's works for the piano. His own compositions include the music to Shakespeare's *Julius Caesar;* a musical ballad entitled *The Minstrel's Curse;* a symphonic poem, *Nirwana,* and numerous songs, choruses and pieces for the piano. His fame as a composer, however, is secondary to that as a pianist and conductor.

BUL'RUSH, a rush-like plant of the marshes, common in Europe and America. The creeping root has astringent properties and was at one time used in medicine, but is now abandoned in favor of more effective remedies. The round, almost leafless stems, from two to ten feet high and bearing a few small, brown flowers near the top, are the most useful part of the plant. The little ark of bulrushes which the mother of Moses made in which to hide him in the marshes indicates the very old history of the plant. The stems are used in making chair bottoms, mats, etc., and in California a common variety is employed in covering wine bottles for packing. See Horsetail Rush.

BULWER-LYTTON, *bul'wer lit'un,* Edward George Earle, first Lord Lytton (1803-1873), an English novelist whose most enduringly popular work was *The Last Days of Pompeii.* Others of his historical novels are *Rienzi* and *The Last of the Barons,* while his "novels of manners," so-called, include *Pelham,* his first successful work, *Ernest Maltravers, The Caxtons* and *My Novel.* Theatrical as these all are in places, and affected in style as well as in sentiment, they were widely read at the time and have continued to have a general vogue because they have stories of interest to tell. Bulwer wrote plays also, and two of them, *The Lady of Lyons* and *Money,* have not yet been dropped from the stage. From another play, *Richelieu,* came the two best-known quotations from Bulwer:

The pen is mightier than the sword

and

In the lexicon of youth * * * * there is no such word as "fail."

Bulwer-Lytton was born in London, studied at Cambridge, where he won honor for facility in verse-making, and after his graduation spent some time in Paris. On his return to England he became estranged from his mother because he made a marriage of which she did not approve, and found himself obliged in some way to provide an income. Thus he was practically forced into literature, and plays and novels followed one another in rapid succession, *Pelham* appearing in 1828, and *Kenelm Chillingly,* his last novel, in the year of his death. Meanwhile, he interested himself in politics, and sat in Parliament from 1831 to 1841 and from 1852 to 1866, attaining considerable influence. In 1866 he was raised to the peerage as Baron Lytton. He was buried in Westminster Abbey.

His son, EDWARD ROBERT BULWER-LYTTON, became a notable literary character, whose *Lucile* will make him long remembered. See LYTTON, EDWARD ROBERT BULWER.

BULYEA, *bul'yay,* GEORGE HEDLEY VICARS (1859-), a Canadian legislator and administrator, first lieutenant-governor of Alberta. He was born at Gagetown, N. B., and received his schooling there and at the University of New Brunswick. He went to Winnipeg in 1882 and the following spring to Qu'Appelle, Sask., where he engaged in business. He was elected to the northwest council in 1894 and for many years was a member of the executive council. In the territorial government he was commissioner of agriculture and of public works (1899-1905), and on the organization of Alberta as a province, September 1, 1905, became lieutenant-governor, a position which he filled with distinction for ten years. In 1915, at the conclusion of his second term as lieutenant-governor, Bulyea became the first chairman of the provincial public utilities board.

BUMBLEBEE. See BEE, subtitle *Bumblebee.*

BUNDESRAT, *boon'des raht,* a German word meaning *council of the confederation,* applied to the Federal council which assists the Reichstag, or Parliament, of the German Empire in its legislative functions. Its members represent the states, considered as units, and not the people, and all the delegates from one state must vote as a unit, according to instructions given them by their government. See GERMANY, subhead *Government.*

BUNGALOW, *bung'ga lo,* a type of house or residence now very popular in the United States and Canada, and to some extent in Europe. All rooms are conveniently arranged on one floor, and spacious verandas, roofed and often screened, are provided. Simple of construction, of wood, stone or brick, yet affording opportunities of rustic and artistic effects in material, roofing and shape, the bungalow, in the minds of many people, makes the ideal home. But in cities the bungalow is frequently more expensive than a two-story dwelling. The one-story plan demands a larger lot to provide sufficient space within and without, and where the price of real estate is high a few feet may add largely to the investment. A modest, comfortable bungalow can be erected for as little as $700.

A NORTH AMERICAN BUNGALOW
A characteristic form, showing in plan the convenient arrangement of living rooms.

TYPES IN INDIA
At left, the native bungalow; at right, that of a higher class native or of a white family.

The bungalow originated in India. Besides those owned privately there are military bungalows built by the government for the use of travelers along the main highways; they are constructed of wood, bamboo or similar materials. Those of the Europeans are generally built of sun-dried bricks, and have thatched or tiled roofs.

BUNION, *bun'yun,* a painful swelling, eventually becoming a distortion of the bone, usually of the joint of the great toe, but in some cases of the little toe. Bunions result from the irritation of the small membranous sac located there and are usually caused by tight shoes. At first the affection is a small swelling, but if not cured it develops into a very painful sore and may permanently deform the toe. To cure a bunion the first thing to do is permanently to remove the pressure by adoption of properly-fitted shoes. Rest and wet dressings will complete the cure, if attended to at the beginning of the trouble. Stubborn cases, however, require a surgical operation.

BUNKER, *bung'ker,* **HILL,** BATTLE OF, the first important battle of the American Revolution. It was decisive, not because either side won a sharp victory but because the brave, steadfast stand of the colonial troops convinced the colonies that the contest with the trained soldiers of England, though unequal, would be by no means hopeless. Had the outcome been a real defeat for the colonists it is possible that resistance to England might have then ceased.

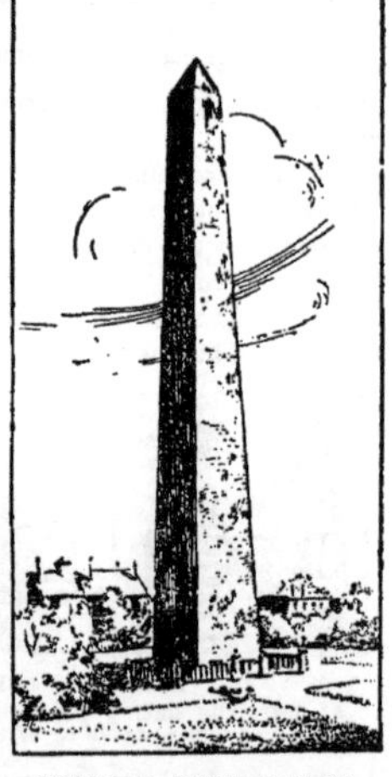

BUNKER HILL MONUMENT

This famous battle took place June 17, 1775, on Breed's Hill and Bunker Hill, Charlestown, Mass., between 1,500 Americans under Colonel Prescott and 3,000 British under General Howe. These forces were but a part of those operating near Boston, for the British troops holding that city numbered about 10,000, while the colonials had no fewer than 15,000 in the neighborhood. Learning that the British were planning to fortify Bunker Hill and so make good their hold on Boston, the Americans quietly occupied the adjoining height of Breed's Hill and threw up an earthwork there during the night of June 16.

At daybreak the British discovered how they had been forestalled, opened fire from their ships in Charlestown harbor and later in the day landed a force to charge the hill. Up the slope in perfect order marched the British regulars, and no opposition met them until they were close to the redoubt, for Putnam had issued his famous command, "Don't one of you fire until you see the whites of their eyes," and it was obeyed. Repulsed by the sudden, fierce fire, the British fell back with great loss, but soon rallied for a second charge, which ended as did the first. Charlestown was burned to the ground by shells which fell into it during the engagement.

Both sides then rallied, and at 4:30 in the afternoon a third charge took place. The Americans had used up all their ammunition, and were in consequence forced to yield after they had lost some of their bravest men, among them General Joseph Warren. All in all, the British had lost 1,054 in killed and wounded and the Americans 450; it was a costly triumph for the former, while for the latter it answered all the purposes of a victory.

Bunker Hill Monument. This battle, so far-reaching in its effects, is commemorated by a monument which was dedicated in 1843. It is a granite shaft 221 feet in height, and stands as nearly as can be ascertained on the spot where General Warren fell. Though it is on Breed's Hill, that height is now commonly known as Bunker Hill. On June 17, 1825, just half a century after the battle, Lafayette laid the corner stone and Webster delivered an address; eighteen years later, at the dedication, Webster again gave one of his memorable orations. The monument cost over $125,000, which sum was raised by popular subscription.

A.MC C.

BUN'SEN, ROBERT WILHELM EBERHARD (1811-1899), one of the discoverers of spectrum analysis (which see) and the inventor of the Bunsen burner, the Bunsen battery and numerous other devices which greatly aided the progress of science during the last half of the nineteenth century. Bunsen was an eminent

German chemist, and for thirty-seven years was professor of chemistry in the University of Heidelberg.

Bunsen Battery, a form of galvanic battery, the cell of which consists of a glass cup in which is placed a cylinder of zinc open on one side, and within this is a porous earthen cup containing a rod or prism of carbon. The glass cup is filled with weak sulphuric acid and the earthen cup with nitric acid. The electric current is developed by the action of the acids on the zinc. The Bunsen battery works quickly, but it is now little used because more convenient batteries have been invented. See ELECTRIC BATTERY.

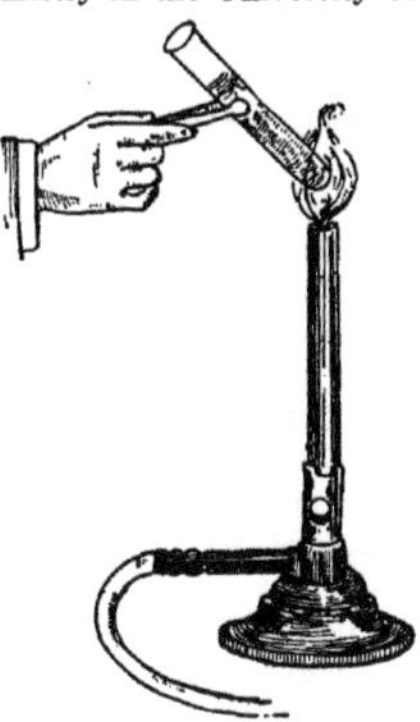

BUNSEN BURNER
Showing, by the hand and test-tube, how it is commonly used.

Bunsen Burner, a form of gas burner especially adapted for heating. It consists of a tube, in which, by means of holes in the side, the gas becomes mixed with air before burning, so that it produces a smokeless flame which gives no light but produces intense heat. It is widely used in laboratories and sometimes in soldering.

BUNTING, the popular name of a seed-eating bird of the finch family. The common bunting, or corn bunting, seen in most cultivated districts of England, does considerable damage to maturing crops, but its food is very largely of weed seeds and it is therefore a friend of the farmer. The *snow bunting,* or *snowbird,* is one of the few birds to be seen in the Arctic regions of Canada. In the United States the *cowbird,* or cow blackbird, is frequently called the *cow bunting.* Many buntings are beautifully colored, and for that reason the *painted bunting,* especially, is valued and sought as a cage bird. See FINCH; BIRDS.

BUN'YAN, JOHN (1628-1688), the author of the most famous and one of the greatest allegories ever written—*The Pilgrim's Progress.* This remarkable book has been translated into seventy-five languages and dialects—more than any other book except the Bible; and has been read with eager interest in every part of the world. Briefly stated, it is the story of the spiritual life of man, the account of Christian's journey through the world and his final triumph in the Celestial City. Religious books which mirror forth truthfully the struggle of the soul were written before, and have been written since; why, then, should this one, the work of an uneducated tinker, have taken such a hold on the affections of people everywhere?

JOHN BUNYAN
Undaunted by environment, he wrote one of the world's greatest books while imprisoned for speaking in public.

The Pilgrim's Progress is not merely a religious dissertation; it is a vivid, dramatic story, with an allegory so plain that it never perplexes or retards the reader who is anxious to know what happens next. For Bunyan did not content himself with simply stating that his hero passed through periods of despondency and doubt; he showed him wallowing in the Slough of Despond and shut up in Doubting Castle by the Giant Despair. Lions and "foul fiends" move through the pages, which are yet so enlivened by humor, kindliness and simple, natural touches that every reader can recognize in the trials of Christian a lifelike picture of his own existence.

Bunyan's Life. The son of a poor tinker, Bunyan was born at Elstow, near Bedford, in England, and after a very brief period spent in school began helping his father at his forge. At the age of sixteen, however, he ran away from home and joined the Parliamentary army in the civil war against Charles I. He was a boy of intense nature and strong passions, and the spirit of the age turned much of his intensity toward religious questions. The fiery preaching of the Puritans terrified him, and all his reckless living and profanity could not deaden his conscience. At last, mainly through the efforts of his wife, he determined to change his manner of life, and after a spiritual struggle which lasted for years, peace came to him. With characteristic zeal he threw himself into Christian work, and as a preacher in the Baptist Church he awakened others to all the fears which had been his in his early days.

But dissenters were not allowed to hold public meetings, and for his violation of this law Bunyan was arrested and placed in Bedford jail. For twelve years he remained there, supporting his family by making shoe laces; and in the intervals of his employment he wrote the work for which he is famous. After his release in 1672 he was allowed to preach when and where he chose, and the closing years of his life seem to have been very happy. In August, 1688, he rode through a hard storm to reconcile a father and son who had had a violent quarrel, and the exposure brought about his death. A statue was erected in his honor in Bedford almost two centuries after his death, but his chief monuments are his written works. These include, besides *The Pilgrim's Progress, The Life and Death of Mr. Badman, The Holy War* and *Grace Abounding*, the last-named a sort of spiritual autobiography. A.MC C.

BUOY, *boi* or *boo'y*, an anchored float designed to aid navigation by indicating the location of shoals and the courses of channels in harbors or rivers. All ships carry charts which show navigators where to expect buoys and what facts to learn from them. Thus, if a captain entering a harbor in the United States or Canada sees a buoy with one side red and the other side black, he knows he must keep the red on his starboard side. He knows, too, that horizontal red and black stripes mark danger spots and green buoys especially dangerous places, while vertical black and white stripes indicate the deepest part of the channel and white buoys show safe anchorage.

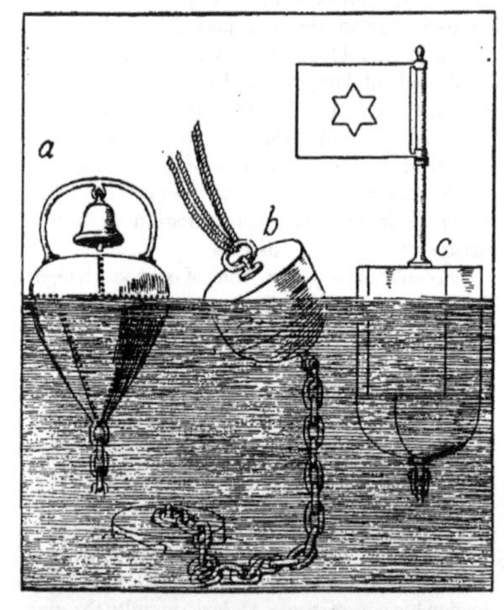

BUOYS

(*a*) Bell buoy; (*b*) can buoy; (*c*) buoy carrying special signal flag.

Among the various types of buoy are the *bell buoy*, which clangs dolefully as it is tossed about; the *whistle buoy*, whose motive power is air compressed by the waves themselves; the *lantern buoy*, a little, floating lighthouse; the *spar buoy*, a wooden pole which is weighted so that it stands upright; the *can buoy*, an iron cylinder with a dome-shaped base; and the *nun* or *nut buoy*, round and tapering at both ends.

BUR'BANK, LUTHER (1849-), an American horticulturist who has won international fame through his successful experiments in plant improvement. He is one of the few men who have devoted their lives to the work of creating new timber trees, fruits, flowers, vegetables, grains and grasses and to the improvement of many familiar species. In reviewing his service to mankind in this field, one recalls the words written by Jonathan Swift two centuries ago:

> And he gave it as his opinion that whoever could make two ears of corn, or two blades of grass, to grow upon a spot where only one grew before, would deserve better of mankind, and do more essential service to his country than the whole race of politicians put together.

In the little town of Lancaster, Mass., where he was born and educated, Burbank learned his first lessons in natural history, for from his early boyhood days he was happiest when studying the trees, birds and flowers. He was unable to go farther in school than the town academy, and when a young man worked in a factory in Worcester, Mass. Though he displayed there a marked aptitude for mechanics and invented a machine that would perform the work of half a dozen men, his real interest lay in nature's creations.

LUTHER BURBANK

He began his real life work as a market gardener and raiser of seeds. It came to him with special force as he busied himself with the cultivation of his vegetable garden that only one of his potato plants bore a seed ball, and it seemed reasonable to him that the offspring of this particular plant ought to show an even greater divergence from the general type. Working out this idea, he produced the famous Burbank potato, of which more than six hundred million bushels have since been raised.

To find a larger and more favorable field for his labors, Burbank went to California, and in the year 1875 settled in a valley about fifty miles north of San Francisco. After a disheartening struggle he finally saved enough money to buy a small tract of land, whereupon he started a nursery of his own, becoming in a few years the owner of a business that netted him $10,000 a year. It was not material prosperity, however, for which he was working, but an ideal which had been uppermost in his thoughts for many years—the creation of new species of plants. In 1893, therefore, he sold his nursery and began his career of experimentation and achievement.

At Santa Rosa, fifty-two miles north of San Francisco, Burbank lives in a beautiful home surrounded by rare trees, shrubs and flowers; his great experiment farm is located at Sebastopol, eight miles distant.

General Methods. Burbank achieves results by carrying on two general lines of work—*crossing* and *selection*. The former process is the uniting of two plants to produce a third, and he brings about this union by placing the pollen of one plant upon the stigma of another, leaving to natural forces the marvel of fertilization (see CROSS-FERTILIZATION). The pollen is collected just before it is ready to fall, and is applied at once to the blossom of another species. Ordinarily he uses his finger tip to place the pollen upon the stigma of the plant to be fertilized, but sometimes it is placed on the stigma with a soft brush.

Selection means the choosing of the best plants and the rejecting of the unfit, or those below certain standards. Thousands of plants must generally be grown in the effort to produce one improved species, and all of these must be examined with painstaking care. Sometimes but one specimen out of several hundred thousand will be approved. When the test is completed the rejected plants are burned, a practice that in the early days of his experiments caused Burbank's neighbors to view him with astonishment.

Important Achievements. It is impossible to give a detailed or even a condensed description of all the very numerous Burbank creations, but a few of special interest should be mentioned. He has originated several varieties of berries, which are of great commercial value. The offspring of the native California dewberry and a Siberian raspberry, which he calls the *Primus* berry, was the "first-known recorded fixed species directly created by man." The *phenomenal* berry, characterized by its enormous size, is the result of crossing the California wild dewberry and the Cuthbert raspberry; the color of this fruit is light crimson. Of still greater interest is the *white blackberry*, a berry of snow-white color and so nearly transparent that its small seeds can be seen.

Burbank's experiments with plums and prunes have revolutionized this industry in California. His success with shipping plums has been especially remarkable, hundreds of carloads being exported each season. One of his most interesting creations is the *plumcot*, the offspring of a Japanese plum and the apricot. The plumcot is delicious in flavor, and the flesh of each fruit is peculiar in color, being yellow, pink, white or crimson. He has also produced a plum which tastes exactly like the Bartlett pear. Among other new fruits is the *pomato*, produced by selection alone from the fruit of the potato. This curious fruit grows upon the potato vine, but in size and general shape resembles a small tomato. Its delicious white flesh suggests several different fruits.

Equally marvelous are the Burbank flower creations. Lilies, roses, petunias, dahlias, poppies and many other beautiful flowers have responded freely to his efforts to create new varieties, but probably none has attracted more general interest than his great *Shasta daisy*, named for his favorite snow-capped peak of the Sierras. The Shasta daisy is the offspring of the English daisy, the wild American daisy and their pure white Japanese cousin. This lovely flower, with its brilliant white petals and golden center, grows from four to six inches in diameter. (A comparison of this flower and the ordinary daisy appears under the title DAISY.) Timber and forest tree culture, too, has engaged his attention, the very rapid-growing *royal* and *paradox* walnuts and the big *sweet chestnut*, which always bears abundantly in six months, being among the most remarkable. His efforts to produce an edible cactus free from thorns are well known; the spineless cactus is fully described in these volumes, under the heading CACTUS.

Mr. Burbank personally placed the stamp of accuracy upon the above article.

BURDETTE, *bur det'*, ROBERT JONES (1844-1914), an American clergyman and humorist, who became famous originally through his paragraphs contributed to the Burlington (Iowa) *Hawkeye*. He was born in Greensboro, Pa., and attended public school at Peoria, Ill. In 1862 he joined the Forty-seventh Illinois Volunteers, serving through the War of Secession. He wrote for several papers after the war and finally became associate editor of the *Hawkeye*, from which he was soon quoted in other newspapers throughout the land. He began to lecture in 1877 and ten years later became a licensed preacher in the Baptist Church. Among his books are the famous *Rise and Fall of the Mustache and Other Hawkeyetems* and *Chimes from a Jester's Bells*.

BUR'DOCK, a coarse, hairy weed with hooked flowers which stick to the clothing of passers-by or to the hair of animals. In this latter respect it is particularly troublesome in the United States and Canada, wherever cows

BURDOCK
Branch, leaves and flowers. At left is a "basket" such as children make with the prickly, hairy flowers.

or sheep are pastured. The burdock is also known as *cockle button, beggars' button, burr-bur* and *stick button*. The plant grows from one foot to three feet high, with large, thin, roundish or heart-shaped leaves. To be destroyed the roots should be grubbed up before seed is ripe. There is a demand, however, for burdock seed, leaves and root for medical purposes. The root and seed are specially prepared and used for blood and skin diseases by some people, though physicians now advise against using anything of this kind for diseases of the blood or skin. The leaves of the burdock are used as a cooling poultice for swellings, etc. About 50,000 pounds of burdock root are imported into America each year. The best is said to come from Belgium.

BUREAU, *bu'ro*, a chest of drawers for clothing. The modern bureau usually has a mirror for toilet purposes, and is frequently called a *dresser*. The name comes from a French word meaning *coarse woolen cloth*, because in olden times writing desks and chests of drawers were covered with sack cloth.

Bureau also means a division in a department of government, as the Bureau of Animal Industry, in the United States Department of Agriculture, and the Bureau of Education in the Department of the Interior.

Bureaucracy means governmental control exercised largely by bureau officials or other really subordinate officers. The term is generally used as suggesting unwarranted official control.

BUREAU OF AMERICAN REPUBLICS. See PAN-AMERICAN UNION.

BURGESSES, *bur'jes es*, HOUSE OF, the first legislative body ever assembled in America. It was in the colony of Virginia, in the year 1619, that the call was issued for this assembly, and two burgesses, or citizens, from each plantation met the governor and his council in the church at Jamestown. In all, there were twenty-seven men present. They enacted a number of needed laws and shortly adjourned, but the influence of the first session determined the colonists to continue its work, and the assembling of the burgesses became an annual affair. James I tried to suppress this movement toward representative government in the colonies, but Charles I sanctioned it in order to gain trade concessions. Throughout all the stirring times preceding the Revolutionary War the house of burgesses, though remaining loyal to England, stood firm for the liberties of the colonies. It was in that assembly in March, 1775, that Patrick Henry delivered his stirring speech which electrified the colonies and strengthened their patriotic impulses; this speech is best remembered for the memorable words—

> Is life so dear or peace so sweet as to be purchased at the price of chains and slavery? Forbid it, Almighty God! I know not what course others may take, but as for me, give me liberty, or give me death!

BURGLARY, *bur'gla ri.* In criminal law burglary is defined as "the breaking and entering by night into the dwelling house of another, with intent to commit a felony." Entering into a house or building through an open doorway with intent to steal does not constitute burglary, but is classed as robbery or larceny. "Breaking" must occur to place the felony in the criminal division of burglary. What constitutes breaking is open to various constructions, but it is usually held that such breaking need not be accompanied by violence. The opening of a window by sliding the catch without damage to glass or framework is sufficient "breaking" to come within the meaning of the law. The usual punishment for burglary is imprisonment for a term not exceeding twenty years. In most countries the killing of a burglar in self-defense or in defense of family or property is not a crime. In England, however, the law regards such killing as manslaughter, but extenuating circumstances may be pleaded to avoid punishment. See FELONY; ROBBERY; LARCENY; MANSLAUGHTER.

BUR'GOMASTER, or **BURGERMEISTER,** *bur' ger my' ster,* the title of the chief magistrate of a city or large town in Germany and the Netherlands. The burgomaster is practically the same officer as the English and American *mayor,* French *maire* and Scottish *provost.* He is a salaried official. To attain the position one must have had some legal training, must possess a thorough grasp of economic science, good business ability, practical sense and a fairly-broad general knowledge. The burgomaster directs policies in finance, commercial enterprise, education, relief of the poor, social reform, and must know how to manage men. He is elected by the municipal body, but in most parts of Germany the government reserves the right to accept or reject the people's choice. A study of the preparation and career of the burgomaster explains why German cities are so well-governed, and carries a lesson which American municipalities have not yet heeded. However, a movement in the direction of greater efficiency in American civil administration is seen in the employment of city managers (see CITY MANAGER).

About the year 1905 a comic opera, *The Burgomaster,* held the attention of the public. Peter Stuyvesant, the Dutch governor of New York, was the leading character, but there was no attempt at character drawing.

BURGOYNE, *bur goin',* JOHN (1722-1792), known to fame as an English general of the Revolutionary War, but also a successful dramatist. After serving in various parts of the world, he was in 1777 appointed commander of an army against the Americans. His taking of Ticonderoga was considered a tragedy by Americans, but his later defeats overbalanced this victory. A part of his army fought a battle at Hubbardton, a detachment of his Hessians was defeated at Bennington, Vt., and on October 17 Burgoyne himself, after a furious battle, was forced to surrender with his whole army at Saratoga, which meant the turning point of the war. So important were the effects of this conflict that it has been termed one of the "fifteen decisive battles of the world." He was coldly received on his return to England and deprived of his command, but upon change of Ministry was appointed commander-in-chief in Ireland. Later he occupied himself mainly with the writing of comedies, including *The Maid of the Oaks, The Lord of the Manor,* and *The Heiress,* a play that still holds the English stage. See REVOLUTIONARY WAR IN AMERICA; FIFTEEN DECISIVE BATTLES.

BURGUNDY, *bur'gun di,* a name which in medieval and early modern times denoted a varying territory, first a kingdom, then a duchy, and finally a province of France which became famous for its red wines. The Burgundians, from whom the name was taken, were a Germanic people who early in the fifth century crossed into Gaul and set up a kingdom there. A century later they were conquered by the Franks, but the Frankish Burgundy was not quite the same in area as the original kingdom. The Treaty of Verdun in 843 split the territory, and by the close of the ninth century there were two Burgundies existing side by side.

BURGUNDY
Limits of the province at time of its greatest extent, about 1477.

In 937 these united to form the kingdom of Arles, but a small portion in the northwest remained independent and took the name of the Duchy of Burgundy. France gradually acquired a hold on this territory, which was governed until 1361 by members of the royal house of Capet. In 1363, the Capetian line having died out, Philip the Bold was made Duke of Burgundy and under his descendants the power of the duchy increased steadily. Much new territory was added, and finally Duke Charles the Bold became strong enough to defy the king himself. It appeared for a time that Charles might be able to set up an independent kingdom and wrest still more territory from France, but in 1477 he was killed in battle against the Swiss, and all his territory but the original duchy passed with his daughter's marriage into the control of the House of Hapsburg. Burgundy proper was seized by Louis XI and annexed to France. This old Burgundy forms the present departments of Côte-d'Or, Saône-et-Loire, Yonne, part of Ain and part of Aube.

BURIAL, *ber' e al,* a method of disposing of the dead, from Anglo-Saxon and German words meaning to *hide in the ground,* or *to conceal.* Different peoples adopt different methods of burial, but all are accompanied by some ceremony. The savage races usually expose bodies to wild animals or birds of prey; the Hindus formerly threw their dead into the Ganges River, although they are learning European ways, and the Egyptians embalm the bodies and preserve them in costly tombs. However, the two most common methods have been placing the dead in the ground or burning, better known as cremation. Both forms were practiced among the Greeks and Romans, though cremation came to be almost the sole method during the latter years of the Roman republic.

The method of burying has varied. In some cases, as with the early Babylonians, the bodies were placed on the surface of the ground and mounds were raised over them. The dead were buried in their garments, and their ornaments, weapons and utensils placed with them. Roman burial ceremonies were extravagantly splendid and long drawn out. The earliest Egyptians placed their dead in tombs, surrounded by articles of the toilet, food, and drink, showing their belief in a material afterlife. The pyramids of the deserts are tombs of the dead (see PYRAMIDS).

Among civilized nations of to-day cemeteries are set apart in which are buried the clothed bodies, in wooden, cement or metal boxes, after embalming. After the introduction of the Christian religion the practice of cremation almost entirely disappeared, because of the belief in the resurrection of the body. It has lately been revived, however, being considered by vast numbers of people a more sanitary method. It is certain that in many cases the hillside cemetery proves a source of contamination to the water supply of town and country. See EMBALMING; CREMATION.

Related Subjects. The following articles deal in part with methods of disposing of the dead; in part with other phases of the subject. But all are of interest in this connection:

Arlington National Cemetery	Mausoleum
Catacombs	Mummy
Coffin	Sarcophagus
Cremation	Taj Mahal
Embalming	Tomb
Epitaph	Towers of Silence
	Westminster Abbey

BURKE, *burk,* EDMUND (1729-1797), an English orator and statesman, accounted the greatest political writer of the eighteenth century. Every American high school student knows of him by reason of his speech *On Conciliation with America,* by which he sought in vain to induce the British government to adopt a conciliatory policy toward its colonies.

EDMUND BURKE
Goldsmith said of him, "He wound himself into his subject like a serpent."

He was born at Dublin, Ireland, and was destined by his lawyer father for the same profession, but he found literature far more to his liking. In London, whither he had gone to study law, he won recognition and several minor political offices by the publication of his *Vindication of Natural Society* and *Origin of Our Ideas of the Sublime and Beautiful,* and finally, in 1766, was elected to Parliament. It was a stirring time in that body, and Burke at once took his place as a foremost figure in all its deliberations—a position he held for thirty years. Authorities do not hesitate to declare him the most influential orator the House of Commons has ever known, and his wonderful powers were always employed on the side of justice and right. No hint of scandal or of political corruption ever attached itself to his name.

His great speech *On Conciliation with America* was delivered in 1775, when he was at the summit of his influence, and might seem to indicate that he was a Liberal, but this was not by any means the case. The French Revolution never had his sympathy, and he resisted firmly attempts at Parliamentary reform; but he was always on the side of movements which were for the betterment of humanity. He opposed the slave trade, and put years of research and labor into his effort to uphold the rights of the people of India as against the greedy officials who oppressed them to enrich themselves. The culmination of this struggle was the impeachment of Warren Hastings (which see), in which Burke took an active part, delivering a wonderful speech.

His writings were numerous, and include in addition to those mentioned above and his published speeches, *Reflections on the Revolution in France* and *Observations on a Pamphlet on the Present State of the Nation.* In all that he wrote there is a richness of imagery and a wealth of figures which shows him to have been a poet in spirit, though the medium through which he expressed himself was prose.

BURKE'S PEERAGE, the name usually applied to a publication entitled *Genealogical and Heraldic Dictionary of the Peerage and Baronetage of the United Kingdom,* first compiled in 1826 by John Burke, an Irish man of letters. It contained the names of all British peers and baronets in alphabetical order, and was the first work of its kind. The publication is still issued annually and is regarded as the best authority on the genealogy of leading British families.

BUR'LAP, a strong, heavy cloth made of jute, flax, hemp or manila. Its natural color is a lifeless tan, and at one time it was used in a very coarse texture without coloring, merely for packing and for coffee bags. It is still used for those purposes; but it has been discovered that burlap can be made a highly ornamental cloth for decoration in the home. It is now sold in many weaves, coarse and fine, and dyed in every color. As covering for walls in place of wall paper, as hangings, as cushion tops—in a large number of ways, embroidered, and stenciled, or plain—burlap is now a favorite fabric for interior decoration. Some of the finer qualities of art burlaps are expensive improvements on the cheap burlap packing cloth.

BURLESQUE, *bur lesk',* a story, poem or theatrical performance which makes a laughing matter of some serious work, by words or actions which are a travesty on the original. Noble thoughts, for instance, are expressed in the most commonplace language; things insignificant are talked of in glowing words, making the thing described seem absurd and ridiculous. And that is why such compositions were given the name *burlesque,* the Italian word for raillery, mockery or jesting. The most famous of the early English writers of burlesque was Chaucer, who ridiculed some of the long-drawn-out tales of the Middle Ages. *Don Quixote,* by Cervantes, a burlesque on absurdly romantic tales of chivalry, is the most famous example of this class of literary work (see DON QUIXOTE). As a form of the drama, burlesque was well known to the Greeks. Molière's comedies are the best-known examples in French literature. Some of the works of Gilbert and Sullivan, particularly *Pinafore* and *The Mikado,* with their burlesque on fads and affectations, are well-known examples. The burlesque in the theaters of to-day, however, is merely a mixture of vaudeville and ballet, and such performances have in many cities lost moral quality.

BUR'LINGAME, ANSON (1820-1870), an American statesman and diplomatist, whose most important achievement was his negotiation, in 1868, of a treaty between the United States and China. In this the latter country for the first time accepted the principles of international law, and really opened its doors to the world. Burlingame was born in New York state and was educated at the University of Michigan and the Harvard Law School. After practicing law in Boston and serving as state senator, he was elected to the national House of Representatives on the Know-Nothing ticket (see KNOW-NOTHINGS). In 1861 President Lincoln appointed Burlingame minister to China, a post which he held until 1867. The following year, while serving as head of an embassy of the Chinese government, he concluded at Washington the epoch-making treaty which is known by his name.

BUR'LINGTON, a town in Halton County, Ontario, thirty miles west of Toronto, at the head of Lake Ontario, on the Grand Trunk Railway. It also receives freight service from the Canadian Pacific Railway, has connection with Hamilton and Oakville by electric line, and is on the Toronto-Hamilton Highway. The Burlington district is celebrated for its fruits, and the town's largest industrial establishments are a cannery, an evaporator and several basket factories. Population in 1911, 1,831; in 1916, about 2,500.

BUR'LINGTON, IOWA, the county seat of Des Moines County, and a railroad, manufacturing and trading center. The population of 24,324 has varied but slightly since 1910. The city, covering an area of about twelve square miles, is situated on bluffs along the west bank of the Mississippi River. *Orchard City* is the descriptive local name. About it lies a rich agricultural country. From the bluffs in Crapo Park, a playground of 100 acres in the southern part of the city, is a superb view of the river and the surrounding country. Near the town are coal fields and limestone quarries.

Burlington is in the southeast corner of the state. Chicago is 206 miles northeast, Des Moines, the state capital, is 166 miles northwest, and down the river 221 miles is Saint Louis. The Chicago, Burlington & Quincy Railroad has large shops here. East of the city the railroad crosses the Mississippi over a fine iron bridge which affords a splendid view up and down the river. Other railroad lines running into Burlington are the Toledo, Peoria & Western, the Chicago, Rock Island & Pacific and the Muscatine North and South. There is an extensive river commerce in raw materials, and modern steamboats are replacing the old side-wheelers that made the river famous in pioneer days.

The important industrial establishments include manufactories of engine works, farm implements and furniture; pearl button and soap-making plants are among the smaller enterprises. A Federal building costing $100,000, constructed in 1890, a large Y. M. C. A. building, a city library, the Tama building, bank buildings and churches are of special architectural note. The city has a commission form of government, adopted in 1910.

History. At the suggestion of Lieutenant Zebulon Pike, the explorer of the Pike's Peak region, a fur-trading post was established on the site of Burlington in 1829. The first permanent settlement, made in 1833, was called Flint Hills, from the Indian word *Shokokon,* but was soon renamed after Burlington, Vt. In 1838 a city charter was granted by the territory of Wisconsin, and Burlington was the meeting place of the territorial legislatures of Wisconsin from 1836 to 1838, and of Iowa from 1838 to 1840. Robert Jones Burdette, the humorist and preacher, once a citizen of Burlington, was locally known as "The *Hawkeye* Man" because of his clever work when associate editor of the *Hawkeye.* T.M.H.

BURLINGTON, VT., one of the leading lumber markets in the United States, the largest city of the state in population, and the county seat of Chittenden County. It is situated on the northwestern border of the state and on Lake Champlain, forty miles northwest of Montpelier, the state capital. The city has a good harbor and is a port of entry. Besides its water transportation it has the service of the Central Vermont and Rutland railways, and electric lines operate north and south. In 1910 the population was 20,468; this had increased to 21,247 in 1914. The city's area is ten square miles.

The country around Burlington is famous for its picturesque mountain scenery, and the city is especially favored by being built on ground which rises 300 feet above the lake. It thus commands a fine view of water, mountain and valley. The city itself is beautiful, with wide, shady streets, handsome homes and fine public buildings; the city hall, county courthouse, Federal building and library cluster around a square in the center of the city. The superior educational advantages of Burlington attract students from a large territory. Here are located the State University of Vermont, state agricultural and medical colleges, Saint Joseph's and Saint Mary's academies, Bishop Hopkins Hall (for girls), Vermont Episcopal Institute (for boys), and the Fletcher-Carnegie, Billings and Burlington Law libraries. Besides the Mary Fletcher Hospital, it has several fine sanitariums, asylums for orphans and homes for the aged and destitute. It is the see of a Roman Catholic archbishop and of an Episcopal bishop. Large cotton and woolen mills are in operation here, and the manufacture of stone and lumber products and proprietary medicines is extensive. Great quantities of lumber are shipped from Canada through this point.

The first settlement on the site of Burlington was made in 1773; it was organized as a town in 1797 and incorporated as a city in 1865. During the War of 1812 it was a military post. The city for some years was the home of Ethan Allen (which see), one of the most conspicuous figures in the history of Vermont; his burial place is in Greenmount Cemetery.

BUR'MA, the largest and most easterly province of British India. To those who read of it, it becomes a land of romance, with its "tinkly temple bells," the mist on its rice fields, and its pagodas, "lookin' eastward to the sea";

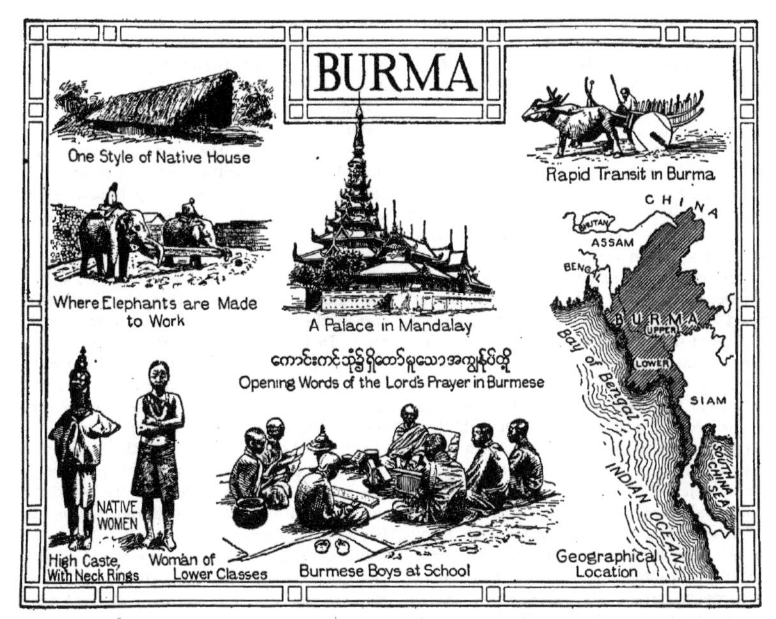

One Style of Native House

Rapid Transit in Burma

Where Elephants are Made to Work

A Palace in Mandalay

Opening Words of the Lord's Prayer in Burmese

High Caste, With Neck Rings

Woman of Lower Classes

Burmese Boys at School

Geographical Location

and it seems not difficult to understand the "ten-year soldier" when he declares that "if you've 'eard the East a-callin', you won't never 'eed nought else." And more prosaic writers admit that these pictures, as Kipling gives them in *On the Road to Mandalay,* are not overdrawn.

The area of this great province on the eastern coast of the Bay of Bengal is 236,738 square miles, somewhat less than that of Saskatchewan or of Texas; but its population of 12,115,217 is twenty-five times that of the province named and three times that of Texas. Assam, Tibet, China and Siam border it on its landward side.

The People and Their Civilization. In many lands of romance a depressing influence is the degradation and misery of the people, but in Burma the traveler does not feel this. The native Burmese, who are of the Mongolian race, are a good-natured, cheerful people, who take life easily because they live in a land so kind that it is only necessary to "tickle her with a hoe and she laughs with a harvest." Their philosophy of life, if it may be called by so formal a name, is to work as little as possible and to spend the rest of their time in frolics and festivities. The betel nut and the "whackin' white cheroot" are their chief indulgences (see BETEL). Men and women alike dress much in silks of bright colors, and a city street on a festival day is a brilliant sight.

In the hill regions to the east live the Shans, a hard-working people whose views of life are far more serious than those of the Burmese. That they have not the artistic eye for dress that distinguishes the Burmese may be seen from the accompanying illustration.

The Irrawaddy River is navigable for hundreds of miles, and is connected with three canals dug since British rule began. Carriage roads have been built in many places, and over 1,500 miles of railway are open, with more under construction. Rangoon, the capital, is joined to Mandalay and Maulmain, the other chief towns, by railroads. See RANGOON; MANDALAY.

The Land and Its Resources. Much of the surface of Burma is hilly or mountainous, parallel ranges running with considerable regularity from north to south. The highest mountains are on the north, where an extension of the great Himalayas shuts off Burma from Tibet. Rivers flow in the valleys running north and south, contributing much to the fertility of the soil by the silt which they spread. Chief of these rivers is the Irrawaddy, which flows nearly through the center and drains three-fourths of the country. The delta plain at its mouth, of great extent and very fertile, is perhaps the most valuable part of the whole country. The eastern districts are drained by a second great river, the Salwin.

Climate. Burma lies approximately in the same latitude as Mexico and Central America, and thus is almost entirely within the tropics, but the differences in elevation cause decided variations in climate. To the north, in the region of great mountains, are districts where frost is not uncommon, and in the central parts of the country there are conditions typical of temperate regions—hot summers and cool winters. Near the coast and throughout the long, narrow projection that runs southward, there is a tropical sameness of climate, with heavy rainfall, which averages 160 inches a year.

Wild Life. Few lands have more luxuriant plants and more of the great animals which civilization is so surely driving out in many places. For in Burma there are still great tracts of unexplored land which are clothed with tropical forests. Here grow teakwood, one of the most valuable products of the country; ironwood, palms of all sorts and the ever-useful bamboo. In and about these jungles range the greatest of game animals, the tiger, the leopard, rhinoceros, deer, crocodile, and most important of all, the elephant. No other part of Asia produces elephants of such great size, and it is from Burma that most of the trained elephants of India are secured. These animals are of the greatest importance to the Burmese, who make of them beasts of burden, training them especially to drag and stack the great logs of teakwood.

Minerals. The Burmese have never paid much attention to their mineral resources, which are believed to be great. The gold which is found in the sand and gravel of the river bottoms they wash out, and in the same labor-saving way, they have made valuable discoveries of rubies, jade and sapphires. The oil wells are a source of wealth to the country, producing over 230,000,000 gallons each year.

Agriculture. Two-thirds of the people of Burma make their living from the soil, which is in places extraordinarily fertile. In the production of rice, which forms the staple food of its people, Burma leads the world, for five-sixths of its cultivated land is given over to rice growing. Each year, except when poor crops cause a famine and much of the product is needed at home, Burma exports about $50,000,000 worth of rice. See RICE.

History and Government. The Burmese claim a long history in the land which they now inhabit, but the early centuries are shrouded in obscurity. From the eleventh century, when its known history began, to the sixteenth, one little kingdom after another made itself powerful and sought to bring the whole territory under its sway. From 1580 to 1750, Pegu, in the south, was supreme, and it was during this period that the Europeans turned their attention to the rich and fertile land. In 1824 an English force entered Burma, but not until 1853 was England able to declare even a part of the country British territory. Another invasion in 1885 resulted in the capture of the king, or "Lord of the White Elephant," as he was called, and the annexation of Burma to British India. Within ten years the risings throughout the country were put down, and from that time Burma has slowly but steadily progressed.

Burma is governed as a province of British India, having a lieutenant-governor and a legislative council. For purposes of local government it is divided into eight districts, each in charge of a commissioner. E.D.F.

No important books on Burma are published in America. Among the best English books is Harmer's *Story of Burma;* it can be purchased from booksellers in large cities.

BURNE-JONES, SIR EDWARD (1833-1898), the greatest of the Pre-Raphaelite painters (see PRE-RAPHAELITES). He was studying at Oxford, for the Church, but, coming under the influence of Rossetti, he and his friend William Morris decided to devote their lives to art. At first his pictures, which show clearly the influence of his master Rossetti but are better in drawing, were not well received, but while he was still a comparatively young man he came to be looked upon as one of the most gifted painters England had ever produced. Whether his subjects were from the Bible, as in his *Christ Crucified Upon the Tree of Life;* from medieval legends, as in *Cophetua and the Beggar Maid* and the *Beguiling of Merlin;* or from mythology, as in *Wine of Circe, Pan and Psyche,* or the *Pygmalion* series, he introduced into his pictures a romantic atmosphere and a poetic feeling which makes an intense appeal to many. Those who care for Burne-Jones's paintings at all are likely to care for them very much indeed. He painted both in water color and in oil, and whatever his medium he always produced a rich, warm color.

His son, Sir PHILIP BURNE-JONES (1861-), has also won distinction as a painter, his best-known work being a striking but repellent *Vampire* based on Kipling's poem of that name. In addition, he also produced excellent portraits of his father, of Watts and of Kipling.

BURNETT, FRANCES ELIZA HODGSON (1849-), a well-known American author who has written many popular novels and a brilliant story for children, *Little Lord Fauntleroy*, published in 1886. The hero, a beautiful boy with long, fair curls, whose mother kept him dressed in velvet suits, with dainty blouse and wide collar, at once became the most popular child character in fiction. His manner of dress gave a new term to children's clothing, and "Fauntleroy suits" became as popular as the book itself. The story was dramatized, and for years was played throughout the United States and Canada. Every child actor of the period who was so fortunate as to possess long curls became a stage Lord Fauntleroy.

FRANCES HODGSON BURNETT

Frances Hodgson was born in Manchester, England, but America claims her as one of its novelists, for she has lived in the United States, with the exception of trips to Europe, since 1865. In 1873 she married Dr. L. M. Burnett, and since that time has used Burnett as a part of her pen name. She was divorced from Dr. Burnett in 1898, and two years later married Dr. Stephen Townsend, who has assisted her in some of her dramatic writing.

Surly Tim's Trouble, published in *Scribner's Magazine* in 1872, brought her to public notice; five years later she wrote *That Lass o' Lowrie's*, a strong and vivid story of life in the English mining districts—and with this book her fame was established. Then followed one novel after another in quick succession. Among the most noteworthy of these are *Haworth's; A Lady of Quality*, one of her most striking and dramatic stories; *A Fair Barbarian; Editha's Burglar; The Shuttle*, based on the international marriage question; *The Dawn of a To-morrow*, a story of the good wrought by a little waif of the slums; *The Secret Garden*, and *T. Tembarom*. In the last, one of the best-liked American novels of recent years, humor and pathos are delightfully blended as the author tells the experiences of an interesting young newspaper reporter and the surprising change in his fortunes.

Among Mrs. Burnett's books for children, *Sara Crewe* probably ranks next to *Lord Fauntleroy* in popularity. A recent story, *Racketty-Packetty House*, was dramatized and played in New York and Chicago by a cast composed entirely of children. *A Lady of Quality* and *The Dawn of a To-morrow* have also been dramatized, and in the latter Miss Eleanor Robson, playing the part of the slum waif named Glad, enjoyed one of the greatest successes of her career. B.M.W.

BURNHAM, *burn'am*, DANIEL HUDSON (1846-1912), an American architect whose name is connected especially with the development of the modern skyscraper and with the movement for the beautifying of the great cities. He was born in Henderson, N. Y., and was educated in Chicago and in Massachusetts. He established himself in Chicago in 1872, soon after its great fire, and the firm of which he was the head began to make a new sky line for the city that rose from its ruins to become the fourth city of the world. Among the Chicago buildings which Burnham designed were the Masonic Temple, for years the highest building in the city, the Great Northern Hotel, the Railway Exchange and the Field department store. The famous Flatiron building, a landmark of New York City, and the handsome Selfridge department store in London, are also products of his genius.

In 1893 Burnham became the architect for the World's Columbian Exposition at Chicago. His work there revealed a richness of imagination and an appreciation of beauty that gave him a world reputation, and he was called upon by various cities to suggest ideas for their improvement. Together with Saint Gaudens, Frederick Olmsted, Jr., C. F. McKim and other eminent artists, he laid the plans for the beautifying of the national capital, the artistic Union Station in that city being entirely his own design. The "Chicago Plan," now being carried out, and destined in time to make Chicago one of the finest cities in the world, was one of his last creations.

BURNHAM, Sherburne Wesley (1838-), an American astronomer, noted for his great achievement in discovering and cataloguing double stars, his accomplishments along those lines exceeding those of any other observer. He was born at Thetford, Vt., and educated in the local academy. He began the study of astronomy while a stenographer, and after his appointment as clerk of the United States circuit court for the northern district of Illinois, he spent his spare time in studying the heavens and became an amateur astronomer of remarkable ability. In 1876 he became connected with the Chicago Observatory. From this position he went to the Lick Observatory, California, and on the opening of the Yerkes Observatory in Wisconsin by the University of Chicago, he was appointed professor of practical astronomy in that institution. He published a catalogue of stars discovered by him from the founding of the Yerkes Observatory to 1900, also a general catalogue of all known double stars visible in the northern hemisphere, in 1907, and *Measures of Proper-Motion Stars,* in 1912.

BURNS, John (1858-), the first laboring man and the first Socialist to hold a seat in the British Cabinet. Nothing better expressed the keynote of his character than the statement he once made about himself: "Came into the world struggling, struggling now, and prospects of continuing." From a place as an unknown ten-year old boy working for a shilling or two a week in a candle factory to the post of a prominent Cabinet minister, entitled to the prefix "Right Honorable" and a salary of £5,000 a year, is a long, toilsome journey.

Burns was right when he said that his life was a struggle. But if he grew up to manhood in comparative poverty, he did not grow up in ignorance. He read much; Robert Owen, Tom Paine and William Cobbett were his favorite authors. He also read some of the works of John Stuart Mill, whose arguments against socialism were so weak, he said, that they converted him to it. After serving for seven years as an engineer's apprentice, Burns worked as gang foreman on the West African coast for a year. The year's earnings were promptly spent on a six-months' tour of France, Germany and Austria, to study economic conditions.

Becoming a firm supporter of socialism, his eloquence made him a conspicuous figure in labor circles. He was arrested in 1878 and in 1886, charged with inciting mobs to violence, but both times proved his innocence. He continued to proclaim the right of free speech on London's streets, and in 1887 was again arrested and imprisoned for six weeks. His lawyer on this occasion was Herbert H. Asquith, a young man about his own age, who later became Prime Minister.

Burns was still working in Hoe's printing-press shops in 1889, when he was elected to the London County Council. In 1892 he was elected to Parliament for Battersea, and has been regularly reëlected since. In the Campbell-Bannerman and Asquith Ministries he sat in the Cabinet as President of the Local Government Board. In 1914 he was for a few months President of the Board of Trade, but resigned in August because he was opposed to British participation in the War of the Nations. To some extent his acceptance of a Cabinet position lessened his prestige among the radicals, but on various occasions he proved his loyalty to union labor and to socialism.

BURNS, Robert (1759-1796), one of the world's greatest writers of verse, the idolized "Bobbie Burns" whom every Scotchman regards with a deep, personal love. Though most o his very best poems were written in the

ROBERT BURNS

The "poet of homely human nature, not half so homely or prosaic as it seems."

—*Kellogg.*

Scottish dialect they belong no more to Scot land than to the world at large, for their tenderness, passion and sweetness have in them a universal appeal.

Burns was the son of a tenant farmer, and was born on January 25, 1759, at Alloway, in Ayrshire, in a little cottage which has become a Mecca to thousands of pilgrims. His father could give him little enough, but whenever he could spare him to go to school he did it gladly, and he encouraged the longings for knowledge which were born in the boy. Most of young Robert's education was gained from reading, to which he earnestly devoted himself. In this way he learned what the best English poets could teach him, and thus he cultivated the instinct for poetry which was a part of his nature.

At an early age he had to begin working on the farm, and by the time he was fifteen he was doing the work of a man. In 1781 he went to Irvine to learn the business of flax-dressing, but the building in which he was working was destroyed by fire, and he was forced to abandon that living. When his father died, Robert took a small farm at Mossgiel with his brother Gilbert, but the venture was not highly successful. Robert, however, began by his poems to attract the attention not only of his neighbors but of educated men of the vicinity, and this seems not strange when it is remembered that *The Cotter's Saturday Night, To a Mouse* and *The Jolly Beggars* were produced at that time.

An unhappy and unsuccessful love affair with Jean Armour of Mossgiel decided him to emigrate to Jamaica, and to obtain money for his passage he published by subscription in 1786 a volume of his poems.

This volume gained the approval of eminent men in Edinburgh, and at their suggestion he gave up his voyage and went to the city to make arrangements for publishing a new edition. The books sold far better than he had dared to hope, and the young man, admired and flattered, was received in the highest society. Scott, then a boy of fifteen, saw him and was deeply impressed. "I never saw," he wrote years later, "such another eye in a human head, though I have seen the most distinguished men in my time."

Returning to the country with about $2,500 which the sale of his books had brought him, Burns took a farm at Ellisland, near Dumfries, and in 1788 married Jean Armour. It was during his residence on this farm that he wrote, in a single day, *Tam O'Shanter*. Again farming was not successful, and Burns accepted the post of exciseman, performing his duties conscientiously. The spectacle of Scotland's greatest poet testing ale and collecting duties on it is a strange one to people of to-day, but those of his own time appear to have seen nothing unusual in it.

In 1791, completely discouraged with farming, Burns moved to Dumfries and relied entirely on his salary as exciseman. He continued to write, increasing his local fame by a number of beautiful songs adapted to old Scottish tunes. But the life in Dumfries was of the wrong sort for a man possessed of as little self-control as was Burns. The idle and the dissipated gathered around him, for his brilliant wit gave a charm to their meetings; while the more respectable classes refused to admit him to their society because of these low associations and his own increasingly dissipated habits. In the winter of 1795 his health began to decline, and in the following summer he died. His wife and four children were made comfortable by the proceeds from a subscription edition of his poems which his friends and admirers at once brought out.

THE BURNS COTTAGE
Where the poet was born.

Burns was one of the most human of all the world's great writers—the things which interested and moved him interest and move every man who keeps himself open to impressions. Honest, proud, friendly and warm-hearted, with a sound understanding and vigorous imagination, he combined with these qualities the high passions which were his ruin. Burns himself felt that justice had never been done him, but he owed the unhappiness and failure of his life fully as much to his own lack of self-control as to outward circumstances. And yet the epigram that "it was Burns' virtues that killed him" has in it much of truth, for his understanding and his better judgment were at war continually with his passions, and the struggle wore out even his strong body.

As to his poetry, there is but one verdict—of its kind it is unsurpassed. The charm of the simple peasant home, the pathos of the daisy cut from its stem, or the field mouse despoiled of its nest—these he not only felt himself to the full but was able to make others feel. And his love songs, such as *My Luv's Like a Red, Red Rose; Highland Mary; Bonnie Doon,* and *O Wert Thou in the Cauld Blast,* are of the very essence of tenderness, and will endure long after more elaborate songs have perished. Of the lines from his poems which made for themselves a place in the common speech, the following may be noted:

> The best laid schemes o' mice and men
> Gang aft a-gley.
>
> Man's inhumanity to man
> Makes countless thousands mourn.
>
> Oh wad some power the giftie gie us
> So see oursel's as others see us!
>
> The rank is but the guinea's stamp,
> The man's the gowd for a' that.

C.W.K.

See AYR for illustration of Burns' Memorial. Consult Lockhart's *Life of Burns;* Shairp's *Life of Burns;* Stevenson's "Robert Burns," in *Familiar Studies of Men and Books.*

BURNS AND SCALDS, *skaldz,* dangerous and painful injuries to any part of the body, caused by excessive heat. Burns are produced by dry heat, scalds by hot water or steam, but the effects and remedies are practically the same in both cases, and the term *burn* is ordinarily used in speaking of any injury of this nature. Serious burns are dangerous because they leave a raw surface exposed to germs; also, where a considerable area of the skin is destroyed and the elimination of waste matter by the skin is interfered with, extra work is thrown upon the kidneys and intestines, and inflammation of these organs may result. Pneumonia is liable to occur when the lungs and bronchial tubes have been irritated by inhalation of hot air or steam.

Burns of a serious character should have the attention of a physician, but whether a physician is called or not, measures should be taken at once to exclude the air from the burned places and to relieve the smarting sensation. Carron oil, consisting of equal parts of lime water and raw linseed oil, is an excellent remedy if it can be procured free from bacteria. Strips of gauze saturated with the oil are applied to the burned places, and bandages are tied on to hold the gauze in place. Lard and baking soda, or olive oil and vaseline, will also prove effective. Whatever grease is used, it must be sterilized by heat and then cooled before being used. If no remedies are at hand immerse the burned parts in water, to exclude the air until proper treatment can be given. A coating of flour or baking soda will also keep the air from reaching the burned parts, but flour and water should never be used together, as the mixture will harden and the patient will suffer intensely when the particles are removed. To offset the effects of shock in case of severe burns, loosen the clothing, keep the patient quiet and lying down, apply warmth and see that he has plenty of fresh air.

W.A.E.

BURN'SIDE, AMBROSE EVERETT (1824-1881), an American soldier who served on the Union side throughout the War of Secession. He was graduated from the Military Academy at West Point in 1847, was engaged in garrison duty for a number of years, and in 1853 resigned from the army service to take up the manufacture of firearms at Bristol, R. I. In 1856 he invented the breech-loading rifle known by his name. When the war began in 1861 Burnside reëntered the army, and as colonel of Rhode Island volunteers took part in the first Battle of Bull Run. In 1862, as commander of the Department of North Carolina, he captured the Confederate garrison on Roanoke Island and was raised to the rank of major-general of volunteers.

He was then transferred to the Army of the Potomac; later he was twice offered the chief command of the Army of Virginia, but declined it. With great loss of life his force held the stone bridge at Antietam, which was the important post of that battle, and when, later, General McClellan was relieved, Burnside took command. After the disastrous Battle of Fredericksburg he was superseded by Hooker and transferred to the Department of the Ohio. During 1864 and 1865 he served under Grant and took part in many important battles.

After the war Burnside was connected with various railroad enterprises, was governor of Rhode Island from 1866 to 1869, and from 1875 until his death he was a United States Senator. His habit of closely shaving his chin and allowing his beard to grow on the sides of his face brought about the use of the term *burnsides,* as applied to sidewhiskers.

BURR, AARON (1756-1836), an American statesman whose talents and energy fitted him to rise high in political life. His ambition, however, and his inability to meet opposition serenely, led him into actions which branded him in the popular mind of his time as a murderer and a traitor.

Burr was born at Newark, N. J., and was graduated in 1772 at Princeton College, of which his father, Aaron Burr, and his grandfather, the celebrated Jonathan Edwards, had been presidents. In the Continental army, which he joined in 1775, he gained a high reputation for courage, rising to the rank of lieutenant-colonel. He resigned from the service in 1778, was admitted to the bar and practiced in Albany and then in New York, quickly becoming a leader in his profession. He served in the state legislature, was attorney-general of New York, and in 1791 was elected to the United States Senate. He has sometimes been called the "first boss of New York state."

AARON BURR
Who might have been President of the United States except for the opposition of one man.

From his very entrance into political life Burr stood as an opponent and rival of Alexander Hamilton. In 1800 he was a candidate for President of the United States, and received the same number of votes as Jefferson; but the House of Representatives, chiefly through the influence of Hamilton, elected Jefferson, and Burr became Vice-President (see Constitution of the United States, Art. II, Sec. 2). This was another grievance against Hamilton, and when in 1804 Burr was defeated in the race for the governorship of New York, he laid that also to Hamilton's influence, probably with good reason. Intensely angry now, he forced a duel upon Hamilton. The two met at Weehawken, N. J., on July 11, 1804, and at the signal Hamilton fired into the air. Burr, however, took careful aim, and his great rival fell mortally wounded.

The outcry was loud, and Burr fled to Georgia, but later returned to Washington and completed his term as Vice-President. His restless ambition, however, would not let him view calmly the ruin of all his political hopes, and he therefore prepared to raise a force for an adventure in the Southwest. Perhaps he meant to conquer Texas and establish there a republic, with himself at its head. Apparently he believed that he might be successful in detaching the Western states from the Union and thereby revenging himself for the slights which he had suffered. Several men of wealth and influence like Harman Blennerhassett (which see) were won over by his promises and Andrew Jackson received him as a friend. The scheme had not progressed far before it became known to the government. His force scattered in the lower Mississippi, his confederate, Wilkinson, turned against him, and he was arrested and tried for treason. Chief Justice John Marshall, however, directed his acquittal. His reputation was ruined, but after some years spent in Europe he boldly returned again to New York in 1812 and opened a law office. Despite his ability, he never regained a large practice and was shunned by society.

Theodosia Burr (1783-1813), his only daughter, was celebrated not only for her beauty and cleverness, but for her unshaken devotion to her father. She was his housekeeper and favorite companion until her marriage in 1801 to Joseph Alston, afterward governor of South Carolina. During her father's trial she worked constantly to arouse public sympathy for him, and on his return from Europe sailed for New York to meet him. Her ship, the *Patriot*, disappeared during the voyage and no one on board was ever heard from again; whether it was wrecked by a storm or captured by pirates was not known. A.B.H.

Consult Todd's *The True Aaron Burr;* Orth's *Five American Politicians: A. Burr.*

BURRELL, *bur'el,* MARTIN (1858-), a Canadian legislator and expert in horticulture, Dominion Minister of Agriculture since 1911. He was born and educated in England, but in 1886 settled on the Niagara peninsula, near Saint Catherine's, Ont. Here he engaged in fruit growing, and was soon known as an authority on horticulture. Removing to British Columbia in 1900, he continued his business of fruit growing, and also became active in local politics. In 1907 the government of British Columbia sent him to England as its fruit commissioner, but a year later Burrell returned to Canada to sit as a Conservative in the House of Commons, of which he was still a member in 1916.

BURROUGHS, *bur'ohz,* JOHN (1837-), one of America's favorite writers on outdoor life, a literary naturalist whose descriptions of birds, bees and flowers are among the most charming in all literature. He was born at Roxbury, N. Y., and spent his boyhood on his father's farm, working, reading and studying. The essay was always his favorite form of literature, and he found especial enjoyment in reading Emerson, Walt Whitman and Matthew Arnold. The first, he says, awakened his religious feelings; the second quickened his interest in human nature; the third taught him to think and write clearly. His first published book, *Walt Whitman as Poet and Person* (1867), was written while he was a clerk in the Treasury Department at Washington. He left this position in 1873, was for several years a national bank examiner, and then, having bought a fruit farm at West Park, near Esopus, on the Hudson, settled down to the congenial life of farmer, student of nature and essayist.

JOHN BURROUGHS

Burroughs' essays are read and loved both in the home and in the school, and are enjoyed equally by the children and their elders. Few other nature writers have quite his gift for making the beauties of outdoor life a reality to the reader. He has the high art of writing so clearly and simply that one forgets his manner of expression and becomes absorbed in the matter. His sympathetic interest in his little friends of the woodlands and his amazing powers of observation are happily revealed in the quaint and suggestive titles of his essays, *Bird Enemies, The Tragedies of the Nests, An Idyl of the Honey-Bee, Winter Neighbors, A Taste of Maine Birch, Winter Sunshine* and others. *Whitman, a Study* and *Literary Values* are representative of his literary essays; *The Light of Day* gives his personal religious views. His experiences on a Western trip with Theodore Roosevelt are interestingly told in *Camping and Tramping with Roosevelt.* He has also written a number of poems, collected under the title *Bird and Bough.*

BUR'WASH, Nathanael (1839-), a Canadian educator and Methodist theologian, from 1887 to 1913 president and chancellor of Victoria College, and for half a century a leader in educational reform in Ontario. Dr. Burwash was graduated from Victoria College in 1859, when it was still at Cobourg, Ont., his birthplace. He continued his theological studies at Yale College and at Garrett Biblical Institute of Northwestern University, Evanston, Ill. He was ordained a Methodist minister in 1864, and two years later was appointed professor of natural science in Victoria College. Later he was dean of the theological faculty for thirteen years, and then president for twenty-six years. He was largely responsible for the federation of Victoria College and the University of Toronto. Dr. Burwash has taken a prominent part in the direction of the Methodist Church in Canada, was Methodist secretary of education from 1874 to 1886, and has been president of the general conference and delegate to the ecumenical conferences at Washington, D. C., in 1901 and at London, England, in 1911. Of his many writings the most significant are *Wesley's Doctrinal Standards; Inductive Studies in Theology; A Manual of Christian Theology,* and *Life and Times of Egerton Ryerson.*

BURYING BEETLE, also called Sexton Beetle, is an insect which in North America reaches a length of about one and a half inches. The name is due to a characteristic unknown among other insect families. It has a keen sense of smell, which guides it promptly to small dead animals, the basis of its domestic economy. Having found a small carrion, it burrows around and under the body until the animal is about five inches below the surface of the ground. In this carrion the female deposits her eggs, and when the larvae (young) hatch in about two weeks they live until mature upon the decaying matter and then begin to repeat the life history of their parents.

There are ten slightly-differing species in America; they are of nearly the same size and most of them are black, with either two red spots or a red band on each of the lateral wing covers.

BUSHEL, the common measure of all bulky articles of commerce, equal to four pecks, or thirty-two quarts. The standard bushel in the United States and Canada contains 2,150.42 cubic inches, being equal to a cylinder eight inches deep and eighteen and one-half inches in diameter, interior measure. In Great Britain an *imperial bushel* is used, having a capacity of 2,218.192 cubic inches. See Weights and Measures.

BUSHMEN, a tribe of African dwarfs inhabiting the Kalahari Desert and the plains in the north of the Cape of Good Hope province. They are a fierce, unsociable and only partly-

HOW THE BUSHMEN LIVE
A native village in the "Bush."

civilized people, leading a wandering life and living by hunting. Cultivation of the land has never appealed to them. They appear to be of separate stock and are not in any way kin to the milder Hottentots. Their skin is a dirty, yellowish-white and their language harsh and guttural, with many curious clicking sounds somewhat similar to those in the Zulu tongues. Their caves show some signs of art, being covered with carvings and rude drawings. They still use poisoned weapons, which make them greatly feared by the neighboring tribes. Their numbers are steadily decreasing, and they have in recent years shown an inclination to accept a few of the advantages of civilization. Little is known about their social affairs, except that each family apparently constitutes a community unto itself. They are nominally subjcet to the laws of the Union of South Africa, but are left to their own devices as long as they do not indulge in robbery.

BUSINESS, *biz'ness,* **COLLEGE,** a school for the training of persons for commercial positions such as those of clerk, bookkeeper or stenographer. The first business instruction of any kind, so far as known, was given by Mr. R. M. Bartlett of Philadelphia, who, in 1846, began to give instruction to a few private pupils in bookkeeping and other commercial subjects. The business college was the outgrowth of his work, and by 1860 all leading cities of the country had one or more of these commercial schools. Since that time their number has greatly increased and commercial departments have been established in many of the public high schools. For a number of years these schools gave only elementary instruction in arithmetic, bookkeeping and penmanship, and had no special text-books, but as they grew in number and patronage a wider range of subjects and special texts were provided.

The growth of commerce and manufactures and the extension of railways, steamship lines, the telegraph and the telephone not only increased the volume of business, but also made commercial transactions much more complex than formerly. From these conditions arose a demand for a more extended business training for youth, and to meet this demand the courses of study in business colleges have been extended until now in the best schools they include commercial arithmetic; a thorough system of accounting, including banking and commission; shorthand and typewriting; commercial law; at least one modern language, usually German or Spanish; political economy, and commercial geography. Many high schools provide commercial courses of one to two years. Schools of commerce having four-year courses of college grade are maintained in many universities and their number is constantly increasing. Most of the other commercial schools are conducted as private enterprises.

Nearly every city of 5,000 people and upwards now has its local business school, in which tuitions range from $25 to $75 for the various courses offered. See ACCOUNTING; BOOKKEEPING; SHORTHAND.

BUST, in sculpture, a figure representing the upper part of the human body, sometimes only the head and neck, but often including parts of the breast and shoulders. This form of sculpture was practiced by the Greeks as early as the sixth century B. C., and is shown in the Hermae, which were heads of the god Hermes mounted on pillars and erected along the roads to serve as guideposts. The Greeks did not make portrait busts of their great men to any extent until the time of Alexander the Great, but from that period there has survived a celebrated series of busts of Alexander and his successors, and also many representations of distinguished poets, philosophers and orators, including Plato, Zeno and Demosthenes. Both marble and bronze were used, the latter more generally than marble.

During the days of the Republic the Romans filled their public places with portrait busts, and the popularity of this form of sculpture continued until the third century of the Christian Era. In the Capitoline Museum and in the Vatican are famous collections of busts of the emperors, and good examples may also be seen in the British Museum, London, and in the Louvre, in Paris. A magnificent private collection of busts, mostly bronzes, belonging to a philosopher of the time of Cicero, which was unearthed at Herculaneum, has been placed in the museum of Naples.

Bust portraiture was a lost art from the sixth to the thirteenth century, but enjoyed a splendid revival through the Italian sculptors of the Renaissance, and there has been no decline in the art since that time. Jean Antoine Houdon (which see), a French sculptor of the eighteenth century, excelled in this field, and at the present time practically every sculptor of note is successful in bust portraiture.

For illustrations of busts, see articles CAESAR, ALEXANDER THE GREAT, HOMER, etc.

BUTLER, BENJAMIN FRANKLIN (1818-1893), an American politician and general, probably of all Northern officers during the War of Secession the one least liked in the South, where he was commonly known as "Beast Butler." He was born at Deerfield, N. H., studied at Waterville College, Maine, and having gained admission to the bar, practiced law at Lowell, Mass., building up a wide reputation. In the state legislature he worked for labor reform, carrying out the same policy which led him in his practice to fight for the interests of the factory workers against the corporations. Shortly after the outbreak of the War of Secession he was made major-general of volunteers and placed in command of the Department of Eastern Virginia, and though he showed no great military ability he came prominently before the public because of his declaration that slaves within the Union lines were "contraband of war" (see CONTRABAND).

BENJAMIN F. BUTLER

During his administration of New Orleans in 1862, to which duty he had been assigned, he issued vigorous repressive orders which won him his unsavory nickname and caused Jefferson Davis to proclaim him an outlaw, to be hanged if captured. He afterward held commands in Virginia and North Carolina, but General Grant removed him in 1864 and he returned to political life. Elected to Congress as a Republican in 1866, he served until 1879, except for two years, and was especially active in the impeachment proceedings against President Johnson. After striving vainly several times to gain the governorship of Massachusetts, he was elected to that office in 1882 by the Democrats, and two years later was the Greenback-Labor candidate for President.

BUTLER, NICHOLAS MURRAY (1862-), an American educator, since 1902 the president of Columbia University. He was born in Elizabeth, N. J., was educated at Columbia College and after graduation took special courses in Berlin and Paris. Following his studies abroad, he was appointed assistant in philosophy at Columbia, and when the institution was reorganized as a university he became the first dean of the faculty of philosophy. He founded and was the first president of the New York College for the Training of Teachers, now the Teachers College of Columbia University, and it was through his influence, while a member of the state board of education of New Jersey, that manual training was introduced into the public schools of that state.

Dr. Butler succeeded Seth Low as president of Columbia University in 1902. He has also taken an active interest in politics, and in 1912 was chosen to succeed James S. Sherman as nominee on the Republican ticket for Vice-President of the United States, Mr. Sherman having died just after the Presidential election, but before the meeting of the Electoral College. He is the editor of the *Educational Review, The Teachers' Professional Library,* the *Great Educators* series and the *Columbia University Contributions to Philosophy and Education,* and has written numerous papers and addresses on educational subjects.

BUTLER, Samuel (1612-1680), an English writer of the reign of Charles II, now remembered chiefly as the author of *Hudibras,* a poem published in three parts, between 1663 and 1678, that holds the Puritans up to mockery and ridicule. It became immensely popular among the frequenters of the London coffee-houses and taverns and at the English court, because of its wit, drollery and sarcastic thrusts at the staid and sober Puritans. Hudibras, from Hugh de Bras, one of the Knights of King Arthur, is the name of the hero of the poem; Butler found the inspiration for the character in a country gentleman whom he had served as attendant. Some of the most familiar proverbs of our common speech have their origin in Butler's work, such as "I smell a rat," "Spare the rod and spoil the child," "Look before you leap." The lines which follow illustrate very well the style and form of wit:

He'd undertake to prove, by force
Of argument, a man's no horse.
He'd prove a buzzard is no fowl,
And that a Lord may be an owl,
A calf an Alderman, a goose a justice
And rooks Committee-men or Trustees.

BUTLER, PA., the county seat of Butler County, is in the west-central part of the state, about thirty miles north of Pittsburgh. It is on Conequenessing Creek and on the Bessemer & Lake Erie, the Buffalo, Rochester & Pittsburgh and the Pennsylvania railroads and has interurban lines to Pittsburgh. The population, which in 1910 was 20,728, was 25,543 in 1914. The area is two and a half square miles.

The city has several parks, a courthouse, a public library and a county hospital. Large deposits of oil, natural gas, coal and iron are found in the vicinity. Important industries include steel-car works, woolen and silk mills, flour mills and manufactories of bottles and plate glass, oil-well machinery, carriages, white lead, pearl buttons and brass and iron beds. Butler was settled about 1798, was made the county seat in 1802, and was incorporated as a borough in 1803. It was named in honor of General Richard Butler, an officer in the War of Independence.

BUTTE, *bute,* a hill standing alone, or a mountain rising abruptly above the surrounding lower country. The term originated from the French word meaning *mound, hillock* or *elevation.* Buttes abound in the Rocky Mountain region. Many of them have been formed by the erosion of ancient plateaus, and are prominent features in the landscape. The term is also applied to high mountains in Canada and England, though it is not so widely used in this respect in the United States. The city of Butte, Montana, is so named because of its nearness to elevations of this nature. See Plateau.

BUTTE, *bute,* Montana, the largest city in the state and the center of the greatest copper-mining district in the world. One-seventh of the world's production and one-fourth of America's output of copper is shipped from Butte. Enormous amounts of zinc, gold and silver, together with copper, comprise an annual production worth nearly $80,000,000. There are 150 mines in operation in and about Butte, and $1,500,000 is the monthly pay-roll to the 12,000 workers in mines and mills. At Anaconda, twenty-six miles distant, is located the Washoe smelter, the world's greatest reduction works.

Butte is the county seat of Silver Bow County. It is in the southwestern part of the state, situated on a broad plateau 5,485 feet above sea level, between the Bitter Root Mountains on the west and the Rocky Mountains on the east. In this thinly-settled state, which is the third largest in the Union, Butte's nearest city neighbors are far distant. Helena, the state capital, is seventy-three miles northeast, and 383 miles west is Spokane. Salt Lake City is 397 miles southeast, Seattle 672 miles northwest, and Chicago about 1,526 miles southeast. These distances are minimized by service over four great transcontinental railroad lines, the Northern Pacific, the Great Northern, the Chicago, Milwaukee & Puget Sound, and the Oregon Short Line, a part of the Union Pacific system. Connecting with Kansas City and Denver is the Chicago, Burlingtou & Quincy Railroad. Butte is the terminus of the Butte, Anaconda & Pacific, the first electrified railroad in America. Electric power is furnished Butte by the Great Falls of the Missouri River, 130 miles distant, Canyon Ferry, Madison Valley and Big Hole, and it is used by the railroads, the mines and the city. As a result, smoke is practically eliminated from the district.

The National Parks Highway connects Butte with the Glacier National Park to the north, and Yellowstone National Park, to the south, is reached over the Montana-Utah Highway. "Butte" generally designates a much larger district than the city itself, and includes five smaller towns. The city covers an area of more than five square miles. During the ten years preceding 1900 the population increased 184.2 per cent. In 1910, including the suburban towns, it had reached 60,000, and in 1914 it was more than 80,000. Although Americans predominate, practically every foreign country is represented, Ireland by the greatest number.

Butte adds to the natural advantages of climate and mineral and agricultural wealth every convenience of a modern city. The water supply is brought from the mountains at an original cost of $4,000,000. The handsome public buildings include the Federal building, library, Masonic Temple, theaters, schools, churches, banks and hospitals. Near the city are a number of pleasure resorts easily reached by electric cars. The Columbia Gardens, owned and operated by ex-Senator William A. Clark, is one of the most popular parks. The Montana State School of Mines, located at Butte, has an attendance of about 100. This school offers advantages to students of engineering. Besides its vast mining activities, Butte operates extensive planing mills, tile factories, iron-works and machine shops.

History. After the gold rush to California, prospectors combed the hills far and near for rich ores. A few placer miners panned gold from Silver Bow Creek, which runs through the city of Butte, as early as 1864. A town site was laid out in 1867 and incorporated in 1876. All provisions came to the camp by ox carts until the completion of the Northern Pacific Railroad in 1883. About that time rich copper deposits were discovered and the life of the city assured.

BUT'TER, an important dairy product made from the fat of milk, known to man for nearly 4,000 years and now one of the most extensively-used foods in temperate regions. It has been said that "bread is the staff of life, but bread and butter is a gold-headed cane."

Butter was formerly made from the milk of goats and sheep, but the market product that has so many familiar uses in the modern household is made only from the milk of cows. Fat occurs in milk in the form of tiny globe-like particles. Formerly it was supposed that each particle was enclosed in a thin skin, or membrane, but this belief is no longer held by scientists, for several reasons. One of these is that by violent agitation of hot milk with an egg beater the fat globules may be divided into smaller ones, and the milk is still normal in appearance. Were each globule surrounded by a membrane this would not be the case. When milk is shaken about, or churned, the liquid globules are solidified and the fat particles cling together, forming the compact mass known as butter.

At least twenty centuries before the Christian Era men made butter by churning milk in skin bags, but in that period it was semi-liquid in form and was always spoken of as being "poured out." Butter was valued by the ancients as a medicine, and as an ointment which they rubbed on the body after bathing, and it was burned in lamps as we now burn oil, but it seems to have been used as a food only to a slight extent. Even at the present time the people of Southern Europe prefer olive oil to butter as a food.

It is said that the Arabs learned how to make butter by accident. They put milk in skin bags, which were carried across the deserts on the backs of camels, and the jolting of their burden during the journey caused the milk to be churned into a butter mass. In India, where on account of the climate it is difficult to keep food sweet for any length of time, the natives make fresh butter every day by shaking milk in bottles.

The Modern Method of Making Butter. The butter made according to these crude methods is, of course, inferior to the product of the modern creamery or dairy farm. The chief processes in butter making of the present day are creaming, or separating, ripening, churning and working.

Separating. Though butter is sometimes made from fresh milk, ordinarily only the cream is used. Cream is a thick, oily substance composed of the globes of fat that rise and gather on the surface of milk. There are two general methods of separating the cream from milk—the gravity and the centrifugal method. The gravity method consists in setting the milk in a cool place in shallow pans, or putting it

HOW YOUTH'S HARDSHIPS ARE DECREASING
Old and new processes of butter making.

in deep cans immersed in cold water. The cream, which is lighter than the water and other substances in the milk, rises to the surface in from eighteen to twenty-four hours, when it is skimmed off the top. If deep cans are used, the skim milk may be drawn off from below.

In creameries (see CREAMERY) and large dairies the cream separator (see SEPARATOR, CREAM) has come into general use. In this device the cream is separated from the milk in a bowl or drum which whirls around at the rate of from 5,000 to 8,000 revolutions a minute. Small hand separators skim from 200 to 1,000 pounds of milk an hour, and the larger machines 4,000 pounds or more, while under favorable conditions only about 0.1 per cent of fat is left in the skim milk.

Ripening. Cream is usually *ripened*, or *soured*, before it is churned; that is, certain germs are allowed to develop in it which give the butter the agreeable flavor and aroma that we associate with a good market product. Sometimes the ripening is left to chance, but in this case there is always the danger that undesirable germs will develop. Hence artificial "starters" of soured skim milk or prepared ferments are frequently added to the cream. The use of improved cultures for the purpose of ripening cream has been greatly extended in the United States through the efforts of the dairy schools and the agricultural experiment stations of the various states. After separating, the cream is cooled to at least 50° F., then warmed to 60° to 70° and held at this temperature until distinctly sour or until it contains about five-tenths per cent acidity. It is then cooled to near 50° and held at that temperature until churned, usually a few hours.

Churning and Working. During the churning process the cream should be kept at a temperature of from 50° to 65° F. As the cream is shaken about, the liquid fat solidifies and the minute granules of fat unite; these collect as they are brought into contact with each other, and when they reach the size of wheat kernels the buttermilk is drained off and the butter is washed.

Finally, the butter is "worked," to remove the buttermilk, to incorporate the necessary salt and to reduce the butter to a compact mass. The amount of salt used is commonly one ounce to the pound, but this depends upon the taste of the consumers. Europeans as a rule salt their butter to a less degree than do Americans. After the product is worked it is made into rectangular prints or packed in boxes or tubs for marketing. In domestic butter making, the churning and working of the butter are usually done by hand labor (see CHURN), but in creameries these operations are performed by machinery.

The Qualities of Butter. The special qualities which determine the relative excellence of butter are flavor, texture, salt and color. Some purchasers require a butter that is mild and delicate; others prefer a high flavor, but whatever the flavor it should be distinctive. Bad flavors are often due to odors of other foods or of decaying substances, these odors being readily absorbed by butter if it is brought into immediate contact with them. The keeping quality of butter depends upon the care exercised in the making. Butter made from sweet as well as from properly-ripened cream, and which is well worked and evenly salted, retains its original quality for a considerable length of time. If the cream is too old and the buttermilk is not fully removed the butter is liable to become rancid. Scrupulous cleanliness should be the rule throughout the entire process. Butter is said to have a good texture if it is hard and firm and will spread easily and does not have a greasy, oily feel and taste.

Butter is naturally golden-yellow in color, this quality varying with the feed and the breed of the cows. If made in the spring and early summer, when the cows are feeding upon the fresh new grass of the pastures, it is usually darker than when made at other seasons, and some artificial butter color is commonly added so there may be a uniform color throughout the year. Annatto is in general use for the purpose, and turmeric, saffron, marigold leaves and carrot juice are also occasionally employed.

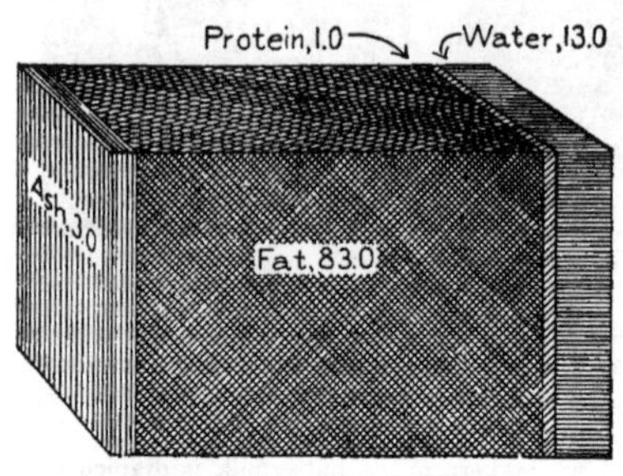

COMPOSITION OF BUTTER

Composition. Butter varies in composition with the conditions under which it is made, but the following figures represent an average grade: fat, about 83 per cent; water, 13 per cent; protein, 1.0 per cent; ash, 3.0 per cent. Of these, water is the most variable. According to the standard fixed by the United States government, butter should contain not less than 82.5 per cent of fat nor more than 16 per cent of water.

Food Value. Butter is highly nutritious, and is one of the most wholesome and most easily digested of all food fats. About 19.7 per cent of the total fat in the daily food of the average person is furnished by butter.

Production. For its size, Denmark is the leading butter-making country of the world, and the quality of the Danish product is unequaled. In quantity the United States is foremost among the butter-making countries, with a total output of nearly 1,700,000,000 pounds a year. About 1,000,000,000 pounds are made on the dairy farms, and the remainder in creameries and city dairies. The value of this output is over $405,000,000. The leading butter states are Wisconsin, Minnesota, Iowa, Illinois, New York, Pennsylvania, Michigan and Ohio. Among the Canadian provinces, Ontario leads in the production of home-made butter and is second in the output of creameries and factories; Quebec produces the largest amount of factory and creamery butter. The yearly output of home-made butter in all the provinces is about 140,000,000 pounds, and of factory and creamery butter, about 65,000,000 pounds. The total value is about $46,000,000.

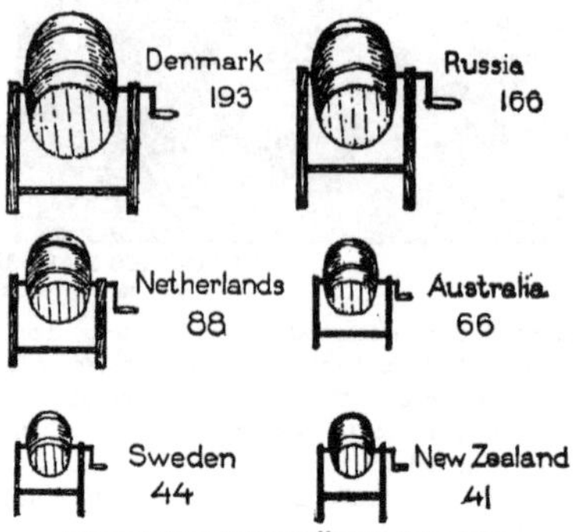

Figures Represent Millions of Pounds

PRINCIPAL BUTTER-EXPORTING COUNTRIES

Adulterants and Imitations. Cottonseed and other oils and various fats have been used as adulterants of butter. The most common manufactured substitute is a fat made of suet, oil, butter, cream and milk, known as *oleomargarine* (which see). Renovated or process butter is a butter of poor quality from which disagreeable odors and flavors have been removed, and to which cream or milk has been added to give it a good appearance. Such a product is much inferior to oleomargarine, and its sale is strictly regulated by law. In the United States all butter of this character must be labeled *renovated.*

A TEST FOR PURITY
Explanation appears in the text.

Tests of Purity. If butter does not smell

sweet it has fermented and has become rancid. Such butter has probably been made from old cream collected for several days, and it contains decomposing particles of fat. There is a characteristic test which pure butter, free from artificial fats, responds to easily. If a bit of butter be melted in a test tube set in warm, not hot, water, and kept at an even temperature for half an hour, it should at the end of that time show clear if pure; if not pure or if it contains artificial fat it will be cloudy. A little pure butter heated in a spoon over a gas jet will simmer evenly and quietly, but will proclaim the presence of oleomargarine by noisy sputtering and popping. E.H.F.

BUTTER'CUP, or **CROWFOOT,** a dazzling yellow wild flower or weed of the roadside and field, found in England, the United States and Canada. From May to September this "little children's dower" brightens the waysides and pasture-lands, but the farmer is not pleased at its appearance; to him it is a troublesome weed. Because of the bitter, burning juice in the plants animals will not eat them. The buttercup grows from one to two and one-half feet high. The leaves usually have three parts and are notched; the flowers, about an inch across, have five smooth, shining petals of yellow. There is also a creeping buttercup, whose stem creeps along the ground and sends out new roots here and there; also a swamp buttercup which loves the moist, shady spots, and a water buttercup whose blossoms float on the water. See **Ranunculus.**

The buttercups, bright-eyed and bold,
Held up their chalices of gold
To catch the sunshine and the dew.
—Dorr, in *Centennial Poem.*

THE STORY OF THE BUTTERFLY

BUTTERFLY, a beautiful insect with wings so dainty and so brightly colored that it is often referred to as a "winged flower." That its name is made up of two common words is perfectly plain, but the reason is not so clear. Probably the "butter" refers to the color of some of the best-known species. Little children in learning to pronounce the long name sometimes get it twisted about and give it a form which seems much more fitting—"flutter-by."

How Butterflies Differ from Moths. Other insects there are which look much like the butterflies; these are the moths. In structure the two are much alike, but there are certain simple differences which will help anyone to distinguish between them. The butterflies are usually bright-colored and have slender bodies, while the moths are in general dusky and thick-bodied; the butterflies love the sunshine and are to be seen flitting about only in the daytime, and rarely on a cloudy day, but the moths are abroad in the twilight. Then, too, the butterfly has a little knob or hook at the end of its feelers, or *antennae,* while the moth has not, but the most noticeable distinction of all is that a butterfly when it alights holds its wings erect, while a moth spreads its wings out flat. An illustration showing fundamental differences is given in the article **Moth.**

Body Structure. The body of a butterfly has three parts—head, thorax and abdomen. The conspicuous parts of the head are the *antennae*, which are not only feelers, but also ears and nose to the butterfly; the eye-clusters,

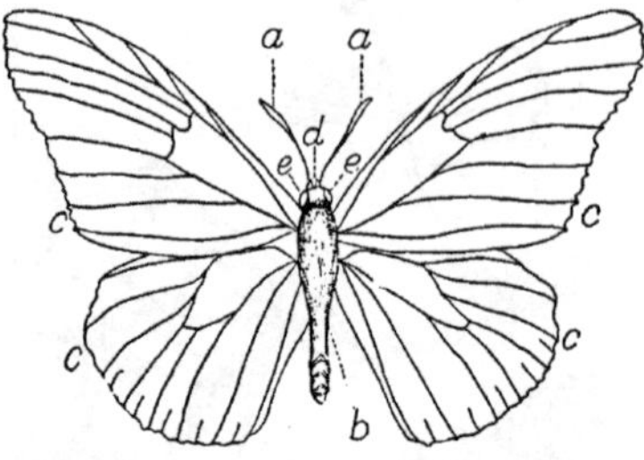

PRINCIPAL PARTS

(*a*) Antennae
(*b*) Body
(*c*) Wings
(*d*) Mandibles
(*e*) Eyes

which in some species are made up of as many as 20,000 tiny eyes, or *facets;* and the long, sucking tube into which the mouth parts have been modified. When not in use this is coiled up like a watchspring. To the thorax are attached the six feeble legs and the four wings.

It is the wings of the butterfly which make it the exquisite, graceful creature that it is. They are large and strong, the front pair usually triangular and the second rounded. Their structure is curious, for they are made up of membranes stretched on a framework of double tubes. The inner tubes are filled with air; the outer ones are veins. Thickly covering the membrane of the wings are tiny scales, arranged in overlapping rows like shingles on a roof. When looked at under a microscope they are found to resemble feathers; thus one is made to marvel at the perfection sometimes displayed in insect life. Minute as they are, these scales are of the utmost importance to the butterfly, as any child has discovered who has carelessly grasped a butterfly by the wings. When released the little insect has not only lost much of its beauty, for all the brilliant coloring is in the scales, but it flies very feebly or not at all. No one who wishes to collect butterflies will ever seize them by the wings, for he knows that in so doing he is ruining his "specimens."

Habits of Life. Butterflies do not, like birds or even some of the lower classes of life, have the "homing" instinct. They make no sort of structure to live in, and seem to have no choice of a home spot except that the eggs must be laid on some substance which will feed the young. Thomas Wentworth Higginson, a lover of butterflies, has written of them—

Birds have their nests; they rear their eager young,
And flit on errands all the livelong day;
Each fieldmouse keeps the homestead whence it sprung;
But thou art Nature's freeman—free to stray
Unfettered through the wood,
Seeking thine airy food,
The sweetness spiced on every blossomed spray.

Their food is chiefly the nectar of flowers, but they are very dainty eaters, seeming to need almost nothing to keep them alive.

One drop of honey gives satiety;
A second draught would drug thee past all mirth.

Their life is very short, lasting with most species but a few days. As soon as the eggs for the next brood are deposited, the insect dies.

In one point besides their wings butterflies resemble birds—in the decided difference in color and even in size between the male and female of the same species. So entirely different are they at times that none but a scientist could tell that they were not of two species rather than one.

There is one curious fact about butterflies which does not seem to accord with their beauty and daintiness. They love an unpleasant odor, especially the smell of decaying matter, and nothing will attract them more quickly than a rotten banana or a spoiled fish. In the tropics, where the natives catch the great

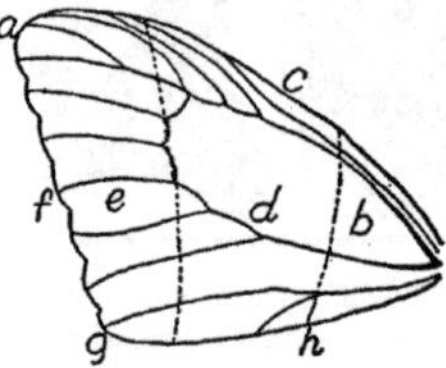

OUTLINE OF WING

(*a*) Apex
(*b*) Base
(*c*) Costal margin
(*d*) Discal area
(*e*) Limbal area
(*f*) Outer margin
(*g*) Inner angle
(*h*) Inner margin

gorgeous-hued specimens for which collectors are willing to pay high prices, they make use of over-ripe fruit to attract the insects to the spot where they lie in wait.

Butterflies share with some other insects the remarkable power of mimicry, which provides protective coloring. Many of the common brown and reddish butterflies that float about in the time of falling leaves look enough like

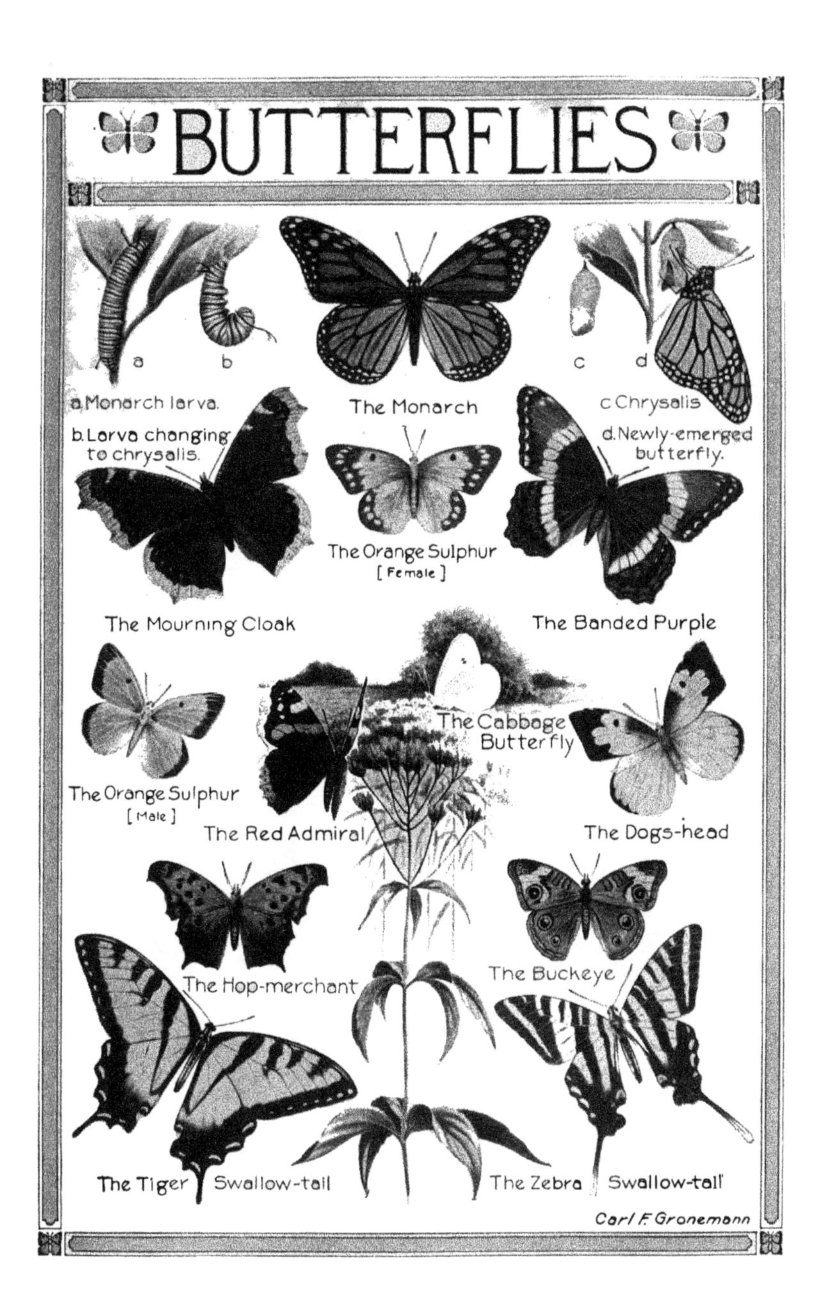
BUTTERFLIES
a
b
c
d
a. Monarch larva.
b. Larva changing to chrysalis.
The Monarch
c Chrysalis
d. Newly-emerged butterfly.
The Orange Sulphur [Female]
The Mourning Cloak
The Banded Purple
The Cabbage Butterfly
The Orange Sulphur [Male]
The Red Admiral
The Dogs-head
The Hop-merchant
The Buckeye
The Tiger Swallow-tail
The Zebra Swallow-tail
Carl F. Gronemann

dead leaves to escape detection, but there is in India a wonderful butterfly which has very special markings that aid in this deception. When its wings are spread they are spotted with purple and orange, and are entirely unleaflike, but when the insect alights and folds its wings the closest observer would have difficulty in distinguishing it from a dead leaf still attached to its twig. A little extension on the hind wing, like the "tail" of a swallowtail butterfly, imitates perfectly the leaf-stalk, and every rib and vein is present. See PROTECTIVE COLORATION.

Life History. The Greeks used the same word, *psyche,* to mean *butterfly* and *soul,* for to them the beautiful insect was the symbol of the soul. Often in their pictures death was

LIFE HISTORY OF A BUTTERFLY
(1) Eggs, highly magnified; (2) caterpillar; (3) chrysalis; (4) butterfly.

shown in the image of a butterfly flitting from the dead man's lips—the soul leaving the body. The appropriateness of this charming conception is found in the life history of the butterfly, whose emergence from an apparent death suggests strikingly the immortality of the soul. For the butterfly undergoes a complete metamorphosis, living in four distinct forms before it has completed its life history: the egg; the *larva,* or caterpillar; the *pupa,* or chrysalis; and the *imago,* or perfect insect. The first man who ever watched a butterfly emerge from a chrysalis, to all appearance dead, must have felt it to be one of the strangest things he ever looked upon; and the knowledge that before this mummy-like stage came the dull, groveling existence of a caterpillar could but have made the wonder greater. Nor does the wonder ever die; each time the transformation is watched it seems as marvelous as before.

The Egg. The butterfly deposits its eggs singly or in clusters upon or near the plants on which the young must feed. The eggs of some species hatch in a few weeks, or even days, but others take months to come to maturity. In cold climates the eggs deposited in the fall are not injured by winter's severity and do not hatch until spring. When they do open there comes out, not a winged creature which looks like the parent butterfly even as much as a scraggly little bird resembles its beautiful parents, but a crawling thing which looks and moves like a worm. See CATERPILLAR.

The Larva. The larva, or caterpillar, the second stage in the development, is familiar. The huge green worms, the cabbage worms and the fuzzy brown, yellow or white caterpillars are by most people as much loathed as the butterflies are admired. Many of them deserve the disgust which they occasion, for they do much harm by destroying vegetation. Most of the harmful caterpillars, however, are the larvae (young) of moths and not of butterflies. A caterpillar does not eat casually and often daintily, as does a butterfly, but applies itself industriously to this, its only task. Trees may be almost stripped of their leaves, the cabbage garden of the truck farmer ruined, but the caterpillar must be fed, for it is storing up fat for the long weeks or months during which it can have no food. The length of time this caterpillar stage endures varies with the locality, the season and the species. In temperate climates it lasts from three to four months, while in the cold regions the period is often ten months, but the caterpillar never fails to know when it is over and to make preparations for the coming pupa stage.

The Pupa. The caterpillars of moths spin for themselves cocoons of silk (see MOTH; COCOON), but those of the butterflies shut themselves up in hard, smooth cases, and are known as *chrysalids.* For the most part these are of a dull color, escaping detection by their resemblance to the objects about them, but some are golden and shiny; and it is these which have suggested the name chrysalis, which means *gold.* Some of these chrysalids hang head downward from twigs or the under side of leaves; others are suspended in a nearly horizontal position by silken cords.

In the pupa state the insect looks dead, but it breathes through small pores and within its

OUTLINE AND QUESTIONS ON THE BUTTERFLY

Outline

I. Description

(1) Insect
(2) Structure
  (a) Head
    1. Antennae
    2. Eye-clusters
    3. Mouth parts
  (b) Thorax
    1. Legs
    2. Wings
      a. Number
      b. Framework
      c. Membrane and scales
  (c) Abdomen
(3) Color
  (a) Variation due to sex
  (b) Protective coloration

II. Life Habits

(1) Lack of "home"
(2) Feeding habits
  (a) Nectar
  (b) Fondness for unpleasant odor
(3) Powers of flight
(4) Migration

III. How Distinguished from Moth

(1) Difference in
  (a) Color
  (b) Shape of body
  (c) Shape of antennae
(2) Flies by day, moth flies by night
(3) Position of wings when at rest

IV. The Life Cycle

(1) The egg
  (a) Where deposited
  (b) Time required for hatching
(2) The larva or caterpillar
  (a) Structure
    1. Head
      a. Eyes
      b. Feelers
      c. Biting jaws
      d. Color and covering
    2. Segments
    3. Legs and leg-stumps
  (b) Self-protection by means of
    1. Concealment
    2. Unpleasant taste
    3. Disagreeable odor
  (c) Food
    1. Voraciousness
    2. Harm to crops
  (d) Molting
  (e) Duration of caterpillar stage
(3) Pupa or chrysalis
  (a) Distinction between chrysalis and cocoon
  (b) Duration of this stage
  (c) Life processes
(4) The perfect insect or imago

V. Classification

(1) Brush-footed butterflies
(2) Blues and coppers
(3) White, sulphur and orange tips
(4) Skippers
(5) Swallowtails

Questions

Explain the allusion in the italicized word in the following quotation from Shakespeare

For men, like butterflies,
Show not their *mealy* wings but to the summer.

Taking the name *caterpillar* in its literal significance, could it be fittingly applied to the smooth green tomato worm?

What is the distinction between a cocoon and a chrysalis?

Why cannot the housekeeper, when she has destroyed all the "dusty millers" flying about her house, feel certain that her furs and flannels are safe?

What proves that butterflies are not dainty in all their tastes?

When birds migrate, it is usually because they cannot find sufficient food for themselves. How does the reason for butterfly migrations differ from this?

How would you construct a cyanide jar?

Outline and Questions on the Butterfly—Continued

What decided difference exists between the feeding habits of the larva, or caterpillar, and those of the imago, or full-grown insect?

Why can you not always be sure when you see a male and a female butterfly that they belong to the same species?

How does the butterfly that has just emerged from the chrysalis differ in appearance from the same insect a few hours later?

Why is it not wise to allow children to make collections of butterflies?

Mention three ways in which the insect when in its larval stage protects itself from the attacks of stronger animals.

Do butterflies have a "home life" as do birds, bees and ants?

Why are chrysalids so named?

What is the best stage at which to begin the study of the life history of the butterfly? Why is this not always possible?

How do the mouth parts of the caterpillar differ from those of the perfect insect?

How many wings has a butterfly?

Are they all the same shape?

What is the sole task of the caterpillar?

In which continent are butterflies most abundant and most gorgeous?

How many legs has a caterpillar? To what part of the body are they attached?

Describe the eyes of a butterfly.

What do the mouth parts look like when not in use?

Do the eggs of all butterflies hatch in about the same length of time?

Which are the largest butterflies, and why are they named as they are?

What explanation is given of the name *butterfly?*

Describe the remarkable manner in which a certain butterfly guards against the attacks of birds or other animals.

To what family does the beautiful "mourning cloak" belong?

What must the amateur collector particularly guard against in mounting his specimens?

What is there peculiar about the way in which a caterpillar grows? How many times does this process take place before the cocoon stage?

If you see a big, gorgeous-winged insect poised on a twig, how can you tell whether it is a butterfly or a moth? If the insect is dead and mounted, can you still tell which of the two it is?

To which group do the white cabbage butterflies belong?

Describe the correct form of net for catching butterflies?

What goes on within the cocoon during its period of quiescence?

Name the various stages in the life of a butterfly.

What is meant by a complete metamorphosis? Name another insect that undergoes the same process. Name one that does not show all the stages.

How do natives in the tropics attract butterflies for capture?

Are these insects as long-lived as ants? As bees?

What must the female find before it lays its eggs?

What double duty do the tubes in the gauzy wings perform?

Is there any kind of caterpillar which pays for the food it eats?

What is meant by "protective coloration"? Do any other insects have it?

Of what did the Greeks make the butterfly representative? Give the myth of Psyche.

horny covering the mysterious life changes continue. The wings, the legs, the body—all are formed during this inactive period, which lasts with some butterflies but a few weeks, while with others it continues through the winter. When it is completed the case splits and out comes the imago, or perfect insect. At first it looks little enough like an airy, bright-hued butterfly, for its wings are soft, colorless, and closely folded. But air and sunshine work a speedy miracle, and very soon the "wingéd blossom, liberated thing" is ready to float away among the "other flowers, still held within the garden's fostering."

As to how far these full-grown butterflies can fly, there is little accurate information. Their flight seems languid, almost lazy, but they move swiftly enough to be difficult to capture when on the wing. A few species, if they do not find at hand the kind of plant food which the larvae must have, migrate in search of it, sometimes for a hundred miles or more. Many of the long flights are not really flights at all—the insects are simply borne along by the wind, as a milkweed seed or a winged maple seed might be.

Classification. Scientists have long names for the different species of butterflies, and any person who wishes to form a collection of any value should make himself familiar with these from some such publication as Holland's *Butterfly Book*. Here, however, a less technical classification will do. All the 650 species which appear in the United States and Canada are grouped in five families.

(1) The greatest is the family of *brush-footed* butterflies. Many of the most conspicuous and beautiful of the temperate-region butterflies are of this family, for instance, the familiar red-brown milkweed butterfly, or monarch, the mourning cloak and the thistle butterfly.

(2) A second great class is the family of so-called *blues and coppers*. These slender-bodied insects have gossamer wings, in shimmering blue or coppery shades. Almost all of them are small, and some, as the *hairstreaks*, have little projections or tails on their hind wings.

(3) The most familiar family contains the *white, sulphur and orange tips*, to which belong the roadside butterflies everywhere to be seen. Varying shades, from the white of the cabbage butterfly to deep sulphur, are to be found, and the size of wing expansion ranges from an inch to two and one-half inches.

(4) The *skippers* make up a large family, but not many species of it are familiarly seen. Brown, blackish or dull gray, these are for the most part inconspicuous and small.

(5) The most beautiful and striking of all are the large *swallowtails*, which are distinguished by the pronounced tails on their hind wings. They are the largest of the butterflies, and show many exquisite colors and markings, black, yellow, greenish-white and even reds and blues appearing on their gauzy wings. Especially admired is the great black and yellow *tiger swallowtail*, which is found almost everywhere in temperate North America.

It is in South America that butterflies most abound, and some of the tropic species are among the most gorgeous of all insects. They have not only brilliant colors but a great wing expanse, some of them measuring eight or nine inches across.

Nature-Study Lessons. Few, if any, among living things are more interesting and more profitable for study than the butterfly in its various stages. Many specimens can be found in every part of the world; a little watchfulness and care will probably enable children to secure several different kinds, either at home or at school, under direction of parents or teacher. The greater the difference, the more valuable the specimens will be for study.

The study of the life history of a butterfly should properly begin with the egg. Unfortunately the pupils may not always be able to find eggs; in that case, they may begin with the second, or caterpillar, stage. Teachers and parents will find that the children will take a lively interest in the development of the caterpillar. A caterpillar may be kept in a glass case, set in a sunny place; if he is fed and given a twig and leaves to which to attach his chrysalis, the observer may soon see him spin himself into his retreat and finally emerge a perfect butterfly. Let the children keep a record of daily observations of any changes they may notice. Not only will they be interested in the caterpillar, but they will, unknown to themselves, be learning how to observe carefully and systematically. Incidentally there will be found many opportunities to teach lessons of kindness.

After the butterfly has been observed it should be allowed to fly away, without being touched with the fingers, for butterflies do no harm, and they make all out-of-doors more beautiful.

Butterfly Collections. The mania to collect is a natural one in children, and butterflies have always been one of the favorite objects of collectors. So far as possible unscientific collections should be discouraged, for it is far better for the children to observe the live insects. If, however, collections are to be made, they should be made as carefully as possible, and the mounted insects should be identified.

First of all, the collector needs a net of tarletan, or mosquito-netting, with a light handle from three to five feet in length, the ring being about a foot in diameter and the bag about eighteen inches deep. When the insects have been caught with this instrument, they are killed by being placed in a wide-mouthed bottle containing cyanide of potash. A few lumps of this substance are placed in the jar, cotton is laid over them, and then a heavy paper, pierced with many holes, is pasted over the cotton. The bottle should be kept closed when not in use, and care should be taken in handling it, as the cyanide is poisonous.

When taken from the poisoning jar the insects should be pinned upon the cork bottom of a tin "field-box," care being taken not to brush the wings. The final mounting may be upon cork or some other substance, such as velvet, as the collector may prefer. Scientists who make of the collecting and mounting of butterflies a business have an elaborate series of instruments, including setting-blocks, drying-racks and special pins and needles, but the equipment described above will be sufficient for amateur collectors. M.S.

Consult Holland's *Butterfly Book;* Hornaday's *Taxidermy and Zoölogical Collecting.*

Related Subjects. In the following articles will be found much that will prove helpful to the reader who is interested in butterflies:

Antennae	Larva
Caterpillar	Metamorphosis
Chrysalis	Moth
Cocoon	Protective Coloration
Insect	

BUTTERMILK, the milky liquid remaining after butter has been made from cream. It contains casein, sugar, a small amount of lactic acid, some butter fat and a few other substances distributed through the liquid, which is largely water. The lactic acid gives buttermilk a slightly sour taste, and the casein, sugar and fat render it of some value as a food. When fresh it constitutes a nourishing drink; but as it absorbs bacteria rapidly when exposed to the air, and soon becomes unwholesome, buttermilk designed for drinking should be kept in closed vessels and on ice. See FOOD, subhead *Chemistry of Food;* MILK.

BUTTERNUT, or **WHITE WALNUT,** a large spreading tree of the walnut family, not quite as attractive or useful as the black walnut. A native of America, it is found in the Northeastern states growing to a height sometimes of eighty feet. It is best suited to cold climates and thrives in deep, rich loam.

BUTTERNUT
Leaves and fruit.

Above the grayish bark are seen the large, hairy, yellow-green leaves and small flowers or yellow-green catkins. Then come the oblong, pointed nuts in spongy, hair-covered ribbed husks. While soft and green the nuts are often preserved as pickles. Later, when dried, they are hard-shelled, sweet and oily, of excellent flavor, but not marketed as much as other nuts. White walnut wood is light brown, soft, coarse-grained, with a satiny luster, and is used in cabinet work and interior finishing of houses. The sap has been used a little in sugar-making, the root-bark in medicine. The bark of the stems has been used for dyeing; the homespun, home-dyed "butternut" uniforms of some regiments in the War of Secession being yet remembered.

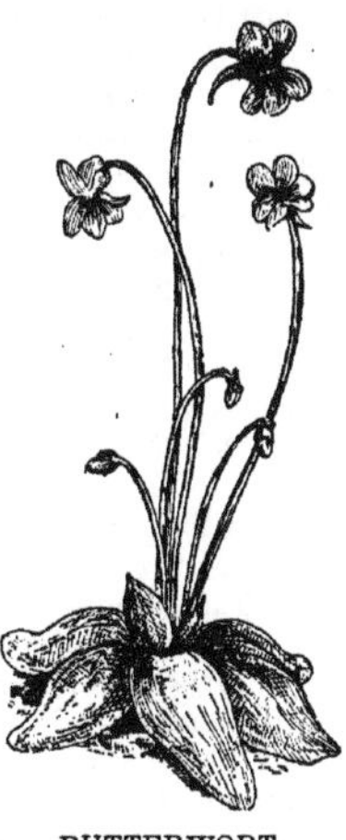

BUTTERWORT

BUTTERWORT, *but'er wert,* one of the carnivorous, or insect-eating, plants, which grows in bogs or soft grounds. The dainty flowers are mostly purple, though on some plants they are yellow. The short, thick leaves secrete a juice which attracts small insects. The edges of the leaf roll over on the insect, which dies and serves as food for the plant. In the north of Sweden, in Lapland and the Alps the leaves are used to curdle milk. In some greenhouses the butterwort is cultivated as a curiosity. Some say the plant was so named because of the buttery feeling of the leaves; others attribute the name to the use of the plant in curdling milk. See Carnivorous Plants.

BUT'TERWORTH, Hezekiah (1830-1905), an American author and editor, whose stories and histories for young people are among the best of juvenile writings. He was born and reared on a Rhode Island farm, received a common school education, and by writing stories earned enough to pay for a course in rhetoric and composition at Brown University. With this preparation he went to Boston with the idea of becoming a great writer, and soon found a minor position on the staff of *The Youth's Companion.* In 1871 he became its editor, and it was truthfully said that *The Companion* carried his name around the world, for his energy, enthusiasm and personality caused the periodical to grow in circulation from 140,000 to 400,000 in the decade between 1877 and 1887. He was its editor until 1894.

Butterworth's published volumes number about sixty. Seventeen of these belong to the *Zigzag* series, books of travel that combine fact and fiction, in which he takes the reader all over the world. These and his biographical and historical tales—*In the Boyhood of Lincoln, The Wampum Belt, In Old New England* and others—are full of charm and interest to young readers. He also wrote numerous ballads, cantatas and hymns, his *Bird with the Broken Wing* being a popular Sunday School song. Butterworth traveled extensively in Europe and in America, and was in constant demand as a lecturer.

BUTTON, *but''n,* a piece of ivory, bone, metal or other material, usually round and employed for the purpose of holding together different parts of wearing apparel, or for ornament. The name is derived from the French word *bouton,* which literally means something *pushed out,* and is appropriate in view of the fact that buttons, to be useful, must be pushed through what are called buttonholes. The original use of buttons was only for ornament. The robes of the Greeks and Romans were kept in place by means of strings, girdles or brooches, and it was not until more complicated garments came into use that buttons were employed as fasteners.

As to the materials of which buttons are made, it has been said that it would be hard to find any material that has not been used in their manufacture. Gold, silver, iron—in fact, metal of all kinds—wood, paper, bone, horn, shell, stone, glass, potatoes, vegetable ivory, and even dried blood are among the materials in daily use as buttons.

Buttons are of two general classes; those which are attached to the garment by means of threads which pass through holes in the buttons themselves, and those which have a shank or loop of metal, or tuft of cloth, by means of which they are affixed. In either case the head of the button may or may not be covered with cloth to match the material to which it is to be attached.

Pearl Buttons. One of the most important branches of the button industry has for many years been the manufacture of pearl buttons. The luster of the inside shell of the pearl oyster has always been considered singularly attractive, and soon after it was found to be easily made into buttons the pearl-button industry grew until it exceeded that of any other kind. In the United States the industry was handicapped because the shells had to be brought from great distances and the cost of production was therefore high. In the latter part of the nineteenth century a substitute for the pearl oyster shell was found in the "nigger-head" mussel of the Mississippi River, and the industry received a fresh impetus. These shells are collected at all times, but those taken during the winter are found most suitable for button making. They are soaked for a few days in clear, fresh water, are then cut to the required sizes by drills, holes are bored, and

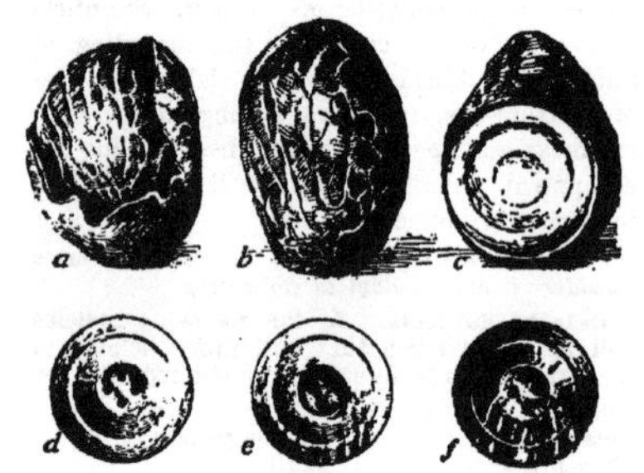

BUTTONS FROM IVORY NUTS

(*a-b*) Ivory nut, with broken shell exposing the nut; (*c*) partly turned piece; (*d*) white button; (*e*) the same, sprayed with shellac; (*f*) sprayed, developed and shellac removed.

when polished, the buttons are ready for market. The finished buttons are sewn to cards containing a certain number, all of the same size. The first button factory on the Mississippi was opened in 1891 and now many towns and villages for a distance of 200 miles along the river banks from Fort Madison south to Sabula have their mussel-fishing or button-making industry.

Vegetable Ivory Buttons. Perhaps the most interesting of all button-making processes is that by which they are manufactured from *vegetable ivory,* as the fruit of corozo nut palm is called. The tree produces clusters of nuts in a husk sometimes as large as an ostrich egg. The nuts are shelled by machinery and the kernels are extracted. The application of heat causes the nut to become as hard as stone, with the appearance of ivory. Vegetable ivory can be dyed to any desired color by chemical processes. The value of vegetable ivory buttons produced in 1912, in the United States alone, exceeded $3,500,000. See IVORY PALM.

Metal Buttons. Except when used for uniforms, metal buttons are employed principally where strength is required rather than ornamentation. They are made by cutting out disks or blanks from sheet metal and then molding or pressing them to the desired shape. Designs or letters are stamped by means of dies. Waterbury, Conn., is the center of the metal button industry in America, its first factory having been established in 1800, fifty years after one had been opened at Philadelphia, Pa. In England the great button center is Birmingham, where the industry attained great importance as long ago as the time of Queen Elizabeth. Canada has button factories in all its large cities, but the greater part of its supply is imported from Britain.

Fraternal and Society Buttons. An important branch of the button industry has recently developed in the United States in the manufacture of club, society, fraternity and political buttons. These are usually made of celluloid with a metal backing. On a disk of metal a picture or design is placed and on top of this a sheet of transparent celluloid, soaked in alcohol and other chemicals, is pressed. The button is then backed by a sheet of metal with a pin for attaching it to the coat, or with a stud to pass through a buttonhole. Fraternity buttons are often made of mosaic design of precious and semi-precious stones and are sometimes very artistically designed.

Most of the buttons made in the United States are used in that country; few are imported and practically none exported. The annual production of the 450 factories engaged in the button industry is valued at over $20,000,000 (1915). F.ST.A.

BUZZARD, *buz'ard,* a species of hawk common both in Europe and North America. In the United States and Southern Canada the name is more commonly applied to the *turkey buzzard,* which feeds on decaying things—things which even a goat would not touch. The common buzzard of Europe feeds upon mice, frogs, toads, worms and insects, and is very sluggish in its habits. The red-tailed and red-shouldered buzzards, commonly known as *hen-hawks,* are not as harmful as is generally supposed. Although they do steal chickens at times, they more than atone for this by ridding the meadows and orchards of destructive pests, such as mice, worms and insects. The defense of hawks in Longfellow's *The Birds of Killingworth* might well apply to these buzzards. Buzzards are known for their wonderful ease and grace in flight; they often remain in the air for hours at a time. Their sight is very keen. When their prey is seen from on high, they descend silently with a graceful swoop and capture the unsuspecting victim. See HAWK; TURKEY BUZZARD.

BY-LAW, a particular or private law made by an incorporated body for the regulation of its affairs. The power to make or change its by-laws belongs to every corporation even without express authority. A by-law which is counter or contrary to a constitutional provision or to a settled rule of law is invalid. By-laws of a municipal corporation are called *ordinances,* and are true laws, for they have the state's authority behind them. In general, however, by-laws are simply agreements or rules as to methods of work among members of an incorporated business organization, or of a society for the promotion of any object.

BY'RON, GEORGE NOEL GORDON, Sixth Lord (1788-1824), an English poet who in his own day was looked upon as second only to Shakespeare. Handsome, passionate, with a carefully-cultivated air of mystery and gloom, he stood as one of the most prominent figures of his day. His ofttimes bitter smile was copied as carefully as was his loose collar and flowing tie, and the word "Byronic" acquired a very definite meaning.

Byron was born in London on January 22, 1788. Until the age of seven he was entirely under the care of his mother, and to her unwise indulgences and equally unwise severities the waywardness that marked his after life may have been partly due. On reaching his seventh year he was sent to the grammar school at Aberdeen, and four years after, in 1798, the death of his granduncle gave him the titles and estates of the family. Mother and son then removed to Newstead Abbey,

GEORGE NOEL GORDON BYRON

No one else—except, perhaps, Wordsworth—who could write so well could also write so ill. "I can never recast anything," he said; "I am like the tiger—if I miss the first spring I go grumbling back to my jungle."

the family seat, near Nottingham. Soon afterward Byron was sent to Harrow, where he distinguished himself by his unsystematic reading, rather than by careful study. In athletic sports he excelled, despite the lameness which had resulted from a childish illness. In 1805 he entered Trinity College, Cambridge. Two years later appeared his first poetic volume, *Hours of Idleness,* which was criticised with unnecessary severity in the *Edinburgh Review.* This criticism roused Byron and drew from him his first really notable effort, the celebrated satire, *English Bards and Scotch Reviewers.*

In 1809, in company with a friend, Byron visited the southern provinces of Spain and voyaged along the shores of the Mediterranean. The result of these travels was *Childe Harold's Pilgrimage,* the first two books of which were published on his return in 1812. The poem was immediately successful, and Byron "awoke one morning and found himself famous." During the next two years *The Giaour, The Bride of Abydos, The Corsair* and *Lara* appeared, and Byron's literary reputation grew steadily.

During these years, however, he was living in the most reckless dissipation. In 1815 he married the daughter of Sir Ralph Milbanke, evidently intending to give up his reckless life; but the marriage proved unhappy, and in about a year Lady Byron left him and refused to return. This rupture gave rise to much popular indignation against Byron, and reports were circulated which caused him to leave England, with an expressed resolution never to return. He visited France, the field of Waterloo, Brussels, the Rhine, Switzerland and the north of Italy; for some time lived at Rome and later at Geneva, where he completed his third book of *Childe Harold.* Not long after appeared *The Prisoner of Chillon; The Dream, and Other Poems;* in 1817 *Manfred,* a tragedy, and *The Lament of Tasso,* all of which helped to make him as popular as a poet as he was unpopular as a man.

From Italy Byron made occasional excursions to the islands of Greece, and at length visited Athens, where he sketched many of the scenes of the fourth and last book of *Childe Harold.* Between 1817 and 1822 appeared, among other poems, five books of *Don Juan* and a number of dramas. While living at Pisa he was for a time intimate with Shelley, one of the few men whom he entirely respected and with whom he was really confidential. Shelley brought out all that was best in his nature, and had the association not been broken by Shelley's tragic death, Byron might in time have found himself and worked out a mode of life which should have been worthy of his powers. Besides his contributions to the *Liberal,* a periodical which he helped Leigh Hunt and Shelley to found at this time, he completed *Don Juan,* with *Werner,* a tragedy, and *The Deformed Transformed,* a fragment. These are the last of Byron's poetical works.

In 1823, troubled perhaps by the consciousness that his life had too long been unworthy of him and driven by an absolute passion for liberty, he threw himself into the struggle for the independence of Greece. In January, 1824, he arrived at Missolonghi, where he was received with the greatest enthusiasm. The air of Missolonghi began to affect his health, and on April 9, 1824, while riding in the rain, he became ill; a fever followed, which ten days later ended fatally.

Byron's natural force and genius were perhaps superior to those of any other Englishman of his time, and won for him in his own day a fame which it is hard for those of a later day to understand. After his death his work was for some time as far underrated as it had been overrated during his life, and it is only within the last few decades that a calm judgment has been passed on his writings. These will live on account of their descriptions of nature, the hatred of sham which they express and their keen insight into the heart of man.

In his own day quotations from Byron were on everyone's lips, and even to-day some of them are everywhere familiar:

No sleep till morn, when Youth and Pleasure meet
To chase the glowing hours with flying feet.

Man!
Thou pendulum betwixt a smile and tear.

A change came o'er the spirit of my dream.

Man's love is of man's life a thing apart;
'Tis woman's whole existence.

The isles of Greece, the isles of Greece!
Where burning Sappho loved and sung.

C.W.K.

BYZANTINE, *bi zan' tin,* **EMPIRE,** called also the Eastern, Greek, or Later Roman Empire, once played a very important part in the history and civilization of Europe, for all through the Dark Ages it stood as a bulwark

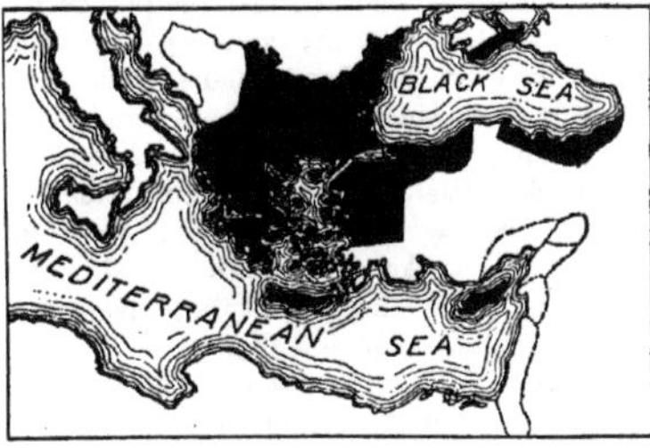

BYZANTINE EMPIRE
Its extent during the reign of Justinian.

against the barbarians and guarded from their inroads the precious legacy of culture which had been left by the ancients.

For almost a thousand years, from the death of Theodosius the Great in A. D. 395 to the fall of Constantinople in 1453, the Byzantine Empire existed as a separate dynasty, with its capital at Constantinople. Theodosius before his death divided his dominions between his two sons, Honorius and Arcadius, and the latter was the first of the Byzantine emperors. He proved a weak ruler, who made few attempts to hold his power, but let it be exercised by ministers. His son, Theodosius II (reigned 408-450), was the next emperor in name, but in reality Pulcheria, his sister, was the ruler. An able ruler she proved to be, carrying on successful war against the Persians and gaining accessions of territory by helping the Western emperor, Valentinian III. Ravages of Attila and the Huns were averted only by the payment of an annual tribute.

The emperors who followed were men of ability, who placed the Empire on a sound basis financially and trained an excellent body of soldiery, so when Justinian came to the throne in 527 he found all things for material advancement ready to his hand. He proved fully able to take advantage of the preparations which these earlier emperors had made, and brought the Empire to the highest point of prosperity and power that it ever attained. His aim was to make his dominions one country in fact as well as in name, and his compilation of laws as well as his conquests were directed toward that end. See JUSTINIAN.

His unfortunate successor, Justin II (reigned 565-578), was harassed on one frontier by the Persians, on the other by the terrible Avars. Most of Italy was lost to the Lombards. The reign of Heraclius (610-641) presented a series of overwhelming reverses, retrieved later by glorious victories. The Persians took Syria, Palestine and Asia Minor, and the invading hordes advanced to a point within sight of Constantinople. Shrewdly gaining time by a humiliating treaty, Heraclius collected his forces and inflicted a defeat upon the Persians at Issus. But a new enemy was gathering strength, whom the exhausted, outlying provinces were too weak to resist—the Mohammedans, with their fanatical missionary zeal. Between 635 and 641 the latter captured Syria, Judea and all the African possessions, but this resulted in good rather than harm to the Empire, which thereby became more truly a unit.

The eighth and ninth centuries saw a peculiar religious struggle—the war of the Iconoclasts (which see). This weakened the Empire at a time when it needed all its strength to oppose its enemies. One ruler tore down images and closed monasteries and convents; the next restored them. Not until the latter half of the ninth century was the controversy finally settled against the Iconoclasts. One of the outstanding figures during these years was the Empress Irene (reigned 797-802), who had formed the ambitious plan of uniting the Eastern and Western empires by marrying Charlemagne. The Bulgarians were very troublesome during the reign of Leo V (813-820), and in the succeeding reign the Saracens gained a firm foothold in Crete and Sicily. See SARACENS.

The Macedonian dynasty (867-1057) established a rule that was on the whole wise and beneficial. The Empire made some distinct gains, as in the reduction of the Bulgarian kingdom to the rank of a province in 1018. But the Seljuk Turks were constantly threatening, and under the weak emperors who followed the Macedonian line they possessed themselves of nearly all of Asia Minor.

The steady advance of the Mohammedan power alarmed all Christian Europe, and during the reign of Alexius Comnenus (1081-1118) began the wonderful movement known as the Crusades (see CRUSADES). As the hosts marched toward Asia Minor by way of Constantinople, the movement necessarily had an important influence on the fortunes of the Byzantine Empire. Alexius wanted help against the Turks, but the vast numbers that responded alarmed him; their depredations within his territory led to serious conflicts, and finally, under later emperors, to open hostility. In 1204 Constantinople was taken by the Crusaders, who established the Latin Empire (1204-1261), with Count Baldwin of Flanders as first emperor. This Latin Empire was never strong, and in 1261 the ruler of Nicaea, Michael Palaeologus, took Constantinople and reëstablished the Greek Empire. His dynasty lasted until the downfall of the Empire in 1453.

The Turks were steadily pressing closer and closer, but by this time it was the Ottoman Turks, who had overthrown the Seljuks. Province after province fell into their hands, and by the beginning of the fifteenth century the emperors were practically vassals of the sultan. In 1453 the Turks, with an army of 400,000 men under Mohammed II, captured Constantinople, and the Byzantine Empire was at an end. They yet held the city at the beginning of the War of the Nations in 1914.

By 1453 the western countries of Europe had developed to a point where they could not easily be overthrown by the Mohammedan conquerors of Constantinople; but had the Byzantine Empire not stood through so many centuries at the gateway of Europe fighting back the barbarous hordes, all Europe might have been swept over. Without it, too, all that was best in the world's past would have been lost, and all that is best in modern civilization retarded for hundreds of years. This is the true significance of the Byzantine Empire.

Byzantine Art. The great Byzantine Empire did not only guard the art and learning of the Greeks and transmit them to Western Europe when its scholars were scattered after the conquest; it developed an art of its own. This dated from the time of Constantine (A.D. 330), and was to a certain extent an endeavor to give expression to the new elements which Christianity had brought into the life of men. From the first the tendency was toward an almost Oriental splendor rather than toward the beautiful simplicity of the Greeks. Richness of material, brilliance of color and decorative effect were the prime aims of the artist. After the Renaissance the Venetian school of painters borrowed many of their principles from Byzantine art, and the result was an unsurpassed richness of color.

In sculpture the aim was the same—toward ornamentation, extravagant costumes, elaboration of details rather than the true proportion of parts and the dignity and correctness of outline for which the Greeks had striven. The naked figure or simple draping were no longer seen, models were not employed, and the figures made little pretense at being true to nature. Despite this artificiality, however, many of the works of the best period, from the sixth to the eleventh century, possessed a considerable beauty and dignity.

One of the favorite branches of art was mosaic work, and in this the artists succeeded in obtaining a characteristically brilliant effect with costly stones. O.B.

BYZANTIUM, *be zan'she um,* the original name of the city of Constantinople (which see). It was changed to Constantinople in A.D. 330 by Constantine the Great, who made it the capital of his empire.

Cc

C, the third letter in the English alphabet, as well as in all other alphabets derived from the Phoenicians. In English it is really an unnecessary letter, except as used in the diagraph *ch,* for its soft sound is represented by *s,* its hard sound by *k.* The Greek *gamma,* with the sound of hard *g,* was transformed by the Romans into two letters, which seem to have been used interchangeably. Thus to this day some authorities write *Caius* Julius Caesar and others *Gaius* Julius Caesar.

The Phoenician letter from which the Greek *gamma* was taken, and through that the modern c, is supposed to have been a picture of a camel's head, which in time was simplified into a symbol much like the figure 7. The Greeks turned this around, and the Romans gradually made its straight lines curved, so that it became the *c* of to-day.

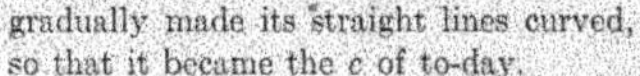

CABAL', an English Ministry which united for a party purpose under the reign of Charles II. The word *cabal,* meaning *secret,* was formed from the initials of the persons taking part in the intrigue, Clifford, Ashley, Buckingham, Arlington and Lauderdale, and was employed as a term of reproach. This faction, though unpopular, was essentially a committee of the Privy Council and proved a forerunner of the modern Cabinet. See PRIVY COUNCIL; CABINET.

Conway Cabal. In American history this name was given to a faction organized in 1777 among a group of officers in the colonial army having for its prime object the displacement of Washington by Gates. Conway joined the "cabal" so earnestly that it was known by his name. Horatio Gates was to be promoted above Washington, with power to supersede him, and Conway was to be inspector-general when the board of war was reconstructed. In the correspondence that followed Washington was abused; affairs became so complicated that Conway resigned and went to France, while Gates and Mifflin were removed from the board of war, the former being assigned to forts on the Hudson. The trickery and meanness of the conspiracy were so apparent that the scheme had the effect of raising rather than lowering him in public esteem.

CABBAGE, *kab'aj,* one of the commonest of garden vegetables, the parent stock from which the Brussels sprouts, the cauliflower and the kohl-rabi (all of which see) have been derived. Originally it was a wild plant growing on the shores of Britain, but the cultivated varieties have spread all over the temperate zones of the world. The leaves only are eaten, and in the common varieties these are compactly folded together into a rounded, dense head sometimes nearly a foot in diameter. The word has come indirectly from the Latin *caput,* meaning *head.* The headed cabbages are of several kinds—the green, the red or

CABBAGE
Showing comparative size of roots and the distance they descend into the ground.

purple and the Savoy, or wrinkled-leaved, varieties. The red cabbage is used chiefly for pickling, while the green is either cooked in various ways and served hot, or eaten raw as a salad. The German *sauerkraut* is made of cabbage salted and pressed in barrels until it ferments slightly. Cabbage is over ninety-seven per cent water, and thus has practically no food value, but it is extremely valuable as a preventive of or remedy for scurvy.

Cabbages are also fed to cattle, and the Channel Islands, the home of a fine breed of cattle, produce a remarkable cow cabbage which often grows to a height of fifteen feet.

Cabbages may be had the entire year, for autumn as well as spring and summer planting yield excellent results.

Cabbage Enemies. There are several fungus growths which attack cabbage plants, often doing great damage; no effective remedy has been found except in a rotation of crops and the planting of the vegetables in fresh soil. Of insect pests the most destructive is the *cabbage worm,* of which there are several species. These are the larvae (young) of the white butterfly; sometimes they practically destroy cabbage gardens by feeding on the leaves and burrowing into the heads of the cabbages. Since it is unsafe to spray cabbages with Paris green, they should be treated with a kerosene emulsion before the heads form.

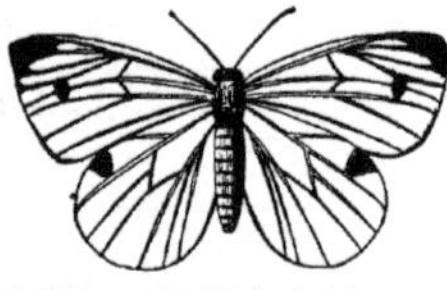

CABBAGE BUTTERFLY

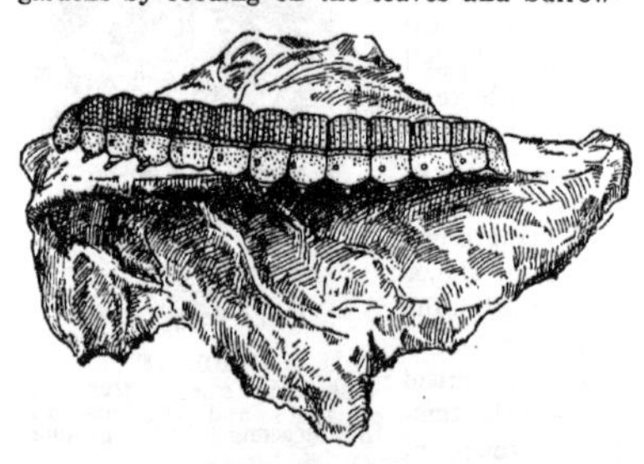

CABBAGE WORM

CABBAGE PALM, a name given to various species of palm trees, whose young, tender leaf-buds resemble the cabbage and when pickled or boiled are used as food. A cabbage palm found in the Southern United States, also called *palmetto,* has fan-shaped leaves and grows to a height of from thirty to fifty feet. In the West Indies it is a tall, graceful tree whose leafy top is sometimes 200 feet above the ground. The cabbage palm is a species of the *areca palm* (see PALM).

CAB'INET, a term generally applied in modern usage to the group of officials, called *ministers,* or *secretaries,* who are heads of the important departments through which the government of a country is carried on, and who form an advisory body to the chief executive, who may be termed a President, a Premier or a Prime Minister.

The Cabinet of the President of the United States. The President is sole head of the executive branch of the government, which is divided into various departments that have been created from time to time by act of Congress. The heads of these departments constitute the President's Cabinet. Though the word *Cabinet* is not mentioned in the Constitution, the establishment of the various subordinate departments is implied in the clause which states that the President may "require the opinion, in writing, of the principal officer in each of the executive departments, upon any subject relating to the duties of their respective offices" (see Art. II, Sec. 2, Clause 1).

Congress is given authority in the Constitution to pass all laws necessary for carrying into effect the powers vested in the government of the United States, and in accordance with this authority the first session of the First Congress established, in 1789, the departments of State (at first called Foreign Affairs), of the Treasury and of War (which then included both military and naval affairs). The secretaries of these departments, and the Attorney-General, who was then an official of the Judicial Department of the government, comprised Washington's first Cabinet. The Attorney-General became a regular Cabinet official in 1814.

The Postoffice Department was organized in 1794, but the Postmaster-General was not included among the Cabinet officials until 1829. The Department of War having been divided in the meantime, the Secretary of the Navy was created a Cabinet officer in 1798. In 1849 the Department of the Interior was established, in 1889 the Department of Agriculture, and in 1903 the Department of Commerce and Labor. This latter department was divided by act of Congress in 1913 and a new Department of Labor was created. Each of these ten departments is described under its appropriate title in these volumes.

The officers who comprise the Cabinet are appointed by the President, but their appointment must be confirmed by the Senate. They do not have seats in Congress, as do the ministers in England, but in addition to their responsibilities as heads of their respective departments, subject always to the overruling authority of the President, they act as advisers to the chief executive. The President and his Cabinet officers work together in close harmony, and there is a mutual interchange of advice and assistance. In 1886, by the Presidential Succession Law (which see), the heads of the executive departments then in existence were made eligible to the Presidency. The greatest prestige is attached to the office of the Secretary of State, but all Cabinet members draw the same salary, $12,000 per year.

The British Cabinet. As now organized, the British Cabinet is composed of a group of men who fill the highest executive offices in the government, and who act as a unit in directing the government. They serve as advisers to the sovereign, are responsible for all of his acts, and as Cabinet members become a committee of the Privy Council (which see). The Cabinet has come to have the character of an executive committee for the party in power, and the members resign from office when that party is defeated.

The official head of the Cabinet is the Prime Minister. The Cabinet officers are appointed by the Crown on the recommendation of the Prime Minister. He himself may hold one or more of these offices, and like the other members of the Cabinet, he has a seat in one of the houses of Parliament.

The number of Cabinet officials varies, but the Cabinet always includes the following ministers: First Lord of the Treasury, Lord Chancellor of England, Lord President of the Privy Council, Lord Privy Seal, five Secretaries of State (for Home Affairs, Foreign Affairs, the Colonies, India and War), Chancellor of the Exchequer and First Lord of the Admiralty. Usually, but not necessarily, there are included the Chancellor of the Duchy of Lancaster, Postmaster-General, First Commissioner of Works, President of the Board of Trade, Chief Secretary for Ireland, Lord Chancellor of Ireland, and the presidents of the Local Government Board, of the Board of Trade, of Agriculture and of Education. See GREAT BRITAIN, subhead *Government.*

In the Dominion of Canada. The Cabinet of the Dominion of Canada is treated in full in the article CANADA, under the subtitle *Dominion Government.*

Consult Bryce's *American Commonwealth*; also any good text-book on civics. Names of the latter may be obtained from school-book publishers.

CABLE, GEORGE WASHINGTON (1844-), one of the best-known of the Southern group of American novelists, whose most popular stories give a remarkable picture of the life and scenery of Louisiana. He was born in New Orleans, began to earn his own living at the age of fourteen, and five years later enlisted in the Confederate army. After the war he joined the staff of the New Orleans *Picayune* and under the pen name of "Drop Shot" became known as the writer of humorous sketches and poems. In 1879 he was writing for *Scribner's Magazine,* and about the same time published *Old Creole Days,* the first of a series of books on which his fame rests. Since 1886 he has made his home in Northampton, Mass., where he founded the Northampton Institute, devoted to the culture and education of wage-earning people.

GEORGE W. CABLE

Cable excels in descriptions of the rivers, swamps and forests of Louisiana, and it has been said that he "knows every mood and whim of the wilderness on the gulf and river." His *Bonaventure, Strong Hearts* and *The Grandissimes* reveal his skill in picturing scenes of river and forest life; *Dr. Sevier, Posson Jone'* and *Old Creole Days* are notable for their vivid descriptions of the life and picturesque beauty of the Old French quarter in New Orleans. His stories have the fault of over-emphasizing the weaknesses of the Creoles, and the author is greater as an artist than as an historian of social conditions. To him, however, American readers owe the interpretation of a phase of Southern life that is little known, and his pen has perpetuated for them scenes of beauty in French New Orleans that are fast disappearing. His other writings include *The Negro Question; John March, Southerner; Bylow Hill;* and *Kincaid's Battery.*

CABLE, SUBMARINE. When you read in your morning newspaper about something that happened across the ocean or on the other side of the world only a few hours earlier, do you ever realize that the telegraph under the sea, which brought the news, is one of the greatest wonders of this wonderful age? Your grandfather probably does, for he can remember when weeks or months were necessary for messages to come from beyond the seas. Of course wireless telegraphy and wireless telephony are even more remarkable, but it is not probable that they will ever make such huge changes in the world as those the cable brought about. Nor is it likely that they will entirely supplant the cable, for wireless systems are helpless during electric storms, while the telegraph beneath the ocean never fails.

In 1864 a New York merchant wishing to buy or sell in London had to wait twenty days or more for an answer to each of his letters. A year later only a few hours were required to exchange cable messages. Thus the cable has made possible an enormous trade between nations, and has brought far-away lands like Australia into close touch with Europe and America.

Cable Problems. To lay a continuous wire along the bottom of the ocean from continent to continent was by no means the only task

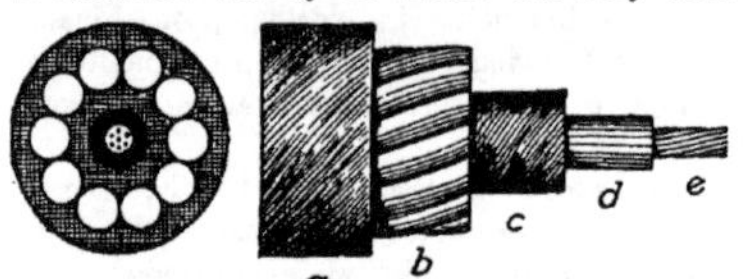

HOW A CABLE IS MADE

(*a*) Outer wrapping of jute, etc.
(*b*) Galvanized wire for protection against anchors, sharp rocks and fishes.
(*c*) Jute wrapping.
(*d*) Gutta percha or rubber.
(*e*) Seven or more copper wires twisted together. This is the part of the cable which carries the message.

The illustration at left shows a cross section of a cable.

The size of an ocean cable is not the same throughout. In the deepest sea a cross-section of the cable is slightly smaller than the one shown above, but in the shallow water its diameter is sometimes three times as great, for there it is subjected to greater dangers.

of the men who gave the world the first cable. The electrical resistance in a wire 2,500 miles long is enormous. (If the reader is not familiar with the meaning of such words as *resistance,* he should refer to the article ELECTRICITY.) But a strong current in a cable is apt to burn the insulation and render the whole length useless, so the current actually employed is so weak that ordinary telegraph instruments cannot be operated by it. A still greater difficulty was found in the fact that the outer coat of protecting wires shown in the illustration makes the cable an electrical condenser (an explanation of which will be found in the article LEYDEN JAR), and stores electricity which resists the passage of the message-carrying current. Thus the first Atlantic cable, completed in 1858, required over an hour to transmit a message of ninety words from Queen Victoria to President Buchanan, and after a few weeks was permanently disabled, presumably by too great strength of current. The illustration explains in detail the construction of the cable now in use.

Instruments. The first man successful in devising a receiver sensitive enough to be operated by the feeble current of the cable was Lord Kelvin. This instrument was the *reflecting galvanometer.* A concave mirror about half the size of a ten-cent piece was hung by a silk thread fastened to a brass ring above and below. A magnetic needle attached to the back of the mirror turned it back and forth, according to the direction of the current in two coils of insulated copper wire, and the direction of this current was controlled by the telegraph key of the operator on the other side of the ocean. The mirror reflected a spot of light on a scale three feet away; if the spot was moved to one side a dot was indicated, if to the other, a dash (see the article TELEGRAPH for an explanation of the Morse alphabet).

Lord Kelvin later invented the *siphon recorder,* which since 1867 has displaced the mirror instrument. The siphon is a small glass tube with its upper end in ink and its lower end upon a strip of paper which passes slowly along while the message is being received. Like the mirror of the older apparatus, the siphon is moved to the right or left by the magnetism of the current, but with a greater degree of accuracy. But because the magnetic action would not be strong enough actually to drag the inked point back and forth on the paper, the siphon is rapidly vibrated up and down by an independent magnet, exactly in the manner of the hammer of an electric bell.

So many improvements have been made in

the apparatus of submarine telegraphy that messages are now sent at the rate of sixty or more words a minute. An early change of importance was the addition of a large tinfoil condenser, which is connected to the receiving instruments and to the ground, and helps to overcome the condensing effect of the cable itself. The automatic transmitter is a device to gain both accuracy and speed; it is operated by a strip of paper in which holes have been punched, representing the dots and dashes of the letters of the alphabet. A still more valnable invention is that which makes it possible to transmit messages in both directions at the same time, as in the *duplex telegraph*. *Quadruplex* or *multiplex* transmission is not yet possible.

Cost of Messages. Though charges for cable service are less than formerly, they are still much higher than for land telegrams. Rates are constantly being changed by the companies, but for regular messages the cost is about twenty-five cents a word from New York to London, Paris, Berlin, Brussels, Amsterdam, Glasgow or Dublin; fifteen cents to Havana; thirty-six cents to the British West Indies; fifty cents to Porto Rico; forty cents to Panama, and sixty-five cents to most of the South American countries. From San Francisco to the Philippines, Japan or China the rate is over a dollar a word, but to Australia not over eighty cents.

Because of the great cost of cable messages it is customary to send them in codes in which a single word may be made to stand for an entire sentence. Recently some of the companies have introduced several varieties of reduced rate messages, which, like telegraph night letters, are designed to keep the lines busy during hours when they would otherwise be idle. Thus a message from New York to London which the company may deliver at any time within twenty-four hours costs but half the regular amount; a letter, not in cipher, to be delivered within twenty-four hours costs seventy-five cents for thirteen words; and a *week end* letter, sent Saturday for delivery Monday, costs one dollar fifteen cents for twenty-five words.

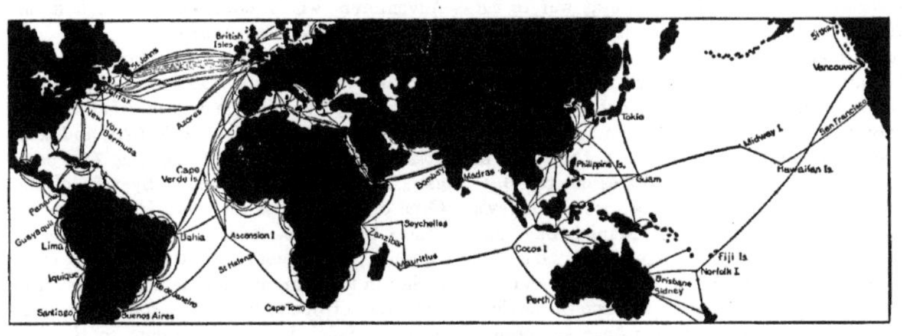

PRINCIPAL OCEAN CABLES IN .1917

The Atlantic Cable. No incident in history gives greater evidence of the power of perseverance than the story of the laying of the first Atlantic cable. When Cyrus W. Field, in 1854, first became interested in the project, the longest cable in operation was between shores less than 100 miles apart. Engineers of the United States navy had previously discovered that the ocean bed between Newfoundland and Ireland was nearly level and composed of soft mud, apparently an ideal resting place for the delicate strand which might link the Old World to the New, but many prominent scientists declared it impossible to lay a cable over two thousand miles long, or to operate it if laid.

Field's first attempt to lay a short cable from Canada to Newfoundland, in 1855, was a failure, but he succeeded in this part of his task the next year. Then followed nine years of discouragement, brightened only by the temporary success of the cable of 1858, which was laid by the British warship *Agamemnon* and the United States warship *Niagara,* after two cables had broken in midocean. When the cable failed after only a few weeks' service, the opposition to Field increased, but again he set about his task of raising capital for the enterprise. The *Great Eastern,* then the largest ship afloat, was chartered in 1865, and after losing a thousand miles of cable on its first trial succeeded in putting in place two cables which permanently joined the two countries, and proved the soundness of Field's ideas. Since then many other cables have been added to these pioneers, as the accompanying map shows. See FIELD, CYRUS W.

The Pacific Cables. Not until 1902 was there direct cable connection between the two shores of the Pacific Ocean. In that year Canada, Australia and New Zealand were joined. The total length of the cable is nearly 8,000 miles, but it is divided into several sections, touching at the Fiji Islands and other less important places. In 1903 a second Pacific cable was put in operation between San Francisco, the Hawaiian Islands and Manila; branches of it have since been extended to Japan and the East Indics. In some of the depths of the Pacific, cable-layers found the waters to be six miles deep.

Cutting Cables in War. International law recognizes the right of a nation at war to cut the cables belonging to its enemies, but most nations agree that neutral cables should not be harmed. A cable connecting a neutral country with the enemy may be cut within territory occupied by the enemy, but some authorities maintain that it should not be cut in the open sea. The United States cut the British cable at Manila in 1898, when Dewey's fleet fought the Battle of Manila Bay, but denied responsibility for damages. C.H.H.

CABOT, *kab'ut,* JOHN (1450-1498) and SEBASTIAN (1476-1557), two well-known navigators, father and son, whose names are prominently connected with the discovery of the continent of North America. They were natives of Italy, but when Columbus made his voyage to the New World they were living at the port of Bristol, England. The wonderful information carried back to Europe by Columbus inspired them to seek a shorter route to the Spice Islands of the East Indies, which they expected to reach by sailing westward on a course that lay to the north of the route followed by Columbus.

John Cabot headed the expedition, for which King Henry VII granted the letters of authority, and it is probable, though not certain, that Sebastian accompanied him. The ship, very similar to the boats used by Columbus, sailed in May, 1497, and on June 24 of the same year Cabot landed on the North American coast in the neighborhood of Cape Breton. The land he claimed for the king of England, convinced that it was a part of Asia. This discovery approaches in importance the great achievement of Columbus in 1492, for the Cabot voyage of 1497 gave England its claim to the mainland of North America, and prepared the way for a long period of exploration and later founding of the English colonies in the New World.

In 1498, John Cabot, possibly accompanied by his son, made a second voyage westward in the attempt to find a shorter route to the East, sailing up the west coast of Greenland until he was blocked by icebergs and passing over Davis Strait to the modern Baffin Land (see NORTHWEST PASSAGE). His homeward route led him as far south as North Carolina. Shortly after his return to England he died. Sebastian Cabot was for several years in the service of the Spanish king, and visited Brazil and the River La Plata in 1525. Later he was appointed chief pilot for England, and in 1554 became governor of the Company of Merchant Adventurers which obtained for England an important trade with Russia.

CABRAL, *kah brahl',* PEDRO ALVAREZ (1460-1526), a Portuguese navigator whose fame rests on one voyage. He landed in what is now Brazil during the winter of 1500-1501, and took possession in the name of the king of Portugal. He had sailed for India by way of the Cape of Good Hope, but was driven from his course by adverse winds and the equatorial current. A Spaniard had reached this coast earlier in the year, and had taken possession in the name of Spain, but Spain could not make good its claim, because according to the line of demarcation the new territory fell within Portuguese limits. Cabral finally reached India, where he made important commercial treaties with native princes. For the immediate effects of Cabral's enterprise, see BRAZIL, subhead *History;* DEMARCATION, LINE OF, with map.

CACAO, *ka ka'o,* the tree from the seeds of which are prepared cocoa, chocolate and cocoa butter. It grows almost everywhere in warm countries, although it is native to tropical America, and is not to be confused with the palm which bears cocoanuts; this is a very different tree. See COCOA and CHOCOLATE, for a discussion of the important products of this tree.

CACTUS, *kack'tus,* a group of interesting plants found in warm, arid regions, whose peculiarities of structure and habits furnish a striking example of adaptation to climatic conditions. The problem of these plants is to secure and retain as much moisture as possible. To prevent loss of moisture by evaporation, they expose as little surface as possible to the sun, and the leaves are greatly reduced. Their fleshy stems also contain a large amount of tissue which is adapted to retaining water. Except for a few African species, the cactus is an American plant, extending southward through Mexico and Central America to Southern South America, and growing abundantly

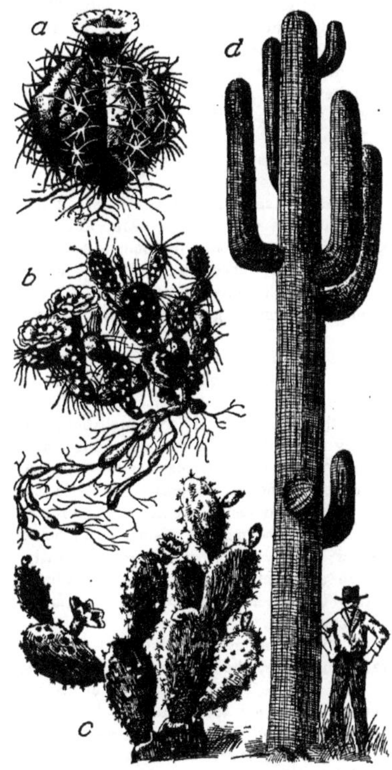

VARIETIES OF CACTUS
(a) Globe Cactus (c) Mexican Opuntia
(b) Hairy Opuntia (d) Giant Cactus

in the United States not far from the Mexican border. About 1,000 forms have been identified. The cactus is the national flower of Mexico and the state flower of Arizona and of New Mexico,

The species vary greatly in form and size. The largest specimens, great numbers of which are found in the drainage basin of the Gulf of California, belong to the *Cereus* group, and are giant plants from fifty to sixty feet in height, with thick, column-like branching stems. Globe-shaped or short cylindrical forms are common among the smaller varieties; sometimes the stems are ribbed, and frequently they bear prominent knob-like growths covered with spines. The flowers are usually large and showy, and the fruit, which stores a great deal of nourishment, is in many cases edible.

The most useful species, belonging to the group known as *Opuntia,* is the *prickly pear,* which has been naturalized in the countries bordering on the Mediterranean Sea under the name of *Indian fig.* In these countries it is a popular food product. Another species of *Opuntia* is cultivated in Mexico to feed the cochineal insect, from whose body a valuable coloring matter is made (see COCHINEAL). Other important species are the *night-blooming cereus* and the *"old-man" cactus;* the latter bears an abundance of white hairs instead of the typical spines, and is well known in greenhouses.

The Spineless Cactus. As a result of several years of experimentation with the prickly pear, Luther Burbank has produced a spineless cactus which is a valuable food plant for both man and animals (see BURBANK, LUTHER). In the process of development the spines were wholly removed from the outside of the plant, and most of the woody fiber from the interior. His spineless cactus grows from eight to sixteen feet high. Its fruit has somewhat the shape of a short, thick cucumber, with flat ends, and it is about two to three inches in diameter. The colors of the fruit are crimson, orange, yellow, purple and white, and its flavors are as numerous as those of the apple. Though produced at one-half the expense of oranges, it sells for about the same price. It may be eaten raw, or made into jams, jellies and syrups. The slabs or stems of the spineless cactus are an excellent stock food, and are estimated to possess about one-half the nutritive value of alfalfa. The joints of the plants make excellent pickles.

Six months after planting, some varieties of the cactus in fairly good soil will produce seventy-five tons of forage to the acre; after the second or third year they will sometimes produce as much as 150 tons to the acre. A cactus leaf ten inches across will sometimes produce thirty to forty full-sized cactus pears. L.B.

CADET, *kay det',* a term applied in the United States, Canada and England to pupils studying at naval and military colleges. The officer of lowest commissioned rank in the United States navy was formerly called a cadet, but the term *midshipman* has now supplanted it. Military cadets are trained in the United States at West Point and in many private military academies; at Sandhurst and Woolwich in England, and at Kingston, Ontario, in Canada. See MILITARY ACADEMY; NAVAL ACADEMY.

Originally the term was applied to the younger son of a noble house, as distinguished from the elder son. The modern words *caddie*, one who carries golf clubs for a player, and *cad*, an offensive or vulgar person, are derived from the word *cadet*.

CADILLAC, *kad e yak'*, ANTOINE DE LA MOTHE (1660-1720), a French nobleman, founder of Detroit, Michigan, and a leading character among the early Frenchmen in America. He served as a captain in Acadia (now Nova Scotia) till recalled to France by Louis XIV, who wanted information about the French colonies, their harbors and their defenses. When he returned to New France he was made commander of Michilimackinac (now Mackinac, Mich.), the second important post in Canada. In 1699 he again went to France to obtain Louis' support in founding a settlement which would become the commercial center of the Northwest.

Louis favored the plan and promised him 200 settlers and six companies of soldiers. But when Cadillac arrived in Montreal, because of the jealousy of the Canadian officials he was able to get only fifty settlers and fifty soldiers. He chose for his town the site of Detroit, and in June, 1701, landed there with his small company. In 1707 he brought the Miami Indians to terms, and founded a post among them in Alabama. When Louis made him governor of Louisiana he punished the Natchez tribe and built a fort in their country and one at Natchitoches as an outpost against the Spaniards. In 1717 the government and trade of Louisiana passed to John Law's Western Company, and Cadillac returned to France, where three years later he died.

CADILLAC, *kad'i lack*, MICH., the county seat of Wexford County, is situated on Little Clam Lake, about fifty miles east of Lake Michigan, toward the northern part of the state. Grand Rapids is ninety-eight miles southwest, Traverse City, on the lake, is forty-two miles north and west, and across the state eighty-five miles southeast is Bay City. Cadillac is on the Grand Rapids & Indiana and the Ann Arbor railways, and is an important supply center for the northern part of the state. The area of the city is nearly four square miles. The population in 1910 was 8,375; in 1914 it was 9,387. Americans predominate, and of the foreign-born, Scandinavians outnumber all others.

Cadillac has extensive lumber interests and a number of allied manufactories. Little Clam Lake and Clam Lake afford fishing for pike, pickerel and perch. Brook trout are taken from the streams of the surrounding country. The noteworthy buildings of the town are the Federal building, erected in 1915 at a cost of $75,000, a Y. M. C. A. costing $50,000, a Carnegie Library costing $25,000, and Mercy Hospital.

Cadillac was settled in 1871. In 1875 it was incorporated as the town of Clam Lake. It received its present name in honor of Cadillac, the French army officer and explorer of Michigan, and was chartered as a city in 1877. The commission form of government was adopted in 1914.

CADIZ, *ka'diz*, one of the most important seaports of Spain, capital of the province of the same name, sixty miles northwest of Gibraltar, on the Atlantic coast. It is well built and strongly fortified, well paved, and, unlike most Spanish cities, is very clean. From the sea the city presents an imposing spectacle, its snow-white buildings apparently rising abruptly from the deeply-blue sea. Nearly every house receives a new coat of whitewash yearly. Though apparently spotlessly clean, Cadiz is far from being a healthful city, owing to imperfect sanitation.

The chief buildings are the great hospital, the customhouse, the old and new cathedrals, the theaters, the bull ring, capable of accommodating 12,000 spectators, and the lighthouse of Saint Sebastian. The Bay of Cadiz, a large basin enclosed by the mainland on one side and a projecting tongue of land on the other, has a good anchorage and is protected by the neighboring hills. It has four forts, two of which form the defense of the grand arsenal, at La Carraca, four miles from Cadiz, where there are large basins and docks.

Cadiz has long been the principal Spanish naval station. Its trade is large, its exports being mainly wine and fruit. The city was founded by the Phoenicians about 1100 B.C., and was one of the chief seats of their commerce in the west of Europe. In the First Punic War it fell into the hands of the Carthaginians, and in the Second Punic War it surrendered to the Romans. In 1587 Sir Francis Drake entered Cadiz harbor and destroyed the warships collected there, a feat which he jokingly described as "singeing the King of Spain's beard." Population in 1910, 67,174.

CADMUS, *kad'mus,* a Greek hero to whom legend gave the honor of having introduced the Phoenician alphabet into Greece. His name has been made proverbial through the familiar—

> "May blessings be upon the head of Cadmus, the Phoenicians, or whoever it was that first invented books."

He was the son of Agenor, king of Phoenicia, and the brother of Europa (which see).

When his sister was carried off by Jupiter in the form of a bull, Cadmus was directed by his father to hunt for her and not to return without her. With his brothers, he set forth on the long quest. One by one the brothers became tired and stopped by the wayside, but Cadmus kept on until told by an oracle that his search was useless. This oracle also directed him to follow a cow which he should shortly meet; and where she should lie down there he was to found a city. He carried out these instructions, and the city which he founded was Thebes in Boeotia. After killing a dragon which guarded a fountain near the site of his proposed city, Cadmus sowed the teeth of the dragon and there sprang up a group of armed men. These men contended with one another until all but five of them fell, and these five became, with Cadmus, the first inhabitants of the new city.

CAEDMON, *kad'mun,* a poet of early England, sometimes called the "Anglo-Saxon Milton." He lived in the latter half of the seventh century, and in his early years toiled as a laborer on the abbey lands at Whitby. What is known of his life is derived from the *Ecclesiastical History* of the Venerable Bede, in which it is told that a vision appeared to Caedmon one night commanding that he sing the praises of God. Thereupon he began to sing verses the words of which he had never heard before. After he awakened he remembered the words he had sung, and added others to them. When the Abbess Hilda, of the monastery at Whitby, learned of his gift, he was admitted into the abbey and made one of the brethren. Several Anglo-Saxon poems are attributed to him, but critics doubt that he was the author of all of them. The most important of these is the so-called *Paraphrase,* giving the story of *Genesis, Exodus* and a part of *Daniel.* Many passages in this remarkable specimen of Old English literature suggest strongly the poetry of Milton.

CAEN, *kahN,* a city in old Lower Normandy, on the banks of the Orne ten miles from the bay at the mouth of the Seine. It is the capital of the department of Calvados, and is 150 miles from Paris. Among its interesting buildings are the castle which William the Conqueror ordered to be built, and several churches in the Norman style, among them Saint Pierre, famous for its spire. It has a university, a public library with more than 100,000 volumes, and a museum. The surrounding district has fertile fields and vineyards, and its quarries produced the stone of which Winchester and Canterbury cathedrals are built. This stone is especially suitable for carving. Its export, and that of Caen's manufactures of lace, woolen and cotton goods, crepe and cutlery, are facilitated by a canal which extends to the sea. The population of the city is nearly 50,000.

CAESAR, *see'z'r,* a title of highest political authority, originally the family name of the first five Roman emperors. So great was the fame of Caius Julius Caesar (which see) that after the death of Nero, the last of that imperial family, the name was adopted by succeeding Roman emperors as a kingly title. This practice continued, and the same title is now found in the German form of *kaiser,* and in the *czar* of Russia.

CAESAR, *see'z'r,* CAIUS JULIUS (100-44 B.C.), one of the greatest men not only in the history of Rome but of all the world. With Alexander the Great and Napoleon he stands as one of those military commanders who built up a great empire by conquest. But he was not only a general; as a statesman he stood preeminent among the men of his time, showing a remarkable grasp of the broad principles of statecraft. As an orator he ranked second only to Cicero. Nor were his abilities as an historian less noteworthy; for his *Commentaries on the Gallic War,* known to every student of Latin, is used everywhere as a textbook not simply because it is written in Latin, but because it presents in unusually simple and vigorous style a straightforward account of events, and is therefore a model of historic writing.

Beginnings of His Rise to Fame. This greatest of the Romans, who "moved through life calm and irresistible like a force of nature," was born of a patrician family, but early showed that his sympathies were strongly in favor of democracy. This does not mean that he believed in government "of the people, by the people and for the people"; the time in which he lived made that impossible; but he did hold firmly to the principle of extending Roman citizen rights to the provincial subjects and to the overthrow of aristocratic privilege.

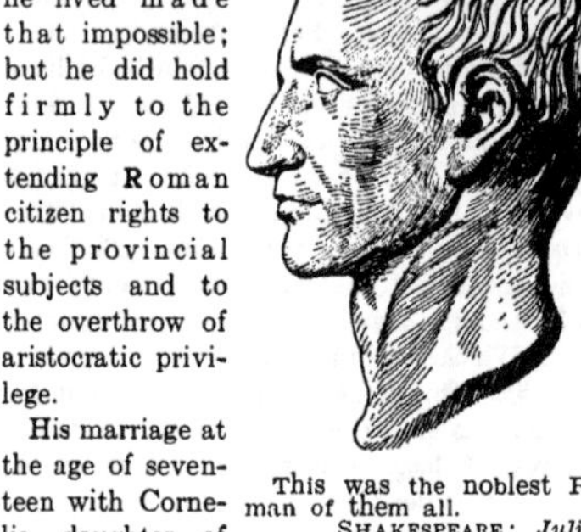

This was the noblest Roman of them all.
SHAKESPEARE: *Julius Caesar.*
The drawing is from a bust in the British Museum.

His marriage at the age of seventeen with Cornelia, daughter of the democratic leader Cinna, strengthened his sympathies, and because he refused to divorce her at the command of the aristocrat, Sulla, he was forced to flee from Rome and give up all his property. On Sulla's death he returned to Rome, and though he left again to study eloquence at Rhodes he was from that time increasingly interested in public affairs, always taking the part of the people. His marriage with Pompeia, a young woman of high social connections who became his second wife in 67 B.C., combined with his personal talents to win him popularity in Rome, and his attempts to confer citizenship on the Latins beyond the Po secured him the sympathies of the Italians. He was elected to various offices, and in all of them won added favor by lavish expenditures and splendid public games.

Connection with Pompey. Though the conspiracy of Catiline brought discredit on all the popular party, including Caesar, his political prospects were not really injured. After a year spent as *propraetor* (governor) in Spain he returned to Rome and formed a coalition with Crassus, a man of enormous wealth and great political ambition, and with Gnaeus Pompey, who had returned from Asia Minor two years before as the great military leader of the day and the idol of the people. Thus was formed the first triumvirate (which see)—not as an organized form of government, but simply as a union to promote the interests of its three members. Through the support which he thus gained, Caesar was elected consul for 59 B.C., and received the command of the province of Gaul for a term of five years, later extended to ten years.

Gallic Campaigns. It was almost an accident that Caesar became a soldier; he was by nature and training a politician rather than a general. But when, in 58 B.C., he went as proconsul to Gaul, it soon became evident that he was a military genius of high order. His courage knew no bounds, and his generalship was so great that in all his nine years in Gaul he only twice lost a battle at which he was present in person. He was adored by his soldiers as perhaps no other man has ever been. In his nine campaigns he reduced the whole province to subjection, and at one time bridged the Rhine that he might pursue the invading Germans whom he had driven back. His campaigns against the Helvetii, the Belgae and the Germans won him immense popularity at Rome, where public thanksgivings were decreed in his honor. Twice during his years in Gaul, in 55 and in 54 B.C., he invaded Britain, but his victories there were little more than nominal, as he left no troops to hold the island.

Civil War. Meanwhile, not everyone at Rome was rejoicing over Caesar's victories; Pompey was becoming ever more jealous and inclining more and more to the aristocratic, or senatorial, party. In 49 B.C. he procured the passage of a decree ordering the disbanding of Caesar's army, but the victorious commander was not prepared to submit, and with his legions crossed the Rubicon—the little stream which separated his provinces of Gaul from Italy. This was virtually a declaration of war, from which there was no turning back; and to this day when an act is based on a momentous decision, it is said the person has "crossed the Rubicon." Little bloodshed was necessary in Caesar's advance upon Rome, for the people everywhere welcomed him, and Pompey, with the Senate and nobles, fled to Greece.

Having made himself in less than three months master of all Italy, Caesar hastened to Spain to overthrow Pompey's legates there. On his return from this expedition he was appointed dictator, an office which he held but eleven days. In January he followed Pompey into Greece and defeated him on the plains of Pharsalia, August 9, 48 B.C. When the news of

From the painting by Gérôme

THE DEATH OF CAESAR.

"And, in his mantle muffling up his face,
Even at the base of Pompey's statue * * * great Caesar fell."

this victory reached Rome, Caesar was appointed dictator for one year, consul for five and tribune for life.

At the Height of His Power. Before returning to Rome Caesar brought to a successful conclusion the war which had been undertaken in order to place Cleopatra on the throne of Egypt, and proved himself not strong enough to resist the charms of the queen who a few years later played so disastrously with the life of Antony. Returning through Pontus, he defeated Pharnaces and informed the Senate of his victory in the laconic dispatch, *"Veni, vidi, Vici"* (I came, I saw, I conquered). He defeated the remaining forces of the party of Pompey at Thapsus in Northern Africa, and Cato killed himself at Utica rather than fall into the hands of this universal conqueror. Now undisputed master of the Roman world, Caesar showed his greatness and magnanimity by pardoning the followers of Pompey. The dictatorship was bestowed upon him for ten years by a grateful people, and his victories were celebrated by magnificent triumphs.

New Jealousies. After his return from defeating the two sons of Pompey in Spain, in the year 45 B.C., fresh honors were conferred upon him. He was made *imperator* for life, and his portrait was stamped upon the coins of the realm. In the correction of the calendar, which had fallen into great confusion, he performed an important service, and he proposed many public improvements, such as founding public libraries, draining the marshes, enlarging the harbor at Ostia and digging a canal across the Isthmus of Corinth. None of these designs, however, was he allowed to carry out, for the aristocrats were still suspicious of him, fearing that he meant to make himself king. At a public festival, indeed, a crown was offered him by Mark Antony, but he refused it—unwillingly, if the reports of historians are to be believed. Antony made reference to this in his masterly funeral oration over Caesar, in the words, according to Shakespeare—

> You all did see that on the Lupercal
> I thrice presented him a kingly crown,
> Which he did thrice refuse: was this ambition?
> Yet Brutus says he was ambitious;
> And, sure, he is an honourable man.
> I speak not to disprove what Brutus spoke,
> But here I am to speak what I do know.

The people had greeted this renunciation with the wildest enthusiasm, but the suspicions of the aristocrats were not quieted, and a plot was formed to kill him.

On March 15, in the year 44 B.C., he was assassinated, receiving over a score of wounds from the daggers of men who had accepted favors at his hands and whom he had believed his friends. Shakespeare in his *Julius Caesar* has made the whole story of the conspiracy and murder as familiar as something that happened yesterday and has described, too, the immediate political result of Caesar's death. The succeeding years showed well how great a disaster it really was to the state, and proved that only under the beneficent rule of one wise man could Rome really prosper.

Character. Any man so outstanding among the men of his own time and of succeeding centuries is certain to be the center of heated argument, and Caesar is no exception to this rule. In his case, however, the disagreement has been almost entirely about his political system, which to some seems a despotism of the worst type, to others a view of empire remarkable for the age in which it was conceived. Almost all historians agree in praising Caesar's kindliness and generosity toward his enemies; his marvelous power of mind which made it possible for him to give an apparently undivided attention to half a dozen matters at the same time, and his irresistible charm of personality. Much of the knowledge of him which has come down to later times is derived from Plutarch's *Lives,* in which Caesar is compared with Alexander the Great; but the Greek historian, according to modern historic judgment, does not always do justice to the most original genius Rome ever produced. W.L.W.

Consult Caesar's *Commentaries on the Gallic War,* which may be had in interlinear translations; Dodge's *Caesar,* in "Great Captains Series"; Ferrero's *Greatness and Decline of Rome,* vol. ii.

Related Subjects. The following articles in these volumes will throw further light upon the story of Caesar or the history of the times in which he lived:

CAFFEINE, *kaf'e in,* or **THEINE,** *the'in,* an important alkaloid whose presence in tea and coffee gives them their stimulating properties (see ALKALOID). Caffeine is an odorless, slightly-bitter solid which crystallizes in slender, silk-like needles. In coffee it occurs in the proportion of from 0.8 to 3.6 per cent, and in tea from 2 to 4 per cent. When taken in moderate quantities it has the stimulating effect of increasing the circulation and arousing one to a greater degree of activity; this is declared by many to be a harmless and gently-stimulating effect. When taken in large quantities, however, it is decidedly injurious, causing nervousness, insomnia, rise of temperature and paralysis of heart action. The person who drinks coffee or tea to excess almost always suffers from nervous disorders. Many authorities to-day believe it would be better for one not to take artificial stimulants of any kind habitually. There is always a depressing reaction of greater or less degree; and the alkaloids are liable to create a need for ever-increasing amounts in order to satisfy the craving for stimulation. See ALKALOID.

CAGLIARI, *ka lyah're,* the capital of Sardinia, a city of 60,000 people, at the southern end of the island. It is connected with nearly every part of Sardinia by rail, and is the port for most of the trade with the mainland. It is 268 miles from Naples and 727 from Gibraltar, and but slightly out of the direct steamer route between those ports.

The city is said to be of Phoenician origin. Among its attractions for the tourist are an amphitheater, botanical gardens, a cathedral, built six hundred years ago by Pisans, and the university, founded by the Spanish king, Philip II, in 1596. Grain, wine, salt from lagoons in the harbor, and goatskins are its chief exports. See SARDINIA.

CAIN, *kane,* the first man in the world to kill another human being, was the eldest son of Adam and Eve. Cain was a tiller of the soil, and Abel, his brother, was a keeper of the sheep. In due time each offered sacrifice to the Lord. Cain offered the fruit of the ground and Abel the firstlings of his flock. "The Lord had respect unto Abel and his offering, but unto Cain and to his offering he had not respect." Cain was angry and killed Abel. As a punishment the Lord pronounced a curse upon Cain and made him a wanderer in the earth. Cain feared that he would be slain, and the Lord placed a mark upon him and commanded that no man should harm him. After this he dwelt in the land of Nod. (See *Genesis* IV.)

CAINE, [THOMAS HENRY] HALL (1853-), an English writer whose novels, rather melodramatic for the most part, have nevertheless a gloomy power which accounts in large measure for their popularity. Caine is of Manx descent—that is, his ancestors came from the Isle of Man; and he had the wisdom to lay the scenes of many of his novels in that little island in the Irish Sea which he knew so well. Most clearly and sympathetically has he drawn his island characters, with their intense natures and their somewhat limited outlook on life, and in this one field his books make a real contribution to the knowledge of various peoples. Caine was born at Runcorn, England, and educated to be an architect, but he found journalism much more to his liking and for six years was a leading writer on the Liverpool *Mercury.* In 1881 he went to London on the invitation of Rossetti, with whom he lived for a year, until the latter's death. His *Recollections of Rossetti,* which picture the poet during this last strange year of his life, is a book of importance.

His first novel, *The Shadow of a Crime,* appeared in 1885, and was followed by *The Deemster, The Bondman, The Manxman, The Christian, The Eternal City, The White Prophet* and *The Woman Thou Gavest Me.* The last four of these, his best-known works, aroused considerable discussion because of the daring with which he handled problems of religion and sex. *The Christian, The Eternal City* and *The White Prophet* also had long and successful runs in dramatic form.

CAIRO, *ki'ro,* the most populous city of Africa, is the capital of Egypt and one of the cosmopolitan cities of the world. Here the people of the East and the West meet, but seldom mingle socially or politically. The streets present a scene of picturesque and vivid coloring, and the ear is assailed by a babel of tongues. Among its inhabitants are Egyptians, Arabs, Nubians, negroes, Turks and representatives of every race in Europe. Cairo might be appropriately called the *city of mosques,* but it received from its founder, Gohar, a general of the Fatimite Caliph, Al Moez, in 973, the name of El Kahira, which means the *victorious city.*

The site is well chosen, on the east bank of the Nile, 130 miles southeast of Alexandria, on the Mediterranean Sea, and eighty-four miles west of Suez by the old caravan route across the desert. It is 5,340 miles from New York, direct through the Strait of Gibraltar, and 2,540 miles from London. On the map of the world it will be found opposite New Orleans, and in the same latitude. The city is divided primarily into two portions, the eastern and the western. These are again

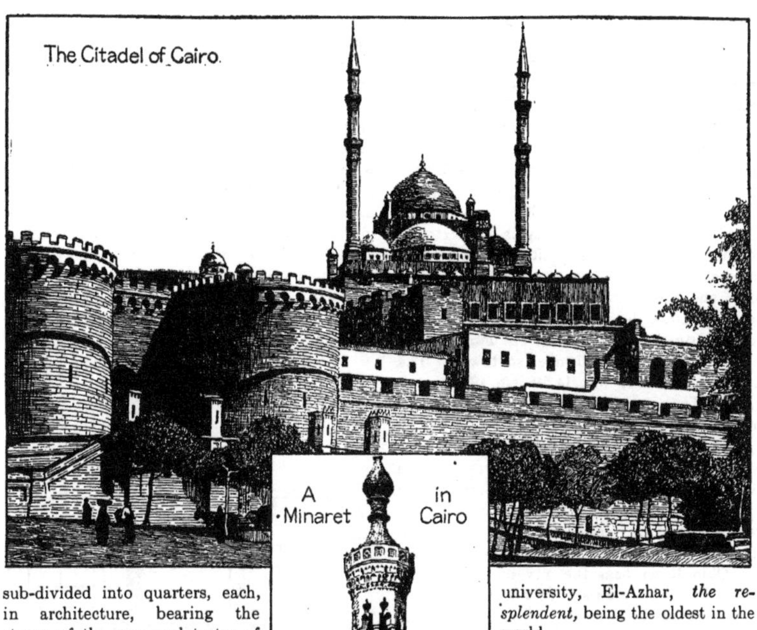
The Citadel of Cairo.

A Minaret in Cairo

sub-divided into quarters, each, in architecture, bearing the stamp of the race and tastes of its inhabitants.

The old portion of the city shows no trace of the advance of Western influences. Crooked, narrow and dirty streets are lined by high stone houses with barred windows. Rearing their turrets and domes above the surrounding dirt and squalor are numerous mosques, some of which are strikingly beautiful and are considered fine examples of the best Arabian architecture. The Gami-ibn-Tulun, erected on what tradition says is the spot where God spoke with Moses, is a magnificent square building surmounted by four minarets and a profusely-ornamented dome. The Gami Sultan Hasan, although comparatively modern, is an imposing building, considered one of the most beautiful of all the mosques of Cairo. As an educational center, Cairo ranks high among Eastern cities, its university, El-Azhar, *the resplendent,* being the oldest in the world.

The modern portion of Cairo is typically French, with broad, well-lighted boulevards, streets lined by well-appointed stores and offices, and, in the center of the European colony, the famous Ezbekia Gardens, covering an area of over twenty acres. Near the garden are the government offices, both British and native, all the important modern buildings and the palace of the khedive, the nominal ruler of Egypt.

Although chiefly noted as a social center, sometimes called an Oriental Paris, Cairo is important as an industrial city. The chief manufactures are textiles, curios to sell to tourists, metal articles, gold and silver work and essences of flowers. Its markets receive ostrich feathers, ivory and hides from the Sudan, shawls from India, tobacco from Turkey; and it is a forwarding center for European manufactured goods. Cairo was formerly an important slave market, but this traffic was finally suppressed in 1877. From 1798 to 1801 the city was held by the French. It was handed over to the Turks by the British, in 1801, and Mehemet Ali seized the reins of government and established a dynasty. In 1915, with the rest of Egypt, Cairo was declared a British possession. Population, 654,500. F.ST.A.

CAIRO, *kay'ro,* ILL., a city in the extreme southern part of the state, situated on an arm of land just above the confluence of the Mississippi and the Ohio rivers. It is the county seat of Alexander County. Chicago is 364 miles northeast, Saint Louis, 148 miles northwest, and Memphis, 170 miles southwest. The population in 1910 was 14,548; in 1914 it was 15,392. The area of the city exceeds two square miles.

Cairo is on the Illinois Central, the Mobile & Ohio, the Saint Louis Southwestern, the Saint Louis, Iron Mountain & Southern and the Cleveland, Cincinnati, Chicago & Saint Louis railroads, and has an an extensive river commerce. The Illinois Central Railway crosses the Ohio River near the city on a fine steel bridge two miles long and fifty feet above high water. It was built in 1888 at a cost of nearly $3,000,000.

Cairo is so situated at the head of all-year navigation of an important river system that it is the natural distributing center and shipping point for a large territory. Six large grain elevators, the Singer Manufacturing Company, creameries, wagon and buggy factories, flour mills, wood-working plants and packing houses are among the important industrial establishments. Quantities of hardwood and cottonwood from the surrounding country are shipped.

Cairo's great problem of recurring river floods has been solved. Levees begun in 1857, many times reconstructed and reinforced, have, since a great flood in 1913, been strengthened sufficiently to withstand any probable height to which the water will rise (see LEVEE). The land is drained and protected from seepage. Among the important buildings are the Federal building, erected in 1871 at a cost of $275,000, the Safford Memorial Library, a government customhouse, a United States marine hospital and the Saint Louis, Iron Mountain & Southern depot. Five miles from Cairo, at Mound City, is a national cemetery.

In Charles Dickens' story, *Martin Chuzzlewit,* Cairo is the city called Eden. Dickens was a losing stockholder in an early financial attempt to create a city on the site of Cairo, and the town was long an unpromising place. Its settlement was a commercial experiment on the part of those foreseeing the great trade possibilities of the location. Three attempts were made, the first in 1818, the second in 1835 and the last in 1851-1854. Cairo was chartered as a city in 1857. The city adopted the commission form of government in 1913. W.F.V.

CAISSON, *kay's'n,* in civil engineering, a water-tight box or casing, used for several different purposes. It is usually constructed of sheet-iron, and so made that it can either be floated or sunk in the water, as desired.

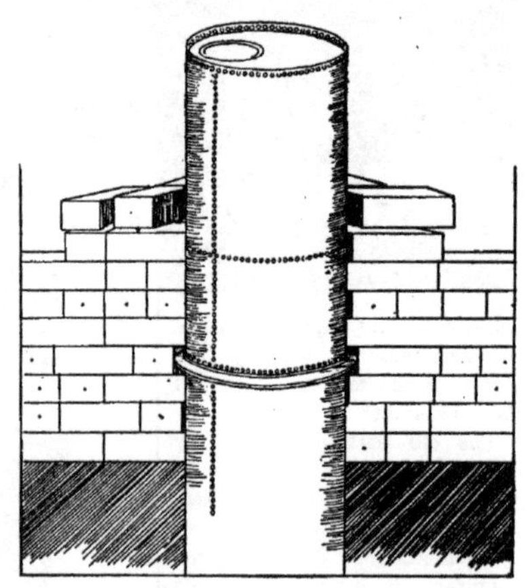

ONE FORM OF CAISSON

Its most general use is for building foundations for bridges, dams, or walls in the water. For this purpose the caisson is sometimes built so that it can be floated to the position where it is desired, and there sunk. A caisson to be used in this way is sometimes made with a wooden bottom which can be separated from the sides. Then when the concrete for the structure is placed inside, the sides can be removed. In other cases the caisson is made without a bottom, and the bottom edges are made so that they will cut into the earth. When the casing is thus sunk into the earth below the water, the water is pumped or forced out, and the earth is taken out by ingenious hoisting devices, the caisson sinking deeper as the work progresses.

For this kind of work it is sometimes necessary to use what is termed a pneumatic caisson. First, a metal caisson open at one end and with cutting edges is sunk into the water, bottom side up. On this is placed another caisson open at the top, and in this the desired masonry is built, thus driving the lower caisson further into the ground. After this is begun compressed air is forced into the lower caisson until the water inside is forced into and out through the earth beneath. Then men are sent into the space inside the caisson through an opening called an air lock, and they continue digging out the earth inside the walls of the casing until the surface is reached on which it is desired to have the foundation rest. The conditions under which these men work are similar to those under which a diver works. The masonry above has grown as the hole below has grown deeper, and finally rests on the rock bottom prepared for it. A similar use of the caisson is common in excavating tunnels under rivers. In this work, however, the caisson is driven horizontally. The great railway tunnels under the Detroit River at Detroit and under the Hudson River at New York were excavated by this method. See TUNNEL.

In sinking foundations for very large buildings a caisson open at both ends, the lower end with cutting edges, is forced into the earth. The caisson sinks lower as the earth inside is removed. When rock bottom is reached, the caisson is filled with concrete. In this manner a concrete pillar resting on solid rock is secured. Most of the "skyscrapers," or very tall buildings, rest upon foundations of this sort.

Caissons, of various forms and in various combinations, are used for raising ships out of the water, and also for floating docks. Cofferdams (which see) are used in preference to caissons where a foundation is to be built in shallow water. See DRY DOCK. F.ST.A.

CALABASH, *kal'a bash,* a name given to a gourd similar to a squash or pumpkin. It is an annual, cultivated in the same manner as a squash. The smooth, hard shells are dried and used as vessels for holding liquids.

Calabash Tree. This must not be confused with the calabash mentioned above. It is an evergreen tree growing in South America, which bears a gourd-like fruit. The shell is very hard and tough and is used for domestic utensils of many kinds. For boiling water it can be placed over a fire many times without being burned. The wood of the tree is used for coach building.

Calabash Pipe, a tobacco pipe made from the stem of a calabash of the squash variety and considered by many smokers to be the mildest and sweetest kind of pipe. It is of comparatively recent origin, the "invention"

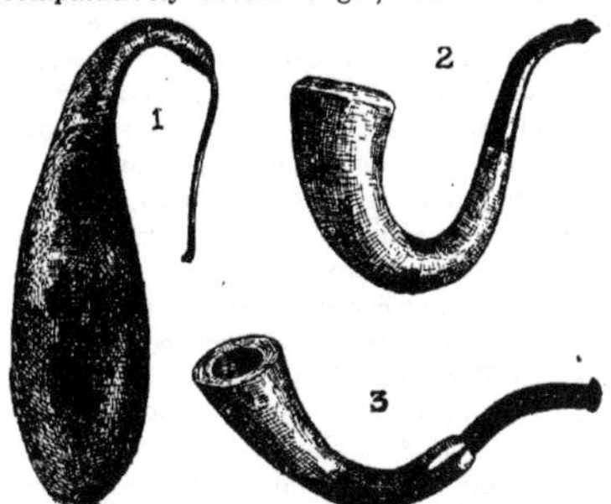

CALABASH: GOURD AND PIPES
(1) Calabash as grown; (2, 3) pipes with stems of differing curves, resulting from varying curves in the gourd.

of a British soldier in South Africa during the Boer War. Having broken the bowl of his pipe he hollowed out the thin end of a calabash gourd and attached a mouthpiece to it. Calabash pipes are now in great demand and are sold at prices ranging from $5 to $25.

CALAIS, *kal ay'* or *kal'is,* a seaport and manufacturing town of Northern France, the nearest continental city to the British Isles. It is important as the chief port for passenger

LOCATION OF CALAIS

traffic between France and England. The town is in the Department of Pas-de-Calais, twenty-five miles southeast of Dover, England, and 185 miles north of Paris.

Calais has an interesting history, dating back to the medieval period of England and France. Old Calais was a celebrated fortress. It was defended by a citadel built in 1560 and by four old forts. A new circle of defenses, strengthened by a deep moat, now surrounds the town and its suburb of Saint Pierre, which are hemmed in by the canal and harbor. Some of the old buildings, including a Gothic cathedral and the town hall, still remain.

In 1346, after the Battle of Crécy, Edward III besieged Calais for nearly a year, when famine forced the town to surrender. Through the entreaty of Queen Philippa, the inhabitants were spared, but were later expelled to make room for English settlers. For two centuries Calais was ruled and largely populated by the English. In 1558, England's war with France ended in the loss of Calais to Francis, Duke of Guise. This was the last of the English possessions on the Continent, and its loss had important effects upon English history (see ENGLAND, subhead *History*). In 1595 it was captured by Spain but was restored to France in 1598 by the Treaty of Vervins. The town has important manufactures of silk and cotton goods and bobbinet lace.

An international tunnel to connect Calais with the English coast has long been considered, and work was actually begun on this project at one time. However, political leaders of both France and England decided that in the event of war between the two nations such a tunnel would be a menace to both, and the work was abandoned. In the War of the Nations (began 1914) the Germans twice tried to take this important seaport, because it would open the way to an invasion of England, but they were unsuccessful. The great German offensive of March, 1918, had this important port as one of its objectives.

CALCIMINE, *kal'si mine,* or cold water paint, has for its basis whiting or carbonate of lime. Calcimine must not be confused with whitewash, which is made from caustic lime. Carbonate of lime, or whiting, will not adhere to a surface; therefore it is necessary to use a binder, which is usually glue, casein or one of the resinous gums. Calcimine is extensively used for inside decorating because of its beauty, cheapness and hygienic qualities.

The United States Department of Agriculture gives the following directions for making calcimine:

Take 16 pounds of dry Paris white (whiting), and pulverize till free of lumps, then mix with one gallon of boiling water. To this add one-half pound of white sizing glue after it has soaked for four hours in one-eighth gallon of cold water. The glue should be dissolved in a glue pot. Any tint desired may be given the calcimine by stirring liquid coloring into the stock. The above recipe will make about two gallons of calcimine weighing 12¾ pounds per gallon. It may be used at once, but is better after standing for half an hour. Ocher, cochineal and logwood are the materials usually used for tinting.

The word is sometimes incorrectly written *kalsomine.*

CALCITE, *kal'site,* a term applied to various minerals, all of which are modifications of crystallized carbonate of calcium. Calcite includes limestone, all the white and most of the colored marbles, chalk and Iceland spar. It is generally white, although shades of gray, yellow, violet, green, red and blue are known. Calcite is found in commercial quantities in Rossie, N. Y., in Copper Mine in the Lake Superior region, in Warsaw, Ill., and also in England and in Androsburg, South Africa. Smaller quantities occur also in other widely-scattered communities.

CALCIUM, *kal'se um,* is never found in a free state, but in its compounds is one of the most abundant and most widely distributed of the chemical elements, forming about 3.5 per cent of the earth's crust. It is a metal whose oxide is quicklime; its hydroxide is slaked lime. Its phosphate forms the main part of the mineral matter of the bones of animals. As a carbonate it appears in calcite, chalk, limestone, marble, coral and shells, and as a sulphate it forms large deposits known as gypsum, alabaster and selenite. Besides, it appears as a constituent in many minerals, such as fluorspar and apatite, and is found in all soils, in the ash of plants, dissolved in sea water and in all springs. The hardness of spring water is almost always due to the calcium compounds dissolved in the water. When quite pure it is a silver-white metal with a high luster, soft enough to be cut with a file, though much harder than lead. It is about one and a half times as heavy as water, and is ductile and malleable. See each mineral named; also DUCTILITY; MALLEABILITY. J.F.S.

CAL'CIUM CAR'BIDE, a hard, brittle, dark gray, crystalline solid. It has a lustrous surface when freshly broken, but quickly tarnishes in the air. The method of making it cheaply was discovered about 1895 by T. L. Willson, a Canadian engineer. It is made on a large scale by heating a mixture of lime and coke in an electric furnace. Its chief use is in making acetylene gas. The process is more fully explained in the article ACETYLENE.

CAL'CULATING MACHINES, machines for performing various arithmetical operations, including addition, subtraction, multiplication and division. These are made in many patterns, the simplest of which is an adding machine only, like the fare register on a street car, which adds but one unit at a time. This contains a series of wheels, usually three, each of which bears numbers from 0 to 9. When the cord which operates the register is pulled the wheel respresenting units moves one step, so as to mark the next higher number. In making a complete revolution this wheel registers ten times, but as the zero appears in the units column, the wheel which represents the tens column is moved forward one figure. Every time the zero in the units column appears in the indicator a small ratchet catches the tens wheel and pulls it forward one figure. When the tens column shows the zero, exactly the same process takes place in the hundreds column.

The typical adding machine used in business to-day is based on the same principle as the fare register. Instead of waiting for the figures in each column to appear, the adding machine may be operated by keys, like a typewriter. For example, to add 350 and 638: the keys, 3 in the hundreds column, 5 in the tens and 0 in the units, must be pushed down like the letters on a typewriter; the machine records them when the operator turns a crank or lever; the number 638 is picked out in the same way, and then by another turn of the lever the machine adds the two numbers at one operation. The process is exactly like that on street car registers, when the register reads 9 or 19 or 29, and another fare is rung up; instead of adding one figure or one column at a time, the machine adds three, four or more by a complicated system of interlocking wheels and ratchets. The cash register (which see) is merely an adding machine of a special form.

On the more complicated forms of calculating machines it is possible not only to add, but also to subtract, multiply and divide. On the simplest machines multiplication is merely repeated addition; on the earliest multiplication machine, to multiply 87 by 5 required four turns of a crank, each turn merely adding 87 to the preceding number. The newest machines are much more complicated, but operate more simply, one or two turns of a crank or pressures on a button being sufficient for almost any problem. Machines are now constructed to figure discounts and interest. See, also, ABACUS; SLIDE RULE.

CALCULUS, *kal'ku lus,* the highest branch of mathematics, studied only in colleges and universities after a thorough preparation in algebra, geometry and trigonometry. It deals with the projectives of variable quantities and with their rate of change. The two problems below are given to indicate briefly the character of the work in calculus. The first is in the practical field of mechanics, the second a theoretical problem in astronomy:

A ball is fired up a hill whose inclination is 15°; the inclination of the piece is 45°, and the velocity of the projectile is 500 ft. per sec.; find the time of flight before it strikes the hill, the path made by the ball, and the distance of the place where it falls from the point of projection.

A particle describes an ellipse under an attraction always directed to one of the foci; it is required to find the law of attraction, the velocity and the periodic time.

CALCUTTA, *kal kut'a,* called by the Hindus KALI GHATA, meaning "the landing place leading to the temple of the goddess Kali." It is a city of temples and palaces and, next to London, is the most populous of all the cities of the British Empire. Capital of the presidency of Bengal, and until 1912 the capital of the great Indian Empire, it stands on the left bank of the Hooghly, an arm of the sacred River Ganges, eighty-six miles from the sea. By mail routes Calcutta is 11,120 miles from New York; from London to Calcutta is 7,400 miles. Like most Oriental cities, Calcutta is divided into two parts, forming practically two cities. One is modern and occupied by Europeans; this section has broad, well-kept streets, fine residences and many beautiful public buildings. The other, the native quarter, has narrow, crooked streets, squalid houses, and nothing but the brilliant coloring affected by the natives in their dress and ornaments to relieve the drab monotony of dirt.

LOCATION MAP

Commerce. From a commercial point of view the city is admirably situated, having at its doors the great waterways of the Ganges and Brahmaputra, from whose fertile valleys it receives vast quantities of produce. The rivers provide cheap means of communication between inland provinces and the sea. The port of Calcutta extends nearly ten miles along the river and is one of the busiest in the world. Over one-third of the entire foreign trade of India passes through the city, which in addition to its almost perfect water transportation has the terminals of three great railway systems. The exports exceed the imports in value, having reached the total of almost $300,000,000 against imports valued at nearly $175,000,000 in 1915. The principal exports are opium, hides, skins, grain, indigo, cotton, silk and jute. The center of the jute industry is at Howrah, on the opposite bank of the Hooghly and connected with Calcutta by a floating bridge 1,530 feet in length. The imports are chiefly manufactured goods, machinery, textiles, salt and liquor.

Buildings and Monuments. Between the river and the fashionable suburbs is the Maidan, or park, the pride of the city, containing Fort William, the largest fortress in India. The park is the parade ground for the wealth and fashion of Calcutta. It contains numerous statues and monuments, among which is a plain tablet indicating the position of the terrible Black Hole of Calcutta (which see). On the north side of the park stands the former residence of the viceroy of India, built by Lord Wellesley, brother of the Duke of Wellington, in 1799 at a cost of $5,000,000. Alongside the park, for a distance of two miles, runs the most famous street in Calcutta, the Chowringhee. The residence of Warren Hastings, one of the most prominent figures in the history of India, is kept by the Indian government as a house of entertainment for Indian princes. The town hall, supreme court, mint and general post office are also notable buildings. In the graveyard of the old cathedral is the tomb of Job Charnock, who founded the city on August 24, 1690.

History. The early history of Calcutta is closely bound up with that of the East India Company (which see). The small settlement established in 1690 gradually grew in importance; neighboring villages were absorbed and formed the nucleus of what was to be not only the center of commercial activity but the political capital of an empire that was to compensate Britain for the loss of its American colonies. In 1756 the town was captured by Surajud Dowlah, native ruler of Bengal; terrible atrocities were committed, but it was later rescued by Clive and Admiral Watson. In 1773 it became the seat of government, but in 1912 Britain restored the capital to Delhi, the original capital. Population in 1911, including suburbs, 1,222,313. See map, accompanying article ASIA.

F.ST.A.

CALDER, *kal'der,* JAMES ALEXANDER (1868-), a Canadian educator and political leader, for many years a leading school official of the Northwest Territories and since 1905 a conspicuous member of the Saskatchewan government. With the exception of Premier Walter Scott he is perhaps the foremost Liberal in the province. He was born in Oxford County, Ont., and was graduated with honors from Manitoba University in 1888. From 1891 to 1894 he was principal of the Moose Jaw high school, then for six years was inspector of schools of the Northwest Territories, and from 1901 to 1905 was deputy commissioner of education for the Territories. In 1905 he was elected to the Saskatchewan Assembly, and was appointed provincial treasurer and minister of education in the Scott ministry. In this position he was chiefly responsible for the organization and development of Saskatchewan's public school system. He continued to sit in the assembly, and was generally recognized as Premier Scott's chief lieutenant. In 1912 he temporarily assumed the duties of minister of railways in addition to his other offices, and since 1913 has been minister of railways.

CALEDONIA, *kal e doh'ni a,* the poetic name for Scotland, but historically the name by which the northern portion of that country and its inhabitants first became known to the Romans. The Roman Agricola invaded the country in 82, defeated the Caledonians in 83, and again in 84 in a battle of which a detailed description is given by the historian Tacitus. Later attempts to subdue the Caledonians were unsuccessful. They were the Scots and Picts of early English history who harassed the Britons after the Romans withdrew.

OLD CALEDONIA

In his *Lay of the Last Minstrel,* Scott has an invocation to Scotland in which the poetic name for the country is used:

> O Caledonia! stern and wild,
> Meet nurse for a poetic child!
> Land of brown heath and shaggy wood,
> Land of the mountain and the flood,
> Land of my sires! what mortal hand
> Can e'er untie the filial band,
> That knits me to thy rugged strand!

CALENDAR, *kal' en dar,* a systematic division and record of time. All calendars must recognize two great natural divisions of time—the day and the year, both of which are solar periods. The month, on the other hand, seems to have been suggested by the periods of the moon. The week is an arbitrary division whose origin is ascribed to various causes.

Among ancient peoples there were many differences in calendars. The Egyptians divided the year into twelve months of thirty days each, and added five days at the end of the year. The year was thus too short by nearly six hours. The Greek year included twelve months of thirty and twenty-nine days, alternately. This arrangement gave a year of only 354 days, 11¼ days short of a solar year. To make up this difference an extra month was added in alternate years, except that every eight or nine years the extra month was omitted.

The old Roman calendar was even more confused. The earliest known system among the Romans was a year of ten lunar months, four of which had 31 days, the remainder only 30. This year of 304 days was too short by about one-sixth. Each year thus began two months earlier in the season than the last, five natural or solar years forming six calendar years. About 700 B.C. January was added to the beginning and February to the end of the year. This made a year of 354 days, as in the ancient Greek calendar. Every second vear a month of 22 or 23 days had to be added to compensate for the 11¼ days lost. About 450 B.C. the months were rearranged in their present order, but March was regarded as the beginning of the year. Owing to the addition of a day for "luck," the calendar year was then 366¼ days long. In a period of twenty-four years, therefore, the calendar would be twenty-four days ahead of true time. A law was passed authorizing the pontiffs to deduct twenty-four days at any time in the last eight years of a period of twenty-four. This power was abused to such an extent for personal ends, such as collecting revenue, that by the time of Julius Caesar the calendar equinox differed from the astronomical equinox by three months.

In 46 B.C., Caesar decreed the reform of the calendar. The periods of the moon were disregarded, and the year was divided into twelve months of 31 and 30 days alternately, except February, with 29 days. February was to have 30 days every fourth year. The seasons were readjusted to the calendar by making the year 46 B.C. fifteen months long, from October 13 to the second following December 31. This year is known as the "year of confusion." The new year then began on January 1. In the reign of Augustus one day was taken from February and added to August, in order that the month named for Augustus might be as long as that named for Julius Caesar. The lengths of the following months were then rearranged to prevent three months of 31 days occurring in succession.

The Julian calendar provided a year of 365¼ days, or 11 minutes 14 seconds longer than the true solar year. This difference led to a gradual change in the calendar date of the equinox, until about 1580 it fell on March 11, ten days earlier than it should have occurred. In 1582, Pope Gregory XIII determined to correct this discrepancy by dropping ten days from October. By this arrangement the day that would have been October 5, 1582, in the Julian calendar became October 15, and the next equinox was thus restored to its proper date. Under the Julian calendar, a single day was gained in about 400 years, because the calendar year was a few minutes longer than the solar year. To correct this discrepancy, the Gregorian calendar omits the additional day in February in century years not divisible by 400. Thus 1600 was a leap year, but 1700, 1800 and 1900 were common years. The year 2000 will be a leap year. The difference between the civil calendar and the astronomical year now averages only 25.95 seconds, which amounts to a single day in 3,330 years. Thus in 4912 the calendar will be one day ahead of the sun.

The Gregorian calendar was adopted almost immediately by the Roman Catholic nations of Europe. The states of Germany retained the old style until 1700, and England did not change until 1752. Russia and the other countries in which the Greek Church is supreme still retain the Julian calendar. In the present century, the old style is thirteen days behind the new style; thus January 14, 1917, in England, was January 1, 1917, in Russia.

For ascertaining any day of the week for any given time within two hundred years from the introduction of the New Style, 1753, to 1952, inclusive.

YEARS 1753 TO 1952									Jan.	Feb.	Mar.	Apr.	May	June	July	Aug.	Sept.	Oct.	Nov.	Dec.
1753g 1754d	1781g 1782d	1800e 1801a	1828q 1829a	1856q 1857a	1884q 1885a	1900g 1901d	1928h 1929d	a	4	7	7	3	5	1	3	6	2	4	7	2
1755e 1756p	1783e 1784p	1802b 1803c	1830b 1831c	1858b 1859c	1886b 1887c	1902e 1903a	1930e 1931a	b	5	1	1	4	6	2	4	7	3	5	1	3
1757c 1758f	1785c 1786f	1804h 1805d	1832h 1833d	1860h 1861d	1888h 1889d	1904k 1905f	1932k 1933f	c	6	2	2	5	7	3	5	1	4	6	2	4
1759g 1760q	1787g 1788q	1806e 1807a	1834e 1835a	1862e 1863a	1890e 1891a	1906g 1907d	1934g 1935d	d	2	5	5	1	3	6	1	4	7	2	5	7
1761a 1762b	1789a 1790b	1808k 1809f	1836k 1837f	1864k 1865f	1892k 1893f	1908l 1909b	1936l 1937b	e	3	6	6	2	4	7	2	5	1	3	6	1
1763c 1764h	1791c 1792h	1810g 1811d	1838g 1839d	1866g 1867d	1894g 1895d	1910c 1911f	1938c 1939f	f	7	3	3	6	1	4	6	2	5	7	3	5
1765d 1766e	1793d 1794e	1812l 1813b	1840l 1841b	1868l 1869b	1896l 1897b	1912m 1913e	1940m 1941e	g	1	4	4	7	2	5	7	3	6	1	4	6
1767a 1768k	1795a 1796k	1814c 1815f	1842c 1843f	1870c 1871f	1898c 1899f	1914a 1915b	1942a 1943b	h	7	3	4	7	2	5	7	3	6	1	4	6
1769f 1770g	1797f 1798g	1816m 1817e	1844m 1845e	1872m 1873e		1916n 1917g	1944n 1945g	k	5	1	2	5	7	3	5	1	4	6	2	4
1771d 1772l	1799d	1818a 1819b	1846a 1847b	1874a 1875b		1918d 1919e	1946d 1947e	l	3	6	7	3	5	1	3	6	2	4	7	2
1773b 1774c		1820n 1821g	1848n 1849g	1876n 1877g		1920p 1921c	1948p 1949c	m	1	4	5	1	3	6	1	4	7	2	5	7
1775f 1776m		1822d 1823e	1850d 1851e	1878d 1879e		1922f 1923g	1950f 1951g	n	6	2	3	6	1	4	6	2	5	7	3	5
1777e 1778a		1824p 1825c	1852p 1853c	1880p 1881c		1924q 1925a	1952q	p	4	7	1	4	6	2	4	7	3	5	1	3
1779b 1780n		1826f 1827g	1854f 1855g	1882f 1883g		1926b 1927c		q	2	5	6	2	4	7	2	5	1	3	6	1

Ecclesiastical Calendar. The Church calendar is regulated partly by the sun's position and partly by the moon's phases. Such days as Christmas, the Feast of the Circumcision and the Nativity of the Blessed Virgin are *fixed days,* originally set according to the solar calendar. Such days as Easter, however, are known as *movable feasts,* their date being determined by the moon's periods. Thus Easter is the first Sunday after the first full moon upon or following the vernal equinox. The other principal movable feasts are Ash Wednesday, Palm Sunday, Good Friday, Ascension and Pentecost.

Hebrew Calendar. Hebrew chronology begins with the Creation, which is supposed to have taken place 3,760 years and 3 months before the beginning of the Christian Era. To the number of years in the Gregorian calendar, 3,761 must be added to find the number of the Hebrew years; thus 1917 in the Gregorian calendar is 1917+3761, or 5678. The Hebrew year ordinarily consists of twelve lunar months: Tisri, Hesvan, Kislev, Tebet, Sebat, Adar, Nisan, Yiar, Sivan, Tamuz, Ab and Elul. These months are alternately 30 and 29 days long. At intervals of nineteen years, however, an extra or *embolismic* month of 29 days, called Veadar, is inserted between Adar and Nisan, and Adar is given 30 days instead of 29, as usual.

Mohammedan Calendar. The Mohammedans reckon time from the Hegira, which occurred in A. D. 622. The Mohammedan year consists of twelve lunar months, or 354 days. As the calendar year is much shorter than the solar

TABLE OF DAYS

[illegible]		[illegible]		[illegible]		[illegible]		[illegible]		[illegible]		[illegible]	
Monday	1	Tuesday	1	Wednesday	1	Thursday	1	Friday	1	Saturday	1	SUNDAY	1
Tuesday	2	Wednesday	2	Thursday	2	Friday	2	Saturday	2	SUNDAY	2	Monday	2
Wednesday	3	Thursday	3	Friday	3	Saturday	3	SUNDAY	3	Monday	3	Tuesday	3
Thursday	4	Friday	4	Saturday	4	SUNDAY	4	Monday	4	Tuesday	4	Wednesday	4
Friday	5	Saturday	5	SUNDAY	5	Monday	5	Tuesday	5	Wednesday	5	Thursday	5
Saturday	6	SUNDAY	6	Monday	6	Tuesday	6	Wednesday	6	Thursday	6	Friday	6
SUNDAY	7	Monday	7	Tuesday	7	Wednesday	7	Thursday	7	Friday	7	Saturday	7
Monday	8	Tuesday	8	Wednesday	8	Thursday	8	Friday	8	Saturday	8	SUNDAY	8
Tuesday	9	Wednesday	9	Thursday	9	Friday	9	Saturday	9	SUNDAY	9	Monday	9
Wednesday	10	Thursday	10	Friday	10	Saturday	10	SUNDAY	10	Monday	10	Tuesday	10
Thursday	11	Friday	11	Saturday	11	SUNDAY	11	Monday	11	Tuesday	11	Wednesday	11
Friday	12	Saturday	12	SUNDAY	12	Monday	12	Tuesday	12	Wednesday	12	Thursday	12
Saturday	13	SUNDAY	13	Monday	13	Tuesday	13	Wednesday	13	Thursday	13	Friday	13
SUNDAY	14	Monday	14	Tuesday	14	Wednesday	14	Thursday	14	Friday	14	Saturday	14
Monday	15	Tuesday	15	Wednesday	15	Thursday	15	Friday	15	Saturday	15	SUNDAY	15
Tuesday	16	Wednesday	16	Thursday	16	Friday	16	Saturday	16	SUNDAY	16	Monday	16
Wednesday	17	Thursday	17	Friday	17	Saturday	17	SUNDAY	17	Monday	17	Tuesday	17
Thursday	18	Friday	18	Saturday	18	SUNDAY	18	Monday	18	Tuesday	18	Wednesday	18
Friday	19	Saturday	19	SUNDAY	19	Monday	19	Tuesday	19	Wednesday	19	Thursday	19
Saturday	20	SUNDAY	20	Monday	20	Tuesday	20	Wednesday	20	Thursday	20	Friday	20
SUNDAY	21	Monday	21	Tuesday	21	Wednesday	21	Thursday	21	Friday	21	Saturday	21
Monday	22	Tuesday	22	Wednesday	22	Thursday	22	Friday	22	Saturday	22	SUNDAY	22
Tuesday	23	Wednesday	23	Thursday	23	Friday	23	Saturday	23	SUNDAY	23	Monday	23
Wednesday	24	Thursday	24	Friday	24	Saturday	24	SUNDAY	24	Monday	24	Tuesday	24
Thursday	25	Friday	25	Saturday	25	SUNDAY	25	Monday	25	Tuesday	25	Wednesday	25
Friday	26	Saturday	26	SUNDAY	26	Monday	26	Tuesday	26	Wednesday	26	Thursday	26
Saturday	27	SUNDAY	27	Monday	27	Tuesday	27	Wednesday	27	Thursday	27	Friday	27
SUNDAY	28	Monday	28	Tuesday	28	Wednesday	28	Thursday	28	Friday	28	Saturday	28
Monday	29	Tuesday	29	Wednesday	29	Thursday	29	Friday	29	Saturday	29	SUNDAY	29
Tuesday	30	Wednesday	30	Thursday	30	Friday	30	Saturday	30	SUNDAY	30	Monday	30
Wednesday	31	Thursday	31	Friday	31	Saturday	31	SUNDAY	31	Monday	31	Tuesday	31

year, the Mohammedan new year constantly retrocedes through the seasons. For example, in 1914 the new year fell on the date corresponding to November 19 in the Gregorian calendar, in 1915 on November 8. In the course of thirty-two and one-half years the Mohammedan new year completes its backward course through the seasons. The Mohammedan calendar also divides the years into cycles of thirty years each. Of each cycle nineteen are regular years of 354 days, and eleven years have an extra day. This method of computation is nearly as accurate as the Gregorian calendar. The Mohammedan calendar, based on the periods of the moon, has an error of one day in about 2,400 years; the Gregorian calendar, based on the revolutions of the earth around the sun, has an error of one day in 3,330 years.

Perpetual Calendar. A perpetual calendar is one which shows the day of the week for any date. In the simple form above, the letters after each year refer to the table of months, while the figures in the table of months refer to the table of days. For example, to find on what day of the week December 25, 1900, fell, look in the table of years for 1900. The letter *g* is attached. Then look for *g* in the table of months; in the parallel line, and under December, is the number 6. The twenty-fifth day in column 6 of the table of days is Tuesday. Christmas, 1900, fell on Tuesday. W.F.Z.

CALGARY, *kal'ga ri,* the largest city in the province of Alberta and the largest city in Canada between Winnipeg and Vancouver. In 1891 Calgary was a community of 3,800 people, and ten years later the population was only 4,800. The town was prosperous, but had as yet given little sign of future greatness. Since 1901, however, it has grown from a frontier town to a great manufacturing and trading city. In 1911 it was the home of 43,704 people, a total which made it the tenth city in size in the Dominion. By 1916 the population had increased to 56,302, making it the eighth city in the Dominion.

Ideal Location. The reason for this tremendous growth is Calgary's location, which is ideal both for manufacturing and for distribution of products. The city is nearly midway between Winnipeg and the Pacific coast, being 811 miles east of Victoria, B. C., and 860 miles west of Winnipeg. It is 194 miles south of Edmonton, the capital of Alberta, 480 miles west of Regina and 478 miles southwest of Saskatoon. The three great Canadian transcontinental railways, the Canadian Pacific, the Canadian Northern and the Grand Trunk Pacific, give Calgary connection with these and many other cities.

Calgary, by its location, is the natural supply station for the rich surrounding stock-raising and farming sections and also for the mining districts of the Rocky Mountains. The country tributary to this city is estimated at 150,000 square miles, an area equal to nearly three times the state of Michigan. The marvelous growth of the Canadian West, added to Calgary's natural advantages, has also brought many manufacturing establishments to the city. Within a short distance is a supply of coal sufficient to supply cheap power and fuel for centuries. The prospects for natural gas are equally favorable. Electrical power for manufacturing purposes is developed on the Bow River a short distance west of the city.

Calgary's site is remarkably attractive. Just to the west of the city are the foothills of the Rocky Mountains, and beyond them are visible the snow-capped peaks of the mountains themselves. Within easy reach of Calgary are Banff and other famous mountain resorts. The city lies on a plateau, at an altitude of 3,437 feet. On one side the plateau is cut by the Bow River, whose cool waters flow from the pass traversed by the main line of the Canadian Pacific Railway. At Calgary the Bow River is joined by one of its tributaries from the southwest, the Elbow River.

Trade and Industry. To Calgary are sent the raw products of the prairies and the mountains—cattle, horses, hogs, sheep, grains, coal, clay, stone, lumber. The city forwards a part of these products to other centers, but an increasing proportion is used in local factories. The value of the manufactured articles in 1916 was estimated at more than $12,000,000, as compared with less than $8,000,000 in 1910 and only $600,000 in 1900. The factories of the city produce flour and other cereal products, biscuits and bread, meat, brick and tile, building materials, cement, harness and other leather goods, soap, carriages and wagons, and lumber and foundry products of all kinds.

Calgary is an important railway divisional point, with large repair shops which add more than $1,000,000 to the city's manufactures. As a distributing center the city's importance is indicated by the presence of nearly two hundred wholesale houses.

Noteworthy Buildings and Other Features. The city is laid out, for the most part, in regular rectangles, with wide streets. In and about the town are a number of pretty recreation spots, including Victoria Park, 103 acres, Reservoir Park, eighty-three acres, and Shaganappi Park, ninety-six acres. Conspicuous among the city's buildings are the Hudson's Bay Company's Stores, built at a cost of $1,000,000; the *Herald* Block; the Canada Life Assurance Building; and the Canadian Pacific's Palliser Hotel, which cost $1,500,000. The city has over thirty public schools, a number of Roman Catholic separate schools, four colleges, the Provincial Normal School and the Provincial Institute of Technology.

Government and History. Since 1909 Calgary has had government by a commission of three members, one of whom is mayor and has charge of the city hall departments; one of public works; and the third, of public utilities. The city owns its street railways, water works, electric light and power plant, municipal market and municipal paving plant. It is one of the chief stations of the Royal Northwest Mounted Police, and is the seat of the Calgary judicial district.

The year 1883 is usually given as the date of the founding of Calgary. More than a century before, in 1752, the French built Fort La Jonquière not far from the present site of the city, and later Fort Bow, or Bow Fort, was a trading post for many years. The building of the Canadian Pacific Railway led to the formation of a settlement here, first known as *Calgarry,* in honor of a small estate of that name in the Hebrides Islands. The word is of Gaelic origin, and means *clear, running water.* Calgary was incorporated as a city in 1894. See colored map, article CANADA. J.MC C.

CALHOUN, *kal hoon',* JOHN CALDWELL (1782-1850), an American statesman, one of the most able and distinguished advocates of the "states' rights" doctrine. For forty years he was prominent in national affairs, and in all that time no word was spoken against his character, no doubt was ever felt as to his great ability. But he had championed what proved an impossible cause, striving for a union which should be outwardly strong but should have as its elements practically independent states. As Webster's slogan, famous the world over, was "Liberty and Union, now and forever, one and inseparable"; as Jackson's was "The Union must and shall be preserved," so Calhoun's was "Liberty dearer than Union."

Calhoun was born on March 18, 1782, near Abbeville, S. C., of Scotch-Irish parents. His early education was meager, for the family was poor, but he studied hard and in 1802 was able to enter the junior class at Yale College. Graduating with honors two years later, he continued his study of the law, was admitted to the bar and began to practice at Abbeville. In the next year (1808) he was elected to the state legislature, and there he proved so noticeably above the average in ability that in 1811 he was sent to Congress. Almost at once he became prominent as an adherent, strangely enough, of Henry Clay, and the two "war hawks," as they were called, did much to bring on the War of 1812 with England. A national bank, a strong navy, a protective tariff—all these Calhoun favored in his early years, for he was then a nationalist rather than a states' rights adherent. As Secretary of War from 1817 to 1825, in the Cabinet of Monroe, he displayed remarkable ability.

JOHN C. CALHOUN

In 1824 by a large majority he was elected Vice-President on the ticket with John Quincy Adams, but by this time his views were beginning to shift, and from the date of his reëlection to the Vice-Presidency in 1828, this time under Jackson, his theories as he held them to his life's end were fairly well established. The agricultural states of the South were smarting under what they called the "tariff of abominations"—a protective measure which favored the manufacturing states of New England at their expense; and Calhoun came forward as their champion with his "South Carolina Exposition," which declared that no state could be bound by a Federal law which it regarded as unconstitutional. This led to a sharp break with Jackson, and the nullification question was for years in the forefront of public notice. Calhoun resigned the Vice-Presidency in 1832 and entered the Senate, and though Clay arranged a compromise which prevented open warfare on the nullification question, Calhoun remained a bitter critic of Jackson's administration. See NULLIFICATION.

From 1832 to 1843 he served in the Senate, in 1844 was Secretary of State under Tyler, and in the next year reëntered the Senate, where he served until his death. At first his advocacy of slavery was merely incidental to his faith in states' rights, but gradually he came to look upon slavery not merely as necessary but as desirable. He was largely responsible for the admission of Texas to the Union and therefore for the swiftly-following Mexican War, but he ardently opposed that conflict. To the last he was active in his efforts for his beloved South, writing a final great speech in 1850 when he was so weak and ill that he had to allow it to be read by a colleague. His *Disquisition on Government* and *Discourse on the Constitution and Government of the United States* were remarkable discussions of constitutional questions. A.MC C.

Consult Von Holst's *Life of Calhoun*; Dodd's *Statesmen of the South*.

CALICO, *kal'i ko,* **AND CALICO PRINTING.** When your grandmother was a girl she thought herself very fortunate to have a dress or two of pretty calico "print." Nowadays we have so many cloths beautiful in texture as well as in color or design that calico is no longer a favorite for gowns. But grandmother's calico cost perhaps seventy-five cents a yard, while to-day from five to ten yards of it may be had for that price, so because of its decreased cost many other uses have been found for it. The cheaper grades of some materials, such as *percale* and *cretonne,* are really calico, though people would not buy them as readily if they were called by that name. In England the word *calico* includes plain white cotton goods, but in the United States and Canada it refers only to the less expensive cloth stamped with designs in color

Printing. Most calico has its pattern on one side only. This is because it is printed, very much like a newspaper. A copper roller on which is engraved the design is pressed tightly against a padded cast-iron cylinder so placed that the cloth passes between the two, like clothes going through a wringer. The color, mixed with starch or flour to keep it from blotting, is transferred from the engraved part of the roller to the cloth; a knife called the *color-doctor* keeps the smooth part of the roller clean. If more than one color is to be printed there must be a roller for each. Sometimes, to give the appearance of a cloth in which the design is woven, calico is printed on both sides, the pattern matching exactly.

Before the days of rotary presses calico was printed by hand work. The design was carved in relief on a wooden block. Anyone who has worked with a rubber stamp knows how hard it is to make an even impression and to set the stamp in exactly the right place each time. Imagine, then, the difficulties of a calico printer of the eighteenth century, with a wooden stamp ten inches long and six inches wide. Yet many very beautiful prints were made, some of them in several colors. It was this process of hand stamping which gave calico its name, for it was originally introduced into Europe from Calicut, in India, where such stamping originated.

Dyed Calico. The pretty blue and white calicoes so often seen in Dutch patterns, and some more expensive cloths of white and one color, are prepared in several ways. Sometimes the fabric is printed with a *mordant* (see DYEING AND DYESTUFFS) and then dyed, the dye taking effect only where the mordant is located. In another process the cloth is first dipped in a mordant, then printed with a chemical which takes away the mordant in the spots which are to remain white, then dyed. Still another method prints the cloth with wax or clay or a chemical through which the dye cannot penetrate. All of these ways of marking the pattern give it a color which will not fade.

The ancient Egyptians were familiar with some of the cloth-printing processes which employ mordants, and the Chinese have from the earliest times known how to print with wood blocks. The Japanese used to form charming patterns by dipping the leaves of trees in dye and pressing them on cotton cloth; they also had a method of painting designs through stencils. Even the Incas, the Indians of Peru, knew how to print cloth before Europeans learned the art from India. K.A.G.

THE STORY OF CALIFORNIA

CALIFORNIA, called the GOLDEN STATE because of the epoch-making discovery of gold in 1848, is one of the Pacific coast states and the second largest in the American Union. Its name, according to most authorities, is taken from that of a fabled island near the equator in the far western seas. Some scholars, however, believe that it is a contraction of the Spanish words *caliente forno*, meaning *hot furnace*, which early travelers used to describe the arid, hot regions of the south. Exceeded only by Texas in area, California has 158,297 square miles, of which 2,645 are water surface; and its 2,377,549 inhabitants (1910) place it twelfth in rank according to population (see subhead, *Population*, below). In this great state Illinois, Iowa and Ohio might be placed, and there would be room to spare, but the population of those three well-settled states is more than five times that of California. Except Texas, it has the greatest north-and-south length of any of the states—750 miles; and its width is about 200 miles.

To the north lies Oregon, to the east are Nevada and Arizona, the latter separated from it by the Colorado River; to the south is that part of Mexico known as Lower California, and to the west stretches the Pacific Ocean. Though there is in the shore line a very decided bend, the eastern boundary line bends, also, so the state is approximately the same width throughout all its great length. The coast line of California is over 1,000 miles in length, and only Florida has a greater stretch of coast. Indeed, there is so little difference between the two in this respect that some authorities are inclined to give the preference to California.

To many who have never visited it California stands as the romance state of the West—the land "flowing with milk and honey" toward which their desires tend; while to those who live there, its "native sons," it is indeed the El Dorado, or golden land, by reason of its climate, its wonderful scenery, its unrivaled fertility and its mineral wealth. The popular name of the state; the term *Golden Gate*, applied to its finest harbor; the golden poppy, which is its state flower; the name El Dorado, given to one of its counties, all point to the position which it holds in the eyes of its admirers.

Surface Features. As might be expected, this vast state has within its borders the widest diversity of surface. Great peaks almost 15,000 feet in height, depressions well below sea level, far-stretching valleys and river lands, deserts and some of the most productive regions in the world—all these are to be found. Nothing seems to be lacking except the stretches of plateau land which the middle Western states have in such plenty. Despite this variety, however, a relief map of California shows a fairly simple, orderly arrangement. To the east and to the west there are high mountain ranges which are joined together toward the ends of the state, and between these ranges is a valley so extensive that it almost merits the name of a plain. The mountains, and consequently the valley, extend in a generally northeasterly and southwesterly direction, almost parallel with the coast.

The Mountains. The Coast Range, as it is called, is in reality a system, made up of a number of smaller ranges. It is in fact a coastal range, running down in many places close to the sea and breaking off abruptly in steep cliffs, rendering the scenery of much of the shore region strikingly picturesque. As a whole, this western mountain system has a height of from 3,500 to 8,000 feet, but outstanding summits, as San Bernardino and San Jacinto, attain altitudes of over 10,000 feet.

To the east, across the great valley, is the far loftier Sierra Nevada, or Snowy Range, the highest and steepest range in the United States. This runs along the eastern border of the state for over 400 miles, and has an average width of about fifty miles. Above its rugged central mass, which has its lowest pass about 5,000 feet above sea level, rise almost a score of peaks above 10,000 feet in height. Chief among these is Mount Whitney, the highest summit in the United States if Alaska be not counted, but its altitude of 14,898 feet is but little greater than that of Fisherman Peak, which is 14,448 feet (see WHITNEY, MOUNT). Other lofty summits are Mount Corcoran, 14,093 feet; Kaweah Peak, 13,752 feet; and Mount Brewer, 13,886 feet. Long stretches lie entirely within the region of eternal snows, and small glaciers are numerous, while of glacial lakes, most of them 8,000 feet above sea level, there are no fewer than 1,500. On the east the Sierra Nevada drops abruptly to the great plain, but to the west deep canyons have been worn by the rushing rivers on their way to join the Sacramento or the San Joaquin. The most famous of these steep valleys, because the most accessible, is Yosemite (which see), but it would have in magnificence of scenery a number of rivals if the others were equally

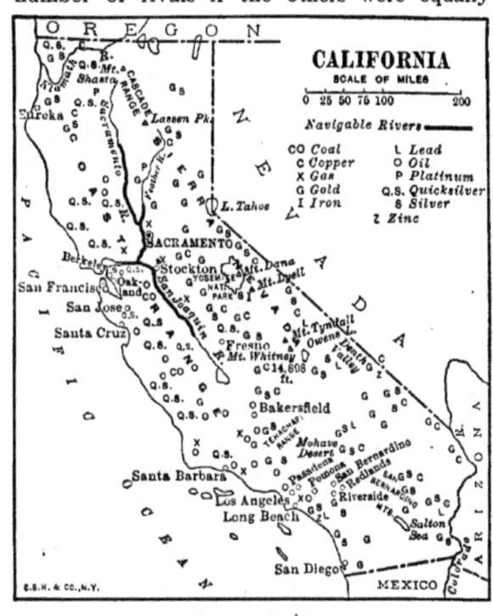

CALIFORNIA

This map shows boundaries, chief rivers, mineral products, leading cities and the highest point of land in the state.

well known. Of recent years the Hetch-Hetchy Valley has attracted wide attention because of the determined attempt on the part of the city of San Francisco to dam up its river to furnish a city water supply. Accounts of the manifold beauties of the region appeared in newspapers and periodicals all over the country, and efforts were made to prevent what seemed to many the vandalism of such a course, but without success (see SAN FRANCISCO).

To the north the Coast Range and the Sierra Nevada are connected by spurs of the Oregon Cascades, which contain a number of prominent peaks, among them Mount Shasta (which see), far famed for its grandeur and beauty. To the south the great border ranges are joined by the Tehachapi Mountains, the northern boundary of that varied region known as Southern California.

Southern California. This comprises not only some of the most productive sections of the state, but in the remote southeast some of the most desolate and arid waste land in all the country, as well. The mountains are not so high as those farther north, but most of the surface is more or less broken, and occasional peaks, as Mount Lowe, have acquired more than local fame. In the western part, toward the sea, the valleys are fertile, and when irrigated with the waters of the mountain streams that flow through them they become wonderful garden spots, producing in great abundance sub-tropical fruits of all kinds. Of this populous seaward region Los Angeles is the metropolis.

The eastern part of Southern California is made up of the Mohave and Colorado deserts, over whose alkaline sands little rain ever falls. Part of this waste region is a high plateau, but near its northern limit, not far to the southeast of Mount Whitney, the highest point in the United States, is the deepest depression in all North America—the fitly-named Death Valley (which see), almost 300 feet below sea level at its lowest point.

The Great Valley. This is the largest valley west of the Rocky Mountains, and the most important. It is 400 miles in length and ninety in width; its area of 18,000 square miles is almost equal to the combined areas of New Hampshire and Vermont. Fairly level throughout most of its extent, this valley has a slight dip toward the center from both north and south, and each section is drained by a great river (see subhead *Waters,* below). Only in one place is the wall of mountains which shuts it in broken through—on the western side, where the two rivers have cut an opening perhaps a mile wide.

The southern part of the valley is not so well watered as is the northern, nor do its waters drain into the Pacific, but into marsh-bordered lakes. Some of this southern valley region is typical Western desert, with its gleaming alkali soil and its dusty sagebrush.

Waters. Naturally in so mountainous a state, the only great rivers are in its one large valley. There are two of these, the Sacramento and the San Joaquin; both rise in the Sierra Nevadas, but hundreds of miles apart. Finding the valley floor, the former flows south and the latter north until they join about sixty miles northeast of San Francisco and make their way to the Golden Gate. At intervals along their course they are joined by tributaries from the east, which flow down from the great Sierras brimming with water from the melting snow. These rapid mountain torrents are not only fine hiding places for the brook trout, beloved of fishermen, but furnish water for irrigation and power for electric plants. At certain seasons of the year both the Sacramento and the San Joaquin overflow their banks and have real flood plains.

The southern part of the valley, beyond the area drained by the San Joaquin, has its mountain streams as well, but these find their way not to the sea but to brackish lakes, of which Tulare and Buena Vista are the most important. West of the Coast Range, which here withdraws from the shore line, the Salinas River drains the west-central part of the state, and in the north are the Klamath and the Eel.

Of the mountain lakes of California, some of which rival the famous Swiss lakes in beauty, the most widely known is Lake Tahoe, on the borderland of Nevada. This largest of the glacial lakes has a length of twenty miles and a depth of 1,500 feet, and its clear, icy waters abound in trout.

Climate. There is nothing of which a Californian is prouder than of his climate, and for the most part with reason. The state extends from about the latitude of Boston to that of Savannah, and might be expected to have a wide variety of climate. So indeed it has, but the differences in latitude have far less to do with this than has altitude or distance from the sea, and the northern end of the state has almost as mild a climate as has the southern. Nowhere save in the high altitudes of the Sierras is anything like the winter cold of the Eastern coast states known, and everywhere except in those localities live stock can remain out-of-doors throughout the year and find plenty of grass for grazing. The valleys are so protected by the mountains that high winds are practically unknown, and there is a large proportion of sunny days. In these inner valleys the range of temperature is greater than on the coast, the summer heat sometimes becoming oppressive. Each part of the state believes in the superiority of its own particular climate, even the intensely hot southeastern desert section claiming for itself great healthfulness because of the lack of moisture in the air; but most visitors agree that it is in the southern section, in the neighborhood of San Diego, that the climate most nearly approaches perfection. There the average temperature is 68° in summer and 54° in winter. A favorite boast of the Californian is that it is necessary to "sleep under blankets" the year round, and this is apparently true, for even in the warmer sections the nights are delightfully cool. Throughout the state, except in the cold mountain regions, flowers bloom out-of-doors at all seasons.

Instead of being divided into a winter of intense cold and a summer of great heat, with transition periods between, the year has but two seasons, a wet and a dry, the former lasting from late October to April, the latter throughout the rest of the year. The rainfall, however, varies decidedly in different localities, decreasing gradually from north to south. In the extreme north it may be as much as fifty-one inches in a year; at San Francisco it is about twenty-three inches, at Los Angeles fifteen inches, and at San Diego ten inches. In the mountainous region and in the Great Valley it is sufficient for nearly all agricultural purposes, but south of the Tehachapi Mountains, even in the most fertile sections, irrigation is necessary to successful soil-cultivation. Dwellers in these irrigated regions believe that they notice an increasing humidity of the air as a result of this continued application of water. Snow is everywhere practically unknown, except high in the mountains.

The climate of California has been one of the chief features in its remarkable development, not only because it has made possible the growing of crops which can be raised in few other localities of the United States, but because it has attracted to the state hundreds of thousands of people. Some go because their health demands it, others just for the delight of living in a climate without extremes of heat and cold.

A feature of the climate of California is its numerous heat and moisture belts, many of them but a few miles in extent. Only a study of local conditions will make clear the causes of some of the peculiar climatic variations.

A north-and-south mountain range, for instance, may force the winds from the ocean to drop their moisture, and a region of heavy timber and luscious grass, like that about Humboldt Bay, is the result. The great interior valleys, much hotter in summer than the coast region, are constantly being cooled by currents of ocean air which forces itself through the passes; but not all of the regions profit alike. Fresno receives comparatively little of the cool air, and thus has a summer climate fitted to the production of the raisin grape, while certain sections farther north feel more of the cooling effect, and are great wine-grape regions. There may be, too, a protecting foothill range which creates a frostless belt and

The illustration is of "General Sherman," the giant sequoia, pronounced by the United States government the largest tree in the world, measured by the number of cubic feet of wood in it. Twenty men with outstretched arms, hand in hand, can just encircle it.

THE KING OF TREES

makes orange-growing possible outside of the regular orange region. This may be seen in Tulare County. Again, as at Watsonville, a sharp rise of ground will condense ocean currents into cool fogs of long duration, and an apple-growing region is created.

Distinctive Plants and Animals. It is the climatic conditions which determine the plant life of a region, and in lesser degree its animal life, and since California is in a sense an "inland island," walled in by mountains and differing from even its neighboring states in climate, it has developed a vegetation largely its own. All the growths except the forest trees, which are somewhat independent of surface moisture, are at their best during the rainy season, and throughout much of the state little green is to be seen during the summer months except in the irrigated sections. The rounded hills covered with a thick mat of sun-dried grass, dotted everywhere with dusty gray-green scrub oaks, are most familiar summer sights, but in the rainy winter everything is a riot of grass and flowers.

In so large a state there are certain to be distinct zones of vegetation. The coast regions of the north have gigantic forests, mostly of cone-bearing trees, which are here found in greater variety and profusion than anywhere else on earth, and the mountains, both north and south, have such forest growths above a certain altitude. Yellow pine, sugar pine, fir, spruce, cedar and hemlock—all are there; but most interesting and distinctive of all is the sequoia, which grows nowhere but in California. Of this evergreen tree there are two species, the redwoods and the big trees, specifically so called. The former occur as magnificent forests over much of the northern region; the latter, the oldest and largest trees on earth, form scattered groves on the western slopes of the Sierra Nevadas. It is difficult for the mind to grasp the age of these forest monarchs—some of them, authorities declare, must have been great trees in the days when Abraham went out from Ur of the Chaldees. See REDWOOD; SEQUOIA.

Other trees which grow in few places elsewhere in North America are the eucalyptus, with its narrow, drooping leaves; the Monterey cypress, a scraggy, picturesque, Japanese-looking tree much used for ornamental purposes in gardens; the madroña, a brilliant splash of color when its thin, bright-red outer bark peels off and shows the vivid green beneath; and the pepper tree, one of the most beautiful trees in the world, with spreading branches, feathery leaves and long clusters of bright-red berries.

To enumerate the flowers which grow in California would be an almost endless task, so profusely do they flourish everywhere. The beautiful California poppy grows wild, but is much grown as a garden plant as well. The poinsettia, with its brilliant red leaves surrounding the insignificant little flower, has come to be widely known, and is grown in greenhouses farther east, where it is popular as a Christmas decoration, but nowhere else does it become so large or so gorgeous as in California. That may be said as well of other plants—of the geranium, for instance, which is elsewhere a "pot plant" or a garden plant of moderate size, but in California attains a vinelike growth and stretches to the very roofs of the bungalows. But the glory of the state is its roses. Roses of every variety, even those perfect kinds which the Easterner looks upon only as hothouse flowers, grow everywhere, and many a house is covered to its roof with thick-blossoming vines.

Animals. In the days before the earliest white men entered California, many animals to-day unknown there roamed in its forests and mountains—the tapir, the wild horse, the lion and even the elephant. And the white men, when they came, found the cougar, the elk, the coyote, and, most characteristic of all, the grizzly bear. To-day the bears are found only in the wildest parts of the state, while the cougar hides in the rocks or the *chaparral,* as the thick, brushy growths are called. Coyotes are still numerous, and in the mountains deer are to be found. Among birds, the quail, so much hunted for food, is most widely known, but the species not found elsewhere are the more interesting. These include the road runner (which see), a peculiar woodpecker and many song birds.

One of the animals which every visitor to California wants to see is the seal, or more properly, sea lion. Some of these sea lions attain a great size, occasional specimens weighing as much as a ton. Various places along the rocky coast are still their haunts, but nowhere can they be observed better than on their great rookeries opposite the Cliff House in San Francisco.

Agriculture. The products of California are as varied as its surface, and few are the temperate or semi-tropical crops which cannot be grown, at least in some sections. For a time wheat was the great product, and on the immense ranches more of this grain was grown than in any other state of the American Union. Later the farms were made smaller, and agriculture was practiced on a more intensive scale. The valleys of the Sacramento and San Joaquin are remarkably fertile, and it is impossible to give in detail the crops that are there brought to a high state of development. Vegetables, grains, forage plants, hops and fruits of all kinds flourish, even orange groves appearing in the foothills to the northernmost limit of the Great Valley.

It is in Southern California that oranges and other tropical fruits are chiefly grown, and the production of them has come to be one of the foremost industries of the state. Two-fifths of the oranges of the world and over three-fourths of those of the United States are grown in this one state, and lemons, grape fruits and citrons are also produced in abundance. Pineapples are grown to a slight extent, but Florida produces ninety-nine per cent of America's output. Almost nowhere else in America are olives produced, although the industry promises success in Arizona, and nowhere else in all the world is there nearly so large an area devoted to the growing of raisin grapes as in California. Almonds and walnuts are also produced in large quantities, and of the orchard fruits, as plums, peaches, apples, pears, cherries, apricots and quinces, there is not one which cannot be successfully grown. Many of these fruits have not the fine flavor of the Eastern varieties, but all are of large size and great beauty. The black cherry, the prune and the apricot, the last-named produced almost exclusively in California, are especially important, and are shipped eastward in great quantities. Berries also flourish, and may be had almost the year round, and grapes are grown in immense quantities for the making of wines.

Other crops include sugar beets, in the production of which California ranks not lower than second among the states; tobacco, a fairly-new experiment, but successful; and cotton, which is grown in the newly-reclaimed Imperial Valley.

Irrigation. In almost every part of the state a certain amount of irrigation is necessary, and in the southern portion it is considerable. In the Sacramento River Valley, also, irrigation is rapidly extending. Statistics show that in all about one-fifth of the crops are produced from irrigated land, but this proportion is bound to grow as new areas are opened up for

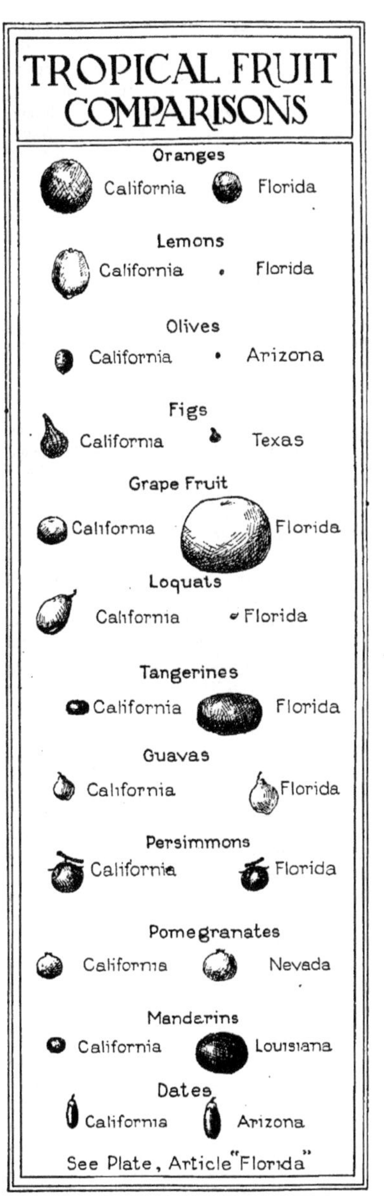

cultivation. Streams are numerous and supply much of the water needed, but many of the irrigation enterprises are private and depend upon wells. As Arizona is to-day profiting much from waters drawn from the Colorado River, so California has profited in recent years. The Imperial Valley, in the southern part of the state, was a parched desert, but investigation proved that its soil was good, and not the alkaline substance of some of the near-by deserts, and water was turned upon it. The result was immediate, and the fertile region thus opened up proved inviting to settlers. At Yuma, Arizona, the United States Reclamation Service has in process of construction a reservoir which will provide irrigation facilities for thousands of acres on the California side

Stock-Raising. In the days of the Mexican occupation of California stock-raising was the chief industry, for the wide grass lands needed no irrigation. Vast herds of cattle and flocks of sheep grazed on the public lands, but with the development of other forms of agriculture this particular branch declined in relative importance, the sheep coming into especial disfavor because the grass which they cropped did not grow again. The raising of stock is still a thriving industry in parts of the state, however, the cattle, horses, sheep, mules and swine having a combined value of about $125,000,000. The poultry business is also profitable, and the country around Petaluma is one of the chief chicken-growing centers in the world.

Mining. It is impossible to think of California without recognizing its mineral wealth, and particularly its gold, for it was the discovery of gold in 1848 that revolutionized the history of the state. In the one year of 1849 more than 40,000 people started out for the new gold fields by the overland route, and many of them perished in the deserts before they reached their goal (see GOLD). But obstacles greater than deserts and gigantic mountain chains are necessary to keep man from the places where gold may be found, and the mining industry grew rapidly. The earliest methods were most primitive, for much of the gold lay near the surface, and a pick, shovel and washing pan were all these pioneers needed. As the shallow gravel streams were exhausted, mining, properly so called, was introduced, and remarkably rich veins of gold were opened up, most of them on the western slope and in the foothills of the Sierra Nevadas. The "mother lode" proved to be one of the richest gold-bearing quartz veins in the world. Between 1850 and 1859 about $55,000,000 worth of gold was produced annually in the state; from that time the decline has been steady, but not sudden, for about $20,000,000 worth was mined in 1915. Since 1897 California and Colorado have been the chief gold-mining states, now one, now the other, holding first place. In all over $1,500,000,000 worth has been taken from the mines—far more than any other one state or country has produced. It is little wonder that California is called the *Golden State.*

Silver, too, is mined, though in much smaller quantities, and the annual yield of copper is valued at about one-fourth the annual yield of gold. California also produces considerable lead and zinc, a little coal, various clays useful in brick and tile making, two-fifths of the world's output of quicksilver, and millions of barrels each year of petroleum. Especially remarkable has been the development in petroleum production, the number of barrels increasing between 1899 and 1913 from less than 2,000,000 to 98,000,000, an output which gave California first rank among the states. More recently Oklahoma has outranked it in the value of its output of petroleum, but not in quantity. In its yield of borax California ranks first, Death Valley being the principal source of supply. Tungsten has been produced in paying quantities in the Mohave Desert, since 1915, and the output is rapidly increasing.

Fisheries. Whaling is no longer the flourishing industry it once was, but it has not died out, and San Francisco is now the chief whaling port in the world. Detailed statistics as to the fisheries, whether coastal or inland, are not available, but it is certain that California is one of the foremost states in the taking and canning of salmon. In addition to this, sturgeon, smelt, halibut, soles, mackerel, cod and bass are caught in the coast waters, as well as the characteristic sand-dab, red snapper and pompano. The tuna, one of the largest of the mackerel tribe, is also plentiful; it has of late years come into increased popularity, and is now being canned as extensively in the state as is salmon. The fishes of the streams include Rocky Mountain trout of large size, black bass and shad. Oysters, mussels, clams, crabs, shrimps and crawfish increase the output of sea food materially.

Manufactures. A state with such immeasurable natural resources has endless possibilities for manufactures, but the almost total

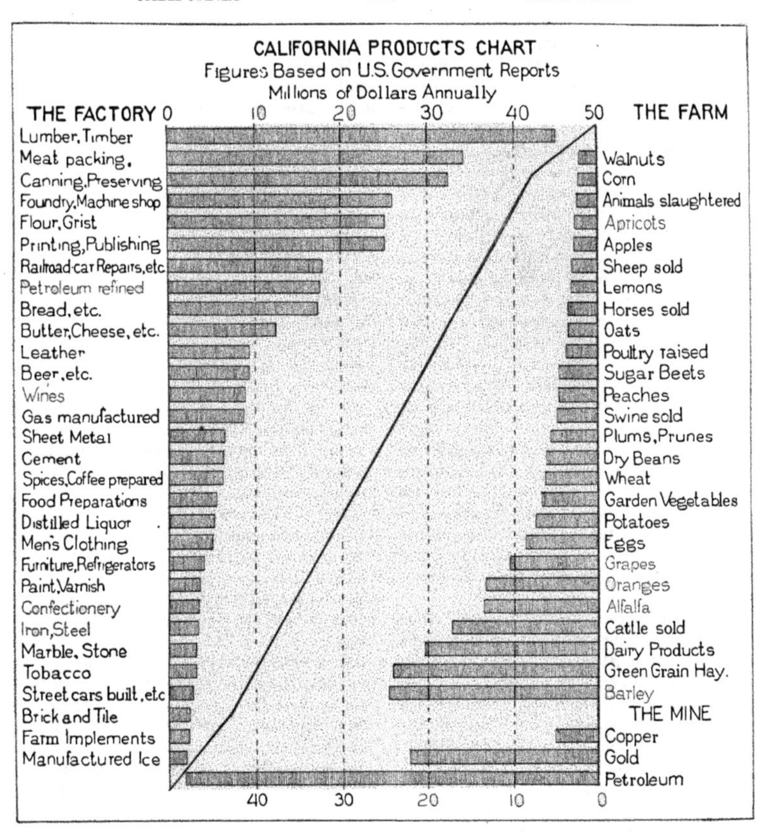

absence of coal and the great distance from any of the large coal fields for a long time retarded factory growth. The mountain streams furnished water power, though this was not everywhere available, but with the discovery of the rich oil fields manufacturing entered on a period of development. Chief of the industries are the cutting of timber and the making of lumber products, in which California ranks sixth among the states. Almost half of the lumber output is California redwood, and because this is particularly well adapted to the making of shingles this has become a branch industry of considerable size. In all, the lumber and timber products for a single year are valued at about $45,000,000.

Other industries, in the order of their approximate value, are meat packing, centered chiefly at San Francisco and Los Angeles; the canning of fruits and vegetables, peaches and asparagus ranking first in value; the making of flour, and the printing and publishing industries. San Francisco has one of the largest iron works in America. California leads all the states in the making of vinous liquors and of olive oil, but these industries cannot rank in the value of their products with those listed above. Sugar refining is becoming increasingly important, the total output of beet sugar being about 317,500,000 pounds a year.

Commerce and Transportation. San Francisco has one of the finest harbors in the world, and within its great bay of about 450 square miles are to be found ships from all countries of the Orient and from many ports of the United States, Western Canada and South America. This chief seaport of the Pacific coast has built up a large foreign trade—a trade amounting to between $80,000,000 and $100,000,000 annually. Just how much the Panama Canal will do ultimately to increase this cannot be predicted, but it is certain that the routing of ships from Eastern United States ports thus made possible, and the greater ease with which the commerce of the Atlantic states of South America may now reach these Western ports, will be of constantly-increasing benefit. Los Angeles and San Diego also have good harbors, that of the former city being at San Pedro, but their commerce does not rival that of San Francisco. There is a considerable coastwise trade, and pleasure steamers do a profitable business between the California ports and those farther north.

Of the inland waters, the Sacramento and San Joaquin rivers alone are of importance for transportation. The former is navigable for 260 miles, the latter for 195. On the southeastern border is the Colorado, one of the great rivers of America, which during all but the driest part of the year is navigable for about 300 miles.

Railway connections with the other states of the Union are furnished by two transcontinental lines, the Atchison, Topeka & Santa Fe and the Southern Pacific, both of which enter the state from the south, while through its center runs a branch of the Southern Pacific, which meets the Union Pacific at Ogden, Utah. The Western Pacific extends from San Francisco to Salt Lake City, where it joins the Denver & Rio Grande. The Southern Pacific also has lines extending from Los Angeles to Portland, Ore., with numerous branches, so now nearly all parts of the state are in easy reach of railway transportation. The total length of lines within the state is over 8,000 miles. The trips from one end of the state to the other afford a series of wonderful views of ocean, gold-fruited orchards, riots of flowers, grass-grown hills and towering, wooded mountains which are nowhere to be surpassed.

California has been very progressive in the construction of electric railways, and about the large towns there is a network of these. Indeed, the development is so rapid that it is almost useless to give statistics. In 1913 there were over 1,600 miles of electric railways in the state, the system that joins Los Angeles to the numerous beautiful towns in its vicinity being especially complete.

Population. In 1845 the white population of California was but little above 5,000; in 1850 it was 92,600, and in 1860, 379,900, for gold had been found. Percentage increase has never again been as great as in those early days, but the growth has been steady, until in 1914 the population was estimated at 2,757,895. At the last official census, in 1910, there were 15.2 inhabitants to the square mile—about half that of the United States as a whole. Somewhat less than half the people live on farms or in strictly rural communities, and the percentage of city and town population is increasing much more rapidly than that of the rural. There were in the state in 1910, 36,248 Chinese, 41,356 Japanese, and 16,371 Indians; statistics show that the Indians are increasing in number rather than diminishing, as is the case in many places. The Indians, peaceable and for the most part self-supporting, live on reservations, most of which are in the south, near the desert country. Their most characteristic employment is basket-making, but many of them are farmers on a small scale. As regards the Chinese and Japanese, the decade between 1900 and 1910 saw a decrease of 9,505 in the number of the former, and an increase of 31,205 in the latter.

Of incorporated towns having a population of 4,000 or more, there were forty-three in 1910, and in these lived about half of the people. The chief cities are Sacramento, the capital; San Francisco, the commercial metropolis; Los Angeles, noteworthy by reason of its phenomenal growth, and now the largest city; Oakland, the chief railway terminal of the San Francisco region; Berkeley, a university town; San Diego, San Jose, Fresno, Stockton, Alameda, Santa Rosa and Pasadena. All of these and others are described in these volumes in their alphabetical order.

Educational Institutions. The state maintains one of the best public school systems in the Union, and has always been known for the high standards required of its teachers, as well as for its liberal pay to teachers. In charge of the system is a state board of education, with a superintendent of public instruction at its head, and funds are provided through a system of state taxation. As in all states which are not thickly populated, the rural school problem has been difficult, but California has solved it quite satisfactorily by means of its well-graded country schools for children of elementary school

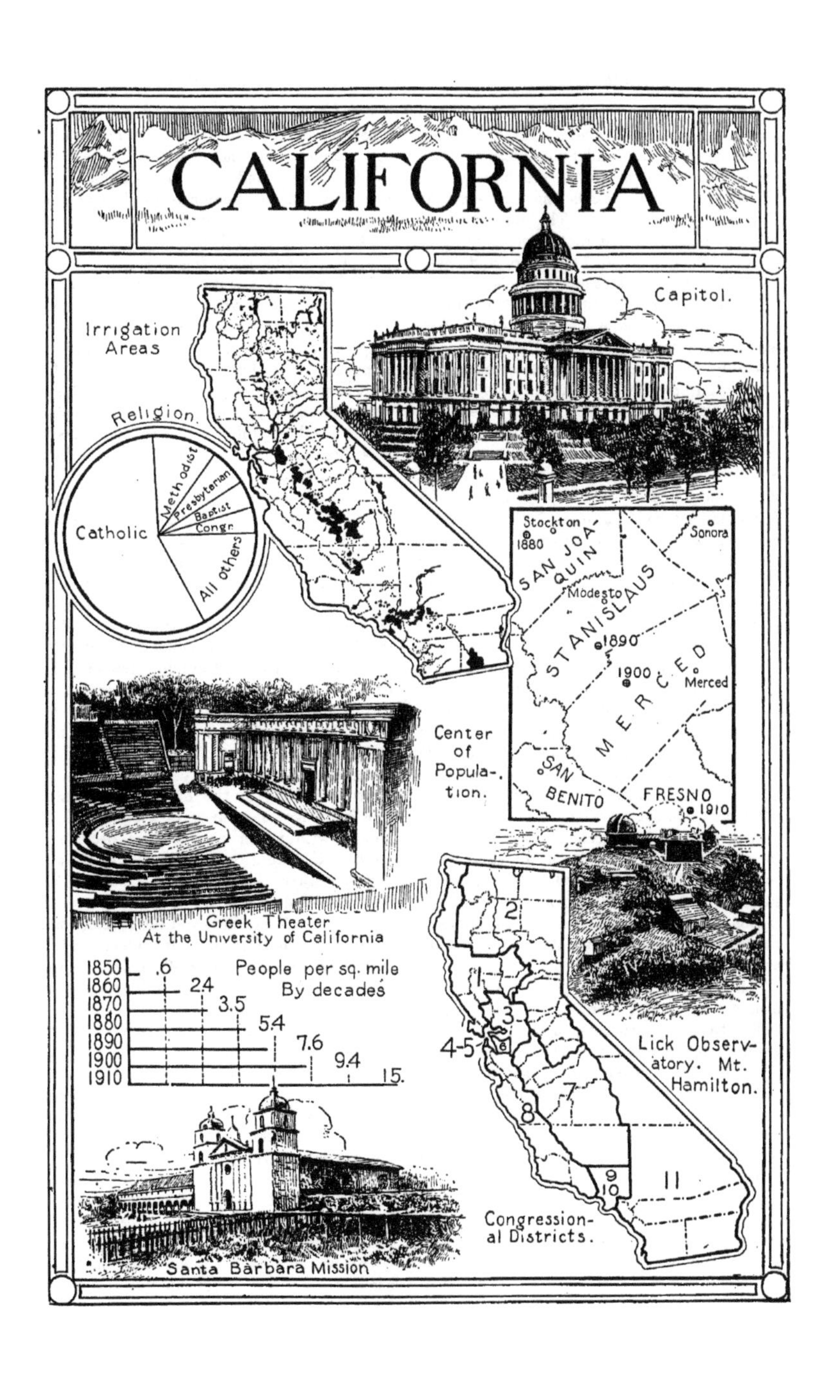
CALIFORNIA
Capitol.
Irrigation Areas
Religion.
Catholic
Methodist
Presbyterian
Baptist
Congr.
All others
Stockton 1880
Sonora
SAN JOAQUIN
Modesto
STANISLAUS
1890
1900
MERCED
Merced
SAN BENITO
FRESNO
1910
Center of Population.
Greek Theater At the University of California
People per sq. mile By decades
1850 .6
1860 2.4
1870 3.5
1880 5.4
1890 7.6
1900 9.4
1910 15.
Lick Observatory. Mt. Hamilton.
1
2
3
4-5
6
7
8
9
10
11
Congressional Districts.
Santa Barbara Mission

age and its union district high schools. At Chico, Fresno, Los Angeles, San Diego, San Francisco, San Jose and Santa Barbara there are normal schools, and at the head of the educational system stands the University of California, at Berkeley. The state has another university of first rank—Leland Stanford Junior (which see), at Palo Alto, and many colleges and denominational schools.

The percentage of illiterates in the state—that is, of people above the age of ten who cannot read or write—is 3.7; it would be considerably lower than this if it were not for the foreign-born population.

State Institutions. For taking care of its defective and dependent classes, the state has established insane asylums at Agnew, Napa, Stockton and Ukiah; a school for the deaf, dumb and blind is at Berkeley, a home for feeble-minded children is at Eldridge, and a home for adult blind at Oakland. The penal institutions include state prisons at San Quentin and Folsom City, a state reform school at Whittier, and the Preston School of Industry at Ione. The state has been progressive in its attempts to avoid that harm to lesser wrongdoers which results from housing them with confirmed criminals, and has established a reformatory in which first offenders are confined.

Government. California is governed under a constitution dating from 1879, but in 1911 twenty-one amendments to this document were adopted. Most of these were concerned with minor matters, but several of much importance provided for the initiative and referendum and for women's right to vote on an equal basis with men. There are certain restrictions to suffrage rights; no Chinaman may vote, nor can one who has been convicted of a crime or who cannot read the constitution and write his own name.

The executive power is vested in a governor, lieutenant-governor, secretary of state, comptroller, treasurer, attorney-general and surveyor-general, each of whom is elected for four years. There is no bar to reëlection. The legislature consists of a senate of forty members, elected for four years, and an assembly of eighty members, elected for two years. The length of sessions is not limited, but pay is allowed members for only sixty days. The judiciary comprises a supreme court, consisting of a chief justice and six associates, elected for a term of twelve years; a superior court for each county, and inferior courts established by the legislature.

Local Government. The units of local government are the fifty-eight counties, which are administered under uniform laws. Provision is also made for the organization of townships. A commission form of government is permissible for all cities and towns, and in 1914 no fewer than fifteen cities had adopted this plan, among them the capital city.

California has a state flag, adopted in 1911. It displays, on a white background, a dark-brown grizzly bear, with a red stripe below and a red star above. The legend is "California Republic." See illustration, in FLAG color plate.

History. Compared with that of the Eastern states, the history of California has been short, but it has been full of interest and romance. The Mexican peninsula of Lower California was discovered in 1533, but not until 1542 did an explorer enter the limits of the present state. This was the Spaniard Cabrillo, who visited the vicinity of Santa Barbara. In 1579 Sir Francis Drake made important coast explorations, sailing as far north as 43°, by reason of which he claimed the territory for England. However, not until almost two centuries later, in 1769, was the first attempt made at settlement, by a Spanish Catholic mission, established at San Diego by the Franciscans. As this was successful, other missions were promptly erected, and by 1823 there were twenty-one, some well to the north of San Francisco. This mission period is one of the most interesting in the history of the state. The buildings themselves, of a style of architecture with which people have been made familiar in recent years through restorations and imitations, were most picturesque, with their surrounding orange groves, grape plantations and cattle ranches; and the religious work which was done from these as centers was little short of marvelous. Over 80,000 Indians were converted to Christianity and taught farming or other civilized pursuits.

The Spaniards, however, had never acknowledged the right of the friars to the land; the former had begun settlements in 1777, which after the Mexican revolution in 1821 increased and expanded. The missions began to decline, and in 1834 were formally disestablished by the Mexican government, the Indian converts being scattered and the valuable buildings left to plunder and decay. Recently there has been a movement in California for the preservation of these old landmarks.

The people of the United States had early begun to feel an interest in this far western region, and in 1826 the first emigrant wagon entered the state. Not until the Mexican War, however, did the United States gain possession of the country, which in 1846 was declared a territory of the Union (see MEXICAN WAR).

Discovery of Gold. Two years later came the discovery of gold, and the rush of "forty-niners" at once began. From all over the world people flocked to the territory, and within twelve years over 260,000 had arrived. Most of them came from the East by wagon caravans, over mountain passes and through canyons which it seems incredible any wagon could ever have crossed. The big covered "prairie schooners," with their motto of "California or bust," were familiar sights to people throughout the Central United States. As was natural, by far the larger proportion of these early arrivals were men without families, and the absence of home life, the sudden rise to wealth or the black despair of failure, and the roughness of existence in a section which had prepared no laws to care for so large an influx, led to widespread disorder, and many of the mining camps were lawless places. Laws of a sort there were, however, each camp making its own, and horse-stealing or nugget-stealing was looked upon as somewhat worse than killing and punished accordingly. Nothing else gives so clearly the impression of this whole reckless, melodramatic period as do the stories of Bret Harte; *The Luck of Roaring Camp* and *The Outcasts of Poker Flat* show not only the roughness but the sentiment and the unexpected heroisms which characterized these "forty-niners."

One of the important things which these early Californians had to do was to establish communication with the rest of the world. Some of the exports could go by boat, and steamship lines were established which made the 19,000-mile toilsome passage around Cape Horn. But even more interesting were the overland routes—the Merchant's Express, which had 2,000 wagons and 20,000 yoke of oxen for freighting across the continent, and the Pony Express, which relayed mail from Missouri to San Francisco in the short space of ten days. It was in connection with this latter business that "Buffalo Bill" established his reputation as a Wild West rider (see CODY, WILLIAM F.). There were stage lines, too, which passed twice a week from Saint Louis to San Francisco, and made it possible to complete the trip from coast to coast in three weeks. In 1869 the Union Pacific Railroad was completed, and the dangerous days of stage travel were over.

MOUNT SHASTA, AS SEEN FROM THE SOUTH

Progress as a State. In 1850 California was admitted to the Union with a no-slavery constitution (see COMPROMISE OF 1850). When the War of Secession broke out the state leaned for a time toward secession, but the Federal party triumphed and the state furnished to the Union cause several companies of volunteers and almost a million and a half dollars. The history of the state since the war has been one of marked economic development. Not only to people from the East, but to those beyond the sea as well, it has seemed a land of promise, and its population has grown steadily. In its development nothing has played a larger part than irrigation (which see). A California International Midwinter Exposition was held in 1894 at San Francisco, and showed an illuminating record of progress.

A number of disasters have at times done more or less damage in different parts of the state. The first was the earthquake of April

18, 1906, which injured several of the coast towns and led in San Francisco to fires which destroyed much of the business section. In February, 1914, serious floods occurred in Los Angeles and neighboring towns as the result of very heavy rainfall, and property to the amount of millions of dollars was lost. Two months later, on April 30, 1914, Lassen Peak, a volcano in Northern California, long believed extinct, became active, and on May 14 it threw out ashes and rocks, with great clouds of smoke and steam. No lives were lost, but some were in very serious danger. Since then from the same peak there have been over a hundred eruptions.

Oriental Immigration. One of the most difficult problems with which California has had to deal has concerned the immigration of the Chinese. They came in great numbers, and while they were for the most part quiet and peaceable and made excellent servants, they never became naturalized and they worked so cheaply that white laborers could not compete with them. Agitation against them began early, and in 1881 a Chinese Exclusion Act (which see) was passed by the Federal government. Since this did not strictly prohibit them from entering the state, supplementary laws have been necessary since that date.

The problem of the Japanese arose later and was more difficult of solution. Since the beginning of the twentieth century it has been acute. Because of treaty arrangements with Japan the United States has not been able to exclude the Japanese, but Japan has been prevailed upon to restrict the emigration of laborers to the United States. A new phase of the question came up when it became apparent that the Japanese held in their possession much of the fruit-growing land, and the state made repeated efforts to prohibit the acquiring of territory by aliens. Upon the insistence of the Federal government, the law finally passed in May, 1913, did not discriminate against the Japanese, but instead safeguarded treaty rights.

Recent Expositions. During 1915 eyes were turned toward California from all parts of the world, for early in that year there was opened at San Francisco the great Panama-Pacific International Exposition (which see), the buildings for which had been two years under construction. At the same time there was opened in San Diego the Panama-California Exposition, which concerned itself more with the southern part of the state, and sought by its architecture and its exhibits to recall the old mission days. Both of these are described in more detail under their titles.

Other Items of Interest. The state has examples of every kind of mountain which is to be found on the globe; there are volcanic mountains, folded mountains, and those craggy, precipitous peaks which are the result of a breaking of the earth's crust. Wind and water, too, have done their work, and many of the hills have been eroded to the rounded form.

In August, 1916, a new national park was created in the Sierra Nevada Mountains. Its special claim to interest may be seen in its name—the Lassen Volcanic National Park.

Even over the arid stretches of the Mohave Desert rain falls occasionally, and when it does the effect is marvelous. Everywhere bright-hued flowers spring up, until the desert is literally carpeted with them. Within a few days, however, they wither and die, and soon all traces of them are lost under the shifting sands.

San Francisco has the only large and well-sheltered harbor in more than one thousand miles of coast.

California produces over ninety-five per cent of the English walnuts grown in the United States. The value of the crop in 1915 was about $4,000,000.

Almost the entire coast of California is bordered by a submarine plateau, which extends in some places but ten, in others several hundred miles out into the Pacific. Once this plateau, over which the water is nowhere very deep, was dry land, a part of the North American continent.

There has been introduced into this state the edible bamboo, the sprouts of which constitute one of the favorite vegetables in China and Japan. The plants thrive remarkably well, and those who have eaten the sprouts pronounce them not inferior to asparagus.

The largest gold nugget ever found in California weighed 195 pounds and was worth $43,534. O.B.

Consult Bancroft's *History of the Pacific States of North America;* Royce's *California,* in "American Commonwealths Series."

Related Subjects. More detailed information about the geography and resources of California may be gained from the following articles:

CITIES AND TOWNS

Alameda	Long Beach
Bakersfield	Los Angeles
Berkeley	Oakland
Eureka	Pasadena
Fresno	Pomona

BEAUTIFUL CALIFORNIA.

Illustrating the semi-tropical vegetation of the "Golden State." The building is the Mechanical and Civil Engineering Building of the University of California, at Berkeley.

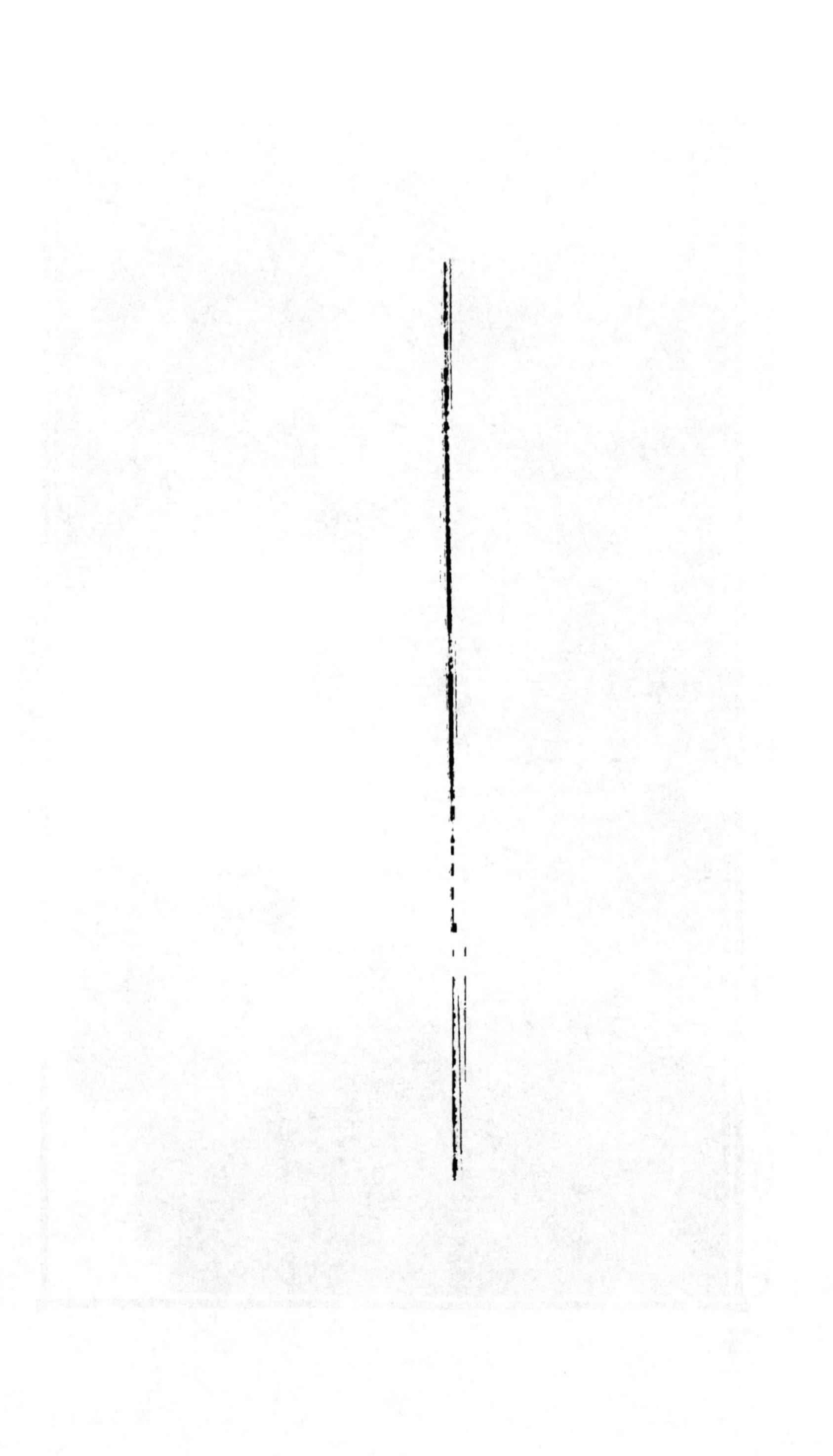

RESEARCH QUESTIONS ON CALIFORNIA

An outline suitable for California will be found with the article "State."

In traveling, during the summer months, from the coast through the great valley and on to the heights of the Sierra Nevadas, what different climates would you encounter?

What is the most famous mountain peak within the state? Which is the loftiest?

Mention four different types of mountains which are found in California.

How is suffrage restricted?

What difference would there be in the climate if the mountains ran at right angles to the coast instead of parallel with it?

How was gold first mined in the state? Why is the same method no longer employed?

Is the number of Chinese increasing or decreasing in the state? The number of Japanese? Of Indians?

What are the chief landmarks left by the early Church occupancy of the territory? What is the present attitude toward them?

What is the steepest mountain range in the United States? Describe briefly the most famous valley in these mountains.

Describe the chief beauties of California's most famous mountain lake

Name four trees which grow in few places besides this state.

Who was the first white man to enter the territory of California? Who first explored the coast line?

If the state were shut off from communication with the rest of the world, what would be its chief lack?

What effect does the Coast Range have on the shore line?

What is the most famous tree of the state? How large does it grow? What is known about the age of these trees?

Where is there an area which was once but a parched desert which has been made literally to "blossom like the rose"?

In what way does the Panama Canal affect California?

Describe the state flag.

How many national parks are there in the state? Why was each set apart?

How do the islands of the northern coast differ from those of the southern?

Of what value are the rapid torrents which flow down from the Sierras?

What is meant by the statement that California is an "inland island"?

Where is Death Valley? Describe conditions there.

How large a proportion of the oranges of the world are grown in this state?

From the chart, which is the most valuable product?

In the making of what products does California surpass all the other states?

From the map, find fifteen names that suggest the early Spanish occupancy of the country.

Has the state any active volcanoes?

Give several reasons why California is called the "Golden State."

How large is the "great valley"?

What is meant by the "climatic belts" of the state? Show how they affect crops.

During the early agricultural stage, what was the chief crop? How does that product rank now?

How many states have a longer coast line than that of California?

Redlands	San José
Riverside	Santa Ana
Sacramento	Santa Barbara
San Bernardino	Santa Cruz
San Diego	Stockton
San Francisco	Vallejo

LEADING PRODUCTS

Alfalfa	Orange
Apricot	Ostrich
Barley	Peach
Cherry	Petroleum
Gold	Plum
Grape	Prune
Grape Fruit	Raisins
Lemon	Walnuts
Lumber	Wine
Olive	

MOUNTAINS

Cascade Range	Whitney, Mount
Coast Range	Rocky
Mount Shasta	Sierra Nevadas

RIVERS

Colorado	San Joaquin
Sacramento	

UNCLASSIFIED

Catalina	Salton Sea
Death Valley	Tahoe, Lake
Golden Gate	Yosemite Valley
Mare Island	

CALIFOR'NIA, Gulf of, an arm of the Pacific Ocean, on the west coast of North America, lying between the peninsula of Lower California and the mainland of Mexico. It was formerly known as the Sea of Cortez, having been first explored by Cortez in 1536. The gulf is about 700 miles long; in width it varies from seventy to 150 miles, and in depth, from 600 to 6,000 feet. The Colorado River is the most important stream flowing into it. Valuable pearl and sponge fisheries are located on the western shore. It contains numerous islands, the most important of which are Angel de la Guarda and Tiburon.

CALIFOR'NIA, Lower, the most westerly section of Mexico, a finger-shaped strip of land extending south and slightly east from the American state of the same name. Though nearly 800 miles long, it is in most places only about one-tenth as wide. It is separated from Mexico proper, except for about fifty miles at its north end, by the Gulf of California, an arm of the Pacific a little wider than the peninsula itself. Its area is 58,328 square miles.

All that has been generally known in the past about Lower California is that it is largely mountain and desert, but it is rich in minerals and contains several towns and a population of over 50,000. Gold, silver, copper, lead and gypsum are found in the mountains, and mining industries are being developed; since the opening of the Panama Canal there has been increased activity in this respect. The peninsula was partly explored by Francisco de Ulloa in 1539, but for over a century it was regarded as an island. It was not until 1842 that its connection with the state of California was discovered.

The territory of Lower California, or Baja California, as it is called by the Mexicans, is self-governing, having a governor and legislature elected by the people, but subject in some respects to the Federal laws of Mexico. During the revolutionary troubles in Mexico beginning in 1912 it maintained an almost complete separation.

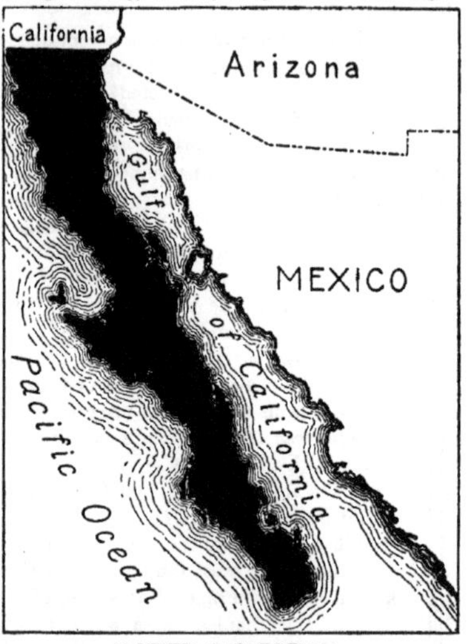

LOWER CALIFORNIA

CALIFOR'NIA, University of, second in enrollment of regular students among the universities of America, a coeducational institution of high standards and of rapidly-increasing growth, is located at Berkeley. Its campus of 520 acres climbs the lower slopes of the beautiful Berkeley Hills, commanding a view of San Francisco Bay and the Golden Gate. When by the Congressional act of 1862 California received its portion of the Federal land grant, it was decided to establish a university to take the place of the College of California, founded in 1855. This university was chartered as a state institution in 1868, was opened for instruction at Oakland a year later, and removed to its present location in 1873.

Through the generosity of Mrs. Phoebe A. Hearst, an international architectural competition was opened in 1896 for the purpose of obtaining plans for a group of buildings worthy of the magnificent university site, and in 1907 the first building erected under the Hearst Plan was completed. This was the white granite Hearst Memorial Mining Building, costing $644,000. Notable among other imposing buildings are the Greek Theater, an open-air structure seating 7,500 people and situated on the side of a hill in a grove of giant eucalyptus trees; the library, costing a million and a quarter dollars; Benjamin Ide Wheeler Hall, costing $730,000, and the Sather Campanile, a 300-foot bell tower costing $200,000. These three buildings are of granite.

The following university departments are maintained at Berkeley: the colleges of arts and sciences, commerce, agriculture, mechanics, mining, civil engineering and chemistry; the schools of architecture, jurisprudence and education, and the University Extension division. On Mount Hamilton, in Santa Clara County, is the Lick Astronomical Department (see LICK OBSERVATORY). In San Francisco are located the school of design, the George Williams Hooper Foundation for Medical Research, and the colleges of law, medicine, dentistry and pharmacy. The Scripps Institution for Biological Research at La Jolla, the Graduate School of Tropical Agriculture at Riverside, and the University Farm at Davis are also parts of the university. Far away in Santiago, Chile, is the D. O. Mills Observatory, a branch of the Lick Astronomical Department.

The buildings and grounds belonging to the University of California were valued in 1915 at nearly $8,000,000, and it has endowment funds amounting to about $5,570,000. A large number of courses are free to students who are residents of the state. The total enrollment of students is over 7,500 (including the summer session, over 11,000), and the faculty numbers nearly 800. The library, increasing at the rate of 20,000 volumes a year and containing in 1916 over 380,000 volumes, possesses the famous historical collection of Hubert Howe Bancroft (which see). The scientific publications of the university are also of importance.

The facts that tuition is entirely free, that student self-government is really successful, that various departments and chairs are devoted wholly to original research, tax-supplied moneys as well as privately-given endowments being so used, are noteworthy characteristics.

Since 1906 the university has set an example in preventive medicine by maintaining a self-supporting infirmary which gives the students all the medical and hospital care they may need in return for an infirmary fee of $6 a year. This means they are kept well and taught to avail themselves of modern scientific medicine. V.H.H.

CALIGULA, *ka lig'u la,* CAIUS CAESAR AUGUSTUS GERMANICUS (A.D. 12-41), a cruel and dissipated ruler who ranks with Nero among the wicked and tyrannical emperors of Rome. He was the youngest son of Germanicus and the nephew of Tiberius, whom he succeeded in the year 37. In the beginning of his reign he made himself very popular by his mildness and his lavish expenditures, but at the end of eight months he was seized with a disorder which permanently affected his brain, and after his recovery his career was marked by crimes and excesses that made his reign, in the words of one historian, "a tissue of follies." Though he degraded the imperial dignity by fighting as a gladiator in the arena, he considered himself a god and caused sacrifices to be offered in his own honor. At last a band of conspirators caused him to be assassinated.

CALIGULA

CALISTHENICS, *kal iss then'iks,* from two Greek words meaning *beautiful* and *strength,* is the art or practice of exercising the body for the purpose of keeping it in health, developing good posture, giving strength to the muscles and grace to the carriage. The term is usually applied to the light systematic exercises that may be performed without any apparatus, or by the use of such light apparatus as Indian clubs, dumb-bells and wands. *Calisthenics* is simply another term for *light gymnastics.* Physical culture, a broader term than either calisthenics or gymnastics, includes the cultivation of the entire physical being, with attention to diet, bathing, mental relaxation, etc., as well as bodily exercise. See GYMNASTICS; PHYSICAL CULTURE.

CALIXTUS, *ka lix'tus,* the name of three Popes. CALIXTUS I was born a slave, but rose by the year 219 to be Bishop of Rome, as the head of the Church was then called. He suffered martyrdom in 224.

CALIXTUS II, Guido of Vienne, Pope from 1119 to 1121, was a son of the Count of Burgundy. In the second year of his reign he expelled the antipope, Gregory VIII, from Rome and two years later concluded with the German emperor, Henry V, the famous Concordat of Worms (see CONCORDAT).

CALIXTUS III, Alonzo Borgia, was Pope from 1455 to 1458. Though aged and feeble, he tried to institute a crusade against the Turks, but failed. Rodrigo Borgia, father of the notorious Caesar and Lucretia Borgia, who became Pope as Alexander VI, was his nephew.

CALLA, *kal'a,* a "flower which is not a flower." Best known of the plants which bear the name is the stately *Ethiopian lily,* or *calla lily,* which came from the banks of the Nile and is grown in America only in greenhouses. Any observer would say that it has a most conspicuous, pure-white flower, but the white funnel is really but an outer leaf, or *spathe,* while the real flowers are tiny unnoticeable things crowded on the club-shaped spadix in the center.

CALLA

Botanists say that not only is this plant not a *lily,* but neither is it a *calla,* for the real calla, or marsh-arum, is a much smaller plant which grows in the bogs of North America and of Europe. Its shining, arrow-shaped leaves and its white spathe make it very ornamental, and the shape if not the color of the latter proclaims it a near relative of Jack-in-the-pulpit. In Lapland the root of this marsh-arum is ground to make bread. See ARUM.

CALLAO, *kahl yah'o,* the chief port of Peru, the shipping point for Lima, the capital, which is seven miles east. Its population in 1915 was about 40,000. The site of the city is a flat, rocky shelf, only eight feet above the sea level. Sea breezes keep the temperature moderate, but the city as a whole is not attractive for residence by reason of its narrow, irregular streets and its lack of good pavements and modern sanitation and adornment. The harbor is excellent. It is a port of entry for European and Japanese steamship lines, and its maritime importance is being greatly advanced by the Panama Canal, which brings it relatively near to the North Atlantic ports. Callao is 1,340 miles from the city of Panama, and, by way of the Panama Canal, is about 6,120 miles from London, about 3,350 miles from New York and about 2,760 miles from New Orleans. It is 1,302 miles from Valparaiso, Chile, and about 3,800 miles from San Francisco.

CALLIOPE, *ka li'o pe,* in Greek mythology, one of the nine goddesses, called MUSES, who presided over music, poetry and science (see MUSES). *Calliope,* whose name indicated the sweetness of her voice, was the Muse of epic poetry. In some myths she is loved by Apollo, and their son is Orpheus, who charmed the trees and rocks and even the fierce Cerberus with his enchanting music.

The same name has been given to a modern musical instrument which produces tones by means of a series of steam whistles. The name of this instrument is commonly and incorrectly pronounced with the accent on the first syllable, as if there were but three syllables. The calliope was at one time a feature of every old-fashioned circus, and is still occasionally heard on pleasure boats. It is one of the least pleasing of the mechanical musical instruments, as the tones it produces are loud and harsh.

CALMS, *kahmz,* REGIONS OF. In the Atlantic and Pacific oceans there are regions along the Tropic of Cancer and the Tropic of Capricorn, where for days and sometimes weeks at a time there is no wind. These regions are known as the *Calms of Cancer* and the *Calms of Capricorn,* respectively. The calms are caused by the mingling of the currents of air coming from the warm and the cool regions. In the regions of calms these currents become the same in temperature, and consequently have an equal pressure. Each counteracts the influence of the other so that there is no current of air or wind. The regions of calms move north and south with the sun, being farther north in the summer in the northern hemisphere and farther south in the winter. On the ocean the sailors call these regions the *doldrums.*

Over the equator there is the region of *Equatorial Calms.* Here the air current is upward and there is no horizontal current. This region also moves north and south with the sun.

In the days of sailing vessels the regions of the calms were usually shunned by sailors, since a ship might lie in them for several weeks awaiting favoring winds. See HORSE LATITUDES; WIND.

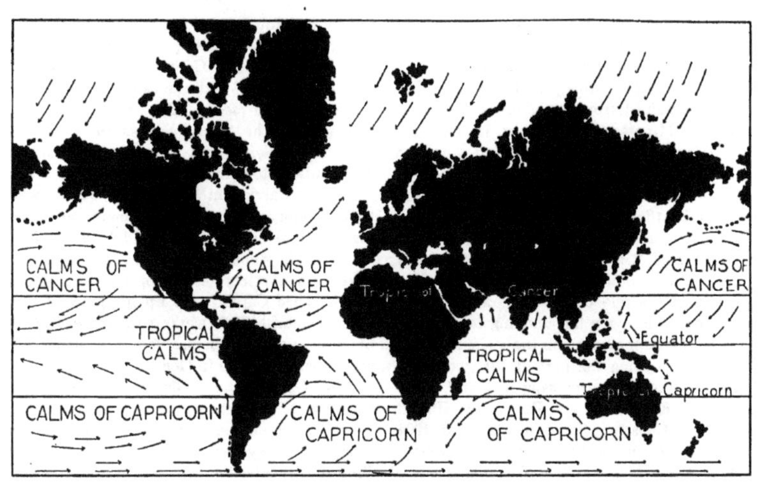

THE REGIONS OF CALMS

CALOMEL, *kal'o mel,* a widely-known drug which has especial popularity in regions where malaria prevails. In the low river lands of the Southern United States, for instance, almost every household has its supply of calomel, and many people take it at intervals whether any particular need for it is felt or not. Besides stimulating the flow of bile, it acts as a powerful cathartic, but there are certain dangers of mouth poisoning or salivation attendant on its use, and there are safer and better purgatives than calomel. The regular dose varies from one-half grain to ten grains. Calomel is a mercury compound, and as it appears on the market is a white, heavy powder which is insoluble in water, ether or alcohol. Most of it is prepared by heating mercurous sulphate with common salt in an iron vessel and cooling the vaporized product to condensation point, but some occurs in nature as *horn quicksilver.* W.A.E.

CALORIE, *kal'o ri,* or **CALORY,** the amount of heat required to raise the temperature of one kilogram of water one degree centigrade (C.), which is about the same amount that is required to heat one pound of water 4° Fahrenheit (F.). The *small,* or *gram,* calorie is the amount of heat required to raise the temperature of one gram of water one degree centigrade. The calorie, therefore, is a unit employed in measuring quantities of heat. *Calorimetry* is the measurement of the quantity of heat a body absorbs or gives off in changing from one temperature to another, or from one state to another; for example, from a solid to a liquid or a liquid to a solid. In freezing, water gives off a certain amount of heat—seventy-nine calories per kilogram, or thirty-six calories per pound—and the same amount of heat is absorbed in changing the ice back to water. The fuel values of foods are expressed in calories. See Food, subhead *Food as Fuel.* J.F.S.

CALUMET, Mich., a city which owes its importance to the vast copper interests of the vicinity. It is about seventy miles northwest of Marquette, and is on the Keweenaw Central, the Mineral Range and the Copper Range railroads. The famous Calumet and Hecla copper mine, near the city, is one of the richest in the world. Population of the city and Laurium and suburbs, all one industrial section, 32,345 in 1910.

CALUMET, *kal'u met,* the "pipe of peace" of the American Indians, used on all ceremonial occasions, but especially when treaties of peace were ratified. The pipe was passed around the circle of warriors and each one took

THE CALUMET

a solemn puff. Native tobacco, with which willow bark or sumac leaves were mixed, was smoked. In the East and Southeast the bowl was made of white stone; in the West, of a red clay obtained from the famous pipestone quarry of Minnesota, mentioned in Longfellow's *Hiawatha.* The long stem, of wood or reed, was decorated with feathers, porcupine quills or women's hair. The name was given this pipe by the French Canadians.

CALVÉ, *kal va',* EMMA, (1866-), the stage name of EMMA DE ROQUER, a celebrated dramatic soprano, born in France, whose interpretation of the rôle of Carmen, in the opera of that name, has won her undying fame. She made her first appearance on the stage at Brussels in 1882, singing the part of Marguerite in *Faust,* and she afterwards sang in grand opera in Paris, in England, Spain, Russia, the United States and Canada. Between 1893 and 1904 she was one of the leading stars at the Metropolitan Opera House in New York City, where she was best liked in the rôle of Carmen and in that of Santuzza in *Cavalleria Rusticana.* Madame Calvé has also appeared with distinction in Massenet's *Sapho,* Thomas's *Hamlet* and Samara's *Flora Mirabilis.* Since 1909 she has devoted her time mainly to concert work. At the height of her fame her rich soprano voice had a range of two and one-half octaves, and was even and clear throughout, but her success was due in no small measure to her extraordinary dramatic ability.

CAL'VIN, JOHN (1509-1564). "The greatest theologian and disciplinarian of the great race of the reformers," and one of the foremost leaders in the history of Christianity, was born at Noyon in Picardy, France. His father was secretary to the bishop of Noyon, and his mother was a beautiful and devout woman. He received his early education in Paris with the children of De Montmor of the nobility. Later he studied in the universities of Orleans, Bourges and Paris, first for the priesthood and then for the law, early distinguishing himself by his industry and remarkable intellectual power. He became dissatisfied with the teachings of the Roman Catholic Church and in 1532 allied himself with the cause of the Reformation (which see). Two years later, at twenty-eight, he published *Institutes of the Christian Religion,* one of the most important contributions to Christian literature of all time.

JOHN CALVIN

In 1536 Calvin entered Geneva on his way to Strassburg, expecting to remain only one night. Here he was found and held by William Farel, whom he had known in Paris. Farel was a celebrated evangelist, and under his influence Calvin gave up his journey and entered upon his life work; with the exception of three years of banishment, he spent the remainder of his life in Geneva.

When Calvin entered Geneva the city was on the verge of political and religious ruin. With Farel's assistance he soon wrought remarkable changes in the government and the people. A Protestant confession of faith was adopted by the city and made binding upon all citizens. Immoral practices were abolished, and those notoriously unworthy were excluded from the holy communion. Calvin's arbitrary rule, however, aroused strong opposition and he was expelled from the city. He then spent three years in Strassburg, teaching theology and writing.

He was recalled to Geneva, where he perfected an organization with himself at the head, which directed the religious and political affairs of the city and controlled the social and individual lives of the people. This rule was established under difficulties and Calvin was involved in numerous controversies. During this time, through Calvin's orders, Michael Servetus, who had written a book on the Trinity, was burned at the stake. This blot upon the career of the great reformer has never been wiped out.

During his supremacy in Geneva, Calvin maintained correspondence with all the great thinkers of Protestantism, and he was consulted upon points of law and theology by leaders throughout Europe. He also published numerous works which have always been considered standard authority upon the subjects they discussed. In 1561, the followers of Calvin separated from the Lutherans, thus forming the first great division in the Protestant Church. Soon after this Calvin died. The principles of his theology are embodied in the creeds of the Presbyterian and reformed Protestant churches.

Calvinism. The chief points in Calvin's creed can be summarized as follows:

(1) The knowledge of God is important in the mind of man.

(2) Creation depends absolutely and continuously upon God, who fosters and guides it by His secret inspiration.

(3) Sin springs from the will of man. When sin enters the soul it brings with it a train of circumstances.

(4) Man was originally a pure being made in the image of his Creator; but he is now fallen and corrupted through his voluntary departure from the good.

(5) Christ is the mediator to redeem man from sin.

(6) Men are saved by an act of absolutely free, unmerited grace on God's part, without regard to good works; men, on the other hand, are never condemned, save on the ground of their own sin.

(7) God, according to Calvin, foreordains or predestines some to surrender to His grace and be saved. He also predestines others, who are no worse, to reject His offer of grace and be lost.

This last article of Calvin's creed has been the source of endless theological discussion (see PREDESTINATION).

Consult Shedd's *Dogmatic Theology;* Hodge's *Systematic Theology.*

CALYCANTHUS, *kal i kan' thus,* a group of North American shrubs, with fragrant bark, leaves and flowers. Their name, from the Greek words for *cup* and *flower,* refers to the cup which surrounds their pistils. There are but four known species, one of which is found in California and bears brown flowers. The others grow wild in the Eastern United States, and have blossoms of a dull purple color. The bark of these plants is known as Carolina, or American, allspice.

CALYPSO, *ka lip' so,* in Greek mythology, a sea nymph who dwelt on a lonely island, on the shores of which Ulysses was shipwrecked. She promised Ulysses immortality if he would remain with her, and succeeded in detaining him for seven years, when he was overcome with longing to see his wife and child. At last Zeus sent the fleet Hermes to Calypso with the message that she must permit Ulysses to depart, and she helped him build the raft on which he sped upon his homeward course. She then died of grief.

CAM, *kam,* in machinery, a simple contrivance for converting a uniform rotary motion into a varied, sliding motion. It is a projecting part of a wheel or other revolving piece, so placed as to give an alternating or varying motion to another piece that comes in contact with it, and is free to move only in a certain direction. The heart-shaped wheel, shown in c in the illustration, is one of the most common forms in use. It is mounted on

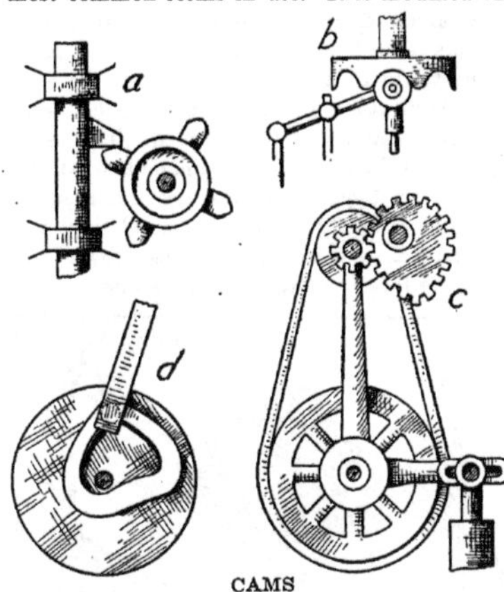

CAMS

(*a*) Cam wheel (*c*) Cam gear-wheel
(*b*) Cylindrical cam (*d*) Face cam

a shaft and imparts an irregular motion to the wheel in which it meshes.

CAMAGÜEY, *kah' mah gway,* long known by the name of PUERTO PRINCIPE, is the largest interior city of Cuba, and had in 1915 a population of 79,166. Midway between the north and south coasts of the island, it lies on a moderate plateau and is bordered by two small streams. It is the principal seat of the old Creole aristocracy, which prides itself on its ancient and purely-Spanish lineage and on its long residence in Cuba. Camagüey was founded in 1515, and to-day it contains many buildings more than a century old. Its cathedral, while much altered and extended from the original structure, has stood for about two hundred years. Some of the bridges rank in age with the buildings of the eighteenth century. The city contains also many modern business houses and residences. It is connected by rail with Havana and with Santiago de Cuba, and by another line with Nuevitas, its port to the north. The raising of live stock has always been an occupation of much importance in the plain about Camagüey, and the city has been a good market for tropical fruits, choice woods, tobacco, sugar and molasses, beeves and honey. See CUBA.

CAMBODIA, *kam bo' di a,* a once powerful kingdom, now part of French Indo-China, is situated on the Gulf of Siam, with Siam, Annam and Cochin-China surrounding it on the north, east and south. It is a densely-wooded territory covering an area of 67,741 square miles, and is therefore considerably larger than Virginia and West Virginia. Although it is in the torrid zone and consists chiefly of low-lying land, the climate is not as hot and unhealthful as might be expected. The navigable river Mekong is to Cambodia what the Nile is to Egypt. It provides communication and annually overflows, inundating and fertilizing a wide territory. Rice, sugar and maize are raised in large quantities, and tropical

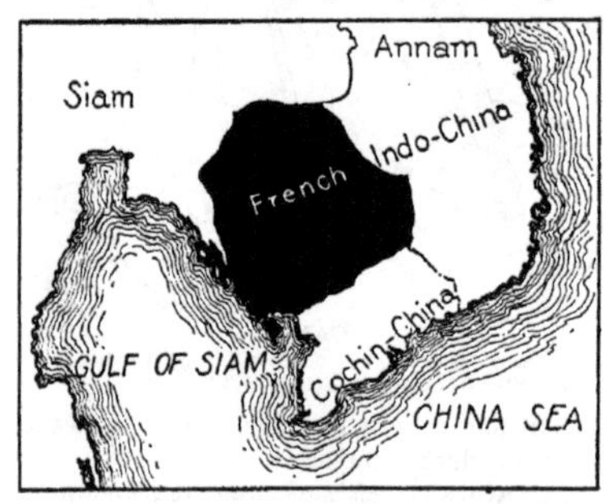

LOCATION OF CAMBODIA

fruits grow in great abundance. Gold, silver, lead, copper and iron are mined. Fishing and agriculture are the principal occupations, and cattle are raised in great numbers. The forests produce eighty kinds of valuable timber, rosewood, ebony and pine being largely exported. Gamboge, a resinous gum obtained from a tree native to Cambodia, is an important article of commerce. Tigers, leopards and elephants roam the forests, and there are also many kinds of bright-plumaged birds.

Although nominally governed by a king, the country is administered by French officials, for in 1863 Cambodia became a French protectorate and in 1884 a French possession. The capital is Pnom-Penh; previous to 1866 Oudong was the capital. In 1907 a treaty was concluded between France and Siam under the terms of which Cambodia acquired certain territory in exchange for other districts ceded to Siam, and the boundaries of the French possessions were clearly defined. Population, about 1,500,000.

CAM'BRIAN PERIOD, the oldest division of geologic time in which well-preserved remains of animal life are found. The rocks formed during this period constitute the CAMBRIAN SYSTEM. They include sandstones, conglomerates, slates and shales. The fossils found in them include those of starfish, sponges and a number of other species resembling some of the shellfish common to-day. The system is usually divided into Lower, Middle and Upper Cambrian. The Cambrian System in order of formation follows the Algonkian. See GEOLOGY; ALGONKIAN SYSTEM; FOSSIL.

CAMBRIDGE, *kame'brij,* MASS., a city famed for its historic, educational and literary associations, and especially as the seat of the first college established in the United States, now Harvard University (which see). Cambridge is one of the county seats of Middlesex County, Lowell being the other, and it is a suburb, and practically a continuation, of Boston, on the opposite side of the Charles River. It is served by the Boston & Maine and the Boston & Albany railroads, and by an extensive system of electric interurban lines. Passenger trains operate from Harvard Square, Cambridge, through a tunnel to the Charles River and thence over a bridge and underground into the Boston subway. A number of bridges span the river between the cities. The area of Cambridge exceeds six square miles. The population, which in 1910 was 104,839, in 1915 was 111,669. It is the fifth city in population in Massachusetts.

Different sections of the city are known as *Old Cambridge,* which contains the grounds of Harvard University; *North Cambridge,* becoming a fine residential section; *East Cambridge* and *Cambridgeport,* which are manufacturing sections. Along the Charles River is a beautiful waterfront park, and a dam east of the city prevents a tide at the mouth of the river. The chief features of the city are Harvard University, Radcliffe College (for women), the Protestant Episcopal Theological Seminary, Andover Theological Seminary and the Massachusetts Institute of Technology. There are a fine city hall, a courthouse, a Y. M. C. A. building and several noteworthy churches, including Christ Church (Protestant Episcopal), built in 1761.

The total value of the manufactured products annually produced in Cambridge is over $44,000,000. Printing is one of the oldest and most important industries, and the well-known Riverside, Athenaeum and University presses are established here. The chief manufactures are foundry and machine-shop products, copper, tin and sheet-iron products, canned meat, furniture, musical instruments, candies, carriages and lumber products.

Cambridge was first settled as Newe Towne in 1630. The present name was adopted in 1638, in honor of Cambridge, England, and a charter was granted in 1846. It has been the home of many famous Americans, and contains several features of historical interest; among these is the Washington Elm under which George Washington stood when he assumed command of the Continental army in 1775. In beautiful Mount Auburn Cemetery are the graves of Longfellow, Lowell, Holmes, Agassiz, Phillips Brooks, John Fiske and many other distinguished Americans. Among the fine old colonial mansions in Cambridge are the homes of Longfellow and Lowell; Craigie House, where Longfellow lived for many years, and which Washington occupied at the time of the Revolution, is now preserved as a memorial of America's favorite poet. M.W.

CAMBRIDGE, OHIO, a shipping point and manufacturing center in Guernsey County, situated about midway between the geographical center and the eastern border line of the state. Marietta is fifty-nine miles south. This region is rich in coal, natural gas, petroleum and pottery clay, and through Cambridge large quantities of coal are shipped over the Baltimore & Ohio and the Pennsylvania railroads, both of which serve the city. Mining is the principal industry, but the city has sheet and tin-plate mills, iron and steel mills, planing mills, a large earthenware plant and a glove factory. The division shops of the Pennsylvania Railroad are also located here. The Carnegie Library, the courthouse and a home for children are buildings worthy of note. Cambridge was settled in 1806 and incorporated in 1831. In 1910 the population was 11,327; in 1914 it was 12,640. The area of the city is two and one-half square miles.

CAMBRIDGE, UNIVERSITY OF, one of the two great universities of England, rivaling Oxford for first place among educational institutions. It is usually regarded as of later foundation than Oxford University, though many have endeavored to establish its claims to seniority. The first Cambridge college was established in 1281 by Hugh de Balsham, bishop of Ely, and named Peterhouse, or Saint Peter's College. The university now comprises seventeen colleges, of which Downing, founded in 1800, is the most modern.

Government of the University. Each of its seventeen colleges constitutes a separate corporation, governed by laws and usages of its own but subject to the laws of the university, just as states and provinces are, in some matters, subject to national laws. The government of the university is vested in a chancellor, the masters or heads of colleges, fellows and students of colleges, and it is incorporated as a society for the study of all the arts and sciences. The legislative senate consists of the chancellor, vice-chancellor and those who have taken the degree of Doctor or Master of Arts. The discipline of each college is maintained by the dean.

Examinations. An examination is required to qualify a student for entrance into the university. The subjects include classics, mathematics, one of the Gospels in Greek and selected Latin subjects. Later there is another examination in the classics and general subjects, followed finally by a more severe examination as a test of qualification for a degree. The students, called undergraduates, number about 3,000, and there are about forty professors in the various departments. There are three terms yearly. The university exchanges professors for a few weeks each year with several American universities.

Revenue. The annual income of the university is about $300,000, derived from fees and endowments. The cost of completing a course, which occupies three years, varies with the tastes and habits of the student, but as a rule an undergraduate would need an allowance of $1,000 to $1,500 a year.

Women's College. No degrees are conferred on women. The students of Girton and Newnham, the chief centers of women's education in England, have full facilities for attending lectures and examinations at the university, receiving certificates on graduation instead of degrees. F.ST.A.

CAMBRIC, *kame'brik,* a thin cloth woven from linen yarn of the finest quality, so called from the town of Cambrai, France, where it is said the cloth was first made. One of the items listed in the private expense record of King Henry VIII, dated 1534, is cambric for his shirts. This fabric is now used generally in making handkerchiefs, collars, cuffs, fine underclothing and fine baby clothes, and is also used as an embroidery linen. Cheap imitations, made of cotton yarn, are also on the market.

CAMBYSES, *kam bi'seez,* II (? -522 B.C.), a son of Cyrus the Great, who became, after the death of his father, in 529 B.C., king of the Medes and Persians. In the beginning of his rule he had his brother Smerdis secretly put to death, because he feared his influence. In the fifth year of his reign he invaded Egypt, conquering the whole kingdom within six months, but of his army of 50,000 men sent to take possession of Ammon, in the Libyan Desert, not one returned. He likewise failed to conquer the Ethiopians. Learning that a usurper who resembled his murdered brother Smerdis had seized the throne, he started back to Persia, but died on the way. According to the Greek historian Herodotus, Cambyses was a cruel ruler whose conduct was almost that of a madman.

CAM'DEN, N. J., the county seat of Camden County, is situated in the southwestern part of the state, on the east bank of the Delaware River and opposite Philadelphia, with which it is connected by several lines of steam ferries. It is the terminus of the Atlantic City and the West Jersey & Seashore railroads and of divisions of the Pennsylvania Railroad. The population in 1910 was 94,538; in 1915 it was 104,349.

Camden occupies an area of seven square miles on nearly level ground, from Cooper River on the north to Newton Creek on the south, a stream which separates it from Gloucester City. Among the prominent buildings and institutions are the county courthouse, city hall, Carnegie and other public libraries, Y. M. C. A. building, club buildings, Homeopathic and Municipal hospitals, Elks' Home, Home for Friendless Children and the West Jersey Orphanage. The city has Forest Hill Park and many small parked squares.

Camden is a railway center of importance, and shipping and manufacturing are extensive industries, the total annual value of manufactured products being over $49,000,000. More than 5,000 men are employed in the shipbuilding industry, and 9,000 by the Victor Talking Machine Company. Among other manufactures are canned soups, foundry products, machinery, cotton and woolen goods, oil cloth, curtains and embroidery, leather, boots and shoes, steel pens, lumber, chemicals and paint. In the suburbs of the city are large truck gardens.

A Quaker settlement along the river at this point, where there was a crossing, was organized in 1773 and named in honor of the Earl of Camden. It became a city in 1828. Walt Whitman made his home at Camden from 1873 until the time of his death in 1892. C.M.C.

CAMEL, *kam'el,* "the ship of the desert," one of the most useful, though one of the most strangely-formed animals. The camel's peculiarities make it valuable to traders in the great deserts of Africa and Asia. In its stomach it has pouches which enable it to store water, and it can therefore go days without drinking. On the back of the *Arabian camel,* the species which makes its home in Arabia, North Africa and India, is a large hump; the *Bactrian camel,* of Central Asia, has two humps. In both species of the animal the humps are of muscle, fat and flesh, and if the camel has to go a long time without food it keeps alive by absorbing the food elements in them. The foot of the camel has only two toes, and its broad base adapts it to traveling in the sand. On the bottoms of its feet, on its knees and on its chest the camel has pads which prevent the sand from wearing away its flesh. In a dust storm it lies down, stretches its long neck on the ground and closes its nostrils. Because of its powerful teeth the camel is able to eat shrubs and hard, dry vegetables. Its sense of smell and its vision are very acute, and it can sight water at a great distance. The average length of life of the camel is forty years.

Its Usefulness to Man. Camels and their riders cross the deserts in caravans, which are sometimes miles in length. The camel is not fully grown until its sixteenth or seventeenth year, but it is trained from its fourth year to kneel, to rise and to carry burdens. A mature camel will carry a load of a thousand pounds. An ordinary camel will carry its pack twenty-five miles a day for three days without water; some will go without water fifty miles a day for five days. A specially-trained camel will carry a rider a hundred miles in a day. Like a pacing horse, it lifts the legs on the same side at the same time.

To the people who live in desert countries the camel is valuable in many other ways than as a beast of burden. The two-humped camel grows a very long winter coat, from the hair of which the natives weave cloth. The Arabs also use camel's hair cloth, especially for tent coverings, and we read in the Bible that John the Baptist "had his raiment of camel's hair." The meat of the young animal resembles veal, and is a favorite with the Arabs. A very durable leather is made from the hide of the camel. Artists in America and Europe use fine camel's hair for their brushes.

The camel is not like the horse, an animal which learns to love and to serve its master intelligently. It is easily governed, but this is because of its dullness, not because of a gentle nature. It is always one of the most interesting animals in the zoölogical garden or circus menagerie. In captivity it is fed on alfalfa, hay, carrots and bread, but very little grain.

Origin. Some authorities believe the first camels or camel-like animals lived in North America, thousands of years ago, when a large part of the continent was desert. When the climate became more humid and forests grew, these animals migrated. Some crossed Siberia into Asia, others went to South America, where their descendants survive as llamas (which see). The llama lives in the mountains, where it has plenty of water, but it still has a stomach like the camel's, capable of storing water. At the time of the rush to California of the gold-seeking "forty-niners" camels were brought to America for use in crossing the desert, but this experiment was not a success. V.L.K.

CAMELLIA, *ka mel' i a,* a beautiful, waxy, rose-like flower which is borne on a plant closely related to those which yield tea. The leaves of this plant are dark green, shining and laurel-like, and serve to set off most effectively the large blossoms. The common camellia, now much grown in greenhouses in America, is native to Japan, but there the flowers are single and usually red, while under cultivation they have doubled and have developed various colors, white, pink, reddish and yellow. For best results, camellias should be planted in a loose, black mold in a cool greenhouse, and given plenty of water. Although they will grow from seeds, the finest specimens are produced by grafting. Dumas' famous novel, translated into English as *Camille,* was called in the original *The Lady of the Camellias.* One species of this flower is known as the *japonica.*

CAMELLIA

CAMEO, *kam' e o,* the general name for all precious stones which are engraved with raised figures, in contrast to *intaglios,* which have the figures sunk below the surface. In a special sense the term *cameo* is applied to a gem

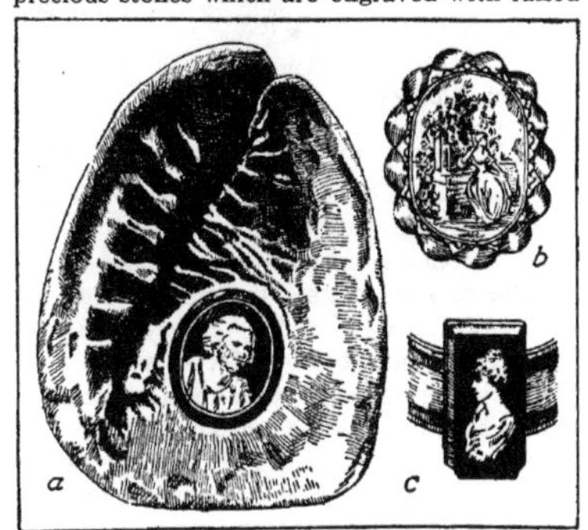

CAMEOS
(*a*) Sardonyx shell with portrait cut in bas-relief.
(*b*) Cameo brooch.
(*c*) Cameo ring.

which has layers of different colors, with the figures so engraved that they appear in one color and the background in another. Onyx, sardonyx and agate are the stones in general use for cameos, but very beautiful artificial specimens are produced from various kinds of shells and fine glass. Shell in particular yields cameos of rare delicacy and loveliness. A famous specimen of an imitation of cameo in glass is the Barberini Vase in the British Museum, made at Rome about the first century before Christ. This has a blue background and figures of a delicate, half-transparent white. Both the Greeks and the Romans excelled in the art of cameo cutting. At the present time dinner rings with cameo settings are highly popular. See PRECIOUS STONES.

CAMERA, *kam'era,* the popular name for any apparatus used for taking pictures by photography. The term is really the Latin word for *chamber,* and is abbreviated from *camera obscura* (which see). The camera obscura works exactly like the human eye. In the eye the crystalline lens reflects upon the sensitive retina any rays of light coming from the outside; in the camera obscura the rays are reflected upon a screen, placed in a darkened chamber, just as the retina is in a darkened chamber of the eye. In the ordinary photographic camera the place of the screen or retina is taken by a sensitized celluloid film or glass plate. The film or plate is previously prepared so that it will preserve the image of any objects reflected on it.

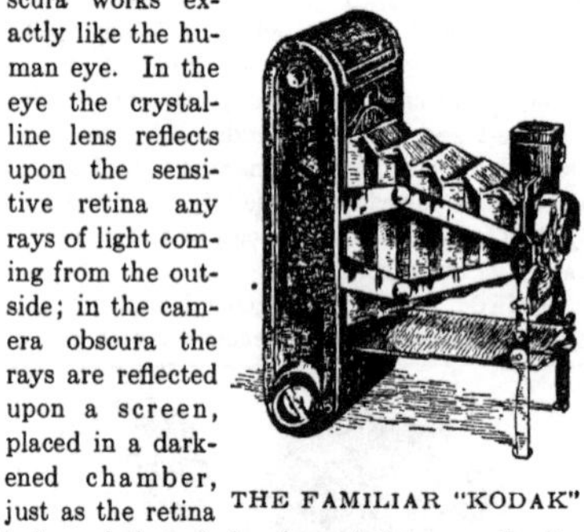

THE FAMILIAR "KODAK"

There are as many different kinds of cameras as there are purposes and personal tastes. The essential parts, however, are few and are easily remembered—the box, the lens, the shutter and the screen. The box is fitted with a telescopic arrangement, made of light-proof cloth, rubber or other material, by which the lens can be placed at a proper distance from the screen (see LENS). At the outer end of this telescope arrangement is a frame in which the lens is set; at the other end, at the back of the box, is the screen. In large cameras, such as those used by professional photographers for taking portraits, a piece of ground glass, in a frame, is used as a screen. The photographer moves the lens backward and forward until the image on the screen is clear. Then he is ready to remove the screen and substitute the sensitized plate. He covers the lens with a cap, which usually contains a shutter. The speed with which the shutter opens and closes can be regulated to a fraction of a second, and the size of the opening is also adjustable. The period during which the shutter is open is called the *exposure,* for the sensitized plate is exposed to the rays of light, which pass through the lens. A large camera is usually mounted on a tripod when in use, to keep it motionless during the exposure.

The smaller, or hand, cameras are exactly the same in principle, though slightly different in details of manufacture and operation. Large cameras, with glass plates, are not easily carried around. The first notable improvement was the substitution of sheets of celluloid film for glass. The sheets were separated by black paper, and when a film was exposed it was withdrawn from the camera together with its protecting black paper. The film "packs," as they were called, were not entirely satisfactory, and the latest cameras all use films in rolls. The entire roll, mounted on a spool, is inserted in the camera, and as it is exposed bit by bit is gradually rolled up on a second spool. The roll of film is protected from the light by black paper.

Small cameras also differ from large ones in the method of adjusting the lens and making the exposure. They have an additional pair of lenses, called the *finder.* These are mounted outside the telescopic arrangement and have no connection with the photographic lens; they are in the same relative

CAMERA USED BY PHOTOGRAPHERS

position, however, and anything visible in the finder will appear in the photograph. In nearly all types of cameras the shutter is controlled by a spring, which may be released by pressure on a lever or on a rubber bulb which in turn releases the lever.

Uses of the Camera. The camera is now used in almost every art and industry. Photoengraving (which see) is perhaps its most important commercial use, but there are dozens of other processes and industries in which it plays a part. Special cameras, both large and small, are made for particular purposes. The solar camera, for example, is used in connection with a telescope to photograph stars and planets not visible to the naked eye (see ASTRONOMY). At the other end of the scale are cameras specially made for scientific research with the microscope. Then there are the *multiplying* camera, in which a number of lenses are used for taking several pictures at one time; the *stereoscopic* camera, a double camera for giving a double picture on one plate; the *copying* camera, used for copying photographs from negatives; and the *cycloramic* camera, which turns on a central pivot and takes a panoramic view at a single exposure. One of the most remarkable kinds of cameras is used for taking moving pictures. This type of camera is fully explained in the article MOVING PICTURE. For further details of the use of cameras, see PHOTOGRAPHY, subhead *Amateur Photography*. K.A.G.

CAM'ERA LUCIDA, *lu'si da,* a simple instrument used for sketching objects seen through a magnifying glass. It consists of a four-sided prism, having two sides at right angles, and the other two at an oblique angle,

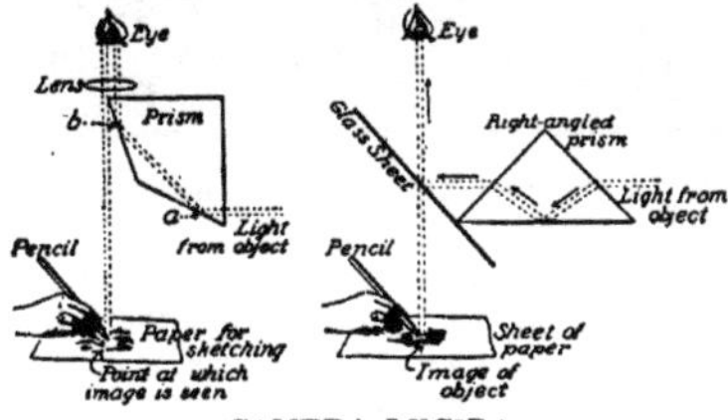

CAMERA LUCIDA

and a magnifying glass, both being attached to a frame in the positions shown, in the illustration at the left. The light from the object enters the prism and is reflected at *a* to the point *b,* where it is reflected to the lens, through which it passes to the eye. The magnified image is thrown upon the paper where the sketch is made. Another form, shown at the right in the illustration, uses a right-angled prism with a sheet of glass for a reflector. But the form first described is in most general use. The camera lucida is used for sketching parts of insects, the cell structure of plants and other very small objects. It is seldom seen outside biological laboratories.

CAM'ERA OBSCURA, *ob sku'ra,* a box arranged for sketching landscapes and other large objects. The box contains a mirror, *a,* set at an angle of 45°; a double-convex lens (see LENS), such as is used in a photographic cam-

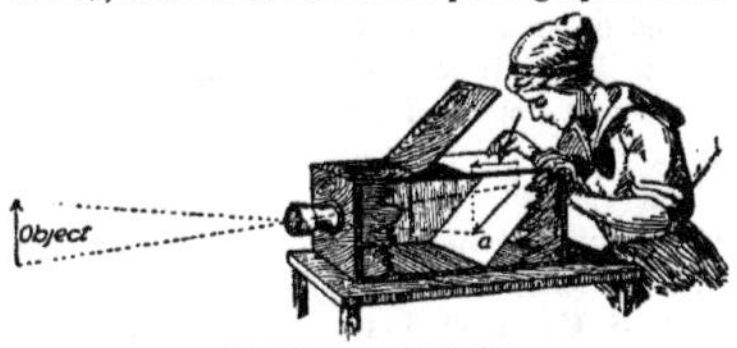

CAMERA OBSCURA

era, is placed in the front end, and the top has a lid which can be raised. Under the lid is a ground glass screen. The image of the object is formed on the mirror and reflected upon the screen where it can easily be sketched with a lead pencil. The instrument should be used in a dark room, or a black cloth should be thrown over it and the head and shoulders of the one making the sketch, otherwise the image will be too dim to admit of sketching.

Before photography became so common, the camera obscura was in general use by artists in making sketches for illustrated papers and magazines, but it is now seldom used except as a toy.

CAMERON, AGNES DEAN (1863-1912), a Canadian educator, traveler and author, known for her study of Canada's natural resources. She was born in Victoria, B. C., where she began teaching at the age of fifteen, and for twenty-five years was closely identified with the public schools of the province. During the last ten years of her service she was principal of the South Park public school, Victoria. In 1908 she traveled 10,000 miles, from Chicago to the Arctic Ocean, by way of the Athabaska, Slave and Mackenzie rivers, returning through the Peace River Valley; this exploit gave her fame as the first woman explorer to reach the Arctic Circle. She was accompanied on this rough adventure only by her niece. Miss Cameron was for several years associate editor of the *Educational Journal of Western Canada,* made frequent appearances on the lecture platform and wrote many articles for newspapers and magazines, as well as two books, *The New North* and *The Outer Trail*. She died in Victoria.

CAMERON, George Frederick (1854-1885), a Canadian lyric poet and journalist whose promising career was cut short by an untimely death. He was born at New Glasgow, Nova Scotia, and was educated at Queen's University, Kingston. After spending several years in the United States he returned to Canada to become editor of the Kingston *News*. Meanwhile he had written a number of poems, strongly marked by an idealistic strain, and won praise for his lyrics from such exacting critics as Matthew Arnold and Swinburne. Cameron's poems were collected after his death in a volume entitled *Lyrics on Freedom, Love and Death,* and selections may be found in Stedman's *Victorian Anthology* and in the *Oxford Book of Canadian Verse*. Among his best poems are *What Matters It; Ah Me! the Mighty Love; True Greatness,* and *The Future*.

CAMILLE, *ka meel'*, the title and the name of the heroine in the English version of a celebrated play by Alexander Dumas the Younger. The play created a sensation when produced in Paris in 1852, and has since enjoyed great popularity in English-speaking countries because of the opportunity it gives for display of highest emotional power. Sarah Bernhardt, Olga Nethersole and Helena Modjeska have all been notable Camilles. Verdi's popular opera *La Traviata,* in which Sembrich and Tetrazzini have starred, is based on the story of Camille. In 1917 Galli-Curci scored a triumph in this part.

CAMORRA, *ka mor'a,* a Neapolitan secret society of criminals which for about a century has been active in extorting petty sums from the common people and larger amounts from the more wealthy, by threats of violence. For a time during the Bourbon rule (before 1860) the members of the society, called *Camorristas,* worked openly, agreeing to commit crimes of any sort for a money payment. Under the present government of Italy they formed a political machine, in complete control of the city of Naples. In 1899 the Crown deprived the municipal government of authority pending an investigation, after which a citizens' league drove the Camorristas from power. It is believed that a final blow at the Camorra was struck in 1912 when six of its members were tried for murder and twenty-five others were accused of being members of a criminal society. It was a notable case; thirty-two lawyers appeared for the defense and over 600 witnesses were examined. The result was a verdict of guilty against a number of Camorristas, who received sentences of imprisonment ranging from four to thirty years.

CAMP. The words *camp* and *camping* suggest many things—woods, prairies, mountains, lakes, rivers, fishing, hunting, rest for tired nerves, out-of-door life. Camping includes everything from spending the night in a tent in your own back-yard to passing several weeks in some secluded spot, miles from the nearest town. The camper may travel on foot, on horseback, in a wagon, or in these modern days, in an automobile. He may use a canoe, a sailboat, a motor launch or a steam-yacht. No matter where or how one goes, the thought of camping should act like magic, should make one restless for the freedom of out-of-doors, and for the strength and vigor which only an outdoor life can give.

There is only one way to get the real flavor of outdoor life; it is to get into the wilds, away from the cities. This does not mean that the boy or girl who has been brought up in a large community cannot learn a great deal about camping without going many miles from home. Much can be learned from books and by practice; many experiments in making shelters and beds, building fires and cooking can be tried before the camping trip begins. In fact, it is wrong for the inexperienced person to start with the idea that only experience in the woods or the mountains can teach him. Experience will teach him confidence and many things which are not found in books, but a little early information will make his experience more pleasant.

Equipment. The amount of equipment to be carried is largely a matter of personal preference. It is desirable to reduce the weight to a minimum, especially if the camping trip includes long walks. If horses or canoes are used the amount may be slightly increased, but the real camper has no desire to make a pack-horse of himself. He quickly learns to get along without some of the things which make his back and shoulders weary when he carries them day after day. It is not a bad plan for the beginner, when he returns from his first long trip, to sort his equipment into three piles—one, of the things he used every day; one, of those he used at least once; and one, of those he did not use at all. On his next trip this third class should be left at home, and the second class, too, if possible. The only possible exception to this arrangement is the first-aid-to-the-injured kit.

Clothing. In the matter of clothing each person must be the judge of what is needed. Any old suit will do for daily wear for a boy or man, although knickerbockers are useful. A woman should wear a short woolen skirt. Some campers wear corduroy because it is warmer and more durable. Woolen shirts are better than cotton, because they dry rapidly if they get wet, and they keep the body at a more even temperature. Woolen or heavy

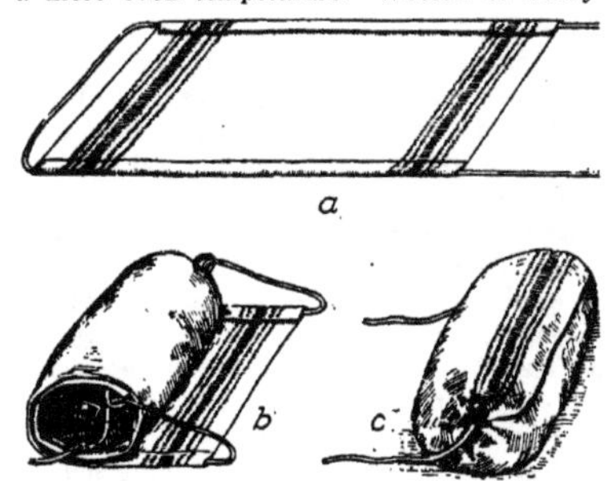

HOW TO FOLD AND ROLL A PACK
(*a*) Blanket with edges folded over rope.
(*b*) Blanket partly rolled, showing method of making loops.
(*c*) Pack rolled up.

leather gloves are useful in nearly any kind of weather. Hats should have small brims, but many campers prefer to use caps, which are less likely to blow off in a high wind or catch on overhanging branches. Shoes should be waterproof, if possible, and should be a trifle easier and larger than ordinary street shoes. Some experienced woodsmen wear moccasins, which allow a better grip on slippery surfaces.

Every camper should have a waterproof canvas bag in which to keep extra clothing and other personal supplies, although few go to this trouble. Four pairs of woolen socks, two flannel shirts, two sets of woolen underwear, one suit of pajamas, one pair of trousers and a woolen sweater will be all the extra clothing needed on a trip through the woods in the spring and autumn. An extra pair of shoes, a pair of moccasins, some thread, needles, scissors, a toothbrush, several towels and a note book can all be kept in the bag.

Camp Supplies. In the amount and kind of supplies there is a wide range. Catalogues of sporting goods are filled with all sorts of devices, some of them really valuable, some of them quite unnecessary. A combination knife, corkscrew and can-opener is most useful, but an experienced camper will make a good pocket knife do the work of all three; a hunting knife and a small ax are also indispensable. Cooking kits are offered in endless combination, but a small empty lard pail, a small frying pan, a tin cup, a knife, fork and spoon are enough for the simplest camping. A supply of nails, heavy twine and wire will often prove useful, and every party should have at least one compass and one waterproof match-box. A magnifying glass, a field-glass and a camera are non-essentials, but will add greatly to the pleasure of a trip.

Food. Camp food is almost entirely a matter of personal choice. Fruits and vegetables can be bought dried or canned; milk is obtainable either condensed in cans or as a powder, and soups are now made in the same fashion. If the camp is not too far from habitation, an occasional supply of fresh meat, eggs, butter, etc., will gratify the appetites of the campers. Coffee, cocoa, bacon and corn-meal cakes are the staples of camp cooking, but bacon is the most important of all.

Proper Location for a Camp. Few people believe that a camp is really a good one unless it is on or near water. The temptation to camp on the very edge of a lake or a stream is always strong, but it should be resisted because the low ground is almost sure to be damp and is frequently infested with mosquitoes and other insects. If there is no high land near the shore, the best place for a camp is some point projecting into the water. Here the currents of air will probably be strong enough to keep insects away.

If a choice can be made between two locations, one near firewood and the other near water, the latter is likely to be more picturesque, but the former involves less work. It is much easier to carry water than wood. If the camp is more or less permanent, involving the use of tents, ground sloping to the south is best, for the open end of the tent can then be placed so that the sun's rays will reach the interior. A camp should never be made, if avoidable, in dense woods, where falling limbs are dangerous, or in a hollow, which will collect water after a rain. Dead wood and heavy underbrush are breeding places for mosquitoes, and they retain dampness.

can be wired to make a firm support, but two upright forked sticks, with a third laid across them, make a satisfactory substitute. The simplest and in some ways the best method is to hang kettles on a long trailer resting in the crotches of forked uprights, as shown in the illustration. Such a trailer can be shifted by one person.

If a fire is being built to provide warmth, one of the best methods is to lay it between two large logs laid parallel. One of the approved ways of building a fire to give heat for a night is illustrated herewith. Two green stakes should be driven into the ground nearly perpendicular. Two heavy logs, for fire-dogs, may be laid at right angles to the line between the stakes. Logs may then be piled against the stakes to any height, and two more stakes driven to hold them in position. As each log burns away the others will drop down, and the fire will burn as long as logs remain above it.

HOW CAMP FIRES SHOULD BE BUILT
At left, night fire; at center and right, fires for cooking.

Camp Fires. The experienced camper starts a fire as soon as the camp is located, and one glance at a fire will tell just how much camping experience the builder has had. There are many ways of building a fire, and as many ways of arranging and supporting the pots and pans. If a high wind is blowing and the camp is somewhat unprotected, it is wise to dig a fire hole, so the live coals will not be blown away. Another simple method is to build the fire in a V-shaped space protected by two logs. The logs should be smoothed off on top, so frying pans and other utensils may be set on them. At the open end of the V the fire should be kept burning briskly, and here the kettle of water may be set to boil. At the closed end of the V should be a bed of live coals, and here the frying and baking should be done. Instead of two logs, two rows of flat stones may be used.

In permanent camps it is customary to make racks on which to hang kettles and pots over the fire. Three pieces of lead pipe, for example,

Building a fire is not difficult in dry weather, but if it is raining the camper may experience trouble. If some dry wood cannot be found under the trees, search should be made for a dead cedar. Cedar splits easily and burns quickly. After the wood is split, some of the smaller pieces should be piled up in the form of a hollow pyramid, with shavings stuffed inside it. Dry birch bark or dead twigs on the lee side of a tree can generally be found, if a cedar is not available; but if nothing else is at hand, it may be necessary to chop into a fallen tree for dry shavings. The fire should be built on the lee side—that is, the side protected from the wind—of a boulder or clump of trees. In the winter it is best not to build a fire under a tree covered with snow, as the heat will melt the snow and the water may put out the fire. The inexperienced camper frequently gathers too little firewood; it is better to gather too much than to leave the cooking at a critical stage or get up before daybreak to hunt for wood to keep the fire going.

As a general rule hardwoods are the best for camp fires. Hardwoods make a slow-burning fuel which yields lasting coals, whereas softwoods give a quick, hot fire that soon dies out. The following woods burn scarcely at all, if they are wet or green: aspen, black ash, balsam, box-elder, sycamore, tamarack and poplar. Chestnut, red oak and red maple burn slowly when green. All soft pines crackle and embers are likely to shoot off. Some of the hardwoods, such as sugar maple and white oak, shoot off long-lived embers and must be watched for some time after the fire has started. The best of firewoods is hickory, either green or dry. It makes a hot fire, and burns down to a bed of coals which keep an even heat for hours.

Shelter. If the weather is clear and fine, no camper will want a shelter. Man has not yet produced anything more restful than a bed of pine boughs under a clear, star-lit sky. But there are many days, even in summer, when some protection is advisable. If the rain is not falling too hard or the wind is not too strong, there is no better shelter than a pine grove. The needles which have fallen to the ground make the softest of beds, and the interlaced branches overhead keep off the rain.

METHOD OF PREPARING A SHELTER

The simplest shelters suitable for a temporary camp are made of brushwood. These can be made in any shape to suit the preferences of the campers, but the lean-to is the only practical brush camp if there are more than three persons in the party. Two crotched sticks should first be driven upright into the ground, about eight feet apart, and a stout sapling laid across them; the sapling may also be laid in the crotch formed by a branch of a standing tree. On this cross-piece should be leaned a number of saplings or poles, the lower ends of which may be secured by sticking them into the ground or by rolling a log against them. On this framework should be laid boughs of hemlock or spruce, or other heavy brush, which should be lapped like shingles, so they will shed the rain. If the shelter is built on sloping ground, a trench should be dug at the back and sides to carry off the rain.

Many campers carry portable tents or waterproof canvas sheets which can be stretched for shelter. These are good, the only objection to them being that they are heavy and are an unnecessary addition to a pack which is probably heavy enough. Tents, however, are essential to comfort, if the camp is to be permanent. For different kinds of tents and their uses, see TENT.

Camp Discipline. Camp life is an intimate association of people, and probably nothing else so fully discloses strength and weakness of character. Especially in a large camp there is likely to be some difference of opinion on all matters. If these differences are merely friendly, no harm is done, but there is always the possibility of serious misunderstanding or quarrels. For this reason the most successful camp is invariably one in which all the members submit cheerfully to an informal discipline established by one or two of the older and more experienced members. For example, if one of the campers wants breakfast at eight and all of the others want it at seven, the advocate of a late breakfast should yield without argument. If most of the party want to spend a day fishing, a minority should not spoil the sport by grumbling about partridges or ducks. If the camp is very large, it may sometimes divide into several parties—one for fishing, one for hunting, one for tramping, etc.

In summer camps for boys and girls discipline means submission to rules which the camp counsellors or masters have adopted. Certain hours of the day should be fixed for meals, for swimming and bathing, for cleaning camp, etc., but some time should also be left free. Each member should do a required share of the work. Shirkers, whether they are young or old, boys or girls, men or women, are not wanted in a real camp. More than anything else cheerfulness is the essential of a good camper; a cheerful "tenderfoot" is infinitely better company than a sulky veteran. W.F.Z.

Consult Hank's *Camp Kits and Camp Life;* Kephart's *Book of Camping and Woodcraft.*

CAMP, WALTER (1859-), one of the leading American authorities on amateur athletics, sometimes called the "dean of American football." The name of Walter Camp is probably familiar to every boy who has ever been on a football team. His interest in athletics dates from his undergraduate days at Yale University, where he played on the university football and baseball teams, rowed on his class crew, won the high hurdles and represented Yale (with Slocum) in the first Intercollegiate Tennis Meet. After his graduation in 1880 he became active in the management of Yale athletics, and for many years was chairman of the athletic committee. He gradually became recognized as the leading authority on football, has been for many years a member of the football rules committee and its secretary, and is editor of the committee's official organ, *Spalding's Football Guide.* Each year he selects an "All-America" football team, composed of the star players of the different colleges, and his selections are generally regarded as authoritative.

Camp's intimate knowledge of schools and athletics has been used to advantage in books written primarily for boys. These are among the most popular of all juvenile books, and include *The Substitute, Jack Hall of Yale, Old Ryerson* and two *Danny Fists* series. He also wrote the *Book of College Sports, American Football, Football Facts and Figures,* and numerous articles published in magazines.

CAMPANILE, *kahm pah ne' la,* from the Italian word for *bell,* is the name given to the old bell towers belonging to the churches of Italy. They were built for the same purpose as the steeples of ordinary churches—to hold the bell which by its ringing should assemble the people; but they differ from steeples in that they are not joined to the churches to which they belong, but form separate buildings. The earliest ones date from the sixth century, but not until three or four centuries later did they become very numerous.

The most famous examples are the campanile of the cathedral at Florence, designed by Giotto in the fourteenth century and faced with red, white and black marble; and the Leaning Tower of Pisa, inclining almost fourteen feet from the perpendicular. Saint Mark's campanile, possibly the most notable of all, was 302 feet high and a landmark of

CAMPANILE OF SAINT MARK'S, VENICE

See article TOWER, noting similarity between the campanile above and the tower comprising the twenty-five upper stories of the Metropolitan Building, New York City. Modern builders have adapted to present-day needs with excellent results a structure which a thousand years ago represented little more than a sentiment.

Venice for over a thousand years, dating from A.D. 900. In 1902 it collapsed; work of restoration began in 1905, and the new campanile, shown in the illustration, was completed in 1912. See PISA.

CAMPANULA, *kam pan' u la,* from the Latin *campana,* meaning *bell,* the name of a group of plants which bear nodding, bell-shaped flowers, white, blue or lilac in color. Probably the favorite species is the slender little *harebell,* also called the *bluebell of Scotland,* which grows in meadows and on rocky hillsides, and is also found high up on the mountain slopes. The more showy varieties of campanula are popular in gardens, and are especially effective as border flowers. Those species that blossom year after year may be planted from young cuttings in the spring or from seeds. They are easily cultivated.

CAMPBELL, *kam'bel,* ALEXANDER (1788-1866), a religious leader and reformer, founder of the Christian (Disciples) Church. He was born in the County of Antrim, Ireland, the son of a minister. His father, Thomas Campbell, emigrated to America in 1807 and settled in Washington, Pa. Two years later, the family, of which Alexander was the eldest, followed him and soon afterward settled in Bethany, W. Va. While a student in Glasgow, Scotland, Alexander Campbell broke with the old religious ties. On arriving in America he found that his father had also withdrawn from the Seceders, a branch of the Presbyterian Church, and had formed an independent organization called the "Christian Association of Washington," whose aim was the promotion of Christian union. *A Declaration and Address* had been drawn up by Thomas Campbell, setting forth the object of the Association. The principles of this document the son heartily espoused, and at once became the champion of the restoration movement which it advocated, namely, a return to the New Testament as the rule of faith and practice.

Because of his gifts as a speaker and his qualities of leadership, Alexander Campbell soon became the recognized head of the movement, and the influence of his teaching spread rapidly through the Ohio Valley. The first organization of his followers was at Brush Run in Washington County, Pa. As they had become immersionists they were admitted into the Red Stone Baptist Association, and continned their affiliation with that body until compelled to withdraw because of doctrinal differences. This separation took place throughout the country where Campbell's influence had gone, between the years 1827 and 1830, resulting in the religious body known to-day as the Disciples of Christ.

To present more effectively the cause he advocated, Campbell began in 1823 the publication of the *Christian Baptist,* which was later merged into the *Millennial Harbinger* and continued its monthly appearance until his death. In 1840, recognizing the need of an educated ministry, he founded Bethany College and became its first president.

In his advocacy of truth he entered the field of public debate, and in platform power he had few rivals. Among his published works, which comprise more than sixty volumes, are his published debates, *The Christian System, Memoirs of Thomas Campbell* and *The Living Oracles.* See DISCIPLES OF CHRIST. T.W.G.

CAMPBELL, SIR, ALEXANDER (1822-1892), a Canadian statesman, one of the leaders in the movement for Confederation, first Postmaster-General of the Dominion and for twenty years the leader of the Conservative party in the Senate. A Scotchman by descent and an Englishman by birth, as a boy of two he was taken to Canada by his parents, who settled first at Lachine and later at Kingston. At Kingston, Campbell became the friend and partner of a young barrister, John A. Macdonald, whose name later became a household word in Canada. Both the partners were active in politics; Campbell was in turn alderman of Kingston, speaker of the legislative council of Upper Canada, and commissioner of Crown lands. At Confederation he was summoned to the Dominion Senate, from which he resigned in 1887, to serve until his death as lieutenant-governor of Ontario. His activity in the Senate was not so conspicuous as that of less able men of his party in the House of Commons, but for twenty years his influence was perhaps second only to that of Sir John Macdonald. Macdonald repeatedly gave him positions in the Cabinet, first as Postmaster-General, from 1867 to 1873, and later as Minister of Militia and Defense, 1880 to 1881, as Minister of Justice, 1881 to 1885, and again as Postmaster-General, 1885 to 1887.

CAMPBELL, THOMAS (1777-1844), a noted English poet, best known to general readers for his stirring lyrics *Hohenlinden, Ye Mariners of England* and *The Battle of the Baltic,* and his ballad of *Lord Ullin's Daughter.* Certain lines from his poems have become so familiar that they are everywhere quoted, without thought of their source. Such are—

> 'Tis distance lends enchantment to the view.
> Like angel visits, few and far between.
> Coming events cast their shadows before.

Campbell was born and educated in Glasgow, and while in the university in that city won a reputation by his poetical translations from the Greek. His *Pleasures of Hope,* published in 1799, made him immediately famous and won him a pension of $1,000 a year. The chief of his later works, besides those mentioned above, were *Gertruae of Wyoming,* a narrative in verse of the Wyoming Valley Massacre in Pennsylvania, and an anthology called *Specimens of British Poets.* From 1820 to 1830 he edited the *New Monthly Magazine,* and in 1826 was made lord rector of the University of Glasgow. He was buried in the Poets' Corner, Westminster Abbey.

CAMPBELL, WILLIAM WILFRED (1861-1918), a Canadian whose lyrical poems have placed him in the front rank of contemporary poets. He was born at Kitchener (then Berlin), Ont., was educated at the University of Toronto and the Episcopal Theological Seminary at Cambridge, Mass., and was for six years an Episcopal clergyman. In 1891 he retired from the Church and secured a position in the civil service at Ottawa. His poems on the lake region have earned for him the title of "The Poet of the Lakes." His first volume was *Lake Lyrics and Other Poems.* His poem entitled *The Mother,* which appeared in April, 1891, is said to have received more notice than any other single poem that ever appeared in the American press. Among his later works are *The Dread Voyage* and *Beyond the Hills of Dreams,* two volumes of verses; *Mordred* and *Hildebrand,* dramas; *A Beautiful Rebel,* an historical novel; and *The Scotsman in Canada.* He also edited the *Oxford Book of Canadian Verse.*

CAMPBELLFORD, *kam'bel ford,* a town in Northumberland County, Ontario, on the Grand Trunk Railway, thirty miles northwest of Belleville and thirty-three miles east of Peterborough. Campbellford is also on the Trent River, and with the completion of the Trent Valley Canal will probably experience a new development. There is an abundance of water power, which is utilized for manufacturing purposes, chiefly in the making of pulp, paper and flour. There are large saw and planing mills, besides a foundry, tannery, shoe factory and bridge works. The electric light and power system is owned by the town. The region tributary to Campbellford is known for fruits and mixed farming. Population in 1911, 3,051; in 1916, about 3,000.

CAMPBELLTON, *kam'bel ton,* N. B., a town in Restigouche County, at the head of deep-water navigation on the Restigouche River, which empties into Chaleur Bay. It is sixteen miles from Dalhousie, at the mouth of the river. The town is of great importance for its lumber, shingle and planing mills, but other industrial establishments, notably machine works and foundries, are worthy of notice. Large quantities of the town's manufactures are shipped by boat, and the Intercolonial Railway and the International Railway of New Brunswick, both operated by the Dominion, receive the remainder. The vicinity of Campbellton is a favorite resort for trout and salmon fishermen. The town was founded about 1810 and was incorporated in 1889. It was practically destroyed by fire in 1910, but was rapidly rebuilt. Population in 1911, 3,817; in 1916, estimated, 4,530. J.T.R.

CAMPEACHY, or **CAMPECHE,** *kahm pay' chay,* a seaport of Mexico, capital of the state of the same name, is situated on the west coast of the peninsula of Yucatan, on the Gulf of Campeachy, at the mouth of the San Francisco River. Shipbuilding and the manufacture of cigars are the chief industries. A considerable trade in campeachy wood, sisal hemp, and wax is maintained, but the harbor is shallow and can be entered only by vessels of light draught. Population, 17,000.

Gulf of Campeachy, the southwest part of the Gulf of Mexico, lying just to the west of the Peninsula of Yucatan. Its land boundaries are the Mexican states of Campeachy, Vera Cruz and Tabasco. It extends into the Gulf of Mexico as far northward as latitude 21°.

CAMP-FIRE GIRLS, an organization for girls, intended to take the place among them that the Boy Scouts takes among boys (see BOY SCOUTS). It lays emphasis upon out-of-door life, but has as its central purpose the awakening of a realization of the beauty and dignity of home-building and home-keeping. Fire, the symbol of the home, of service and romance, is the chosen emblem, and the three degrees of membership are known as Wood-Gatherer, Fire-Maker and Torch-Bearer. To become a Wood-Gatherer any girl between the ages of ten and twenty has but to learn and repeat the prime law of the organization as follows:

Seek beauty.
Give service.
Pursue knowledge.
Be trustworthy.
Hold on to health.
Glorify work.
Be happy.

To pass on to the degree of Fire-Maker a girl must learn the Fire-Maker's song, a sort of chant which runs as follows:

As fuel is brought to the fire,
So I purpose to bring
My strength,
My ambition,
My heart's desire,
My joy
And my sorrow
To the fire
Of humankind.
For I will tend
As my fathers have tended,
And my father's fathers
Since time began,
The fire that is called
The love of man for man,
The love of man for God.

COUNCIL OF THE CAMP-FIRE GIRLS
The Guardian of the Fire elevates the Wood-Gatherer to the rank of Fire-Maker.

She must also prepare herself for certain definite things, namely:

To help prepare and serve, together with the other candidates, at least two meals for meetings of the Camp-Fire.

To mend a pair of stockings, a knitted undergarment and hem an article having a hem at least one yard in length.

To keep a written, classified account of all money received and spent for at least one month.

To tie a square knot five times in succession correctly and without hesitation.

To sleep with open window or out-of-doors for at least one month.

To take an average of at least half an hour daily outdoor exercise for not less than a month.

To refrain from soda water, chewing gum and candy between meals for at least one month.

To name the chief causes of infant mortality in summer. Tell how and to what extent it has been reduced in one American community.

To know what to do in the following emergencies: clothing on fire; person in deep water who cannot swim; open cut; frosted foot; fainting.

To know the principles of elementary bandaging and how to use surgeon's plaster.

To know what a girl of her age needs to know about herself.

To commit to memory any good poem or song not less than twenty-five lines in length. Know the words of *America*.

To know the career of some woman who has done much for her country or state.

For a Fire-Maker to become a Torch-Bearer is comparatively easy, and depends on the winning of certain "honors." The head of each Camp-Fire organization is the Guardian of the Fire, who must send to the national headquarters in New York City for a license before she is permitted to serve. The Camp-Fire circle was organized in 1911 by Luther H. Gulick and his wife, and has spread rapidly.

A local Camp-Fire may be formed at any time by a group of girls who have agreed to comply with the rules of the organization and have sent in their names and that of the older woman whom they wish as guardian to the national headquarters. This local Camp-Fire chooses for itself a special name, and holds weekly meetings, at which the guardian must be present. The national organization requires no fee, but a local Camp-Fire may decide to pay regular dues, which are then expended as the chapter desires. Very interesting ceremonials are suggested for the weekly meetings and for the more formal Council-Fire which takes place once a month, and information as to these may be obtained from the national office. J.H.B.

CAMPHOR, *kam'fer,* a whitish, semi-transparent gum, with a sharp aromatic taste and characteristic odor, obtained from the wood and bark of a group of trees belonging to the laurel family. The camphor of commerce comes from a tree extensively cultivated in Japan and on the island of Formosa, and is extracted from the chipped wood by steam distillation. The gum is drained and pressed to free it of volatile oil, and the remaining mass is then purified. Spirits of camphor, the most common liquid form of this substance, is a mixture of ten parts camphor, seventy parts alcohol and twenty parts water. Applied externally it acts as a counter-irritant (see BLISTER) and to a slight extent as an anesthetic. As it has antiseptic properties, it is also used in liniments. Taken internally it acts as a nerve stimulant and is used in cases of hysteria, inflammation of the large intestine and cholera. In large doses it is a poison, causing convulsions resulting in death.

CAMPHOR

Camphor gum is used in the manufacture of celluloid and of certain explosives. It is sometimes placed in furs and woolen garments to protect them from moths, though the cheaper substance, naphthalene, is much more commonly employed for this purpose. Camphor has been made artificially from turpentine, but not economically enough to compete with the natural product. The word *camphor* is applied to a class of substances similar to and including the camphor of laurel. Among these are menthol (peppermint camphor), thymol (thyme camphor) and borneol (from a magnificent tree native to Borneo and the adjacent islands of Sumatra and Labuan). J.F.S.

CAMPANINI, *kahm pah nee'ne,* CLEOFONTE (1860-), an orchestra conductor, prominently identified with the notable present-day musical ventures in Europe and America. He was born at Parma, Italy, and was educated as a violinist. When twenty-three years of age, while conductor of the Parma Opera House, he was offered an engagement at the Metropolitan Opera House, New York City. Under his baton Madame Sembrich made her New York debut in *La Sonnambula.* From 1903 to 1906 he conducted orchestras at Milan, Rome, Naples and Venice. At Milan he produced the first version of Puccini's *Madame Butterfly,* which was not favorably received; later he successfully directed this same opera at Covent Garden, London, and in America's leading cities. As conductor of the Manhattan Opera House, New York City, he introduced *Louise, Samson and Delilah, Thais* and *The Damnation of Faust.* From 1910 to 1913 he was director of the Chicago Grand Opera Company, and in 1913 he succeeded to the position of general manager.

CAMPUS MARTIUS, *kam'pus mar'shus,* a large open space outside the walls of ancient Rome, set apart for military combats and athletic exercises, and sacred to the god Mars, for whom it was named. It lay between the Pincian, the Quirinal and the Capitoline hills and the Tiber River, and in early times had an area of about 320 acres. Later the field was made smaller by the erection of numerous private and public buildings, and towards the end of the republic the Campus Martius became a suburban pleasure ground for the Romans, with gardens, shady walks, baths and theaters. Agrippa erected there the old Pantheon, and Augustus built a magnificent tomb for himself. The site of the original Campus Martius is now occupied by a thickly-settled portion of the modern business city. See map, in article ROME.

CAM'ROSE, a town in the Canadian province of Alberta, 175 miles northeast of Calgary and forty-seven miles southeast of Edmonton. Since 1906, when it was incorporated, it has had a remarkable development. Then it was a little settlement at the end of a branch line of the Canadian Pacific Railway; now it is an important railway center, with branch lines of the Canadian Northern and the Grand Trunk Pacific, as well as the Canadian Pacific, radiating in all directions. From an uninhabited spot on the plains it grew to a population of 1,586 in 1911, and to over 2,000 in 1916. Its growing importance has won distinct recognition, for it is now the headquarters for one-fourth of the provincial telephone system, and also has one of the two provincial normal schools. The normal school was opened in January, 1915, in a splendid new building erected at a cost of $250,000.

The Camrose district is famous for its rich farms and its coal deposits. Mixed farming in the region brought wholesale and farm implement houses to the town. The coal deposits, a good grade of semi-bituminous, are in many places near the surface and are easily mined. Valuable deposits are known to exist inside the town limits. Camrose owns its water works and electric light and power plant, which are operated at a low cost because of the cheapness of coal. The town is under the single-tax system, there being no licenses or taxes on buildings or improvements of any kind (see

Single Tax). A large part of the population of the town and the surrounding country is Norwegian, and the town is the seat of the Norwegian Lutheran College. J.D.S.

CANAAN, *ka'nan.* See Palestine.

CANAANITES, *ka'nan ites,* in general, the name given to the heathen nations dwelling in Palestine west of the Jordan before the conquest of the country by the Israelites. At the time of the Israelitish invasion these nations were the Hittites, Jebusites, Hivites and Amorites. According to the Old Testament the Canaanites were descendants of Ham, the second son of Noah; but as a matter of fact only a part of the nations dwelling in Palestine at the time of the Israelitish invasion were Canaanites. The Canaanites were gradually subdued by the Israelites, and in Solomon's time they all paid tribute.

In language, government, morals and religion these people were different from the Israelites, the principal feature of their religion being the worship of Baal and Asherah, his consort, who was called "the happy." The symbol of Asherah was the stem of a tree, though this was sometimes carved into an image. The symbol of Baal was probably a cone, and represented the rays of the sun. It was undoubtedly the mingling of these symbols in large numbers which constituted the groves of Baal, so frequently mentioned in the Old Testament. See Palestine.

CANADA, Dominion of. A little more than a century and a half ago, in 1763, at the close of the French and Indian War, Great Britain formally acquired ownership of the French possessions in North America, then called New France. At that time there were a few scattered settlements along the Saint Lawrence and its tributaries, but even Montreal and Quebec, the largest, were small villages. Canada in those days was but a small part of the present province of Quebec. There were a few settlements in Nova Scotia and on Prince Edward Island, but with these exceptions the rest of the great area now included in the Dominion was the home of Indians and wild beasts. The population of Canada was not more than 70,000.

For about fifteen years this new British possession was practically a colony of Frenchmen, with a few British officials. During and after the Revolutionary War, however, a new element appeared, and by 1783 there were thousands of English-speaking settlers. They had left the rebellious English colonies, and had moved northward to establish new homes for themselves. These United Empire Loyalists formed the first important groups of English-speaking Canadians, and on the foundation-stones which they helped to lay has risen a great self-governing nation which remains a loyal, integral part of the British Empire.

Practically the whole of Canada's present population is descended from Europeans, and its present area is slightly larger than the whole of Europe. Canada, except the southern part of Ontario, Quebec and the Maritime Provinces, lies north of the forty-ninth parallel of latitude. In Europe, north of this parallel, are the richest and most populous nations. If Canada had centered under its one government all the peoples, the industries and the resources of Europe north of this parallel, it would be the greatest nation in the western hemisphere and probably one of the most populous and the richest in the world. It would boast of a population of 250,000,000 people.

In Europe north of the forty-ninth parallel are the British Isles, Belgium, the Netherlands, Denmark, Norway, Sweden, nine-tenths of Germany, nearly all of Russia, the northern third of Austria-Hungary and all of France north of Paris. Near the sixtieth parallel are three great capital cities, Petrograd, Christiania and Stockholm. Moved directly westward, Petrograd would lie on the east shore of Hudson Bay, about as far north as the northern end of Labrador. Then Stockholm would be in the middle of Hudson Bay and Christiania would be on the west shore near Fort Churchill. London, Berlin and Vienna would all fall within the Dominion (see second map following).

In proportion to the population which Canada would have if Europe were transplanted upon it, its present population is but a tiny nation. The census of 1911 gave the Dominion a total of 7,206,643 inhabitants, about the same as Rumania, one of the smaller European nations. Bulgaria has about 4,300,000 people, Serbia and Greece together have over 6,000,000, while Montenegro adds 516,000. Thus the Balkan states, though only a small spot on

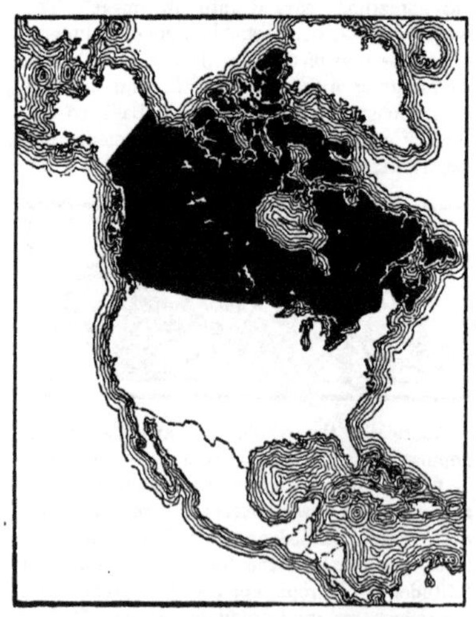

THE DOMINION OF CANADA
Showing geographical position in North America, showing Newfoundland and Labrador, also.

the map of the world, have a population two and a half times that of the Dominion.

These comparisons can merely suggest the vast possibilities which lie within the Dominion. Its natural wealth has already given it a remarkable development, but its resources are still to a great extent undeveloped. What these resources are, how they have been and are being used, how the people who develop them live and govern themselves, how they have struggled in the past and are still struggling to make the history of Canada a noble one—all this is the story of Canada.

Area. The Dominion of Canada, with a total area of 3,729,665 square miles, occupies slightly less than one-half of the North American continent. With the exception of Alaska, Newfoundland and Labrador, it includes all of North America north of the United States. The greatest length of the Dominion, from east to west, is about 2,700 miles; from north to south its greatest extent is 1,600 miles. Canada is the largest country in the world, Russia and China excepted. Its vast area is divided into provinces and territories as follows:

	SQUARE MILES
Alberta	255,285
British Columbia	355,855
Manitoba	251,832
New Brunswick	27,985
Nova Scotia	21,428
Ontario	407,262
Prince Edward Island	2,184
Quebec	706,834
Saskatchewan	251,700
Northwest Territories	1,242,224
Yukon	207,076
Total	3,729,665

This total does not include Hudson Bay, which covers 443,750 square miles, and the Gulf of Saint Lawrence, which covers 101,562 square miles. It does, however, include those portions of the Great Lakes which lie within the Canadian boundaries, and all other water surface in the Dominion. The total area under water is 125,755 square miles. The areas of the provinces as given in the table above were established in 1912. By the boundary revision of that year the area of Manitoba was increased by 178,100 square miles, Ontario by 146,400 square miles and Quebec by 354,961. The additions to Quebec, Ontario and Manitoba were formerly part of the Northwest Territories.

The People of Canada

Their Number. From a population of 70,000, scattered along the rivers, in 1760, Canada has grown to a nation which spreads from ocean to ocean and at the last census (1911) included 7,206,643 persons. It is an interesting fact that in the forty years from 1871 to 1911 the population of Canada increased approximately as much as in the whole preceding century. More than one-half of the increase since 1871 came in the decade from 1901 to 1911. In 1791 the total population was nearly 150,000, of which only about 25,000 were in Upper Canada (Ontario). In 1871, at the first census of the Dominion, Ontario had nearly one-half of the total, which was then 3,689,257. During each decade until 1901 the population increased by about 500,000; but from 1901 to 1911 it jumped from 5,371,315 to 7,206,643, an increase

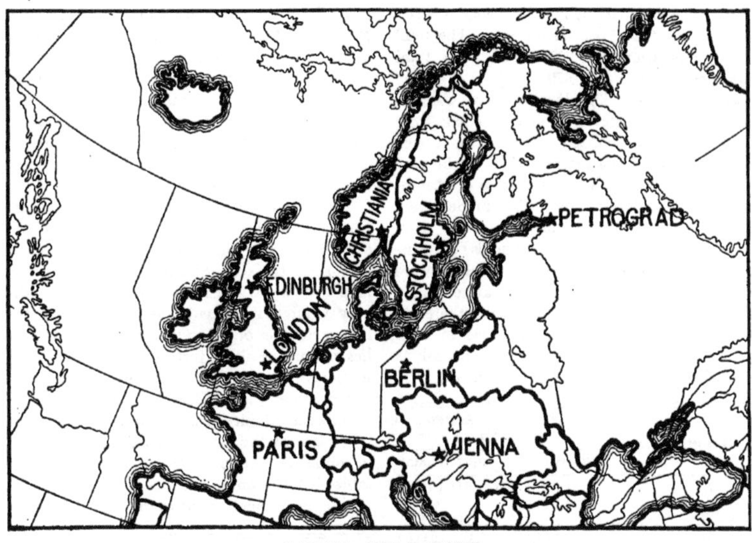

CANADA AND EUROPE
A map of comparative locations in like latitudes. Stockholm, for illustration, is as far north as the center of Hudson Bay; Paris is just south of the latitude of the international boundary.

nearly four times the normal. The table below shows the distribution of the population by provinces, the density of population and the number of males and females. For the growth of population in the provinces, see the article on each province, in these volumes.

POPULATION OF THE DOMINION OF CANADA

PROVINCES	TOTAL, 1911	DENSITY OF POPULATION PER SQ. MI.	MALE	FEMALE
Alberta	374,663	1.46	223,989	150,674
British Columbia	392,480	1.10	251,619	140,861
Manitoba	455,614	6.18	250,056	205,558
New Brunswick	351,889	12.61	179,867	172,022
Nova Scotia	492,338	22.98	251,019	241,319
Ontario	2,523,274	9.67	1,299,290	1,223,984
Prince Edward Island	93,728	42.91	47,069	46,659
Quebec	2,003,232	5.69	1,011,502	991,730
Saskatchewan	492,432	1.95	291,730	200,702
Yukon	8,512	0.04	6,508	2,004
Northwest Territories	18,481	0.01	9,346	9,135
Totals for Canada	7,206,643	1.93	3,821,995	3,384,648

Urban and Rural Communities. Fully as noteworthy as the increase in population as a whole has been the marked growth of cities and towns. In 1901 the urban population of the Dominion was only 37.6 per cent of the total, while in 1911 it had risen to 45.5 per cent. In ten years the urban population increased four times as much as the rural, and in Ontario and the three Maritime Provinces the rural population is actually decreasing. The relative growth of cities has been greatest in the Western provinces, where many flourishing towns have risen "over-night." Calgary's population from 1901 to 1911 increased 994 per cent; Vancouver's, 272 per cent, and Winnipeg's, 221 per cent. Many other cities, both in the east and west, increased forty or fifty per cent or more. About 2,354,000 people, or 32.7 per cent of the total population, live in towns or cities having more than 5,000 inhabitants. The total urban population is 3,280,964 and the total rural population is 3,925,679.

Their Origins. The Dominion of Canada, like the United States, includes people from all parts of the civilized world. This varied stock has come more recently than in the case of the United States, and the process of absorption is comparatively slow. The French were the first settlers, and to this day, in the province of Quebec, there are hundreds of communities which are as French as they were 150 years ago. French is still the native tongue of nearly all Quebec. In Nova Scotia the early settlers were mostly Scotch, whereas in New Brunswick, Ontario and the eastern townships of Quebec the first important settlements were made by United Empire Loyalists from New England. The Western provinces also have their share of Scotch, Irish, English and French, as well as a larger proportion of other nationalities. The table in column one shows the birth or descent of the most numerous foreign elements. For the purposes of this table, which is based on the census of 1911, the original Canadians, the Indians, are included.

NATIONALITIES	INHABITANTS CLASSIFIED ACCORDING TO	
	ANCESTRY (a)	BIRTHPLACES
English	1,823,150	510,674
Irish	1,050,384	92,874
Scotch	997,880	169,391
Total British	3,896,985	6,453,104 (b)
Austro-Hungarian	129,103	121,430
Chinese	27,774	27,083
Dutch	54,986	3,808
French	2,054,890	17,619
German	393,320	39,577
Indian	105,492	
Italian	45,411	34,739
Japanese	9,021	8,425
Jewish	75,681	
Polish	33,365	
Russian	58,639	100,971 (c)
Scandinavian	107,535	49,194
Other Foreign	214,441	350,693 (d)
Total	7,206,643	7,206,643

(a) Canadian citizens born in the United States are classified in this table according to the country from which their ancestors came.
(b) Includes 5,619,682 born in Canada.
(c) Includes Polish and Russian Jews.
(d) Includes 303,680 born in the United States.

Their Religions. The Dominion has no State Church, although in Quebec the Roman Catholic Church has certain privileges which it has held since the days of French rule. The Roman Catholic Church was for a long time the only Church in Canada, and it still has the largest number of adherents, the Roman Catholics comprising about forty per cent of the population. The Roman Catholics are strongest in Quebec and are of least relative importance in Alberta, Manitoba and Ontario. The Presbyterians, Methodists and Anglicans rank next in order in the number of members. The table below shows the membership of the leading denominations at the census of 1911:

RELIGIONS	TOTAL MEMBERSHIP	PER CENT OF TOTAL POPULATION
Anglicans	1,043,017	14.47
Baptists	382,666	5.31
Christians	16,773	0.23
Congregationalists	34,054	0.47
Greek Church	88,507	1.23
Jews	74,564	1.03
Lutherans	229,864	3.19
Mennonites	44,611	0.62
Methodists	1,079,892	14.98
Mormons	15,971	0.22
Presbyterians	1,115,324	15.48
Roman Catholics	2,833,041	39.31
Salvation Army	18,834	0.26
Others (including Pagans and No-Religion)	229,525	3.18
Total	7,206,643	

Physical Characteristics of the Dominion

The physical features of Canada are comparatively simple and easily understood. More than one-half of the total surface slopes gently towards Hudson Bay. To the east and west are higher lands, rising in the west to some of the highest mountains in North America. The interior is not so much a valley as a great trough, irregular in shape, but considerably wider at the north than at the south. Although the whole of this interior might be expected to drain into Hudson Bay, two low heights of land are sufficient to turn Canada's largest rivers, the Mackenzie and the Saint Lawrence, into smaller troughs of their own; the Mackenzie flows northwest to the Arctic Ocean and the Saint Lawrence northeast to the Atlantic Ocean.

For purposes of study, then, the surface may be divided into these three regions: (1) The hilly eastern half, including the Laurentian plateau and the extension of the Appalachian system east of the Saint Lawrence Valley;

IN THE CANADIAN ROCKIES

The Three Sisters, Canmore

Giant Steps in Paradise Valley

Mount Assiniboine

(2) the interior plains, characterized by their lack of trees; (3) the mountains between the plains and the Pacific coast, including the Rocky Mountains and a number of smaller ranges. Each of these sections is described in detail below.

1. **Eastern Canada.** In this part is included all the territory from Hudson Bay to the Labrador coast, as well as an area west of Hudson Bay (see LAURENTIAN PLATEAU). In this section the highest points, in Labrador, are about 8,000 feet above the level of the Atlantic Ocean. From Labrador westward the altitude declines quickly, and the vast interior of Northern Quebec seldom exceeds 2,000 feet in altitude. The interior is a succession of low ridges of hard rocks, sometimes covered with a few trees, but as often bare. Between these ridges are lakes, swamps and rapidly-flowing rivers. With the exception of the Saint Lawrence, all the important rivers flow into Hudson Bay, which also receives several large rivers from the plains. The Hudson Bay basin is bounded at the south by a ridge, hardly noticeable; this is known as "the height of land."

South of this ridge the rivers are tributary to the Saint Lawrence, whose valley was the most important physical feature in determining the early course of Canadian history. Without the Saint Lawrence River, French Canada would probably have clung to the Atlantic shores; but with it the early explorers, trappers, traders and missionaries found an easy way to the interior. It is not astonishing that the Saint Lawrence Valley, including the fertile plains of Southern Ontario, has always been the richest and most populous part of Canada. Though there are some deposits of petroleum, natural gas and salt, the plains have no metals, and the wealth of this section is based on its fertile soil and temperate climate. The Saint Lawrence Valley is essentially a farming section, whereas the extensive areas of Northern Ontario and Quebec are mainly unsuited for cultivation.

As the Saint Lawrence approaches its mouth the valley becomes narrower and is shut in by the rocky Gaspé Peninsula. This peninsula and the Maritime Provinces of New Brunswick, Nova Scotia and Prince Edward Island comprise the *Acadian* region. Taken as a unit this region is a continuation of the Appalachian highlands, which extend from Alabama northward parallel to the Atlantic coast of the United States. The coast line of these provinces is long and irregular, both on the Atlantic side and on the Gulf of Saint Lawrence. Coal and some metals are found in New Brunswick and Nova Scotia, and in the Gaspé Peninsula, which belongs to Quebec, are copper and the largest deposits of asbestos in Canada. The surface of this section was once almost completely covered with forests, and large tracts of timber still remain.

Besides the Saint Lawrence basin and the Acadian region, there is one other important feature of Eastern Canada's topography—the Niagara escarpment (meaning a steep slope). This is a line of cliffs forming a break in the Saint Lawrence basin and it runs from Queenston Heights west to the head of Lake Ontario, near Hamilton, then northwest until it forms the Bruce Peninsula, shutting off Georgian Bay from Lake Huron. The Niagara escarpment causes falls on the rivers which plunge over it, Niagara Falls (which see) being the greatest and most famous.

2. **The** Interior Plains. Between Eastern Canada and the mountains on the west is a vast region of plains, about 700 miles wide from east to west at the international boundary and gradually narrowing to 400 at the Arctic Ocean. These plains are a continuation of the great interior plains of the United States, and in a general way the two regions of plains are similar, but the Canadian plains are more broken and have more timber. The international boundary, latitude 49° N., nearly coincides with the watershed which divides the drainage into Hudson Bay from the great Mississippi system. From the international boundary the plains slope gradually northward and from the Rocky Mountains gradually eastward. The general slope, therefore, is from southwest to northeast.

Some interesting estimates of the degree of slope were made by George M. Dawson, while he was director of the Dominion Geological Survey. A line drawn northeast from the intersection of the international boundary and the mountains to a point on the prairie north of Lake Winnipeg would have an average drop of 5.38 feet per mile. A line drawn straight east to the valley of the Red River would have an average slope of 4.48 feet per mile. Southwestern Alberta has an altitude of 4,500 feet, whereas the lowest point in the Red River Valley is only 800 feet above sea level.

A considerable part of this difference in altitude is due to two escarpments, or sharp bluffs, which break the surface and really form three separate lands. The first level is the valley

AREA AND POPULATION

Political Divisions	Area, Square Miles: Land	Water	Total	Population, 1911: Rural	Urban	Total	Seat of Government	Population
Canada, Dominion of	3,603,910	125,725	3,729,665	3,925,502	3,281,141	7,206,643	Ottawa	87,062
Provinces:								
Alberta	252,925	2,360	255,285	232,726	141,397	374,673	Edmonton	24,900
British Columbia	353,416	2,439	355,855	188,796	203,684	392,480	Victoria	31,660
Manitoba	231,926	19,906	251,832	255,249	200,365	455,614	Winnipeg	136,035
New Brunswick	27,911	74	27,985	252,342	99,547	351,889	Fredericton	7,208
Nova Scotia	21,068	360	21,428	306,210	186,128	492,338	Halifax	46,619
Ontario	365,880	41,382	407,262	1,194,785	1,328,489	2,523,274	Toronto	376,538
Prince Edward Island	2,184		2,184	78,758	14,970	93,728	Charlottetown	11,203
Quebec	690,865	15,969	706,834	1,032,618	970,614	2,003,232	Quebec	78,190
Saskatchewan	243,382	8,318	251,700	361,067	131,365	492,432	Regina	30,213
Territories:								
Yukon	206,427	649	207,076	4,647	3,865	8,512	Dawson	3,013
Northwest Territories	1 207,926	34,298	1,242,224	18,481		18,481		

LEADING INDUSTRIES

Political Divisions	Value of Products: Agriculture, 1911	Manufactures, 1911	Mining, 1914	Fisheries, 1915
Canada, Dominion of	$725,301,375	$1,165,975,639	$128,475,499	$35,860,708
Alberta	48,124,564	18,788,825	12,773,669	94,134
British Columbia	16,982,193	65,204,236	24,202,924	14,553,320
Manitoba	68,218,308	53,673,609	2,428,902	742,925
New Brunswick	20,360,596	35,422,302	1,034,706	4,737,145
Nova Scotia	24,171,381	52,706,184	17,514,786	9,166,851
Ontario	296,595,793	579,810,225	52,147,973	3,341,182
Prince Edward Island	11,553,780	3,136,470		933,682
Quebec	133,329,871	350,901,656	12,259,637	2,076,851
Saskatchewan	105,964,889	6,332,132	710,840	165,888
Yukon			5,402,062	63,730

PRINCIPAL RIVERS

Name	Length Miles	Drainage Basin Sq. Mi.	Name	Length Miles	Drainage Basin Sq. Mi.	Name	Length Miles	Drainage Basin Sq. Mi.
Abitibi	340	11,300	George	365	20,000	Red Deer	385	18,300
Albany	610	59,800	Hamilton	350	29,100	Rupert	380	15,700
Assiniboine	450	52,600	Kazan	455	32,700	Saguenay	405	35,900
Athabaska	765	58,900	Koksoak	535	62,400	St. John	390	21,500
Attawapiskat	465	18,700	Kootenay	400	15,500	St. Lawrence	1,900	†309,500
Backs	605	47,500	Lewes	338	35,000	St. Maurice	325	16,200
Big	520	26,300	Liard	550	100,700	Saskatchewan	1,205	158,800
Bow	315	11,100	Mackenzie	2,525	682,000	Severn	420	38,600
Churchill	1,000	115,500	Miramichi	135	5,400	Skeena	335	19,300
Columbia *	465	39,300	Moose	340	42,100	Slave	265	180,000
Coppermine	525	29,100	Nelson	390	370,800	S. Saskatchewan	865	65,500
Dubawnt	580	58,500	N. Saskatchewan	760	54,700	Stewart	320	21,900
Eastmain	375	25,500	Nottaway	400	29,800	Stikine	335	20,300
English	330	20,600	Ottawa	685	56,700	Thompson	270	21,800
Fraser	695	91,700	Peace	1,065	117,100	Winisk	295	24,100
French	180	8,000	Pelly	330	21,300	Winnipeg	475	44,000
Gatineau	240	9,100	Red	545	63,400	Yukon *	655	145,800

* Partly in Canada.
† Not including the Great Lakes.

LAKES

Name	Area Sq. M.	Elevation Feet	Name	Area Sq. M.	Elevation Feet	Name	Area Sq. M.	Elevation Feet
Aberdeen	514	130	Great Bear	11,821	391	Nipigon	1,730	852
Abitibi	331	900	Great Slave	10,719	520	Nipissing	330	640
Athabaska	2,842	690	Huron *	14,331	582	Ontario *	3,727	247
Atlin	343	2,200	Kaminuriak	368	320	Payne	747	
Aylmer (N. W. T.)	612	795	Kaniapiskau	441	1,850	Rainy *	260	1,109
Baker	1,029	30	Kootenay	220		Reindeer	2,436	1,150
Bras d'Or	360	†	Lesser Slave	480	1,890	St Clair *	257	578
Claire	404	700	MacDougall	318		St. John	350	
Clearwater	478	790	Mackay	980	700	Seul, Lac	392	1,153
Clinton-Colden	674	1,221	Maguse	490	350	South Indian	1,531	800
Cree	407	1,530	Manitoba	1,817	700	Superior *	11,178	603
Dubawnt	1,654	500	Martre, Lac la	1,225	450	Winnipeg	9,459	714
Erie *	5,019	573	Melville	1,298		Winnipegosis	2,086	831
Etawney	625	750	Michikamau	613	1,650	Wollaston	906	1,300
Garry	980		Mistassini	975	1.350	Woods, L. of the *	1,325	
Gras, Lac de	674	700	Moose	552	812	Yathkyed	858	300

* Partly in Canada. † Sea level.

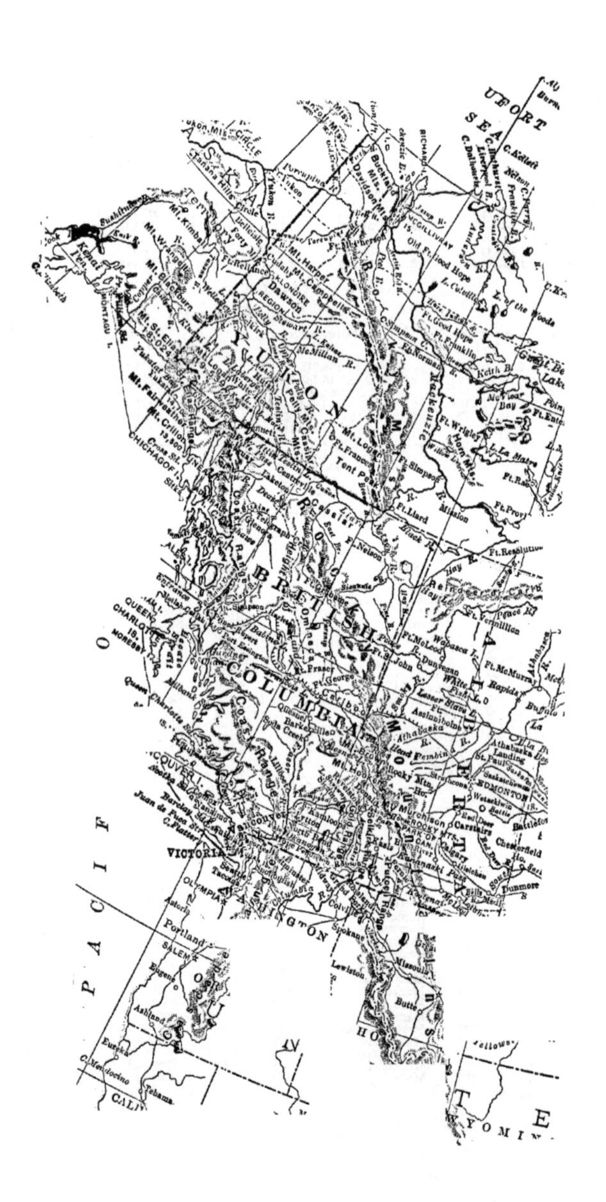
BRITISH COLUMBIA
YUKON
VICTORIA
Vancouver
Portland
Spokane
Butte
Dawson
KLONDIKE REGION
EDMONTON
Calgary

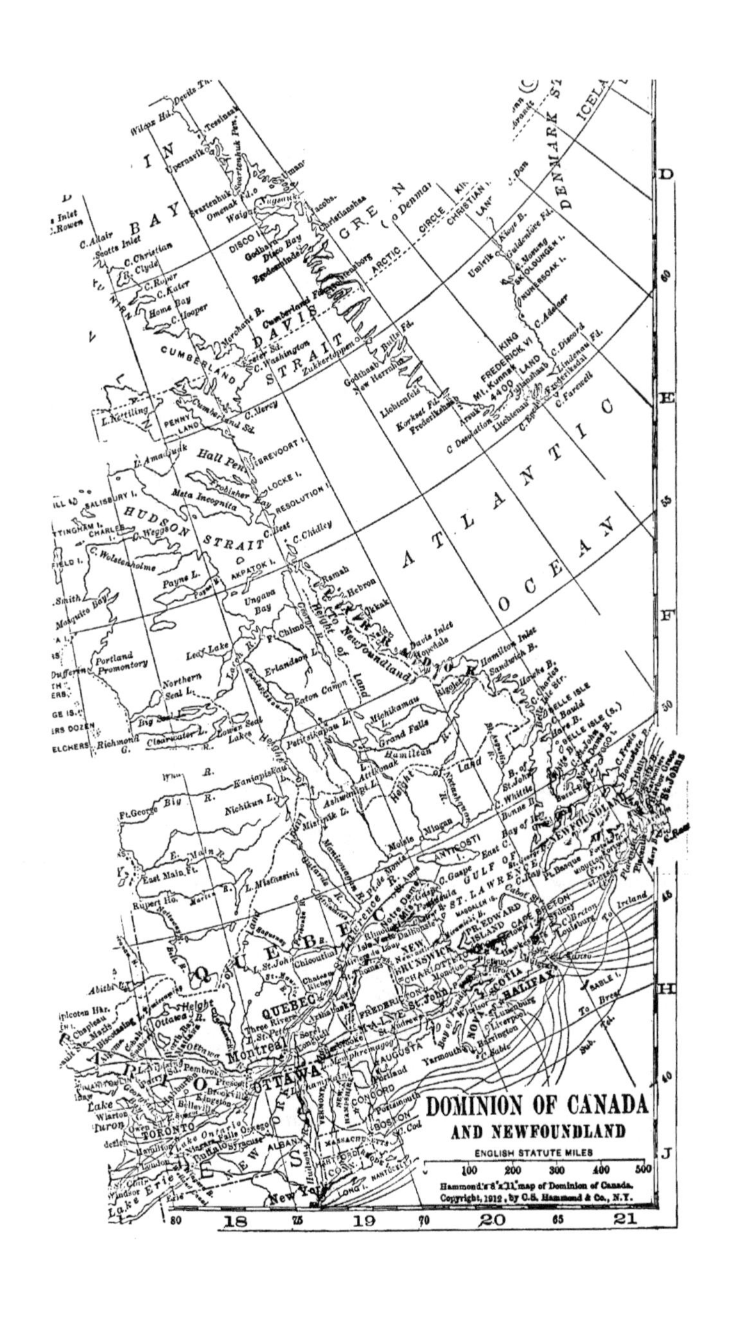
DOMINION OF CANADA
AND NEWFOUNDLAND
ENGLISH STATUTE MILES
100 200 300 400 500
Hammond's 8"x11" map of Dominion of Canada.
Copyright, 1912, by C.S. Hammond & Co., N.Y.
ATLANTIC OCEAN
HUDSON STRAIT
DAVIS STRAIT
GULF OF ST. LAWRENCE
QUEBEC
OTTAWA
Montreal
TORONTO
HALIFAX
NEWFOUNDLAND
St. Johns
NOVA SCOTIA
NEW BRUNSWICK
FREDERICTON
CHARLOTTETOWN
ANTICOSTI
Lake Ontario
Lake Erie
Lake Huron
New York
BOSTON
CONCORD
AUGUSTA
ALBANY
Ungava Bay
Hamilton Inlet
Grand Falls
Belle Isle
Cape Breton
Sydney
Yarmouth
Liverpool
Cumberland Sd.
Frobisher Bay
Resolution I.
Cape Farewell
ARCTIC CIRCLE
DENMARK ST.
To Ireland
To Brest
D
E
F
H
J
18
19
20
21

PROGRESS IN POPULATION MATERIAL, INDUSTRIES, ETC.

	1871	1901	1911
Area	3,272,500	3,729,665	3,729,665
Population	3,485,761	5,371,315	7,206,643
Population per sq. mile	1.06	1.16	1.29
Net debt	$ 77,706,517	268,480,004	340,042,052
Gross debt	$115,492,683	354,732,433	474,941,478
Assets	$ 37,786,165	86,252,429	134,899,435
Chartered banks and branches, No	204	758	2,376
Capital	$ 37,095,340	67,035,615	103,009,256
P. O. and Government Savings Banks:			
Amount on deposit	$ 4,569,297	56,048,957	58,094,331
Agriculture:			
Farms and farm property		$1 787,102,630	4,231,840,636
Farm products		$ 363,126,384	782,886,494
Farm animals		$ 275,167,627	631,103,420
Manufactures:			
Establishments, number	41,257	14,650	19,218
Value of products	$221,617,773	481,053,375	1,165,975,639
Revenue, Consolidated Fund	$ 19,335,560	52,514,701	117,780,410
Customs	$ 11,841,105	28,425,284	71,838,088
Excise	$ 4,295,945	10,318,266	16,869,837
Public Works	$ 1,146,240	5,770,071	10,818,834
Post Office	$ 612,631	3,441,505	9,146,952
Expenditure, Consolidated Fund	$ 15,623,081	46,866,368	87,774,198
Charges on debt	$ 6,013,625	13,490,153	14,116,044
Railways and Canals	$ 921,355	6,558,375	11,758,269
Public Works	$ 675,575	3,659,441	9,216,300
Post Office	$ 815,471	3,931,446	7,954,223
Provincial Subsidies	$ 2,624,940	4,250,607	9,092,472
Expenditure, Capital Account	$ 3,670,397	7,693,857	30,852,963
Exports, total	$ 74,173,618	196,487,632	297,196,365
Per capita	$ 21.08	36 37	41 52
To Great Britain	$ 24,345,240	105,328,956	136,965,111
To United States	$ 30,815,185	72,382,230	119,396,801
To other countries	$ 19,013,193	18,776,446	40,834,453
Imports, total	$ 96,192,971	190,415,525	472,247,540
Per capita	$ 27.31	35.24	65 97
From Great Britain	$ 49,168,170	43,018,164	109,936,462
From United States	$ 29,022,387	110,485,008	284,934,739
From other countries	$ 8,756,925	27,734,816	67,080,117
Duty collected	$ 11,843,656	29,106,980	73,312,368
Aggregate trade	$170,266,589	386,903,157	769,443,905
Sea-going vessels, inwards, tonnage	2,521,573	7,514,732	11,919,339
Sea-going vessels, outwards, tonnage	2,594,520	7,028,330	10,377,847
Vessels employed on inland waters, entered, tonnage	4,055,198	5,720,575	13,286,102
Vessels employed on inland waters, cleared, tonnage	3,054,797	5,766,171	11,846,257
Vessels employed in coasting trade, tonnage		34,444,796	66,627,934
Mineral production, total		$ 65,804,611	103,220,994
Metallic minerals		$ 41,939,500	46,105,423
Non-metallic minerals		$ 23,865,111	57,115,571
Railways:			
Mileage	2,695	18,140	25,400
Capital and Bonded Debt	$257,035,188 *	816,110,837	1,528,689,201
Earnings	$ 19,358,085 *	72,898,749	188,733,494
Working expenses	$ 15,775,532 *	50,368,726	131,034,785
Passengers carried, number	5,541,814 *	18,385,722	37,097,718
Freight, tons	6,331,757 *	36,999,371	79,884,282
Government Railways, working expenses	$ 442,993	5,739,052	10,037,879
Government Railways, revenue	$ 565,714	5,213,381	10,249,394
Post Offices, number	3,943	9,834	13,811
Postal revenue	$ 1,079,767	4,641,608	9,146,952
Postal expenditure	$ 1,271,006	5,153,622	7,954,223
Money orders issued, number	120,521	1,151,024	4,840,896
Value of orders issued in Canada	$ 4,546,434	17,956,258	70,614,862
Value of orders paid in Canada	$ 4,067,735	14,324,289	54,297,619
Immigrants arrived		49,149	311,084

* In 1876

MOUNTAINS

Name	Location	Altitude Feet	Name	Location	Altitude Feet	Name	Location	Altitude Feet
Alberta	Alberta	12,000	Dome	Alberta	11,650	Robson	Br. Columbia	13,068
Alexander	Alberta *	11,650	Fairweather	Br. Columbia	15,292	St. Elias	Yukon	18,024
Assiniboine	Alberta *	11,860	Forbes	Alberta	12,100	Sir Sandford	Br. Columbia	11,590
Athabaska	Alberta *	11,900	Goodsir	Br. Columbia	11,676	Saskatchewan	Alberta	11,500
Augusta	Yukon	13,918	Hubbard	Yukon	12,064	Stutfield Peak	Alberta	11,400
Bryce	Alberta *	11,686	Hungabee	Br. Columbia	11,447	Temple	Alberta	11,626
Chown	Alberta *	11,500	Lefroy	Alberta *	11,220	"The Twins"	Alberta	11,800
Columbia	Alberta *	12,740	Lituya	Br. Columbia	11,745	Vancouver	Yukon	15,660
Cook	Yukon	13,758	Logan	Yukon	19,539	Victoria	Alberta *	11,355
Crillon	Br. Columbia	12,750	Lyell	Alberta *	11,463	Woolley	Alberta	12,000
Diadem Pk.	Alberta	11,500	Newton	Yukon	13,774			

* Alberta—Br. Columbia boundary.

of the Red River, together with the rich prairies west of Winnipeg. On this lowest level is a series of lakes, including Manitoba, Winnipeg, Winnipegosis and many smaller ones, which empty by rapid rivers into Hudson Bay. Lake Winnipeg and its vicinity were once covered by a glacial lake which covered an area greater than all of the present Great Lakes combined (see GLACIAL PERIOD).

Overlooking this flat prairie region is the first escarpment, which raises the average altitude from 800 feet to 1,600 feet. This level, too, has many lakes, whose surplus waters are carried off by the Churchill, English, Dubawnt and other rivers, northeast to Hudson Bay. The southern part of this level is drained by the Saskatchewan River, which rises at the foot of the Rocky Mountains and carries waters from all three levels into Lake Winnipeg and thus into Hudson Bay.

The second level rises gradually until it reaches the Missouri Coteau, sometimes called the *Grand Coteau des Prairies,* in Central Saskatchewan. This second escarpment represents a rise to an average level of 3,000 feet, slowly sloping upward to a maximum of 4,500 feet in Southwestern Alberta. Throughout the plains region there are few hills, and an elevation of a few hundred feet is so unusual that it is usually called a "mountain." A characteristic of the region is the deep valleys which the rivers have worn for themselves in the soft rock. These are usually narrow and sometimes reach a depth of 100 or 200 feet, so that they are miniature canyons.

The southern part of the plains is treeless, except in the river beds. If the distance is not too great the course of the rivers can be followed by the narrow green band of trees on the banks. There are numerous lakes on the second and third levels, but in the southern part the climate is so dry that many of them are strongly alkaline and some of them dry up entirely during the summer. The northern part of the plains, north of the North Saskatchewan River, is separated from the Saskatchewan drainage system by a line of highlands which acts as a watershed. This northern system includes the Athabaska and the Peace rivers, together with all the other great rivers and the lakes which are tributary to the Mackenzie, the greatest river system in Canada. The Saskatchewan and the Red River, though smaller than the Mackenzie, are economically more important. They were formerly the main avenues of travel, and their valleys are now the centers of population and wealth in the Canadian West.

3. The Mountains. The third of the great physical divisions is the mountain belt which extends over practically the whole of British Columbia, the Yukon and the western part of Alberta. This does not mean that all of British Columbia and the Yukon is mountainous, for there are long stretches of flat or rolling country which separate the ranges. The entire region, however, is a part of the Cordilleran belt which extends along the western coast of both North and South America. While its physical characteristics are somewhat confused and complicated, its dominating features are the Rocky Mountains and the Coast Ranges. All the ranges run approximately northwest in parallel lines, there being only a few smaller cross-ranges.

In these ranges there is an abundance of magnificent scenery. The eastern slopes are often gradual, but on the west the mountains rise abruptly from green valleys to snow-covered peaks. The perpetual snow and ice on the summits of these peaks is due to their high latitude rather than their altitude, for only a few peaks rise over 10,000 or 11,000 feet. Among the loftiest of these are Robson, Alberta, Assiniboine, Columbia, Murchison, Fairweather, Vancouver and Logan. Mount Logan, in the Yukon Territory, has an altitude of 19,539 feet, the highest point in Canada. For a detailed account of the physical geography of the mountain region, see ALBERTA; BRITISH COLUMBIA; YUKON TERRITORY.

Climate. The climate of a country so vast in extent as Canada cannot be briefly summarized. Even in the limits of a single province there are frequently astonishing variations, due primarily to differences in latitude and altitude, but also to the presence of large bodies of water, sheltering mountain ranges and other local conditions. Canada extends from latitude 42° N. almost to the North Pole, a distance considerably more than one-fourth of the total distance from the North to the South Pole. Southern Ontario, the southernmost part of the Dominion, is in the latitude of Rome, Constantinople and Peking. Between the extremes of polar cold and the mildness of the temperate zone are many steps which can only be indicated roughly. Taken as a whole, Canada has bracing weather, both summer and winter, with plenty of sunshine. The winters are severe and are accompanied by a heavy fall of snow. Spring is short, as is also summer,

Meteorological Service
CANADA
AVERAGE DATE
OF THE
LAST FROST IN SPRING
R.F. Stupart, Dir.
PACIFIC OCEAN
ATLANTIC OCEAN
HUDSON BAY
DAVIS STRAIT

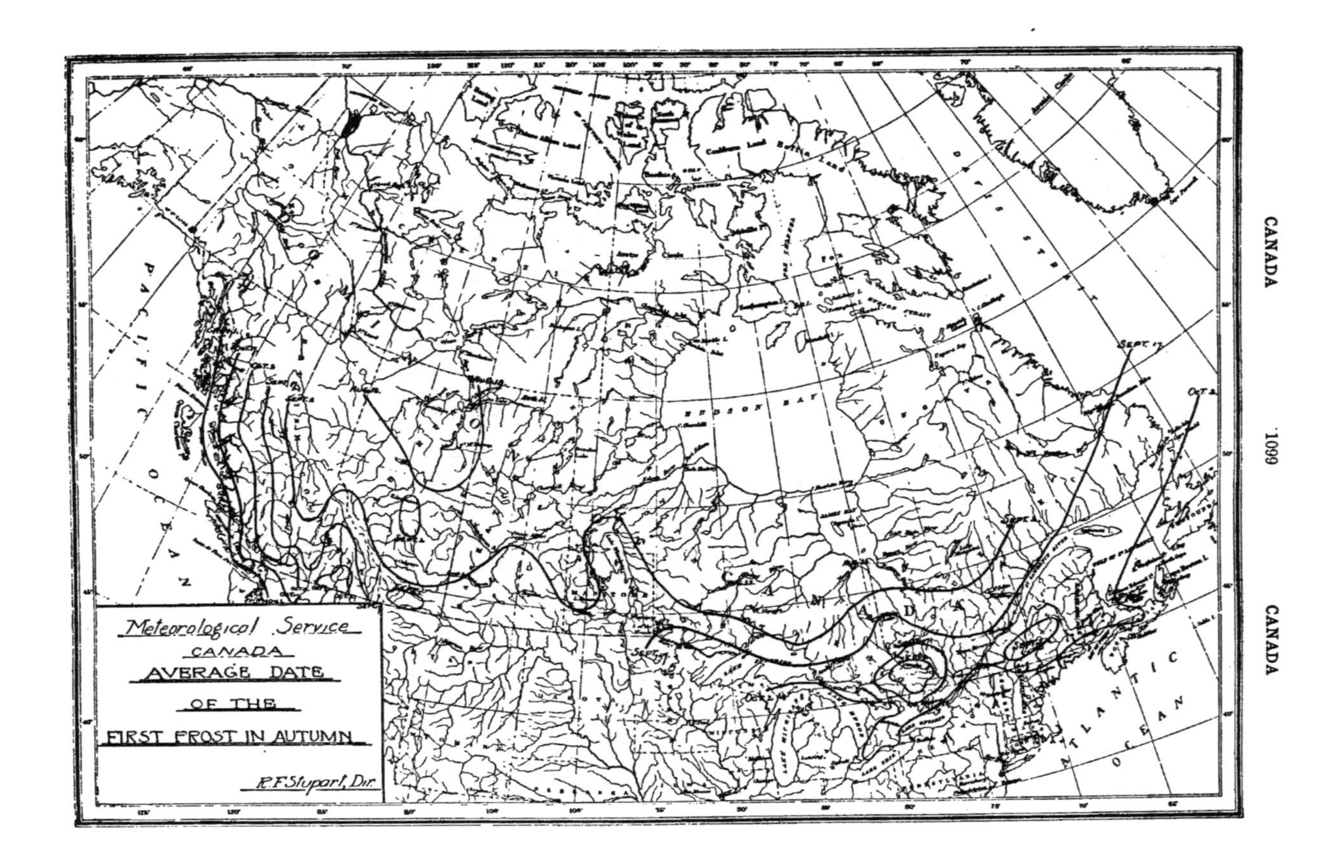
Meteorological Service
CANADA
AVERAGE DATE
OF THE
FIRST FROST IN AUTUMN
R.F. Stupart, Dir.
PACIFIC OCEAN
ATLANTIC OCEAN
HUDSON BAY
DAVIS STRAIT
SEPT. 17
OCT. 2

and the delightful autumn, with its Indian summer, is regarded by many as the best season of the year. For details of the climate in different localities, see the articles on the separate provinces.

On the Atlantic coast and in the Hudson Bay region the climate is distinctly Arctic or sub-Arctic. The summers are short and warm, while the winters are long and very cold. The Labrador peninsula is not very inviting to settlers. The cold Labrador Current, laden with icebergs, which sweeps southward through Davis Strait past the shores of Greenland, keeps the entrance to Hudson Bay frozen for eight or nine months each year and carries icebergs and pack ice along the Labrador coast sometimes until midsummer.

South of the Gulf of Saint Lawrence are the Maritime Provinces, whose climate is affected by the warmer currents from the Gulf of Mexico. The Maritime Provinces, especially on the coasts, have frequent fogs and an excessive amount of rainfall, sometimes as high as fifty or fifty-five inches a year. The Saint Lawrence section has a bracing climate; the winters are cold and dry, with much snow and occasionally short periods of exceedingly cold weather. The winter air is invigorating, and the Saint Lawrence Valley is the scene of many outdoor winter sports, such as snow shoeing, tobogganing and skiing. The summers are warm and pleasant, without much humidity, but at the necessary time there is usually an abundance of rain for the crops. Northern Ontario, like the Labrador peninsula, has severe winters, with much more snow than the southern part. Even in winter, however, there is a great deal of sunshine, and the cool, clear summers have made the region a favorite resort.

The climate of the western half of the Dominion, including the mountain and prairie sections, shows even greater variety of climate than the eastern half. The prevailing winds are from the west, and reach the coast after a passage over the warm Pacific Ocean, whose waters are 20° warmer than those of the Atlantic. When these westerly winds strike the cold Coast Range they lose much of their warmth and moisture. Along the coast there is consequently a warm, rainy climate. Flowers bloom the year around, and fruits and vegetables almost reach perfection. Somewhat lightened, the winds sweep eastward to the Rocky Mountains, whose western slope has much snow and rain and abundant vegetation. The intervening interior plateau, however, gets little rain, and the growth of vegetation depends largely on irrigation.

By the time the winds cross the Rocky Mountains they have lost nearly all of their strength. A local wind, called the *chinook* (which see), has marked influence on the climate, cooling the air in summer and warming it in winter. A noteworthy result of the chinook wind is the open character of the winters; the snow is quickly melted and evaporated by the warm, dry winds. The prairie provinces as a whole have a distinctly continental climate with short, warm summers and very cold, dry winters (see CLIMATE). Both in summer and in winter there is abundant sunshine. The growing season for crops is short, but as the sun is above the horizon for nearly twenty hours out of the twenty-four, nearly all staple grains and vegetables mature far north in the Peace River Valley, and even beyond. There is always in these extreme northern regions, however, the fear that frost will come before the crops are matured.

Plant and Animal Life

Native vegetation in Canada is less varied than might be expected from the great extent of the country and the wide differences in climate. The entire northern and northeastern part of the Dominion has an Arctic or sub-Arctic type of plant life. The vegetation is sparse, and includes only mosses, lichens and a few hardy herbs and willows. A large area between Hudson Bay and the Arctic Ocean is almost a desert; this section is usually known as the *barren grounds,* or *tundras.* South of this region appears an entirely different type of vegetation. The Saint Lawrence Valley and the Maritime Provinces were once covered with great forests, including both hard and soft woods. Large areas have been cleared or burned, and others have been deforested from natural causes. There are still, however, large areas of standing timber which constitute one of the great resources of Eastern Canada. The most important species are white spruce, white pine, balsam fir and hemlock, among the cone-bearing trees, and birch, maple, basswood, elm, oak and ash, among the hardwoods. This section also has many wild flowers and wild fruits.

SOME ANIMALS OF CANADA

Polar Bear

Otter

Caribou

Virginian Deer

Grizzly Bear

Beaver

Fox

Musk-ox

Lynx

Moose

The prairies of Manitoba, Saskatchewan and Alberta are almost treeless. Manitoba has small areas of spruce, fir, cedar, poplar, paper birch, ash and other trees, but they are of little importance except locally. In Southern Saskatchewan and Alberta the only native trees are poplar, willow and cottonwood, and even these are confined to the banks of the rivers. North of the Saskatchewan River the vegetation becomes more plentiful, and from Hudson Bay to the Rocky Mountains is a broad belt of spruce, tamarack and poplar. The trees are not as large as those of the Eastern sections, nor are they as important commercially. A few spring flowers, notably the crocus, flourish on the prairies, but the dry summer heat seems to prevent the growth of any later flowers, even of the hardy daisy.

The greatest forests remaining in Canada are in the Pacific, or mountain, belt. There are thousands of square miles of virgin timber, giant trees, many of them 200 to 300 feet high. Spruce, hemlock and cedar are most common in these regions.

The Animals of Canada. The animal life of Canada, like the plant life, may be divided into several belts, or regions. Roughly considered, these belts are *circumpolar;* that is, the animals found there are of the same species as those found in Europe and Asia at the same distance from the North Pole. There are no animals which are distinctively Canadian and not found elsewhere. The beaver, to be sure, is so common and so characteristic that it has become the national emblem, but the beaver is also found in other countries, though in ever-decreasing numbers.

Of the larger animals there is still a great variety in the unsettled regions. The musk-ox and the caribou are common in the Hudson Bay region and farther south in winter, and the woodland caribou is found in all the provinces except Prince Edward Island. The stately moose ranges the forests, and a few bison, the American buffaloes, roam over the plains, once the home of countless thousands of these characteristically American animals. The Virginian deer and the black-tailed deer are still plentiful in all Southern Canada, but the wapiti, or American elk, which once wandered in great bands from Quebec to the Pacific and from the Peace River far southward into the United States, has been almost exterminated. Only a few small bands still remain on the prairies. The pronghorn antelope is another native of the plains.

The black bear is common in nearly all parts of Canada, except along the Arctic shores, where the polar bear has his haunts. In the Rockies and the other mountains in the West there are many grizzly and brown bears. More characteristic of the mountains perhaps are the bighorn, or Rocky Mountain sheep, and the Rocky Mountain goat, whose agility and surefootedness, even on the sharpest peaks and the most precipitous slopes, protect them from wholesale slaughter. Among the other large animals still to be found in various parts of the Dominion are the timber wolf, the coyote, the puma (or cougar) and the red fox. Silver fox, lynx, beaver, otter, marten, fisher, mink and skunk are the most important and numerous of the fur-bearing animals. Hares, rabbits and squirrels are plentiful in many parts. All the fur-bearing and game animals are now protected by law from hunters during stated seasons, but their number, nevertheless, seems to be steadily decreasing, and the fur trade is becoming of less and less importance.

Birds. Although Canada has a few birds which are unknown in regions to the south, most of Canada's birds are migratory and are well known in all temperate climates. Canada is their breeding-ground, but when the cold weather approaches they flock to the warmer southland (see BIRDS, subtitle *Migration of Birds*). The game birds, or wild fowl, are particularly numerous in the west, where their breeding-grounds extend from Southern Manitoba and the Western prairies even to the Arctic Ocean. Besides many ducks and geese, there are gulls, cormorants, albatrosses, fulmars, petrels and other sea-birds. Golden eagles, bald-headed eagles, owls, hawks, ravens and crows are common. In parts of Ontario the wild turkey and quail are occasionally seen, and in British Columbia the California quail and the mountain partridge are found. There are many varieties of grouse, including the prairie chicken and the so-called partridge. The Canada jay, the waxwing, grosbeak, snow bunting and sometimes the raven remain in Canada throughout the winter. Songbirds are found everywhere, especially in regions which are still well-wooded; robins, orioles, thrushes and catbirds sing as gayly in Canada as anywhere else, and the English sparrow, though only recently introduced, is already a nuisance in many towns. One of the prettiest of the birds and also the smallest is the ruby-throated humming bird, which is everywhere, even in the mountains.

Canada's Natural Resources

The Dominion of Canada is essentially an agricultural country. Its greatest natural resource is the great belt of fertile land which extends with scarcely a break from the Atlantic to the Pacific. This belt is the southern part of the Dominion, including the rich river valleys of British Columbia, the fertile plains of the prairie provinces and the productive valley of the Saint Lawrence. With the exception of the barren strip around Lake Superior this belt is the best-developed section of Canada.

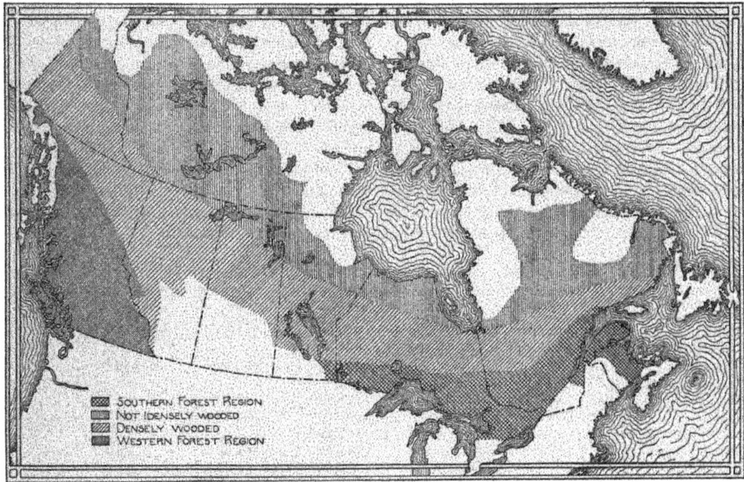

FOREST AREAS OF CANADA

North of this fertile belt is a strip of prairie extending almost entirely across the continent. Less fertile and less suited for cultivation than the southern belt, it is yet an important area, capable of great agricultural development. The raising of cattle and sheep is comparatively a minor branch of industry, but can be easily developed in this region and in the higher areas near the foothills of the Rocky Mountains. The northern part of Canada is unsuited to agricultural development, but it has great mineral deposits whose extent is scarcely appreciated. Gold, silver, copper, iron, lead and coal are found in large areas, even as far south as the United States boundary. The coal and iron deposits of the Maritime Provinces are among the greatest in North America.

The forests constitute another great resource. It has been estimated that the forest area of Canada is about 500,000,000 acres, from fifty to sixty per cent of which are covered with timber of merchantable size. At the present time the lumbering industry is important in all the provinces except Prince Edward Island and the prairie provinces, but even in the latter attempts are constantly being made to utilize the timber now available and provide for an increase in the future supply.

It has been said that the Canadian fisheries are the greatest in the world. This statement is no exaggeration if account is taken of the possible extent of the fishing industry rather than its present development. On both the Atlantic and Pacific coasts the fisheries, with proper regulation, are capable of almost indefinite expansion, and the inland, fresh-water fisheries are also only beginning to yield their wealth. In connection with the forests and fisheries the subject of conservation (which see) has naturally been a leading issue. The mineral wealth of a country cannot be increased by man, but the forests and fisheries can be replenished and even increased by care. Forests and fisheries, properly cared for, are inexhaustible, and for this reason conservation of these resources becomes an issue which every nation must always face.

This brief summary merely indicates the principal lines of development on which the Dominion's future depends. Unlike many lesser countries, it has a variety of resources. A smaller country might be suitable for farming, but might be totally lacking in mineral wealth or in fisheries. Canada is fortunate in that it has all the natural resources—including water power to help in their development—on which the economic prosperity of a nation depends. The extent to which these resources have been utilized is told below in detail.

Agriculture

Agriculture is necessarily one of the first pursuits in which men engage, and to this day fully one-half of the total population of the Dominion are directly engaged in or dependent for their living on the cultivation of the soil. Thousands of people, too, are employed in the trade and transportation of food products, in milling, in meat packing and in other industries which arise as a result of agriculture. The first settlers made their little clearings in the Saint Lawrence Valley and in the Maritime Provinces, and a little farming was done in British Columbia even before its admission to the Dominion, but it was not until the end of the nineteenth century that the fertile lands of the prairie provinces were opened. To-day in both these grains. Grain may be raised successfully far beyond the present northern limits of grain-growing areas. Good wheat has been raised as far north as Fort Vermilion, on the Peace River, about 800 miles north of the United States boundary, and at Fort Simpson, on the Mackenzie River, 1,000 miles north of that boundary. The danger from frost, however, is present in many sections during late summer, and it is extremely unlikely that grains will be raised on a large scale in the north.

The following table shows the production of the important field crops by provinces for 1915, as well as the totals for 1914 and for the average period, 1910-1915.

PRODUCTION IN BUSHELS OF PRINCIPAL FIELD CROPS, 1915.

PROVINCE	WHEAT	OATS	BARLEY	RYE	FLAX	CORN*
Alberta...........	51,355,000	107,741,000	6,984,000	463,000	1,124,000	5,700
British Columbia..	525,000	4,390,600	106,900			5,400
Manitoba..........	96,425,000	69,471,000	17,763,000	155,000	374,000	60,000
New Brunswick....	267,000	5,559,000	48,000			1,000
Nova Scotia.......	247,000	3,487,000	128,400	4,500		2,300
Ontario...........	18,912,000	122,810,000	15,369,000	1,551,000	62,000	16,991,000
Prince Edward Is.	623,000	6,832,500	106,800			3,400
Quebec............	1,411,000	42,182,000	2,255,000	145,000	7,000	801,000
Saskatchewan.....	195,168,000	157,628,000	10,570,000	75,600	9,061,000	8,300
Total, 1915........	376,303,600	520,103,000	53,331,000	2,394,000	10,628,000	17,897,000
Total, 1914........	161,280,000	313,078,000	36,201,000	2,016,000	7,175,200	17,175,000
Average, 1910-1915.	196,026,000	343,612,000	41,436,000	2,155,000	13,033,000	19,087,000

* Including both fodder corn and corn for husking.

agriculture represents a capital investment for the Dominion of about five billion dollars. Of this total one-fourth is in Ontario, about one-fifth in Saskatchewan and one-tenth each in Alberta and Manitoba.

At first the plains of Alberta, Saskatchewan and Manitoba were regarded as useful only for grazing, but once the immense possibilities of this region were realized it became one of the greatest grain-growing areas in the world. The development of the west has come almost entirely since the beginning of the twentieth century. For many years Ontario was the largest producer of wheat and oats in Canada, but Saskatchewan since 1911 has ranked first

Agriculture has naturally reached its highest development in the Dominion in the sections which have been cultivated the longest time; these are Prince Edward Island, a few sheltered valleys in Nova Scotia and New Brunswick, the valley of the Saint Lawrence above Quebec and the southern peninsula of Ontario. It is true, however, that these sections are better known for the production of a few crops than for the value of their total output. Apples are cultivated extensively in the Annapolis valley in Nova Scotia and the Saint John valley in New Brunswick. Potatoes and turnips receive special attention in the Maritime Provinces. Ontario's annual crop of 7,000,000 bushels or

The Normal Annual Precipitation in Canada.

The Meteorological Service, Toronto. R. F. Stupart, Dir.

more is nearly seventy per cent of the total apple crop of the Dominion. Southern Ontario, especially the Niagara peninsula, is famous for its peaches, grapes and small fruits. Berries, in fact, flourish anywhere in Southern Canada, even on the Western prairies. British Columbia has a smaller area of land suited for cultivation, but its variety of soil and climate adapts it for many fruits and vegetables which do not flourish elsewhere.

In the older sections, and even on the more recently settled prairies, there is an increasing tendency toward diversification of crops and the raising of cattle and other live stock. For some years, at least, there were many farms on which wheat was the only crop; if the wheat crop was even a partial failure there was serious hardship for the farmer. Nowadays many farmers have cattle, horses and swine, and perhaps a few sheep. The raising of beef for export is an important branch of the industry. In Eastern Canada, and to a lesser degree in the West, dairying and poultry-raising have had a noteworthy development.

Live Stock. Horses, cattle and other live stock are raised in large numbers. Heavy draught horses are raised chiefly in Ontario, but also to an increasing extent in the Western provinces. Thoroughbreds, including carriage and saddle horses, are bred in many sections. Large numbers of horses of all kinds were sent to Europe early in the War of the Nations for use by the allied armies. The raising of live stock is noticeably decreasing in the Eastern provinces, but is increasing rapidly in the West. For example, the number of horses in Saskatchewan increased from 332,000 in 1910 to 609,000 in 1914, and in Alberta from 294,000 to 519,000. During the same period the number of swine in Saskatchewan increased from 125,000 to 454,000, and in Alberta from 143,000 to 397,000. The live-stock industry is gradually shifting its center from east to west, but in the process is not growing very fast as a whole. The table below shows the 1915 totals for the Dominion.

	NUMBER	VALUE
Horses	2,947,000	$371,430,000
Milch Cows	2,673,000	153,632,000
Other Cattle........	3,363,000	143,498,000
Sheep	2,058,000	14,550,000
Swine	3,434,000	42,418,000

No branch of agriculture has developed more rapidly than dairying and its allied industry, poultry-raising. The annual production of butter increased from 141,400,000 pounds in 1900 to 201,800,000 pounds in 1910, and is still increasing steadily. Quebec produces nearly two-thirds of the total, and Ontario about one-sixth. Ontario has about one-half of the $15,000,000 worth of poultry in the Dominion, and Alberta, British Columbia, Manitoba and Quebec together have nearly all of the other half. The egg production is over 125,000,000 dozen a year.

Irrigation. The development of Western Canada, while due in the first instance to the construction of transcontinental railways, has been greatly stimulated by irrigation. This is especially true in British Columbia, and to an increasing extent in Alberta and Saskatchewan. Irrigation on a small scale was practiced on the prairies when these districts were first settled, but the irrigated areas were small, and the works were crude. There was at that time no law regulating the use of water for irrigation, and every farmer and rancher took as much as he needed or wanted. The confusion which resulted was untangled by a Dominion law of 1894, and its amendments. This law applied only to the Northwest, but in British Columbia and the Eastern provinces irrigation was already under provincial control. The development of irrigation has been marked, especially since 1910. The best-known and the largest of all the irrigation projects is that of the Canadian Pacific Railway, in the region between Calgary and Medicine Hat. The vicinities of Lethbridge and Moose Jaw are also irrigation centers. In British Columbia irrigation is most extensive in the south, in the Okanagan, Thompson and Columbia valleys. For further details, see each of the provinces named.

How the Government Helps the Farmer. The Dominion government, through the Department of Agriculture, helps the farmer, the rancher and everybody whose living is derived from the soil, in many ways. In 1887 the Department established the central experimental farm at Ottawa and branch farms at Nappan, N. S., Brandon, Man., Indian Head, Sask., and Agassiz, B. C. Since then fourteen additional branch stations have been established; at all of which information may be had as to the best methods of preparing the soil, the most profitable crops to raise, and thousands of other facts which the farmer wants to know. The-health-of-animals branch of the Department of Agriculture not only enforces the laws relating to quarantine and inspection, but is constantly carrying on valuable experiments relating to animal diseases. Through the efforts of this branch Canada is now entirely free from rinderpest, pleuro-pneumonia and the foot-and-mouth disease. The government has also provided a national registration system for pedigreed live stock, and the effect on the cattle has been noteworthy.

The activities of the government in relation to agriculture are almost numberless. It furnishes information as to the operation of dairying factories and maintains model creameries and cheese factories. Inspectors give information about the best ways to pack fruit, butter and cheese. The government also offers bounties, or subsidies, under certain conditions, to persons who provide cold-storage facilities. One of its most important services is in the extension of markets. Commercial agents are stationed at various foreign cities throughout the world to study foreign markets and the best methods of introducing Canadian products or furthering their sale.

Fisheries of Canada

The waters in and about Canada contain the principal food fishes in greater abundance than do the waters in any other part of the world. It is no exaggeration to say that Canada has the most extensive fisheries in the world. It is true, however, that the total value of the annual catch in the United States is far in excess of that in Canada, but a large part of this excess is due to the fact that fish caught off Canadian coasts and landed in United States ports are credited to the United States. The Canadian fisheries are naturally divided into three groups—the Atlantic, the Pacific and the inland fisheries.

Atlantic Fisheries. The Atlantic fisheries may be divided into two classes, the *deep-sea* fisheries and the *inshore,* or *coastal,* fisheries. The deep-sea fishing grounds are off the "Banks," twenty to ninety miles from the coast of Newfoundland, and the fishing is usually carried on from trawlers of forty to 100 tons. Trawling with hook and line, with herring and squid as bait, is the customary method. Cod, haddock, hake and halibut are the principal varieties caught. In the inshore, or coastal, fisheries smaller boats are used, with crews of two or three men, using nets, hand-lines and trawls. The inshore fisheries extend along the coast line of Quebec and the Maritime Provinces, a total length of more than 5,000 miles. On this coast are many caves and natural harbors, in most of which fish can be taken with little effort. In addition to the fishes already mentioned, the principal varieties caught inshore are herring, mackerel, shad, smelt, flounder and sardine. Lobsters are taken in large numbers, and excellent oyster beds exist in many parts of the Gulf of Saint Lawrence.

Pacific Fisheries. The fisheries on the Pacific coast are wholly in British Columbia. For many years they were of little value, then they rose slowly to second place, and finally to an undisputed leadership. Here, as on the Atlantic coast, trawling is the usual method; small dories are used for setting and hauling the lines. The irregular coast of British Columbia, with thousands of inlets, bays and islands, makes a shore line over 7,000 miles long, which probably is better stocked with fish than any other part of the world. Herring exist in abundance, but the most important catch is salmon (which see). North of Queen Charlotte Sound considerable attention is being given to the halibut fisheries.

Inland or Fresh-water Fisheries. It has been estimated that the Dominion has about 220,000 square miles of fresh-water area, practically the whole of which is well stocked with fish. The Canadian part of the Great Lakes is only

VALUE OF THE CATCH

PROVINCES	1912	1914	CAPITAL INVESTED 1914
British Columbia	$13,667,125	$13,891,398	$12,489,613
Nova Scotia	9,367,550	8,297,626	7,110,210
New Brunswick	4,886,157	4,308,707	3,600,547
Ontario	2,205,436	2,674,685	1,506,581
Quebec	1,868,136	1,850,427	1,445,871
Prince Edward Island	1,196,396	1,280,447	948,667
Manitoba	1,113,486	606,272	303,927
Saskatchewan	139,436	148,602	30,941
Alberta	111,825	81,319	15,878
Yukon	102,325	68,265	11,798
Total	$34,667,872	$33,207,748	$27,464,033

one-fifth of the area of Canada's fresh-water lakes. The most important fresh-water fishes are whitefish, trout, pickerel, pike and sturgeon.

The fresh-water fisheries also include many spots in which game fish abound. Canada, in fact, is the fisherman's paradise, for every part of the Dominion except Southern Alberta and Saskatchewan offers him unending opportunity. Trout, salmon, pike, pickerel, muskelunge and bass are the most numerous of the game fishes.

Value of the Fisheries. The market value of all kinds of fish and fish products taken by Canadian fishermen averages $33,000,000 a year; the biggest catch ever taken was in 1912, when the total was $34,667,000. The table on page 1107 shows the catch and the capital invested by provinces, according to the reports of the Department of Marine and Fisheries, both for the largest year and for 1914. The latter year, it will be noted, fell far behind.

Mining

It has long been known that Canada is rich in mineral deposits, but only in recent years has the approximate extent of this resource been determined. As explained above (see PHYSICAL CHARACTERISTICS), the Eastern and Western sections of the Dominion are geologically a continuation of the mountain systems of the United States, and it is now known that the minerals found in the Appalachian chain and the Rocky Mountains and Coast Ranges exist in the corresponding parts of Canada. There is a third mineral-bearing area, the Lake Superior region, almost midway between the Atlantic and Pacific coasts. While the existence of these mineral deposits was known for years, peculiar conditions prevented development. In the first place, without transportation facilities it was impossible to mine profitably. Then the mining regions had a sparse population, and, except in Nova Scotia, extended over a large area. The severity of the climate doubtless discouraged some prospectors, but the chief difficulty was that the great coal supplies, necessary for smelting, were in Nova Scotia, while the richest ore deposits lay 2,000 miles or more away.

Coal is by far the most valuable mineral product at present, although gold held this honor at one time. Nova Scotia for many years has produced an average of 6,000,000 to 7,000,000 tons a year, approximately half of the total for the Dominion. British Columbia, with an average output of 2,000,000 to 3,000,000 tons, and Alberta, with an output which has several times touched 4,000,000 tons, are the only other important producers. Most of the coal of these provinces is bituminous, or soft coal, of fairly good quality, but small pockets of anthracite have been opened near Calgary and on Queen Charlotte Island. Anthracite has also been found in Northern Ontario, along the Albany River Small deposits of bituminous coal exist in New Brunswick, but in the other provinces only lignite of various grades is found. The total coal production of the Dominion now averages 14,000,000 tons a year.

Gold. The gold production of Canada reached its highest point in 1900 and 1901, when the placer mines (see GOLD) of the Klondike were being worked to their fullest capacity. Gold has been mined in small quantities in Nova Scotia almost from the days of the pioneer settlers, and small deposits are known in Quebec. Placer mining in British Columbia was also an early industry, but the gold output had never reached a total value of more than $1,500,000 until 1895. In that year it was just over $2,000,000; in 1897, the year of the great Klondike strike, it was over $6,000,000, and by 1900 it reached $27,000,000. During the Klondike excitement the introduction of lode and hydraulic mining in British Columbia increased the output of that province from an average of less than $500,000 to one of $5,000,000. After 1900 the output of the Dominion slowly declined until the opening of the Porcupine district in Ontario in 1911, when it quickly rose from its low average of $9,000,000 to a new level of $15,000,000 or $16,000,000 a year. See KLONDIKE; PORCUPINE.

Silver and Lead. During the nineteenth century silver was mined in small quantities in Ontario and Quebec and to a trifling extent in British Columbia. The deposits were not known to be large or valuable, and little was done to develop them until the end of the century. British Columbia's output increased from 77,000 ounces in 1892 to 5,472,000 ounces in 1897, but now averages only 3,000,000 ounces, worth about $1,750,000. Most of Canada's silver production is in Ontario, where the metal was discovered in the Cobalt district in 1904 and 1905. That district alone now yields over 95 per cent of the output of the Dominion. In nearly all parts of the Dominion where silver is mined it is found in combination with lead; the production of this latter mineral is about 37,000,000 pounds a year. Among the minor mineral products are arsenic, brick-clay, corundum and graphite.

Other Minerals. Canada is the world's greatest producer of nickel and asbestos. Its only rival in the production of nickel is New Caledonia, an island near Australia, which is now in second place, and in asbestos it has no rival. Nickel is mined in Ontario, in the Sudbury district, northeast of Lake Huron, and the asbestos comes from the eastern part of Quebec, Thetford being the center of the industry. Iron is found in every province, but its production has been unimportant until recent years. Upon the outbreak of the War of the Nations the production of Canadian iron increased enormously, due to war demands. Large amounts of copper are extracted each year in Ontario and British Columbia, and smaller quantities in Quebec. The great war gave such an impetus to mineral development that its effects in this direction were destined to be permanently beneficial. The table below summarizes the output of the important minerals of the Dominion, the figures of production being annual averages:

Mineral	Quantity	Unit	Value
Asbestos	130,000	tons	$ 3,700,000
Cement, Portland	8,000,000	bbls.	10,000,000
Coal	15,000,000	tons	37,000,000
Copper	75,000,000	lbs.	10,000,000
Gypsum	600,000	tons	1,200,000
Iron			
Lead	37,000,000	lbs.	1,700,000
Lime	7,000,000	bu.	1,500,000
Natural Gas	20,000,000	cu. ft.	3,300,000
Nickel	47,000,000	lbs.	14,000,000
Pig Iron	80,000	tons	1,000,000
Sand and Gravel			2,500,000
Silver	30,000,000	oz.	19,000,000
Granite			2,000,000
Limestone			3,000,000
Total mineral production, annual average			$130,000,000

Manufactures

Canada has great natural resources and abundant power to develop them, but the expansion of the manufacturing industries has become a feature only in recent years. Between 1900 and 1910, for example, the output of Canadian factories increased 142 per cent, from $481,000,000 to $1,165,000,000. Móre recently, since the beginning of the War of the Nations, Canada has begun to manufacture on a large scale certain articles, such as shells and bullets, which it had previously made only on a small scale or not at all. War has also increased the production of pig iron and steel, and of some prepared foods which can be sent to the men in the trenches, but some of this increase has been at the expense of the regular industries of peace. Foundries, iron works and mills which formerly made bridges and structural steel began in 1914 to turn out munitions.

The country was well suited to the development of this new industry, but when the war broke out Great Britain naturally looked to the United States, with its greater industries, for immediate supplies. Before long, however, Canadian industries were also able to respond to the unprecedented demand. At first there was considerable confusion in the allotment of orders among Canadian manufacturers, but the Shell Committee and its successor, the Imperial Munitions Board, did remarkable work in the organization of the new industry. It was estimated that the total of all war orders placed in Canada in 1915 was about $600,000,000, and for 1916 was probably greater. Nearly half of the total amount was spent for shrapnel, but a considerable percentage was for high-explosive shells, rifles and cartridges, harness, clothing and foodstuffs. The effect of these war orders on Canadian industry was tremendous; the output of Canadian manufactures was increased fifty per cent in a single year. Some of this increase was due to the stimulation of existing branches of manufactures, but most of it came from the new industries. The boom created in the iron and steel industry by the war exerted an influence in almost every other branch of industrial activity.

Manufacturing has naturally developed most in those sections which have offered the best opportunities. In Nova Scotia coal is plentiful, and in Ontario, Quebec and to a less degree in New Brunswick there is abundant water power. Large sections of these provinces are manufacturing communities. In Ontario the electric power derived from Niagara Falls is distributed to towns at a distance. The forests supply raw material for furniture and all kinds of wood products. An abundant supply of hemlock has made these provinces the leaders in the leather industry. Cotton goods are made chiefly in Quebec and woolens in Ontario. In British Columbia the manufacture of lum-

STATISTICS OF MANUFACTURES (1910)

PROVINCES	NUMBER OF ESTABLISHMENTS	EMPLOYEES	CAPITAL	VALUE OF PRODUCTS
Alberta	290	6,890	$ 29,518,346	$ 18,788,825
British Columbia	651	33,312	123,027,521	65,204,236
Manitoba	439	17,325	47,941,540	53,673,609
New Brunswick	1,158	24,755	36,125,012	35,422,302
Nova Scotia	1,480	28,795	79,596,341	52,706,184
Ontario	8,001	238,817	595,394,608	579,810,225
Prince Edward Island	442	3,762	2,013,365	3,136,470
Quebec	6,584	158,207	326,946,925	350,901,656
Saskatchewan	173	3,250	7,019,951	6,332,132
Canada	19,218	515,203	$1,247,583,609	$1,165,975,639

ber products, smelting of ore and the packing and preserving of fish are most important and have raised the province to third place in manufactures.

In the prairie provinces, however, manufacturers have had little inducement, and the manufacturing industries are comparatively small and few, the most important ones being dependent on agriculture for raw materials. At Winnipeg, Edmonton, Calgary and other cities meat packing is a growing industry; the making of flour and grist-mill products, bread and other bakery goods and car repairing are the other important industries. For details of manufactures, see each province. The table above summarizes the manufacturing industries of the Dominion in 1910, the date of latest available statistics.

Transportation

Canals. Canada's commercial and industrial development was due in the first place to its facilities for navigation by water, especially the Great Lakes and the Saint Lawrence River. The early settlers made the Saint Lawrence a great highway; their descendants have continued to use it. More than that, they have improved the channel and have built a system of canals, so that vessels drawing not more than fourteen feet of water may pass up the Saint Lawrence through the Great Lakes to the head of Lake Superior or the foot of Lake Michigan. From the Strait of Belle Isle to Port Arthur is 2,233 statute miles; to Duluth, 2,357 miles; and to Chicago, 2,289 miles. Ocean-going vessels ascend the Saint Lawrence River as far as Montreal, 1,003 miles from the Strait of Belle Isle. From Montreal westward smaller vessels pass through nine canals, in order, as follows: Lachine, Soulanges, Cornwall, Farran's Point, Rapide Plat, Galop, Welland and Sault Ste. Marie, usually called the *Soo*. The total length of these canals is seventy-four miles, and the total lockage, or height overcome by locks, is 553 feet. All of these canals except the last two are usually called the Saint Lawrence canals; they enable ships to avoid the rapids in the river. The Welland Canal connects Lake Erie and Lake Ontario, and the Sault Ste. Marie Canal con-

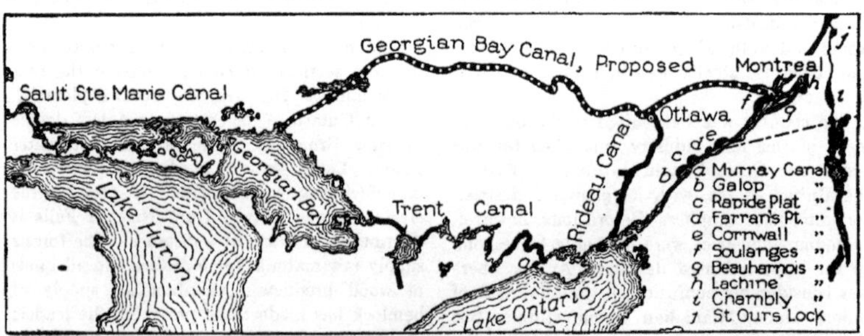

CANALS OF THE DOMINION

The above map does not include the Welland Canal, which is shown in a map with the article bearing that title. Special maps also show the Georgian Bay and Sault Sainte Marie canals, in articles of the same name.

nects Lake Huron and Lake Superior (see WELLAND CANAL; SAULT STE. MARIE CANAL).

Besides these canals on the main line of water communication, there are a number of others, which may be called branches. In actual operation, however, these other canals serve a distinctly local traffic. The Murray

LIFT LOCK AT PETERBOROUGH
The largest of its kind in the world.

Canal, from the Bay of Quinte to Lake Ontario, is used only by coasting vessels and is not on the through line from Montreal to Lake Superior. The Ottawa-Rideau system, connecting Kingston, on Lake Ontario, with Montreal by way of Ottawa, is 248 miles long, but is a canalized river for a large part of the distance. Another important system extends from Sorel, Que., to Chambly, then to Saint Johns, through the Chambly Canal and up the Richelieu River to Lake Champlain. The Trent Canal is a series of short canals or channels between Lake Huron and Lake Ontario. The hydraulic lift lock on this canal at Peterborough has a lift of sixty-five feet and is the largest lock of its kind in the world. A similar lock at Kirkfield has a lift of fifty feet, five inches. The Trent Canal is used as yet only for local traffic. Saint Peter's Canal, on Cape Breton Island, connects Saint Peter's Bay with the Bras d'Or Lakes and practically cuts the island into two.

In addition to these canals plans have been drawn for a number of new canals. The largest and most important of these is the Georgian Bay Ship Canal (see GEORGIAN BAY, for illustration), from the northeast corner of Georgian Bay to the Ottawa River, and then along the Ottawa Valley to Montreal. This canal will enable ocean-going vessels to steam directly from Montreal to Lake Huron and will shorten the water route between Montreal and ports on Lake Michigan and Lake Superior by nearly 300 miles. It is estimated that the construction of the canal will cost $125,000,000, which is about the total already spent on all the other canals.

Railroads. Canada, with more than 30,000 miles of railway, has a greater mileage in proportion to population than any other country except the Australian Commonwealth. Long ago men realized that the development of Canada could not come without intercolonial and transcontinental railways. Not only were railways needed for an economic reason, but for a political one, to serve as a bond of union between provinces so widely separated. The need of railways as an economic and political bond explains why the government has always been willing to grant liberal subsidies to private railway companies, and has itself built and now operates two of Canada's great railway systems. See colored map, with this article.

The first railroad in Canada was built in 1836 from La Prairie, near Montreal, to Saint Johns, a distance of sixteen miles. Ten years later another short line was completed from Montreal to Lachine. Canada's railway system had its real beginning in 1851, when Parliament authorized the construction of a line from Quebec to the western limit of Upper Canada and also a branch from Quebec to Portland, Maine. This line, the Grand Trunk Railway, was completed in 1856, raising Canada's mileage to 1,414.

Since 1856, when the main line of the Grand Trunk was completed from Quebec to Sarnia, Ont., on the Saint Clair River, the Grand Trunk has gradually increased its mileage to 3,100 in Canada alone, besides numerous connections, either owned or leased, to cities in the United States. The main line extends from Portland, Me., through Montreal to Chicago. Three parallel lines extend across the Ontario peninsula, and these are connected by branch lines, one of which extends to Lake Timiskaming. The road was long famous for the old suspension bridge over Niagara Falls, and for the tubular bridge over the Saint Lawrence at Montreal, and the modern steel structures which have replaced the original bridges are noteworthy examples of engineering skill.

Intercolonial Railway. The Intercolonial was the first of the great railway systems to be owned and operated by the Dominion government. It was planned in the days before Confederation—hence its name—and after Confederation was completed by the Dominion. It was opened to traffic in 1876, and now has a mileage of over 1,450. The main line extends from Moncton, N. B., to Montreal, by way of Levis and the south bank of the Saint Lawrence River.

Canadian Pacific Railway. The construction of a transcontinental railway was one of the conditions on which British Columbia entered the Confederation. From the first the railway was a political issue, and politics prevented progress. The government began the actual work of construction, but in 1881 turned over the task to a group of capitalists, including George Stephen, later Lord Mount Stephen, and Donald A. Smith, later Lord Strathcona and Mount Royal (see subtitle *History*, below). The road was finished in 1885, and its mileage is now a total of 12,000. The main line extends from Montreal, Quebec, to Vancouver, B. C. In addition to its main line and branches, the "C. P. R.," as it is generally known, has leased or has traffic rights over several roads which give it connection with ports in the Maritime Provinces and with Boston.

Canadian Northern Railway. In 1896 the firm of Mackenzie, Mann & Company, contractors, built a 100-mile railway known as the Lake Manitoba Railway and Coal Company's line. Mackenzie, Mann & Company owned the road. From this small beginning has grown the Canadian Northern Railway, the third of the great transcontinental systems. This road has opened to settlement a great area in the Northwest, north of the Canadian Pacific and the Grand Trunk Pacific. The mileage of the Canadian Northern is about 6,600. Through trains from Quebec to the Pacific coast were first operated in 1915.

Grand Trunk Pacific. This railway system was also opened to through traffic in 1915. The Grand Trunk Pacific Railway Company was incorporated by Parliament in 1903. It agreed with the government to construct a railway from Winnipeg to Prince Rupert, B. C., on the Pacific Ocean, and to operate a line from Prince Rupert to Moncton, N. B. The section east of Winnipeg was built by the Dominion government and is technically known as the National Transcontinental Railway, whereas the line west of Winnipeg is the Grand Trunk Pacific. Under the original agreement, the Grand Trunk Pacific was to operate the entire system from Moncton to Prince Rupert, but on the completion of construction, it refused to operate the lines east of Winnipeg. This section, therefore, is now operated as a part of the Canadian Government Railways. A number of important branches have also been built by the government, the most important being the Hudson Bay Railway, extending from Pas, about 180 miles northeast of Prince Albert, to Port Nelson on Hudson Bay. At Pas a branch of the Canadian Northern Railway furnishes connection with Hudson Bay Junction. This branch is intended to open a new route to Europe for the grain of the Northwest provinces. See HUDSON BAY.

Facts and Figures about Canadian Railroads. The total railway mileage of the Dominion (about 30,000) is divided among the provinces approximately as follows:

Ontario	9,250
Quebec	4,000
Manitoba	4,000
Saskatchewan	5,000
Alberta	2,500
British Columbia	2,000
New Brunswick	2,000
Nova Scotia	1,400
Prince Edward Island	280
Yukon	100

The Canadian railways also own about 225 miles of line in the United States, crossing United States territory in passing from one Canadian point to another and used only for Canadian traffic.

Since Confederation the Dominion has spent about $325,000,000 for the construction of government railways, over $200,000,000 for operating expenses and nearly $200,000,000 in subsidies to other than government roads; this makes a grand total of $725,000,000. All the railways combined carry about 50,000,000 passengers and 100,000,000 tons of freight a year. The railways employ about 160,000 men, and their gross earnings are about $250,000,000 a year.

Commerce and Shipping. The regulation of Canada's foreign trade is the legal right of the government of Great Britain. As Canada grew and became more important in the Empire, this right fell into disuse. First, the British government grew into the habit of consulting the Dominion's interests and of making no treaties without its approval. Now the Dominion authorities often conduct their own negotiations, and the resulting treaties are subject to a merely nominal approval by the home government.

Canadian ships carry a considerable part of Canada's products to foreign shores and bring back the products of other lands. The greatest part of Canadian commerce, however, is still carried in British bottoms. Both on the Atlantic and on the Pacific the Dominion has many ports from which ocean liners plough their way across the seas to Europe, Asia, Australia and Africa. Halifax and Sydney, N. S., Saint John, N. B., Quebec and Montreal are the chief ports on the Atlantic side; Vancouver, Victoria and Prince Rupert on the Pacific. Each of these ports is a great railway terminal.

Most of Canada's trade is with Great Britain and the United States. Canada in normal years exports more goods to Great Britain than to the United States, but it imports nearly twice as much from its neighbor as from the mother country. Practically the entire export trade is in raw materials, the products of farms, fisheries and mines. Farm products make up about one-third of the total value, and wheat alone is more than one-fifth. Of the imports nearly three-fourths are manufactured goods, including farm implements and machinery, automobiles, various iron and steel products, cotton and woolen goods, sugar and coal. Canada exports a small quantity of coal to districts near the mining regions, but imports six times as much in other sections. The total foreign trade amounts approximately to $700,000,000 a year, of which imports constitute $400,000,000 and exports the remainder. These figures are exclusive of the vastly-increased volume of exports of munitions of war to the allies during the War of the Nations, of which no exact statistics are obtainable.

Education

By the British North America Act of 1867 (which see), the entire control of education is left to the individual provinces. All legislation on educational matters is within the jurisdiction of the provincial legislatures. The constitution makes only a single condition, that the privileges of denominational and separate schools in Ontario and Quebec shall not be denied. These had existed for so long at the passing of the Act that it was felt that any change would be an injustice.

Broadly considered, the educational systems of the provinces are double ones; there are, first, the non-denominational public schools in Protestant communities, and second, the Roman Catholic schools, chiefly in the French and Irish communities, in which the teachings of the Roman Catholic Church form the basis of education. The question of "separate schools" has long been perplexing and has caused much bitterness in politics. The entire question, while it occasionally figures in national politics, is essentially provincial, and is treated more at length in the articles on the separate provinces in these volumes, under the subhead *Education.*

In all the provinces the expenses of the educational system are paid from the public revenues. Each province usually contributes liberally to all its schools, but the local districts bear the chief burden. Public elementary education is free throughout the Dominion, except in parts of Quebec, where parents or guardians pay certain small fees. All the provinces except Quebec have laws making education compulsory. Each province has a department, or ministry, of education, which enforces throughout the province the laws for uniform training of teachers, uniform text-books and uniform examinations and grading of pupils. There are secondary schools, high schools, or collegiate institutes, colleges and universities in all of the provinces, and in all of them advanced methods are employed.

In the application of modern educational theories to its school problems, Canada has not been backward. Nature study, domestic science, manual training, agriculture and technical or vocational training of all kinds are receiving much attention. Much of the work done is due to Sir William C. Macdonald, whose liberality provided several large endowment funds for these new branches of school work (see MACDONALD, SIR WILLIAM C.). In 1899 he provided funds to establish manual training centers in connection with the regular work of the public schools. The popularity of the new system led the local authorities in nearly every case to adopt manual training as a regular part of the curriculum. Later he provided a fund for a school garden at each of five rural schools in each of five provinces. The garden and the nature study with it were treated as a part of the regular school work. School gardens are no longer experiments, but an integral part of public school work. Progress along these lines has been made in all the provinces, but more especially in Ontario (which see, subhead *Education*), the most populous of the provinces.

For statistics of illiteracy in Canada, see ILLITERACY.

The Dominion Government

Character of the Government. The British North America Act of 1867, together with the amendments of 1871 and 1886, constitutes the supreme law of the Dominion. This Act was passed by the British Parliament to establish a union between the "Provinces of Canada, Nova Scotia and New Brunswick," with a constitution "similar in principle to that of the United Kingdom." It specifies the organization and method of government. It assigns to the provinces and to the Federal government their respective powers. It is noteworthy that any powers not specifically granted to the provinces are reserved to the Dominion. This is the opposite plan from that adopted by the United States, whose constitution specifies the powers to be exercised by the national government and leaves all others to the states.

The constitution of Canada was the first attempt to ȧdapt British principles of government to a Federal union. The government bears striking resemblances both to that of Great Britain and to that of the United States. It furnishes perhaps the most remarkable example in all history of various types of government in combination. The Dominion of Canada is a dependency, and has a monarchical form of government, for the hereditary sovereign is the supreme head of the state. Yet it is practically a self-governing democracy, for the entire body of citizens has a voice in government. The sovereign exercises his powers through a Parliament elected by the people and through a Ministry responsible to it. The country has both representative and responsible, or Parliamentary, government. The United States has representative, but not responsible, government, for the elected representatives have no direct control over the executive. The Dominion is a Federal government; that is, it includes a number of provinces which possess distinct powers for provincial purposes. Yet each of these provinces is a part of a greater whole and its individual interests are subordinated to the general good. To sum up, therefore, these paradoxical facts, the Dominion is dependent, yet self-governing; monarchical yet democratic; a group of states, yet a single union.

Division of Functions. The three functions of government are to enforce law, to make law and to interpret law. These functions are exercised by the three branches of the Dominion government, the executive, the legislative and the judicial. The Governor-General, as representative of the Crown, is the chief executive and his assent is necessary to every law. The details of departmental affairs are supervised by a Ministry, or Cabinet, whose members are appointed by the Governor-General, but are members of and responsible to Parliament. Parliament is the law-making body, and by the authority of the British North America Act it has also established the courts which interpret the law.

The Governor-General. The king or queen of the United Kingdom of Great Britain and Ireland appoints the Governor-General as the personal representative of the Crown. The Governor-General, therefore, has a double responsibility, for he is not only the governor of a great nation but the guardian of imperial interests. As the Crown's representative he reports to the Secretary of State for the Colonies in London on all matters of interest or importance to the Empire, and official communications between the Dominion and the imperial government pass through him. He opens and closes the sessions of Parliament and assents to or reserves all bills passed by Parliament. He *reserves* a bill by refusing his assent to it and forwarding it to England for the consideration of the Privy Council (which see). All his official acts are in the name of the "Governor-General in Council," with the advice and consent of the Ministry. If he should feel it impossible to agree with his constitutional advisers, he has the right to dismiss them and appoint others to their places. This course has never been taken since Confederation, for it is contrary to the spirit of responsible government. The Governor-General receives a salary of $50,000 a year, and usually holds his office for five years.

The Ministry. By the terms of the British North America Act the executive council which advises the Governor-General is called the "King's Privy Council for Canada." Some of the members of the council act as the heads of executive departments; they comprise the "Ministry," "Cabinet," or "Government." When the Governor-General appoints a minister he first designates him as a member of the Privy Council and then as the head of a department. Privy councillors retain their rank after they resign from office, but this is an honor which carries no responsibilities with it. A minister who has been defeated for reëlection remains a member of the council and is technically one of the Governor-General's advisers, but he is hardly ever called upon for his opinion. Councillors are sometimes chosen as "ministers without portfolios" in order to place them in the Cabinet, without requiring them to assume executive duties.

The members of the Cabinet hold office only as long as they are supported by a majority in the House of Commons. They are responsible directly to the House, which is the direct representative of the people. This system is called responsible government. Most of the Cabinet members sit in the House of Commons, but three or four are usually Senators. All ministers must be members of Parliament, and the government, therefore, is in the hands of what is really a committee, made up of members of the two Houses of Parliament. The number of ministers usually varies from sixteen to twenty, although there is no fixed number. Ministers in charge of departments receive a salary of $7,000 a year in addition to their sessional allowance of $2,500 as members of Parliament. The permanent head of a department is the deputy-minister, who holds office under the civil service regulations (see CIVIL SERVICE IN CANADA). The duties of the various ministers are summarized below:

1. The President of the Council presides over the meetings of the Ministry. He has no executive duties, except such as relate to the work of the council as a whole. This office is usually, though not necessarily, held by the Premier (see subhead, below).

2. The Minister of Justice and Attorney-General of the Dominion is the legal adviser of all the government departments. The administration of justice, including the control of the Royal Northwest Mounted Police and of prisons, is in his hands. He also reviews all the laws passed by the provincial legislatures.

3. The Minister of Finance has charge of the Dominion finances. He presents the annual budget to Parliament, explains the government's financial policy as regards the raising and expenditure of revenue, and is responsible for the collection and distribution of funds. See BUDGET.

4. The Minister of Trade and Commerce executes all laws relating to commerce, industry and allied subjects which are not definitely assigned to some other department. He is also in charge of the census and statistics branch, which was formerly a part of the Department of Agriculture.

5. The Minister of Agriculture, besides the division of industry which gives him his title, has charge of public health, copyrights, trademarks and patents.

6. The Minister of Marine and Fisheries has supervision of the ocean and inland fisheries, of the lighthouse and life-saving service, of the examination of ships' captains and mates, harbors, piers and docks and practically the entire field of fisheries and navigation. The Minister of Marine and Fisheries also acts as Minister of Naval Service.

7. The Minister of Militia and Defense is responsible for the administration of all military affairs, including the military college at Kingston, Ontario. He acts as president of the militia council, which is composed of the Minister, the Deputy Minister, the Chief of the General Staff, and three other officers of the army; this council advises the Minister of Militia.

8. The Minister of the Interior is in charge of the government of the Northwest Territories, the Indians, public lands, forestry branch and the geological survey.

9. The Minister of Railways and Canals is responsible for the management of the Intercolonial Railway, owned by the Dominion government, and for a general supervision of the government canals. He also has some duties in connection with general problems of transportation.

10. The Minister of Public Works has charge of the construction and maintenance of all public works and buildings, except railways and canals.

11. The Postmaster-General controls the management of the Postoffice Department.

12. The Minister of Customs manages the collection of customs duties.

13. The Minister of Inland Revenue has charge of weights and measures and of all excise and other internal taxes.

14. The Minister of Labor acts as arbitrator in labor troubles, and under specified conditions may intervene to end strikes. He may also investigate labor conditions generally and issue reports on them.

15. The Minister of Mines investigates the mineral resources and conditions of the mining industry and issues reports of his findings. He has comparatively few duties and usually holds another position in the Ministry.

16. The Secretary of State registers all documents under the great seal of the Dominion, has charge of public printing and of all official correspondence between the Dominion and provincial governments.

17. The Secretary of State for External Affairs has charge of relations with the British and foreign governments. This office is usually held by the Premier.

Premier. The Premier is literally *first* member of the Ministry. He is chosen by the Governor-General to form a Ministry, although the formal power of appointing ministers is vested in the Governor-General. Legally no such person as Premier exists, but the name is given to the head of the Ministry; usually he acts as president of the Privy Council. The Premier is the head of the Cabinet, or Ministry, he is the leader of the party in power, and he is, under a system of responsible government such as that in Great Britain and Canada, the head of the legislative branch of the government. He is also, in practice, the executive

THE DOMINION PARLIAMENT BUILDING

As it stood before the fire referred to in the text. Rebuilding operations were in progress before the end of 1916.

head of the government, for the Governor-General almost never acts without his advice and consent. Communication between the Cabinet and the Governor-General takes place through the Premier, although the individual ministers may communicate with the Governor-General on any important matters relating to their departments.

If the Premier dies or resigns, the Cabinet is thereby dissolved, and the ministers hold office only until a new Premier is appointed and their successors are named. If the government, that is, the Ministry, is defeated on some important question in Parliament, there are two courses open to the Premier and his ministers. They may resign, whereupon the Governor-General selects a new Premier, usually the leader of the opposition, or they may convince the Governor-General that they still hold public confidence and that the vote does not represent popular sentiment. In this second case the Governor-General may dissolve Parliament and call for a general election.

Parliament. The legislative functions are exercised by a Parliament of two houses. The upper house is the Senate; the lower house is the House of Commons. In nearly all respects the rights and powers of the two houses are exactly the same, but any bills for raising revenue or appropriating funds must originate in the House of Commons. The upper house may reject such a bill as a whole or it may accept it, but it is not allowed to amend it. This differs from the practice in the United States, where the Senate may amend financial bills, although it may not originate them. The powers of Parliament are listed in the British North America Act under twenty-nine heads; the provinces are not allowed to legislate on these subjects:

1. Public debt and property.
2. Regulation of trade and commerce.
3. Raising of money by any system of taxation.
4. Borrowing of money on the public credit.
5. Postal Service.
6. Census and statistics.

7. Military and naval service and defense.
8. Salaries and allowances of all officers of the Dominion.
9. Beacons, buoys, lighthouses and Sable Island.
10. Navigation and shipping.
11. Quarantine and marine hospitals.
12. Seacoast and inland fisheries.
13. Inter-provincial or international ferries.
14. Currency and coinage.
15. Banks and banking.
16. Savings banks.
17. Weights and measures.
18. Bills of exchange and promissory notes.
19. Interest.
20. Legal tender.
21. Bankruptcy and insolvency.
22. Patents.
23. Copyrights.
24. Indians and their lands.
25. Naturalization.
26. Marriage and divorce.
27. Criminal law, except the organization of courts.
28. Penitentiaries.
29. Powers especially stated as not belonging to the provinces.

Membership. The Senate was originally composed of seventy-two members—twenty-four each from Ontario and Quebec and twelve each from New Brunswick and Nova Scotia. The theory of this division was that the Senators should represent the interests of a province rather than of a local district. Since 1867 the admission of other provinces has made some readjustments necessary, and the total membership is now eighty-seven. The Senators are appointed by the Governor-General, on the advice of the Privy Council. The political party in power, therefore, controls the nomination of new Senators. A Senator must be thirty years old, and must own $4,000 worth of real or personal property in the province for which he is appointed. The appointment is for life, but a Senator may resign, and his seat automatically becomes vacant if he becomes a bankrupt, is convicted of treason or other crimes, transfers his allegiance to some other power, or loses the property or residence qualification.

The House of Commons represents the body of voters more directly than does the Senate, for its membership is constantly changing. No property qualification is required, but a member must be a British subject either by birth or naturalization. He need not reside in the district for which he is elected; this practice is like that of Great Britain but unlike that of the United States. The British North America Act fixes the representation of Quebec at sixty-five members, and each of the other provinces is allowed a number of members which bears the same relation to its population as sixty-five is to the population of Quebec. Quebec, by the census of 1911, has one member of the House for 30,810 inhabitants. The population of any province divided by 30,810 gives the number of representatives to which it is entitled in the House. The total number of members, in accordance with the Representation Act of 1914, is 234, divided as follows: Alberta, 12; British Columbia, 13; Manitoba, 15; New Brunswick, 11; Nova Scotia, 16; Ontario, 82; Prince Edward Island, 3; Quebec, 65; Saskatchewan, 16; Yukon Territory, 1. As this act could not go into effect until the dissolution of the Twelfth Parliament, the number of members early in 1917 was still 221.

Each Senator and member of the Commons receives an allowance of ten cents per mile for traveling expenses, and a sessional allowance of $2,500, provided that the session lasts over thirty days.

Officers of Parliament. Each house has a presiding officer, called the *Speaker*. In the House of Commons he is elected by the members, but in the upper house he is appointed by the Governor-General. In each house there are a number of permanent officers who are not members. The chief clerk supervises the keeping of the journals, the translation of public documents, etc. All debates and proceedings in both houses are recorded by official reporters. French or English may be used in addressing either house, but both must be used in all laws and records. The sergeant-at-arms has charge of messengers and pages, and supervises the care of the parliamentary chamber and office-rooms. The Senate also has a "gentleman usher of the black rod," whose duty it is to summon the Commons to the Senate chamber to hear the Governor-General's address at the beginning and end of the session.

How a Bill Becomes a Law. Bills may originate in either house, except that financial measures must come from the House of Commons. To become a law a bill must pass three readings in both houses, and must then be approved by the Governor-General. The voting is by yeas and nays; those in favor of a bill say *yea* in unison, those opposed, *nay*. If the vote seems close, a roll-call or division is demanded. After a bill has passed one house it is sent to the other for consideration. Here it goes through the same process, during which it may be amended, unless it is a bill for raising money. If amendments are added the amended bill must be sent back to the house in which it originated. Each house must agree to any amendments proposed by the other house; if this becomes impossible, the bill is dropped for that session.

Committees of Parliament. The bills introduced into Parliament at each session may total several hundreds or even thousands. It is impossible for the houses as a whole to give proper consideration to more than a few of these, but it is possible to designate small groups of members to consider all the bills on certain subjects. Such groups, or *standing committees,* are named by the Speaker in each house. It is their duty to consider in detail the provisions of any act which may be referred to them. For example, a bill may be introduced to make certain changes in the banking laws. By vote of the house the bill will be referred to the standing committee on banking. After the committee has considered the proposed law, it reports to the whole house with the suggestion that the bill be passed, be amended in certain respects, or be rejected. The committee's recommendation is usually, though not necessarily, accepted. In addition to the standing committees, *special committees* may be appointed from time to time to consider bills which do not fall within the sphere of any standing committee.

The Courts. The British North America Act did not provide Dominion courts, but it authorized Parliament to provide for the "constitution, maintenance and organization of a general court of appeal for Canada, and for the establishment of any additional courts for the better administration of the laws of Canada." The creation of a Supreme Court was delayed until 1876, and since that time two other Federal courts, the Exchequer Court and the Admiralty Court, have been established. These three are the only regular Dominion courts. In one sense, however, the courts of the provinces are Dominion courts, for the judges are appointed and paid by the Dominion government, acting through the Governor-General in Council; the procedure and jurisdiction of these courts is fixed by the provinces.

FIRST PARLIAMENT BUILDING, TORONTO (1796-1813)

Supreme Court. This is the highest court in Canada, and has appellate jurisdiction on all civil and criminal cases. In unusual cases, where the question at issue is extremely grave or is of importance not merely to Canada but to other parts of the British Empire, appeals from the decision of the Supreme Court are allowed to the Privy Council in London; but the court is made, as far as possible, the final authority on legal controversies arising in the Dominion. Besides the cases which have arisen in the lower courts and have been appealed, the Supreme Court may be consulted by the Governor-General in Council on important questions relating to provincial or Dominion legislation, education or "any other matter of a constitutional nature on which it is necessary to obtain a judicial opinion." Besides the Chief Justice there are five puisne judges, all appointed for life. *Puisne* is an old form of the word *puny,* and is pronounced in the same way; it designates these judges as *associate,* or *lower,* judges. The Chief Justice receives an annual salary of $10,000; the puisne judges each receive $9,000.

Exchequer Court. The functions of this court were originally exercised by the Supreme Court, but in 1887 a separate court was established. It has original jurisdiction in all claims or actions against the Crown, and all cases affecting the property or other interests of the Crown. It also has jurisdiction in cases regarding copyrights, trade-marks and patent cases, and in time of war it is a prize court. The court has a single judge, appointed for life at a yearly salary of $8,000.

Admiralty Court. This is not technically a separate court, but a division of the Exchequer Court. The duties of the latter are so heavy that special local judges have been appointed to hear cases in any way relating to shipping, navigation, trade and commerce in Canadian waters. Admiralty business may be brought before the Exchequer Court at Ottawa, or before one of the local judges in admiralty who hold court at Toronto, Quebec, Halifax, Saint John, N. B., Victoria, Charlottetown and Dawson. Only a judge of a superior or of a county court, or a barrister of not less than seven years' standing is eligible for appointment, by the Governor-General, as a local judge in admiralty.

Provincial Government. The details of provincial government are given in the articles on each province elsewhere in these volumes, and only the characteristics common to them all need be discussed here. The system is modeled on that of the Dominion. The lieutenant-governor, as representative of the Crown, occupies the same relative place as that of the Governor-General in the Dominion. The men who are in active charge of government comprise the *executive council;* they are members of the legislature and are responsible to it. In all of the provinces except Quebec and Nova Scotia the legislature is composed of a single elected house. In these two provinces there is also an upper house, the legislative council, whose members are appointed by the lieutenant-governor for life. The provincial courts have sole control over all matters affecting civil and property rights, and therefore come into closer contact with all classes of people than do the Dominion courts. The Dominion Parliament makes laws relating to crime and punishment, but the trial takes place in provincial courts. This division of powers was caused by the fact that Quebec in 1867 still used the French civil code, whereas the other provinces all used the English code. It was therefore necessary to leave civil procedure to the provinces, but criminal procedure is determined by the Dominion.

By the British North America Act the provinces are given entire control of a specified number of subjects, and their jurisdiction is limited to these, as follows:

(1) Amendment of the provincial constitution, except as regards the office of lieutenant-governor.

(2) Direct taxation to raise money for provincial purposes.

(3) Borrowing money on the credit of the province.

(4) Provincial offices, and the appointment and payment of provincial officers.

(5) Reform and penal institutions, also hospitals, asylums and charitable institutions.

(6) Municipal institutions.

(7) Shop, saloon, auctioneer's and other licenses.

(8) Public works, except such as are interprovincial in character.

(9) Guarantee of civil rights and property.

(10) Administration of justice.

(11) Matters of local or private nature.

(12) Education. G.H.L.

History of Canada

Age of Discovery and Exploration. The sagas of the Northmen recite the deeds of one Leif Ericson, "a large man and strong, of noble aspect," who is said to have reached the shores of North America, probably the Labrador coast, in the year 1000. The sagas further tell how he and his men sailed southward to a land of many large trees, which may have been Nova Scotia. That the brave Vikings reached Greenland is certain, but whether or not they ever reached the shores of Canada cannot be definitely known, for they left no traces of their visits. See VIKINGS; ERIC THE RED.

The first voyager known to have reached the mainland of North America was John Cabot, who discovered the bleak Labrador coast in 1497 (see CABOT). Like the other early navigators, Cabot was looking for a northwest passage to India, but he also hoped to "bring back so many fish that England will have no more business with Ireland." This part of his ambition was realized, for soon afterwards fishermen from Europe began to pay yearly visits to the Newfoundland Banks. The voyages of John Cabot and his son Sebastian did not disclose a new way to India, but they brought fishermen and fur-traders to America. In 1534 a brave seaman of Saint Malo, France, sailed westward to take possession of all lands he should find for King Francis I of France. This was Jacques Cartier (which see). He entered the Gulf of Saint Lawrence and coasted along the shores of Labrador, which he reported was barren enough "to be the land allotted of God to Cain." Cartier's experiences, however, aroused great interest in France. He made a second and a third voyage, discovered the Saint Lawrence River, and sailed up it as far as Hochelaga, an Indian village on the present site of Montreal. In connection with Cartier's third voyage an attempt was made to found a permanent settlement about ten miles up the river from Hochelaga, but after a terrible winter the colony was abandoned.

For more than half a century France thereafter neglected America. The failure of the colony, the cold winters and the barren coast discouraged the hardiest of adventurers. More than that, France was suffering from civil war, and brave men were needed to fight at home. Not until the beginning of the seventeenth century did the French again turn to the work of exploring and settling this new land which they claimed.

Coming of Fur-trader, Missionary and Colonist. Although the French as a nation paid no attention to this unexplored land across the seas, fishermen from France, as well as from Spain, Portugal and England, continued to visit the Grand Banks. On one occasion there were 150 French ships off the Banks, and it was a rare day when at least one ship did not sail from France bound for the fishing-grounds. Gradually the men of the sea recognized the profits to be made on land. Some of them built huts near the shore and began to exchange knives, hatchets, liquor and trinkets of all kinds for the furs which the Indians had secured.

The work of exploration was renewed in 1603 by Samuel de Champlain, the "Father of New France," and in the next year he took part in founding the settlement of Acadia (which see). A company was organized to colonize the land now included in New Brunswick and Nova Scotia; it was given a monopoly of the fur trade, but one of the conditions attached to the monopoly was that the natives should be converted to Christianity. Acadia led an uncertain existence for a century and a half. It was alternately held by the French and English, and more than once it was all but wiped out.

The first permanent settlement was made at Quebec in 1608 by Champlain (see CHAMPLAIN, SAMUEL DE). Until his death in 1635 he labored in every possible way to make the colony a success. He encouraged exploration of the interior, stimulated the fur trade, sent missionaries to the Indians, took part in Indian wars and at the same time encouraged settlement. The pioneer work of the pious Franciscan friars (Récollets) and the Jesuits who followed them inspired a wealthy French nobleman to found a new colony, primarily to provide a home for a new order of nurses. In 1642 Sieur de Maisonneuve, who had been sent to found the colony, ascended the Saint Lawrence River to the island on which Montreal now stands. The new colony barely managed to exist.

In 1627 the Company of New France, generally called the Company of the Hundred Associates, was given a monoply of the fur trade throughout the French possessions. Far from fulfilling its contract to add 4,000 colonists within fifteen years, the company occasionally discouraged settlement. But until 1663 New France remained under the control of the company. In that year Louis XIV dissolved the company, and made New France a royal colony. He chartered a new company, the Company of the West, to enjoy a monopoly of trade; it was no more successful than its predecessor, and its charter was annulled in 1674.

Worse than the failure of France to give proper support to the colony were the troubles within the colony itself. The Iroquois were frequently on the warpath and made life miscrable for the colonists. There was also much quarreling among the traders, among the priests, and finally between the bishop and the governor. Monseigneur de Laval, the first bishop of Canada, was strong enough to secure the dismissal of two governors with whom he had quarreled, but the later governors were little inclined to take advice from the priests.

New France under Royal Government. For a short time the new royal government promised a great future for New France. The colony had been on the verge of disaster, but it was now flourishing. In 1665 two thousand colonists arrived, with horses and sheep. Three able officials arrived in the same year—Sieur de Courcelle, the governor; Jean Baptiste Talon, the intendant, or treasurer; and Marquis de Tracy, a general. De Tracy's first attacks on the Iroquois failed, but as he had been sent out to destroy or subdue them, he tried again, this time with success. The Iroquois sued for peace, and for twenty years the colonists were free from Indian attacks. Many of the soldiers who served under de Tracy settled in Canada, and annual shipments of colonists continued to arrive. Courcelle was recalled in 1672, and was succeeded by Frontenac, the greatest governor since Champlain. Frontenac ruled with a firm hand, he held the Iroquois in check, but he frequently quarreled with the intendant and other officials, who resented his attempts to regulate the fur trade. The king, in disgust, finally recalled both Frontenac and Duchesneau, the intendant, in 1682.

Not the least of Frontenac's services to New France was the opening of the West. The Jesuits had heretofore given most of their attention to the Hurons, but the practical destruction of this nation by the Iroquois left them without a field. Frontenac encouraged them to go into the unexplored West, and he gave aid to Marquette and Joliet, to La Salle and Tonty. These men discovered the Mississippi and traced it to its mouth, thus giving France a claim on the great interior valley which it finally surrendered to the United States in 1803. Posts were established at Mackinac, Niagara and other points, and New France seemed embarked on a new era of prosperity. See MARQUETTE; JOLIET; LA SALLE.

The French, however, were not without ambitious rivals, for in the north the Hudson's Bay Company (which see) was taking away the trade of the northern Indians, and on the southeast Albany was becoming a great center for marketing furs. Frontenac's successors, moreover, were unable to handle the Iroquois Indians as Frontenac had done. Attempts to conquer or conciliate resulted in enraging them, and for several years they put an end to the French fur trade. Their raids had a terrible climax at Lachine, which they destroyed to a house and to a man. In this crisis, King Louis XIV turned again to Frontenac, who seemed the only man, in spite of his seventy years, to save the colony from destruction.

The Struggle for New France. The reappointment of Frontenac coincided with the beginning of the struggle between France and England—France to keep New France, England to conquer it. There had been wars between them before this time, and when the French in Canada were not fighting both the Iroquois and the English, they were fighting the Iroquois. But it was now clear that the French were planning a great American empire, while William III of England was beginning to organize Europe against French aggression. In 1689 the real struggle began, and for three-quarters of a century it continued, with occasional periods of peace or truce, during which the nations made further preparations to renew the conflict.

It was the ultimate aim of the French to hold the English to the Atlantic seaboard, or even to drive them out of North America. Frontenac at once began attacks on the New England colonies, and for years the colonies suffered the horrors of border warfare. The success of their raids encouraged the French, but infuriated the English, who made an attempt with a small fleet to capture Quebec and Montreal. The Peace of Ryswick in 1697 brought a temporary halt to the hostilities, and in the next year Frontenac died. Peace did not last long; in the next years three more wars were fought (for details, see FRENCH AND INDIAN WARS). After the fall of Quebec, in 1759, the triumph of England became almost inevitable, but the French made desperate attempts to regain Quebec and hold Montreal. When no aid came to them from France, the French army in America was driven to surrender in 1760. By the Treaty of Paris in 1763 France surrendered its possessions in America to England.

Government and Society under French Rule. Subject to the approval of the king, the royal governor of New France was practically an absolute ruler. He commanded the army, and was authorized to conduct all negotiations

BISHOP'S PALACE, QUEBEC
Where the first Parliament of Lower Canada met, in 1792.

with the Indians and with foreign powers. The strong governors, like Frontenac, stretched these powers to cover almost all problems. Second in authority to the governor was the *intendant,* who controlled the finances and the administration of justice. Incidentally he was a sort of spy on the governor. The governor was generally a noble of high rank; the intendant was usually a lawyer and of humble birth. The bishop ruled the Church, and also sat in the supreme council, which issued the official decrees in most matters of civil government. There was a bishop's court at Quebec for the trial of offenses against the Church, and there were local judges for civil cases at Quebec, Three Rivers and Montreal. Appeal was allowed from all courts to the council and from the council to the intendant, who could try any case. The people were never consulted on government matters; the spirit of government was well expressed by one intendant, who wrote that "it is of the greatest consequence that the people should not be left at liberty to speak their minds."

Paternalism and absolutism ruled in society as in government. Richelieu, in chartering the Company of the Hundred Associates, had introduced into Canada a feudal system much like that of France. His object seems to have been to establish a Canadian aristocracy and also to find an easy method of dividing the land among settlers. The land was granted by the king to *seigniors,* who parceled it out to *censitaires,* who held their land on the payment of *cens* or quit-rent. This rental was very low, usually about one cent an acre. The actual tillers of the soil were the *habitants,* the lowest class of society.

The Canadian aristocracy, as a whole, was a penniless lot. The rich nobles of France were too fond of court life to exchange it for the colonial frontier. The nobles in Canada, therefore, were mostly officers of regiments which had served in Canada or merchants and farmers who had become prosperous. The more aristocratic of them, too proud to engage in farming or ordinary trade, were always heavily in debt, but many of them found a way to sustenance if not to wealth through the fur trade. These exiled gentlemen became the explorers and defenders of New France.

Not all of the men in New France, however, submitted to the feudal system. Love of adventure, disgust with farming, ambition for wealth, all combined to lead many of the younger men into the woods. Here they hunted and traded for furs, and many of them lived with the Indians and took Indian wives. These *coureurs de bois,* "runners of the woods," as they were called, were brave and hardy, but they caused much trouble for the governor. They freely broke the regulations for the fur trade, and any attempt to punish them was likely to make them outlaws and enemies of their country. After months in the distant forests, where they secured the choicest furs, they would return to the settlements for a few weeks, to spend in revelry and dissipation all the money they had received for the pelts. When their money was gone they would wave an *adieu* to civilization and plunge again into the wilderness.

After Britain's Conquest. The change from French to British rule was made without much disturbance. Some of the Indians, led by Pontiac, a chief of the Ottawas, rose against their new masters; but the French as a whole were glad to return to peaceful occupations. A few of the seigniors, merchants and higher officials—perhaps 400 in all—returned to France, but the habitants were quickly reconciled to British rule. For three years the conquered territory was under military rule, until George III reorganized the government by proclamation in 1763.

Prince Edward Island and Cape Breton were added to Nova Scotia, Anticosti was given to Newfoundland, and the main colony was organized as a separate province under the new name of Quebec.

In 1763 there were 60,000 French and only 500 English in Canada. The home government felt that the incoming English settlers would soon absorb the French element, but in this opinion they were mistaken. Fortunately the governor, General James Murray, tried to satisfy the majority of the population. He had no high opinion of the Englishmen in Quebec, whom he called "men of mean education, traders, mechanics, publicans, followers of the army." Discontent remained, however, and the necessary reforms in the government were made by the Quebec Act (which see) in 1774. This Act met some bitter opposition in England because of the great concessions it made to the French-Canadians, but the wisdom of its authors was soon to be proved, for in 1775 and again in 1812 the French-Canadians refused the chance to rebel against England and even helped to fight England's battles.

At the outbreak of the Revolutionary War the New England colonies tried hard to secure the coöperation of Quebec, and when their efforts failed they made vigorous attacks on the loyal colony (for the military events of this period, see REVOLUTIONARY WAR IN AMERICA). The war brought one great gain to Canada. Some 40,000 colonists who remained loyal to the king and the United Empire abandoned their homes in the United States and crossed the border into the loyal colony. Many of them left valuable estates or gave up influential and remunerative positions as ministers, judges and other officials. From a life of comparative ease they plunged again into a wilderness. Many settled in Nova Scotia, some in Eastern Quebec, but most of them made homes for themselves in the unoccupied lands which now constitute New Brunswick and Southern Ontario. To this day their descendants are the most important single element in these provinces.

The Struggle for Responsible Government. As the number of English-speaking settlers increased it became evident that the Quebec Act did not provide a satisfactory government. In 1791, therefore, the Constitutional Act divided the colony into Upper Canada and Lower Canada. Upper Canada had a population of 20,000, mostly United Empire loyalists; Lower Canada had six times as many people, mostly French-speaking. For fifty years Upper and Lower Canada remained separate provinces, each with a governor, an executive council corresponding roughly to a Cabinet, a legislative council and an assembly. The members of the assembly were elected by the people, and in theory the government was representative and responsible to the people. In practice, however, the governor and the officials appointed by him were in absolute control. For half a century the history of Upper and Lower Canada is a struggle for complete representative government (for details, see ONTARIO; QUEBEC; subhead *History* under each article). A similar struggle was going on in New Brunswick and Nova Scotia.

Bitter political strife was temporarily pushed aside by the War of 1812 (which see), when the Canadians, both French and English, rallied in defense of their country. Nor did politics in the East prevent the exploration and development of the West. There were the two great fur-trading companies, the Hudson's Bay Company and the Northwest Company, which later joined forces. Settlements were made in the Red River Valley (see MANITOBA, subhead *History*), and the explorations of Sir Alexander Mackenzie, Simon Fraser and David Thompson opened up this wilderness and laid the basis for a great Dominion extending from the Atlantic to the Pacific.

War and expansion had no effect on the political issues in the provinces. In Upper and Lower Canada the radicals actually resorted to open rebellion to secure the reformation of the government. Under Louis J. Papineau in Lower Canada and William Lyon Mackenzie in Upper Canada, the radicals took arms, but at the first contact with the militia the rebels were dispersed. The rebellion was doomed to failure, for only the most impulsive and thoughtless of the people had been carried away by the eloquence of a few leaders. The great mass of the people were not merely loyal, but were disgusted with the violent methods of the reformers.

The rebellion of 1837 naturally brought discredit on all the reformers, even Robert Baldwin and the other moderate Liberals. But it had one great result: It called the British government's attention to the critical state of affairs in the Canadas and led to the Act of Union. The Earl of Durham was appointed Governor-General of British North America and High Commissioner to investigate the abuses in the government, and as a result of his famous report, issued in 1839, Upper and Lower Canada were reunited by act of the British Parliament.

The operation of the Act of Union disappointed the reformers. The governor and the legislative council appointed by the Crown still maintained control. In the legislative assembly the two provinces elected an equal number of representatives, but the rapid growth of Upper Canada soon caused a demand for representation by population. Meanwhile the first three governors, Sydenham, Bagot and Metcalfe, refused to recognize the responsibility of the executive council, or Ministry, but Lord Elgin in 1848 admitted the principle of responsible government, and since that date Canada has been practically self-governing. The Rebellion Losses Bill, which was aimed to compensate the citizens of Upper and Lower Canada for their losses in the rebellion of 1837, was bitterly opposed by many Conservatives, because, they said, it put a premium on rebellion. But when the supporters of the bill were victorious in the general elections, Lord Elgin forced the Conservative Ministry to resign and called into office the leaders of the reformers, Lafontaine and Baldwin. It is noteworthy that as long as the Act of Union was in force the Ministry always included the leader of the dominant party in each province.

In Nova Scotia and New Brunswick there had been similar struggles for responsible government. The British government finally instructed the governor of Canada to rule in accordance with the well-understood wishes of the people "and summon to the Ministry those who held the general confidence and esteem of the province," and within a year or two the Maritime Provinces compelled their governors to adhere to the same principles.

During the next fifteen years a succession of Ministries carried on the government. There was always some friction between Upper and Lower Canada, and in each province there were local issues which further divided the great political parties. As a result no Ministry ever had a large majority in the assembly, and none stayed in office very long. The most prominent in this changing array of ministers were Allan MacNab, Robert Baldwin, Sir Louis Lafontaine, Sir Etienne Taché, Sir John A. Macdonald, George Brown, Sir Francis Hincks, John S. Macdonald and Thomas D'Arcy McGee.

In spite of those constant changes three important reforms were enacted. The Municipal Corporation Act of 1849, of which Baldwin was the author, provided a system of municipal government for Upper Canada. In 1854 seigniorial tenure was abolished in Lower Canada, and the clergy reserves were secularized in Upper Canada (for details, see subheads *History* under ONTARIO; QUEBEC). Another event of great importance in 1854 was the negotiation of a commercial treaty between Canada and the United States; it provided for free trade between the countries and remained in force for ten years.

Confederation. The idea of confederation, of a Federal union in which each province should retain control of its local affairs while transferring its general powers to a central government, was not a new one in 1864. It had been suggested early in the nineteenth century, and the Earl of Durham had been one of its ardent supporters. It was also gaining ground both in Upper and in Lower Canada, chiefly because real government was rapidly becoming impossible there under the Act of Union. Finally, in 1864, after one Ministry had succeeded another in an apparently endless chain, George Brown proposed a coalition Ministry, including men of all parties, to work for a union.

While the statesmen in the two Canadas were thus beginning to put their heads together, the Maritime Provinces became alarmed at the possibility of war between Great Britain and the United States (see TRENT AFFAIR); they considered the advisability of a union and called a conference of delegates to meet at Charlottetown, P. E. I. When the Canadians heard of this coming conference, they asked and received permission to send a delegation, which included John A. Macdonald, Georges E. Cartier and George Brown. The proposed union of the Maritime Provinces was overshadowed by the greater plan of union of all the provinces, and the Charlottetown Conference adjourned after deciding to hold a second conference at Quebec a month later. It was at the Quebec Conference (which see) that a series of seventy-two resolutions embodying a plan of government was adopted.

The action of the conference was received with joy in Great Britain and in Upper and Lower Canada. Newfoundland and Prince Edward Island rejected the plan entirely, and New Brunswick and Nova Scotia accepted it only after much debate and delay. Delegates were sent to England to present the resolutions to the home government, and in March, 1867, Parliament passed the British North America Act providing for a union of Canada, Nova Scotia and New Brunswick as one Dominion of Canada. The Act, except in a few minor details, followed the plan of the resolutions adopted by the Quebec Conference.

Expansion of the Dominion. The British North America Act provided for the division of Canada into two provinces, Quebec and Ontario, previously known as Lower and Upper Canada, respectively. It also made provision for the addition of other parts of British North America to the new Dominion. During the first session of the Dominion Parliament a resolution was passed calling for the annexation of Rupert's Land and the Northwest to Canada. One of the arguments in favor of this change was that the Hudson's Bay Company was interested in its own trade rather than in the development of the West, and also because of the possibility of aggression by the expanding United States. Under pressure from the British government the company finally surrendered its control of the Northwest; it received £300,000 and one-twentieth of all the land lying south of the north branch of the Saskatchewan River and west of Lake Winnipeg, and it also retained its posts and its trading privileges.

YORK FACTORY
For many years one of the main posts of the Hudson's Bay Company. At the mouth of the Nelson River, opposite Port Nelson.

Red River Rebellion. The Hudson's Bay Company formally surrendered its territorial rights to the British government on November 19, 1869. In anticipation of the further transfer to the Canadian government, Hon. William McDougall had already been appointed lieutenant-governor of Rupert's Land. The news that he was on his way to Fort Garry (Winnipeg) was the signal for an uprising of the Metis, or half-breeds, in the Red River Valley. The Metis, ten thousand of them, had not been consulted about the change in government, and they were aroused by the possibility that the new government would place restrictions upon them and would perhaps even drive them from their homes. A number of Fenians and Americans, agitating for annexation to the United States, added to the excitement, but the French half-breeds, led by Louis Riel, formed the storm center. Riel was a brilliant man, eloquent and magnetic, but vain and self-seeking to an extraordinary degree. He organized a "provincial government," forbade McDougall to enter Rupert's Land, and, to terrorize his opponents, executed Thomas Scott, a young Orangeman from Ontario. Hitherto the rebellion had created little interest in Eastern Canada, but this cold-blooded execution raised a storm of indignation. An expeditionary force was quickly gathered and placed under the command of Colonel (later Lord) Garnet Wolseley. Before the expedition reached Fort Garry the rebels dispersed, and Riel fled across the border to the United States. (See RIEL, LOUIS.)

New Provinces. While the troops were on the way westward, the Dominion Parliament passed an act (July 15, 1870) creating the province of Manitoba. The arrival of the soldiers at Fort Garry, whose name was at this time changed to Winnipeg, was followed almost at once by that of Sir Adams Archibald, the first provisional governor.

In 1871 British Columbia finally agreed to enter the Dominion, subject to a number of conditions, the most important of which was that a transcontinental railway should be begun within two years and completed within ten years. The failure of the government to fulfil its contract nearly led to the withdrawal of British Columbia from the Dominion, but with the driving of the last spike in 1885 British Columbia was firmly bound to the other proviuces.

Prince Edward Island, which had been one of the most vigorous opponents of Confederation, was the next to join, and in 1873 became a province. The Dominion government assumed the heavy provincial debt and also bought the rights of certain absentee landowners. The Dominion now included all of British North America except Newfoundland. To remove any possible doubt as to the Canadian jurisdiction of the unoccupied Northwest, in 1878 an imperial order in council was issued, annexing to the Dominion all British possessions in North America except Newfoundland, which remains to this day a separate colony.

For more than twenty-five years the organization of the Dominion remained unchanged, but by the end of the nineteenth century there was a strong demand for better government in the Northwest Territories. This vast section was growing rapidly in population, and the loose territorial government was unsuited to the new conditions. Finally, in 1905, Parliament created two new provinces, Alberta and Saskatchewan, including in them the old districts of Alberta, Assiniboia, Saskatchewan and Athabaska. Thus a solid row of provinces stretched from ocean to ocean.

In 1912 a further change was made in the provincial boundaries of Manitoba, Ontario and Quebec. The greater part of the old district of Keewatin was divided between Manitoba and Ontario, and all of Ungava was added to Quebec. The Northwest Territories (which see) now include only the districts of Franklin, Mackenzie and part of Keewatin. Maps showing old and new areas appear with the articles on these provinces.

National Problems. The unification and expansion of the Dominion were not accomplished without many difficulties. The organization of a Federal government brought in its train a series of national problems, but many of the people were still too much absorbed in local issues to get a proper perspective of the new questions which had to be answered. Lord Monck, the first Governor-General, called on Sir John A. Macdonald, who had been foremost in the Confederation movement, to form the first Ministry. Sir John, or "John A.," as he was popularly known, summoued to his aid able men of all parties, among them Sir A. T. Galt, Sir Georges E. Cartier, Sir Alexander Campbell and William McDougall.

At the first general election, in August, 1867, when this Ministry appealed to the country for a vote of confidence, it was overwhelmingly defeated in Nova Scotia. For about two years Nova Scotia agitated for the repeal of the British North America Act, even sending a delegation to appeal to the British Parliament. The Imperial government refused absolutely to allow Nova Scotia to withdraw from the union, and a compromise was finally effected, by increasing the amount of the annual subsidy to be paid to the province by the Dominion.

The first years of the Macdonald Ministry were occupied chiefly in organizing the new government. Postal rates were unified and the postoffice placed under the control of a single department, the tariff was systematized, and a civil service and a militia were established. A national banking system was created by an act of 1871, and the Intercolonial Railway was opened to traffic in 1876. The most serious difficulties arose in connection with the extension of the Dominion to the west. The Red River Rebellion (see above), while easily suppressed, caused great anxiety, and the Pacific Railway was a political issue for ten years.

Foreign Policies. Without doubt the most notable achievement of Macdonald in dealing with foreign relations was the settlement of a number of long-standing disputes between the United States on the one hand and Canada and Great Britain on the other. At the request of the Canadian government Great Britain asked the United States to make a peaceful settlement of the questions in dispute. A joint commission was appointed, with Macdonald as Canada's representative, which met at Washington, the capital of the United States, in February, 1871. The United States claimed that Great Britain should pay for the damage inflicted by the *Alabama,* a Confederate cruiser which was fitted out in England (see ALABAMA, THE). This claim the commission submitted to arbitration. The Canadian claim for damages from the Fenian raids was withdrawn by the Dominion at the request of the Imperial government, but in return for this surrender Great Britain guaranteed a loan for the construction of Canadian railways. The Saint Lawrence River and the Great Lakes were opened to ships of both countries, and the Oregon boundary was at last definitely settled. The fisheries dispute was turned over to a commission which held sessions at Halifax, N. S., and granted the United States the unrestricted use of Canadian waters for ten years in return for a cash payment of $5,500,000. The Treaty of Washington in this way disposed of several troublesome problems.

Fall of the Conservatives. When British Columbia entered the Dominion in 1871 it was with the understanding that a transcontinental railway would be begun at once. In 1872 two companies of Canadian capitalists, one headed by Sir Hugh Allan, the other by Hon. David Macpherson, sought a charter for the construction of this railway. While negotiations were under way to unite the two companies, the charge was made in the House of Commons that Macdonald and certain other members had received and used money furnished by Sir Hugh Allan to influence voters in the elections of 1872. While it did not appear that Macdonald or any other member of the Ministry had profited personally from these transactions, it was generally admitted that Allan had paid out large sums of money with the understanding that his company would be awarded the contract. Public feeling became very intense, and the Macdonald Ministry resigned in October, 1873, before the charges could be brought formally. The Governor-General, then the Earl of Dufferin, called on Alexander Mackenzie, the leader of the Liberal opposition, to form a new Ministry, which received an overwhelming majority at the general elections.

Mackenzie and Liberal Rule. The new Premier, Mackenzie, soon announced that it would be impossible to build the railroad as had been planned. He proposed that the Dominion should itself undertake the task and build the road a little at a time, as its finances permitted. This suggestion was unsatisfactory to British Columbia, which forced the government to agree to build a wagon road and telegraph line at once and to complete the railway by 1890. During the years that the Mackenzie Ministry held office a number of important laws were passed. One established the Dominion Supreme Court, another adopted the Australian ballot (which see) and a third organized the Northwest Territories. Immediately after the Red River Rebellion the Northwest began to fill with settlers, who needed some organized government. A new feature of this government, which has since been copied in Australia and elsewhere, was the Royal Northwest Mounted Police (which see).

Though the Liberal Ministry was responsible for these and other advances, it was not very popular. It was handicapped in its policies by a strongly Conservative Senate and a powerful opposition even in the House of Commons. Canada, moreover, was passing through a period of business depression, which was unjustly laid at the doors of the Liberals. The factories of the United States, which were then suffering from a similar depressed condition, tried to flood Canadian markets with goods they could not sell at home. The opposition, seizing their chance, began to cry for a protective tariff, and "Canada for the Canadians." This became Sir John Macdonald's "national policy," on which he fought the elections of 1878 and was returned to power.

Thirteen Years of Conservative Government. From 1878 until his death in 1891 Macdonald was at the helm. A protective tariff was established, and the Canadian Pacific Railway was completed. Macdonald rejected Mackenzie's plan of piecemeal construction, and awarded the contract for the entire job to a new Canadian Pacific Railway Company, of which Lord Mount Stephen (then George Stephen) and Lord Strathcona (then Donald A. Smith) were prominent members. The work of construction was pushed so rapidly that the last spike was driven in November, 1885, five years before the date fixed in the contract.

The construction of the railway caused a second rebellion of the half-breeds led by Louis Riel. After the Red River Rebellion each of the half-breeds or Metis had been given 240 acres of land, but as the white settlers began to appear they gave up their lands and settled farther west, on the banks of the Saskatchewan. The coming of the railway, with the possibility that they would again be forced to move, led the half-breeds into open rebellion which was quickly suppressed. The rebellion left its mark on the whole Dominion rather than on the Northwest alone, for every province was interested in ending it. Common dangers "strengthened in the hearts of Canadians the union which Confederation had brought about."

While Macdonald lived he held his party together, and at succeeding elections was returned to power; the Liberals' attempts to make capital out of the tariff issue met with little success. Macdonald's successor as Premier was Sir John J. C. Abbott, who was followed in turn by Sir John Thompson and Sir Mackenzie Bowell. During Thompson's tenure of office the Bering Sea controversy was settled, but the succeeding Ministries were noteworthy chiefly for the dissensions among the Conservatives. In 1896 Sir Charles Tupper assumed office and reorganized the Ministry, but the country had lost confidence in the Conservatives and at the following elections returned the Liberals to power.

The Laurier Ministry. Sir Wilfrid Laurier, the new Premier, was the first French-Canadian to hold that office. During his administration of fifteen years, from 1896 to 1911, Canada showed a remarkable economic development, and correspondingly, a development of national self-confidence and national unity. But this growth of national unity and individuality was no more marked than the strengthening of the ties which bind Canada to the Empire. The power of these ties has been illustrated in many ways. The outbreak of the war in South Africa in 1899 gave the Dominion an opportunity to show its loyalty by sending three contingents of troops. The laying of the Pacific cable from Canada to Australia, the establishment of penny postage throughout the Empire, the granting of preferential tariffs to British goods imported into the Dominion, and the celebration of the Quebec Tercentenary are merely examples of the workings of this national spirit. In 1905 the British garrison in Halifax was replaced by Canadian troops, and Esquimalt, the naval station in British Columbia, was placed under Dominion control. Canada is now solely responsible for the defense of the Dominion, and it has also become bound, at least in a measure, to make the British Empire as self-sustaining in defense as in commerce and industry.

Internal Development. The great economic development of the Dominion during the last years of the nineteenth century and the first decade of the twentieth is conspicuous. This has been in part the result of a moderate tariff which provided low duties on manufactured goods and certain food products and raw materials used by the Dominion, but at the same time provided some protection to Canadian industries. The discovery of gold in the Yukon and the organization of Alberta and Saskatchewan as provinces led thousands of settlers westward. It was nothing but a stroke of genius which led the government to advertise the rich lands of the west, and much of the credit for the growth of the prairie belongs to Sir Clifford Sifton, Laurier's Minister of the Interior, who conceived the plan. The construction of two new transcontinental railways —the Grand Trunk Pacific and the Canadian Northern, in addition to the Canadian Pacific— opened up a vast territory to settlement and stimulated national pride. The west is being steadily developed with an eye to its future possibilities, not merely being exploited for temporary gain.

Foreign Relations. From time to time new disputes have arisen between the Dominion and the United States, or old ones have been revived. Laurier proposed to settle all disputes by the formation of a Joint High Commission. Such a commission was appointed in 1898, the Canadian representatives being Sir Wilfrid Laurier, Sir Louis H. Davies and Sir Richard Cartwright. The negotiations, unfortunately, were broken off by a sharp disagreement over the Alaska boundary (see ALASKA, subhead *History*); this dispute was not settled until 1903. Also in 1903 the many questions arising from the joint use of the Great Lakes and other boundary waters led to the appointment of a commission to arbitrate these disputes, and in 1908 the two countries reached an agreement on an accurate system of marking the international boundary. At the same time the fisheries question was being arbitrated by the Hague Tribunal, which upheld the claims of Canada and Newfoundland on all important points.

Another step of great importance was the negotiation of a commercial treaty with France in 1907. The negotiations were conducted by two Canadians and the treaty was ratified by the Dominion. Equally striking was the free hand given to Canada to negotiate with Japan in the same year. The result was an agreement that Japan would restrict emigration to Canada, and the Dominion government would proteet the Japanese in Canada from violence and aggression.

Reciprocity and Liberal Defeat. For many years the tariff relations between the Dominion and the United States have been unsatisfactory, and attempts have been made from time to time to readjust them. In 1910 an American delegation visited Canada to urge reciprocity (which see), and in 1911 a treaty was formally presented for approval. The treaty was promptly ratified by Congress, but in the Dominion Parliament it met bitter opposition. The debate in Parliament continued from January to May, 1911, and after an interval, for ten days in July. Parliament was suddenly dissolved on July 29, and the issue presented to the country. The general election on September 21 gave the Conservatives a large majority in the House of Commons. The Laurier Ministry promptly resigned, and on October 10 Robert L. Borden (now Sir Robert) formed a new Ministry.

The Borden Ministry. One of the first measures introduced and carried by the new Ministry greatly extended the boundaries of Manitoba, Ontario and Quebec. Of more vital importance, however, was the change, or attempted change, in the naval policy. Several months after a conference in London, in June, 1912, between Canadian and British ministers, the Conservative government decided to present to the Dominion Parliament certain emergency proposals in regard to the navy. The Borden policy, in brief, provided for the immediate construction of three dreadnaughts, to

ROYAL MINT, OTTAWA

Established by royal decree, but supported by the Dominion. Here all Canadian money is produced, and also some of the gold sovereigns of the mother country. Each of the latter so coined bears on the reverse side a small *C*. The mint was opened in 1909, and is in charge of a deputy mint master from London.

be built in Great Britain at a cost of $35,000,000, to form a part of the British navy and to be under the control of the British admiralty. In spite of the Premier's repeated statements that this was an emergency policy, formulated in the face of grave danger, the Liberals insisted that the principle of a strictly Canadian navy should not be departed from. Both sides agreed as to the necessity for a naval contribution, but differed as to the method by which this was to be accomplished. The government's proposals, though carried in the Commons, were defeated in the Senate.

Minor Questions of Policy. The debates on the naval policy overshadowed all other issues during 1912 and 1913. Early in 1914 Premier Borden announced a plan for the addition of nine Senators and the redistribution of seats in the House of Commons. The budget for 1914-1915, in spite of the strenuous efforts of the Liberals to secure the reduction or abolition of the duties on foodstuffs, failed to provide any downward revision of the tariff. Another problem which threatened serious consequences concerned the granting of financial aid to the Canadian railways, none of which could have been built without the assistance of the government. The Liberals, under Sir Wilfrid Laurier's leadership, opposed the continuance of such aid, and particularly objected to a new arrangement, made with the Canadian Northern Railway, by which the government, in return for aid granted by it, acquired $40,000,000 of the $100,000,000 capital stock. All of these issues sank into insignificance in the autumn of 1914, on the outbreak of the War of the Nations, when Canada at once proffered its aid to the mother country. Its part in the struggle was a most honorable one. G.H.L.

Canada and the War

The Nation's Loyalty and Patriotism. Within three hours after Great Britain had declared war a call was issued for a special session of the Dominion Parliament. At the beginning of the session, on August 18, the Duke of Connaught, Governor-General, boasted that "the spirit which animates Canada inspires His Majesty's dominions throughout the world, and we may be assured that united action to repel the common danger will not fail to strengthen the ties that bind together these vast dominions." Sir Wilfrid Laurier, the leader of the Liberal opposition, instead of offering the usual objections to the government's policy, promised the whole-hearted support of his party in the fight "for freedom against oppression, for democracy against autocracy, for civilization against reversion to that barbarism in which the supreme law, the only law, is the law of might." Parliament responded to these appeals by appropriating $50,000,000 for war expenditures, the amount to be raised by additional duties and taxes.

Everywhere throughout the Dominion its citizens were giving evidence of the spirit which animated the people. A special regiment, the Princess Patricia's Light Infantry, was equipped by one man, $500,000 was donated by another, $100,000 for a battery of quick-firers was offered by a third, one million bags of flour were donated to Great Britain by the Dominion government, and flour, cheese, horses, tinned salmon and other commodities were offered by the provincial governments in large quantities.

Patriotism alone, however, wins no battles. The Canadian government at once placed at the disposal of the British government the cruisers *Niobe* and *Rainbow,* and on August 5 purchased two submarines, just completed for Chile, and held these ready for the defense of the western coast. It also offered to send an expeditionary force of 20,000 men or more. Within three weeks 32,000 men had volunteered and were in training at Camp Valcartier, while 150,000 more had volunteered. The first Canadian contingent reached Plymouth, England, on October 8, 1914, and before the end of the year a part of it was on the fighting line. In April, 1915, in the second Battle of Ypres, the Canadian division, though greatly outnumbered, held its ground at Saint Julien, and thereby saved the allies from a great disaster. Men accustomed to civilian life, men drawn from the bar, the universities, the shops and counting houses, from every activity, were plunged into the most scientific, the most bloody and the most devastating war in the history of mankind. And these men proved their right to stand side by side with the bravest veterans. The Canadians made good. In the second Battle of Ypres, or the battle of Saint Julien, as the Canadians' part in it is known—

> "The Canadian officers, and indeed men, had the Wellington touch—the touch of the man who never lost an English gun which he did not recover. What Wellington did in the Pyrenees the Canadians did on the flattest plain in Europe. Within four hours they turned around and advanced. They endured every sort of difficulty, mental or mechanical. They met men terribly sick, half-blind and weak in the limbs from the fumes let loose by the Germans. They had to face shrapnel, rifle fire, machine-gun fire, in a country absolutely defenseless. They were wounded by every sort of engine—by the bayonet, by gas, by bits of metal of every shape and size. But they charged home—against all the rules of war as understood in German textbooks."

Meanwhile those who were left at home cheerfully assumed the burdens thrust on them by war. The aim of the government was to raise an army first of 250,000, then gradually increasing to 500,000 men. In the work of organization the most conspicuous man was Major-General Sir Sam Hughes, and to the front in France General Sir Arthur Currie was despatched as commander in chief of Canada's overseas forces. In 1917 the call for men became so pressing that the government passed a conscription act, following the example of the United States. This was patriotically received in all the provinces excepting Quebec, where in the spring of 1918 draft riots occurred. Another remarkable evidence of Canadian loyalty is the Canadian Patriotic Fund, over $15,000,000, raised by voluntary subscriptions, and used in taking care of the wives and children of those who have gone to the front. See, also, WAR OF THE NATIONS. G.H.L.

Military Affairs

Canadian Militia. The constitution of Canada, based on the British North America Act of 1867 (which see), authorizes the Federal government of Canada to provide and maintain a militia force for the defense of the country and the maintenance of law and order within its borders.

Composition. By the Militia Act (Chapter 41, Revised Statutes of Canada, 1906), all male inhabitants of Canada between the ages of eighteen and sixty, if they are able-bodied and not exempt or disqualified by law, and are British subjects, shall be liable to serve in the militia, although the Governor-General may require all the male inhabitants of Canada capable of bearing arms to serve in a case of grave emergency.

Active Militia. The general system under which the militia is organized at present is based upon the voluntary offering for service of male citizens between the ages of eighteen and forty-five, and the militia is divided into two sections, the *active* and the *reserve.* The active militia consists of (1) Corps raised by voluntary enlistment; (2) Corps raised by ballot.

Minister of Militia. The command-in-chief of the militia is vested in His Majesty the King, but the Parliamentary head of the militia forces of the country is the Minister of Militia and Defense, one of the Cabinet Ministers of the Dominion government, who is responsible for the administration of military affairs, including the initiative in all matters involving the expenditure of money. The Minister is advised by a militia council, of which he is president, with the Deputy Minister of Militia and Defense as vice-president. As constituted at present, the militia council, in addition to the Minister, consists of six members, and an official of the department acting as Secretary. The members are—

The Deputy Minister.
The Chief of the General Staff.
The Adjutant-General.
The Quartermaster-General.
The Master-General of the Ordnance.
The Accountant and Paymaster-General.

Military Districts. The Dominion, for purpose of military administration, is divided into ten military districts, each administered by a staff officer, who is responsible for the maintenance and efficiency of the militia units within the boundaries of his district.

Permanent and Non-Permanent Forces. The active militia is divided into the *permanent* and *non-permanent* forces, the permanent force (not exceeding 5,000) being enrolled for continuous service and furnishing schools of instruction for the active militia and instructors for the same. The non-permanent militia consists of units of various arms, namely, cavalry, artillery, engineers, infantry and rifles, army service corps, army medical corps, with details of various other administrative departments necessary for the administration of a military force in the field.

Officers and Men. Officers are appointed by His Majesty, represented by the Governor-General-in-Council, and are nominated in the first instance by the officers commanding units, who are responsible for the quality and standing of the candidates put forward. These may be either men already serving in the units of the active militia, or citizens who have not hitherto served but who are obliged to qualify at classes of instruction before being confirmed in their rank. The men are enrolled voluntarily for periods of three years. An officer holds office during pleasure.

Strength. Under the above general system, the strength of the active militia during the year 1914 was 70,064, with 15,067 horses. The personnel was made up of 5,379 officers and 64,685 N. C. O.'s (non-commissioned officers) and men.

Overseas Expeditionary Force. Upon the outbreak of the War of the Nations, in 1914, the Canadians at once began to coöperate with the mother country, and the general system of the militia was utilized for the creation of an expeditionary overseas force. Provisional units were officered and at the beginning manned largely from the units of the active militia. These latter maintained their skeleton organization in their various centers throughout the war and continued to provide recruiting centers for the expeditionary units. In less than two years after hostilities began more than 350,000 men were under arms. See subtitle *Canada and the War.*

The Cadet System of Military Training. Some of the universities, colleges, academies and high schools in Canada have conducted cadet work for more than a century. It has been conducted as a part of the regular work of the public schools in Toronto for over forty years. Cadet work is recognized and encouraged by the Dominion government, and by the governments of all the provinces of Canada, as a very effective agency for developing a boy's physical and executive powers; for making obedience to law an essential habit; for revealing his duties as a citizen; for training him to defend his country and the fundamental principles of Christian democracy, if they are threatened; for arousing genuine patriotism—not an arrogant or offensive consciousness of national pride; for training a boy in habits of neatness, definiteness and dignified behavior, and for making him conscious of his value as an individual when coöperating with his fellows, reverencing guiding laws for the achievement of a common aim.

Cadets are not soldiers. They take no pledge to give military service, but they are prepared more effectively and much more cheaply to do military service than it is possible to prepare them in any other way. Those who were cadets as boys at school are more quickly and more thoroughly trained for efficient work in defense of home and justice and liberty when they are men. Comparatively few Canadians object in any way to cadet training in the schools. A statement was issued in 1912 by leading ministers of Protestant and Roman Catholic churches in Canada strongly approving cadet work for national, physical and ethical reasons.

The following statement shows the rapid increase in cadet work in the provinces of Canada from March 31, 1913, to March 31, 1916:

and second, to the very active interest of the Minister of Militia, Sir Sam Hughes. The War of the Nations undoubtedly added to the recognition of the value of cadet work during the latter part of the period, but the increase was much greater during the two years preceding the war than it was after the war began. In 1916 there were 654 companies, with a total of 26,160 cadets in universities, colleges, academies and high schools. There were 638 companies with 25,520 cadets in public, separate and other preparatory schools. "Separate" schools are elementary schools conducted by religious minorities. They are national schools which are under the regulation of the education departments of the provinces of the Dominion in which they exist.

There were in the same year 128 companies with 5,120 cadets in municipalities, which are not conducted by boards of education, and eight companies with 320 cadets in government institutions. Lord Strathcona in 1911 made a grant of $500,000 to the Militia Department of Canada. The interest on this fund is divided annually among the provinces of Canada proportionally on the basis of population, and is used in accordance with the regulations of the deed of trust issued by Lord Strathcona. There is a local Strathcona Trust in each province, consisting of military representatives appointed by the Militia Department of the Dominion, and educational representatives appointed by the Militia Department of the provinces.

PROVINCES	COMPANIES			CADETS		
	1913	1916	Increase	1913	1916	Increase
Maritime Provinces..	87	103	16	3,480	4,120	640
Quebec.............	282	451	169	11,280	18,040	6,760
Ontario.............	220	437	217	8,800	17,480	8,680
Manitoba...........	57	160	103	2,280	6,400	4,120
Saskatchewan.......	20	68	48	800	2,720	1,920
Alberta.............	73	131	58	2,020	5,240	3,220
British Columbia....	20	78	58	800	3,120	2,320
In all Canada.......	759	1428	669	30,300	57,120	26,820

The increase made yearly from March, 1912, to March, 1916, is as follows:

DATES	COMPANIES		CADETS	
	Total	Increase	Total	Increase
Year ending March, 1912....................	506		20,240	
" " " 1913....................	759	253	30,300	10,060
" " " 1914....................	1,117	358	44,680	14,380
" " " 1915....................	1,322	205	52,880	8,200
" " " 1916....................	1,428	106	57,120	4,240

The rapid increase in the number of cadets during the four years named is due mainly to two causes: first, the Strathcona Trust Fund,

The income from the Strathcona Trust Fund must be used as follows: fifty per cent for physical training, thirty-five per cent for military drill and fifteen per cent for rifle shooting. University cadets and the cadets of colleges and all secondary schools are inspected by officers of the Militia Department of the Dominion, and the Strathcona grants are paid according to the reports of these officers. The cadets in public and separate schools are inspected by the same cadet inspectors who represent the Militia Department, and the payments to these schools are made on the joint reports of the school inspectors of the Education Department, and the cadet inspectors of the Militia Department. Payments to the schools are made by the treasurer of the Strathcona Trust for each province, on authority of certificates issued by the education departments of the different provinces.

New teachers receive training in physical work at normal schools and other institutions for training teachers; special classes are conducted by the Militia Department for teachers already engaged in teaching, at a time suitable for the local authorities, wherever thirty or more teachers unite to take a course; and certificates of two grades are issued to those who pass the required examination at the close of the course.

Special summer schools are conducted under the direction of the Militia Department for teachers who wish to qualify for certificates in the department of physical training or military drill. The Militia Department makes a grant of one dollar per cadet to regular teachers who hold certificates as cadet instructors in order to supplement their salaries for teaching the regular school subjects. The Militia Department pays one dollar a year to school boards for each cadet in proper uniform on the day of the annual inspection. This grant provides a fund for the purchase of uniforms. The government supplies cadet companies with rifles, hats or caps, and belts, taking a bond from each school board for the value of the supplies issued. Ammunition for indoor and outdoor marksmanship practice is supplied free by the Militia Department, and matches for cadets are conducted at the annual meetings of the Dominion and the provincial rifle associations. Cadet camps are conducted annually in each military district throughout Canada, on the military camp grounds.

Among the cadet companies of the Dominion in 1916 were twenty-one squadrons of mounted cadets, and forty-four other companies, or squadrons, affiliated with militia regiments. In Ontario the Strathcona Trust offers three gold medals for competition in each county—one for the best marksman in the high schools, one for the best in the public or separate schools, and one for the best among former pupils under eighteen years of age. After eighteen a young man may join the regular militia.

Canadian Military Camps. When it was found in 1914 that Canada would have to train large armies the government decided to establish three large camps in addition to the camps already in existence throughout the Dominion—Valcartier Camp in the east, Camp Borden in the center, and Camp Hughes in the west:

Camp Valcartier. Camp Valcartier is situated in the province of Quebec sixteen miles northwest of the historic city of Quebec. It contains 12,700 acres, practically twenty square miles. It is nearly six miles in length and averages three and one-third miles in width. It lies in a valley almost surrounded by mountains, and is admirably adapted to the training of all classes of soldiers. The soil is a light sandy loam several feet deep, so that it is easily drained. The Jacques Cartier River flows through the camp. The river provides an adequate supply of excellent water, fine facilities for bathing, and opportunities for engineering work in the construction of pontoon bridges.

There are nearly three miles of rifle ranges, in all fifteen hundred ranges, a larger number than in any other camp in the world. The mountains make a perfect background for the ranges, and also provide unsurpassed opportunities for artillery practice.

The part of the camp occupied by the tents of the soldiers, the offices of the commanding officer and his staff, and the buildings for all kinds of stores, is well lighted by electric lights, and has a system of water supply that brings water in pipes to every building and tent, and excellent streets and roads with good drainage and sewage systems. All requisite engineering services have been provided for a camp of 35,000 men.

The Canadian Northern Railway connects Camp Valcartier with the city of Quebec, so that soldiers may be easily sent to the camp or transferred from it to the transport ships when ready for overseas service.

Camp Hughes. Camp Hughes is in the province of Manitoba eighteen miles east of Brandon and one hundred and fourteen miles west of Winnipeg. It contains 102,000 acres, practically 160 square miles, and is a forest reserve owned by the Dominion government. There are five lakes on the reserve. The portion actually used for camp purposes contains about thirty square miles, and is fitted to provide accommodation for 40,000 men. The camp is situated on the Canadian Pacific Railway. The tented city is well supplied with good streets, water works, sewers, electric lights, and excellent buildings for the headquarters of the officers, the hospital, dental work, store rooms, and other buildings. The arrangements for rifle practice, trenching, artillery practice and field operations are very complete. There is a parade ground for reviews that is unsurpassed in any country.

In all the military camps of Canada the Young Men's Christian Association conducts special work for the soldiers. They have large tents or buildings in which meetings are held, and free reading rooms and writing rooms are kept open. The Y. M. C. A. officers in most of the camps direct the sports and entertainment of the men.

The large camps afford the best possible opportunities for training armies under conditions most like the conditions of actual warfare.

Camp Borden. Camp Borden is in the province of Ontario sixty-one miles north of Toronto. It contains 20,000 acres, a little more than thirty-one square miles, and is more than seven miles long and over four miles wide. Two branches of the Nottawasaja River run through the camp through deep gorges in which beautiful lakes have been formed by damming the rivers. These lakes are used for swimming by the soldiers. There are 500 shower baths among the tents. Excellent water for drinking and other purposes is pumped from deep flowing artesian wells into high tanks and flows through the great camp to every tent. There are fifteen miles of water mains and seventeen miles of electric lighting in the camp. The main streets of the camp are made of concrete, and there are ten miles of sewers with a modern clarification tank system.

The bake ovens have a capacity of twenty tons per day. There are 180 field kitchens.

The buildings for officers' headquarters, hospital, dental clinics, store houses, etc., are large substantial structures. The hospital is 137 by 43 feet, is fitted with two operating rooms, a dispensary, laboratories and X-ray room, and has ample accommodation for medical stores, linen, etc. The Canadian army was the first in the world to have a dental unit—independent of the medical unit. The dental building is 152 feet long by 25 feet wide, and has twenty-four chairs and the necessary laboratories.

A stadium for sports, lectures and entertainments seating 20,000 has been erected. There are miles of targets for rifle practice. The camp at present has accommodations for 40,000 men. S.H.

Consult Parkman's books relating to France in America, particularly *Pioneers of France in the New World, La Salle and the Discovery of the Great West,* for early history; *The Chronicles of Canada,* edited by G. M. Wrong and H. H. Langton; *Makers of Canada,* a series of biographical studies of great Canadians; Pope's *Sir John A. Macdonald;* Willson's *Life of Lord Strathcona and Mount Royal;* Tupper's *Recollections of Sixty Years;* Washburn's *Trails, Trappers and Tenderfeet in Western Canada;* Maxwell's *Canada of To-day.* The government publishes a helpful volume, *The Canada Year Book,* which may be secured on application to the Census and Statistics Office, Ottawa.

Related Subjects. The following index will simplify reference to the many topics in these volumes which have to do, directly or indirectly, with Canada. The articles on the provinces have also detailed lists of related subjects.

CANALS

Georgian Bay Ship Canal
Rideau
Sault Sainte Marie
Trent
Welland

CITIES AND TOWNS

See lists under different provinces.

DISTRICTS

Assiniboia
Athabaska
Franklin
Keewatin
Klondike
Labrador
Mackenzie
Ungava

GOVERNMENT

Cabinet
Governor-General
Lieutenant-Governor
Parliament
Premier
Province
Royal Northwest Mounted Police
Territory

GULFS AND BAYS

Baffin's Bay
Belle Isle, Strait of
Fundy, Bay of
Georgian Bay
Hudson Bay
James Bay
Juan de Fuca, Strait of
Passamaquoddy Bay
Puget Sound
Saint Lawrence, Gulf of

HISTORY

Acadia
Aix-la-Chapelle, Treaty of
British North America Act
Clayton-Bulwer Treaty
Dominion Day
Empire Day
Erie, Battle of Lake
Flag
Fort Niagara
French and Indian Wars
Hudson's Bay Company
Jay Treaty
Louisburg, Sieges of
Lundy's Lane, Battle of
Maple Leaf, The
Northwest Company
Paris, Treaties of
Quebec, Siege of
Quebec Act
Quebec Resolutions
Quebec Tercentenary
Queenstown Heights, Battle of
Rebellion of 1837
Red River Rebellion
Revolutionary War in America
Rupert's Land
Saskatchewan Rebellion
Thames River, Battle of the
Union, Act of
United Empire Loyalists
War of the Nations
Webster-Ashburton Treaty

The following is a list of those who, whether as soldiers, statesmen or administrators, have had a part in the making of Canadian history. Their biographies appear in these volumes:

Abbott, Sir John J. C.
Aberdeen, John C. Gordon, Earl of
Amherst, Baron
Archibald, Sir Adams
Argyll, Ninth Duke of
Arthur, Sir George
Aylesworth, Sir Allen
Bagot, Sir Charles
Baldwin, Robert
Beck, Sir Adam
Blake, Edward
Blondin, Pierre Edouard
Borden, Sir Frederick
Borden, Sir Robert L.
Bourassa, Henri
Bowell, Sir Mackenzie
Bowser, William J.
Brock, Sir Isaac
Brodeur, Louis Philippe
Brown, George
Bulyea, George H. V.
Burrell, Martin
Calder, James Alexander
Campbell, Sir Alexander
Carleton, Sir Guy
Cartier, Sir Georges E.
Carter-Cotton, Francis
Cartwright, Sir Richard J.
Casgrain, Thomas Chase
Chapleau, Sir Joseph
Cochrane, Francis

OUTLINE AND QUESTIONS ON CANADA

Outline

I. Location and Size

(1) Latitude
(2) Longitude
(3) Position as compared with European nations
(4) Boundaries
(5) Size
 (a) East and west length, 2,700 miles
 (b) Greatest north and south extent, 1,600 miles
 (c) Actual area, 3,729,665 square miles
 (d) Area compared with that of other countries
 (e) Area of separate provinces

II. The People

(1) Population
 (a) Percentage of increase
(2) Proportion of urban and rural population
(3) Density
(4) Origins of the people
(5) Religion
 (a) Predominance of Roman Catholics
(6) Education
 (a) Provincial control
 (b) "Separate schools"
 (c) How expenses are met
 (d) Adoption of modern theories
 (e) Illiteracy statistics

III. The Land

1) Surface regions
 (a) Eastern Canada
  1. Highlands to the east
  2. Saint Lawrence Valley
  3. Acadian region
  4. Niagara escarpment
 (b) Interior plains
  1. Prairie region
  2. First escarpment
  3. Missouri Coteau
 (c) The western mountains
  1. Rocky Mountains
  2. Coast ranges
  3. Outstanding peaks
(2) Drainage
 (a) River systems
  1. Those draining into Hudson Bay
  2. Those draining into Arctic Ocean
  3. Rivers of Pacific slope
 (b) Lakes
(3) Climate
 (a) Conditions due to latitude
 (b) Variations due to surface and presence of large bodies of water
 (c) Average temperatures
  1. Atlantic coast and Hudson Bay region
  2. Maritime provinces
  3. Saint Lawrence section
  4. Prairie provinces
  5. Western coast

IV. Plant and Animal Life

(1) Plants
 (a) Dependence on climate
 (b) Plant regions
(2) Animals
 (a) Game animals
 (b) Birds

V. Industries

(1) Agriculture
 (a) Cereals
  1. Wheat
  2. Oats
  3. Barley
  4. Rye
  5. Corn
 (b) Fruits
  1. Apples
  2. Peaches
  3. Grapes
  4. Berries
 (c) Other crops
  1. Alfalfa
  2. Flax
  3. Hay
  4. Potatoes
 (d) Stock-growing
 (e) Dairying and poultry-raising
 (f) Irrigation
(2) Fisheries
 (a) On the Atlantic coast
  1. Deep-sea fishing
  2. Inshore fishing
 (b) On the Pacific coast
 (c) Fresh-water fisheries
 (d) Value of annual catch
(3) Mining
 (a) Coal
 (b) Gold
 (c) Silver and lead
 (d) Other minerals
 (e) Effect of war on output
(4) Manufacturing
 (a) Natural location of districts
 (b) Leading products

Outline and Questions on Canada—Continued

VI. Transportation and Commerce

(1) Navigable rivers
(2) Canals
(3) Railroads
  (a) Intercolonial Railway
  (b) Canadian Pacific
    1. Part in development of the country
  (c) Canadian Northern
  (d) Grand Trunk Pacific
(4) Foreign commerce
  (a) With what countries
  (b) Value of imports and exports

VII. Government

(1) General character
  (a) Resemblances to that of Great Britain
  (b) Resemblances to that of United States
(2) Departments
  (a) Executive
    1. Governor-General
    2. The Ministry
  (b) Legislative
    1. Senate
      a. Membership
      b. Powers
    2. House of Commons
      a. Membership
      b. Powers
    3. Officers of Parliament
  (c) Judicial
    1. Supreme Court
    2. Exchequer Court
    3. Admiralty Court
(3) Provincial governments
(4) Military system

VIII. History

(1) Discovery and exploration
(2) Colonization
(3) Struggle between France and England for control .
(4) Opening of the West
(5) Struggle for responsible government
(6) Act of Union
(7) Confederation
(8) Period of expansion
  (a) New provinces established
(9) National problems
(10) Recent development
(11) Foreign relations
(12) Canada and the War of the Nations

Questions

What remarkable percentage of increase in population did one Alberta town have between 1901 and 1911?

How many men did Canada send to the aid of England during the first two years of the War of the Nations?

Would the shooting season for wild fowl be earlier or later in Western Canada than in Texas? Why?

What is the most valuable mineral product of the Dominion? Has it always been so? Which province furnishes most of it?

How does the constitution of the Dominion resemble that of the United States? How does it differ from it?

Who was "the Father of New France"? What did he do to win the title?

Has the largest province the largest population? In which province are there the most people to the square mile?

How do the great interior plains of Canada differ from those of the United States?

What animal has been made the national emblem of Canada?

Where are the mineral-bearing regions located? What relation do those of the United States bear to them?

How are the public schools supported?

What was the first resource of this vast land which attracted the Frenchmen? The second?

What are *Metis?* What part did they play in the Red River Rebellion?

Show by statistics that the increase in population was not an even, steady growth.

Outline and Questions on Canada—Continued

What peculiarity of surface is it which causes Niagara Falls?

Where are the forested regions of the Dominion?

What is meant by "separate schools"?

What were Cabot's two objects in making the voyage which resulted in the discovery of Labrador?

When did the idea of Confederation originate? When was it put into actual operation?

How large a water area is included in Canada?

How was the "overseas expeditionary force" organized?

Why is there so much rain on the Pacific coast and so little just over the mountains?

What are the Newfoundland "Banks"? What has been their importance in the development of the country?

How does Canada's trade with Great Britain in normal times compare with that with the United States?

In what way do the legislative departments of Quebec and Nova Scotia differ from those of the other provinces?

What is meant by "responsible" government? How was it gained in Canada?

How many countries in the world are larger than the Dominion?

Which is the largest province? Has it always been so?

What effect did the Saint Lawrence have on the development of the country?

Why have the Maritime Provinces a warmer, more humid climate than regions in the same latitude farther west?

Mention four ways in which the government helps the farmer.

How is Canada's foreign trade regulated now?

What is meant by the word *puisne?* How is it used in connection with the judicial system?

When were Upper and Lower Canada constituted? What was the difference in the character of their population?

What nation of Europe has about the same population as Canada?

What is the "height of land"? What effect does it have?

Why is the coast of Labrador colder than the western coast in the same latitude?

Name two great factors in the development of the West.

How does the Dominion rank among the countries of the world as regards railway mileage in proportion to population?

Trace the course of procedure by which a bill becomes a law.

When Canada came into the possession of England, what was the proportion of English to French inhabitants?

What part of Canada lies in the same latitude as London? As Paris? As Stockholm?

What effects did the great glacier have on the surface of Canada?

To what part of the country did Longfellow refer in his line, "In the *Acadian* land, on the shores of the Basin of Minas"?

Where is the most magnificent scenery to be found?

What has been the progress of the live-stock industry?

Name three canals that serve different parts of the country, and tell why each one is important.

How large a population has the part of Europe which in latitude corresponds to Canada?

Colborne, Sir John
Connaught, H. R. H., the Duke of
Costigan, John
Craig, Sir James
Crothers, T. W.
Davies, Sir Louis Henry
Denison, George Taylor
Denonville, Marquis de
Derby, Earl of
Devonshire, Duke of
Doherty, Charles J.
Dorion, Sir Antoine
Douglas, Sir James
Drummond, Sir George Gordon
Dufferin and Ava, Marquis of
Dunsmuir, James
Durham, Earl of
Fielding, William Stevens
Fisher, Sydney Arthur
Fitzpatrick, Sir Charles
Foster, Sir George Eulas
Frontenac, Comte de
Galt, Sir Alexander T.
Gouin, Sir Lomer
Graham, George Perry
Greenway, Thomas
Grey, Earl
Haggart, John Graham
Haldimand, Sir Frederick
Hardy, Arthur Sturgis
Harvey, Sir John
Haultain, Frederick
Hazen, John Douglas
Head, Sir Edmund W.
Head, Sir Francis Bond
Hearst, William H.
Hincks, Sir Francis
Howe, Joseph
Hughes, Sir Sam
Jette, Sir Louis A.
Joly de Lotbinière, Sir Henri
Jones, Alfred Gilpin
Kemp, Albert Edward
King, William L. Mackenzie
Lafontaine, Sir Louis
Lansdowne, Marquis of
Laurier, Sir Wilfrid
Lemieux, Rodolphe
Lisgar, Baron
Lougheed, Sir James A.
McBride, Sir Richard
Macdonald, Sir Hugh
Macdonald, Sir John A.
Macdonald, John Sandfield
McDougall, William
McGee, Thomas D'Arcy
Mackenzie, Alexander
Mackenzie, William Lyon
MacNab, Sir Allan
Macpherson, Sir David
Maisonneuve, Sieur de
Metcalfe, Baron
Middleton, Sir Frederick
Minto, Earl of
Monck, Viscount
Montcalm, Marquis de
Monts, Sieur de
Morin, Auguste Norbert
Mount Stephen, Baron
Mowat, Sir Oliver
Mulock, Sir William
Murray, George Henry
Nelson, Wolfred
Norris, T. C.
Oliver, Frank
Otter, Sir William D.
Papineau, Louis J.
Perley, Sir George H.
Pope, Sir Joseph
Pugsley, William
Reid, John D.
Robinson, Sir John B.
Roblin, Sir Rodmond
Roche, William J.
Rogers, Robert
Ross, Sir George
Rowell, N. W.
Scott, Walter
Secord, Laura
Sifton, Arthur L.
Sifton, Sir Clifford
Simcoe, John Graves
Simpson, Sir George
Strathcona and Mount Royal, Baron
Sydenham and Toronto, Baron
Taché, Sir Etienne P.
Thompson, Sir John
Tilley, Sir Samuel
Tupper, Sir Charles
Tupper, Sir Charles H.
Vaudreuil-Cavagnal, Marquis
White, Sir William
Whitney, Sir James P.
Wilmot, Lemuel Allan
Wolfe, James

ISLANDS

Anticosti
Belle Isle
Cape Breton
Magdalen
Manitoulin
Queen Charlotte Islands
Sable
Thousand Islands
Vancouver

LAKES

Athabaska
Bras d'Or
Champlain
Erie
Great Bear
Great Lakes
Great Slave
Huron
Lake of the Woods
Manitoba
Memphremagog
Muskoka Lakes
Nipigon
Nipissing
Ontario
Rainy
Saint Clair
Simcoe
Superior
Winnipeg
Winnipegosis

LEADING PRODUCTS

Alfalfa
Apple
Asbestos
Barley
Cattle
Coal
Cod
Copper
Corn
Fish
Flax
Fruits
Fur
Gold
Hay
Herring
Horse
Iron
Lumber
Nickel
Oats
Potato
Rye
Salmon
Silver
Wheat

MOUNTAINS

Assiniboine, Mount
Athabaska, Mount
Cascade Range
Columbia, Mount
Hooker, Mount
Laurentian Plateau
Logan, Mount
Robson, Mount
Rocky
Saint Elias
Selkirk

POLITICAL DIVISIONS

Alberta
British Columbia
Manitoba
New Brunswick
Northwest Territories
Nova Scotia
Ontario
Prince Edward Island
Quebec
Saskatchewan
Yukon

RIVERS

Albany
Assiniboine
Athabaska
Chaudière
Churchill
Columbia
Fraser
Gatineau
Hamilton
Kootenay
Mackenzie
Miramichi
Montmorency
Moose
Nelson
Ottawa
Peace
Red River of the North
Restigouche
Saguenay
Saint John
Saint Lawrence
Saskatchewan
Skeena
Stikine
Yukon

CANADA BALSAM, *bawl'sam,* a resinous substance obtained from the Balm of Gilead fir, common in Canada and the United States. In odor it resembles turpentine, and it has a bitter taste. It is used in medicine and in making varnishes, and because of its almost perfect transparency it is valuable in mounting objects on glass for examination through a microscope. It is also of great value as a cement for joining the lenses of eyeglasses, spectacles and optical instruments.

CANADA GOOSE, the common wild goose of the North American continent, which nests in the region stretching from the northern limit of tree growth in the lower Yukon valley as far south as Indiana. No member of its race has more interesting migratory habits than this bird of passage. When autumn heralds

CANADA GOOSE

the approach of winter vast numbers from the fur-bearing sections of Canada assemble along the shores of Hudson Bay, and as the increasing cold drives them to seek a sunnier home, great flocks, arranged in long, converging lines, with an old gander at the head, begin their southward flight.

These living wedges of feathered travelers fly high in the air, the loud, hoarse "honk" of the leader and the answering calls of his followers often being the only sounds that break the silence of the solitary regions through which they pass. Early in October they reach the coasts of the Eastern and Middle states, and through the winter are found in various warmer parts of the Union, even to the most southern portion of Florida. Their food is chiefly grass and berries.

The Canada goose is about thirty-six inches long, and wears a grayish-brown coat. The head, neck and tail are black, the under parts gray, and there is a broad white patch on the throat. The nest, which is usually placed on the ground, is made loosely of twigs and grass, and holds from five to seven pale green or white eggs. As this bird is much sought by hunters, because of the delicacy of its flesh, it is included in the list of game birds which are protected by law during the greater part of the year. The Canada goose has been introduced into Europe. B.M.W.

CANADA THISTLE, a variety of thistle that more than all others of its family deserves the name of *pest,* for it is one of the most troublesome weeds in existence. A native of Europe, it reached Canada and the United States through the mingling of its seed with imported grains, and now grows in wild profusion from Newfoundland south to Virginia and west to the great plains, a most unwelcome visitor in the farmer's fields. Nature has given it the best possible weapons with which to maintain a successful battle for existence against its enemy, man. Within the tubes of its small purplish flowers is formed a sweet nectar, which attracts countless bees and butterflies, wasps, flies and beetles. The pollen grains have a slight stickiness and are carried away with the greatest ease on the hairs of the insects, to be later deposited on the stigma of another flower. The plant also is well adapted to fertilization by means of its seeds and its horizontal rootstocks, every portion of which can produce a new plant. For this reason, partial uprooting may work harm, for several plants may spring up where only one grew.

CANADA THISTLE

Fields where this thistle appears can be cleared of it only by the most thorough cultivation, for the rootstocks must be entirely rooted out, and before the plant goes to seed. If the weed gets into the field of a plant that is not cultivated, as oats or grass, it will gain such headway that the land will have to be cultivated the next season and planted to a different grain. Sometimes the farmer turns a field infested by thistles into a sheep pasture and lets the sheep aid him in ridding the land of the troublesome visitor.

The plant is slender and branching, and grows from one to three feet in height. The prickly leaves are long, sword-shaped and deeply notched, and grow close together on the stem. See THISTLE; SEEDS, subtitle *Seed Dispersal.*

CANADIAN LITERATURE. The literature of Canada springs from two great roots which can never form a single tree. Each root, however, has already brought forth a sturdy sapling which is making room for itself among the literatures of the world. So long as French is spoken in Canada, the French-Canadians will have a literature which not only records their own achievements and voices their own aspirations, but also shows the influence of France. The highest honor which can be awarded to the work of a French-Canadian author is the laurel crown of the French Academy.

The Canadian who writes in English is less dependent on the traditions of the mother country. It is true, he has the splendid heritage of all English literature, but he also has a separate history which encourages independence. British rule has turned the literary eye of French Canada back to France itself, but it has given English Canada a new national consciousness which is reflected in its literature.

French Canada. The literature of French Canada divides itself naturally into two periods, the year 1763 being the dividing line. The books written before that year were the work of explorers and missionaries, and naturally deal with discovery and travel. The men who first attempted to plant European civilization on American soil wrote vivid accounts of their struggles. The chronicles of Cartier's voyages, the narratives of Champlain and Hennepin, the histories of Lescarbot, Sagard, Le Clerq and Charlevoix form a body of literature of the highest rank. The volumes of the *Jesuit Relations,* which contain the reports of the missionaries, must not be omitted from this group.

Totally different in character, but equally important, were the Breton and Norman folk-songs, which, in the course of time, acquired some of the spirit of their new homes. They were once transmitted orally from generation to generation, and it is only in recent times that some of them have been put into print. To this day they recall the France of two or three centuries ago, when the Quebec habitant and the French peasant led lives of much the same character.

Patriotism and Histories. The beginning of British rule over French Canada in 1763 caused a struggle between two civilizations which has not yet entirely disappeared, and created a new type of French-Canadian literature which is distinguished chiefly by race-patriotism. Its first important products were the speeches of Louis J. Papineau. These speeches, purely political in purpose, rank first, in comparison with written works, both in popular esteem and in point of time. Soon after appeared the histories of Canada by Michel Bibaud and François X. Garneau, the latter still regarded by the French as their standard authority. Garneau's work has considerable distinction of style and has had a great influence, but it is marred by an excessive patriotism which prevents it from being authoritative. It remains valuable, however, as "the first great literary stimulus to racial self-respect."

Of later historians there is a long list, beginning with Benjamin Sulte, Abbé Casgrain and Sir James Le Moine, each of whom is treated in alphabetical order in these volumes. Others whose names are familiar for histories and monographs on historical or economic subjects are Thomas Chapais, Etienne Parent, Laurent David and Alfred D. DeCelles. Among journalists and publicists must be mentioned Henri Bourassa, also famous as an orator, Joseph Charles Taché, Hector Fabre and Abbé J. A. Damours.

Fiction and Poetry. In fiction and poetry the same patriotic strain is everywhere apparent. The novelist and the poet are inspired by the events of French-Canadian history, and when the keynote is not love of Canada it is love of the great mother Church, or of France, the mother country. In fiction the first work of importance is Philippe de Gaspé's *Les Anciens Canadiens* (1863), a rambling tale with a familiar plot, but valuable as a faithful, vivid picture of the *ancien regime,* of the days of seigneurs, voyageurs, coureurs de bois, Indians, sailors and soldiers. About the same time appeared numerous poems from the pen of Octave Cremazie, who became a national poet, not because he was great, but because he celebrated the occasions which "touched some lasting aspiration of his race."

Later novelists are few, and little known outsid their own communities. A classic of its kind is *Jean Rivard,* by A. Gerin-Lajoie; it is the French-Canadian adaptation of the cry, "Back to the land." Among later novels which have won fame for their writers are Choquette's *Claude Paysan,* an artistic story of a habitant's hopeless love; Mme. Laura Conan's *L'Oublie,* crowned by the French Academy; and Napoleon Bourassa's *Jacques et Marie,* a tale of simple people who bring to mind Longfellow's *Evangeline.* Of poets there is a long list, including Leon Le May, who translated *Evangeline* into French verse; Benjamin Sulte, better known as an historian; P. J. O. Chauveau, Louis Fiset and Alfred Garneau, three minor writers whose work shows delicate poetic touches; and above all, Louis H. Frechette, who, of all French-Canadian poets, has probably the best claim to greatness.

English Canada. The literature of English-speaking Canada begins at a later date but with a character similar to that of French Canada. The first writers were men like Samuel Hearne, Sir Alexander Mackenzie and Alexander Henry, who wrote accounts of their own travels and discoveries. This period of exploration coincided roughly with the rapid settlement and development of New Brunswick and Upper Canada (Ontario) by the United Empire Loyalists. For a generation or two these settlers were too busy to write books, and it was not until about 1830 that English Canada paid much attention to literature.

The struggle for responsible government, both before and after the Union of 1841, was accompanied by a flood of controversial political literature, much of which is now of little value. The speeches and writings of Bishop Strachan, Sir John Beverly Robinson, William Lyon Mackenzie, Egerton Ryerson and Joseph Howe, however, are brilliant pieces of work as literature, regardless of the great influence they had on the events of the day. The chief literary figure of this period is Thomas Chandler Haliburton, who wrote under the pen-name of "Sam Slick." Haliburton in private life was a distinguished lawyer and judge, but to posterity he is still the first Canadian humorist. One of his distinctive gifts was his aptitude for short, pithy sayings. "Circumstances alter cases" is perhaps the best-known quotation from his books.

Influence of Political Life. With the coming of Confederation there was a distinct growth in literary activity in the Dominion. A Canadian national spirit first asserted itself, and perhaps for the first time there were signs of a truly national literature. Just pride in home and native land and the appreciation of the pioneers' sacrifices inspired the poet as well as the novelist and the historian. The speeches which had for their purpose the arousing of a spirit of unity still remain as a distinct type of literature, and many of the later speeches of Sir Charles Tupper, Sir John A. Macdonald, Alexander Mackenzie, Sir Alexander T. Galt, Edward Blake and Sir Wilfrid Laurier have an enduring quality which makes them good literature. The speeches of Laurier are especially noteworthy as combining the "emotional appeal of the French-Canadian with the reasoned presentment of constitutional precedents and principles" which are so characteristic of English orators and statesmen.

History and Allied Subjects. The formal histories, except those written by Francis Parkman, who was not a Canadian, do not rise to great heights. The standard *History of Canada* is by William Kingsford; it is an accurate record, but has little charm of style. John C. Dent's *The Last Forty Years* (1841-1881) and his *Story of the Upper Canada Rebellion* are well written and clear. Hannay's *History of Acadia* is a valuable work, and Col. William Wood's *The Fight for Canada* is praised as one of the best historical books written by a Canadian. The history of the Northwest is a special field which has been ably treated by Alexander Begg, George Bryce, George M. Adam, Archbishop Taché and others. *From Ocean to Ocean,* by Rev. George Monro Grant, is one of the most interesting travel-books of its kind. Other important books on the Northwest are Agnes C. Laut's *Conquest of the Northwest* and Beckles Willson's *The Great Company,* a history of the Hudson's Bay Company. The books on history and government written by Sir John G. Bourinot are standard authorities.

Biography has been devoted mainly to political leaders and statesmen, especially those of the period immediately preceding and following

Confederation. Among the best-known biographers and essayists are Sir William Dawson, the geologist, Sir Daniel Wilson, T. Arnold Haultain, John Castell Hopkins and W. D. Le Sueur. Most distinguished of them all is Goldwin Smith, a unique figure in Canadian literature. As author, teacher and lecturer his influence for progress was world-wide. His views were often those of a small minority, but his lofty principles won him the respect and admiration of all.

Fiction. For many years Canada lagged behind the rest of the English-speaking world in fiction. It had authoritative historians and famous poets long before any of its novelists were well known. With two exceptions there were no Canadian novelists until after the Dominion was formed; these were John Galt, the founder of Guelph, Ont., who wrote *Lawrie Todd, or the Settlers in the Woods* (1830), a vivid account of frontier life, and Major John Richardson, who wrote *Wacousta, or the Prophecy* (1832), an exciting tale of Pontiac's conspiracy. The next novel of importance did not appear until 1877, William Kirby's *The Golden Dog, a Legend of Quebec.* James De Mille wrote about thirty novels, including *Helene's Household,* a story of Rome in the first century of the Christian Era.

Included in the list of recent authors are Sara Jeanette Duncan Cotes, Grant Allen, Robert Barr, Lily Dougall, Catherine Parr Traill, Norman Duncan, William Douw Lighthall, Charles G. D. Roberts, Arthur Stringer, Ernest Thompson Seton and Margaret M. Saunders, whose autobiography of a dog, *Beautiful Joe,* is a children's classic. A novelist of international reputation is Charles W. Gordon ("Ralph Connor"), and perhaps even better known is Sir Gilbert Parker. Worthy of special mention are the humorous sketches of Stephen B. Leacock, sometimes called the successor of "Sam Slick." The most notable of these authors are given places in these volumes in their alphabetical order.

Poetry. Perhaps the most remarkable long poem ever written by a Canadian is the tragedy *Saul,* by Charles Heavysege. He also wrote numerous short lyrics, one of which, called *Night,* begins with these lines:

'Tis solemn darkness; the sublime of shade;
Night by no stars nor rising moon relieved;
The awful blank of nothingness arrayed
O'er which my eyeballs roll in vain, deceived.

Among other early poets were Robert Sweeney, Oliver Goldsmith, a collateral descendant of the author of the *Vicar of Wakefield,* and Susanna Moodie. The Canadian Goldsmith wrote *The Rising Village,* which presents a picture of a prosperous community, in contrast to *The Deserted Village* written by his famous kinsman. Mrs. Moodie's work, in poetry as in prose, gives vivid pictures of pioneer life in Upper Canada. *The Canadian Herd-Boy (A Song of the Backwoods)* indicates the character of much of her work. The follow ing familiar stanza illustrates its style:

Through the deep woods at peep of day,
The careless herd-boy wends his way,
By piny ridge and forest stream,
To summon home his roving team;
Cobos! Cobos! from distant dell
Sly echo wafts the cattle-bell.

At a later date the number of Canadian poets became larger, and works of many of them are familiar not only in Canada but throughout the English-speaking world. Charles Sangster, sometimes called the "Canadian Wordsworth," was inspired mainly by Canadian scenery and history. *The Rapid,* from which the following lines are quoted, is one of his best-known poems:

All peacefully gliding,
The waters dividing,
The indolent batteau moved slowly along,
The rowers, light-hearted,
From sorrow long parted,
Beguiled the dull moments with laughter and song.

But the last stanza tells the sad end of the story:

Fast downward they're dashing,
Each fearless eye flashing,
Though danger awaits them on every side;
Yon rock—see it frowning!
They strike—they are drowning!
But downward they speed with the merciless tide;
No voice cheers the rapid that angrily, angrily
Shivers their bark in its maddening play;
Gaily they entered it—heedlessly, recklessly,
Mingling their lives with its treacherous spray!

Another poet of distinction is Charles Mair, who was the founder of the Canadian classical-nature school. His poetic drama, *Tecumseh,* is noteworthy for its insight into Indian character and for its striking descriptive passages:

There was a time on this fair continent
When all things throve in spacious peacefulness.
The prosperous forests unmolested stood,
For where the stalwart oak grew there it lived
Long ages, and then died among its kind.
The hoary pines—those ancients of the earth—
Brimful of legends of the early world,
Stood thick on their own mountains unsubdued.

OUTLINE ON CANADIAN LITERATURE

I. FRENCH-CANADIAN LITERATURE

Explorers, Missionaries and Their Chief Works

Jacques Cartier. 1494-1557
Bref Récit de la Navigation . . . . de Canada

Samuel Champlain. 1567-1635
Des Sauvages: ou Voyage de Samuel Champlain

Marc Lescarbot. 1570-1630
Histoire de Nouvelle France

Louis Hennepin. 1640-1706
Description de la Louisiane

Pierre Charlevoix. 1682-1761
Histoire . . . . de la Nouvelle France

"The Jesuit Relations"

Unwritten Folk Songs

Historians and Their Chief Works

François X. Garneau. 1809-1866
Histoire de Canada

Sir James M. Le Moine. 1825-1912
Chronicles of the Saint Lawrence
Quebec, Past and Present

Henri R. Casgrain. 1831-1904
Pelerinage au Pays d' Evangeline

Benjamin Sulte. 1841-
Histoire des Canadiens-Français

Henri Bourassa. 1868-
The French-Canadian in the British Empire

Poets and Novelists and Their Chief Works

Philippe de Gaspé. 1784-1871
Les Anciens Canadiens

Pierre J. O. Chauveau. 1820-1890
Donnaconna

Antoine Gerin-Lajoie. 1824-1882
Jean Rivard

Octave Cremazie
Les Morts
Le Drapeau de Carillon

Alfred Garneau. 1836-1904
Poésies

Leon Le May. 1837-
Poemes Couronées
Evangeline

Louis H. Fréchette. 1839-1908
Mes Loisirs
La Voix d'un Exile
Les Oiseaux de Neige

Benjamin Sulte. 1841-
Les Laurentiennes

II. ENGLISH-CANADIAN LITERATURE

Explorers and Pioneers

Alexander Henry. 1739-1824
Travels and Adventures in Canada

Samuel Hearne. 1745-1792
Journey from Prince of Wales Fort to the Northern Ocean

Sir Alexander Mackenzie. 1755-1820
Voyages on the River Saint Lawrence and Through the Continent of North America

Earl of Selkirk. 1771-1820
Sketch of the Fur Trade in North America

Alexander Ross. 1783-1856
Adventures of the First Settlers on the Oregon River
The Red River Settlement

Political and Controversial Writers

Bishop John Strachan. 1778-1867
Sir John B. Robinson. 1791-1863
William Lyon Mackenzie. 1795-1861
Egerton Ryerson. 1803-1882
Joseph Howe. 1804-1873
Sir Francis Hincks. 1805-1885

Historians

William Kingsford. 1819-1898
History of Canada

Goldwin Smith. 1823-1910
History of the United States
The Political Destiny of Canada

Sir John G. Bourinot. 1837-1902
Manual of Constitutional History
Canada Under British Rule

John C. Dent. 1841-1888
Last Forty Years
The Canadian Portrait Gallery

James Hannay. 1842-1910
History of Acadia
History of the War of 1812

George Bryce. 1844-
Remarkable History of the Hudson's Bay Company
Short History of the Canadian People

Arthur G. Doughty. 1860-
Quebec Under Two Flags
The Siege of Quebec

Beckles Willson. 1869-
The Great Company
Life of Lord Strathcona

Agnes C. Laut. 1872-
The Lords of the North
Canada, Empire of the North

Novelists

John Galt. 1779-1839
The Annals of the Parish

Major John Richardson. 1787-1865
Wacousta

Thomas C. Haliburton. 1796-1865
The Clockmaker (Sam Slick)
The Old Judge
Wise Saws and Modern Instances

Outline on Canadian Literature—Continued

Catherine Parr Traill. 1802-1899
Backwoods of Canada
Susanna Moodie. 1803-1885
Mark Hurdlestone, the Gold Worshiper
William Kirby. 1817-1906
The Golden Dog
James De Mille. 1837-1880
Andy O'Hara
A Castle in Spain
Grant (C. G. B.) Allen. 1848-1899
The Devil's Die
The Woman *Who Did*
Robert Barr. 1850-1912
A Woman Intervenes
In the Midst of Alarms
W. B. Basil King. 1859-
The Wild Olive
The Inner Shrine
William D. Lighthall. 1857-
The False Chevalier
Charles G. D. Roberts. 1860-
Barbara Ladd
The Young Acadian
The Backwoodsman
Ernest Thompson Seton. 1860-
Wild Animals I Have Known
Biography of a Grizzly
Charles W. Gordon ("Ralph Connor"). 1860-
Black Rock
The Sky Pilot
Margaret M. Saunders. 1861-
Beautiful Joe
Sir Gilbert Parker. 1862-
Pierre and His People
The Battle of the Strong
The Weavers
Sara Jeanette Duncan Cotes. 1862-
An American Girl in London
The Burnt Offering
Stephen B. Leacock. 1869-
Nonsense Novels
Behind the Beyond
Norman Duncan. 1871-1916
Doctor Luke of the Labrador
The Cruise of the Shining Light
Arthur Stringer. 1874-
The Wire Tappers
The Hand of Peril

Poets and Dramatists

Charles Heavysege. 1816-1876
Saul
Charles Sangster. 1822-1893
The Saint Lawrence and the Saguenay
Hesperus and Other Poems
Charles Mair. 1838-
Tecumseh
Isabella V. Crawford. 1851-1887
Old Spookses' Pass
William Henry Drummond. 1854-1907
The Habitant
Johnnie Courteau
George F. Cameron. 1854-1885
Standing on Tiptoe
William D. Lighthall. 1857-
Thoughts, Moods and Ideals
William Wilfred Campbell. 1861-
Beyond the Hills of Dreams
The Dread *Voyage*
W. Bliss Carman. 1861-
Low Tide on *Grand Pré*
Pipes of Pan
Archibald Lampman. 1861-1899
Among the Millet
Lyrics of Earth
Frederick G. Scott. 1861-
The Hymn of Empire *and Other* Poems
Jean Blewett. 1862-
Out of the Depths
E. Pauline Johnson. 1862-1913
As Red Men Die
The Song *My Cradle* Sings
Duncan Campbell Scott. 1862
The Magic House and Other Poems
Arthur Stringer. 1874-
Hephaestus and *Other Poems*
Robert W. Service. 1876-
The Spell *of* the *Yukon*
Songs of a *Sourdough*

Miscellaneous Authors

Sir Daniel Wilson. 1816-1892
Prehistoric Man
Caliban, the Missing Link
Sir John W. Dawson. 1820-1899
Acadian Geology
The Story of the Earth and Man
Thomas D. McGee. 1825-1868
History of the Irish Settlers in America
Sir Sandford Fleming. 1827-1915
The Intercolonial: A History
Time and Its Notation
George Monro Grant. 1835-1902
From Ocean to Ocean
Nathanael Burwash. 1839-
Life of Egerton Ryerson
George Taylor Denison. 1839-
History of Cavalry
Bégin, Louis Nazaire. 1840-
The Infallibility of the Sovereign Pontiffs
Sir John Murray. 1841-1914
Narrative of the Cruise *of H. M. S. Challenger*
Sir William Osler. 1849-
The Principles and Practice of Medicine
John W. Bengough. 1851-
Caricature History of Canadian Politics
James Bonar. 1852-
Malthus and His Work
James Mavor. 1854-
Wages, Theories and Statistics
Agnes Dean Cameron. 1863-1912
The New North

Among the popular poets are Isabella Valancey Crawford, George Frederick Cameron, William Wilfred Campbell, Bliss Carman, Archibald Lampman, Duncan Campbell Scott, Robert W. Service, Emily Pauline Johnson, Jean Blewett and Helen M. Johnson. One of Bliss Carman's best poems is *Low Tide on Grand Pré*, which begins with these lines:

> The sun goes down, and over all
> These barren reaches by the tide
> Such unelusive glories fall,
> I almost dream they yet will bide
> Until the coming of the tide.

The poems of Robert W. Service have received both praise and censure. His debt to Rudyard Kipling is evident, but he deserves great credit for harmonizing the form and the content of his poems. The form he chooses has a rhythm and freedom which are characteristic of life in the Yukon as he describes it. *The Law of the Yukon* is one of his best-known poems:

> This is the law of the Yukon, and ever she makes it plain:
> "Send not your foolish and feeble; send me your strong and your sane:
> Strong for the red rage of battle; sane, for I harry them sore;
> Send me men girt for the combat, men who are grit to the core;
> Swift as the panther in triumph, fierce as the bear in defeat,
> Sired of bulldog parent, steeled in the furnace heat."

Authors who have won distinction in other fields are also occasional writers of verse; among them may be mentioned Sir Gilbert Parker, Charles G. D. Roberts, William D. Lighthall and William Kirby. Unique among all Canadian verse are the poems of William Henry Drummond, who wrote in a quaint French-English dialect, which is well illustrated in *Leetle Bateese*, the story of a "regular imp" of a five-year-old, who chased the hens and scared the cows, and then was—

> Too sleepy for sayin' de prayer to-night?
> Never min', I s'pose it'll be all right.
> Say dem to-morrow—ah! dere he go!
> Fas' asleep in a minute or so—
> An' he'll stay lak dat till de rooster crow,
> Leetle Bateese!
>
> Den wake up right away tout de suite
> Lookin' for somet'ing more to eat,
> Makin' me t'ink of dem long leg crane—
> Soon as dey swaller, dey start again;
> I wonder your stomach don't get no pain,
> Leetle Bateese!

But in spite of all the mischief he causes and the worry he gives his poor grandfather, the old man says:

> But leetle Bateese! please don't forget
> We're rader you're stayin' de small boy yet;
> So chase de chicken an' mak' dem scare,
> An' do w'at you lak wit' your ole gran'père,
> For w'en you're beeg feller he won't be dere—
> Leetle Bateese!

Relating to Literature in English. The reader who wishes a complete view of all literature in English should consult the following subjects, and the further references printed at the end of each:

American Literature — Drama
Biography — English Literature
Essay — Novel
History — Poetry
Literature — Prose

W.F.Z.

CANADIAN RIVER, a river that rises in the northeastern part of New Mexico and flows easterly through the "Panhandle" of Texas and Oklahoma, forming the most important tributary of the Arkansas. Its length is 900 miles. Its course closely follows a part of the former boundary between Oklahoma and Indian Territory. The quantity of water it carries varies greatly at different seasons; sometimes it is almost dry; at other times it cannot be forded at all, and is not dependable for navigation.

CANAL. As water highways for the transportation of people and their possessions, and as channels excavated for purposes of drainage and irrigation, canals have always played an important part in the progress of mankind. The first artificial watercourses were probably irrigation ditches, but the many obstructions to travel in navigable rivers and the delays and dangers experienced in journeying over rough trails and roadways and across deserts and mountains by means of animal power, led to the building of canals for navigation many hundreds of years before railroads solved the problem of rapid transportation. This article discusses only the canal for navigation. Drainage and irrigation channels are described under the headings DRAINAGE and IRRIGATION. The Chicago Drainage Canal (see CHICAGO DRAINAGE CANAL) is an important example of a channel excavated for the purpose of carrying away the sewage of a city.

In Egypt, Assyria, India and China navigation canals were in operation long before the Christian Era, and Nebuchadnezzar, the great Babylonian ruler of the sixth century B. C., restored a canal that classic writers say was originally built eleven centuries before his time. This was the royal canal of Babylon, connecting the Tigris and the Euphrates rivers. It is an interesting tradition that a predecessor of the present Suez Canal (which see), joining the Nile River to the Red Sea, was begun about 600 B. C. by an Egyptian king; this is said to have been destroyed in A. D. 767 by a Mohammedan caliph. In the thirteenth century the Chinese constructed the most important work of its kind after the beginning of the Christian Era—the Grand Canal, connecting the Yangtse-kiang and the Pei-ho. This canal is 650 miles long and from five to six feet deep.

In comparison, canal building in Europe developed somewhat tardily, the present magnificent systems dating from about the twelfth century. The invention of the canal lock, however, in 1481, the honor of which is claimed both by Italy and Holland, gave new impetus to the construction of artificial watercourses, and Europe has a canal mileage at the present time of about 13,300. The total for the world is about 26,000, representing an expenditure of over two billion dollars, and over half of this is in a few countries of the continent of Europe. Statistics for the world's important canals may be found in the table at the end of this article.

Details of Construction. In railroad building the track may run up or down grade, but the course of a canal must consist of one or more level sections, or *reaches*. The Cape Cod Canal, which cuts through a narrow strip of land where the Cape and the Massachusetts mainland join, is an example of a waterway which connects two points on a single water level (see CAPE COD CANAL). Canals built over a route of different levels, like the Erie, consist of several reaches, the adjoining extremities of which are usually connected by locks. A lock is a chamber with stone or concrete side walls and water-tight gates at each end. In passing from a lower to a higher level a vessel goes through the gates at the lower end, and floats into the chamber. The lower gates are then closed and the valves in the sides or bottom of the lock are opened, allowing the water from the higher level to flow in. When the water in the lock has reached the level of that above the upper gates, these gates are opened and the boat continues its journey.

The dimensions of a canal are determined by the size of the vessels which are expected to use it. It must be wide enough at the surface and bottom to permit any two boats to pass without touching each other, and the depth should be at least one and one-half feet greater than the draft of the largest vessels that navigate it. The canal bed is always made flat. When the channel is excavated through soft earth the sides slope outward from the bottom, and the harder the material the steeper the banks. It is customary to cut the banks perpendicular, or nearly so, when the canal runs through rocks. Power drills and explosives are used to break up rock.

Embankments or aqueducts are built to carry canals across valleys; culverts are provided to carry streams beneath them, and these waterways are crossed by bridges wherever they are intercepted by ordinary traffic routes. The aqueduct usually takes the form of a masonry-arch bridge, the top of which is made into a channel or trough to conduct the water. Steel is now used to some extent in constructing these troughs.

Canals of the United States. The advantages that would result from building canals in a new and sparsely-settled country were foreseen by George Washington and other statesmen of the early national period. A canal around the rapids of the Connecticut River, at South Hadley, Mass., the first artificial waterway built in America, was completed in 1793, but the first important work of this kind was the Erie Canal, across New York state, begun in 1817. Its completion in 1825 was a significant event in the economic history of the country; it made a city of the town of Buffalo, at its western terminus, and was largely responsible for the early commercial supremacy of New York City. The subsequent improvements made on this canal and on other canals constructed later in New York state are fully described under the headings ERIE CANAL and NEW YORK STATE BARGE CANAL.

The enlargement of these waterways by the state of New York, which necessitated an

FIRST LOCK BUILT ON THE ERIE CANAL

expenditure of over $150,000,000, represents a revival of interest in canal building. During the era of railroad expansion following the War of Secession, public interest in artificial waterways suffered a noticeable decline. Nearly all of the important canals in the country were opened before the war. Among these are the Chesapeake and Ohio, between Washington and Cumberland, Md. (1850), the Illinois and Michigan, between Chicago and La Salle, Ill. (1848), and the system of locks about the rapids of the Saint Mary's River, known as the Sault Sainte Marie Canal (1855). The Hennepin, or Illinois and Mississippi, Canal, extending from the Illinois River near Hennepin to the Mississippi near Rock Island, was begun in 1892 and completed in 1908, but it was almost the only boat canal started during the period between the close of the war and the beginning of the twentieth century. The above canals are described in these volumes under their respective titles.

Since 1900 there have been constructed a third great lock (the Davis) on the Sault Sainte Marie Canal, and a ship canal across Cape Cod, connecting Buzzard's and Barnstable bays; both of these projects were completed in 1914. The latter was financed by a private corporation. The new lock at Sault Sainte Marie is the largest in the world, 1,350 feet long between gates, 80 feet wide and 24.5 feet deep. The construction of a fourth lock of similar dimensions has been authorized by Congress, and work on this was well under way by the end of 1915.

Other projects completed in 1915 were the Lake Washington Canal extending from Puget Sound to Lake Washington; and the Dalles-Celilo Canal, around the Dalles Rapids, which opens the Columbia River to light-draft boats up stream as far as Priest Rapids, and to Lewiston, on the Snake River, in Idaho. Work also progressed on the Ohio River project, which, when completed, will provide a navigable waterway from Pittsburgh to Cairo, Ill. Up to 1915 thirty-one out of the fifty-three proposed locks and dams had been finished or were under construction. Extensive improvements have also been made on the Louisville and Portland Canal around the falls of the Ohio. In the same year, 1915, the Illinois legislature passed a bill providing for the construction of a canal between Joliet and Utica, a distance of sixty-five miles, and a new canal board was appointed by the governor of Pennsylvania to complete estimates for a proposed canal between the Pittsburgh district and Lake Erie. In 1917 all sections of the New York Barge

CLIMBING A HILL

Here are four locks which show very clearly the way in which a steamer is lifted from one canal level to another. At the start the water in the first lock was at the same level as that in the foreground; the boat steamed into it, the gates were closed, and the lock was filled with water from great pipes which enter it at the bottom. This brought the level of the water in the first lock even with that in the second, and the steamer is now entering the second look. In coming down hill the process would be reversed.

Canal were ready for service, so the first two decades of the twentieth century have a very creditable record in the matter of canal building.

To the United States also is due the honor of completing a project that has held the interest of the nations for centuries—the great cut across the isthmus that joins the two American continents. The Panama Canal, opened the same year as the Cape Cod Canal and the new Davis lock, is accessible to all countries on an

Canals of the United States

CANALS	Cost of Construction*	Length, Miles	Depth, Feet†	LOCATION
Albemarle and Chesapeake	$1,641,363	11½	12	Norfolk, Va., to Albemarle Sound, N. C.
Augusta	1,500,000	7	10	Savannah River, Ga., to Augusta, Ga.
Beaufort	502,078	20	10	Beaufort Inlet, N. C., to Pamlico Sound.
Black River	3,581,954	35	4	Rome, N. Y., to Lyons Falls, N. Y.
Black Rock Channel	3,000,000	3¼	22	Connects Lake Erie and Niagara River at Buffalo, N. Y.
Brazos River	255,000	32	5	Brazos River to Matagorda Bay, Tex.
Cape Cod (ship canal)	12,000,000	8–13	25–30	Connects Buzzard's Bay and Barnstable Bay.
‡Cayuga and Seneca	2,232,632	23	7	Montezuma, N. Y., to Cayuga and Seneca Lakes, N.Y.
‡Champlain	4,044,000	61	6	Whitehall, N. Y., to Watervliet, N. Y.
Channel	450,000	32	5	Between Apalachicola River & St. Andrews Bay, Fla.
Chesapeake and Delaware	4,000,000	14	9	Connects Chesapeake and Delaware Bays.
Chesapeake and Ohio	11,290,327	185	6	Cumberland, Md., to Washington, D. C.
Colbert Shoals	2,350,000	8	7	Colbert Shoals, Tennessee River, Ala.
Company	90,000	23	4½	Miss. Riv. at New Orleans, La., to Bayou Black.
Dalles-Celilo	4,800,000	8½	7	Columbia River, from Big Eddy to Celilo Falls, Ore.
Delaware and Raritan	4,888,749	66	7	New Brunswick, N. J., to Bordentown, N. J.
Delaware Division	2,433,350	60	6	Easton, Pa., to Bristol, Pa.
‡Erie	52,540,800	339	7	Albany, N. Y., to Buffalo, N. Y.
Estherville-Minim Creek	174,619	5	6	Winyah Bay, S. C., to Santee River.
Fairfield	50,000	4½	5	Alligator River to Lake Mattamuskeet, N. C.
‖Florida East Coast	3,500,000	350	5	Mayport, Fla., to Miami.
Galveston and Brazos	340,000	38	5	Oyster Bay, Tex., to Brazos River, Tex.
Harlem River (ship canal)	2,700,000	8	15	Connects Hudson River (via Spuyten Duyvil Creek) and Long Island Sound
Illinois and Michigan	6,339,098	96	5	Chicago, Ill., to La Salle, Ill.
Illinois & Mississippi (Hennepin)	7,320,000	75	7	Illinois River to Miss. River near Rock Island, Ill.
Lake Drummond	2,800,000	22	9	Connects Chesapeake Bay with Albemarle Sound.
Lake Landing	25,000	4	5	Lake Mattamuskeet to Wysocking Bay, N. C.
Lake Washington-Puget Sound	5,000,000	6½	36	Connects Lake Washington and Puget Sound.
Lehigh Coal and Navigation Co.	4,455,000	108	6	Coalport, Pa., to Easton, Pa.
Lewes	356,000	11½	6	Connects Rehoboth and Delaware Bays.
Louisville and Portland	5,716,686	2½	9	At Falls of Ohio River, Louisville, Ky.
Mattamuskeet Out Fall	500,000	7	10	Hyde County, N. C.
Miami and Erie	8,062,680	274	5½	Cincinnati, Ohio, to Toledo, Ohio.
Miami and South New River	...	...	...	Lake Okeechobee to Miami, Fla.
Morris	5,100,000	103	5	Jersey City, N. J., to Phillipsburg, N. J.
Muscle Shoals and Elk R. Shoals	3,156,919	16	5	Big Muscle Shoals, to Elk River Shoals, Tenn.
North New River	...	...	...	Lake Okeechobee to Ft. Lauderdale, Fla
N. J. Coastal Inland Waterway	450,000	114	6	Cape May to Bay Head, N. J.
Ohio	(a)4,695,204	70	4	Cleveland, Ohio, to Dresden, Ohio.
‡Oswego	5,239,526	23	12	Oswego, N. Y., to Syracuse, N. Y.
Panama	375,000,000	50	41–45	From Colon to Panama across the Isthmus of Panama.
Pennsylvania	7,731,750	193	6	Columbia, Northumberland, Wilkes-Barre, Pa.
Portage Lake and Lake Superior	1,725,000	25	20	From Keweenaw Bay to Lake Superior.
Port Arthur (ship canal)	...	7	26	Port Arthur, Tex., to Gulf of Mexico.
Sabine-Neches	1,081,000	16	26	Port Arthur Canal to mouth Sabine River, Tex.
Salem	...	2	5–8	Salem River to Delaware River.
Santa Fe	70,000	10	5	Waldo, Fla., to Melrose, Fla.
Sault Ste. Marie (ship canal)	4,000,000	1⅓	18	Connects Lakes Superior and Huron at St. Mary's Riv.
Schuylkill Navigation Co.	12,461,600	108	6¼	Mill Creek, Pa., to Philadelphia, Pa.
Sturgeon Bay and Lake Michigan	287,000	1¼	20	Between Green Bay and Lake Michigan.
St. Clair Flats	1,180,000	3	20	Canal through delta at mouth of St. Clair River.
St. Mary's Falls	9,400,000	1⅓	18½	Connects Lakes Superior and Huron at Sault Ste. Marie, Mich.
St. Mary's Falls (parallel canal)	9,475,000	1½	24½	Connects Lake Superior and Huron.
‖West Palm Beach	...	12	...	...

Canals of Canada

CANALS	Cost of Construction*	Length, Miles	Depth, Feet†	LOCATION
Chambly	$728,999	12	6½	This canal overcomes the rapids between Chambly and St. Johns.
Cornwall	7,242,804	11	14	Cornwall to Dickinson's Landing.
Lachine	13,404,970	8½	14–18	Montreal to Lachine.
Rideau	5,531,332	133¼	5	Connects River Ottawa with Lake Ontario.
Sault Ste. Marie (ship canal)	5,000,000	1⅓	18¼	Connects Lake Superior and Huron at St. Mary's Riv
Soulanges	8,000,000	14	15	Cascade Point to Coteau Landing.
‖Trent	13,611,000	236	6–8⅓	Connects Lake Ontario and Lake Huron via Trent Riv
Welland (ship canal)	▲29,250,951	26¾	14–25	Connects Lake Ontario and Lake Erie.
††Williamsburg	10,490,184	12¼	9–14	Along St. Lawrence River.

* And improvements. † Navigable depth. ‖ Under construction. ▲ Not including cost of improvements and changes in locks, etc., now under way and involving an additional cost of about $20,000,000. †† Consisting of the Farran's Point, Rapide Plat and Galop Canals. (a) Original cost of canal extending from Cleveland to Portsmouth, 317 miles, but now abandoned between Portsmouth and Dresden. ‡ The Erie, Oswego, Champlain and Cayuga and Seneca canals have been enlarged by the state of New York to a depth of 12 feet, and they now comprise the New York State Barge Canal.

SOME FOREIGN CANALS	Length, Miles	Depth, Feet	Bottom Width, Feet	Estimated Cost
Suez — Mediterranean and Red Sea	103	35	108	$127,000,000
Kronstadt — Petrograd	16	20½	...	10,000,000
Manchester — Manchester and Liverpool	35½	28	120	85,000,000
Kaiser Wilhelm (Kiel Canal) — Baltic and North Seas	61	36	72	40,000,000
Elbe and Trave	41	10	72	6,000,000
Berlin-Stettin (Hohenzollern Canal)	136	9.8	32–39	12,500,000

ON THE NEW YORK STATE BARGE CANAL
A view of three of a series of locks, showing how canal navigation is carried over a low watershed.

equal basis, and is destined to influence the economic history of the entire world (see PANAMA CANAL).

About 4,500 miles of canals have been constructed in the United States, but not much more than half of this mileage is now in actual use. See list of names below.

Canals of Canada. The comprehensive canal system of the Dominion of Canada is fully described under the heading CANADA, subtitle *Transportation.* See, also, the special articles RIDEAU CANAL; SAINT LAWRENCE RIVER; SAULT SAINTE MARIE CANAL; TRENT CANAL; WELLAND CANAL.

Canals of Europe. Because of the character of the country, the "low countries" (Netherlands) led the other parts of Europe in canal building, and by 1250 an extensive network of artificial waterways had spread over that region. At the present time Holland alone has over 2,400 miles of canals, and Belgium about 1,345. The great canal system of France, now aggregating over 3,000 miles, was begun early in the seventeenth century. This statement means a good deal when it is realized that France is only two-thirds as large as Texas, and it contains more than half of the mileage in canals boasted by the entire United States. In 1666 the French government began the construction of one of the most celebrated engineering projects of the century, the Languedoc Canal, connecting the Mediterranean Sea and the Bay of Biscay, and 148 miles in length. It was first opened in 1668, and since then has shared with the other waterways of the country in the great sums appropriated by the government for improvement and maintenance. No tolls on French state waterways have been levied since 1888.

Germany's most important canal, the Kiel, or Kaiser Wilhelm (see KAISER WILHELM CANAL), provides a waterway between the Baltic and the North seas. It was completed in 1895, but was later rebuilt and was again opened to navigation in June, 1914. The commercial importance of this watercourse is unquestioned, and its strategic value was demonstrated in the War of the Nations. Another notable watercourse is the new Hohenzollern Canal, also opened in 1914. It joins the Oder and Spree rivers, providing waterway communication between Berlin and Stettin on the North Sea. In the spring of 1915 a canal between the Vistula and Oder rivers was opened, 182 miles in length. Since 1875 the government has been engaged in the systematic reorganization and development of its waterways; rivers have been dredged, that is, canalized, to make them navigable, old canals have been rebuilt and new ones constructed. Of about 8,000 miles of navigable waterway, approximately 3,000 miles are canals.

The Russian system of artificial waterways was begun by Peter the Great. There are now in the country about 500 miles of canals and over 550 miles of natural watercourses canalized. One of the most recent projects is the canal between Kronstadt, on the Gulf of Finland, and Petrograd, on the Neva River. It is sixteen miles in length and 20.5 feet deep and cost $10,000,000. Within recent years until the outbreak of the War of the Nations the government has been appropriating between $8,000,000 and $12,000,000 a year for waterway development. The Danube River, which crosses Austria-Hungary, is the most important watercourse in that country, but there is also a canal system, local in character, which adds about 1,700 miles to the natural waterways of the dual monarchy. In Greece is the important Corinth Canal (see CORINTH), across the isthmus which joins the Peloponnesus to the northern part of the country. It is one of the world's famous short canals.

The most important canal in the British Isles, and one of the most remarkable in the world, is that between Manchester and Eastham, on the Mersey River, six miles from Liverpool (see MANCHESTER SHIP CANAL). It was opened for traffic in 1894, and has since been enlarged and improved. Other notable British canals are the Grand Canal, 165 miles long, between Dublin and Ballinasloe, on the River Shannon; the Caledonian Canal, extending across Scotland from Inverness to Fort William; and the canal between the firths of Forth and Clyde, thirty-five miles in length. A royal commission on canals and waterways was appointed in 1912 to investigate the matter of modernizing and nationalizing the canal system of Great Britain. There are in the United Kingdom about 4,700 miles of artificial waterways, 4,000 of this mileage belonging to England and Wales. C.H.H.

Related Subjects. Of the canals of the world, the following are treated in separate articles in these volumes:

Cape Cod	Manchester Ship Canal
Chesapeake and Ohio	New York State Barge Canal
Chicago Drainage	Nicaragua
Corinth	Panama
Erie	Rideau
Georgian Bay Ship Canal	Sault Sainte Marie
Hennepin	Suez
Illinois and Michigan	Trent Canal
Kaiser Wilhelm	Welland

CANARY, *ka na' ri,* a beautiful little bird of the finch family, the best loved of all the cage birds because of its remarkable gift of song and its cheerfulness and friendly characteristics. It was first found in the Canary Islands and Madeira, acquired its name from the former, and was taken to Europe about the beginning of the sixteenth century. Since then it has found its way into households in various parts of the world, and the centuries of breeding and domestication have brought about marked changes in its appearance. In a state of nature the bird has a dull-green plumage streaked with darker shades, while the canary in captivity is usually a bright yellow, though sometimes orange, reddish or pale yellow. The wild canary is not more than five and one-half inches long; its tame brother is sometimes eight inches in length. The topknots of some

CANARIES

Below, usual type of canary; above, American goldfinch, or wild canary.

and the long, slender shapes of others are all the results of breeding. The Scotch Fancy canary, with its long, slender, curved body, bent almost to a semi-circle, is one of the strangest of these results.

In the Harz Mountains and other parts of Germany and in the British Isles and Belgium, the raising of canaries is an important industry, and large prices are paid for the highest type of singing birds. The Harz Mountain canaries are the most famous of all. In the United States and Canada the birds cost from $1 up to $75, but sometimes $150 has not been considered too high a price to pay for an especially fine singer. The birds require a clean cage, good seed, some green food, lime and plenty of cold water. Beyond this they need little care and thrive almost anywhere. Their average length of life is close to twenty years. In America, the name *wild canary* is often given to the American *goldfinch,* or thistle bird, which, though entirely different, does somewhat resemble the captive canaries.

Among several excellent books on the subject may be mentioned *Canaries and Cage Birds* by W. A. Blakston; and *The Canary Book,* by R. L. Wallace.

CANARY ISLANDS, a group of islands in the Atlantic Ocean about sixty miles west of the northwest coast of Africa, covering a total area of 2,807 square miles. They were discovered by the Spaniards in 1630 and were named *Canaria,* a word derived from the Latin *canis,* meaning *dog,* for at that time a large, fierce breed of dogs, now extinct, lived there. The islands are thirteen in number, the most important of which are Teneriffe, Grand Canary, Palma, Hierro, Gomera, Lanzarote and Fuerteventura. The remaining six are uninhabited. All are rugged, volcanic and mountainous, frequently presenting precipitous cliffs to the sea. The most notable peak is that of Teneriffe, which rises to a height of 12,182 feet. The Canaries are not reached directly from America, but by way of European ports. They belong to Spain, and are governed as a province.

The climate is mild and healthful and the soil is so fertile that the islands in ancient times earned the name of Fortunate Islands. There are no rivers of note, but there are numerous streams. These contain no fish except eels; in the surrounding seas, also, fish are scarce. All kinds of domestic animals have been introduced and thrive well. Agriculture is primitive, but the fertile soil produces large quantities of grain, fruit, vegetables and flowers.

Little is known of the Guanches, the tribe who originally inhabited these islands. They were for the most part exterminated by their conquerors, the remnant being absorbed by intermarriage. The present inhabitants are darker than the northern Spaniards, and are well formed and hardy. Population, estimated, 1910, 475,000.

CANCELLATION, *kan sel la'shun,* is a "short cut" in mathematics. Its purpose is to avoid long and difficult multiplications and divisions, both in whole numbers and in fractions. It is based upon these principles:

(1) Both dividend and divisor may be divided by the same number, and the quotient is unchanged; or, a common factor may be dropped from dividend and divisor and the quotient remains unchanged.

(2) The numerator and denominator may be divided by the same number and the value of the fraction remains unchanged; or, a common factor may be dropped from numerator and denominator and the value of the fraction remains unchanged.

In division, the most common use of cancellation is the division of both terms by 10, 100, and so forth. For illustration:

(1) $6500 \div 1300$. Divide each number by 100 by striking off two zeros in each; then the problem becomes $65 \div 13 = 5$.

(2) $75000 \div 250$. Upon dividing both terms by 10 the problem becomes $7500 \div 25 = 300$.

But it is also used in the division of both terms by other factors than 10, 100, etc.; for example:

$2200 \div 55 = \text{quotient}$

Divide dividend and divisor by 11, and the problem becomes

$200 \div 5 = q$

$q = 40$

See what common factor is taken out in the following:

(3) $650 \div 39 = q$
$50 \div 3 = 16\frac{2}{3}$. (Factor taken out is 13.)

(4) $960 \div 36 = q$
$80 \div 3 = 26\frac{2}{3}$. (Factor taken out is 12.)

(5) $225000 \div 450$
$22500 \div 45$
$1500 \div 3 = 500$. (Factors taken out are 10 and 15.)

(6) $108000 \div 7200$
$1080 \div 72$
$30 \div 2 = 15$. (Factors taken out are 100 and 36.)

(7) $99180 \div 360$
$9918 \div 36$
$1102 \div 4 = 275\frac{1}{2}$. (Factors taken out are 10 and 9.)

Taking out of factors other than 10, 100, etc., is not so general as it should be. Children's attention should be drawn to it (see FACTORING). Here is an illustration to show

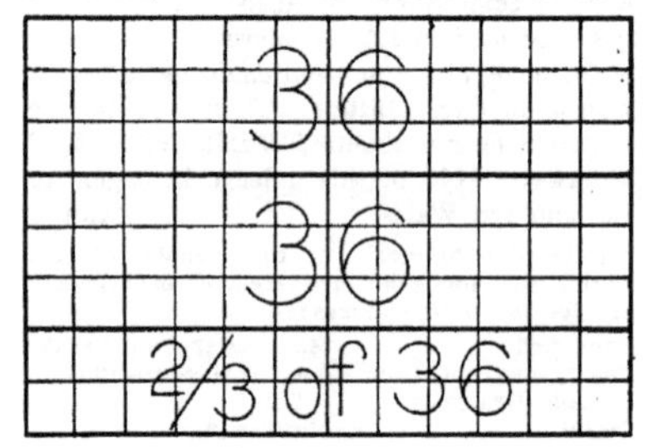

the common factor and why it can be dropped:

$96 \div 36 = \text{quotient}.$

This reads, "How many 36's in 96?" 96 is made up of eight 12's; 36 is made up of three 12's, as shown in the diagram; then the problem is seen to become, "How many 3 rows in 8 rows?" and then, "How many 3's in 8?"

This put into arithmetical form appears:

$$(8 \times 12) \div (3 \times 12) \text{ or } \frac{8 \times \overset{1}{\cancel{12}}}{3 \times \underset{1}{\cancel{12}}} = \frac{8}{3} = 2\frac{2}{3}$$

Cancellation is used very much more freely when division is written in *fraction* form than when it appears with the division sign (÷);

$$\text{for example, } 13500 \div 270 = \frac{\overset{150}{\cancel{13500}}}{\underset{3}{\cancel{270}}} = \frac{150}{3} = 50.$$

(a) Divide both terms by 10
(b) Divide both terms by 9
(c) Divide both terms by 3

Let us look at cancellation as used in a concrete problem:

"How many blocks of ice 36 inches by 24 inches can be packed in a layer in an ice house 60 feet wide and 90 feet long?"

$$\text{Solution: Number of blocks} = \frac{\overset{10}{\cancel{60}} \times 90 \times \overset{\overset{1}{\cancel{6}}}{\cancel{144}}}{\underset{\underset{1}{\cancel{6}}}{\cancel{36}} \times \underset{1}{\cancel{24}}}$$

(a) Divide both terms by 24
(b) Divide both terms by 6
(c) Divide both terms by 6

and the division appears

$$\frac{10 \times 90 \times 1}{1 \times 1} = 900$$

Cancellation shortens the work in multiplication and division of fractions; for example, $\frac{2}{3} \times \frac{7}{8} \times \frac{12}{17} \times \frac{34}{35}$. Let this appear $\frac{2 \times 7 \times 12 \times 34}{3 \times 8 \times 17 \times 35}$. The numerator of the new fraction is the product of 2, 7, 12, 34, while the new denominator is the product of 3, 8, 17, 35. We may divide the numerator and denominator by the same number without changing the value of the fraction:

(a) Divide both terms by 3: (3 below and 12 above);
(b) Divide both terms by 7: (35 below and 7 above);
(c) Divide both terms by 2: (2 above and 8 below);
(d) Divide both terms by 17: (34 above and 17 below);
(e) Divide both terms by 4: (4 below and 4 above); and the product appears as below:

$$\frac{\overset{1}{\cancel{2}} \times \overset{1}{\cancel{7}} \times \overset{\overset{1}{\cancel{4}}}{\cancel{12}} \times \overset{2}{\cancel{34}}}{\underset{1}{\cancel{3}} \times \underset{\underset{1}{\cancel{4}}}{\cancel{8}} \times \underset{1}{\cancel{17}} \times \underset{5}{\cancel{35}}} = \frac{2}{5}$$

An illustration in mixed numbers follows:

$$2\frac{1}{2} \times 8\frac{1}{3} \times 7\frac{1}{5} = \frac{5}{2} \times \frac{25}{3} \times \frac{36}{5}$$

(a) Divide both terms by 5: (5 above and 5 below);
(b) Divide both terms by 3: (36 above and 3 below);
(c) Divide both terms by 2: (12 above and 2 below); and we have—

$$\frac{\overset{1}{\cancel{5}}}{\underset{1}{\cancel{2}}} \times \frac{25}{\underset{1}{\cancel{3}}} \times \frac{\overset{\overset{6}{\cancel{12}}}{\cancel{36}}}{\underset{1}{\cancel{5}}} = \frac{25 \times 6}{1} = 150$$

A problem in division of fractions becomes a problem in multiplication of fractions, and so cancellation is helpful in division of fractions. For example:

$$(1) \quad \frac{12}{13} \div \frac{9}{26} = \frac{\overset{4}{\cancel{12}}}{\underset{1}{\cancel{13}}} \times \frac{\overset{2}{\cancel{26}}}{\underset{3}{\cancel{9}}} = \frac{8}{3}$$

$$(2) \quad 8\frac{5}{9} \div 7\frac{1}{3} = \frac{\overset{7}{\cancel{77}}}{\underset{3}{\cancel{9}}} \times \frac{\overset{1}{\cancel{3}}}{\underset{2}{\cancel{22}}} = \frac{7}{6}$$

A.H.

CANCER, *kan'ser*, the common name of a dangerous malignant tumor that may grow in the human body or in the bodies of other vertebrate animals. The name comes from a Greek word meaning *crab*, and was applied to the tumor because the enlarged veins around the swelling resemble the claws of a crab. Though there are many different forms of cancer, certain characteristics are common to all of them. Cancerous growths are always composed of a fibrous framework enclosing a mass of cells and a milky-white cancer juice. They have no definite limits, but involve surrounding tissues, and, because of their tendency to spread by means of the veins and lymphatics, they often cause growths of similar character in other parts of the body.

Classification. A convenient mode of classification divides cancers into three groups—*scirrhus, encephaloid* and *epithelial cancer*. In the first class are those which have a large proportion of fibrous elements, which make them very hard. They do not grow rapidly, but are difficult to check because of their tendency to spread and to ulcerate. Scirrhus most commonly attacks the female breast, but sometimes internal organs are affected. In encephaloid the cell elements are found in greater proportion than the fibrous, making this form of cancer very soft. Because of its rapid growth it is sometimes known as *acute* cancer. The internal organs or the limbs are its most common seats. Epithelial cancer occurs most frequently in the skin and mucous membranes, or where these unite, as on the lips. Though this form of cancer does not spread so rapidly as the other forms nor cause secondary growths to the same extent, it has the same tendency to affect surrounding glands. All forms tend to recur after removal.

Causes. Cancer may be started by local irritation, as by the stem of a pipe on the lip, gallstones in the gall bladder and the rubbing of a corset steel on the breast. Continued irritation is the usual excitant of the cancer, but a single accident may cause it. Heredity seems to have some influence, though this may be simply heredity of the habits that tend to develop cancer, such as smoking or always being well-fed. Careful research and observation have established the following facts: Nearly one-half of all cancers occur in the stomach and intestines; in other words, in that part of the body in which food is stored which has not always been disinfected by heat. Outside surfaces of the body protected against dirt and injuries are almost entirely free from cancerous growths, while these growths do attack those parts not protected by clothing and therefore exposed to dirt and injuries. Preventive measures would therefore include the avoidance of all possible irritation of the tissues; eating only those foods which are absolutely clean or have been sterilized by cooking; practicing moderation in choosing one's diet, and observing hygienic rules of living.

Treatment. Statistics show that cancer is one of the most frequent causes of death among persons over forty years of age, and that beyond a certain stage the disease is incurable. Institutions for the study and treatment of this scourge of humanity are found in every civilized country, and the problem is engaging the attention of the most skilled investigators in the medical and scientific world. The most common method of treatment is removal by knife, and as yet, so far as known, this is the most reliable method.

In recent years the X-rays and radium forms of treatment have been attracting wide attention. Within certain limits both have been found of value in checking cancerous growths, but the ultimate cure is still an unattained goal. In regard to their use, about which many false hopes have been cherished, the following points should be remembered: their curative effects are practically limited to superficial cancers of the skin, to superficial growths of mucous membrane which are not true cancers and to some deeper-lying tumors not very malignant. That large group of cancers known as *generalized,* that is, disseminated through the body, cannot be benefited by radium or the X-ray, nor can those which are not accessible to these methods of treatment. Others which are accessible sometimes grow so rapidly they offset the curative effects. It should also be remembered that the supply of radium is limited, and its fabulous cost puts it out of the reach of the great majority (see RADIUM).

If the victims of cancer could be brought to a realization of the fatal danger that lies in delay, death statistics for this disease would tell a different story. Any lesions on the lips, sores in the mouth, lumps on the breast and other irritations liable to lead to cancer that do not disappear in a few weeks should have the attention of a physician and be removed. Thorough examination and proper treatment at an early stage of the disease are absolutely essential. Patent medicines that deaden pain without affecting the cause of the irritation are worse than useless, for the temporary relief they afford deceives the victim and permits the incipient cancer to make dangerous headway.

W.A.E.

CANCER (the crab), the fourth sign of the zodiac, entered by the sun on or about the twenty-first of June and quitted a month later. The symbol is ♋. The constellation of Cancer is no longer in the sign of Cancer, but at present occupies the place of the sign of Leo (see ZODIAC; PRECESSION OF THE EQUINOXES). The Tropic of Cancer is the name given to the northern tropic (see TROPICS). According to mythology Cancer is the crab that attacked Hercules when he was destroying the Hydra, a monster with a hundred heads.

CANDELABRUM, *kan de la' brum,* from the Latin *candela,* properly refers to a candle holder, but in late years the name has also been given to any ornamental support of a lamp. Workers in metals since the ancient era have displayed their genius in devising graceful forms. The Etruscans probably excelled in the art. The most beautiful examples have been found in the buried cities of Herculaneum and Pompeii.

CANDIA, *kan' di a,* the capital of the island of Crete (which see) before it became the possession of Greece in 1912. It is an ancient city; it is on the site of the ancient Heracleum, which was the seaport of Cnosus. The modern Candia dates from the ninth century, when it was founded by the Saracens. Genoa owned it and fortified it in the twelfth century, and later it was an outpost of Venice. For later history, see CRETE. Population in 1910, 22,683.

CAN'DLEBERRY, or **BAY'BERRY,** or **WAX MYRTLE,** *mer' t'l,* a shrub from whose wax-covered berries the popular bayberry candles are made. These burn with a pleasant, piny odor, and are connected by tradition with Christmas, the old belief being that on that day—

> A bayberry candle burned to the socket
> Brings health to the body, joy to the heart,
> And gold to the pocket.

The candleberry shrub, which grows from four to eighteen feet in height, is to be found all along the eastern coast of North America, but is much more abundant in the South. The berries with their coating of wax are boiled, and the greenish-white tallow is skimmed from the surface. From a bushel of berries four or five pounds of wax may be obtained.

CANDLEFISH, a salt-water fish, from twelve to fifteen inches long, belonging to the smelt family. It derives its name from a peculiar custom among the Indians of Alaska, whose coasts it frequents. The Indians literally make a candle of it by drying it and forcing through it the pith of a rush or a strip of bark as a wick. When the wick is ignited it burns freely. The extreme oiliness of the fish, strangely enough, is not unpleasant to the taste, for the oil has a fine flavor. When fried the flesh is considered superior to that of the trout. The oil is sometimes extracted and is used as a substitute for cod-liver oil.

CANDY, *kan'di*, a name given to almost any sweetmeat which has sugar as its main ingredient. Different methods of making, with the addition of various flavorings, fruits, nuts or other ingredients, produce almost innumerable kinds of candy, and the demand is so great that new and tempting combinations are being constantly evolved. The United States uses more candy than any other nation in the world, the output of its factories averaging over $135,000,000 each year. The range of prices at which it is sold is wide. Some of it, containing a large proportion of glucose, is put on the market at ten cents a pound, while other varieties bring as high as a dollar or more a pound, in part because of the maker's name.

If the candy that is eaten each year were evenly distributed among the inhabitants of the country, it would not be at all harmful. Physicians generally agree that a moderate amount of pure candy, eaten immediately after a meal, is thoroughly wholesome, for the system needs a certain proportion of sugar. If eaten constantly, however, at intervals during the day, even the best candy will do harm in many ways. Especially important is it that children should not be allowed to eat much candy, and that such as they do have should be pure. Constant visits to the candy shop with pennies and the purchase there of the cheap, brightly-colored candy so attractive to children's eyes, work great harm.

Story of the Candy Industry. So prominent a place has candy made for itself in social as well as industrial life, so almost necessary has it come to seem to many people, it is difficult to imagine that a comparatively short time ago there was no such thing. But if children before the beginning of the nineteenth century had candy at all, they had it in the form of little sugar pills or pellets given them by the doctors, for the first candies were made by physicians and druggists. The purpose was to conceal the unpleasant taste of medicines, and there are still on the market to-day many drugs which are mixed or coated with sugar to make them palatable. Some of these, as peppermint, hoarhound or wintergreen drops of various kinds, have passed from the medicine class and are now chiefly eaten as candy.

The popularity of the doctors' sugar-coated pellets suggested to certain enterprising men in England the idea that sweetmeats without medicine might be even more popular, and with their earliest products the candy industry began. It spread in time to all civilized countries, but it was only with the invention of machinery which made manufacture far easier, cheaper and more rapid that it became a great business.

As in all industries, there have been constant improvements in method. Manufacturers are now compelled by the laws of most states and provinces to keep their factories clean, light and airy. Then, too, the harmful coal-tar preparations which were once used to color candies are generally forbidden by pure-food laws, so a person may to-day buy colored candies without the fear that they are poisonous. Even green-colored candy, formerly avoided because it was supposed to contain arsenic, is now harmless, for in almost all cases the dye used is simply spinach juice. See ADULTERATION IN FOODSTUFFS AND CLOTHING.

The Making of Candy. There are so many kinds of candies that an enumeration of them and of the methods of manufacture would be an impossibility, but they divide themselves naturally into certain classes. All cream candies, for instance, have as their basis what is called a *fondant*. This French term is commonly used because France was for a long time the only country which could make these creams in perfection. The fondant is made, in the factories, by adding to a large quantity of sugar a small amount of glucose, and boiling the whole with water until a thick, clear syrup is obtained. This is poured upon huge marble slabs, allowed to cool, and then worked with long paddles until it becomes white, creamy and smooth. Innumerable kinds of candy are made with this fondant as a basis—chocolates, bonbons, patties, nut- and fruit-nougats, cocoanut creams, and so on.

Fondant is easily made in the home by

boiling together two cups of granulated sugar, one cup of water and a very little cream of tartar. When the syrup has become thick it is allowed to cool, then beaten with a spoon or paddle until it has a creamy consistency. A little practice will enable one to produce candies which are the equal of the best sold in the shops.

The process of molding candies in the great factories is interesting. Shallow boxes are filled with cornstarch, into which are stamped by machinery little hollows of the shape wanted for the candy. Into these the cream candy is run from a tank, and after it has hardened the candy is dumped with the starch into a hopper which allows the starch to drop through a sieve and brushes from the finished candies all the starch that has clung to them.

Hard candies are made of sugar and water boiled to a heavier syrup than is the fondant. It is colored and flavored, then allowed to harden without being stirred or beaten. Much of the stick candy is still rolled out by hand. Caramels of the best grade are made of sugar and pure cream, boiled until the product is of a proper consistency and cooled on marble slabs. The candied fruits, nuts and flowers which have become so popular in recent years are made by dipping them into a syrup which has been boiled until it is just on the point of recrystallizing into sugar. Chocolate creams, too, are a dipped candy, the fondant centers being thrown into the chocolate and lifted out with a wire spoon. Perhaps there is no candy which is more popular than the various kinds of chocolates, and the demand for them has caused a vast increase in the chocolate business (see CHOCOLATE). A.MC C.

CAN'DYTUFT, a group of plants belonging to the mustard family, that take their name from Candia, the old English name for the island of Crete, from which the seeds were carried to England in the sixteenth century. Three species of candytuft are well-known garden flowers, easily cultivated in ordinary soil. *Purple candytuft* grows to a height of about one foot, and bears its flowers in flat-topped clusters, the blossoms in the center of the cluster being smaller than those on the edge. The four petals, pink or pale purple in color, show an irregularity that is typical of the candytuft group, for the two outer petals are much longer than the others. A garden bed of this species, with the shades of pink and purple melting into each other on the nodding flower heads, is a most attractive sight.

CANDYTUFT

Bitter candytuft, grown in gardens and also found as a weed in Western and Central Europe, is a small plant six to twelve inches high, whose erect, branching stem bears white, sometimes purple-tinted, flower clusters. The root, stem, leaves and seeds of this species are said to have medicinal properties. *Evergreen candytuft,* a lover of warm climates, is valued for its abundant, pure white flowers. In a favorable climate these flowers remain in bloom throughout the winter.

CANE, *kane,* the name applied to various plants which have a reed-like stalk or stem, including the sugar cane and bamboo among the grasses, and the rattans among the palms. The stems of plants of this type are of untold value to man, because of their lightness, strength and flexibility. In tropical countries of the East, for instance, where the bamboo flourishes, the natives use the plant to build their houses, and much of their furniture is also of this wood. Batavia, in Java; Sarawak, in Borneo; Singapore and Penang, in the Straits Settlements, and Calcutta, in India, are centers of a flourishing trade in canes and rattans, the Western European countries and Canada and the United States importing large quantities. In these latter countries fishing rods, walking sticks, umbrella handles, baskets, chair seats and many other articles made from cane plants are familiar objects.

Because so many walking sticks have been made of the wood of these plants, the term *cane* is now applied without discrimination to any walking stick. See BAMBOO; SUGAR CANE; RATTAN, with which are illustrations.

CANE SUGAR. See SUGAR, subhead *Cane Sugar.*

CANIS MAJOR, *ka'nis ma'jer,* meaning the GREAT DOG, is a constellation in the southern hemisphere containing Sirius, the Dog Star, the brightest star in the heavens. The dog is the faithful companion of the mighty hunter Orion, whom he eternally watches over in the sky. The three familiar stars forming the belt of Orion point directly down to the nose of the dog formed by the star Sirius, one of the nearest to earth, but so distant that it takes eight years for its light to reach us. See ASTRONOMY.

CANKER SORE, *kang'ker sore,* a small ulcer or collection of ulcers, which forms in the mouth, usually on the tongue. Ulcers are quite common in children and are usually due to indigestion. They can be removed by the use of pulverized alum or silver nitrate, but the latter remedy should be used with caution, for it may cause a sore as painful as the one it removes.

CANKERWORM, *kang'ker werm,* the caterpillar of either of two moths, the females of both of which species are wingless. The eggs of one species hatch in the spring, those of the other in the fall or early winter, and the caterpillars may be distinguished by the number of white stripes across their backs, the spring caterpillars having eight and the autumn ones six. These caterpillars are very destructive to the foliage of fruit trees and shade trees, often eating the entire foliage of an orchard or grove in a few days. When the worms are disturbed they drop from the leaves and hang suspended by a silken thread. Since they must climb the trees to reach the foliage, one of the best means of preventing their ravages is to fasten bands of tarred paper or some other sticky substance around the trees early in the spring. Shade trees may be sprayed with Paris green, but this should be used with caution on fruit trees. See CATERPILLAR; MOTH.

CANNA, *kan'a,* one of the most brilliant and ornamental of garden plants. Its development has been achieved within recent years, and now no plant is more popular for large beds or as backgrounds for lower-growing flowers. For the canna is tall—frequently as tall as a man; and even the so-called dwarf varieties are four feet in height. The large leaves, dark-green or bronze, spread about the central stock, at the end of which appear the gorgeous red, yellow or orange flowers. Cannas are very easily grown, few plants repaying more generously the care expended upon them. They need a rich, warm soil and plenty of moisture; and the blossoms as they wilt should be picked, to prevent seeding. At the close of the flowering season the big root is dug up, kept from freezing through the winter and replanted in the spring.

CANNA

CANNIBAL, *kan'i bal,* a person who eats human flesh. Spanish discoverers at the time of Columbus found that the custom of eating their fellow beings existed among the Caribs, a West Indian tribe, and from their name came the modern word *cannibal.* Man-eating races were mentioned, however, by the Greek historian Herodotus and other ancients, and reference is made to the practice in the writings of Marco Polo (which see), an Italian adventurer of the thirteenth century. Cannibalism until recently prevailed among the savages of West and Central Africa, New Guinea, the Fiji Islands, Australia and New Zealand, and is still practiced in Sumatra and other islands of the East Indies. The early North American Indians were accustomed to eat their prisoners in time of war, and even so highly civilized a race as the Aztecs of Mexico consumed the human victims offered as sacrifices to their god of war. The heart of a brave man was regarded as an especially choice morsel.

Among some races cannibalism began as a religious rite, and in the course of time there developed a natural appetite for human flesh. Others have believed that the virtues of the deceased passed on to the one who ate of his flesh, and so children often consumed their dead parents. In some cases men have eaten human beings simply because there was lack of other food supply. Even highly-civilized beings may be driven to eat of human flesh by extreme hunger, as is shown in the records of sieges and shipwrecks. Cannibalism and cannibals are mentioned frequently in stories of adventure and travel, a well-known instance being the rescue of Friday from the cannibals by Robinson Crusoe. Other stories are told of mariners adrift at sea who cast lots to see which one should be killed and give his body to satisfy the hunger of his companions. Even in cases of dire necessity, however, such a proceeding is not held to be justifiable, and there are cases on record of the trial, condemnation and punishment of rescued sailors who had during their apparently hopeless hours resorted to this course.

CANNING AND PRESERVING. See FOOD PRODUCTS, PRESERVATION OF.

CANNING CLUBS are one evidence of the established popularity of "team work" in various modern activities. The spirit of business organization, arising from the desire to secure a higher grade of efficiency, has made itself felt in both the home and the school, and as one result the domestic art of canning fruits and vegetables has been placed on a systematic basis. There are in many communities leagues of housewives and their daughters, who put up fruits and vegetables according to the most modern methods, who read bulletins and other printed matter giving recipes, suggestions and instruction, and who send their produce in a spirit of friendly rivalry to local, district or county fairs and festivals. A particularly-helpful feature of such clubs is the custom of holding demonstrations and meetings for discussion and interchange of views and experiences. School canning clubs for the girls are similarly organized, and these are as much in favor as the boys' corn clubs. (The corn-club movement is fully described in these volumes in an appropriate subhead under the heading CORN.)

The benefits, both direct and indirect, derived from organizing canning clubs are not hard to find. The housewife who wishes to have a bountiful supply of wholesome, nutritious foods, put up in an attractive form and in such a way that they will keep through the winter, finds that it pays to learn the best and quickest methods of canning. Working systematically and enjoying the advantages of coöperation and of club discussions, she finds herself growing more and more efficient and the work increasingly interesting. Where mothers and daughters belong to a club there results a closer bond between them in the home activities, and the girls are provided with a useful and helpful occupation in their spare hours. School clubs and home clubs alike teach lessons in economy and thrift, and in many cases they prove to be a bond of union between the home and the school.

Equipment. Club members find that best results are obtained when the utensils and other equipment are carefully chosen and arranged in an orderly way. The systematic worker has a clock in a convenient position, for each step of the process is timed, and within easy reach there is a thermometer and a set of graduated scales. She also does not forget her note book, in which records are carefully written as the work proceeds. Pans and tins, a canning boiler, plenty of clean cloths and towels, an abundant supply of pure, clean water, sharp coring and paring knives, jars, tops, rubbers, labels, etc., are of course essential. Tin cans with solder-hemmed caps are used for putting up many fruits and vegetables. Glass jars, tops and rubbers are kept in a pan of warm water on the back of the stove until ready for use. For canning all kinds of fruits and vegetables, steam-pressure canners have been found most successful. Steam under pressure raises the heat to about 250° F., and makes possible complete sterilization. A wash boiler, or other vessel with a close-fitting top, can be turned into a home canner by making a false bottom with lifting handles. Such a device is adequate for canning on a small scale, and is known as a hot-water bath outfit.

Definitions and Methods. The successful canner must be familiar with several terms that are peculiar to this household art. The more important of these are here listed:

Scalding. Dipping fruit in boiling water for the purpose of removing the skin.

Blanching. Boiling quickly in water. This is done by placing fruit or vegetables in a cloth bag and lowering it into the water. Purposes—to eliminate acids and other bitter substances; to reduce bulk of vegetable greens.

Cold dipping. Plunging food products into

HOME CANNING CLUBS. Above: Out-of-doors hot-water bath outfit. Below: Demonstration conducted by government expert.

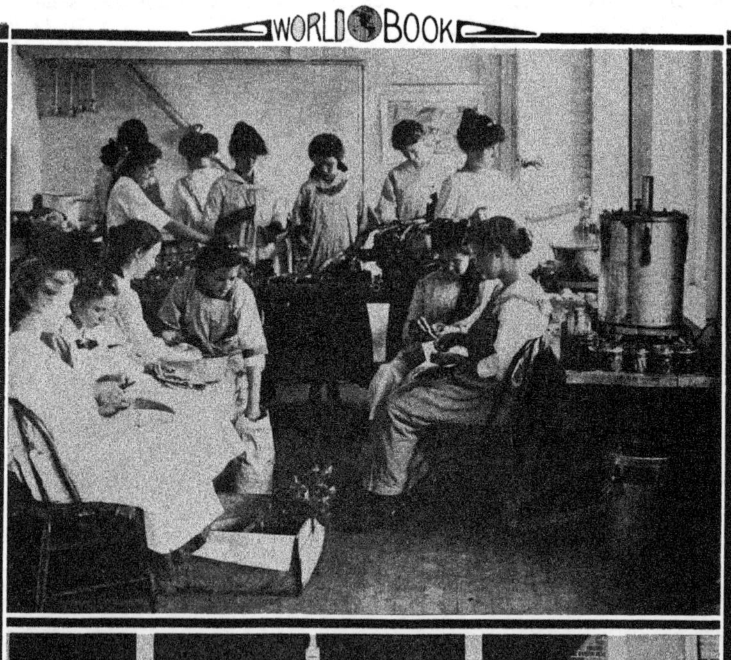

cold water immediately after blanching, to harden the pulp under the skin and to make it easier to handle products in packing.

Sterilizing, also known as *processing.* Boiling fruits or vegetables for a certain period after completely sealing the container. The object is to destroy germs.

Tinning the steel. Putting the hot steel used for capping tins in sal ammoniac and solder, turning the steel several times until it is smooth and bright, and then dipping it in soldering "flux."

Capping. Soldering the solder-hemmed tops on the cans with the capping steel.

Tipping. Sealing the air hole or vent in the center of the tin cap.

Flux. Prepared by adding to muriatic acid as much zinc as will dissolve, and then adding water equal in amount to acid. The fluid is used to clean the steel and to wipe surfaces to be soldered. When applied to tin it adds a coating of zinc, to which the solder will readily adhere. Should be applied carefully and none be permitted to run inside the can.

It is not possible here to give all of the standard recipes used by club members in canning the different varieties of fruits and vegetables, but the following directions illustrate the methods employed:

For canning soft fruits, such as berries, cherries, etc. Can the fruit the day it is picked. Grade; then rinse by pouring water over fruit through a strainer. Cull, seed and stem. Pack at once in glass jars or tin cans. Add thin, boiling hot syrup. Place rubber and top in place and partially tighten. (Cap and tip tin cans.) Sterilize in hot-water bath outfit sixteen minutes, or in steam-pressure outfit under five pounds steam ten minutes. Remove jars. Tighten covers and invert to cool. Wrap in paper and store.

For canning tomatoes. Grade according to size, ripeness and color. Scald in hot water so as to loosen skins. Dip quickly into cold water and then take out. Core and skin. Fill container with whole tomatoes, adding one level teaspoonful of salt to a quart. Place in position rubber and cap. Partially seal, but not tight. (Cap and tip tin cans.) Sterilize twenty-two minutes in hot-water bath outfit, or fifteen minutes in steam-pressure outfit under five pounds steam. Remove jars, tighten covers, invert to cool. Wrap in paper and store.

Club Demonstrations. Meetings for all the club members, at which the standard methods of canning are demonstrated and principles and terms are explained, are of great practical value. Those conducted in the school, attended both by pupils and their mothers, are of special interest. The teacher or leader makes all arrangements before the demonstration begins, placing in order the canning outfit, tables, supplies, food products, etc., and seeing that the fire is in readiness. The club members are then called to order, preliminary instructions are given, and each pupil is assigned a task. By varying the work, each member is able to participate in all the canning processes, from preparing the containers and fruit to labeling the cans and entering the records. Then follow the clearing up and putting away of the apparatus. To add variety and interest contests are sometimes introduced; different members compete to see who is quickest and most efficient in canning a specified amount of fruit, or in scalding, filling, labeling, etc.

The work in canning can be easily correlated with other school subjects—geography, history, physiology, reading, composition and arithmetic—and the pupils who cultivate their own plots of ground and can the produce which they themselves raise are learning practical lessons in agriculture.

How to Organize a Club. The United States Department of Agriculture is making a special effort to interest homes and schools in canning-club work, and to this end is coöperating with state agricultural colleges. In some states there is an agent in charge of boys' and girls' club work at the state college of agriculture, from whom necessary instruction and information concerning the organizing of clubs may be obtained. If there is no state agent application should be made to the States Relation Service, Office of Extension Work North and West, Department of Agriculture, Washington, D. C. Report blanks, bulletins and other necessary material will be mailed on request. Club members are required to attend canning demonstrations and special meetings called by state, district or local leaders, to make exhibits of a specified number of canned products at some fair, to attend and take part in the programs of the local club meetings, and to fill out the regular canning report blanks and forward them to the local leader, state agent or chief of the extension work at Washington. B.M.W.

Related Subjects. More or less directly connected with the above theme are the following articles, which will be found in their alphabetical order in these volumes:

Boys' and Girls' Clubs
Food Products, Preservation of
Gardening
School Garden

CANNON, *kan'un,* a term usually applied to big guns, as distinguished from any kind of small firearms. This name was first used about the beginning of the fifteenth century and is derived from the Latin word *canna,* which means a *tube* or *reed.* It is applied to any form of firearm which is fired from a fixed or movable carriage and which fires a projectile of greater caliber than one and one-half inches. In the article ARTILLERY (which see) a description is given of the various kinds and sizes of modern cannons or guns, as well as of the methods of manufacture. See, also, MACHINE GUN.

CANNON, *kan'un,* JOSEPH GURNEY (1836-), an American legislator and politician, except for two intervals of two years each a member of the Federal House of Representatives since 1873, and its Speaker from 1903 to 1911. As Speaker he perfected the system which he had helped Thomas B. Reed to install in 1890, the system which made the Speaker in practice the absolute master of the House. Through this system he exercised an influence on legislation which can with difficulty be over-estimated.

JOSEPH G. CANNON
For more than a generation a leader in the American Congress.

Cannon was for years one of the most picturesque figures in public life. Very tall, thin and angular, yet wiry and seemingly weather-beaten, his was a figure to delight the caricaturist, and in the days of his greatest power his features, more or less exaggerated, were familiar to every newspaper reader in America. He is famous as a wit and a story-teller, yet on occasion he can be most uncommunicative. His humor, his unconventionality, his bluntness and his ability to make personal friends of his bitterest political opponents are well known. On his eightieth birthday, in 1916, the House of Representatives celebrated in his honor, as evidence of the esteem in which he is held by the members. Though he was born in the South, at Guilford, N. C., his home for over half a century has been at Danville, Ill.; there everybody knows him as "Uncle Joe." This title accompanied him to Congress, and as "Uncle Joe" Cannon he is known to millions who have never seen him.

Cannon was admitted to the Illinois bar in 1858, and only three years later was elected to his first public office, state's attorney of Vermilion County, of which Danville is the county seat. This position he filled with ability until 1868. He became a member of the House of Representatives in 1873, and was reëlected without interruption until 1891. During this period he gradually became one of the leading Republicans in that body, and from 1889 to 1891 was chairman of the committee on appropriations. Like many other Republicans he was defeated for reëlection for the term 1891 to 1893, but in the latter year again took his seat. He was again chairman of the committee on appropriations from 1897 to 1903, and was then chosen Speaker.

Cannon perfected the system by which the Speaker controlled legislation, and he ruled, it must be said, with no uncertain hand. He appointed all committees, and he himself was chairman of the committee on rules, which determined the methods of transacting business in the House. A movement to change the rules, and especially to remove the Speaker from the committee on rules, began soon after Cannon became Speaker, and was finally successful in 1910. Thereafter the Speaker was not a member of the committee on rules, and that committee itself is chosen by a caucus of Representatives, not by the Speaker. While it is true that Cannon's conduct as Speaker was chiefly responsible for the revolt, he was merely carrying out a system which he believed to be right. That there was little personal feeling against him was shown by the speedy failure of an attempt, made at the same time, to remove him from the Speaker's chair. When Congress met in 1911 the Democrats were in the majority and elected Champ Clark as Speaker. Cannon failed of reëlection in 1912, and was not in Congress from 1913 to 1915, when he was again elected. W.F.Z.

CANOE, *ka noo',* **AND CANOEING.** A canoe is a long, narrow boat, of light weight, designed to be propelled through the water by paddles. Canoes are made in many sizes and patterns, each of which is best for a certain purpose. A canoe, for example, which will ride easily through river rapids may be too heavy and slow for use in quiet lagoons and lakes. For all ordinary purposes, however, a fourteen-foot canoe, with a two-foot beam, is very serviceable. Its weight should be as little as possible, the average being from forty to sixty pounds. As a general principle the lighter the canoe the more expensive it is. One advantage of a light canoe is that it may be easily carried by one man from one lake or stream across a stretch of land to another watercourse. The best canoes cost as much as $100, or even more; these are strong and large enough to hold a load of 500 or 600 pounds, and if given ordinary care, will last for years. For the boy or girl who wants to canoe as a sport there are cheaper makes, some priced as low as $25.

ON NORTHERN STREAMS

The Art of Canoeing. To the person who has never tried canoeing, it seems the simplest thing in the world. A canoe is sharp both at the bow and stern, and can be propelled backward and forward with ease. It is very light, and responds immediately to the slightest touch of the paddle in water. But this responsiveness also makes canoeing difficult, for a poor paddler will have trouble in keeping his course even in smooth water. A canoe is easily upset, as it is usually flat-bottomed and has no keel.

Special canoes, which are so made that a sail can be hoisted in the bow, often have a detachable or collapsible keel. Even with a keel, however, a canoe requires careful management.

Getting In and Out of a Canoe. The first step in canoeing is to learn how to get in and out. It is always safest to get in over the end; a heavy weight suddenly thrown on the side of a canoe is almost sure to upset it. If it is impossible for any reason to get in at the bow or stern, a step over the sides directly into the bottom is reasonably safe, especially if there is another person already in the canoe to help keep its balance.

How to Paddle. Paddling requires practice and skill, but it is not difficult except in rough water. The best place for the paddler is at the stern, where the force of his stroke not only drives the boat forward but steers it. The paddle may be driven into the water on either side of the canoe, and good paddlers take turns. If there is a second paddler in the canoe, he or she should sit in the bow, and the bow paddle should be on the left side when the stern paddle is on the right. If the paddle is dipped in on the right side, the right hand should grip the paddle lightly near the blade, and guide it into the water. The left hand should hold the handle firmly and should supply the drive. The blade should enter the water with the flat side at right angles to the course of the canoe, and should make little or no splash. As the paddle is drawn back through the water, the blade should approach the stern until finally the flat side of the blade is parallel to the course of the canoe.

When the paddle has reached this point, it acts as a rudder, and a slight turn of the wrist to the right or left will turn the boat in the opposite direction. Steering is perhaps the most difficult part of canoeing, especially in rough water. In smooth water a long steady stroke is easiest, but in rough water a short quick one is usually best; it will keep the canoe in its course, whereas a slow stroke would give the wind time to turn its bow around between strokes.

Distribution of Weight. A canoe offers least resistance to the water if all the weight is in the stern. The bow is then high in the water or even out of water, and the canoe barely skims the surface. This plan is good on a calm day, but if there is any wind blowing, a sudden gust may turn the bow around in spite of everything the paddler in the stern can do. If a heavy sea is running the weight should be evenly distributed, so that the bow and the stern set equally low in the water.

A "DUGOUT"
Eastern American style, hollowed out of a log.

Kinds of Canoes. *Primitive.* The earliest canoes were made by tying together thin strips of wood and stretching a skin over them; the result looked like an umbrella upside down.

The next stage was the use of tree trunks, which were hollowed out by fire or by crude knives. Such canoes, called *dugouts,* are still common in the interior of Africa. The natives of North America have a variety of canoes, including the bone-frame covered with skin,

BRITISH COLUMBIA CEDAR DUGOUT

Some of the early boats of this character were a hundred feet long.

which the Eskimos still use, and the familiar birch-bark canoe. The birch-bark canoe is still used by the poorer Indians in many parts of the United States and Canada, but most Indians use the modern cedar boat, which is stronger, lighter and more easily paddled

Modern. The best modern canoes have cedar frames, covered with a heavy canvas. The ordinary canoe is open or undecked. Sailing canoes usually have a fore- and after-deck, leaving only a well in the middle of the boat for the sailor or canoeist. Such canoes have water-tight compartments in the bow and stern, and collapsible center boards for use when the sail is up. In small canoes the space below decks is wasted, but in larger ones it often provides sleeping-quarters.

Motor Attachments. It is now possible to buy an adjustable motor and propeller which can be fastened to the stern of a canoe or row-boat in a few minutes. Such motor attachments are made under various trade names, but they are all constructed on the principle of the larger electric motors used in regular power boats. The detachable motors weigh from sixty to one hundred pounds, and cost from $50 to $100. W.C.

CANON, *kan'un,* **LAW,** a body of Church law which regulates many of the doctrines and other affairs of the Roman Catholic Church. It is drawn from the opinions of the ancient fathers of the Church, the epistles and bulls of Popes and the decrees of Church councils, to which are added certain maxims of the civil law and the teachings of the Bible.

CANONIZATION, *kan un i za'shun,* in the Roman Catholic Church, an impressive ceremony by which duly-qualified deceased persons are declared saints. One of the necessary qualifications is the candidate's performance of at least two miracles. The power to canonize is vested in the Pope, who conducts a rigid investigation into the former mode of life and genuineness of the miracles attributed to the prospective saint. This examination is a step toward canonization, which takes place many years after beatification, or announcement that the person is one of the "blessed." The object of the further delay is to allow sufficient time to collect additional proof of the fitness of the candidate. This being established, a day (usually the anniversary of the death of the new saint) is set aside on which to honor his memory, his name is placed in the list of the saints, and churches and altars are consecrated to him. G.W.M.

CANORA, *ka no'ra,* a town in Saskatchewan, on the main line of the Canadian Northern Railway and on the Regina-Hudson Bay branch of the Grand Trunk Pacific. It is in the eastern part of the province, about thirty-five miles from the Manitoba boundary, 193 miles northeast of Regina and twenty-six miles north of Yorktown, which is on the Canadian Pacific Railway. The completion of the Hudson Bay Railway will make Canora an important junction point. It is already a center for the shipment of grain. Canora owns an electric light plant and a waterworks system, the cost of the latter being about $100,000. A hospital and a race track are conspicuous features. Population in 1911, 435; in 1916, about 1,200.

CANOVA, *ka no'va,* ANTONIO (1757-1822), one of the greatest Italian sculptors of his time, whose special work was to bring new life to the declining art of sculpture in Italy. At the Academy of Venice he had a brilliant career, and in 1779 he was sent by the senate of Venice to Rome, where he produced his *Theseus Vanquishing the Minotaur.* In 1783 Canova undertook the execution of the tomb of Pope Clement XIV in the Church of the Apostles, a work inferior to his second and perhaps his best public monument, the tomb of Pope Clement XIII in Saint Peter's. *Psyche and Butterfly, Hebe,* the colossal *Hercules Hurling Lichas into the Sea,* the *Pugilists* and the group *Cupid and Psyche* are among his more noted works. He executed in Rome his *Perseus with the Head of Medusa.*

He was summoned to Paris three times by

Napoleon, and in 1812 he completed a colossal marble statue of the emperor, who is represented as a god. Canova's influence on the art of his time gives him a permanent place among great sculptors. When he began his work, Italian sculpture, having departed a long way from the strength, beauty and idealism of Michelangelo, was weak and affected; Canova brought it back to the lofty standards of the Renaissance.

CANTALOUPE, *kan' ta loop,* a favorite variety of *muskmelon* (see MELON).

CANTATA, *kan tah' ta,* a story set to music, sung by a chorus and including solos, duets, trios and quartets. The cantata is a shorter composition than the oratorio or opera. It differs further from the oratorio in that the latter always has a sacred theme; the cantata may be based upon any subject, religious or secular. However, there is little difference between a short oratorio and a religious cantata, and the great church cantatas of Bach are sometimes classed as oratorios. The opera is more elaborate than the cantata, and is further distinguished from it in being presented with scenery and by means of acting.

Dudley Buck (which see) was one of the greatest American composers of cantatas, among his best-known works being *The Golden Legend, Light of Asia, Voyage of Columbus* and the melodious *King Olaf's Christmas.* Lewis Carroll's charming nonsense verses, *The Walrus and the Carpenter,* have been set to music by an English composer, Percy E. Fletcher, forming a cantata that is a favorite with children. One of the world's most charming cantatas is that of *Queen Esther.* See OPERA; ORATORIO; MUSIC.

CANTEEN, *kan teen',* in military life, a place set apart in an army post where the soldiers may buy clothing, food and other necessities, and enjoy gymnasium, reading and recreation privileges. In the United States army the term *post exchange* has superseded *canteen.* Previous to 1901 the post exchange was permitted to sell beer and wine to the troops, but in that year the sale of all intoxicating beverages was abolished by act of Congress. The system of selling liquor to the soldiers through the post exchange is generally known as the *canteen system,* and it went out of existence largely through the opposition of the temperance societies, particularly the W. C. T. U. The advisability of restoring the canteen system is much debated. Those favoring it say that the soldiers, obliged to buy liquor outside the post, obtain an inferior quality, and are furthermore subjected to greater temptations than when permitted to make such purchases in the well-regulated exchange, and that drunkenness and disorder are more likely to occur under the present system. In the British army the canteen is in two departments, the wet and the dry, the former carrying ale, porter and mineral water; the latter, groceries.

In both the United States army and the armies of Europe the name canteen is applied to a flask or bottle, holding about two pints, in which the soldier carries his liquid refreshment while on the march. This is made of metal, leather or wood. In the British army the corresponding vessel is called the water-bottle, while the canteen is a tin vessel that serves the purpose of pan, dish and plate.

CANTERBURY, *kan' ter ber i,* a city in the county of Kent, England, sixty-two miles east of London, famous for its cathedral and for its close association with the country's religious and scholastic development. Since the year

CANTERBURY CATHEDRAL
The seat of the Archbishop of Canterbury, the ruling official of the Church of England.

597, when Saint Augustine came with missionaries from Rome, Canterbury has been the central point of the Church of England. The present cathedral dates from the eleventh century, with additions made 400 years later, and is a magnificent specimen of Gothic architecture. Here was situated the shrine of Thomas à Becket, to which multitudes of pilgrims came annually and which added greatly to the sanctity of the cathedral. Near the site of the shrine is the tomb of Edward the Black Prince, with the helmet, shield and equipment he wore in battle.

There are numerous other churches and educational institutions, including a grammar school founded by Henry VIII. The ancient walls surrounding the city can still be traced, and many relics point to the fact that it was an important military post under the Romans. The trade of the city is chiefly agricultural, and it has tanneries, breweries and other manufactories. Population, 24,889.

Archbishop of Canterbury. The primate and ruling head of the Established Church of England has since the year 599 been styled the Archbishop of Canterbury. The holder of this office has always been chosen for his learning, piety and ability to rule. The most famous archbishops of olden days were Saint Augustine, Lanfranc, Saint Dunstan, Thomas à Becket, Cranmer and Laud. The archbishop is granted great privileges of rank, and from his hands the ruler of Great Britain receives anointment and the crown in Westminster Abbey. The present archbishop, Randall Thomas Davidson, was appointed in 1903.

CANTERBURY TALES, the great production of the first renowned English poet, Geoffrey Chaucer (which see). His plan for this work was most ambitious, and although he died before he had carried it to completion, there remains enough to give it a place as one of the remarkable works of English literature. In the delightful *Prologue,* known to every high-school student, there is described a company of men and women who are about to set out on a pilgrimage to the shrine of Thomas à Becket. These are of every rank of life, and Chaucer showed himself in his descriptions of them a master character-painter. To while away the time on the journey they plan to tell stories, each person to contribute two. These romantic tales, most of which are in verse, are now little read except by students of early English poetry, but the *Prologue* repays study by any person. It is a little difficult to read, for the spelling is different from that of to-day, but editions with notes are many and inexpensive, and the difficulty soon passes with practice. The following quotations, which are very well known, need no interpreter:

He was a veray parfit gentil knight.

And gladly wolde he lerne, and gladly teche.

CANTON, *kan'ton,* the name of a political division in some countries of Europe, derived from the Italian *cantone,* meaning a corner or *angle.* In Switzerland each of the twenty-two states comprising the Swiss republic is known as a canton (see SWITZERLAND). The French canton is a division of the political unit known as *arrondissement,* and is the seat of a justice of the peace. There are 2,908 cantons in France, each consisting of twelve *communes* (see COMMUNE).

CANTON, *can ton',* a very important commercial center of the Orient, a Chinese city of great antiquity, and until the middle of the nineteenth century entirely removed from the influences of modern civilization. Like all Eastern seaports, it is divided into two parts—an old city, distinctly Oriental, harboring all the manners, customs and prejudices of the East, and a modern part, devoted to commerce and forming a neutral territory on which East and West meet to trade. The city is admirably situated on the eastern bank of the Pearl River, eighty miles from the sea, in the province of Kwang-Tung. The ancient part is surrounded by a fortified wall six miles in circumference, with twelve gates for traffic. These gates are shut at night and carefully guarded. Two water gates protect the river east and west of the city. The river is the only home of many thousands of the inhabitants of Canton, who occupy boats and rafts moored to the banks.

According to Western ideas the term "clean" can be applied to very few Asiatic cities, but Canton almost deserves that description. The streets of the old city are narrow, with houses of only two stories, the ground floor generally being a store or warehouse. The assortment of merchandise found in these unpretentious stores is almost endless. The products of every quarter of the globe are here collected, modern inventions from the West lying by the side of Eastern curiosities thousands of years old. The Chinese do not use milk either fresh or in the form of butter or cheese, but grocery stores contain such food as horse, cat and dog flesh, hawks, owls and edible birds' nests.

The new city is cosmopolitan. Chinese merchants have assimilated Western ideas and are expert, courteous and industrious. While the old city retains its pagodas, mosques and temples, the new quarter contains churches, libraries, schools and business premises of the most modern architecture. The officials of the city and province reside in the old town; representatives of foreign countries and their staffs occupy a special quarter of the new city outside the walls.

The industries of Canton are varied, embracing the manufacture of silks, cotton goods,

FLOATING HOMES OF THOUSANDS OF PEOPLE

In certain portions of Canton the river is so crowded with boats which are the permanent homes of many people that a person might walk upon them from one side of the river to the other.

porcelain, glass, paper, sugar, lacquered ware, firecrackers and metal goods. Until the close of the seventeenth century commerce with China was carried on under great difficulties. All foreigners were "barbarians" or "devils," to be kept out of the country at all cost. However, Canton was one of the first ports to admit foreign trade. It restricted the number of merchants to deal with the foreigners, and made those dealers responsible for all customs dues on what they imported. The Chinese War of 1840 nominally threw open the port of Canton, but it was not until 1861 that the port and city became actually open to foreigners. For many years Canton held a monopoly of the tea trade with England, but it suffered the loss of a great portion of this on the opening of other Chinese ports in 1860, and is now exceeded in importance as a tea center by Shanghai and Hongkong. The city is about 7,500 miles from the Pacific coast cities of America. Population, in 1915, about 900,000. F.ST.A.

CAN'TON, Ill., a city of Fulton County, situated in a fertile agricultural district in the western part of the state, twenty-eight miles west of Peoria. Railroad transportation is provided by the Toledo, Peoria & Western and the Chicago, Burlington & Quincy railroads. Canton has coal-mining interests and large manufactories of agricultural implements, cigars and cigar boxes, tile, brick, lumber and foundry products. Its prominent features include a public library, parks and a hospital. The place was settled about 1832 and was first incorporated in 1849. In 1910 it had a population of 10,453; in 1914, 12,438.

CANTON, Ohio, a progressive city, noted for the variety of its industries. It is the county seat of Stark County, and is situated in the northeastern part of the state, on Nimishillen Creek. Massillon is eight miles west, Cleveland is fifty-nine miles north and west and Pittsburgh is 101 miles southeast. The Pennsylvania, the Baltimore & Ohio and the Wheeling & Lake Erie railways meet here and electric lines operate north, east and west from the city. Canton was settled in 1805, was incorporated as a village in 1822 and chartered as a city in 1854. Its rapid growth is shown by the increase of population from 50,217 in 1910 to 59,139 in 1915; the foreign element, in which Austrians predominate, comprises only eight per cent of this number.

The district in which Canton is located is well adapted to growing wheat, corn, oats and fruit, and rich deposits of coal, limestone and pottery clay are found in the vicinity. The output of its 220 industries was valued at $52,000,000 in 1915. One of the largest sheet-metal works in the United States is located here. Pottery, tiles, a variety of bricks, iron bridges, steel products, agricultural implements, watches and enameled ware are among the leading manufactures. Large quantities of coal and grain are shipped from here. Canton has a $100,000 Federal building, a city hall, an auditorium which seats 4,200, a county courthouse, a county workhouse, a $250,000 Y. M. C. A. building and a Carnegie Library with 30,000 volumes. The park reservations cover 168 acres.

Canton was the home of William McKinley, twenty-fourth President of the United States; his home and imposing tomb and two monuments erected to the heroes of the Spanish-American War are features of interest in the city.

CANUTE, *ka nute'* (about 994-1035), a king of England, Denmark and Norway with whose name there is associated a beautiful medieval legend. According to it the great king, wearied of the flattery of his courtiers, who would have him believe they thought him to be divine, determined to give them a lesson. He had his throne placed near the seashore just before time for high water, and as the tide came up he commanded it in stern tones to go back. Then turning to his amazed courtiers he told them to cease to call divine a king whose word had no power over the water at his feet.

Canute was not always such a devout man, for when he became king of England in 1014 on the death of his father, Sweyn, he began his reign by devastating the country and putting out of his way all who opposed him. Not until 1017 was he accepted as king by the whole island, and thereafter there seemed to be a change in his character, for he ruled wisely and humanely, restoring old English customs and giving important posts to English subjects who proved worthy. At the death of his brother in 1018 he gained Denmark, two years later conquered Norway, and in 1031 forced Malcolm of Scotland to acknowledge his supremacy.

CAN'VAS, a strong, coarse cloth, originally made of hemp, as shown by the name, which is derived from the Latin *cannabis*, meaning *hemp*. Ordinary canvas is now made from cotton, though better grades are made of flax. The weight and fineness of canvas depend largely on the purpose for which it is to be used. The best grades, light and thin, are favorite materials for men's and women's summer clothing; these are called by the trade name of *duck*. Duck is usually made of flax, that is, linen, as also are certain heavier grades which are used by artists for paintings. The cloth used by artists varies in thickness according to the size of the picture—the larger the picture the heavier the cloth. Canvas is also used to a considerable extent for awnings, tents, tarpaulins and sails. The sail cloth used on large vessels is usually made of linen of a good quality, as it is subject to hard wear, but smaller sails are most frequently made of cotton or of mixed goods.

CAN'VASBACK, a large fresh-water duck, native of North America, which is highly prized for food on account of its delicate flavor. It breeds in the northern parts of the United States and in Canada, and in the winter migrates to the valley of the Mississippi and the marshy lands of the Atlantic coast. Its food consists chiefly of the roots of the wild celery, and wherever that plant is plentiful, the canvasback, if unmolested, may be found. Its plumage is black, white, chestnut brown and slate color. The *red head duck* is often substituted for the canvasback on the market, owing to similarity of color, especially on the head. The canvasback was being hunted to extinction before the passage of game laws by states and provinces for its protection. See GAME, subhead *Game Laws*; also color plate, in article BIRDS.

CANYON, or **CAÑON,** *kan'yun*, a deep, narrow river valley or gorge with steep sides which in some instances are nearly perpendicular. Canyons are worn to their great depths by the erosion of countless thousands of years. The name comes from the Spanish word *cañon*, meaning *funnel*. Canyons form some of the grandest and most awe-inspiring scenery in the world. The Grand Canyon of the Colorado in Arizona is in some places over a mile deep. Next in size is the Grand Canyon of the Yellowstone, in Yellowstone National Park. This canyon is 1,500 feet deep, and its walls are formed of rocks, colored red, brown, black and gray. The Royal Gorge in Colorado is one of the most remarkable narrow gorges in the world. Its walls are perpendicular for almost its entire length. Many small canyons notable for their beauty occur in the Canadian Rockies.

Related Subjects. Investigation of this topic will not be complete without reference to the following articles:

Erosion
Grand Canyon of the Colorado
Royal Gorge
Yellowstone National Park

CAPE BRETON ISLAND, *bret'un,* or *brit' un,* a rocky, irregularly-shaped island which is part of the province of Nova Scotia, Canada. The island takes its name from a cape at the eastern end, a name which always re-

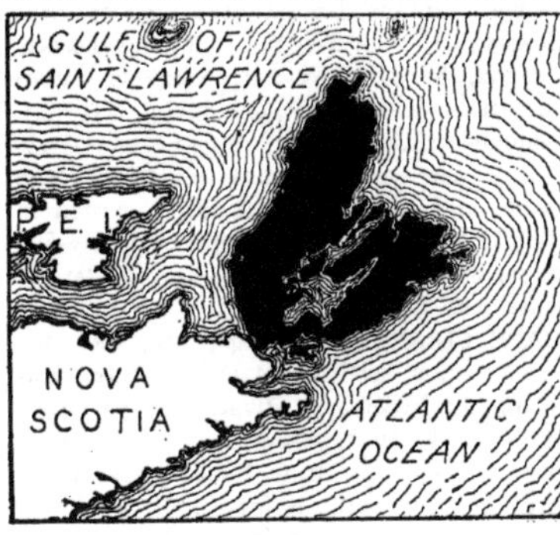

LOCATION MAP

minded many of the good Acadian peasants of their distant home across the seas. Cape Breton Island is geologically a continuation of the Nova Scotia peninsula, and is separated from it only by the narrow Strait of Canso, which connects Saint George's Bay, to the north, with Chedabucto Bay, to the south.

Like the adjoining mainland, Cape Breton Island is noted for its forests, its mines, its fisheries and its picturesque landscapes. The timber, consisting chiefly of pine, hemlock, spruce, birch, oak and maple, is locally valuable, but has long since become an insignificant factor in the Canadian supply. The coal fields, however, are among the most important in the whole Dominion. The important fields are three: the Sydney field, in the northeast, with the splendid harbor of Sydney as its central point; the Inverness field, along the west shore of the island; and the Richmond field, just north of Chedabucto Bay. These fields are the largest contributors to the province's annual output of about 8,000,000 tons, more than half of the present total for the Dominion.

Next in importance to the industries dependent on coal and iron are the fisheries, which employ about 7,000 men and support about one-third of the population of the island. Cod, mackerel, herring and whitefish are the big catches. Compared with New Brunswick and the mainland of Nova Scotia, there is little game fishing, but many tourists are attracted to the island by its picturesque scenery and delightful summer weather. Potatoes, wheat and other grains are raised in small quantities, but most foodstuffs are brought from the mainland.

Like the mainland of Nova Scotia, the island was originally a French possession. The English seized it in 1745, restored it to France three years later, recaptured it in 1758, and finally gained permanent possession in 1763. At first it was a separate colony, but later was united with Nova Scotia.

The island has an area of 3,120 square miles and is practically divided from north to south into two separate parts by the chain of Bras d'Or Lakes and the canal on Saint Peter's Isthmus. It is divided into four districts or counties—Cape Breton, Inverness, Victoria and Richmond. The principal towns are Sydney, having a population of about 20,000, Dominion, Sydney Mines, North Sydney, Glace Bay and Inverness, each of which is described elsewhere in these volumes. Louisburg, once a great fortress and the most important point on the island, is now an incorporated town. The total population of the island, including a few hundred Indians, was 122,084 in 1911. Most of the people are descendants of Scotch Highlanders, but about 15,000 are French Acadians. A.H.M.

CAPE COD CANAL, a sea-level waterway without locks, built across the narrow strip of land where Cape Cod joins the Massachusetts mainland, and opened to navigation in 1914. It connects Buzzard's Bay with Barnstable Bay, an extension of Cape Cod Bay, at Sandwich, Mass., and from shore to shore is eight miles in length. The total length of the channel, however, excavated from a thirty-foot depth in either bay, is thirteen miles. The canal has a uniform depth of twenty-five feet at average low water, and with the exception of one bend runs straight across the Cape. The approaches in both bays are from 250 to 350 feet wide at the bottom. To protect the entrance on the Cape Cod Bay side a massive breakwater 3,000 feet long has been built, and a lighthouse has been constructed on the end of this breakwater. The canal throughout the greater part of the course is 100 feet wide at the bottom, but there are three passing points where it increases to 250 feet. It is illuminated throughout its course by electricity, according to a street-lighting plan, and is crossed by modern railway and highway bridges. The expense of operation is met by tolls.

The Cape Cod Canal means a saving of seventy miles of distance for vessels engaged in the carrying trade between Boston and New York and Southern ports. Previous to the opening of the canal ships carrying an aver-

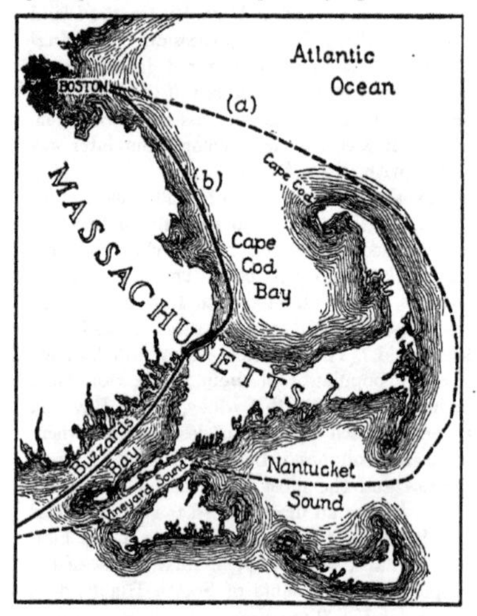

CAPE COD CANAL
(*a*) Route before opening of canal; (*b*) present route, through canal.

age of 25,000,000 tons of freight a year were forced to make the trip around the Cape by way of Nantucket Sound, a distance of 334 miles, as against 264 miles through the canal. The former route is difficult and dangerous because of the shoals in the Sound, the frequent fogs that obscure the signals from the lighthouse and the reefs on the shores of the Cape. Between 1843 and 1903 over 2,100 vessels were wrecked in the region of the Nantucket shoals, and about 700 persons perished. The construction of the canal has therefore meant a saving of lives and property as well as of time and distance.

The agitation for such a canal is not a recent movement, for the project was considered as early as 1697. During the War of the American Revolution General Washington, finding it impossible to send his troops to New York over a sea route, said, "The interior barrier (across Cape Cod) should be cut in order to give greater security to navigation and against the enemy." The route was first surveyed in 1791, and several companies were formed to build the canal before its construction was made a possibility by the decision of Mr. August Belmont to finance the undertaking. Work was begun in 1909, and the canal was completed in 1914, the cost of construction amounting to $12,000,000. Though it is possible that the United States government will some day take over the canal and increase its depth, the waterway is even now of importance from a military standpoint. It can accommodate the smaller cruisers of the navy, most of the auxiliary ships and the entire torpedo and submarine fleet, as it existed in 1917. If increased to a depth of thirty-five feet its draft would be ample for the largest battleships. See CANAL. B.M.W.

CAPE GIRARDEAU, *je rahr' doh,* Mo., one of the earliest settlements of the state, situated on the Mississippi River, in Cape Girardeau County, on the southeastern border of the state. By rail Saint Louis is 131 miles northwest; Cairo, Ill., is fifty miles southeast, and Memphis is 170 miles south and west. The Frisco Line, built to the city in 1902, and the Cape Girardeau & Northern Railway, constructed in 1905, serve the city. Several boat lines connect with the important river points, and an electric line extends to Jackson, Mo. In 1910 the population was 8,475, and by 1914 it had increased to 10,033. The area is two and one-half square miles.

Though Cape Girardeau is principally a manufacturing city, it derives a large revenue from the fertile, well-cultivated district in which it is situated, and its good river and railway transportation facilities make it a shipping point of considerable importance. Good timber, especially cypress, abounds in the vicinity; timber, lime, limestone, flour and mineral paints are the leading articles of commerce. Shoes lead in the manufactures of the city, 770 people being employed in the industry. Other important manufactured products are ice cream, cement, tobacco products, interior finish, pressed brick and a variety of lumber products.

The notable buildings are a $100,000 Federal building constructed in 1910, the courthouse, a $200,000 office building constructed in 1908, and Saint Francis Hospital, completed in 1915 at a cost of $195,000. For its size the city has exceptional educational advantages. Besides its public schools it has one of Missouri's five state normal schools, with six buildings and a campus of fifty-five acres (see MISSOURI, subhead *Education*), the Loretto Convent, a library, Saint Vincent's College for men and Saint Vincent's Convent, the last two being the oldest Roman Catholic schools west of the Mississippi River.

Cape Girardeau was settled by the French in 1765, and named for Ensign Girardot, commander of the first military post on the Mississippi River. This was the first settlement of English-speaking people in the upper territory of what was afterwards known as the Louisiana Purchase. The home of Don Louis Lorimier, the Spanish commandant who issued many grants, is still standing. F.J.M.

CAPE OF GOOD HOPE, PROVINCE OF THE, the oldest of the federated provinces now forming the Union of South Africa, occupying the southern extremity of the African continent. In 1910, when the Union was formed,

LOCATION MAP
Area in black shows the Province of the Cape of Good Hope and Natal.
(*a*) German Southwest Africa, captured by the British in 1915
(*b*) Bechuanaland Protectorate
(*c*) Transvaal
(*d*) Orange Free State
(*e*) Basutoland

the province was known as Cape Colony. It is bounded on all sides by British possessions, consisting of Natal, Orange Free State, Bechuanaland, the Transvaal and German Southwest Africa, the latter annexed in 1915, after its capture by the British in the War of the Nations. The area of the province is about 277,000 square miles, or 12,000 miles larger than the area of Texas and slightly more than that of the great Canadian province of Alberta.

The surface of the country consists for the most part of undulating prairie, or *veldt*, as it is called, with many wide stretches of flat, barren desert. In parts, however, there are mountainous regions, the ground rising in plateaus from the sea until it culminates in the Drakensberg range of mountains. For a considerable part of the year the highest peaks of these mountains, which attain a height of nearly 10,000 feet, are covered by snow. The coast is rugged and bare, with few natural harbors. Algoa Bay, Mossel Bay, Table Bay and False Bay are the most important indentations on the south and west shores. On the east the coast line is almost unbroken. The only important river is the Orange River, which forms part of the northern boundary of the province (see ORANGE RIVER). There are numerous small streams, many of which are dry in the summer.

Climate. The climate of the Cape of Good Hope is exceptionally pleasant and healthful. Rains are more abundant in the east than in the western portions, which are sandy and arid. The variation in temperature is slight. In the higher inland regions the nights are cold, frost is frequent, and snow falls occasionally, but from sunrise to sunset the heat is intense. The average annual temperature in Cape Town, the capital, is over 62° F. Darkness succeeds sunset very rapidly, the change in temperature at that time being more marked than at any other. The benefit of the general climate to those suffering from pulmonary complaints is becoming more apparent.

Agriculture. Like all South African countries, the Cape of Good Hope suffers from lack of water. The soil is fertile and will produce large crops of cereals, fruit and flowers when water is plentiful. So unreliable is the water supply, however, that agriculture is not as important an industry as it would be under more favorable circumstances. Corn, called *mealies* throughout South Africa, is extensively grown, and in a good season will yield 100 bags for every bag sown. Fruit is grown near cities and wherever irrigation is possible. Grapes are grown in many parts and the wine of the country was formerly famous, but careless cultivation and lack of attention to the vines have reduced the quality of the grapes and the wine produced is now of poor quality.

Sheep, goats and cattle are the principal sources of wealth. Vast flocks of Angora goats are kept by natives and Boer farmers, and produce mohair of very fine quality. The Cape of Good Hope is one of the world's most important wool-exporting countries, the value of the wool crop being exceeded only by the export of diamonds.

Mineral Wealth. Although the diamond is the most important source of wealth, it by no means represents the country's only mineral resource. Coal is extensively mined, and gold, iron, lead and copper are found in quantities sufficient to justify mining. The Cape of Good Hope is the most famous diamond country in the world and exports stones to the value of over $30,000,000 every year. Kimberley, in the district of Griqualand West, is the center of the diamond industry, controlled by the De Beer's Company. In less than forty years the mines of Kimberley have produced diamonds to the value of over $500,000,000. This was the result of the accidental finding of a diamond in the bank of the Vaal River in 1867 by a Boer farmer who did not even know that diamonds were valuable.

Inhabitants. The total population of the province was 2,564,965 in 1910. Of these 19,763 were Malays, 415,282 of mixed races, 582,377 Europeans and the remainder Kaffirs, Hottentots, Bechuanas, Zulus and Basutos. The original inhabitants are generally supposed to have been Bushmen, small numbers of whom still exist. They are small, almost pygmies, living by hunting and with no knowledge of agriculture. They were conquered by more warlike tribes from the north, belonging to Bantu stock. The colored races have freely intermingled and have proved quicker and more intelligent in adopting civilized pursuits than the more northerly tribes. The white inhabitants are chiefly Dutch and English. See BUSHMEN.

Ostrich Farming. An industry that is steadily growing in importance is ostrich farming. This is a source of considerable revenue, as hundreds of thousands of birds are owned, and feathers to the value of more than $6,000,000 are annually exported, under normal conditions in the world. The ostrich farms are mostly in the southwestern sections. Raising alfalfa is an important part of ostrich farming and has proved peculiarly adapted to the soil and climate.

Transportation. There are no navigable rivers in South Africa, and the question of transportation has always been one of great importance. All the larger towns are now connected by railroads, and Cape Town is the terminus of the Cape-to-Cairo Railway, which runs through the province from north to south. There are many districts, however, which are far from the railroads, the nearest market being distant several days' journey by ox wagon. The roads have been greatly improved during recent years, but travel in the rainy season becomes difficult. The chief vehicles used are light, two-wheeled, hooded carriages called Cape carts, and the heavy farm wagons are drawn by spans of six or eight mules or sixteen oxen. In calculating distances in the country the time taken for the journey is usually stated, six miles an hour being the usual rate of travel; thus a place thirty-six miles away would be "six hours" distant.

Cities. Cape Town, the capital, a beautiful city at the foot of Table Bay, is the most important port of South Africa, with excellent docks and railway facilities. It had in 1910 a population of 149,461, including suburbs. Port Elizabeth, with 30,688, is the center of the ostrich-feather industries. Kimberley, the world's diamond market, had 29,525 inhabitants in 1911. East London, an important and thriving port, has a population of about 21,000. Other towns include Graaff-Reinet, Cradock, Worcester, Oudtshoorn, Uitenhage, Beaconsfield, Paarl and King William's Town. The more important cities are described elsewhere in these volumes.

Animal Life. The settling of the country has made such changes that animals that were plentiful only a few years ago have now nearly disappeared. Lions are occasionally seen in the extreme north, and the hippopotamus is found in some parts of the Orange River country. Antelopes, hyenas, jackals, baboons and monkeys abound. There is also a wild dog, known as the Cape hunting dog, which, however, is fast being exterminated.

History. Towards the end of the fifteenth century Vasco da Gama opened up a sea route to India by sailing round the Cape of Good Hope. Although many vessels sought shelter in South African waters, no attempt was made to occupy any territory on that continent for nearly two centuries. The first settlers were Dutch, driven out of their own land in search of religious freedom. They were followed by French Huguenots, and from 1652 to 1806 the colony thrived under Dutch rule. In 1806, however, the British seized the territory and declared the Dutch, or Boers, as they were called, subjects of the British government. In time this led to great disaffection, and the Boers, rather than submit to the British, migrated, or "trekked," in large numbers and formed the colonies of the Orange Free State and the Transvaal.

During the South African War (1899-1902) the Cape of Good Hope was invaded by Boers from the northern provinces, and many of the burghers rose in sympathy with them. When peace was declared the burghers of the Cape of Good Hope who had taken up arms against the British were punished by disfranchisement and fines. Dr. Jameson, leader of the Jameson Raid, was made premier of the province in 1904, and was instrumental in bringing about the union of the South African provinces, which were finally proclaimed as the Union of South Africa in 1910. E.D.F.

Related Subjects. In connection with the above, the reader is referred to the following articles:

Boer	Jameson, Leander Starr
Bushmen	Ostrich
Diamond	South African War
Gama, Vasco da	Union of South Africa

CAPER, *ka'per*, the unopened flower bud of the caper bush, which, pickled in vinegar and salt, is widely used in making a sauce for meats, especially boiled mutton. The sharp, biting taste of the caper is not usually pleasing when first tried, but, like that of the olive, becomes palatable on further acquaintance. Physicians advise against the use of such strong condiments, for they are apt to destroy the taste for wholesome food.

CAPER
Flower, bud and leaves.

The caper bush is a low, trailing shrub, native to the countries bordering on the Mediterranean Sea, where it climbs over walls and in and out of crevices. It is grown from cuttings in greenhouses in Great Britain, the Northern United States and Canada, and from seed in the Southern United States. The plant begins to bloom early in the summer, and the flowers appear until winter. The buds are gathered each morning, placed at once in salt and vinegar and later put on the market, the finer specimens in bottles, the coarser in small barrels. The flower buds of the marsh marigold and nasturtium are often pickled and eaten as a substitute for capers, but the practice is not to be commended.

CAPETIAN, *ka pe'shan,* **DYNASTY,** a famous line of French kings, which took its name from the first of the house, Hugh Capet. These kings ruled from 987 to 1328, and in nearly every case throughout this period son followed father in regular succession. Under the Capetians the French monarchy grew steadily in power and influence, and the greater part of the English domain in France was recovered from England and added to the French possessions. This was also the period of the Crusades, three Capetian kings being themselves at the head of crusading armies. The most important political event of the period was the admission of representatives of the common people, known as the Third Estate, to the National Assembly, which occurred in 1302 during the reign of Philip the Fair. The last Capetian king, Charles IV, was succeeded in 1328 by Philip VI, of the House of Valois.

CAPE-TO-CAIRO, *ki'ro,* **RAILWAY,** a trunk line of railway under construction through the entire continent of Africa, from Cairo and Alexandria to Cape Town. The project was first considered by Cecil Rhodes with a view to opening up the vast resources of the continent and was first undertaken as a private enterprise. It was intended to push the line from both ends, running south from Cairo and north from Cape Town, the two ends to meet somewhere in the vicinity of the great Central African lakes. The total length when finished will be about 5,700 miles, making it the longest continuous line in the world.

In 1889 the first rails were laid at Kimberley, that town already having rail connection with Cape Town. Through Rhodesia the roadbed was quickly constructed, little ballasting being necessary. At one time the construction was so rapid that 492 miles of track were laid in 500 working days.

Difficulties Experienced. It was not until the banks of the Zambezi River were approached that serious engineering problems presented themselves. The crossing of the Zambezi River was made by means of a cantilever bridge with a clear span of 500 feet across the gorge immediately below the Victoria Falls. This bridge, the highest in the world, 420 feet above the water, was completed

HIGHEST BRIDGE IN THE WORLD
On the Cape-to-Cairo Railway, over the Zambezi River.

in eighteen months. It was built out from both banks and joined in the middle, a strong net being suspended beneath it to catch any workman who might accidentally fall.

Farther north difficulties increased, and the fever-stricken districts through which the line passed rendered the task of construction one of great difficulty and danger. White men were necessary to superintend the labor, which was chiefly that of unskilled natives. Large rewards were offered to competent men who would brave the climate, but the response was discouraging. Hundreds of healthy men, after a few months in the jungles, returned to civilization with health permanently impaired. Still the work was pushed on and plans were matured for its completion.

During the early stages of construction the natives resented the appearance of the railroad, and clashes between various tribes and the laborers were frequent. Gradually, however, the natives were persuaded that the railroad was not intended to work them injury. Toleration took the place of opposition, and finally great numbers volunteered to work.

The Northern Portion. The Egyptian government undertook the construction of the line from Cairo southward. In 1896 work was projected southward from Wadi Halfa, the point to which the line had been brought by the khedive of Egypt. About 1,400 miles of track have been laid. At Atbara the line crosses the River Nile by means of a steel bridge 1,000 feet in length. The material for this bridge was made in the United States in less than a year, no English firm being willing to guarantee delivery in less than two years.

Object of the Railway. When the line is completed it will form the main artery of continental communication, with branches extending east and west. Thus it is probable that the products of the interior will find ready outlet to the coast, colonization will be accelerated and regions now uninhabitable will be turned into productive, healthful provinces. Cecil Rhodes dreamed of a British South

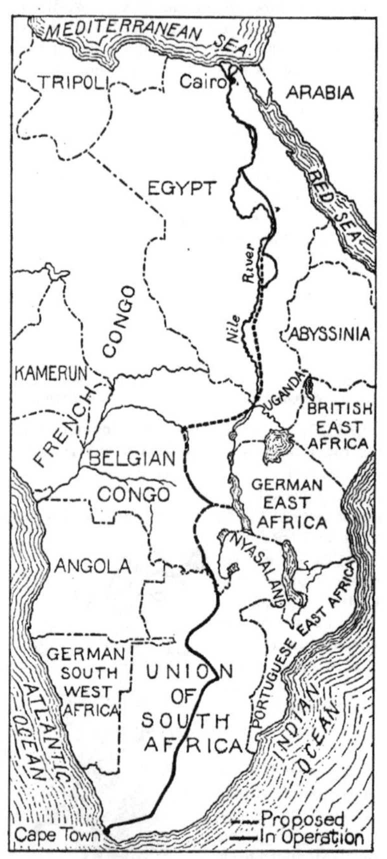

CAPE-TO-CAIRO RAILROAD

Africa which should form an integral part of the Empire; he also hoped that some day a line from the Cape to Cairo would run through none but British territory. That part of Africa which is already settled forms a very small portion of the whole continent, and the railroad will undoubtedly open up territory capable of absorbing a great proportion of the surplus population of European countries.

Other Facts of Interest. No final estimate of the cost is yet possible. In some parts the cost of actual construction is very slight, while in others it is abnormally heavy. All material has to be conveyed great distances. To feed the workmen is a task of tremendous difficulty, and cost of medical attention is also one of the large items of expense. In Rhodesia the cost of construction was about $22,500 per mile; north of the Zambesi it was in many places considerably more. Just after the bridge over the Zambesi was completed $40,000,000 had been expended on the southern portion; a similar sum had probably been expended on the northern end up to 1914, when the War of the Nations brought work to a halt, largely because of the demand for money in Europe. The time taken for completion of the railroad is considerably in excess of that first estimated, but it is practically assured that by the year 1925 the Cape of Good Hope will be connected with the Mediterranean Sea by rail.

The trunk line and most of the branches are now in the hands of the governments of Egypt and the Union of South Africa. A portion of the route is now covered by water and there remains only a gap of 600 miles not connected by combined rail and water travel. The all-rail route is intended entirely to eliminate water transportation, with its necessary transshipments. F.ST.A.

CAPE TOWN, the capital of the Cape of Good Hope Province, in the Union of South Africa. It is beautifully situated on the lower slopes of Table Mountain, overlooking Table Bay, on the South Atlantic Ocean, less than a hundred miles northwest of Cape Agulhas, the southernmost point of the African continent. It is a city of many attractions; the houses and public buildings are in modern European style, with the exception of a few typical old Dutch dwellings, reminders of the days before the great "trek," when the Dutch emigrated farther north to establish themselves independently. The streets are gay with a mass of color, Malays, Hindus, Zulus, Kaffirs and whites mingling in the crowded thoroughfares. The city is well illuminated, has electric street cars, telegraph and telephone service and good railroad connections with the other provinces of the Union. The harbor is excellent and is protected by a breakwater over 4,000 feet long, and the docks cover an area of over sixteen acres. Its commerce is very extensive, as it is one of the principal gateways to South Africa and is the depot of supplies for a vast territory.

The city has numerous parks and gardens noted for their beautiful flowers and trees and many fine buildings, among which are the Houses of Parliament, the Supreme Court, the Museum and a cathedral and some remarkable mosques. The Cape Observatory, one of the most important astronomical institutions in the world, is situated here. The climate is healthful, though dust storms are frequent and sometimes serious. The suburbs are noted for their beauty, and the surrounding country is fertile and rich in natural resources. Population, excluding suburbs, about 67,000, of whom 29,933 are Europeans.

CAPE VERDE, *vurd,* **ISLANDS,** a group of ten volcanic islands and four islets in the Atlantic Ocean, west of Africa, belonging to Portugal. They are named after Cape Verde, or Green Cape, on the African coast, 300 miles distant. Their area is 1,480 square miles. The islands produce rice, maize, coffee, tobacco, sugar cane, nuts and various fruits. Most of the inhabitants are negroes or of mixed race. The chief town is Praia, a seaport on São Thiago, or Santiago, the largest island. Porto Grande, on São Vicente, is a coaling station for steamers and has the best harbor in the group. Until 1854 slavery was a settled institution in the islands, and it was not until 1876 that all slaves were declared free. Population in 1910, 150,000.

CAPILLARIES, *kap'i la riz.* When you prick through your skin with the point of even the finest needle, blood oozes out. This is because the needle has entered a minute blood vessel. These small blood vessels are the capillaries, and they connect the arteries with the veins. Some of them are so small that only one blood corpuscle at a time can pass through. They are largest in the marrow of the bones and smallest in the brain. The capillaries are distributed all through the body and are so numerous in the skin that they form a complete network; you cannot penetrate the skin anywhere without opening a capillary.

The walls of the capillaries are very thin, and the nutritive material in the blood passes through them into the tissues. The waste matter is also absorbed through them into the blood. In the capillaries of the lungs the blood receives oxygen and gives off carbonic acid gas. See BLOOD; CIRCULATION OF THE BLOOD; ARTERIES; VEINS. W.A.E.

CAPILLARITY, *kap i lar' i ti.* If you drop ink on your paper, and hold the corner of a piece of blotting paper in it the ink will rise into the blotter, or if you hold a piece of crayon over it the crayon will absorb the ink.

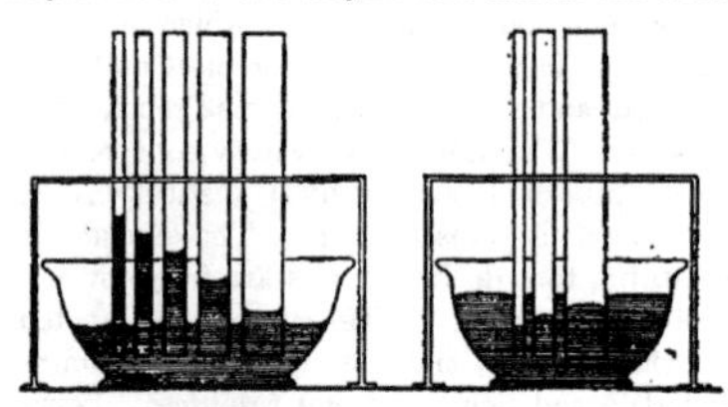

FIG. 1 FIG. 2

Again, if you take a number of straws of different sizes and place them in a glass of water, the water will rise in the straws above the surface of the water in the glass (see Fig. 1). If a little red ink is added to the water, its height in the straws is more easily seen. This tendency of liquids to rise in small tubes and into the hairlike pores of porous bodies is called *capillarity;* the word is from *capillary,* which means *hairlike.*

There are many illustrations of capillarity in nature. By it water is drawn up into a plant through the tiny rootlets; wood absorbs moisture and thereby swells; and the capillaries of the blood vessels carry the blood to every part of the body. We also make use of capillarity in various ways. It is capillarity that draws the oil up in lamp wicks; it causes water to enter the pores of sugar and other substances which it dissolves.

But this sort of capillarity exists only between liquids and solids which have attraction for each other. Among those which repel the action is directly opposite. If a small glass tube is placed in an upright position in a glass of mercury, the mercury in the tube will be lower than that in the glass, and its surface will be convex (see Fig. 2). A greased rod or tube placed in water will affeet the water in the same way. Any boy or girl can perform many interesting experiments to illustrate capillarity.

CAP'ITAL, as usually defined, is that part of the accumulated wealth of an individual or community that is available for the further production of wealth. Capital, therefore, is not necessarily money, although it may be represented by money. As popularly used, it means only the wealth which is used in business. A manufacturer, for example, does not regard his home as capital, but if he owns his factory building he considers this as an important part of his capital. So, too, his stock of raw materials is capital, but the electric automobile which is used by his wife is not capital.

Theories of Capital. The business man, therefore, regards as capital only the wealth which yields him an income. This is slightly different from the view of the classical or orthodox economist. The classical economists, including John Stuart Mill, regard as capital only the products of past industry which are themselves used as a means of further production. This definition excludes such things as lands, which are used in production but are not themselves the products of labor.

Many modern economists defend this latter definition of capital on the grounds of social policy. Capital which is the result of past industry is a saving and represents a sacrifice; some producer has refrained from using up his product. Other producers create things merely for the enjoyment these give, and use them up. The man who saves and sacrifices should receive a reward, which is *interest.* If a man receives interest or profit on land or some commodity whose production involves no sacrifice on his part, he receives a profit which he has not earned. This is the argument which forms the basis of the single tax (which see).

The Socialist has still another theory of capital. He defines capital as any form of wealth which yields an income without any exertion on the part of its possessor. The tools of a carpenter or a shoemaker, in this sense, are not capital. Capital, the Socialist argues, is a monopoly of the essential means of production; the laborer, who creates wealth, instead of receiving the profit of his labor merely receives a living wage. The Socialist believes that capital will finally have universal control of production; when this stage is reached, he asserts, there will be a revolution, either sudden or gradual, by which privately-owned capital will be taken over by the community and cease to be the basis for a claim to a share in the wealth produced by the workers.

Fixed and Circulating Capital. The best-known division of capital is into two classes, *fixed* and *circulating.* Fixed capital includes all forms which are capable of use more or less indefinitely, such as buildings, machinery and tools. Circulating capital, on the other hand, includes those forms which are used up in a single operation; the raw materials of manufacture, for example, can be used only once before they are transformed or absorbed into some finished product. This distinction was once important, because economists of the eighteenth century and the early nineteenth century believed that wages of laborers were paid out of circulating capital (see WAGES, for an explanation of the Wages Fund theory). Recent economists are inclined to pay little attention to this classification, but divide capital into active and passive classes. Active capital, such as machines, gives or creates utility; passive capital, such as raw material, has utility imparted to it.

The accumulation and distribution of capital present one of the greatest problems of economic science. Nearly all economic reform, including public ownership of great public utilities and regulation of trusts, have as their object the control, in whole or in part, of the means of production.

Since inventive genius is busy all the time devising new agencies for aiding production, the mass of capital in relation to a given volume of product or in relation to a given force of factory workers continually grows, the rôle of capital in industry becomes greater and greater, the share of the product deducted on account of the ownership of such capital naturally keeps pace with the growth of the means of production, and the question as to the adjustment of interest between workers and owners of capital thrusts itself into the foreground as a vital social question. E.A.R.

Related Subjects. Closely associated with this topic are the following, to which the reader is directed

Interest	Stock, Capital
Rent	Trusts
Single Tax	Wages
Socialism	

CAPITAL (in architecture). See ARCHITECTURE; also COLUMN, for illustration.

CAPITAL PUNISHMENT, the severest punishment that can be prescribed by the courts—that of death. The word *capital* means *pertaining to the head,* and was applied to criminal law from the fact that beheading was the early form of execution. The manner of inflicting the death penalty has greatly varied; the methods of olden times were often barbarous and cruel, but public sentiment in modern days discountenances any but the most humane means for the infliction of this penalty.

Capital Punishment Offenses. In the United States capital punishment is inflicted for murder and treason, and in military law for sedition, violence, gross neglect of duty, desertion, or disobedience in vital cases to lawful commands. Each state or province has jurisdiction over its own territory, and therefore the laws governing crime differ. The court before which the trial is held pronounces the death sentence, but in many jurisdictions the laws require a warrant from the chief executive before the execution may take place.

In Canada the death sentence may be imposed for treason, murder, rape, piracy with violence and "upon subjects of a friendly power who levy war on the king in Canada."

How Penalties Are Enforced. Sentence of death is executed by hanging in most parts of the United States, Canada, Great Britain, Scotland, Ireland, Austria and Japan. In France and Germany the sword and the guillotine are the usual means; strangulation by the garrote is employed in Spain.

Electrocution. The method by electrocution was first adopted by the state of New York, becoming effective on January 1, 1889. Since that time the states of New Jersey, Pennsylvania, Massachusetts, Ohio, North Carolina, South Carolina, Kentucky, Virginia, Arkansas and Vermont have adopted this method.

Where Death Penalty Is Abolished. Among the states of the Union, Michigan abolished the death penalty in 1846 except for treason, and Wisconsin wholly abolished it in 1853. Maine abolished it for the second time in 1887, and it has also been discontinued in Rhode Island, Minnesota, Kansas, North Dakota, South Dakota and Oregon, imprisonment for life being substituted as a penalty for murder in the first degree. Alaska abolished the death penalty in 1915, and Tennessee in 1916. In the European countries Belgium has had no legal execution since 1863. Switzerland abolished it in 1874 and though the right of restoring it was given to each canton, or state, only seven out of the twenty-two exercised this power. Rumania abolished it in 1864, Holland in 1870, and Portugal has likewise discontinued it. Since 1868 in England all executions are required by law to take place privately, and this plan has also been adopted widely in the United States and Canada. A.E.

CAPITALS OF THE UNITED STATES. The national legislative body of the United States, since the Declaration of Independence, has met in session and passed laws in nine different cities. The Continental Congress, which passed the famous resolution of Richard Henry

Lee and adopted the subsequent Declaration, sat in Philadelphia from September 5, 1774, to December, 1776.

During the course of the Revolution the dangerous proximity of the British troops made necessary frequent changes of headquarters for the national governing body, and Congress met in the following cities:

Baltimore, Dec. 20, 1776, to Mar., 1777.
Philadelphia, Mar. 4, 1777, to Sept., 1777.
Lancaster, Pa., Sept. 27, 1777, to Sept. 30, 1777.
York, Pa., Sept. 30, 1777, to July, 1778.
Philadelphia, July 2, 1778, to June 30, 1783.
Princeton, N. J., June 30, 1783, to Nov. 20, 1783.
Annapolis, Md., Nov. 26, 1783, to Nov. 30, 1784.
Trenton, N. J., Nov. 30, 1784, to Jan., 1785.
New York, Jan. 11, 1785, to June, 1790.

For the next ten years Philadelphia was the seat of government, but that city was regarded only as a temporary location, as the Constitution, drafted in 1787, authorized Congress to exercise exclusive legislation over some district, not exceeding 100 square miles, which should be ceded by particular states and be acceptable to Congress as the seat of government. President Washington favored a site for the capital city somewhere on the Potomac, and he was authorized by Congress to select a site. Maryland made a cession of 69¼ square miles, and Virginia one of 30¾ square miles, on opposite sides of the river (see DISTRICT OF COLUMBIA). The city of Washington, which was laid out on the Maryland side of the Potomac, became officially the national capital in 1800. The Virginia cession was subsequently returned to that state. B.M.W.

CAPITOL. Historically the name *capitol* is applied to the Capitoline Mount, the smallest and most famous of the seven hills on which Rome was built; sometimes the name is applied to the great temple of Jupiter, which stood on the southern of the two summits of the Capitoline. The northern summit was the site of the citadel of the ancient city. The temple to Jupiter, in which was carried on the worship of three great deities, Jupiter, Juno and Minerva, was the center of the state religion. The edifice was begun by one of the legendary kings of Rome, and was later destroyed and rebuilt again and again, each time with greater magnificence. The final structure, that of Domitian, lasted until the tenth century. Other important structures on the Capitoline Hill were the temple of Jupiter Tonans (Jupiter of Thunder), erected by the Emperor Augustus; and the Tabularium, a magnificent library built in 73 B. C., for the housing of the public records. See ROME.

The modern application of the word relates to a pretentious building in the seat of government of a country, province or state, from which the government administers the laws. The *city* is thus the *capital city;* the *building* is the *capitol.*

CAPRI, *kah pre',* known in ancient times as *Capreae,* is a beautiful island belonging to Italy, in the Bay of Naples, celebrated for its delightful climate and invigorating air. It yearly attracts over 30,000 tourists. Capri has an area of five and three-fourths square miles, and its highest elevation, Mount Solaro, rises 1,920 feet above the sea. The islanders produce white and red wine, oil and fruits, and also engage in fishing. The towns are the capital, Capri, in the eastern part, and Anacapri, in the west, built on a plateau 880 feet above the sea and reached by a winding road cut through the rock. Population of the island, about 6,800.

CAPRICORNUS, *kap ri kor' nus,* meaning *the goat,* is applied to a constellation of the southern hemisphere and the tenth sign of the zodiac, marking the winter solstice, about December 21. The symbol of Capricornus is ♑, representing the horns of the animal, and it generally appears in art as a figure with the forepart of a goat and the hindpart of a fish. According to tradition, the goat represents the god Pan, who had the head of a man with the legs of a goat. By some it is supposed to represent Amalthea, the goat who suckled the infant Hercules. The form *Capricorn* is the name of the southern tropic. See ZODIAC; TROPICS.

CAPSICUM, *kap' si kum,* an annual shrubby plant with wheel-shaped flowers and many-seeded berries. Although chiefly natives of the East and West Indies, China, Brazil and Egypt, these plants have spread to other tropical or subtropical countries. Capsicum is cultivated for its fleshy fruit, which at times reaches the size of an orange; it is variously colored and very sharp to the taste. The fruit or pod

CAPSICUM

is used for pickles and sauces. Some kinds are especially cultivated for medicinal uses, mainly as a counter-irritant. Dried or pulverized, these fruits are used in cases of neuralgia and rheumatism and as an ingredient of a gargle in extreme cases of sore throat.

CAPTAIN, *kap' tin,* the title given to a commissioned officer in the army and navy, taking rank next above a lieutenant. In the army a captain commands a company and is assisted by first and second lieutenants. In the navy he commands a battleship. The naval officer takes rank one step above military officers bearing the same title; hence a naval lieutenant is equal in rank to a military captain and a naval captain is of the same rank as a military colonel.

The pay of a captain varies in different countries. In the United States the military captain's pay is from $1,800 to $2,800 per year, depending on length of service; a naval captain receives $4,500. A British army captain draws $1,029; French, $676; German, $1,096; Russian, $350; Austrian, $600; Italian, $760. Naval captains draw pay as follows: England, $2,443; Germany, $1,631 to $2,231; France, $1,902 to $2,296; Russia, $2,871; Italy, $1,621. The pay of captains in Canadian forces is considerably higher than that of officers in similar rank in the permanent British forces, and is based on practically the same scale as that of United States officers.

CAPUCHINS, *kap' u chinz,* or *kap u sheenz',* an order of friars, founded in 1525, which is a branch of the Order of Saint Francis, the name being taken from the pointed hood, or *capouch,* which is a part of their habit. They are clothed in brown or gray, and wear sandals instead of shoes. The rules of the Order prescribe that the members shall live by begging, and that no gold, silver or silk shall be used about their altars. The Capuchins are characterized by their devout piety and simple sermons, and, with the Jesuits, were the most effective preachers and missionaries of the Church in the sixteenth century. They are now most numerous in Austria. In the United States they have monasteries in the dioceses of Milwaukee and Green Bay, Wis., in New York City and in Leavenworth, Kan. See FRIAR. G.W.M.

CARABAO, *kah rah bah' o,* the name given to a variety of the Asiatic water buffalo found in the Philippine Islands, where it is domesticated and highly valued as a beast of burden. It is very slow in its movements and will not work during the heat of the day, but will travel through bogs and marshes through which no other animal could pass. Like its Indian relative, it loves the water, and is a

THE CARABAO AT WORK

good swimmer. In its wild state it is fearless, and if wounded becomes dangerous, charging with great speed and ferocity. It is a little smaller than the Indian water buffalo, which is often six feet high at the shoulder and has a spread of horns exceeding its height. In color the carabao is a slaty bluish-black, and it becomes almost hairless when aged.

Strange as it may appear to Western people, the carabao dislikes white people, for it never has become accustomed to their odor; but it exhibits something bordering on affection for its dirty, often ill-smelling, native master. See BUFFALO, for illustration of water buffalo.

CARACAS, *kah rah' kahs,* the principal city and the capital of the Federal republic of Venezuela and also of the Federal district of Caracas, faces the north bank of the River Guayra, in latitude 10° 30′ north, about 3,000 feet above sea level. Its port is La Guayra, on the Caribbean Sea, six miles distant. This port, by way of Saint Thomas, D. W. I., is 4,250 miles from London, about 2,070 miles from New Orleans and 1,435 miles from New York. The population of Caracas in 1915 was about 85,000.

Two principal avenues, crossing at right angles, divide the city into quarters, in all of which the streets are numbered on a model plan, rendered practicable by their symmetry. The streets are narrow, for the purpose of securing shade; and the buildings are not tall, for fear of earthquakes. Street-car lines, electric lights and power, adequate water supply, good sanitation, telephone service, social advantages, literary activity and educational facilities render the city desirable for residence, though it is not, in a large sense, a trading or manufacturing place. There are numerous public squares, or little parks, with gardens and statuary. The Plaza de Bolivar and the park Independencia (on the western hill) are prominent attractions. Among the public buildings are the capitol, the university, the cathedral, the national library, the opera house, the Masonic Temple, the Pantheon and the archbishop's palace.

Caracas was founded in 1567, and named Santiago de Leon de Caracas. It was almost destroyed by an earthquake in 1812. The city is famous as the birthplace of Simon Bolivar, the greatest of South American heroes, and as the cradle of South American independence. It contains a statue of James Monroe, fifth President of the United States, whose memory is revered in Venezuela because of the Monroe Doctrine (which see). F.ST.A.

CAR'AT, a weight of 3.17 troy grains, used by jewelers in weighing precious stones and pearls. The term is derived from the Arabic *carat*, meaning a *bean* or *seed*. In ancient times the seeds of the coral and carob trees were used as weights for precious stones, which were described as being of so many "beans weight" or "carats." The carat is divided into 4 *carat grains*, which, in turn, are divided into 2, 4, 8 or 16 parts for more accurate measurements. The term is also used to express the amount of gold in an alloy, a carat being $1/24$ of the total weight. So, if $18/24$ of an alloy is pure gold, it is said to be *18 carats fine*, and when it is *24 carats fine* it is pure, or *solid gold*. See ALLOY; DIAMOND.

CARAVAGGIO, *kah ra vah' jo*, MICHELANGELO MERISI DA (1565-1609), a celebrated Italian painter, founder and leader of the naturalistic school of painting in Italy. The name Caravaggio, by which he is generally known, is the name of his birthplace. After studying in Milan and Venice, he went to Rome, where he made great progress in his art and became widely known as the painter who followed nature and disregarded tradition. His fiery temper and quarrelsome disposition proved a greater hindrance to him than the opposition of his rivals, and he was forced to leave Rome after killing a comrade in a brawl. Thereafter he painted in Naples, in Malta and in various cities of Sicily. In 1609, having obtained pardon for his crime from the Pope, he started back to Rome but was waylaid on the road and fatally wounded.

Caravaggio was a pioneer in the movement for naturalness in painting and his influence extended far beyond his native land. He painted two groups of pictures—scenes from everyday life and religious subjects. Of the former, good examples are his *Card Players, The Gipsy Fortune Teller* and *Love Conquered.* The figures in his religious canvases are painted with such boldness and realism that they were highly displeasing to the people of his time, who preferred idealistic treatment of the saints. His masterpiece, *The Burial of Christ*, now in the Vatican, is one of the great pictures of the world. Another well-known canvas is *The Supper at Emmaus.*

CAR'AVAN, a Persian word meaning *people* or *army*, is taken over into the English language to mean a very special group of people—one of the large companies which travel together across the deserts of Asia or Africa. For the most part these travelers are merchants who dare not set out alone for fear of robbers and the beasts of the desert, but there are also each year, at certain holy seasons, caravans of pilgrims who journey from Cairo or Damascus to Mecca (see MOHAMMEDANISM). Whatever the purpose, camels are used as the means of conveyance, for only camels could travel for days or even weeks across the desert lands (see CAMEL).

Centuries ago, when the cities about the eastern Mediterranean shores were the great trade centers of the world, immense caravans journeyed to them every year from more easterly parts of Asia, bringing the rugs and spices of Persia, the embroideries, tea and silks of China, the jewels and shawls of India, and sometimes there were in one of these great trains from 1,000 to 5,000 camels. The caravan might thus be several miles in length, and the camping spots populous villages. To-day throughout much of the Orient railways have made the old-time caravans unnecessary, but no railroad has as yet made its way into the interior of the deserts to which the camel easily penetrates.

The thought of these great caravans, moving noiselessly across the sands, without haste yet without rest, has always been a fascinating one to Western minds, and allusions to them in literature are numerous. Bryant in his *Thanatopsis* speaks of —

The innumerable caravan which moves
To that mysterious realm where each shall take
His chamber in the silent halls of death.

Milnes, with less figurative meaning, writes:

While o'er the neighboring bridge the caravan
Winds slowly in one line interminable
Of camel after camel.

CAR'AVEL, the name given to the type of ship used by Columbus on his first voyage across the Atlantic, and in general by the Spanish and Portuguese seamen of the fifteenth and sixteenth centuries to vessels undertaking long voyages. The caravel was a ship of 200 or 300 tons, having a deep hull and high decks fore and aft, and carrying a double tower at the

A CARAVEL

stern and a single one at the bow. The vessel had four masts. Faithful reproductions of the three ships of the fleet of Columbus, the *Nina, Pinta* and *Santa Maria,* were constructed in Spain, towed to the United States, and exhibited at the World's Columbian Exposition in Chicago in 1893. Afterwards they were a permanent exhibit in the lagoons of Jackson Park, the site of the fair grounds.

The word is also applied to a small fishing vessel of ten or fifteen tons used in Spain, Portugal and the Azores, and to a large Turkish warship.

CAR'AWAY, a plant often cultivated for its fruit or seeds, which have a spicy fragrance and a warm, biting taste. These seeds are used in cakes, candies, breads, some varieties of cheese, and also in medicine. By distillation of these seeds a light, easily-evaporated oil is obtained. The plant is a native of Europe and of Asia, and has finely-cut leaves and white flowers. It is a biennial (see BIENNIALS).

CARBIDES, *kahr'bydz,* compounds formed by the union of carbon with metals. Those of most interest are the carbide of iron and the carbide of calcium. The presence of carbon in iron in varying proportions changes its properties, the difference between pig iron, wrought iron and steel being due largely to the different proportions of carbon and carbide of iron in each. Calcium carbide is of interest as the source of acetylene gas and of calcium cyanamide (see ACETYLENE). It is a grayish, coarse-grained solid, from which the gas is set free when it is placed in water. Calcium carbide is manufactured at Niagara Falls, N. Y., and at Sault Sainte Marie, Mich. J.F.S.

CARBOHYDRATES, *karbohy'drates.* When you sit down in the morning and enjoy your breakfast of oatmeal or other cereal, relish your piece of bread and butter and sip your coffee, to which you have added milk and sugar, you are introducing into your body a number of substances that belong to the group known as *carbohydrates.* The carbohydrates are chemical compounds that are found in great quantities in the animal and vegetable world, and are composed of carbon, oxygen and hydrogen. Among the carbohydrates are starch, various kinds of sugar, and cellulose. All vegetable foods are rich in carbohydrates. Starch forms a great proportion of the nourishing part of the cereals, wheat, oats, corn, rye and rice; these contain, besides, a good proportion of *fats* and *proteins,* the other chief groups which form the nourishing part of the various articles of food (see FAT; PROTEINS). Granulated sugar is a pure carbohydrate. Molasses, honey and fruits contain large proportions of carbohydrates in the form of special sugars. Milk also contains a carbohydrate known as *lactose,* or *sugar of milk.*

The carbohydrates are among the most necessary and most important substances in our daily food. The starches and sugars found in all vegetable foods are easily digested. Like the fats, they serve as fuel to keep the body warm and to provide energy for its movements. Neither fats nor carbohydrates, however, can be converted into muscle, as can proteins. It is impossible to live on carbohydrates alone, or even on carbohydrates and fats together. A certain amount of protein is necessary in the diet, to provide for the growth of the body, and to repair the waste of muscular tissue, which is always going on. See FOOD. J.F.S.

CARBOLIC, *karbol'ik,* **ACID,** or **PHENOL,** *fee'nol,* or *fee'nohl,* an acid obtained from coal tar, having a peculiar and not unpleasant odor. When pure, carbolic acid is in the form of white crystals; when long exposed to the light, however, these crystals turn red. Carbolic acid is easily melted. It is only slightly soluble in cold water. It is a powerful disinfectant, and is employed for cleansing surgical instruments, walls, floors and utensils used about those afflicted with contagious diseases. It was formerly in general use to protect wounds in surgery, but under modern methods it is not commonly so employed. Taken into the stomach, except in very weak solutions, it acts as a powerful poison and causes death if relief is not prompt (see ANTIDOTE, for treatment). In many localities its sale is prohibited except upon a physician's prescription. J.F.S.

CAR'BON. We all know what *coal* is, and sometimes we see *charcoal.* We are familiar with the so-called "lead" in our pencils, which is not lead at all, but *graphite,* and occasionally we see a *diamond.* All of these substances are different forms of the element carbon, but the diamond is crystallized in one way, graphite in another, and charcoal is not crystallized at all; in other words, its atoms are not arranged in any definite order. Coal contains much carbon in combination with other elements. Some hard coals, or *anthracites,* contain over ninety per cent of carbon. The coke left in the retorts when coal is heated to make gas is, like wood charcoal, an impure amorphous (uncrystallized) carbon. The purest amorphous carbon is made by heating sugar in a loosely-covered crucible until gas ceases to come off.

Carbon is one of the chemical elements, and it exists in the three forms mentioned. It forms more compounds than any of the other elements. Its compounds are found in every plant and animal, and the branch of chemistry that treats of these compounds is called *organic chemistry.* With oxygen it forms carbonic oxide, or carbon monoxide, and carbonic-acid gas (which see). With hydrogen, it forms an extensive class of compounds known as hydrocarbons, which differ widely in their chemical and physical characteristics. The various carbonates occur very abundantly in the earth's crust. J.F.S.

Related Subjects. A broader view of this topic may be secured by reading the following articles in these volumes:

Boneblack	Diamond
Carbides	Gas
Carbonates	Graphite
Charcoal	Hydrocarbons
Coke	Lampblack

CARBONATES, *kar'bon ates,* salts of carbonic acid, such as sodium carbonate and calcium carbonate. Calcium carbonate, or carbonate of lime, is the most abundant of the natural carbonates. It exists as calcite, limestone and marble, chalk and marl, and is the chief constituent of egg-shells, oyster-shells, pearls and coral, and an important constituent of bones. Whiting is fine calcium carbonate, and putty is the same material mixed with linseed oil. The common carbonate of soda, or sal soda, is a well-known washing powder, and bicarbonate of soda is used in cooking and for numerous other purposes. Carbonate of copper, known to the mineralogist as *cerussite,* is a valuable ore from which copper is obtained, and one of the iron ores, *siderite,* is a carbonate of iron. Some carbonates are used in dyeing and others in medicine. When a carbonate is treated with an acid, carbonic-acid gas (which see) is set free. J.F.S.

CARBONDALE, PA., a coal-mining city of Lackawanna County, in the northeastern part of the state. It is sixteen miles northeast of Scranton, thirty-five miles northeast of Wilkes-Barre and 199 miles northwest of New York City. It is served by the Delaware & Hudson, the Erie, and the New York, Ontario & Western railroads. The area of the city is about three square miles. The population in 1910 was 17,040; it had increased to 18,532 in 1914.

Carbondale is situated in the Lackawanna Valley. The Catskill Mountains in New York may be clearly seen to the northeast. The hills surrounding the city form one of the richest anthracite coal sections of the state, and the town takes its name from this great natural resource. Coal mining and shipping are the principal industries; industrial plants of importance are silk and knitting mills, foundries and machine shops, ice refrigerating plants, car shops and glass factories.

In the center of the city is Memorial Park. The prominent public buildings are the municipal building, the post office, state hospital for the criminal insane (located at Fairview, five miles distant), four banks, an emergency hospital, a private hospital and a city library.

The coal mines were opened in 1824, and the settlement at Carbondale, begun that year, was incorporated in 1851. The city adopted the commission form of government in the year 1913. R.M.S.

CARBON DISULPHIDE, *di sul'fide,* or **CARBON BISULPHIDE,** a compound of carbon and sulphur, which is known as a heavy, colorless liquid. When pure it has rather a pleasant odor, but ordinarily, owing to the presence of impurities, it has a very unpleasant smell. It readily dissolves india rubber, gutta-percha, sulphur, phosphorus and resins. Carbon disulphide is sometimes used to kill moths and other insects, as well as burrowing animals, such as moles and woodchucks, and small animals in buildings, as rats and mice in mills. Another important use is in the manufacture of artificial silk from wood pulp. Carbon disulphide is manufactured by heating its elements, carbon and sulphur, together. The form of carbon commonly used is coke, and the temperature required is a bright red heat. Electric furnaces are now used almost exclusively for the purpose. J.F.S.

CARBONIC-ACID GAS, or **CARBON DIOXIDE,** *dy ox'ide,* or *dy ox'id,* a gas formed when charcoal, wood, coal, oil or almost any material containing the element *carbon* burns in a free supply of air. The name carbon dioxide signifies that the molecules of the gas consist of one atom of carbon combined with two atoms of oxygen. The name *carbonic-acid gas* has reference to the faintly-acid character of the liquid obtained by dissolving the gas in water. Under the misleading name of *soda-water*, this solution is familiar as an effervescent drink. Carbonic-acid gas is colorless and has only a very faint odor. It is about one and a half times as heavy as air, and a common lecture experiment is to pour it from vessel to vessel like water. When so poured upon a lighted candle it extinguishes the flame, for carbon dioxide does not support combustion. Indeed the gas is often used to extinguish fires (see FIRE DEPARTMENT, subhead *Fire Extinguisher*).

Carbon dioxide exists in the atmosphere in the small proportion of three parts in 10,000. Air containing ten times this proportion of carbon dioxide can be breathed without danger, but in pure carbon dioxide or in air containing a large proportion of it men and animals are quickly suffocated. Since they suffer much the same as in drowning, it is thought that the gas is not an active poison and that the evil effects are really due to lack of free oxygen, for animals cannot separate the oxygen of the compound from the carbon.

Interchange Between Plants and Animals. All animals, including man, give off carbonic-acid gas in breathing; it is also produced by fires, and from these sources it is constantly poured into the atmosphere. But what the animals give off the plants absorb, for to them it is a life-sustaining element, as is oxygen to animals, so that the relative proportion of the gas in the atmosphere does not change. Under the action of sunlight the leaves of plants absorb carbonic-acid gas through their pores and liberate part of the oxygen, returning it to the air. Thus plants and animals supply the atmosphere in a measure with what each group needs to enable it to live and grow.

Choke Damp. Carbonic-acid gas is produced in fermentation and decay. Being heavier than the air, it is liable to remain for a long time in wells and silos, as well as in caves and mines, in sufficient quantities to suffocate one entering these places. A lighted lantern lowered into the suspected place shows at once whether or not the gas is present. If the light is extinguished the place is unsafe. This gas is formed in mine explosions and often accumulates in poorly-ventilated chambers, where many miners lose their lives by suffocation. See MINING.

Liquid Carbon Dioxide. Carbon dioxide, when subjected to a pressure of about 450 pounds to the square inch and a temperature of 5° F. below zero, is easily changed to liquid. This acid is manufactured on a large scale by forcing the gas into steel cylinders by means of a powerful pump, until the pressure becomes sufficient to change the gas into liquid. The large quantities of carbon dioxide produced in the process of brewing are now saved and used in this way. Liquid carbon dioxide is also made directly in factories established for the purpose. It is used in the making of soda water, beer, champagne, mineral waters and other effervescent drinks, to which it imparts a sparkling appearance and biting taste. J.F.S.

CARBONIFEROUS, *kar bon if'er ous,* **PERIOD,** that period of geologic time during which many of the world's coal beds were formed. A more recent name given by American geologists is *Pennsylvania Period.* The rocks formed during this period are also known as the *Carboniferous System.* They lie between the Devonian System below and the Permian System above (see GEOLOGY), and include the coal measures—that is, the layers of rock between which the coal is found. The rocks include coarse masses or clusters known as millstone grit, limestone, sandstone and shales. Some of these, particularly the limestones and sandstones, are valuable building stone. In some localities the rocks of the system contain valuable deposits of iron ore.

Extent. During the Carboniferous Period it is probable that all of North America east of the Rocky Mountains was above the sea, and the Carboniferous System is found at various places from the eastern slope of the Appalachian Mountains westward as far as Kansas. The great coal fields of the Appalachian field extend from the northern part of Pennsylvania to Alabama, and reappear in Nova Scotia. There are large coal fields in Ohio, Indiana and Illinois, and also in Iowa, Missouri, Kansas and Texas and in the provinces of Nova Scotia and New Brunswick, Canada. These locations are shown in maps accompanying the article COAL. Modern research leads to the opinion that the eastern and western coal fields were formerly connected but were later separated by erosion; that is, the waters which rushed down the Mississippi Valley carried away the coal and thus separated these fields. The coal measures cover only a small part of the Carboniferous System. In Europe coal measures occur in the British Isles, the basin of the Rhine and in other places. Everywhere they are of great thickness. There are extensive coal deposits in China and also in Australia.

Life. Our greatest scientific interest now is in the plant life of the period, because the coal was formed mostly from these plants. Fossils found in the coal measures show that the plants were similar to our ferns, horsetails and club mosses, but that they were very much larger than any of these forms now in existence, even in tropical regions. Ferns and horsetails grew to the size of trees and club mosses were of gigantic proportions. Not only were the plants of unusual size, but vegetation was more luxuriant than any now known, and growth must have been more rapid than even in the most thriving tropical regions of the present time.

Animals resembled salamanders, frogs and other forms that lived partially on land and partially in water. Insects were numerous, as also were various kinds of fish, mollusks and crustaceans. No remains of large animals have been found. See FOSSIL.

Duration. No system of rock of any time enables the geologist to tell how long the Carboniferous Period continued. It is estimated that a vigorous growth of vegetation will yield annually one ton of dried vegetable matter to the acre, or 640 tons to the square mile. If this annual growth could be continued for 1,000 years and all the vegetable matter compressed to the density of coal it would form a seam of coal about six inches thick. At this rate it would take at least 2,000 years to produce a layer of coal one foot thick. But on some of the coal fields the combined thickness of the seams is from 100 to 250 feet. Between the layers of coal are layers of rock whose formation may have required as much time as that of the coal. But as vegetation grew more rapidly than at present, that condition would shorten to some extent the time required. Thus, taking all conditions into consideration, geologists estimate the duration of the Carboniferous Period at from 2,000,000 to 5,000,000 years, a length of time which must be taken for granted rather than fully comprehended. See COAL; DEVONIAN PERIOD. W.F.R.

CARBON MONOXIDE, *mon oks'id,* a colorless gas, formed when carbon dioxide is passed through heated carbon. In it the carbon is combined with only half as much oxygen as in carbon dioxide. Its molecule consists of one atom of each element, whence the prefix *mon,* meaning *one.* It is very poisonous, and burns with a blue flame. The little blue flame seen in anthracite or charcoal fires is caused by the burning of carbon monoxide. It is found in ordinary illuminating gas, especially in "water gas" (see GAS), which, because of it, is dangerous to inhale in even the minutest quantities. Carbon monoxide sometimes escapes from coal stoves and furnaces and causes stupefaction and death to whole families. One volume of it in 100,000 volumes of air produces symptoms of poisoning, and one volume in 750 of air causes death in half an hour. Although itself odorless and tasteless, its escape from a stove or gas pipe into the air can usually be detected by the odor of the other gases which escape with it. J.F.S.

CARBORUNDUM, *kar bo run'dum,* a very hard substance used in place of emery and sandpaper as an abrasive (for polishing). Carborundum is made by mixing in proper proportions coke, sand, sawdust and a small quantity of salt, and subjecting the mixture to intense heat in an electric furnace constructed especially for the purpose. It requires about thirty-six hours to complete the process. When it comes from the furnace, carborundum is a mass of coarse crystals that reflect nearly all the colors of the rainbow, but when ground to powder it is gray. The finest powder is used for polishing; coarser grades are used in place of emery. Hones of various degrees of fineness are made from it, the finest being used for sharpening razors and surgical instruments. Carborundum is a chemical compound of carbon and silicon. It is manufactured in large quantities at Niagara, N. Y. See, also, CORUNDUM; EMERY. J.F.S.

CARBUNCLE, *kar'bung k'l,* the name now applied to the crimson and scarlet varieties of garnet when cut with a smooth, rounded surface curving outwards. The term comes from a Latin word meaning *little coal;* the ancients gave the name to any red, fiery stone that glowed in the darkness like a burning coal. Certain ancient Jewish writers held that the ark of Noah received its light from carbuncles and other precious gems.' The carbunele is named in *Exodus* XXVIII as one of the gems in the breastplate of Aaron, the high priest of the Children of Israel. See GARNET; PRECIOUS STONES; HIGH PRIEST.

CARBURETOR, *kar' bu ret ur.* In an internal combustion engine using a liquid fuel, a carburetor is a device for converting the liquid into a mist or vapor and for mixing the vapor with air in such proportions that the mixture will burn instantly and without smoke. The function of the carburetor, in other words, is to insure perfect combustion. It must mix the fuel with air under wide variations of engine speed and power, must vaporize the mixture under variations of temperature, must respond at once to a demand for greater or less speed, and must do all these things in spite of variations in the quality of the gasoline or other fuel used. The carburetor is, therefore, a very delicate instrument, but at the same time its method of operation is simple and easily understood.

The process may be more quickly understood if the meaning of the word carburetor is known. To *carburate* means to saturate or combine chemically with carbon; that is, the carburetor combines air with carbon, which is not pure, but may be in the form of gasoline, kerosene or other liquid. The process of mixing the air and carbon is called *carburation.* The modern carburetor usually has three inlets and one outlet. One of the inlets is for the fuel, another is for air, and there is usually an additional inlet, called the auxiliary or secondary inlet, also for air. The only outlet is to the engine, and it usually has the throttle valve attached to it.

The Process of Carburation. The fuel enters the carburetor sometimes by gravity from a tank situated at a higher level; but usually it is forced into the carburetor by air pressure or automatically by combustion which creates a pressure greater than the pressure of air in the storage tank. An excessive flow of gasoline into the carburetor is prevented by a float in a chamber through which the gasoline flows. As the level of gasoline in the carburetor reaches the desired point, the float rises and finally closes a valve in the supply pipe.

From the float chamber a passage leads to a jet nozzle, through which the gasoline is sprayed into the mixing chamber. The nozzle is about one-sixteenth of an inch above the level of the liquid in the float chamber when the air pressure at the jet and in the float chamber are the same. Under these conditions there is no flow of liquid. However, when the piston of .the engine is drawn back, the air pressure at the nozzle and in the air pipe is reduced. As the pressure at the nozzle is then less than the pressure in the float chamber, the liquid is at once forced through the nozzle into the mixing chamber. When the engine has been at rest it is necessary to make this initial suction at the nozzle by "turning the engine over," or "cranking" it, which may be done by hand or by a mechanical appliance called a "starter."

The air inlet, which is usually protected with a fine screen to keep out dirt, also opens into the mixing chamber. Under normal conditions the air inlet is adjusted so that a sufficient supply of air enters to secure perfect combustion. However, if the engine is going at high speed, the liquid, being heavier, tends to flow into the mixing chamber more rapidly than needed. To offset this increase, most carburetors are

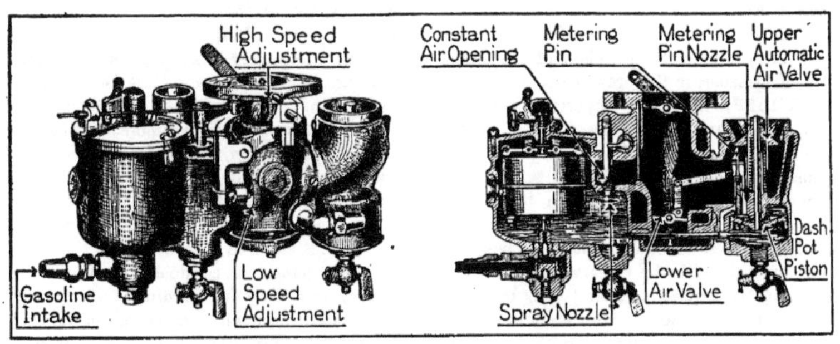

ONE FORM OF CARBURETOR
Showing exterior appearance and cross-section through center.

provided with an auxiliary air inlet. The valve on this inlet is held closed by a spring until the pressure inside the mixing chamber is dangerously low, when the valve opens in response to the pressure of air from the outside.

The only outlet from the carburetor leads to the cylinder of the engine. The gas, mixed with air, enters the cylinder, and is there exploded by an electric spark. For further details, see GAS ENGINE. W.F.Z.

CARDAMOM, *kahr' da mum,* the dried fruits and seeds of different species of reed-like herbs called cardamoms. They have a spicy taste and are used in making sauces, curries and cordials. They are an excellent relief in cases of colic and, on account of their flavor, they are often used with other medicines. Those recognized in America as *true* or *official cardamoms* and known in commerce as *Malabar cardamoms* are the product of a plant of the mountains of Malabar, in the province of Madras.

CAR'DIFF, the most important port and commercial center of Wales, capital of the county of Glamorgan. It existed before the Roman conquest of Britain in 55 B.C., and as early as the Norman conquest in 1066 was a town of considerable importance. The city is 170 miles west of London, on the River Taff at its junction with the estuary of the Severn, and near the largest coal and iron mines in Great Britain. Its docks afford accommodation for the largest vessels afloat, and cover an area of more than 160 acres. The city derives considerable revenue from harbor dues, but the docks are the property of the Marquis of Bute, to whose family the town owes much of its prosperity. The chief exports are coal and iron and steel manufactures, the value of coal annually exported exceeding $117,000,000. In normal times the exports of Cardiff exceed in weight those of London by some thousands of tons, but London, Liverpool and Newcastle rank above it in the weight of incoming products.

The town is well laid out, has good street car systems, electric lighting and excellent railroad facilities. The building of greatest interest is the castle, dating from the eleventh century, now restored and occasionally occupied by the Marquis of Bute. There are also many schools, a university college, town hall, public library, markets and public baths. Population in 1911, 184,663.

CAR'DINAL, in the Roman Catholic Church, a counselor of the Pope and a dignitary next in rank to him. The name, which comes from the Latin word for *hinge* and consequently denotes something of great importance, was first given to priests who were permanently attached to a church. Later its use was restricted to those priests who held prominent positions. Then, as the new administration of the affairs of the Church became more burdensome, the Pope called upon bishops to assist him, and these were designated as cardinal-bishops. Finally the cardinal became a counselor of the Pope and gave his entire time to his office. Under the direction of the Pope the cardinals conduct the administration of the Church.

Cardinals are appointed by the Pope, and while it is intended to have a representative of each prominent nation in the College of Cardinals, by far the greater number of them have always been Italians. The first American cardinal, McCloskey, was appointed in 1875. There are living at present (1917) three Americans who have been raised to that rank; these are James Gibbons of Baltimore, William H. O'Connell of Boston, and John M. Farley of New York. Diomede Falconio was appointed a cardinal while in America and by some is ranked as an American cardinal, but in 1913 he was transferred to Rome.

Taken together, the cardinals form the *Sacred College,* or College of Cardinals, which is an incorporated body with an income of its own. The number may vary, but it was fixed at seventy by Sixtus V in 1586. On the death of the Pope the Sacred College elects his successor, from its own membership. The official symbols of a cardinal are the *biretta,* or red cap; the sapphire ring; the purple cassock; the miter of white silk, and the red hat. This last is placed on the head of the newly-made cardinal at the time of his appointment, by the Pope himself, but is then laid away and never worn again. At the cardinal's funeral it is placed on his casket. G.W.M.

CARDINAL BIRD, or REDBIRD, one of the most beautiful wild birds of North America, so-named from the rosy red plumage of the male bird. The cardinals belong to the finch family, and are also known as the *crested redbird,* the *Virginia redbird* and the *Virginia nightingale.* James Lane Allen has given these birds another name in his sympathetic and tender book, *The Kentucky Cardinal,* in which he pleads for their protection from the sportsman's gun. They are about eight inches in length, and bear on the head a conspicuous crest that gives them a rather distinguished appearance. The males are bright vermilion above, paler beneath, with tints of gray on the back, and black forehead and throat. The females are modestly attired in olive-gray and buff.

The cardinals are found chiefly in Southeastern Canada and in the Eastern United States from New York to Florida, and are also common in Mexico and the Bermuda Islands. The migratory birds of the cooler sections travel southward in flocks on the approach of winter, but a few stragglers usually linger in the swamps of Pennsylvania nearly all winter. Occasionally a cardinal is seen as far north as Massachusetts; there are a few in the province of Ontario. South of the Ohio River these birds are permanent residents of the localities where they are found. They usually nest in a thicket of brambles or in a low tree, their little home being loosely made of twigs, grass, weed-stems, etc., and lined with grass or roots. There are three to five dull white or bluish eggs, with spots of reddish-brown and lilac, and two broods are usually raised each season. They feed on insects, worms and the seeds of small fruits, and are useful destroyers of insect pests. The male bird never neglects his mate during the breeding season, but keeps her and the young well supplied with food.

CARDINAL BIRDS

The cardinal is one of the sweetest songsters of America, its songs reminding one of the nightingale's music, though lacking the haunting melancholy of the latter's notes. Easily tamed, it is popular as a cage bird, and large numbers are shipped to Europe under the name of *Virginia nightingale*.

CARDINAL FLOWER, or **INDIAN PINK,** a large, intensely-red and very showy flower. It is a native of America, growing to a height of two or three feet in low, swampy places, or on muddy banks of streams. It is a straight-growing, attractive flower, worthy of cultivation in moist borders. In England it is much prized and cultivated. A blue-flowered species is called *blue cardinal flower*. All species are members of the lobelia family. See LOBELIA.

CARDINAL FLOWER

CARDS, PLAYING, oblong pieces of cardboard bearing certain spots and figures which are used in playing games of chance or skill. They are considered by some to be an innocent recreation of the fireside and by others the most widespread gambling device the world has ever seen.

The set of cards commonly used is known as a *deck*, or *pack*, and consists of fifty-two cards containing four suits—spades and clubs, which are printed in black, and diamonds and hearts, printed in red. A suit is composed of thirteen cards consisting of king, queen, knave (or jack), known as face cards, and ten spot-cards, sometimes called pip-cards, ranging in number of spots from one, or the ace, to ten. The natural rank in the suit sometimes places the king highest, and so on down, the ace being the lowest; but in most games this rank is changed, the ace being the highest, the king second, etc. A great variety of games may be played with cards. There are round games, as in hearts or poker, in which any number of persons may join; four persons may play euchre, whist, bridge, etc.; two may play pinochle, cribbage or bezique, and one person may find diversion in the game of solitaire. Rules for many games appear in these volumes.

Playing cards are of ancient origin. The course that card playing took in its European diffusion shows that it probably came from the Orient, for it was first found in the Eastern and Southern countries. Historical trace of cards was found earliest in Italy, then in Germany, France and Spain, in the order named.

CARDSTON, a town in the southwest corner of the province of Alberta, fourteen miles north of the United States boundary and sixty-seven miles southwest of Lethbridge. It is on a branch of the Canadian Pacific Railway and is near Saint Mary's River, one of whose tributaries, Lee's Creek, flows through the town. Cardston is the center of a prosperous dry-farming region, in which grains are the leading products. There is also considerable stock raising and some dairy farming. The most conspicuous structure in the town is the $500,000 Mormon Temple, which gives the town its popular name, the "Temple City." A large part of the population of Cardston and the district is Mormon. There is fine scenery, good fishing and big-game hunting in the mountains west and south of Cardston. Population in 1911, 1,207; in 1916, estimated, 1,800. J.T.B.

CARIBBEAN, *kair i be' an,* **SEA,** a body of water which might be called a mammoth lagoon. It begins at the north coasts of South America and Panama and the eastern shores of Central America and Yucatan, and is almost

LOCATION MAP

enclosed by the West Indies. Through it all ships must pass proceeding to or from the Atlantic end of the Panama Canal. From Trinidad on the east to Yucatan on the west it is over 1,700 miles long; its greatest width is from Haiti and Jamaica to Panama, about 700 miles. In parts the sea is of great depth, soundings of 16,000 feet (nearly three miles) having been found off the coast of Cuba and near the coast of Venezuela. The broadest entrance to it is the Yucatan Channel, 120 miles wide. In days of adventure this sea was the "Spanish main."

CARIBOU, *kair' a boo,* or *kair a boo',* an Indian name for the American reindeer, which is now rarely found south of Canada, but formerly was common as far south as Wyoming. There are three species found in Canada, known

CARIBOU

as the *Newfoundland* caribou, the *barren ground* and the *woodland.* These distinctions are made more on account of the districts in which they are found than because of any important differences in the animals themselves. Caribou roam about in the summer, constantly changing their ground to escape the flies and insects that become a terrible pest, but in winter they gather together in herds, feeding on winter berries and the leaves of shrubs. Their large, hairy hoofs enable them to travel easily in the snow. They have large antlers, one branch of which extends over the forehead in front .

Caribou are carefully protected by game laws in Canada and the United States and are hunted only in the winter. When snow is on the ground their tracks are easily followed. To be successful, the hunter must approach the game against the wind, for if the caribou gets the scent of man it flees at such a pace even over deep snow that pursuit is hopeless. Huntsmen generally use a sporting rifle, firing a cartridge containing a soft, lead-nosed bullet. The steel bullet used in warfare would not disable a caribou unless it entered the heart or brain. The caribou is said to fall a victim to wolves less frequently than does the moose, although the latter is stronger and a more courageous fighter. The greater speed and ability to travel over loose snow enables the caribou to outdistance his pursuers, while the moose, sinking deep in the snow, must stand at bay and trust to hoof and horn to drive off the wolves. See REINDEER.

CARICATURE, *kar'i ka ture,* comes from an old Italian word meaning to *overload* or *exaggerate.* It thus relates to any representation, whether written or in picture form, in which the peculiarities of a person or object are so exaggerated as to appear ridiculous. Written caricatures include such satires as *Don Quixote,* which laughed out of existence the absurd romances of chivalry, or as Swift's *Gulliver's Travels,* which represented the politicians of England as giants or as pygmies. But far more commonly the word is used to mean grotesque pictures, such as those with which modern newspapers and periodicals have made everyone familiar.

Its Necessary Elements. A successful caricature must have in it more than a suggestion of fact. To represent a strenuous man of action in the garb and the hesitating state of mind of Hamlet, or a notedly grave and thoughtful man as a clown is wide of the mark. A keen sense of perception and true humor are as much a part of the caricaturist's equipment as the ability to draw; if he possesses all these he may be a molder of public opinion as truly as is an editor. The downfall of the "Tammany tiger" (see TAMMANY) was brought about once through Thomas Nast's clever series of caricatures, and many a politician has been made in pictures to appear so ridiculous as to be obliged to give up all claims to office. In a sense the public man is unfortunate who possesses certain strong facial characteristics, for these the artist is certain to magnify in his pictures. Simply two rows of shining teeth in a drawing are sufficient to suggest Theodore Roosevelt; the bald head, slight fringe and strong mouth of William J. Bryan serve the cartoonist well. Not all *cartoons* are caricatures, but most of the former which have the element of humor rely for it largely upon the use of caricature.

Although it has taken on new importance in recent years, the art of caricaturing is by no means new. The Assyrians and Egyptians made use of it, as did the Greeks and Romans and the great painters of the Renaissance. Especially famous are the caricatures drawn by Leonardo da Vinci. In England the art really began with Hogarth, one of the greatest caricaturists the world has ever seen, and was continued by Rawlinson, Cruikshank, Leech, Tenniel, Du Maurier and others. American caricaturists who have gained wide celebrity are, in addition to Nast, mentioned above, Davenport, Outcault, Opper, McCutcheon and Briggs. Nast and Davenport put venom into their pictures when they desired to pillory an alleged enemy of the public. Outcault favored children's escapades and was the creator of the wonderfully-popular "Buster Brown." McCutcheon believes there is full opportunity for the exercise of his powers in pictures which cause one to smile, with never a suspicion of a sting, or which convey some lesson or warning against wrong conditions in society. Briggs portrays the common American boy, with his joys and sorrows, and also reveals with human touch the complex elements in older boys.

A list of the caricaturists who are treated in these volumes is given with the article CARTOON.

CARLETON, *kahrl'ton,* SIR GUY (1724-1808), a British soldier and colonial governor who rendered good service to England in the French and Indian and the Revolutionary wars. Appointed governor of Quebec in 1775, he later took supreme command of the British forces in Canada, successfully repelled the American attacks in the early years of the Revolution, led in the capture of Crown Point and was raised to the rank of lieutenant-general. In 1777 he was superseded by Burgoyne, but at the close of the war succeeded Sir Henry Clinton as commander-in-chief. For his services he was created Baron Dorchester by the king and was granted a pension of £1,000 a year. From 1786 to 1796 he was again governor of Quebec, proving a popular administrator.

CARLETON, WILL (1845-1912), an American poet whose writings are widely read and enjoyed because of his simple and natural treatment of home life and the joys and sorrows that are common to humanity. He was born near Hudson, Mich., educated at Hillsdale College, in his native state, and soon after his graduation in 1869 began to write home-life poems. He first won public approval by his *Over the Hills to the Poorhouse.* He was called to the lecture platform and continued in this work until shortly before his death. In his lectures he was accustomed to read and recite selections from his own writings. One lecture, two hours in length, was entirely in verse. During the last five years of his life he edited the literary magazine *Everywhere.* Among his volumes of poems, which he began to publish in 1871, are *Farm Ballads, Farm Legends, Farm Festivals, City Festivals* and *Poems for Young Americans.*

Carleton's poetry is simply written, enlivened by humor and bits of homely philosophy, and graced by touches of pathos. The following lines, from *The First Settler's Story,* are typical of his style and his manner of philosophizing:

Boys flying kites haul in their white-winged birds;
You can't do that way when you're flying words;
"Careful with fire," is good advice we know,
"Careful with words," is ten times doubly so.
Thoughts unexpressed may sometimes fall back dead;
But God Himself can't kill them when they're said.

CARLETON PLACE, a town located in Lanark County, Ontario, twenty-eight miles southwest of Ottawa and forty-five miles northwest of Brockville. It is on a small tributary of the Ottawa River, and is near several small lakes which provide excellent fishing for sportsmen. The town is a division point on the main line of the Canadian Pacific Railway, which maintains large repair shops there. Foundry products, woolen goods, gloves and flour are the leading manufactures. Carleton Place was founded in 1818, and was named for a small town of the same name not far from Glasgow, in Lanark County, Scotland. Population in 1911, 3,621; in 1916, about 3,900.

CARLISLE, *kahr lyle',* PA., county seat of Cumberland County, in the southeastern part of the state, nineteen miles southwest of Harrisburg. It is an attractive town, favorably situated in the fertile agricultural Cumberland Valley, and is served by the Cumberland Valley and the Philadelphia & Reading railroads and electric interurban lines. The population in 1910 was 10,303; it was 10,589 in 1914.

It is the seat of the United States Indian Industrial and Training School (which see), and of Dickinson College (non-sectarian), which includes the Metzger Institute for girls. James Buchanan (President) and R. B. Taney (Chief Justice) were graduates of Dickinson College. In the mountains near Carlisle is Mount Holly Springs, a popular summer resort. The prominent industrial plants include boot and shoe factories, machine shops and manufactories of railway frogs and switches, axles, carpets, ribbon, hosiery and paper-boxes. The combined manufactured products have an annual value of about $3,415,000.

Carlisle was organized as a town in 1751 and was incorporated as a borough in 1872. Major John André was one of a number of British prisoners held here during the War of Independence, and at the time of the Whisky Rebellion George Washington made Carlisle his headquarters.

CARLISLE, *kahr lyle',* **INDIAN SCHOOL,** the popular name of the UNITED STATES INDIAN INDUSTRIAL AND TRAINING SCHOOL (which see).

CARLSBAD, *karls' baht,* or **KARLSBAD,** a free, royal city of Bohemia, seventy miles northwest of Prague, and one of the most celebrated of European watering places. Tens of thousands of visitors in normal times are annually attracted to its hot mineral springs. The chief ingredients of the water are carbonate of soda, sulphate of soda and common salt. The Sprudel, the most famous of the nineteen springs, has a temperature of 165° and discharges about 2,000 quarts of water a minute. The waters are clear and odorless, and are sometimes beneficial in cases of gout, kidney troubles, rheumatism and stomach complaints. The salt obtained from the water is exported in large quantities. The discovery of the hot springs is attributed to Charles IV in 1347, and they were known as *Charles's Bath,* as the king bathed in the waters and believed in their curative properties. The town was raised to the rank of a free royal city in 1707. Population, about 15,500.

CARLYLE, *kahr lyle',* THOMAS (1795-1881), an eminent Scottish writer, in whose work breathes an intense hatred of shams and a belief that work and duty, not happiness, should be the aim of life. Although pessimistic, bitter and unhappy, he yet stands as one of the men who helped to raise the tone of society in his day.

THOMAS CARLYLE

Carlyle was born at Ecclefechan, in Dumfriesshire, and was educated for the Church. Finding during his years at the University of Edinburgh that general reading was far more to his taste than theological studies, he gave up the idea of becoming a minister, and for a time was a teacher. This troubled his independent spirit, however, and in 1818 he removed to Edinburgh and began to support himself, frugally enough, by literary work. His career as an author may be said to have begun with the publication in the *London Magazine* of his *Life of Schiller,* which was enlarged and printed separately in 1825, attracting much favorable attention. In 1824 appeared his translation of Goethe's *Wilhelm Meister,* and in the next year his *Specimens of German Romance.*

Sartor Resartus (The Tailor Re-tailored), which appeared in 1834, won for its author immediate fame and remains one of the works which give him high rank. The grotesque and the sublime, the pathetic and the humorous, crowd the pages of this book, but through it all there shows Carlyle's love for sincerity at any cost. In 1837 appeared the *French Revolution,* a vivid dramatic picture of that historic movement. His *Chartism, Past and Present* and *Heroes and Hero-Worship* held the audience which his earlier work had gained for him, and *Oliver Cromwell's Letters and Speeches, with Elucidations and a Connecting Narrative* roused new enthusiasm among students. The largest and most laborious work of his life, *Frederick the Great,* appeared between 1858 and 1865, and after this little came from his pen.

In 1866, having been elected lord rector of Edinburgh University, he delivered an installation address to the students on *The Choice of Books.* While he was still in Scotland the news reached him that his wife had died suddenly in London, and from his grief he never recovered. Remorse for his treatment of her hung like a cloud over the rest of his life, which was passed in close seclusion. He died at Chelsea in 1881, and in the years immediately following, his *Reminiscences* and *Life,* as well as the *Letters of Jane Welsh Carlyle,* were published by James Anthony Froude. Some of the revelations contained in these works greatly injured Carlyle's reputation for a time, but gradually the bitterness has been forgotten, while the sincerity and true greatness of the man have come once more to the fore.

Of no author is it more true that "the style is the man" than of Carlyle, for his disjointed, rugged sentences and his fiery appeals really give a very true picture of him. The following quotations give some of the notes of his philosophy:

> Clever men are good, but they are not the best.
>
> There is no life of a man, faithfully recorded, but is a heroic poem of its sort, rhymed or unrhymed.
>
> Happy the people whose annals are blank in history-books.
>
> The true university of these days is a collection of books.

A. MC C.

CAR'MAN, Albert (1833-), a Canadian clergyman and educator, for more than thirty years general superintendent of the Methodist Episcopal Church in the Dominion. Dr. Carman was born at Iroquois, Dundas County, Ont., where he attended the grammar school. In 1855 he was graduated from Victoria College, then at Cobourg, and two years later was appointed professor of mathematics in Albert College, Belleville. Through his efforts Albert College was affiliated with the University of Toronto in 1860, and in 1868 was given a charter as a university. Dr. Carman, who had been principal of Albert College for a year, became the first chancellor of the university. In 1859 he was ordained a minister of the Methodist Church, and from 1874 to 1883 was its bishop. In the latter year, on the union of the various Methodist denominations in Canada, he became general superintendent of the united Church. He was an early advocate of prohibition, and always took an active interest in educational matters, being at various times a senator of Victoria University and the University of Toronto, and a governor of Wesleyan Theological College at Montreal. Alma College for Women, at Saint Thomas, Ont., was founded largely through his efforts.

ALBERT CARMAN

CARMAN, [William] Bliss (1861-), the foremost lyric poet of Canada. He was born in Fredericton, New Brunswick, and enjoys a common ancestry with Ralph Waldo Emerson. In addition to his natural abilities and his heritage of culture, Carman had excellent educational advantages, including courses at the University of New Brunswick, Harvard and the University of Edinburgh. In 1893, when his first volume of published verse, *Low Tide on Grand Pré,* appeared, he was already known as a magazine writer and as office editor of the New York *Independent;* this volume of verse brought him to the public notice as a young poet of striking promise—a promise which his later work has fulfilled.

Carman's depth and richness of imagination and his gift for expressing his emotions in beautiful and fitting language are suggested by the following lines from *At the Granite Gate:*

> And the lone wood-bird—Hark!
> The whippoorwill, night-long,
> Threshing the summer dark
> With his dim flail of song,
> Shall be the lyric lift,
> When all my senses creep,
> To bear me through the rift
> In the blue range of sleep.

Among Carman's volumes of poems are the series of *Songs from Vagabondia* (with Richard Hovey), *Pipes of Pan, Ballads of Lost Haven* and *A Winter Holiday*. He is also a writer of graceful prose, and is the author of several volumes of essays, including *Kinship of Nature, Friendship of Art* and *The Making of Personality*. None of his later works is considered superior to his first published volume.

CARMEN, a world-famous opera by the French composer, Georges Bizet, which has enjoyed extraordinary popularity since its first production in Paris in 1875. It is based on a novel of the same name, the work of the French novelist, Prosper Mérimée. The heroine, Carmen, is a Spanish girl of fiery temperament and wonderful fascination, whose heartless treatment of her lover, Don José, drives him to kill her in a frenzied outburst of jealousy. The opera is full of color and dramatic interest, and the music is richly melodious. The famous *Toreador* song and the *Habanera,* sung by Carmen, are enduringly popular, as are the many stirring choruses, the music of the overture, and the exquisite melodies of the third act, written for the secondary soprano rôle. Adelina Patti and Emma Calvé are among the famous prima donnas who have sung the leading part in this opera; the latter proved to be the greatest Carmen of all time. Among later artists Mary Garden and Geraldine Farrar have interpreted the part with notable success.

CARMINE, *kahr'min,* a beautiful red coloring matter made from the dried bodies of the cochineal, an insect native to Mexico and Central America. To obtain good results in making carmine it is necessary to have a clear sunny day, as the bright, pretty red cannot be obtained except in the sunlight. Carmine is used in making artificial flowers, water colors, rouge and red ink, in silk dyeing and in miniature painting. See COCHINEAL.

CARNATION, *kar na'shun,* one of the most popular of flowers. It has been cultivated from very ancient times for its clove-like fragrance and beauty, for perfumes and for decoration. It was brought from Southern Europe long ago as a wild flower of lilac hue. Under cultivation it now grows in many forms, and in tints dainty or bright, to please the taste of all. It has been called the "winter flower," blooming

. . . while the hollyhock,
The pink, and the carnation vie
With lupin and with lavender,
To decorate the fading year.
—MOIR, in *The Birth of the Flowers.*

chiefly in the winter months, from October till the end of March. Carnations are usually raised from "layers" or "cuttings." They need good turfy loam mixed with a little manure and leaf mold, and some sharp sand to keep it loose.

This flower is subject to several diseases, the worst being rust. This can be cured by spraying with a solution of sulphide of potassium. When large blossoms and a long stem are desired, remove all but the bud at the very end.

CARNEGIE, *kahr neg'i,* ANDREW (1835-), a famous and very practical philanthropist, is an American of Scotch birth. His gifts to the public have exceeded in amount those of any other man in the history of the world. From Dunfermline, where he was born on November 25, 1835, he emigrated to the United States at the age of thirteen. In his first position, that of weaver's assistant in a cotton factory at Allegheny, Pa., he earned about a dollar a week, but twenty years later his industry and ability had made him a wealthy man. As a messenger boy in a Pittsburgh telegraph office he had learned telegraphy, and after becoming an operator advanced rapidly to the rank of division superintendent for the Pennsylvania Railroad. Through friendship with the inventor he had entered the organization of the Woodruff Sleeping Car Company, the most successful forerunner of the Pullman Company, and had made a moderate fortune. By cautious investment in Pennsylvania oil lands he increased his wealth enormously. After the War of Secession he entered the iron business, and soon became one of the industrial leaders of America.

ANDREW CARNEGIE

In 1868 he introduced the Bessemer process into the American steel industry. The great steel works which he established at Homestead, Pa., and elsewhere, grew rapidly, and in 1899 he consolidated all his interests in the Carnegie Steel Company, at that time one of the greatest industrial institutions ever established. When in 1901 it was merged with the United States Steel Corporation (the "steel trust"), he retired from business with a fortune estimated at a half-billion dollars. Since then he has devoted his time largely to philanthropic and literary interests.

In 1912 Mr. Carnegie announced that with the gift of $125,000,000 to a corporation which will carry out those of his philanthropic schemes which he had not already endowed, his personal fortune was reduced to $25,000,000, about one-twentieth of its former amount. He expects to die a comparatively poor man.

The scope of his public spirit has been world-wide. Besides the five institutions which receive special attention below, his gifts include $11,000,000 to the Carnegie Institute of Technology, Pittsburgh; $10,000,000 to Scotch universities, including Saint Andrews and Aberdeen, for both of which he has been lord rector; $5,000,000 as a benefit fund for employees of the Carnegie Steel Company; a $2,500,000 trust for Dunfermline, his native town; $2,000,000 to the Carnegie Church Peace Union, which aims to enlist all churches on behalf of permanent peace; $1,750,000 for the Peace Palace at The Hague; and $850,000 for the grounds and buildings of the Pan-American Union at Washington.

As a writer he has expressed himself vigorously in denunciation of war, and his works, *The Gospel of Wealth, The Empire of Business* and *Problems of To-day,* have had international reading. He was made a Commander of the Legion of Honor of France in 1907, and in 1911 received the peace medal of the Fourth International Congress of American States.

Carnegie Endowment for International Peace. This is probably the most practical of the many organizations striving to banish war from the world. It does not rival or supplant any other institution, but endeavors to coöperate with and aid all peace-promoting organizations at home and abroad. In addition it is active in educating nations to a greater friendship with each other. The income from its fund of ten million dollars, given by Mr. Carnegie in 1910, is administered by a board of trustees of which Elihu Root was the first president. It has three active divisions, those of economics and history, of international law, and of intercourse and education. The first two are primarily for research, the last for spreading the information which they gain and for promoting international good will. It was proposed to establish an Academy of International Law at The Hague, to which advanced students from all nations would come. However, the outbreak of the War of the Nations in 1914 caused this and some other schemes to be suspended for a time.

Carnegie Foundation for the Advancement of Teaching. The annual proceeds of the fund of $15,000,000 given to this organization by Mr. Carnegie in 1905 and 1908 are distributed in pensions to teachers in the United States, Canada and Newfoundland retiring from the faculties of universities, colleges and technical schools, and to their widows. By dealing only with those schools which will bring their admission requirements and standards of teaching up to a specified level, the administrators of the fund have beneficially influenced the quality of higher education. Some opposition has been met from strictly denominational schools, which are not included in the benefits of the foundation, and from others which object to certain requirements as to their government. An educational research fund of $1,250,000 was added by Mr. Carnegie in 1913.

Carnegie Hero Funds. These gifts of Mr. Carnegie are for the financial support of those incapacitated for work, either temporarily or permanently, in heroic attempts to save human life, and for the aid of widows and orphans of

CARNEGIE HERO-FUND MEDAL
(Panel designed to receive inscription.)

heroes. Medals of three classes are given. The original fund of $5,000,000 set aside in 1904 for the United States, Canada and Newfoundland has been supplemented by similar gifts for Great Britain and Ireland, France, Germany, Switzerland, Belgium, Netherlands, Sweden, Norway, Italy and Denmark. Military bravery is not recognized.

Carnegie Institution. This organization, founded in 1902 to encourage in the broadest and most liberal manner investigation, research and discovery and the application of knowledge to the improvement of mankind, has an

CARNEGIE INSTITUTION

endowment of $22,000,000. The institution offers no regular class work and no degrees. Its administration building is in Washington, D. C. The President of the United States, the Vice-President, the Speaker of the House of Representatives, the secretary of the Smithsonian Institution and the president of the National Academy of Sciences are *ex-officio* members of the board, and the United States government guarantees the free use of its public records, museums and libraries to all persons connected with the Institution.

The organization has shown extraordinary breadth in its work of advancing scholarship and has aided research on subjects as widely different as bacteria and the historical sources of Browning's poem, *The Ring and the Book.* It has already brought to the attention of the world a number of scholarly men and has rendered immense assistance to universities. The value of future activities of the organization would be hard to estimate, for it possesses not only breadth of vision and good will, but ample funds for every need.

Carnegie Libraries. Over fifty million dollars has been given by Mr. Carnegie to libraries in English-speaking countries, and it is largely as the result of his beneficence that the public library has become a prominent institution in every community of importance in the United States and Canada. Cities which have not accepted his aid have at least been stimulated to better their libraries by the examples of their neighbors. Professional training for librarians has been encouraged. The type of architecture now almost standard for American public libraries owes its prominence to the activities of the Carnegie commission. C.H.H.

CARNEGIE, Pa., a suburb of Pittsburgh, noted for its extensive steel-manufacturing industries. The borough was formed by the consolidation of Mansfield and Chartiers in 1894, and named for Andrew Carnegie, the American steel manufacturer and philanthropist. The population increased from 10,009 in 1910 to 11,150 in 1914. About sixty per cent are American, the remainder consisting of Poles, Italians, Russians and Greeks.

Carnegie is situated in the beautiful Chartiers valley, in Allegheny County, in the southwestern part of the state, about eight miles southwest of Pittsburgh. The Pennsylvania, the Wabash and the New York Central railways serve the city. Electric lines communicate with Pittsburgh and adjacent towns. Carnegie is an important coal-mining center; natural gas and oil also are found in the vicinity. There are extensive iron, steel, lead and glass plants and manufactories of ploughs, stoves and tinware. The most notable structures are the $100,000 Federal building, completed in 1916, the Carnegie Free Library, the Masonic Temple and Saint Paul's Orphan Asylum.

CARNELIAN, *kahr neel' yan,* or **CORNELIAN**, a kind of stone (chalcedony), usually of a clear, rich, reddish color, but sometimes yellow, brown or white. It takes an excellent polish and is used in jewelry for seal rings, bracelets, necklaces and other ornamental articles. One of the first stones used for carving, engraving and ornamental purposes, it has also been prized as a charm. Goethe says:

"Carnelian is a talisman,
It brings good luck to child and man."

CARNIVOROUS, *kar niv' o rus,* **ANIMALS** are flesh-eating animals, natives of all parts of the world excepting Australia and New Zealand. This order of animals includes those of many sizes, from the tiny ermine, which can be hidden in a pocket, to bears which weigh as much as 2,000 pounds. Excepting the bears, all walk on the under surface of their toes. Many have sharp claws, heavy hair and strong, agile limbs. But all of them have large, strong teeth, with sharp, cutting edges, so they can cut and tear flesh food with ease. The animals in the zoölogical gardens in the city parks are nearly all flesh-eaters, or carnivorous—the lion, tiger, leopard, wild cat, bear, wolf, badger, seal, sea lion, walrus and many others. Many carnivorous animals, however, eat vegetable food, as well. The polar bear is very fond of grass, and some bears eat berries, nuts and the honey of the bumblebee. Although many carnivorous animals are beasts of prey, they are as well an aid to mankind, for they feed on mice, gophers and insects. They are hunted mainly for sport, but some are valuable for their fur. From this flesh-eating class of animals have come our two most valuable pets—the cat and the dog. See FUR-BEARING ANIMALS, and articles relating to those named above.

Related Subjects. To the student who wishes to gain an idea of the number and variety of these flesh-eating mammals, the following list will be useful:

Aard-wolf
Badger
Bear
Bloodhound
Bulldog
Cat
Cheeta
Civet
Collie
Coyote
Dachshund
Dingo
Dog
Ermine
Eskimo Dog
Ferret
Fox
Foxhound
Fox Terrier
Glutton
Great Dane
Greyhound
Hound
Hyena
Ichneumon
Jackal
Jaguar
Leopard
Lion
Lynx
Marten
Mastiff
Mink
Mongoose
Newfoundland Dog
Ocelot
Otter
Ounce
Panther
Pointer
Polecat
Poodle
Pug
Puma
Raccoon
Ratel
Retriever
Sable
Saint Bernard Dog
Scotch Terrier
Setter
Shepherd Dog
Skunk
Skye Terrier
Spaniel
Spectacled Bear
Spitz
Staghound
Tasmanian Wolf
Terrier
Tiger
Vampire Bat
Weasel
Wild Cat
Wolf

CARNIVOROUS PLANTS, the name given an interesting group of plants which lure insects into their grasp, then devour them. The flowers of such plants are usually so colored that from a distance, to the fly or insect, they appear to be decaying meat—a feast in store for them. But, instead, the adventurers are caught in the sticky hairs, where they die and are held until digested; or they slip down the inside of the flower into a watery trap. Most carnivorous plants live in moist places, where they obtain no nitrogen from the soil, so it must be supplied by insects. Their structure is so curious and their action so effective, it almost seems they have intelligence. See BOTANY, for illustrations.

Related Subjects. The articles on the following topics describe in more detail the methods by which these plants trap their food:

Bladderwort
Butterwort
Pitcher Plants
Sundew
Venus's Flytrap

CARNOT, *kahr no',* MARIE FRANÇOIS SADI (1837-1894), a French statesman, President of the French republic from 1887 to 1894. He was educated as an engineer and advanced rapidly in his profession, until he was appointed prefect of the lower Seine during the siege of Paris, in 1871. When, after the close of the Franco-German War, the French people reorganized the government and established a republic, Carnot took an active part in the nation's councils, and held public office without interruption until his election as President of France to succeed Jules Grevy (which see). He commanded the highest respect for his character and ability, and the Panama Canal scandals, which occurred during his term of office, did not reflect on his honor. During a celebration given for him at Lyons he was killed by an anarchist.

CAROB, *kair' ob,* a useful pod-bearing plant found in a wild state in all countries bordering the Mediterranean. It grows somewhat like an apple tree. The foliage is dark green and evergreen; the flowers, insignificant. The pods, brown, leathery and from four to ten inches long, are the most valuable part of the plant, for they contain a sticky pulp which tastes like manna. They are used as forage for horses and cattle, and sometimes as food for man. The brown, shiny, bean-like seeds are of no value. Carobs have been called *locust* and *Saint John's bread,* from the belief that they were eaten by John the Baptist in the wilderness. They are also thought to be the husks referred to in the parable of the Prodigal Son. The carob is not the same as the American locust, although it is somewhat similar and is an excellent shade tree.

CAROL, *kair'ol,* a song of rejoicing which forms a part of the celebration of the great festivals of the Christian Church, but associated most commonly with Christmas. The first Christmas carol, recorded in *Luke* II, 13-14, was sung by the heavenly chorus of angels on the plains of Bethlehem, and some of the most beautiful carols ever written, such as Nahum Tate's *While Shepherds Watched Their Flocks by Night,* and Charles Wesley's *Hark! the Herald Angels Sing,* were inspired by this old sweet story. Carol singing was very popular in Europe in the Middle Ages, and still is a special feature of the Christmas season in England, where bands of men and boys go about the streets for several nights before Christmas, singing in the open air.

Dinah M. Craik's familiar Christmas song, a special favorite in England, has the simplicity and picturesque charm of many of the older carols:

God rest ye, merry gentlemen; let nothing you dismay,
For Jesus Christ, our Saviour, was born on Christmas Day.
The dawn rose red o'er Bethlehem, the stars shone through the gray,
When Jesus Christ, our Saviour, was born on Christmas Day.

God rest ye, little children; let nothing you affright;
For Jesus Christ, your Saviour, was born this happy night;
Along the hills of Galilee the white flocks sleeping lay,
When Christ, the Child of Nazareth, was born on Christmas Day.

CAROLINGIANS, *kar o lin'je anz,* the second line of Frankish kings, whose name is derived from that of the most illustrious member of the house, Charlemagne, or Charles the Great. The Carolingians were the immediate successors of the Merovingian line (see MEROVINGIANS), the declining strength of the latter house being the occasion of their rise to royal power. In the seventh century the weak Merovingian kings gave the real governing power to officers called mayors of the palace, the most famous of whom was Charles Martel. His son, Pippin the Short, was crowned king of the Franks in 751, which marks the formal beginning of the Carolingian dynasty. The height of its fame and power was reached in the reign of Charlemagne, son of Pippin, whose conquests expanded the kingdom into a great empire. See CHARLEMAGNE.

Charlemagne's son divided his empire among his three sons, and at his death (840) his son, Charles the Bald, became king of the part of his territory which corresponds to modern France. The successors of Charles were weak and incompetent, and the dynasty came to an end with Louis V, who died in 987. The Carolingian was succeeded by the Capetian dynasty (which see).

CARP, *kahrp,* a fresh-water fish, originally from Asia, but now found in most parts of the northern hemisphere. In Europe it is much prized as a delicacy, and brings a high price. There are three varieties in European waters,

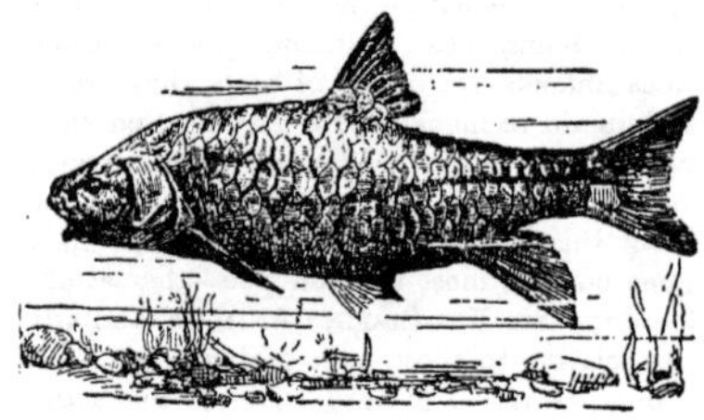

CARP

scale carp, evenly covered with scales; *mirror carp,* with larger, irregular scales; and *leather carp,* with only a few scales, much of the skin being bare. Carp feed on all sorts of material, both animal and vegetable. In seeking food they root like pigs, and so produce much muddy water. This has made their presence objectionable in many places.

The carp was introduced into the United States by the United States Fish Commission in 1877. It has multiplied enormously, especially in the streams of the Mississippi Valley, and has become one of the important food fishes. The quality of its flesh is not high, and most of the American carp are consumed by the poorer classes in the large cities. The fact that the carp's flesh remains firm a long time after being removed from water is of advantage in shipping it. The best market carp weigh four or five pounds; larger ones bring a lower price. Specimens weighing twenty-five pounds or more are sometimes taken.

The carp is a close relative of the goldfish, as well as of most of the small fishes of the lakes and streams that we call "minnows."

CARPATHIAN, *kahr pa' thi an,* **MOUNTAINS,** an important mountain system in Europe, extending in a semi-circle from the Danube River near Pressburg to Orsova, also on the southern bank of the same stream. It is the eastern extension of the great Central Europe mountain system. The mountains are not covered with perpetual snow, even in their highest altitudes, and glaciers are entirely absent. They are noted for the richness of their deposits of gold, lead, silver and copper, and many mines have been worked for centuries. The lower slopes are covered with forests of fir, oak and beech, in which roam wolves, lynxes and bears. The ranges of which the system is composed all lack the lofty grandeur of the Alps, the highest point, Gersdorfeispitze, being only 8,737 feet above sea level.

During the War of the Nations fierce fighting took place each year in the passes over the ranges, and it is estimated that as many as 500,000 men were killed and buried in the mountain chain, which was alternately held by Russians and Austrians.

CARPENTER, FRANK GEORGE (1855-), an American traveler, newspaper writer and authority in geography, whose letter-writing tours for various American journals have taken him into nearly every quarter of the globe. He was born at Mansfield, O., and was educated at the University of Wooster in that state. Beginning his career as a legislative correspondent for the *Cleveland Leader* in 1879, he was not long in broadening his field, and in 1888-1889 made a trip around the world for a newspaper syndicate. Since then he has visited almost every part of Europe, Asia, Africa and South America. He reported the Mexican revolution of 1913, and has written also of life and conditions in Cuba, Central America, the Panama region and Canada.

Carpenter's *Geographical Readers,* with their vivid descriptions of the author's personal observations, written in an interesting and popular style, are favorites in the common schools. His industrial series, comprising *How the World is Fed, How the World is Clothed* and *How the World is Housed,* are for children's reading; they are easy to understand and are full of interesting information.

CARPENTER'S HALL. See PHILADELPHIA, subhead *Historic Buildings.*

CARPENTRY, *kahr'pen tri.* What boy or man is there who does not truly like to work with hammer and nails and saw? Most men are too busy at other things to be able often to indulge their tastes in this direction, but when they have the opportunity they thoroughly enjoy a little carpentry. As for boys, the desire to fashion things, large or small, useful or ornamental or neither, is a part of their very being.

How a Carpenter Builds a House. It will be seen from the following description of a carpenter's work that he is something more than a mere driver of nails. In a small way he is an architect and an engineer, and along certain lines a mathematician. Of course he must also be skilful with his tools, but a first-class carpenter is even more active with his brains than with his hands.

The Frame. A wooden house is built around a skeleton called the frame, of which there are two types, the *braced* frame and *balloon* frame. The former is a complete structure of itself; the latter depends upon the boards of the walls and floors for bracing against winds and other stresses. The braced frame is very seldom built in the United States and Canada, and only the balloon frame is described in this article.

In Fig. 1 is shown the frame of a small one-story gable-roof house, and the following paragraphs tell exactly how it is constructed. By carefully studying both the pictures and the text, anyone with a taste for carpentry should be able to build correctly any simple type of house. One not familiar with the names for different kinds of lumber ought also to read the article on that subject.

In the following paragraphs a description of the parts is first given in order that a clear idea may be had of their relation; afterwards is told the usual order in which they are put together:

Sills. These are marked *a a* in Fig. 1. In the present instance they are 4"x4", and the exact length of the sides of the house. (Explanation: ' indicates *foot* or *feet;* " indicates *inches.*) For a longer house it would be necessary to join and spike two timbers. Fig. 2 shows how this may be safely done. The sills are joined at corners as in Fig. 3, and spiked. If the house is built upon a concrete foundation, a bed of lime mortar is spread over the top of the latter and the sills set in it with

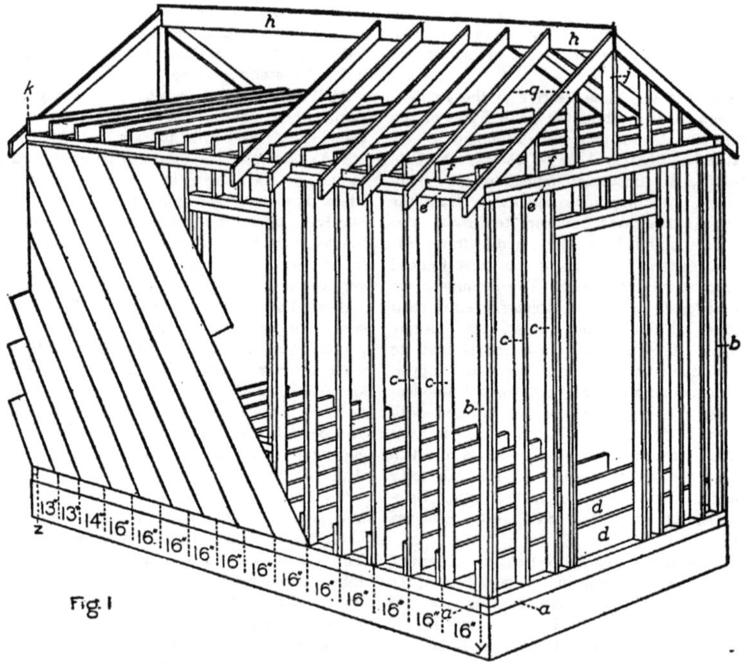

Fig. 1

their edges an inch from the outside. If the building will be subject to high winds, bolts are imbedded in the concrete while it is being made, and the sills fastened as in Fig. 4.

Corner posts. The uprights, *b b* in Fig. 1, are of two 2"x4", spiked together. They should be

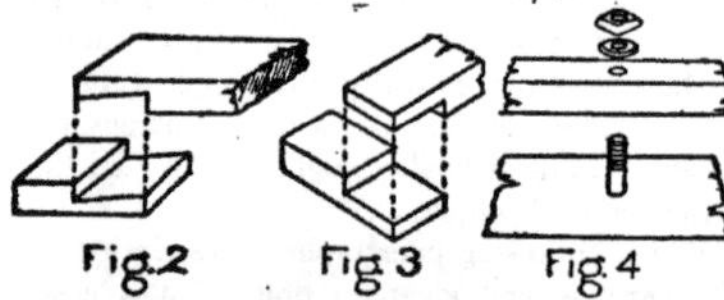

Fig. 2 Fig. 3 Fig. 4

about 8" longer than the distance desired between ceiling and floor. Thus, a house of this type with a ceiling 7' 4" high has an 8-foot post.

Studs. These are indicated by *c c* in Fig. 1, and are 2"x4". The distance between them must be so arranged that if the house is plastered the ends of the laths, which are 48 inches long, can be nailed to them; therefore they are usually placed either 16" or 12" from the center of one to the center of another. Study of the spacings along the side (*y* to *z*) will show how to space the other walls. It will be noticed that the first center is measured from the edge of the corner post and that the unavoidable spaces of odd size are all made at one end of the wall, so that few laths need to be cut.

Joists. The floor beams, or joists, *d d* in Fig. 1, are 2"x8". In houses of longer span the larger dimension of the joists should be increased one-half inch for every foot of length beyond twelve. They are set beside the studs and corner posts, and spiked to them as well as to the sills. To make their tops level they are notched on the under side, as in Fig. 5. Additional strength may be gained by bridging, as in Fig. 6.

Plates. The horizontal timbers *e e* in Fig. 1 are double 2"x4", like the corner posts. They are joined at the corners in a manner similar to the sills (Fig. 3). Gable houses are sometimes built without the plate at front and back, the studs continuing to the rafters.

Ceiling beams. These are shown at *f f* in Fig. 1, and are 2"x4". For longer spans they should be 2"x6". Because laths are nailed to them they are directly above the studs, with an additional one close to the plates at each end. Their projecting corners (k) will have to be sawed off before the roof is put on.

Rafters. These timbers, marked *g g* in Fig. 1, are spaced about 2 feet apart, the outside edge of the first pair being flush with the outside of the plate. Directions for measuring and cutting rafters are given below. The ridge board *h* is an inch thick and about 6 inches wide, and cut as shown in Fig. 14.

King post. This is marked *j* in Fig. 1. The

short studs beside it are cut as explained below, under *Cutting the Rafters*. They are spaced 2 feet *on centers*.

Partitions. The frame of a partition is of 2"x4" vertical studs resting on a 2"x4" and

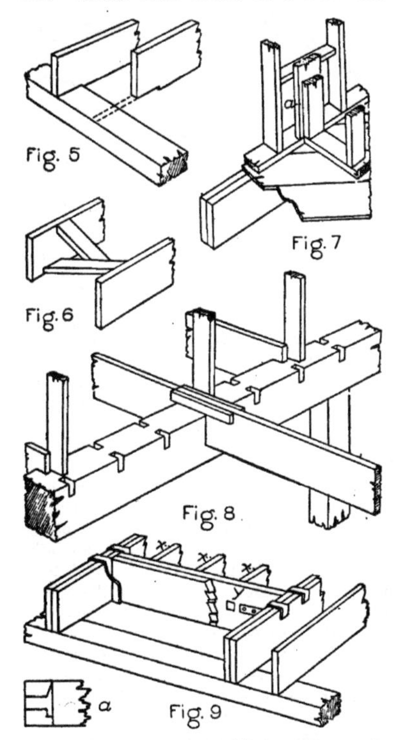

crowned by a 2"x4". A partition running parallel with the joists should be set on two 2"x8", as in Fig. 7. In a small house a partition at right angles to the studs may safely rest on them, but in a building of larger span, posts and a beam should be added, as in Fig. 8. Unless there is a cement floor to the cellar the posts should be set in the ground a few feet. Care must be taken that there is a space at the edge of every partition, to which laths can be nailed. Fig. 7 (*a*) shows how this may be effected.

Window openings. These are framed with double 2"x4"'s at top and side, as in Fig. 10. In some cases a stud may form part of a side, as at *a;* but usually at least one side must be like *b*. An opening should be about 2 inches longer than the finished window frame, and about 7 inches wider (unless the window is hung without weights, in which case the opening is only 2 inches wider).

Door openings. These should be 3 inches higher than door frames and 6 inches wider. They are framed above and at the sides like windows.

Openings for chimneys, cellarways, etc. Horizontal openings are framed as in Fig. 9. The joists *x x* should be fastened to the header *y*

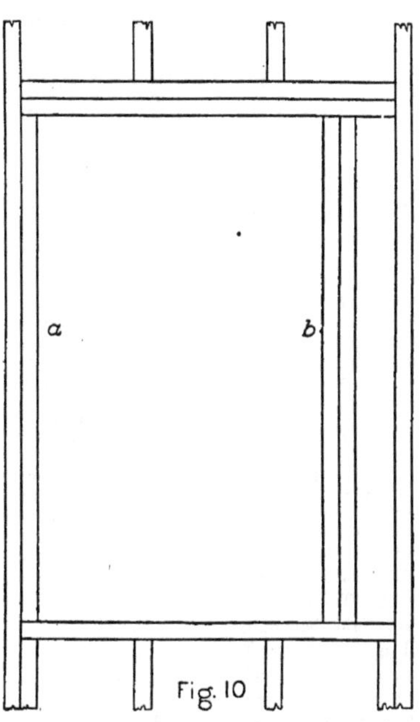

with *mortise and tenon* joints, as shown at *a*, in addition to being spiked. None of the joists should come within 4 inches of the brick work of a chimney.

The Outside. The boards which cover the frame are called *sheathing;* they are usually put on diagonally, because in this form they give stronger bracing to the frame. They are sawed to the proper length after being nailed, as in Fig. 1. The best sheathing is shiplap, but common boards of the best grade are suitable where warmth is not important.

Building paper. One or two thicknesses of paper are put on over the sheathing, and these should be carefully brought around all corners and edges of openings, to keep out wind. See Fig. 16.

Siding. The outer covering is laid horizontally, commencing at the bottom. Beneath the siding are the two pieces *a* and *b* in Fig. 12, which form a *water table*.

Roofing. The roof is usually covered with common boards, laid two inches apart except at the eaves, where they are close together.

Corner boards. These finishing boards are

nailed vertically over the siding. If one corner board is 4 inches wide, the other should be 5, as in Fig. 11.

Shingles. Each row of shingles covers all of the row below it except 4 to 6 inches. The row at the eaves should slightly overhang the last board, and be double, as in Fig. 15. As each row is laid a board is temporarily nailed to the roof, with its upper edge marking the location of the next row. Care should be taken to see that no joint between two shingles is directly above one in the row below. Very broad shingles should be split. Two nails placed about two inches above the exposed portion will hold the ordinary shingle. Fig. 14 shows a method of making the ridge water-tight.

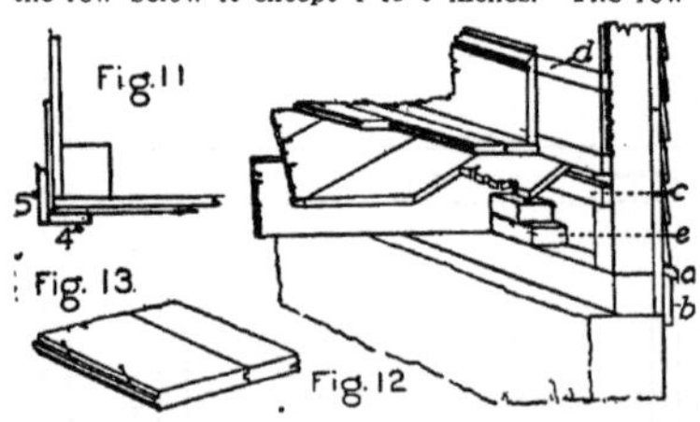

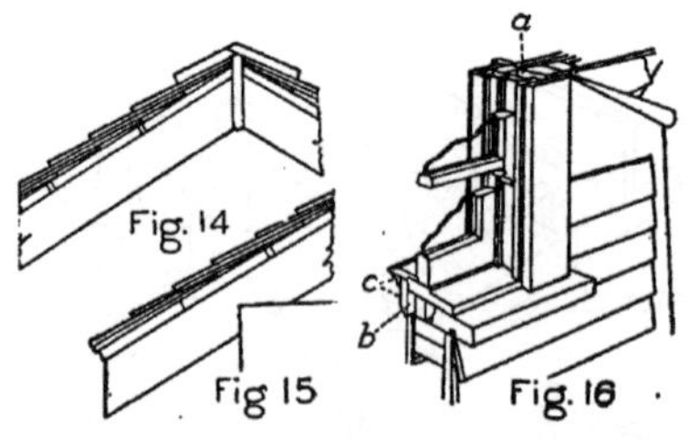

Eaves. The edge of the roof may be finished in many ways, which can best be learned by examining finished houses. Metal eave-troughs, or gutters, may be attached either before or after shingling.

The Inside. *Floors* are usually double, as in Fig. 12. The lower layer is of plain or matched boards laid diagonally so that their shrinkage will not cause cracks in the upper layer of finished flooring. The latter is *blind nailed,* as in Fig. 13. Where the end of a diagonally cut board will be unsupported, a small strip *c* (Fig. 12) should be fastened.

Walls. If plaster is to be applied, walls and ceilings must first be covered with laths set horizontally about ¼-inch apart. Remember always to have a nailing strip for the end of every lath. In place of plaster the walls may be finished with composition board or with V-joint lumber, blind-nailed.

In cold climates sheathing is put inside the studs as well as outside. In this case vertical nailing strips must be added before the laths are attached.

Baseboards are attached as shown in Fig. 12. With V-joint they may be omitted and the angles filled with *quarter-round,* which is like a quarter of a small cylinder.

Windows. All the details of a window are shown in Fig. 16. Most of the parts come from the mill already cut in proper sizes. In the picture, *a* is the space for the pulley weights, *b* is a plaster ground like *d* in Fig. 12, and *c c* are trimmings. A good rule for windows is that there should be a square foot of glass for every 100 cubic feet of interior space to be lighted. Before commencing a building it will be well to find out if the size of window desired is in stock at the local planing mill.

Doors. Both frames and doors are usually purchased ready made. A common size of door is 2′ 8″x6′ 8″x1⅜″. Inside doors are set high enough to swing over carpets, and a *saddle* is put underneath them in the door frame. The details of door hanging can best be learned by examining doors already in use.

Filling. Where cold winds are frequent it is advisable to fill spaces between joists with old bricks or odds and ends of 2″x4″, as at *e* in Fig. 12. Openings between rafters at the plate may also be blocked.

Cutting the Rafters. Because of the diagonal cuts in rafters, their measurement is more difficult than that of other parts of the frame. In actual cutting the steel square is used, but before building commences the length of rafters can best be estimated by a graphic drawing, as in Fig. 17. On a large sheet of paper draw a line *ac* to represent the plate, using a convenient scale, say ¾ of an inch to the foot. If the roof is to be ¼ pitch (see ROOF), the perpendicular line *bd* at the center will be ¼ *ac*. Now draw the lines *ba* and *bc;* their measurement gives the length of the rafters from ridge to edge of plate. If the rafters are to be notched to the plate for half their thickness, this will be the measurement along their center, as will be seen from Fig. 18. Do not forget to add to the length of the rafter thus found the distance which you wish it to project beyond the plate. The distance *ad* is called the *run* of the rafter; the distance *bd* its *rise*.

A steel square has a scale of inches on each outer edge, besides scales of various other sorts and tables which cannot be explained here. To measure a common rafter like the one in the house illustrated, the inch scales are used. In the present instance the rafter has a *run* of 6 feet and a *rise* of 3 feet, and in one foot of run will rise 6 inches. Lay the square on the edge of a 2″x4″, with the 6″ mark of one arm and the 12″ mark of the other as at *a* in Fig. 18. Mark along the line *p* for the *plumb cut,* but in cutting the rafter later, remember to allow a half-inch for the ridge-board. The dotted lines are added to show what part this rafter will occupy in the finished frame. Now mark where the 12″ line rests and move the square so that the 6″ mark rests on this point, as in position *b*. In the sixth position of the square (*c*) the 12″ mark will be at a point directly over the edge of the plate. Move the square to the seventh position and mark the plumb cut of the notch. The foot cut (*f*) will be at right angles to it.

In cutting the studs between the plate and the gable, remember that a stud one-third of the distance between the edge of the plate and the king post will be one-third as high as the latter. To cut the upper edge of one of these studs mark a rise of one inch, one-half its width.

Construction. In erecting a house, make frequent tests with a level. Set the sills in place first, level them and be sure that their corners are square. Next lay and level the joists; to make certain that their tops are all at the same height lay a long board across them to hold the level. Set the corner posts in place, using the level to make sure that they are perpendicular. They may be held in position by temporary braces. Some carpenters nail the plate to the studs before putting the latter in place, but this makes a heavy wall to lift, and for amateurs it may be better to nail the plate to the corner posts first, setting up the studs one by one. The window and door openings are not sawed until after the frame of studs is all up, and it is often convenient to lay the rough flooring first, so that the carpenters can walk about. Rafters should not be erected until the ceiling beams are spiked in place, or their weight will spread the walls apart. Enough of the sheathing to hold the frame in the position in which its members have been found to be level and perpendicular should also precede the rafters. The end rafters and the ridge board, which should be leveled with special care, are put in place before the intermediate rafters. Window and door frames are set before siding. The order of construction of the rest of the house is a matter of convenience.

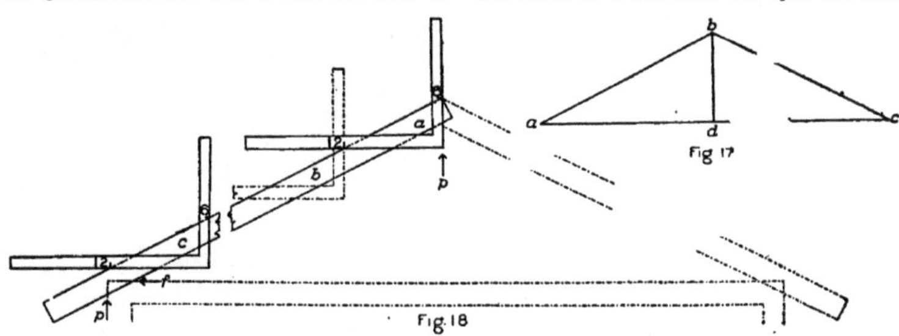

Fig 17

Fig. 18

To mark the positions of the siding cover a piece of strong twine with chalk, stretch it along the line for the top of a piece of siding and snap it against the wall; it will leave a distinct mark on the building paper.

Nails should be large enough for strength, but never of a size that will split the wood. For names of different sizes of nails see the article on that subject.

The only hardware necessary for the building will be locks and hinges. Small pieces of tin called *flashing* must be placed where the chimney is built through the roof, to prevent the leaking of water into the house. C.H.H.

CARPETBAGGERS, *kahr'pet bag erz.* After the close of the War of Secession nearly all the whites in the Southern states were deprived by Congress of the privilege of voting, because they had taken part in the rebellion against the government. This condition was taken advantage of by unscrupulous politicians and adventurers from the North, who took up a temporary residence in the Southern states that they might control the negro vote and be elected to office. The name comes from the old-fashioned traveling bag, which was made of carpet with leather mountings. The carpet-bag suggested the temporary character of the residence of these Northern adventurers. The state governments administered under those conditions were of the worst sort imaginable. Enormous taxes were levied and the money was frequently spent in extravagance and speculation, leaving the states burdened with debt. See RECONSTRUCTION.

The name is also sometimes applied to those Northern politicians who, before the war, took up their residence in the South with a view to representing those states in Congress.

CARPET BEETLE, sometimes called BUFFALO BUG or BUFFALO MOTH, a troublesome beetle about one-eighth of an inch long, marked with black, white and red. The larva (young) is a short, brown, hairy grub that feeds on carpets and woolen clothing, hence the name. This beetle and its young are difficult to remove. The best preventive is use of rugs instead of carpets on hardwood floors. Spraying a carpet with benzine and then airing, and putting tarred paper underneath when laying again, will kill beetles. Pyrethrum, powder and naphtha balls are also helpful. See BEETLE.

CARPETS AND RUGS, textile coverings for the floor. The Latin word *carpita,* from which the name *carpet* is derived, really means *rug,* but at the present day a carpet is regarded as distinct from a rug. The difference lies in the fact that a carpet is used to cover the entire floor surface of a room, while a rug covers only a part. Also, while most rugs are woven all in one piece, carpets are usually made up of strips of varying width sewed together.

Antiquity of Carpets. Carpet making and rug weaving by hand have for centuries been Oriental arts, and it was not until the nineteenth century that machinery was employed to meet the growing demand for floor coverings. In ancient times, even in the remote days when our ancestors dwelt in caves and lived by hunting, it is probable that their rude dwellings were made more habitable by the use of skins for rugs and for hangings at entrances. Couches were no doubt made of skins, and it would be only natural that if the supply were sufficient more should be placed on the ground or floor of the cave. Thus it is certain that the first floor covering was a rug, and not a carpet, in the modern sense of the word. Gradually the use of skins was extended to coverings for seats, and as soon as the art of weaving was discovered textiles began to take their place.

That carpet making is of very ancient origin is proved by the fact that there are still in existence carpets known to have been made nearly 1,500 years before the Christian Era. The palaces of the Pharaohs of Egypt were decorated with carpets and rugs, and these were also used in the temples.

How Carpets and Rugs Are Made. The methods employed in weaving Oriental carpets are the same now as they were centuries ago. On a wooden framework, its size depending on the length and width of the carpet or rug to be made, are stretched threads of hemp, cotton, wool, or silk, to form the *warp,* or foundation. To these threads are knotted tufts

HOW ORIENTAL RUGS ARE MADE

of wool, silk, camel or goat hair, or mixtures of those materials. The ends of the knotted pieces are allowed to protrude, all on the same side of the warp. After a row of such pieces has been added, a thread of the same material as the warp is run in alternately over and above the warp threads. The knots and weft thread are pressed tightly together by means of a blunt comb. Row after row of knots and weft thread are added, until the desired size is obtained. On the number of knots in a square inch largely depends the value of the carpet. In some carpets the number may be 200, or even fewer, while in others there may be as many as 750 knots to the square inch. Such weaving is, of course, a very slow process, and often the work of several weavers for more than a year is required to complete one carpet.

In Eastern countries most of the weaving is done in the homes of the people, and the patterns woven are sometimes handed down in one family for generation after generation. Throughout the East the method of weaving is the same, with variations only in the form of knot tied and in the patterns used. Rugs and carpets produced in Mohammedan countries have patterns of geometrical design only, as the laws of the Koran forbid the reproduction of the image or likeness of any living thing.

Modern Carpets. In all Western countries the steam power loom has entirely superseded the weaving of carpets and rugs by hand. The origin of carpet weaving in the United States is found in the rag-carpet industry, which flourished as recently as the early part of the nineteenth century. Careful housewives preserved their rags, which they cut into long, narrow strips and then sewed together. These strips were then made into carpets by local or traveling carpet makers. Later, women in the home learned the art, and the majority of men and women living to-day remember this familiar home process.

In 1791 the first carpet-weaving factory was erected in Philadelphia. Since that date the industry has grown until at the present time the United States produces and uses more carpets than any other country in the world. In 1841 Erastus Bigelow perfected the first power loom in the United States. This greatly reduced the cost of production and made it possible to do in one day by machinery what had previously required many weeks to do by hand. The Bigelow company manufactures two of the finest domestic rugs on the market —the Anglo-Persian and Anglo-Indian. Both are imitations of Oriental weaves.

Varieties of Oriental Rugs. The art of making Oriental rugs is so ancient that it is difficult definitely to trace to its origin. It is not improbable that Egypt improved the crude work of its predecessors, for the similarity of the forms and designs found in modern Oriental rugs to forms and inscriptions of Egyptian architecture carries the art of weaving back to that ancient civilization. Rugs are written pages. In their chaotic designs is a symbol language, the key of which, through the passing of centuries, is all but lost. The Ispahan rug, with its gorgeous field of *Ispahan* red and varied floral designs, with a fine tracing of puzzling, fascinating lines, inaugurated a new era of rug making. This rug is hardly in existence to-day, only a few perfect specimens and many fragments being preserved in private collections.

Oriental rugs of the present time are classified into six distinct groups—Persian, Turkish, Turkestan, Caucasian, Chinese and Indian. Each group may be subdivided, the names of the rugs being derived from the various districts and towns in which they are made. Persia leads in the production of beautiful and costly rugs, and there are about nineteen varieties of this class. The Kermanshah, Tabriz, Shiraz, Saruk, Sehna, Kurdistan and Serape are among the finest weaves. Of the Turkestan family, the Bokhara, Khiva and Samarkand are probably the best known. Ghiordes, Anatolian, Kiz-Kllm and Hamidie are prominent names among those of the Turkish group, and Karabagh, Kazak, Cashmere and Daghestan are familiar patterns in the Caucasian division. Chinese rugs form a distinct type, odd geometrical designs or unique varieties of the lotus flower being distributed on harmonious grounds of blue, red or yellow. Some fine Indian rugs come from Tanjore and Benares. These are usually distinguished by a medallion center.

One of the most pronounced patterns in Oriental rugs is the *prayer rug,* which always accompanies the devout Mohammedan. Upon this rug he performs his devotions, his face and the point of the pattern being directed toward Mecca. Eastern rugs are the most durable and most valuable of all rugs; some have been in use for 300 years and their price runs into thousands of dollars. Large rugs of fine weave have commanded a price as high as $45,000; a rug nine by twelve feet may command a price of $2,000, and one about two by three feet may cost from $50 to $150.

Modern Makers. Possibly the rug most familiar at the present time is the *Brussels,* sold either as a rug, or "art square," as it is sometimes called, or as a carpet. Originally made in Brussels, it was copied in England, whence it was introduced into America. The *Axminster,* named after a small town in England where it was first manufactured, is a comparatively inexpensive and attractive copy of the Turkish designs. The *Wilton,* somewhat more expensive than the Brussels or Axminster, is also made in Oriental patterns. Imitation Brussels and Wilton carpets and rugs are sold as Tapestry Brussels, attractive in color and design but lacking in the wearing qualities of the genuine Turkish articles or the better qualities of American rugs. A variety of Axminster with the nap or pile on both sides is now extensively manufactured in the United States and is sometimes called a *Smyrna.* Some cheaper forms of carpet are manufactured of undyed material, and a pattern is afterwards stamped by a printing machine similar to that used in calico printing.

It is only within comparatively recent years that the use of carpets has become universal. Modern methods of manufacture have placed floor coverings within the reach of families in the most moderate circumstances, and there are few homes in Canada or the United States which are not ornamented with carpets or rugs of some kind. The difficulties met in cleaning the carpet have led to greater use of rugs, which can be easily moved for the removal of dust. The advent of the vacuum cleaner, however, will doubtless serve to extend the use of carpets. J.S.C.

CARRACCI, or **CARACCI,** *kah rah' che,* a celebrated family of Italian painters, the three leading members of which founded in Bologna, in 1582, a famous academy of painting, "the academy of those on the right road." Their school is known in the history of art as the *Eclectic School,* which means "a bringing together of the best points of various systems"; their object was to do away with the uninspired imitation of Raphael and Michelangelo which then prevailed, and to unite the special excellencies of the great masters of the Renaissance—the drawing of Michelangelo, the color of Titian, the grace and symmetry of Raphael and the light and shade of Correggio.

Ludovico (1555-1619), the eldest of the three Carracci, was born in Bologna. Later, in Florence, he came under the influence of Andrea del Sarto; in Parma, of Correggio; in Venice, of Veronese and Tintoretto. On his return to Bologna he became associated with his two cousins, AGOSTINO (1557-1602) and ANNIBALE (1560-1609), in carrying out the program of their famous school, which may be regarded as the first modern academy of art.

Among Ludovico's important canvases are *Sermon of John the Baptist,* in the Gallery of Bologna, and *Conversion of Saint Paul,* in the Munich Gallery. He also painted a number of sacred frescoes. Annibale's *Madonna Appearing to Saint Luke and Saint Catharine* and *The Resurrection,* both in the Louvre, are among his greatest achievements; his *Three Marys,* in Castle Howard, Yorkshire, England, is a wonderful example of pathos in art. The galleries of Paris, Petrograd, Madrid, Florence and Rome contain specimens of his landscapes. Agostino was an engraver of first rank as well as a painter, and his engraving of Tintoretto's *Crucifixion* was preferred by that master to the original. A celebrated canvas is his *Last Communion of Saint Jerome,* now in Bologna. The famous fresco decorations of the gallery of the Farnese Palace, in Rome, are the joint work of Agostino and Annibale.

CARRANZA, *kah rahn' zah,* VENUSTIANO (1860?-), a Mexican leader and general who was recognized by the United States and other nations, in 1915, as President of Mexico, although he styled himself "First Chief." Until the successful revolt against Diaz in 1911, Carranza was one of the great landowners of Mexico, and he had occupied a high judicial position. He was known as a high-class Mexican, educated and wealthy; his estates in Northern Mexico included thousands of acres. This aristocrat was a friend of President Madero, who appointed him governor of the state of Coahuila. Carranza refused to recognize the provisional government formed by General Huerta, and in March, 1912, after Madero's death, was acclaimed First Chief of the Constitutionalists, the name by which the former adherents of Madero now called themselves.

VENUSTIANO CARRANZA

Carranza's career for the next few years is a part of Mexican history, which is chiefly a record of civil war. In the districts over which he held sway he preserved a fair measure of order, although several times he showed a complete disregard of the rights and wishes of neutrals, including the United States and its representatives. He proved himself, however, a capable soldier, and his armies succeeded in defeating and scattering those of Villa, his chief rival. During the summer of 1915 Carranza became in fact ruler of all Mexico excepting sections in the north and in the south. This condition was accepted by the United States and the six leading South American republics as indicating Carranza's ability to bring peace to his country; therefore his government was formally recognized on October 19, 1915.

Early in 1916 Francisco Villa, former leader but then a bandit, attacked Columbus, New Mexico, killed a number of people, fired the town and escaped southward into the Mexican desert. A United States cavalry force, in pursuit, had at first the coöperation of the Carranza forces, but within a month met opposition from his soldiers. It developed that Carranza's leadership was in the balance; his power increased, however, but his authority was continually menaced by active banditry, particularly by Villa. See MEXICO, subtitle *History;* VILLA, FRANCISCO.

CARRARA, *kah rah'rah,* **MARBLE,** the fine, white, crystalline limestone from which the most famous statues of the world have been made. This sugar-grained material is valued not only for its pure white beauty, but also for its smoothness when polished. The mountains which yield this valued marble surround the city of Carrara in Northern Italy and reach a height of over 5,000 feet. Many sculptors work there to save the expense of shipping the marble. Although the Carrara quarries have been worked for 2,000 years, they having furnished the material for the Pantheon at Rome, the supply is still practically inexhaustible. These quarries employ about 10,000 men. Illustrations showing the quarries at Carrara appear with the article MARBLE.

CARREL, *kair'el,* ALEXIS (1873-), a French biologist and surgeon, famous for his experiments concerning the nature and processes of animal life. Dr. Carrel was born in France, received his professional training there, and first won distinction as a surgeon at the University of Lyons. But it was in the United States, to which he went in 1905, that he carried on the experiments which made him world-famous and won for him the Nobel Prize in medicine for 1912 (see NOBEL PRIZE). As fellow of the Rockefeller Institution for Medical Research, a position he held from 1909 to 1914, he made some of the most important discoveries in the history of medicine and biology. He kept alive certain organs of dead animals, and thus proved that many organs, such as the heart or the stomach, may live and carry on their functions even after the death of the body. Thus he proved a difference between general death, which ends the life of a distinctive animal or person, and elemental death, which ends the life of its tissues. These discoveries suggested the possibility that dead or defective organs could be replaced by sound, live ones. Dr. Carrel successfully transplanted veins and arteries, kidneys and even blood-vessels, all of which were kept alive in cold storage.

Not only has he kept separate organs alive, but he has kept alive separate cells apart from the organs of which they were once parts. He has demonstrated that these separate cells will grow independently of each other if they are given proper nourishment, and also that their growth can be stimulated from two to forty times by adding what he calls *tissue juice.* This juice is derived from the tissues of animals, and Dr. Carrel found that the juice of a certain animal produced quicker growth when applied to another animal of the same species than when applied to cells of any other species. These experiments and discoveries may result in new methods for treating injuries and diseases, but Dr. Carrel's work along these lines came to a temporary halt in 1914, when he returned to his native land to serve as surgeon in its army in the War of the Nations. In France, however, his work among the wounded led to his discovery of a new method of sterilizing deep wounds. This consists in the continuous irrigation of infected wounds with a solution of bleaching powder neutralized with soda mixed with boracic acid. Washing continues for several days until the wound is found to be disinfected. See WAR OF THE NATIONS.

Among his numerous scientific treatises are: *The Preservation of Tissues and Its Application to Surgery; The Surgery of Blood Vessels; Results of a Replantation of the Thigh;* and *The Transplantation of Veins and Organs.*

CARRIAGE, *kair'ij,* a word derived from the Latin *carrus,* meaning a *chariot,* and now applied to a wheeled vehicle for the conveyance of passengers, as distinct from those used for transportation of freight. The first vehicle used in ancient times was probably somewhat in the form of a sleigh light enough to be drawn over the ground. Then came the narrow, two-wheeled war chariot, from which have been gradually evolved the light, springed vehicles of the present day. The heavy, broad-rimmed wheels of the chariot have given place to narrow, rubber-covered wheels, running on steel axles with ball bearings. Carriages have assumed a great variety of shapes as the result of individual fancy or the necessities of existing roads. From the very earliest ages the use of carriages was regarded as a luxury in Europe and from time to time laws were passed restricting their use. At the present time a tax must be paid on all vehicles in England used for other than purely commercial purposes.

Carriages may now be divided into two classes—those with two wheels and those with four. The carriage- and coach-making industries have suffered greatly since the introduction of dependable, low-priced automobiles, the use of the latter having greatly increased in all cities and having almost driven the

carriage out of many rural districts. The different leading varieties of carriages will be found described in these volumes under the names by which they have become known.

CARRIER PIGEON, *pij'un,* or **HOMING PIGEON,** a variety of domestic pigeon, which, because of wonderful flying powers and its love of home, can be trained to carry messages. Carrier pigeons are larger than the

CARRIER PIGEON

doves, and they have long wings, a large mass of naked skin at the base of the beak, and a circle of naked skin around the eyes. Their speed is marvelous, and the distance they can fly without rest seems almost unbelievable. An American homing pigeon is known to have made a journey of 1,040 miles without stopping. It is thought that the pigeon was first used as a carrier by the Chinese, but the date is unknown. When Joshua invaded Palestine, however, this means was used to communicate with the camps on either side of the Jordan.

Carrier pigeons are trained for service in war. During the War of the Nations (which see) they carried thousands of messages, flying at a height of about one-half mile. Seldom was one killed, owing to swiftness of flight and small size. Messages are placed in small capsules, tied just above the foot.

CAR'ROLL, Charles (1737-1832), an American statesman of Revolutionary fame, who gave freely of his wealth to further the patriot cause, and was influential throughout the war and during the early national period. He was born in Annapolis, Md., but received his education abroad. On his return to Maryland in 1765 he settled on an estate in Frederick County, the name of which appears in all his signatures, for, to distinguish himself from others who had his name, he always signed himself, "Charles Carroll of Carrollton." In 1776 he was elected to the Continental Congress from Maryland, and was one of the signers of the Declaration of Independence; it is an interesting fact in connection with this that he outlived all of the other signers. In 1776 the Continental Congress sent him, with Samuel Chase and Benjamin Franklin, on a fruitless mission to Canada in the hope of persuading the Canadians to join cause with the Americans. In 1789 he was elected as the first Senator from Maryland under the Constitution of the United States, serving until 1792. Carroll's last public act was performed on July 4, 1828, when he turned the first spadeful of earth in the building of the Baltimore & Ohio Railroad—the first passenger railway in the United States.

CARROLL, Lewis. See Dodgson, Charles.

CARROT, *kair'ut,* a familiar garden and field vegetable of the parsley family, cultivated for its root, which is both a table and a stock food. The plant is a biennial; that is, its entire period of growth extends through two

THE CARROT
Illustration shows plant stalk, leaves, flowers, seed and root.

seasons (see Biennial). The roots are yellow, white or reddish, and are slender and tapering, those raised for table use being smaller and finer-grained than the stock varieties. Carrots are grown from the seed. Those for the table may be sown in the spring as soon as the weather is settled; the rows should be from one to two feet apart, and the growing plants kept from two to three inches apart in the row. They thrive best in a rich soil containing sand and clay, and need little attention.

Carrots for stock are planted in April or May and grow well into the fall before maturing. The roots are an excellent food for farm animals, especially for dairy cattle, as they make a good quality of milk for butter. For the table, carrots are cooked in cream, boiled and seasoned in various ways, used in soups, etc. They rank well among nutritious vegetables, as their sugar proportion is about 9 per cent. Other ingredients are water, 88.2 per cent; protein, 1.1; fat, .4; fiber, 1.1; ash, 1.0 (see FOOD, subhead *Chemistry of Food*). The loss of sugar resulting from boiling may be lessened by cooking the whole carrot quickly or cutting it into large pieces. Like many garden vegetables, carrots may be stored through the winter. If dried thoroughly and packed in a box between layers of sand or moss they will keep for months.

In Germany, the carrot, cut into small pieces and thoroughly dried, is used as a substitute for coffee, and in the United States it has been used to adulterate coffee. The plant contains a coloring matter that is sometimes used to color butter. The wild carrot is a troublesome weed with a woody root.

CAR'SON, CHRISTOPHER (1809-1868), a famous American frontiersman, better known as KIT CARSON. No romantic hero of fiction has probably aroused more interest in lovers of adventure than has this cool-headed, daring trapper and hunter. He was born in Madison County, Kentucky, but was early taken to Missouri, where for a time he was apprenticed to a saddler. In 1826 he began his adventurous life by accompanying a party of hunters to New Mexico. Later he went several times to the Pacific coast and acted as hunter for Western army garrisons. He was with Fremont in two expeditions across the Rocky Mountains and occasionally helped Western ranchers to drive cattle and sheep for long distances through the wild country. At one time Carson alone took a drove of more than fifty mules and horses for a distance of 500 miles through an almost uninhabited region.

Appointed United States agent to the Utah and Apache Indians in 1854, he performed notable service for the government through his friendship with influential chiefs, and during the War of Secession, as a scout in the Southwest, he acted with great energy and skill in behalf of the Union. At the close of the war he was brevetted brigadier-general. Many of his thrilling adventures as scout, guide, hunter, trapper and Indian fighter were almost incredible, and in cunning and resourcefulness, as well as in woodcraft, he rivaled, if he did not excel, the most expert Indians.

CARSON, SIR EDWARD (1854-), a British statesman, noted for his violent opposition to Home Rule for Ireland. Carson was born in Dublin, was graduated from Trinity College there, and in 1892 was elected to the British Parliament as member for Dublin University, at the same time becoming Solicitor-General for Ireland. During the years that followed he was a conspicuous figure in all the debates on questions relating to Ireland. His opposition to the Irish Nationalists, led by John Redmond, naturally allied him with the Conservatives, or Unionists, and he was Solicitor-General in the Salisbury and Balfour Ministries from 1900 to 1906. When Arthur J. Balfour retired from the leadership of this party, Carson was one of the three men who were considered as his successor. In Parliament he was recognized as a brilliant debater, absolutely fearless but inclined to extremes, and he was passed by in favor of Andrew Bonar Law, who was chosen leader of the Conservatives in the House of Commons.

SIR EDWARD CARSON

In 1912 Carson was the leader of the Ulstermen in their resistance to Home Rule. The introduction of a new bill to establish Home Rule in Ireland was followed by threats of revolution in Ulster, whose Protestant Orangemen declined to be ruled, so they said, by the Roman Catholics. Even before the bill was presented to Parliament Carson organized a massmeeting of nearly 100,000 armed men, who paraded before him and Bonar Law, as a warning to the government. In the House of Commons Carson was bitter in his denunciation of the bill, and in Ireland he became "general" of the Ulster volunteers. With the Orange battle-flag used by William III at the Battle of the Boyne waving over his head, he was the first of 350,000 Ulstermen who signed the solemn covenant to resist, under arms if necessary, the establishment of Home Rule.

This violent opposition led to a proposal for compromise, but before any agreement could be reached the outbreak of the War of the Nations in 1914 indefinitely postponed a decision. When the Asquith Ministry was reorganized, in May, 1915, Carson was appointed Attorney-General, but he held office for a few months only, resigning in October because he was not in sympathy with the views of the Ministry on the campaign in the Dardanelles. See Home Rule.

CARSON CITY, Nev., the state capital and county seat of Ormsby County, is thirty-two miles south of Reno and 250 miles northeast of San Francisco, on the Virginia & Truckee Railroad. Northeast twenty-one miles is Virginia City, home of the Comstock, a famous silver mine. The population in 1910 was 2,448; in 1914 it was 3,500.

Carson City is situated in Eagle Valley, at the eastern base of the Sierra Nevada Mountains. Lake Tahoe, on the state boundary, is fourteen miles west. The city is the base of supplies for the surrounding mining and agricultural country. The tall green shafts of many Lombardy poplars are a picturesque feature of the town. The important public buildings are the state capitol, a Federal building erected in 1888 at a cost of $150,000, the United States mint, built in 1869 at a cost of $300,000, and a state library, one of the finest in the West. At Carson City is the state orphans' home; two miles southeast is the state prison, and three miles south is a United States government Indian school. The city is abundantly supplied with water from natural springs.

Carson City was named for "Kit" Carson, the famous scout and guide. It was settled as a trading post in 1851, and in the early days miners and lumber-jacks, adventurers and fortune seekers comprised the life of the camp. The place was laid out as a town in 1858, and became the capital of the state in 1861. It was chartered as a city in 1875. E.D.V.

CARTAGENA, *kahr ta je'na,* a Spanish seaport in the province of Murcia, with one of the largest and safest harbors on the Mediterranean Sea. It is a naval and military station, the arsenal containing barracks, docks, hospitals and machine shops. Lead smelting is largely carried on, and there are in the neighborhood rich mines of excellent iron, copper and zinc. Esparto grass is grown in the vicinity and is used for making shoes, ropes, mats and a kind of coarse cloth. Lead, iron ore, oranges and other fruits are exported.

Cartagena was founded by the Carthaginians under Hasdrubal about 243 B.C., and was called New Carthage. It was taken by Scipio Africanus (210 B.C.) and was long an important Roman town. It was destroyed by the Goths and revived in the time of Philip II. Population in 1910, 96,983.

CARTE BLANCHE, *kahrt blahNsh,* a French term meaning *white paper,* is a blank paper signed by one in authority and entrusted to another to fill in as he pleases. Thus in 1649, Charles II, king of England, tried to save his father's life by sending a *carte blanche* from The Hague to the Parliament to be filled up with any terms which they would accept as the price of his safety. The term is now used to mean, in substance, "Do as you please."

CARTER, Mrs. Leslie (1862-), one of the leading actresses on the American stage, whose ability to portray powerful emotion has given her the name of "the American Bernhardt." She made her first appearance as an actress in 1890 in David Belasco's *Ugly Duckling,* at the Broadway Theater, New York. Though she showed promise, her performance in the play was not remarkable, and after a season in *Miss Helyett,* a musical comedy, she retired from the stage for a period of study. For eighteen months she was trained and coached by Belasco, appearing at the end of that time, October, 1895, as Maryland Calvert in *The Heart of Maryland.* Her success in this play was immediate and extraordinary, and she repeated her triumph three years later in London.

Mrs. Carter was equally successful in *Du Barry,* by Belasco and John Luther Long, a play that she used for three years, and in *Zaza,* Belasco's English version of a French drama. In 1906 she severed her connection with Belasco and became her own manager. Mrs. Carter's interpretation of any rôle that calls for emotional acting of a strong and varied character is generally regarded by critics as powerful and impressive.

CARTER-COTTON, Francis L. (1847-), a Canadian journalist and legislator, one of the foremost Conservatives in British Columbia and a member of the provincial legislature almost continuously since 1890. He was born in Yorkshire, but as a young man settled in Vancouver, which remained his home. In 1886 he founded the Victoria *News-Advertiser,* which became under his editorial direction one of the most influential newspapers in the province. Elected to the legislature in 1890, he served continuously until 1900; he was again elected in 1903, and has served since without interruption. Carter-Cotton was early recognized as an authority on financial questions, and was provincial minister of finance from 1898 to 1900. Sir Richard McBride appointed him chief commissioner of lands and works, a position he held from 1904 to 1910. His interest in higher education led him in 1906 to endow a chair of mathematics in McGill University College of British Columbia, and in 1912 he was elected first chancellor of the University of British Columbia.

CARTHAGE, *kahr'thayj,* one of the most celebrated cities of ancient times, in the third century before Christ the great commercial rival of Rome. It was situated in North Africa, near the site of the modern Tunis, on

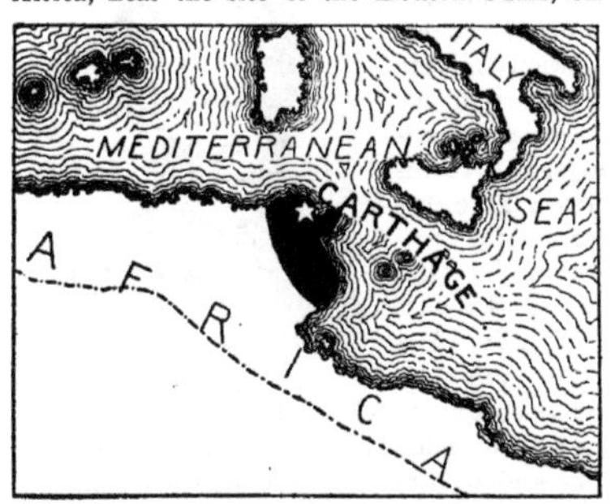

ANCIENT CARTHAGE

a peninsula extending into a small bay of the Mediterranean Sea. Carthage is famed in legend and in history, and two stories are told of its origin. In classic myths it was founded by Dido, daughter of a Phoenician king of Tyre, and under her kindly rule became a rich and flourishing city. How Aeneas and his company of Trojan exiles were driven by storms to seek her hospitality, how the Carthaginian queen entertained the hero for many months, and on his departure threw herself in grief and despair upon the blazing funeral pile—all is vividly told in Vergil's story of the wanderings of Aeneas. (See DIDO; AENEID.)

The historical account of the city's founding is that merchants from the neighboring colony of Utica and from the mother city of Tyre, in Phoenicia, established a trading post on the site about 850 B.C., and called it Karthadshat, or "New City." Because of its splendid location on the Mediterranean shore, Carthage became one of the greatest commercial centers of antiquity, and its people early began to extend their dominions by colonization and by conquest. In the third century B.C., when it was at the height of its power, Carthage was a magnificent city of about 700,000 people, holding sway over the northern coast of Africa from the Greater Syrtis (the modern Gulf of Sidra) to the Strait of Gibraltar, and over Sardinia and nearly all of Sicily, while tribute was collected from the natives of Southern Spain and of Corsica. All of the coasts and islands of the Mediterranean were visited by Carthaginian merchantmen, and these hardy traders even ventured to the Azores, Britain and the Baltic Sea.

Historians and excavators have learned many interesting facts about the great city, whose site is now marked by two or three small hamlets and a few ruins. Across the peninsula on which the city was built was a triple wall of towers. The sides were likewise defended by walls, and two harbors, connected by a canal, served for the navy and for merchant vessels. The military harbor was circular in shape and provided with docks sixteen feet in width, which were large enough to hold 220 vessels. Less than a mile north of the harbors rose the hill of Byrsa, the citadel of the city. Excavators have unearthed the ruins of the ancient walls and of public buildings and tombs.

Nothing is known of Carthaginian history before the sixth century B.C. The first wars of importance were fought with the Greeks in the fifth century over the control of Sicily, which was abandoned by the Greeks in 275 B.C. The conquest of the southern part of Italy by the Romans brought Rome and Carthage into close contact, and as the Romans viewed the commercial supremacy of the Carthaginians as a grave menace to their own rising glory, war followed inevitably. The first conflict began in 264 B.C. (see PUNIC WARS). Two other wars were fought, and in 146 B.C. Carthage was captured and destroyed after a desperate siege of two years. The burning of the city continued more than two weeks.

The northern coast of Africa became a Roman province, and Carthage, rebuilt by the Emperor Augustus in 29 B.C., was accounted one of the finest cities in the Roman Empire in the second and third centuries of the Christian Era. The Vandal king, Genseric, made it his capital in 439, and nearly a century later it was wrested from the Vandals by Belisarius, the great general of the Emperor Justinian. In 647 Carthage was destroyed by the Arabs, and was never rebuilt. B.M.W.

CARTHAGE, Mo., the county seat of Jasper County, is a progressive industrial city situated in the southern part of the state, near the western border, about one-half mile south of Spring River. Joplin is twenty miles southwest, and Springfield is ninety-six miles northeast. The city is served by the Saint Louis & San Francisco Railway, constructed to the city in 1870; the Missouri Pacific Railway, built to this point in 1874; and the Carthage & Western Railway, built here in 1906. The area of the city is a little less than five square miles. Its population in 1910 was 9,483, and in 1914, 9,510.

Carthage, locally called the *Queen City*, is located in a district rich in lead, zinc, cobalt, marble and limestone. There are ten quarries in the vicinity, employing about 300 people, the annual pay-roll amounting to $260,000. Shoe factories, flour mills, machine shops, bed-spring factories and foundries are prominent industrial enterprises. The most noteworthy buildings are a post office, erected in 1911 at a cost of $75,000, a $125,000 high school, erected in 1910, and a $25,000 Carnegie Library, erected in 1909. Carthage College, a business college and a school of telegraphy supplement the public means of education. The city has three parks.

Carthage was settled in 1833 and became the county seat in 1842. It was incorporated as a city in 1873. H.L.M.

CARTHUSIANS, *kahr thu'zhanz*, an Order of monks which takes its name from the solitary village of Chartreuse, in the French Alps, where in the eleventh century Saint Bruno of Cologne, with six companions, established the first Carthusian monastery. These hermit monks wore rude garments, ate vegetables and coarse bread, slept on beds of straw, and were rigid in their observance of fasting, seasons of prayer and night watching. The modern Carthusians live with almost the same austerity. The members of the Order were from the beginning well educated and given to hospitality and charity. At one time they had the finest convents in the world, of which La Grande Chartreuse, in France, and the Certosa di Pavia, south of Milan, are among the most celebrated. At the present there are fewer than twenty Carthusian monasteries on the European continent.

In England the name *Charterhouse* was applied to the Carthusian monasteries. The most famous of these, established in London in the fourteenth century, became in the course of time an almshouse for old men and a free grammar school for poor boys. Thackeray attended this school in his boyhood, and has made it the background for affecting scenes in *The Newcomes*. The only English Carthusian monastery at the present time is situated eight miles northwest of Brighton, near Steyning.

Chartreuse, a strong liquor made in three colors, originated with the monks of La Grande Chartreuse. G.W.M.

CARTIER, *kahr tya'*, SIR GEORGES ETIENNE (1814-1873), a Canadian statesman, Prime Minister of Canada from 1857 to 1862, one of the "Fathers of Confederation," the man who was chiefly responsible for Quebec's entrance into the Confederation in 1867. He also deserves recognition as one of the first to see the need of railways, and to his tireless energy was due the completion of the Grand Trunk Railway. For twenty years he was Quebec's unquestioned political leader, and with Lafontaine and Laurier he stands among the greatest French-Canadian statesmen. At the same time he used to describe himself as "an Englishman speaking French," and the British population in Quebec always felt that he represented them as much, if not more, than did the men of their own blood.

SIR GEORGES E. CARTIER

Cartier was born in the village of Saint Antoine, Quebec, on September 6, 1814. After graduating in law from the College of Saint Sulpice, in Montreal, he began to practice his profession in that city in 1835. Scarcely had he begun to make headway when he was involved in the rebellion of 1837, led by Louis Papineau. His participation in this revolt kept him out of public life for a decade, but in 1848 he entered Parliament as member for his native county.

During the next twenty-five years he was one of the leaders in Canadian affairs. He was soon recognized as the leader of the more liberal wing of the Conservative party, and in 1855 became provincial secretary. Two years later he was appointed Attorney-General for Lower Canada, and from 1857 to 1862 was joint Premier with Sir John A. Macdonald.

During the next five years Cartier spoke and wrote constantly in favor of Confederation, the formation of a Dominion, and in the face of great opposition carried the province of Quebec with him. He took part in the conferences which preceded the passage of the British North America Act, and in 1868 was rewarded by a baronetcy of the United Kingdom. In the first Dominion Ministry he was Minister of Militia and Defense, but in 1872 a religious issue caused his rejection by his Quebec constituents. He was lâter elected, however, in Manitoba, but the mortification of the defeat at home was bitter. His health was already poor, and he died on May 20, 1873.

CARTIER, JACQUES (1494?-1557?), a bold and daring navigator whose explorations in America under the flag of France gave that country the basis for its claim to the great domain of Canada. In the year 1534, Cartier was commissioned by Francis I to head an expedition to North America for the purpose of adding new lands to the French dominions, and to discover a passage to China (see NORTHWEST PASSAGE). In early summer the little fleet of two small vessels reached the coast of Newfoundland and passed through the Strait of Belle Isle into the Gulf of Saint Lawrence. After sailing along the shores of New Brunswick, Cartier made a landing at Cape Gaspé, on the eastern coast of the present province of Quebec, and took formal possession of the country in the name of Francis I. The following May he made a second trip to the New World; on this voyage he discovered the Saint Lawrence River, giving it the name of the saint on whose feast day he first sighted its waters. He journeyed up the river to a small Indian village over which towered a great hill called by him Mount Royal, the present site of the city of Montreal. On a third visit, in 1541, he built a fortified post near the site of Quebec, but the fort was soon abandoned.

CARTILAGE, *kahr'ti laj,* or **GRISTLE.** If you look at the end of the breast-bone of a chicken, you will notice that it is not bone at all, but a white, elastic substance that can easily be bent in any direction. Further observation will show that this substance gradually changes into bone, and that it is difficult to tell just where the bone begins. This white elastic substance is cartilage, or gristle. Cartilage is found in the body of every animal that has a backbone. It is of two kinds, temporary and permanent. Temporary cartilage gradually changes to bone before the skeleton is mature. The tip of the breast-bone of a fowl over a year old, for example, is hard bone; in the chicken it was cartilage. There is also more cartilage in the skeleton of a child than in that of an adult; in the latter most of the temporary cartilage has turned to bone. Permanent cartilage is found around the joints and does not change to bone. The external ear is of this kind of cartilage. It also occurs in the nose and the eyelids. It is white, tough and flexible, is a sort of soft padding between the harder bones, and is useful in holding the bones in position at the joints and enabling the joints to bend. See BONE.

CARTOON'. As most commonly used today, the word *cartoon* refers to a humorous or satiric picture intended to hold up to ridicule some public man, party or movement. Practically all of the great daily papers, as

"THE MASTER MUSICIAN IS DEAD"
The above picture presents a cartoon of more serious aspect than usual in illustrations of that class. It is by John T. McCutcheon, and reproduction is by permission.

well as a great many magazines, have such sketches, and they play a large part in recording history. In this sense the word means practically what caricature meant earlier, and a brief history of the art of caricature, or cartoon-making, is given under that title. Cartoons in recent years have undergone a decided change of spirit. They are much better-natured—far less bitter. Rarely indeed, except in sensational journals, is the old, venomous, biting style of cartoon to be seen; most of those shown in the daily papers are merely humorous or entertaining, and have no touch of satire.

The great success of clever cartoonists should interest young students to study carefully the elements that enter into the pictures of their favorite artists. The technique, the elimination of all unnecessary lines, the selection of the essential characteristics, the certain emphasis, the appeal to the sense of the humorous by just the right line—all these enlist our keenest interest, study and criticism. But the intelligence back of the pencil, the knowledge of history, the understanding of present conditions in government, the appreciation of individual characteristics prove that the successful cartoonist must be a student as well as an artist, with imagination and with charity.

The power of a cartoonist of ability can scarcely be over-estimated, and his responsibility is also great. The follies and wickedness of life are legitimate subjects for caricature, but honest effort, ideal enterprise, heroic adventure into untried paths, however futile they may appear, should not be the objects of attack by the ambitious cartoonists.

Any one who has an obedient pencil, a touch of the true cartoonist's genius, should cultivate the gift as one of highest value, and make of his art an element of helpfulness in the world.

As the article CARICATURE relates, Leonardo da Vinci made caricatures, but he also made cartoons. In those days the two were entirely different, for originally a cartoon was a drawing made on heavy paper and used as a model for a large picture in fresco, tapestry or oil. It was made exactly the size of the picture intended, and the design was transferred to the canvas or wall by means of tracing or pin-pricking. The process was much like that by which school children produce maps, for carbon was smeared on the back of the drawing. Some of the old cartoons or designs of the master painters have a real value of their own, aside from that of the finished painting. Most famous of all which have survived are seven cartoons of Raphael for the Vatican tapestries, which are preserved in the South Kensington Museum in London. Modern painters seldom use the cartoon method, though it has not been entirely abandoned.

Related Subjects. Of the famous cartoonists or caricaturists, the following are treated in these volumes:

Bengough, John Wilson
Briggs, Clare A.
Cruikshank, George
Du Maurier, George Louis
Gibson, Charles Dana
Hogarth, William
McCutcheon, John T.
Nast, Thomas
Opper, Frederick B.
Outcault, Richard F.
Tenniel, John

CARTOUCHE, *kahr toosh'*, a word applied to various oval ornaments, scrolls or shields used in heraldry and art. A special use of the term occurs in Egyptian archaeology, the cartouche being an oblong frame with oval ends, which enclosed the names of kings and queens engraved on monuments. The scroll-shaped volute of the capital crowning the shaft of an Ionian column (see COLUMN) is an example of the use of the cartouche in architecture. In heraldry the term denotes an oval shield with a convex surface, on which were borne the arms of the Pope and Church officials of noble birth.

CARTOUCHE
The name of Khufu (Cheops).

The modern use of the word is borrowed from the Italian *cartoccio,* meaning *roll of paper*. Originally a cartouche was a roll of parchment, paper or other material which held the charge of powder and shot of a firearm, that is, a cartridge. The word *cartridge* is itself a corruption of *cartouche*.

CARTRIDGE, *kahr'trij,* a corruption of the French word *cartouche,* the name given to a case of paper or metal holding a charge of explosive powder and a bullet or a number of pellets, to be fired from firearms. The cartridge used in modern rifles and machine guns is a cylinder made from brass and contains a percussion cap at the base. The explosive is inserted in the cylinder and the bullet is placed on top of it, the brass being crimped or pressed against the bullet to hold it firmly in place. Cartridges for shotguns are usually made of stout paper in several thicknesses, with a brass base to give additional strength and hold the percussion cap. A *blank cartridge* is a cartridge with a charge of explosive, but without ball or shot. See AMMUNITION; PROJECTILE.

CART'WRIGHT, EDMUND (1743-1823), the English inventor whose crude machine for cotton-weaving, constructed in 1785, laid the foundation for the magnificent power looms of the present day. He was born in Nottinghamshire, educated at Oxford, and became a clergyman of the Church of England. When past forty years of age he began to take an interest in machinery, an interest which made him one of the chief promoters of the prosperity of the English people. Misguided workmen, who thought that machinery would take away their only means of earning a living, burned down the mill in which four hundred of his looms were set up, but Cartwright was not discouraged. He invented a wool-carding machine, a device for making ropes, and a steam engine in which alcohol was used as fuel, besides helping Robert Fulton in his experiments on steamboats. Having spent his own income in bringing out his inventions, which yielded him little in return, Cartwright was granted $50,000 by Parliament in 1809. See LOOM; WEAVING.

CARTWRIGHT, PETER (1785-1872), a Methodist circuit-rider, or itinerant preacher, one of the most influential and most picturesque figures of the pioneer days of Illinois. He was born in Amherst County, Virginia, and when five years old was taken by his family to Logan County, Kentucky. Peter grew up a wild and reckless boy, given to gambling and other dissipations, but he was converted at about the age of sixteen and shortly afterwards became a local preacher. In 1803 he was accepted into the regular ministry, and three years later Bishop Asbury, the first Methodist bishop in America, ordained him an elder.

In 1824 Cartwright settled in Sangamon County, Illinois, then a struggling pioneer region inhabited by a few resolute and enterprising settlers. The spirit of the man made itself felt in the life and development of that community for half a century, and it has been truthfully said of him that he had as much to do with the upbuilding of Central Illinois as any other man of his time. Not only was he a powerful and tireless preacher, but he interested himself in the spread of education and in the political affairs of the state. When the capital was at Vandalia, Cartwright was a Democratic member of the state legislature, and at one time was Abraham Lincoln's unsuccessful opponent in a campaign for election to Congress.

Many tales have come down about his eccentric habits, his gifts for story-telling and his fearlessness in dealing with the unruly members of the community. In some cases, it is said, he found the "arm of the flesh" a better weapon than moral suasion. His life was a powerful influence for righteousness, and his memory is still revered. He lies buried in the village of Pleasant Plains, Sangamon County, the scene of some of his most active labors. Among several well-known writings are his pamphlet entitled *Controversy with the Devil* and *The Autobiography of the Rev. Peter Cartwright.*

CARTWRIGHT, SIR RICHARD JOHN (1835-1912), a Canadian statesman and financier, recognized as one of the foremost authorities on tariff and financial problems, and for many years second only to Sir Wilfrid Laurier among the leaders of the Liberal party. He was for forty years a prominent figure in the Legislative Assembly of Canada, and its successor, the Dominion House of Commons, and then for nearly a decade more was the leading Liberal in the Senate.

Sir Richard was born on December 4, 1835, at Kingston, Ont., where his family was prominent. Young Richard was sent to Trinity College, Dublin, Ireland, and on his return began the study of law. He never practiced, for the call of politics was stronger than the law. Unlike some boys of well-to-do parents, he was trained to business habits, and later in life was director and president of one of the largest banks in Canada. After 1863, however, when he first entered the Legislative Assembly for Canada, he gave most of his time to public affairs.

At the beginning of his public career Cartwright was a Conservative, and during the first years of Sir John A. Macdonald's Ministry was one of that statesman's supporters. About 1870, however, there were some sharp differences of opinion between them, especially on banking reform, and the Pacific railways scandal of 1873 put an end to their relations. It is no small tribute to his popularity that he was reëlected to Parliament as a Liberal by the same district which had previously chosen him as a Conservative. From 1873 to 1878 he was Minister of Finance in the Mackenzie Ministry, and then for sixteen years, while the Liberals were in opposition, he was the chief spokesman against the financial and trade policy of the Conservatives.

In 1896, when Laurier became Premier, he made Cartwright Minister of Trade and Commerce. Laurier advocated a mild protective tariff, while Cartwright was an outspoken free-trader, but the latter's Parliamentary experience, his quickness in debate and his knowledge of economic questions made him a great asset to the Liberals. In 1897 he represented the

Dominion at Washington in negotiations for better commercial relations with the United States, and in 1898 he served on the Joint High Commission for settling all existing disputes between the United States and Great Britain relating to Canada. On several occasions, during the absence of Laurier, Cartwright was acting Premier. He was knighted by Queen Victoria in 1879, and in 1902 was given the honor of membership in the Imperial Privy Council. From 1904 until his death he was one of the Liberal leaders in the Senate. His *Reminiscences,* completed in the year of his death, is a valuable record of Canadian politics. W.F.Z.

CARUSO, *ka roo' zo,* Enrico (1873-), an Italian operatic tenor, one of the most popular grand opera stars of modern times. He was born in Naples, where he began singing in churches at the age of eleven. Ten years later he sang his first part in opera, and in 1896, as Alfredo in *La Traviata,* attracted much favorable notice. Through four seasons he sang the leading tenor rôles at the municipal opera house of Milan, where he took the part of Jean in the first Italian performance of Massenet's *Sapho.* His appearances in the leading cities of Russia and Germany, in Buenos Aires, London, Rome and Lisbon were highly successful, and his triumphs were repeated in 1904 when he began his first American season at the Metropolitan Opera House in New York.

ENRICO CARUSO
In costume.

Caruso's extraordinary success is due to the very unusual power, sweetness and range of his voice. He has always been able to attract immense audiences, which accounts for the enormous sums paid for his services. He made a contract with the New York Metropolitan Company by which his salary was nearly $200,000 a year for four years. He is paid about $40,000 a year for singing into phonographs, and the "records" thus made bring prices ranging as high as $7 each. In 1913 he sang in Vienna in three performances, receiving for each of these about $5,000. Caruso's most successful rôles are Rhadames in *Aïda,* Manrico in *Il Trovatore,* Turiddu in *Cavalleria Rusticana* and Johnson in *The Girl of the Golden West.*

CARVER, John (1575-1621), a leader of the Pilgrim Fathers, chosen during the *Mayflower* voyage to be first governor of Plymouth Colony. He was born in England and because of the severe religious persecutions went to Leyden, in Holland, then a refuge for the Puritans. He was an elder in the church and one of those who secured the original patent for the new colony in America. His election was unanimous, and he justified the choice of the colonists by his firmness, prudence and ever-ready courage. Four months after the arrival in Massachusetts, however, he died from the effects of a sunstroke.

CARVING, according to the most general use of the term, is the art of fashioning ornamental or natural-appearing figures in ivory and wood. Work in stone or marble usually comes under the head of *sculpture* (which see), but the term *carving* is properly applied to the smaller figures cut in stone or marble, such as leafage, scroll work, statuettes, etc. The oldest known examples of wood carving are Egyptian. This beautiful art came into extensive use early in the Christian Era, and was long popular in the decoration of the churches of Central Europe, especially in Germany, where the shrines and altars were sometimes adorned with whole scenes from the lives of the saints. Among the Swiss peasants, wood carving is a regularly-organized occupation.

Wood is admirably adapted to the representation of those forms which have life and movement, and a good piece of wood carving has a freshness and vigor that carved ivory and stone or marble do not possess. The art is one that boys and girls can learn without great difficulty and it is being offered now quite generally as a course of study in art and manual training schools. See Wood Carving.

That ivory carving was practiced among the ancients is known from the specimens found in Egyptian tombs and in the ruins of Nineveh; ivory was also used by the Greeks and Romans for various ornamental purposes. Among the most interesting specimens of the early period are the wax writing tablets used by the Roman consuls, the outsides of which were made of ivory ornamented with beautifully-carved figures. Ivory was used extensively in the Middle Ages in the ornamentation of the churches and for other purposes. Coffins were often covered with carved ivory plates, and among the gifts received by Charlemagne were two richly-carved doors. Ivory adapts itself readily to the carver's art because of its elasticity and evenness of texture, and at the present time it is used quite generally for toilet articles, ornaments of various kinds, chessmen, knife handles and other objects.

CARY, *ka'ri,* ALICE **(1820-1871)** and PHOEBE **(1824-1871),** two American poets, sisters and life-long companions, whose graceful and picturesque verses were greatly admired in their own day, and are still read and enjoyed. They were born and brought up on a farm in the valley of the Miami, in Ohio. Their only education was that afforded by the little country school in the district, and for years they depended for reading matter on their home library—a shelf of about half a dozen books. Though their educational advantages were so limited, the sisters constantly studied and wrote, holding fast to the idea that some day others in the great world outside would read their poems.

When Phoebe was fourteen and Alice eighteen they began to see their names in print, and in the course of the next few years they became widely known as writers of charming little poems that found circulation through various papers and magazines. Whittier wrote them a letter of encouragement and appreciation, and when in 1849 Horace Greeley visited them in their home they determined to try their fortunes in New York. In 1851 they settled in the city, and there the sisters lived and wrote together for the next twenty years, mingling with a brilliant group that included Horace Greeley, Bayard Taylor, Richard and Elizabeth Stoddard, Justin McCarthy and Thomas Bailey Aldrich. In this circle they themselves were held in great esteem.

There is great similarity in their choice of subjects, though they show differences in poetic temperament. The poetry of each may be divided into certain groups, as ballads and narrative poems, religious poems and hymns, poems of nature and of sentiment, and poems for children. Alice was more delicately imaginative than her sister, but less dramatic, and Phoebe possessed more wit. In their nature poems, Alice saw the soul in nature, and loved to interpret its moods, while Phoebe associated nature with human experiences. They loved children equally well, but Alice showed in her poems more of the motherly affection for them; Phoebe regarded them as comrades.

One of Alice's finest achievements is her *Pictures of Memory,* beginning—

> Among the beautiful pictures
> That hang on Memory's wall
> Is one of a dim old forest,
> That seemeth best of all.

Edgar Allan Poe said of this poem that in rhythm it was one of the most perfect lyrics in the English language. Of all of Phoebe Cary's poems none made a wider appeal than the hymn *Nearer Home,* which begins—

> One sweetly solemn thought
> Comes to me o'er and o'er—
> I am nearer home to-day
> Than I ever have been before;
>
> Nearer my Father's house,
> Where the many mansions be;
> Nearer the great white throne,
> Nearer the crystal sea.

B.M.W.

CARYATIDES, *kar i at' i deez,* or **CARYATIDS,** the name applied in Greek architecture to the figures of women dressed in long robes and standing upright in graceful positions, when used as columns to support a roof. The most celebrated of these figures appear on the southwest porch of the Erechtheum, a temple on the Acropolis of Athens. The corresponding male figures are called *Atlantes.* The name is the plural of the Greek *Caryatis,* meaning *a woman of Caryae* (in Laconia). In this city was a temple to the goddess Artemis, and during her annual festivals there virgins were accustomed to dance in her honor. Some authorities say that these dancing maidens suggested to architects the idea of using their images as columns, thus explaining the origin of the term *Caryatides.*

CARYATIDES
In the porch of the Erechtheum.

CASABA, *kah sah' ba,* or **CASABA MELON,** one of the largest varieties of muskmelons, not extensively known in America until the present century. It was introduced from Cassaba, near Smyrna in Asia Minor, and is sometimes known as the *Persian melon*. Its color both inside and out is yellow. The flesh is of excellent quality. The outside is divided like other melons by longitudinal channels, but is smooth, lacking the network characteristic of American muskmelons. See color plate, with article FRUITS.

CASABIANCA, *kah zah byahng' ka,* the boy who "stood on the burning deck, whence all but him had fled." Mrs. Hemans' famous poem, which begins with the lines quoted, tells how he remained steadfast until the vessel was destroyed by an explosion, rather than leave without his father's express command. And the father, meanwhile, lay too deeply wounded to hear his son's cry.

Casabianca was a real boy, only ten years old at the time he met his death. The father was captain of the *Orient,* the flagship of Napoleon's fleet, and during the Battle of Abukir Bay in 1798 was compelled to assume command of the whole fleet because the admiral had been killed. The death of the brave boy roused much sympathy and admiration, and Mrs. Hemans expressed the general feeling in the closing words of her poem:

> But the noblest thing that perished there
> Was that young faithful heart.

The poem has often been parodied and thus has been deprived of some of its appeal, but the boy's deed itself has never lost its power to thrill.

CASCADE, *kass kayd',* **RANGE,** a mountain chain in the western part of North America, extending from Northern California through Oregon and Washington to British Columbia and Alaska. It is usually called a northward continuation of the Sierra Nevada, from which it is separated by a series of deep valleys near Mount Shasta. This peak, rising 14,380 feet above the sea, is the loftiest in the range and is one of the most magnificent mountains in North America. Mount Rainier, or Tacoma, whose summit is 14,408 feet above the sea, Mount Adams, 12,470 feet, and Mount Hood, 11,225 feet, are other beautiful peaks. Many of the peaks of the range are extinct volcanoes.

As a mass the Cascade Range is comparatively low and broad, though rugged in outline and topped by snow-clad peaks. Particularly in Oregon and Washington, where the Columbia and the Klamath rivers have cut deep canyons through the mountains, the scenery is magnificent. The slopes, particularly on the west, are heavily wooded and seem to form a great dark cloak for bright green pastures, shimmering blue lakes and glistening white peaks. The Cascade Range is the home of the Douglas fir, a tree which often reaches a height of 200 feet or more and is justly considered one of the most beautiful trees in the world. Practically the whole of the mountain forest area in Oregon and Washington is now included in national forest reserves.

The Cascade Range is generally considered to end at the International Boundary. Its natural continuation in British Columbia would be the nearest mountains to the Pacific, which are called the Coast Range. This is comparatively low, seldom reaching above 8,000 feet, but is snowy and contains numerous glaciers. It is cut into deeply by splendid fiords, and the larger rivers, such as the Fraser, the Skeena and the Stikine, have cut profound and picturesque canyons and valleys through the range. The Canadian Pacific Railway follows the Fraser canyon to the sea, while the Grand Trunk Pacific uses the Skeena valley 500 miles to the northwest.

Cascade Tunnel. The Great Northern Railway has cut a tunnel under the crest of this range at a point about fifty miles southeast of Everett and fifty-five miles east of Seattle. The tunnel is sixteen feet wide, twenty-one and a half feet high and 13,416 feet, or 2.6 miles, long. This is one of the most noted engineering works of its kind on the North American continent. A similar tunnel on the Northern Pacific Railway, about fifty miles directly southeast of Seattle, is 9,850 feet long. See map, NORTH AMERICA. A.P.C.

CASCARA
(*a*) Male flower
(*b*) Female flower
(*c*) Fruit

CAS'CARA, a laxative drug from the bark of the *cascara sagrada,* known as *chittim - wood bark,* or *sacred bark.* The plant grows as a large shrub or a small tree in the Northern United States and Southern Canada and south as far as Northern California. The dried pieces of bark are broken up and packed tightly and usually kept a year before being used. Merely an extract is used in medicine, never the crude bark. Because of the value of cascara as a laxative, many patent medicines with names sounding like it have been put on the market.

CASE, a term which denotes the relation of a noun or a pronoun to the other words in the sentence. In Latin case is always indicated by *inflecting,* or changing the ending of words, but in English most case-endings have been dropped and case is commonly indicated by the *order* in which the words are used; that is, the place which the word occupies in the sentence. To illustrate: In the sentence, "Arthur gave Ernest Mary's ball," all except one of the words are nouns, each used in a different way. *Arthur* is used as the subject of the sentence, *ball* is the direct object of the verb *gave, Ernest* is the indirect object, and *Mary's* is used to modify *ball* and indicates ownership. If such a sentence were written in Latin each of these different uses would be indicated by changing the endings of the words themselves. In Old English, too, nouns were inflected to show their case, but in modern English they are inflected only in one case, the possessive.

Order Determines Case. But, although the inflected forms of nouns have disappeared, the names of the different cases have been retained and these are used, in sentence analysis, for indicating the relationships of nouns to other words. For example, the only way in which the noun *Arthur* is inflected is by adding *'s* to form the possessive. Taking the sentence used above, let us see how changing the position of the various words will alter their case and therefore the meaning of the sentence. For example, in the sentence, "Arthur gave Ernest Mary's ball," *Arthur* is the subject of the sentence, therefore is said to be in the *nominative* case. If the sentence is changed to read, "Ernest gave Arthur Mary's ball," *Arthur,* simply through its changed position, becomes the indirect object of the verb and is then said to be in the *objective,* or *accusative,* case. Or, the sentence might read, "Ernest gave away Arthur's ball," in which event *Arthur's* would be in the *possessive,* or *genitive,* case.

There are, then, three cases, in English: the *nominative,* the *possessive* (or genitive), the *dative* and the *objective* (or accusative).

Uses of the Nominative. A noun is said to be in the nominative case when it is used as the *subject of a verb,* as, "*Miriam* called at the house one evening in midsummer"; when it is used as the *predicate complement* of a verb, as, "Edison is a great *inventor*"; when it is used *in apposition* with another nominative word, as, "That is Peary, the famous *explorer*"; when it is used as the *person or thing addressed,* as, "*Agatha,* where have you been?" or when it is used independently as the subject of a participial clause, in what is called the *nominative absolute,* as, "The *moon* having risen, we started."

Uses of the Objective. A noun is said to be in the objective case when it is used as the direct or indirect object of a verb, two uses which can be illustrated in the same sentence: "Max threw Peter the *ball.*" *Ball* is the object of the verb *threw; Peter* is the indirect object, in reality the object of the preposition *to,* understood. Both are said to be in the objective case. A noun is also in the objective case when it is the object of a preposition, as, "Max threw the ball against the *wall.*" Here *wall* is the object of the preposition *against.*

The subject of an infinitive must also be in the objective case. In Latin the infinitive construction is used very freely and the subject of the infinitive is always in the accusative case, which corresponds to the English objective case. In English the infinitive construction with the subject expressed is possible only in conjunction with a transitive verb. In the sentence, "I believed him to be honest," *him* is the subject of the infinitive *to be honest,* and the entire phrase is the object of the verb *believed.* Where the infinitive *to be* takes a complement, the case of that complement must be objective, to agree with the case of the infinitive's subject, as in the sentence, "They thought *me* to be *him.*"

A noun or a pronoun in apposition with another noun or pronoun in the objective case is also put in the objective; as, "I have engaged Johnson, the *carpenter*"; "I have engaged Johnson, *him* whom you recommended"; "They were amazed to see her again, *her* whom they had thought dead."

Forming the Possessive, or Genitive. A noun is said to be in this case when it denotes ownership; as, *Paul's* umbrella; *Stanley's* skates. Possession may also be indicated by means of a prepositional phrase, as, the banks *of the river.*

Singular Possessives. The possessive case is formed in the singular by adding to the nominative form of the noun the letter *s,* preceded by an apostrophe ('). In old English the possessive was formed by adding *es* to the noun, as in Chaucer's phrase, "lordes werre" (lord's war). As this gave many possessives the same form as the plural, gradually the *e* was dropped and the apostrophe substituted to indicate the omission, so we now write it *lord's*. If a word ending in an *s* sound becomes awkward to pronounce through the addition of another *s*, the apostrophe alone is added; as, *Moses'* law, for *goodness'* sake. In some words, however, the *s* is almost always added; as, *Burns's* poems, the *empress's* command. There is no fixed rule for the formation of such possessives, and the problem is often avoided altogether by using the prepositional phrase with *of* where it does not affect the meaning; as, in the law *of Moses,* the poems *of Burns.* The tendency is to indicate possession by means of the preposition *of* in most cases where the object is without life; as, the clasp *of the necklace,* not the *necklace's* clasp; the cause *of the war,* not the *war's* cause.

Plural Possessives. In forming plural possessives the apostrophe only is added if the noun ends in *s;* as, in *ladies'* gloves, *birds'* nests. When the plural form of the noun does not end in *s* the possessive is formed as in the singular, by adding both the apostrophe and the *s,* as, *men's* clothes, *women's* hats, *children's* toys.

Compound Possessives. Only the last word in a compound word is inflected, as, *sister-in-law's.* The same rule is followed if joint possession is to be indicated, as, *Thurber and Brown's* studio, *William and Mary's* uncle. But where the possession is not common, the sign of the possessive must be added to each noun; as, *Grant's* and *Lee's* forces; the poem is neither *Bryant's* nor *Whittier's.*

Double Possessive. There is a curious idiomatic double possessive in common use which is a mixture of two constructions, the inflected possessive and the possessive indicated by *of;* as, a play *of Shakespeare's;* a cousin *of Philip's.* In some cases this double form is absolutely necessary to bring out the meaning; for instance, "this picture *of my uncle*" does not convey at all the same meaning as "this picture *of my uncle's.*" The same is true of expressions like "this introduction *of Shaw,*" or "this introduction *of Shaw's.*"

For a detailed account of the cases and inflections of pronouns, see PRONOUN. L.M.B.

CASEIN, *ka'se in,* the substance in milk which forms the curd in making cheese. Casein alone will not form curd, but needs the action of some acid. In its natural state it bears a close resemblance to the white of an egg. It also occurs in beans and peas and is one of their most valuable food elements. It is a compound of nitrogen, carbon, hydrogen, oxygen and sulphur. See MILK; FOOD, subhead *Chemistry of Food.*

CASGRAIN, *kas graN',* HENRI RAYMOND (1831-1904), a Canadian historian, known as Abbé Casgrain. He was born at Rivière Ouelle, Que., studied theology at Quebec Seminary, and was ordained a priest in 1856. After teaching for three years, he became a priest at the Basilica in Quebec, where he remained until 1873. He is chiefly known for the historical studies of French Canada to which he devoted the remainder of his life. His first book, *Légendes Canadiennes,* appeared in 1861. Of his later books the most valuable for students are *Biographies Canadiennes; Un Pélerinage au Pays d'Evangéline;* and *Montcalm et Lévis.* Casgrain's aim was to portray the early French-Canadians as they really were, and he succeeded in presenting vivid narratives, but he was sometimes misled by his racial sympathy to exaggerate the commendable features of their history and to minimize their faults or weaknesses.

CASGRAIN, THOMAS CHASE (1852-), a Canadian barrister and statesman, Postmaster-General of the Dominion and since 1912 a member of the International Waterways Commission which may consider and determine all questions relating to the Great Lakes and other waterways on the boundary between the United States and the Dominion. Casgrain was born in Detroit, Mich., but his parents' home was in Quebec, where he attended the seminary and Laval University. Called to the bar in 1877, he soon rose to eminence in his profession. In 1885 he was junior counsel for the Crown at the trial of Louis Riel for treason. He sat as a Conservative in the Quebec legislature from 1886 to 1890 and again from 1892 to 1896. During the latter period he was also attorney-general of the province of Quebec, and secured a law against corrupt practices at elections. For eight years (until 1904) he was then a member of the Dominion House of Commons. In 1914 he was again elected to the House of Commons and became Postmaster-General in the Borden Ministry.

CASH'MERE GOAT, a variety of goat remarkable for its fine, silky fleece, which is used to make the famous, costly cashmere, or kashmir, shawls. A full-grown goat yields only eight ounces of the valuable down, which underlies the long hair. To make a shawl a yard and a half square the fleece of ten goats is required; it takes about a year to weave one shawl, the work all being done by hand. As the year-old sheep, or merino, furnishes the finest wool, it can be seen why the shawls of

CASHMERE GOAT

the highest quality are very expensive. From $25 to $100 is paid for those of good quality, but some of the most exquisite patterns and most perfect workmanship have been priced as high as $1,500. This is the more remarkable when it is remembered that labor in the Orient can be bought for a few cents per day.

The native homes of the cashmere goats are Tibet and India, but they have been raised in France and Germany. Those which live in the higher altitudes are deep yellow; those lower are yellowish-white, and those still farther down pure white. The colder the region where the goat pastures, the heavier the fleece. The flesh of the cashmere goat is suitable for food, and when well cared for the animal gives a rich milk. See GOAT.

CASH REGISTER, a mechanical device for calculating and registering sales and money received. It is now in almost universal use in retail stores. The first practical cash register was invented in 1879 by James Ritty of Dayton, Ohio, and from that time to the present that city has been the world's center for the manufacture of such machines. Since the first cash register was patented great improvements have been made and the device is now a very complicated piece of machinery, the working of which, however, is simple.

It consists of a metal box with drawers for money and papers, and a mechanism which works somewhat similarly to that of a typewriter. Keys marked with figures, when pressed, operate levers which force into view tickets with legends showing the amounts of money received from a sale, and also cause that amount to be registered on a meter. Each sum registered is added to the previous registry. The total sum registered may be seen at any time by examining the meter, which is usually kept locked. Cash registers are made with separate departments for each salesman, and some are made to deliver automatically a receipt for amount registered. These machines cost the merchants who use them from $40 to $750, according to size and accessories.

CASINO, *ka se'no,* a game of cards played usually by two persons, though more may play if desired. A full pack of fifty-two ordinary playing cards is used. Four cards are dealt to each player, and four are placed face up on the table. Each player in turn takes from the faced cards on the table any card of similar value to one in his hand, or any combination of cards making up that value. For instance, a player holding a nine may pick up a nine or a five and four, or six and three. If no card can be picked up which will match cards in his hand, the player lays a card face up on the table.

A variation is made by building. Thus, a player may add a four to a three already on the table, provided he has a seven with which to pick them up. His opponent may, however, have a seven and may himself pick up the built cards and deprive his opponent of them. The cards are dealt four by four until all have been played. The last to pick up a trick then claims the remaining cards on the table. Score is taken as follows: for little casino (two of spades), 1; big casino (ten of diamonds), 2; each ace, 1; the highest number of cards held by a player, 3; greatest number of spades held by one player, 1. The game is usually twenty-one points. The book *Hoyle's Games* gives rules covering every possible playing condition.

CASPIAN, *kas'pi an,* **SEA,** the largest inland sea in the world, situated between Europe and Asia. Its greatest length is 760 miles, its breadth varies from 100 to 280 miles, and it covers nearly 170,000 square miles, an area larger than the areas of Illinois, Wisconsin and Iowa combined. It lies ninety-six feet below the level of the sea. On three sides it is surrounded by Russian territory, and on the east by Persia. About 850 square miles of its area is occupied by islands. There is no ebb or flow of tides in this sea, and its water is less salt than that of the ocean, because of the many rivers that flow into it. The Volga, Ural, Terek and Kura are the most important of its supply streams, and these form important links in the commercial communications between Europe and Asia. Astrakhan, at the mouth of the Volga, is its most important port. The development of the oil wells in the Baku territory has recently added considerably to the commercial value of this inland sea. See ASTRAKHAN; BAKU; and for location on map, see EUROPE.

CASS, LEWIS (1782-1866), an American statesman, soldier and diplomat, whose name is connected with important events in American history from the War of 1812 to the War of Secession. He studied at the famous Phillips Academy in his native town of Exeter, N. H., completed a law course in Marietta, O., and later became prominent in Ohio politics. On the outbreak of the War of 1812 he was made colonel of Ohio volunteers, becoming brigadier-general of the regular army after a year of seivice. In 1813 General Cass was appointed governor of the Territory of Michigan, then a great unsettled tract of country on the western frontier. During his long term of office, which continued until 1830, he made treaties with the Indians, built roads, organized townships and counties and laid the foundations of civilization in the wilderness under his control.

LEWIS CASS
Michigan's greatest statesman in the early period.

Cass became Jackson's Secretary of War in 1831. He was ambassador to France between 1836 and 1842, resigning in the latter year because of his disapproval of the Webster-Ashburton Treaty (which see). In 1845 the state of Michigan elected him to the Senate of the United States, and he represented the state in that body until 1857, except for a brief period in 1848 while he was campaigning for the Presidency on the Democratic ticket. A letter of his, dated 1847, contains the first definite statement of the doctrine of "squatter sovereignty." During the anti-slavery struggle he favored Clay's compromise measures, upheld the Fugitive Slave Law, and voted for the Kansas-Nebraska Bill. See SQUATTER SOVEREIGNTY; FUGITIVE SLAVE LAWS; KANSAS-NEBRASKA BILL.

Cass was appointed Secretary of State by Buchanan, and resigned in 1860 because of the President's refusal to strengthen the forts at Charleston, S. C. He was one of the statesmen of the "old school," deserving of honor because of his work in building up and Americanizing an important section of his country.

CASSANDRA, *kă san' dra,* in classical mythology, the unhappy prophetess of Greek legend who was doomed to utter her prophecies to unbelieving ears. She was the daughter of the Trojan king and queen, Priam and Hecuba, and was loved by Apollo, from whom she received the power to foretell the future. When she would not return his love the angry god decreed that none should believe her words. Again and again she warned her countrymen not to keep the stolen wife of Menelaus, the beautiful Helen, and she vainly begged them not to take the Wooden Horse within the walls of Troy (see TROY). In some of the Grecian myths she is carried away to Greece by Agamemnon and there murdered by his wife, Clytemnestra. This story is told by Aeschylus in his tragedy, *Agamemnon.*

The unhappy lot of Cassandra is the subject of a fine poem by Schiller, a stanza of which is here given:

And men my prophet wail deride!
  The solemn sorrow dies in scorn;
And lonely in the waste I hide
  The tortured heart that would forewarn.
Amid the happy, unregarded,
  Mock'd by their fearful joy, I trod;
Oh, dark to me the lot awarded,
  Thou evil Pythian god!

CASSAVA, *ka sah' va,* an important food crop, widely cultivated in tropical countries but native of South America. It is a bushy shrub, with broad, shining, hand-shaped leaves. There are two forms, bitter and sweet, both cultivated for their starchy roots in the West Indies, Africa and tropical America. Excellent crops are produced from stem cuttings on light, sandy, well-drained soils prepared as for corn. Bitter cassava contains a poisonous juice, which when pressed out and boiled becomes a delicious sauce called *casareep.* The tapioca of commerce is made from cassava. The only state in which the cultivation of sweet cassava has attracted attention is Florida, but it is not an important industry there, owing to the high cost of labor in its production compared with labor conditions elsewhere in its range.

CASSEL, or **KASSEL**, *kahs'el,* one of the most picturesque towns of Germany, capital of the province of Hesse-Nassau. It is situated on both banks of the Fulda River, ninety-one miles northeast of Frankfort-on-the-Main. The city possesses one of the finest collections of pictures in Europe, with splendid examples of the works of Frans Hals, Rembrandt and Van Dyck. It is also an important musical center and has a fine opera house, built by Jerome Bonaparte. The commerce is extensive and the chief articles of manufacture are engines, mathematical and scientific instruments, porcelain, knives and chemicals. There are large lithographing plants, where some of the world's finest lithography is produced. Population in 1910, 153,196.

CASSIA, *kash'a,* a large group of pod-bearing plants found in tropical regions. The cassias consist of trees, shrubs and herbs. The leaflets of several kinds form the well-known drug called *senna.* The pods contain a thick pulp which belongs to the sugar class of laxatives; both leaves and flowers are also used as medicines. The bark of an entirely different plant of the laurel family is commonly called *cassia bark.* Having the flavor of cinnamon, and being cheaper, it is often substituted for it. The cassia of the Bible was probably cassia bark.

CASSIOPEIA, *kas i o pe'ya,* one of the most conspicuous constellations in the northern hemisphere, sometimes called the "Lady in her Chair." It is about the same distance from the North Pole as the Great Bear, and contains a large number of stars; five of the brightest are arranged somewhat in the form of a *W.* According to legend, Cassiopeia was the mother of Andromeda, who was chained to a rock on the seashore to appease the anger of the gods and was rescued from a monster by Perseus. See ANDROMEDA.

THE CONSTELLATION OF CASSIOPEIA

CASSOWARY, *kas'o wa ri,* a large, shy but inquisitive bird belonging to the same family as the ostrich and emu. It is a native of New Guinea and stands about five feet high. Its peculiar hairlike feathers are black in both sexes. The head and neck are bare and bluish in color, and its head is crowned by a bony crest of brilliant blue, scarlet and purple. The wings are so short that the bird is unable to fly, but its legs are powerful and it can run with nearly the speed of a horse. Disliking the sunshine, this timid bird builds its rough nest of leaves and grass in dense, wooded places, and comes out in the morning and evening for its meal of fallen fruit, bulbs or insects. The plumage of the cassowary is used by the natives for head-ornaments, mats and rugs, and the flesh is considered delicious.

CASSOWARY

CAST, a reproduction of a statue or other work of art in a mold. The model to be copied is covered with plaster, so applied that it may be removed in sections. When these sections are carefully put together they form a mold which is filled with liquid plaster. When the plaster is dry the mold is removed, uncovering a copy of the original model. Most of the famous statues of the world have been copied in this way, thus furnishing models for museums and schools.

CASTE, *kast,* a word from the Portuguese *casta,* meaning *family, strain, breed* or *race.* In a general sense it means a hereditary division of society on the basis of occupation, wealth, religion, etc. Specifically, however, it applies to the classes into which the Hindus are divided by religious laws. It is probable that caste was originally grounded on a difference of descent and mode of living, and that the separate castes were originally separate races. The caste system prevails principally in India, but it is known to exist or to have existed in many other regions.

In India there are four castes: the Brahmans, or those highest in authority, usually priests; the military order; the husbandman or trader caste, which is divided into sub-castes, according to occupation; and the servant caste. All those below the fourth are called Pariahs, or outcasts. This system has kept the people of India contented in the castes to which they were born. They can never rise above their caste, but can lose caste by wrong-doing, and to regain it must perform certain religious rites. In other countries the only true social divisions are those marked by degrees of ability and by achievements.

CASTILE, *kas teel',* **AND ARAGON,** *air'a gon,* or *ah'rah gohn,* formerly two separate and powerful kingdoms of Spain, which were united

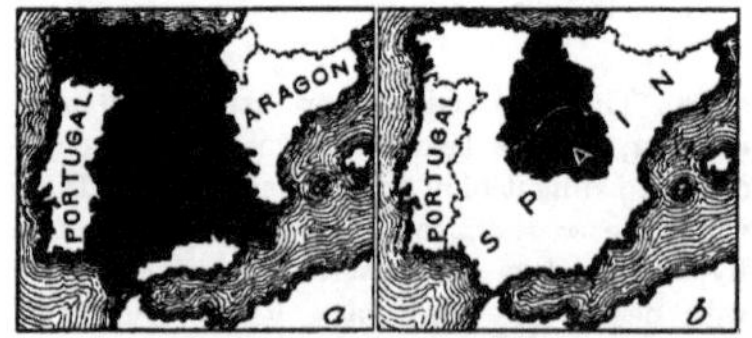

CASTILE (IN BLACK)
(*a*) Old Castile; (*b*) the modern province.

in 1469 by the marriage of Ferdinand of Aragon and Isabella of Castile. These monarchs are the ones whom Columbus interested in his proposed voyages, and thus they earned a place in American history. The combined territories formed the nucleus of the modern kingdom of Spain. Aragon extended over the northeastern part of the peninsula, while Castile occupied the greater part of what is now Spain, extending from the Bay of Biscay south-

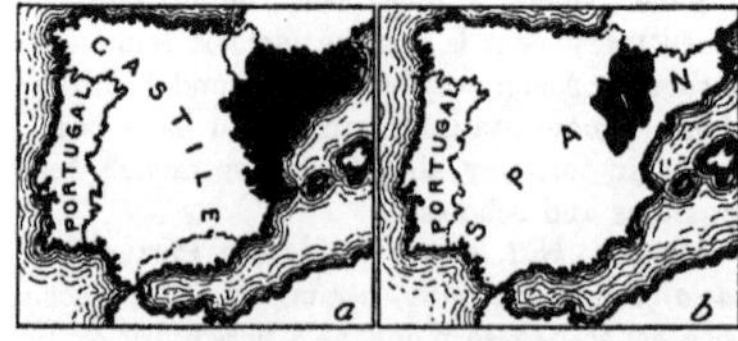

ARAGON (IN BLACK)
(*a*) In the time of Isabella; (*b*) present-day province.

ward. In the great struggle against the Moors, Castile had always played a prominent part, and the additional strength derived from its union with Aragon made possible the death blow to Moorish dominion in Europe (see MOORS). The language of Castile is still the literary language of Spain and to a great extent of the Spanish-speaking countries of South America.

Under the present system of government Castile is divided into Old and New Castile and again subdivided into provinces, the most important of which is Madrid, in which the capital of Spain is situated. Aragon is now divided into the provinces of Saragossa, Huesca and Teruel.

CASTLE, *kas''l,* a word derived from the Latin *castellum,* meaning *a fort,* applied to an edifice serving as a residence and a place of defense. The Romans erected permanent fortified camps in their colonies and conquered territories, surrounding them with walls and by

CASTLE
The fortress of the medieval period.

(1) Moat	(6) Rampart
(2) Drawbridge	(7) Portcullis
(3) Wicket	(8) Donjon, or Keep
(4) Merlon	(9) Turret
(5) Embrasures	(10) Escutcheon

other means making them strong enough to resist attack. The castle was a natural evolution from such camps and originated in the desire for a safe place in which to dwell and in which to store valuable possessions. In the days of feudalism there was almost constantly a state of war. Each feudal leader gathered his retainers round him and was prepared to protect his possessions by force of arms, or to take from his neighbors what he wanted by the same means. The feudal barons of Europe, the robber chiefs of the Rhine and the nobles of France, especially, built for themselves castles, some so strong that they could resist any attack. The Crusaders taught Europe much concerning the art of war and incidentally how to build castles.

Sometimes a castle was built to defend a

town and to be the residence of the feudal chief, who was paid by the town for protection. A village would perhaps spring up around a castle already built, and its inhabitants would be more secure, feeling that they had a stronghold to which to retire in time of danger. In building a castle every advantage was taken of position. Sometimes the castle would be perched upon almost inaccessible crags. If placed on level ground the entire building was surrounded by a *moat*—a deep ditch filled with water. Across the moat was a *drawbridge* which could be raised or lowered at will. The outer wall was of great height and thickness, strengthened at intervals with towers in which were loopholes through which missiles could be discharged. The main entrance was protected by a *portcullis,* an armored gate or door, which could be raised to allow free passage or lowered to completely bar the way. An inner wall surrounded the main court in which was the *donjon,* or keep, the strongest part of the castle, and the residence of the chief and his family. The keep was usually fifty or sixty feet high, with thick walls, sometimes honeycombed with passages leading to all parts of the castle and, in many cases, beyond the outer walls.

The invention of gunpowder rendered the most powerful castle useless as a protection, and though many were built at a later date they were no longer regarded as strongholds. In many parts of Europe the ruins of castles may still be seen, and some few have been kept in a habitable state. For the most part, however, they are now merely objects of curiosity to tourists, to be inspected on payment of a small fee.

A graphic description of an attack on a castle is contained in Sir Walter Scott's *Ivanhoe.* F.ST.A.

CAS'TOR AND POL'LUX, the twin sons of Zeus and Leda, and the heroes of some of the most picturesque stories in Greek and Roman mythology. They are often called the *Dioscuri,* which means *sons of Zeus.* The "illustrious twins," as Horace speaks of them, were champions of the manly sports, Castor favoring especially the art of horsemanship and Pollux that of boxing. Helen of Troy was their sister. One of their exploits was their invasion of Attica to rescue her from Theseus, who had carried her off to Athens. This story is a variation of the Grecian myth concerning Helen (see TROY). They also shared in the dangers of the Calydonian hunt, in which the greatest heroes of Greece engaged in order to rid the fields of Calydon of a ravenous boar. They sailed on the Argonautic expedition (see ARGONAUTS) and were afterwards honored as patrons of voyagers.

Castor, who was mortal, was slain in battle, and Zeus, to comfort the grieving Pollux, permitted him to share his immortality with his brother. Thus the brothers lived one day in Olympus, and the next in Hades. According to another story, Zeus placed them among the stars as Gemini, the Twins.

In art the brothers are usually represented with snow-white steeds. The Romans believed that they rode on their horses at the front of battle during the fight at Lake Regillus. Macaulay, in his *Lays of Ancient Rome,* has a famous description of them:

> So like they were, no mortal
> Might one from other know;
> White as snow their armor was,
> Their steeds were white as snow.
> Never on earthly anvil
> Did such rare armor gleam,
> And never did such gallant steeds
> Drink of an earthly stream.

CASTOR OIL, a well-known purgative, obtained from the seeds of the castor oil plant. This is a native of Africa, but is now cultivated in all the warmer regions of the globe. The oil is obtained from the seeds by bruising and pressing. When pure and fresh, it should be clear, colorless and sticky. The disagreeable taste, so familiar to all, can be partially overcome by use in capsules or coffee, with orange peel or as a "sandwich" between two layers of fruit juice. Castor oil is also used to oil machinery, for making sticky fly paper, and in the East Indies as lamp oil. Castor oil plants are no longer considered a paying crop, other products having like uses now taking their place in the commercial world. They are, however, cultivated in gardens for ornament, even in temperate regions.

CASTOR OIL PLANT

Showing, also, flower and seeds.

SOME OF THE CATS

CAT, the purring, fireside member of the cat family, which includes, also, the wildcat, cheeta, lynx, tiger, lion, etc. The domestic cat is about twelve to sixteen inches long, rather large and broad of head, slender-bodied, with short, powerful jaw, sharp teeth and claws and strong muscles. It belongs to a family of animals that is better armed than any other for the destruction of animal life. As a destroyer of mice and rats the cat is unexcelled, and although a bird enemy, too, its real use to mankind is often under-estimated. With eye-sight adapted to darkness; and with patience and cunning, the cat watches for prey which is a pest to man. One of the first things a settler in a new country needs is a cat. Mice are usually so numerous there that without this little nightwatchman it would be difficult to keep necessary stores of food. In many cities the worth of cats is officially recognized; a definite sum is furnished yearly for the maintenance of cats in many post offices and public buildings, to keep away the ever-destructive mice. The cat becomes more strongly attached to places than to people, and will often desert families who have moved, to go back to the old home. Cats enjoy petting, and show jealousy if neglected for other pets.

It is believed that the cat was originally tamed from the wild state and domesticated in Egypt. In that original "granary of the world," the cat was loved and honored. From the Egyptian cat-headed moon-goddess Pasht, it is believed, has come the expression *pussy*.

Peculiarities of the Cat. There are a number of things about the cat family, not seen in other animals, which are of especial interest. The most conspicuous of these are the claws. The forefeet of cats have five toes each, the hind feet four, and every toe is fitted with a sharp, hooked claw so joined to the tip bone of the toe that it can be pulled in or thrown out at will. This makes it possible for the cat to walk softly on the fleshy part of its feet, or to scratch and tear when the need arises.

The eyes of cats hold peculiar interest. The color varies in different kinds from yellow to orange, from blue to emerald green. The pupils can be wonderfully expanded or contracted, sometimes being just a vertical slit with a pin hole at each end through which the light enters. Because of this power cats see well in the dark.

To show their emotions, when pleased and contented, cats cause a vibration deep in the throat, which from the sound is called a "purr." When angry they hump their backs high in the middle, raise their hair and tails and glare and spit and howl.

To drink, cats lap up milk or water with the tongue. When they catch mice or other animals they seem to delight in playing with the victim, holding it a short time, then letting it free, and again with a quick movement catching the creature and shaking it; this routine continues until they finally put an end to the life of the unfortunate subject of their playfulness.

Objections to Cats in Families. In spite of the usefulness of cats, some people will not keep them as pets, for it is true they have treacherous instincts. One minute a cat may be purring happily; the next, its claws may be out, scratching and tearing. So a cat is not always a good pet for a child. Then, too, cats delight in killing birds. It is charged that cats are unclean animals and carriers of disease; the truth, however, is that if a cat is given a fair chance in life it keeps itself one of the cleanest of pets.

Cat Aristocrats. Among the various breeds or races of cats now carefully bred for the cat shows, two of the most curious are the tailless *Manx* cat, or rabbit-cat, and the *Persian* cat, with its long silky fur. The *tortoise shell,* the color a mixture of black, white and brownish or fawn color; the large *Angora,* and the blue, or *Carthusian* and *Maltese* cats, with long, soft, grayish-blue fur, are the species most admired by the lovers of cats. The name *tabby* is derived from a street in Bagdad-Attab —celebrated for its manufacture of watered or moiré silks, in England called *taffeta.* The word *tabby-cat* does not refer to the sex of the cat, but to the fur, which is usually yellow, marked with orange, red, brown or black.

The wild members of the cat family, danger-

Outline for Essays on the Cat

Why I Love My Cat

(1) Its beauties
- (a) Soft fur
- (b) Markings
- (c) Bright eyes
- (d) Graceful tail
- (e) Long "whiskers"

(2) Its cleverness
- (a) How it found its way home when stolen
- (b) Its ability to catch mice
- (c) How it drove away the dogs that would have harmed its kittens

(3) Its affectionate ways
- (a) Likes to rub against people
- (b) Its satisfied purr
- (c) Its fondness for curling up in my lap

A Cat's Story

(1) Where I was born

(2) My brothers and sisters
- (a) Number
- (b) Description

(3) Why I was given away

(4) My new home
- (a) The place
- (b) The people

(5) Why I like girls better than boys
- (a) They feed me
- (b) They are not rough with me

(6) My little mistress
- (a) Description
- (b) Her treatment of me

(7) What I like best
- (a) The warm corner
- (b) My favorite food

ous to man and other animals, referred to at the beginning of this article, are described under their respective titles. M.S.

Consult Williams's *The Cat, Its Care and Management;* Champion's *Everybody's Cat Book.*

CATACOMBS, *kat'a kohmz,* subterranean galleries and caves forming the usual burial places of the early Christians who did not practice cremation. They bought plots of land, either as communities or families, and under the surface excavated cemeteries, or crypts. The term *catacomb* is said to have been applied originally to the district near Rome which contained the chapel of Saint Sebastian, in the vaults of which, according to tradition, the bodies of Saint Peter and Saint Paul were first deposited.

CHRIST AS THE GOOD SHEPHERD
A symbol of the love and faith of the early Christians.

The catacombs of the early Christians of Rome consisted of long, narrow galleries, usually about eight feet high and five feet wide, branching off in all directions, forming a perfect maze of corridors. When one story

CRYPT OF SAINT CECILIA, ROME

was no longer sufficient, staircases were made, and a second line of galleries was dug out beneath. The graves, or *loculi,* to receive the bodies, were cut into the walls of the gallery, one above another. They were closed laterally by a slab, on which there was occasionally a brief inscription or a symbol, such as a dove, an anchor or a palm branch, and sometimes all of these. Some of the inscriptions and epitaphs were beautifully carved, some merely scratched upon the slab, and others were painted in red and black. In later times beautiful frescoes were common, in which are indicated the Christian faith and devotion.

It is now regarded as certain that in times of persecution the early Christians frequently took refuge in the catacombs, since burial places had the right of protection by law, and they also gathered there to celebrate in secret the ceremonies of their religion. The practice of burial in the catacombs completely died out in the beginning of the fifth century, and the existence of such subterranean burial grounds was forgotten. Six centuries later excavations laid bare what were then supposed to be the ruins of previous cities.

The term *catacombs* has also been applied to certain ancient subterranean quarries in Paris, which have been used since 1786 as burial places. It is said that 6,000,000 bodies lie in these catacombs, where the bones are arranged in fanciful designs along the sides of the passages. F.ST.A.

CATALEPSY, *kat'a lep si,* a word derived from the Greek *katalepsis,* meaning a *seizure.* It is now applied to a condition in which a person suddenly becomes unconscious and remains rigidly fixed in the attitude assumed when the attack commenced. The seizure may terminate quickly or it may continue for some time. The action of the heart and lungs continues, and the pulse and temperature remain natural. This condition is generally the consequence of some other disease. Catalepsy has often been mistaken for death, and cases are on record in which full preparation for burial had been made before the mistake was discovered. The assertion has been made that occasionally persons in a cataleptic state are buried alive, but there is slight evidence in support of this.

CATALINA, *kat a le'na,* or **SANTA CATALINA,** a favorite California resort and scene of annual fishing tournaments, is one of the Santa Barbara Islands, near Los Angeles. It is from one to four miles wide and about

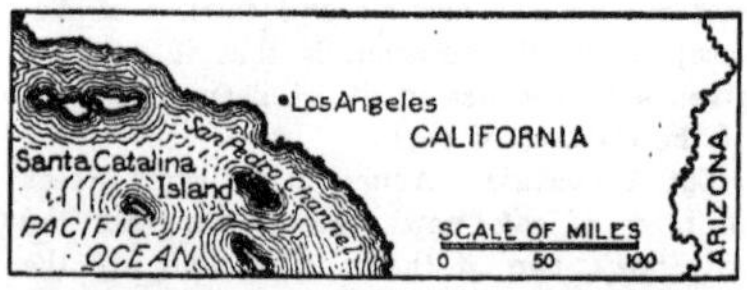

twenty-two miles long. The waters around this hilly, wooded island abound in fish, and following the policy of conservation of natural resources, by act of the California legislature

Catalina has been made a fish reservation. Formerly the canning of tuna was rapidly exhausting the supply of that fish, and as many as 150 fishing launches were seen there each day. Tourists visit Catalina in constantly increasing numbers. They are now invited to view the wonderful marine gardens in glass-bottomed boats; the sight is a never-ending source of wonder and delight, for one sees a jumbled rainbow of color, a gracefully-moving mass of curious shapes—starfish, octopus, long-armed jelly-fish, angel-fish, sea cucumber, crab and squid—of varied colors, moving in the clear blue water, in and out among the ribbon-like olive and amber leaves of kelp.

CATALPA, *ka tal'pa,* or **INDIAN BEAN,** a rapidly-growing shade and ornamental tree. Some species are natives of Asia; others belong to the United States and Southern Canada. Hardy varieties are cultivated in cool temperate climates. The leaves are large, heart-shaped and pointed; the flowers, trumpet-shaped, are white and tinged in the throat with purple and violet. They are followed by pods containing winged seeds. These pods, often nearly a foot long, hang from the leafless twigs until spring. Catalpa wood is used for fence posts and railroad ties.

CATANIA, *kah tah'nya,* a beautiful city of Sicily, capital of the province of the same name, situated on the lower slopes of Mount Etna, fifty-nine miles southwest of Messina. It has been repeatedly visited by violent earthquakes; one of the worst of these was in 1693, when it was almost entirely destroyed; and it has been partially laid in ruins by lava from eruptions of Mount Etna. The city was one of the most flourishing of Greek cities in Sicily and was important under the Romans. Catania manufactures silk and other fabrics, besides lava ware and amber ware. It exports grain, fruits, sulphur and wine. Population in 1911, 211,699.

CATAPULT, *kat'a pult,* a machine which works on the same principle as a crossbow, used by the ancients to throw heavy missiles, chiefly stones or bars of iron. The toy weapon of this same nature now used by boys and called a slingshot in the United States and Canada is described under the title SLING. In the trench fighting prevailing in the War of the Nations various forms of the ancient catapult were revived and found effective, throwing poison-gas missiles straighter and for a greater distance than they could be hurled by hand. Some of the missiles were bombs or grenades, sometimes of primitive manufacture, consisting merely of tin cans filled with gunpowder or dynamite ignited by means of a fuse. In some cases a stick served the purpose of a catapult; one end was stuck deep into the ground and to the other was attached

THE CATAPULT

At the left, the machine projected a number of arrows at one time; that at the right threw a large stone when the spring was released. Such instruments of destruction were common for hundreds of years.

a loop of cloth, leather or rope. The bomb was placed in the loop, the stick was bent as far as possible and suddenly released; it sprang back into an upright position and hurled the bomb into the hostile trenches.

CATARACT, *kat'a rakt,* a disease of the crystalline lens of the eye or its capsule, which produces in these structures a loss of transparency which affects the vision in varying degrees, from slight impairment to blindness.

The origin of the term *cataract* is a matter of dispute. It is generally accepted, however, that it is derived from a Greek word meaning a *waterfall,* this name being applied to the disease under the belief formerly held that the opacity, or lack of transparency, was caused by an opaque liquid falling from above downward in front of the crystalline lens.

The normal crystalline lens of the eye is a bi-convex body enclosed in a sac called its *capsule.* Both the lens itself and this capsule are transparent; the function of the lens is to focus rays of light and cause them to form a perfect image on the retina.

There are many varieties of cataracts, such as the *congenital* (existing from birth), the *senile* (occurring in the aged), the *traumatic* (due to injuries), and cataracts produced by diseases of the eye itself or diseases of the body generally, such as diabetes, gout, kidney and diseases of the blood vessels. The variety of cataract most commonly met with is the senile. The cause of this variety is not definitely known. It is certain, however, that old age predisposes to its development, and that proper care of the eyes, regulation of the demands made upon them, wearing of glasses fitted by a competent oculist, if necessary, and observance of the laws of general health all greatly diminish the tendency to its development.

The development of the disease is usually slow, without pain or inflammation. The vision is affected in greater or less degree, depending upon the location, number, form, size and density of the non-transparent areas the cataract produces in the lens. When the greater portion of the lens has become affected, the pupil, which is normally black in color, assumes a grayish or white appearance, and such a cataract is designated as a *ripe,* or *mature,* cataract; if only a small portion of the lens is affected, the cataract is designated as *unripe,* or *immature.*

The vision of an eye affected with a mature cataract is reduced to blindness, while that of an immature cataractous eye varies in proportion as the cataract is developed. If the non-transparent spots or opacity of the cataract be located in the center of the lens, vision may be considerably affected, even though the opacity itself be small in size, owing to to the fact that an opacity in such a situation interferes greatly with the entrance of the light rays to the retina of the eye; if the spots or opacities be located at some distance from the center of the lens, vision may be only slightly affected.

Interference with vision is the chief, and in the majority of cases the only, disturbance a cataract causes a patient. These visual disturbances manifest themselves in different ways, such as fogginess of vision, seeing double or multiplying objects, floating specks before the eyes, etc. Often the development in old people of the change in vision called popularly "second-sight," permitting them to read comfortably without reading glasses which had previously been nécessary, is an early manifestation of a beginning cataract.

There is no drug or medical treatment which will "dissolve," "absorb," or "take off" a cataraet. The various advertised remedies claiming such powers are snares and deceits, and the exploiters of them are quacks and imposters. In the early stages of a beginning senile cataraet, the patient may often be made comfortable by proper regulation of the demands made upon his eyes and the wearing of properly-fitted glasses, which require, as a rule, frequent changing. After the cataract has reached maturity, or is ripe, a surgical operation is the only scientifically-recognized method of restoring vision. The operation is usually successful.

Each variety of cataract enumerated requires a special management and treatment suited to its particular type. See EYE. R.J.T.

CATARACT. See WATERFALL.

CATARRH, *ka tahr',* in the broadest sense, the name applied to inflammation of any mucous membrane. Such a condition is accompanied by profuse discharge of the natural secretion and is designated according to the seat of irritation, as gastric catarrh, intestinal catarrh, catarrh of the bladder, etc. More frequently the term is applied to chronic irritation of the nasal passages. Acute attacks of nasal catarrh have symptoms similar to those of a severe cold—spasms of sneezing, watery discharge from nose and eyes, feeling of weight about the forehead, etc. (see COLD). Neglect of chronic catarrhal trouble is sometimes responsible for impaired sight or hearing. This condition should be treated by a competent physician. W.A.E.

CATAWBA, *ka taw'ba,* a formerly powerful tribe of Indians who lived along the Catawba River in North and South Carolina. When first known they were the most important tribe south of the Ohio River, next to the Cherokee. They were agriculturists, but were constantly at war with the Iroquois and other northern tribes. They favored the British in the French and Indian War, and aided the Americans in the Revolution. War, disease and dissipation caused their numbers rapidly to decrease, until now but very few remain, and these live on a reservation in York County, South Carolina. For their customs and habits, see INDIANS, AMERICAN.

CATBIRD, a common American bird of the mocking bird family, so named because one of its calls sounds like the mewing of a cat. It is a creature of moods and attitudes. Sometimes it is seen proudly perched, tail erect singing the sweetest of songs, often in imitation of other birds; at other times it will be seen head and tail down, with a dejected appearance. It is found throughout the Northern and Middle states and Southern Canada, in thickets and shrubberies, where it lives an active existence, chiefly in the pursuit

CATBIRD

of insects. Its plumage is a deep slate color above and lighter below, with a reddish-brown patch beneath the tail. In winter it migrates to the extreme southern part of the United States, or even to Mexico and Central America.

CATECHU, *kat' e choo,* a resinlike substance used in tanning, dyeing, calico printing and in medicine. It is obtained chiefly from the wood of various acacia trees native to India, the best quality being the product of trees about one foot in diameter. The bark is removed for use in tanning, and the heartwood is cut up into pieces and boiled in water until a tarry substance results. After this has partially hardened, rough blocks or balls of it are wrapped in large leaves, and in this form it is placed on the market. The neutral dyes which this catechu, or *cutch,* produces are very valuable. In medicine it is a valuable astringent.

CATERPILLAR, a wormlike creature which bears somewhat the same relation to a butterfly that an ugly, rough bulb bears to a stately lily; for from this crawling thing, which is to most people one of the most loathsome of living objects, there develops in time the most beautiful of all the insects (see BUTTERFLY; MOTH). The curious name probably means *hairy cat;* it thus would really seem to belong to the woolly caterpillars only, but it is applied as well to the hairless ones.

The Story of a Caterpillar. *Molting.* When the egg of a butterfly hatches, it is a tiny caterpillar which crawls out; and as the butterfly has taken care to place the egg near some plant, it finds its food near at hand. At once it begins to eat and to grow, but its skin does not grow with it as does the skin of most animals. Soon this becomes too tight, and the caterpillar prepares to throw it off. A split appears on the upper part, near the head end, and the caterpillar then wriggles itself out, appearing in a new soft skin which has been folded in under the old one. This, too, is in turn outgrown, and the process is repeated four or five times before the caterpillar is fully grown. In the temperate regions most species remain in the caterpillar state for two or three months, but in cold climates, where they pass the winter in a condition of inactivity, longer intervals elapse between successive molts. Some species of the Arctic regions take from two to three years to pass from the egg to the butterfly state.

Structure. A caterpillar has usually twelve rings or segments, not including the head, and to each of the first three of these is attached a pair of five-jointed legs. These develop later into the legs of the perfect insect, but the feet or leg-stumps on the abdomen are not really legs and are shed with the last skin. Occasionally, as in the so-called measuring worm, there is but one pair of legs on the abdomen, and the larva moves by drawing these hind legs up to the front pair, thus forming a loop or arch of its body.

The head has six eye-spots on each side, a pair of short, jointed feelers, by which the caterpillar guides itself in moving, and strong, biting jaws which are very different from the sucking mouth-parts of the butterfly. The body may be naked or covered with hairs, bristles or spines, which in caterpillars living an exposed life are often brightly colored. On the skin of the hairless species there frequently appear lines, eye-spots or ringed spots of bright colors contrasting with that of the body, and many a caterpillar might by reason of its color be considered really beautiful if it were not for its wormlike appearance. For more details of the part played by the caterpillar in the development of a moth or butterfly, see articles on those insects.

Protection. Like most living things, many caterpillars have ways of protecting themselves from attack. Some of them make nests for themselves of rolled leaves or of spun silk; some burrow into plants. In some species there are present glands which secrete an unpleasant fluid, while with others it is a sickening taste which saves them from being eaten by other small animals. But in spite of these devices a very, very small proportion of the caterpillars that are hatched ever come to the cocoon-building stage, for not only do larger animals eat them, but tiny parasites burrow in their bodies and drain their life.

Feeding of Caterpillars. Mature butterflies do no harm, and the cocoon, or chrysalis, is to all appearance a lifeless thing needing no food, but caterpillars are heavy eaters, and it is this which makes them a serious menace. Some few, as the silkworms (see SILK), are valuable, but most of them do nothing to pay for what they devour. One of the things that makes farming hard work is the constant fight that must be waged against caterpillars, for roots, leaves, flowers and fruit have all their particular enemies among these larvae (young). Sometimes, in years when caterpillars are especially numerous, great fields are made bare of vegetation, and orchard, forest and shade trees are stripped of their leaves. Exceedingly troublesome are such pests as the cabbage worm, the cotton worm, the army worm and the cutworm. In most countries the governmental departments of agriculture have made special studies of these plant enemies, and they willingly furnish to the farmers information relating to the best means of fighting them. C.H.H.

Related Subjects. The reader is referred to the following articles in these volumes, important in the study of the caterpillar:

Butterfly	Moth
Molting	Tent Caterpillar

Consult the books referred to under articles BUTTERFLY; MOTH.

CATFISH, a group of smooth-skinned fishes with protective spines on the fins and "whiskers", around the mouth. On being taken from the water they sometimes make a curious, purring noise, hence the name. Most catfish live in fresh water, but some tropical species live in the ocean. Catfish are especially abundant in South America and Africa, but a number of species are found throughout the United States and Canada, and are there often known as *horned pout* and *bullhead.* The bullhead has been called "the friend of the poor," because it can be caught easily with any bait, hook or tackle. The largest specimens of catfish have weighed 150 pounds, but the average weight is about thirty-five pounds. The flesh is sweet and nutritious. The species known as the bullhead averages in American lakes and rivers from one to two pounds. In the Mississippi Valley the blue and channel catfish are very abundant.

CATFISH

CATHARINE, *kath'erin,* the name of two rulers of Russia who exercised great influence.

Catharine I (1684-1727), wife of Peter the Great, and after his death empress. Born of poor parents, who died when she was three years old, she never received an education. In 1701 she married a dragoon of the Marienburg garrison. When the town was taken by the Russians in 1702 she fell into the hands of a Russian officer, who passed her on to another, and finally she was taken by Peter the Great. She acquired a wonderful influence over him, and in 1712 he married her. In 1724 she was crowned at Moscow, and on the death of her husband a palace intrigue made her his successor. Though immoral and unlearned, she had charm and an active mind, and did much for Peter and for Russia.

Catharine II (1729-1796), a firm, talented empress of Russia, but an unprincipled woman with two ruling passions—love and ambition. So profoundly did she influence her country that she became known after her death as CATHARINE THE GREAT. In 1745 she was married to Peter, nephew of the Empress Elizabeth. Peter came to the throne on the death of Elizabeth in 1762. But Catharine, whose married life was not happy, won over the Imperial Guards, with the assistance of the Orloff family and others, and after Peter had reigned for a few months he was deposed, thrown into prison and afterwards killed. Catharine was then proclaimed empress.

CATHARINE THE GREAT

On the death of Augustus III of Poland she caused one of her favorites to be placed on the throne, and by this she profited in successive partitions of that country. By the war with the Turks, which occupied a considerable part of her reign, she conquered the Crimea and opened the Black Sea to the Russian navy. Her dream, however, of driving the Turks from Europe and restoring the Byzantine Empire was not to be fulfilled. She improved the administration of justice, and the condition of the serfs, constructed canals, founded the Russian Academy and in a variety of ways contributed to the enlightenment and prosperity of the country. Her enthusiasm for reform, however, was checked by the events of the French Revolution, for she felt that that great upheaval was caused by giving the common people too exalted an idea of liberty. Yet with all her faults, Catharine II was one of the most remarkable sovereigns of modern times. M.S.

CATHARINE DE' MEDICI, *kath'er in day may'de che* (1514-1589), one of the most unscrupulous queens in history, the wife of Henry II of France and mother of three French kings. The daughter of Duke Lorenzo, one of the famous Italian family of the Medici, she came to France schooled in the selfish political principles of sixteenth-century Italy, and her whole course was directed by her unbounded personal ambitions. Catharine began to interfere in state affairs in the reign of her son Francis II, and on his death in 1560 she took the entire government into her hands, for the new king, Charles IX, was only a boy of ten.

She deliberately encouraged the hostility between Roman Catholics and Protestants, playing off one party against the other and taking care that neither should obtain the balance of power. Alarmed by the growing strength of the Protestant faction, led by the Prince of Condé and Admiral Coligny, she planned the Massacre of Saint Bartholomew's Day (August 24, 1572), to which she persuaded her weak son Charles IX to give his consent. Her course of intrigue continued through the reign of her son Henry III, and she died with scarcely a friend to mourn her loss. See SAINT BARTHOLOMEW'S DAY, MASSACRE OF.

CATHARINE OF ARAGON, *air'a gon* (1485-1536), the first queen of Henry VIII of England, who quite without any fault of her own figured in a famous divorce suit that played an important part in the English Reformation. Catharine was the daughter of Ferdinand and Isabella, the Spanish monarchs who started Columbus on his voyage to the New World in 1492. At the age of sixteen she became the wife of Arthur, Prince of Wales, son of Henry VII of England. Five months after the marriage Arthur died, and the king, because he wished to keep Catharine's dowry, arranged for her a marriage with his second son, Henry, then a boy of twelve (see HENRY VIII). As the two were already so closely related by her first marriage, a dispensation from the Pope had to be obtained to make this marriage legal.

Henry and Catharine were married in 1509, immediately after he came to the throne, and lived together for twenty years. Five children were born to them, only one of whom, a daughter named Mary, lived. Henry's disappointment in not having a son to succeed him was bitter, and he pretended to see in this a sign of Heaven's displeasure because he had married his brother's widow. His love for Anne Boleyn, the queen's beautiful maid of honor, fixed him in his determination to have the first marriage annulled, and he accordingly obtained a divorce from Catharine, and married Anne. This course of action brought about a separation of the English Church from the Church of Rome, and was the beginning of the Reformation in England.

Catharine lived in retirement after her divorce, and though persecuted by agents of the king, refused to the last to sign away her rights or those of her daughter Mary.

CATHAY, a name formerly applied to China, but now used only in poetical allusions to that country. The northern provinces of China were conquered about A.D. 907 by the hordes of "Khitai," and Marco Polo, in writing about the country, called it by their name. Tennyson in *Locksley Hall* writes, "Better fifty years of Europe than a cycle of Cathay."

CATHEDRAL, *ka the'dral.* To almost anyone this word brings an immediate and a vivid picture—the view of a great, gray stone building with towers pointing skyward, with pointed arches and elaborately-carved rose windows. It is not a new building, this picture cathedral, but has century-old legends clustering about it as thickly as the ivy which shrouds it. All of these picturesque details, however, are not essential to the idea of a cathedral, which may be but a tiny church if it have one important thing beneath its roof. This is the chair or throne of the bishop, for *cathedra* means *seat,* and any church structure which

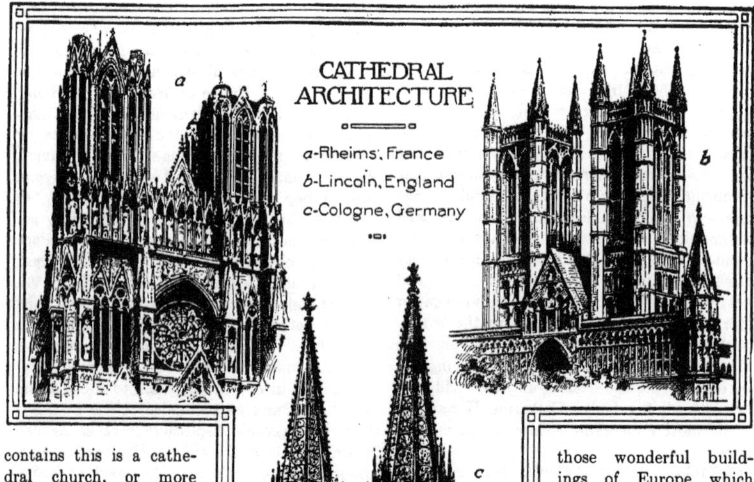

contains this is a cathedral church, or more briefly, a cathedral. In any bishop's province, or diocese, there can be but one such church, and because this is likely to be the largest and wealthiest one, the idea of stately magnificence has grown up naturally about the word.

America, like Europe, has its cathedral in every bishop's province, and some of these are beautiful structures. The cathedral of Notre Dame in Montreal, Canada, is one of the largest religious buildings in all North America; in Albany, N. Y., is the Roman Catholic Cathedral of the Immaculate Conception, a fine example of Gothic architecture; Saint Paul, Minn., and St. Louis, Mo., have Roman Catholic cathedrals, and New York City has the splendid Saint Patrick's Roman Catholic Cathedral, which cost about $2,500,000 and is one of the most magnificent of all church buildings in the United States. In process of construction, however, is an even greater one, the Protestant Episcopal Cathedral of Saint John the Divine, in New York City, which is to cost about $10,000,000.

How the Great Cathedrals Were Built. The glory of the word, however, comes not from these comparatively new structures, but from those wonderful buildings of Europe which date from the Middle Ages. When Peter the Hermit (which see) preached the First Crusade, he roused that religious fervor which is mainly responsible for the erection of the great cathedrals. Some might go on this crusade and on those which followed, but some must stay at home, and the latter sought a way to show that their devotion was as great as that of those who marched to fight the infidel. Great gifts were made by the wealthy, and the returning crusaders brought back rich trophies, but a large part of the vast sums needed came from the poor peasants at home. They denied themselves, they saved, they even worked for days at a time helping to drag the great stones which went into these holy buildings.

A tale is told of one woman who was so poor that she could give nothing at all, but she walked day by day beside the oxen which drew the loads, and fed them wisps of grass. Thus the cathedrals which constitute the very finest examples of the art of the Middle Ages are monuments not only to the victory of the Cross over the infidels, not only to the artistic sense and exquisite workmanship of the times, but to the extraordinary devotion of the people. Sometimes a cathedral would be centuries in building, and the fervor must often have flagged, but the religious purpose was kept always in view, and it is said that in no structures ever erected is there so little evidence of the tendency to slight work because it will not be seen; the most minute carving in the farthest and darkest corner is as carefully done as the sculpture over the great central doors.

The Most Famous European Cathedrals. In general, a great cathedral was built in the most conspicuous place in a city, and stood as a landmark for miles around, its towers far over-topping the roofs of the town. No special style of architecture was prescribed for cathedrals, but most of those in England and many of those on the Continent are Gothic in style and cross-shaped in arrangement, having chapterhouse, cloister, chapels and crypt connected with them.

Among the most noted on the Continent are the cathedral of Milan; the great cathedral at Florence, with its huge dome and magnificent campanile, the whole constituting one of the finest examples of the Italian-Gothic style; the wonderful Cologne cathedral; the Cathedral of Notre Dame at Paris, with its celebrated gargoyles; and the cathedrals of Amiens and Rheims, which suffered great damage in 1914 and 1915, during the early stages of the War of the Nations. All over the world there was a feeling of the profoundest regret at the destruction visited on these buildings, not only because of their beauty but because of what they stand for in religious history.

In England there are the cathedrals of Canterbury, York, Salisbury, Lincoln and Exeter, and, perhaps the most famous of all, Saint Paul's in London; Scotland has but two, those of Glasgow and Kirkwall. Detailed description of these is impossible here, but many of them are described under their titles or under the names of the cities in which they are located. For the reader interested in art few more inspiring subjects could be suggested than these cathedrals which date from the Middle Ages, for they are practically an epitome of the art of the times in which they were built. A.MC C.

Consult Pratt's *Cathedral Churches of England;* Wilson's *Cathedrals of France;* Bell's *Handbook to Continental Churches.*

CATHODE, *kath'ode,* **RAYS,** the rays springing from the negative or cathode electrode in a glass tube or other vessel from which the air has been exhausted and into whose opposite ends electrodes have been sealed (see CROOKES TUBES). The rays can be produced by connecting the electrodes with an induction coil (which see) or with an electric machine. They fill the tube with a bluish light, and if a piece of platinum is placed in front of the cathode electrode it becomes red hot. These rays are of interest because it was by experimenting with them that the Roentgen or X-rays were discovered. See ROENTGEN RAYS.

CATHOLIC UNIVERSITY OF AMERICA, located at Washington, D. C., is the highest educational institution in the United States under the direction of the Roman Catholic Church. It was founded for the purpose of giving Roman Catholics an opportunity to study higher courses under the guidance of their own Church, and is purely a post-graduate university. The founding of such an institution was advocated by Archbishop John Spalding for several years before Pope Leo XIII granted the constitution in the year 1887. The department of theology began its first sessions in 1889, and the other departments, philosophy, letters and sciences, were added as the buildings were completed.

The school of sacred sciences is divided into four departments—biblical, dogmatic, moral and historical. To the faculty of philosophy belong the departments of philosophy proper, experimental psychology and kindred branches. There are in addition courses in literature, philology, and physical, biological and social sciences. In 1911 a Teachers' College was opened for Roman Catholic Sisters who are in the teaching profession. The institution has property and investments valued at nearly $2,700,000, and a library of about 100,000 volumes. The tuition is $125 a year. About 700 students are enrolled, and they are under the guidance of a faculty of about eighty professors and instructors. Since its foundation Cardinal Gibbons has been chancellor of the institution. G.W.M.

CATILINE, *kat'i line* (108-62 B. C.), the celebrated leader of a band of Roman conspirators, whose plots against the republic were discovered and thwarted by Cicero. Catiline was of patrician birth, and his full name was LUCIUS SERGIUS CATILINA. In his youth he joined the party of Sulla, and revealed himself as a cruel and greedy supporter of that dictator. Possessed of great physical strength, a lawless nature and unscrupulous daring, he soon came into prominence politically, serving as quaestor in 77 B.C., praetor in 68 B.C. and governor of Africa the following year.

On his return to Rome in 66 B.C. he planned to gain the consulship, but was impeached for misgoverning his province and was thus disqualified. Disappointed in his hopes and burdened with debts, he plotted with several other dissolute young nobles to murder the chief men of the state and to seize the power for themselves. At this time Cicero, the great orator, was consul. Having learned of the conspiracy, he placed the city in a state of defense, and in the senate chamber, with Catiline himself present, exposed the whole affair in his famous *First Oration Against Catiline*. The guilty leader, stunned by the eloquence of the consul, attempted a feeble reply, and then fled from the chamber, with cries of "traitor" and "parricide" ringing in his ears.

He escaped from Rome during the night and hurried to his camp in Etruria. Many of his followers deserted when they heard the news of the suppression of the conspiracy and of the execution of the leaders who had remained in Rome. In 62 B.C., Catiline's force was defeated near Pistoria in Etruria, by a Roman army under Antonius, and he himself was slain while fighting with desperate courage.

Related Subjects. The story of Catiline will be considerably extended by reference to the following titles:

Cicero	Quaestor
Consul	Sulla
Praetor	

CATKINS
(a) Willow (c) Quaking asp
(b) Seaside alder (d) American white birch

CAT'KIN, a tail-like form of blossoms consisting of flowers of one sex, without petals, covered with modified leaves which shield them until pollination. The pussy willow catkins, demure little heralds of spring, are small and oval, grayish-white and silky. Those of the shade-tree willow are usually yellow and appear before or with the leaves. The drooping catkins of the poplar, the bright yellow early spring catkins of the birch, the sweet, heavy fragrance from those of the chestnut in June and July, are familiar to the lover of trees.

CAT'NIP, or CATMINT, a weed of the mint family found near barns and homes, and widely scattered throughout North America and Europe. It has much the same fascination for cats as has valerian root; because they eat it so greedily it received its name. This plant grows erect, to a height of two or three feet, with rose-tinged, whitish flowers, and downy, heart-shaped leaves, green above and whitish below. The catnip tea brewed from the leaves is still a popular home spring "tonic" in many rural districts.

CATNIP
Leaves and flowers.

CATO, *ka'toh,* in ancient history, the name of an honored Roman family, two members of which rose to fame. One of these was a type of the primitive Roman of the early republic; the other died for his principles in the declining days of the republic.

Marcus Porcius (234-149 B.C.), surnamed THE ELDER, THE WISE and THE CENSOR, was the son of a peasant farmer of Latium. His life on the small estate which he inherited from his father and diligently cultivated with his own hands taught him to revere the simple manners and customs of his Roman forefathers, and this was the keynote of his character. At the age of seventeen he fought under Fabius Maximus in the Second Punic War, and took part in the crushing defeat of the Carthaginians at Zama, in 202 B.C. Meanwhile he had removed to Rome, where he advanced rapidly in rank, serving as quaestor, aedile and praetor, and in 195 B.C. he gained the consulship.

Cato came prominently into public life at a time when Rome was being influenced by Greek ideas and customs, and throughout his entire career he was an unflinching and uncompromising foe of these new ideas, which he believed were a corrupting element in the state. In 184 B.C. he was elected censor, and so severe and unsparing was he in the performance of his duties that the title "the Censor" became his distinctive surname, and has been borne by him through all the centuries of history.

Cato was sent to Carthage in 157 B.C. on a mission for the state. Impressed by the prosperity of that city, he returned home to warn his countrymen that Rome had still a dangerous rival, and thereafter he concluded every speech with the famous exhortation, "Carthage must be destroyed." In the year of his death the great city was razed to the ground by Roman might. Cato was equally great as a writer and as a statesman, and he was the first important author who wrote Latin prose. Of his many works only one, a treatise on agriculture, has survived.

Marcus Porcius (95-46 B.C.) is known in history as CATO OF UTICA, from the place of his death, to distinguish him from his great-grandfather, Cato the Censor. He fought with distinction in the war against Spartacus, the gladiator, served as military tribune in Macedonia and was made quaestor in 65 B.C. In the latter office he was generally commended for the reforms he introduced in the management of the public treasury, and was chosen tribune of the people in 63 B.C. As tribune he delivered a speech accusing Caesar of having had a share in Cataline's plot.

Cato's opposition to Caesar was due to his inability to see that the declining state needed the service of such a genius as the great Roman leader. He placed himself on the side of Pompey when the latter broke with Caesar, and as soon as the news of Pompey's defeat at Pharsalia reached him, he departed to North Africa. There he was placed in command of the defense of Utica, but, convinced that the cause of the republic was hopeless and preferring death to surrender, he took his own life with the sword. B.M.W.

Related Topics. A broader knowledge of the life and times of these men will be gained from the following articles:

Catiline	Punic Wars
Carthage	Quaestor
Censor	Rome, subhead *History*
Consul	Spartacus
Pompey	Tribune
Praetor	

CATS'KILL MOUNTAINS, a rugged and beautiful range of mountains lying west of and nearly parallel to the Hudson River in New York state, the southern end of the range being about 100 miles by rail from New York

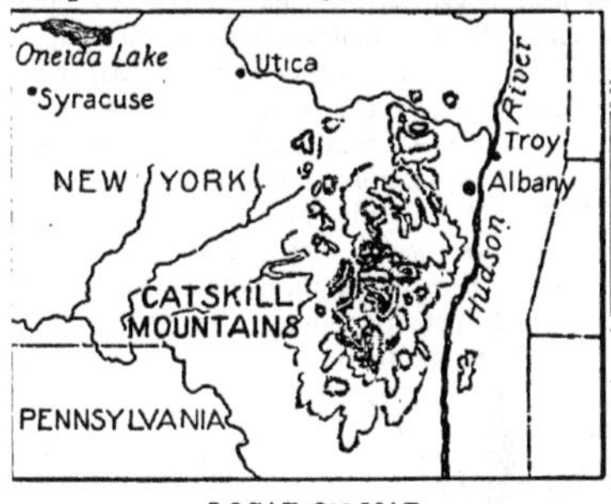

LOCATION MAP

City. The rock formation shows that these mountains were once a river plateau, but the rains and frosts, and probably glaciers, have carved these rocks of the ancient sea into many interesting shapes. And now, covered with trees and woodland plants and flowers, with waterfalls and little rivers here and there, they furnish ideal spots for a large number of summer resorts and sanitariums. This wild, interesting scenery is reached by three routes, the views from each approach being well worth a long journey. The Catskills are fifty miles long and thirty miles wide, the two highest peaks being Slide Mountain, 4,250 feet, and Hunter Mountain, 4,025 feet. Beginning in these mountains is the ninety-two-mile Catskill Aqueduct which furnishes 500,000,000 gallons of water a day to the city of New York. Here, too, is the Dunderberg, the scene of Washington Irving's *Rip Van Winkle*, a masterful piece of fiction of the days of King George the Third. See AQUEDUCT. T.E.F.

CATSUP, *kat'sup.* See KETCHUP.

CATT, CARRIE CHAPMAN, one of the most zealous and successful of the American leaders in the movement for woman suffrage. She was born at Ripon, Wis., and educated in the Iowa State Industrial College, later studying law.

In 1884 she married Leo Chapman and in 1890, four years after Mr. Chapman's death, was married to George W. Catt, a civil engineer.

Mrs. Catt taught school and advanced to the position of superintendent of schools in Mason City, Iowa, before she decided to devote all of her talents and energy to the equal suffrage cause. In 1890 she became state lecturer and organizer of the suffrage movement in Iowa, and since that time has lectured in nearly every state in the Union and in Canada and in almost every country of Europe. She has also served as president of the National American Woman Suffrage Association and of the International Woman Suffrage Alliance. To Mrs. Catt is due a large part of the credit for the progress being made in the suffrage movement in America. See WOMAN SUFFRAGE.

CAT-TAIL, a wild plant of the swamps and marshes, useful and decorative. There are two species found throughout the United States and Southern Canada. Sometimes only a few plants are found in one spot; in other

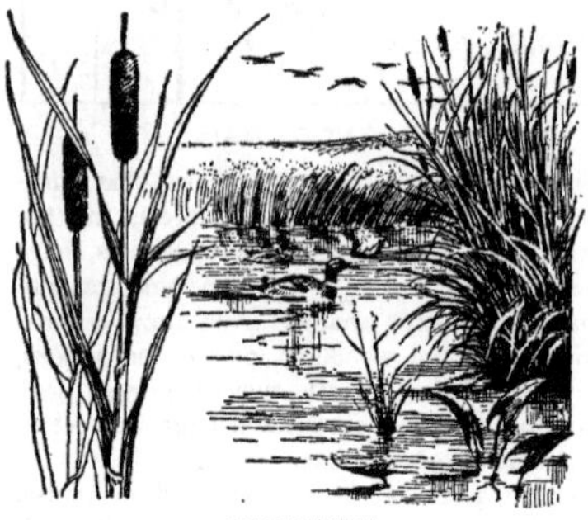

CAT-TAILS

places acres of marsh are covered with their waving green leaves and handsome, seal-brown top growths of dense, pollenlike material and oval spikes of flowers. And here, daintily perched at the tip-top point of a flower stalk, we will find the little marsh wren, singing its song, or from the hidden nests in the lower leaves hear the chatter of bittern, rails and grebes. The larger species of cat-tails, or *bulrushes,* as they are sometimes called, grow to a height of five or six feet, have long, broad leaves, and the yellow male flowers right above the brown female ones. The smaller cat-tails have narrow leaves, and the male and female flowers are divided by a short space of bare stalk.

The roots of cat-tails are rich in starch and are eaten by the Cossacks of Russia. In England, too, they are eaten under the name of *Cossack asparagus.* The silky down of cat-tails is used in dressing wounds and for upholstering. The pollen of this family of plants is very inflammable, and in some places in Europe and India is used as tinder. Soaked in kerosene, the cat-tail serves the small boy as a torch on a festive occasion.

CATTEGAT, *kat'te gat,* or **KATTEGAT,** meaning *cat's throat* in the language of Scandinavia, is the name of a strait forming one of the connecting links between the Baltic and the North seas. It is situated between Sweden

LOCATION MAP

and Denmark, and is 150 miles long and ninety miles wide. Navigation is attended with danger, on account of its numerous shoals and frequent storms. The fisheries are important, herring especially, that abundant product of the North Sea being caught in great numbers.

CATTERMOLE, GEORGE (1800-1868), an English water-color painter and illustrator whose work is associated with that of some of the most famous novelists of the nineteenth century. Scott's *Waverly Novels* were illustrated by him, as were a number of the works of Dickens. With the latter writer, especially, he was on very intimate terms, and reference to him in the *Letters* of Dickens is frequent. His chief fame, however, was due not to his illustrative work but to his water colors, which won for him honors in various exhibits. His knowledge of color and his feeling for the romantic and picturesque made him especially successful in painting scenes of chivalry.

CATTLE, *kat''l,* cud-chewing animals that have been used for domestic purposes since at least 2,000 years before Christ. They are valuable for their milk, hides and meat. Formerly cattle were used to a large extent as beasts of burden.

The domesticated cattle of the world are now of two species, the cattle found in Europe and America and the humped cattle of India, called *zebus.* The zebu is more extensively used as a beast of burden than are the cattle of Western countries. The cattle of Europe and America have descended from three species, two of which were domesticated and bred by the ancient lake dwellers; the other species is known to have existed in Scandinavia. The lines of descent from these prehistoric species to the cattle of the present are not clearly marked, and the origin of most of the breeds of the present time is rather obscure.

Cattle in Europe and America have been brought to a much higher state of usefulness than in any other part of the world. According to differences in environment, the natural divides between valleys, the artificial boundaries between countries and between different parts of the same country in its division into counties and townships, cattle have been classed into breeds, but the breed distinctions are not great enough to call for a fine zoölogical classification.

The Familiar Breeds. The common breeds are the Holstein-Friesian, Guernsey, Jersey, Ayrshire, Brown Swiss, Dutch Belted, Kerry, Dexter, French Canadian, Shorthorn, Hereford, Aberdeen-Angus, Galloway, Red Polled and Devon. Below will be shown the classification of these fifteen breeds into types according to the greatest usefulness. The breeds are distinguished one from another mainly through color differences, horned characteristics and differences in type. The horned characteristics may be discussed here, and the type and colors left until the individual breeds are described.

With the exception of the Aberdeen-Angus, Galloway and Red Polled breeds, cattle are horned. The horns of the various breeds vary all the way from the short, blunt horns about twelve inches long, of the Shorthorns and Herefords, that stick straight out from the sides of the head, to the long, beautiful, upcurving horns of the Ayrshires. At one time the native cattle of Texas were characterized by long horns which were so wide-spreading that the name *Texas longhorn* came into use. The horns on some of these cattle were ten feet from tip to tip. These were inferior range cattle which quickly gave way to better types, due to the introduction of superior beef blood. A few of these longhorns are still found in Mexico and other parts of Southern North America, and they are yet plentiful in South America.

Hornless Breeds. Breeds like the Aberdeen-Angus and Galloway are said to be *polled,* or, having the horns removed, though occasionally cows and bulls appear without horns in all breeds. This has led to the development of polled strains within the Jersey, Holstein-Friesian, Shorthorn and Hereford breeds. Herds of cattle naturally polled, or those that have been artificially dehorned, are much quieter and may be herded together like sheep. This practice of artificially dehorning cattle has become more and more common because cattle kept quiet are easier to handle and yield more milk or lay on fat more quickly.

Dehorning Cattle. Cattle may be dehorned in two ways. The most humane way and the easiest way is to kill the horns on young calves, and this should be done before the calves are three weeks old. The hair around the small buttons is clipped close. The little horn button is then rubbed with a stick of caustic potash which has been moistened in water. The horn must be rubbed hard until blood appears, so that every vestige of the horn be destroyed, for if all of the young horn tissue is not destroyed, small, unsightly, misshapen nubs of horns will grow. Two precautions must be taken; the potash should be wrapped in dry paper where it is held in the fingers, so the skin may not be injured, and

SOME MEMBERS OF THE OX FAMILY

one should be careful that none of the liquid runs down into the eyes of the calf, because blindness would ensue. Horn tissue is harder than skin tissue, and anything which would destroy the horn would destroy any skin tissue or tissues of similar nature with which it might come in contact. Cattle thus dehorned while young will have a much more sightly head than those from which the horns have been clipped when older.

The second way of dehorning applies to older animals and consists merely of clipping off the horns close to the head with suitable clippers, or of sawing them off. This method is very painful to the animal, but is humane in that it makes the animals more docile and less dangerous to others of their own kind and to human beings. The one precaution to be observed in this second method is to make sure that the horns are clipped enough so that no future growth may take place.

Two Types of Cattle. To-day cattle are bred for meat and for milk. This has led to more or less clearly-defined forms of animals for these purposes. Those which have been bred for great production of milk, cheese and butter are said to be the *dairy type;* those which produce barely enough milk for the support of their young and which are of a bodily form to yield a large amount of the highest priced cuts of meat, are classed as the *beef type.* Between the two there is a type known as the *dual purpose type.*

The Dairy Type. The dairy type of animal is characterized by a general outline of body that is "wedge-shaped from before backwards." This is brought about by a large, full udder, a large abdomen, and by the fact that in the dairy cow the hips and pelvic arch are usually somewhat higher than the shoulders and withers. The dairy cow is spare in form, with no superfluous flesh. Her joints are prominent, and her general form shows a looseness and openness not to be found in the beef animal.

A dairy cow reaches full growth at about five years of age. It is best to breed her when she is fifteen to eighteen months old, so her first calf will be born before she is twenty-seven months old. Her average useful life extends over a period of about five years, that is, she will not live on the average beyond eight years, although individual cows sometimes continue to produce until twelve or fourteen years of age. A cow should produce a calf every twelve months. She should be milked about ten and one-half months out of every twelve and be allowed to rest about six weeks previous to the birth of the next calf. During this resting period she should be well fed so she may store up fat and energy for use in her next productive period. The time between the birth of one calf and the next, when a cow is milking, is known as the lactation period.

Cows should be milked twice a day, as a rule. When an individual yields more than forty pounds, which is about twenty quarts, it may be best to milk her three times a day. A cow which is inclined to be docile and turn most of her food into milk, showing no tendeney to lay on fat, is said to have a good "dairy temperament." This does not mean a sluggish temperament; she must be bright and full of nervous force and energy, but must be quiet.

The important dairy breeds are the Holstein-Friesian, the Guernsey, the Jersey and the Ayrshire. The Holstein-Friesian breed originated in Holland, where the natural conservatism of the Dutch people has caused them to prefer this one breed of cattle for upwards of 2,000 years. This breed has found greatest development in the United States.

The Guernsey and Jersey breeds do not produce so great a quantity of milk, but make up for less quantity by a greater content of butter fat. The Guernsey and Jersey have been purely bred for upwards of 200 years on the islands of the same names in the English Channel, off the coast of France.

The Ayrshire breed, which originated in County Ayr, Scotland, is of about the same size as the Jersey, but is of a more closely knit, compact type. These animals produce a good quantity of milk of excellent quality, and have a reputation as better grazers than the other breeds.

Holstein-Friesian cattle are black and white. The color of the Jersey in general is solid fawn, varying through all the shades from light to dark. White is allowable and may occur in well-defined patches. The Guernsey cow is generally larger than the Jersey, and perhaps a little coarser. The color is yellowish, brownish or reddish fawn. This is wholly unlike the fawn of the Jersey, and is not likely to be mistaken after a few individuals of each of the two breeds have been seen. White markings are more common with Guernseys than with Jerseys. Ayrshire cattle are red and white, although occasionally a brown-and-white animal may appear.

Mature Holstein-Friesian cows should weigh 1,000 to 1,200 pounds. They yield the most milk of any of the dairy breeds; a good average yield for a mature Holstein-Friesian cow would be thirty pounds of milk per day for 300 days, or 9,000 pounds of milk in one year, and this milk will test about 3.5 per cent butter fat. A mature Guernsey cow should weigh 1,000 pounds and yield 6,000 pounds of milk in one year, testing five to six per cent butter fat. A mature Jersey cow will weigh from 700 to 900 pounds, and will produce in one year 6,000 to 7,000 pounds of milk, testing 4.5 to 5.5 per cent of butter fat. A mature Ayrshire cow will weigh about the same as the Jersey, and will produce 7,000 to 8,000 pounds of milk, testing four to five per cent butter fat.

The following are the best yearly records up to January 1, 1916, of the leading cow in each of the four great dairy breeds:

BREED	COW	LBS. MILK	LBS. BUTTER FAT	LBS. BUTTER 85 PER CENT FAT
Holstein-Friesian	Banostine Belle Dekol	27404	1058	1245
Guernsey	May Rilma	19763	1073	1262
Jersey	Jacoba Irene	17253	952	1118
Ayrshire	Lily of Willowmoor	22106	888	1045

Brown Swiss, Dutch Belted, Kerry, Dexter and French Canadian are minor breeds of cattle classed as dairy breeds.

Care and Feeding of Dairy Cattle. The foods used in feeding dairy cattle are divided into two classes, *coarse foods*, or *roughage*, and *concentrates*. Examples of coarse foods are alfalfa, clover hay, corn fodder and corn silage. The concentrates, those foods containing a high percentage of digestible matter, ordinarily used in feeding, are divided into three groups, according to the amount of protein that they contain. Protein is the chemical compound in the food from which is built the hair, hoofs, horns and lean meat of animals. The casein, or curd, of milk is nearly pure protein. The protein in the milk and the other protein materials in the body cannot be built from anything except the protein in the foods, therefore the importance of the protein.

If good foods are used, with plenty of protein in them, the feeding will be good. There are three protein food groups. The *low protein group* contains corn, oats, wheat, rye, barley, buckwheat, hominy chop, dried beet pulp and corn and cob meal. The grains, corn, oats, etc., all should be ground for dairy cows. These contain less than twelve per cent of protein in their total. The *medium protein group,* between twelve per cent and twenty-five per cent, contains wheat bran, mixed wheat feed, standard wheat middlings, flour wheat middlings, buckwheat feed, pea meal, cull beans and some others. The *high protein group,* twenty-five per cent and over, is made up of malt sprouts, linseed oil meal, cottonseed meal, gluten feed, brewers' dried grains, distillers' dried grains, buckwheat middlings and a few others. In order thoroughly to understand the feeding of animals one must be familiar with all of these feeds and their relative values.

The food that an animal eats in twenty-four hours is called a *ration.* With all the clover hay and corn silage that she will eat, or the equivalent amount of other coarse foods, it would be good practice to feed a cow a mixture of equal parts of three concentrates, choosing one from each group noted above. This is a simple and effective way of combining foods into a ration. For example, an efficient ration might be clover hay, corn silage and a grain mixture of corn meal, wheat bran and cottonseed meal, equal parts by weight. Feed her one pound of the grain mixture to three or three and a half pounds of milk yielded daily; one of the foods in the grain mixture should be a bulky food, like wheat bran. The combinations of grains that should be used will depend on the relative prices of the concentrates available. Corn silage, roots or some such succulent food should always be provided. Three or more concentrates should always be used in the grain mixture.

In addition to being well and generously fed, dairy cattle must always be provided with a comfortable, light, well-ventilated stable, with well-bedded stalls. The feeder must see that his animals have access to a clean supply of fresh water twice a day, and salt should be fed to the extent of one ounce a day. The cows may have this salt mixed with their grain every day or receive the equivalent amount twice a week.

Young stock, bulls and cows not giving milk may be fed good coarse foods and enough of the following mixture of concentrates to keep them growing well: equal parts by weight of wheat bran, ground oats and corn meal, with ten pounds of oil meal mixed with every hundred pounds of the other three foods.

Calves intended for the dairy herd should be taught to drink when two or three days old, but during these days a calf should be left with its mother. Then it should be separated from the mother and put in a clean, well-bedded pen by itself. To teach it to drink, use a small pail into which it can easily get its head. Put about two quarts of warm milk fresh from the cow into the pail, back the calf up into the corner of the pen and straddle its neck. Hold the pail in the left hand and with the right hand hold its nose down into the milk with two fingers slightly into its mouth. Hold these two fingers a little apart so that the milk will be drawn in between them. Gradually withdraw the fingers. After a few attempts the calf will usually drink properly as soon as it gets a good taste of the milk. It should get whole milk for a week or ten days, and then should gradually be turned to skimmed milk. Change from the whole milk to the skimmed milk at the rate of one pound per day, and increase the amount of skimmed milk as the appetite of the calf demands it.

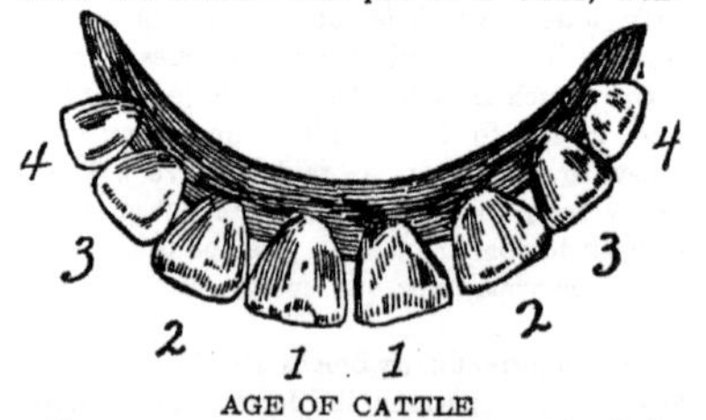

AGE OF CATTLE

The age may be told as follows: The teeth marked *1* appear when the animal is eighteen months old; those marked *2*, at the age of twenty-seven months; those marked *3*, at the age of three years, and those marked *4* at three years nine months.

MODEL COW BARN.

Dairy farming on a large scale, under nearly ideal conditions. Photograph from the Haskell Institute, Lawrence, Kansas.

Milk should always be fed at a temperature of 90° to 100°.

The greatest trouble in feeding calves comes from cold milk, dirty pails and cold, dirty pens. No one should ever feed a calf out of a pail he would not be willing to drink out of himself. As soon as the calf will eat good clover hay, it should receive all it will eat. A good grain mixture is the one indicated above for bulls and young stock. The main thing of importance in caring for young animals at all times is to keep them growing well.

The Beef Type. The beef type is characterized by a form of body of the shape of a parallelogram when viewed from the side. The animal is closely knit, with a small, fine bone, heavily and evenly fleshed all over, in direct contrast to the openness of frame and angular characteristics of the dairy type.

The best-known beef breeds are the Shorthorn, the Hereford, the Aberdeen-Angus and the Galloway. The *Shorthorn* breed, formerly called *Durham,* originated in Northeast England, in the counties of Durham and York, and has attained great prominence as a beef breed. The Herefords originated in County Hereford, in Southwest England, but have not been as uniform in their development, lacking, as a breed, in the hindquarters. The Aberdeen-Angus and Galloway breeds have not attained so great prominence as the other breeds, but are both good for beef production.

Shorthorn cattle are red, red and white, pure white, or a mingling of red and white called *roan.* Hereford cattle are red, with a white face, white legs below the knees, white breast, white brush, white on the top of the neck and white along the bottom of the abdomen. Aberdeen-Angus and the Galloway cattle are black and are polled. Galloway cattle have longer hair than the other breeds, and their hides when tanned with the hair on make the best overcoats and robes.

Feeding Beef Cattle. In growing and feeding beef cattle, two systems are in general practice: (1) Through the grain-growing regions, the cattle are bred and fattened on the same farm; (2) the cattle are bred and grown for a year or two on the range and then shipped into the grain-growing country to be fattened and finished for market. In the first of these systems, the animals to-day are grown on mother's milk and the best of corn, just as rapidly as possible. When they are marketed at six to eight months of age by this method they are called *baby beeves;* they are sold at a live weight of 450 to 500 pounds and bring the highest prices on the market. Good alfalfa or clover hay, with plenty of corn and some oil meal after the calves have been weaned, is the typical ration in producing this kind of meat. On farms where baby beef raising is not practiced, the cattle are kept growing well with good hay (clover hay, mostly), corn and a little of some concentrate such as cottonseed meal, gluten food, or oil meal, to provide more protein than is found in a strictly corn and hay ration.

In modern farm beef raising the cattle are turned to market at a weight of from 1,000 to 1,200 pounds at about two years of age. Formerly beeves were kept on roughage with very little grain until three years old, when they were fattened and turned to market at about 2,000 pounds live weight. It is now considered more economical to keep the animals growing and market not later than at two years of age. For example, a popular method, assuming that the calves are born in the spring, gives them milk and pasture the first summer, a light grain ration the first winter, pasture the second summer, a light grain ration the second winter; then they are put on a heavy grain ration with pasture the third summer. This finishes them for the fall market, and this method utilizes pastures and roughage to the greatest advantage; the method is modified to fit the conditions on the individual farm and to fit the market, and the animal reaches the market when a few months over two years old.

The beef-breeding cows reach about the same age as the dairy cows. The cow is considered to have passed a useful life if she has produced five calves. The cows are kept as long as they will breed regularly, and are then sold for what they will bring.

The Great Cattle Ranges. There is a part of the country through the western part of North and South America adjoining and including the mountain ranges which is low priced and generally unfit for farm purposes. This is partly due to the nature of the country, but mostly to the lack of rainfall. However, these lands produce nutritious grasses, which are available to animals that can travel over relatively-long distances for water and sparse vegetation. This range country lent itself to the breeding and rearing of cattle and thus developed the great cattle ranches. Here the yearly cost of feeding a steer is small, so animals are usually kept until two or three years old and then shipped to be finished on farms. The range develops the frames, but the animals are sold as feeders relatively thin in flesh. Some farmers prefer two-year-old feeders weighing 1,100 pounds, because of the greater *spread*. By spread is meant the increase per

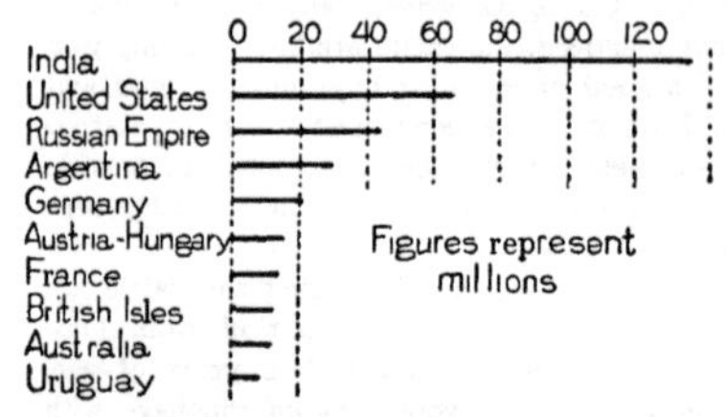

COUNTRIES RAISING THE MOST CATTLE

pound that the finished steer sells for over the cost per pound of the feeder. Thus, if an 1,100 pound feeder were sold at an increase of $1.50 per hundred weight over the purchase price, there would be a gain of $16.50, whereas the gain on the original weight of an 800 pound feeder would only be $12. Thus it might be more advantageous to fatten the larger animal, even though it cost more per pound of gain.

The great cattle ranges of Western North America are being rapidly given over to sheep ranching and to grain farming. The beef business is becoming more a small farm business instead of great cattle ranch business.

Something of the same type of cattle raising has developed on range land in other countries, particularly in Argentina, South America. With the increase in price of meat in the world's markets and the better methods of refrigeration, Argentine beef is being sold on the English, United States and Canadian markets. The United States is no longer exporting beef; there is not a sufficient amount in America, and the supply for home consumption must be augmented from other countries. The effect of the shipment of beef from Argentina to the United States will be to keep the prices of meat lower to the consumer, and of course the farmer in the United States will lose where the city consumer will gain. The importation of meats from Argentina into the United States increased rapidly during 1914-1915. The exports of beef from Argentina to all countries increased from 830,213 quarters in 1905 to 4,356,254 quarters in 1915, an increase of 3,526,041 quarters in ten years. E.G.S.

Related Subjects. The following articles will be found to contain much additional information which will be of interest to readers of this article:

Beef	Meat and Meat Packing
Butter	Milk
Cheese	Ruminants
Cow	Zebu
Dairying	
Horn	

Consult Eckle's *Dairy Cattle and Milk Production; Bulletin 75*, Bureau of Animal Industry (Washington, D. C.), also *Bulletin 34*, from same source; *Breeds of Dairy Cattle*, Farmer's Bulletin 106, United States Department of Agriculture.

CATULLUS, *ka tul'us*, CAIUS VALERIUS (about 87-about 54 B. C.), a famous Roman lyric poet, whose verses are remarkable expressions of his personal feelings and experiences. He formed a deep attachment for a woman who appears in numberless poems under the name of Lesbia, and his love for her was the theme of some of his finest lyrics. His joys and sorrows, his despair when he found Lesbia unfaithful to him, his affection for his friends and his hatred for his enemies are all expressed in his poetry without the slightest restraint.

Among the longer poems of Catullus the most remarkable are his *Nuptials of Peleus and Thetis;* his two marriage poems, forerunners of the marriage songs of Spenser, Jonson and Herrick; and a weird poem, *Attis*, which suggests an Oriental influence and is unlike anything else in Roman literature.

The following translation of a passage in one of his songs is suggestive of his poetic imagination:

> Suns may set and rise; we, when our short day has closed, must sleep on during one eternal night.

Catullus was a master of epigram, and his poetry contains numerous expressions like the following:

> What woman says to fond lover should be written on air or the swift water.

CAUCASUS, *kaw'ka sus*, or **CAUCASIA,** *kaw kay'shi a*, a region in the southeastern part of the Russian Empire, divided into two portions by the Caucasus Mountains, which extend from the Black Sea to the Caspian Sea. The portion to the north of the mountains is called Cis-Caucasia, or *on this side of the mountains;* that to the south, Trans-Caucasia, or *across the mountains*. The combined area of the two sections is 184,603 square miles, or nearly equal to that of Ohio, Indiana, Michigan and Wisconsin combined. Politically the two territories are divided into thirteen provinces, under the rule of Russia. The Caucasus is rich in all natural resources, and its oil wells are second in value only to those of the United States. The principal portion of the oil trade is carried on through the port of Baku, on the Caspian, and Batum, on the Black Sea. Population, about 10,500,000.

The Caucasus Mountains. This system extends from west to east for a distance of about 900 miles and forms a natural barrier between Asia and Europe. The inhabitants of the region have been regarded as typical of the white, or Caucasian, races of the world, hence the name. The mountains are bold and rugged, many peaks being higher than the highest of the Alps, though lacking the scenic effects found in the latter. They are formed principally of granite and schists, the only peaks showing distinct volcanic origin being Elbruz, 18,470 feet above sea level, and Kasbek, 16,546 feet. In the east numerous passes cross the mountains, but the western end of the range is impassable. The lower slopes are covered with forests of oak, birch, fir, ash, beech and elm, in which are found deer, goats, wolves, lynxes, foxes and other wild animals, besides numerous birds of prey. The rivers Kuban and Rion drain the western slopes, emptying their waters into the Black Sea. In the east the Kura and Terek flow into the Caspian Sea.

CAUCUS, *kaw' kus,* a meeting of members of a political party, called for the purpose of nominating candidates for office, to decide on a policy, or both. The caucus is the most democratic means of party government, and was the chief feature of the American political system for over a century. In a party caucus all the recognized members of the party were allowed to vote and speak. The voters of a town, a village, a legislative district or other small governmental unit usually held a caucus, and from these smaller units delegates were sent to a similar caucus or convention of the county or state. Originally the caucus was an excellent method of party government, but as the population of local units increased, the way opened for corruption and government by bosses. A few powerful leaders would make a "slate" of candidates they wished nominated, and it was seldom that the opposition could unite to overthrow this list, even if it desired to do so. The abuses which gradually arose made some other method inevitable, and the local caucus has now been supplanted nearly everywhere by the direct primary (which see).

The caucus is still used in legislative bodies, including Congress and many of the Parliaments of the world. In the United States it has become a recognized feature of party government. The members of a single party in either house of Congress meet in secret session to decide on a policy and the party's candidate for Speaker or other offices. The decision of such a party caucus is regarded as binding on all the members of the party, whether they are present or not. Until 1824 the members of each political party in the Congress of the United States held a caucus for the purpose of nominating candidates for President and Vice-President.

The system of party caucuses greatly simplifies procedure in a legislative body; if a party which has a majority of members can decide on a definite policy it can outvote the minority without delay. On the other hand there is always the danger that the minority in a caucus will be overruled, and that the majority of the caucus is really only a minority in the legislative body. Once it was a rare occurrence for any legislator to refuse to abide by the decision of his party caucus, but in more recent days of *insurgents* and *bolters* and shifting political allegiance the caucus has lost some of its importance. W.F.Z.

CAULIFLOWER, *kaw' li flou er,* a garden variety of cabbage, which, instead of a bud-like head of leaves, is a very tight head of flowers. More delicately flavored than cabbage, it is highly esteemed as a table vege-

CAULIFLOWER

table, and is served with cream or butter sauce, or is pickled. The cauliflower requires fertile soil, well drained but moist, and must be sheltered from the sun to keep the heads white and attractive for the market; this is often secured by tying the leaves over the head as soon as it appears. In food value it does not differ from cabbage, both being low, though both are of value in furnishing bulk in one's dietary, and in supplying various mineral salts. See FOOD.

CAUSTIC, *kaws'tik,* a word derived from the Greek *kaustikos,* meaning *burning,* and in surgery applied to various substances which destroy animal tissues. They are used chiefly to prevent the action of poison and to burn away the infection from sores and bites. One of the most familiar forms of caustic is found in nitrate of silver, called *lunar caustic,* generally used in the form of a pointed stick about the size of a lead pencil. Potassium hydroxide, another burning substance, is called *caustic potash.* A familiar caustic, widely used for domestic purposes, is *lye,* or *caustic soda,* valuable as a cleanser when used in small quantities dissolved in water, but a very destructive chemical if used carelessly. Carbolic acid, a very poisonous disinfectant and employed also in dressing wounds, is classed as a caustic.

CAVALIERI, *kah vah lya're,* LINA (1874-), an Italian operatic soprano, one of the most beautiful women on the operatic stage. In private life she is MADAME LUCIEN MURATORE, wife of a well-known operatic tenor. Cavalieri was born at Rome, and began her career as a singer at cafés and concert halls. Possessed of a determination to advance, and equipped with ability, a strong will and a fighting spirit which recognized no obstacles in her path to success, the obscure Roman cabaret singer found herself in time an admired and recognized star in the very center of the world of opera. After a brief appearance on the lyric stage, she made her professional debut in 1900 as Nedda in *Pagliacci,* at the Royal Theater, Lisbon. Since then she has played principal rôles in a repertory of noted operas, including *La Bohéme, La Traviata, Rigoletto, Mignon* and *Fedora.*

LINA CAVALIERI

CAVALIERS, *kav a leerz'* (horsemen), a name applied during the Civil War in England to the gaily-dressed troops devoted to the cause of Charles I, as opposed to *Roundheads,* the name given to the adherents of Cromwell and the Parliamentary cause (see COMMONWEALTH OF ENGLAND). As now more generally used, the name applies to a gay dancing partner or woman's escort.

CAVALLERIA RUSTICANA, *kah vahl la ree'a roo ste kah'na,* a popular one-act opera written in Italian, the music of which was composed by Pietro Mascagni. Its first presentation, at Rome in 1890, brought that composer from obscurity to fame. The words of the opera were written by two of Mascagni's friends, who based their work on a tale of Sicilian life, by Giovanni Verga. Turiddu, the handsome young lover of the story, goes away to the wars, and in his absence his sweetheart, Lola, marries Alfio, the village carter. When Turiddu returns home he consoles himself with a beautiful village maiden, Santuzza, but the fickle Lola soon wins back his love. In despair, Santuzza tells Alfio of her unhappiness, and the jealous husband challenges Turiddu to a duel with knives, and kills him.

Though these events all occur in a single act, at one point in the play the stage is empty, and during the interval the orchestra plays the popular *Intermezzo,* the favorite melody in the opera. The swift movement, dramatic interest and attractive music of *Cavalleria Rusticana* have given it enduring popularity; its success has been far greater than that of any of Mascagni's other operas. The soprano rôle, that of Santuzza, was one of the triumphs of Emma Calvé (which see.)

CAVALRY, *kav'al ri,* a body of mounted troops forming an important branch of all modern armies. In the days when armed knights rode to battle surrounded by their relatives, the knights and their esquires were the only mounted men, and battles depended a great deal on their personal prowess. There was no concerted action; each knight fought by himself, and battles developed into a series of individual combats. There came about a gradual change in tactics and mounted men were organized into troops, who as a body fought together, and when charging an enemy sought to deliver a tremendous blow at opposing troops by combined weight of horses and men. Not only did cavalry become useful on account of quickness of movement, but under skilful and dashing leadership often proved capable of turning the tide of battle and inflicting decisive defeats.

The original form of cavalry was similar to that of mounted infantry. Horses served a good purpose by conveying men quickly to parts of the battlefield where they were needed; when there they dismounted and fought as infantry. Cromwell proved, as no other leader before him had done, the value of concerted action and the staggering weight of a blow delivered by charging cavalry. The Cavaliers, or soldiers of the king, were individually as brave as Cromwell's troopers and probably more expert with arms and horses, but they lacked the unity of action that swept the "Roundheads" to victory. Napoleon carried the use of cavalry to a high state of perfection and developed the finest cavalry leaders the world had ever known. Waterloo proved, however, that veteran infantry will withstand the most furious cavalry onslaught.

Modern cavalry regiments are armed with swords or sabers, and carbines and pistols, or with lances. The conditions of fighting, however, have greatly changed, and spectacular cavalry charges are almost things of the past. Cases may occur in which small numbers of mounted men acting as patrols or scouts may be opposed to each other. But if a charge in the old style were attempted in warfare such as prevailed in the War of the Nations, beginning in 1914, men and horses would be destroyed by machine-gun fire before they could get within striking distance of the enemy. The cavalry of the future will doubtless revert to the mounted infantry type, use their horses merely as a means of transportation, and fight on foot from one position to another.

The Cossacks of Russia formed the most terrible cavalry regiments of Europe under the old system of fighting, and they sustained their century-old reputation as valiant warriors in the War of the Nations. In the same gigantic conflict the Uhlans of the German army, armed with their long, sharp-pointed poles, were fully as dauntless and brave. See ARMY. L.R.G.

CAVE, or **CAVERN**, a hollow opening in the crust of the earth, usually caused by the action of water in the more or less soluble strata of limestone rocks. In volcanic rocks caves were sometimes formed by the action of gases while the rock was in a plastic state, leaving hollows or air holes similar to those formed in bread while baking. To geologists caves are of great interest, especially so when found to contain animal remains, as often happens when the caves occur in limestone formations. The remains are sometimes those of long-extinct animals, and at other times they consist of fossils of animals that still exist. In the limestone caves of England fossilized bones of hyenas, reindeer, bears and wolves are frequently found. Human remains, relics of prehistoric man, have also been found in caves, the walls of which bear traces of rude carvings and drawings. See STONE AGE; BRONZE AGE.

An interesting feature is found in the stalactites and stalagmites which abound in limestone caves. In many cases the stalactites, hanging from the roof, have been joined by stalagmites, resulting in a pillar or column apparently supporting the roof. (See STALACTITE.) These stalactites often assume fantastic shapes and sometimes are of great beauty. In one cave in Yorkshire, England, there is a row of eight stalactites, shaped like icicles, producing, when gently struck, the effect of a peal of eight bells.

Some caves resemble mere burrows; others are of great extent, with vast, vaulted chambers and branching passages miles in length. The Mammoth Cave in Kentucky has more than 150 miles of passages. The cave of Frederickshall in Norway is noted for its great depth, which is said to exceed 11,000 feet. The Wyandotte Cave, in Indiana, and the Luray Caverns in Virginia are celebrated for the beauty of their stalactites. One of the most remarkable caves in the world is Fingal's Cave, on the island of Staffa off the coast of Scotland. The walls and roof are formed by columns of basalt, each perfectly shaped and finished as if by the hand of man. See FINGAL'S CAVE; LURAY CAVERNS; MAMMOTH CAVE; WYANDOTTE CAVE.

CAVE DWELLERS. In various parts of the world there have been discovered interesting caves of a former age, containing bones of animals and human beings, remains of tools and weapons and other debris. The examination and study of these have led scientists to believe that ages ago such caves were the habitation of primitive men and their families. To these rude householders of a remote past students of archaeology apply the name *cave dwellers*, or *cave men*, and the caverns in which they lived are classified according to various geologic periods. The earliest cave dwellers of Europe of whom there is much detailed knowledge ranged over England and Wales, France, Belgium, Germany, Hungary and Switzerland. In their dwellings have been found implements of warfare made of bone and unpolished stone, and pieces of bones of animals slain with these crude weapons. There are piles of awls, lance-heads, hammers and saws, arrow-heads and harpoons and needles made of bone. The animal remains are those of the reindeer, horse, ibex, bison, antelope, musk sheep, cave bear and lion.

This group of cave men lived by hunting and fishing and wore garments of skin sewed together with the sinews of the reindeer or strips of intestine. They used stone lamps filled with fat to illuminate their dwellings and had a knowledge of fire, but possessed no domestic animals. Though they knew nothing of spinning or of the art of pottery making, they were skilful in drawing animal figures, as evidenced by numerous pictures engraved on stone and ivory. Among the primitive races of today the Eskimos are nearest in habit to these ancient cave men.

There are other groups of caves in Europe showing varying degrees of civilization, some of which were occupied well into the historic era.. Some of the later caverns are believed to have been inhabited by men of refinement who were forced to take refuge in them during periods of invasion. In Southwestern North America there are remains of lodging places excavated in cliffs, but the races inhabiting these are known as cliff dwellers (which see).

CAVIAR, *kav'i ahr,* or *ka vyahr',* or **CAVIARE,** a table delicacy prepared from the roe of the sturgeon and other large fish. The word *caviar* is of Turkish or Tartar origin, but the delicacy might almost be called the national luxury of the Russians. Caviar has a peculiar piquant taste which in popular fancy can be truly appreciated only by refined palates. Thus it seems to have been considered in Shakespeare's time, for Hamlet is made to say of a certain play that it "pleased not the million; 'twas caviare to the general." Its high price even to-day prevents it from becoming an article of general diet, and it is usually found only at the tables of the well-to-do or of the higher grade of restaurants.

Caviar is prepared by freeing the eggs from the tissue which holds them together. The eggs are then carefully washed and rubbed with salt, after which they are dried and packed in kegs, bottles or cans. The best caviar comes from the shores of the Caspian Sea, but the abundance of sturgeon in the Great Lakes and their tributary regions has given rise to its manufacture in the United States and Canada.

CAVE DWELLINGS IN MEXICO
These were similar in many respects to the cliff-dwellings of another period.

CAVITE, *kah ve' ta,* the principal United States naval station in the Philippine Islands, and capital of the province of the same name. It is situated on Luzon Island, eight miles southwest of Manila. Near here, in the Spanish-American War, through which the United States came into possession of the Philippines, the Spanish fleet was first attacked by Admiral Dewey, on May 1, 1898, and the city was occupied by United States marines immediately after the engagement. It is important as a coaling station and has an arsenal, dry docks and repair shops, but owing to the nearness of Manila will probably not become a large city. The leading native industries are the manufacture of tobacco and hemp. Population, about 4,500.

CAVOUR, *ka voor'*, COUNT CAMILLO BENSO DI (1810-1861), a far-sighted Italian statesman whose efforts to unite the Italian people under a single ruler give him a place among the "nation-makers" of the world. He entered the Parliament of Sardinia in 1848, when the whole Italian peninsula was aflame with the spirit of patriotism. In 1852 he became the Prime Minister of Victor Emmanuel, king of Sardinia, and in this office Cavour worked with one aim in view—the union of Italy under a central government which should be independent of Austria. His first step was to make Sardinia an ally of England and France in the Crimean War, thus securing for his country the recognition of the European powers in the peace congress at the close of the war. He then forced Austria into a war with Sardinia, which resulted in a victory for the Italian state and the annexation of the greater part of Lombardy. With the help of Garibaldi, Cavour was able, by the beginning of 1861, to unite all Italy except Venice and Rome, and he lived to see the meeting of the first Italian Parliament. See ITALY, subhead *History;* GARIBALDI, GIUSEPPE.

CAWEIN, *kay wine'*, MADISON JULIUS (1865-1914), one of the popular recent poets of America, distinguished especially for his sympathetic treatment of nature, which may be seen from these lines in one of his introductory poems:

> If the wind and the brook and the bird would
> teach
> My heart their beautiful parts of speech,
> And the natural art that they say them with,
> My soul would sing of beauty and myth
> In a rhyme and a metre that none before
> Have sung in their love, or dreamed in their lore,
> And the world would be richer one poet the more.

Cawein was born in Louisville, Ky., and was educated in the high school there. His first poems in book-form were published in 1887, and from that time on volume followed volume with rapidity. The quality was not always uniform, but his best work was certain to win the attention of enthusiastic readers, and a volume of his verse was one of the events of the literary year. His death, which was looked upon as a distinct artistic loss, called forth a number of poetic tributes, one of which contains these lines:

> Here, here the pain is sharpest! For he walked
> As one of these and they knew naught of fear,
> But told him daily happenings and talked
> Their lovely secrets in his listening ear!
> Yet we do bid them grieve, and tell their grief;
> Else were they thankless, else were all untrue;
> O wind and stream, O bee and bird and leaf,
> Mourn for your poet, with a long adieu!

CAWNPORE, *kawn'pohr,* or **KANPUR,** the scene of the blackest atrocities of the Sepoy Rebellion, but now a flourishing commercial center of India. It is noted for its manufactures of cotton, harness, shoes and other leather goods, and is admirably situated at the junction of four great Indian railway systems. It is 628 miles northwest of Calcutta and forty miles southwest of Lucknow, on the south bank of the sacred River Ganges, in the United Provinces of Agra and Oudh. In 1857 the European garrison was destroyed by the rebels under Nana Sahib, who indiscriminately murdered men, women and children. The bodies of 200 victims were cast down a well, now marked by a memorial stone of white marble. Population in 1911, 178,557. See SEPOY REBELLION.

CAX'TON, WILLIAM (1422-1491), a celebrated English printer, whose introduction of the art of printing into England laid the foundation for the literary glories of the Elizabethan Era. In his youth he was apprenticed to a London merchant, and after the death of his master went over to Bruges in Belgium, where he established a business of his own. There he prospered in his trade and also mastered several languages. In the course of time he made a translation into English of a popular romance about the story of Troy, and in order to obtain several copies of it he learned the newly-discovered art of printing. This work was printed at Bruges or Cologne about 1474, and is the first book printed in the English tongue.

In 1476 Caxton returned to England and set up a printing shop in one of the almshouses near Westminster Abbey. In 1477 he published the first book printed in England, entitled *Dictes and Notable Wise Sayings of the Philosophers*. From that time until his death he was busily engaged in writing, translating and printing, including among his publications many of Chaucer's stories and Malory's tales of King Arthur. His services to English literature cannot be overestimated; his numerous translations helped to fix the literary form of the language, and his introduction of printing is counted among the chief events that brought about the revival of learning in England.

CAYENNE, *ka en'*, whose name is everywhere known through its association with red pepper, is the capital of French Guiana, the last remnant of continental America now held by France. It is on the Isle of Cayenne, near the mouth of the Cayenne River, in latitude 5° north; it is about 3,800 miles from France and about 2,500 miles from New York. It had a population of 13,527 in 1915. The harbor is shallow, and moderately-large ships anchor six miles away to avoid grounding. There are two quays, but no docks for repairs. The city is square in plan, clean and well built of wood and brick. There is a monthly exchange of mails with the island of Martinique.

Cayenne was founded by the Dutch in 1604, but was taken by the French in 1676. In the first decade of the present century the imprisonment of Capt. Alfred Dreyfus, a French soldier, on Devil's Island, near Cayenne, on a charge of treason widely deemed false, attracted the attention of the world. The island on which he was confined forms part of the penal colony for Asiatic and African criminals.

CAYUGA, *ka yoo' ga,* a tribe of Indians formerly living on the shores of Cayuga Lake, New York. The name means *swamp dwellers.* The Cayuga were of the Iroquoian family and belonged to the Five Nations (which see). They were the last to join this confederation and were known as the *Youngest Brother.* At the beginning of the Revolutionary War a large part of them removed to Canada and never returned; the others became scattered among other tribes. About 800 of them now live on Grand River reservation in Ontario. For customs and habits, see INDIANS, AMERICAN.

CAYUGA LAKE, a beautiful lake, one of the system known as the *Finger Lakes,* from their long, tapering shape. It is situated a little west of the center of the state of New York, is thirty-eight miles long and from one to three and one-half miles wide. It discharges its waters into Lake Ontario, through the Seneca and Oswego rivers. The principal towns on its banks are Cayuga, Ithaca and Aurora. The lake is much frequented by tourists and sportsmen, and Ithaca is particularly attractive in the summer on account of the boat races which take place on the waters of the lake, between crews from Cornell University at Ithaca and other colleges and universities. Its southern end extends down to Ithaca's city-limits.

CECILIA, *se sil' yah,* SAINT, the patron saint of music, a favorite subject for the songs of poets and the brush of artists. She suffered martyrdom about A.D. 230 and received the martyr's crown. In the Roman Catholic Church her festival (November 22) is celebrated with beautiful music. Her story forms one of Chaucer's *Canterbury Tales,* and Dryden in his *Alexander's Feast,* and Pope in his *Ode on Saint Cecilia's Day* have sung her praises. Raphael, Domenichino, Dolce and Mignard have represented her in celebrated paintings. The Church of Saint Cecilia in Rome is one of the most beautiful places of worship.

CECROPS, *se' krops,* a character who figures in Greek tradition as the first king of Attica and the builder of the famous citadel of Athens, named Cecropia in his honor. Various stories sprang up about him; he was said to have introduced marriage, the burial of the dead, and writing and other arts. In the myths he is a half-snake, half-man, who came from Crete or from Egypt and founded the city of Athens. When Athena and Poseidon disputed as to which should have the honor of naming the city, Cecrops decided in favor of Athena.

CEDAR, *se' der,* an important ornamental and commercial species of cone-bearing evergreen trees belonging to the pine family. They are distinguished by their horizontal, wide-spreading branches and fine, compact leaves.

THE CEDARS
Trees and cones.

Their reddish wood is fragrant and very durable. Chips of cedar are sprinkled among furs and woolens as a protection against moths; sometimes entire chests made of this wood are used for storage of winter garments. The inner bark of red cedar furnished the Indian many useful articles, such as blankets, ropes and the walls of his home. The white cedar is used for boats, telegraph poles, railroad ties, etc. The wood of lead pencils is nearly always cedar.

Cedars of Lebanon. The stately, strong, beautiful cedar of Lebanon, a favorite with poets and painters and frequently mentioned in the Bible, is now rarely seen. To construct the wonderful Temple of Solomon, King Hiram of Tyre furnished wood from the cedars of Lebanon. The gigantic task was done by 80,000 hewers and must have taken a very long time. The most celebrated group of cedars of Leba-

CEDAR OF LEBANON
One of the two specimens in California, believed to have been standing fully three thousand years at the date of Christ's birth.

non was located not far from the village of Tripoli until 1916, when the German army destroyed them. The circumference of the largest varied from about eighteen to forty-seven feet. Two fine specimens of the cedar of Lebanon are located in Santa Rosa, Cal.; one of these, illustrated herewith in a drawing from a photograph, is believed by David Starr Jordan to be close to 5,000 years old.

CEDAR RAPIDS, IOWA, a manufacturing city and a distributing point of importance in the state. It is situated in Linn County, in the eastern part of the state, about midway between the northern and southern boundary lines. Cedar River divides the city into two sections, known as the East Side and the West Side. Dubuque is seventy-nine miles northeast, Chicago is 219 miles east and north, and Saint Paul is 250 miles northwest. Saint Louis, Kansas City, Omaha and Saint Paul are about equally distant from Cedar Rapids. Four large railway systems serve the city—the Chicago & North Western; Chicago, Milwaukee & Saint Paul; Chicago, Rock Island & Pacific, and the Illinois Central; electric lines extend to Iowa City and Lisbon. The first settlement was made in 1845, and named for the rapids in the river. The city was incorporated in 1856; it adopted the commission form of government in 1908. The population increased from 32,811 in 1910 to 35,858 in 1914; the state census of 1915 claimed 40,667. The area exceeds twelve square miles.

Parks and Boulevards. Cedar Rapids is a blend of hill, dale and winding river, with all the aspects of a modern, progressive city. The two sections of the town are united by nine concrete bridges, and their limits enclose miles of well-made streets, many beautiful churches, numerous handsome residences, public playgrounds and parks; the largest and most attractive of the latter is Beaver Park (seventy acres) with its zoo; Ellis, Daniels, Riverside and Sinclair parks are all of considerable size, and several districts have been landscaped for residential parks.

Buildings and Institutions. All the city government buildings are located on an island in Cedar River, in the heart of the city. Among the noteworthy architectural features are the Cedar Rapids high school, the $100,000 public library, the American Trust building, and the Westminster Presbyterian, Saint Paul's Methodist and Immaculate Conception (Roman Catholic) churches. In addition to its public library and schools, it has Coe College (Presbyterian), founded in 1881, and a Masonic library, the only exclusively Masonic library in the United States. There are Y. M. C. A. and Y. W. C. A. buildings, Knights of Pythias and Elks homes and homes for friendless children and aged people. The international headquarters of the Order of Railway Conductors of America is also located here.

Commerce and Industry. Owing to its location in a rich agricultural district, Cedar Rapids is a trade center of importance and has a large number of wholesale jobbing and commission houses. The rapids in the river furnish abundant power, but electric power is used by the greater number of the city's 190 factories; the value of their output in 1915 was estimated at $46,100,000. So extensive is the product of cereals that the city is locally called *The Cereal City;* two of its cereal mills are among the largest in the United States. It also claims to have the largest independent starch works and independent meat-packing plant in the Union. Pumps, windmills, iron pipe and wire fence are also important manufactures. Railroad shops give employment to hundreds of men. J.W.

CELEBES, *sel'e beez,* one of the most important of the Dutch East India Islands, situated between Borneo and the Moluccas. It is wonderfully rich in natural resources and covers an area of 72,000 square miles, a little more than that of the state of Oklahoma. Gold is found in the valleys, and the northern portion abounds in sulphur and copper, with occasional valuable deposits of tin. Diamonds in small quantities and other precious stones in somewhat greater abundance are found, and there are extensive coal fields. The climate is tropical, and for Europeans decidedly unhealthful.

Flowers and fruits grow in great profusion; cereals and vegetables are cultivated. Coffee and spice are the principal sources of wealth, but the trade in trepang is important (see TREPANG). Wild animals are numerous, including deer, buffalo, goats, baboons and the peculiar babirussa, a wild hog with double tusks growing between the eyes and snout (see illustration, in article BABIRUSSA). The inhabitants, mostly of Malay stock, are intelligent, industrions and capable of a fairly-high degree of civilization. In the interior there still remain remnants of barbaric tribes. The capital and most important port is Macassar, through which nearly all the foreign trade passes. The island has been in possession of the Dutch since 1660, except for a short period in 1811, when it was held by the British. Population, estimated at about 2,000,000; of these, only 2,500 are whites, in government positions or in trade.

CELERY, *sel' er i,* a wholesome vegetable of the parsley family, native to the temperate parts of Europe, but now extensively grown in the United States and Canada. Wild celery is bitter, tough and woody, and to some people

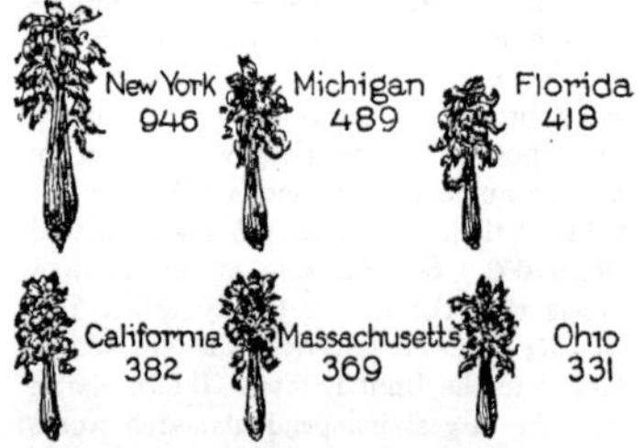

VALUE OF THE ANNUAL CELERY CROP
For average years.

poisonous, but in its cultivated form it is a crisp, tender, edible stalk, and is eaten either raw or cooked. Rich, mellow, sandy loam is the best for celery raising.

Merely the raising of celery is not the only important part of the industry. To make it white, crisp, tender and marketable it must be "blanched." It is the *etiolation,* the scientific name for the destruction of the coloring

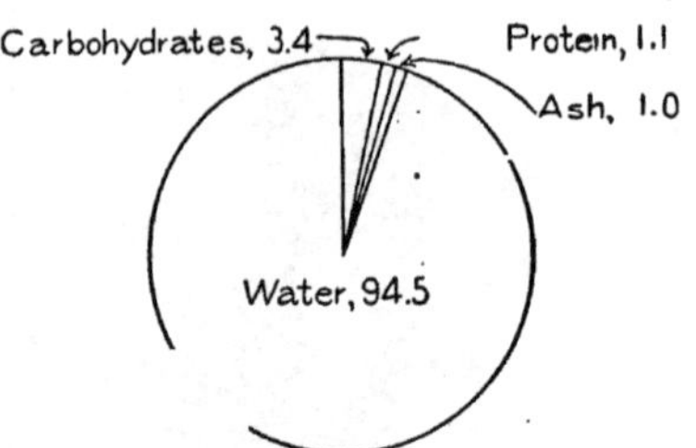

FOOD VALUE OF CELERY

matter in the plant tissue, the "blanching" or "whitening" of the stalks, which means success or failure of a crop. This is done by allowing the plant to grow in the dark. The method adopted depends on the time the crop is to be used. The most common method of blanching on a small scale, and that which produces celery of the finest flavor, is to heap soil about the stalks. For plants for early shipment blanching boards are often placed on edge along each side of a row of plants.

In Michigan, California and New York, thousands of acres are devoted to the celery industry. Kalamazoo, Michigan, is known as "the celery city"; almost all of that state's large celery crop is grown in the vicinity and shipped from that city. S.L.A.

CELESTINE, *sel' es tin,* the name of five Popes, of whom the most noteworthy were Celestine III and Celestine V.

Celestine III, Pope from 1191 to 1198, came to the high office at the age of eighty-five. While he was a thoroughly conscientious man, he did not possess quite the force of character which the troubled times required. The Emperor Henry VI immediately forced Celestine to crown him, and was strong enough not only to refuse the Pope the tribute which previous emperors had paid, but to appoint German bishops himself. Celestine excommunicated Leopold of Austria for his imprisonment of Richard the Lion-Heart and attempted to punish John of England for his treason during Richard's absence, but John refused to recognize his jurisdiction.

Celestine V became Pope in July, 1294, at the age of seventy-nine. Previous to that time he had won wide fame by the severity of the penances which, as a monk, he had imposed upon himself, and he had founded a monastic Order known as the Celestines. One of his first acts was to issue a decree declaring that any Pope had the right to abdicate, and in December of the same year in which he was elected he took advantage of this right and resigned his high office. His reason was that he wished to return to his devotions and severities. Boniface VIII, his successor, feared that a strong clerical party might grow up about Celestine, so he had the latter placed in prison, where he died.

CELL, *sel,* in biology the "start in life" of every living thing—the unit of structure. Not much more than one-thousandth of an inch in diameter, a cell is a soft mass of living, jelly-like matter, albuminous, like the white of an egg. This is called the *protoplasm.* In it, as a central organ, is a roundish structure, like the yolk of an egg, called the *nucleus.* The nucleus is generally more solid than the rest and sometimes has within it a smaller body called the *nucleolus,* and all this is usually surrounded by a cell-wall. The protoplasm possesses life-giving properties; the duty of the nucleus is to govern the reproduction of the cells. The cell multiplies by the division of the whole cell into two. This process begins at the nucleus, when the cell reaches a certain size. Then the two parts grow to the size of the first, and each repeats the process. This is continued, cell on cell, until there results the completed animal or plant. The lowest forms of plants or animals, however, are always a single cell (see PROTOPLASM).

Cells, singly, are nearly spherical in outline, but when pressed by other cells, they may take on modified forms—spindle-shaped, cylindrical, star-shaped, or they may appear as polygons.

CELLINI, *chel le' ne,* BENVENUTO (1500-1571), the most distinguished goldsmith of the Italian Renaissance, famed also as a sculptor, engraver and writer. In his youth he began to study the art of the goldsmith in his native city of Florence, but was driven out of that place as the result of a duel. Because he was quick-tempered and quarrelsome he had to move frequently, so he lived at various times in Rome, Florence and other Italian cities, and at one time was in France. His whole life was a series of adventures.

Under the patronage of Pope Clement VII, Cellini became the greatest worker in metals of his time. Later he was invited to the court of Francis I of France, and for that ruler modeled the famous bronze figure, *Nymph of Fontainebleau,* now in the Louvre. The best example of his work as a sculptor is the bronze statue of *Perseus with the Head of Medusa,* a gruesome subject, which still adorns one of the museums of Florence. Richness of decoration and elaboration of detail characterize nearly all of his productions, and these qualities give to his work as a goldsmith marvelous beauty. His famous *Autobiography,* begun in his fifty-ninth year, is one of the masterpieces of the world's literature.

CELLULOID, *sel' u loid,* an attractive artificial substance extensively used as a substitute for ivory, bone, hard rubber and coral. Having such a close resemblance to these in hardness, elasticity and texture, sometimes only experts can detect it. Celluloid is composed of cellulose, or vegetable fibrine, reduced by acids to a substance resembling soluble cotton (see GUNCOTTON); camphor is then added, and the compound is molded by heat and pressure to any desired shape. The manufacture of this much-used substance is simple and interesting, but sometimes injurious to the workmen, through the acids used. The finished product is not as safe as the things it resembles, for the touch of a lighted match or a spark will set it aflame. It explodes with a flash, like gunpowder. Celluloid is used chiefly for buttons, handles for knives, forks and umbrellas, billiard balls, backs of brushes, piano keys, napkin rings, opera-glass frames, pipe-stems, films for cameras and other small articles. It can be variously colored.

CELLULOSE, *sel' yu lohs,* a substance found in all plants except a few of the lowest kinds. It forms the walls of the plant cells which contain the living protoplasm. Cotton fibers are composed almost entirely of cellulose, and so are hornets' nests. It is also abundant in flax fibers (and therefore in linen), and in straw. Wood is largely cellulose, and the wood pulp used in making cheap white paper is nearly pure cellulose.

Cellulose contains the same proportions of its elements, *carbon, hydrogen* and *oxygen,* as does starch. Like starch, also, it can be converted into the sugar *glucose* by treatment with sulphuric acid. It is classed among the *carbohydrates.* Cellulose swells when wet, and that prepared from corn pith is sometimes used as a lining in warships to prevent leakage when a shell pierces the side of the vessel. It will not dissolve in hot water (as starch does), nor even in fairly-strong acids or alkalies. In the chemical laboratory paper is used for filtering almost any liquid, and this filter paper is pure cellulose.

Very strong acids, however, act upon cellulose. *Parchment paper* is made by dipping paper into concentrated sulphuric acid for a moment and then into water. The cellulose swells on the surface and becomes hardened. Strong alkalies also swell cellulose, a fact which is made use of in making *mercerized cotton.* A mixture of nitric and sulphuric acids converts cellulose into the explosive nitro-cellulose, known as *pyroxylin* or *guncotton.* Nitrocellulose is soluble in a mixture of ether and alcohol, producing the compound collodion. When collodion is used to dress wounds, the alcohol and ether evaporate and a thin layer of nitrocellulose is left which protects the wound from infection.

Artificial silk, or *lustracellulose,* is made by dissolving cellulose in suitable solvents and forcing the liquids through minute holes into some medium which immediately reforms the cellulose in fine threads which shine like silk. In one process collodion is used. Here the threads are formed by the evaporation of the alcohol and ether. These threads are nitrocellulose at first, but this is then changed back into cellulose by chemical treatment. In the viscose process, the cellulose (usually wood pulp) is swelled with the alkali and caustic soda, and then treated with carbon disulphide. The product, called viscose, is dissolved in a little water and then is forced through the holes into a strong salt solution. The threads formed have only to be heated and washed to change them back to cellulose.

Cellulose is present in all foods of vegetable origin. It is especially abundant in the stalks and leaves of plants, and so we find a large proportion of it in such foods as celery, lettuce and spinach. Only a little of the cellulose is digested by man (a much larger proportion by herbivorous animals, such as the horse and cow), but, nevertheless, it is practically indispensable to the diet. It affords bulk and stimulates the peristaltic (churning) motions of the bowels.

J.F.S.

Related Subjects. Reference to the following topics will be helpful:

Carbohydrate
Cotton
Explosives
Guncotton
Protoplasm

CELTS, *selts,* or *kelts,* or **CELTIC PEOPLE,** a division of the Aryan branch of the human family, supposed to have been the first Aryan settlers in Europe. The onslaughts of the Teutons, Slavonians and other races drove them westward, and at the beginning of the historic period they were the principal inhabitants of Britain, Ireland, France, Belgium, Switzerland and Northern Italy. About 235 B.C. a Celtic tribe conquered and settled that portion of Asia Minor to which the name Galatia was given. The Celts were a restless, energetic people, and held their place in Western Europe until conquered by the Romans, who generally called them Gauls. Of the Celtic peoples surviving at the present time, there are two divisions, the *Gaelic,* or *Gadhelic,* comprising the Scotch Highlanders, the Irish and the Manx; and the *Cymric,* comprising the Welsh and the Bretons (the people of Brittany). The Cornish dialect, formerly belonging to the Cymric group, became extinct only recently.

CEMENT, *se ment',* or *sem' ent.* Pieces of paper may be stuck together with mucilage and paste, pieces of wood and other articles may be joined with glue, and paper may be fastened to walls with paste. We hold bricks in a wall with mortar and fasten metallic fixtures to glass lamps with plaster of Paris. All preparations used for sticking articles together are known as *cements.* A cement must be soft and pliable, so it will fill all spaces and crevices where it is used. It must harden or "set" within a reasonable time after it is applied. Most cements are mixed with water, and they harden on exposure to the air or by a chemical change in the substance. Glue and some other cements are softened by heating, and these harden on cooling. All cements must have the property of adhering so firmly to the objects to which they are applied that on becoming solid they will hold them together.

Portland Cement. The most important cement is that prepared from certain varieties of limestone and used extensively for building purposes. The varieties of this cement usually recognized are *natural rock* cement, made direct from a limestone containing a good proportion of clay; *Portland cement,* and *slag cement.* When the term *cement* is used without qualification, Portland cement is meant. This cement was first made in England and took its name from its resemblance to a limestone obtained from Portland, on the English Channel. There are two processes in the manufacture, the wet and the dry. The cement must contain certain proportions of lime, clay and silica. When the natural rock contains these substances in proper proportion the process of manufacture consists in quarrying the rock, breaking it into small pieces, which are thoroughly dried in kilns, then grinding to a powder. This powder is then burned to a clinker in kilns constructed especially for the purpose, coal dust or crude petroleum being used for a fuel. The clinker is ground to a fine powder, which is packed in sacks or barrels for the market.

When clay and limestone need to be mixed before burning, the wet process is used. Each ingredient is ground separately; then they are wet and mixed in proper proportions. The mixture is made into bricks, which are dried, ground and burned, and again ground to a powder, as in the dry process.

Hydraulic cement is cement which hardens under water. Since nearly all Portland cement is hydraulic, the term is now but little used. *Slag cement* is made from the slag of smelting furnaces where steel is made. Only certain varieties of slag are suitable for the purpose, and it is but little used in the United States and Canada.

Cement is manufactured in large quantities in the United States, Canada, England and Germany. The yearly output in the United States is about 93,000,000 barrels, and in Canada it is about 8,600,000 barrels. The leading states in the order of their production are Pennsylvania, Indiana, California, New York, Illinois. See CONCRETE. W.F.R.

Consult Meade's *Cement, Its Manufacture and Uses;* Taylor's *Practical Cement Testing.*

CENIS, *see nee',* an important mountain in the Alpine system, between the Italian province of Turin and Savoy in France, with an altitude of 11,755 feet. Over this mountain, to enable him to conduct a military campaign against Italy, Napoleon constructed a road forty miles in length, and this for years thereafter formed the main route between the two countries. A railway tunnel now penetrates the mountain at heights varying from 3,775 to 4,245 feet. Work on its construction was commenced in 1857, and it was completed in 1872. The total cost was about $15,000,000 and so difficult was the task of boring through the rock that the greatest ingenuity was called for on the part of the engineers. The power-drill and air compressor now used in all tunneling operations were first tried and found successful in the Mont Cenis tunnel.

CENOZOIC, *se no zo'ik,* **ERA,** the division of geologic time which extends from the Mesozoic Era to the present. Through its different periods plants and animals steadily approached forms of life as they now exist. In this era was developed the ancestor of man, though the exact time of his origin is unknown. See GEOLOGY, for diagram showing the place of the Cenozoic Era; also FOSSIL; MESOZOIC ERA.

CENSER, *sen'sur,* a vessel in which incense is burned during religious ceremonies. Among the ancient Jews it was used to offer perfumes in sacrifices, that for the tabernacle being of brass, that for the temple, of gold. Censers of various forms are still used in the Roman Catholic Church at mass, vespers and other offices, as well as in some Anglican and other churches. A censer is usually of ornamental form, supported by chains with which it is swung to and fro by hand, the perfumed smoke issuing from the perforated top. In Shakespeare's time the term was applied to a bottle perforated and ornamented at the top and used for sprinkling perfume, or to a pan for burning any odoriferous substance.

CENSER

A characteristic type of the eighteenth century.

CENSORS, *sen'sorz,* in ancient Rome, two officials whose chief duty was to take the census of the citizens and record the amount of property owned by each. Among other functions, they kept a strict watch over public and private morals, and could reduce the rank of a person or take away his right to vote if they found him guilty of immoral conduct. From the name of these Roman officials comes the English word *censorious,* which means *fault-finding.* The censorship was created about 444 B.C., and for a century was open only to the patricians, but in 351 B.C. the plebeians secured the right of election to the office, and from their ranks came the most distinguished of all the censors, Cato the Elder (see CATO). During the empire the emperors exercised the powers of the censor under the name *prefect of morals*. See PATRICIAN; PLEBEIAN.

Modern Censors. In modern usage *censor* is applied to an officer who examines books, pamphlets, newspaper articles, etc., before they are published, to determine whether or not they contain objectionable matter, and to a similar officer who passes on plays before they are publicly presented. In democratic countries like the United States and Canada, where liberty of the press and freedom of speech are zealously defended, any sort of censorship is generally unpopular and only the stern necessities of war would excuse it. Censors usually have considerable power in such countries as Russia, where popular freedom is limited. For the censorship of moving picture displays, see MOVING PICTURE, subhead *National Board of Censorship.*

CENSUS, *sen'sus,* an official counting of all the people of a country or section of country, which may also include statistics relating to age, place of birth, occupation, etc., and data on the various industries. In fact, the word has come to be applied to any gathering of statistics which is carried on by means of direct questioning; that is, by a government enumerator who goes from house to house and obtains his information by personal interviews.

The first reliable census of a modern European nation was taken by Sweden in 1749. Most of the countries of Europe at the present time order an official counting of their population every ten years. In Germany, where the system was founded by Frederick William I of Prussia, the census is taken every five years.

The National Census of the United States. In accordance with the Constitution, the first census of the United States was taken within three years after the assembling of the first Congress, and every tenth year since then has been a census year. The first census was completed in 1790; the results were collected in a pamphlet of fifty-six pages, and the total cost was $44,377. Seventeen marshals and 200 assistants, who carried their quill pens and ink-horns in their saddle-bags as they made their toilsome journeys through the sparsely-settled country, had the work in charge.

The records that have survived from the early period show that entries were made on blanks of every conceivable size and form, for the government was too poor in those days to furnish its census-takers with uniformly-ruled blanks. An enumerator of the second census (1800), who ran out of paper before he finished his work, made entries on the back of an old periodical containing an essay by Benjamin Franklin on the *Art of Procuring Pleasant Dreams.* There are other census relics quite as interesting.

The census of 1790 made individual records only for heads of families, the members of the families being grouped as free whites, males sixteen years and over, males under sixteen years, females, other free persons, and slaves. Slowly the census broadened its field, and in 1850, after considerable experimenting, the service was put on a really informational basis. In that year census-takers began to record every inhabitant by name, and for the first time there was a complete classification of the people according to age, sex, color, defects, etc. There were, besides, special schedules for agriculture and for manufacturing and mechanical industries.

The work was taken out of the hands of the marshals in 1880 and placed in charge of a census office at Washington. In 1902 the permanent Bureau of the Census, now a division of the Department of Commerce, was created by act of Congress.

The thirteenth census of the United States, that of 1910, the first undertaken after the creation of the permanent bureau, required the services of 55,000 census-takers and 3,000 clerks. The country was divided into 320 districts, each under the control of a supervisor, who had charge of the enumerators in his district. There were four main lines of inquiry—population, agriculture, manufactures and mines and quarries. The facts collected by the army of census-takers are tabulated at the Washington office, by means of remarkably ingenious electrical machines. The census reports fill over a dozen large volumes which are for sale at cost price by the Superintendent of Documents of the Government Printing Office, Washington, D. C. Abstracts of the census for any state may be purchased in cloth binding for one dollar; extracts of population, agriculture, or the like, may be had in pamphlet form at prices ranging from five cents to fifteen cents.

State Census. The various states authorize a census every ten years; this is taken about midway between two national census years. The state census follows the main lines of the national census, though in less detail, and special attention is paid to collecting statistics on state industries.

Dominion of Canada. The first Federal census of the Dominion of Canada was taken in 1871, and since then an official enumeration has been taken every ten years. The work is under the supervision of the Department of Trade and Commerce. The fifth census, taken in 1911, required the services of 264 commissioners and 9,703 enumerators. Because of the large immigrant population, the birthplace of each parent and details concerning nationality, naturalization and date of immigration are recorded. Complete statistics are collected in regard to industries, products, occupations and religion, and tables are also compiled showing the number of infirm and illiterate residents of the country. B.M.W.

Consult Merriam's *America's Census-Taking from the First Census;* Wright and Hunt's *History and Growth of the United States Census.*

CENT, *sent,* the name given to a copper coin representing the one-hundredth part of a dollar, the smallest in value in the United States monetary system. The word is derived from the Latin *centum,* meaning *one hundred,* and in slightly-differing forms is applied to a small coin of most countries. The word *cent* has been in use in the United States since 1786, when Congress passed an act making the dollar of one hundred cents the basis of the coinage system. The composition and weight of the cent have been changed several times, but the coin was finally standardized in 1873, when an act was passed authorizing the coinage of one-cent pieces containing ninety-five per cent of copper and five per cent of tin and zinc, and weighing forty-eight grains. A cent is legal tender for any amount not over twenty-five cents in any one payment; if fifty pennies are offered in payment of a fifty-cent debt the creditor may legally refuse to accept the proffer. Two-cent and three-cent coins, which have been issued at various times, are now very rare.

The British system of coinage has been found cumbersome, with its pounds, shillings and pence, so Canada long ago adopted the decimal system of the United States. The British penny is equal in value to two cents, United States or Canadian; the half-penny is equivalent to one cent. Americans, in speaking of their own coins, use the terms *cent* and *penny* interchangeably. Italy, France, Chile, Brazil, Peru and in fact all countries not under British jurisdiction, have their coinage based on the decimal system, the *cent, centavo, centesimo* or *centime* being in each case a small coin which represents one-hundredth of a coin of larger denomination.

CENTAUR, *sen'tawr,* a mythical creature, half man, half horse, supposed by the ancient Greeks to live in Thessaly. Although wild and lawless, the centaurs were represented as capable of good attributes, for Chiron (which see) was one of the wisest teachers of the great Greek heroes. At one time, when a certain king was being married, the centaurs appeared at the celebration and tried to carry off the bride. The battle which ensued was one of the favorite subjects in Greek art. An interesting explanation of the centaur myth is that the early Greeks, totally unacquainted with horseback riding, saw occasional riders come out of Thessaly and fancied that man and horse were one being. See CENTAURUS.

THE CENTAUR CHIRON AND CUPID

CENTAURUS, *sen tawr'us,* a constellation of the southern hemisphere which contains the earth's nearest neighbor among the stars. It has been calculated that this star, Alpha Centauri, is only 4.4 light years from the earth (see

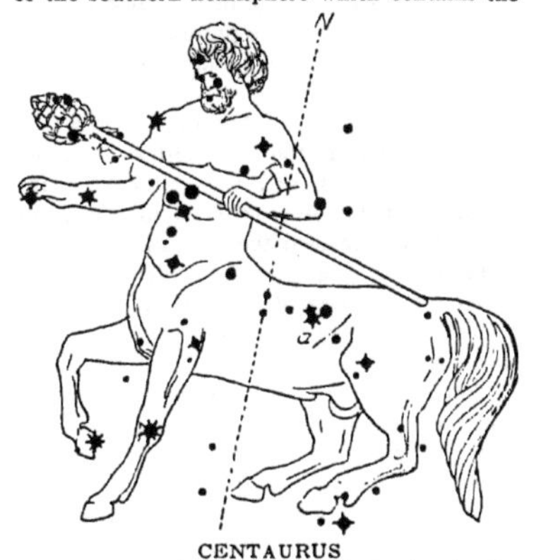

CENTAURUS
(a) Alpha, the third brightest star in the heavens. (See, also, maps of the stars, in article ASTRONOMY.)

ASTRONOMY, for explanation). It must be remembered, however, that light travels a distance of 186,330 miles a second; the distance of 4.4 light years is therefore too great to be expressed in figures that would convey any meaning. Centaurus is represented in mythology as half man, half horse, chief of the centaurs, a teacher of all the manly and noble arts who was accidentally killed by Hercules and placed among the stars by Jupiter. See CENTAUR.

CENTENNIAL EXPOSITION, *sen ten' ni al ex po zish' un,* a world's fair, the first international exhibition of arts, manufactures and products of the earth held in America. This exposition was planned to commemorate the hundredth anniversary of the Declaration of Independence, and was held in the summer of 1876, in Fairmount Park, Philadelphia, the city in which the Declaration was written. Its site covered 236 acres, within which about 200 buildings were erected. The main building was nearly 2,000 feet long and 464 feet wide. Other important buildings were Machinery Hall, Agricultural Hall, Horticultural Hall and Memorial Hall. The last-named, used as an art gallery, was constructed in permanent form of granite, glass and iron, and is now the Pennsylvania Museum of Art. Nearly fifty foreign governments were represented in the exhibits. Prizes were awarded contributors, and the judges were some of the most famous men of science and in the professions. Nearly 10,000,000 people paid admission to the grounds.

The exhibition was important because it showed Americans the superiority of some European products, and thus stimulated increased effort for improvement in American goods. It also opened the eyes of Europeans to the fact that a manufacturing and commercial nation was developing in America which threatened European supremacy in those fields. At this exhibition the Bell telephone was first exhibited, and there the grace and beauty of articles of Japanese manufacture became apparent to the people of the United States for the first time.

Since the Centennial, other international exhibitions have been held in the United States, each contributing a permanent building to the city in which it was held. See WORLD'S COLUMBIAN EXPOSITION; PANAMA-PACIFIC EXPOSITION; ALASKA-YUKON-PACIFIC EXPOSITION.

CENTIGRADE, *sen' ti grayd,* a type of thermometer which is graduated on the scale of one hundred. In the Fahrenheit thermometer, the freezing point is 32° above zero and the boiling point 212°; in the Centigrade the freezing point is zero and the boiling point 100°. The two are made precisely alike, and the length of tube between freezing point and boiling point does not differ; but on the Fahrenheit that length of tube is divided into 180 degrees and on the Centigrade into 100 degrees. If, therefore, it is desired to translate Centigrade readings into Fahrenheit, the number of degrees Centigrade must be multiplied by $\frac{180}{100}$, or $\frac{9}{5}$, and the product increased by 32. To change Fahrenheit to Centigrade, on the other hand, multiply the Fahrenheit reading by $\frac{5}{9}$ and subtract 32. See THERMOMETER.

CEN'TIMETER, a measure of distance equivalent to 0.3937 of an inch. It is the hundredth part of a meter, from which the metric system derives its name. The term centimeter is usually expressed by the letters *cm.* See METRIC SYSTEM; METER.

CENTIPEDE, *sen' ti peed,* a wormlike creature, with a body consisting of numerous rings or segments, each of which bears a small pair of legs. There are so many legs it was once popularly believed there were a hundred,

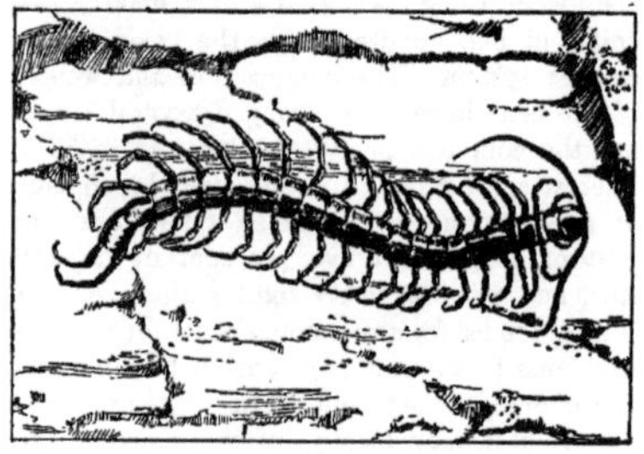

A CENTIPEDE

and so it got its name. But really no centipede has yet been found to possess more than thirty-one pairs of legs. Its body is flattened, with feelers usually long and many-jointed. Commonly found under stones and dead wood, it is a creepy, crawling thing much despised and generally considered poisonous. However, the common centipede is quite harmless; as a matter of fact, it eats earthworms and insects which would be harmful to vegetation. Though some species of tropical countries inflict severe and often dangerous bites, they, too, eat destructive insects such as cockroaches and beetles. Some of these tropical species grow to a length of eighteen inches.

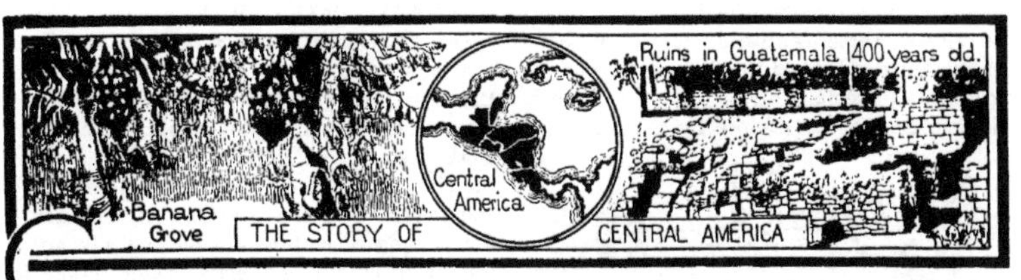

CENTRAL AMERICA, the tropical land mass connecting North America and South America. Authorities differ as to the exact definition of the term, some preferring to include in it the southernmost states of Mexico; but as most commonly understood, Central America consists of the republics of Guatemala, Honduras, Salvador, Nicaragua, Costa Rica and Panama, and the colony of British Honduras. Lying between the Atlantic and the Pacific, it stretches from Mexico to Colombia, and has an area of about 181,500 square miles. When Panama was a part of Colombia it was classed as South American territory.

The Central American states, described under their titles, are in a very backward condition, and the entire area attracted only casual attention from the rest of the world until plans began to be formed for connecting the Atlantic and Pacific oceans by a great ship canal. Then, since the narrowest strip of dividing land is in Central America, eyes were turned in that direction, and at least one state, Panama, has been one of the centers of world interest for years.

Physical Features. In general, Central America is mountainous, and its western range, though broken at intervals by little east and west ranges, really is continuous with the great mountain systems of North America and South America. Volcanoes are numerous, though for the most part inactive, and ruins show where more than one city has been destroyed by them. The highest peak, Acatenango, rises 13,800 feet above the sea. On the west the coast in most places drops abruptly to the water, presenting a gloomy and frowning appearance, but on the east hot, moist plains slope gently back throughout most of its length.

Everywhere the climate is warm, and in the lowlands most unhealthful. This accounts in a measure for the backwardness of the states, for Europeans and Americans have not cared to risk their lives to establish industries, but the wonders that have been accomplished in the Canal Zone in Panama show what may some time be done to make even the damp coast regions healthful, by the extermination of the fever-carrying mosquitoes.

History. Interest in Central America lies in the far-distant past and the immediate present, for the remote past of much of this territory was noteworthy. An intermediate period is colorless. Here, long before the first white man visited the country, there grew up a civilization unmatched elsewhere in the western hemisphere. The Maya Indians had built for themselves great cities and monuments which are fit to rank with the Pyramids of Egypt; they had worked out a calendar, and had invented a method of writing. All this, scholars believe, happened as long ago as the fifth or sixth century A. D., and by the time Europeans reached the coasts the civilization as well as the cities had fallen into decay and the Indians had forgotten the meaning of progress. Of recent years much interest has been shown in the remains of this old culture.

TYPICAL VILLAGE STREET

The modern history of Central America begins in 1502, when Columbus first sighted its shores. A dozen years later a Spanish adventurer conquered a part of the present Costa Rica, and in 1524 Pedro de Alvarado succeeded in gaining control over most of Guatemala and Salvador. The great Spanish conqueror, Cortez, entered the country in 1525, and overcame the tribes that had held out until then. For almost three centuries from that date Central America formed the Spanish captaincy-general of Guatemala, but in 1821 the Spanish rule was brought to a close by a revolution.

A NATIVE HUT

The newly-independent territory was at once divided into the five states of Costa Rica, Guatemala, Honduras, Salvador and Nicaragua, which for a brief time were a part of the empire of Mexico. In 1823 they freed themselves, and became the Republic of the United States of Central America, but the union lasted only sixteen years. Since that time various efforts have been made to form another union, sometimes by force of arms, but the governments in the single states have always been too unstable for any one of them to assume a firm leadership. By colonization Great Britain gained a hold in the country during the nineteenth century, establishing British Honduras, and in 1903 another state, Panama, was added to the number of republics.

Almost incessant wrangling within and among the states has retarded progress, but the Washington Peace Conference of 1907 was the beginning of a period of diminishing interstate friction. A Central-American Court was established, and it has been very successful in arbitrating troubles between states. It has also outlined interstate regulations as to roads, telegraph systems and commerce. O.B.

Related Subjects. The following articles, if read in connection with the above, will give a more detailed knowledge of Central America:

Belize	Nicaragua
Bluefields	Nicaragua, Lake
British Honduras	Nicaragua Canal
Colon	Panama
Cortez, Hernando	Panama, Isthmus of
Costa Rica	Panama, Republic of
Guatemala	Panama Canal
Honduras	Salvador
Leon	San José
Maya	San Salvador
Mosquito Coast	Tegucigalpa

CENTRAL FALLS, R. I., a city in Providence County, in the northeastern section of the state near the border line, and on the Blackstone River. Providence, the capital, is four and one-half miles south. The New York, New Haven & Hartford Railroad enters the city, and electric lines reach adjacent cities and towns. The city is extensively engaged in the making of cotton, woolen and silk goods, leather, haircloth and glass, water-power for manufacture being provided by the river. A chocolate mill was established here in 1780, and the place was known for a long time as Chocolateville; in 1871 it became a part of the city of Lincoln, and it was incorporated independently in 1895. Since that time its growth has been rapid. According to the Federal census the population increased from 22,754 in 1910 to 24,707 in 1914.

CENTRA'LIA, Ill., a city in Marion County, in the south-central part of the state, sixty miles east of Saint Louis, Mo. The settlement made here in 1853 was first chartered as a city six years later. In 1910 the population was 9,680; in 1914 it was 10,938. The area is three square miles. It is served by the Illinois Central; the Chicago, Burlington & Quincy; the Illinois Southern, and the Southern railroads.

The city has parks, a city hall, a Carnegie Library and a hospital. It is near large deposits of coal, and coal mining is one of the principal industries. Fruits, especially apples and strawberries, are shipped in large quantities. There are railroad machine and repair shops, canning factories, glass works, marble yards, a zinc smelter and manufactories of boxes and crates, shirts and overalls.

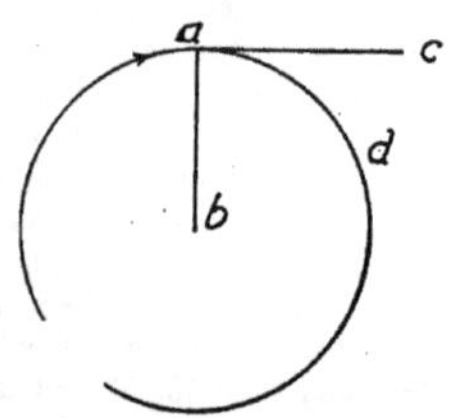

CENTRIFUGAL FORCE

Illustrating a stone at (*a*) tied to a string (*ab*). When stone is at (*a*), revolving around (*b*), it is traveling towards (*c*); but the string is pulling towards (*b*), hence the stone must take the resultant path (*ad*). See COMPOSITION OF FORCES.

CENTRIF'UGAL, *sen trif'u gal,* **FORCE.** Mud flies from the rim of a rapidly rotating carriage wheel, and water flies from a swiftly rotating grindstone. If a string be tied to the handle of a pail filled with water the latter can be swung over the head without spilling the water. The force which produces these results is called centrifugal force, a name which means *flying from the center*. The tendency of every body in motion is to move in a straight line, and this tendency causes the development of centrifugal force in bodies moving in a circular or curved path. (See CENTRIPETAL FORCE.)

CENTRIPETAL, *sen trip' e tal,* **FORCE.** This is the force which counteracts centrifugal force (see above). In the wheel and the grindstone it is the force of cohesion (which see) that keeps these bodies from flying to pieces and their fragments from scattering in straight lines. In case of the pail of water it is the string. Should the string break or should a person lose his grip on it the pail would fly off in a straight line, but it would soon change its direction owing to the force of gravity, and come to the earth in a curved path. Gravity is the great centripetal force of the earth. Were it not for this, the rapid rotation of our planet as the equator is approached would cause all bodies on its surface to fly off into space in straight lines.

CENTURY PLANT, or **AGAVE,** *a gay' ve,* an interesting plant of the deserts, belonging to the aloe family, now often cultivated as a decorative house or garden plant. These plants are generally large, and have a massive tuft of fleshy, spiny-edged leaves, sometimes striped or margined. They live for many years—ten to seventy—before flowering. This long delay gives them the common name of *century plant,* although it is misleading. Because of its name many people believe it lives much longer than a century and blossoms but once in a hundred years. When the time for flowering approaches, a tall stem springs from the center of the tuft of leaves and grows very rapidly until it reaches a height of fifteen, twenty or even forty feet, and bears, toward the end, a large number of flowers. Then the leaves, having given their store of nourishment to the production of this large stalk and its flowers droop and die, but the roots soon send up new shoots for the new generation.

The sap of some of the aloes, when fermented, yields an intoxicating beverage resembling cider, called by the Mexicans *pulque.* From others is produced the familiar *mescal,* another intoxicating drink of the Mexicans. The leaves are used as fodder, and their fibers are also formed into thread, cord and ropes. An extract from the leaves is used as a substitute for soap. Slices of the withered flower-stem are used as razor-strops.

CEPHALOPODA, *sef a lop' o dah,* the name of a class of mollusks which includes the cuttlefish, the octopus, the pearly nautilus and the squid, all of which are described under their titles. The name means *head-footed,* and is given these animals because they have a number of arms, or feet, around the mouth. The arms usually have little cups, or "suckers," on the under side, which enable the animal to fasten itself securely to its prey.

THE CENTURY PLANT

CERAMIC, *se ram' ik,* **ART.** See POTTERY.

CERBERUS, *sur' be rus,* in Greek and Roman myths, Pluto's three-headed dog, the grim guardian of the entrance to the underworld. His jaws dripped with foam, from which sprang the deadly nightshade (see NIGHTSHADE). Early and late he kept his watch, allowing no living being to enter the gates of Hades, nor any spirit to pass out of them. Orpheus, however, searching for his wife Eurydice, played such melting strains on his lyre that Cerberus was won over and permitted him, though a mortal, to enter the realms of Pluto.

REPRESENTATIONS OF CERBERUS

(a) Hercules capturing Cerberus. (From a vase painting.)

(b) A bronze Cerberus dating from ancient times, now in the Vatican Museum, Rome.

The last of the twelve labors of Hercules (which see) was the descent into Hades to secure the savage Cerberus. Hercules speedily accomplished this fearful task, but when he brought the triple-headed monster to Eurystheus, at whose command the deed was performed, the latter fled in terror, and taking refuge in a huge jar would not come out of his hiding place until Hercules had carried the dog back to Hades.

CEREALS, *se're alz.* See GRAINS.

CERES, *se'reez,* in ancient mythology, a Roman goddess who protected and watched over the fruits of the earth, and especially the grains. She was the daughter of Saturn and Rhea and mother of Proserpina (in Greek, Persephone). According to the interesting myth, when her daughter was stolen and carried off to Hades, Ceres neglected the earth during her search for Proserpina, and all vegetation died. The Romans built temples and celebrated the festival of the *Cerealia,* in honor of Ceres. The sacrifices made to her consisted of pigs and cows. Ceres was usually represented in full attire, holding ears of corn and a lighted torch, and with poppies, her sacred flower. The Greeks worshiped the goddess under the name *Demeter*. The earliest planetoid known was named Ceres, according to the custom of giving stars and constellations names taken from mythology. See PROSERPINA.

CEREUS, *se're us.* See CACTUS.

CERRO GORDO, *ser'ro gor'do,* BATTLE OF, in the war between Mexico and the United States, one of the most important battles fought by the Americans on their triumphal march from Vera Cruz to the City of Mexico. Cerro Gordo is a mountain pass near Jalapa, sixty miles northwest of Vera Cruz. There a force of 12,000 Mexicans, under Santa Anna, was attacked and totally routed by 8,500 Americans under General Scott, on April 18, 1847. The Mexicans lost about 1,000 in killed and wounded; the Americans sixty-three killed and 368 wounded. This victory cleared the way for the Americans nearly to the capital of Mexico, which was captured on September 14.

CERVANTES SAAVEDRA, *ther vahn'tays sah ah vay'drah,* MIGUEL DE (1547-1616), universally acknowledged to be the greatest of Spanish novelists, rich in imagination and keen of wit—the author of *Don Quixote,* a book translated into nearly all the languages of the world. This book (which is described in detail under its title) was written to ridicule the extravagantly sentimental stories of chivalry that the people of that day were so fond of reading, but in itself it is a masterpiece of satirical writing, containing a wealth of humor and revealing sure insight into human nature. The early life of the author is not known in much detail. It is certain that he was born in Alcala de Henares, of pure Castilian stock, but his education is largely a matter of conjecture. Early in his career he joined the army, for there is record of the gallant part he played in the Battle of Lepanto (1570), at which the Turkish power in Europe received its death blow. There is further record of his experiences as a soldier and of his capture in 1575 by a band of Algerine pirates. For five years he bore with praiseworthy fortitude the hardships of prison life, being finally ransomed through the efforts of friends and relatives.

CERVANTES

Returning to Spain, Cervantes married and began seriously to devote himself to literature. He published some sonnets in 1583, and two years later a pastoral novel, *Galatea,* gained for him considerable notice. Following this appeared a long list of plays—from twenty to thirty or more—but he was not successful as a dramatist, and for a time lived in very straitened circumstances. Only a few verses came from his pen between the period of dramatic writing and his emergence as the author of *Don Quixote*. The first part of this book, which was destined to win him undying

CERVANTES' HOME
It was here that *Don Quixote* was written.

fame, was published in 1605; ten years later part two appeared. Though the work was received with every evidence of popular favor, it brought the author scant financial returns. Second only to his masterpiece is a little volume of twelve *Exemplary Tales,* published in 1613. Even had there been no *Don Quixote,* these stories would have given him high rank as a novelist. He continued his literary work until the last year of his life, among his later writings being a review of contemporary poets (in rhyme), entitled *Journey to Parnassus.* M.S.

CETACEA, *se ta' she ah,* or **CETACEANS,** an order of marine animals—the whales, dolphins and porpoises—surpassing in size all others in existence. They are true mammals, with warm blood, and they breathe by means of lungs (see MAMMAL). But in becoming used to life in the water, their bodies, outwardly and inwardly, have assumed much of the form and structure of fish. The body ends in a tail which is expanded into two horizontal lobes. There are no hind limbs, and the fore limbs are broad paddles, or flippers, enclosed in a continuous sheath of thick skin. The fishlike appearance is further increased by a fin on the back, but this is a simple fold of skin and does not contain bony spines.

The whale and its allies have no teeth in the full-grown state, but, instead, have triangular plates of baleen, or whalebone, which are developed on ridges across the palate. The nostrils open directly upward on the top of the head and are closed by valves of skin, which are under the control of the animal. When a cetacean comes to the surface to breathe, it blows the air out violently, and the vapor it contains, becoming condensed into a cloud, resembles a column of water and spray.

Related Subjects. For a detailed discussion of the various cetaceans, the student may consult the following articles:

Dolphin	Narwhal
Dugong	Porpoise
Grampus	Rorqual
Manatee	Whale

CETTINJE, *tset' en yay,* the capital of the kingdom of Montenegro, the smallest of the Balkan states. It is situated in a barren and stony region, 2,100 feet above the Adriatic Sea, about ten miles inland, and is little more than a village. The royal palace is unpretentious, only one story in height, and the public buildings have no claim to architectural beauty. The trade is unimportant, and there are practically no manufactures. The city was founded in the fifteenth century and has repeatedly been conquered and sacked by the Turks. Population, about 3,500.

CEVENNES, *sa ven',* a mountain system of Southern France, running northeast from the most northerly hills of the Pyrenees for a distance of 320 miles. The highest peak in the whole system is Mezene, 5,755 feet above sea level, but the average altitude is not more than 3,000 feet. The lower slopes of the mountains are very fertile, and large crops of grain, fruit and vegetables are raised. In many localities mulberry trees are cultivated for their leaves, upon which silkworms feed. The mountains have been the scene of much warfare, and at various times have furnished shelter for the persecuted Waldenses, Albigenses and Camisards. In the Vosges Mountains, a northern extension of the Cevennes system, some of the fiercest fighting of the War of the Nations took place.

CEYLON, *se lon',* a large island in the Indian Ocean, often called the "Pearl of the Orient" because of the richness of its soil and its natural beauty. The romance of it has been felt by thousands who know little more about the island than is contained in those lines of Heber's famous hymn—

> . . . . the spicy breezes
> Blow soft o'er Ceylon's isle,
> Where every prospect pleases
> And only man is vile.

It lies fifty-five miles southeast of the extreme southern point of the peninsula of India, being separated from it by the shallow Palk

Strait. The island is pear-shaped, 271 miles in extent from north to south and 137 miles in greatest width, with an area of 25,281 square miles. It is therefore about five-sixths as large as Ireland or the state of South Carolina. The surface is very mountainous in the in-

LOCATION MAP

terior, with a low-lying coast line of coral reef. The most noted mountain is Adams Peak (7,420 feet), to which Buddhists make religious pilgrimages.

People and History. Ceylon, like India, is a very ancient land, and contains some of the most remarkable ruins in the world. It is also one of the chief centers of Buddhism. The native inhabitants are the Singhalese, who have lived here many centuries. They are Buddhists and a gentle and peaceable, but effeminate, people. The men have the curious custom of dressing like women. They wear skirts and earrings and have long hair, which they do up on their heads with immense tortoise-shell combs.

Europe entered into Ceylon's history by the invasion of the Portuguese in 1505, of the Dutch in 1636, and of the British in 1795. In 1831 the entire island consented to British rule and became a crown colony of the Empire, having a governor appointed in London; each of its nine provinces was placed under a government agent. It is now one of England's most prosperous colonies. Its population is over 3,500,000, more than half Singhalese; others are Hindus, Mohammedans and Europeans. Colombo is the capital and one of its three great harbors, the others being Galle and Trincomalee, the latter one of the finest in the world. There are 600 miles of railroad on the island. See COLOMBO.

Vegetable and Animal Life. Ceylon is a tropical island, situated between 6° and 10° north of the equator. Its climate is hot on the coast, but cool and delightful in the uplands, where white residents build their homes. Agriculture is the chief occupation. The cocoanut palm, breadfruit, plantain, cinnamon, mango and bamboo abound, while there are plantations of tea, rice, coffee and tobacco. Over 457,000 acres of land are devoted to the tea plantations, while more than 644,000 acres are cultivated for rice. The bo-tree, banyan, ebony and satinwood trees grow in the forests, while the cinchona tree is cultivated for its bark, from which quinine is made. There are gold, graphite and precious stones in the mountains and pearl fisheries on the coast.

The most important agricultural feature is the cultivation of the tea plant. Its growth has been truly marvelous. In 1873 the export of tea was only twenty-three pounds; at the beginning of the War of the Nations in 1914 it was nearly 200,000,000 pounds. More than one thousand plantations, chiefly in the mountains, employ about 400,000 people, most of whom come over from India, where employment is difficult to find. See TEA.

Animal life is very abundant, there being innumerable varieties of birds and insects and many larger animals. Of the latter, the elephants are the largest and most numerous, several thousands of them roaming wild through the forests. Many, however, are tamed and make valuable beasts of burden. Many of the tame elephants in India come from Ceylon; the remainder from Burma. The Ceylon elephant does not have the long ivory tusks for which the African elephant is hunted. K.A.G.

No American publisher has issued a book relating to Ceylon. The administrative offices at Colombo issue *The Blue Book of Ceylon* annually; it may be secured by application. Humphrey's *Travels East of Suez,* published in 1915, may be ordered through city booksellers.

CHAD, LAKE, also spelled TCHAD, a large lake lying between the Sudan and the Sahara Desert of North Africa. Within recent years it has greatly shrunk in size, but it is still a very large volume of water, covering an area of nearly 40,000 square miles—about as large as the state of Ohio. In extremely dry seasons it has been known to cover only about half that area, and in certain sections towns normally on its shores may be twenty miles away from the water. Although the lake is supposed to have been known to the ancients and is referred to by Ptolemy, it was first visited by white people in 1823. There are many islands, most of them inhabited by bands of pirates who live by pillaging the native tribes living on the shores. The territory at the eastern end of the lake is fertile; that on the west is a desolate sand-strewn desert. The waters of the lake and the rivers Shari and Yeu, which flow into it, abound with crocodiles, fish of many kinds and turtles.

CHADWICK, George Whitfield (1854-), considered by critics the most important American musical composer, with the exception of MacDowell. Born in Lowell, Mass., he received his early musical education in America, but later studied with the best European masters. Returning to America in 1880, he entered the New England Conservatory as instructor, and later became director. Since 1897 he has conducted the annual music festival at Worcester, Mass. Chadwick's most important compositions are the oratorio *Judith* and the music for the *Columbian Ode,* sung at the opening of the World's Fair in Chicago.

CHAFFINCH, *chaf'inch,* a handsome European songster belonging to the finch family, so named because of its fondness for chaff or grain. Its loud, clear notes make it one of the most popular cage birds in Europe, especially in Germany. The chaffinch can be trained to sing distinct melodies, and almost to utter human words. The distinguishing features of the male are a bluish-gray crown, chestnut back and black wings marked by conspicuous white bars. In England, where flocks spend the winter, the chaffinch is a common visitor of the gardens, shrubberies and hedgerows. The nest is made chiefly of moss and wool, and contains four or five purplish-buff eggs spotted with purplish-red. The bird feeds principally on insects and their larvae (young), on grains and on the seeds of various weeds.

The chaffinch has not been introduced into America, although some effort has been made in that direction.

CHAIN, *chane,* a series of links of metal joined together and forming a flexible band to serve useful purposes or worn simply for ornament. The use of chains is of very ancient origin. They are mentioned frequently in the Bible, and in ancient times were worn as ornaments or as signs of office. The simplest form of chain, whether large or small, is made of a series of links, each made from one piece of metal bent to the looped shape, with the ends welded together. Iron, steel, brass and bronze are used where strength is required; gold, silver, platinum and various alloys are used for ornamental purposes. The links of chains are sometimes made of several pieces, as in the case of bicycle chains. When great strength is required, as in cables, each link is sometimes reënforced by a brace, or stud, of metal joining the sides of the link to prevent it from being torn asunder or compressed (see illustration). As a rule all very large chains are made by hand, each link being shaped and forged separately. Small chains are made by machinery.

THREE STYLES OF CHAINS
At top, twisted links; in center, straight links; at bottom, stud links.

Chain, a term applied to a definite measurement in surveying, and to the instrument with which that measurement is made. The length of the surveyors' chain is 66 feet; the actual chain consists of a series of 100 links, each 7.92 inches in length. Ten square chains, or 100,000 square links, make one acre. In surveying, a steel band or tape has supplanted the more cumbersome chain, but the term *chain* is still applied. The width of country roads and some village streets is 66 feet, this width conveniently resulting from the length of a chain. See Surveying.

CHALCEDONY, *kal sed'o ni,* a beautiful variety of quartz, first found abundantly near Chalcedon in Bithynia; because of this fact the name was given. A little is now found in California and Colorado, also in rock cavities in Iceland and Scotland. The common chalcedony, also called *white agate,* resembles milk diluted with water, is semi-transparent and more or less clouded with circles and spots. There are several other kinds, appearing in different colors and formations, such as agate, onyx, chrysoprase, sard, carnelian and sardonyx, each with its special beauties. When polished, chalcedony is valued for brooches, necklaces, ornamental boxes, cups, etc., and many pieces contain interesting vegetable fossils. In the Book of *Revelation* the third foundation of the wall of the Holy City is described as being made of chalcedony. The famous agatized wood of the fossil forest in Chalcedony Park, Arizona, is caused by a chalcedony deposit from water replacing the woody fibers.

CHALDEA, *kal de'ah,* the name anciently applied to the extreme southern district of Babylonia, lying to the west of the mouths of the Tigris and Euphrates rivers, along the Persian Gulf. "Ur of the Chaldees" was the home of Abraham, from which Jehovah called him to begin a new life in the land of Canaan (*Genesis* XI, 31). It is supposed that the Chaldeans were a Semitic people who came from Arabia at a very early date and settled in the neighborhood of Ur. From there they began to make war upon the other inhabitants of Babylonia, and at various periods Chaldean princes sat upon the Babylonian throne. About 626 B.C. Nabopolassar founded the Chaldean (or New Babylonian) Empire, and he and his famous son, Nebuchadnezzar, made Babylonia a world power.

The last Chaldean king ruled about 556 B.C., but during this period the union between the Babylonians and Chaldeans had become so complete that in later history the terms Chaldean and Babylonian are used without distinction between the two. In the book of *Daniel* Chaldean is applied not only to the Babylonian people but to a class of magicians. See BABYLONIA.

CHALK, *chawk.* The mark made on the blackboard with white crayon contains millions of fragments of the shells of tiny creatures which ages and ages ago lived on the bottom of the sea. As these little animals died they

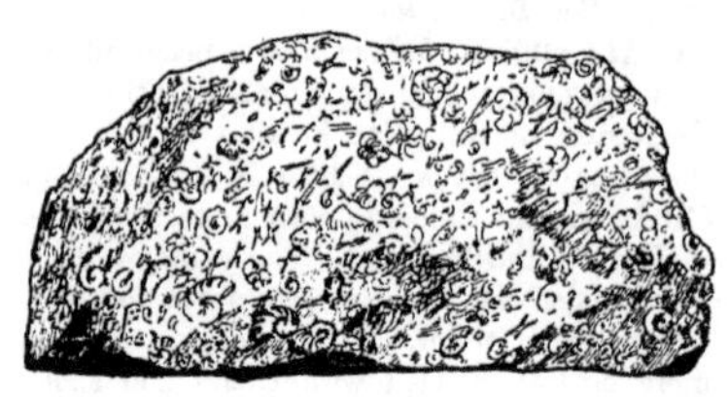

CHALK

left thin skeletons which in the course of ages turned to a soft, white limestone that we know as *chalk.*

When seen through the microscope, a piece of chalk shows hundreds of these tiny shells of different forms and sizes. Some resemble snail shells; others are circular and beautifully marked, while others resemble needles. The needle-like objects came from sponges. Chalk, then, is nearly pure limestone, composed almost entirely of these shells.

The great white cliffs in France and England on each side of the Strait of Dover are composed of chalk, and the cliffs on the English side gave the name of "Albion" to England centuries ago, for *Albion* is believed to have been derived from Gaelic words meaning *white* and *hill.* There are extensive beds of chalk also under the city of London. In the United States it is found in Arkansas, Texas, Iowa, Montana and a few other states. Some of the beds are over 600 feet thick, so the period during which the chalk was forming must have been very long (see CRETACEOUS SYSTEM).

Uses. The use with which we are most familiar is for marking on blackboards. The crayons are made by grinding the chalk, mixing the powder with some substance which will hold it together, and shaping the paste in molds. For this reason if crayon dust be examined with the microscope no fragments of shells may be found. When ground and mixed with water chalk forms a whitewash; *whiting* is ground chalk from which the grit has been removed. In both England and America chalk is used to some extent in the manufacture of cement (which see). As a rock it is of no value for building purposes, because it is easily crushed.

CHAL'LENER, FREDERICK SPROSTON (1869-), a Canadian painter, noted for his landscapes and mural decorations. Though born and educated in England, he has made Canada his home since 1883. He studied at the Ontario School of Art, and also in Italy, Egypt and Syria. He is a member of the Ontario Society of Artists and of the Royal Canadian Academy of Art. Among his best pictures are *Workers of the Fields* and *A Song at Twilight* (both in the National Gallery, Ottawa), *When the Lights Are Low* and *A Quiet Old Road* (in the Provincial Art Gallery, Toronto). Equally famous are his mural paintings in the Russell Theater, Ottawa; the King Edward Hotel, Toronto; the Royal Alexandra Hotel, Winnipeg, and the Hotel Macdonald, Edmonton.

CHALLENGER EXPEDITION. One of the greatest journeys of purely scientific investigation for studying the conditions of the oceans, and the one most fruitful in results, was the undertaking known as the "Challenger Expedition." At the suggestion of the Royal Society of London the British government fitted out the ship *Challenger*, of 2,300 tons, for a long cruise around the world. The vessel sailed from Portsmouth, England, in December, 1872, and did not return to England until May, 1876. During the four years the *Challenger* cruised in all directions in the Atlantic, Southern and Pacific oceans, traveling nearly 68,000 nautical miles and making investigations at 362 stations.

During the journey deep-sea soundings were continually made; samples of water at various depths were taken, in order to study its chemical composition and its temperature; and a great number of specimens of the animal life as well as of the vegetation of the oceans were collected. The currents of the oceans were studied and careful atmospheric and meteorological observations were recorded. A great number of geological specimens, taken both from the bottom of the sea and from the coasts, were also gathered. The results of the observations made and the study of the materials collected are embodied in the *Challenger Report*, which was published in fifty volumes in London under the editorship of Sir John Murray.

CHAM'BERLAIN, the family name of a father and son conspicuous in English political life beginning soon after the middle of the nineteenth century.

Joseph Chamberlain (1836-1914) was one of the leading English statesmen of his day and was frequently called the greatest Colonial Secretary England has ever had. He began public life as a Radical; he ended it as a Unionist, exactly the opposite. Twice he deserted his political chief—once Gladstone, the Liberal, once Balfour, the Conservative—and both times his withdrawal caused the defeat of the Ministry and the division of the party of which he had previously been a member. These changes of party represented a gradual development in a statesmanship which was at first local, then national, and lastly imperial. The key to Chamberlain's political career is that he began as a reformer of local conditions in Birmingham and as he grew older he devoted his attention less to Birmingham, even less to England, and most to the Empire of which they were parts. His life, therefore, falls into three periods.

JOSEPH CHAMBERLAIN

As a Local Leader. Chamberlain was born on July 8, 1836, in London, where his father was a prosperous business man. At sixteen he began work in his father's office, but two years later went to Birmingham to assist in the management of a screw factory in which his father had an interest. He invented a screw with a point so hard that it could be driven into wood without the necessity of first boring a hole, and largely through his efforts the business was amazingly successful. At the age of thirty-eight Chamberlain retired from active business with a fortune. Meanwhile, he had become prominent in the political as well as the business life of Birmingham, and in 1873 was elected mayor.

A National Figure. Chamberlain's work in Birmingham gave him a national reputation which was properly recognized by his election to the House of Commons in 1876, and only four years elapsed before his ability won him a place in Gladstone's Cabinet as President of the Board of Trade. In 1886 Gladstone appointed him President of the Local Government Board, but his opposition to Gladstone's Home Rule Bill led him to resign after two months. With other Liberals who opposed this measure, he then organized the Liberal Unionist party, and succeeded in overthrowing the Ministry. In 1888 he was one of the three delegates sent to the United States to settle the Canadian fisheries' dispute, but the most important result of the visit, so far as Chamberlain himself was concerned, was his marriage to Miss Mary Endicott, the daughter of President Cleveland's Secretary of War.

During these years the Liberal Unionists were drifting further away from the Liberals, and in 1895 their party existence practically came to an end when Chamberlain accepted the post of Secretary of State for the Colonies in Lord Salisbury's Cabinet.

"Think Imperially!" The keynote of Chamberlain's life from 1895 until his death was expressed in this appeal to his countrymen. It was as Colonial Secretary, an office which he held for eight years, that his most important work was done. After 1895 the "economic necessities of a world-wide empire" were his first care. He determined that the colonies, instead of being alternately neglected and exploited, should be steadily encouraged and given coöperation. The colonies' appreciation of his attitude was shown during the South African War, when Canada, New Zealand and Australia proved eager to send troops to the aid of the mother country. It was during Chamberlain's term in office that the Australian colonies were united into a commonwealth.

In 1903 Chamberlain introduced tariff reform as an issue in British politics, by proposing to give the colonies a preference in trade. Balfour, as Prime Minister and leader of the party, tried to keep the tariff out of politics. Chamberlain refused to compromise, resigned from the Cabinet, and finally in 1905 forced the issue before the people. The Liberals won a sweeping victory, which was generally interpreted as the deathblow to tariff reform. In 1906 Chamberlain's health began to fail, but he continued to advise the tariff reformers and sat in Parliament until the year of his death, though his active leadership was at an end. He died on July 2, 1914.

[Joseph] Austen Chamberlain (1863-), the oldest son of Joseph Chamberlain, had already won great honors before his father died. He was educated at Rugby School and at the University of Cambridge, and entered Parliament in 1892. From 1895 to 1905 he was a member of the Balfour Ministry, in which he held the posts of Civil Lord of the Admiralty, Financial Secretary to the Treasury, Postmaster-General and finally Chancellor of the Exchequer. This gradual promotion was partly due, perhaps, to Balfour's attempts to keep the tariff reformers from destroying the Unionist party, but it was also in recognition of his ability. After 1906, when his father retired from active leadership, Austen Chamberlain was the acknowledged champion of the tariff reformers. From 1892 until 1914 he represented East Worcestershire in Parliament, but in 1914, after the death of his father, he was elected by Birmingham West, the constituency which his father had represented for twenty-nine years. In May of the next year he became Secretary of State for India in the coalition Cabinet headed by Asquith, which the War of the Nations made necessary. W.F.Z.

CHAMBER OF COMMERCE, an organization of traders and merchants for their mutual benefit, or for the wider purpose of promoting the business and commercial interests of their community. In the United States, Canada and England membership in these local organizations is voluntary, and their usefulness depends on the energy and ability of the members. The fundamental purpose of such bodies is to increase the prosperity of the community, and incidentally of its individual business interests. To this end a chamber of commerce may investigate general business conditions at home and abroad, transportation facilities and their possible improvement, extension of credit and any other business factors. The recommendations of such a body frequently influence local, state or provincial, and occasionally even national, legislation. One of the commonest activities of such chambers is the distribution of printed matter in which the advantages of the city or district are set forth in terms to attract new industry.

In Canada the functions of a chamber of commerce are exercised by the chartered boards of trade. These are corporations formed under a Dominion statute, and must include at least thirty persons who are "merchants, traders, brokers, mechanics, manufacturers, managers of banks, or insurance agents resident in the district." This corporation may elect twelve of its members to form a board of arbitration. Any three members of the board may arbitrate any cases submitted to it. These are usually disputes of local importance, but sometimes of wider interest. In some of the European countries, notably in France and Germany, the boards have a semi-official character, and exercise certain administrative functions which belong to the municipality in the United States and Canada.

Among the most active chambers of commerce in the United States are those of New York, Philadelphia, Chicago, Saint Louis, New Orleans, Atlanta, San Francisco and Portland (Oregon). Important Canadian boards of trade include those of Vancouver, Saskatoon, Regina, Toronto, Montreal, Winnipeg, Calgary and Halifax.

CHAMBER OF COMMERCE OF THE UNITED STATES, an organization in which membership is open to local chambers of commerce and other associations of business men. It was organized at a national commercial conference called by President Taft and held in Washington, D. C., on April 22 and 23, 1912. Its purpose, roughly defined, is to do nationally what the individual chamber of commerce does locally (see CHAMBER OF COMMERCE).

It studies and encourages the organization of associations of business men, and puts the results of its investigations at the service of organizations which desire to add to their efficiency. It analyzes the statistics of commerce and production, both at home and abroad, watches dangers which might retard commercial development, and makes note of opportunities which might result in expansion. One of its objects is to keep a close watch on congressional legislation affecting the commercial interests of the country. In a general way it aims to do for the commercial interests of the nation what the American Federation of Labor does for labor. It should be noted, however, that the methods of the two organizations are quite different; the Federation of Labor maintains agents in Washington and operates through a central organization, whereas the Chamber of Commerce maintains no lobby and operates through its constituent members and their influence upon the members of Congress.

Unlike the chambers of commerce in France, Germany and other European countries, the Chamber of Commerce of the United States has no official relation to the government. The government pays no part of its expenses, nor is an arbitrary tax levied for its support, a condition which exists in some European countries. The chamber, however, acts voluntarily as an adviser, with respect to appropriations, executive order and legislation, and since its organization has exercised a considerable influence in the framing of paragraphs of new laws that relate directly to commercial and industrial operations.

Membership. The membership of this national Chamber of Commerce includes organizations and individual persons or firms. Every commercial or manufacturers' association, not organized for private purposes, is eligible to membership. Such associations include organizations whose membership is confined to a single trade or group of trades, and also those local or state organizations whose chief purpose is the development of the commercial and industrial interests of a community. Individual persons or firms which belong to any association already a member of the chamber of commerce of the United States are eligible to individual membership. The number of individual members is limited to 5,000, but there is no limit to the number of organization members. The membership in 1916 included about 700 associations and 3,600 individuals. The national headquarters are at Washington, D. C.

Control of Its Policy. One feature of the work of the Chamber of Commerce of the United States is unique—the method by which its policy is framed. The board of directors has no right to commit the chamber to any project or policy. Expression of the chamber's opinion upon any public question can be made only after a referendum has been taken and the vote of the organization members recorded. The right to vote is restricted to organizations; individual members are required to express their opinions through their respective local organizations. This is a unique procedure, one which has never been used to obtain an expression of public opinion either in the United States or in any other country. While it was first regarded as a complicated piece of machinery, it has been found that legislative bodies are susceptible to a statement of this actual vote when they would never be influenced by the action of a board of directors. H.W.

CHAMBERS, ROBERT WILLIAM (1865-), a popular American novelist and writer of short stories, born in Brooklyn, N. Y. Before he began his career as an author he studied art in the Julien Academy at Paris, and for a time made illustrations for *Life*, *Truth*, *Vogue* and other New York periodicals. In 1893 he published *In the Quarter*, and his literary activity since that time has continued without interruption. Among the best-known of more than twenty of his stories are *Iole*, *The Fighting Chance*, *The Firing Line*, *Ailsa Page*, *The Common Law*, *The Business of Life*, *Athalie* and *The Girl Philippa*.

Chambers has undeniably the gift of writing an interesting story. He belongs, however, to that group of American novelists who have chosen as their themes some of the baser aspects of modern life, and his books have an unwholesome tone that makes them dangerous reading for young people.

CHAMBERSBURG, PA., an industrial borough in Franklin County, of which it is the county seat. It is situated in the extreme southern part of the state, about midway between the eastern and western boundary lines, on Conococheague and Falling creeks. Harrisburg is fifty miles northeast by rail. The Cumberland Valley and Western Maryland railroads afford transportation, and electric lines extend to adjacent cities. The population was 11,800 in 1910; it had increased to 12,192 in 1914. The area of the borough exceeds three square miles.

Chambersburg is located in a cultivated and thickly-settled section of the Cumberland Valley, where marble, limestone and freestone abound. It is largely engaged in making hosiery, dresses, paper, iron, foundry products, silk and condensed milk, and also contains the shops of the Cumberland Valley road. Besides the public schools, there are Wilson College (Presbyterian), for girls, organized in 1870, Penn Hall Preparatory School, also for girls, and a public library. It has a hospital, and homes for children and the aged. Beyond the city are Pen-Mar, Mont Alto and Wolf Lake parks.

On the present site of Chambersburg a settlement was made in 1730 by Benjamin Chambers, an Irish emigrant in whose honor it was named; it was known for a long time as Falling Springs. In July, 1864, during the War of Secession, General McCausland entered the town and demanded $100,000 tribute in gold. Upon its refusal to grant his request the place was burned, the loss sustained being estimated at $1,000,000. After the war the town was quickly rebuilt. J.S.C.

CHAMELEON, *ka me' le un,* a lizard which has the remarkable power of changing color within a few moments whenever it desires to

CHAMELEON
Small illustration shows detail of "hand."

do so. Although native to Asia, Africa and Southern Europe, chameleons are also found in the Southwestern United States and the West Indies. The best-known species has a naked body six or seven inches long, but the species in the United States is hardly more than half the latter length. The head is large, the tail long, round and slender. The skin of the body is loose, and hangs in folds at the throat. The body is covered with very tiny scales or granules. The animal changes its color, either in accordance with its surroundings or through anger, fear, temperature or light. From a bronze-brown it can change to emerald green, and again back to brown. Its power of fasting and habit of inflating itself gave rise to the fable that it lives on air, but in reality it feeds upon insects, taking its prey by rapid movements of an exceptionally long, sticky tongue. In general habit chameleons are dull and sluggish, except in the use of the eyes and tongue.

CHAMINADE, *shah me nahd',* CECILE LOUISE STEPHANIE (1861-), one of the best-known women musicians of modern times, was born in Paris. When only eight years of age she composed sacred music that won the praise of Bizet, the composer of *Carmen.* She studied for several years under excellent teachers, began a successful career as a pianist at the age of eighteen, and became in time well known as a music conductor. Chaminade's fame, however, rests chiefly on her compositions, which include such familiar instrumental pieces as *The Scarf Dance, The Flatterer, Morning,* etc., and many charming, melodious songs. Among the latter are *Madrigal, Rosamunde, Berceuse* and *The Silver Ring.* As a composer she is distinctly original, and her compositions are regarded as valuable exercises for the piano student.

CHAMOIS, *sham' mi,* a shy, goatlike antelope, famed for its fleetness and its keenness of scent. It lives in the high mountains of Europe and Western Asia, and was once very common in the Alps. In the summer it is found near the snow line; in the winter, lower down, in the forests. It is a rather small animal, with a brownish winter coat that changes to fawn color in summer and gray in the spring. Its head is pale yellow, marked by a black band surrounding the eyes and extending from the nose to the ears. Its horns, which are about six or seven inches long, are round and almost smooth, and they grow straight upward until near the tip, where they suddenly end in a sharp hook that is bent backward. The tail is black.

During the feeding time, which is in the morning, one animal is always standing on guard in some prominent place, for the pur-

CHAMOIS

pose of warning the rest of approaching danger. The pursuit of chamois is difficult and dangerous, as they live in the steepest, roughest mountains, and are so quick and light they can easily jump across a ravine fifteen feet wide.

Though the flesh is highly prized as food, the chief value of a chamois lies in its skin, which is used to make the very soft, warm, flexible leather known as chamois skin. Most of the skin now sold as such, however, comes from the skin of sheep, but it lacks the velvety softness of the genuine.

CHAMOMILE, or **CAMOMILE,** *kam' o mile,* a well-known daisylike plant belonging to the Composite family (which see). It is a perennial and has slender, trailing stems. The flower of the chamomile of commerce is white, with a yellow center. Both leaves and flowers are bitter and aromatic. The fragrance is due to the presence of an oil, of a light blue color when first extracted. Both the leaves and the flowers are used as poultices, as in cases of earache, and also medicinally in the form of tea. The flowers are sometimes used in place of hops. Chamomile is cultivated in gardens in the United States and Southern Canada. The *golden marguerite,* with its beautiful bright or pale-yellow flowers, is a variety of chamomile, as is also the common, troublesome, ill-smelling *mayweed,* with its small, white, yellow-centered flowers.

CHAMPAGNE, *sham pane',* an expensive French wine, white or red, sparkling or "dry," sweet or acid, most of which finds a market in the United States. It is made chiefly in the department of Marne, in the former province of Champagne, although a similar wine is made elsewhere. Much domestic champagne is consumed in America, most of which is made in California.

The best qualities are made almost exclusively from black grapes. The creaming or slightly-sparkling champagnes are more highly valued and are higher in price than the full-frothing wines. The small quantity of alcohol which the latter contain nearly all escapes from the froth as it rises to the surface, carrying with it the fragrance and leaving the liquor nearly tasteless. The property of creaming, or frothing, possessed by these wines is due to the fact that they are partly fermented in the bottle, carbonic acid being thereby produced. Because this fermenting takes place under pressure, the bottles used must be of the strongest quality. Keeping champagne cool prevents too much frothing, and that is one reason why it is usually served from a bucket of ice. See WINE.

CHAMPAIGN, ILL., a city of Champaign County, situated in the east-central part of the state, eighty miles east and north of Springfield and 128 miles south and west of Chicago. The neighboring city of Urbana, which joins it on the east, contains the University of Illinois (see ILLINOIS, UNIVERSITY OF). Champaign is served by the Cleveland, Cincinnati, Chicago & Saint Louis, the Illinois Central and the Wabash railroads, and by the Illinois Traction System. The area is nearly four square miles. In 1910 the population was 12,421; in 1914 it was 13,835.

Champaign has several parks, a Federal building erected in 1906 at a cost of $80,000, a Masonic Temple, the Burnham Library and Burnham Hospital and the University Armory, a building which cost $300,000. The city is surrounded by a rich agricultural section, for which it and Urbana are trade centers. It has railroad shops of the Illinois Central, an ice and cold-storage plant, tool and textile factories and foundries. Founded in 1855, Champaign was chartered as a city in 1860 and rechartered in 1883.

CHAMPLAIN, *sham plane',* a lake 125 miles long and from one to fifteen miles wide, lying between the states of New York and Vermont, with its northern end in Quebec. It covers an area of about 600 square miles, contains

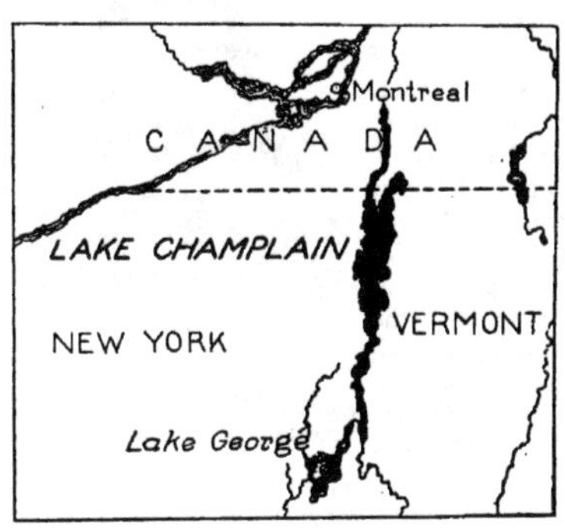

LAKE CHAMPLAIN

many islands and is a beautiful and popular summer resort. Salmon, trout and sturgeon abound, and the lake is navigated by large excursion steamers. Numerous small streams flow into it from the south, and the surplus waters are carried into the Saint Lawrence by the Richelieu River.

On September 11, 1814, a decisive naval engagement was fought between British and United States vessels in the harbor of Plattsburg, on Lake Champlain. The forces were almost evenly balanced, any superiority existing being on the side of the British. After severe fighting and heavy losses on both sides the British were defeated.

CHAMPLAIN, Samuel de (1567-1635), a French explorer and colonial pioneer, the founder of Quebec and known in history as the "Father of New France." He was the real creator of the French dominion in America. Parkman, the great historian of the French in America, sketches him in these words:

"Of the pioneers of the North American forests, his name stands foremost on the list. It was he who struck the deepest and boldest strokes into the heart of their pristine barbarism. His character belonged partly to the past, partly to the present. The *preux chevalier,* the crusader, the romance-loving explorer, the curious knowledge-seeking traveler, the practical navigator, all found their share in him."

Champlain was born in Brouage, a little town on the Bay of Biscay. His father, a ship captain, trained him in the principles of navigation, but the boy entered the army. His seaman's training stood him in good stead, however, in 1599, when he was offered the command of one of several vessels about to sail to the West Indies. During the next two years he visited all the principal ports of Mexico and the West Indies, and even traveled inland to Mexico City. His account of this voyage, which brought him to the notice of King Henry IV of France, is noteworthy for one of the earliest suggestions, if not the first, for a canal across the Isthmus of Panama.

CHAMPLAIN
Who laid the foundations for a vast French domain in America.

His first voyage to Canada was made in 1603, when he explored the Saint Lawrence River as far as the Lachine Rapids. The next year he was back again, this time exploring the coast as far south as Cape Cod. On this second voyage his patron, Sieur de Monts, accompanied him and founded Port Royal (Annapolis). In 1608, having persuaded De Monts to establish a settlement on the Saint Lawrence River, Champlain founded the settlement of Quebec, to which he gave its present name. On his previous explorations he had already established friendly relations with the Algonquin and the Huron Indians, and in 1609 he joined them in a successful raid against the Iroquois. In this expedition he discovered the beautiful lake which has since borne his name. Champlain's help at this time won for the French the lasting friendship of the Algonquins, but also the hatred of the Iroquois, who were forced to make friends first with the Dutch and then with the English. This was the beginning and perhaps the cause of much murderous border warfare.

After this exciting adventure Champlain returned to France to tell his story and win greater support for his colony. From 1613 until 1629 he crossed the Atlantic every year. He was lieutenant-governor of the colony, but he was more—he was the very life of New France. Yet he was not able to strengthen and protect Quebec as much as was necessary, and in 1629 was compelled to surrender his settlement to an English fleet. Taken a prisoner to England, he was soon released, and after Canada was restored to France in 1632 he returned to Quebec as lieutenant-governor. He died on Christmas day, 1635. M.S.

CHAMPS ELYSÉES, *shahN za le za',* a Paris boulevard, one of the most beautiful in Europe, extending from the Place de la Concorde to the Place de l'Etoile. It is nearly 300 feet wide, double the width of most American boulevards, and 1¼ miles in length. At the end near the Place de l'Etoile is the famous Arc de Triomphe erected to celebrate the victories of Napoleon (see Arch of Triumph). The boulevard is lined with trees and contains scores of beautiful buildings. There are many cafes, before which those Frenchmen known as *boulevardiers* love to sit and partake of refreshments while watching the passing stream of vehicles and pedestrians. See Paris, subhead *Boulevards, Avenues, Park and Gardens.*

CHANCELLOR, *chan'sel er,* a word meaning originally *doorkeeper,* now used to designate various important officers of the government. In Germany, for instance, the chief administrator, in England known as the *Prime Minister,* is called the *Chancellor,* Bismarck having been the first to hold the title.

In England the *Lord High Chancellor* is not an administrative but a judicial officer, the highest in the kingdom. He is the adviser of the Crown and the Keeper of the Great Seal, the official sign of royal authority. Only the royal family itself and the Archbishop of Canterbury take rank above him, and he receives a salary equivalent to $50,000 a year. He is a member of the Cabinet and the presiding officer of the House of Lords. His duties are very numerous, chief among them being the supreme judgeship of the Court of Chancery. He is the official guardian of all infants, as well as of people of unsound mind. The *Chancellor of the Exchequer* is the British Minister of Finance, and a member of the Cabinet.

In the United States and Canada the term has no official meaning, but is sometimes used instead of *president* as the title of the head of a university.

CHANCELLORSVILLE, *chan'sel ers vil,* BATTLE OF, one of the most important engagements of the War of Secession, in which the Confederates won a decisive victory, but suffered the loss of "Stonewall" Jackson, their greatest general next to Lee. The battle was fought on May 2-4, 1863, at Chancellorsville, Va., eleven miles west of Fredericksburg. The Federal Army of the Potomac, numbering about 130,000, under command of General Hooker, had for some months been lying entrenched on the north side of the Rappahannock River. In April Hooker began a general movement against Lee's Army of Northern Virginia, stationed at Fredericksburg, on the opposite side of the river. Lee's entire force numbered about 60,000.

Hooker led the bulk of his army across the river and took a position at Chancellorsville, while the rest of the troops, under General John Sedgwick, were ordered to cross below Fredericksburg, so Lee was between two sections of the Union forces. The Confederate commander, however, outgeneraled Hooker by getting the latter's army between two of his own wings, and the Union forces were completely defeated in the fight of May 2 and 3. "Stonewall" Jackson, who was Lee's "right arm" in the first day's fight, was shot by mistake by some of his own men while reconnoitering in the darkness on the night of May 2. Eight days later he died. On May 4 Lee attacked and defeated Sedgwick, and the entire Union army withdrew to the opposite side of the Rappahannock. The Union loss was about 17,300; the Confederate, about 12,460. Lee was so encouraged by this victory that he immediately planned an invasion of the north, which ended disastrously at Gettysburg, two months later.

CHANCERY, *chan'ser i,* COURT OF. The court of chancery was formerly the highest court of England, and second in authority only to Parliament. At present it is a division of the High Court of Justice. It is presided over by the Lord High Chancellor, and from this circumstance it derived its name. The purpose of the court is to settle cases which do not fall under the common law. The chancellor is regarded as the personal representative of the king, and as such possesses extraordinary authority. In the United States, courts of equity have taken the place of courts of chancery.

CHAN'NEL ISLANDS, a group of islands in the English Channel, ten miles from the coast of France, representing all that remains to England of its once great possessions in France. Their combined area is seventy-five

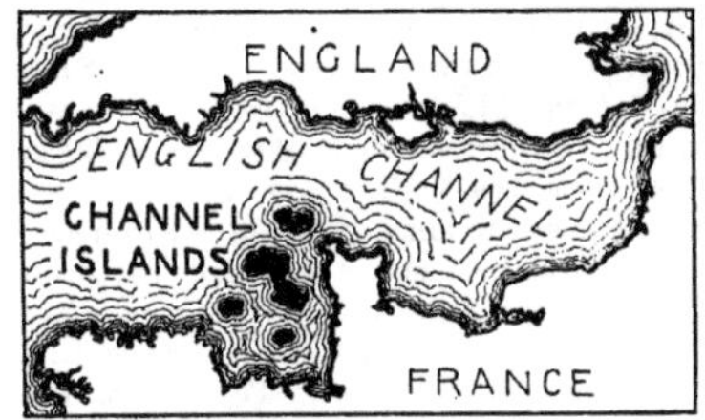

LOCATION MAP

square miles, five square miles more than the area of the District of Columbia. Although politically English, the islanders are typically French in manners and customs, and they pride themselves on belonging to the race which conquered England in the days of William I, the Conqueror. Jersey, Guernsey, Alderney and Sark are the only inhabited islands, but there are numerous rocks and islets, many of which are submerged at high tide. The climate is mild and healthful, and flowers and vegetables are grown in great quantities, reaching the London markets several weeks before the English crops. Stone for building purposes is exported, and the islands are famous for their dairy cattle, the Jerseys, Guernseys and Alderneys. Although strongly fortified, the islands find their greatest protection in the dangerous tides and currents, which render navigation for large vessels extremely difficult.

CHANNING, *chan'ing,* WILLIAM ELLERY (1780-1842), one of the most famous American preachers, whose influence is still felt in social and political reforms, through his memory and his writings. He was born at Newport, R. I., and studied at Harvard College. His first appointment as a pastor was in 1803, when he was placed in charge of the congregation of the Federal Street Church in Boston. At first his sermons did not show a strong denominational spirit, but gradually he became a decided Unitarian and taught the doctrines of that denomination with great zeal and success. Noble and fearless, he was a strong advocate of temperance, international peace and freedom. His most popular essays are those on *National Literature*, *John Milton* and *Self-Culture*. Coleridge said of him, "He has the love of wisdom and the wisdom of love." There is a bronze statue of him in Boston.

CHANUTE, *cha nute'*, KAN., a city in Neosho County, in the southeastern part of the state, about 120 miles southwest of Kansas City and forty miles west and south of Fort Scott. It is on the Atchison, Topeka & Santa Fe and the Missouri, Kansas & Texas railroads. The population in 1910 was 9,272; in 1914 it was 11,429. The area of the city exceeds two square miles. Chanute is surrounded by an extensive natural-gas and petroleum field, and gas is exclusively used for manufactories. The industrial plants include large railroad shops of the Santa Fe, oil refineries, drilling-tool works, brick and cement plants, smelters and flour mills. Chanute was settled in 1872 and incorporated the following year. In 1912 it adopted the commission form of government.

CHAPARRAL, *chap a ral'*, dense, tangled brushwood, consisting of low, thorny shrubs, brambles and briars, that grows on the dry soil of Texas, Arizona, New Mexico and Mexico. The word is derived from the Spanish *chaparro*, meaning *evergreen oak*, and was first used in the United States about 1846, during the Mexican War. References to chaparral occur in the writings of Bayard Taylor, Robert Louis Stevenson, Helen Hunt Jackson, Stewart Edward White and others who have written of the Western country. Mrs. Jackson's description of this shrubby plant in her *Glimpses of Three Coasts* is often quoted:

> Nobody will ever, by pencil or brush or pen, fairly render the beauty of the mysterious, undefined, undefinable chaparral.

CHAPLAIN, *chap'lin*, a clergyman attached to an army or navy, performing for the soldiers or sailors under his charge the duties a minister performs for his congregation.

United States army chaplains are appointed by the President, with the advice and consent of the Senate, the Secretary of War making assignments and transfers. There are no restrictions as to denomination. Each regiment of cavalry, infantry and field artillery has its chaplain; one is assigned to the corps of engineers and to the Military Academy, and there is a specified number, varying from time to time, for the coast artillery corps. The number allowed to the navy bears a definite relation to the total membership of the navy and marine corps.

The rank, pay and allowances of a chaplain in the United States army, after seven years' service, are those of a captain of infantry; until then his grade is that of a first lieutenant. Unusual ability is recognized by advancement to the rank of major, though there may be among the chaplains no more than fifteen majors at any one time. A chaplain in the navy begins as an acting-chaplain, with the rank of a junior-grade lieutenant, and after three years becomes chaplain, progressing through the various grades of lieutenant, lieutenant-commander, commander and captain. In the Canadian service some chaplains are honorary captains, others honorary majors. See RANK.

France has no chaplains in its military service, the office having been abolished when Church and State were separated, but in the War of the Nations many hundreds of priests enlisted as privates and throughout the war served as chaplains when not on the firing line. In England there are chaplains of two classes—permanent and occasional—the latter being appointed for temporary service in special districts.

CHAPLEAU, *shap lo'*, SIR JOSEPH ADOLPHE (1840-1898), a Canadian statesman, one of the foremost criminal lawyers and political orators of his time. He was born at Ste. Therese de Blainville, Quebec, on November 9, 1840. He was twenty-one when he was called to the bar of Lower Canada, as Quebec was then called, and in a short time was recognized as a great criminal lawyer. In 1867 he was elected to the Quebec Assembly, and in 1873 became solicitor-general of the province. At short intervals he next became provincial secretary, leader of the Conservative opposition, and finally premier of Quebec from 1879 to 1882. He was Secretary of State in the Dominion Ministries of Sir John A. Macdonald and Sir John J. C. Abbott from 1882 to 1892, when he held the post of Minister of Customs for six months. During the last six years of his life he was lieutenant-governor of Quebec.

CHAPMAN, George (1557 or 1559-1634), the poet and dramatist of Shakespeare's day who is remembered chiefly as having been the first to translate into English verse Homer's immortal epics, the *Iliad* and the *Odyssey*. Such critics as Pope, Lamb and Coleridge greatly admired these translations for their lofty language and swiftness of action; and they inspired one of the finest sonnets Keats ever wrote—*On First Looking into Chapman's Homer*—in which occur these oft-quoted lines:

Oft of one wide expanse had I been told
That deep-browed Homer ruled as his demesne;
Yet did I never breathe its pure serene
Till I heard Chapman speak out loud and bold:
Then felt I like some watcher of the skies
When a new planet swims into his ken.

Chapman was born near the town of Hitchin, in Hertfordshire, and learned his Greek at Oxford. When he was about thirty-five he published his first long poem, *The Shadow of Night,* and in 1598 his first play, a comedy which bore the quaint title of *The Blinde Begger of Alexandria, Most Pleasantly Discoursing His Variable Humours.* The *Iliad* and *Odyssey* translations were published in instalments, appearing at intervals throughout a period of nearly twenty years. It was not until 1611 that the entire twenty-four books of the *Iliad* were completed, and not until 1616 that the *Odyssey* was published in its entirety.

All this time, however, Chapman was writing successful plays, among the most popular being the comedies of *Al Fooles, The Widow's Tears* and *Monsieur d'Olive,* and the tragedy of *Bussy d'Ambois.* A play called *Eastward Hoe,* written by Chapman in collaboration with Ben Jonson and John Marston, led the Stuart king, James I, to send him to prison because of a satirical remark about the Scotch, and the play was ordered reprinted with the offending passage omitted. As a writer for the stage, however, Chapman did not equal the other dramatists of the Elizabethan period, either in his handling of plot or of character.

He also wrote a number of long poems, made some translations from Latin literature, and completed the paraphrase called *Hero and Leander* which Christopher Marlowe had begun and left unfinished at his death.

CHAPULTEPEC, *chah pool te pec',* Battle of, the last battle of the war between the United States and Mexico before the capture of the City of Mexico. The scene of the engagement was the small hill of Chapultepec, which, crowned by an imposing fortified castle, guarded the gates to the capital city. On September 12, 1847, the American force under General Scott began an attack on the hill, which continued the following day and resulted in the retreat of the Mexicans to the capital. On the 14th the Americans entered the city and the war was practically at an end. About 7,500 Americans and 4,000 Mexicans engaged in the three-days' fight, the loss on each side being over 800.

CHARADE, *sha rade',* a popular form of riddle, the answer to which is a word of several syllables, each of which alone is in itself a word. Each syllable, taken as a word, is described, and finally a puzzling definition of the whole word is given. The following is an example: "Some one threw my first and second at me, and it hit my third. It did not hurt me, for it was only a branch of my whole." The answer is *Mistletoe.* A girl, sitting under a high table, would suggest the word *misunder stand.*

A pleasing charade requiring more thought is in the form of a rhyme, as—

"My first is a circle, my second a cross;
If you meet with my whole, look out for a toss."

The answer is *Ox.* Then, too, charades are often presented in the form of little plays, each syllable representing a scene. They are then called *acting* charades. This form of amusement is much in vogue on social occasions. It is thought that the name was derived from a French word meaning *idle talk,* which in its turn was derived from Spanish words meaning *speech and actions of a clown.*

CHARCOAL, *char' kole,* is the familiar brittle, coal-like material produced when wood burns incompletely, and hence often found in the ashes of a wood fire. The origin of the word *char* is doubtful, but some authorities derive it from the Anglo-Saxon *cearcian,* meaning *to crackle.* Charcoal was formerly made in large quantities by cutting down trees, piling the logs into mounds or pyramids, covering these with earth and setting the wood on fire. The earth restricted the draught, or supply of air, to the fire and thus kept the wood from burning completely to ashes. In countries where hardwood is plentiful, charcoal is still made in ways similar to that described. Sometimes the heating is carried out in closed iron retorts, and the escaping gases are cooled so as to condense the acetic acid, wood alcohol and acetone which they contain.

Although *hardwood charcoal* is the most common variety, almost any plant or animal material can be charred. *Sugar charcoal* is the purest form of the element which chemists call *carbon*. When this burns it leaves no ashes at all. Commercial charcoal is carbon mixed with the impurities which remain as ashes when the charcoal is burned. After wood charcoal the next most common commercial varieties are *lampblack* and *animal charcoal*, or *boneblack*. Boneblack is made by charring bones; lampblack, by burning oil and letting its yellow flame strike against a cold metal cylinder which turns slowly so as not to become heated in any one part.

Uses of Charcoal. Wood charcoal is used as a fuel. It gives a smokeless fire. It was formerly the only fuel used in the smelting of iron ores, but for this purpose it has been almost completely replaced by coke, a form of carbon made from coal in much the same way as charcoal is made from wood. Large quantities are still used in the old-fashioned black gunpowder, which is a mixture of charcoal, sulphur and saltpeter. For military purposes this kind of gunpowder has now been largely replaced by other explosives, which have the double advantage of being much more powerful and of yielding little or no smoke. Charcoal gunpowder, however, is cheaper than these smokeless powders, and is therefore commonly used in blasting rocks, in clearing land of tree stumps, and in loosening soil in some places, so that the roots of trees and plants can grow to greater depths than would otherwise be possible.

Charcoal has the property of absorbing large quantities of gases. Boxwood charcoal will absorb ninety times its own volume of ammonia gas, and cocoanut charcoal 170 times its own volume. Charcoal is sometimes used to sweeten the air of rooms. Lampblack is much used in paints and in printing and drawing inks. Carbon inks, such as India ink and printing inks, do not fade like ordinary writing inks. Animal charcoal is largely used in the sugar refinery and in the distillery. Black as it is, it has the power of removing the color from crude sugar, syrups and crude liquors, leaving them as clear and colorless as water.

In drawing, charcoal pencils, which are merely charred twigs, are much used for rapid, sketchy work, when the object is not only softness of finish but the attainment of the maximum of effect with as few lines as possible. See CARBON. J.F.S.

CHARD, SWISS CHARD or SEA KALE, a valuable but not extensively-cultivated vegetable. It is a form of common garden beet, but its roots are small and woody. The center rib of the leaf is the desirable part and is prepared much the same as asparagus. The leaf itself is also cooked for greens. Chard should be cultivated much the same as beets, requiring, however, a little richer soil. Covering the plants with straw in the fall will aid early growth and help blanch the stems. It is a vegetable which will repay care and cultivation.

SWISS CHARD

CHARGÉ D' AFFAIRES, *shahr zha' da fair'*, a French phrase meaning *charged with affairs*, now used generally to indicate a diplomatic agent of inferior rank sent by one country to another. He takes rank after ambassadors, ministers and resident ministers, and is given his credentials not by the ruler of his state but by the minister of foreign affairs. Nor is he accredited to the ruler of the state to which he goes, but to the minister of foreign affairs. When two nations are on the verge of a break and ambassadors and ministers have been withdrawn, special *chargés d'affaires* may be appointed to carry on the necessary communication. At any time, also, that an ambassador is absent from his post a member of his staff is made *chargé d'affaires*. See DIPLOMACY.

CHARGE OF THE LIGHT BRIGADE, a stirring, patriotic poem by Alfred Tennyson, written to celebrate the memory of the English brigade of light cavalry whose heroic charge against the Russian center, in the Battle of Balaklava, has won it undying fame. This battle, one of the most important engagements of the Crimean War (see CRIMEA), was fought on October 25, 1854, with the Turkish, French and English forces contending against the Russians. Through a mistake in issuing orders, the English cavalry brigade under Lord Cardigan, numbering about 600 men, was commanded to charge the Russian guns at the end of a long valley. Though they knew "someone had blundered," they rode to the attack at the word of command, while, in the language of the poet—

> Cannon to right of them
> Cannon to left of them
> Cannon in front of them
> Volley'd and thunder'd.
>
> Storm'd at with shot and shell,
> Boldly they rode and well,
> Into the jaws of Death,
> Into the mouth of Hell
> Rode the six hundred.

Only a remnant of the brave company returned from the ride into the "jaws of Death." A French officer who witnessed the charge said, "It is magnificent, but it is not war." Yet that splendid example of devotion to duty has been an inspiration to the world through all the years that have passed, and whoever reads the story of the "Charge of the Light Brigade" feels as Tennyson did when he wrote the closing words of the poem:

> When can their glory fade?
> O the wild charge they made!
> All the world wonder'd.
> Honor the charge they made!
> Honor the Light Brigade,
> Noble six hundred.

CHAR'IOT, the original of all modern wheeled vehicles. The name comes from the Latin *carrus,* from which also *car* and *carriage* are derived. The chariot of ancient times had two wheels surmounted by a boxlike body in which the driver stood, and was probably first used in war. Two or four horses were used and in many cases the axles of the wheels were armed with scythelike blades with which to mow down the ranks of the enemy. The ancient Britons used chariots both in war and for state occasions, and the conquering Romans took back home with them many of these vehicles and used them in their triumphal processions. Egyptians, Assyrians, Greeks and Romans vied with each other in the magnificence of their chariots, which were built for display and not for speed.

The reins of the harness were sufficiently long to be tied round the waist of the driver, leaving his hands free for the use of weapons. The wheels had four, sometimes eight, spokes, and were cumbersome and heavy. Many noted groups of statuary exist which depict a chariot drawn by two horses urged on at full speed by a warrior whose spear and quiver of arrows are ready to his hand. In olden days chariot races were common; what is regarded as the finest description of such an event is found in Lew Wallace's historical novel *Ben Hur.*

Arms on armour clashing bray'd
Horrible discord, and the madding wheels
Of brazen chariots ray'd; dire was the noise
Of conflict. —MILTON: *Paradise Lost.*

CHARITY, *char'i ti.* In the word *charity* are summed up the acts of mercy that man performs for the relief of his fellow creatures who are suffering from poverty, sickness or other ills. Charity is a practical working out of the doctrine of the Brotherhood of Man; it is an expression of man's love for humanity, and offers a common meeting-ground for all those who find it "more blessed to give than to receive," regardless of their faith or creed. In the words of Pope (from the *Essay on Man*)—

> For modes of faith let graceless zealots fight
> His can't be wrong whose life is in the right.
> In faith and hope the world will disagree,
> But all mankind's concern is charity.

Individual charity, the kind that Christ told about in his parable of the Good Samaritan, where a wayfarer saw another in trouble and "had compassion on him," has always existed and always will exist as long as there is suffering in the world. In modern times, however, charity has come to be especially identified with organization, and with groups of individuals who are working together for permanent and not temporary results.

Regulated Charity. Relief work of an organized character had its beginning in the early Christian Church, and the churches are still active agents in the field of charity. To leave this field entirely to religious bodies, however, would be a serious mistake; for the churches are under the control of many denominations, and hence cannot bring about universal coöperation of effort. Organized charity has grown out of the feeling that general coöperation in charity work is a necessity.

As the first step in this direction came the formation of relief societies whose purpose was to do away with haphazard methods of giving and to place the work on a systematic basis. Then came associations for improving the condition of the poor, whose aim was to make the aid rendered of permanent value. Relief societies now are maintained in almost countless numbers in various parts of the world, including in their work the care of destitute, neglected and delinquent children, friendless boys and girls, impoverished families—in fact, people of every description who are in need of a helping hand.

Bureaus of Charities. The final step in systematized charity was the formation of the Charity Organization Society, which exists under various names in different cities, such as Associated Charities, United Charities, Boards and Bureaus of Charities. The first of these societies, and the one on which the others have been modeled, was started in London in 1869, receiving the support of such eminent men as Gladstone and Ruskin. Its founders stated that its main object was "cure, as distinguished from the mere alleviation of distress." Hardly less important was the aim to bring about such coöperation between existing relief societies as would do away with any overlapping of their fields of effort.

Societies of this character are now maintained in about 150 cities of the United States, in nearly 100 cities of Great Britain and in all the larger Canadian and Australian cities. Various charitable organizations similar to these are also found on the continent of Europe. The first American Charity Organization Society was founded in Buffalo, N. Y., in 1877.

All of these societies work on certain fundamental principles. First of all, they investigate all cases that come to their attention. A record is made for each family and placed on file for reference. All possible information is obtained, and this is placed at the disposal of individuals or relief societies that are interested. In this way the charity organization society makes possible coöperation among all the philanthropic agencies of the city.

Personal service is another important feature of the work of these bureaus. Not only do they see that help is given a needy family, but a friendly interest is kept up by means of a band of voluntary workers called friendly visitors. These organizations also seek to bring about social reforms, and to interest the community in establishing playgrounds, public baths, swimming-pools, comfort stations and better sanitary conditions in general. The recent movement in great cities for better housing conditions in the poorer districts is largely due to the broad work of the Charity Organization Society and interested individuals.

While the charity bureaus often must give help outright, especially when times are hard and winters are severe, as far as possible they encourage self-help. Therefore, laundries, sewing rooms and employment bureaus are established according to local conditions. The carrying on of the many activities of these bureans calls for men and women of education and high ability, and some societies have training schools where their workers are especially prepared for social service.

The charity bureaus are supported by voluntary contributions, and are administered by boards of directors chosen from among the contributors. Actively at the head of the work is the superintendent, who has a corps of assistants. The larger cities are divided into districts, each of which is managed by a district superintendent. W.B.G.

Authoritative information concerning the work of the charity bureaus of America may be obtained from the periodical *Charities*, published by the Charity Organization Society of New York City. On the general subject of charity, consult Warner's *American Charities;* Henderson's *Dependent, Defective and Delinquent Classes;* Devine's *The Practice of Charity.*

CHARITY, Sisters of, also written Sisters of Mercy, is the name given to a number of Orders of women in the Roman Catholic Church which are devoted to the care and education of the sick, the poor, the aged or the orphaned. Each Order is known by its special gown or habit, usually loose robes of black, relieved at the throat and about the face by a touch of white. The members of all the Orders are forbidden to marry. The first organization was established in France by Saint Vincent de Paul in 1629. The Order was approved by the Pope, and it spread rapidly wherever the Roman Catholic Church was found. It is one of the strongest, best-known and generally appreciated organizations within the Church. Because of their self-sacrificing lives and their systematic devotion to assisting the needy, the members have been spared persecution many times during religious conflicts. They have been saved by opposing forces when cities in which they were established were besieged and nearly destroyed. There are now a number of Orders in America which are popularly known as the Sisters of Charity. The first of these was founded in Maryland in 1809, by Mother Seton, and its branches are numerous. G.W.M.

CHARLEMAGNE, *shahr' le mane* (742-814), the first of the Holy Roman emperors and the only ruler of whose name *the Great* has been made a real part—for *Charlemagne* means

CHARLEMAGNE

The inscription declares that "Charles the Great ruled as emperor for fourteen years"; the sword and the orb represent respectively his might and his divine right; while the emblems above, the eagle of Germany and the fleur-de-lis of France, indicate that his empire marked the beginning of those two great states. The original painting of the above was by Albert Dürer; it is now in the National Museum at Nuremberg, Germany.

literally *Charles the Great*. His influence on the history of Europe is hard to over-estimate, for he lived just at the close of the Dark Ages, and by his enlightened measures did much to hasten the dawn of a newer and better civilization.

He was the son of Pepin the Short and the grandson of the famous Charles Martel. On his father's death in 768 he became joint king of the Franks with his brother Carloman, but three years later Carloman died, and Charlemagne was recognized as sole king of the Franks. Desiderius, king of the Lombards, already angered because Charlemagne had married his daughter and divorced her, supported the claims of Carloman's children to their father's part of the kingdom, and against him Charlemagne undertook his first campaign. This being victoriously ended, he seized all the Lombard possessions and placed on his own head the famous Iron Crown of Lombardy. In 774, before leaving Italy, Charlemagne visited Rome and formally approved the donation of certain lands made by his father to the Pope. This is looked upon as the beginning of the Papal claims to temporal power, which caused so much disturbance in Europe throughout medieval times.

Campaigns. From this time on his long reign was filled with wars; it is said that he made, in all, fifty-two campaigns. Lombards, Saracens and Saxons especially were time after time forced to defend themselves against him,

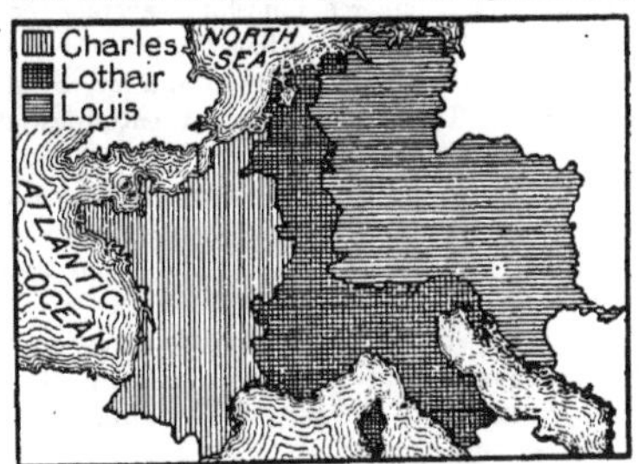

CHARLEMAGNE'S EMPIRE
As divided in 843.

usually in vain. Yet despite his success Charlemagne was not a great warrior. His genius lay rather in organization, and this helped him not only to win his victories but to weld his great empire with its unrelated peoples into something approaching unity. The religious motive was often strong in his wars. For this reason he undertook in 777 an expedition against the Saracens in Spain, and it was on his return march that his rear guard under Roland was attacked and cut to pieces by the wild peoples of the Pyrenees in the famous

Pass of Roncesvalles. He was determined, too, to establish Christianity among the Saxons, and for almost thirty years waged intermittent war against them. During the struggle, after one of numerous revolts, Charlemagne had 4,500 Saxon prisoners put to death at one time —all in an effort to force the Saxons to become Christians. In time they yielded to these forceful methods; Saxony became a part of Charlemagne's empire, and most of the Saxon leaders of the old regime were put to death.

Holy Roman Empire. In 800 Pope Leo III called Charlemagne to Rome to ask his aid in a struggle against a hostile faction. After Charlemagne was victorious the Pope rewarded him by placing upon his head a crown of gold and proclaiming him emperor of the Romans, the successor of Augustus and Constantine. Thus was established the Holy Roman Empire, that curious monarchy which played so large a part in the history of medieval Europe.

Importance in History. It is not only or chiefly as a conqueror that Charlemagne was an important world figure. He was as well a statesman who bound together his empire and prevented the great nobles from becoming too powerful by his *missi dominici,* or officials appointed by him and responsible to him. He protected commerce, punishing severely the robbers who had made perilous the life of traveling merchants, and encouraged and improved agriculture. Then, too, he was an enthusiastic patron of learning. He formed at his court a school for nobles and their sons, with Alcuin as teacher, and he himself learned to read Latin and even Greek, though he could not write legibly.

His great empire, which included not only modern France but Germany, Holland, Belgium, Switzerland, Hungary, most of Italy and a part of Spain, was left to his son, Louis I, but the son was not as strong as the father, and the carefully built structure was in time torn apart. A.MC C.

Consult Davis's *Charlemagne,* in "Heroes of the Nations" Series; Prutz's *Age of Charlemagne;* Mombert's *History of Charles the Great.*

Related Subjects. For additional information connected with the life and work of Charlemagne, see

Charles (France)
Charles Martel
Franks
Holy Roman Empire
Iron Crown
Subhead under *Crown*
Pepin

CHARLEROI, *shar le roi',* Pa., a thriving borough in Washington County, in the southwestern part of the state, on the Monongahela River, forty miles southeast of Pittsburgh. Railway transportation is provided by the Pennsylvania Railroad. Charleroi was settled in 1890 and was incorporated the following year; its growth is due to the mining industry and to manufactures, the principal products being various kinds of glass and shovels. In 1910 the population was 9,615; in 1914 it was 11,185. The area is less than one square mile.

CHARLES [England], the name of two English sovereigns of the royal Scottish House of Stuart, both of whom were firm believers in the doctrine of the "divine right of kings." The life of the first of the two was a sacrifice to this belief.

Charles I (1600-1649), son of James I, persisted in a course of tyranny throughout his reign that ended in his execution and the establishment of the Commonwealth of England. He came to the throne of England in 1625; within the next four years he convened three Parliaments and dissolved each of them because they refused to submit to his arbitrary ways. To the famous Petition of Right, drawn up by the third Parliament, he at first agreed, but speedily violated its most important clauses by attempting to raise money by unlawful taxes and loans. Between 1629 and 1640 Charles governed England without a Parliament, using the infamous courts of the Star Chamber and High Commission to make his various methods of raising money seem legal.

CHARLES I
The famous triple portrait, by Van Dyck.

In 1639 the king's attempt to force Scotland to use English forms of worship led to a rebellion, and he was obliged to call a Parliament in order to have money voted to crush the insurrection. In 1640 the famous Long Parliament assembled (so-called because it remained in session twelve years), but Charles succeeded no better with this assembly than with the others, and civil war began when he attempted to seize five of its leading members. The king had on his side the nobility, gentry and clergy, while the Puritans and the people of the great trading towns supported Parliament. In the course of the struggle the "man of the hour," Oliver Cromwell, came into prominence, and his great victories at Marston Moor (1644) and Naseby (1645) marked the ruin of the king's cause. In 1646 Charles escaped to Scotland, but was delivered up to the English Parliament. In 1649 he was tried, condemned as a public enemy of the nation and beheaded. The private life of this unfortunate king was blameless.

Charles II (1630-1685), son of Charles I, was the first of the restored Stuart line. In 1651 he was proclaimed king by the Scotch, but his army was defeated by Cromwell at Worcester, and he fled to France. The death of Cromwell in 1658 and the popular dissatisfaction with the Commonwealth as a form of government opened the way for his return, and in 1660 he was crowned as Charles II of England. His first Parliament gave him all the privileges which earlier assemblies had fought to keep his father from enjoying. Among the important events of his reign were a war with the Dutch, the great plague and fire of London, the Rye House Plot and the passage by Parliament in 1679 of the famous Habeas Corpus Act.

The court of Charles II was accounted the most immoral in all English history, and the evil life of the king and his associates was reflected in the literature of the Restoration Period.

B.M.W.

Related Subjects. The reader is referred to the following articles in these volumes:

Commonwealth of England	Long Parliament
Cromwell, Oliver	Naseby, Battle of
Divine Right of Kings	Petition of Right
Habeas Corpus	Restoration, The
Hampden, John	Rye House Plot
	Star Chamber

CHARLES OF FRANCE

CHARLES [France], the name of ten sovereigns who have worn the crown of France. The first was Charles the Bald, youngest son of Charlemagne's son Louis, who received the western portion of his father's empire when it was divided by the Treaty of Verdun in 843. The kingdom over which he ruled until 877 was the nucleus of modern France, and he is therefore known as Charles I of France. Charles II, surnamed THE FAT, ruled from 885 until 887, when his subjects, wearied by his cowardly method of defending the country from the attacks of the Northmen, deposed him.

Charles III, called THE SIMPLE, came to the throne in 893. During his reign the territory later known as Normandy was ceded to the Northmen, and Lorraine was conquered. Imprisoned during a revolt of his subjects, he died in captivity in 929. Charles IV, known as THE FAIR, was king of France from 1322 to 1328, the last of the Capetian line (see CAPETIAN DYNASTY). His rule was marked by the strengthening of the royal power and the suppression of the lawless nobles in the kingdom.

Charles V, surnamed THE WISE (1337-1380), was born in the same year in which the Hundred Years' War (which see) began. When his father, John the Good, was taken captive by the English at the Battle of Poitiers, in 1356, Charles ruled in his stead and was crowned king in 1364. He fought England for several years, wresting from his enemies nearly all that they had won from his father, and was equally successful in establishing order in his own kingdom. Charles was a patron of art and literature, and laid the foundations of the National Library of France (see BIBLIOTHEQUE NATIONALE). The famous prison known as the Bastille (which see) was built by him to keep the lawless citizens of Paris in order.

Charles VI (1368-1422), son of Charles V, was a boy of twelve when his father died. Four of his uncles divided the kingly power among them, and their personal ambitions soon brought the country to a state of great disorder. Finally, in 1388, Charles took the governing power into his own hands and ruled wisely until 1392. In that year he suffered from an attack of insanity, and when it became evident that his mind was permanently weakened his uncles regained their power.

The rivalry between two of these, the Duke of Burgundy and the Duke of Orleans, split the country into two warring factions. Henry V of England, making the weakness of France serve his own purposes, invaded the country and in 1415 won a great victory at Agincourt (which see). Five years later the Peace of Troyes was signed, by which Charles VI acknowledged Henry V as his successor and disinherited his own son. When the king died in 1422, nearly all of France was under the control of the English.

Charles VII (1403-1461), who succeeded his father Charles VI in 1422, fell heir to a crown that was claimed by the English for their king, Henry VI. With nearly all of his realm in the hands of the foreign foe, the young king looked on helplessly while the English continued their conquests, and when Orleans was besieged in 1428 the outlook for France was dark indeed. In 1429 came another terrible defeat, but in that year the deliverer of France appeared—the heroic Joan of Arc (which see). Inspired by her faith and enthusiasm, the French raised the siege of Orleans, and on July 17, 1429, Charles was crowned at Rheims. In the years that followed the French drove the English from all their holdings in France except Calais.

As soon as Charles knew that his claim to the throne was secure, he began to reorganize the government, and in the course of time peace and prosperity returned to France. He was, however, a timid and irresolute ruler, and it is to his lasting discredit that he made no effort to save Joan of Arc from her terrible fate.

Charles VIII (1470-1498) succeeded his father, Louis XI, in 1483, when he was only thirteen years of age. For the next eight years the kingdom was wisely governed by the boy king's sister, Anne of Beaujeu. In 1491 he married Anne, Duchess of Brittany, thereby adding the duchy to the French realm. Charles became king in fact as well as in name at the age of twenty-one, and his reign is memorable because of his invasion of Italy in 1494. This was an epoch-making event in European history, for it was the beginning of four centuries of interference by the Northern nations in the affairs of Italy. Charles accomplished the conquest of the kingdom of Naples in 1495, but a league was formed against him and his efforts came to nothing.

Charles IX (1550-1574), son of Henry II and Catharine de' Medici (which see), succeeded his elder brother Francis II at the age of ten. Even after he was declared of age, his mother, who had acted as regent, was the real sovereign of the nation. His reign was one of the unhappiest in French history, disturbed continually by civil wars, intrigues and strife between the Roman Catholics and Protestants. Though not vicious, the young king was weak and easily influenced, and so was persuaded by his mother to permit the greatest outrage of his entire reign, the massacre of Saint Bartholomew's Day (August 24, 1572). Charles himself suffered terrible remorse for having given his consent to the massacre, and died two years later. See SAINT BARTHOLOMEW'S DAY, MASSACRE OF.

Charles X (1759-1836), younger brother of Louis XVI and Louis XVIII, and the last sovereign of the older Bourbon line of kings, was a striking example of the old saying, "A Bourbon never learns anything and never forgets anything" (see BOURBONS). Succeeding his brother, Louis XVIII, in the year 1824, he began at once to revive the old despotic rule which had driven the French people to the terrible Revolution of 1789 (see FRENCH REVOLUTION). All liberal measures were disregarded, the clergy was restored to power, the constitution was ignored and laws were changed merely by the king's proclamation. In 1830 the people of Paris rose in revolt, and in August of that year Charles abdicated in favor of his grandson, Henry of Bordeaux. The French, however, chose Louis Philippe, Duke of Orleans, as their king. Charles escaped to England, and afterwards took up his residence in Austria, where he died. See FRANCE, subtitle *History*. B.M.W.

CHARLES, in the history of the HOLY ROMAN EMPIRE, the name of seven monarchs who bore the title *Holy Roman emperor*. In theory the Holy Roman emperors were successors of Charlemagne, but in fact they ruled over the German dominions and Italy. Excepting Charlemagne (Charles the Great), Charles V and Charles VI were the most important of the emperors who bore the name of Charles.

Charles V (1500-1558) was one of the most powerful sovereigns of the sixteenth century. Heir to the rich and populous provinces of the Netherlands and to the dominions of Spain and the Austrian House of Hapsburg, he became king of Spain as Charles I in 1516, and was crowned Emperor Charles V in 1520 as successor to Maximilian I. His reign was greatly disturbed by wars with Francis I of France and Solyman the Magnificent, sultan of Turkey. In his second war with Francis I an imperial army plundered Rome and took the Pope prisoner. Charles and Francis ended their struggles in 1544, but in the meantime the great Reformation movement had developed in the German dominions of the emperor.

Had Charles been able at the beginning of his reign to turn his attention to religious matters in Germany, he might have prevented the growth of Protestantism during his lifetime. When, in 1546, the year of Luther's death, he began serious efforts to suppress the movement, he found the Protestants too strong for him, and by the Peace of Augsburg (1555) it was agreed that the people of each German state should adopt the religion, whether Protestant or Roman Catholic, of the ruling prince of that state. Charles began, however, the persecution of the Protestants in Spain and the Netherlands that were continued by his son Philip II of Spain.

Wearied by his years of warfare and saddened by his failure to make all of his subjects think alike in matters of religion, the emperor in 1555 and 1556 gave up to his son, Philip, the crowns of the Netherlands and Spain, and to his brother, Ferdinand, his imperial authority.

Charles VI (1685-1740), the last of the direct male line of the House of Hapsburg, and the second son of the Emperor Leopold I, was Holy Roman emperor from 1711 to 1740. In 1700, on the death of Charles II of Spain, Charles of Hapsburg claimed the Spanish throne as the rival of Philip of Anjou. This brought on the War of the Spanish Succession, in which Great Britain and Holland aided Charles. When he became emperor of Germany in 1711, Charles was forced by his allies to give up his claim to the Spanish crown, but was permitted to retain the Spanish possessions in the Netherlands and in Italy. In 1713 he published the Pragmatic Sanction, by which his daughter Maria Theresa was to inherit all the possessions of the House of Austria. Charles spent more than twenty years of his reign trying to win the consent of the European powers to the "Pragmatic Sanction."

B.M.W.

Related Subjects. The reader is referred to the following important articles:

Charlemagne	Reformation
Hapsburg, House of	Spain (*History*)
Holy Roman Empire	Succession Wars
Netherlands (*History*)	Utrecht, Peace of
Pragmatic Sanction	

CHARLES [Sweden] the name of several Swedish monarchs.

Charles IX (1550-1611), third son of Gustavus Vasa (see GUSTAVUS I), began his rule as regent of the kingdom in 1592, on the death of his brother John. In this position he gave his support to the establishment of Protestantism in Sweden. He was crowned king in 1604, and during his reign engaged in wars with Poland, Russia and Denmark. Charles was the founder of the University of Gothenburg and the author of a rhymed history of his war with Poland.

Charles X, GUSTAVUS (1622-1660), who reigned from 1654 to 1660, was the nephew of the great Gustavus II Adolphus (which see) and successor of Queen Christina. Soon after his accession he invaded Poland, and having forced Frederick William, elector of Brandenburg, to give him aid, defeated the Poles in a famous battle at Warsaw (1656). During a war with Denmark he secured for his own kingdom the Danish provinces of Scania and Holland, and laid siege to Copenhagen. The Dutch then came to the help of the Danes, and Frederick William turned against Charles, so successfully that the Swedish forces were defeated both on land and on sea.

Charles XI (1655-1694) succeeded his father Charles X in 1660, but the kingdom was ruled by his mother, Hedwig, until the boy king had reached the age of seventeen. His reign began with wars against the Germans, the Dutch and the Danes. After the restoration of peace Charles began a period of reform. He diminished the power of the nobles, cut down the public debt, reorganized the army and navy and brought them to a high degree of excellence, and by his wise management of the public revenues put the finances of the kingdom on a firm basis.

Charles **XII** (1682-1718), one of the most remarkable kings of the middle period, was the eldest son of Charles XI, whom he succeeded in 1697. At that time Sweden was one of the great European powers, and the Baltic Sea was practically a Swedish lake. The growing power of the Scandinavian kingdom to the north was jealously watched by three European sovereigns—Frederick IV of Denmark, Augustus of Poland and Peter the Great of Russia. When the young king ascended the Swedish throne these rulers decided that the time was ripe for them to strike for the control of the Baltic, and in 1700 the War of the North began.

Charles threw himself into the conflict with a reckless daring that has won for him the name of "Madman of the North." Though he won several brilliant victories, in the end he overestimated his strength and made a foolhardy invasion of Russia. At Pultowa (1709) his army was nearly wiped out by the forces of Peter the Great, and he fled southward to Turkey. After spending five years in fruitless plots and schemes for revenge, which led to his imprisonment by the Turks, he escaped to Stralsund, a Swedish possession in Prussia. For a year he conducted a brilliant defense of the place, yielding finally to a combined force of Danes, Saxons, Prussians and Russians. Soon after this he invaded Norway, and was killed while besieging Frederikshald. B.M.W.

CHARLES I (1887-), emperor of Austria, who is also Charles IV, king of Hungary. He succeeded his great-uncle, Francis Joseph, on November 21, 1916. Before that date he was the Archduke Karl Franz Joseph. The young emperor is a man of simple tastes. Before he became heir to the throne through the assassination of his uncle, the Archduke Francis Ferdinand, there were many at the Viennese court who had never seen him. In the War of the Nations he served at the front as nominal head of the army until German officers assumed the Austrian commands. His portrait appears on page 505.

CHARLES EDWARD, the YOUNG PRETENDER. See STUART, CHARLES EDWARD.

CHARLES MARTEL, *sharl mar tel'* (about 688-741), a famous leader of the Franks, who won his title of *Martel,* meaning *the hammer,* by his celebrated defeat of the Arabs on the plains of Tours, in A. D. 732. It was this battle which saved the Christian civilization of Western Europe from being overwhelmed by the power of Mohammedanism. Under the last Merovingian kings Charles held the position of mayor of the palace, but exercised real kingly authority. He thus prepared the way for his son Pepin (which see).

CHARLESTON, S. C., the largest city of the state and one of the most conspicuous historical cities in the South. Its population increased from 48,833 in 1910 to 60,427 in 1915. Charleston is the county seat of Charleston County, and is situated on the southeastern coast of the state, on a tongue of land between the Ashley and Cooper rivers. These two rivers unite immediately below the town to form the spacious harbor which communicates with the Atlantic Ocean at Sullivan's Island, about seven miles below. Savannah is 115 miles southwest; Columbia, the state capital, is 129 miles northwest. The city is served by the Atlantic Coast Line, Charleston & Western Carolina, Seaboard Air Line and Southern railways. The last named was the first railroad in the United States constructed to be operated by steam locomotives (1830); it extended from Charleston to Hamburg, and was called the South Carolina Railroad. Charleston is also a port of call for several lines of steamers. It is popularly called the *Plumbline Port to Panama,* as it occupies the same meridian of longitude (80°) as does this great gateway to the commercial world. The area of Charleston is about three and a quarter square miles.

Though Charleston occupies a foremost rank commercially, it still retains many aspects peculiar to the Southern states. The stately mansions and lofty piazzas, the many gardens with their profusion of magnolias, camellias, jessamine and azaleas, the wide streets bordered with live-oaks and other shade trees, stamp it as one of the old leisure-loving cities of the South. It has nine miles of water front, and the harbor has been so improved by the construction of jetties as to admit large vessels. The defenses include Fort Sumter, Fort Johnson and Fort Moultrie, on Sullivan's Island, where the Federal government has expended over $500,000 to render the artillery post one of the best-equipped in the United States. Several million dollars have been expended in improving the navy yard, on Cooper River.

Parks and Boulevards. Charleston claims to have the most beautiful gardens in the United States. Of the many parks, Hampton is the largest (318 acres). White Point Gardens, Marion Square, Hampstead Wall, Colonial and Washington parks are among the other recreation grounds of the city. Legare Street and the Battery, a broad promenade about 500 yards long, are part of the residential districts. King Street is the principal retail thoroughfare; Rutledge and Ashley avenues extend the length of the city from north to south. In the vicinity are several attractive resorts easily reached by street railway. The Isle of Palms is about ten miles north of the harbor.

Institutions and Public Buildings. Among the most noteworthy buildings are the white marble United States customhouse, constructed at a cost of $3,400,000; a $450,000 post office and a $75,000 immigration depot. The Charleston Museum, Gibbes Memorial Art Building, the city hall, Hibernian Hall, Exchange, Citadel, Old Powder Magazine and the City Market are all buildings of architectural, historical or commercial interest. The most conspicuous of Charleston's churches are Saint Michael's (Episcopal), with its chimes and lofty tower dating from 1761, and the Roman Catholic Cathedral. Besides the public school system the city has the College of Charleston, dating from 1788, which has a museum; the Avery Normal Institute (colored), and the Medical College of South Carolina. There are also numerous academies and business schools. The library of Charleston is maintained by subscription and is the third oldest in the United States, having been established in 1743. Among the benevolent institutions are the Home for Mothers, Widows and Daughters of Confederate Soldiers; the Euston Home and the Charleston Orphan Home.

Commerce and Industry. Charleston is a great distributing point and wholesale jobbing center for the Southeast. It is the only coal-export port on the South Atlantic coast, and ships coal here for Cuba and South America. It is also the great fuel-oil distributing point for this section, having two companies with tank capacities of 700,000 gallons each. An abundance of sea-island cotton is grown in the vicinity; this and rice, fruits, lumber, naval stores and fertilizer form the chief exports. The industries are numerous and varied, textile, fertilizer, lumber and machinery plants being the most important establishments.

History. Charleston is one of the oldest cities in the United States. An English settlement was made here in 1670, and named *Charles Town*, for King Charles II. A company of Huguenots joined the settlement in 1685. In 1775 it was the third seaport in importance in America, and was the first Southern city to join the Revolution. In 1776 the provincial congress of South Carolina met in Charles Town and in the same year the first independent state constitution was adopted. In 1783 Charleston was incorporated, and until 1790 it was the capital of the state. The city was visited by the greatest earthquake known in the history of the United States, in August, 1886; more than $8,000,000 worth of property was destroyed, three-fourths of the homes were demolished or injured and many people were killed. For forty years the city struggled to regain the position it held previous to the War of Secession; the progress it has made during the last decade is remarkable. The South Carolina Inter-State and West Indian Exposition, held in 1901, was an important factor in the growth of the city. S.S.R.

CHARLESTON, W. Va., the capital of the state and the county seat of Kanawha County, is situated in the middle western part of the state, about midway between the northern and southern borders, and on the Great Kanawha River, at its junction with Elk River. Huntington is fifty miles west, Wheeling is 272 miles north and east, and Cincinnati is 211 miles west and north. Railway transportation is provided by the Chesapeake & Ohio, Kanawha & Michigan, Kanawha & West Virginia and the Coal & Coke railways, and there is steamboat connection with all important ports on the Mississippi and Ohio rivers. Electric lines operate west from the city. Charleston was built around a fort that was erected in 1786. It was incorporated as a town in 1794 and as a city in 1870, and since that year has been the capital of the state with the exception of ten years when Wheeling was the seat of government (1875-1885). The population increased from 22,996 in 1910 to 27,703 in 1914, and the city now ranks third in the state.

Charleston lies between high hills a mile apart. A beautiful boulevard extends for miles along the river bank; lined by trees with overhanging branches, and bordered with elegant homes, with wide, well-kept lawns, this is the most exclusive residential district. Everywhere there is an abundance of flowers, foliage and magnificent old trees. The Capitol, a $300,000 Federal building, a customhouse, county courthouse, Y. M. C. A. building, a public library, a convention hall and a monument to "Stonewall" Jackson are features of interest.

The vicinity of Charleston is rich in deposits of bituminous coal, salt, iron, oil and timber, and because of its fine shipping facilities by rail and water, for the Kanawha River has an excellent system of locks and dams, the city is the distributing point for all these products. An abundance of natural gas promotes interest in manufactures. Here are large railroad repair shops, boat-building yards and one of the largest ax factories in the world. There are also glass, color and veneer works and lumber mills. S.P.P.

CHARLEVOIX, *shahr le vwah'*, Pierre Francis Xavier de (1682-1761), a French missionary and traveler, whose writings form a valuable part of the early literature of Canada. Charlevoix became a Jesuit in 1698, when he was sixteen years old, and from 1705 to 1709 taught at Quebec. Later he returned to France, but in 1720 was back again in America, this time to gather information about the "Western Sea," which was then supposed to lie only a short distance west of the Mississippi River. He ascended the Saint Lawrence River and the Great Lakes as far as Michilimackinac (now Mackinac), then reached the Mississippi by way of Lake Michigan and the Illinois River, and descended the Mississippi to New Orleans in a small boat. His knowledge of New France is displayed in his best-known work, *Histoire et description generale de la Nouvelle France* (that is, *History and General Description of New France*).

Charlevoix, Mich., the county seat of Charlevoix County, named for Pierre Charlevoix, is 210 miles northeast of Grand Rapids, on the Pere Marquette Railroad. It is best known as a summer resort. Population in 1910, 2,420.

CHARLOTTE, *shar'lot,* N. C., a commercial and industrial center of importance in the state, and the county seat of Mecklenburg County. It is situated on Sugar Creek, near the southern state line, about midway between the eastern and western borders. Raleigh, the capital, is 174 miles northeast. The city has the service of the Southern, the Piedmont & Northern, the Norfolk Southern and the Seaboard Air Line railways. The population, which was 34,014 in 1910, had increased to 37,951 in 1914. The city's area is thirteen square miles.

Charlotte is the trade center for an agricultural and cotton-growing section, and the kindred cotton industries claim its chief interest; these are cotton-weaving and the manufacture of cotton-mill machinery, cottonseed oil and other by-products. In this locality there are more than 400 cotton mills, operating 5,000,000 spindles. Fertilizers, belting, saddlery, harness, drugs, cement and various kinds of machinery are also made here. Gold deposits are found in this section of the state, and a branch mint was established here in 1838; at the beginning of the War of Secession it was closed, but was reopened as an assay office in 1869.

Charlotte has a Federal building, county courthouse, an auditorium, Y. M. C. A. and Y. W. C. A. buildings, many handsome churches and fine business buildings. The number of its benevolent institutions is above the average for cities of its size. In addition to its private and public schools, there are the Presbyterian College, Saint Mary's Seminary, North Carolina Medical College, Elizabeth College and Conservatory of Music for young women, a Carnegie Library and a library for colored people. Biddle University (Presbyterian) for colored students is outside the city limits.

The place was settled in 1750, was incorporated in 1768 and in 1774 became the county seat. Here on May 20, 1775, the Mecklenburg Declaration of Independence (which see) was adopted, and a monument has been erected to its signers. In September, 1780, Lord Cornwallis occupied the city and during his stay pronounced it a "hornet's nest," a name since then adopted by the city as its emblem. Charlotte was also the headquarters of General Gates in 1780, and here in the War of Secession the full Confederate Cabinet met for the last time. M.W.

CHARLOTTENBURG, *shahr lot'en boork,* a western suburb of Berlin, capital of Germany, named after Queen Charlotte, whose husband, King Frederick I, here erected for her a magnificent castle. In the gardens of the castle is a royal tomb in which rest the remains of Frederick William III, Queen Louisa, Empress Augusta and Emperor William I. Charlottenburg is a beautiful residential district, containing many fine mansions, and is connected with Berlin by a magnificent avenue named the Charlottenburger Chaussee. Along the banks of the River Spree, which flows through the suburb, are many flourishing factories and industrial institutions. . Population in 1910, 305,181.

CHARLOTTETOWN, *shahr'lot town,* the capital and largest city of Prince Edward Island and the county town of Queen's County. It is on the south shore of the island and has a beautiful harbor, almost completely landlocked, formed by the confluence of three small streams, the Hillsborough, York and the Elliott rivers. These are also known as the East, North and West rivers, respectively. Its shipping facilities have given Charlottetown considerable commerce, chiefly in agricultural products and fish, and it has regular steamship connection with Boston, Halifax and ports on the Saint Lawrence River. The Prince Edward Island Railway provides connection with all important points on the island. Population in 1911, 11,203; in 1916, about 12,000.

The manufacturing enterprises of Charlottetown are varied, including pork packing, foundries and machine shops and railway repair shops. Of less importance among its products are aërated water, condensed milk, tobacco, soap and canned lobsters. In the neighborhood of the city are numerous fox farms and several Karakul sheep farms. The Parliament buildings, the Government House, the provincial insane asylum and the law courts building are among the conspicuous structures of the city. Prince of Wales College, founded in 1860, is housed in a building erected at a cost of $60,000. Saint Dunstan's College, the normal school and a large private school for boys should also be mentioned among the educational institutions. Victoria Park, with an area of sixty acres, is one of the attractive features of the city. Charlottetown was settled in 1786, and was named in honor of Queen Charlotte, wife of George III. The conference which led to the Quebec Conference, and thus to the organization of the Dominion, was held at Charlottetown in 1864.

CHARON, *ka'ron,* the ragged old ferryman of the Lower World, a character in Greek mythology. He is represented as the son of Erebus and Night, bent and old, with matted beard and tattered garments. Gloomily, with

CHARON
The boatman of the Styx. A detail from the painting by Neide.

one oar, he ferried the shades of the dead across the rivers Styx and Acheron to the realm of Hades. But the mythological story tells us that only those would he take who had had a proper burial, and in whose mouths was placed an obolus, the coin Charon exacted as his fee. All others were compelled to wander wearily on the shores of the river for a century; after that time Charon would take them without charge to their final resting place.

Charon appears frequently in literature and art. Homer does not mention him, but he is pictured in Vergil's *Aeneid.* The hero Aeneas is ferried across to Hades in the boat which had previously carried only shades. Though Charon for a long time refused to perform this service, he was finally persuaded to do it. The great painting by Polygnotus, *Odysseus in the Lower World,* shows this ancient ferryman. On some early Etruscan monuments he appears as an ugly, animal-faced demon of death, with tusks and pointed ears, carrying snakes or a large hammer. One of the best of the paintings illustrating the myth of Charon is by Neide.

CHART, a map or drawing made for a particular purpose, in which accuracy of detail is the chief requirement. The one possibly in most common use is the *mariner's,* or *hydrographic,* chart. This shows a seacoast with every detail of rock, shoal, depth, sounding, bank, channel, bay and harbor so exactly located that a ship may be guided safely by it through the most dangerous seas. The *topographic* chart, also common, shows with similar accuracy the details of any land surface and is mainly for the guidance of military men and surveyors. *Climatic* charts present by outline and diagram the rainfall, temperature and direction of the winds of certain localities. These are prepared daily by the United States Weather Bureau and are designed to be of aid to navigation, by giving warning of storms, and also for the information of all people whose activities may depend upon weather conditions. There are also *celestial* charts, on which stars and constellations are correctly shown, and *heliographic* charts, which locate the spots on the sun's surface. A great variety of *educational* charts are used in teaching.

CHARTER, a written instrument or contract given by a government as evidence of certain political or business privileges granted. The Great Charter (see MAGNA CHARTA) granted by King John is the world's most historic charter, as it conferred on the English-speaking race privileges which are the foundation of the liberty of Britain, Canada, Australia and the United States. Charters are granted by states to banks, corporations and associations, authorizing them to conduct their business within specified limits. A state or province by charter authorizes the organization of a village or city government; the charter sets forth the powers and obligations of such a government. See CORPORATION.

CHARTER OAK, an historic tree which is said to have concealed the charter of Connecticut for two years. It stood where now is Charter Oak Place, in the present city of Hartford. Its age was computed at nearly a thousand years, when, finally, it was blown down in August, 1856. A white marble monument now marks the spot. James II found Connecticut's charter a barrier to his plan to make that community a part of his New England. So, in 1687, at his command, Sir Edmond Andros, the governor-general of New England, went to Hartford and demanded the

THE CHARTER OAK

The tree that is said to have been the hiding place of the Connecticut charter.

delivery of the charter. Appearing to submit, the colonists went to the council chamber to carry out the ceremony, but while there the lights were snuffed out and the document was carried to a hiding place in the hollow of a tree, where it remained until the deposition of Andros. Early reports of this incident referred to the tree as an elm. Some people declared that the paper was hidden in the home of a prominent colonist, but about 1789 the belief became settled that this oak had concealed the famous charter.

CHARTER OAK MONUMENT

Erected on Charter Oak Place, in the city of Hartford. It was here that the famous tree stood.

CHARTISM, *char'tiz'm,* in England, grew out of the oppressive conditions under which workingmen lived and was a movement which attempted radical reform. The Reform Bill of 1832 had bettered matters somewhat, but had not silenced the discontent among the laboring classes, which by 1838 had become acute. From that date until 1848 the Chartist movement was at its height. A formal demand, known as the *National People's Charter,* called for six reforms: (1) universal suffrage; (2)

equal electoral districts; (3) vote by ballot; (4) annual Parliaments; (5) no property qualifications for members of Parliament; (6) salaries for members of Parliament.

Monster meetings were held at which hundreds of thousands were present, and huge petitions were presented to Parliament. Directly, the movement accomplished nothing, though it did leave an influence on the trend of thought of the times. The repeal of the Corn Laws brought improved conditions, and after 1848 the movement gradually died out. See CORN LAWS.

CHASE, SALMON PORTLAND (1808-1873), an eminent American statesman and jurist, who as Chief Justice of the United States Supreme Court presided over the impeachment trial of President Johnson. His greatest fame, however, was achieved in the Cabinet of President Lincoln. He was born in Cornish, N. H., and was educated at Dartmouth College. After studying law in Washington, D. C., he began to practice in Ohio, where he took part in the defense of so many runaway slaves that the slave-holders of Kentucky nicknamed him "the attorney-general of fugitive slaves." He became the recognized leader of the anti-slavery movement in Ohio, and throughout a term of office as United States Senator, from 1849 to 1855, he vigorously opposed the extension of slavery into the new territories and the passage of the Kansas-Nebraska Bill (which see). The Liberty party in 1843 and the Free-Soil party in 1848 had called upon him to prepare their national platforms (see LIBERTY PARTY; FREE-SOIL PARTY).

Chase was elected governor of Ohio in 1855 and again in 1857. He had by that time joined the new Republican party, and in 1860 was one of the candidates for the Presidential nomination. Failing to secure this honor, he accepted the office of Secretary of the Treasury under Lincoln. His career as a Cabinet officer marks him as one of the great Secretaries for during the perilous years of the War of Secession the national credit was maintained, funds were secured to carry on the struggle and a new national banking system was created. Differences with Lincoln regarding the President's war policy caused him to resign in 1864, and in the same year Lincoln appointed him to succeed Chief Justice Taney as head of the Supreme Court. His dignified and impartial conduct of the impeachment trial of President Johnson was highly praised.

There is an excellent life of Chief Justice Chase written by Albert Bushnell Hart for the *American Statesmen Series.*

CHAT, a popular name of "the clown among birds." The chats are small, lively birds of the warbler family. During the mating season the males perform many extraordinary twists and turns in the air. Their song is a mixture of whistles, wails, clucks and chuckles, which gave them their name. In the United States and Canada the yellow-breasted, or polyglot, chat is a larger species, olive-green above and white below, with a yellow breast. It builds its nest in briary thickets.

CHATHAM, *chat'am,* a town of Northumberland County, New Brunswick, situated on a series of terraces rising from the south bank of the Miramichi River, twenty-five miles above its entrance into Mirimichi Bay. Its harbor is excellent, and provides anchorage for vessels drawing twenty-five feet of water. Chatham is the largest town in New Brunswick, on the eastern, or Gulf of Saint Lawrence, shore, and is the fourth town in size in the province. It is on the Intercolonial Railway, 112 miles northeast of Fredericton, the capital of the province. The Miramichi River is a great logging stream, and lumber is Chatham's chief interest. There are several lumber and planing mills, and opposite the town, on the north bank of the river, is a large pulp mill. These mills have a combined working force of about 500 men, one-tenth of the total population of Chatham. The fisheries in the vicinity are also noteworthy, smelts, sardines, bass and oysters being most important. Chatham is the seat of a Roman Catholic bishop, convent and hospital. The town was founded in 1816, and incorporated in 1896. Population in 1911, 4,666; in 1916, estimated, 5,000. J.A.F.

CHATHAM, the county town of Kent County, Ontario, is forty-seven miles east of Detroit, sixty-four miles southwest of London and 180 miles southwest of Toronto. It is frequently called the *Maple City,* from the large number of maple trees in its parks and on its boulevards. Chatham's geographical position, midway between Lake Saint Clair and Lake Erie, has made it an important railway center. The Canadian Pacific, Grand Trunk, Wabash and Pere Marquette railways enter the city, and the Michigan Central runs seven miles to the south. Connection with the latter is by the Chatham, Wallaceburg & Lake Erie, an electric railway. Chatham also has excellent facilities for shipment by water, for the Thames River, which flows through the city, is navigable for lake boats up to this point. The town was founded in 1812, and was incorporated as a city in 1895. Population in 1911, 10,770; in 1916, 12,863.

The trade of the city is largely in agricultural products and supplies. There are also numerous factories and mills, the leading manufactures being motors, well machinery, boilers, wagons and carriages, wheels and axles, woolen goods, flour and beet sugar. Light and power are supplied from Niagara Falls, and for lighting and heating there is also an abundance of natural gas, piped from the Tilbury oil fields, about fifteen miles southwest. The educational institutions, in addition to the public schools, include a collegiate institute, the Ursuline convent school, and the Canada Business College, all three of more than local fame. Harrison Hall contains the city and county offices, but the most conspicuous structure in the city is the government armory erected in 1905 at a cost of $500,000. W.M.F.

CHATHAM, Earl of. See Pitt, William.

CHATTAHOOCHEE, *chat a hoo'che,* a large, muddy river which forms for a considerable distance the boundary between Georgia and Alabama. The name means *pictured rocks,* and was given it by the Creek Indians because of the vari-colored rocky banks between which it flows. It rises in Northern Georgia, in the Blue Ridge, flows southwest and then south, and after its junction with the Flint River receives the name of Apalachicola. For two-fifths of its entire course of 500 miles it is navigable for large steamboats, and Columbus, Ga., at the head of navigation, receives much of its vast supply of cotton and sends out many of its textiles, cottonseed products and machines over its waters. It also furnishes water power to the city, by reason of its descent of 120 feet in three miles. One of the most famous poems of Sidney Lanier is the *Song of the Chattahoochee,* which sings of the course of the river—

Out of the hills of Habersham,
Down the valleys of Hall,

and on to the place where

The dry fields burn, and the mills are to turn,
And a myriad flowers mortally yearn.

CHATTANOOGA, *chat a noo'ga,* Tenn., the county seat of Hamilton County, is an important railroad center and a rapidly-developing industrial city in the southeastern part of the state, near the Georgia line. The city is 150 miles southeast of Nashville and 140 miles northwest of Atlanta. It is on the Tennessee River, navigable to this point for eight months of the year, and on the Alabama Great Southern; the Central of Georgia; the Queen & Crescent; the Nashville, Chattanooga & Saint Louis; the Southern; the Tennessee, Alabama & Georgia, and the Western & Atlantic railroads. The rapid increase in population is noted in the census reports; in 1900 it had 30,154; in 1910, 44,604; and in 1915, 48,476. The area is six and a half square miles, and close about the city are a dozen populous suburbs.

Chattanooga is beautifully situated on the south bank of the winding Tennessee River, in a great natural amphitheater surrounded by historic and picturesque hills. The various points of interest are made accessible by boulevards and electric lines. Southwest of the city is Lookout Mountain, its summit rising 2,126 feet above sea level and commanding a magnificent view into seven states. East and south is Missionary Ridge; a short distance southeast in Georgia is the field of Chickamauga, now a national military park fifteen square miles in extent. Throughout the grounds monuments and historical tablets have been erected by the various states in honor of their soldier dead. Fort Oglethorpe, a brigade post of several thousand acres, adjoins the park; a regiment of cavalry is stationed here. During the Spanish-American War in 1898 Fort Oglethorpe was a mobilization camp; 60-000 soldiers were encamped at one time on the Chickamauga battlefield. Immediately southeast of the city is one of the largest national cemeteries, containing 13,322 graves. Warner, Boynton, Montague, East Lake and Houston parks cover 150 acres. Signal Mountain, north of the city, is a popular pleasure resort.

Buildings and Institutions. Prominent buildings of the city are the Federal building, erected at a cost of $350,000; the city hall, costing $200,000; the county courthouse, which cost $500,000; a Carnegie Library, armory, museum, Y. M. C. A. and Y. W. C. A. buildings, the Masonic Temple and the Terminal Station. Among a number of noteworthy bank, hotel and church buildings are the Temple Court, James and Hamilton National banks, the Patten hotel and the First Presbyterian church.

Educational institutions are the University of Chattanooga (Methodist Episcopal), the Chattanooga College of Law, Baylor School, McCallie School and the Girls' Preparatory School. There are also a number of private schools, a musical conservatory, the Baroness Erlanger Hospital, and the Vine Street Orphans' Home.

Industry. The geographical location of Chattanooga, a natural gateway between the hills, has made the city an important railway center. It is the southern headquarters of the Interstate Commerce Commission. Water commerce consists of cotton, grain, coal, iron ore and manufactured products. Chattanooga is, however, of greatest importance as a manufacturing center. Power for more than 280 industries is furnished by a dam and power plant at Hale's Bar on the Tennessee River, completed in 1913 at a cost of $9,000,000, and several similar projects on the Oconee River. The city has more than $50,000,000 invested in manufactories, 12,000 men are employed, and the value of the annual product exceeds $65,000,000. Among many large industrial plants are manufactories of iron and steel (more pig iron is smelted here than in any other Southern city), textiles, wood-working products, leather goods, tanning extracts, fertilizers, machinery, stoves, medicines, bricks and tiles.

History. In the early days, river voyagers landed here to avoid the rapids of the Tennessee River, and the locality, settled about 1835, was known as Ross's Landing. Ross was the name of a Cherokee chief whose people were moved West by the government in 1838. In 1851 the town was incorporated as Chattanooga and became a city in 1866. During the War of Secession it was one of the most important strategic points in the Confederacy and the struggle for its possession led to some of the severest battles of the war (see CHICKAMAUGA, BATTLE OF). In the course of the fighting the city was almost destroyed. Immediately succeeding the war the manufacturing of iron was begun, to restore the ruined railroads; a long period of development and prosperity followed, and the city has become one of the important industrial centers of the South. In 1911 the commission form of government was adopted. H.F.W.

CHATTANOOGA, BATTLE OF, a decisive battle of the War of Secession, which in reality consisted of three engagements, all fought in the vicinity of Chattanooga, Tenn., on November 23 to 25, 1863. General Bragg, with a Confederate army of 40,000, had defeated the Federals at Chickamauga, and had then encamped before Chattanooga, his lines extending from Lookout Mountain along Missionary Ridge for about twelve miles. The Federal General Rosecrans had been succeeded by Thomas, and Grant had just been placed in supreme command of the army of 60,000 in that region. A campaign all along the line was mapped out; Sherman was instructed to attack the Confed-

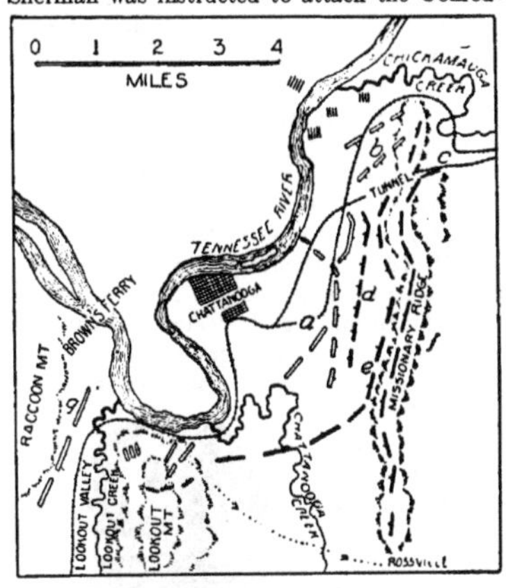

MAP OF BATTLEFIELD

erates on the extreme right and to advance along Missionary Ridge toward their center; Thomas was to attack in the center, and Hooker was to strike at their extreme left and drive them from Lookout Mountain.

Thomas, on the twenty-third, met with the first success, driving back Bragg's advance guards, and on the next day Sherman was successful until stopped by a strongly fortified gap in the mountain ridge. The most famous part of the engagement, however, was Hooker's attack on Lookout Mountain—the "Battle above the Clouds"—by which the Confederates were dislodged from their position there. On the morning of the twenty-fifth Thomas's troops, ordered to make an attack on the Confederate center, exceeded their commands by rushing up Missionary Ridge under constant and heavy fire and driving the Confederates in confusion from the summit. This remarkable and spectacular feat ended the battle. The Confederates lost in killed and wounded 6,687, the Federals, 5,815.

CHATTEL, *chat''l,* a term closely akin to the word *capital,* used in law to mean almost the same thing as the phrase *personal property* (which see). There is, however, a slight difference technically, chattels being only such personal property as can be physically delivered. Thus money in the hand is a chattel, but a claim for money due is not.

Chattels may be *personal* or *real*, the former being all such movable articles as furniture, money or clothes. A *chattel real*, on the other hand, is any interest in land less than actual ownership, as a lease or a mortgage. Growing crops also come under this title. The term *goods* is narrower than *chattel*, meaning practically the same as *chattels personal*, and the commonly used expression *goods and chattels* is thus a mere repetition for emphasis, as the first word adds nothing to the meaning. See MORTGAGE, subhead *Chattel Mortgage*.

CHAUCER, *chaw'sur,* GEOFFREY (about 1340-1400), the first great poet of England, known as the "Father of English Poetry." Writers of verse there had been before him, and some of these rose above the average low rank, but he was the first to show that poetry, masterly in technique as well as in content, could be written in the shifting, developing English language. The changes which have taken place in the language since his day have prevented his wide popularity, for it is not now possible to read his works without considerable study; but those who do give them attention find themselves well repaid. Critics do not hesitate to give Chaucer a rank among English poets secondary only to that of Shakespeare and Milton.

GEOFFREY CHAUCER

Little is known of the boyhood or education of Chaucer, but it is certain that during the English invasion of France in 1359-1360 he was imprisoned, was ransomed by the king and became a squire in the royal service. That he was an efficient servant is shown by the fact that he was sent on several important missions to France and to Italy, and the effect of these journeys is evident in his works, which show strong traces of French and Italian influence. In 1374 he was made comptroller of customs for London, and in 1386 was elected to Parliament. At times during the latter part of his life, when the political party to which he belonged was not in power, he was very poor, and not until a year before his death was he given permanent financial relief by the king. Chaucer was a diplomat, a business man and a poet of high rank; that he could be all these shows that he must have been a man of the greatest range of ability.

His first works were translations or at least adaptations from the French, but later the Italian writers became his models, and under their domination he produced such poems as *Troylus and Cryseyde*, one of his most beautiful works; *Legende of Good Women*, *Palamon and Arcite* and *The Parlement of Foules*. In his third and greatest period he was thoroughly English in his theme and in his treatment of it, though the plan of his greatest work, the *Canterbury Tales* (which see), was one which had been used before in Italy by Boccaccio. The dramatic ability shown in his descriptions of characters in this remarkable work has led many to speculate as to what Chaucer might have become in an age when the drama was the chief form of literature; but he lived in a story-loving age, and the ability to tell a story was perhaps the greatest of all his gifts. A.MC C.

Consult MacKaye's *Complete Poetical Works of Chaucer: Now First Put into Modern English;* Hammond's *Chaucer: A Bibliographical Manual.*

CHAUDIÈRE, *sho dyair',* a river of the province of Quebec, Canada, famed for its beautiful falls, which attract many visitors. Its steep, rocky banks and the many little wooded islands which obstruct its channel are most picturesque. The Chaudière has its source in a number of small streams which flow into Lake Megantic near the border of Maine and only a few miles from the source of the Kennebec. Issuing from Lake Megantic, the Chaudière flows northward and then northwestward in a wide curve, and after a course of 120 miles empties into the Saint Lawrence about seven miles above the city of Quebec. The falls, which are two and one-half miles from its mouth, make it of little value for navigation.

CHAUFFEUR, *sho'fer,* from a French word signifying *to make hot,* was in its original meaning used to denote a stoker of a steamship or a locomotive. As automobiles became common the term was adopted not only in France but also in English-speaking countries to refer to the man who looked after the fuel and the machinery. To-day the word has become well established in English to signify the professional driver of any motor vehicle. In the United States and Canada chauffeurs receive wages ranging from $50 to $100 per month. The hours of service may be long every day, for a chauffeur must be at the call of his employer day and night.

CHAUTAUQUA, *sha tawk' wa,* a name given to a remarkable system of popular education, which is the evolution of a Sunday-school assembly held at Chautauqua Lake, New York, in the summer of 1874, for the instruction of

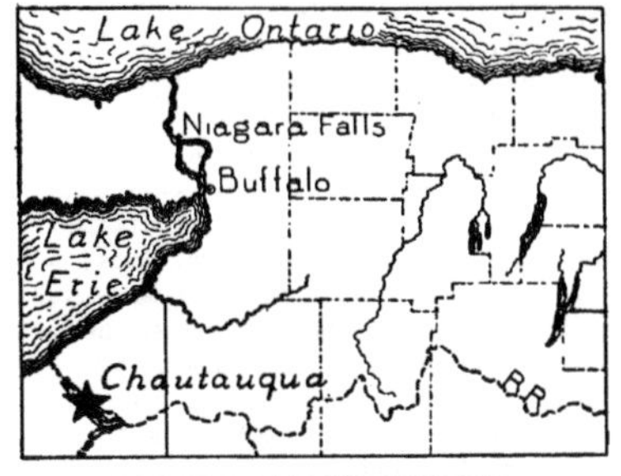

LOCATION OF CHAUTAUQUA

Sunday-school teachers. The movement was popular from the first, and has increased from year to year in scope until it has grown to large proportions. It now has fifteen departments in its summer schools, and an assembly attended by 40,000 to 50,000 persons annually; there is also a home reading circle with thousands of members, and it has property on Chautauqua Lake worth $1,250,000, with over 600 cottages and public buildings for its summer population.

Chautauqua Institution. The plan of the founders of the movement—Lewis Miller of Akron, O., and Rev. (afterwards Bishop) John H. Vincent—was for religious instruction only, but the scope of the work soon broadened until it aimed at an education that should be at once intellectual, ethical and spiritual. In 1879 a group of schools was established with graded courses of study covering four years, in which literature, art, history and pedagogy were taught. From 1883 to 1893 Dr. William Rainey Harper, president of the University of Chicago from 1891 to 1906, gave his summers to Chautauqua work, and during this time a complete system of summer schools was established. This system now includes fifteen distinct schools, embracing courses in English, European and ancient literature, history, pedagogy and nearly all the arts and sciences. George Vincent, president of the University of Minnesota, son of Bishop Vincent, became president in 1907, with about 200 leading educators as instructors. In 1915 his title was changed to chancellor, Arthur E. Bestor becoming president. Bishop Vincent became chancellor emeritus. The sessions of the schools are held during the months of July and August. An important feature of the Chautauqua movement has always been the popular exercises of the summer assembly. They consist of talks on interesting topics, lectures by noted speakers from all over the world, concerts and various recreations. These are all free to Chautauqua visitors.

Chautauqua Literary and Scientific Circle, the name given to the home reading course, is the best-known branch of the Chautauqua work. Each course consists of four years of reading, known as American, English, European and Classical years, and includes history, art, travel, literature and science. The work of each year is complete in itself, and each member of the Circle reads the same books. In addition, there are eighty-eight courses for those who wish to specialize. The books used are specially prepared for the courses, with required readings in *The Independent,* the present current events element of the course, which in 1914 absorbed *The Chautauquan,* the former official organ. There are also a monthly bulletin, *The Round Table,* and a membership book of hints and helps for home study. Diplomas are awarded for completing the four-years' reading course. For some years the plan included teaching by correspondence, but this has been discontinued.

The Chautauqua Literary and Scientific Circle was organized in 1878 with the idea that it would meet a recognized want with persons who had been denied a liberal education, and would appeal to old and young alike. It proved amazingly popular, 7,000 enrolling the first year. Over 300,000 have joined the Circle; 60,000 of them have completed at least one course. The idea pleased the English so well that they have patterned their British Home Reading Union after it. Branches of the Chautauqua system have been established in Japan and South Africa. The work has also proved as popular in Canada as in the United States.

Local "Chautauquas." The idea of the Chautauqua Assembly spread through America, and local "Chautauquas" sprang up everywhere. These assemblies employ popular lecturers and other entertainers and hold sessions of several days, which are largely attended. There are 4,000 of these "Chautauquas." A.B.

CHAUVEAU, *sho vo',* PIERRE JOSEPH OLIVIER (1820-1890), a Canadian statesman, educator and man of letters, one of the most talented French-Canadians of his generation, at one time premier of Quebec but now remembered chiefly for numerous graceful poems from his pen. He was a barrister by profession, but as a young man was best known for the poems and clever letters on political and social topics which he contributed to newspapers. From 1844 to 1855 he was a member of the assembly, and during the last three and a half years of this period was a member of the Hincks-Morin ministry, first as solicitor-general for Lower Canada and later as provincial secretary. Then for twelve years he was chief superintendent of education for Lower Canada, a position in which he wielded great influence. In August, 1867, a month after Confederation, he was called on to form a ministry, and until 1873 was premier of Quebec. For a year he was then Speaker of the Dominion Senate, but he resigned from that body in January, 1874. In spite of this varied activity, he found considerable time for writing. Besides his occasional poems his most noteworthy book is a novel entitled *Charles Guérin,* a clever tale which was very popular in its day.

CHECK, or **CHEQUE,** a simple order written by anyone who has money deposited in a bank, instructing the bank to pay a specified sum to a person named or to the bearer. Once a California lumberman was buying a section of timber land; in order not to lose his option he was obliged to make a hurried payment, and, picking up a shingle, he wrote on it, *Blank National Bank, San Francisco, pay* J. *H. Sullivan ten thousand dollars,* then added his name and the date. This order was just as much a check as though it had been given on the printed forms of his bank, and as such the bank treated it.

Very nearly all the business of the United States and Canada is carried on by means of checks, though in other countries they are less popular. The checking system, with its assistant, the clearing house, makes it possible for a community to transact many times as much business with a given amount of currency as it otherwise could carry on. Thus in New York City checks to the value of several hundred million dollars are drawn daily, yet only a score of millions in actual cash changes hands.

A Checking Account. When you deposit money in a checking account you are required to sign your name on a card, which the bank preserves so that if any other person attempts to get money by writing your name the forgery may be detected. The teller gives you a *bank book* and a *check book.* The first shows the amount of your deposit, and every time you add money to your account you take the book with you so the teller may enter the figures; in small banks once a month you give it to the bookkeeper to record the money the bank has paid out for you, but this system is being superseded by a monthly "statement" from the bank. The check book contains blank checks; when you wish to pay out money you fill one of them as shown in the illustration.

For each check there is a stub on which is space for a memorandum of the particulars of the check and for addition and subtraction of amounts deposited and withdrawn. It is a wise plan to number your checks and their stubs, so that at the end of the month, when the bank returns to you your paid checks, you can quickly discover which ones are still unpaid. The balance shown in your bank book should be greater than that shown in your check book by the sum of the checks outstanding; if it is not, either you or the bookkeeper has made an error. If you wish to give a check when you do not have your check book with you, it is permissible to take another bank's blank, cross out the name and substitute that of your own bank, but this is poor policy because you may forget to record the amount in your check book.

Your signature on a check must always be like the copy given the bank; for instance, if the latter is J*ohn A. Low* you must not sign J. *A. Low.*

Checks in the United States generally carry the words *Pay to the* o*rder of*..............; in Canada the form is *Pay to* .... *or Bearer.* In the latter case it is customary to run the pen through the word *Bearer* and substitute *Order,* for otherwise if the check is lost anyone who finds it may cash it. This word *order* means that by endorsing (see below) the check the owner may order the money paid to another. If you yourself wish to draw cash from your account, you may write a check payable to *Self* or to *Cash.*

An advisable step is to write the purpose of each check on its face. Thus if you are paying a bill you may write *In full of account to date,* and when your creditor endorses and cashes the check it becomes a receipt. A check does not, however, constitute a payment until the bank *honors* it, that is, pays out the money for which it calls; so it is never wise to give a receipt for an account paid by check unless the manner of its payment is stated on the receipt.

Endorsing a Check. If you receive a check you may get cash for it, transfer it to another person, or deposit it to the credit of your account in the bank. If you are cashing it you merely sign your name on the back, across the left end. This form is called *endorsement in blank,* and you should not execute it until you reach the bank, for if you lose the endorsed check anyone can present it for payment. If you are making the check over to another person the correct form of endorsement is *Pay to the order of (name),* followed immediately by your signature. This is called *endorsement in full;* it obliges the man to whom you transfer it to add his endorsement, thus admitting that he has received the value named, whereas a check endorsed in blank may go through a dozen hands and receive no signatures except that of the last possessor.

Your signature to an endorsement should read exactly like your name as written on the face of the check by the drawer, even if he has misspelled it, but in the latter event you must write your correct bank signature immediately beneath the other. When you endorse a check you become responsible for its payment if it proves to be worthless, so too much care cannot be exercised.

Worthless Checks. Many people are careless about keeping account of the checks they give out, and occasionally issue one for more money than they have in the bank. If you receive and dispose of a check for ten dollars and the drawer has only nine dollars and ninety cents in the bank, the check will be returned to you marked *Insufficient Funds,* or in Canada, *N. S. F.,* which stands for *Not Sufficient Funds.* Since a bank honors checks in the order in which they are presented, not the order in which they are made out, it is always wise to dispose of a check the same day it is received.

If a check is returned to you for insufficient or no funds and there are endorsements on it above yours, you must *protest* the check at once if you wish to hold the endorsers responsible. Protesting consists in giving a formal, legal, sworn notice of non-payment. A *post-dated* check, one issued before the date it bears, is not due until that date and cannot be protested till the next day.

Certified Checks. If someone has given you a check and you doubt its worth, it is a good plan to take it to his bank and have it marked *certified* by the bookkeeper before depositing it in your own bank. The certification makes the bank responsible for payment. Sometimes you may wish to have a check of your own certified, especially if you are sending it some distance. A certified check should not be confused with a cashier's check, which is a bank's own order to pay.

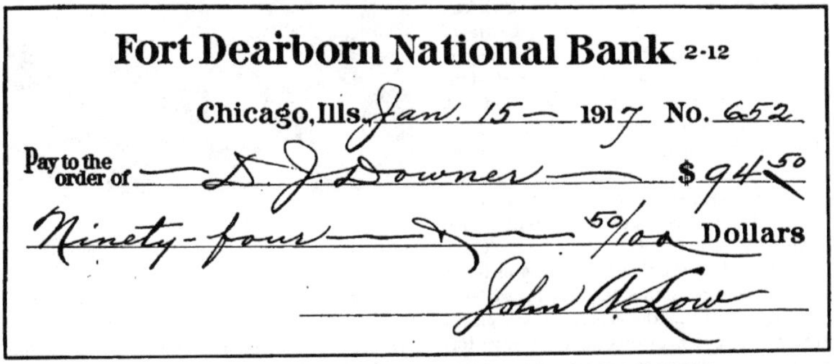

Fort Dearborn National Bank 2-12

Chicago, Ills. Jan. 15 1917 No. 652

Pay to the order of D. J. Downer $94 50/100

Ninety-four 50/100 Dollars

John A. Low

THERE IS SLIGHT CHANCE TO ALTER A CHECK THUS MADE

Exchange. In Canada it is customary for a bank to charge a fee for accepting a check payable in another town, even if it is drawn on a branch of the same bank. This fee varies according to the amount of the check and the influence of the person presenting it. When sending a check to a person in another place you should add the presumable amount of the exchange to the amount you are paying. Never write *Forty Dollars and exchange,* for a check must bear a definite amount. In the United States since 1916 no exchange is charged by banks which are members of the Federal reserve system.

Stopping Payment. If after you have given out a check you wish for any reason to prevent its payment you may do so by giving written notice to the bank and releasing the bank from responsibility for error. This is the proper course to follow with a lost check; if a second one is then issued it should be plainly marked *Duplicate,* in red ink.

Protecting a Check. If a forged check is cashed the bank is the loser, but the loss on a *raised* check (one on which the amount has been fraudulently increased), even though the signature be genuine, must be met by its maker. It is therefore wise to use extreme care, in making out a check, to leave no blank spaces in or after the statement of amount. A good form to follow is shown in the illustration. There are a number of patent "protectors" on the market, with which the amount may be indelibly indicated. C.H.H.

CHECKERS, *check'erz,* or **DRAUGHTS,** *drafts,* a game for young or old, a battle of wooden men on a cardboard field, the players being the generals. The board has sixty-four alternating black and white, or black and red, squares, either black or white squares being used as the "line of march," or the spots upon which the "men" move. Each player is given a set of twelve men, small round pieces of wood or bone. The two sets are of different colors, or "uniforms," usually black and white. These men are placed on the first three rows of black or white squares on each side of the board, leaving two open rows in the center. Each player in turn moves one man diagonally, always forward, to the next square, except when "jump-

BOARD SET FOR PLAY

ing," as explained below. In the remainder of this article it is assumed that the men are placed on white squares.

The object of the game is to move the men of one side so skilfully that the progress of the other men is blocked, or to capture all of the enemy's men. If a man is moved next to an enemy's man and an open space is left behind him, the opposing man may jump over to the next open square and so capture a man

A TEST PLAY

In the illustration a game has nearly reached its close. The white plays next and should win the game.

and get farther into the enemy's lines. More than one may be captured at a time if there are alternate men and open spaces in a forward line. If a man of one side gets across the board to the rear line of squares of the other side, he is *crowned,* or made a *king.* That is, the enemy gives up one of the men he has captured and puts him on top of the man to be *kinged.* A king may move either backward or forward, but always on the white squares, one square at a time except when making a capture, so he has the advantage over all other men. When one side has captured all the men of the other side the game is won, or if a blockade is caused on the board where all the men of one side are hemmed in by the other and any move means capture, the game is won.

There are other rules in the game of checkers, which some players observe and some do not. For instance, if one side fails to capture a man of the enemy, either through oversight or because it would place his man in a dangerous position, the opponent may compel him to capture the man, or may remove the delinquent soldier from the board, and then has the privilege of the next move.

Checkers is a very ancient game, known by the Egyptians, Greeks and Romans. It was played in Europe in the sixteenth century. An old form of checkers is known in China as "the game of circumvention." M.S.

CHEESE, *cheez,* an important food, made chiefly from the "curds" of milk, the product of a flourishing industry in every grazing and agricultural section.

Process. The simplest variety is the cottage cheese, or Dutch cheese, made by many housewives. To make this, the milk is allowed to curdle and is then subjected to a very gentle heat, as great heat toughens the curd. The whey is then drained off, the curd is salted, and if desired, cream is mixed with it. By far the larger part of the commercial cheese is made in factories, and the process, though differing in details, is practically the same in its essentials for the various kinds. This process includes curdling the milk with acid or rennet, separating the whey from the curd, grinding and salting the curd and packing it in molds of various sizes and shapes. These are then subjected to pressure, that all the whey may be forced out. Sometimes all the butter fat of the milk is left in it; in this case the cheese is known as full-cream; sometimes but a part is left, and half-skim cheese results. Full-skim cheese, which contains no butter fat, is in general hard, tasteless and horny, and in some places its manufacture is forbidden. All except the cottage variety are the better for being ripened, that is, kept for several months in a cool place.

Kinds. Some kinds of cheese are hard, some are soft, according to the method of ripening or the amount of water which is allowed to remain in them. Certain kinds, chiefly made in Europe, are famous and in great demand. These include *Roquefort,* a soft cheese which has been allowed to ripen until a harmless blue mold has formed through it; *Edam,* a hard, yellow cheese sold in red-painted balls; *Parmesan* and *Gorgonzola,* hard cheese; *Swiss,* a hard cheese which is somewhat porous and filled with Swiss "eyes," and *Neuchâtel, Camembert* and *Limburger,* all soft cheese. The United States makes mostly Cheddar cheese, commonly known as American cream cheese, nine-tenths of its huge product being of that variety.

Amount Produced. Canada and the United States are among the greatest cheese-producing countries in the world. In Canada the standard has been kept high by the passage of laws forbidding the sale of skim-milk cheese, and the result has been great popularity for Canadian varieties in other countries. Almost 200,000,000 pounds are exported every year, Ontario alone exporting more than the entire United States sends abroad.

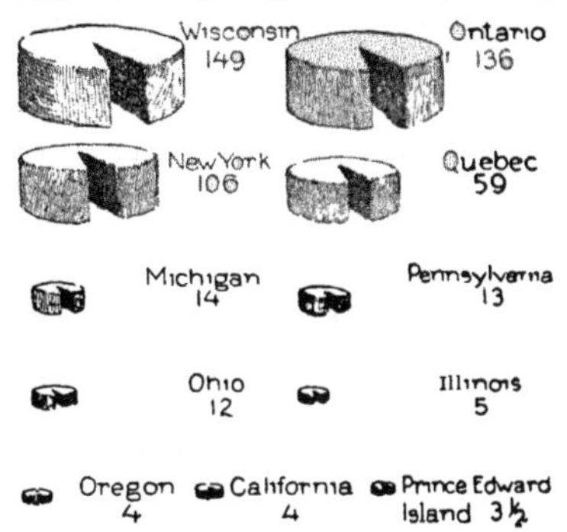

WHERE CHEESE IS MADE

An average production of five years is represented in the totals.

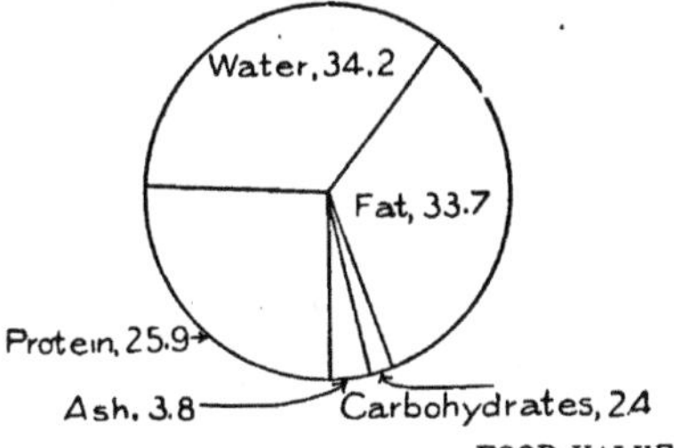

FOOD VALUE OF CHEESE

At left, full-cream cheese; at right, cottage cheese.

In the United States the production of cheese, including that made on the farms, amounts to about 300,000,000 pounds a year. About one-half of this is now produced in Wisconsin. The exports, approximately 17,000,000 pounds, are greatly overbalanced by the 30,000,000 or 40,000,000 pounds imported from Europe in normal years. Many foreign brands of cheese are now made with considerable success in the United States.

Food Value. Cheese long had the name of being a very indigestible substance, and later a saying gained currency to the effect that "cheese digests everything but itself." But experiments have proved conclusively that by most people cheese is easily digested. Occasionally there is a person who cannot eat it, but there is scarcely a food, however wholesome, of which the same may not be said. There are also highly nutritive qualities in cheese, which contains a large percentage of tissue-building and of energy-forming substances. As a heat producer, cream cheese ranks high, and should therefore be used more in winter than in summer. Its fuel value averages 2,000 calories per pound, or almost three times that of an equal weight of eggs. Just because of this high fuel value, cheese should not be eaten in large quantities. See CALORIE; FOOD, subhead *Chemistry of Food.* E.H.F.

Consult reports of state or provincial Agricultural Experiment Stations (or Farm), also Yearbook of U. S. Department of Agriculture.

CHEETA, CHEETAH, *chee'ta,* or **HUNTING LEOPARD,** *lep'ard,* a large cat of the African jungles, three or four feet high, and

THE CHEETA

about the length of a leopard. Its limbs are so slender and its body so long that it is the quickest animal known for running short distances. Because of this fact it chases its prey, and does not crouch and steal upon it, like most of the cats. Tawny-colored, black-spotted, excepting on the throat, the skin of the cheeta is valued for wearing apparel by the chiefs of African tribes.

The cheeta is also well known in India, where it is tamed and trained for hunting. Like a falcon, it is held in leash and kept blindfolded until the game is seen. Then it is loosed and it makes a quick dash for the animal, which it holds down until the hunters come. The Crusaders introduced the cheeta into Europe for this purpose in the fourteenth century, but it is no longer so employed.

CHELSEA, *chel'se,* MASS., in Suffolk County, is a residential suburb of Boston, three miles northeast of the city and connected with it by the Boston & Maine Railroad, electric lines and steam ferries. The Mystic River, which separates Chelsea and Charlestown, a part of Boston, is crossed by a long bridge. The area of the city is two square miles. In 1910 the population was 32,452; in 1915 it was 43,426, including a large element of Jews, Irish and Armenians.

Chelsea has a Federal building, erected in 1908 at a cost of $125,000, a courthouse, city hall, state armory, Carnegie Library, United States marine and naval hospitals, a soldiers' home and Ye Old Pratt House, a Revolutionary tavern. Although principally a residential city, it has important manufactures of rubber and elastic goods, foundry and machine-shop products, stoves and furnaces, tiles and pottery, mucilage and paste, shoes, woolens, brass goods, wireless apparatus, lithographs, etc.

The city was settled in 1626 as Winnisimmet. A part of Boston from 1634 to 1638, it was then incorporated as the town of Chelsea, and became a city in 1857. The city suffered a $17,000,000 property loss by fire in 1908.

CHELTENHAM, *chel't'n am,* an English watering place, popularly known as "Asia Minor" because of the numbers of returned Anglo-Indians whom its mineral springs attract. It is in Gloucestershire, eight miles northeast of Gloucester and 109 miles west by north of London, and had in 1911 a population of 50,035. Its waters, efficacious in diseases of the stomach and the liver, are its chief but not its only claim to fame, for its educational institutions, from grammar school to college, have more than a local reputation. Cheltenham, despite its age of more than eleven centuries, is distinctly a modern town, well laid out and adorned with beautiful thoroughfares.

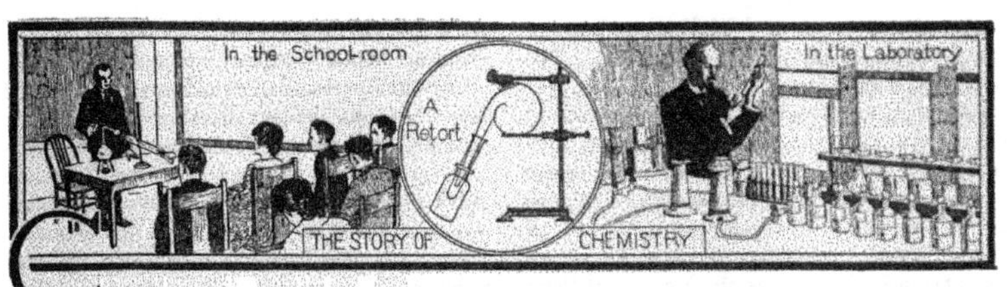

CHEMISTRY, *kem' is tri,* one of the most wonderful of the sciences, which deals not with the appearance or the value of matter, but with its composition. That is, it seeks to discover just what every substance in the world, whether seen or unseen, is made of, and what are the relations of the component parts to each other. Perhaps the briefest definition which can be given of it is that it is "the science of the composition of substances." It is chemistry that has made clear the wonderful fact that the flashing diamond, the gritty, black charcoal and the soft, lead-like graphite of which pencils are made are all composed of one substance—carbon; it is chemistry that has proved that oxygen and ozone are simply two forms of the same substance, and that the rusting of iron is essentially the same kind of process as the burning of wood.

Importance in Every-Day Life. Chemistry is not simply an attempt to reduce matter to first principles, on the part of men who cannot be content to take things just as they seem. It is of the utmost practical importance, and has in recent years worked wonders in connection with many of the commonest pursuits. Even before they were thoroughly understood, many of the principles of chemistry were constantly used in such processes as dyeing, soap-making, glass and pottery manufacture, but their use was not scientifically worked out. At present, however, most great manufacturing concerns have their trained chemists whose business it is to watch constantly processes and products, both their own and, so far as possible, their competitors', in the effort to discover new useful substances, new uses for substances already known, improved methods and means of decreasing expense. Soap factories, steel plants, mills, mines and packing houses all have their laboratories, which are considered as essential as the business offices. The insistence in recent years upon purity in all food products has given to chemistry a new commercial importance, for nothing less than minute analysis can unfailingly determine adulteration.

Growth of the Science. It might seem as though, in the development of sciences, chemistry would have been the very last one to appear, for much of that with which it deals cannot be handled and is invisible, and could never force itself upon the attention of anyone. For instance, water has always been one of the central substances about which man's life has grown up, and man has therefore needed to have considerable knowledge of water. But if he knew where it was to be found; that it would quench thirst, put out fires and help all living things to grow; that it would not run uphill unless forced, and had a tendency to "seek its own level," he had enough practical facts to live by. What mattered it to him whether water was an individual substance or a compound of other substances?

Alchemy. But there was one substance in which, by reason of their greed, men early became especially interested. That was gold. If they could just find out how gold was made, they could have plenty of the precious metal without all the labor and expense of mining it. And thus many centuries ago men began to study into the composition of substances that they might find something which would turn less valuable metals into gold. This study became known as *alchemy,* probably from *Chemia,* an old name of Egypt, the country where the study first grew up. See ALCHEMY.

Beginnings of a True Science. Needless to say, these alchemists, or philosophers, as they called themselves, never succeeded in making gold, but they did something quite as valuable, in leading the way to the science of chemistry: In their experiments they inevitably discovered many things for which they were not looking—properties of matter, new substances and new ways of making old ones, and above all, the healing properties of drugs. Medicine in its modern sense grew up side by side with chemistry.

Strange theories were formed from the half-known facts as they emerged, and one of these theories, common in the early years of the sixteenth century, concerned itself with the relation between medicine and chemistry. The body, said these early chemists, is made up of various chemicals—and then each proceeded to make for himself a list of these body substances. If one of the chemicals was present in excess, they argued, disease was certain to result, and many illnesses were labeled as growing out of too much or too little of some one substance. Paracelsus, the greatest of these doctor-chemists, really effected many cures and made discoveries that are of the utmost value to modern pharmacy. But men knew too little of anatomy and physiology, as well as of chemistry itself, to carry this really helpful phase of the science very far.

Finally there arose men who realized that if this study of substances and their composition was to become a real science it must be carried on for its own sake and not by reason of its relation to gold-making or to healing. From that time on, progress was comparatively rapid. Robert Boyle, for instance, who lived in the seventeenth century, held as theory much that has now been proven fact. Thus he knew much about the doctrine of elements, so fundamental to the science, and announced the difference between a chemical compound and a mixture (see below, *Chemical Compounds*). Other men who later added greatly to the growth of the science, and whose work is discussed under their names in these volumes, were Bunsen, Sir Humphry Davy and Remsen. Other modern chemists have added vastly to the stores of knowledge left by these men, but they have altered little the main principles laid down by them. Thus the central doctrines of *elements*, substances which cannot be decomposed, have never been disturbed.

Elements. When an electric current is passed through water, two gases, oxygen and hydrogen, are obtained, which are substances entirely different from water. If we try by any means known to science to-day to obtain either from oxygen or from hydrogen any simpler substance we will not succeed. Neither will we succeed if we try to decompose iron, or gold, or carbon. Such substances which cannot be decomposed into other simpler substances or cannot be transformed into one another are named *elements*. Elements are substances whose molecules contain only one kind of *atoms*. Oxygen contains nothing else but oxygen; gold contains nothing but gold. Up to the present chemists have found that there are in nature eighty-three elements. All the thousands and thousands of other materials that are in the world are either chemical *compounds* of these eighty-three elements, or mixtures made up of more than one element or compound. Air, for instance, is a mixture of the two elements oxygen and nitrogen; salt is a chemical compound of the elements sodium and chlorine; and milk is a mixture of several chemical compounds.

The following list contains the name of each element, its symbol (see subhead below, *Chemical Symbols*), and its atomic weight:

Name	Symbol	Atomic Weight
Aluminium	Al	27.1
Antimony	Sb	120.2
Argon	A	39.88
Arsenic	As	74.96
Barium	Ba	137.37
Bismuth	Bi	208.0
Boron	B	11.0
Bromine	Br	79.92
Cadmium	Cd	112.4
Caesium	Cs	132.81
Calcium	Ca	40.07
Carbon	C	12.005
Cerium	Ce	140.25
Chlorine	Cl	35.46
Chromium	Cr	52.0
Cobalt	Co	58.97
Columbium	Cb	93.5
Copper	Cu	63.57
Dysprosium	Dy	162.5
Erbium	Er	167.7
Europium	Eu	152.0
Fluorine	F	19.0
Gadolinium	Gd	157.3
Gallium	Ga	69.9
Germanium	Ge	72.5
Glucinum	Gl	9.1
Gold	Au	197.2
Helium	He	4.0
Holmium	Ho	163.5
Hydrogen	H	1.008
Indium	In	114.8
Iodine	I	126.92
Iridium	Ir	193.1
Iron	Fe	55.84
Krypton	Kr	82.92
Lanthanum	La	139.0
Lead	Pb	207.2
Lithium	Li	6.94
Lutecium	Lu	175.0
Magnesium	Mg	24.32
Manganese	Mn	54.93
Mercury	Hg	200.6
Molybdenum	Mo	96.0
Neodymium	Nd	144.3
Neon	Ne	20.2
Nickel	Ni	58.68
Niton	Nt	222.4
Nitrogen	N	14.01
Osmium	Os	190.9
Oxygen	O	16.0
Palladium	Pd	106.7
Phosphorus	P	31.04
Platinum	Pt	195.2
Potassium	K	39.1

Name	Symbol	Atomic Weight
Praseodymium	Pr	140.9
Radium	Ra	226.0
Rhodium	Rh	102.9
Rubidium	Rb	85.45
Ruthenium	Ru	101.7
Samarium	Sa	150.4
Scandium	Sc	44.1
Selenium	Se	79.2
Silicon	Si	28.3
Silver	Ag	107.88
Sodium	Na	23.0
Strontium	Sr	87.63
Sulphur	S	32.06
Tantalum	Ta	181.5
Tellurium	Te	127.5
Terbium	Tb	159.2
Thallium	Tl	204.0
Thorium	Th	232.4
Thulium	Tm	168.5
Tin	Sn	118.7
Titanium	Ti	48.1
Tungsten	W	184.0
Uranium	U	238.2
Vanadium	V	51.2
Xenon	Xe	130.2
Ytterbium	Yb	173.5
Yttrium	Y	88.7
Zinc	Zn	65.37
Zirconium	Zr	90.6

Only one-half of the elements in this list are found in abundance in nature; the others are rare. Thus, the earth's crust in the main is composed of the elements oxygen, silicon, aluminium, iron, calcium, magnesium, potassium and sodium. Over ninety-eight per cent of the weight of the earth's crust is composed of the above-mentioned elements. The chief elements of the atmosphere are, as we have stated, oxygen and nitrogen; those of water are oxygen and hydrogen.

Theories about the Elements. We have defined elements as substances which are not related to each other and which cannot be transformed into one another. But let us state here that since men began to think about the "nature of things," as the Latin poet Lucretius expresses it, they have always been inclined to believe in the unity of matter. That is, they were inclined to think that all the substances found in nature consist essentially and ultimately of one and the same substance, and that all of them are made of a single primordial matter. The ancient Greek philosophers believed that the universe and all that is in it were formed by the combination of four substances, namely, fire, air, water and earth. They considered these as elements. The belief in the unity of matter and in the transmutability of the elements into one another forms the basis of the teachings of alchemy.

Modern chemistry is based, of course, on other principles, and uses positive methods in research and investigation. This explains the wonderful advance this science has made during the last hundred years; but the mind of man will always find a special fascination in trying to speculate about the origin of things or the ultimate cause of things. There are yet to-day many eminent men of science who believe that all the elements are composed of the same matter.

The Electron Theory of Atomic Structure. This view of the ultimate unity of the elements will be strengthened if the latest theory about the structure of atoms proves to be correct. According to this theory the atoms of all substances are made up of a great number of *electrons* (see ELECTRON). The elements vary from each other because their atoms are composed of different numbers of electrons and because these are grouped together in a number of different ways. We know that the newly-discovered element called *radium* has the power of emitting continuously streams of free electrons. We also know that radium is changing into the element *helium,* which is an example of the actual transformation of one element into another. But before any definite conclusions can be reached on this as well as on other similar subjects a great amount of scientific work will have to be done.

It is worth mentioning here that the spectroscope (which see) has shown that we find in the sun and in the stars the same elements as on earth, which indicates that the highest temperatures existing anywhere in nature are not capable of decomposing the elements.

Chemical Compounds. The subject of compounds in chemistry is a very interesting one, for a chemical compound is a very different thing from many of the substances which are usually thought of as compounds. For instance, one who eats a piece of cake can say at once, "There is sugar in this cake; there is butter; there are flour and vanilla; there are eggs." It is one body—a piece of cake; but it is also very evidently made up of many substances. That is, it is a compound, in the ordinary sense of the word.

There are, however, other things which do not announce so clearly their composite character. No one on tasting common salt would think of it as a compound, and water seems not in the least like *two* things, yet both of these every-day substances are compounds. They differ, however, very decidedly from such a mixture as the cake, where each article put in keeps its own properties, except as these are modified by heat. An easily-tried experiment will show more clearly than description can do the difference between the two kinds of composite bodies.

If a small quantity of very fine iron-filings be mixed thoroughly with a small quantity of powdered sulphur, the iron remains iron and the sulphur remains sulphur. They may be distinguished from each other when looked at through a microscope, and a magnet held over the mixture will quickly draw out the iron, leaving the sulphur. But if the mixture is placed in an iron spoon (or a glass test tube) and held over a hot flame, something curious results. Something is formed which is neither iron nor sulphur—which is not like either iron or sulphur. The new substance may be pounded to a powder, but no magnet, however strong, can now draw out the iron, for the simple reason that, as iron, it is not there. The new substance is just as real and has just as distinct properties of its own as had the two elements which combined to make it, but there is one difference. Any person who knows the proper chemical means for decomposing the new substance could reduce it again to iron and sulphur, while no means known to man could have divided either of the original elements.

The iron and sulphur before they were heated formed what is known as a *mechanical mixture,* each keeping its own properties; after they were heated they formed a *chemical compound.* Many of the very commonest things, which seem as simple as anything could well be, are in reality chemical compounds, for water and salt, spoken of above, are of this nature. Air, on the other hand, is a mere mechanical mixture of gases. The piece of cake we spoke of is a mechanical mixture of the chemical compounds, water, salt, sugar, starch, fats, proteins, etc.

Atoms. There are definite ways in which chemical compounds are made up. All matter which exists in the world, according to chemists, is made up of inconceivably tiny particles called *atoms.* These are far too small for the most powerful microscope ever to discover, and no atom can be further divided. Now when a certain number of atoms of one element are brought close to atoms of another element, various things may happen. They may remain exactly as they have been, neither substance showing the slightest interest in the other; one atom of one kind may seize upon one or more atoms of the other kind and unite with them to form a tiny particle of a new substance—a *chemical compound,* in which each of the original atoms loses its identity; or both kinds of atoms may wait until some force, as heat or electricity, puts them in such a condition that they can unite.

Atoms which will thus unite with each other, either with or without aid, are said to have a *chemical affinity* for each other, and unless two substances have such affinity they cannot be forced to unite. No amount of mixing or melting or heating will make of them anything but a mechanical mixture. In the experiment described above, the sulphur and iron filings united to form a new substance with properties of its own, not just because they were melted together but because they also have a chemical affinity for each other.

In the very simplest form of a chemical compound, one atom of one substance unites with one atom of another. But often one atom of one element will seize upon two or three or even four of another, or two atoms of one may combine with three of another. It is easier for some elements to enter into combination than for others, because some elements are gases and some are solids, and the latter are much more dependent on outside forces to make it possible for them to unite with substances for which they have even the strongest chemical affinity.

Chemical Symbols. Each element has its own name, and also a "nickname," or abbreviation, which is known as its *symbol.* Usually this symbol is the first letter of its name, as *O* for oxygen, *N* for nitrogen and *H* for hydrogen. Some of the very common chemical compounds have simple names—names which give no hint that the substance is a compound, as *water* and *salt.* But this is not enough for the chemists, who have devised a system of naming all compounds which not only shows that they are such, but gives the elements of which they are composed, as well, and the number of atoms of each which enters into the combination. By this simple system, the symbols of the elements which make up a substance are written together as a *formula,* thus—*NaCl. Na* stands for sodium, the Latin name for which is *natrium,* and *Cl* for chlorine, and the substance declares itself at once as a compound of sodium and chlorine—a compound for which the common name is salt.

OUTLINE AND QUESTIONS ON CHEMISTRY

Outline

I. What It Is

(1) "The science of the composition of substances"
(2) Its wonderful achievements

II. Its Development

(1) Reasons for the beginning of such a science
(2) Alchemy
  (a) Its purpose
  (b) Its methods
(3) The real science
  (a) Its connection with medicine
  (b) Discovery of fundamental principles

III. Branches

(1) Organic chemistry
  (a) The carbon compounds
  (b) Not necessarily a study of living organisms
(2) Inorganic chemistry
  (a) Lack of sharp distinctions
(3) Special classifications
  (a) Physiological chemistry
  (b) Agricultural chemistry
  (c) Electro-chemistry
  (d) Industrial chemistry

IV. Subject Matter

(1) Elements
  (a) What an element is
  (b) Number
  (c) Occurrence in nature
  (d) Old theories about elements
  (e) The electron theory
(2) Chemical compounds
  (a) Distinguished from mechanical mixture
  (b) Atoms
  (c) Chemical affinity
  (d) Combination affected by outside forces
    1. Heat
    2. Electricity
    3. Light
    4. Mechanical force
  (e) Laws of combination

V. Chemical Symbols

(1) Method of naming elements
(2) Method of naming compounds

Questions

Why is the study of chemistry of increasing importance to-day?

From the point of view of chemistry, what is the difference between air and water?

Mention two substances which are really compounds but seem like simple substances.

In the formula H_2SO_4, what do the small figures indicate? What can you tell about the composition of the substance for which the formula stands?

What curious theory did old-time chemists hold as to the origin of disease?

What is meant by a "chemical compound"?

Give the chemical symbols for two inorganic substances commonly present on the dinner-table.

What extraordinary relationships has chemistry discovered between substances that are apparently widely separated?

Why can brass be decomposed, while copper, which seems not unlike it, cannot be?

If you heat iron filings and sulphur in a test tube, can you then draw out the iron with a magnet?

What element, in some compound or other, is present in every living thing, so far as is known?

Why should a soap-factory have a chemist among its employees?

How large a proportion of the earth's crust is made up of eight elements? What are they?

In this instance one atom of sodium combines with one atom of chlorine, but in cases where the number of atoms is not thus equal, figures must be used. These figures are made small and are written to the right of and below the letters, thus—O_3, which means three atoms of oxygen, combined into one molecule of ozone. Water is made up of hydrogen and oxygen in the ratio of two atoms to one, two atoms of hydrogen uniting with one of oxygen, and the formula is therefore H_2O.

Branches of Chemistry. The two great branches into which chemistry is divided are commonly known as *organic* and *inorganic* chemistry. The names may not be the best that could be devised, but they have been used so long that there is no thought of change.

Organic chemistry is that division which treats of the *carbon* compounds. Hundreds of compounds of this element are found in living organisms—plants and animals—and indeed no living thing is known which does not contain carbon compounds. In the early days of chemical study it was believed that all the so-called organic substances existed in living plants and animals only—in other words, that they could be produced only in living organisms. But in 1828 a chemist produced in his laboratory an organic compound, called *urea*, from its elements, and later others were produced artificially until the theory of a *vital principle* was given up. This branch is thus better described as the chemistry of the carbon compounds.

Inorganic chemistry treats of those compounds which do not have carbon in their make-up. The dividing line is not, however, quite sharp, because carbon itself and some carbon compounds, especially those which are found as minerals, are commonly discussed in books on inorganic chemistry.

Other special classifications exist, according to the differing purposes of chemical study, as follows:

Biochemistry, or *physiological chemistry*, treats of the chemical changes which take place in living plants and animals.

Agricultural chemistry deals with the problems of the farm and farm products. Although of comparatively recent development, it has already assumed great importance.

Industrial chemistry treats of the application of chemical knowledge to the manufacturing of products. These two branches are divisions of *applied* chemistry. J.F.S.

Related Subjects. The articles in these volumes which have to do with chemistry are numerous. To make reference to them easy, the following index is given, which lists all of those closely related to the subject, except the elements. A list of those is given in the article above, and all of the important elements are treated in separate articles. The reader who takes time to study the text-matter of this list will have a good foundation knowledge.

CHEMISTRY, *kem'is tri,* **OF FOOD.** See FOOD, subhead *Chemistry of Food.*

CHEMNITZ, *kem'nitz,* one of the chief commercial and industrial cities of Germany. It is in Saxony, on the Chemnitz River, thirty-eight miles west-southwest of Dresden. The town owns its water, gas and electric plants, and conducts municipal pawnshops. In this "Saxon Manchester," as the town is called, the greatest industries are the manufacture of machinery, in which over 20,000 men are employed, and the making of textile fabrics, chiefly cottons. A large part of its calicoes, ginghams and other printed goods is sent to the United States. Chemnitz had in the early years of the nineteenth century a growth rivaling that of the western cities of America. In 1900 its population was 207,000; in 1910, 287,340.

CHEOPS, *ke'ops,* an Egyptian king of the fourth dynasty, builder of the famous Great Pyramid at Gizeh, near Cairo. He lived about 2900 B. C. According to Herodotus, the "Father of History," he was an oppressive ruler who stopped at nothing to secure funds to complete his pyramid, even sacrificing his daughter's honor. But others believe he was considered a wise and powerful king. The Egyptians called him Khufu, and the pyramid "the glory of Khufu." It took twenty years and 100,000 men working constantly to complete this work of wonder. See PYRAMIDS.

CHEQUE, *check,* the English and Canadian form of the word CHECK (which see).

CHERBOURG, *sher boor',* a stronghold of Northern France, chiefly known to Americans as the European terminus of a great trans-Atlantic steamship line. It is on the English Channel, at the mouth of the Divette River, 232 miles west-northwest of Paris. Its commercial port, which has an outer harbor and an inner basin, is commodious and unusually secure, but its features of greatest importance are its naval port and its fortifications. The port is cut out of solid rock, and has three great basins with a combined area of fifty-five acres. Outside of both ports, and protecting them from the north, is a vast breakwater accounted the most remarkable work of its kind in the world. It is over two miles in length, 650 feet wide at its base and thirty feet at its summit, and encloses an area of 3,700 acres. Where its two branches meet there is a great fort, and smaller forts crown the ends. Two strong fortifications back of the town protect it from the landward side as well.

Cherbourg has numerous industries, but neither they nor its buildings differ materially from those of other cities of its size. Population in 1911, 43,731.

CHER'OKEE. When Fernando De Soto made his famous expedition to the Mississippi River in 1539-1540 he passed through the territory of an Indian tribe more nearly civilized than any other that he had found. They were described as living in log houses and tilling the land. These Indians were the Cherokee, a branch of the Iroquoian family and one of the most important tribes east of the Mississippi. They formerly occupied all the mountain region of the present states of West Virginia, North and South Carolina, Tennessee, Alabama and Georgia. The name is said to mean *cave people* in the Choctaw language, because there were so many caves in the region they occupied. Before the Revolutionary War they were in frequent conflict with the settlers of the Southern colonies, and during the war they favored the British.

In 1785 the Cherokee made a treaty with the United States. Their independence was acknowledged, and their title to the lands they occupied was confirmed. From that time the tribe made rapid progress in civilization. In 1820 the Cherokee Nation was organized, and a few years later a constitution modeled after that of the United States was adopted. Schools were established, and Sequoya, a half-blood, invented an alphabet which soon enabled them to read in their own language. In 1827 they established the first Indian printing press in the United States, and the next year began the publication of the *Cherokee Phoenix,* the first Indian newspaper. Meanwhile, gold had been discovered in their territory, the inhabitants of Georgia were clamoring for their land, and notwithstanding the treaty which the United States, under President Jackson, had made with the Cherokee, they were compelled by force to give up their homes and remove to Indian Territory, where some of their number had previously gone. The journey was made on foot, and many lost their lives through hardship and suffering. During this time of trial, their great leader, John Ross, accompanied them and shared their sufferings. Some of the Cherokee refused to be removed from their ancestral home, but fled to the mountains. Their descendants, numbering about 1,700, live in Western North Carolina.

Arriving at their destination, they reëstablished the government and made Tahlequah the capital. Just as they had regained something of their former prosperity the War of Secession broke out, and again disaster was brought upon them. Some sided with the South, others with the North, and the nation furnished soldiers to both the Confederate and the Federal armies. During the war the country was ravaged by hostile forces; at the close of the war the Cherokee made a new treaty with the United States, which freed the slaves they had held and adopted them as citizens. Schools were reëstablished and industries were revived. Prosperity continued and the Cherokee became a civilized and thriving people. In 1892 they sold their western territorial extension, and in 1906 they gave up their government and became citizens of the United States. Many of them are among the most prosperous, influential and cultured citizens of Oklahoma. They number about 20,000. See FIVE CIVILIZED TRIBES; INDIANS, AMERICAN. E.S.C.

CHERRY, *cher'i,* a small, plumlike fruit of many varieties, sweet and sour, cultivated and wild. It has been called the "home fruit," because it has not been grown extensively for commerce. It requires more work in harvesting than any other fruit and is not easily shipped long distances, but cherries always

Sweet is the air with the budding haws, and the valley stretching for miles below
Is white with blossoming cherry-trees, as if just covered with lightest snow.

The illustration shows form of tree, appearance of flowers, detail of leaves and the ripened fruit.

find a ready market. Either wild or cultivated, the trees are ornamental, with their satiny brown bark against the oval, dark-green leaves, dotted here and there with bunches of light, bright-red or purplish-black fruit, feasts for birds as well as food for man.

Before the leaves come, the wealth of white blossoms make the trees seem "covered with

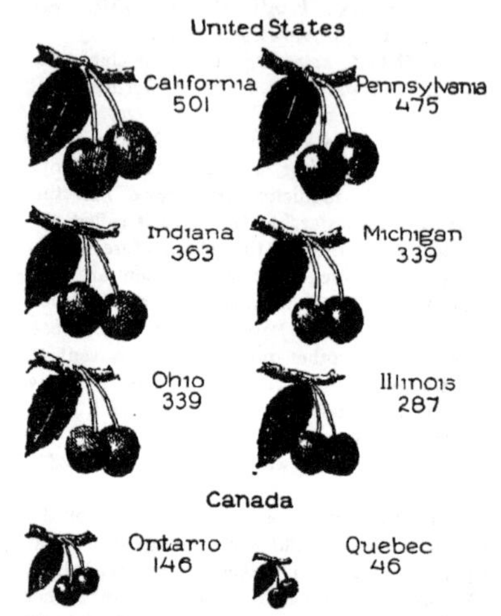

THE CHERRIES THAT GROW IN A YEAR

The figures represent an average of crops for five years.

lightest snow." And who has not heard of and in his mind's eye seen the graceful branches of dainty pink cherry blossoms of Japan at Cherry Festival time—trees so covered with soft bloom it would seem some fleecy pink clouds must have dropped down at sunset time!

Cherry wood, especially that from the black cherry, is fine-grained and beautiful, very valuable for the manufacture of fine furniture. Luther Burbank has done much work with cherries to make the fruit more valuable commercially, and has recently developed a wonderful sour cherry—a boon to canners, for, when picked, it leaves its stone on the tree (see BURBANK, LUTHER). In the eastern part of America sour cherries form an important item in the canning business, but on the Pacific coast the sweet cherry industry is more highly developed.

Choke Cherry. Sometimes the little wild choke cherry tree, with its puckery, bitter, "choking" fruit, is mistaken for a young, wild, black cherry tree, with its much-prized fruit. But a taste of one choke cherry will teach one to watch carefully for its much broader leaves and the disagreeable odor of leaves and bark. Yet the birds strip the fruit from the choke cherry trees, and so their seed has been widely distributed.

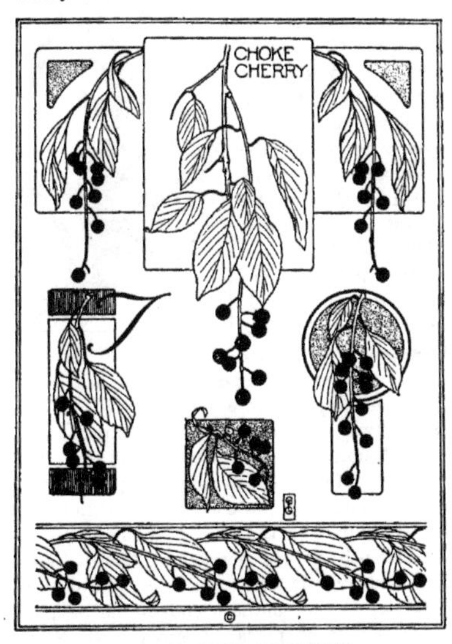

DESIGNS FOR BOOKLET
Suggested designs for cover page of a school booklet devoted to the cherry.

Cherry Brandies, Cordials and Remedies. Opposed to the choke cherry we have the wild black cherry. The tree grows large and spreading. The juice of its bitter-sweet, purple-black fruit forms the basis of all old-fashioned home remedies for "that tired feeling"; and wild cherry extracts often appear in doctors' prescriptions as a tonic. The cherry brandies and cordials of commerce and of home production are made from the little, wild, black cherry. Cherry bounce, the old-fashioned beverage of story books, is also made from this cherry of many uses. M.S.

CHERRY LAUREL, *law'rel,* a well-known ornamental evergreen shrub, a native of Asia Minor, but now cultivated in Europe and America. It is commonly called *laurel,* but it must not be confused with the sweet bay or other tree species of laurel. The leaves, which are thick and leathery, and sometimes the scentless flowers, yield the poison laurel-water which is like the oil of bitter almonds. The stone-fruit is round and shiny.

CHERSONESUS, *kur so ne'sus,* from the Greek *chersos,* meaning *dry land,* and *nesos,* meaning *island,* is a name which the ancient Greeks applied to several peninsulas. Three of the most important of these were the Thracian Chersonesus, northwest of the Hellespont, corresponding to the peninsula of the Dardanelles; the Tauric Chersonesus, the peninsula formed by the Black Sea and the Sea of Azov, now called the Crimea; and the Cimbrian Chersonesus, the modern Jutland.

CHER'UB, or ÇHERUBIM, *cher'u bim.* A cherub is one of an order of angelic beings, ranking next to the order of *seraphim.* The form *cherubim* is the plural of *cherub.* The cherubim are believed to excel in knowledge, the word *cherub* being derived from the Hebrew word *to know.* In art they are usually represented by heads, with one, two or three pairs of wings, and in the earliest religious paintings their faces are thoughtful and intelligent. The early painters also held strictly to a prescribed color scheme when representing cherubim in a Glory of Angels, a Glory being a portrayal of the several orders of angels in circles. The inner circle, that of the seraphim, is red, the symbol of love; the second, that of cherubim, is blue, emblem of light and knowledge. This law of color was observed in the oldest pictures, in illuminated manuscripts and in stained glass. Later artists, however, gave themselves more freedom in representing angelic beings, a change noticeable in such celebrated paintings as Raphael's *Sistine Madonna* and Perugino's *Coronation of the Virgin.*

In the Raphael picture the Madonna is descending from clouds composed of heads of thousands of cherubim, which are shown in a golden-tinted background. In the Perugino picture the floating cherubim have wings of various colors, blending in an exquisite harmony of tones. The aspect of serious meditation noticeable in the cherubic faces painted by the more reverent artists is beautifully illustrated in the two famous cherubim at the base of the *Sistine Madonna,* and in the cherubim in Perugino's *Assumption of the Virgin.* See MADONNA.

CHERUBINI, *ka roo be'ne,* MARIA LUIGI CARLO ZENOBIO SALVATORE (1760-1842), an Italian musical composer, excelling especially in sacred music. He was born at Florence, and commenced his musical studies at the age of six, under his father's instructions. At nine he began to study under eminent masters and soon showed a genius for composition. Before he was sixteen he had produced his creditable *Mass and Credo in D,* and a *Te Deum* for male voices, which is still often sung. His fame first became general in 1805, when he went to Vienna to compose an opera for the New Imperial Opera House. That production, *Faniska,* won him many friends, notably Hadyn and Beethoven, who pronounced him the greatest composer of sacred music of the age. After 1809 he wrote sacred music almost exclusively. He made several visits to London, being appointed at one time composer to the king, and later superintendent of the king's chapel. In 1821 he became director of the Paris Conservatory, and during his administration of more than twenty years he brought it to a high standard of excellence. His masterpiece is the opera, *Les deux Journées* ("The Water Carrier").

CHESAPEAKE, *ches'a peek,* **AND OHIO CANAL,** a waterway along the north side of the Potomac River from Georgetown, D. C., now a part of the city of Washington, to Cumberland, Md. This canal has an interesting history, for as far back as 1774 it was an idea of Washington's to make the Potomac navigable from tidewater to the Alleghanies. The scheme was interrupted by the Revolutionary War, but in 1784 a company was formed to revive it; Washington was the organization's head until he became President of the United States. The project was abandoued in 1820, but was later taken up and completed in 1850 at a cost of over $11,000,000. The canal is 184 miles long, sixty feet wide and six feet deep, with seventy-four locks having a total lift of 609 feet. Comparatively little traffic passes through it. See CANAL.

CHESAPEAKE, *ches'a peek,* **BAY,** a large inlet on the Atlantic coast extending northward through the states of Virginia and Maryland, dividing the latter into two parts, called respectively, near the Bay, the Eastern and the Western Shore. The channel at the entrance is twelve miles wide, with Cape Henry and Cape Charles on either side. The bay is 200 miles long, from four to forty miles wide, and has a depth of from thirty to sixty feet in the channel. The coast is very irregular, having many bays and inlets and large estuaries at the mouths of the numerous rivers which empty into it. The most important of the latter are the James, the York, the Rappahannock, the Rapidan and the Potomac, on the west; the Susquehanna, on the north; and the Elk, the Chester and the Choptank, on the east.

The shores are low and marshy and abound in wild waterfowl, while the shallow, brackish waters contain vast natural beds of the famous Chesapeake oysters. Oyster beds are also planted scientifically, and the oyster trade of the Maryland and Virginia beds is the largest in the world. (The details of the oyster industry are given in these volumes under the title OYSTER.) As the bay is navigable for deep-sea vessels nearly its entire length, it has a large foreign as well as coastwise trade. The most important port is Baltimore, which is situated on the west shore in Maryland, on the Patapsco River; in the value of its commerce Baltimore is the sixth port in the United States and it is the seventh city in population (1916). Norfolk and Portsmouth, in the eastern part of Virginia, at the southern end of the bay, are other important ports. Norfolk, with Portsmouth, is the largest naval station in the United States and one of the largest coaling stations in the world. The United States Naval Academy is at Annapolis, on the west shore of the bay, in Maryland, not far from Washington, D. C. The Indians called the bay the *Great Salt Water.*

LOCATION MAP

CHESS, the name of a very interesting and fascinating game. It is in no sense a game of chance, but is the most intellectual of all games of skill, for it not only trains the power of observation but is a mental contest which brings forth such qualities as foresight, resource, imagination and ingenuity on the part of the players. Chess has often been compared to a game of strategy as played by two opposing generals on the battle field, and it resembles war in the sense that it consists of attack and defense and has a definite object in view, towards which all the moves of the game lead.

The name in all the European languages is derived from the Persian word *shah,* which means *king,* and indicates the aim of the game, which is the surrender of the king.

The Board and the Pieces. The game of chess is played by two persons on a board which is divided into sixty-four squares, arranged in eight rows of eight squares each, and colored alternately white and black. The same board is used in the game of checkers. Each player has a set of sixteen men; one set is colored white and the other black. Eight of each set are of the lowest grade and are named *pawns;* the other eight are of various grades and are named *pieces.* The pieces on each side consist of *a king, queen,* two *bishops,* two *knights* and two *rooks,* or *castles.* The board must be placed so that each player shall have a white square at his right hand.

The Position of the Pieces. At the beginning of the game all the men are arranged upon the two rows of squares next to the players, the pieces on the first, or nearest, row, and the pawns on the row immediately in front of the pieces. The king and queen occupy the two central squares facing the corresponding pieces on the opposite side. The rule to be remembered is that the queen *always* occupies her own color, which means that the white queen is set on the light square and the black queen on the black square. The two bishops occupy the squares next to the king and queen; the two knights, the squares next

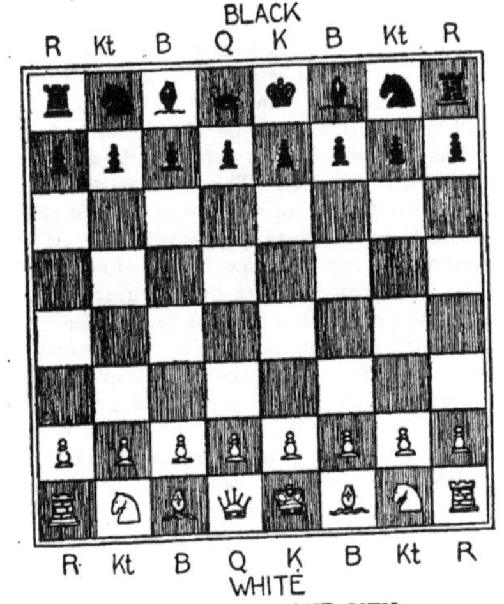

CHESSBOARD AND MEN
The position of the pieces at the beginning of the game.

the bishops; the castles occupy the last, or corner, squares. The illustration shows how the men are arranged when the game starts. The men standing on the king's or queen's side of the board are named respectively *king's* and *queen's men.* Thus, king's bishop or knight is the bishop or knight on the side of the king. The pawns are named from the

BLACK

QR1 / QR8	[illegible]	QB1 / QB8	[illegible]	K1 / K8	[illegible]	KKt1 / KKt8	[illegible]
[illegible]	QKt2 / QKt7	[illegible]	Q2 / Q 7	[illegible]	KB2 / KB7	[illegible]	KR2 / KR7
QR3 / QR6	[illegible]	QB3 / QB6	[illegible]	K3 / K 6	[illegible]	KKt3 / KKt6	[illegible]
[illegible]	QKt4 / QKt 5	[illegible]	Q4 / Q 5	[illegible]	KB4 / KB5	[illegible]	KR4 / KR5
QR5 / QR4	[illegible]	QB5 / QB4	[illegible]	K 5 / K 4	[illegible]	KKt5 / KKt4	[illegible]
[illegible]	QKt6 / QKt3	[illegible]	Q6 / Q 3	[illegible]	KB6 / KB3	[illegible]	KR6 / KR 3
QR7 / QR2	[illegible]	QB7 / QB2	[illegible]	K7 / K 2	[illegible]	KKt7 / KKt2	[illegible]
[illegible]	QKt8 / QKt 1	[illegible]	Q8 / Q 1	[illegible]	KB8 / KB 1	[illegible]	KR8 / KR1

WHITE

NAMES OF THE SQUARES

pieces in front of which they stand, such as *king's pawn, queen's castle's pawn,* and so on. The names of the men are abbreviated, as follows: King, *K;* King's Bishop, *KB;* King's Knight, *KKt;* King's Castle or Rook, *KC* or *KR;* Queen, *Q;* Queen's Bishop, *QB;* Queen's Knight, *QKt;* Queen's Castle, *QC* or *QR;* Pawns, P.

The Moves of the Pieces. In chess a man captures by occupying the position held by the captured man, who is then removed from the board. In this the game differs from checkers, where the piece played is set one square beyond the man "jumped." The *pawn* moves straight forward one square at a time, with two exceptions: when it is moved first, in which case it may be advanced either one or two squares, at the discretion of the player; and when it captures a man, at which time it always moves diagonally one square, to the position of the captured man. A pawn never moves backward. A piece or another pawn directly in front of it stops its progress. When a pawn reaches the eighth row, or the extreme limit of the board, it may be exchanged for any piece previously lost which the player chooses. As a rule the queen, the most valuable piece, is chosen, if during the game that piece has been lost. This is called *queening a pawn.*

The rook, or *castle,* moves for any distance

in a straight line either forward, backward or sidewise, but *not diagonally*.

The *bishop* moves any distance either backward or forward, but *only diagonally*. It must be noted that a bishop always moves on squares of the same color.

The *queen* is the most powerful piece on the board; she can move any distance in any straight line, either forward, backward, sidewise or diagonally, as far as her path is clear. It is of course understood that one of her own men stops her progress, but she may capture an opponent exposed to direct approach.

The *king* is at once the weakest and most valuable piece on the board. As regards direction, he is as free as the queen, but for distance he is limited to *one square* at a time. Standing on any central square, he commands the eight squares around him, and no more.

Castling. Besides his ordinary move the king has another, by special privilege, in which the castle participates. Once in the game, if the squares between king and castle are clear, if neither king nor castle has been moved, if the king has not been attacked by any hostile man and if no hostile man has commanded the square over which the king has to pass, the king's or queen's castle can be placed next to the king and the king can be moved over

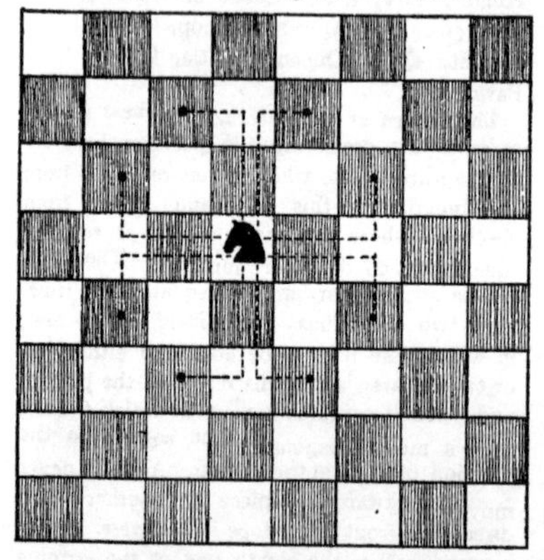

THE MOVES OF THE KNIGHT
The dotted lines show possible direction and distance in any one move; the dot shows where any of these moves will place him.

the castle to the adjoining square. This move is called *castling.*

The *knight,* unlike the other pieces, has a peculiar move. He moves over two squares at a time, one of which is diagonal and the other is straight. He may move in any direction and he can leap around any man occupying a square intermediate to that to which he intends to go. The knight always moves to a square of a different color. The knight, like the king, when on a central square on the board, commands eight squares, which are at two squares' distance, as shown in the third illustration.

The Value of the Pieces. If the pawn is taken as the standard of unity the relative value of the pieces is as follows: pawn, 1; bishop or knight, 3; rook, 5; queen, 9. The knight or the bishop is usually known as a "minor piece." The value of the pieces also depends upon the state of the game. Thus, at the end of the game a pawn is much more valuable than at the beginning, and a knight is generally stronger than a bishop; on the other hand, two bishops at the end are more valuable than two knights.

Check and Checkmate. The definite aim in chess is to force the surrender of the opposing king. The king in chess cannot be taken; he can only be in such a position that if it were any other piece he would be taken. When a piece or pawn attacks him, he is said to be *in check;* that is, he is in such a position that the next opposing move would capture him, and the opponent is bound to give notice by saying "check." When the king is in check all other plans must be abandoned and all other men sacrificed, if necessary, to save him from that situation. This is done either by removing him to an adjacent square not commanded by any man of the adversary, or by interposing one of his own men, and so screening him from check, or by capturing the attacking man. When the king can no longer be defended on being checked by the adversary, he is *checkmated,* and the game is over.

When neither of the players is able to checkmate the other, the result is a *drawn game.* When the player having the superior force, by oversight or want of skill blocks his opponent's king so that he cannot move without going into check and none of his other men can be moved, such a situation is known as *stalemate,* and the game is considered a draw.

Notation. The rows of squares running straight up and down the board are called *files;* those running from side to side are called *lines.* Each of the sixty-four squares of the chess board has a name and two numbers, as is shown in the second illustration. Each square is named after the piece which occupies it at the beginning of the game, and is called the *king's square* or the *queen's square,* and so on; the whole file has the same name. But each player counts from his own side, and it is easily seen that row number 1 for him is row number 8 for his adversary, and row number 2 for him is row number 7 for his adversary, and so on. Other signs used in chess books or in the explanation of chess problems are: (—), *to;* (*x*), *takes.*

Opening, Middle and End. A game of chess can be divided into three parts: the opening, the middle and the end. In the opening each player seeks to move his pieces in such a way as to secure the best strategic position for the actual battle which develops in the middle game. The various openings of a game are explained in all books of chess, and any player who wishes to gain proficiency must master the openings. A few broad principles governing the opening are to play forth the minor pieces early, to castle the king in good time and not expect to establish a strong attack with half of one's forces at home.

The actual battle takes place in the middle game and results in the capture of such a number of pawns and pieces as usually decides which side will eventually win the game. It is during the middle game, where such an endless variety of situations is to be found, that the players have the opportunity to display all their ingenuity and power of combination. A few simple hints which ought to guide a player during any part of the game are to try always to perceive the motive of the adversary before making the next move; to look over the board to see whether he cannot make a better move than the one he intended to make; to be careful not to play into his opponent's hand by being tempted to capture a piece which is only intended as a bait.

The Scholar's Mate. We give below, as an example, a short game which has been practiced upon young and inexperienced players and which never fails to cause such a player the greatest astonishment. It is called the *scholar's mate,* and in this game checkmate is given in the first few opening moves. The movements can be followed on the diagram in the second illustration:

WHITE	BLACK
1. P-K 4	1. P-K 4
2. KB-QB 4	2. KB-QB 4
3. Q-KR 4	3. KKt-KB 3
4. Q-KB 7 and checkmate	

History. The game of chess, which is the most cosmopolitan game and is played now in every part of the world, originated in Asia. It seems probable that it was invented in India, and from this place was introduced into Persia. The Arabs conquered Persia in the seventh century, learned the game and introduced it into all the countries they conquered afterwards. In this way chess reached Spain, whence it spread all over Europe. Benjamin Franklin popularized the game in the United States. O.B.

CHEST, or **THORAX,** *tho'raks,* the boxlike portion of the human body that lies between the neck and the abdomen. It is shaped somewhat like a cone, with the narrower end upward. The ribs, which are attached to the breast bone in front and to the spinal column behind, form its sides. The only opening at the top is the windpipe, which leads to the throat. The bottom is formed by a layer of soft muscle known as the diaphragm, through which arteries and veins lead downward. Within the chest are the heart, the terminals of the great arteries and veins, the lungs, the windpipe, the bronchi, the œsophagus and the thoracic duct.

In the act of breathing the muscles which connect and cover the ribs cause them to be drawn upward and outward, while the diaphragm flattens downward. Thus the chest can be increased in size in every direction, and it is so constructed that this expansion occurs without injury to the delicate organs within it. Ordinarily when people breathe they extend the chest from one to two inches, but a very deep breath may expand it easily three inches. In physical examinations for life insurance a three-inch expansion is expected.

CHESTER, an old cathedral city of England, on the north bank of the River Dee, about twenty-two miles from the sea and sixteen miles southeast of Liverpool. The town is one of the oldest in England and is considered one of the most picturesque. Its history dates authentically from the third Roman invasion, A. D. 80, being identified with Deva, an important Roman military station. Chester has a number of striking antique features. It is still surrounded by its ancient wall, about two miles in circumference, eight feet thick and pierced by four gates. A promenade extends along the top. Another unique feature is called the "Rows," and consists of an open but covered gallery running along the second story of the fronts of houses on the two principal streets.

These galleries are reached by steps ascending from the street and are lined with shops. The streets in this section were excavated by the Romans eight to ten feet below the ground level, in the solid rock. There are also a number of the half-timbered houses of the sixteenth century on these streets.

CHESTER CATHEDRAL

The cathedral is one of the oldest in England and was once the church of Saint Werburgh's Abbey, one of the richest abbeys in the kingdom (see ABBEY). Its grammar school was founded by Henry VIII. There are also an old Saxon church dating back to the eleventh century and a castle built by William the Conqueror. The Grosvenor Bridge over the Dee, a stone arch 200 feet long, is one of the finest bridges in Europe. Chester, though proud of its antiquities, has introduced all modern improvements. It is an important river port and trade center, and has a number of large industries outside its walls. Population in 1911, 39,028.

CHESTER, GEORGE RANDOLPH (1869-), a popular American writer, best known as the author of *Get-Rich-Quick Wallingford*, a collection of stories written in a breezy, entertaining style, and enlivened by humor that is typically American. The best literary quality of their author is his ability to present types of character that are met in real life. Chester was born in Ohio, left home at an early age to make his own way in the world, and after holding a number of positions of a varied sort he became a reporter on the Detroit *News*. From this position he advanced to that of Sunday editor of the Cincinnati *Enquirer*, and he soon became a regular contributor to leading magazines. Besides several series of Wallingford tales, his collected stories include *Cordelia Blossom*, a character sketch of a woman who understood politics as well as society; *Five Thousand an Hour*, the tale of an effort to acquire honestly a million dollars within a specified time; and *The Making of Bobby Burnit*, the history of a rich young man who developed from a business novice to a man of first importance in his city.

CHESTER, PA., the oldest city in the state, situated in Delaware County, in the extreme southeastern corner of the state, and on the Delaware River. Philadelphia is fifteen miles northeast and Wilmington, Del., is fourteen miles southwest. The city has fine transportation service through the Pennsylvania, the Baltimore & Ohio and the Philadelphia & Reading railroads and the Southern Pennsylvania Traction Company. In recent years the population has grown rapidly; it increased from 38,537 in 1910 to 40,474 in 1914.

Formerly shipbuilding was the chief industry in Chester; several vessels of the United States navy were built in its immense shipyards, which are classed with the largest in the United States. But its good harbor and exceptional transportation facilities by water and rail have given variety to industry, and have made the city the trade center for a very prosperous section. Manufacturing interests are largely centered in silk, cotton and woolen goods, ship and railway machinery, dyestuffs and building materials; over 8,000 people are employed in its 300 factories. Chester has a Federal building, a city hall, erected in 1724, the Deshong Memorial Art building and grounds, three parks, two free libraries and two hospitals. The home of William Penn, founder of Pennsylvania colony, is a feature of historical interest. The locality is the seat of Pennsylvania Military College, Crozier Theological Seminary (Baptist) and Swarthmore College.

Chester was settled by Swedes in 1643 and was known as Upland until 1682, when the name was changed to its present one by William Penn. It was laid out in 1700, was incorporated as a borough in 1701 and became a city in 1866. Here, in 1777, Washington reassembled his troops after the Battle of the Brandywine.

CHES'TERFIELD, PHILIP DORMER STANHOPE, Earl of (1694-1773), an English writer and statesman whose political career has been overshadowed by the remarkable grace and polish of his manners. His name has become a synonym for elegance of demeanor, and to say that a man has the manners of a Chesterfield is to pay the highest possible compliment to his good breeding. Chesterfield's letters to his son, in which he gave him advice in matters of etiquette, are famous and are justly admired for their literary excellence.

Lord Chesterfield succeeded his father, the third earl, in 1726. Two events stand out prominently in his political career—his appointment as ambassador to The Hague, in 1728, and as Lord-Lieutenant of Ireland, in 1745. Both positions he filled with ability. As a member of the House of Lords he was an active and bitter opponent of Walpole (see WALPOLE, HORACE). Chesterfield was made Secretary of State in 1746; two years later he resigned from office and retired to private life.

CHES'TERTON, GILBERT KEITH (1874-), an English poet, essayist and novelist, one of the most original and forceful of the modern group of British writers. The outstanding feature of his work is an extreme fondness for paradox. He was educated at Saint Paul's School and later attended the classes of the Slade Art School. His first important publication, a volume of poems collected under the title of *The White Knight,* appeared in 1900, just after the outbreak of the South African War. During the next three years Chesterton became widely known through his brilliant anti-imperialistic articles in the *Speaker* and the London *Daily News,* and at the close of the war he was asked by John Morley to write a sketch on Browning for the *English Men of Letters* series. His discussion of Browning and one of Dickens, which appeared later, are illuminating and sympathetic literary criticisms.

Chesterton's philosophy, that of a man violently opposed to the philosophy of the modern age, is strikingly set forth in such volumes of essays as *Heretics* and *Orthodoxy.* His fiction includes two collections of ingenious detective stories, *The Innocence of Father Brown* and *The Wisdom of Father Brown,* and the novels *Manalive* and *The Flying Inn.* His vigor and intellectual power make him a stimulating force in literature, and he is one of the most influential writers in England at the present time.

CHESTNUT, *ches'nut,* a stately tree of ornamental and commercial value, of which there are five known species, two of them American. The most important American species grow as high as 100 feet, thriving best on high, gravelly or sandy land, or on mountain sides. The largest known chestnut tree is the "Chestnut of a Hundred Horsemen" near the foot of Mount Etna, which at one time is said to have sheltered one hundred men on horseback.

The chestnut tree is a joy to the eye the year round; in the spring appear the well-shaped, glossy, dark-green leaves; then come yellow, fragrant catkins, and in autumn leaves of pure gold with borrowed summer sunshine.

The chestnuts, lavish of their long-hid gold,
To the faint Summer, beggared now and old,
Pour back the sunshine hoarded 'neath her favoring eye.
LOWELL: *An Indian-Summer Reverie.*

And last we find it standing "knee-deep" in its own yellow leaves, and scattered all about are the velvet-lined burs, turned brown with frost, yielding their store of smooth, brown nuts. American chestnuts have the finest flavor, but those of Spain and Italy are the largest, and in those countries they form a staple of food among the peasants.

Ordinarily chestnut trees bear nuts only after the tenth or twelfth year, but Luther

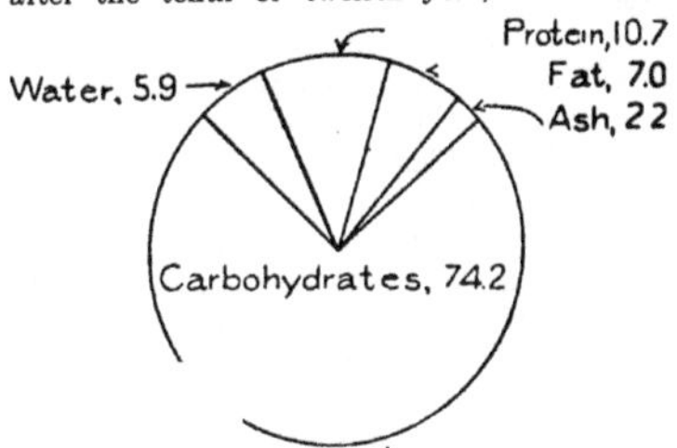

FOOD VALUE OF CHESTNUTS

Burbank has produced a chestnut seedling which bears nuts in eighteen months. It is also a hardy seedling, and might possibly be the means of restoring chestnut trees speedily in the Eastern United States, where most of them have been killed by a fungus blight.

This trouble was first noticed in 1904 near New York, and has been spreading rapidly, there being no known cure.

The well-known *chinquapin* is a miniature species of chestnut. Chestnut bark is valuable for tanning. The timber is used for woodwork, furniture, railroad ties, fence posts and fuel. The nuts are eaten raw or boiled, baked or roasted, and sometimes dried and ground into flour. The horse chestnut is an entirely different tree. M.S.

CHEVIOT, *chev'i ut,* **HILLS,** a low mountain range lying partly in Northumberland, England, and partly in Roxburghshire, Scotland, forming about thirty-five miles of the boundary line between the two countries. The hills extend from the River Tweed on the northeast to the sources of the Liddel on the southwest. They are well grouped, smooth in contour and covered with grass, and are the grazing ground of the famous Cheviot sheep. The region is also noted for its grouse. During the Border wars the hills were the scene of much of the romance and history of those troublous times, and they will always be associated with the old English ballad, *Chevy Chase*.

CHEWING GUM. See subhead, in article GUM.

CHEYENNE, *shi en'* or *she en'*, a tribe of Indians belonging to the Algonquian family. When first known to white men they lived in Minnesota, north of the Minnesota River, but were gradually driven westward into the present states of South Dakota, North Dakota and Montana. They have always been a strong, brave people, who have held women in high regard. When first known they lived in villages and cultivated the land, but later they became the most skilful and daring Indian riders of the Plains. In 1832 a portion of the tribe removed to Arkansas and became known as the Southern Cheyenne. Their treatment by the whites was unjust, and from 1860 to 1878 they were actively engaged in border warfare. In 1876 they took a prominent part in the battle in which General Custer and all his men were slain. They have since become friendly. The southern branch is located in Oklahoma, where they live upon farms and have become civilized and have been admitted to citizenship. The northern branch is on a reservation in Montana. In all they number about 3,500. For customs and habits see INDIANS, AMERICAN; see, also, ALGONQUIAN INDIANS.

CHEYENNE, *shi en'*, WYO., the capital of the state and county seat of Laramie County, situated in the extreme southeastern part of the state, on Crow Creek. Denver is 106 miles south, Omaha is 516 miles east, and Salt Lake City is 520 miles west. The Union Pacific, the Colorado & Southern and the Chicago, Burlington & Quincy railways serve the city. Cheyenne was settled in 1867, when the Union Pacific Railway was constructed to this point; in 1869 it was chosen as the capital. In 1913 the commission form of government was adopted. The water works, which cost $2,000,000, are owned by the city. The population in 1916 was 11,320, which included the soldiers at the military post, Fort D. A. Russell.

Cheyenne is located on the eastern slope of the Laramie range of mountains, 6,050 feet above sea level, with Crow Creek, a tributary of the South Platte River, winding half way around it. The surrounding country is a picturesque mountainous region, with many waterfalls, lakes and springs, and this natural beauty extends into the city through its several parks. Cheyenne is the home of the original Frontier Days' Wild West Show, which annually attracts large numbers of visitors and contestants. The buildings and grounds for this entertainment, including the city lakes and land adjoining, cover 160 acres.

Industrially, the vicinity is largely interested in stock-raising, and from the city great numbers of beef cattle are shipped to Eastern markets. The kindred industry of meat-packing is extensive, and the creameries supply a large section around Cheyenne. Here are located the large shops of the Union Pacific Railroad, which employ about 1,000 men. Iron is mined in the vicinity.

Besides the state capitol, the city has a $400,000 Federal building, the governor's mansion, soldiers' and sailors' home, county hospital, a Carnegie Library, and a Roman Catholic convent. Fort D. A. Russell, three miles from the city, is one of the largest forts in the United States; its construction and equipment cost $7,000,000. R.M.L.

CHIAROSCURO, *ke ah ro skoo'ro.* One of the most difficult things to master in painting is the handling of light and shade, or *chiaroscuro*, as it is called, from Italian words meaning *light* and *dark*. Unless objects in the light stand out and those in the shadow are properly subordinated, perspective seems to be lacking. Correggio and Rembrandt rank as the great masters of chiaroscuro.

CHICAGO, the "metropolis of the Northwest," is the largest city of Illinois, the second largest of the western hemisphere and the fourth largest in the world. The three cities which surpass it—London, Paris and New York—were, even the youngest of them, great cities while the site of Chicago was still but a wilderness of marsh and forest roamed over by the Indians. It is not alone for its size that Chicago is notable, but also for the rapidity of its growth and for the restless energy which has characterized every step of its advance. William Vaughn Moody, a poet who spent many years in Chicago, described with a true poetic insight its dominant spirit in the lines—

And yonder where, gigantic, wilful, young,
Chicago sitteth at the northwest gates,
With restless, violent hands and casual tongue,
Moulding her mighty fates.

Size and Location. Chicago is a rambling city spread over 194 square miles, with 4,685 miles of streets, one-half more than the total mileage of roads in the state of Delaware. Its lake frontage, which through that part of its length not occupied by railroad tracks is one of the most charming outlooks possessed by any large city, is twenty-four miles, and with compactly-built suburbs, over thirty miles; the greatest east and west extent of the city, to suburban limits, is ten and one-half miles.

Chicago is in Cook County, of which it is the county seat. It lies along the southwestern shore of Lake Michigan, on a plain but fifteen feet or thereabouts above the level of the lake, or 596 feet above sea level, and much of the land along the shore has been built up from a flat beach.

Though it is called a Western city, and is Western in spirit, Chicago is in reality well to the east of the center of the country; it is 2,274 miles from San Francisco, and but 911 from New York. Its marvelous growth in population and commercial and industrial importance has been largely due to its position at the head of Lake Michigan, where it formed for many years the only outlet for the products of the Middle West.

Plan of the City. The original plan of the city was influenced very decidedly by the Chicago River, a little stream, but a very important one. It is formed by two branches, one from the northwest and one from the southwest, which unite about a mile from the lake, meeting the lake midway between the northern and southern limits of the city. This stream is sluggish and unpicturesque, and has for much of its history been filthy and disease-breeding, but commercially it is more important than any other river of its length in the world. Not nearly so large a commerce, in tonnage, passes through the Suez Canal in a year as through this little river.

By the Chicago River and its branches the city is divided into three well-recognized districts, or "sides": the South Side, including all the territory south of the river; the North Side, including all that to the north; and the West Side, much the greatest in area, to the west of the branches. Three great tunnels, recently reconstructed, and no fewer than sixty-six bridges connect the various parts of the city with each other. In the main the streets are regularly laid out, crossing each other at right angles, and most of them are broad.

The Business Section. One feature very characteristic of Chicago is its "Loop," or business district, which is crowded into an area little more than one and one-half square miles in extent. Not all of its great business houses are within that space, but the larger proportion, and by far the more important, of them are there. In other cities a man may have to travel miles to consult his dentist, his oculist and his physician, buy his clothing and lunch at his favorite restaurant; in Chicago

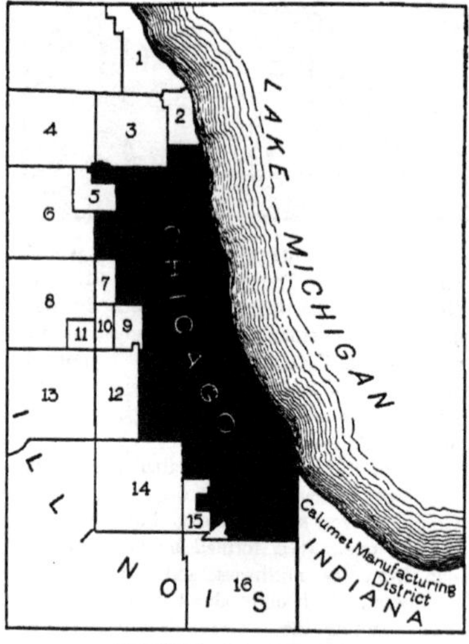

SUBURBAN DISTRICTS

1. New Trier Township
2. Evanston
3. Niles
4. Maine
5. Norwood Park
6. Leyden
7. Oak Park
8. Proviso
9. Cicero
10. Berwyn
11. Riverside
12. Stickney
13. Lyons
14. Worth
15. Calumet
16. Thornton

he can do it all within a very few blocks. This has its advantages, but it also has its disadvantages. The crush in the streets and the din from the cars, wagons, elevated trains and automobiles are by no means soothing to the hardened resident, while to the stranger they are nerve-racking. During comparatively late years the noise and the crowding have been greatly lessened by the construction of tunnels, thirty feet below the surface, through which most of the heavy freight is carried.

Within the business district State Street stands as the center of the retail trade. Department stores have been brought to a high state of efficiency in Chicago, and the group on State Street is the largest in the world. The great retail establishment of Marshall Field & Company, covering an entire block, is unmatched elsewhere the world over in size and equipment. Fifth Avenue is the center for the wholesale dry goods trade; La Salle Street is the financial district, or "Wall Street of Chicago," and South Water Street, a little north of the "Loop," is a succession of produce markets, to which the wagons of the truck farmers and those from the outlying markets throng in the early hours, before the rest of the city is awake.

The most notable street of the downtown district, and for a mile one of the finest vistas in the world, is Michigan Avenue, the first street west of the lake. With the grassy stretch of Grant Park to the east and many of the most substantial and striking buildings of the city on the west, and with its beautiful lighting system, it is probably unexcelled. In accordance with its "city beautiful" idea, Chicago has been devoting much attention to the beautifying of its lake front. Colonnades, fountains, pillared terraces and ornamental bridges have been added, and the plans for the future include athletic grounds, a stadium, concrete steps leading to the water's edge, and statuary among the trees.

The Blackstone Hotel, in its architectural features much resembling some of the great hotels of New York; the Auditorium, with its large hotel and magnificent theater; the McCormick Building; the tile-faced Railway Exchange; the People's Gas Building, with its eighteen gigantic one-piece granite pillars; the University Club and Monroe buildings; the Tower Building, the top of whose pinnacle is the highest point in the city, and the Michigan Boulevard Building constitute an imposing row and give a most picturesque skyline to this city which, in its flat surroundings, is all too barren of the picturesque. The only building which stands on the east side of Michigan Avenue between Park Row and Randolph Street, a distance of one mile, is the Art Institute, a beautiful structure in Renaissance style. It contains permanent art collections of great value, and there is seldom a time when a special exhibit of note is not housed there. There are, moreover, a valuable art library, a lecture hall, and a well-attended school of art instruction. Immediately to the south of this building stands one of the city's newest ornaments, Lorado Taft's remarkable fountain, "The Spirit of the Great Lakes."

Other Notable Buildings. No building in Chicago may be over 200 feet in height. Chief of the great structures, in point of size, is the combined city hall and courthouse, occupying the square bounded by Randolph, Clark, Washington and La Salle streets. It is a magnificent structure of steel and granite, with great columns and allegorical figures in bas-relief, the whole erected at a cost of about $10,000,000. The square formed by Adams, Dearborn, Jackson and Clark streets is occupied by another

"SPIRIT OF THE GREAT LAKES"

Lakes Superior, Michigan, Huron, Erie and Ontario are artistically represented by five female figures, built upon a rocky base at relative elevations. A stream of sparkling water rising in the basin held by "Superior" overflows into the shell held by "Michigan," and so on from shell to shell until "Ontario" surrenders her pleasant guardianship over the unpolluted waters of the greatest fresh-water lakes in the world to the rough keeping of the turbulent "Saint Lawrence." In ideal conception this is one of Lorado Taft's masterpieces.

fine structure, the Federal Building, which is in the shape of a rectangle surmounted by a cross and a crowning dome. The Tacoma, far surpassed to-day by many other office buildings, is noteworthy as the first steel-frame "sky-scraper" ever erected; the Masonic Temple is notable as the highest building in the city (354 feet in height), and the Continental and Commercial Bank structure as having the largest floor space, excepting two, in the United States.

A detailed enumeration of the office buildings would be both profitless and uninteresting; suffice it to say that they must needs be numerous and large to house the vast army of people that is poured into the Loop district every morning. Some of these buildings are notable because of special architectural excellence, some because of artistic interior decorations, and some because of their luxurious appointments. Perhaps the most impressive thing about all of them is the care and effort that has been required to make them firm and safe. The soil that underlies the city is an unstable mixture of sand, gravel and blue clay, and it is necessary to sink great shafts of steel and concrete down to bed rock from fifty to 110 feet below the surface in order to make the foundations secure.

Of hotels of all sorts Chicago has a large number, both in the downtown section and in the outlying residence districts. There are also many theaters, ranging in size from the Auditorium, which seats 3,747 people, to the Little Theater, with a seating capacity of ninety-nine. In all, there are over 600 theaters, including moving picture houses.

Libraries. Of the scores of libraries in Chicago the largest and most popular is the Public Library, which dates from the years immediately following the great fire of 1871. It has over a half million volumes, and the annual circulation is over 3,000,000 volumes. One of the most beautiful and complete library buildings in the country houses this collection, and there are almost a score of branches in different parts of the city, several of which are contained in fine buildings. The other two large libraries are the Newberry and the John Crerar, the former occupying an imposing granite building on the North Side, the latter housed in temporary quarters on Wabash Avenue. These are both reference libraries, and their books

THE ART INSTITUTE

are not for circulation. The Newberry collections are especially valuable on such subjects as literature, history, music and genealogy, while the John Crerar specializes in the natural, physical and social sciences.

Local Transportation. In so widely-scattered a city with its business so centralized, transportation is a big problem, and one which has been met in three ways. First, there are the electric street railways, which have over 1,000 miles of track, and connect all parts of the city. Authorities do not hesitate to say that no city in the United States possesses a better street railway system. In the management of the surface lines the city is a partner, receiving fifty-five per cent of the net profits. In 1915 this sum amounted to over $20,000,000, the accumulation of years, and it is being held to finance at no distant day a system of subway transportation.

There are four elevated roads, two to the West Side, one to the South and one to the North. In the downtown district these form a "loop" about the main business section, enclosing the streets from Lake to Van Buren and from Wabash to Wells Street, and this it is which gives the popular name "Loop" district to this section. The Loop encircles the great retail and wholesale stores. The elevated roads are clean, safe and rapid.

In addition to these purely local lines, most of the great railways entering the city have suburban divisions. In all, it is estimated that all the lines collect daily an average of almost 1,500,000 fares.

The Park System. Chicago has two popular names; it is called the *Windy City* and the *Garden City*, the latter name having been given to it because of its parks. To-day it is very far from being the first city of the country in its proportion of park area to population, but it has an unusually well-planned system of beautiful parks, in total area over 4,600 acres. Of the two score, or thereabouts, of parks, seven are of considerable extent. Lincoln Park, on the North Side, has an area of 317 acres, but is being largely added to by the creation of new land on the lake shore at its northern limit. This is the favorite park of the children, who are attracted not so much by the beautiful, shaded drives, the conservatory or the lagoon, as by the zoölogical garden. About 1,700 animals, one of the finest collections in the country, are housed here, some of them in buildings which are models in their way. Most noteworthy of the statues with which Lincoln Park is liberally adorned is the equestrian statue of Grant and the famous *Lincoln* by Saint Gaudens.

On the South Side the most important parks are Jackson, with 542 acres, and Washington, with 371 acres. The former, stretching for one and one-third miles along the lake, is the old site of the World's Columbian Exposition, of which a few buildings have been preserved as memorials. Beautiful drives, lagoons for boating, a rose garden and excellent golf and tennis facilities have made this one of the city's most popular parks. Housed in the former Fine Arts Building of the Exposition is the Field Columbian Museum, a valuable collection in the fields of natural history and ethnology, but this institution by 1920 is to have a magnificent permanent home in Grant Park in the downtown district. A mile west of Jackson Park, and connected with it by the boulevard remembered as the Midway Plaisance of the World's Fair, is Washington Park, especially noted for its effective landscape gardening. The third large park on the South Side is Marquette, one of the newer playgrounds, with an area of 322 acres, much frequented by reason of its long golf course. Downtown, between Michigan Avenue and the lake, is a more recently laid out area known as Grant Park, once famous as Lake Front Park. It is in this park that most out-of-door exhibitions, such as the aviation meets, are held.

The West Side parks are Humboldt, 205 acres; Garfield, 187 acres, noted for its conservatory, the largest in the country; and Douglas, 182 acres. Connecting the various parks is a splendid system of boulevards, aggregating over seventy miles and forming one of the finest drives in America. Most of these are lined with beautiful homes and contain central grass plots decorated with trees and flowers. The people of Chicago also have the benefit of the natural parks or woodland regions which have been purchased and opened up by the Cook County forest reserve board. These lie in a semi-circle about the city, and are all easy of access.

Playgrounds and Beaches. One of the things of which Chicago has most reason to be proud is its system of small parks. These are so located as to be accessible to the people who need them most—those in the thickly-settled districts; and they contain practically all that visitors of any age can demand for pleasure or relaxation. There are gymnasiums with trained instructors, swimming pools, fully-equipped playgrounds for children of various ages, sand piles, wading pools, skating ponds, reading rooms and club rooms, all free. In the summer season thousands seek the bathing beaches which may be found at intervals along the lake front from the northern section of the city to the south end. Many of these are controlled by the city, and the new Clarendon Avenue Beach, opened to the public in 1916, is one of the best-equipped municipal beaches in the world.

How the City Gets Its Water. Lake Michigan furnishes an inexhaustible store of water. To bring into the city and distribute the 575,000,000 gallons or more used daily, an intricate system of cribs, lake and land tunnels and pumping stations has been constructed. Indeed, Chicago might well have the title *city of tunnels* added to its other nicknames, so extensively is the land beneath it honeycombed. From two to four miles out in the lake there are five cribs, with which connect nine tunnels well below the bottom of the lake, and these in their turn convey the water to ten main land tunnels. Some of the lake tunnels are fourteen feet in diameter.

The most important thing about drinking water is that it shall be pure, and of course it cannot be if impure matter in great quantity is dumped into the lake. Despite this fact all the sewage of the city for a long time found its way into the lake, but by 1875 it became clear that some other method of sewage disposal must be found if the health of the city were not to suffer. Attempts were made to use the old Illinois and Michigan Canal, but this proved inadequate, and between 1892 and 1900 a new canal, one of the finest sanitary works in all the world, was built (see CHICAGO DRAINAGE CANAL). By means of this the vast volume of sewage of the city finds its way through the Illinois River to the Mississippi, and so to the Gulf of Mexico. Chicago River no longer in reality empties into Lake Michigan; the Drainage Canal meets it, and the current, reversed, flows from the lake westward through the city.

The World's Greatest Railway Center. Chicago stands supreme as a railroad center. No railroad runs through Chicago, for every train that enters the city reaches a terminal; and the thirty-four lines terminating there have a combined mileage which is half that of all the railroad systems of the United States. It is believed that the number of railways centering in the city will never be increased, as there is no room for another roadbed, except at such enormous expense as to be prohibitive. Some of those already entering the city are unpleasantly crowded in the hours when local traffic is heavy. Six large stations, all but one of them in the downtown district, accommodate the passenger service, and by 1922 a great new Union Station will have been completed, rivaling any found elsewhere in the world. Each road has its extensive freight depots in various parts of the city; a belt line extends almost around three sides of the city, connecting the different roads and forming a complete freight transfer system.

The entrance into the city of so many great railways made necessary very dangerous grade crossings, but beginning in 1892 these were in large measure done away with by the elevation of tracks, at a cost of a million dollars a mile. To-day Chicago has within its limits more than twice as many miles of elevated track as have all the other cities of the United States together.

Commerce and Industries. Naturally a city that is the greatest railroad center in the world might be supposed to have a large rail commerce. It has that, and more. Chicago is also one of the greatest of inland ports, lines of steamers, both freight and passenger, connecting with all the other important lake ports. Over 6,000 ships a year enter and leave its harbor, and these deposit 8,500,000 tons of freight and bear away an equal amount. The city is a sort of clearing house; it does not keep all that is brought into it, but reships much of it. The iron which constitutes over fifty per cent of the weight of its lake imports it makes use of in its great suburban steel mills, but much of the lumber and grain that arrives is shipped again, Chicago ranking as the greatest grain market and the greatest lumber market in the world. It is also first in its export of packing-house products. Among the cities of the United States only New York surpasses Chicago in the volume and value of its trade.

Docking facilities for years were inadequate, and partly to remedy this condition a great Municipal Pier, near the mouth of the river, was completed in 1916. It is built of concrete and steel and extends over half a mile into Lake Michigan. At its farther end is space used for a recreation center 660 feet long and 300 feet wide, which has easily accommodated 100,000 people in a single day. The pier cost $4,500,000.

With the coal fields of Illinois so near and the raw materials from the great Middle West so easily available, Chicago has become an important manufacturing center. Over 350,000 people are employed in its various establish-

CHICAGO'S MUNICIPAL PIER

There are two great warehouses, one above the other, on each side, each almost a half mile in length. The recreation pier, 660 feet long, is at the extreme right in the illustration.

ments, and the total value of the products is over a billion and a quarter dollars annually. Largest of these industries is that of slaughtering and meat-packing, carried on at the Union Stockyards, by far the greatest establishment of its kind in the world. To quote the popular statement, "Every part of the animal is used but the squeal," and to-day the by-products, made from parts that were formerly thrown away, reach a value of scores of millions of dollars each year. Iron and steel products, machine shop and foundry products, men's clothing, railroad cars and lumber products are manufactured in vast quantities, while the bakery products reach a value of about $27,-000,000 a year. Printing and publishing is an important industry, but in this regard Chicago does not rank with Boston, New York or Philadelphia.

Schools and Other Institutions. Chicago has a complete system of public schools ranging from the kindergarten through the grammar grades and high schools to the Chicago Normal College, with its three practice schools for teachers. In the twenty-two high schools and 279 grammar schools there are enrolled almost 320,000 pupils, and the teaching force numbers over 7,000. The regular school term is ten months, and during half of that time night schools are also conducted, their enrollment averaging about 26,000. There are schools for the blind, the deaf and the crippled, and in certain schools special classes are held for subnormal children. Many of the high schools and more than half of the grammar schools include manual training in their courses, and domestic science teaching is becoming increasingly important.

CHICAGO'S SCHOOLS IN 1844

Corner of State and Madison streets, now the most crowded street corner in the world.

Of institutions of higher learning the University of Chicago is the most prominent. Northwestern University, America's greatest Methodist institution, located at Evanston, has its professional departments of law, medicine, dentistry and the college of commerce in the city, and there are, in addition, Saint Ignatius College, Loyola University, Armour Institute of Technology, Lewis Institute, several medical schools and excellent theological schools. The Art Institute, which has in attendance upon its classes about 2,500 students each year, has been mentioned above. Few other art schools in the country offer as complete courses.

There are about 1,000 churches of all denominations, and a large number of hospitals, the most noted being the Cook County Hospital, Michael Reese, Saint Luke's, the Presbyterian, Wesley, Mercy and the Augustana. In Hull House (which see) the city has one of the best-known social settlements in the world, with Miss Jane Addams at its head; others which have won a wide reputation are Chicago Commons, Northwestern University Settlement and the University of Chicago Settlement. The United Charities and the Jewish Aid Society maintain corps of trained investigators whose duty it is to discover the needs of the poor and unfortunate and to see that aid is furnished them. There are also smaller charitable organizations, many of which have specialized in some particular field.

Administration. A mayor, elected for a term of four years and paid $18,000 per year, the highest salary of any municipal officer in America, is the chief executive, and he is assisted by a council of one chamber, composed of seventy aldermen, two from each of

OUTLINE AND QUESTIONS ON CHICAGO

Outline

I. Position and Size

(1) Latitude, 41° 53′ 6″ north
(2) Longitude, 87° 38′ 1″ west
(3) Situation on Lake Michigan
(4) Distance from other large cities
(5) Area
(6) Population
(7) Rapid growth

II. Description

(1) Plan of city
 (a) Determined by Chicago River
 (b) The business section or "Loop"
 (c) Important streets
(2) Notable buildings
 (a) Public
 (b) Office buildings
 (c) Hotels
 (d) Theaters
(3) The park system
 (a) "The garden city"
 (b) North Side parks
 (c) South Side parks
 (d) West Side parks
 (e) Playgrounds and beaches
 (f) The Outer Park plan
 (g) The "City Beautiful"
(4) Educational institutions
(5) Churches
(6) Charitable institutions

III. Water Supply and Sewage

(1) Cribs
(2) Tunnels
(3) Amount of water used daily
(4) Drainage Canal

IV. Transportation

(1) Railway systems
 (a) Greatest railway center
 (b) Elevation within city
(2) Local transportation
 (a) Street railways
 (b) Elevated roads

V. Commerce and Industry

(1) Rail commerce
(2) Lake commerce
(3) Docks
(4) Manufactures

VI. The People

(1) Nationalities represented

VII. Government and History

(1) Departments of government
(2) Revenue
(3) History
 (a) Settlement
 (b) Growth to 1870
 (c) The great fire
 (d) Later growth

Questions

What is the meaning of the name *Chicago*?
What was the first "skyscraper" ever built?
How many gallons of water are used in the city each day?
What is the largest foreign-born element in the city?
Show that Chicago is not geographically a Western city.
How does it rank as to size among the world's cities?
What was the greatest calamity that ever befell the city?
What engineering project has changed the course of a river?
Why has the city no "skyscrapers" as tall as those of New York?
What is the "Loop"? Why is it so called?
Why is it unlikely that any more railroads will ever terminate in Chicago?
What gave Chicago a larger registration of voters in 1916 than any other city in the country?
Mention three pieces of statuary of which the city may feel justly proud.
What provision does the city make for the recreation of its people?
What is the greatest industry? How many cities surpass it in this?

CHICAGO IN 1830

The stockade is the second Fort Dearborn, built after the destruction of the first, which is pictured under the title FORT DEARBORN. The John Kinzie house is at the right of the picture.

thirty-five wards. Certain department heads, as the chief of police and the fire chief, are appointed by the mayor and go out of office with him, but throughout the departments themselves civil service methods prevail. The total revenue and expenditures of the city amount to about $60,000,000 annually.

The People. Chicago has a greatly-varied population, about seventy-seven per cent of its inhabitants being foreign born or of foreign parentage. By far the most numerous of these adopted citizens are the Germans, of whom there are over half a million. That is, Chicago is a larger German city than is Frankfort, one of the most important cities of the German Empire. Austrians rank next in number, then the Irish, Russians, Swedes and Italians, in that order. Newspapers are published regularly in at least ten languages, and within the confines of the city the church service is given in at least a score. The total population of Chicago in 1910 was 2,185,283; in 1913 it was estimated at 2,388,500; and in 1916 by the Federal census authorities at 2,497,722.

History. Interest in the history of Chicago centers in its growth, remarkable even among American cities. Other cities have had "booms," but Chicago's expansion has been continuous. Attempts have been made to prove that the name *Chicago* is from an Indian word meaning *mighty*, or that it has some poetic or high moral significance, but the general opinion is that it is a form of the Indian name for the everywhere-present wild onion. The first white visitors to the site were Marquette and Joliet, who stopped there in 1673, and other great French explorers followed them. In 1779 a negro from San Domingo built a cabin on the north bank of the Chicago River, and in 1804 this came into the possession of John Kinzie, the first white man to make his home on the site of the city. The Federal government in 1804 built Fort Dearborn (which see) on the south bank of the river, and though this was abandoned when the Indian massacre of 1812 occurred, it was rebuilt four years later. In 1830 maps were made, definitely marking out the town of Chicago, which had a total area of three-eighths of a square mile and contained twenty-seven voters. When incorporated, three years later, the town had a slightly-increased area and a population of 550, while its tax-levy reached the total of $48.90. The first city water works, constructed in 1834, consisted of a well that cost $95.

From this time on the growth was steady, if not particularly rapid. The Illinois and Michigan Canal, begun in 1836 and completed in 1848, and the Chicago & Galena Union Railroad, which later was the nucleus of the great Chicago & North Western system, brought the little city into touch with the territory to the west, the territory upon which its prosperity was to depend; and the population increased from 4,480 in 1840 to almost 300,000 in 1870. The city's first and greatest calamity occurred in 1871; a terrible fire broke out on October 8 on the West Side, extended north and west and raged for two days and nights, destroying property valued at $196,000,000 and rendering 100,000 persons homeless. With wonderful rapidity the city was rebuilt, the old wooden structures being replaced in large measure by those of brick and stone.

In its later history Chicago has suffered much from labor troubles. Out of these grew the Haymarket Riot of 1886, in which seven policemen were killed. Serious strikes have occurred at intervals in the Stockyards, but most noteworthy of these movements were the railway strike in 1894, put down only with the aid of Federal troops, and the teamsters' strikes of 1904-1905. An event of more pleasing character was the World's Columbian Exposition (which see) of 1893, the greatest world's fair held up to that time. On December 30, 1903, there occurred in the Iroquois Theater a fire in which 572 lives were lost, and as a result of this disaster theaters not only in Chicago but all over the world have been built and equipped with more thought of safety. The city has always been a favorite meeting place for conventions, and among others the national conventions at which Lincoln, Grant, Garfield, Blaine, Cleveland, Harrison, Bryan, Roosevelt, Taft and Hughes were nominated for President of the United States were held there. E.D.F.

CHICAGO, UNIVERSITY OF, one of the most prominent institutions of higher education in the United States. It is located in Chicago and centered on the Midway Plaisance, which connects Jackson and Washington parks, and has over a score of buildings in the Gothic style which are unsurpassed on any campus in the country. The campus itself, except for the buildings, has little of the picturesqueness of those in many smaller towns or in less crowded areas.

HARPER MEMORIAL LIBRARY
This building illustrates the prevailing style of architecture on the university campus.

Its rapid growth is one of the things about the university which its students celebrate in song; it is the newest of the great universities, though in its antecedents it dates from the middle of the nineteenth century. The old University of Chicago, a Baptist school of college rank, was opened in 1857, but was compelled through lack of funds to surrender its charter in 1886. Four years later, largely through the efforts of the American Baptist Educational Society, the new university was opened, and though it is in no sense a denominational institution and exacts no religious tests of students or teachers, its charter provides that the president and two-thirds of the trustees must be Baptists.

Organization. The university thus chartered in 1890 is organized into five departments: (1) schools and colleges, including the four-year undergraduate courses, as well as the graduate schools; (2) university extension, which by correspondence courses and lectures attempts to reach those not in attendance on its classes; (3) university libraries, laboratories and museums; (4) the university press; and (5) the affiliated and coöperating schools.

In arranging its courses the university mapped out a plan for itself differing from that of any other American university. The scholastic year is not the usual period of nine months, but is divided into four quarters, each of which is subdivided into two terms. The summer quarter, at most universities a vacation period, is at the University of Chicago the busiest term of the year; teachers flock to it from all parts of the country, for in three summer courses they can complete an ordinary college year of work. Since the courses are arranged by quarters, a student may drop out for three months at any time without interfering with his work. The four years of undergraduate work are divided into the junior colleges, or freshman and sophomore years, and the senior colleges, the junior and senior years. At the end of the first period a certificate is issued, and at the end of the second the bachelor's degree is given. Especial emphasis is laid on graduate work, and everything possible is done to encourage students in original research.

Growth. Though many benefactors have given liberally to the upbuilding and equipment of the university, its growth has been largely due to the bequests of John D. Rockefeller, who at various times contributed sums totaling about $35,000,000. William Rainey Harper, president from its foundation to 1906, developed a policy which attracted students from every part of the Union. His successor, Harry Pratt Judson, has carried out approximately the same policy, and the result has been continued growth in attendance and in financial resources. The libraries of the university contain almost 600,000 volumes; the total assets of the institution are over $31,000,000; the faculty numbers about 400, and the total registration averages over 6,000. It must be borne in mind, however, that perhaps a third of this number are in attendance for the summer quarter only.

Late in 1916 plans were announced whereby the university would in the future possess the greatest medical department in America. A fund of $15,000,000 is to be available for the purpose. Rush Medical College, long affiliated with the university, and the Presbyterian Hospital, on Chicago's West Side, will be the nucleus of a great post-graduate department, while the undergraduate school will be built along the Midway on land already owned by the university.

CHICAGO DRAINAGE CANAL, officially known as the CHICAGO SANITARY AND SHIP CANAL, is a great sanitary project to provide pure water for the millions of that city. Ages ago the Great Lakes found an outlet to the ocean by way of the Illinois and Mississippi rivers, and the channel through which this great stream flowed forms the valley of the Illinois River. When Chicago discovered that it must protect Lake Michigan from the insanitary effect of its sewage, the city engineers turned their attention to the water course of the past ages. A brief examination showed that a canal connecting the lake with the Desplaines River could be constructed without engineering difficulty. The necessary legislation was obtained and the great channel, commonly known as the Chicago Drainage Canal, was begun September 3, 1892, and completed in January, 1900, at a cost of about $50,000,000.

The canal proper is twenty-eight miles long, and varies in width in different sections from 110 feet at the bottom and 198 feet at the water line in the narrowest section to 202 feet and 290 feet in the widest section. The sections cut through rock have a width of 160 feet at the bottom and 162 feet at the top. The depth of the cut varies from thirty to thirty-six feet; the depth of water is never less than twenty-two feet, and is usually about twenty-four feet six inches. By means of the controlling works at Lockport, twenty-nine miles inland, consisting of flood gates and a beartrap dam, the depth and flow of water are easily regulated. Ordinarily the flow is about 300,000 cubic feet per minute, but the full capacity of the canal is 600,000 cubic feet per minute.

The Chicago Drainage Canal is one of the greatest engineering works in the world. It has changed the course of the Chicago River and made it an outlet of Lake Michigan, when formerly it flowed into the lake; it is the only river in the world whose flow is away from its mouth. In connection with the construction of the canal the entire sewage system of Chicago had to be changed. Formerly all sewers emptied into the lake; now they empty into the canal, and the water supply of the city has been saved from pollution. In the near future the canal will doubtless form a link in a deep waterway between the Great Lakes and the Gulf of Mexico. See CANAL. W.F.R.

CHICAGO HEIGHTS, ILL., a manufacturing suburb of Chicago, distant twenty-seven miles south from the center of the city. It is at the southernmost boundary of Cook County, and occupies the highest land within the county lines. The Chicago & Eastern Illinois, the Michigan Central and the Elgin, Joliet & Eastern railroads serve it, and it is connected with Chicago and neighboring towns by electric railway. During the first decade of the nineteenth century its population almost trebled, for while it had in 1900 but 5,100 inhabitants, the number had grown by 1910 to 14,525. Its manufacturing plants, which include factories for the making of chemicals, iron and steel products, glass, pianos, automobiles, brick, clothing and many other articles, are grouped on the east side of the city, where are to be found, also, the homes of most of the factory workers. The municipality owns the water works. Settled in 1835, the town had a slow growth, and not until 1900 was it chartered as a city.

CHICKADEE, *chik'a dee.* See TITMOUSE.

CHICKAHOMINY, *chick a hom'i ni,* a small river in Southeastern Virginia, about ninety miles long, famed for the numerous battles that were fought on or near its banks during the War of Secession. The Chickahominy rises sixteen miles northwest of Richmond and flows in a southeasterly direction, entering the James River about twenty-two miles below the village of City Point. During a part of its course it flows through a wooded swamp that in the wet season becomes almost impassable. The river was therefore an important military barrier. On or near it were fought, in 1862, the battles of Fair Oaks, Mechanicsville, Gaines's Mill (or Cold Harbor), Savage's Station, Frazier's Farm and Malvern Hill; and in 1864, the second Battle of Cold Harbor.

CHICKAMAUGA, *chik a maw'ga,* BATTLE OF, one of the most desperate engagements of the

War of Secession, in which the generalship of General Thomas saved the Union army from total rout, and won for him his title, "The Rock of Chickamauga." The battle was fought near Chickamauga Creek, in Georgia, on September 19 and 20, 1863, between a Federal force of 55,000, commanded by General Rosecrans, and a Confederate army of 70,000, under General Braxton Bragg. The Confederate general had led his troops in pretended retreat southward from Chattanooga, with Rosecrans following. The latter, however, perceiving that the retreat was only a pretense, drew up his forces in battle line, and by sending Thomas to the extreme left prevented Bragg from shutting off the Union army from Chattanooga.

The battle began on the morning of September 19, and continued throughout the day without decisive results. On the second day, by reason of a misunderstanding of orders issued by Rosecrans, a division was withdrawn from the Union right, and through the gap in the battle line Longstreet charged with his Confederate troops, driving the Federal right and center back towards Chattanooga in great disorder. Thomas, commanding the left, then had to face an attack from the entire Confederate army, and he held his ground throughout the day without flinching, retiring that night only when ordered to do so by Rosecrans. The Federal loss was over 16,000; the Confederate, about 17,800.

Chickamauga National Military Park. The National Military Park on the site of the Battle of Chickamauga covers an area of fifteen square miles. It was dedicated on September 19-21, 1895, and was the first battlefield to be completely set apart to commemorate the engagement fought within its boundaries. In order that visitors may clearly understand the movements and positions of the armies, monuments and other historical guide marks have been erected at various points on the field, and there are also several lofty observation towers.

CHICK'ASAW, an important tribe of Muskhogean Indians who formerly lived in the northern part of Mississippi and the western part of Tennessee. They were closely related to the Choctaw. These Indians were discovered by Fernando De Soto in 1540. When he attempted to make some of them work for him he was attacked by them and lost several men. The Chickasaw were a powerful and warlike tribe and almost constantly at war with other tribes. They were opposed to the French, and favored the English in the struggles of those nations for possession of territory in America.

The Chickasaw were always friendly to the United States. As civilization advanced they began to remove westward, and in 1834 sold all their remaining land east of the Mississippi and removed to the western part of Indian Territory, where they bought land of the Choctaw, and became the Chickasaw Nation. They are now citizens of the United States and of Oklahoma, civilized and prosperous. See FIVE CIVILIZED TRIBES. For customs and habits, see INDIANS, AMERICAN.

CHICKASHA, *chik'a shaw,* OKLA., a shipping point of importance in its territory. It is the county seat of Grady County, and is situated southwest of the geographical center of the state, on the Washita River. Oklahoma City is forty-one miles northeast. The Atchison, Topeka & Santa Fe, the Chicago, Rock Island & Pacific and the Frisco railroads serve the city. In 1895 the first settlement was made, and the city was incorporated in 1897; it was named for the Chickasaw tribe of Indians. There was an increase in population from 10,320 in 1910 to 13,873 in 1914. The area exceeds three square miles.

Chickasha is located in the fertile valley of the Washita River, which produces an abundance of Indian corn, cotton, vegetables and fruits. Stock raising and cotton growing are the leading industries; the stock-feeding pens here are among the largest in the United States, and the city is one of the greatest shipping points in the state for hogs and cattle. Cottonseed oil mills and railway machine-shops are the largest employers of labor. Chickasha has a $175,000 Federal building, a $100,000 bank building and a fine theater. Besides its public schools it has the State Industrial School, Saint Joseph's Academy, Oklahoma College, for women, and a Carnegie Library. A.E.

CHICKEN POX, a contagious disease common among children, characterized by an eruption somewhat like that of smallpox. The two diseases, however, are otherwise very different; chicken pox is rarely dangerous, and smallpox vaccination is not effective in preventing it. Fever is usually present twenty-four hours before the appearance of the eruption, and there may be vomiting, restlessness and slight pains in the legs and back.

Red pimples break out first upon the face, scalp and neck, later upon the limbs and back. They come in "crops," new blotches appearing while the older ones are maturing. In from twelve to twenty-four hours these pimples are filled with a thin fluid, which is not liable to become pus if kept from infection. By the fourth or fifth day crusts form, which fall off a few days later. The fever ranges from 100° F. to 102° F., falling to normal after the first two or three days.

The body of the patient should be sponged each day and the crusts be kept oiled. Scars will not form if rules of cleanliness are observed and scratching is prevented. The patient must be kept quarantined until all the crusts have disappeared, as the disease is very contagious. W.A.E.

CHICLE, *chik''l,* or *chik'le,* a gumlike, milky juice of a tree, used in the manufacture of chewing gum. The tree which produces chicle is the naseberry or *sapodilla,* a native of central and tropical South America, but also cultivated in Mexico. Most of the chicle used in America is obtained from Yucatan and Southern Mexico. See GUM, subhead *Chewing Gum.*

CHICOPEE, *chik'o pe,* MASS., a manufacturing city of Hampden County, situated on the east bank of the Connecticut River at the mouth of the Chicopee River, in the southwestern part of the state. Springfield is three miles south, and Hartford, Conn., is twenty-nine miles south. The Boston & Maine Railway serves the city, and electric lines connect with Holyoke and Springfield. Chicopee was founded in 1640, became a town in 1848 and was incorporated as a city in 1890. The villages of Chicopee Falls, Fairview and Willimansett are included within the city limits. Its area is over twenty-four square miles; its population in 1915 was 30,138, an increase of 4,737 since 1910.

The growth of Chicopee is due to the abundant water power available for manufacturing. Among its industrial plants are the Ames Sword Company, the largest factory of its kind in the United States, the Fisk Rubber Company and the New England Westinghouse Company. There are also manufactories of cotton and knit goods, automobiles, agricultural implements, carpets, athletic goods, bronze statuary, counting scales and other commodities. The massive bronze door of the Senate wing of the national Capitol, and the equestrian statue of Washington, in one of Boston's public gardens, are products of Chicopee. Educational advantages are afforded by a fine public school system, the Academy of Our Lady of the Elms and a library. J.E.H.

CHICORY, *chik'o ri,* an herb, native of Europe and Asia, but now cultivated and found wild in the United States and Southern Canada. It has a fleshy root, spreading branches, coarse leaves, like those of the dandelion, and bright blue, sometimes pink or white, flowers. The leaves when blanched are used as a salad, and the roots, when young and tender, are also used as food. The long, fleshy, milky root has for years been dried, roasted and ground and used for adulterating coffee, but pure food laws now make this impossible, unless its presence is stated on the label. It may easily be detected in coffee by putting a spoonful of the mixture into a glass of clear, cold water; the coffee will float on the surface and the chicory will separate and discolor the water as it precipitates. Chicory was formerly called *succory.*

CHICORY

CHICOUTIMI, *she koo te me',* the county town of Chicoutimi County, Quebec, at the confluence of the Saguenay and Chicoutimi rivers, on the Canadian Northern and the Roberval-Saguenay railways, 227 miles north of the city of Quebec. Chicoutimi is perhaps best known for the beauty of its surroundings, and is visited annually by hundreds of tourists who take the boat trip up the Saguenay River. The Saguenay valley is also a famous resort for hunters and fishermen. But the town is important as a lumbering center, and its pulp mills export about 60,000 tons of dry pulp a year. There are also lumber mills, tanneries, sash and door factories and other industrial establishments which utilize the products of the forests. Chicoutimi is the seat of a Roman Catholic bishop and has a beautiful cathedral which cost $350,000, a seminary costing $600,000, a large normal school and a well-equipped public hospital. Chicoutimi was founded in

1847 and was incorporated as a town in 1879. Population in 1911, 5,880; in 1916, about 6,500.

CHIFFON, *shif'on,* a very soft, thin, gauzy material used for veils, trimmings, ruches and various dainty garments for women. The finest qualities are silky and are made in all colors, but chiefly delicate tints of blue, green, pink, and cream and white are used. Cotton chiffons are also made, and dainty chiffon ribbons are popularly used by florists to decorate bouquets of flowers and plants to be sent as gifts. Chiffon lace is chiffon embroidered with silk.

The name chiffon is French and means *rag* or *flimsy cloth,* and in that language is used to suggest anything decorative worn by women.

CHIHUAHUA, *che wah'wah,* the capital of the state of Chihuahua, and the most important commercial city in Northern Mexico. It is built on a table-land surrounded by mountains 4,650 feet above sea level, and is on the Chihuahua River. The city is 700 miles north of Mexico City and 225 miles south of El Paso, Tex. The state of Chihuahua once produced more silver than any other district in Mexico, and the city was the center of large mining industries; the silver mines are not yet exhausted. When not torn by revolutions the city supports prosperous textile, iron and other industries. Compared with some other Mexican cities, it is well laid out, clean, healthful and well supplied with water by a fine aqueduct built in the seventeenth century. One of its chief features is a fine cathedral, completed in 1789. Chihuahua was founded by the Spaniards in 1539. In the eighteenth century its population was 70,000; now it is about 40,000. In the latter part of 1916 it was the scene of desperate fighting between the forces of Villa and of Carranza. The former attacked the city repeatedly and finally succeeded in capturing it, but evacuated it the next day. Later, he held it again for a short time.

CHILBLAIN, *chil'blane,* a stinging, burning, itching inflammation appearing in cold weather, usually on the feet, but sometimes on the face or hands. It is not frostbite, as very generally believed, but is caused by too sudden changes of temperature when the circulation of blood is poor. Because tight shoes prevent the warm blood from reaching the ends of the toes, the tender skin, sometimes moist, cannot then well resist the shock of sudden change to heat or cold, and soreness results. Application of tincture of iodine, ichthyol or tincture of camphor will bring relief. Then with plenty of outdoor exercise and frequent changing of stockings, a cure will be effected. Keeping the feet dry and warm and wearing roomy shoes will prevent chilblains. W.A.E.

CHILD, THE. Parents of the twentieth century take better care of their children than did the parents of preceding generations, simply because they know better how to do it. The rearing of children, together with every other field of human endeavor, has been invaded by the scientific spirit. By studying children individually and in groups, statistics and tables have been gathered which enable us to know with some accuracy what we have a right to expect of the average child and how to go about to secure it. The facts so far assembled are not final, by any means; as more children are studied, modifications and changes are bound to be made. But the account of the development of the child which follows represents some of the knowledge which is now at hand.

The Development of the Child. In the beginning the human body consists of a single cell. When this cell begins to grow it divides into two cells; each of these divides in its turn into two more, and so on, until that ever-new marvel, a tiny human being, is formed. By weighing and measuring hundreds of thousands of babies it has been found that the average baby boy, at birth, weighs about 7.3 pounds; the average baby girl, 7.1 pounds. The boy baby should be about 19.68 inches tall; the girl baby, 19.48 inches. In six months the average baby doubles its weight; in a year he trebles it. So a year-old child should weigh about twenty-one pounds. By the sixth year the average boy weighs about forty-five pounds, the average girl about forty-three, and the boy should be just a trifle over, the girl just a trifle under, forty-four inches in height.

The first table below shows a child's increase in weight and height from the age of six to sixteen years:

AGE	BOYS			GIRLS		
	AVERAGE IN LBS.	ANNUAL INCREASE	PER CENT OF INCREASE	AVERAGE IN LBS.	ANNUAL INCREASE	PER CENT OF INCREASE
6½	45.2	...	...	43.4	...	...
7½	49.5	4.3	9.5	47.7	4.3	9.9
8½	54.5	5.0	10.1	42.5	4.8	10.0
9½	59.6	5.1	9.3	57.4	4.9	9.3
10½	65.4	5.8	9.7	62.9	5.5	9.6
11½	70.7	5.3	8.1	69.5	6.6	10.5
12½	76.9	6.2	8.7	78.7	9.2	13.2
13½	84.8	7.9	10.3	88.7	10.0	12.7
14½	95.2	10.4	12.3	98.3	9.6	11.9
15½	107.4	12.2	12.8	106.7	8.4	8.5
16½	121.0	13.6	12.7	112.3	5.6	5.2

The second table, giving heights in inches, is made up from the measurements of American-born children in three American cities; in common with Canadian children, they are a little taller and heavier than the average English, Irish, German or Scandinavian child:

Years	6	7	8	9	10	11	12	13	14	15	16	17
Boys	44.10	46.21	48.16	50.09	52.21	54.01	55.78	58.17	61.08	62.96	65.58	66.29
Girls	43.66	45.94	48.07	49.61	51.78	53.79	57.16	58.75	60.32	61.39	61.72	61.99

The Development of the Senses. In a remarkable book called the *Biography of a Baby*, Miss M. W. Shinn describes the state of a new-born baby thus:

> She took in with vague comfort the gentle light that fell on her eyes, seeing without any sort of attention or comprehension the moving blur of darkness that varied it. She felt motions and changes; she felt the action of her own muscles, and after the first three or four days disagreeable shocks of sound now and then broke through the silence or perhaps through an unnoticed jumble of faint noises. She felt touches on her body from time to time, but without the least sense of the place of the touch. * * * From time to time sensations of hunger and thirst, and once or twice of pain made themselves felt through all the others, and mounted until they became distressing; from time to time a feeling of heightened comfort flowed over her as hunger or thirst was satisfied. * * * For the rest she lay empty-minded, neither consciously comfortable nor uncomfortable, yet on the whole pervaded with a dull sense of well-being. Of the people about her, of her mother's face, of her own existence, of desire or fear, she knew nothing.

The preceding paragraph from Miss Shinn's book is only an imaginative way of saying that a baby, although it is not born with its eyes closed, like a kitten, does not *see;* it does not *hear*, or *smell;* it does not *think;* it *feels* only vaguely and unconsciously. And yet, in this animal-like little being all the elements of the future man or woman are present, and its growth and development during the first years of its life are truly marvelous. We have no way of measuring accurately how fast or how far a child progresses in these first years, but we do know that he is learning to use all of his senses, that he has a constant and insistent desire to touch, taste and handle everything around him; that he is pleased with bright and beautiful colors; that he is alert to pleasant sounds and sensitive to harsh ones; that he acquires very positive likes and dislikes about the food he eats; and that he develops a liking for pleasant odors and a distaste for those that are unpleasant.

This is exactly as it should be, for a child lives by his senses. They are his only way at first of acquiring knowledge of any sort. They furnish all the material his mind has to work with. If his senses are not satisfied, his mind will starve; if they are not developed, his mind will not develop. It is important, therefore, that from the second month on, when the senses begin to be active, plenty of material be furnished for stimulating and developing each sense.

Smell and Taste. It is practically impossible to test a baby's sense of smell, but it is quite probable that this sense does not develop rapidly because the clean surroundings of the well-cared-for baby are practically odorless.

Tests made on a new-born baby seem to prove, however, that the sense of taste is active from the first—that there is a dislike for sour and bitter things and a liking for sweets. It is very desirable that this sense of taste should be wisely developed, because a child's enjoyment of simple and wholesome food depends largely on it. As soon as a child begins eating solid food, he should be encouraged to like the things which are good for him and to dislike those which are unwholesome. It may be mentioned, too, that children should be encouraged to be thirsty, for their bodies need a great deal of water. Every baby should be given plenty of water to drink, and older children should be encouraged to drink large amounts of it.

Hearing. A new-born baby is deaf, usually because the inner ear is full of mucus, and it remains deaf for several days. But if loud noises are not heard by a baby by the end of the fourth week, he should be taken to a physician. Ordinarily, after three or four days the baby becomes very sensitive to sound and starts and trembles if a door is slammed or some one speaks loudly. A sneeze or a whistle will also cause a violent reaction. Music and sound are such important factors in the growth of children that no child ought to be brought up without having the opportunity to hear soft, sweet sounds. His interest in such sounds should be encouraged and stimulated, and early training in music should be begun. The kindergarten admirably provides both the music and the rhythm in which children delight.

Dermal Senses. By dermal senses we mean the sensations in the skin. Babies quickly note the difference between things warm and cold. A baby a week old will cry if he is put into a bath that is a few degrees colder than the one to which he is accustomed. This should be remembered by the person who prepares his bath. The hand of an adult is not sensitive enough for testing the temperature of the water. A thermometer should be used, or, failing that, the elbow. The mucous membrane of a baby's mouth and throat is much more tender than that of a grown person. Food which is merely warm to an adult will seem disagreeably hot to a baby or small child. Anyone who has observed children will realize how indifferent the average child is as to whether his food is more than warm, and a mother's admonition, "Now eat your soup while it is nice and hot," is usually enough to make the child push his plate away and wait until it cools.

Touch. Up to the third month the average baby has done nothing but aimlessly grasp with his hands, which he holds habitually with the thumb inside the palm. But after the second month he may be given every sort of object to handle which will not do him injury. As he grows older the more objects he has—hard and soft, rough and smooth—to play with, the faster will his sense of touch develop. The ordinary toys babies are given may be supplemented by the many objects the ordinary household provides—clothespins, empty spools, napkin rings, spoons, etc. If the baby cannot handle the objects he sees his knowledge of them will be imperfect.

Sight. The eyes of a new-born baby are closed most of the time. The reason that some babies are so wakeful at night is undoubtedly due to the fact that the darkness is pleasanter to their eyes than daylight. They prefer to sleep when it is light and to lie awake in the dark. The eyes of a tiny baby will close if a light is brought near them, but after a few days he will turn his head towards a window or a light, and after a few weeks light will give him pleasure. By the end of two weeks the eyes, which do not at first coördinate, will begin to follow objects and at the end of eight or nine weeks a baby will stare at an object for minutes at a time. By the seventh month he will distinguish faces by staring at strangers and smiling at friends, will turn his head towards a person leaving the room and follow with his eyes objects dropped from his hand.

Muscular Control. At birth a child has no power to make voluntary movements of any sort. When he moves an arm or a leg, when his eyes close at a bright light or when he starts at a loud sound, the movement is a total surprise to him—something he can neither prevent nor repeat. But gradually all of his vague feelings become more distinct by being repeated, and as the connective fibers grow in his brain, the various feelings become associated with one another. The wonderful change in a baby usually occurs when he is about six months old and is due to his discovery that he can move this way or that as he pleases and can direct his movements with his eyes. Immediately he begins doing what he sees other people do. He begins to imitate sounds, facial expressions and movements of all sorts.

The age at which children begin walking varies so greatly that it is impossible to give any date for it. But since a baby learns by imitation, he is likely to begin walking at an earlier age if there are other children in the family. Some babies learn to walk before they are a year old; others do not walk until nearly the second year. Of course the baby kicks and practices creeping before he begins to walk; otherwise he would not have sufficient muscular strength to master the art. Walking has a marked effect on most babies. They get a new view of things when they can see the world from a standing position, and as a rule they actually sleep better, eat more and become better-natured and happier.

In order that growth in muscular control may develop properly, children should be encouraged to be active, to use all the large muscles of their bodies. All children should be free to run around, to romp and to play as much as they wish. When they are a little older they will be greatly assisted in learning control of the smaller muscles by having plenty of tools to work with, and they should be encouraged to make their own toys, playhouses, doll clothes and other things.

Language. Tears, smiles, cries and gestures are the baby's first means of expressing his emotions. A baby cries from the first; he will shed tears any time after the twenty-third day, and he sometimes smiles in the second week. By the fourth month he will stretch out his hands toward the thing he wants, and still later he will put his hands together as if he wanted to grasp an object. Between the eighth and the twelfth month he begins pointing at the thing he wants. He will begin in the sixth month to express affection through imitating the kisses, pats and hugs of other people and begin using a real gesture language. He will tug at his mother's dress if he is hungry, will stretch out his arms to be taken up, and learns to wave "bye-bye." A little later all sorts of coaxing and begging gestures will appear.

Even after he begins to speak he will supplement his words with gestures, just as many savages do. A baby's greatest difficulty at first is learning to articulate. Once this ability has been acquired his progress in learning to talk will be very rapid. Another obstacle is learning to walk. While he is doing this a baby acquires no more speech and may even go backward, but afterwards the learning and understanding of words is very rapid. His progress at this stage will be greatly influenced by the people around him. It is only by observing the language used by a baby and noting his mistakes that an adult begins to realize what an immensely-complicated thing is speech. Surely the fact that most children by the fifth year have obtained a good working knowledge of the mother tongue would alone justify the claim that these are the years of most importance, the years of greatest development.

The Kindergarten Child. Froebel, the man who conceived the idea of the kindergarten, and Maria Montessori, one of the great child educators of to-day, both set out with the idea of helping the child under six to develop to the height of his powers. The necessity for the normal development of the senses has already been shown. This development is bound to go on, whether it is encouraged or not, but if it is systematically fostered and stimulated the child will be better equipped than if he has to acquire everything in a haphazard fashion. Such a system of child training as is furnished by the kindergarten and the Montessori school goes still further. It not only helps a child to develop the senses, but it also trains him to associate his sensations with the spoken symbols, so that everything he learns is made more usable. It also helps him to acquire muscular control, teaches him to use the large muscles of his body and arms and legs and the smaller muscles of the hands and fingers. And hand in hand with this training goes the development of all the mental powers, imagination and reason, memory and perception (see PSYCHOLOGY).

The School Child. Let us suppose that the child up to the age of six has lived in an environment which has developed brain and body to its fullest capacity. Bubbling over with energy, alert, imaginative, eager and curious, expressing himself spontaneously and exuberantly on every occasion, the six-year-old comes to the public school. Here every sense he has begun developing, every interest he has displayed, should be made use of; his curiosity must be stimulated and satisfied; his energy directed. He should go on acquiring more discrimination as to colors, more delicacy of touch, more sensitiveness of hearing, greater muscular control and a larger appreciation of everything beautiful. And he must also go on learning to express himself more clearly and accurately, both in spoken and in written language.

In this development the school, the home and the playground are almost equally important. It is the duty of every parent and every teacher to see that all three are forces for progress and not for retrogression. Under the headings EDUCATION; CHILD STUDY, and other related topics referred to at the close of this article, this phase of a normal child's education is treated in greater detail in these volumes. Space will be given here only to a brief treatment of some of the conditions which must be guarded against.

Physically-Defective Children. It is rapidly becoming the practice to have physical inspection in all public schools. This is of prime importance, because it has been found in many cases that children who are considered obstinate, stupid or positively bad are partly blind or deaf, or are the victims of serious nervous trouble. The eyes and the ears are the principal channels through which knowledge comes, so the child who cannot hear and see perfectly is seriously handicapped. He may not know of his trouble, unless there is actual pain, and for this reason his parents and teachers should be alert for signs. Defective eyesight can be discovered by noting a child's position when he is reading or writing. If his eyes are either more or less than a foot from the book he is reading, he should be given special tests with a set of cards, which can be bought for ten cents, to determine what is the trouble. Nearsight, farsight and astigmatism are the most common ailments (see EYE; BLINDNESS; ASTIGMATISM).

By first determining, by means of a watch, how far a normal child can hear, the standard for testing the child suspected of deafness may be fixed (see EAR). If a child is dull or does not pay attention, or if he asks constantly to have things repeated, he should at once be tested for ear trouble. And it should be remembered that the purpose of testing children in these ways is always to discover whether a doctor's care is needed.

Fatigue. Complete fatigue, or nervous exhaustion, is almost as difficult to recover from as a severe illness. For this reason children must be watched carefully and guarded against overwork, too long hours of work, too great worry over their tasks, not enough work, or work that has not sufficient variety, for all these conditions bring about a state of fatigue which is likely to result in serious harm. The great trouble with our public schools is that the classes are large and the teacher has not the time to give every child sufficient individual attention. This, then, must be the duty of parents. It is essential that they be on the lookout for signs of nervous or bodily fatigue.

In order to avoid excessive fatigue a child must be interested in his work, and he must find a great deal of variety in it. His hours of work must not be too long; he must not do much outside work; he must get plenty of play, plenty of sleep and plenty of good, nourishing food. It is the duty of parents, wherever possible, to coöperate with the teacher in securing the best working conditions within their power for the child—light, well-ventilated school rooms, a comfortable desk and seat, adequate teaching equipment and well-kept, spacious playgrounds. The comfortable seat and desk are of vital importance, because the body of a growing child is very plastic, and a wrong sitting position held for several hours out of every day will change and deform his body.

Signs of Fatigue. The signs of fatigue are inattention, restlessness and irritability. Tests have shown that a person who is very tired is also not as sensitive to touch, that his eye cannot distinguish colors as well as when he is rested and that his muscular control is impaired, for he will be more clumsy and awkward in moving about. He is more likely, too, to be impertinent and undisciplined than when he is rested. A good night's sleep and plenty of wholesome food ought always to restore a child's good temper and energy. If it does not, then the conditions under which he works and plays must be changed.

The Exceptional Child. There is a large class of children who are constituted differently from the average child and for whom no provision is made in the public schools. There is the exceptionally-bright child; there is the eccentric child, who has marked individuality without being either inventive or original; there are the feeble-minded child, the backward child and the wayward child. Of course all children vary a little from the average, because there is actually no such individual person as the *average* child. It is simply a term given to a composite of all the statistics on children. Up to a certain point this variation from the average has no significance, but beyond it we have the abnormal or exceptional child who is so great a problem in the schools. Institutions are now solving the problem of feeble-minded children and those difficult to manage, and in smaller classes and by special instruction, the problem of the exceptional and the backward child. All these exceptional children need an unusual amount of care. H.K.S.

Related Subjects. The following articles in these volumes, relating to children or to children's activities, will be of interest in connection with this topic:

CHILD LABOR, a term relating to the employment of children in industry. Children have been thus employed from the earliest days of recorded history, but the problem of child labor, as it is commonly understood, has developed with the modern factory system. In every country in which manufacturing industries have reached a high state of development competition is keen and effort is constantly being made to keep the cost of production low. In such countries child labor is a vital social and economic issue.

The Development of the System. Under the conditions of labor which preceded the factory system, the employment of children was regarded as a part of their education. Either as apprentices or in the workshops of their parents they learned a trade and "habits of steady industry." While there were many cases of abuse under this system, there was a close personal relation between the master and the child, which usually checked the master's indifference to the child's good. The factory system is characterized by two features which did not exist under any preceding system of labor: first, the employment of workmen in large numbers has tended to destroy any personal relations between master and workman; second, the operation of automatic machinery frequently requires quickness and deftness rather than physical strength. In England, where child labor first became a social menace, the demand for children to work in textile mills was supplied by a vicious system, using pauper children collected from the poorhouses. These children received as pay only their food and lodging. As competition became more intense, the working and living conditions of the children became worse, until they constituted a form of slavery. Children five years of age were sometimes found in the mills. Hours of work were unregulated, and a day of twelve hours or "from sunrise to sunset" was not uncommon.

Such conditions existed in England during the first quarter of the nineteenth century. In the United States child labor did not involve great numbers of children until the period of industrial expansion which followed the War of Secession. In Belgium, Germany and Italy it began to trouble economists and sociologists about 1875 to 1880, and in Canada the problem is even more recent.

The Regulation of Child Labor. The first law regulating child labor, in the modern sense, was passed by the British Parliament in 1802. It applied to cotton mills only, forbade work between 9 P. M. and 6 A. M., limited the working day to twelve hours and required elementary instruction for apprentices. An important act of 1819 prohibited the employment of children under nine years of age in establishments for the preparation and spinning of cotton. These early statutes were weakened, however, by failure to provide for their enforcement. Step by step greater protection was given to the child, until now the minimum age for full-time work is thirteen, and fourteen unless the child possesses certain physical and educational qualifications.

Following the lead of England, Germany passed its first law regulating child labor in 1839, and nearly all European countries now give the child some degree of protection. The following regulations apply to factories: In Germany, thirteen is the age limit; children from thirteen to fourteen may be employed only six hours a day, and children from fourteen to sixteen not over ten hours a day. Thirteen is also the general age limit in France, though a child of twelve may begin work if he has a prescribed educational and medical certificate of fitness; ten years is the limit in Spain, and fourteen in Switzerland and Norway; in several other European countries the age limit is twelve years. The enforcement of the laws in Europe is generally very strict.

In the United States and Canada the problem of regulation is newer. Here, as in other countries, the tendency of many industries is to overwork and underpay their child-workers, and little attention is paid to their educational, physical or moral well-being. All students of social welfare recognize the fact that child labor is an evil whose influence extends to succeeding generations, and that it must be controlled by legislation. The Dominion of Canada has not as yet enacted a national child-labor law, all regulation being in the hands of the provincial governments. The laws of Ontario are typical of the progress which has been made. No child under fourteen years of age is allowed to work in a factory, but children between ages of twelve and fourteen may be employed during July, August, September and October in canning and preserving factories. Ten hours is the maximum working day, and six o'clock is the earliest hour at which work may be begun. There are various other regulations looking to the safety and health of children.

In the United States there is naturally a variety in the details of state child-labor laws. These laws all fix a minimum age below which children must not be employed in one or more specified occupations. Many of them regulate the length of the working day and prohibit night work. Most states require children to procure certificates showing their age and extent of schooling, and in these states employers who hire children without such certificates are liable to a penalty. An education minimum and a certificate of physical fitness are in a few states required in addition to a documentary proof of the child's age.

In more than half of the states the age limit for factory work is fourteen; it is twelve in a few states, fifteen in Michigan, fifteen for boys and sixteen for girls in Ohio and sixteen in Montana. In the factories of California, New York, Ohio, Massachusetts and several other states children under sixteen may work only eight hours. Many states fix a sixteen- or eighteen-year limit upon employment in specified dangerous or hazardous occupations.

The enactment by Congress of a national child-labor law has been agitated by various organizations and individuals for many years, and such a law was passed in 1916. The bill, known as the Keating-Owen Act, was signed by President Wilson in September. However, it was declared unconstitutional in 1918. Its provisions, in brief, were as follows:

> One year after approval of the act it shall be illegal for the products of any mine or quarry that employs children under sixteen years of age to enter interstate commerce, or for the products of any mill, cannery, workshop or manufacturing establishment that employs children under fourteen years of age to enter interstate commerce. Products of industrial establishments that employ children between the ages of fourteen and sixteen years more than eight hours a day, six days a week, or earlier than six o'clock in the morning or later than seven o'clock in the evening, are also excluded. The bill further provides that products sold within a state within thirty days of their manufacture are also debarred from interstate commerce.

CHILDREN, Societies for. In all civilized countries there are organizations having for their purpose the protection and care of children who have become orphans, or who, for other reasons, have been deprived of suitable homes. In America the most widely known of these are the Society for the Prevention of Cruelty to Children, Saint Vincent's Aid Society, the Jewish Relief Association, the Children's Aid Society and the American Humane Association. The purpose of these societies is to protect children from evil associates and from cruelty on the part of those who employ them or have the care of them. Home-finding societies, whose purpose it is to place orphans in suitable homes, are formed in many provinces, states and large cities. Juvenile courts (which see) have jurisdiction over all cases of dependent and delinquent children, and those conducted according to the most advanced methods exercise the right of jurisdiction in regard to placing children outside the home. The administration of these courts is for purposes of guardianship, education and protection, not for trial and punishment. The Children's Bureau (which see), in the Department of Labor, was organized by the United States government to conduct investigation relating to the welfare of children, and to publish reports of the same. H.K.S.

CHILDREN'S BUREAU, a bureau of the United States government, authorized in April, 1912, now under the Department of Labor. Its duty is to investigate and report upon all matters pertaining to child life and children's welfare, especially the questions of birth rate, infant mortality, orphanage, desertion, juvenile courts, employment, and all legislation affecting children.

Previous to the opening of this bureau, the United States government had been spending large sums of money each year in gathering facts and distributing information regarding horses, cattle, hogs and the like, but had not created any service distinctly to study the welfare of children. This new work is not intended to relieve the states of the responsibility of dealing with child problems, but rather to gather such information from all parts of the Union as will best enable the states to promote the interest of the nation's greatest asset, child life. The first director of the new bureau was Miss. Julia Lathrop, for many years a co-worker with Miss Jane Addams at Hull House, Chicago, and long a member of the Illinois Board of Charities. See LATHROP, JULIA C. J.L.

CHILDS, GEORGE WILLIAM (1829-1894), one of the most notable of American publishers and philanthropists. He was born in Baltimore, but began his business career in Philadelphia, becoming a partner in the publishing house of Childs & Peterson in 1849. In 1864 he purchased the Philadelphia *Public Ledger*, one of the earliest of the low-priced daily papers. Under his management it became very influential and made its owner a wealthy man. Mr. Childs' charities, both public and private, were boundless. Among the most noted of his public gifts were a memorial fountain at Stratford-on-Avon in England, a monument over the grave of Edgar Allan Poe, the presentation of a printers' cemetery—"Woodlawn"—in Philadelphia, and a subscription which made possible the endowment of the home for union printers at Colorado Springs. His private benefactions were equally large, and included among many others the educating of 800 boys and girls and the pensioning of many old literary workers. He wrote two books, *The Recollections of General Grant* and *Personal Recollections.*

End of Volume Two

Lightning Source UK Ltd.
Milton Keynes UK
UKHW02n0431160218
317658UK00015B/282/P

9 780483 608009